ANTENNA THEORY

ANTENNA THEORY
ANALYSIS AND DESIGN

FOURTH EDITION

Constantine A. Balanis

Cover Image: Courtesy NASA/JPL-Caltech

Copyright © 2016 by John Wiley & Sons, Inc. All rights reserved

Published by John Wiley & Sons, Inc., Hoboken, New Jersey
Published simultaneously in Canada

No part of this publication may be reproduced, stored in a retrieval system, or transmitted in any form or by any means, electronic, mechanical, photocopying, recording, scanning, or otherwise, except as permitted under Section 107 or 108 of the 1976 United States Copyright Act, without either the prior written permission of the Publisher, or authorization through payment of the appropriate per-copy fee to the Copyright Clearance Center, Inc., 222 Rosewood Drive, Danvers, MA 01923, (978) 750-8400, fax (978) 750-4470, or on the web at www.copyright.com. Requests to the Publisher for permission should be addressed to the Permissions Department, John Wiley & Sons, Inc., 111 River Street, Hoboken, NJ 07030, (201) 748-6011, fax (201) 748-6008, or online at http://www.wiley.com/go/permission.

Limit of Liability/Disclaimer of Warranty: While the publisher and author have used their best efforts in preparing this book, they make no representations or warranties with respect to the accuracy or completeness of the contents of this book and specifically disclaim any implied warranties of merchantability or fitness for a particular purpose. No warranty may be created or extended by sales representatives or written sales materials. The advice and strategies contained herein may not be suitable for your situation. You should consult with a professional where appropriate. Neither the publisher nor author shall be liable for any loss of profit or any other commercial damages, including but not limited to special, incidental, consequential, or other damages.

For general information on our other products and services or for technical support, please contact our Customer Care Department within the United States at (800) 762-2974, outside the United States at (317) 572-3993 or fax (317) 572-4002.

Wiley also publishes its books in a variety of electronic formats. Some content that appears in print may not be available in electronic formats. For more information about Wiley products, visit our web site at www.wiley.com.

Library of Congress Cataloging-in-Publication Data:

Balanis, Constantine A., 1938–
Modern antenna handbook / Constantine A. Balanis.—4th ed.
   p. cm.
   Includes index.
   ISBN 978-1-118-64206-1 (cloth)
   1. Antennas (Electronics)   I. Title.
   TK7871.6.B354 2016
   621.382′4—dc22   2016050162

Printed in the United States of America

SKY10030333_100421

To the memory of my parents, uncle and aunt

Στη μνήμη των γονέων, του θείου και της θείας μου

Contents

Preface — xiii

About the Companion Website — xix

1 Antennas — 1

- 1.1 Introduction — 1
- 1.2 Types of Antennas — 3
- 1.3 Radiation Mechanism — 7
- 1.4 Current Distribution on a Thin Wire Antenna — 15
- 1.5 Historical Advancement — 18
- 1.6 Multimedia — 21
- References — 22

2 Fundamental Parameters and Figures-of-Merit of Antennas — 25

- 2.1 Introduction — 25
- 2.2 Radiation Pattern — 25
- 2.3 Radiation Power Density — 35
- 2.4 Radiation Intensity — 37
- 2.5 Beamwidth — 40
- 2.6 Directivity — 41
- 2.7 Numerical Techniques — 55
- 2.8 Antenna Efficiency — 60
- 2.9 Gain, Realized Gain — 61
- 2.10 Beam Efficiency — 65
- 2.11 Bandwidth — 65
- 2.12 Polarization — 66
- 2.13 Input Impedance — 75
- 2.14 Antenna Radiation Efficiency — 79
- 2.15 Antenna Vector Effective Length and Equivalent Areas — 81
- 2.16 Maximum Directivity and Maximum Effective Area — 86
- 2.17 Friis Transmission Equation and Radar Range Equation — 88
- 2.18 Antenna Temperature — 96
- 2.19 Multimedia — 100
- References — 103
- Problems — 105

3 Radiation Integrals and Auxiliary Potential Functions 127

- 3.1 Introduction 127
- 3.2 The Vector Potential **A** for an Electric Current Source **J** 128
- 3.3 The Vector Potential **F** for **A** Magnetic Current Source **M** 130
- 3.4 Electric and Magnetic Fields for Electric (**J**) and Magnetic (**M**) Current Sources 131
- 3.5 Solution of the Inhomogeneous Vector Potential Wave Equation 132
- 3.6 Far-Field Radiation 136
- 3.7 Duality Theorem 137
- 3.8 Reciprocity and Reaction Theorems 138
- References 143
- Problems 143

4 Linear Wire Antennas 145

- 4.1 Introduction 145
- 4.2 Infinitesimal Dipole 145
- 4.3 Small Dipole 155
- 4.4 Region Separation 158
- 4.5 Finite Length Dipole 164
- 4.6 Half-Wavelength Dipole 176
- 4.7 Linear Elements Near or On Infinite Perfect Electric Conductors (PEC), Perfect Magnetic Conductors (PMC) and Electromagnetic Band-Gap (EBG) Surfaces 179
- 4.8 Ground Effects 203
- 4.9 Computer Codes 216
- 4.10 Multimedia 216
- References 218
- Problems 220

5 Loop Antennas 235

- 5.1 Introduction 235
- 5.2 Small Circular Loop 236
- 5.3 Circular Loop of Constant Current 250
- 5.4 Circular Loop with Nonuniform Current 259
- 5.5 Ground and Earth Curvature Effects for Circular Loops 268
- 5.6 Polygonal Loop Antennas 269
- 5.7 Ferrite Loop 270
- 5.8 Mobile Communication Systems Applications 272
- 5.9 Multimedia 272
- References 275
- Problems 277

6 Arrays: Linear, Planar, and Circular 285

- 6.1 Introduction 285
- 6.2 Two-Element Array 286
- 6.3 N-Element Linear Array: Uniform Amplitude and Spacing 293
- 6.4 N-Element Linear Array: Directivity 312
- 6.5 Design Procedure 318
- 6.6 N-Element Linear Array: Three-Dimensional Characteristics 319
- 6.7 Rectangular-to-Polar Graphical Solution 322

6.8	N-Element Linear Array: Uniform Spacing, Nonuniform Amplitude	323
6.9	Superdirectivity	345
6.10	Planar Array	348
6.11	Design Considerations	360
6.12	Circular Array	363
6.13	Multimedia	367
	References	367
	Problems	368

7 Antenna Synthesis and Continuous Sources — 385

7.1	Introduction	385
7.2	Continuous Sources	386
7.3	Schelkunoff Polynomial Method	387
7.4	Fourier Transform Method	392
7.5	Woodward-Lawson Method	398
7.6	Taylor Line-Source (Tschebyscheff-Error)	404
7.7	Taylor Line-Source (One-Parameter)	408
7.8	Triangular, Cosine, and Cosine-Squared Amplitude Distributions	415
7.9	Line-Source Phase Distributions	416
7.10	Continuous Aperture Sources	417
7.11	Multimedia	420
	References	420
	Problems	421

8 Integral Equations, Moment Method, and Self and Mutual Impedances — 431

8.1	Introduction	431
8.2	Integral Equation Method	432
8.3	Finite Diameter Wires	439
8.4	Moment Method Solution	448
8.5	Self-Impedance	455
8.6	Mutual Impedance Between Linear Elements	463
8.7	Mutual Coupling in Arrays	474
8.8	Multimedia	480
	References	480
	Problems	482

9 Broadband Dipoles and Matching Techniques — 485

9.1	Introduction	485
9.2	Biconical Antenna	487
9.3	Triangular Sheet, Flexible and Conformal Bow-Tie, and Wire Simulation	492
9.4	Vivaldi Antenna	496
9.5	Cylindrical Dipole	500
9.6	Folded Dipole	505
9.7	Discone and Conical Skirt Monopole	512
9.8	Matching Techniques	513
9.9	Multimedia	523
	References	524
	Problems	525

10 Traveling Wave and Broadband Antennas — 533

- **10.1** Introduction — 533
- **10.2** Traveling Wave Antennas — 533
- **10.3** Broadband Antennas — 549
- **10.4** Multimedia — 580
  - References — 580
  - Problems — 582

11 Frequency Independent Antennas, Antenna Miniaturization, and Fractal Antennas — 591

- **11.1** Introduction — 591
- **11.2** Theory — 592
- **11.3** Equiangular Spiral Antennas — 593
- **11.4** Log-Periodic Antennas — 598
- **11.5** Fundamental Limits of Electrically Small Antennas — 614
- **11.6** Antenna Miniaturization — 619
- **11.7** Fractal Antennas — 627
- **11.8** Multimedia — 633
  - References — 633
  - Problems — 635

12 Aperture Antennas — 639

- **12.1** Introduction — 639
- **12.2** Field Equivalence Principle: Huygens' Principle — 639
- **12.3** Radiation Equations — 645
- **12.4** Directivity — 648
- **12.5** Rectangular Apertures — 648
- **12.6** Circular Apertures — 667
- **12.7** Design Considerations — 675
- **12.8** Babinet's Principle — 680
- **12.9** Fourier Transforms in Aperture Antenna Theory — 684
- **12.10** Ground Plane Edge Effects: The Geometrical Theory of Diffraction — 702
- **12.11** Multimedia — 707
  - References — 707
  - Problems — 709

13 Horn Antennas — 719

- **13.1** Introduction — 719
- **13.2** E-Plane Sectoral Horn — 719
- **13.3** H-Plane Sectoral Horn — 733
- **13.4** Pyramidal Horn — 743
- **13.5** Conical Horn — 756
- **13.6** Corrugated Horn — 761
- **13.7** Aperture-Matched Horns — 766
- **13.8** Multimode Horns — 769
- **13.9** Dielectric-Loaded Horns — 771
- **13.10** Phase Center — 773
- **13.11** Multimedia — 774
  - References — 775
  - Problems — 778

14 Microstrip and Mobile Communications Antennas   783

- **14.1** Introduction   783
- **14.2** Rectangular Patch   788
- **14.3** Circular Patch   815
- **14.4** Quality Factor, Bandwidth, and Efficiency   823
- **14.5** Input Impedance   826
- **14.6** Coupling   827
- **14.7** Circular Polarization   830
- **14.8** Arrays and Feed Networks   832
- **14.9** Antennas for Mobile Communications   837
- **14.10** Dielectric Resonator Antennas   847
- **14.11** Multimedia   858
- References   862
- Problems   867

15 Reflector Antennas   875

- **15.1** Introduction   875
- **15.2** Plane Reflector   875
- **15.3** Corner Reflector   876
- **15.4** Parabolic Reflector   884
- **15.5** Spherical Reflector   920
- **15.6** Multimedia   923
- References   923
- Problems   925

16 Smart Antennas   931

- **16.1** Introduction   931
- **16.2** Smart-Antenna Analogy   931
- **16.3** Cellular Radio Systems Evolution   933
- **16.4** Signal Propagation   939
- **16.5** Smart Antennas' Benefits   942
- **16.6** Smart Antennas' Drawbacks   943
- **16.7** Antenna   943
- **16.8** Antenna Beamforming   946
- **16.9** Mobile Ad hoc Networks (MANETs)   960
- **16.10** Smart-Antenna System Design, Simulation, and Results   964
- **16.11** Beamforming, Diversity Combining, Rayleigh-Fading, and Trellis-Coded Modulation   972
- **16.12** Other Geometries   975
- **16.13** Multimedia   976
- References   976
- Problems   980

17 Antenna Measurements   981

- **17.1** Introduction   981
- **17.2** Antenna Ranges   982
- **17.3** Radiation Patterns   1000
- **17.4** Gain Measurements   1003
- **17.5** Directivity Measurements   1010

17.6	Radiation Efficiency	1012
17.7	Impedance Measurements	1012
17.8	Current Measurements	1014
17.9	Polarization Measurements	1014
17.10	Scale Model Measurements	1019
	References	1024

Appendix I: $f(x) = \dfrac{\sin(x)}{x}$ 1027

Appendix II: $f_N(x) = \left| \dfrac{\sin(Nx)}{N \sin(x)} \right|$ $N = 1, 3, 5, 10, 20$ 1029

Appendix III: Cosine and Sine Integrals   1031

Appendix IV: Fresnel Integrals   1033

Appendix V: Bessel Functions   1035

Appendix VI: Identities   1041

Appendix VII: Vector Analysis   1045

Appendix VIII: Method of Stationary Phase   1055

Appendix IX: Television, Radio, Telephone, and Radar Frequency Spectrums   1061

Index   1065

Preface

The fourth edition of *Antenna Theory* is designed to meet the needs of electrical engineering and physics students at the senior undergraduate and beginning graduate levels, and those of practicing engineers. The text presumes that the students have knowledge of basic undergraduate electromagnetic theory, including Maxwell's equations and the wave equation, introductory physics, and differential and integral calculus. Mathematical techniques required for understanding some advanced topics in the later chapters are incorporated in the individual chapters or are included as appendices.

The book, since its first edition in 1982 and subsequent two editions in 1997 and 2005, has been a pacesetter and trail blazer in updating the contents to keep abreast with advancements in antenna technology. This has been accomplished by:

- Introducing new topics
- Originating innovative features and multimedia to animate, visualize, illustrate and display radiation characteristics
- Providing design equations, procedures and associate software

This edition is no exception, as many new topics and features have been added. In particular:

- New sections have been introduced on:
  1. Flexible and conformal bowtie
  2. Vivaldi antenna
  3. Antenna miniaturization
  4. Antennas for mobile communications
  5. Dielectric resonator antennas
  6. Scale modeling
- Additional MATLAB and JAVA programs have been developed.
- Color and gray scale figures and illustrations have been developed to clearly display and visualize antenna radiation characteristics.
- A companion website has been structured by the publisher which houses the MATLAB programs, JAVA-based applets and animations, Power Point notes, and JAVA-based interactive questionnaires. A solutions manual is available only for the instructors that adopt the book as a classroom text.
- Over 100 additional end-of-chapter problems have been included.

While incorporating the above new topics and features in the current edition, the book maintained all of the attractive features of the first three additions, especially the:

- Three-dimensional graphs to display the radiation characteristics of antennas. This feature was hailed, at the time of its introduction, as innovative and first of its kind addition in a textbook on antennas.

- Advanced topics, such as a chapter on *Smart Antennas* and a section on *Fractal Antennas*.
- Multimedia:
  1. Power Point notes
  2. MATLAB programs
  3. FORTRAN programs
  4. JAVA-based animations
  5. JAVA-based applets
  6. JAVA-based end-of-the-chapter questionnaires

The book's main objective is to introduce, in a unified manner, the fundamental principles of antenna theory and to apply them to the analysis, design, and measurements of antennas. Because there are so many methods of analysis and design and a plethora of antenna structures, applications are made to some of the most basic and practical configurations, such as linear dipoles; loops; arrays; broadband, and frequency-independent antennas; aperture antennas; horn antennas; microstrip antennas; and reflector antennas.

A tutorial chapter on Smart Antennas is included to introduce the student in a technology that will advance antenna theory and design, and revolutionize wireless communications. It is based on antenna theory, digital signal processing, networks and communications. MATLAB simulation software has also been included, as well as a plethora of references for additional reading.

Introductory material on analytical methods, such as the Moment Method and Fourier transform (spectral) technique, is also included. These techniques, together with the fundamental principles of antenna theory, can be used to analyze and design almost any antenna configuration. A chapter on antenna measurements introduces state-of-the-art methods used in the measurements of the most basic antenna characteristics (pattern, gain, directivity, radiation efficiency, impedance, current, and polarization) and updates progress made in antenna instrumentation, antenna range design, and scale modeling. Techniques and systems used in near- to far-field measurements and transformations are also discussed.

A sufficient number of topics have been covered, some for the first time in an undergraduate text, so that the book will serve not only as a text but also as a reference for the practicing and design engineer and even the amateur radio buff. These include design procedures, and associated computer programs, for Yagi–Uda and log-periodic arrays, horns, and microstrip patches; synthesis techniques using the Schelkunoff, Fourier transform, Woodward–Lawson, Tschebyscheff, and Taylor methods; radiation characteristics of corrugated, aperture-matched, and multimode horns; analysis and design of rectangular and circular microstrip patches; and matching techniques such as the binomial and Tschebyscheff. Also new sections have been introduced on flexible & conformal bowtie and Vivaldi antennas in Chapter 9, antenna miniaturization in Chapter 11 and expanded scale modeling in Chapter 17.

Chapter 14 has been expanded to include antennas for Mobile Communications. In particular, this new section includes basic concepts and design equations for the Planar Inverted-F Antenna (PIFA), Slot Antenna, Inverted-F Antenna (IFA), Multiband U-type Slot Antenna, and Dielectric Resonator Antennas (DRAs). These are popular internal antennas for mobile devices (smart phones, laptops, pads, tablets, etc.). A MATLAB computer program, referred to as **DRA_Analysis_Design**, has been developed to analyze the resonant frequencies of Rectangular, Cylindrical, Hemicylindrical, and Hemispherical DRAs using TE and TM modal cavity techniques by modeling the walls as PMCs. Hybrid modes are used to analyze and determine the resonant frequencies and quality factor (Q) of the Cylindrical DRA. The MATLAB program **DRA_Analysis_Design** has the capability, using a nonlinear solver, to design (i.e., find the Q, range of values for the dielectric constant, and finally the dimensions of the Cylindrical DRA) once the hybrid mode ($TE_{01\delta}$, $TM_{01\delta}$ or $HE_{11\delta}$), fractional bandwidth (BW, in %), VSWR and resonant frequency (f_r, in GHz) are specified. A detailed procedure to follow the design is outlined in Section 14.10.4.

The text contains sufficient mathematical detail to enable the average undergraduate electrical engineering and physics students to follow, without difficulty, the flow of analysis and design. A certain amount of analytical detail, rigor, and thoroughness allows many of the topics to be traced to their origin. My experiences as a student, engineer, and teacher have shown that a text for this course must not be a book of unrelated formulas, and it must not resemble a "cookbook." This book begins with the most elementary material, develops underlying concepts needed for sequential topics, and progresses to more advanced methods and system configurations. Each chapter is subdivided into sections or subsections whose individual headings clearly identify the antenna characteristic(s) discussed, examined, or illustrated.

A distinguished feature of this book is its three-dimensional graphical illustrations from the first edition, which have been expanded and supplemented in the second, third and fourth editions. In the past, antenna texts have displayed the three-dimensional energy radiated by an antenna by a number of separate two-dimensional patterns. With the advent and revolutionary advances in digital computations and graphical displays, an additional dimension has been introduced for the first time in an undergraduate antenna text by displaying the radiated energy of a given radiator by a single three-dimensional graphical illustration. Such an image, formed by the graphical capabilities of the computer and available at most computational facilities, gives a clear view of the energy radiated in all space surrounding the antenna. In this fourth edition, almost all of the three-dimensional amplitude radiation patterns, along with many two-dimensional graphs, are depicted in color and gray-scale. This is a new and pacesetting feature adopted, on a large scale, in this edition. It is hoped that this will lead to a better understanding of the underlying principles of radiation and provide a clearer visualization of the pattern formation in all space.

In addition, there is an abundance of general graphical illustrations, design data, references, and an expanded list of end-of-the chapter problems. Many of the principles are illustrated with examples, graphical illustrations, and physical arguments. Although students are often convinced that they understand the principles, difficulties arise when they attempt to use them. An example, especially a graphical illustration, can often better illuminate those principles. As they say, "a picture is worth a thousand words."

Numerical techniques and computer solutions are illustrated and encouraged. A number of MATLAB computer programs are included in the publisher's website for the book. Each program is interactive and prompts the user to enter the data in a sequential manner. Some of these programs are translations of the FORTRAN ones that were included in the first and second editions. However, many new ones have been developed. Every chapter, other than Chapters 3 and 17, has at least one MATLAB computer program; some have as many as four. The outputs of the MATLAB programs include graphical illustrations and tabulated results. For completeness, the FORTRAN computer programs are also included, although nowadays there is not as much interest in them. The computer programs can be used for analysis and design. Some of them are more of the design type while some of the others are of the analysis type. Associated with each program there is a READ ME file, which summarizes the respective program.

The purpose of the Power Point Lecture Notes is to provide the instructors a copy of the text figures and some of the most important equations of each chapter. They can be used by the instructors in their lectures but may be supplemented with additional narratives. The students can use them to listen to the instructors' lectures, without having to take detailed notes, but can supplement them in the margins with annotations from the lectures. Each instructor will use the notes in a different way.

The Interactive Questionnaires are intended as reviews of the material in each chapter. The student can use them to review for tests, exams, and so on. For each question, there are three possible answers, but only one is correct. If the reader chooses one of them and it the correct answer, it will so indicate. However, if the chosen answer is the wrong one, the program will automatically indicate the correct answer. An explanation button is provided, which gives a short narrative on the correct answer or indicates where in the book the correct answer can be found.

The Animations can be used to illustrate some of the radiation characteristics, such as amplitude patterns, of some antenna types, like line sources, dipoles, loops, arrays, and horns. The Applets cover more chapters and can be used to examine some of the radiation characteristics (such as amplitude patterns, impedance, bandwidth, etc.) of some of the antennas. This can be accomplished very rapidly without having to resort to the MATLAB programs, which are more detailed.

For course use, the text is intended primarily for a two-semester (or two- or three-quarter) sequence in antenna theory. The first course should be given at the senior undergraduate level, and should cover most of the material in Chapters 1 through 7, and some sections of Chapters 14, 16 and 17. The material in Chapters 8 through 16 should be covered in detail in a beginning graduate-level course. Selected chapters and sections from the book can be covered in a single semester, without loss of continuity. However, it is essential that most of the material in Chapters 2 through 6 be covered in the first course and before proceeding to any more advanced topics. To cover all the material of the text in the proposed time frame would be, in some cases, an ambitious and challenging task. Sufficient topics have been included, however, to make the text complete and to give the teacher the flexibility to emphasize, deemphasize, or omit sections or chapters. Some of the chapters and sections can be omitted without loss of continuity.

In the entire book, an $e^{j\omega t}$ time variation is assumed, and it is suppressed. The International System of Units, which is an expanded form of the rationalized MKS system, is used in the text. In some cases, the units of length are in meters (or centimeters) and in feet (or inches). Numbers in parentheses () refer to equations, whereas those in brackets [] refer to references. For emphasis, the most important equations, once they are derived, are boxed. In some of the basic chapters, the most important equations are summarized in tables.

I will like to acknowledge the invaluable suggestions from all those that contributed to the first three additions of the book, too numerous to mention here. Their names and contributions are stated in the respective editions. It is my pleasure to acknowledge the suggestions of the reviewers for the fourth edition: Dr. Stuart A. Long of the University of Houston, Dr. Leo Kempel of Michigan State University, and Dr. Cynthia M. Furse of the University of Utah. There have been other contributors to this edition, and their contributions are valued and acknowledged. Many graduate and undergraduate students at Arizona State University have written and verified most of the MATLAB computer programs; some of these programs were translated from FORTRAN, which appeared in the first three editions and updated for the fourth edition. However some new MATLAB and JAVA programs have been created, which are included for the first time in the fourth edition. I am indebted to Dr. Alix Rivera-Albino who developed with special care all of the color and gray scale figures and illustrations for the fourth edition and contributed to the manuscript and figures for the Vivaldi and mobile antennas. The author also acknowledges Dr. Razib S. Shishir of Intel, formerly of Arizona State University, for the JAVA-based software for the third edition, including the Interactive Questionnaires, Applets and Animations. These have been supplemented with additional ones for the fourth edition. Many thanks to Dr. Stuart A. Long, from the University of Houston, for reviewing the section on DRAs and Dr. Christos Christodoulou, from the University of New Mexico, for reviewing the manuscript on antennas for mobile devices, Dr. Peter J. Bevelacqua of Google for material related to planar antennas for mobile units, Dr. Arnold Mckinley of University College London (formerly with the Australian National University) for information and computer program related to nonuniform loop antennas, Dr. Steven R. Best of Mitre Corporation for figures on the folded spherical helix, Dr. Edward J. Rothwell, from Michigan State University, for antenna miniaturization information, Dr. Seong-Ook Park of the Korea Advanced Institute of Science and Technology (KAIST), for the photo and permission of the U-slot antenna, and Dr. Yahia Antar and Dr. Jawad Y. Siddiqui, both from the Royal Military College of Canada, for information related to cylindrical DRAs. I would also like to thank Craig R. Birtcher, and my graduate students Dr. Ahmet C. Durgun (now with Intel), Dr. Nafati Aboserwal (now at the University of Oklahoma), Sivaseetharaman Pandi, Mikal Askarian Amiri, Wengang Chen, Saud Saeed and Anuj Modi, all of Arizona State University, for proofreading of the manuscript and many other contributions to the fourth edition.

Special thanks to the companies that contributed photos, illustrations and copyright permissions for the third edition. However, other companies, Samsung, Microsoft and HTC have provided updated photos of their respective smart phones for the fourth edition.

During my 50+ year professional career, I have made many friends and professional colleagues. The list is too long to be included here, as I fear that I may omit someone. Thank you for your friendship, collegiality and comradery. I will like to recognize George C. Barber, Dennis DeCarlo and the entire membership (members, government agencies and companies) of the Advanced Helicopter Electromagnetics (AHE) Program for the 25 years of interest and support. It has been an unprecedented professional partnership and collaboration. To all my teachers and mentors, thank you. You have been my role models and inspiration.

This journey got started in the middle to the late 1970s, at the early stages of my academic career. Many may speculate why I have chosen to remain as the sole author and steward for so many years, dating back to first edition in 1982 and then through the subsequent three editions of this book and two editions of the *Advanced Engineering Electromagnetics* book. I wanted, as long as I was able to accomplish the tasks, to have the books manifest my own fingerprint and reflect my personal philosophy, methodology and pedagogy. Also I wanted the manuscript to display continuity and consistency, and to control my own destiny, in terms of material to be included and excluded, revisions, deadlines and timelines. Finally, I wanted to be responsible for the contents of the book. In the words of Frank Sinatra, 'I did it my way.' Each edition presented its own challenges, but each time I cherished and looked forward to the mission and venture.

I am also grateful to the staff of John Wiley & Sons, Inc., especially Brett Kurzman, Editor, Alex Castro, Editorial Assistant, and Danielle LaCourciere, Production Editor for this edition. Special thanks to Shikha Sharma, from Aptara, Inc., for supervising the typesetting of the book. Finally I must pay tribute and homage to my family (Helen, Renie, Stephanie, Bill, Pete and Ellie) for their unconditional support, patience, sacrifice, and understanding for the many years of neglect during the completion of all four editions of this book and two editions of the *Advanced Engineering Electromagnetics*. Each edition has been a pleasant experience although a daunting task.

Constantine A. Balanis
Arizona State University
Tempe, AZ

About the Companion Website

There is a student companion website that contains:

- PowerPoint Viewgraphs
- MATLAB Programs
- JAVA Applets
- Animations
- End-of-Chapter Interactive Questionnaires

To access material on the student companion site, simply visit
http://bcs.wiley.com/he-bcs/Books?action=index&bcsId=11764&itemId=1118642066.

This book is also accompanied by a password protected companion website for instructors only. This website contains:

- Power Point Viewgraphs
- MATLAB Programs
- JAVA Applets
- Animations
- End-of-Chapter Interactive Questionnaires
- Solutions Manual

Visit the link below for instructions on how to obtain a password.
http://bcs.wiley.com/he-bcs/Books?action=index&itemId=1118642066&bcsId=9777.

CHAPTER 1

Antennas

1.1 INTRODUCTION

An antenna is defined by Webster's Dictionary as "a usually metallic device (as a rod or wire) for radiating or receiving radio waves." The *IEEE Standard Definitions of Terms for Antennas* (IEEE Std 145–1983)* defines the antenna or aerial as "a means for radiating or receiving radio waves." In other words the antenna is the transitional structure between free-space and a guiding device, as shown in Figure 1.1. The guiding device or transmission line may take the form of a coaxial line or a hollow pipe (waveguide), and it is used to transport electromagnetic energy from the transmitting source to the antenna, or from the antenna to the receiver. In the former case, we have a *transmitting* antenna and in the latter a *receiving* antenna.

A transmission-line Thevenin equivalent of the antenna system of Figure 1.1 in the transmitting mode is shown in Figure 1.2 where the source is represented by an ideal generator with impedance Z_g, the transmission line is represented by a line with characteristic impedance Z_c, and the antenna is represented by a load Z_A [$Z_A = (R_L + R_r) + jX_A$] connected to the transmission line. The Thevenin and Norton circuit equivalents of the antenna are also shown in Figure 2.27. The load resistance R_L is used to represent the conduction and dielectric losses associated with the antenna structure while R_r, referred to as the radiation resistance, is used to represent radiation by the antenna. The reactance X_A is used to represent the imaginary part of the impedance associated with radiation by the antenna. This is discussed more in detail in Sections 2.13 and 2.14. Under ideal conditions, energy generated by the source should be totally transferred to the radiation resistance R_r, which is used to represent radiation by the antenna. However, in a practical system there are conduction-dielectric losses due to the lossy nature of the transmission line and the antenna, as well as those due to reflections (mismatch) losses at the interface between the line and the antenna. Taking into account the internal impedance of the source and neglecting line and reflection (mismatch) losses, maximum power is delivered to the antenna under conjugate matching. This is discussed in Section 2.13.

The reflected waves from the interface create, along with the traveling waves from the source toward the antenna, constructive and destructive interference patterns, referred to as standing waves, inside the transmission line which represent pockets of energy concentrations and storage, typical of resonant devices. A typical standing wave pattern is shown dashed in Figure 1.2, while another is exhibited in Figure 1.15. If the antenna system is not properly designed, the transmission line

---

IEEE Transactions on Antennas and Propagation, vols. AP-17, No. 3, May 1969; AP-22, No. 1, January 1974; and AP-31, No. 6, Part II, November 1983.

Antenna Theory: Analysis and Design Fourth Edition. Constantine A. Balanis.
© 2016 John Wiley & Sons, Inc. Published 2016 by John Wiley & Sons, Inc.
Companion Website: www.wiley.com/go/antennatheory4e

2 ANTENNAS

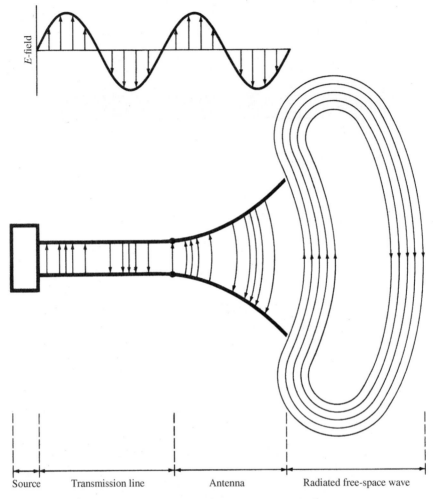

Figure 1.1 Antenna as a transition device.

could act to a large degree as an energy storage element instead of as a wave guiding and energy transporting device. If the maximum field intensities of the standing wave are sufficiently large, they can cause arching inside the transmission lines.

The losses due to the line, antenna, and the standing waves are undesirable. The losses due to the line can be minimized by selecting low-loss lines while those of the antenna can be decreased by

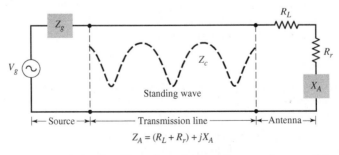

Figure 1.2 Transmission-line Thevenin equivalent of antenna in transmitting mode.

reducing the loss resistance represented by R_L in Figure 1.2. The standing waves can be reduced, and the energy storage capacity of the line minimized, by matching the impedance of the antenna (load) to the characteristic impedance of the line. This is the same as matching loads to transmission lines, where the load here is the antenna, and is discussed more in detail in Section 9.7. An equivalent similar to that of Figure 1.2 is used to represent the antenna system in the receiving mode where the source is replaced by a receiver. All other parts of the transmission-line equivalent remain the same. The radiation resistance R_r is used to represent in the receiving mode the transfer of energy from the free-space wave to the antenna. This is discussed in Section 2.13 and represented by the Thevenin and Norton circuit equivalents of Figure 2.27.

In addition to receiving or transmitting energy, an antenna in an advanced wireless system is usually required to *optimize* or *accentuate* the radiation energy in some directions and suppress it in others. *Thus the antenna must also serve as a directional device in addition to a probing device*. It must then take various forms to meet the particular need at hand, and it may be a piece of conducting wire, an aperture, a patch, an assembly of elements (array), a reflector, a lens, and so forth.

For wireless communication systems, the antenna is one of the most critical components. A good design of the antenna can relax system requirements and improve overall system performance. A typical example is the TV for which the overall broadcast reception can be improved by utilizing a high-performance antenna. The antenna serves to a communication system the same purpose that eyes and eyeglasses serve to a human.

The field of antennas is vigorous and dynamic, and over the last 60 years antenna technology has been an indispensable partner of the communications revolution. Many major advances that occurred during this period are in common use today; however, many more issues and challenges are facing us today, especially since the demands for system performances are even greater. Many of the major advances in antenna technology that have been completed in the 1970s through the early 1990s, those that were under way in the early 1990s, and signals of future discoveries and breakthroughs were captured in a special issue of the *Proceedings of the IEEE* (Vol. 80, No. 1, January 1992) devoted to Antennas. The introductory paper of this special issue [1] provides a carefully structured, elegant discussion of the fundamental principles of radiating elements and has been written as an introduction for the nonspecialist and a review for the expert.

1.2 TYPES OF ANTENNAS

We will now introduce and briefly discuss some forms of the various antenna types in order to get a glance as to what will be encountered in the remainder of the book.

1.2.1 Wire Antennas

Wire antennas are familiar to the layman because they are seen virtually everywhere—on automobiles, buildings, ships, aircraft, spacecraft, and so on. There are various shapes of wire antennas such as a straight wire (dipole), loop, and helix which are shown in Figure 1.3. Loop antennas need not only be circular. They may take the form of a rectangle, square, ellipse, or any other configuration. The circular loop is the most common because of its simplicity in construction. Dipoles are discussed in more detail in Chapter 4, loops in Chapter 5, and helices in Chapter 10.

1.2.2 Aperture Antennas

Aperture antennas may be more familiar to the layman today than in the past because of the increasing demand for more sophisticated forms of antennas and the utilization of higher frequencies. Some forms of aperture antennas are shown in Figure 1.4. Antennas of this type are very useful for aircraft and spacecraft applications, because they can be very conveniently flush-mounted on the skin of

4 ANTENNAS

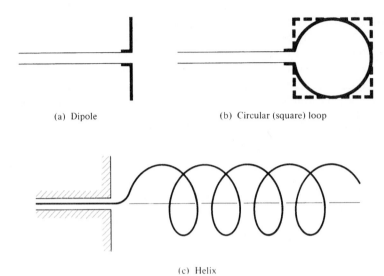

(a) Dipole

(b) Circular (square) loop

(c) Helix

Figure 1.3 Wire antenna configurations.

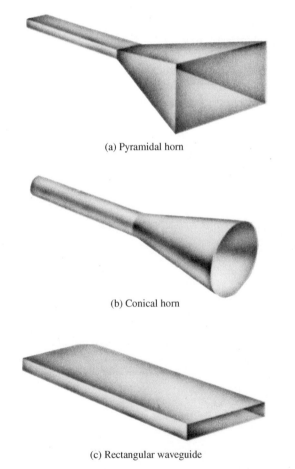

(a) Pyramidal horn

(b) Conical horn

(c) Rectangular waveguide

Figure 1.4 Aperture antenna configurations.

the aircraft or spacecraft. In addition, they can be covered with a dielectric material to protect them from hazardous conditions of the environment. Waveguide apertures are discussed in more detail in Chapter 12 while horns are examined in Chapter 13.

1.2.3 Microstrip Antennas

Microstrip antennas became very popular in the 1970s primarily for spaceborne applications. Today they are used for government and commercial applications. These antennas consist of a metallic patch on a grounded substrate. The metallic patch can take many different configurations, as shown in Figure 14.2. However, the rectangular and circular patches, shown in Figure 1.5, are the most popular because of ease of analysis and fabrication, and their attractive radiation characteristics, especially low cross-polarization radiation. The microstrip antennas are low profile, comfortable to planar and nonplanar surfaces, simple and inexpensive to fabricate using modern printed-circuit technology, mechanically robust when mounted on rigid surfaces, compatible with MMIC designs, and very versatile in terms of resonant frequency, polarization, pattern, and impedance. These antennas can be mounted on the surface of high-performance aircraft, spacecraft, satellites, missiles, cars, and even mobile devices. They are discussed in more detail in Chapter 14.

1.2.4 Array Antennas

Many applications require radiation characteristics that may not be achievable by a single element. It may, however, be possible that an aggregate of radiating elements in an electrical and geometrical

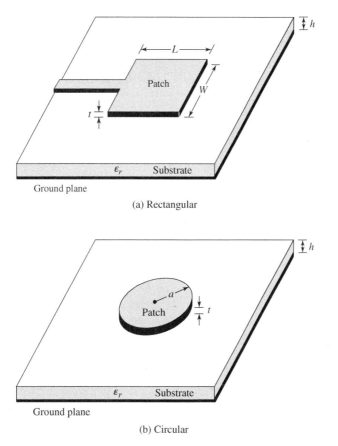

Figure 1.5 Rectangular and circular microstrip (patch) antennas.

6 ANTENNAS

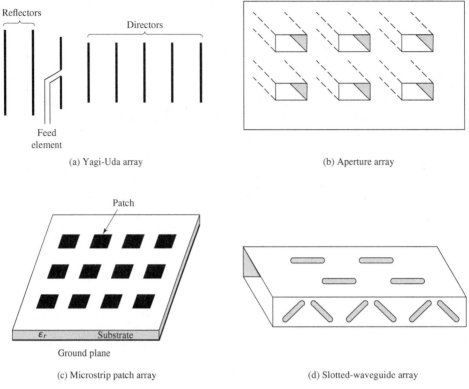

Figure 1.6 Typical wire, aperture, and microstrip array configurations.

arrangement (*an array*) will result in the desired radiation characteristics. The arrangement of the array may be such that the radiation from the elements adds up to give a radiation maximum in a particular direction or directions, minimum in others, or otherwise as desired. Typical examples of arrays are shown in Figure 1.6. Usually the term *array* is reserved for an arrangement in which the individual radiators are separate as shown in Figures 1.6(a–c). However the same term is also used to describe an assembly of radiators mounted on a continuous structure, shown in Figure 1.6(d).

1.2.5 Reflector Antennas

The success in the exploration of outer space has resulted in the advancement of antenna theory. Because of the need to communicate over great distances, sophisticated forms of antennas had to be used in order to transmit and receive signals that had to travel millions of miles. A very common antenna form for such an application is a parabolic reflector shown in Figures 1.7(a) and (b). Antennas of this type have been built with diameters of 305 m or even larger. Such large dimensions are needed to achieve the high gain required to transmit or receive signals after millions of miles of travel. Another form of a reflector, although not as common as the parabolic, is the corner reflector, shown in Figure 1.7(c). These antennas are examined in detail in Chapter 15.

1.2.6 Lens Antennas

Lenses are primarily used to collimate incident divergent energy to prevent it from spreading in undesired directions. By properly shaping the geometrical configuration and choosing the appropriate material of the lenses, they can transform various forms of divergent energy into plane waves. They can be used in most of the same applications as are the parabolic reflectors, especially at

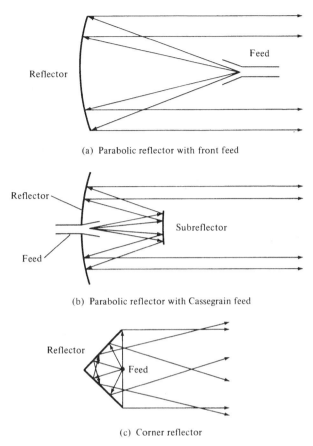

Figure 1.7 Typical reflector configurations.

higher frequencies. Their dimensions and weight become exceedingly large at lower frequencies. Lens antennas are classified according to the material from which they are constructed, or according to their geometrical shape. Some forms are shown in Figure 1.8 [2].

In summary, an ideal antenna is one that will radiate all the power delivered to it from the transmitter in a desired direction or directions. In practice, however, such ideal performances cannot be achieved but may be closely approached. Various types of antennas are available and each type can take different forms in order to achieve the desired radiation characteristics for the particular application. Throughout the book, the radiation characteristics of most of these antennas are discussed in detail.

1.3 RADIATION MECHANISM

One of the first questions that may be asked concerning antennas would be "how is radiation accomplished?" In other words, how are the electromagnetic fields generated by the source, contained and guided within the transmission line and antenna, and finally "detached" from the antenna to form a free-space wave? The best explanation may be given by an illustration. However, let us first examine some basic sources of radiation.

1.3.1 Single Wire

Conducting wires are material whose prominent characteristic is the motion of electric charges and the creation of current. Let us assume that an electric volume charge density, represented by q_v

8 ANTENNAS

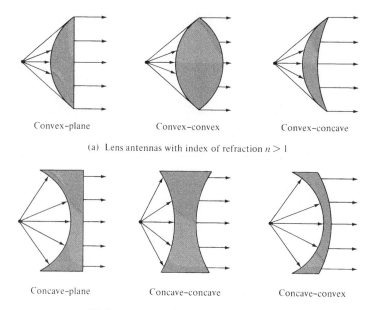

(a) Lens antennas with index of refraction $n > 1$

(b) Lens antennas with index of refraction $n < 1$

Figure 1.8 Typical lens antenna configurations. (SOURCE: L. V. Blake, *Antennas*, Wiley, New York, 1966).

(coulombs/m³), is distributed uniformly in a circular wire of cross-sectional area A and volume V, as shown in Figure 1.9. The total charge Q within volume V is moving in the z direction with a uniform velocity v_z (meters/sec). It can be shown that the current density J_z (amperes/m²) over the cross section of the wire is given by [3]

$$J_z = q_v v_z \qquad (1\text{-}1\text{a})$$

If the wire is made of an ideal electric conductor, the current density J_s (amperes/m) resides on the surface of the wire and it is given by

$$J_s = q_s v_z \qquad (1\text{-}1\text{b})$$

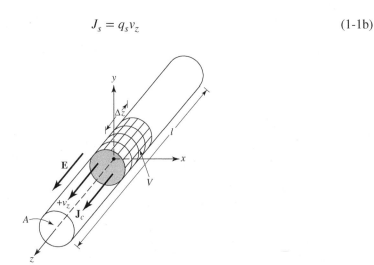

Figure 1.9 Charge uniformly distributed in a circular cross section cylinder wire.

where q_s (coulombs/m^2) is the surface charge density. If the wire is very thin (ideally zero radius), then the current in the wire can be represented by

$$I_z = q_l v_z \tag{1-1c}$$

where q_l (coulombs/m) is the charge per unit length.

Instead of examining all three current densities, we will primarily concentrate on the very thin wire. The conclusions apply to all three. If the current is time varying, then the derivative of the current of (1-1c) can be written as

$$\frac{dI_z}{dt} = q_l \frac{dv_z}{dt} = q_l a_z \tag{1-2}$$

where $dv_z/dt = a_z$ (meters/sec^2) is the acceleration. If the wire is of length l, then (1-2) can be written as

$$l\frac{dI_z}{dt} = lq_l \frac{dv_z}{dt} = lq_l a_z \tag{1-3}$$

Equation (1-3) is the basic relation between current and charge, and it also serves as the fundamental relation of electromagnetic radiation [4], [5]. It simply states that *to create radiation, there must be a time-varying current or an acceleration (or deceleration) of charge*. We usually refer to currents in time-harmonic applications while charge is most often mentioned in transients. To create charge acceleration (or deceleration) the wire must be curved, bent, discontinuous, or terminated [1], [4]. Periodic charge acceleration (or deceleration) or time-varying current is also created when charge is oscillating in a time-harmonic motion, as shown in Figure 1.17 for a $\lambda/2$ dipole. Therefore:

1. If a charge is not moving, current is not created and there is no radiation.
2. If charge is moving with a uniform velocity:
   a. There is no radiation if the wire is straight, and infinite in extent.
   b. There is radiation if the wire is curved, bent, discontinuous, terminated, or truncated, as shown in Figure 1.10.
3. If charge is oscillating in a time-motion, it radiates even if the wire is straight.

A qualitative understanding of the radiation mechanism may be obtained by considering a pulse source attached to an open-ended conducting wire, which may be connected to the ground through a discrete load at its open end, as shown in Figure 1.10(d). When the wire is initially energized, the charges (free electrons) in the wire are set in motion by the electrical lines of force created by the source. When charges are accelerated in the source-end of the wire and decelerated (negative acceleration with respect to original motion) during reflection from its end, it is suggested that radiated fields are produced at each end and along the remaining part of the wire, [1], [4]. *Stronger radiation with a more broad frequency spectrum occurs if the pulses are of shorter or more compact duration while continuous time-harmonic oscillating charge produces, ideally, radiation of single frequency determined by the frequency of oscillation.* The acceleration of the charges is accomplished by the external source in which forces set the charges in motion and produce the associated field radiated. The deceleration of the charges at the end of the wire is accomplished by the internal (self) forces associated with the induced field due to the buildup of charge concentration at the ends of the wire. The internal forces receive energy from the charge buildup as its velocity is reduced to zero at the ends of the wire. Therefore, charge acceleration due to an exciting electric field and deceleration due

10 ANTENNAS

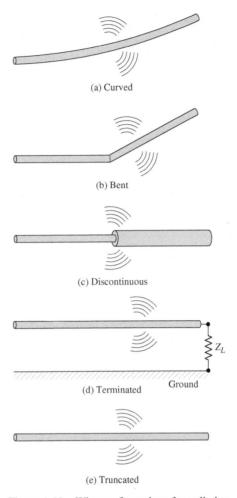

Figure 1.10 Wire configurations for radiation.

to impedance discontinuities or smooth curves of the wire are mechanisms responsible for electromagnetic radiation. While both current density ($\mathbf{J}_c$) and charge density (q_v) appear as source terms in Maxwell's equation, charge is viewed as a more fundamental quantity, especially for transient fields. Even though this interpretation of radiation is primarily used for transients, it can be used to explain steady state radiation [4].

1.3.2 Two-Wires

Let us consider a voltage source connected to a two-conductor transmission line which is connected to an antenna. This is shown in Figure 1.11(a). Applying a voltage across the two-conductor transmission line creates an electric field between the conductors. The electric field has associated with it electric lines of force which are tangent to the electric field at each point and their strength is proportional to the electric field intensity. The electric lines of force have a tendency to act on the free electrons (easily detachable from the atoms) associated with each conductor and force them to be displaced. The movement of the charges creates a current that in turn creates a magnetic field intensity. Associated with the magnetic field intensity are magnetic lines of force which are tangent to the magnetic field.

RADIATION MECHANISM

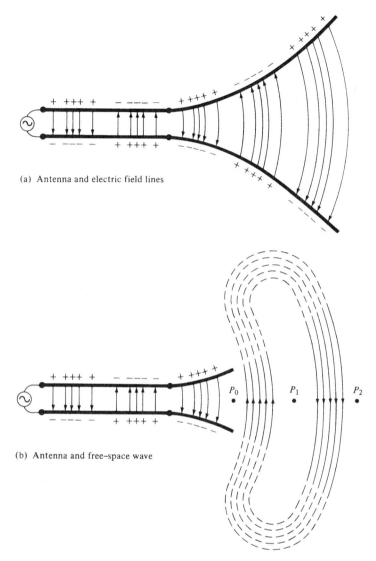

Figure 1.11 Source, transmission line, antenna, and detachment of electric field lines.

We have accepted that electric field lines start on positive charges and end on negative charges. They also can start on a positive charge and end at infinity, start at infinity and end on a negative charge, or form closed loops neither starting or ending on any charge. Magnetic field lines always form closed loops encircling current-carrying conductors because physically there are no magnetic charges. In some mathematical formulations, it is often convenient to introduce equivalent magnetic charges and magnetic currents to draw a parallel between solutions involving electric and magnetic sources.

The electric field lines drawn between the two conductors help to exhibit the distribution of charge. If we assume that the voltage source is sinusoidal, we expect the electric field between the conductors to also be sinusoidal with a period equal to that of the applied source. The relative magnitude of the electric field intensity is indicated by the density (bunching) of the lines of force with the arrows showing the relative direction (positive or negative). The creation of time-varying electric and magnetic fields between the conductors forms electromagnetic waves which travel along the transmission line, as shown in Figure 1.11(a). The electromagnetic waves enter the antenna and have associated

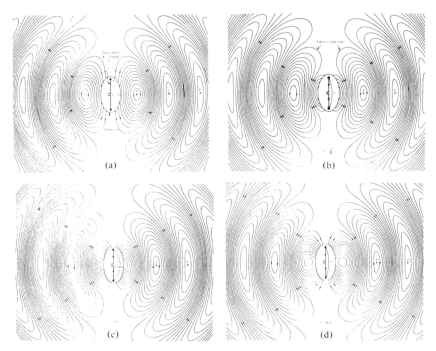

Figure 1.12 Electric field lines of free-space wave for a $\lambda/2$ antenna at $t = 0$, $T/8$, $T/4$, and $3T/8$. (SOURCE: J. D. Kraus, *Electromagnetics*, 4th ed., McGraw-Hill, New York, 1992. Reprinted with permission of J. D. Kraus and John D. Cowan, Jr.).

with them electric charges and corresponding currents. If we remove part of the antenna structure, as shown in Figure 1.11(b), free-space waves can be formed by "connecting" the open ends of the electric lines (shown dashed). The free-space waves are also periodic but a constant phase point P_0 moves outwardly with the speed of light and travels a distance of $\lambda/2$ (to P_1) in the time of one-half of a period. It has been shown [6] that close to the antenna the constant phase point P_0 moves faster than the speed of light but approaches the speed of light at points far away from the antenna (analogous to phase velocity inside a rectangular waveguide). Figure 1.12 displays the creation and travel of free-space waves by a prolate spheroid with $\lambda/2$ interfocal distance where λ is the wavelength. The free-space waves of a center-fed $\lambda/2$ dipole, except in the immediate vicinity of the antenna, are essentially the same as those of the prolate spheroid.

The question still unanswered is how the guided waves are detached from the antenna to create the free-space waves that are indicated as closed loops in Figures 1.11 and 1.12. Before we attempt to explain that, let us draw a parallel between the guided and free-space waves, and water waves [7] created by the dropping of a pebble in a calm body of water or initiated in some other manner. Once the disturbance in the water has been initiated, water waves are created which begin to travel outwardly. If the disturbance has been removed, the waves do not stop or extinguish themselves but continue their course of travel. If the disturbance persists, new waves are continuously created which lag in their travel behind the others. The same is true with the electromagnetic waves created by an electric disturbance. If the initial electric disturbance by the source is of a short duration, the created electromagnetic waves travel inside the transmission line, then into the antenna, and finally are radiated as free-space waves, even if the electric source has ceased to exist (as was with the water waves and their generating disturbance). If the electric disturbance is of a continuous nature, electromagnetic waves exist continuously and follow in their travel behind the others. This is shown in Figure 1.13 for a biconical antenna. When the electromagnetic waves are within the transmission line and antenna, their existence is associated with the presence of the charges inside the conductors. However, when the waves are radiated, they form closed loops and there are no charges to sustain

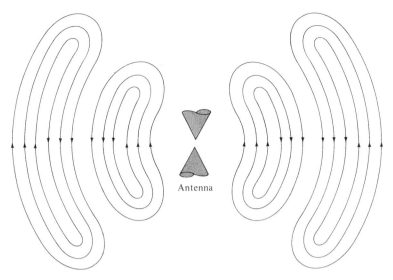

Figure 1.13 Electric field lines of free-space wave for biconical antenna.

their existence. *This leads us to conclude that electric charges are required to excite the fields but are not needed to sustain them and may exist in their absence. This is in direct analogy with water waves.*

1.3.3 Dipole

Now let us attempt to explain the mechanism by which the electric lines of force are detached from the antenna to form the free-space waves. This will again be illustrated by an example of a small dipole antenna where the time of travel is negligible. This is only necessary to give a better physical interpretation of the detachment of the lines of force. Although a somewhat simplified mechanism, it does allow one to visualize the creation of the free-space waves. Figure 1.14(a) displays the lines of force created between the arms of a small center-fed dipole in the first quarter of the period during which time the charge has reached its maximum value (assuming a sinusoidal time variation) and the lines have traveled outwardly a radial distance $\lambda/4$. For this example, let us assume that the number of lines formed are three. During the next quarter of the period, the original three lines travel an additional $\lambda/4$ (a total of $\lambda/2$ from the initial point) and the charge density on the conductors begins to diminish. This can be thought of as being accomplished by introducing opposite charges which at the end of the first half of the period have neutralized the charges on the conductors. The lines of force created by the opposite charges are three and travel a distance $\lambda/4$ during the second quarter of the first half, and they are shown dashed in Figure 1.14(b). The end result is that there are three lines of force pointed upward in the first $\lambda/4$ distance and the same number of lines directed downward in the second $\lambda/4$. Since there is no net charge on the antenna, then the lines of force must have been forced to detach themselves from the conductors and to unite together to form closed loops. This is shown in Figure 1.14(c). In the remaining second half of the period, the same procedure is followed but in the opposite direction. After that, the process is repeated and continues indefinitely and electric field patterns, similar to those of Figure 1.12, are formed.

1.3.4 Computer Animation-Visualization of Radiation Problems

A difficulty that students usually confront is that the subject of electromagnetics is rather abstract, and it is hard to visualize electromagnetic wave propagation and interaction. With today's advanced numerical and computational methods, and animation and visualization software and hardware, this dilemma can, to a large extent, be minimized. To address this problem, we have developed and

14 ANTENNAS

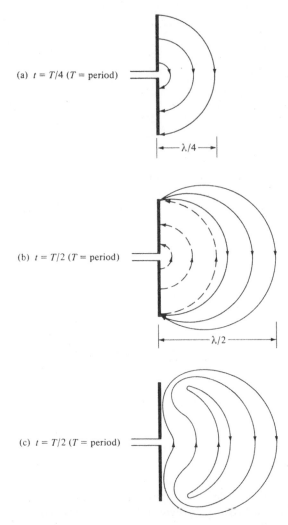

Figure 1.14 Formation and detachment of electric field lines for short dipole.

included in this chapter computer programs to animate and visualize three radiation mechanisms. Descriptions of the computer programs are found in the website created by the publisher for this book. Each problem is solved using the Finite-Difference Time-Domain (FD-TD) method [8]–[10], a method which solves Maxwell's equations as a function of time in discrete time steps at discrete points in space. A picture of the fields can then be taken at each time step to create a video which can be viewed as a function of time. Other animation and visualization software, referred to as *applets*, are included in the book website.

The three radiation problems that are animated and can be visualized using the computer program of this chapter and included in the book website are:

a. *Infinite length line source (two-dimensional) excited by a single Gaussian pulse and radiating in an unbounded medium.*
b. *Infinite length line source (two-dimensional) excited by a single Gaussian pulse and radiating inside a perfectly electric conducting (PEC) square cylinder.*
c. *E-plane sectoral horn (two-dimensional form of Figure 13.2) excited by a continuous cosinusoidal voltage source and radiating in an unbounded medium.*

In order to animate and then visualize each of the three radiation problems, the user needs *MATLAB* [11] and the *MATLAB M-file*, found in the publisher's website for the book, to produce the corresponding FD-TD solution of each radiation problem. For each radiation problem, the *M-File* executed in *MATLAB* produces a video by taking a picture of the computational domain every third time step. The video is viewed as a function of time as the wave travels in the computational space.

A. Infinite Line Source in an Unbounded Medium (**tm_open**)
The first FD-TD solution is that of an infinite length line source excited by a single time-derivative Gaussian pulse, with a duration of approximately 0.4 nanoseconds, in a two-dimensional TM^z-computational domain. The unbounded medium is simulated using a six-layer Berenger Perfectly Matched Layer (PML) Absorbing Boundary Condition (ABC) [9], [10] to truncate the computational space at a finite distance without, in principle, creating any reflections. Thus, the pulse travels radially outward creating a *traveling* type of a wavefront. The outward moving wavefronts are easily identified using the coloring scheme for the intensity (or gray scale for black and white monitors) when viewing the video. The video is created by the *MATLAB M-File* which produces the FD-TD solution by taking a picture of the computational domain every third time step. Each time step is 5 picoseconds while each FD-TD cell is 3 mm on a side. The video is 37 frames long covering 185 picoseconds of elapsed time. The entire computational space is 15.3 cm by 15.3 cm and is modeled by 2500 square FD-TD cells (50 × 50), including 6 cells to implement the PML ABC.

B. Infinite Line Source in a PEC Square Cylinder (**tm_box**)
This problem is simulated similarly as that of the line source in an unbounded medium, including the characteristics of the pulse. The major difference is that the computational domain of this problem is truncated by PEC walls; *therefore there is no need for PML ABC*. For this problem the pulse travels in an outward direction and is reflected when it reaches the walls of the cylinder. The reflected pulse, along with the radially outward traveling pulse, interfere constructively and destructively with each other and create a *standing* type of a wavefront. The peaks and valleys of the modified wavefront can be easily identified when viewing the video, using the colored or gray scale intensity schemes. Sufficient time is allowed in the video to permit the pulse to travel from the source to the walls of the cylinder, return back to the source, and then return back to the walls of the cylinder. Each time step is 5 picoseconds and each FD-TD cell is 3 mm on a side. The video is 70 frames long covering 350 picoseconds of elapsed time. The square cylinder, and thus the computational space, has a cross section of 15.3 cm by 15.3 cm and is modeled using an area 50 by 50 FD-TD cells.

C. E-Plane Sectoral Horn in an Unbounded Medium (**te_horn**)
The *E*-plane sectoral horn is excited by a cosinusoidal voltage (CW) of 9.84 GHz in a TE^z computational domain, instead of the Gaussian pulse excitation of the previous two problems. The unbounded medium is implemented using an eight-layer Berenger PML ABC. The computational space is 25.4 cm by 25.4 cm and is modeled using 100 by 100 FD-TD cells (each square cell being 2.54 mm on a side). The video is 70 frames long covering 296 picoseconds of elapsed time and is created by taking a picture every third frame. Each time step is 4.23 picoseconds in duration. The horn has a total flare angle of 52° and its flared section is 2.62 cm long, is fed by a parallel plate 1 cm wide and 4.06 cm long, and has an aperture of 3.56 cm.

1.4 CURRENT DISTRIBUTION ON A THIN WIRE ANTENNA

In the preceding section we discussed the movement of the free electrons on the conductors representing the transmission line and the antenna. In order to illustrate the creation of the current distribution on a linear dipole, and its subsequent radiation, let us first begin with the geometry of a

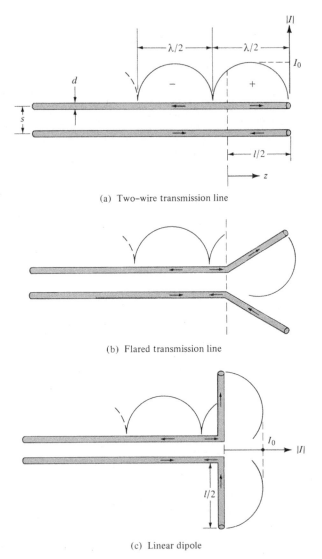

Figure 1.15 Current distribution on a lossless two-wire transmission line, flared transmission line, and linear dipole.

lossless two-wire transmission line, as shown in Figure 1.15(a). The movement of the charges creates a traveling wave current, of magnitude $I_0/2$, along each of the wires. When the current arrives at the end of each of the wires, it undergoes a complete reflection (equal magnitude and 180° phase reversal). The reflected traveling wave, when combined with the incident traveling wave, forms in each wire a pure standing wave pattern of sinusoidal form as shown in Figure 1.15(a). The current in each wire undergoes a 180° phase reversal between adjoining half-cycles. This is indicated in Figure 1.15(a) by the reversal of the arrow direction. Radiation from each wire individually occurs because of the time-varying nature of the current and the termination of the wire.

For the two-wire balanced (symmetrical) transmission line, the current in a half-cycle of one wire is of the same magnitude but 180° out-of-phase from that in the corresponding half-cycle of the other wire. If in addition the spacing between the two wires is very small ($s \ll \lambda$), the fields radiated by the current of each wire are essentially cancelled by those of the other. The net result is an almost ideal (and desired) nonradiating transmission line.

As the section of the transmission line between $0 \leq z \leq l/2$ begins to flare, as shown in Figure 1.15(b), it can be assumed that the current distribution is essentially unaltered in form in each of the wires. However, because the two wires of the flared section are not necessarily close to each other, the fields radiated by one do not necessarily cancel those of the other. Therefore, ideally, there is a net radiation by the transmission-line system.

Ultimately the flared section of the transmission line can take the form shown in Figure 1.15(c). This is the geometry of the widely used dipole antenna. Because of the standing wave current pattern, it is also classified as a standing wave antenna (as contrasted to the traveling wave antennas which will be discussed in detail in Chapter 10). If $l < \lambda$, the phase of the current standing wave pattern in each arm is the same throughout its length. In addition, spatially it is oriented in the same direction as that of the other arm as shown in Figure 1.15(c). Thus the fields radiated by the two arms of the dipole (vertical parts of a flared transmission line) will primarily reinforce each other toward most directions of observation (the phase due to the relative position of each small part of each arm must also be included for a complete description of the radiation pattern formation).

If the diameter of each wire is very small ($d \ll \lambda$), the ideal standing wave pattern of the current along the arms of the dipole is sinusoidal with a null at the end. However, its overall form depends on the length of each arm. For center-fed dipoles with $l \ll \lambda, l = \lambda/2, \lambda/2 < l < \lambda$ and $\lambda < l < 3\lambda/2$, the current patterns are illustrated in Figures 1.16(a–d). The current pattern of a very small dipole (usually $\lambda/50 < l \leq \lambda/10$) can be approximated by a triangular distribution since $\sin(kl/2) \simeq kl/2$ when $kl/2$ is very small. This is illustrated in Figure 1.16(a).

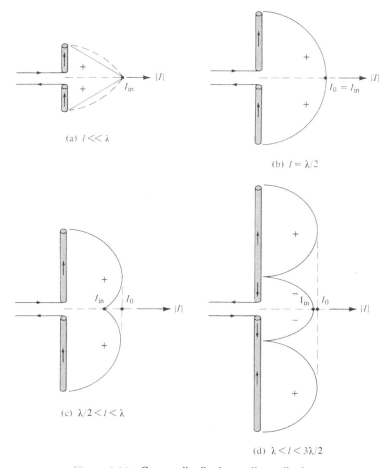

Figure 1.16 Current distribution on linear dipoles.

18 ANTENNAS

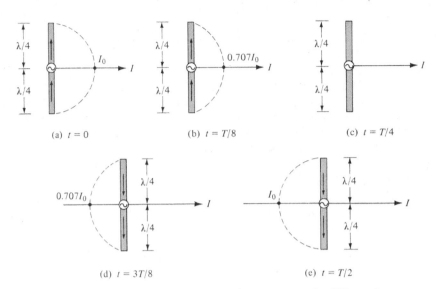

Figure 1.17 Current distribution on a $\lambda/2$ wire antenna for different times.

Because of its cyclical spatial variations, the current standing wave pattern of a dipole longer than $\lambda (l > \lambda)$ undergoes 180° phase reversals between adjoining half-cycles. Therefore the current in all parts of the dipole does not have the same phase. This is demonstrated graphically in Figure 1.16(d) for $\lambda < l < 3\lambda/2$. In turn, the fields radiated by some parts of the dipole will not reinforce those of the others. As a result, significant interference and cancelling effects will be noted in the formation of the total radiation pattern. See Figure 4.11 for the pattern of a $\lambda/2$ dipole and Figure 4.7 for that of a 1.25λ dipole.

For a time-harmonic varying system of radian frequency $\omega = 2\pi f$, the current standing wave patterns of Figure 1.16 represent the maximum current excitation for any time. The current variations, as a function of time, on a $\lambda/2$ center-fed dipole, are shown in Figure 1.17 for $0 \le t \le T/2$ where T is the period. These variations can be obtained by multiplying the current standing wave pattern of Figure 1.16(b) by $\cos(\omega t)$.

1.5 HISTORICAL ADVANCEMENT

The history of antennas [12] dates back to James Clerk Maxwell who unified the theories of electricity and magnetism, and eloquently represented their relations through a set of profound equations best known as *Maxwell's Equations*. His work was first published in 1873 [13]. He also showed that light was electromagnetic and that both light and electromagnetic waves travel by wave disturbances of the same speed. In 1886, Professor Heinrich Rudolph Hertz demonstrated the first wireless electromagnetic system. He was able to produce in his laboratory at a wavelength of 4 m a spark in the gap of a transmitting $\lambda/2$ dipole which was then detected as a spark in the gap of a nearby loop. It was not until 1901 that Guglielmo Marconi was able to send signals over large distances. He performed, in 1901, the first transatlantic transmission from Poldhu in Cornwall, England, to St. John's Newfoundland. His transmitting antenna consisted of 50 vertical wires in the form of a fan connected to ground through a spark transmitter. The wires were supported horizontally by a guyed wire between two 60-m wooden poles. The receiving antenna at St. John's was a 200-m wire pulled and supported by a kite. This was the dawn of the antenna era.

From Marconi's inception through the 1940s, antenna technology was primarily centered on wire related radiating elements and frequencies up to about UHF. It was not until World War II that modern antenna technology was launched and new elements (such as waveguide apertures, horns, reflectors) were primarily introduced. Much of this work is captured in the book by Silver [14]. A contributing factor to this new era was the invention of microwave sources (such as the klystron and magnetron) with frequencies of 1 GHz and above.

While World War II launched a new era in antennas, advances made in computer architecture and technology during the 1960s through the 1990s have had a major impact on the advance of modern antenna technology, and they are expected to have an even greater influence on antenna engineering into the twenty-first century. Beginning primarily in the early 1960s, numerical methods were introduced that allowed previously intractable complex antenna system configurations to be analyzed and designed very accurately. In addition, asymptotic methods for both low frequencies (e.g., Moment Method (MM), Finite-Difference, Finite-Element) and high frequencies (e.g., Geometrical and Physical Theories of Diffraction) were introduced, contributing significantly to the maturity of the antenna field. While in the past antenna design may have been considered a secondary issue in overall system design, today it plays a critical role. In fact, many system successes rely on the design and performance of the antenna. Also, while in the first half of this century antenna technology may have been considered almost a "cut and try" operation, today it is truly an engineering science. Analysis and design methods are such that antenna system performance can be predicted with remarkable accuracy. In fact, many antenna designs proceed directly from the initial design stage to the prototype without intermediate testing. The level of confidence has increased tremendously.

The widespread interest in antennas is reflected by the large number of books written on the subject [15]. These have been classified under four categories: Fundamental, Handbooks, Measurements, and Specialized. This is an outstanding collection of books, and it reflects the popularity of the antenna subject, especially since the 1950s. Because of space limitations, only a partial list is included here [2], [5], [7], [16]–[39], including the first, second and third editions of this book in 1982, 1997, 2005. Some of other books are now out of print.

1.5.1 Antenna Elements

Prior to World War II most antenna elements were of the wire type (long wires, dipoles, helices, rhombuses, fans, etc.), and they were used either as single elements or in arrays. During and after World War II, many other radiators, some of which may have been known for some and others of which were relatively new, were put into service. This created a need for better understanding and optimization of their radiation characteristics. Many of these antennas were of the aperture type (such as open-ended waveguides, slots, horns, reflectors, lenses), and they have been used for communication, radar, remote sensing, and deep space applications both on airborne and earth-based platforms. Many of these operate in the microwave region and are discussed in Chapters 12, 13, 15 and in [40].

Prior to the 1950s, antennas with broadband pattern and impedance characteristics had bandwidths not much greater than about 2:1. In the 1950s, a breakthrough in antenna evolution was created which extended the maximum bandwidth to as great as 40:1 or more. Because the geometries of these antennas are specified by angles instead of linear dimensions, they have ideally an infinite bandwidth. Therefore, they are referred to as *frequency independent*. These antennas are primarily used in the 10–10,000 MHz region in a variety of applications including TV, point-to-point communications, feeds for reflectors and lenses, and many others. This class of antennas is discussed in more detail in Chapter 11 and in [41].

It was not until almost 20 years later that a fundamental new radiating element, which has received a lot of attention and many applications since its inception, was introduced. This occurred in the early 1970s when the microstrip or patch antennas was reported. This element is simple, lightweight, inexpensive, low profile, and conformal to the surface. These antennas are discussed in more detail in Chapter 14 and in [42].

Major advances in millimeter wave antennas have been made in recent years, including integrated antennas where active and passive circuits are combined with the radiating elements in one compact unit (monolithic form). These antennas are discussed in [43].

Specific radiation pattern requirements usually cannot be achieved by single antenna elements, because single elements usually have relatively wide radiation patterns and low values of directivity. To design antennas with very large directivities, it is usually necessary to increase the electrical size of the antenna. This can be accomplished by enlarging the electrical dimensions of the chosen single element. However, mechanical problems are usually associated with very large elements. An alternative way to achieve large directivities, without increasing the size of the individual elements, is to use multiple single elements to form an *array*. An array is a sampled version of a very large single element. In an array, the mechanical problems of large single elements are traded for the electrical problems associated with the feed networks of arrays. However, with today's solid-state technology, very efficient and low-cost feed networks can be designed.

Arrays are the most versatile of antenna systems. They find wide applications not only in many spaceborne systems, but in many earthbound missions as well. In most cases, the elements of an array are identical; this is not necessary, but it is often more convenient, simpler, and more practical. With arrays, it is practical not only to synthesize almost any desired amplitude radiation pattern, but the main lobe can be scanned by controlling the relative phase excitation between the elements. This is most convenient for applications where the antenna system is not readily accessible, especially for spaceborne missions. The beamwidth of the main lobe along with the side lobe level can be controlled by the relative amplitude excitation (distribution) between the elements of the array. In fact, there is a trade-off between the beamwidth and the side lobe level based on the amplitude distribution. Analysis, design, and synthesis of arrays are discussed in Chapters 6 and 7. However, advances in array technology are reported in [44]–[48].

A new antenna array design referred to as *smart antenna*, based on basic technology of the 1970s and 1980s, is sparking interest especially for wireless applications. This antenna design, which combines antenna technology with that of digital signal processing (DSP), is discussed in some detail in Chapter 16.

1.5.2 Methods of Analysis

There is plethora of antenna elements, many of which exhibit intricate configurations. To analyze each as a boundary-value problem and obtain solutions in closed form, the antenna structure must be described by an orthogonal curvilinear coordinate system. This places severe restrictions on the type and number of antenna systems that can be analyzed using such a procedure. Therefore, other exact or approximate methods are often pursued. Two methods that in the last four decades have been preeminent in the analysis of many previously intractable antenna problems are the *Integral Equation* (IE) method and the *Geometrical Theory of Diffraction* (GTD).

The Integral Equation method casts the solution to the antenna problem in the form of an integral (hence its name) where the unknown, usually the induced current density, is part of the integrand. Numerical techniques, such as the Moment Method (MM), are then used to solve for the unknown. Once the current density is found, the radiation integrals of Chapter 3 are used to find the fields radiated and other systems parameters. This method is most convenient for wire-type antennas and more efficient for structures that are small electrically. One of the first objectives of this method is to formulate the IE for the problem at hand. In general, there are two type of IE's. One is the *Electric Field Integral Equation (EFIE)*, and it is based on the boundary condition of the total tangential electric field. The other is the *Magnetic Field Integral Equation (MFIE)*, and it is based on the boundary condition that expresses the total electric current density induced on the surface in terms of the incident magnetic field. The MFIE is only valid for closed surfaces. For some problems, it is more convenient to formulate an EFIE, while for others it is more appropriate to use an MFIE. Advances, applications, and numerical issues of these methods are addressed in Chapter 8 and in [3] and [49].

When the dimensions of the radiating system are many wavelengths, low-frequency methods are not as computationally efficient. However, high-frequency asymptotic techniques can be used to analyze many problems that are otherwise mathematically intractable. One such method that has received considerable attention and application over the years is the GTD/UTD, which is an extension of geometrical optics (GO), and it overcomes some of the limitations of GO by introducing a diffraction mechanism. The Geometrical/Uniform Theory of Diffraction is briefly discussed in Section 12.10. However, a detailed treatment is found in Chapter 13 of [3] while recent advances and applications are found in [50] and [51].

For structures that are not convenient to analyze by either of the two methods, a combination of the two is often used. Such a technique is referred to as a *hybrid method*, and it is described in detail in [52]. Another method, which has received a lot of attention in scattering, is the Finite-Difference Time-Domain (FDTD). This method has also been applied to antenna radiation problems [53]–[57]. A method that has gained a lot of momentum in its application to antenna problems is the Finite Element Method [58]–[63].

1.5.3 Some Future Challenges

Antenna engineering has enjoyed a very successful period during the 1940s–1990s. Responsible for its success have been the introduction and technological advances of some new elements of radiation, such as aperture antennas, reflectors, frequency independent antennas, and microstrip antennas. Excitement has been created by the advancement, utilization, and proliferation of Computational ElectoMagentics (CEM) software that provides students, engineers, and scientists with versatile and indispensable tools for modeling, visualizing, animating, and interpreting EM phenomena and characteristics. In addition, with such tools, electrically large structures that are complex and may otherwise be intractable can be designed and analyzed to gain insight into the performance of systems in order to advance and improve their efficiency. Today, antenna engineering is a science based on fundamental principles.

Although a certain level of maturity has been attained, there are many challenging opportunities and problems to be solved. Phased array architecture integrating monolithic MIC technology is still a most challenging problem. Integration of new materials, such as *metamaterials* [64], *artificial magnetic conductors* and *soft/hard surfaces* [65], into antenna technology offers many opportunities to control, discipline, harness and manipulate the EM waves to design devices with desired and funtional characteristics, and improved performance. Computational electromagnetics using supercomputing and parallel computing capabilities will model complex electromagnetic wave interactions, in both the frequency and time domains. Innovative antenna designs, such as those using *smart antennas* [66], and *multifunction, multiband, ultra wide hand, reconfigurable antennas and antenna systems* [67], to perform complex and demanding system functions remain a challenge. New basic elements are always welcomed and offer refreshing opportunities. New applications include, but are not limited to nanotechnology, wireless communications, direct broadcast satellite systems, global positioning satellites (GPS), high-accuracy airborne navigation, security systems, global weather, earth resource systems, and others. Because of the many new applications, the lower portion of the EM spectrum has been saturated and the designs have been pushed to higher frequencies, including the millimeter wave and terahertz frequency bands.

1.6 MULTIMEDIA

In the publisher's website for this book, the following multimedia resources related to this chapter are included:

a. **Java**-based **interactive** questionnaire with answers.
b. Three **Matlab**-based **animation-visualization** programs designated
   - **tm_open**

- **tm_box**
- **te_horn**

which are described in detail in Section 1.3.4 and the corresponding READ ME file in the book website.

c. **Power Point (PPT)** viewgraphs.

REFERENCES

1. C. A. Balanis, "Antenna Theory: A Review," *Proc. IEEE*, Vol. 80, No. 1, pp. 7–23, January 1992.
2. L. V. Blake, *Antennas*, Wiley, New York, 1966, p. 289.
3. C. A. Balanis, *Advanced Engineering Electromagnetics*, Second edition, Wiley, New York, 2012.
4. E. K. Miller and J. A. Landt, "Direct Time-Domain Techniques for Transient Radiation and Scattering from Wires," *Proc. IEEE*, Vol. 68, No. 11, pp. 1396–1423, November 1980.
5. J. D. Kraus, *Antennas*, McGraw-Hill, New York, 1988.
6. J. D. Kraus, *Electromagnetics*, McGraw-Hill, New York, 1992, pp. 761–763.
7. S. A. Schelkunoff and H. T. Friis, *Antenna Theory and Practice*, Wiley, New York, 1952.
8. K. S. Yee, "Numerical Solution of Initial Boundary Value Problems Involving Maxwell's Equations in Isotropic Media," *IEEE Trans. Antennas Propagat.*, Vol. AP-14, No. 3, pp. 302–307, May 1966.
9. J. P. Berenger, "A Perfectly Matched Layer for the Absorption of Electromagnetic Waves," *J. Comput. Phys.*, Vol. 114, pp. 185–200, October 1994.
10. W. V. Andrew, C. A. Balanis, and P. A. Tirkas, "A Comparison of the Berenger Perfectly Matched Layer and the Lindman Higher-Order ABC's for the FDTD Method," *IEEE Microwave Guided Wave Lett.*, Vol. 5, No. 6, pp. 192–194, June 1995.
11. *The Student Edition of MATLAB*: Version 4: User's Guide, The MATH WORKS, Inc., Prentice-Hall, Inc., Englewood Cliffs, NJ, 1995.
12. J. D. Kraus, "Antennas since Hertz and Marconi," *IEEE Trans. Antennas Propagat.*, Vol. AP-33, No. 2, pp. 131–137, February 1985.
13. J. C. Maxwell, *A Treatise on Electricity and Magnetism*, Oxford University Press, London, UK, 1873, 1904.
14. S. Silver (Ed.), *Microwave Antenna Theory and Design*, MIT Radiation Lab. Series, Vol. 12, McGraw-Hill, New York, 1949.
15. S. Stutzman, "Bibliography for Antennas," *IEEE Antennas Propagat. Mag.*, Vol. 32, pp. 54–57, August 1990.
16. J. Aharoni, *Antennae*, Oxford University Press, London, UK, 1946.
17. S. A. Schelkunoff, *Advanced Antenna Theory*, Wiley, New York, 1952.
18. E. A. Laport, *Radio Antenna Engineering*, McGraw-Hill, New York, 1952.
19. C. H. Walter, *Traveling Wave Antennas*, McGraw-Hill, New York, 1968.
20. E. Wolff, *Antenna Analysis*, Wiley, New York, 1966 (first edition), Artech House, Norwood, MA, 1988 (second edition).
21. W. L. Weeks, *Antenna Engineering*, McGraw-Hill, New York, 1968.
22. E. Jordan and K. Balmain, *Electromagnetic Waves and Radiating Systems*, Prentice-Hall, New York, 1968.
23. R. E. Collin and F. J. Zucker (Eds.), *Antenna Theory*, Parts 1 and 2, McGraw-Hill, New York, 1969.
24. W. V. T. Rusch and P. D. Potter, *Analysis of Reflector Antennas*, Academic Press, New York, 1970.
25. W. L. Stutzman and G. A. Thiele, *Antenna Theory and Design*, Wiley, New York, 1998.
26. R. S. Elliot, *Antenna Theory and Design*, Prentice-Hall, New York, 1981.
27. K. F. Lee, *Principles of Antenna Theory*, Wiley, New York, 1984.
28. R. E. Collin, *Antennas and Radiowave Propagation*, McGraw-Hill, New York, 1985.
29. T. A. Milligan, *Modern Antenna Design*, McGraw-Hill, New York, 1985.

30. J. R. Wait, *Introduction to Antennas and Propagation*, IEE, Hithin Herts, UK, 1966.
31. F. R. Connor, *Antennas*, Edward Arnold, London, 1989.
32. K. Chang (Ed.), *Handbook of Microwave and Optical Components*, Vol. I, Wiley-Interscience, New York, 1989, Chapters 10–13.
33. R. C. Johnson and H. Jasik, *Antenna Engineering Handbook*, McGraw-Hill, New York, 1984.
34. R. C. Hansen (Ed.), *Microwave Scanning Antennas*, Vols. I–III, Academic Press, New York, 1964 (reprinted by Peninsula Publishing, Los Altos, CA).
35. A. W. Rudge, K. Milne, A. D. Olver, and P. Knight (Eds.), *The Handbook of Antenna Design*, Vols. 1 and 2, Peter Peregrinus, London, 1982.
36. Y. T. Lo and S. W. Lee (Eds.), *Antenna Handbook: Theory, Applications, and Design*, Van Nostrand Reinhold, New York, 1988.
37. I. J. Bahl and P. Bhartia, *Microstrip Antennas*, Artech House, Norwood, MA, 1980.
38. J. R. James and P. S. Hall (Eds.), *Handbook of Microstrip Antennas*, Vols. I and II, Peter Peregrinus, 1989.
39. D. M. Pozar, *Antenna Design Using Personal Computers*, Artech House, Norwood, MA, 1985.
40. W. V. T. Rusch, "The Current State of the Reflector Antenna Art-Entering the 1990s," *Proc. IEEE*, Vol. 80, No. 1, pp. 113–126, January 1992.
41. P. E. Mayes, "Frequency-Independent Antennas and Broad-Band Derivatives Thereof," *Proc. IEEE*, Vol. 80, No. 1, pp. 103–112, January 1992.
42. D. M. Pozar, "Microstrip Antennas," *Proc. IEEE*, Vol. 80, No. 1, pp. 79–91, January 1992.
43. F. K. Schwering, "Millimeter Wave Antennas," *Proc. IEEE*, Vol. 80, No. 1, pp. 92–102, January 1992.
44. W. H. Kummer, "Basic Array Theory," *Proc. IEEE*, Vol. 80, No. 1, pp. 127–140, January 1992.
45. R. C. Hansen, "Array Pattern Control and Synthesis," *Proc. IEEE*, Vol. 80, No. 1, pp. 141–151, January 1992.
46. W. F. Gabriel, "Adaptive Processing Array Systems," *Proc. IEEE*, Vol. 80, No. 1, pp. 152–162, January 1992.
47. R. J. Mailloux, "Antenna Array Architecture," *Proc. IEEE*, Vol. 80, No. 1, pp. 163–172, January 1992.
48. R. Tang and R. W. Burns, "Array Technology," *Proc. IEEE*, Vol. 80, No. 1, pp. 173–182, January 1992.
49. E. K. Miller and G. J. Burke, "Low-Frequency Computational Electromagnetics for Antenna Analysis," *Proc. IEEE*, Vol. 80, No. 1, pp. 24–43, January 1992.
50. P. H. Pathak, "High-Frequency Techniques for Antenna Analysis," *Proc. IEEE*, Vol. 80, No. 1, pp. 44–65, January 1992.
51. R. J. Marhefka and W. D. Burnside, "Antennas on Complex Platforms," *Proc. IEEE*, Vol. 80, No. 1, pp. 204–208, January 1992.
52. G. A. Thiele, "Overview of Selected Hybrid Methods in Radiating System Analysis," *Proc. IEEE*, Vol. 80, No. 1, pp. 66–78, January 1992.
53. J. C. Maloney, G. S. Smith, and W. R. Scott Jr., "Accurate Computation of the Radiation from Simple Antennas using the Finite-Difference Time-Domain Method," *IEEE Trans. Antennas Propagat.*, Vol. 38, No. 7, pp. 1059–1068, July 1990.
54. D. S. Katz, M. J. Piket-May, A. Taflove, and K. R. Umashankar, "FDTD Analysis of Electromagnetic Wave Radiation from Systems Containing Horn Antennas," *IEEE Trans. Antennas Propagat.*, Vol. 39, No. 8, pp. 1203–1212, August 1991.
55. P. A. Tirkas and C. A. Balanis, "Finite-Difference Time-Domain Techniques for Antenna Radiation," *IEEE Trans. Antennas Propagat.*, Vol. 40, No. 3, pp. 334–340, March 1992.
56. P. A. Tirkas and C. A. Balanis, "Contour Path FDTD Method for Analysis of Pyramidal Horns With Composite Inner E-Plane Walls," *IEEE Trans. Antennas Propagat.*, Vol. 42, No. 11, pp. 1476–1483, November 1994.
57. A. Taflove, *Advances in Computational Electrodynamics: The finite-Difference Time-Domain Method*, Artech House, Boston, 1998.
58. J. Volakis, A. Chatterjee, and L. C. Kempel, *Finite Element Method for Electromagnetics*, IEEE Press, New York, 1998.

59. J. Jin, *The Finite Element Method in Electromagnetics*, Wiley, New York, 2014.
60. J. M. Jin and J. L. Volakis, "Scattering and Radiation Analysis of Three-Dimensional Cavity Arrays Via a Hybrid Finite-Element Method," *IEEE Trans. Antennas Propagat.*, Vol. 41, No. 11, pp. 1580–1586, November 1993.
61. D. T. McGrath and V. P. Pyati, "Phased Array Antenna Analysis with Hybrid Finite Element Method," *IEEE Trans. Antennas Propagat.*, Vol. 42, No. 12, pp. 1625–1630, December 1994.
62. W. Sun and C. A. Balanis, "Vector One-Way Wave Absorbing Boundary Condition for FEM Applications," *IEEE Trans. Antennas Propagat.*, Vol. 42, No. 6, pp. 872–878, June 1994.
63. E. W. Lucas and T. P. Fontana, "A 3-D Hybrid Finite Element/Boundary Element Method for the Unified Radiation and Scattering Analysis of General Infinite Periodic Arrays," *IEEE Trans. Antennas Propagat.*, Vol. 43, No. 2, pp. 145–153, February 1995.
64. *IEEE Trans. Antennas Propagat.*, Special Issue on *Metamaterials*, Vol. 51, No. 10, October 2003.
65. *IEEE Trans. Antennas Propagat.*, Special Issue on *Artificial Magnetic Conductors, Soft/Hard Surfaces and Other Complex Surfaces*, Vol. 53, No. 1, Jan. 2005.
66. *IEEE Trans. Antennas Propagat.*, Special Issue on *Wireless Information Technology and Networks*, Vol. 50, No. 5, May 2002.
67. *IEEE Trans. Antennas Propagat.*, Special Issue on *Multifunction Antennas and Antenna Systems*, Vol. 54, No. 1, Jan. 2006.

CHAPTER 2

Fundamental Parameters and Figures-of-Merit of Antennas

2.1 INTRODUCTION

To describe the performance of an antenna, definitions of various parameters are necessary. Some of the parameters are interrelated and not all of them need be specified for complete description of the antenna performance. Parameter definitions will be given in this chapter. Many of those in quotation marks are from the *IEEE Standard Definitions of Terms for Antennas* [IEEE Std 145-1993. Reaffirmed 2004(R2004)]. This is a revision of the IEEE Std 145-1983.*

2.2 RADIATION PATTERN

An antenna *radiation pattern* or *antenna pattern* is defined as "a mathematical function or a graphical representation of the radiation properties of the antenna as a function of space coordinates. In most cases, the radiation pattern is determined in the far-field region and is represented as a function of the directional coordinates. Radiation properties include power flux density, radiation intensity, field strength, directivity, phase or polarization." The radiation property of most concern is the two- or three-dimensional spatial distribution of radiated energy as a function of the observer's position along a path or surface of constant radius. A convenient set of coordinates is shown in Figure 2.1. A trace of the received electric (magnetic) field at a constant radius is called the amplitude field *pattern*. On the other hand, a graph of the spatial variation of the power density along a constant radius is called an amplitude *power pattern*.

Often the *field* and *power* patterns are normalized with respect to their maximum value, yielding *normalized field* and *power patterns*. Also, the power pattern is usually plotted on a logarithmic scale or more commonly in decibels (dB). This scale is usually desirable because a logarithmic scale can accentuate in more details those parts of the pattern that have very low values, which later we will refer to as minor lobes. For an antenna, the

a. *field* pattern (*in linear scale*) typically represents a plot of the magnitude of the electric or magnetic field as a function of the angular space.

IEEE Transactions on Antennas and Propagation, Vols. AP-17, No. 3, May 1969; Vol. AP-22, No. 1, January 1974; and Vol. AP-31, No. 6, Part II, November 1983.

Antenna Theory: Analysis and Design, Fourth Edition. Constantine A. Balanis.
© 2016 John Wiley & Sons, Inc. Published 2016 by John Wiley & Sons, Inc.
Companion Website: www.wiley.com/go/antennatheory4e

26 FUNDAMENTAL PARAMETERS AND FIGURES-OF-MERIT OF ANTENNAS

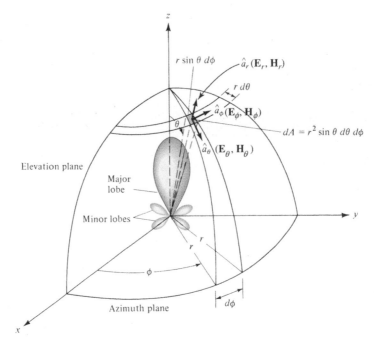

Figure 2.1   Coordinate system for antenna analysis.

b. *power* pattern (*in linear scale*) typically represents a plot of the square of the magnitude of the electric or magnetic field as a function of the angular space.
c. *power* pattern (*in dB*) represents the magnitude of the electric or magnetic field, in decibels, as a function of the angular space.

To demonstrate this, the two-dimensional normalized field pattern (*plotted in linear scale*), power pattern (*plotted in linear scale*), and power pattern (*plotted on a logarithmic dB scale*) of a 10-element linear antenna array of isotropic sources, with a spacing of $d = 0.25\lambda$ between the elements, are shown in Figure 2.2. *In this and subsequent patterns, the plus (+) and minus (−) signs in the lobes indicate the relative polarization (positive or negative) of the amplitude between the various lobes, which changes (alternates) as the nulls are crossed.* To find the points where the pattern achieves its half-power (−3 dB points), relative to the maximum value of the pattern, you set the value of the

a. field pattern at 0.707 value of its maximum, as shown in Figure 2.2(a)
b. power pattern (in a linear scale) at its 0.5 value of its maximum, as shown in Figure 2.2(b)
c. power pattern (in dB) at −3 dB value of its maximum, as shown in Figure 2.2(c).

All three patterns yield the same angular separation between the two half-power points, 38.64°, on their respective patterns, *referred to as HPBW* and illustrated in Figure 2.2. This is discussed in detail in Section 2.5.

In practice, the three-dimensional pattern is measured and recorded in a series of two-dimensional patterns. However, for most practical applications, a few plots of the pattern as a function of θ for some particular values of ϕ, plus a few plots as a function of ϕ for some particular values of θ, give most of the useful and needed information.

2.2.1   Radiation Pattern Lobes

Various parts of a radiation pattern are referred to as *lobes*, which may be subclassified into *major* or *main*, *minor*, *side*, and *back* lobes.

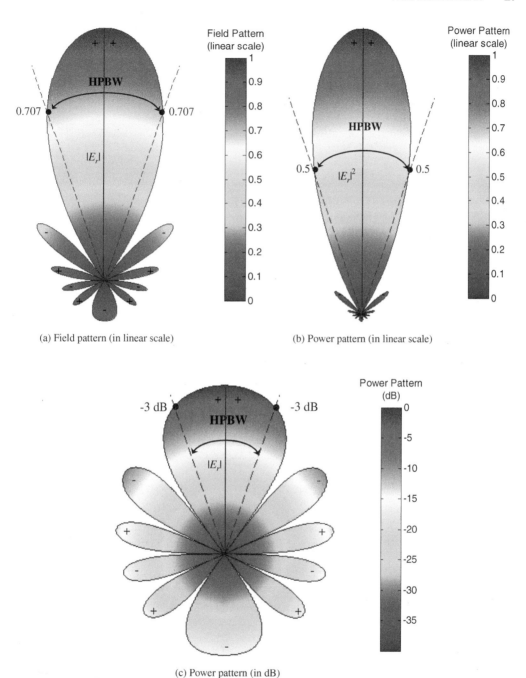

Figure 2.2 Two-dimensional normalized *field* pattern (*linear scale*), *power* pattern (*linear scale*), and *power* pattern (*in dB*) of a 10-element linear array with a spacing of $d = 0.25\lambda$.

A *radiation lobe* is a "portion of the radiation pattern bounded by regions of relatively weak radiation intensity." Figure 2.3(a) demonstrates a symmetrical three-dimensional polar pattern with a number of radiation lobes. Some are of greater radiation intensity than others, but all are classified as lobes. Figure 2.3(b) illustrates a linear two-dimensional pattern [one plane of Figure 2.3(a)] where the same pattern characteristics are indicated.

28 FUNDAMENTAL PARAMETERS AND FIGURES-OF-MERIT OF ANTENNAS

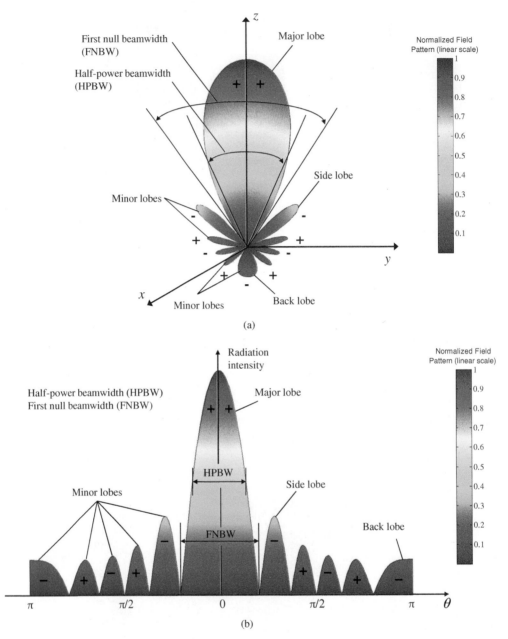

Figure 2.3 (a) Radiation lobes and beamwidths of an antenna amplitude pattern in polar form. (b) Linear plot of power pattern and its associated lobes and beamwidths.

MATLAB-based computer programs, designated as **Polar** and **Spherical**, have been developed and are included in the publisher's website for this book. These programs can be used to plot the two-dimensional patterns, both polar and semipolar (*in linear* and *dB scales*), in polar form and spherical three-dimensional patterns (*in linear* and *dB scales*). A description of these programs is found in the publisher's website for this book. Other programs that have been developed for plotting rectangular and polar plots are those of [1]–[3].

A *major lobe* (also called main beam) is defined as "the radiation lobe containing the direction of maximum radiation." In Figure 2.3 the major lobe is pointing in the $\theta = 0$ direction. In some

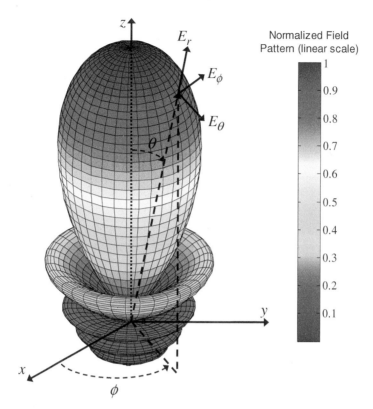

Figure 2.4 Normalized three-dimensional amplitude *field* pattern (*in linear scale*) of a 10-element linear array antenna with a uniform spacing of $d = 0.25\lambda$ and progressive phase shift $\beta = -0.6\pi$ between the elements.

antennas, such as split-beam antennas, there may exist more than one major lobe. A *minor lobe* is any lobe except a major lobe. In Figures 2.3(a) and (b) all the lobes with the exception of the major can be classified as minor lobes. A *side lobe* is "a radiation lobe in any direction other than the intended lobe." (Usually a side lobe is adjacent to the main lobe and occupies the hemisphere in the direction of the main beam.) A *back lobe* is "a radiation lobe whose axis makes an angle of approximately 180° with respect to the beam of an antenna." Usually it refers to a minor lobe that occupies the hemisphere in a direction opposite to that of the major (main) lobe.

Minor lobes usually represent radiation in undesired directions, and they should be minimized. Side lobes are normally the largest of the minor lobes. The level of minor lobes is usually expressed as a ratio of the power density in the lobe in question to that of the major lobe. This ratio is often termed the side lobe ratio or side lobe level. Side lobe levels of −20 dB or smaller are usually not desirable in most applications. Attainment of a side lobe level smaller than −30 dB usually requires very careful design and construction. In most radar systems, low side lobe ratios are very important to minimize false target indications through the side lobes.

A normalized three-dimensional far-field amplitude pattern, plotted on a linear scale, of a 10-element linear antenna array of isotropic sources with a spacing of $d = 0.25\lambda$ and progressive phase shift $\beta = -0.6\pi$, between the elements is shown in Figure 2.4. It is evident that this pattern has one major lobe, five minor lobes and one back lobe. The level of the side lobe is about −9 dB relative to the maximum. A detailed presentation of arrays is found in Chapter 6. For an amplitude pattern of an antenna, there would be, in general, three electric-field components (E_r, E_θ, E_ϕ) at each observation point on the surface of a sphere of constant radius $r = r_c$, as shown in Figure 2.1. In the far field, the radial E_r component for all antennas is zero or vanishingly small compared to either one, or

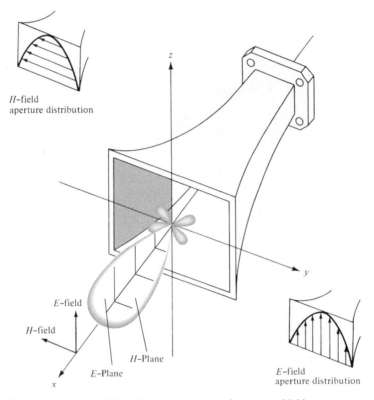

Figure 2.5   Principal E- and H-plane patterns for a pyramidal horn antenna.

both, of the other two components (see Section 3.6 of Chapter 3). Some antennas, depending on their geometry and also observation distance, may have only one, two, or all three components. In general, the magnitude of the total electric field would be $|\mathbf{E}| = \sqrt{|E_r|^2 + |E_\theta|^2 + |E_\phi|^2}$. The radial distance in Figure 2.4, and similar ones, represents the magnitude of $|\mathbf{E}|$.

2.2.2 Isotropic, Directional, and Omnidirectional Patterns

An *isotropic* radiator is defined as "a hypothetical lossless antenna having equal radiation in all directions." Although it is ideal and not physically realizable, it is often taken as a reference for expressing the directive properties of actual antennas. A *directional* antenna is one "having the property of radiating or receiving electromagnetic waves more effectively in some directions than in others. This term is usually applied to an antenna whose maximum directivity is significantly greater than that of a half-wave dipole." Examples of antennas with directional radiation patterns are shown in Figures 2.5 and 2.6. It is seen that the pattern in Figure 2.6 is nondirectional in the azimuth plane [$f(\phi), \theta = \pi/2$] and directional in the elevation plane [$g(\theta), \phi = $ constant]. This type of a pattern is designated as *omnidirectional*, and it is defined as one "having an essentially nondirectional pattern in a given plane (in this case in azimuth) and a directional pattern in any orthogonal plane (in this case in elevation)." An *omnidirectional* pattern is then a special type of a *directional* pattern.

2.2.3 Principal Patterns

For a linearly polarized antenna, performance is often described in terms of its principal E- and H-plane patterns. The *E-plane* is defined as "the plane containing the electric-field vector and the

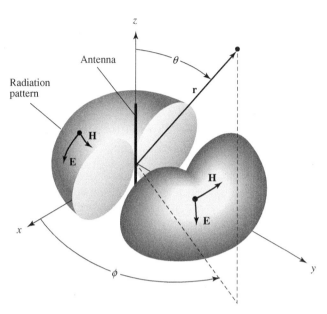

Figure 2.6 Omnidirectional antenna pattern.

direction of maximum radiation," and the *H-plane* as "the plane containing the magnetic-field vector and the direction of maximum radiation." Although it is very difficult to illustrate the principal patterns without considering a specific example, it is the usual practice to orient most antennas so that at least one of the principal plane patterns coincide with one of the geometrical principal planes. An illustration is shown in Figure 2.5. For this example, the x-z plane (elevation plane; $\phi = 0$) is the principal *E*-plane and the x-y plane (azimuthal plane; $\theta = \pi/2$) is the principal *H*-plane. Other coordinate orientations can be selected.

The omnidirectional pattern of Figure 2.6 has an infinite number of principal E-planes (elevation planes; $\phi = \phi_c$) and one principal *H*-plane (azimuthal plane; $\theta = 90°$).

2.2.4 Field Regions

The space surrounding an antenna is usually subdivided into three regions: (a) reactive near-field, (b) radiating near-field (Fresnel) and (c) far-field (Fraunhofer) regions as shown in Figure 2.7. These regions are so designated to identify the field structure in each. Although no abrupt changes in the field configurations are noted as the boundaries are crossed, there are distinct differences among them. The boundaries separating these regions are not unique, although various criteria have been established and are commonly used to identify the regions.

Reactive near-field region is defined as "that portion of the near-field region immediately surrounding the antenna wherein the reactive field predominates." For most antennas, the outer boundary of this region is commonly taken to exist at a distance $R < 0.62\sqrt{D^3/\lambda}$ from the antenna surface, where λ is the wavelength and D is the largest dimension of the antenna. "For a very short dipole, or equivalent radiator, the outer boundary is commonly taken to exist at a distance $\lambda/2\pi$ from the antenna surface."

Radiating near-field (Fresnel) region is defined as "that region of the field of an antenna between the reactive near-field region and the far-field region wherein radiation fields predominate and wherein the angular field distribution is dependent upon the distance from the antenna. If the antenna has a maximum dimension that is not large compared to the wavelength, this region may not exist. For an antenna focused at infinity, the radiating near-field region is sometimes referred to

32 FUNDAMENTAL PARAMETERS AND FIGURES-OF-MERIT OF ANTENNAS

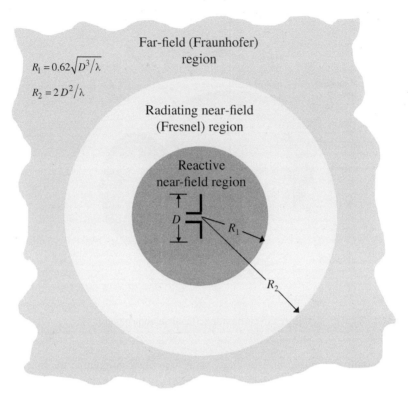

Figure 2.7 Field regions of an antenna.

as the Fresnel region on the basis of analogy to optical terminology. If the antenna has a maximum overall dimension which is very small compared to the wavelength, this field region may not exist." The inner boundary is taken to be the distance $R \geq 0.62\sqrt{D^3/\lambda}$ and the outer boundary the distance $R < 2D^2/\lambda$ where D is the largest* dimension of the antenna. This criterion is based on a maximum phase error of $\pi/8$. In this region the field pattern is, in general, a function of the radial distance and the radial field component may be appreciable.

Far-field (Fraunhofer) region is defined as "that region of the field of an antenna where the angular field distribution is essentially independent of the distance from the antenna. If the antenna has a maximum[†] overall dimension D, the far-field region is commonly taken to exist at distances greater than $2D^2/\lambda$ from the antenna, λ being the wavelength. The far-field patterns of certain antennas, such as multibeam reflector antennas, are sensitive to variations in phase over their apertures. For these antennas $2D^2/\lambda$ may be inadequate. In physical media, if the antenna has a maximum overall dimension, D, which is large compared to $\pi/|\gamma|$, the far-field region can be taken to begin approximately at a distance equal to $|\gamma|D^2/\pi$ from the antenna, γ being the propagation constant in the medium. For an antenna focused at infinity, the far-field region is sometimes referred to as the Fraunhofer region on the basis of analogy to optical terminology." In this region, the field components are essentially transverse and the angular distribution is independent of the radial distance where the measurements are made. The inner boundary is taken to be the radial distance $R = 2D^2/\lambda$ and the outer one at infinity.

The amplitude pattern of an antenna, as the observation distance is varied from the reactive near field to the far field, changes in shape because of variations of the fields, both magnitude and phase. A typical progression of the shape of an antenna, with the largest dimension D, is shown in Figure 2.8.

---

*To be valid, D must also be large compared to the wavelength ($D > \lambda$).
†To be valid, D must also be large compared to the wavelength ($D > \lambda$).

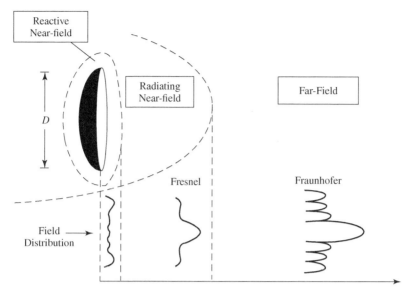

Figure 2.8 Typical changes of antenna amplitude pattern shape from reactive near field toward the far field. (SOURCE: Y. Rahmat-Samii, L. I. Williams, and R. G. Yoccarino, The UCLA Bi-polar Planar-Near-Field Antenna Measurement and Diagnostics Range," *IEEE Antennas & Propagation Magazine*, Vol. 37, No. 6, December 1995 © 1995 IEEE).

It is apparent that in the reactive near-field region the pattern is more spread out and nearly uniform, with slight variations. As the observation is moved to the radiating near-field region (Fresnel), the pattern begins to smooth and form lobes. In the far-field region (Fraunhofer), the pattern is well formed, usually consisting of few minor lobes and one, or more, major lobes.

To illustrate the pattern variation as a function of radial distance beyond the minimum $2D^2/\lambda$ far-field distance, in Figure 2.9 we have included three patterns of a parabolic reflector calculated at distances of $R = 2D^2/\lambda, 4D^2/\lambda$, and infinity [4]. It is observed that the patterns are almost identical, except for some differences in the pattern structure around the first null and at a level below 25 dB. Because infinite distances are not realizable in practice, the most commonly used criterion for minimum distance of far-field observations is $2D^2/\lambda$.

2.2.5 Radian and Steradian

The measure of a plane angle is a radian. One *radian* is defined as the plane angle with its vertex at the center of a circle of radius r that is subtended by an arc whose length is r. A graphical illustration is shown in Figure 2.10(a). Since the circumference of a circle of radius r is $C = 2\pi r$, there are 2π rad $(2\pi r/r)$ in a full circle.

The measure of a solid angle is a steradian. One *steradian* is defined as the solid angle with its vertex at the center of a sphere of radius r that is subtended by a spherical surface area equal to that of a square with each side of length r. A graphical illustration is shown in Figure 2.10(b). Since the area of a sphere of radius r is $A = 4\pi r^2$, there are 4π sr $(4\pi r^2/r^2)$ in a closed sphere.

The infinitesimal area dA on the surface of a sphere of radius r, shown in Figure 2.1, is given by

$$dA = r^2 \sin\theta \, d\theta \, d\phi \quad (\text{m}^2) \tag{2-1}$$

Therefore, the element of solid angle $d\Omega$ of a sphere can be written as

$$d\Omega = \frac{dA}{r^2} = \sin\theta \, d\theta \, d\phi \quad (\text{sr}) \tag{2-2}$$

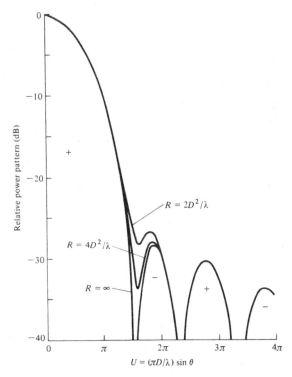

Figure 2.9 Calculated radiation patterns of a paraboloid antenna for different distances from the antenna. (SOURCE: J. S. Hollis, T. J. Lyon, and L. Clayton, Jr. (eds.), *Microwave Antenna Measurements*, Scientific-Atlanta, Inc., July 1970).

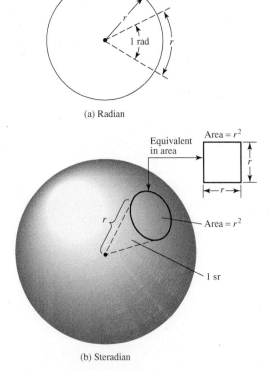

Figure 2.10 Geometrical arrangements for defining a radian and a steradian.

Example 2.1

For a sphere of radius r, find the solid angle Ω_A (in square radians or steradians) of a spherical cap on the surface of the sphere over the north-pole region defined by spherical angles of $0 \leq \theta \leq 30°, 0 \leq \phi \leq 360°$. Refer to Figures 2.1 and 2.10. Do this

a. exactly.
b. using $\Omega_A \approx \Delta\Theta_1 \cdot \Delta\Theta_2$, where $\Delta\Theta_1$ and $\Delta\Theta_2$ are two perpendicular angular separations of the spherical cap passing through the north pole.

Compare the two.
Solution:

a. Using (2-2), we can write that

$$\Omega_A = \int_0^{360°} \int_0^{30°} d\Omega = \int_0^{2\pi} \int_0^{\pi/6} \sin\theta \, d\theta \, d\phi = \int_0^{2\pi} d\phi \int_0^{\pi/6} \sin\theta \, d\theta$$

$$= 2\pi[-\cos\theta]\Big|_0^{\pi/6} = 2\pi[-0.867 + 1] = 2\pi(0.133) = 0.83566$$

b. $\Omega_A \approx \Delta\Theta_1 \cdot \Delta\Theta_2 \underset{=}{\overset{\Delta\Theta_1 = \Delta\Theta_2}{\frown}} (\Delta\Theta_1)^2 = \dfrac{\pi}{3}\left(\dfrac{\pi}{3}\right) = \dfrac{\pi^2}{9} = 1.09662$

It is apparent that the approximate beam solid angle is about 31.23% in error.

2.3 RADIATION POWER DENSITY

Electromagnetic waves are used to transport information through a wireless medium or a guiding structure, from one point to the other. It is then natural to assume that power and energy are associated with electromagnetic fields. The quantity used to describe the power associated with an electromagnetic wave is the instantaneous Poynting vector defined as

$$\mathscr{W} = \mathscr{E} \times \mathscr{H} \tag{2-3}$$

$\mathscr{W}$ = instantaneous Poynting vector (W/m^2)
$\mathscr{E}$ = instantaneous electric-field intensity (V/m)
$\mathscr{H}$ = instantaneous magnetic-field intensity (A/m)

Note that script letters are used to denote instantaneous fields and quantities, while roman letters are used to represent their complex counterparts.

Since the Poynting vector is a power density, the total power crossing a closed surface can be obtained by integrating the normal component of the Poynting vector over the entire surface. In equation form

$$\mathscr{P} = \oiint_S \mathscr{W} \cdot d\mathbf{s} = \oiint_S \mathscr{W} \cdot \hat{\mathbf{n}} \, da \tag{2-4}$$

$\mathscr{P}$ = instantaneous total power   (W)

$\hat{\mathbf{n}}$ = unit vector normal to the surface

da = infinitesimal area of the closed surface   (m²)

For applications of time-varying fields, it is often more desirable to find the average power density which is obtained by integrating the instantaneous Poynting vector over one period and dividing by the period. For time-harmonic variations of the form $e^{j\omega t}$, we define the complex fields **E** and **H** which are related to their instantaneous counterparts $\mathscr{E}$ and $\mathscr{H}$ by

$$\mathscr{E}(x, y, z; t) = \text{Re}[\mathbf{E}(x, y, z)e^{j\omega t}] \tag{2-5}$$

$$\mathscr{H}(x, y, z; t) = \text{Re}[\mathbf{H}(x, y, z)e^{j\omega t}] \tag{2-6}$$

Using the definitions of (2-5) and (2-6) and the identity $\text{Re}[\mathbf{E}e^{j\omega t}] = \frac{1}{2}[\mathbf{E}e^{j\omega t} + \mathbf{E}^*e^{-j\omega t}]$, (2-3) can be written as

$$\mathscr{W} = \mathscr{E} \times \mathscr{H} = \frac{1}{2}\text{Re}[\mathbf{E} \times \mathbf{H}^*] + \frac{1}{2}\text{Re}[\mathbf{E} \times \mathbf{H}e^{j2\omega t}] \tag{2-7}$$

The first term of (2-7) is not a function of time, and the time variations of the second are twice the given frequency. The time average Poynting vector (average power density) can be written as

$$\mathbf{W}_{av}(x, y, z) = [\mathscr{W}(x, y, z; t)]_{av} = \frac{1}{2}\text{Re}[\mathbf{E} \times \mathbf{H}^*] \quad (\text{W/m}^2) \tag{2-8}$$

The $\frac{1}{2}$ factor appears in (2-7) and (2-8) because the **E** and **H** fields represent peak values, and it should be omitted for RMS values.

A close observation of (2-8) may raise a question. If the real part of $(\mathbf{E} \times \mathbf{H}^*)/2$ represents the average (real) power density, what does the imaginary part of the same quantity represent? At this point it will be very natural to assume that the imaginary part must represent the reactive (stored) power density associated with the electromagnetic fields. In later chapters, it will be shown that the power density associated with the electromagnetic fields of an antenna in its far-field region is predominately real and will be referred to as *radiation density*.

Based upon the definition of (2-8), the average power radiated by an antenna (radiated power) can be written as

$$\begin{aligned} P_{rad} = P_{av} &= \oiint_S \mathbf{W}_{rad} \cdot d\mathbf{s} = \oiint_S \mathbf{W}_{av} \cdot \hat{\mathbf{n}} \, da \\ &= \frac{1}{2} \oiint_S \text{Re}(\mathbf{E} \times \mathbf{H}^*) \cdot d\mathbf{s} \end{aligned} \tag{2-9}$$

The power pattern of the antenna, whose definition was discussed in Section 2.2, is just a measure, as a function of direction, of the average power density radiated by the antenna. The observations are usually made on a large sphere of constant radius extending into the far field. In practice, absolute power patterns are usually not desired. However, the performance of the antenna is measured in terms of the gain (to be discussed in a subsequent section) and in terms of relative power patterns. Three-dimensional patterns cannot be measured, but they can be constructed with a number of two-dimensional cuts.

Example 2.2

The radial component of the radiated power density of an antenna is given by

$$\mathbf{W}_{rad} = \hat{\mathbf{a}}_r W_r = \hat{\mathbf{a}}_r A_0 \frac{\sin\theta}{r^2} \quad (W/m^2)$$

where A_0 is the peak value of the power density, θ is the usual spherical coordinate, and $\hat{\mathbf{a}}_r$ is the radial unit vector. Determine the total radiated power.

Solution: For a closed surface, a sphere of radius r is chosen. To find the total radiated power, the radial component of the power density is integrated over its surface. Thus

$$P_{rad} = \oiint_S \mathbf{W}_{rad} \cdot \hat{\mathbf{n}} \, da$$

$$= \int_0^{2\pi} \int_0^{\pi} \left(\hat{\mathbf{a}}_r A_0 \frac{\sin\theta}{r^2}\right) \cdot (\hat{\mathbf{a}}_r r^2 \sin\theta \, d\theta \, d\phi) = \pi^2 A_0 \quad (W)$$

A three-dimensional normalized plot of the average power density at a distance of $r = 1$ m is shown in Figure 2.6.

An isotropic radiator is an ideal source that radiates equally in all directions. Although it does not exist in practice, it provides a convenient isotropic reference with which to compare other antennas. Because of its symmetric radiation, its Poynting vector will not be a function of the spherical coordinate angles θ and ϕ. In addition, it will have only a radial component. Thus the total power radiated by it is given by

$$P_{rad} = \oiint_S \mathbf{W}_0 \cdot d\mathbf{s} = \int_0^{2\pi} \int_0^{\pi} [\hat{\mathbf{a}}_r W_0(r)] \cdot [\hat{\mathbf{a}}_r r^2 \sin\theta \, d\theta \, d\phi] = 4\pi r^2 W_0 \quad (2\text{-}10)$$

and the power density by

$$\mathbf{W}_0 = \hat{\mathbf{a}}_r W_0 = \hat{\mathbf{a}}_r \left(\frac{P_{rad}}{4\pi r^2}\right) \quad (W/m^2) \quad (2\text{-}11)$$

which is uniformly distributed over the surface of a sphere of radius r.

2.4 RADIATION INTENSITY

Radiation intensity in a given direction is defined as "the power radiated from an antenna per unit solid angle." The radiation intensity is a far-field parameter, and it can be obtained by simply multiplying the radiation density by the square of the distance. In mathematical form it is expressed as

$$U = r^2 W_{rad} \quad (2\text{-}12)$$

where

U = radiation intensity (W/unit solid angle)

W_{rad} = radiation density (W/m^2)

The radiation intensity is also related to the far-zone electric field of an antenna, referring to Figure 2.4, by

$$U(\theta, \phi) = \frac{r^2}{2\eta}|\mathbf{E}(r, \theta, \phi)|^2 \simeq \frac{r^2}{2\eta}\left[|E_\theta(r, \theta, \phi)|^2 + |E_\phi(r, \theta, \phi)|^2\right]$$
$$\simeq \frac{1}{2\eta}\left[|E_\theta^\circ(\theta, \phi)|^2 + |E_\phi^\circ(\theta, \phi)|^2\right] \quad (2\text{-}12a)$$

where

$\mathbf{E}(r, \theta, \phi)$ = far-zone electric-field intensity of the antenna = $\mathbf{E}^\circ(\theta, \phi)\dfrac{e^{-jkr}}{r}$

E_θ, E_ϕ = far-zone electric-field components of the antenna

η = intrinsic impedance of the medium

The radial electric-field component (E_r) is assumed, if present, to be small in the far zone. Thus the power pattern is also a measure of the radiation intensity.

The total power is obtained by integrating the radiation intensity, as given by (2-12), over the entire solid angle of 4π. Thus

$$P_{\text{rad}} = \oiint_\Omega U \, d\Omega = \int_0^{2\pi} \int_0^\pi U \sin\theta \, d\theta \, d\phi \quad (2\text{-}13)$$

where $d\Omega$ = element of solid angle = $\sin\theta \, d\theta \, d\phi$.

Example 2.3

For the problem of Example 2.2, find the total radiated power using (2-13).
Solution: Using (2-12)

$$U = r^2 W_{\text{rad}} = A_0 \sin\theta$$

and by (2-13)

$$P_{\text{rad}} = \int_0^{2\pi} \int_0^\pi U \sin\theta \, d\theta \, d\phi = A_0 \int_0^{2\pi} \int_0^\pi \sin^2\theta \, d\theta \, d\phi = \pi^2 A_0$$

which is the same as that obtained in Example 2.2. A three-dimensional plot of the relative radiation intensity is also represented by Figure 2.6.

For an isotropic source U will be independent of the angles θ and ϕ, as was the case for W_{rad}. Thus (2-13) can be written as

$$P_{\text{rad}} = \oiint_\Omega U_0 \, d\Omega = U_0 \oiint_\Omega d\Omega = 4\pi U_0 \quad (2\text{-}14)$$

or the radiation intensity of an isotropic source as

$$U_0 = \frac{P_{\text{rad}}}{4\pi} \quad (2\text{-}15)$$

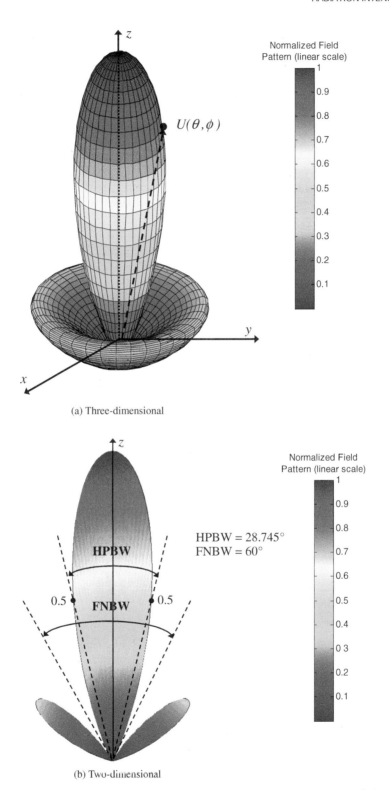

Figure 2.11 Three- and two-dimensional power patterns (in linear scale) of $U(\theta) = \cos^2(\theta)\cos^2(3\theta)$.

2.5 BEAMWIDTH

Associated with the pattern of an antenna is a parameter designated as *beamwidth*. The *beamwidth* of a pattern is defined as the angular separation between two identical points on opposite side of the pattern maximum. In an antenna pattern, there are a number of beamwidths. One of the most widely used beamwidths is the *Half-Power Beamwidth* (*HPBW*), which is defined by IEEE as: "In a plane containing the direction of the maximum of a beam, the angle between the two directions in which the radiation intensity is one-half value of the beam." This is demonstrated in Figure 2.2. Another important beamwidth is the angular separation between the first nulls of the pattern, and it is referred to as the *First-Null Beamwidth* (*FNBW*). Both the *HPBW* and *FNBW* are demonstrated for the pattern in Figure 2.11 for the pattern of Example 2.4. Other beamwidths are those where the pattern is −10 dB from the maximum, or any other value. However, in practice, the term *beamwidth*, with no other identification, usually refers to *HPBW*.

The beamwidth of an antenna is a very important figure of merit and often is used as a trade-off between it and the side lobe level; that is, as the beamwidth decreases, the side lobe increases and vice versa. In addition, the beamwidth of the antenna is also used to describe the resolution capabilities of the antenna to distinguish between two adjacent radiating sources or radar targets. The most common resolution criterion states that *the resolution capability of an antenna to distinguish between two sources is equal to half the first-null beamwidth (FNBW/2), which is usually used to approximate the half-power beamwidth (HPBW)* [5], [6]. That is, two sources separated by angular distances equal or greater than FNBW/2 ≈ HPBW of an antenna with a uniform distribution can be resolved. If the separation is smaller, then the antenna will tend to smooth the angular separation distance.

Example 2.4

The normalized radiation intensity of an antenna is represented by

$$U(\theta) = \cos^2(\theta)\cos^2(3\theta), \quad (0 \leq \theta \leq 90°, \quad 0° \leq \phi \leq 360°)$$

The three- and two-dimensional plots of this, plotted in a linear scale, are shown in Figure 2.11. Find the

a. half-power beamwidth HPBW (*in radians and degrees*)
b. first-null beamwidth FNBW (*in radians and degrees*)

Solution:

a. Since the $U(\theta)$ represents the *power* pattern, to find the half-power beamwidth you set the function equal to half of its maximum, or

$$U(\theta)|_{\theta=\theta_h} = \cos^2(\theta)\cos^2(3\theta)|_{\theta=\theta_h} = 0.5 \Rightarrow \cos\theta_h \cos 3\theta_h = 0.707$$

$$\theta_h = \cos^{-1}\left(\frac{0.707}{\cos 3\theta_h}\right)$$

Since this is an equation with transcendental functions, it can be solved iteratively. After a few iterations, it is found that

$$\theta_h \approx 0.25 \text{ radians} = 14.325°$$

Since the function $U(\theta)$ is symmetrical about the maximum at $\theta = 0$, then the HPBW is

$$\text{HPBW} = 2\theta_h \approx 0.50 \text{ radians} = 28.65°$$

b. To find the first-null beamwidth (FNBW), you set the $U(\theta)$ equal to zero, or

$$U(\theta)|_{\theta=\theta_n} = \cos^2(\theta)\cos^2(3\theta)|_{\theta=\theta_n} = 0$$

This leads to two solutions for θ_n.

$$\cos\theta_n = 0 \Rightarrow \theta_n = \cos^{-1}(0) = \frac{\pi}{2} \text{ radians} = 90°$$

$$\cos 3\theta_n = 0 \Rightarrow \theta_n = \frac{1}{3}\cos^{-1}(0) = \frac{\pi}{6} \text{ radians} = 30°$$

The one with the smallest value leads to the FNBW. Again, because of the symmetry of the pattern, the FNBW is

$$\text{FNBW} = 2\theta_n = \frac{\pi}{3} \text{ radians} = 60°$$

2.6 DIRECTIVITY

In the 1983 version of the *IEEE Standard Definitions of Terms for Antennas*, there has been a substantive change in the definition of *directivity*, compared to the definition of the 1973 version. Basically the term *directivity* in the new 1983 version has been used to replace the term *directive gain* of the old 1973 version. In the new 1983 version the term *directive gain* has been deprecated. According to the authors of the new 1983 standards, "this change brings this standard in line with common usage among antenna engineers and with other international standards, notably those of the International Electrotechnical Commission (IEC)." Therefore *directivity of an antenna* defined as "the ratio of the radiation intensity in a given direction from the antenna to the radiation intensity averaged over all directions. The average radiation intensity is equal to the total power radiated by the antenna divided by 4π. If the direction is not specified, the direction of maximum radiation intensity is implied." Stated more simply, the directivity of a nonisotropic source is equal to the ratio of its radiation intensity in a given direction over that of an isotropic source. In mathematical form, using (2-15), it can be written as

$$D = \frac{U}{U_0} = \frac{4\pi U}{P_{\text{rad}}} \tag{2-16}$$

If the direction is not specified, it implies the direction of maximum radiation intensity (maximum directivity) expressed as

$$D_{\text{max}} = D_0 = \frac{U|_{\text{max}}}{U_0} = \frac{U_{\text{max}}}{U_0} = \frac{4\pi U_{\text{max}}}{P_{\text{rad}}} \tag{2-16a}$$

42 FUNDAMENTAL PARAMETERS AND FIGURES-OF-MERIT OF ANTENNAS

D = directivity   (dimensionless)
D_0 = maximum directivity   (dimensionless)
U = radiation intensity   (W/unit solid angle)
U_{max} = maximum radiation intensity   (W/unit solid angle)
U_0 = radiation intensity of isotropic source   (W/unit solid angle)
P_{rad} = total radiated power   (W)

For an isotropic source, it is very obvious from (2-16) or (2-16a) that the directivity is unity since U, U_{max}, and U_0 are all equal to each other.

For antennas with orthogonal polarization components, we define the *partial directivity of an antenna for a given polarization in a given direction* as "that part of the radiation intensity corresponding to a given polarization divided by the total radiation intensity averaged over all directions." With this definition for the partial directivity, then in a given direction "the total directivity is the sum of the partial directivities for any two orthogonal polarizations." For a spherical coordinate system, the total maximum directivity D_0 for the orthogonal θ and ϕ components of an antenna can be written as

$$D_0 = D_\theta + D_\phi \qquad (2\text{-}17)$$

while the partial directivities D_θ and D_ϕ are expressed as

$$D_\theta = \frac{4\pi U_\theta}{(P_{rad})_\theta + (P_{rad})_\phi} \qquad (2\text{-}17a)$$

$$D_\phi = \frac{4\pi U_\phi}{(P_{rad})_\theta + (P_{rad})_\phi} \qquad (2\text{-}17b)$$

where

U_θ = radiation intensity in a given direction contained in θ field component
U_ϕ = radiation intensity in a given direction contained in ϕ field component
$(P_{rad})_\theta$ = radiated power in all directions contained in θ field component
$(P_{rad})_\phi$ = radiated power in all directions contained in ϕ field component

---

Example 2.5

As an illustration, find the maximum directivity of the antenna whose radiation intensity is that of Example 2.2. Write an expression for the directivity as a function of the directional angles θ and ϕ.

Solution: The radiation intensity is given by

$$U = r^2 W_{rad} = A_0 \sin\theta$$

The maximum radiation is directed along $\theta = \pi/2$. Thus

$$U_{max} = A_0$$

In Example 2.2 it was found that

$$P_{rad} = \pi^2 A_0$$

Using (2-16a), we find that the maximum directivity is equal to

$$D_0 = \frac{4\pi U_{max}}{P_{rad}} = \frac{4}{\pi} = 1.27$$

Since the radiation intensity is only a function of θ, the directivity as a function of the directional angles is represented by

$$D = D_0 \sin\theta = 1.27 \sin\theta$$

Before proceeding with a more general discussion of directivity, it may be proper at this time to consider another example, compute its directivity, compare it with that of the previous example, and comment on what it actually represents. This may give the reader a better understanding and appreciation of the directivity.

Example 2.6

The radial component of the radiated power density of an infinitesimal linear dipole of length $l \ll \lambda$ is given by

$$\mathbf{W}_{av} = \hat{\mathbf{a}}_r W_r = \hat{\mathbf{a}}_r A_0 \frac{\sin^2\theta}{r^2} \quad (W/m^2)$$

where A_0 is the peak value of the power density, θ is the usual spherical coordinate, and $\hat{\mathbf{a}}_r$ is the radial unit vector. Determine the maximum directivity of the antenna and express the directivity as a function of the directional angles θ and ϕ.

Solution: The radiation intensity is given by

$$U = r^2 W_r = A_0 \sin^2\theta$$

The maximum radiation is directed along $\theta = \pi/2$. Thus

$$U_{max} = A_0$$

The total radiated power is given by

$$P_{rad} = \oiint_\Omega U\, d\Omega = A_0 \int_0^{2\pi} \int_0^\pi \sin^2\theta \sin\theta\, d\theta\, d\phi = A_0 \left(\frac{8\pi}{3}\right)$$

Using (2-16a), we find that the maximum directivity is equal to

$$D_0 = \frac{4\pi U_{max}}{P_{rad}} = \frac{4\pi A_0}{\frac{8\pi}{3}(A_0)} = \frac{3}{2}$$

which is greater than 1.27 found in Example 2.5. Thus the directivity is represented by

$$D = D_0 \sin^2 \theta = 1.5 \sin^2 \theta$$

At this time it will be proper to comment on the results of Examples 2.5 and 2.6. To better understand the discussion, we have plotted in Figure 2.12 the relative radiation intensities of Example 2.5 ($U = A_0 \sin \theta$) and Example 2.6 ($U = A_0 \sin^2 \theta$) where A_0 was set equal to unity. We see that both patterns are omnidirectional but that of Example 2.6 has more directional characteristics (is narrower) in the elevation plane. Since the directivity is a "figure of merit" describing how well the radiator directs energy in a certain direction, it should be convincing from Figure 2.12 that the directivity of Example 2.6 should be higher than that of Example 2.5.

To demonstrate the significance of directivity, let us consider another example; in particular let us examine the directivity of a half-wavelength dipole ($l = \lambda/2$), which is derived in Section 4.6 of Chapter 4 and can be approximated by

$$D = D_0 \sin^3 \theta = 1.67 \sin^3 \theta \qquad (2\text{-}18)$$

since it can be shown that [see Figure 4.12(b)]

$$\sin^3 \theta \simeq \left[\frac{\cos\left(\frac{\pi}{2} \cos \theta\right)}{\sin \theta} \right]^2 \qquad (2\text{-}18a)$$

where θ is measured from the axis along the length of the dipole. The values represented by (2-18) and those of an isotropic source ($D = 1$) are plotted two- and three-dimensionally in Figure 2.13(a,b). For the three-dimensional graphical representation of Figure 2.13(b), at each observation point only the largest value of the two directivities is plotted. It is apparent that when

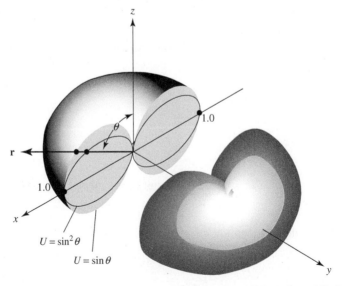

Figure 2.12 Three-dimensional radiation intensity patterns. (SOURCE: P. Lorrain and D. R. Corson, *Electromagnetic Fields and Waves*, 2nd ed., W. H. Freeman and Co. Copyright © 1970).

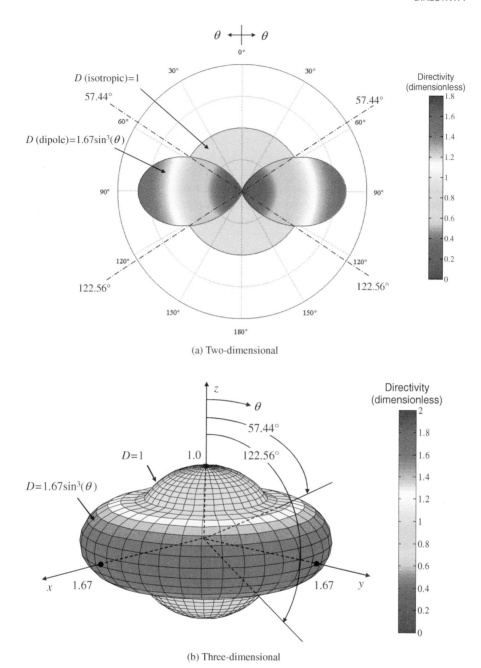

Figure 2.13 Two- and three-dimensional directivity patterns of a λ/2 dipole. (SOURCE: C. A. Balanis, "Antenna Theory: A Review." *Proc. IEEE*, Vol. 80, No. 1. January 1992. © 1992 IEEE).

$\sin^{-1}(1/1.67)^{1/3} = 57.44° < \theta < 122.56°$, the dipole radiator has greater directivity (greater intensity concentration) in those directions than that of an isotropic source. Outside this range of angles, the isotropic radiator has higher directivity (more intense radiation). The maximum directivity of the dipole (relative to the isotropic radiator) occurs when $\theta = \pi/2$, and it is 1.67 (or 2.23 dB) more intense than that of the isotropic radiator (with the same radiated power).

46 FUNDAMENTAL PARAMETERS AND FIGURES-OF-MERIT OF ANTENNAS

The three-dimensional pattern of Figure 2.13(b), and similar ones, are included throughout the book to represent the three-dimensional radiation characteristics of antennas. These patterns are plotted to visualize the three-dimensional radiation pattern of the antenna. These three-dimensional programs, along with the others, can be used effectively toward the design and synthesis of antennas, especially arrays, as demonstrated in [7] and [8]. A MATLAB-based program, designated as **Spherical**, is also included in the publisher's website to produce these and similar three-dimensional plots.

The directivity of an isotropic source is unity since its power is radiated equally well in all directions. *For all other sources, the maximum directivity will always be greater than unity, and it is a relative "figure of merit" which gives an indication of the directional properties of the antenna as compared with those of an isotropic source.* In equation form, this is indicated in (2-16a). The directivity can be smaller than unity; in fact it can be equal to zero. For Examples 2.5 and 2.6, the directivity is equal to zero in the $\theta = 0$ direction. *The values of directivity will be equal to or greater than zero and equal to or less than the maximum directivity ($0 \leq D \leq D_0$).*

A more general expression for the directivity can be developed to include sources with radiation patterns that may be functions of both spherical coordinate angles θ and ϕ. In the previous examples we considered intensities that were represented by only one coordinate angle θ, in order not to obscure the fundamental concepts by the mathematical details. So it may now be proper, since the basic definitions have been illustrated by simple examples, to formulate the more general expressions.

Let the radiation intensity of an antenna be of the form

$$U = B_0 F(\theta, \phi) \simeq \frac{1}{2\eta} \left[|E_\theta^0(\theta, \phi)|^2 + |E_\phi^0(\theta, \phi)|^2 \right] \qquad (2\text{-}19)$$

where B_0 is a constant, and E_θ^0 and E_ϕ^0 are the antenna's far-zone electric-field components. The maximum value of (2-19) is given by

$$U_{\max} = B_0 F(\theta, \phi)|_{\max} = B_0 F_{\max}(\theta, \phi) \qquad (2\text{-}19a)$$

The total radiated power is found using

$$P_{\text{rad}} = \oiint_\Omega U(\theta, \phi) \, d\Omega = B_0 \int_0^{2\pi} \int_0^{\pi} F(\theta, \phi) \sin\theta \, d\theta \, d\phi \qquad (2\text{-}20)$$

We now write the general expression for the directivity and maximum directivity using (2-16) and (2-16a), respectively, as

$$\boxed{D(\theta, \phi) = 4\pi \frac{F(\theta, \phi)}{\int_0^{2\pi} \int_0^{\pi} F(\theta, \phi) \sin\theta \, d\theta \, d\phi}} \qquad (2\text{-}21)$$

$$\boxed{D_0 = 4\pi \frac{F(\theta, \phi)|_{\max}}{\int_0^{2\pi} \int_0^{\pi} F(\theta, \phi) \sin\theta \, d\theta \, d\phi}} \qquad (2\text{-}22)$$

Equation (2-22) can also be written as

$$D_0 = \frac{4\pi}{\left[\int_0^{2\pi}\int_0^{\pi} F(\theta,\phi)\sin\theta\,d\theta\,d\phi\right]/F(\theta,\phi)|_{\max}} = \frac{4\pi}{\Omega_A} \quad (2\text{-}23)$$

where Ω_A is the beam solid angle, and it is given by

$$\Omega_A = \frac{1}{F(\theta,\phi)|_{\max}}\int_0^{2\pi}\int_0^{\pi} F(\theta,\phi)\sin\theta\,d\theta\,d\phi = \int_0^{2\pi}\int_0^{\pi} F_n(\theta,\phi)\sin\theta\,d\theta\,d\phi \quad (2\text{-}24)$$

$$F_n(\theta,\phi) = \frac{F(\theta,\phi)}{F(\theta,\phi)|_{\max}} \quad (2\text{-}25)$$

Dividing by $F(\theta,\phi)|_{\max}$ merely normalizes the radiation intensity $F(\theta,\phi)$, and it makes its maximum value unity.

The beam solid angle Ω_A is defined as the solid angle through which all the power of the antenna would flow if its radiation intensity is constant (and equal to the maximum value of U) for all angles within Ω_A.

2.6.1 Directional Patterns

Instead of using the exact expression of (2-23) to compute the directivity, it is often convenient to derive simpler expressions, even if they are approximate, to compute the directivity. These can also be used for design purposes. For antennas with one narrow major lobe and very negligible minor lobes, the beam solid angle is approximately equal to the product of the half-power beamwidths in two perpendicular planes [5] shown in Figure 2.14(a). For a rotationally symmetric pattern, the half-power beamwidths in any two perpendicular planes are the same, as illustrated in Figure 2.14(b).

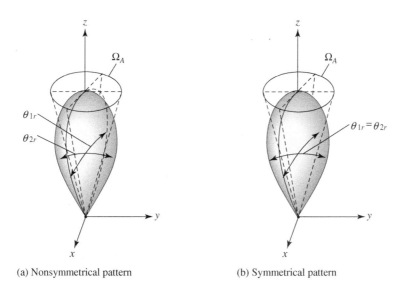

(a) Nonsymmetrical pattern   (b) Symmetrical pattern

Figure 2.14 Beam solid angles for nonsymmetrical and symmetrical radiation patterns.

48 FUNDAMENTAL PARAMETERS AND FIGURES-OF-MERIT OF ANTENNAS

With this approximation, (2-23) can be approximated by

$$D_0 = \frac{4\pi}{\Omega_A} \simeq \frac{4\pi}{\Theta_{1r}\Theta_{2r}} \qquad (2\text{-}26)$$

The beam solid angle Ω_A has been approximated by

$$\Omega_A \simeq \Theta_{1r}\Theta_{2r} \qquad (2\text{-}26a)$$

where

Θ_{1r} = half-power beamwidth in one plane   (rad)

Θ_{2r} = half-power beamwidth in a plane at a right angle to the other   (rad)

If the beamwidths are known in degrees, (2-26) can be written as

$$D_0 \simeq \frac{4\pi(180/\pi)^2}{\Theta_{1d}\Theta_{2d}} = \frac{41{,}253}{\Theta_{1d}\Theta_{2d}} \qquad (2\text{-}27)$$

where

Θ_{1d} = half-power beamwidth in one plane   (degrees)

Θ_{2d} = half-power beamwidth in a plane at a right angle to the other   (degrees)

For planar arrays, a better approximation to (2-27) is [9]

$$D_0 \simeq \frac{32{,}400}{\Omega_A(\text{degrees})^2} = \frac{32{,}400}{\Theta_{1d}\Theta_{2d}} \qquad (2\text{-}27a)$$

The validity of (2-26) and (2-27) is based on a pattern that has only one major lobe and any minor lobes, if present, should be of very low intensity. For a pattern with two identical major lobes, the value of the maximum directivity using (2-26) or (2-27) will be twice its actual value. For patterns with significant minor lobes, the values of maximum directivity obtained using (2-26) or (2-27), which neglect any minor lobes, will usually be too high.

Example 2.7

The radiation intensity of the major lobe of many antennas can be adequately represented by

$$U = B_0 \cos^4 \theta$$

where B_0 is the maximum radiation intensity. The radiation intensity exists only in the upper hemisphere ($0 \leq \theta \leq \pi/2, 0 \leq \phi \leq 2\pi$), and it is shown in Figure 2.15.
Find the

a. beam solid angle; exact and approximate.
b. maximum directivity; exact using (2-23) and approximate using (2-26).

DIRECTIVITY   49

Solution: The half-power point of the pattern occurs at $\theta = 32.765°$. Thus the beamwidth in the θ direction is 65.53° or

$$\Theta_{1r} = 1.1437 \text{ rads}$$

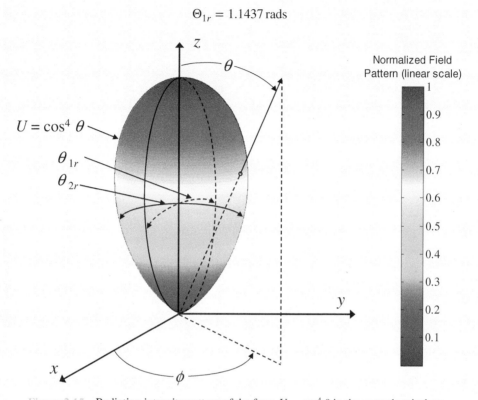

Figure 2.15 Radiation intensity pattern of the form $U = \cos^4 \theta$ in the upper hemisphere.

Since the pattern is independent of the ϕ coordinate, the beamwidth in the other plane is also equal to

$$\Theta_{2r} = 1.1437 \text{ rads}$$

a. *Beam solid angle Ω_A*:
   Exact: Using (2-24), (2-25)

$$\Omega_A = \int_0^{360°} \int_0^{90°} \cos^4 \theta \, d\Omega = \int_0^{2\pi} \int_0^{\pi/2} \cos^4 \theta \sin \theta \, d\theta \, d\phi$$

$$= \int_0^{2\pi} d\phi \int_0^{\pi/2} \cos^4 \theta \sin \theta \, d\theta$$

$$= 2\pi \int_0^{\pi/2} \cos^4 \theta \sin \theta \, d\theta = \frac{2\pi}{5} \text{ steradians}$$

Approximate: Using (2-26a)

$$\Omega_A \approx \Theta_{1r} \Theta_{2r} = 1.1437(1.1437) = (1.1437)^2 = 1.308 \text{ steradians}$$

b. *Directivity D_0:*

Exact: $D_0 = \dfrac{4\pi}{\Omega_A} = \dfrac{4\pi(5)}{2\pi} = 10$ (dimensionless) = 10 dB

The same exact answer is obtained using (2-16a).

Approximate: $D_0 \approx \dfrac{4\pi}{\Omega_A} = \dfrac{4\pi}{1.308} = 9.61$ (dimensionless) = 9.83 dB

The exact maximum directivity is 10 and its approximate value, using (2-26), is 9.61. Even better approximations can be obtained if the patterns have much narrower beamwidths, which will be demonstrated later in this section.

Many times it is desirable to express the directivity in decibels (dB) instead of dimensionless quantities. The expressions for converting the dimensionless quantities of directivity and maximum directivity to decibels (dB) are

$$D(\text{dB}) = 10\log_{10}[D(\text{dimensionless})] \tag{2-28a}$$

$$D_0(\text{dB}) = 10\log_{10}[D_0(\text{dimensionless})] \tag{2-28b}$$

It has also been proposed [10] that the maximum directivity of an antenna can also be obtained approximately by using the formula

$$\frac{1}{D_0} = \frac{1}{2}\left(\frac{1}{D_1} + \frac{1}{D_2}\right) \tag{2-29}$$

where

$$D_1 \simeq \frac{1}{\left[\dfrac{1}{2\ln 2}\displaystyle\int_0^{\Theta_{1r}/2} \sin\theta\, d\theta\right]} \simeq \frac{16\ln 2}{\Theta_{1r}^2} \tag{2-29a}$$

$$D_2 \simeq \frac{1}{\left[\dfrac{1}{2\ln 2}\displaystyle\int_0^{\Theta_{2r}/2} \sin\theta\, d\theta\right]} \simeq \frac{16\ln 2}{\Theta_{2r}^2} \tag{2-29b}$$

Θ_{1r} and Θ_{2r} are the half-power beamwidths (in radians) of the *E*- and *H*-planes, respectively. The formula of (2-29) will be referred to as the arithmetic mean of the maximum directivity. Using (2-29a) and (2-29b) we can write (2-29) as

$$\frac{1}{D_0} \simeq \frac{1}{2\ln 2}\left(\frac{\Theta_{1r}^2}{16} + \frac{\Theta_{2r}^2}{16}\right) = \frac{\Theta_{1r}^2 + \Theta_{2r}^2}{32\ln 2} \tag{2-30}$$

or

$$D_0 \simeq \frac{32\ln 2}{\Theta_{1r}^2 + \Theta_{2r}^2} = \frac{22.181}{\Theta_{1r}^2 + \Theta_{2r}^2} \tag{2-30a}$$

$$D_0 \simeq \frac{22.181(180/\pi)^2}{\Theta_{1d}^2 + \Theta_{2d}^2} = \frac{72{,}815}{\Theta_{1d}^2 + \Theta_{2d}^2} \tag{2-30b}$$

DIRECTIVITY 51

TABLE 2.1 Comparison of Exact and Approximate Values of Maximum Directivity for $U = \cos^n \theta$ Power Patterns

n	Exact Equation (2-22)	Kraus Equation (2-26)	Kraus % Error	Tai and Pereira Equation (2-30a)	Tai and Pereira % Error
1	4	2.86	−28.50	2.53	−36.75
2	6	5.09	−15.27	4.49	−25.17
3	8	7.35	−8.12	6.48	−19.00
4	10	9.61	−3.90	8.48	−15.20
5	12	11.87	−1.08	10.47	−12.75
6	14	14.13	+0.93	12.46	−11.00
7	16	16.39	+2.48	14.47	−9.56
8	18	18.66	+3.68	16.47	−8.50
9	20	20.93	+4.64	18.47	−7.65
10	22	23.19	+5.41	20.47	−6.96
11.28	24.56	26.08	+6.24	23.02	−6.24
15	32	34.52	+7.88	30.46	−4.81
20	42	45.89	+9.26	40.46	−3.67

where Θ_{1d} and Θ_{2d} are the half-power beamwidths in degrees. Equation (2-30a) is to be contrasted with (2-26) while (2-30b) should be compared with (2-27).

In order to make an evaluation and comparison of the accuracies of (2-26) and (2-30a), examples whose radiation intensities (power patterns) can be represented by

$$U(\theta, \phi) = \begin{cases} B_0 \cos^n(\theta) & 0 \leq \theta \leq \pi/2, \quad 0 \leq \phi \leq 2\pi \\ 0 & \text{elsewhere} \end{cases} \quad (2\text{-}31)$$

where $n = 1 - 10$, 11.28, 15, and 20 are considered. The maximum directivities were computed using (2-26) and (2-30a) and compared with the exact values as obtained using (2-23). The results are shown in Table 2.1. From the comparisons it is evident that the error due to Tai & Pereira's formula is always negative (i.e., it predicts lower values of maximum directivity than the exact ones) and monotonically decreases as n increases (the pattern becomes more narrow). However, the error due to Kraus' formula is negative for small values of n and positive for large values of n. For small values of n the error due to Kraus' formula is negative and positive for large values of n; the error is zero when $n = 5.497 \simeq 5.5$ (half-power beamwidth of $56.35°$). In addition, for symmetrically rotational patterns the absolute error due to the two approximate formulas is identical when $n = 11.28$, which corresponds to a half-power beamwidth of $39.77°$. From these observations we conclude that, Kraus' formula is more accurate for small values of n (broader patterns) while Tai & Pereira's is more accurate for large values of n (narrower patterns). Based on absolute error and symmetrically rotational patterns, Kraus' formula leads to smaller error for $n < 11.28$ (half-power beamwidth greater than $39.77°$) while Tai & Pereira's leads to smaller error for $n > 11.28$ (half-power beamwidth smaller than $39.77°$). The results are shown plotted in Figure 2.16 for $0 < n \leq 450$.

2.6.2 Omnidirectional Patterns

Some antennas (such as dipoles, loops, broadside arrays) exhibit omnidirectional patterns, as illustrated by the three-dimensional patterns in Figure 2.17 (a,b). As single-lobe directional patterns can be approximated by (2-31), omnidirectional patterns can often be approximated by

$$U = |\sin^n(\theta)| \quad 0 \leq \theta \leq \pi, \quad 0 \leq \phi \leq 2\pi \quad (2\text{-}32)$$

52 FUNDAMENTAL PARAMETERS AND FIGURES-OF-MERIT OF ANTENNAS

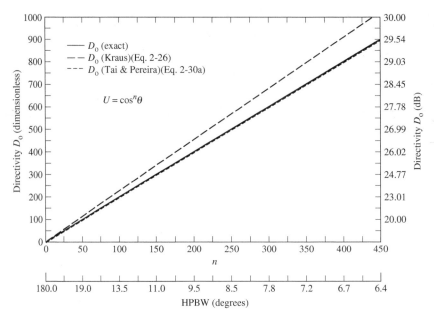

Figure 2.16 Comparison of exact and approximate values of directivity for directional $U = \cos^n \theta$ power patterns.

where n represents both integer and noninteger values. The directivity of antennas with patterns represented by (2-32) can be determined in closed form using the definition of (2-16a). However, as was done for the single-lobe patterns of Figure 2.14, approximate directivity formulas have been derived [11], [12] for antennas with omnidirectional patterns similar to the ones shown in Figure 2.17 whose main lobe is approximated by (2-32). The approximate directivity formula for an omnidirectional pattern as a function of the pattern half-power beamwidth (in degrees), which is reported by McDonald in [11], was derived based on the array factor of a broadside collinear array [see Section 6.4.1 and (6-38a)] and is given by

$$D_0 \simeq \frac{101}{\text{HPBW (degrees)} - 0.0027 \, [\text{HPBW (degrees)}]^2} \quad (2\text{-}33a)$$

However, that reported by Pozar in [12] is derived based on the exact values obtained using (2-32) and then representing the data in closed-form using curve-fitting, and it is given by

$$D_0 \simeq -172.4 + 191\sqrt{0.818 + 1/\text{HPBW (degrees)}} \quad (2\text{-}33b)$$

The approximate formula of (2-33a) should, in general, be more accurate for omnidirectional patterns with minor lobes, as shown in Figure 2.17(a), while (2-33b) should be more accurate for omnidirectional patterns with minor lobes of very low intensity (ideally no minor lobes), as shown in Figure 2.17(b).

The approximate formulas of (2-33a) and (2-33b) can be used to design omnidirectional antennas with specified radiation pattern characteristics. To facilitate this procedure, the directivity of antennas with omnidirectional patterns approximated by (2-32) is plotted in Figure 2.18 versus n and the

DIRECTIVITY 53

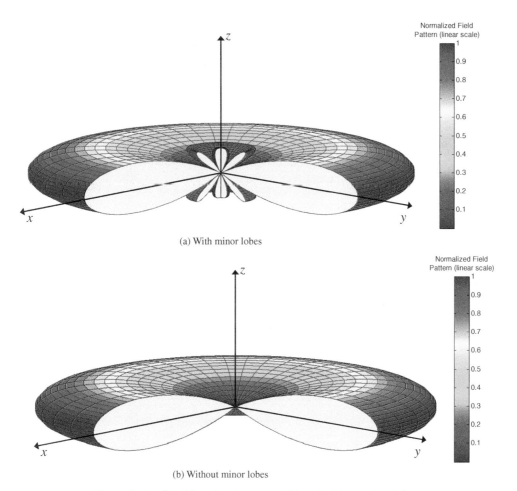

(a) With minor lobes

(b) Without minor lobes

Figure 2.17 Omnidirectional patterns with and without minor lobes.

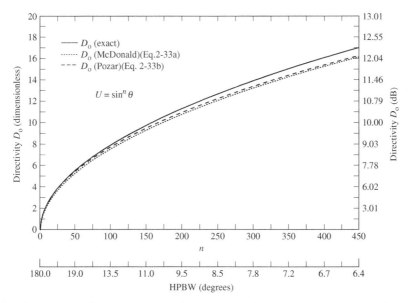

Figure 2.18 Comparison of exact and approximate values of directivity for omnidirectional $U = \sin^n \theta$ power patterns.

half-power beamwidth (in degrees). Three curves are plotted in Figure 2.18; one using (2-16a) and referred as *exact*, one using (2-33a) and denoted as *McDonald*, and the third using (2-33b) and denoted as *Pozar*. Thus, the curves of Figure 2.18 can be used for design purposes, as follows:

a. Specify the desired directivity and determine the value of n and half-power beamwidth of the omnidirectional antenna pattern, or
b. Specify the desired value of n or half-power beamwidth and determine the directivity of the omnidirectional antenna pattern.

To demonstrate the procedure, an example is taken.

Example 2.8

Design an antenna with omnidirectional amplitude pattern with a half-power beamwidth of 90°. Express its radiation intensity by $U = \sin^n \theta$. Determine the value of n and attempt to identify elements that exhibit such a pattern. Determine the directivity of the antenna using (2-16a), (2-33a), and (2-33b).

Solution: Since the half-power beamwidth is 90°, the angle at which the half-power point occurs is $\theta = 45°$. Thus

$$U(\theta = 45°) = 0.5 = \sin^n(45°) = (0.707)^n$$

or

$$n = 2$$

Therefore, the radiation intensity of the omnidirectional antenna is represented by $U = \sin^2 \theta$. An infinitesimal dipole (see Chapter 4) or a small circular loop (see Chapter 5) are two antennas which possess such a pattern.

Using the definition of (2-16a), the exact directivity is

$$U_{max} = 1$$

$$P_{rad} = \int_0^{2\pi} \int_0^{\pi} \sin^2 \theta \, \sin \theta \, d\theta \, d\phi = \frac{8\pi}{3}$$

$$D_0 = \frac{4\pi}{8\pi/3} = \frac{3}{2} = 1.761 \text{ dB}$$

Since the half-power beamwidth is equal to 90°, then the directivity based on (2-33a) is equal to

$$D_0 = \frac{101}{90 - 0.0027(90)^2} = 1.4825 = 1.71 \text{ dB}$$

while that based on (2-33b) is equal to

$$D_0 = -172.4 + 191\sqrt{0.818 + 1/90} = 1.516 = 1.807 \text{ dB}$$

The value of n and the three values of the directivity can also be obtained using Figure 2.18, although they may not be as accurate as those given above because they have to be taken off the graph. However, the curves can be used for other problems.

2.7 NUMERICAL TECHNIQUES

For most practical antennas, their radiation patterns are so complex that closed-form mathematical expressions are not available. Even in those cases where expressions are available, their form is so complex that integration to find the radiated power, required to compute the maximum directivity, cannot be performed. Instead of using the approximate expressions of Kraus, Tai and Pereira, McDonald, or Pozar alternate and more accurate techniques may be desirable. With the high-speed computer systems now available, the answer may be to apply numerical methods.

Let us assume that the radiation intensity of a given antenna is separable, and it is given by

$$U = B_0 f(\theta) g(\phi) \tag{2-34}$$

where B_0 is a constant. The directivity for such a system is given by

$$D_0 = \frac{4\pi U_{max}}{P_{rad}} \tag{2-35}$$

where

$$P_{rad} = B_0 \int_0^{2\pi} \left\{ \int_0^{\pi} f(\theta) g(\phi) \sin\theta \, d\theta \right\} d\phi \tag{2-36}$$

which can also be written as

$$P_{rad} = B_0 \int_0^{2\pi} g(\phi) \left\{ \int_0^{\pi} f(\theta) \sin\theta \, d\theta \right\} d\phi \tag{2-37}$$

If the integrations in (2-37) cannot be performed analytically, then from integral calculus we can write a series approximation

$$\int_0^{\pi} f(\theta) \sin\theta \, d\theta = \sum_{i=1}^{N} [f(\theta_i) \sin\theta_i] \Delta\theta_i \tag{2-38}$$

For N uniform divisions over the π interval,

$$\Delta\theta_i = \frac{\pi}{N} \tag{2-38a}$$

Referring to Figure 2.19, θ_i can take many different forms. Two schemes are shown in Figure 2.19 such that

$$\theta_i = i\left(\frac{\pi}{N}\right), \quad i = 1, 2, 3, \ldots, N \tag{2-38b}$$

or

$$\theta_i = \frac{\pi}{2N} + (i-1)\frac{\pi}{N}, \quad i = 1, 2, 3, \ldots, N \tag{2-38c}$$

In the former case, θ_i is taken at the trailing edge of each division; in the latter case, θ_i is selected at the middle of each division. The scheme that is more desirable will depend upon the problem under investigation. Many other schemes are available.

56   FUNDAMENTAL PARAMETERS AND FIGURES-OF-MERIT OF ANTENNAS

In a similar manner, we can write for the ϕ variations that

$$\int_0^{2\pi} g(\phi)\,d\phi = \sum_{j=1}^{M} g(\phi_j)\Delta\phi_j \tag{2-39}$$

where for M uniform divisions

$$\Delta\phi_j = \frac{2\pi}{M} \tag{2-39a}$$

Again referring to Figure 2.19

$$\phi_j = j\left(\frac{2\pi}{M}\right), \quad j = 1, 2, 3, \ldots, M \tag{2-39b}$$

or

$$\phi_j = \frac{2\pi}{2M} + (j-1)\frac{2\pi}{M}, \quad j = 1, 2, 3, \ldots, M \tag{2-39c}$$

Combining (2-38), (2-38a), (2-39), and (2-39a) we can write (2-37) as

$$P_{\text{rad}} = B_0\left(\frac{\pi}{N}\right)\left(\frac{2\pi}{M}\right)\sum_{j=1}^{M}\left\{g(\phi_j)\left[\sum_{i=1}^{N} f(\theta_i)\sin\theta_i\right]\right\} \tag{2-40}$$

The double summation of (2-40) is performed by adding for each value of $j(j = 1, 2, 3, \ldots, M)$ all values of $i(i = 1, 2, 3, \ldots, N)$. In a computer program flowchart, this can be performed by a loop within a loop. Physically, (2-40) can be interpreted by referring to Figure 2.19. It simply states that

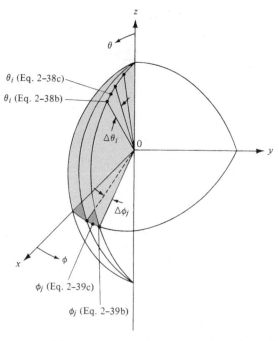

Figure 2.19 Digitization scheme of pattern in spherical coordinates.

for each value of $g(\phi)$ at the azimuthal angle $\phi = \phi_j$, the values of $f(\theta) \sin \theta$ are added for all values of $\theta = \theta_i (i = 1, 2, 3, \ldots, N)$. The values of θ_i and ϕ_j can be determined by using either of the forms as given by (2-38b) or (2-38c) and (2-39b) or (2-39c).

Since the θ and ϕ variations are separable, (2-40) can also be written as

$$P_{\text{rad}} = B_0 \left(\frac{\pi}{N}\right) \left(\frac{2\pi}{M}\right) \left[\sum_{j=1}^{M} g(\phi_j)\right] \left[\sum_{i=1}^{N} f(\theta_i) \sin \theta_i\right] \quad (2\text{-}41)$$

in which case each summation can be performed separately.

If the θ and ϕ variations are not separable, and the radiation intensity is given by

$$U = B_0 F(\theta, \phi) \quad (2\text{-}42)$$

the digital form of the radiated power can be written as

$$P_{\text{rad}} = B_0 \left(\frac{\pi}{N}\right) \left(\frac{2\pi}{M}\right) \sum_{j=1}^{M} \left[\sum_{i=1}^{N} F(\theta_i, \phi_j) \sin \theta_i\right] \quad (2\text{-}43)$$

θ_i and ϕ_j take different forms, two of which were introduced and are shown pictorially in Figure 2.19. The evaluation and physical interpretation of (2-43) is similar to that of (2-40).

To examine the accuracy of the technique, two examples will be considered.

Example 2.9(a)

The radiation intensity of an antenna is given by

$$U(\theta, \phi) = \begin{cases} B_0 \sin \theta \sin^2 \phi, & 0 \leq \theta \leq \pi, \quad 0 \leq \phi \leq \pi \\ 0 & \text{elsewhere} \end{cases}$$

The three-dimensional pattern of $U(\theta, \phi)$ is shown in Figure 2.20.

Determine the maximum directivity numerically by using (2-41) with θ_i and ϕ_j of (2-38b) and (2-39b), respectively. Compare it with the exact value.

Solution: Let us divide the θ and ϕ intervals each into 18 equals segments ($N = M = 18$). Since $0 \leq \phi \leq \pi$, then $\Delta\phi_j = \pi/M$ and (2-41) reduces to

$$P_{\text{rad}} = B_0 \left(\frac{\pi}{18}\right)^2 \left[\sum_{j=1}^{18} \sin^2 \phi_j\right] \left[\sum_{i=1}^{18} \sin^2 \theta_i\right]$$

with

$$\theta_i = i\left(\frac{\pi}{18}\right) = i(10°), \quad i = 1, 2, 3, \ldots, 18$$

$$\phi_j = j\left(\frac{\pi}{18}\right) = j(10°), \quad j = 1, 2, 3, \ldots, 18$$

58 FUNDAMENTAL PARAMETERS AND FIGURES-OF-MERIT OF ANTENNAS

$$U = \begin{cases} B_0 \sin(\theta)\sin^2(\phi) & \begin{array}{l} 0 \leq \theta \leq \pi \\ 0 \leq \phi \leq \pi \end{array} \\ 0 & \text{Elsewhere} \end{cases}$$

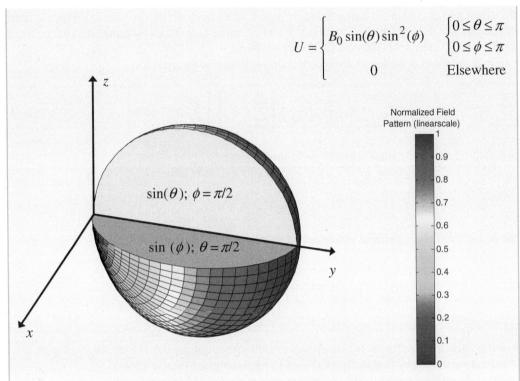

Figure 2.20 Three-dimensional pattern of the radiation of Examples 2.9(a,b).

Thus

$$P_{\text{rad}} = B_0 \left(\frac{\pi}{18}\right)^2 [\sin^2(10°) + \sin^2(20°) + \cdots + \sin^2(180°)]^2$$

$$P_{\text{rad}} = B_0 \left(\frac{\pi}{18}\right)^2 (9)^2 = B_0 \left(\frac{\pi^2}{4}\right)$$

and

$$D_0 = \frac{4\pi U_{\max}}{P_{\text{rad}}} = \frac{4\pi}{\pi^2/4} = \frac{16}{\pi} = 5.0929$$

The exact value is given by

$$P_{\text{rad}} = B_0 \int_0^\pi \sin^2\phi\, d\phi \int_0^\pi \sin^2\theta\, d\theta = \frac{\pi}{2}\left(\frac{\pi}{2}\right) B_0 = \frac{\pi^2}{4} B_0$$

and

$$D_0 = \frac{4\pi U_{\max}}{P_{\text{rad}}} = \frac{4\pi}{\pi^2/4} = \frac{16}{\pi} = 5.0929$$

which is the same as the value obtained numerically!

Example 2.9(b)

Given the same radiation intensity as that in Example 2.9(a), determine the directivity using (2-41) with θ_i and ϕ_j of (2-38c) and (2-39c).

Solution: Again using 18 divisions in each interval, we can write (2-41) as

$$P_{rad} = B_0 \left(\frac{\pi}{18}\right)^2 \left[\sum_{j=1}^{18} \sin^2 \phi_j\right] \left[\sum_{i=1}^{18} \sin^2 \theta_i\right]$$

with

$$\theta_i = \frac{\pi}{36} + (i-1)\frac{\pi}{18} = 5° + (i-1)10°, \quad i = 1, 2, 3, \ldots, 18$$

$$\phi_j = \frac{\pi}{36} + (j-1)\frac{\pi}{18} = 5° + (j-1)10°, \quad j = 1, 2, 3, \ldots, 18$$

Because of the symmetry of the divisions about the $\theta = \pi/2$ and $\phi = \pi/2$ angles, we can write

$$P_{rad} = B_0 \left(\frac{\pi}{18}\right)^2 \left[2\sum_{j=1}^{9} \sin^2 \phi_j\right] \left[2\sum_{i=1}^{9} \sin^2 \theta_i\right]$$

$$P_{rad} = B_0 \left(\frac{\pi}{18}\right)^2 4[\sin^2(5°) + \sin^2(15°) + \cdots + \sin^2(85°)]^2$$

$$P_{rad} = B_0 \left(\frac{\pi}{18}\right)^2 4(4.5)^2 = B_0 \left(\frac{\pi}{18}\right)^2 (81) = B_0 \left(\frac{\pi^2}{4}\right)$$

which is identical to that of the previous example. Thus

$$D_0 = \frac{4\pi U_{max}}{P_{rad}} = \frac{4\pi}{\pi^2/4} = \frac{16}{\pi} = 5.0929$$

which again is equal to the exact value!

It is interesting to note that decreasing the number of divisions (M and/or N) to 9, 6, 4, and even 2 leads to the same answer, which also happens to be the exact value! To demonstrate as to why the number of divisions does not affect the answer for this pattern, let us refer to Figure 2.21 where we have plotted the $\sin^2 \phi$ function and divided the $0° \leq \phi \leq 180°$ interval into six divisions. The exact value of the directivity uses the area under the solid curve. Doing the problem numerically, we find the area under the rectangles, which is shown shaded. Because of the symmetrical nature of the function, it can be shown that the shaded area in section #1 (included in the numerical evaluation) is equal to the blank area in section #1′ (left out by the numerical method). The same is true for the areas in sections #2 and #2′, and #3 and #3′. Thus, there is a one-to-one compensation. Similar justification is applicable for the other number of divisions.

It should be emphasized that all functions, even though they may contain some symmetry, do not give the same answers independent of the number of divisions. As a matter of fact, in most cases the answer only approaches the exact value as the number of divisions is increased to a large number.

60 FUNDAMENTAL PARAMETERS AND FIGURES-OF-MERIT OF ANTENNAS

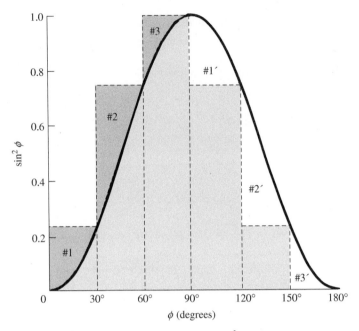

Figure 2.21 Digitized form of $\sin^2 \phi$ function.

A MATLAB and FORTRAN computer program called **Directivity** has been developed to compute the maximum directivity of any antenna whose radiation intensity is $U = F(\theta, \phi)$ based on the formulation of (2-43). The intensity function F does not have to be a function of both θ and ϕ. The numerical evaluations are made at the trailing edge, as defined by (2-38b) and (2-39b). The program is included in the publisher's website for this book. It contains a *subroutine* for which the intensity factor $U = F(\theta, \phi)$ for the required application must be specified by the user. As an illustration, the antenna intensity $U = \sin \theta \sin^2 \phi$ has been inserted in the subroutine. In addition, the upper and lower limits of θ and ϕ must be specified for each application of the same pattern.

2.8 ANTENNA EFFICIENCY

Associated with an antenna are a number of efficiencies and can be defined using Figure 2.22. The total antenna efficiency e_0 is used to take into account losses at the input terminals and within the structure of the antenna. Such losses may be due, referring to Figure 2.22(b), to

1. reflections because of the mismatch between the transmission line and the antenna
2. I^2R losses (conduction and dielectric)

In general, the overall efficiency can be written as

$$e_0 = e_r e_c e_d \tag{2-44}$$

where

e_0 = total efficiency   (dimensionless)

e_r = reflection (mismatch) efficiency = $(1 - |\Gamma|^2)$ (dimensionless)

e_c = conduction efficiency (dimensionless)

e_d = dielectric efficiency (dimensionless)

Γ = voltage reflection coefficient at the input terminals of the antenna
   $[\Gamma = (Z_{in} - Z_0)/(Z_{in} + Z_0)$ where Z_{in} = antenna input impedance, Z_0 = characteristic impedance of the transmission line]

VSWR = voltage standing wave ratio = $\dfrac{1 + |\Gamma|}{1 - |\Gamma|}$

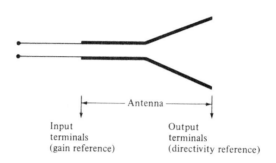

(a) Antenna reference terminals

(b) Reflection, conduction, and dielectric losses

Figure 2.22 Reference terminals and losses of an antenna.

Usually e_c and e_d are very difficult to compute, but they can be determined experimentally. Even by measurements they cannot be separated, and it is usually more convenient to write (2-44) as

$$e_0 = e_r e_{cd} = e_{cd}(1 - |\Gamma|^2) \qquad (2\text{-}45)$$

where $e_{cd} = e_c e_d$ = antenna radiation efficiency, which is used to relate the gain and directivity.

2.9 GAIN, REALIZED GAIN

Another useful figure-of-merit describing the performance of an antenna is the *gain*. Although the gain of the antenna is closely related to the directivity, it is a measure that takes into account the efficiency of the antenna as well as its directional capabilities. Remember that directivity is a measure that describes only the directional properties of the antenna, and it is therefore controlled only by the pattern.

Gain of an antenna (in a given direction) is defined as "the ratio of the intensity, in a given direction, to the radiation intensity that would be obtained if the power accepted by the antenna were radiated isotropically. The radiation intensity corresponding to the isotropically radiated power is equal to the power accepted (input) by the antenna divided by 4π." In equation form this can be expressed as

$$\text{Gain} = 4\pi \frac{\text{radiation intensity}}{\text{total input (accepted) power}} = 4\pi \frac{U(\theta,\phi)}{P_{in}} \quad \text{(dimensionless)} \quad (2\text{-}46)$$

In most cases we deal with *relative gain*, which is defined as "the ratio of the power gain in a given direction to the power gain of a reference antenna in its referenced direction." The power input must be the same for both antennas. The reference antenna is usually a dipole, horn, or any other antenna whose gain can be calculated or it is known. In most cases, however, the reference antenna is a *lossless isotropic source*. Thus

$$G = \frac{4\pi U(\theta,\phi)}{P_{in}(\text{lossless isotropic source})} \quad \text{(dimensionless)} \quad (2\text{-}46a)$$

When the direction is not stated, the power gain is usually taken in the direction of maximum radiation.

Referring to Figure 2.22(a), we can write that the total radiated power (P_{rad}) is related to the total input power (P_{in}) by

$$P_{rad} = e_{cd} P_{in} \quad (2\text{-}47)$$

where e_{cd} is the antenna radiation efficiency (dimensionless) which is defined in (2-44), (2-45) and Section 2.14 by (2-90). According to the IEEE Standards, "gain does not include losses arising from impedance mismatches (reflection losses) and polarization mismatches (losses)."

In this edition of the book we define two gains; one, referred to as *gain* (G), and the other, referred to as *realized gain* (G_{re}), that also takes into account the reflection/mismatch losses represented in both (2-44) and (2-45).

Using (2-47) reduces (2-46a) to

$$G(\theta,\phi) = e_{cd}\left[4\pi \frac{U(\theta,\phi)}{P_{rad}}\right] \quad (2\text{-}48)$$

which is related to the directivity of (2-16) and (2-21) by

$$G(\theta,\phi) = e_{cd} D(\theta,\phi) \quad (2\text{-}49)$$

In a similar manner, the maximum value of the gain is related to the maximum directivity of (2-16a) and (2-23) by

$$G_0 = G(\theta,\phi)|_{max} = e_{cd} D(\theta,\phi)|_{max} = e_{cd} D_0 \quad (2\text{-}49a)$$

While (2-47) does take into account the losses of the antenna element itself, *it does not take into account the losses when the antenna element is connected to a transmission line*, as shown in Figure 2.22. These connection losses are usually referred to as *reflections (mismatch) losses*, and they are taken into account by introducing a reflection (mismatch) efficiency e_r, which is related to the reflection coefficient as represented in (2-45) or $e_r = (1 - |\Gamma|^2)$. Thus, we can introduce a *realized gain* G_{re} that takes into account the reflection/mismatch losses (due to the connection of the antenna element to the transmission line), and it can be written as

$$G_{re}(\theta, \phi) = e_r G(\theta, \phi) = (1 - |\Gamma|^2) G(\theta, \phi)$$
$$= e_r e_{cd} D(\theta, \phi) = e_o D(\theta, \phi) \qquad (2\text{-}49b)$$

where e_o is the overall efficiency as defined in (2-44), (2-45). Similarly, the *maximum realized* gain G_{re0} of (2-49a) is related to the maximum directivity D_0 by

$$G_{re0} = G_{re}(\theta, \phi)|_{max} = e_r G(\theta, \phi)|_{max} = (1 - |\Gamma|^2) G(\theta, \phi)|_{max}$$
$$= e_r e_{cd} D(\theta, \phi)|_{max} = e_o D(\theta, \phi)|_{max} = e_o D_0 \qquad (2\text{-}49c)$$

If the antenna is matched to the transmission line, that is, the antenna input impedance Z_{in} is equal to the characteristic impedance Z_c of the line ($|\Gamma| = 0$), then the two gains are equal ($G_{re} = G$).

As was done with the directivity, we can define the *partial gain of an antenna for a given polarization in a given direction* as "that part of the radiation intensity corresponding to a given polarization divided by the total radiation intensity that would be obtained if the power accepted by the antenna were radiated isotropically." With this definition for the partial gain, then, in a given direction, "the total gain is the sum of the partial gains for any two orthogonal polarizations." For a spherical coordinate system, the total maximum gain G_0 for the orthogonal θ and ϕ components of an antenna can be written, in a similar form as was the maximum directivity in (2-17)–(2-17b), as

$$G_0 = G_\theta + G_\phi \qquad (2\text{-}50)$$

while the partial gains G_θ and G_ϕ are expressed as

$$G_\theta = \frac{4\pi U_\theta}{P_{in}} \qquad (2\text{-}50a)$$

$$G_\phi = \frac{4\pi U_\phi}{P_{in}} \qquad (2\text{-}50b)$$

where
  U_θ = radiation intensity in a given direction contained in E_θ field component
  U_ϕ = radiation intensity in a given direction contained in E_ϕ field component
  P_{in} = total input (accepted) power

For many practical antennas an approximate formula for the gain, corresponding to (2-27) or (2-27a) for the directivity, is

$$G_0 \simeq \frac{30{,}000}{\Theta_{1d}\Theta_{2d}} \qquad (2\text{-}51)$$

In practice, whenever the term "gain" is used, it usually refers to the *maximum gain* as defined by (2-49a) or (2-49c).

Usually the gain is given in terms of decibels, instead of the dimensionless quantity of (2-49a). The conversion formula is

$$G_0(\text{dB}) = 10 \log_{10}[e_{cd} D_0 \text{ (dimensionless)}] \qquad (2\text{-}52)$$

64   FUNDAMENTAL PARAMETERS AND FIGURES-OF-MERIT OF ANTENNAS

Example 2.10

A lossless resonant half-wavelength dipole antenna, with input impedance of 73 ohms, is connected to a transmission line whose characteristic impedance is 50 ohms. Assuming that the pattern of the antenna is given approximately by

$$U = B_0 \sin^3 \theta$$

find the maximum realized gain of this antenna.

Solution: Let us first compute the maximum directivity of the antenna. For this

$$U|_{max} = U_{max} = B_0$$

$$P_{rad} = \int_0^{2\pi} \int_0^{\pi} U(\theta, \phi) \sin\theta\, d\theta\, d\phi = 2\pi B_0 \int_0^{\pi} \sin^4\theta\, d\theta = B_0 \left(\frac{3\pi^2}{4}\right)$$

$$D_0 = 4\pi \frac{U_{max}}{P_{rad}} = \frac{16}{3\pi} = 1.697$$

Since the antenna was stated to be lossless, then the radiation efficiency $e_{cd} = 1$.

Thus, the total maximum gain is equal to

$$G_0 = e_{cd} D_0 = 1(1.697) = 1.697$$

$$G_0(\text{dB}) = 10 \log_{10}(1.697) = 2.297$$

which is identical to the directivity because the antenna is lossless.

There is another loss factor which is not taken into account in the gain. That is the loss due to reflection or mismatch losses between the antenna (load) and the transmission line. This loss is accounted for by the reflection efficiency of (2-44) or (2-45), and it is equal to

$$e_r = (1 - |\Gamma|^2) = \left(1 - \left|\frac{73 - 50}{73 + 50}\right|^2\right) = 0.965$$

$$e_r(\text{dB}) = 10 \log_{10}(0.965) = -0.155$$

Therefore the overall efficiency is

$$e_0 = e_r e_{cd} = 0.965$$

$$e_0(\text{dB}) = -0.155$$

Thus, the overall losses are equal to 0.155 dB. The maximum realized gain is equal to

$$G_{re0} = e_0 D_0 = 0.965(1.697) = 1.6376$$

$$G_{re0}(\text{dB}) = 10 \log_{10}(1.6376) = 2.142$$

The gain in dB can also be obtained by converting the directivity and radiation efficiency in dB and then adding them. Thus,

$$e_{cd}(\text{dB}) = 10 \log_{10}(1.0) = 0$$

$$D_0(\text{dB}) = 10 \log_{10}(1.697) = 2.297$$

$$G_0(\text{dB}) = e_{cd}(\text{dB}) + D_0(\text{dB}) = 2.297$$

which is the same as obtained previously. The same procedure can be used for the realized gain.

2.10 BEAM EFFICIENCY

Another parameter that is frequently used to judge the quality of transmitting and receiving antennas is the *beam efficiency*. For an antenna with its major lobe directed along the z-axis ($\theta = 0$), as shown in Figure 2.1(a), the beam efficiency (BE) is defined by

$$\text{BE} = \frac{\text{power transmitted (received) within cone angle } \theta_1}{\text{power transmitted (received) by the antenna}} \text{(dimensionless)} \quad (2\text{-}53)$$

where θ_1 is the half-angle of the cone within which the percentage of the total power is to be found. Equation (2-53) can be written as

$$\text{BE} = \frac{\int_0^{2\pi} \int_0^{\theta_1} U(\theta, \phi) \sin\theta \, d\theta \, d\phi}{\int_0^{2\pi} \int_0^{\pi} U(\theta, \phi) \sin\theta \, d\theta \, d\phi} \quad (2\text{-}54)$$

If θ_1 is chosen as the angle where the first null or minimum occurs (see Figure 2.1), then the beam efficiency will indicate the amount of power in the major lobe compared to the total power. A very high beam efficiency (between the nulls or minima), usually in the high 90s, is necessary for antennas used in radiometry, astronomy, radar, and other applications where received signals through the minor lobes must be minimized. The beam efficiencies of some typical rectangular and circular aperture antennas will be discussed in Chapter 12.

2.11 BANDWIDTH

The *bandwidth* of an antenna is defined as "the range of frequencies within which the performance of the antenna, with respect to some characteristic, conforms to a specified standard." The bandwidth can be considered to be the range of frequencies, on either side of a center frequency (usually the resonance frequency for a dipole), where the antenna characteristics (such as input impedance, pattern, beamwidth, polarization, side lobe level, gain, beam direction, radiation efficiency) are within an acceptable value of those at the center frequency.

- For *broadband antennas*, the bandwidth is usually expressed as the ratio of the upper-to-lower frequencies of acceptable operation. For example, a 10:1 bandwidth indicates that the upper frequency is 10 times greater than the lower.
- For *narrowband antennas*, the bandwidth is expressed as a percentage of the frequency difference (upper minus lower) over the center frequency of the bandwidth. For example, a 5% bandwidth indicates that the frequency range of acceptable operation is 5% of the bandwidth center frequency.

Because the characteristics (input impedance, pattern, gain, polarization, etc.) of an antenna do not necessarily vary in the same manner or are even critically affected by the frequency, there is no unique characterization of the bandwidth. The specifications are set in each case to meet the needs of the particular application. Usually there is a distinction made between pattern and input impedance variations. Accordingly *pattern bandwidth* and *impedance bandwidth* are used to emphasize this distinction. Associated with pattern bandwidth are gain, side lobe level, beamwidth, polarization, and beam direction while input impedance and radiation efficiency are related to impedance bandwidth. For example, the pattern of a linear dipole with overall length less than a half-wavelength ($l < \lambda/2$)

66   FUNDAMENTAL PARAMETERS AND FIGURES-OF-MERIT OF ANTENNAS

is basically insensitive to frequency. The limiting factor for this antenna is its impedance, and its bandwidth can be formulated in terms of the Q. The Q of antennas or arrays with dimensions large compared to the wavelength, excluding superdirective designs, is near unity. Therefore the bandwidth is usually formulated in terms of beamwidth, side lobe level, and pattern characteristics. For intermediate length antennas, the bandwidth may be limited by either pattern or impedance variations, depending upon the particular application. For these antennas, a 2:1 bandwidth indicates a good design. For others, large bandwidths are needed. Antennas with very large bandwidths (like 40:1 or greater) have been designed in recent years. These are known as *frequency independent* antennas, and they are discussed in Chapter 11.

The above discussion presumes that the coupling networks (transformers, baluns, etc.) and/or the dimensions of the antenna are not altered in any manner as the frequency is changed. It is possible to increase the acceptable frequency range of a narrowband antenna if proper adjustments can be made on the critical dimensions of the antenna and/or on the coupling networks as the frequency is changed. Although not an easy or possible task in general, there are applications where this can be accomplished. The most common examples are the old whip antenna of a car radio and the "rabbit ears" of a television. Both usually have adjustable lengths which can be used to tune the antenna for better reception. Antennas of this type, whose adjustments are made to modify their radiation characteristics, are often referred to as *reconfigurable antennas*.

2.12 POLARIZATION

Polarization of an antenna in a given direction is defined as "the polarization of the wave transmitted (radiated) by the antenna. *Note:* When the direction is not stated, the polarization is taken to be the polarization in the direction of maximum gain." In practice, polarization of the radiated energy varies with the direction from the center of the antenna, so that different parts of the pattern may have different polarizations.

Polarization of a radiated wave is defined as "that property of an electromagnetic wave describing the time-varying direction and relative magnitude of the electric-field vector; specifically, the figure traced as a function of time by the extremity of the vector at a fixed location in space, and the sense in which it is traced, *as observed along the direction of propagation*." Polarization then is the curve traced by the end point of the arrow (vector) representing the instantaneous electric field. The field must be observed along the direction of propagation. A typical trace as a function of time is shown in Figures 2.23(a) and (b). Animation of the these two traces can be performed using the MATLAB computer program **Polarization_Diagram_Ellipse_Animation** found in the publisher's website for this book.

The polarization of a wave can be defined in terms of a wave *radiated (transmitted)* or *received* by an antenna in a given direction. The polarization of a wave *radiated* by an antenna in a specified direction at a point in the far field is defined as "the polarization of the (locally) plane wave which is used to represent the radiated wave at that point. At any point in the far field of an antenna the radiated wave can be represented by a plane wave whose electric-field strength is the same as that of the wave and whose direction of propagation is in the radial direction from the antenna. As the radial distance approaches infinity, the radius of curvature of the radiated wave's phase front also approaches infinity and thus in any specified direction the wave appears locally as a plane wave." This is a far-field characteristic of waves radiated by all practical antennas, and it is illustrated analytically in Section 3.6 of Chapter 3. The polarization of a wave *received* by an antenna is defined as the "polarization of a plane wave, incident from a given direction and having a given power flux density, which results in maximum available power at the antenna terminals."

Polarization may be classified as linear, circular, or elliptical. If the vector that describes the electric field at a point in space as a function of time is always directed along a line, the field is said to be *linearly* polarized. In general, however, the figure that the electric field traces is an ellipse, and the field is said to be elliptically polarized. Linear and circular polarizations are special cases of elliptical,

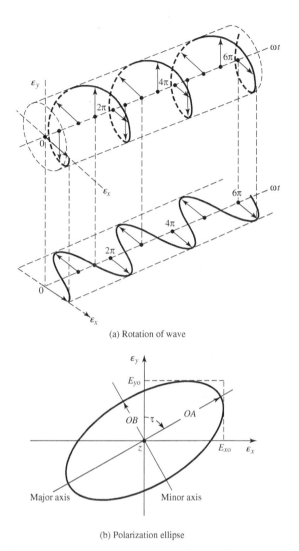

Figure 2.23 Rotation of a plane electromagnetic wave and its polarization ellipse at $z = 0$ as a function of time.

and they can be obtained when the ellipse becomes a straight line or a circle, respectively. The figure of the electric field is traced in a *clockwise* (CW) or *counterclockwise* (CCW) sense. *Clockwise* rotation of the electric-field vector is also designated as *right-hand polarization* and *counterclockwise* as *left-hand polarization*.

In general, the polarization characteristics of an antenna can be represented by its *polarization pattern* whose one definition is "the spatial distribution of the polarizations of a field vector excited (radiated) by an antenna taken over its radiation sphere. When describing the polarizations over the radiation sphere, or portion of it, reference lines shall be specified over the sphere, in order to measure the tilt angles (see tilt angle) of the polarization ellipses and the direction of polarization for linear polarizations. An obvious choice, though by no means the only one, is a family of lines tangent at each point on the sphere to either the θ or ϕ coordinate line associated with a spherical coordinate system of the radiation sphere. At each point on the radiation sphere the polarization is usually resolved into a pair of orthogonal polarizations, the *co-polarization* and *cross polarization*. To accomplish this, the co-polarization must be specified at each point on the radiation sphere." "*Co-polarization* represents the polarization the antenna is intended to radiate (receive) while *cross-polarization* represents the polarization orthogonal to a specified polarization, which is usually the co-polarization."

"For certain linearly polarized antennas, it is common practice to define the co-polarization in the following manner: First specify the orientation of the co-polar electric-field vector at a pole of the radiation sphere. Then, for all other directions of interest (points on the radiation sphere), require that the angle that the co-polar electric-field vector makes with each great circle line through the pole remain constant over that circle, the angle being that at the pole."

"In practice, the axis of the antenna's main beam should be directed along the polar axis of the radiation sphere. The antenna is then appropriately oriented about this axis to align the direction of its polarization with that of the defined co-polarization at the pole." "This manner of defining co-polarization can be extended to the case of elliptical polarization by defining the constant angles using the major axes of the polarization ellipses rather than the co-polar electric-field vector. The sense of polarization (rotation) must also be specified."

The polarization of the wave radiated by the antenna can also be represented on the Poincaré sphere [13]–[16]. Each point on the Poincaré sphere represents a unique polarization. The north pole represents left circular polarization, the south pole represents right circular, and points along the equator represent linear polarization of different tilt angles. All other points on the Poincaré sphere represent elliptical polarization. For details, see Figure 17.24 of Chapter 17.

The polarization of an antenna is measured using techniques described in Chapter 17.

2.12.1 Linear, Circular, and Elliptical Polarizations

The instantaneous field of a plane wave, traveling in the negative z direction, can be written as

$$\boldsymbol{\mathscr{E}}(z;t) = \hat{\mathbf{a}}_x \mathscr{E}_x(z;t) + \hat{\mathbf{a}}_y \mathscr{E}_y(z;t) \tag{2-55}$$

According to (2-5), the instantaneous components are related to their complex counterparts by

$$\mathscr{E}_x(z;t) = \mathrm{Re}[E_x^- e^{j(\omega t + kz)}] = \mathrm{Re}[E_{xo} e^{j(\omega t + kz + \phi_x)}]$$
$$= E_{xo} \cos(\omega t + kz + \phi_x) \tag{2-56}$$
$$\mathscr{E}_y(z;t) = \mathrm{Re}[E_y^- e^{j(\omega t + kz)}] = \mathrm{Re}[E_{yo} e^{j(\omega t + kz + \phi_y)}]$$
$$= E_{yo} \cos(\omega t + kz + \phi_y) \tag{2-57}$$

where E_{xo} and E_{yo} are, respectively, the maximum magnitudes of the x and y components.

A. Linear Polarization

For the wave to have linear polarization, the time-phase difference between the two components must be

$$\Delta\phi = \phi_y - \phi_x = n\pi, \quad n = 0, 1, 2, 3, \ldots \tag{2-58}$$

B. Circular Polarization

Circular polarization can be achieved *only* when the magnitudes of the two components are the same *and* the time-phase difference between them is odd multiples of $\pi/2$. That is,

$$|\mathscr{E}_x| = |\mathscr{E}_y| \Rightarrow E_{xo} = E_{yo} \tag{2-59}$$

$$\Delta\phi = \phi_y - \phi_x = \begin{cases} +(\tfrac{1}{2} + 2n)\pi, n = 0, 1, 2, \ldots & \text{for CW} \\ -(\tfrac{1}{2} + 2n)\pi, n = 0, 1, 2, \ldots & \text{for CCW} \end{cases} \tag{2-60}\tag{2-61}$$

If the direction of wave propagation is reversed (i.e., $+z$ direction), the phases in (2-60) and (2-61) for CW and CCW rotation must be interchanged.

C. Elliptical Polarization

Elliptical polarization can be attained *only* when the time-phase difference between the two components is odd multiples of $\pi/2$ *and* their magnitudes are not the same *or* when the time-phase difference between the two components is not equal to multiples of $\pi/2$ (irrespective of their magnitudes). That is,

$$|\mathscr{E}_x| \neq |\mathscr{E}_y| \Rightarrow E_{xo} \neq E_{yo}$$

$$\text{when } \Delta\phi = \phi_y - \phi_x = \begin{cases} +(\tfrac{1}{2} + 2n)\pi & \text{for CW} \quad (2\text{-}62a) \\ -(\tfrac{1}{2} + 2n)\pi & \text{for CCW} \quad (2\text{-}62b) \end{cases}$$
$$n = 0, 1, 2, \ldots$$

or

$$\Delta\phi = \phi_y - \phi_x \neq \pm\frac{n}{2}\pi = \begin{cases} > 0 & \text{for CW} \quad (2\text{-}63) \\ < 0 & \text{for CCW} \quad (2\text{-}64) \end{cases}$$
$$n = 0, 1, 2, 3, \ldots$$

For elliptical polarization, the curve traced at a given position as a function of time is, in general, a tilted ellipse, as shown in Figure 2.23(b). The ratio of the major axis to the minor axis is referred to as the axial ratio (AR), and it is equal to

$$\text{AR} = \frac{\text{major axis}}{\text{minor axis}} = \frac{OA}{OB}, \quad 1 \leq \text{AR} \leq \infty \tag{2-65}$$

where

$$OA = \left[\tfrac{1}{2}\{E_{xo}^2 + E_{yo}^2 + [E_{xo}^4 + E_{yo}^4 + 2E_{xo}^2 E_{yo}^2 \cos(2\Delta\phi)]^{1/2}\}\right]^{1/2} \tag{2-66}$$

$$OB = \left[\tfrac{1}{2}\{E_{xo}^2 + E_{yo}^2 - [E_{xo}^4 + E_{yo}^4 + 2E_{xo}^2 E_{yo}^2 \cos(2\Delta\phi)]^{1/2}\}\right]^{1/2} \tag{2-67}$$

The tilt of the ellipse, *relative to the y axis*, is represented by the angle τ given by

$$\tau = \frac{\pi}{2} - \frac{1}{2}\tan^{-1}\left[\frac{2E_{xo}E_{yo}}{E_{xo}^2 - E_{yo}^2}\cos(\Delta\phi)\right] \tag{2-68}$$

When the ellipse is aligned with the principal axes $[\tau = n\pi/2, n = 0, 1, 2, \ldots]$, the major (minor) axis is equal to $E_{xo}(E_{yo})$ or $E_{yo}(E_{xo})$ and the axial ratio is equal to E_{xo}/E_{yo} or E_{yo}/E_{xo}.

SUMMARY

We will summarize the preceding discussion on polarization by stating the general characteristics, and the *necessary and sufficient* conditions that the wave must have in order to possess *linear, circular* or *elliptical* polarization.

Linear Polarization A *time-harmonic wave is linearly polarized, at a given point in space, if the electric-field (or magnetic-field) vector at that point is always oriented along the same straight line at every instant of time.* This is accomplished if the field vector (electric or magnetic) possesses:

a. Only one component, or
b. Two orthogonal linear components that are in time phase or 180° (or multiples of 180°) out-of-phase.

Circular Polarization A *time-harmonic wave is circularly polarized, at a given point in space, if the electric (or magnetic) field vector at that point traces a circle as a function of time.*

The *necessary and sufficient* conditions to accomplish this are if the field vector (electric or magnetic) possesses *all* of the following:

a. The field must have two orthogonal linear components, and
b. The two components must have the same magnitude, and
c. The two components must have a time-phase difference of odd multiples of 90°.

The sense of rotation is always determined by rotating the phase-leading component toward the phase-lagging component and observing the field rotation as the wave is viewed as it travels away from the observer. If the rotation is clockwise, the wave is right-hand (or clockwise) circularly polarized; if the rotation is counterclockwise, the wave is left-hand (or counterclockwise) circularly polarized. The rotation of the phase-leading component toward the phase-lagging component should be performed along the angular separation between the two components that is less than 180°. Phases equal to or greater than 0° and less than 180° should be considered leading whereas those equal to or greater than 180° and less than 360° should be considered lagging.

Elliptical Polarization A *time-harmonic wave is elliptically polarized if the tip of the field vector (electric or magnetic) traces an elliptical locus in space. At various instants of time the field vector changes continuously with time at such a manner as to describe an elliptical locus. It is right-hand (clockwise) elliptically polarized if the field vector rotates clockwise, and it is left-hand (counterclockwise) elliptically polarized if the field vector of the ellipse rotates counterclockwise* [13]. The sense of rotation is determined using the same rules as for the circular polarization. In addition to the sense of rotation, elliptically polarized waves are also specified by their *axial ratio* whose magnitude is the ratio of the major to the minor axis.

A wave is elliptically polarized if it is not linearly or circularly polarized. Although linear and circular polarizations are special cases of elliptical, usually in practice elliptical polarization refers to other than linear or circular. The *necessary and sufficient* conditions to accomplish this are if the field vector (electric or magnetic) possesses *all* of the following:

a. The field must have two orthogonal linear components, and
b. The two components can be of the same or different magnitude.
c. (1) If the two components are not of the same magnitude, the time-phase difference between the two components must not be 0° or multiples of 180° (because it will then be linear). (2) If the two components are of the same magnitude, the time-phase difference between the two components must not be odd multiples of 90° (because it will then be circular).

If the wave is elliptically polarized with two components not of the same magnitude but with odd multiples of 90° time-phase difference, the polarization ellipse will not be tilted but it will be aligned with the principal axes of the field components. The major axis of the ellipse will align with the axis of the field component which is larger of the two, while the minor axis of the ellipse will align with the axis of the field component which is smaller of the two.

2.12.2 Polarization Loss Factor and Efficiency

In general, the polarization of the receiving antenna will not be the same as the polarization of the incoming (incident) wave. This is commonly stated as "polarization mismatch." The amount of power extracted by the antenna from the incoming signal will not be maximum because of the polarization loss. Assuming that the electric field of the incoming wave can be written as

$$\mathbf{E}_i = \hat{\boldsymbol{\rho}}_w E_i \qquad (2\text{-}69)$$

where $\hat{\boldsymbol{\rho}}_w$ is the unit vector of the wave, and the polarization of the electric field of the receiving antenna can be expressed as

$$\mathbf{E}_a = \hat{\boldsymbol{\rho}}_a E_a \qquad (2\text{-}70)$$

where $\hat{\boldsymbol{\rho}}_a$ is its unit vector (polarization vector), the polarization loss can be taken into account by introducing a *polarization loss factor* (PLF). It is defined, based on the polarization of the antenna in its transmitting mode, as

$$\text{PLF} = |\hat{\boldsymbol{\rho}}_w \cdot \hat{\boldsymbol{\rho}}_a|^2 = |\cos \psi_p|^2 \text{ (dimensionless)} \qquad (2\text{-}71)$$

where ψ_p is the angle between the two unit vectors. The relative alignment of the polarization of the incoming wave and of the antenna is shown in Figure 2.24. If the antenna is polarization matched, its PLF will be unity and the antenna will extract maximum power from the incoming wave.

Another figure of merit that is used to describe the polarization characteristics of a wave and that of an antenna is the *polarization efficiency* (*polarization mismatch* or *loss factor*) which is defined as "the ratio of the power received by an antenna from a given plane wave of arbitrary polarization to the power that would be received by the same antenna from a plane wave of the same power flux density and direction of propagation, whose state of polarization has been adjusted for a maximum received power." This is similar to the PLF and it is expressed as

$$p_e = \frac{|\boldsymbol{\ell}_e \cdot \mathbf{E}^{\text{inc}}|^2}{|\boldsymbol{\ell}_e|^2 |\mathbf{E}^{\text{inc}}|^2} \qquad (2\text{-}71\text{a})$$

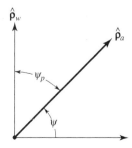

Figure 2.24 Polarization unit vectors of incident wave ($\hat{\boldsymbol{\rho}}_w$) and antenna ($\hat{\boldsymbol{\rho}}_a$), and polarization loss factor (PLF).

72 FUNDAMENTAL PARAMETERS AND FIGURES-OF-MERIT OF ANTENNAS

where

ℓ_e = vector effective length of the antenna

$\mathbf{E}^{inc}$ = incident electric field

The vector effective length ℓ_e of the antenna has not yet been defined, and it is introduced in Section 2.15. It is a vector that describes the polarization characteristics of the antenna. Both the PLF and p_e lead to the same answers.

The conjugate (*) is not used in (2-71) or (2-71a) so that a right-hand circularly polarized incident wave (when viewed in its direction of propagation) is matched to right-hand circularly polarized receiving antenna (when its polarization is determined in the transmitting mode). Similarly, a left-hand circularly polarized wave will be matched to a left-hand circularly polarized antenna.

To illustrate the principle of polarization mismatch, two examples are considered.

Example 2.11

The electric field of a linearly polarized electromagnetic wave given by

$$\mathbf{E}_i = \hat{\mathbf{a}}_x E_0(x, y) e^{-jkz}$$

is incident upon a linearly polarized antenna whose electric-field polarization is expressed as

$$\mathbf{E}_a \simeq (\hat{\mathbf{a}}_x + \hat{\mathbf{a}}_y) E(r, \theta, \phi)$$

Find the polarization loss factor (PLF).

Solution: For the incident wave

$$\hat{\boldsymbol{\rho}}_w = \hat{\mathbf{a}}_x$$

and for the antenna

$$\hat{\boldsymbol{\rho}}_a = \frac{1}{\sqrt{2}} (\hat{\mathbf{a}}_x + \hat{\mathbf{a}}_y)$$

The PLF is then equal to

$$\text{PLF} = |\hat{\boldsymbol{\rho}}_w \cdot \hat{\boldsymbol{\rho}}_a|^2 = |\hat{\mathbf{a}}_x \cdot \frac{1}{\sqrt{2}} (\hat{\mathbf{a}}_x + \hat{\mathbf{a}}_y)|^2 = \frac{1}{2}$$

which in dB is equal to

$$\text{PLF (dB)} = 10 \log_{10} \text{PLF (dimensionless)} = 10 \log_{10}(0.5) = -3$$

Even though in Example 2.11 both the incoming wave and the antenna are linearly polarized, there is a 3-dB loss in extracted power because the polarization of the incoming wave is not aligned with the polarization of the antenna. If the polarization of the incoming wave is orthogonal to the polarization of the antenna, then there will be no power extracted by the antenna from the incoming wave and the PLF will be zero or $-\infty$ dB. In Figures 2.25(a,b) we illustrate the polarization loss factors (PLF) of two types of antennas: wires and apertures.

We now want to consider an example where the polarization of the antenna and the incoming wave are described in terms of complex polarization vectors.

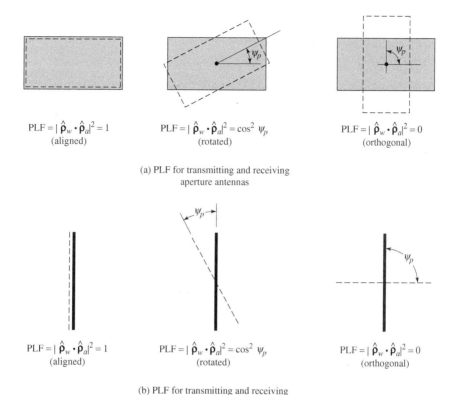

Figure 2.25 Polarization loss factors (PLF) for aperture and linear wire antennas.

Example 2.12

A right-hand (clockwise) circularly polarized wave radiated by an antenna, placed at some distance away from the origin of a spherical coordinate system, is traveling in the inward radial direction at an angle (θ, ϕ) and it is impinging upon a right-hand circularly polarized receiving antenna placed at the origin (see Figures 2.1 and 17.23 for the geometry of the coordinate system). The polarization of the receiving antenna is defined in the transmitting mode, as desired by the definition of the IEEE. Assuming the polarization of the incident wave is represented by

$$\mathbf{E}_w = (\hat{\mathbf{a}}_\theta + j\hat{\mathbf{a}}_\phi)E(r,\theta,\phi)$$

Determine the polarization loss factor (PLF).

Solution: The polarization of the incident right-hand circularly polarized wave traveling along the $-r$ radial direction is described by the unit vector

$$\hat{\boldsymbol{\rho}}_w = \left(\frac{\hat{\mathbf{a}}_\theta + j\hat{\mathbf{a}}_\phi}{\sqrt{2}}\right)$$

while that of the receiving antenna, in the transmitting mode, is represented by the unit vector

$$\hat{\boldsymbol{\rho}}_a = \left(\frac{\hat{\mathbf{a}}_\theta - j\hat{\mathbf{a}}_\phi}{\sqrt{2}}\right)$$

74 FUNDAMENTAL PARAMETERS AND FIGURES-OF-MERIT OF ANTENNAS

Therefore the polarization loss factor is

$$\text{PLF} = |\hat{\boldsymbol{\rho}}_w \cdot \hat{\boldsymbol{\rho}}_a|^2 = \frac{1}{4}|1+1|^2 = 1 = 0 \text{ dB}$$

Since the polarization of the incoming wave matches (including the sense of rotation) the polarization of the receiving antenna, there should not be any losses. Obviously the answer matches the expectation.

Based upon the definitions of the wave transmitted and received by an antenna, the polarization of an antenna in the *receiving* mode is related to that in the *transmitting* mode as follows:

1. "In the same plane of polarization, the polarization ellipses have the same axial ratio, the same sense of polarization (rotation) and the same spatial orientation.
2. "Since their senses of polarization and spatial orientation are specified by viewing their polarization ellipses in the respective directions in which they are propagating, one should note that:
   a. Although their senses of polarization are the same, they would appear to be opposite if both waves were viewed in the same direction.
   b. Their tilt angles are such that they are the negative of one another with respect to a common reference."

Since the polarization of an antenna will almost always be defined in its transmitting mode, according to the IEEE Std 145-1993, "the receiving polarization may be used to specify the polarization characteristic of a nonreciprocal antenna which may transmit and receive arbitrarily different polarizations."

The polarization loss must always be taken into account in the link calculations design of a communication system because in some cases it may be a very critical factor. Link calculations of communication systems for outer space explorations are very stringent because of limitations in spacecraft weight. In such cases, power is a limiting consideration. The design must properly take into account all loss factors to ensure a successful operation of the system.

An antenna that is elliptically polarized is that composed of two crossed dipoles, as shown in Figure 2.26. The two crossed dipoles provide the two orthogonal field components that are not

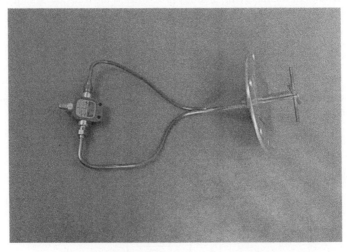

Figure 2.26 Geometry of elliptically polarized cross-dipole antenna.

necessarily of the same field intensity toward all observation angles. If the two dipoles are identical, the field intensity of each along zenith (perpendicular to the plane of the two dipoles) would be of the same intensity. Also, if the two dipoles were fed with a 90° degree time-phase difference (phase quadrature), the polarization along zenith would be circular and elliptical toward other directions. One way to obtain the 90° time-phase difference $\Delta\phi$ between the two orthogonal field components, radiated respectively by the two dipoles, is by feeding one of the two dipoles with a transmission line which is $\lambda/4$ longer or shorter than that of the other $[\Delta\phi = k\Delta\ell = (2\pi/\lambda)(\lambda/4) = \pi/2]$. One of the lengths (longer or shorter) will provide right-hand (CW) rotation while the other will provide left-hand (CCW) rotation.

A MATLAB computer program **Polarization_Propag** is included at the end of the chapter, and it computes the Poinaré sphere angles and the polarization of the wave radiated by an antenna and traveling in an infinite homogeneous medium.

2.13 INPUT IMPEDANCE

Input impedance is defined as "the impedance presented by an antenna at its terminals or the ratio of the voltage to current at a pair of terminals or the ratio of the appropriate components of the electric to magnetic fields at a point." In this section we are primarily interested in the input impedance at a pair of terminals which are the input terminals of the antenna. In Figure 2.27(a) these terminals

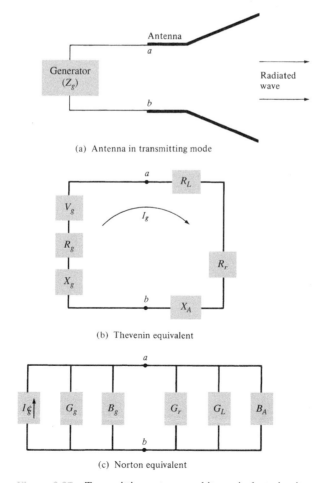

Figure 2.27 Transmitting antenna and its equivalent circuits.

76 FUNDAMENTAL PARAMETERS AND FIGURES-OF-MERIT OF ANTENNAS

are designated as $a - b$. The ratio of the voltage to current at these terminals, with no load attached, defines the impedance of the antenna as

$$Z_A = R_A + jX_A \qquad (2\text{-}72)$$

where

Z_A = antenna impedance at terminals a–b (ohms)
R_A = antenna resistance at terminals a–b (ohms)
X_A = antenna reactance at terminals a–b (ohms)

In general the resistive part of (2-72) consists of two components; that is

$$R_A = R_r + R_L \qquad (2\text{-}73)$$

where

R_r = radiation resistance of the antenna
R_L = loss resistance of the antenna

The radiation resistance will be considered in more detail in later chapters, and it will be illustrated with examples.

If we assume that the antenna is attached to a generator with internal impedance

$$Z_g = R_g + jX_g \qquad (2\text{-}74)$$

where

R_g = resistance of generator impedance (ohms)
X_g = reactance of generator impedance (ohms)

and the antenna is used in the transmitting mode, we can represent the antenna and generator by an equivalent circuit* shown in Figure 2.27(b). To find the amount of power delivered to R_r for radiation and the amount dissipated in R_L as heat ($I^2 R_L/2$), we first find the current developed within the loop which is given by

$$I_g = \frac{V_g}{Z_t} = \frac{V_g}{Z_A + Z_g} = \frac{V_g}{(R_r + R_L + R_g) + j(X_A + X_g)} \quad (A) \qquad (2\text{-}75)$$

and its magnitude by

$$|I_g| = \frac{|V_g|}{[(R_r + R_L + R_g)^2 + (X_A + X_g)^2]^{1/2}} \qquad (2\text{-}75a)$$

where V_g is the peak generator voltage. The power delivered to the antenna for radiation is given by

$$P_r = \frac{1}{2}|I_g|^2 R_r = \frac{|V_g|^2}{2}\left[\frac{R_r}{(R_r + R_L + R_g)^2 + (X_A + X_g)^2}\right] \quad (W) \qquad (2\text{-}76)$$

---

*This circuit can be used to represent small and simple antennas. It cannot be used for antennas with lossy dielectric or antennas over lossy ground because their loss resistance cannot be represented in series with the radiation resistance.

INPUT IMPEDANCE

and that dissipated as heat by

$$P_L = \frac{1}{2}|I_g|^2 R_L = \frac{|V_g|^2}{2}\left[\frac{R_L}{(R_r + R_L + R_g)^2 + (X_A + X_g)^2}\right] \quad (W) \tag{2-77}$$

The remaining power is dissipated as heat on the internal resistance R_g of the generator, and it is given by

$$P_g = \frac{|V_g|^2}{2}\left[\frac{R_g}{(R_r + R_L + R_g)^2 + (X_A + X_g)^2}\right] \quad (W) \tag{2-78}$$

The maximum power delivered to the antenna occurs when we have conjugate matching; that is when

$$R_r + R_L = R_g \tag{2-79}$$

$$X_A = -X_g \tag{2-80}$$

For this case

$$P_r = \frac{|V_g|^2}{2}\left[\frac{R_r}{4(R_r + R_L)^2}\right] = \frac{|V_g|^2}{8}\left[\frac{R_r}{(R_r + R_L)^2}\right] \tag{2-81}$$

$$P_L = \frac{|V_g|^2}{8}\left[\frac{R_L}{(R_r + R_L)^2}\right] \tag{2-82}$$

$$P_g = \frac{|V_g|^2}{8}\left[\frac{R_g}{(R_r + R_L)^2}\right] = \frac{|V_g|^2}{8}\left[\frac{1}{R_r + R_L}\right] = \frac{|V_g|^2}{8R_g} \tag{2-83}$$

From (2-81)–(2-83), it is clear that

$$P_g = P_r + P_L = \frac{|V_g|^2}{8}\left[\frac{R_g}{(R_r + R_L)^2}\right] = \frac{|V_g|^2}{8}\left[\frac{R_r + R_L}{(R_r + R_L)^2}\right] \tag{2-84}$$

The power supplied by the generator during conjugate matching is

$$P_s = \frac{1}{2}V_g I_g^* = \frac{1}{2}V_g\left[\frac{V_g^*}{2(R_r + R_L)}\right] = \frac{|V_g|^2}{4}\left[\frac{1}{R_r + R_L}\right] \quad (W) \tag{2-85}$$

Of the power that is provided by the generator, half is dissipated as heat in the internal resistance (R_g) of the generator and the other half is delivered to the antenna. This only happens when we have *conjugate matching*. Of the power that is delivered to the antenna, part is radiated through the mechanism provided by the radiation resistance and the other is dissipated as heat which influences part of the overall efficiency of the antenna. If the antenna is lossless and matched to the transmission line ($e_o = 1$), then half of the total power supplied by the generator is radiated by the antenna during conjugate matching, and the other half is dissipated as heat in the generator. Thus, to radiate half of the available power through R_r you must dissipate the other half as heat in the generator through R_g. These two powers are, respectively,

78   FUNDAMENTAL PARAMETERS AND FIGURES-OF-MERIT OF ANTENNAS

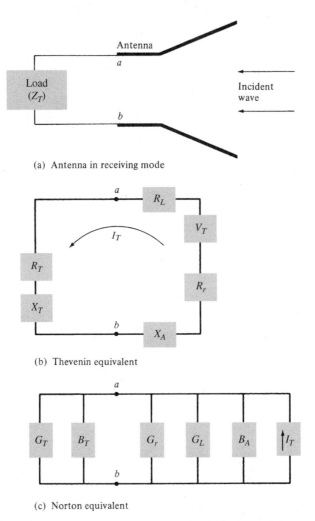

Figure 2.28   Antenna and its equivalent circuits in the receiving mode.

analogous to the power transferred to the load and the power scattered by the antenna in the receiving mode. In Figure 2.27 it is assumed that the generator is directly connected to the antenna. If there is a transmission line between the two, which is usually the case, then Z_g represents the equivalent impedance of the generator transferred to the input terminals of the antenna using the impedance transfer equation. If, in addition, the transmission line is lossy, then the available power to be radiated by the antenna will be reduced by the losses of the transmission line. Figure 2.27(c) illustrates the Norton equivalent of the antenna and its source in the transmitting mode.

The use of the antenna in the receiving mode is shown in Figure 2.28(a). The incident wave impinges upon the antenna, and it induces a voltage V_T which is analogous to V_g of the transmitting mode. The Thevenin equivalent circuit of the antenna and its load is shown in Figure 2.28(b) and the Norton equivalent in Figure 2.28(c). The discussion for the antenna and its load in the receiving mode parallels that for the transmitting mode, and it will not be repeated here in detail. Some of the results will be summarized in order to discuss some subtle points. Following a procedure similar to that for the antenna in the transmitting mode, it can be shown using Figure 2.28 that in the receiving mode under conjugate matching ($R_r + R_L = R_T$ and $X_A = -X_T$) the powers delivered to R_T, R_r, and

R_L are given, respectively, by

$$P_T = \frac{|V_T|^2}{8}\left[\frac{R_T}{(R_r+R_L)^2}\right] = \frac{|V_T|^2}{8}\left(\frac{1}{R_r+R_L}\right) = \frac{|V_T|^2}{8R_T} \qquad (2\text{-}86)$$

$$P_r = \frac{|V_T|^2}{2}\left[\frac{R_r}{4(R_r+R_L)^2}\right] = \frac{|V_T|^2}{8}\left[\frac{R_r}{(R_r+R_L)^2}\right] \qquad (2\text{-}87)$$

$$P_L = \frac{|V_T|^2}{8}\left[\frac{R_L}{(R_r+R_L)^2}\right] \qquad (2\text{-}88)$$

while the *induced* (*collected* or *captured*) is

$$P_c = \frac{1}{2}V_T I_T^* = \frac{1}{2}V_T\left[\frac{V_T^*}{2(R_r+R_L)}\right] = \frac{|V_T|^2}{4}\left(\frac{1}{R_r+R_L}\right) \qquad (2\text{-}89)$$

These are analogous, respectively, to (2-81)–(2-83) and (2-85). The power P_r of (2-87) delivered to R_r is referred to as *scattered* (or *reradiated*) power. It is clear through (2-86)–(2-89) that under conjugate matching of the total power collected or captured [P_c of (2-89)] half is delivered to the load R_T [P_T of (2-86)] and the other half is scattered or reradiated through R_r [P_r of (2-87)] and dissipated as heat through R_L [P_L of (2-88)]. If the losses are zero ($R_L = 0$), then half of the captured power is delivered to the load and the other half is scattered. This indicates that in order to deliver half of the power to the load you must scatter the other half. This becomes important when discussing effective equivalent areas and aperture efficiencies, especially for high directivity aperture antennas such as waveguides, horns, and reflectors with aperture efficiencies as high as 80 to 90%. Aperture efficiency (ε_{ap}) is defined by (2-100) and is the ratio of the maximum effective area to the physical area. The effective area is used to determine the power delivered to the load, which under conjugate matching is only one-half of that intercepted; the other half is scattered and dissipated as heat. For a lossless antenna ($R_L = 0$) under conjugate matching, the maximum value of the effective area is equal to the physical area ($\varepsilon_{ap} = 1$) and the scattering area is also equal to the physical area. Thus half of the power is delivered to the load and the other half is scattered. Using (2-86) to (2-89) we conclude that even though the aperture efficiencies are higher than 50% (they can be as large as 100%) all of the power that is captured by the antenna is not delivered to the load but it includes that which is scattered plus dissipated as heat by the antenna. The most that can be delivered to the load is only half of that captured and that is only under conjugate matching and lossless transmission line.

The input impedance of an antenna is generally a function of frequency. Thus the antenna will be matched to the interconnecting transmission line and other associated equipment only within a bandwidth. In addition, the input impedance of the antenna depends on many factors including its geometry, its method of excitation, and its proximity to surrounding objects. Because of their complex geometries, only a limited number of practical antennas have been investigated analytically. For many others, the input impedance has been determined experimentally.

2.14 ANTENNA RADIATION EFFICIENCY

The antenna efficiency that takes into account the reflection, conduction, and dielectric losses was discussed in Section 2.8. The conduction and dielectric losses of an antenna are very difficult to compute and in most cases they are measured. Even with measurements, they are difficult to separate and they are usually lumped together to form the e_{cd} efficiency. The resistance R_L is used to represent the conduction-dielectric losses.

80 FUNDAMENTAL PARAMETERS AND FIGURES-OF-MERIT OF ANTENNAS

The *conduction-dielectric efficiency* e_{cd} is defined as *the ratio of the power delivered to the radiation resistance R_r to the power delivered to R_r and R_L*. Using (2-76) and (2-77), the radiation efficiency can be written as

$$\boxed{e_{cd} = \left[\frac{R_r}{R_L + R_r}\right]} \quad \text{(dimensionless)} \tag{2-90}$$

For a metal rod of length l and uniform cross-sectional area A, the dc resistance is given by

$$R_{dc} = \frac{1}{\sigma}\frac{l}{A} \quad \text{(ohms)} \tag{2-90a}$$

If the skin depth $\delta[\delta = \sqrt{2/(\omega\mu_0\sigma)}]$ of the metal is very small compared to the smallest diagonal of the cross section of the rod, the current is confined to a thin layer near the conductor surface. Therefore the high-frequency resistance can be written, based on a *uniform current distribution*, as

$$R_{hf} = \frac{l}{P}R_s = \frac{l}{P}\sqrt{\frac{\omega\mu_0}{2\sigma}} \quad \text{(ohms)} \tag{2-90b}$$

where P is the perimeter of the cross section of the rod ($P = C = 2\pi b$ for a circular wire of radius b), R_s is the conductor surface resistance, ω is the angular frequency, μ_0 is the permeability of free-space, and σ is the conductivity of the metal.

Example 2.13

A resonant half-wavelength dipole is made out of copper ($\sigma = 5.7 \times 10^7$ S/m) wire. Determine the conduction-dielectric (radiation) efficiency of the dipole antenna at $f = 100$ MHz if the radius of the wire b is $3 \times 10^{-4}\lambda$, and the radiation resistance of the $\lambda/2$ dipole is 73 ohms.

Solution: At $f = 10^8$ Hz

$$\lambda = \frac{v}{f} = \frac{3 \times 10^8}{10^8} = 3 \text{ m}$$

$$l = \frac{\lambda}{2} = \frac{3}{2}\text{m}$$

$$C = 2\pi b = 2\pi(3 \times 10^{-4})\lambda = 6\pi \times 10^{-4}\lambda$$

For a $\lambda/2$ dipole with a sinusoidal current distribution $R_L = \frac{1}{2}R_{hf}$ where R_{hf} is given by (2-90b). See Problem 2.52. Therefore,

$$R_L = \frac{1}{2}R_{hf} = \frac{0.25}{6\pi \times 10^{-4}}\sqrt{\frac{\pi(10^8)(4\pi \times 10^{-7})}{5.7 \times 10^7}} = 0.349 \text{ ohms}$$

Thus,

$$e_{cd}(\text{dimensionless}) = \frac{73}{73 + 0.349} = 0.9952 = 99.52\%$$

$$e_{cd}(\text{dB}) = 10\log_{10}(0.9952) = -0.021$$

2.15 ANTENNA VECTOR EFFECTIVE LENGTH AND EQUIVALENT AREAS

An antenna in the receiving mode, whether it is in the form of a wire, horn, aperture, array, dielectric rod, etc., is used to capture (collect) electromagnetic waves and to extract power from them, as shown in Figures 2.29(a) and (b). For each antenna, an equivalent length and a number of equivalent areas can then be defined.

These equivalent quantities are used to describe the receiving characteristics of an antenna, whether it be a linear or an aperture type, when a wave is incident upon the antenna.

2.15.1 Vector Effective Length

The effective length of an antenna, whether it be a linear or an aperture antenna, is a quantity that is used to determine the voltage induced on the open-circuit terminals of the antenna when a wave impinges upon it. The vector effective length ℓ_e for an antenna is usually a complex vector quantity

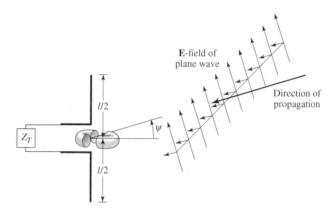

(a) Dipole antenna in receiving mode

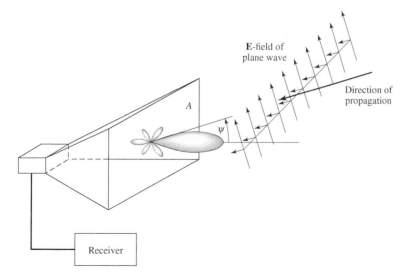

(b) Aperture antenna in receiving mode

Figure 2.29 Uniform plane wave incident upon dipole and aperture antennas.

82 FUNDAMENTAL PARAMETERS AND FIGURES-OF-MERIT OF ANTENNAS

represented by

$$\boldsymbol{\ell}_e(\theta,\phi) = \hat{\mathbf{a}}_\theta l_\theta(\theta,\phi) + \hat{\mathbf{a}}_\phi l_\phi(\theta,\phi) \tag{2-91}$$

It should be noted that it is also referred to as the *effective height*. It is a far-field quantity and it is related to the *far-zone* field $\mathbf{E}_a$ radiated by the antenna, with current I_{in} in its terminals, by [13]–[18]

$$\mathbf{E}_a = \hat{\mathbf{a}}_\theta E_\theta + \hat{\mathbf{a}}_\phi E_\phi = -j\eta \frac{kI_{in}}{4\pi r} \boldsymbol{\ell}_e e^{-jkr} \tag{2-92}$$

The effective length represents the antenna in its transmitting and receiving modes, and it is particularly useful in relating the open-circuit voltage V_{oc} of receiving antennas. This relation can be expressed as

$$V_{oc} = \mathbf{E}^i \cdot \boldsymbol{\ell}_e \tag{2-93}$$

where

V_{oc} = open-circuit voltage at antenna terminals
$\mathbf{E}^i$ = incident electric field
$\boldsymbol{\ell}_e$ = vector effective length

In (2-93) V_{oc} can be thought of as the voltage induced in a linear antenna of length $\boldsymbol{\ell}_e$ when $\boldsymbol{\ell}_e$ and $\mathbf{E}^i$ are linearly polarized [19], [20]. From the relation of (2-93) the *effective length of a linearly polarized antenna receiving a plane wave in a given direction* is defined as "the ratio of the magnitude of the open-circuit voltage developed at the terminals of the antenna to the magnitude of the electric-field strength in the direction of the antenna polarization. Alternatively, the effective length is the length of a thin straight conductor oriented perpendicular to the given direction and parallel to the antenna polarization, having a uniform current equal to that at the antenna terminals and producing the same far-field strength as the antenna in that direction."

In addition, as shown in Section 2.12.2, the antenna vector effective length is used to determine the polarization efficiency of the antenna. To illustrate the usefulness of the vector effective length, let us consider an example.

Example 2.14

The far-zone field radiated by a small dipole of length $l < \lambda/10$ and with a triangular current distribution, as shown in Figure 4.4, is derived in Section 4.3 of Chapter 4 and it is given by (4-36a), or

$$\mathbf{E}_a = \hat{\mathbf{a}}_\theta j\eta \frac{kI_{in} l e^{-jkr}}{8\pi r} \sin\theta$$

Determine the vector effective length of the antenna.
  Solution: According to (2-92), the vector effective length is

$$\boldsymbol{\ell}_e = -\hat{\mathbf{a}}_\theta \frac{l}{2} \sin\theta$$

This indicates, as it should, that the effective length is a function of the direction angle θ, and its maximum occurs when $\theta = 90°$. This tells us that the maximum open-circuit voltage at the dipole terminals occurs when the incident direction of the wave of Figure 2.29(a) impinging upon the small dipole antenna is normal to the axis (length) of the dipole ($\theta = 90°$). This is expected since the dipole has a radiation pattern whose maximum is in the $\theta = 90°$. In addition, the effective length of the dipole to produce the same output open-circuit voltage is only half (50%) of its physical length if it were replaced by a thin conductor having a uniform current distribution (it can be shown that the maximum effective length of an element with an ideal uniform current distribution is equal to its physical length).

2.15.2 Antenna Equivalent Areas

With each antenna, we can associate a number of equivalent areas. These are used to describe the power capturing characteristics of the antenna when a wave impinges on it. One of these equivalent areas is the *effective area (aperture)*, which in a given direction is defined as "the ratio of the available power at the terminals of a receiving antenna to the power flux density of a plane wave incident on the antenna from that direction, the wave being polarization-matched to the antenna. If the direction is not specified, the direction of maximum radiation intensity is implied." In equation form it is written as

$$A_e = \frac{P_T}{W_i} = \frac{|I_T|^2 R_T / 2}{W_i} \quad (2\text{-}94)$$

where

A_e = effective area (effective aperture) (m^2)
P_T = power delivered to the load (W)
W_i = power density of incident wave (W/m^2)

The effective aperture is the area which when multiplied by the incident power density gives the power delivered to the load. Using the equivalent of Figure 2.28, we can write (2-94) as

$$A_e = \frac{|V_T|^2}{2W_i} \left[\frac{R_T}{(R_r + R_L + R_T)^2 + (X_A + X_T)^2} \right] \quad (2\text{-}95)$$

Under conditions of maximum power transfer (*conjugate matching*), $R_r + R_L = R_T$ and $X_A = -X_T$, the effective area of (2-95) reduces to the maximum effective aperture given by

$$A_{em} = \frac{|V_T|^2}{8W_i} \left[\frac{R_T}{(R_L + R_r)^2} \right] = \frac{|V_T|^2}{8W_i} \left[\frac{1}{R_r + R_L} \right] \quad (2\text{-}96)$$

When (2-96) is multiplied by the incident power density, it leads to the maximum power delivered to the load of (2-86).

All of the power that is intercepted, collected, or captured by an antenna is not delivered to the load, as we have seen using the equivalent circuit of Figure 2.28. In fact, under conjugate matching only half of the captured power is delivered to the load; the other half is scattered and dissipated as heat. Therefore to account for the scattered and dissipated power we need to define, in addition to the effective area, the *scattering, loss* and *capture* equivalent areas. In equation form these can be defined similarly to (2-94)–(2-96) for the effective area.

The *scattering area* is defined as the equivalent area when multiplied by the incident power density is equal to the scattered or reradiated power. Under conjugate matching this is written, similar to (2-96), as

$$A_s = \frac{|V_T|^2}{8W_i} \left[\frac{R_r}{(R_L + R_r)^2} \right] \quad (2\text{-}97)$$

which when multiplied by the incident power density gives the scattering power of (2-87).

The *loss area* is defined as the equivalent area, which when multiplied by the incident power density leads to the power dissipated as heat through R_L. Under conjugate matching this is written, similar to (2-96), as

$$A_L = \frac{|V_T|^2}{8W_i} \left[\frac{R_L}{(R_L + R_r)^2} \right] \quad (2\text{-}98)$$

which when multiplied by the incident power density gives the dissipated power of (2-88).

Finally the *capture area* is defined as the equivalent area, which when multiplied by the incident power density leads to the total power captured, collected, or intercepted by the antenna. Under conjugate matching this is written, similar to (2-96), as

$$A_c = \frac{|V_T|^2}{8W_i} \left[\frac{R_T + R_r + R_L}{(R_L + R_r)^2} \right] \quad (2\text{-}99)$$

When (2-99) is multiplied by the incident power density, it leads to the captured power of (2-89). In general, the total capture area is equal to the sum of the other three, or

Capture Area = Effective Area + Scattering Area + Loss Area

This is apparent under conjugate matching using (2-96)–(2-99). However, it holds even under non-conjugate matching conditions.

Now that the equivalent areas have been defined, let us introduce the *aperture efficiency* ε_{ap} of an antenna, which is defined as the ratio of the maximum effective area A_{em} of the antenna to its physical area A_p, or

$$\varepsilon_{ap} = \frac{A_{em}}{A_p} = \frac{\text{maximum effective area}}{\text{physical area}} \quad (2\text{-}100)$$

For aperture type antennas, such as waveguides, horns, and reflectors, the maximum effective area cannot exceed the physical area but it can equal it ($A_{em} \leq A_p$ or $0 \leq \varepsilon_{ap} \leq 1$). Therefore the maximum value of the aperture efficiency cannot exceed unity (100%). For a lossless antenna ($R_L = 0$) the maximum value of the scattering area is also equal to the physical area. Therefore even though the aperture efficiency is greater than 50%, for a lossless antenna under conjugate matching only half of the captured power is delivered to the load and the other half is scattered.

We can also introduce a *partial effective area* of an antenna *for a given polarization* in a given direction, which is defined as "the ratio of the available power at the terminals of a receiving antenna to the power flux density of a plane wave incident on the antenna from that direction and with a specified polarization differing from the receiving polarization of the antenna."

The effective area of an antenna is not necessarily the same as the physical aperture. It will be shown in later chapters that aperture antennas with uniform amplitude and phase field distributions have maximum effective areas equal to the physical areas; they are smaller for nonuniform field distributions. In addition, the maximum effective area of wire antennas is greater than the physical area (if taken as the area of a cross section of the wire when split lengthwise along its diameter). Thus the wire antenna can capture much more power than is intercepted by its physical size! This should not come as a surprise. If the wire antenna would only capture the power incident on its physical size, it would be almost useless. So electrically, the wire antenna looks much bigger than its physical stature.

To illustrate the concept of effective area, especially as applied to a wire antenna, let us consider an example. In later chapters, we will consider examples of aperture antennas.

Example 2.15

A uniform plane wave is incident upon a very short lossless dipole ($l \ll \lambda$), as shown in Figure 2.29(a). Find the maximum effective area assuming that the radiation resistance of the dipole is $R_r = 80(\pi l/\lambda)^2$, and the incident field is linearly polarized along the axis of the dipole.

Solution: For $R_L = 0$, the maximum effective area of (2-96) reduces to

$$A_{em} = \frac{|V_T|^2}{8W_i}\left[\frac{1}{R_r}\right]$$

Since the dipole is very short, the induced current can be assumed to be constant and of uniform phase. The induced voltage is

$$V_T = El$$

where

V_T = induced voltage on the dipole
E = electric field of incident wave
l = length of dipole

For a uniform plane wave, the incident power density can be written as

$$W_i = \frac{E^2}{2\eta}$$

where η is the intrinsic impedance of the medium ($\simeq 120\pi$ ohms for a free-space medium). Thus

$$A_{em} = \frac{(El)^2}{8(E^2/2\eta)(80\pi^2 l^2/\lambda^2)} = \frac{3\lambda^2}{8\pi} = 0.119\lambda^2$$

The above value is only valid for a lossless antenna (the losses of a short dipole are usually significant). If the loss resistance is equal to the radiation resistance ($R_L = R_r$) and the sum of the two is equal to the load (receiver) resistance ($R_T = R_r + R_L = 2R_r$), then the effective area is only one-half of the maximum effective area given above.

Let us now examine the significance of the effective area. From Example 2.15, the maximum effective area of a short dipole with $l \ll \lambda$ was equal to $A_{em} = 0.119\lambda^2$. Typical antennas that fall under this category are dipoles whose lengths are $l \leq \lambda/50$. For demonstration, let us assume that $l = \lambda/50$. Because $A_{em} = 0.119\lambda^2 = lw_e = (\lambda/50)w_e$, the maximum effective electrical width of this dipole is $w_e = 5.95\lambda$. Typical physical diameters (widths) of wires used for dipoles may be

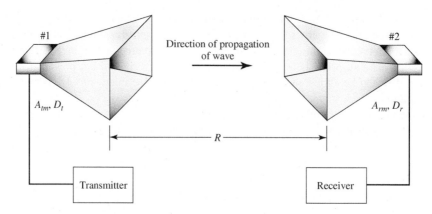

Figure 2.30 Two antennas separated by a distance R.

about $w_p = \lambda/300$. Thus the maximum effective width w_e is about 1,785 times larger than its physical width.

2.16 MAXIMUM DIRECTIVITY AND MAXIMUM EFFECTIVE AREA

To derive the relationship between directivity and maximum effective area, the geometrical arrangement of Figure 2.30 is chosen. Antenna 1 is used as a transmitter and 2 as a receiver. The effective areas and directivities of each are designated as A_t, A_r and D_t, D_r. If antenna 1 were isotropic, its radiated power density at a distance R would be

$$W_0 = \frac{P_t}{4\pi R^2} \tag{2-101}$$

where P_t is the total radiated power. Because of the directive properties of the antenna, its actual density is

$$W_t = W_0 D_t = \frac{P_t D_t}{4\pi R^2} \tag{2-102}$$

The power collected (received) by the antenna and transferred to the load would be

$$P_r = W_t A_r = \frac{P_t D_t A_r}{4\pi R^2} \tag{2-103}$$

or

$$D_t A_r = \frac{P_r}{P_t}(4\pi R^2) \tag{2-103a}$$

If antenna 2 is used as a transmitter, 1 as a receiver, and the intervening medium is linear, passive, and isotropic, we can write that

$$D_r A_t = \frac{P_r}{P_t}(4\pi R^2) \tag{2-104}$$

Equating (2-103a) and (2-104) reduces to

$$\frac{D_t}{A_t} = \frac{D_r}{A_r} \quad (2\text{-}105)$$

Increasing the directivity of an antenna increases its effective area in direct proportion. Thus, (2-105) can be written as

$$\frac{D_{0t}}{A_{tm}} = \frac{D_{0r}}{A_{rm}} \quad (2\text{-}106)$$

where A_{tm} and A_{rm} (D_{0t} and D_{0r}) are the *maximum* effective areas (directivities) of antennas 1 and 2, respectively.

If antenna 1 is isotropic, then $D_{0t} = 1$ and its maximum effective area can be expressed as

$$A_{tm} = \frac{A_{rm}}{D_{0r}} \quad (2\text{-}107)$$

Equation (2-107) states that the maximum effective area of an isotropic source is equal to the ratio of the maximum effective area to the maximum directivity of any other source. For example, let the other antenna be a very short ($l \ll \lambda$) dipole whose effective area ($0.119\lambda^2$ from Example 2.15) and maximum directivity (1.5) are known.

The maximum effective area of the isotropic source is then equal to

$$A_{tm} = \frac{A_{rm}}{D_{0r}} = \frac{0.119\lambda^2}{1.5} = \frac{\lambda^2}{4\pi} \quad (2\text{-}108)$$

Using (2-108), we can write (2-107) as

$$A_{rm} = D_{0r}A_{tm} = D_{0r}\left(\frac{\lambda^2}{4\pi}\right) \quad (2\text{-}109)$$

In general then, the *maximum effective aperture* (A_{em}) of any antenna is related to its maximum directivity (D_0) by

$$\boxed{A_{em} = \frac{\lambda^2}{4\pi}D_0} \quad (2\text{-}110)$$

Thus, when (2-110) is multiplied by the power density of the incident wave it leads to the maximum power that can be delivered to the load. This assumes that there are no conduction-dielectric losses (radiation efficiency e_{cd} is unity), the antenna is matched to the load (reflection efficiency e_r is unity), and the polarization of the impinging wave matches that of the antenna (polarization loss factor PLF and polarization efficiency p_e are unity). If there are losses associated with an antenna, its maximum effective aperture of (2-110) must be modified to account for conduction-dielectric losses (radiation efficiency). Thus,

$$A_{em} = e_{cd}\left(\frac{\lambda^2}{4\pi}\right)D_0 \quad (2\text{-}111)$$

The maximum value of (2-111) assumes that the antenna is matched to the load and the incoming wave is polarization-matched to the antenna. If reflection and polarization losses are also included, then the maximum effective area of (2-111) is represented by

$$A_{em} = e_0 \left(\frac{\lambda^2}{4\pi}\right) D_0 |\hat{\boldsymbol{\rho}}_w \cdot \hat{\boldsymbol{\rho}}_a|^2$$
$$= e_{cd}(1 - |\Gamma|^2) \left(\frac{\lambda^2}{4\pi}\right) D_0 |\hat{\boldsymbol{\rho}}_w \cdot \hat{\boldsymbol{\rho}}_a|^2 \quad (2\text{-}112)$$

2.17 FRIIS TRANSMISSION EQUATION AND RADAR RANGE EQUATION

The analysis and design of radar and communications systems often require the use of the *Friis Transmission Equation* and the *Radar Range Equation.* Because of the importance [21] of the two equations, a few pages will be devoted for their derivation.

2.17.1 Friis Transmission Equation

The Friis Transmission Equation relates the power received to the power transmitted between two antennas separated by a distance $R > 2D^2/\lambda$, where D is the largest dimension of either antenna. Referring to Figure 2.31, let us assume that the transmitting antenna is initially isotropic. If the input power at the terminals of the transmitting antenna is P_t, then its isotropic power density W_0 at distance R from the antenna is

$$W_0 = e_t \frac{P_t}{4\pi R^2} \quad (2\text{-}113)$$

where e_t is the radiation efficiency of the transmitting antenna. For a nonisotropic transmitting antenna, the power density of (2-113) in the direction θ_t, ϕ_t can be written as

$$W_t = \frac{P_t G_t(\theta_t, \phi_t)}{4\pi R^2} = e_t \frac{P_t D_t(\theta_t, \phi_t)}{4\pi R^2} \quad (2\text{-}114)$$

where $G_t(\theta_t, \phi_t)$ is the gain and $D_t(\theta_t, \phi_t)$ is the directivity of the transmitting antenna in the direction θ_t, ϕ_t. Since the effective area A_r of the receiving antenna is related to its efficiency e_r and directivity D_r by

$$A_r = e_r D_r(\theta_r, \phi_r) \left(\frac{\lambda^2}{4\pi}\right) \quad (2\text{-}115)$$

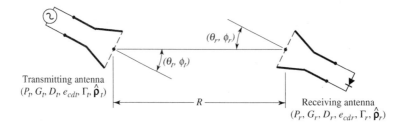

Figure 2.31 Geometrical orientation of transmitting and receiving antennas for Friis transmission equation.

the amount of power P_r collected by the receiving antenna can be written, using (2-114) and (2-115), as

$$P_r = e_r D_r(\theta_r, \phi_r) \frac{\lambda^2}{4\pi} W_t = e_t e_r \frac{\lambda^2 D_t(\theta_t, \phi_t) D_r(\theta_r, \phi_r) P_t}{(4\pi R)^2} |\hat{\rho}_t \cdot \hat{\rho}_r|^2 \qquad (2\text{-}116)$$

or the ratio of the received to the input power as

$$\frac{P_r}{P_t} = e_t e_r \frac{\lambda^2 D_t(\theta_t, \phi_t) D_r(\theta_r, \phi_r)}{(4\pi R)^2} \qquad (2\text{-}117)$$

The power received based on (2-117) assumes that the transmitting and receiving antennas are matched to their respective lines or loads (reflection efficiencies are unity) and the polarization of the receiving antenna is polarization-matched to the impinging wave (polarization loss factor and polarization efficiency are unity). If these two factors are also included, then the ratio of the received to the input power of (2-117) is represented by

$$\frac{P_r}{P_t} = e_{cdt} e_{cdr} (1 - |\Gamma_t|^2)(1 - |\Gamma_r|^2) \left(\frac{\lambda}{4\pi R}\right)^2 D_t(\theta_t, \phi_t) D_r(\theta_r, \phi_r) |\hat{\rho}_t \cdot \hat{\rho}_r|^2 \qquad (2\text{-}118)$$

Example 2.16

Two *lossless* X-band (8.2–12.4 GHz) horn antennas are separated by a distance of 100λ. The reflection coefficients at the terminals of the transmitting and receiving antennas are 0.1 and 0.2, respectively. The maximum directivities of the transmitting and receiving antennas (over isotropic) are 16 dB and 20 dB, respectively. Assuming that the input power in the lossless transmission line connected to the transmitting antenna is 2 W, and the antennas are aligned for maximum radiation between them and are polarization-matched, find the power delivered to the load of the receiver.

Solution: For this problem

$$e_{cdt} = e_{cdr} = 1 \text{ because the antennas are lossless.}$$

$$|\hat{\rho}_t \cdot \hat{\rho}_r|^2 = 1 \text{ because the antennas are polarization-matched}$$

$$\left.\begin{array}{l} D_t = D_{0t} \\ D_r = D_{0r} \end{array}\right\} \text{because the antennas are aligned for maximum radiation between them}$$

$$D_{0t} = 16 \text{ dB} \Rightarrow 39.81 \text{ (dimensionless)}$$

$$D_{0r} = 20 \text{ dB} \Rightarrow 100 \text{ (dimensionless)}$$

Using (2-118), we can write

$$P_r = [1 - (0.1)^2][1 - (0.2)^2][\lambda/4\pi(100\lambda)]^2 (39.81)(100)(2)$$

$$= 4.777 \text{ mW}$$

90   FUNDAMENTAL PARAMETERS AND FIGURES-OF-MERIT OF ANTENNAS

For reflection and polarization-matched antennas aligned for maximum directional radiation and reception, (2-118) reduces to

$$\frac{P_r}{P_t} = \left(\frac{\lambda}{4\pi R}\right)^2 G_{0t} G_{0r} \qquad (2\text{-}119)$$

Equations (2-117), (2-118), or (2-119) are known as the *Friis Transmission Equation*, and it relates the power P_r (delivered to the receiver load) to the input power of the transmitting antenna P_t. The term $(\lambda/4\pi R)^2$ is called the *free-space loss factor*, and it takes into account the losses due to the spherical spreading of the energy by the antenna.

2.17.2 Radar Range Equation

Now let us assume that the transmitted power is incident upon a target, as shown in Figure 2.32. We now introduce a quantity known as the *radar cross section* or *echo area* (σ) of a target which is defined as *the area intercepting that amount of power which, when scattered isotropically, produces at the receiver a density which is equal to that scattered by the actual target* [13]. In equation form

$$\lim_{R\to\infty}\left[\frac{\sigma W_i}{4\pi R^2}\right] = W_s \qquad (2\text{-}120)$$

or

$$\begin{aligned}\sigma &= \lim_{R\to\infty}\left[4\pi R^2 \frac{W_s}{W_i}\right] = \lim_{R\to\infty}\left[4\pi R^2 \frac{|\mathbf{E}^s|^2}{|\mathbf{E}^i|^2}\right] \\ &= \lim_{R\to\infty}\left[4\pi R^2 \frac{|\mathbf{H}^s|^2}{|\mathbf{H}^i|^2}\right]\end{aligned} \qquad (2\text{-}120a)$$

where

σ = radar cross section or echo area   (m^2)
R = observation distance from target   (m)
W_i = incident power density   (W/m^2)
W_s = scattered power density   (W/m^2)
$\mathbf{E}^i$ ($\mathbf{E}^s$) = incident (scattered) electric field   (V/m)
$\mathbf{H}^i$ ($\mathbf{H}^s$) = incident (scattered) magnetic field   (A/m)

Any of the definitions in (2-120a) can be used to derive the radar cross section of any antenna or target. For some polarization one of the definitions based either on the power density, electric field, or magnetic field may simplify the derivation, although all should give the same answers [13].

Using the definition of radar cross section, we can consider that the transmitted power incident upon the target is initially captured and then it is reradiated isotropically, insofar as the receiver is concerned. The amount of captured power P_c is obtained by multiplying the incident power density of (2-114) by the radar cross section σ, or

$$P_c = \sigma W_t = \sigma \frac{P_t G_t(\theta_t, \phi_t)}{4\pi R_1^2} = e_t \sigma \frac{P_t D_t(\theta_t, \phi_t)}{4\pi R_1^2} \qquad (2\text{-}121)$$

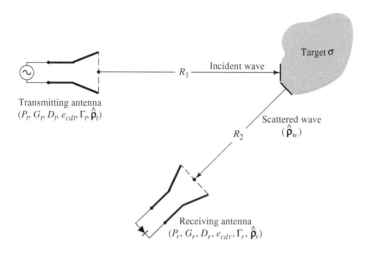

Figure 2.32 Geometrical arrangement of transmitter, target, and receiver for radar range equation.

The power captured by the target is reradiated isotropically, and the scattered power density can be written as

$$W_s = \frac{P_c}{4\pi R_2^2} = e_{cdt}\sigma \frac{P_t D_t(\theta_t, \phi_t)}{(4\pi R_1 R_2)^2} \tag{2-122}$$

The amount of power delivered to the receiver load is given by

$$P_r = A_r W_s = e_{cdt} e_{cdr} \sigma \frac{P_t D_t(\theta_t, \phi_t) D_r(\theta_r, \phi_r)}{4\pi} \left(\frac{\lambda}{4\pi R_1 R_2}\right)^2 \tag{2-123}$$

where A_r is the effective area of the receiving antenna as defined by (2-115).

Equation (2-123) can be written as the ratio of the received power to the input power, or

$$\boxed{\frac{P_r}{P_t} = e_{cdt} e_{cdr} \sigma \frac{D_t(\theta_t, \phi_t) D_r(\theta_r, \phi_r)}{4\pi} \left(\frac{\lambda}{4\pi R_1 R_2}\right)^2} \tag{2-124}$$

Expression (2-124) is used to relate the received power to the input power, and it takes into account only conduction-dielectric losses (radiation efficiency) of the transmitting and receiving antennas. It does not include reflection losses (reflection efficiency) and polarization losses (polarization loss factor or polarization efficiency). If these two losses are also included, then (2-124) must be expressed as

$$\boxed{\begin{aligned}\frac{P_r}{P_t} &= e_{cdt} e_{cdr} (1 - |\Gamma_t|^2)(1 - |\Gamma_r|^2)\sigma \frac{D_t(\theta_t, \phi_t) D_r(\theta_r, \phi_r)}{4\pi} \\ &\quad \times \left(\frac{\lambda}{4\pi R_1 R_2}\right)^2 |\hat{\rho}_w \cdot \hat{\rho}_r|^2 \end{aligned}} \tag{2-125}$$

where

$\hat{\rho}_w$ = polarization unit vector of the scattered waves

$\hat{\rho}_r$ = polarization unit vector of the receiving antenna

For polarization-matched antennas aligned for maximum directional radiation and reception, (2-125) reduces to

$$\boxed{\frac{P_r}{P_t} = \sigma \frac{G_{0t} G_{0r}}{4\pi} \left[\frac{\lambda}{4\pi R_1 R_2}\right]^2} \qquad (2\text{-}126)$$

Equation (2-124), or (2-125) or (2-126) is known as the *Radar Range Equation*. It relates the power P_r (delivered to the receiver load) to the input power P_t transmitted by an antenna, after it has been scattered by a target with a radar cross section (echo area) of σ.

2.17.3 Antenna Radar Cross Section

The radar cross section, usually referred to as RCS, is a far-field parameter, which is used to characterize the scattering properties of a radar target. For a target, there is *monostatic* or *backscattering RCS* when the transmitter and receiver of Figure 2.32 are at the same location, and a *bistatic RCS* when the transmitter and receiver are not at the same location. In designing low-observable or low-profile (stealth) targets, it is the parameter that you attempt to minimize. For complex targets (such as aircraft, spacecraft, missiles, ships, tanks, automobiles) it is a complex parameter to derive. In general, the RCS of a target is a function of the polarization of the incident wave, the angle of incidence, the angle of observation, the geometry of the target, the electrical properties of the target, and the frequency of operation. The units of RCS of three-dimensional targets are meters squared (m^2) or for normalized values decibels per squared meter (dBsm) or RCS per squared wavelength in decibels (RCS/λ^2 in dB). Representative values of some typical targets are shown in Table 2.2 [22]. Although the frequency was not stated [22], these numbers could be representative at X-band.

The RCS of a target can be controlled using primarily two basic methods: *shaping* and the use of *materials*. Shaping is used to attempt to direct the scattered energy toward directions other than the desired. However, for many targets shaping has to be compromised in order to meet other requirements, such as aerodynamic specifications for flying targets. Materials is used to trap the incident energy within the target and to dissipate part of the energy as heat or to direct it toward directions other than the desired. Usually both methods, shaping and materials, are used together in order to optimize the performance of a radar target. One of the "golden rules" to observe in order to

TABLE 2.2 RCS of Some Typical Targets

Object	Typical RCSs [22]	
	RCS (m^2)	RCS (dBsm)
Pickup truck	200	23
Automobile	100	20
Jumbo jet airliner	100	20
Large bomber *or* commercial jet	40	16
Cabin cruiser boat	10	10
Large fighter aircraft	6	7.78
Small fighter aircraft *or* four-passenger jet	2	3
Adult male	1	0
Conventional winged missile	0.5	−3
Bird	0.01	−20
Insect	0.00001	−50
Advanced tactical fighter	0.000001	−60

achieve low RCS is to "*round corners, avoid flat and concave surfaces, and use material treatment in flare spots.*"

There are many methods of analysis to predict the RCS of a target [13], [22]–[33]. Some of them are full-wave methods, others are designated as asymptotic methods, either low-frequency or high-frequency, and some are considered as numerical methods. The methods of analysis are often contingent upon the shape, size, and material composition of the target. Some targets, because of their geometrical complexity, are often simplified and are decomposed into a number of basic shapes (such as strips, plates, cylinders, cones, wedges) which when put together represent a very good replica of the actual target. This has been used extensively and proved to be a very good approach. The topic is very extensive to be treated here in any detail, and the reader is referred to the literature [13], [22]–[33]. There is a plethora of references but because of space limitations, only a limited number is included here to get the reader started on the subject.

Antennas individually are radar targets which many exhibit large radar cross section. In many applications, antennas are mounted on the surface of other complex targets (such as aircraft, spacecraft, satellites, missiles, automobiles), and become part of the overall radar target. In such configurations, many antennas, especially aperture types (such as waveguides, horns) become large contributors to the total RCS, monostatic or bistatic, of the target. Therefore, in designing low-observable targets, the antenna type, location and contributions become an important consideration of the overall design.

The scattering and transmitting (radiation) characteristics of an antenna are related [34]–[36]. There are various methods which can be used to analyze the fields scattered by an antenna. The summary here parallels that in [23], [37]–[40]. In general, the electric field scattered by an antenna with a load impedance Z_L can be expressed by

$$\mathbf{E}^s(Z_L) = \mathbf{E}^s(0) - \frac{I_s}{I_t} \frac{Z_L}{Z_L + Z_A} \mathbf{E}^t \quad (2\text{-}127)$$

where

$\mathbf{E}^s(Z_L)$ = electric field scattered by antenna with a load Z_L

$\mathbf{E}^s(0)$ = electric field scattered by short-circuited antenna ($Z_L = 0$)

I_s = short-circuited current induced by the incident field on the antenna with $Z_L = 0$

I_t = antenna current in transmitting mode

$Z_A = R_A + jX_A$ = antenna input impedance

$\mathbf{E}^t$ = electric field radiated by the antenna in transmitting mode

Green [37] expressed the field scattered by an antenna terminated with a load Z_L in a more convenient form which allows it to be separated into the *structural* and *antenna mode* scattering terms [23], [37]–[40]. This is accomplished by assuming that the antenna is loaded with a conjugate-matched impedance ($Z_L = Z_A^*$).

- The *structural* scattering term is introduced by the currents induced on the surface of the antenna by the incident field when the antenna is conjugate-matched, and it is independent of the load impedance.
- The *antenna mode* scattering term is only a function of the radiation characteristics of the antenna, and its scattering pattern is the square of the antenna radiation pattern.

The antenna mode depends on the power absorbed by the load of a lossless antenna and the power that is radiated by the antenna due to a load mismatch. This term vanishes when the antenna is conjugate-matched.

In general, the field scattered by an antenna loaded with an impedance Z_L is related to the field radiated by the antenna in the transmitting mode in three different ways.

- *First*, the field scattered by an antenna when it is loaded with an impedance Z_L is equal to the field scattered by the antenna when it is short-circuited ($Z_L = 0$) minus a term related to the antenna reflection coefficient and the field transmitted by the antenna.
- *Second*, the field scattered by an antenna when it is terminated with an impedance Z_L is equal to the field scattered by the antenna when it is conjugate-matched with an impedance Z_A^* minus the field transmitted (radiated) times the conjugate reflection coefficient.
- *Third*, the field scattered by the antenna when it is terminated with an impedance Z_L is equal to the field scattered by the antenna when it is matched with an impedance Z_A minus the field transmitted (radiated) times the reflection coefficient weighted by the ratio of two currents (I_m/I_t, I_m = scattering current when antenna is matched with an impedance Z_L, I_t = antenna current in the transmitting mode).

It can be shown that the total radar cross section of an antenna terminated with a load Z_L can be written as [40]

$$\sigma = |\sqrt{\sigma^s} - (1 + \Gamma_A)\sqrt{\sigma^a} e^{j\phi_r}|^2 \qquad (2\text{-}128)$$

where

σ = total RCS with antenna terminated with Z_L

σ^s = RCS due to structural term

σ^a = RCS due to antenna mode term

ϕ_r = relative phase between the structural and antenna mode terms

If the antenna is short-circuited ($\Gamma_A = -1$), then according to (2-128)

$$\sigma_{\text{short}} = \sigma^s \qquad (2\text{-}129)$$

If the antenna is open-circuited ($\Gamma_A = +1$), then according to (2-128)

$$\sigma_{\text{open}} = |\sqrt{\sigma^s} - 2\sqrt{\sigma^a} e^{j\phi_r}|^2 = \sigma_{\text{residual}} \qquad (2\text{-}130)$$

Lastly, if the antenna is matched $Z_L = Z_A (\Gamma_A = 0)$, then according to (2-128)

$$\sigma_{\text{match}} = |\sqrt{\sigma^s} - \sqrt{\sigma^a} e^{j\phi_r}|^2 \qquad (2\text{-}131)$$

Therefore, under matched conditions, according to (2-131), the range of values (minimum to maximum) of the radar cross section is

$$|\sqrt{\sigma^s} - \sqrt{\sigma^a}| \leq \sigma \leq |\sqrt{\sigma^s} + \sqrt{\sigma^a}| \qquad (2\text{-}132)$$

The minimum value occurs when the two RCSs are in phase while the maximum occurs when they are out of phase.

Example 2.17

The structural RCS of a resonant wire dipole is in phase and its magnitude is slightly greater than four times that of the antenna mode. Relate the short-circuited, open-circuited, and matched RCSs to that of the antenna mode.

Solution: Using (2-129)

$$\sigma_{\text{short}} = 4\sigma_{\text{antenna}}$$

Using (2-130)

$$\sigma_{\text{open}} = 2\sigma_{\text{antenna}}(0) = 0 \text{ or very small}$$

The matched value is obtained using (2-131), or

$$\sigma_{\text{match}} = \sigma_{\text{antenna}}$$

To produce a zero RCS, (2-128) must vanish. This is accomplished if

$$\text{Re}(\Gamma_A) = -1 + \cos\phi_r \sqrt{\sigma^s/\sigma^a} \qquad (2\text{-}133\text{a})$$

$$\text{Im}(\Gamma_A) = -\sin\phi_r \sqrt{\sigma^s/\sigma^a} \qquad (2\text{-}133\text{b})$$

Assuming positive values of resistances, the real value of Γ_A cannot be greater than unity. Therefore there are some cases where the RCS cannot be reduced to zero by choosing Z_L. Because Z_A can be complex, there is no limit on the imaginary part of Γ_A.

In general, the structural and antenna mode scattering terms are very difficult to predict and usually require that the antenna is solved as a boundary-value problem. However, these two terms have been obtained experimentally utilizing the Smith chart [37]–[39].

For a monostatic system the receiving and transmitting antennas are collocated. In addition, if the antennas are identical ($G_{0r} = G_{0t} = G_0$) and are polarization-matched ($\hat{\rho}_r = \hat{\rho}_t = 1$), the total radar cross section of the antenna for backscattering can be written as

$$\sigma = \frac{\lambda_0^2}{4\pi} G_0^2 |A - \Gamma^*|^2 \qquad (2\text{-}134)$$

where A is a complex parameter independent of the load.

If the antenna is a thin dipole, then $A \simeq 1$ and (2-134) reduces to

$$\sigma \simeq \frac{\lambda_0^2}{4\pi} G_0^2 |1 - \Gamma^*|^2 = \frac{\lambda_0^2}{4\pi} G_0^2 \left|1 - \frac{Z_L - Z_A^*}{Z_L + Z_A}\right|^2$$

$$= \frac{\lambda_0^2}{4\pi} G_0^2 \left|\frac{2R_A}{Z_L + Z_A}\right|^2 \qquad (2\text{-}135)$$

If in addition we assume that the dipole is resonant and its length is $l = \lambda_0/2$ and is short-circuited ($Z_L = 0$), then the normalized radar cross section of (2-135) is equal to

$$\frac{\sigma}{\lambda_0^2} \simeq \frac{G_0^2}{\pi} = \frac{(1.643)^2}{\pi} = 0.8593 \simeq 0.86 \qquad (2\text{-}136)$$

96 FUNDAMENTAL PARAMETERS AND FIGURES-OF-MERIT OF ANTENNAS

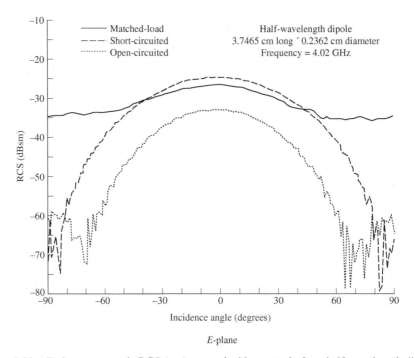

Figure 2.33 E-plane monostatic RCS ($\sigma_{\theta\theta}$) versus incidence angle for a half-wavelength dipole.

which agrees with experimental corresponding maximum monostatic value of Figure 2.33 and those reported in the literature [41], [42].

Shown in Figure 2.33 is the measured E-plane monostatic RCS of a half-wavelength dipole when it is matched to a load, short-circuited (straight wire) and open-circuited (gap at the feed). The aspect angle is measured from the normal to the wire. As expected, the RCS is a function of the observation (aspect) angle. Also it is apparent that there are appreciable differences between the three responses. For the short-circuited case, the maximum value is approximately −24 dBsm which closely agrees with the computed value of −22.5 dBsm using (2-136). Similar responses for the monostatic RCS of a pyramidal horn are shown in Figure 2.34(a) for the E-plane and in Figure 2.34(b) for the H-plane. The antenna is a commercial X-band (8.2-12.4 GHz) 20-dB standard gain horn with aperture dimension of 9.2 cm by 12.4 cm. The length of the horn is 25.6 cm. As for the dipole, there are differences between the three responses for each plane. It is seen that the short-circuited response exhibits the largest return.

Antenna RCS from scale model measurements [43] and microstrip patches [44], [45] have been reported.

2.18 ANTENNA TEMPERATURE

Every object with a physical temperature above absolute zero (0 K = −273°C) radiates energy [6]. The amount of energy radiated is usually represented by an equivalent temperature T_B, better known as brightness temperature, and it is defined as

$$T_B(\theta, \phi) = \varepsilon(\theta, \phi) T_m = (1 - |\Gamma|^2) T_m \qquad (2\text{-}137)$$

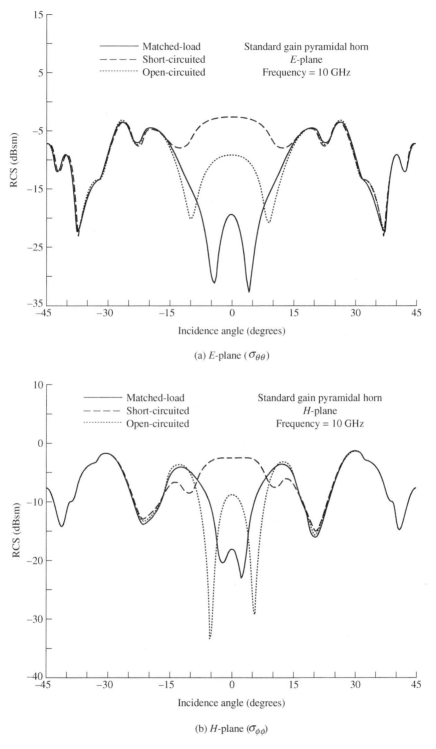

Figure 2.34 *E*- and *H*-plane monostatic RCS versus incidence angle for a pyramidal horn antenna.

where

T_B = brightness temperature (equivalent temperature; K)
ε = emissivity (dimensionless)
T_m = molecular (physical) temperature (K)
$\Gamma(\theta, \phi)$ = reflection coefficient of the surface for the polarization of the wave

Since the values of emissivity are $0 \leq \varepsilon \leq 1$, the maximum value the brightness temperature can achieve is equal to the molecular temperature. Usually the emissivity is a function of the frequency of operation, polarization of the emitted energy, and molecular structure of the object. Some of the better natural emitters of energy at microwave frequencies are (a) the ground with equivalent temperature of about 300 K and (b) the sky with equivalent temperature of about 5 K when looking toward zenith and about 100–150 K toward the horizon.

The brightness temperature emitted by the different sources is intercepted by antennas, and it appears at their terminals as an antenna temperature. The temperature appearing at the terminals of an antenna is that given by (2-137), after it is weighted by the gain pattern of the antenna. In equation form, this can be written as

$$T_A = \frac{\int_0^{2\pi} \int_0^{\pi} T_B(\theta, \phi) G(\theta, \phi) \sin\theta \, d\theta \, d\phi}{\int_0^{2\pi} \int_0^{\pi} G(\theta, \phi) \sin\theta \, d\theta \, d\phi} \tag{2-138}$$

where

T_A = antenna temperature (effective noise temperature of the antenna radiation resistance; K)
$G(\theta, \phi)$ = gain (power) pattern of the antenna

Assuming no losses or other contributions between the antenna and the receiver, the noise power transferred to the receiver is given by

$$P_r = kT_A \Delta f \tag{2-139}$$

where

P_r = antenna noise power   (W)
k = Boltzmann's constant   (1.38×10^{-23} J/K)
T_A = antenna temperature   (K)
Δf = bandwidth   (Hz)

If the antenna and transmission line are maintained at certain physical temperatures, and the transmission line between the antenna and receiver is lossy, the antenna temperature T_A as seen by the receiver through (2-139) must be modified to include the other contributions and the line losses. If the antenna itself is maintained at a certain physical temperature T_p and a transmission line of length l, constant physical temperature T_0 throughout its length, and uniform attenuation of α (Np/unit length) is used to connect an antenna to a receiver, as shown in Figure 2.35, the effective antenna temperature at the receiver terminals is given by

$$T_a = T_A e^{-2\alpha l} + T_{AP} e^{-2\alpha l} + T_0(1 - e^{-2\alpha l}) \tag{2-140}$$

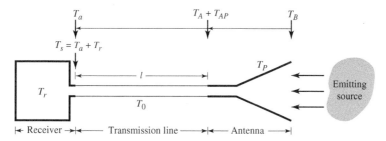

Figure 2.35 Antenna, transmission line, and receiver arrangement for system noise power calculation.

where

$$T_{AP} = \left(\frac{1}{e_A} - 1\right) T_p \qquad (2\text{-}140a)$$

T_a = antenna temperature at the receiver terminals (K)
T_A = antenna noise temperature at the antenna terminals (2-138) (K)
T_{AP} = antenna temperature at the antenna terminals due to physical temperature (2-140a) (K)
T_p = antenna physical temperature (K)
α = attenuation coefficient of transmission line (Np/m)
e_A = thermal efficiency of antenna (dimensionless)
l = length of transmission line (m)
T_0 = physical temperature of the transmission line (K)

The antenna noise power of (2-139) must also be modified and written as

$$P_r = kT_a \Delta f \qquad (2\text{-}141)$$

where T_a is the antenna temperature at the receiver input as given by (2-140).

If the receiver itself has a certain noise temperature T_r (due to thermal noise in the receiver components), the *system noise power at the receiver terminals* is given by

$$P_s = k(T_a + T_r)\Delta f = kT_s \Delta f \qquad (2\text{-}142)$$

where
P_s = system noise power (at receiver terminals)
T_a = antenna noise temperature (at receiver terminals)
T_r = receiver noise temperature (at receiver terminals)
$T_s = T_a + T_r$ = effective system noise temperature (at receiver terminals)

A graphical relation of all the parameters is shown in Figure 2.35. The effective system noise temperature T_s of radio astronomy antennas and receivers varies from very few degrees (typically $\simeq$ 10 K) to thousands of Kelvins depending upon the type of antenna, receiver, and frequency of operation. Antenna temperature changes at the antenna terminals, due to variations in the target emissions, may be as small as a fraction of one degree. To detect such changes, the receiver must be very sensitive and be able to differentiate changes of a fraction of a degree.

Example 2.18

The effective antenna temperature of a target at the input terminals of the antenna is 150 K. Assuming that the antenna is maintained at a thermal temperature of 300 K and has a thermal efficiency of 99% and it is connected to a receiver through an X-band (8.2–12.4 GHz) rectangular waveguide of 10 m (loss of waveguide = 0.13 dB/m) and at a temperature of 300 K, find the effective antenna temperature at the receiver terminals.

Solution: We first convert the attenuation coefficient from dB to Np by $\alpha(\text{dB/m}) = 20(\log_{10} e)\alpha(\text{Np/m}) = 20(0.434)\alpha(\text{Np/m}) = 8.68\alpha(\text{Np/m})$. Thus $\alpha(\text{Np/m}) = \alpha(\text{dB/m})/8.68 = 0.13/8.68 = 0.0149$. The effective antenna temperature at the receiver terminals can be written, using (2-140a) and (2-140), as

$$T_{AP} = 300\left(\frac{1}{0.99} - 1\right) = 3.03$$

$$T_a = 150e^{-0.149(2)} + 3.03e^{-0.149(2)} + 300[1 - e^{-0.149(2)}]$$

$$= 111.345 + 2.249 + 77.31 = 190.904 \text{ K}$$

The results of the above example illustrate that the antenna temperature at the input terminals of the antenna and at the terminals of the receiver can differ by quite a few degrees. For a smaller transmission line or a transmission line with much smaller losses, the difference can be reduced appreciably and can be as small as a fraction of a degree.

A summary of the pertinent parameters and associated formulas and equation numbers for this chapter are listed in Table 2.3.

2.19 MULTIMEDIA

In the website created by the publisher for this book, the following multimedia resources are included for the review, understanding, and visualization of the material of this chapter:

a. **Java**-based **interactive questionnaire**, with answers.
b. **Java**-based **applet** for computing and displaying graphically the directivity of an antenna.
c. **Matlab** and **Fortran** computer program, designated **Directivity**, for computing the directivity of an antenna. A description of this program is in the READ ME file in the publisher's website for this book.
d. **Matlab** plotting computer programs:
   - **2-D Polar** (designated as **Polar**). This program can be used to plot the two-dimensional patterns, in both polar and semipolar form (*in linear or dB scale*), of an antenna.
   - **3-D Spherical**. This program (designated as **Spherical**) can be used to plot the three-dimensional pattern (*in linear or dB scale*) of an antenna in spherical form.
   - **Polarization_Diagram_Ellipse_Animation:** Animates the 3-D polarization diagram of a rotating electric field vector [Figure 2.23(a)]. It also animates the 2-D polarization ellipse [Figure 2.23(b)] for linear, circular and elliptical polarized waves, and sense of rotation. It also computes the axial ratio (AR).
   - **Polarization_Propag:** Computes the Poincaré sphere angles, and thus the polarization wave travelling in an infinite homogeneous medium.

   A description of these programs is in the corresponding READ ME files in the publisher's website for this book.
e. **Power Point (PPT)** viewgraphs, in multicolor.

TABLE 2.3 Summary of Important Parameters and Associated Formulas and Equation Numbers

Parameter	Formula	Equation Number
Infinitesimal area of sphere	$dA = r^2 \sin\theta \, d\theta \, d\phi$	(2-1)
Elemental solid angle of sphere	$d\Omega = \sin\theta \, d\theta \, d\phi$	(2-2)
Average power density	$\mathbf{W}_{av} = \dfrac{1}{2}\mathrm{Re}[\mathbf{E}\times\mathbf{H}^*]$	(2-8)
Radiated power/average radiated power	$P_{rad} = P_{av} = \oiint_S \mathbf{W}_{av}\cdot d\mathbf{s} = \dfrac{1}{2}\oiint_S \mathrm{Re}[\mathbf{E}\times\mathbf{H}^*]\cdot d\mathbf{s}$	(2-9)
Radiation density of isotropic radiator	$W_0 = \dfrac{P_{rad}}{4\pi r^2}$	(2-11)
Radiation intensity (far field)	$U = r^2 W_{rad} = B_0 F(\theta,\phi) \simeq \dfrac{r^2}{2\eta}$	(2-12),
	$\times \left[|E_\theta(r,\theta,\phi)|^2 + |E_\phi(r,\theta,\phi)|^2\right]$	(2-12a)
Directivity $D(\theta,\phi)$	$D = \dfrac{U}{U_0} = \dfrac{4\pi U}{P_{rad}} = \dfrac{4\pi}{\Omega_A}$	(2-16), (2-23)
Beam solid angle Ω_A	$\Omega_A = \displaystyle\int_0^{2\pi}\int_0^\pi F_n(\theta,\phi)\sin\theta \, d\theta \, d\phi$	(2-24)
	$F_n(\theta,\phi) = \dfrac{F(\theta,\phi)}{|F(\theta,\phi)|_{\max}}$	(2-25)
Maximum directivity D_0	$D_{\max} = D_0 = \dfrac{U_{\max}}{U_0} = \dfrac{4\pi U_{\max}}{P_{rad}}$	(2-16a)
Partial directivities D_θ, D_ϕ	$D_0 = D_\theta + D_\phi$	(2-17)
	$D_\theta = \dfrac{4\pi U_\theta}{P_{rad}} = \dfrac{4\pi U_\theta}{(P_{rad})_\theta + (P_{rad})_\phi}$	(2-17a)
	$D_\phi = \dfrac{4\pi U_\phi}{P_{rad}} = \dfrac{4\pi U_\phi}{(P_{rad})_\theta + (P_{rad})_\phi}$	(2-17b)
Approximate maximum directivity (*one main lobe pattern*)	$D_0 \simeq \dfrac{4\pi}{\Theta_{1r}\Theta_{2r}} = \dfrac{41{,}253}{\Theta_{1d}\Theta_{2d}}$ (Kraus)	(2-26), (2-27)
	$D_0 \simeq \dfrac{32\ln 2}{\Theta_{1r}^2 + \Theta_{2r}^2} = \dfrac{22.181}{\Theta_{1r}^2 + \Theta_{2r}^2} = \dfrac{72{,}815}{\Theta_{1d}^2 + \Theta_{2d}^2}$ (Tai-Pereira)	(2-30), (2-30a), (2-30b)
Approximate maximum directivity (*omnidirectional pattern*)	$D_0 \simeq \dfrac{101}{\mathrm{HPBW(degrees)} - 0.0027[\mathrm{HPBW(degrees)}]^2}$ (McDonald)	(2-33a)
	$D_0 \simeq -172.4 + 191\sqrt{0.818 + \dfrac{1}{\mathrm{HPBW(degrees)}}}$ (Pozar)	(2-33b)

(continued overleaf)

102 FUNDAMENTAL PARAMETERS AND FIGURES-OF-MERIT OF ANTENNAS

TABLE 2.3 *(continued)*

Parameter	Formula	Equation Number
Gain $G(\theta, \phi)$	$G = \dfrac{4\pi U(\theta, \phi)}{P_{in}} = e_{cd}\left[\dfrac{4\pi U(\theta, \phi)}{P_{rad}}\right] = e_{cd}D(\theta, \phi)$ $P_{rad} = e_{cd}P_{in}$	(2-46), (2-47) (2-49)
Antenna radiation efficiency e_{cd}	$e_{cd} = \dfrac{R_r}{R_r + R_L}$	(2-90)
Loss resistance R_L *(straight wire/uniform current)*	$R_L = R_{hf} = \dfrac{l}{P}\sqrt{\dfrac{\omega\mu_0}{2\sigma}}$	(2-90b)
Loss resistance R_L *(straight wire/ $\lambda/2$ dipole)*	$R_L = \dfrac{l}{2P}\sqrt{\dfrac{\omega\mu_0}{2\sigma}}$	
Maximum gain G_0	$G_0 = e_{cd}D_{max} = e_{cd}D_0$	(2-49a)
Partial gains G_θ, G_ϕ	$G_0 = G_\theta + G_\phi$ $G_\theta = \dfrac{4\pi U_\theta}{P_{in}}, \quad G_\phi = \dfrac{4\pi U_\phi}{P_{in}}$	(2-50) (2-50a), (2-50b)
Realized gain G_{re}	$G_{re} = e_r G(\theta, \phi) = e_r e_{cd} D(\theta, \phi) = (1 - |\Gamma|^2)e_{cd}D(\theta, \phi)$ $= e_0 D(\theta, \phi)$	(2-49a) (2-49b)
Total antenna efficiency e_0 Reflection efficiency e_r	$e_0 = e_r e_c e_d = e_r e_{cd} = (1 - |\Gamma|^2)e_{cd}$ $e_r = (1 - |\Gamma|^2)$	(2-52) (2-45)
Beam efficiency BE	$BE = \dfrac{\int_0^{2\pi}\int_0^{\theta_1} U(\theta, \phi)\sin\theta\, d\theta\, d\phi}{\int_0^{2\pi}\int_0^{\pi} U(\theta, \phi)\sin\theta\, d\theta\, d\phi}$	(2-54)
Polarization loss factor (PLF)	$PLF = |\hat{\boldsymbol{\rho}}_w \cdot \hat{\boldsymbol{\rho}}_a|^2$	(2-71)
Vector effective length $\boldsymbol{\ell}_e(\theta, \phi)$	$\boldsymbol{\ell}_e(\theta, \phi) = \hat{\mathbf{a}}_\theta l_\theta(\theta, \phi) + \hat{\mathbf{a}}_\phi l_\phi(\theta, \phi)$	(2-91)
Polarization efficiency p_e	$p_e = \dfrac{|\boldsymbol{\ell}_e \cdot \mathbf{E}^{inc}|^2}{|\boldsymbol{\ell}_e|^2 |\mathbf{E}^{inc}|^2}$	(2-71a)
Antenna impedance Z_A	$Z_A = R_A + jX_A = (R_r + R_L) + jX_A$	(2-72), (2-73)
Maximum effective area A_{em}	$A_{em} = \dfrac{|V_T|^2}{8W_i}\left[\dfrac{1}{R_r + R_L}\right] = e_{cd}\left(\dfrac{\lambda^2}{4\pi}\right)D_0|\hat{\boldsymbol{\rho}}_w \cdot \hat{\boldsymbol{\rho}}_a|^2$ $= \left(\dfrac{\lambda^2}{4\pi}\right)G_0|\hat{\boldsymbol{\rho}}_w \cdot \hat{\boldsymbol{\rho}}_a|^2$	(2-96), (2-111), (2-112)
Aperture efficiency ε_{ap}	$\varepsilon_{ap} = \dfrac{A_{em}}{A_p} = \dfrac{\text{maximum effective area}}{\text{physical area}}$	(2-100)
Friis transmission equation	$\dfrac{P_r}{P_t} = \left(\dfrac{\lambda}{4\pi R}\right)^2 G_{0t}G_{0r}|\hat{\boldsymbol{\rho}}_t \cdot \hat{\boldsymbol{\rho}}_r|^2$	(2-118), (2-119)
Radar range equation	$\dfrac{P_r}{P_t} = \sigma\dfrac{G_{0t}G_{0r}}{4\pi}\left[\dfrac{\lambda}{4\pi R_1 R_2}\right]^2 |\hat{\boldsymbol{\rho}}_w \cdot \hat{\boldsymbol{\rho}}_r|^2$	(2-125), (2-126)

TABLE 2.3 (*continued*)

Parameter	Formula	Equation Number
Radar cross section (RCS) (m²)	$\sigma = \lim_{R \to \infty} \left[4\pi R^2 \dfrac{W_s}{W_i} \right] = \lim_{R \to \infty} \left[4\pi R^2 \dfrac{|\mathbf{E}^s|^2}{|\mathbf{E}^i|^2} \right]$ $= \lim_{R \to \infty} \left[4\pi R^2 \dfrac{|\mathbf{H}^s|^2}{|\mathbf{H}^i|^2} \right]$	(2-120a)
Brightness temperature $T_B(\theta, \phi)$ (K)	$T_B(\theta, \phi) = \varepsilon(\theta, \phi) T_m = (1 - |\Gamma|^2) T_m$	(2-144)
Antenna temperature T_A (K)	$T_A = \dfrac{\int_0^{2\pi} \int_0^{\pi} T_B(\theta, \phi) G(\theta, \phi) \sin\theta \, d\theta \, d\phi}{\int_{0\pi}^{2\pi} \int_0^{\pi} G(\theta, \phi) \sin\theta \, d\theta \, d\phi}$	(2-145)

REFERENCES

1. A. Z. Elsherbeni and C. D. Taylor Jr., "Antenna Pattern Plotter," Copyright © 1995, Electrical Engineering Department, The University of Mississippi, University, MS.
2. W. R. Scott Jr., "A General Program for Plotting Three-dimensional Antenna Patterns," *IEEE Antennas and Propagation Society Newsletter*, pp. 6–11, December 1989.
3. A. Z. Elsherbeni and C. D. Taylor Jr., "Interactive Antenna Pattern Visualization," *Software Book in Electromagnetics*, Vol. II, Chapter 8, CAEME Center for Multimedia Education, University of Utah, pp. 367–410, 1995.
4. J. S. Hollis, T. J. Lyon, and L. Clayton Jr. (eds.), *Microwave Antenna Measurements*, Scientific-Atlanta, Inc., July 1970.
5. J. D. Kraus, *Antennas*, McGraw-Hill, New York, 1988.
6. J. D. Kraus, *Radio Astronomy*, McGraw-Hill Book Co., 1966.
7. A. Z. Elsherbeni and P. H. Ginn. "Interactive Analysis of Antenna Arrays," *Software Book in Electromagnetics*, Vol. II, Chapter 6, CAEME Center for Multimedia Education, University of Utah, pp. 337–366, 1995.
8. J. Romeu and R. Pujol, "Array," *Software Book in Electromagnetics*, Vol. II, Chapter 12, CAEME Center for Multimedia Education, University of Utah, pp. 467–481, 1995.
9. R. S. Elliott, "Beamwidth and Directivity of Large Scanning Arrays," Last of Two Parts, *The Microwave Journal*, pp. 74–82, January 1964.
10. C.-T. Tai and C. S. Pereira, "An Approximate Formula for Calculating the Directivity of an Antenna," *IEEE Trans. Antennas Propagat.*, Vol. AP-24, No. 2, pp. 235–236, March 1976.
11. N. A. McDonald, "Approximate Relationship Between Directivity and Beamwidth for Broadside Collinear Arrays," *IEEE Trans. Antennas Propagat.*, Vol. AP-26, No. 2, pp. 340–341, March 1978.
12. D. M. Pozar, "Directivity of Omnidirectional Antennas," *IEEE Antennas Propagat. Mag.*, Vol. 35, No. 5, pp. 50–51, October 1993.
13. C. A. Balanis, *Advanced Engineering Electromagnetics*, Second Edition, John Wiley and Sons, New York, 2012.
14. H. Poincaré, *Theorie Mathematique de la Limiere*, Georges Carre, Paris, France, 1892.
15. G. A. Deschamps, "Part II—Geometrical Representation of the Polarization of a Plane Electromagnetic Wave," *Proc. IRE*, Vol. 39, pp. 540–544, May 1951.
16. E. F. Bolinder, "Geometrical Analysis of Partially Polarized Electromagnetic Waves," *IEEE Trans. Antennas Propagat.*, Vol. AP-15, No. 1, pp. 37–40, January 1967.
17. G. A. Deschamps and P. E. Mast, "Poincaré Sphere Representation of Partially Polarized Fields," *IEEE Trans. Antennas Propagat.*, Vol. AP-21, No. 4, pp. 474–478, July 1973.

18. G. Sinclair, "The Transmission and Reflection of Elliptically Polarized Waves," *Proc. IRE*, Vol. 38, pp. 148–151, February 1950.
19. C. A. Balanis, "Antenna Theory: A Review," *Proc. IEEE*, Vol. 80, No. 1, pp. 7–23, January 1992.
20. R. E. Collin, *Antennas and Radiowave Propagation*, McGraw-Hill Book Co., New York, 1985.
21. M. I. Skolnik, *Radar Systems*, Chapter 2, McGraw-Hill Book Co., New York, 1962.
22. J. A. Adam, "How to Design an "Invisible' Aircraft," *IEEE Spectrum*, pp. 26–31, April 1988.
23. G. T. Ruck, D. E. Barrick, W. D. Stuart, and C. K. Krichbaum, *Radar Cross Section Handbook*, Vols. 1, 2, Plenum Press, New York, 1970.
24. M. I. Skolnik (Ed.), *Radar Handbook*, Chapter 27, Section 6, McGraw-Hill Book Co., New York, pp. 27-19–27-40, 1970.
25. J. W. Crispin, Jr. and K. M. Siegel, *Methods of Radar Cross Section Analysis*, Academic Press, Inc., New York, 1968.
26. J. J. Bowman, T. B. A. Senior, and P. L. Uslenghi (Eds.), *Electromagnetic and Acoustic Scattering by Simple Shapes*, Amsterdam, The Netherland: North-Holland, 1969.
27. E. F. Knott, M. T. Turley, and J. F. Shaeffer, *Radar Cross Section*, Artech House, Inc., Norwood, MA, 1985.
28. A. K. Bhattacharya and D. L. Sengupta, *Radar Cross Section Analysis and Control*, Artech House, Inc., Norwood, MA, 1991.
29. A. F. Maffett, *Topics for a Statistical Description of Radar Cross Section*, John Wiley and Sons, New York, 1989.
30. Special issue, *Proc. IEEE*, Vol. 53, No. 8, August 1965.
31. Special issue, *Proc. IEEE*, Vol. 77, No. 5, May 1989.
32. Special issue, *IEEE Trans. Antennas Propagat.*, Vol. 37, No. 5, May 1989.
33. W. R. Stone (ed.), *Radar Cross Sections of Complex Objects*, IEEE Press, New York, 1989.
34. A. F. Stevenson, "Relations Between the Transmitting and Receiving Properties of Antennas," *Q. Appl. Math.*, pp. 369–384, January 1948.
35. R. F. Harrington, "Theory of Loaded Scatterers," *Proc. IEE* (British), Vol. 111, pp. 617–623, April 1964.
36. R. E. Collin, "The Receiving Antenna," in *Antenna Theory, Part I*, (R. E. Collin and F. J. Zucker, Eds.), McGraw-Hill Book Co., 1969.
37. R. B. Green, "The Effect of Antenna Installations on the Echo Area of an Object," Report No. 1109-3, ElectroScience Laboratory, Ohio State University, Columbus, OH, September 1961.
38. R. B. Green "Scattering from Conjugate-Matched Antennas," *IEEE Trans. Antennas Propagat.*, Vol. AP-14, No. 1, pp. 17–21, January 1966.
39. R. J. Garbacz, "The Determination of Antenna Parameters by Scattering Cross-Section Measurements, III. Antenna Scattering Cross Section," Report No. 1223-10, Antenna Laboratory, Ohio State University, November 1962.
40. R. C. Hansen, "Relationships Between Antennas as Scatterers and as Radiators," *Proc. IEEE*, Vol. 77, No. 5, pp. 659–662, May 1989.
41. S. H. Dike and D. D. King, "Absorption Gain and Backscattering Cross Section of the Cylindrical Antenna," *Proc. IRE*, Vol. 40, 1952.
42. J. Sevick, "Experimental and Theoretical Results on the Backscattering Cross Section of Coupled Antennas," Tech. Report No. 150, Cruft Laboratory, Harvard University, May 1952.
43. D. L. Moffatt, "Determination of Antenna Scattering Properties From Model Measurements," Report No. 1223-12, Antenna Laboratory, Ohio State University, January 1964.
44. J. T. Aberle, Analysis of Probe-Fed Circular Microstrip Antennas, PhD Dissertation, University of Mass., Amherst, MA, 1989.
45. J. T. Aberle, D. M. Pozar, and C. R. Birtcher, "Evaluation of Input Impedance and Radar Cross Section of Probe-Fed Microstrip Patch Elements Using an Accurate Feed Model," *IEEE Trans. Antennas Propagat.*, Vol. 39, No. 12, pp. 1691–1696, December 1991.

PROBLEMS

2.1. An antenna has a beam solid angle that is equivalent to a *trapezoidal* patch (*patch with 4 sides, 2 of which are parallel to each other*) on the surface of a sphere of radius r. The angular space of the patch on the surface of the sphere extends between $\pi/6 \leq \theta \leq \pi/3 (30° \leq \theta \leq 60°)$ in latitude and $\pi/4 \leq \phi \leq \pi/3 (45° \leq \phi \leq 60°)$ in longitude. Find the following:
(a) Equivalent *beam solid angle* [which is equal to number of *square radians/steradians* **or** $(degrees)^2$] of the patch [in *square radians/steradians* **and** in $(degrees)^2$].
  - Exact.
  - Approximate using $\Omega_A = \Delta\Theta \cdot \Delta\Phi = (\theta_2 - \theta_1) \cdot (\phi_2 - \phi_1)$. Compare with the exact.

(b) Corresponding antenna *maximum directivities* of part *a* (*dimensionless* **and** *in dB*).

2.2. Derive (2-7) given the definitions of (2-5) and (2-6)

2.3. A hypothetical isotropic antenna is radiating in free-space. At a distance of 100 m from the antenna, the total electric field (E_θ) is measured to be 5 V/m. Find the
(a) power density (W_{rad})   (b) power radiated (P_{rad})

2.4. Find the half-power beamwidth (HPBW) and first-null beamwidth (FNBW), *in radians and degrees*, for the following normalized radiation intensities:

(a) $U(\theta) = \cos\theta$ (b) $U(\theta) = \cos^2\theta$
(c) $U(\theta) = \cos(2\theta)$ (d) $U(\theta) = \cos^2(2\theta)$ $(0 \leq \theta \leq 90°, 0 \leq \phi \leq 360°)$
(e) $U(\theta) = \cos(3\theta)$ (f) $U(\theta) = \cos^2(3\theta)$

2.5. Find the half-power beamwidth (HPBW) and first-null beamwidth (FNBW), *in radians and degrees*, for the following normalized radiation intensities:

(a) $U(\theta) = \cos\theta \cos(2\theta)$
(b) $U(\theta) = \cos^2\theta \cos^2(2\theta)$
(c) $U(\theta) = \cos(\theta)\cos(3\theta)$
(d) $U(\theta) = \cos^2(\theta)\cos^2(3\theta)$ $(0 \leq \theta \leq 90°, 0 \leq \phi \leq 360°)$
(e) $U(\theta) = \cos(2\theta)\cos(3\theta)$
(f) $U(\theta) = \cos^2(2\theta)\cos^2(3\theta)$

2.6. The maximum radiation intensity of a 90% efficiency antenna is 200 mW/unit solid angle. Find the directivity and gain (dimensionless and in dB) when the
(a) input power is 125.66 mW
(b) radiated power is 125.66 mW

2.7. The power radiated by a lossless antenna is 10 watts. The directional characteristics of the antenna are represented by the radiation intensity of

(a) $U = B_o \cos^2\theta$ (watts/unit solid angle)
(b) $U = B_o \cos^3\theta$ $(0 \leq \theta \leq \pi/2, 0 \leq \phi \leq 2\pi)$

For each, find the
(a) maximum power density (in watts/square meter) at a distance of 1,000 m (assume far-field distance). Specify the angle where this occurs.
(b) exact and approximate beam solid angle Ω_A.
(c) directivity, exact and approximate, of the antenna (dimensionless and in dB).
(d) gain, exact and approximate, of the antenna (dimensionless and in dB).

2.8. The *approximate far zone normalized electric field* radiated by a *resonant* linear dipole antenna used in wireless mobile units, positioned symmetrically at the origin along the z-axis, is given by

$$\mathbf{E}_a \simeq \hat{\mathbf{a}}_\theta E_a \sin^{1.5}\theta \frac{e^{-jkr}}{r}, \quad \begin{array}{l} 0° \leq \theta \leq 180° \\ 0° \leq \theta \leq 360° \end{array}$$

where E_a is a constant and r is the spherical radial distance measured from the origin of the coordinate system. Determine the:
(a) *Exact maximum directivity* (*dimensionless* and *in dB*).
(b) *Half-power beamwidth* (*in degrees*).
(c) *Approximate maximum directivity* (*dimensionless* and *in dB*). Indicate which approximate formula you are using and why?
(d) *Approximate maximum directivity* (*dimensionless* and *in dB*) using another approximate formula. Indicate which other formula you are using and why?
(e) Maximum directivity (dimensionless and in dB) using the computer program **Directivity**.

2.9. You are an antenna engineer and you are asked to design a high directivity/gain antenna for a space-borne communication system operating at 10 GHz. The specifications of the antenna are such that its pattern consists basically of *one major lobe* and, for simplicity, *no minor lobes* (*if there are any minor lobes they are of such very low intensity and you can assume they are negligible/zero*). Also it is desired that the pattern is symmetrical in the azimuthal plane. In order to meet the desired objectives, the main lobe of the pattern should have a *half-power beamwidth* of *10 degrees*. In order to expedite the design, it is assumed that the major lobe of the normalized radiation intensity of the antenna is approximated by

$$U(\theta, \phi) = \cos^n(\theta)$$

and it exists only in the upper hemisphere ($0 \leq \theta \leq \pi/2, 0 \leq \phi \leq 2\pi$). Determine the:
(a) Value of *n* (*not necessarily an integer*) to meet the specifications of the major lobe. *Keep 5 significant figures in your calculations.*
(b) *Exact* maximum directivity of the antenna (*dimensionless* and *in dB*).
(c) *Approximate* maximum directivity of the antenna *based on Kraus'* formula (*dimensionless* and *in dB*).
(d) *Approximate* maximum directivity of the antenna based on *Tai & Pereira's* formula (*dimensionless* and *in dB*).

2.10. In target-search ground-mapping radars it is desirable to have echo power received from a target, of constant cross section, to be independent of its range. For one such application, the desirable radiation intensity of the antenna is given by

$$U(\theta, \phi) = \left\{ \begin{array}{ll} 1 & 0° \leq \theta < 20° \\ 0.342\,\mathrm{csc}(\theta) & 20° \leq \theta < 60° \\ 0 & 60° \leq \theta \leq 180° \end{array} \right\} \quad 0° \leq \phi \leq 360°$$

Find the directivity (in dB) using the exact formula.

2.11. A beam antenna has half-power beamwidths of 30° and 35° in perpendicular planes intersecting at the maximum of the mainbeam. Find its approximate maximum effective aperture (in λ^2) using:

(a) Kraus'

(b) Tai and Pereira's formulas.

The minor lobes are very small and can be neglected.

2.12. The normalized radiation intensity of a given antenna is given by

(a) $U = \sin\theta \sin\phi$ 　　　　　　(b) $U = \sin\theta \sin^2\phi$

(c) $U = \sin\theta \sin^3\phi$ 　　　　　　(d) $U = \sin^2\theta \sin\phi$

(e) $U = \sin^2\theta \sin^2\phi$ 　　　　　(f) $U = \sin^2\theta \sin^3\phi$

The intensity exists only in the $0 \leq \theta \leq \pi, 0 \leq \phi \leq \pi$ region, and it is zero elsewhere. Find the

(a) exact directivity (*dimensionless* and *in dB*).

(b) azimuthal and elevation plane half-power beamwidths (in degrees).

2.13. The *normalized radiation intensity* radiated by an antenna is given by

$$U(\theta, \phi) = \begin{cases} \sin\theta \cos^2\phi & \begin{array}{l} 0° \leq \theta \leq 180° \\ 90° \leq \theta \leq 270° \end{array} \\ 0 & \text{Elsewhere} \end{cases}$$

The maximum of the radiation intensity occurs towards $\theta = 90°$ and $\phi = 180°$. Find the:

(a) *Exact* maximum directivity (*dimensionless* and *in dB*).

(b) Half-power beamwidth (*in degrees*) in the principal azimuth (horizontal) plane.

(c) Half-power beamwidth (*in degrees*) in the principal elevation (vertical) plane.

(d) Maximum directivity (*dimensionless* and *in dB*) using an *appropriate approximate method* that you know. *Indicate which one you are using*.

2.14. Find the directivity (dimensionless and in dB) for the antenna of Problem 2.12 using

(a) Kraus' approximate formula (2-26)

(b) Tai and Pereira's approximate formula (2-30a)

2.15. For Problem 2.10, determine the approximate directivity (in dB) using

(a) Kraus' formula 　　　(b) Tai and Pereira's formula.

2.16. The normalized radiation intensity of an antenna is rotationally symmetric in ϕ, and it is represented by

$$U = \begin{cases} 1 & 0° \leq \theta < 30° \\ 0.5 & 30° \leq \theta < 60° \\ 0.1 & 60° \leq \theta < 90° \\ 0 & 90° \leq \theta \leq 180° \end{cases}$$

(a) What is the directivity (above isotropic) of the antenna (in dB)?

(b) What is the directivity (above an infinitesimal dipole) of the antenna (in dB)?

2.17. The radiation intensity of an antenna is given by

$$U(\theta, \phi) = \cos^4 \theta \sin^2 \phi$$

for $0 \leq \theta \leq \pi/2$ and $0 \leq \phi \leq 2\pi$ (i.e., in the upper half-space). It is zero in the lower half-space.
Find the
(a) exact directivity (dimensionless and in dB)
(b) elevation plane half-power beamwidth (in degrees)

2.18. The normalized radiation intensity of an antenna is symmetric, and it can be approximated by

$$U(\theta) = \begin{cases} 1 & 0° \leq \theta < 30° \\ \dfrac{\cos(\theta)}{0.866} & 30° \leq \theta < 90° \\ 0 & 90° \leq \theta \leq 180° \end{cases}$$

and it is independent of ϕ. Find the
(a) exact directivity by integrating the function
(b) approximate directivity using Kraus' formula

2.19. The maximum gain of a horn antenna is +20 dB, while the gain of its first sidelobe is −15 dB. What is the difference in gain between the maximum and first sidelobe:
(a) in dB   (b) as a ratio of the field intensities.

2.20. The normalized radiation intensity of an antenna is approximated by

$$U = \sin \theta$$

where $0 \leq \theta \leq \pi$, and $0 \leq \phi \leq 2\pi$. Determine the directivity using the
(a) exact formula
(b) formulas of (2-33a) by McDonald and (2-33b) by Pozar
(c) computer program **Directivity** of this chapter.

2.21. Repeat Problem 2.20 for a $\lambda/2$ dipole whose normalized intensity is approximated by

$$U \simeq \sin^3 \theta$$

Compare the value with that of (4-91) or 1.643 (2.156 dB).

2.22. The radiation intensity of a circular loop of radius a and of constant current is given by

$$U = J_1^2(ka \sin \theta), \quad 0 \leq \theta \leq \pi \text{ and } 0 \leq \phi \leq 2\pi$$

where $J_1(x)$ is the Bessel function of order 1. For a loop with radii of $a = \lambda/10$ and $\lambda/20$, determine the directivity using the:
(a) formulas (2-33a) by McDonald and (2-33b) by Pozar.
(b) computer program **Directivity** of this chapter.
Compare the answers with that of a very small loop represented by 1.5 or 1.76 dB.

2.23. Find the directivity (dimensionless and in dB) for the antenna of Problem 2.12 using numerical techniques with 10° uniform divisions and with the field evaluated at the
(a) midpoint (b) trailing edge of each division.

2.24. Compute the directivity values of Problem 2.12 using the computer program **Directivity** of this chapter.

2.25. The far-zone electric-field intensity (array factor) of an end-fire two-element array antenna, placed along the z-axis and radiating into free-space, is given by

$$E = \cos\left[\frac{\pi}{4}(\cos\theta - 1)\right]\frac{e^{-jkr}}{r}, \quad 0 \leq \theta \leq \pi$$

Find the directivity using
(a) Kraus' approximate formula
(b) the computer program **Directivity** of this chapter.

2.26. Repeat Problem 2.25 when

$$E = \cos\left[\frac{\pi}{4}(\cos\theta + 1)\right]\frac{e^{-jkr}}{r}, \quad 0 \leq \theta \leq \pi$$

2.27. The radiation intensity is represented by

$$U = \begin{cases} U_0 \sin(\pi \sin\theta), & 0 \leq \theta \leq \pi/2 \text{ and } 0 \leq \phi \leq 2\pi \\ 0 & \text{elsewhere} \end{cases}$$

Find the directivity
(a) exactly
(b) using the computer program **Directivity** of this chapter.

2.28. The approximate far-zone electric field radiated by a *very thin wire* circular loop of radius a, positioned symmetrically about the z-axis and with its area parallel to the xy-plane, is given by

$$E_\phi \simeq C_0 \sin^{1.5}\theta \frac{e^{jkr}}{r}$$

where C_0 is a constant. Determine the:
(a) Exact directivity (*dimensionless* and *in dB*).
(b) Approximate directivity (*dimensionless* and *in dB*) using an approximate but appropriate formula (*state the formula you are using*).

2.29. The radiation intensity of an aperture antenna, mounted on an infinite ground plane with z perpendicular to the aperture, is rotationally symmetric (not a function of ϕ), and it is given by

$$U = \left[\frac{\sin(\pi \sin\theta)}{\pi \sin\theta}\right]^2$$

Find the approximate directivity (dimensionless and in dB) using
(a) numerical integration. Use the **Directivity** computer program of this chapter.
(b) Kraus' formula (c) Tai and Pereira's formula.

2.30. The normalized far-zone field pattern of an antenna is given by

$$E = \begin{cases} (\sin\theta \cos^2\phi)^{1/2} & 0 \le \theta \le \pi \text{ and } 0 \le \phi \le \pi/2, 3\pi/2 \le \phi \le 2\pi \\ 0 & \text{elsewhere} \end{cases}$$

Find the directivity using
(a) the exact expression   (b) Kraus' approximate formula
(c) Tai and Pereira's approximate formula
(d) the computer program **Directivity** of this chapter

2.31. The normalized field pattern of the main beam of a conical horn antenna, mounted on an infinite ground plane with z perpendicular to the aperture, is given by

$$\frac{J_1(ka\sin\theta)}{\sin\theta}$$

where a is its radius at the aperture. Assuming that $a = \lambda$, find the
(a) half-power beamwidth
(b) directivity using Kraus' approximate formula

2.32. A base station cellular communication systems *lossless* antenna has a *maximum gain* of 16 dB (above isotropic) at 1,900 MHz. Assuming the *input power* to the antenna is 8 watts, what is the *maximum* radiated power density (in watts/cm^2) at a distance of 100 meters? This will determine the safe level for human exposure to electromagnetic radiation.

2.33. A uniform plane wave, of a form similar to (2-55), is traveling in the positive z-direction. Find the polarization (linear, circular, or elliptical), sense of rotation (CW or CCW), axial ratio (AR), and tilt angle τ (in degrees) when
(a) $E_x = E_y, \Delta\phi = \phi_y - \phi_x = 0$ (b) $E_x \ne E_y, \Delta\phi = \phi_y - \phi_x = 0$
(c) $E_x = E_y, \Delta\phi = \phi_y - \phi_x = \pi/2$ (d) $E_x = E_y, \Delta\phi = \phi_y - \phi_x = -\pi/2$
(e) $E_x = E_y, \Delta\phi = \phi_y - \phi_x = \pi/4$ (f) $E_x = E_y, \Delta\phi = \phi_y - \phi_x = -\pi/4$
(g) $E_x = 0.5E_y, \Delta\phi = \phi_y - \phi_x = \pi/2$ (h) $E_x = 0.5E_y, \Delta\phi = \phi_y - \phi_x = -\pi/2$
In all cases, justify the answer.

2.34. Derive (2-66), (2-67), and (2-68).

2.35. Write a general expression for the polarization loss factor (PLF) of two linearly polarized antennas if
(a) both lie in the same plane   (b) both do not lie in the same plane

2.36. A linearly polarized wave traveling in the positive z-direction is incident upon a circularly polarized antenna. Find the polarization loss factor PLF (dimensionless and in dB) when the antenna is (based upon its transmission mode operation)
(a) right-handed (CW)   (b) left-handed (CCW)

2.37. A 300 MHz uniform plane wave, traveling along the *x*-axis *in the negative x-direction*, whose electric field is given by

$$\mathbf{E}_w = E_o(j\hat{\mathbf{a}}_y + 3\hat{\mathbf{a}}_z)e^{+jkx}$$

where E_o is a real constant, impinges upon a dipole antenna that is placed at the origin and whose electric field radiated toward the *x-axis in the positive x-direction* is given by

$$\mathbf{E}_a = E_a(\hat{\mathbf{a}}_y + 2\hat{\mathbf{a}}_z)e^{-jkx}$$

where E_a is a real constant. Determine the following:
(a) Polarization of the incident wave (*including axial ratio and sense of rotation, if any*). You must justify (state why?).
(b) Polarization of the antenna (*including axial ratio and sense of rotation, if any*). You must justify (state why?).
(c) Polarization loss factor (*dimensionless and in dB*).

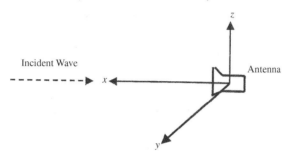

2.38. The electric field of a uniform plane wave traveling along the *negative z*-direction is given by

$$\mathbf{E}^i_w = (\hat{\mathbf{a}}_x + j\hat{\mathbf{a}}_y)E_0 e^{+jkz}$$

and is incident upon a receiving antenna placed at the origin and *whose radiated electric field, toward the incident wave*, is given by

$$\mathbf{E}_a = (\hat{\mathbf{a}}_x + 2\hat{\mathbf{a}}_y)E_1 \frac{e^{-jkr}}{r}$$

where E_0 and E_1 are constants.
Determine the following:
(a) Polarization of the *incident wave*, and why?
(b) Sense of rotation of the *incident wave*.
(c) Polarization *of the antenna*, and why?
(d) Sense of rotation of the *antenna polarization*.
(e) Losses (*dimensionless and in dB*) due to polarization mismatch between the incident wave and the antenna.

2.39. A spherical wave travelling in free-space along the negative y-axis, and whose electric field is given by

$$\mathbf{E}_w = (4\hat{\mathbf{a}}_z + j2\hat{\mathbf{a}}_x)E_w \frac{e^{+jky}}{y}$$

where E_w is a constant, impinges upon a $\lambda/2$ dipole antennas positioned at the origin of the coordinate system whose normalized electric field antenna radiation characteristics along the y-direction are given by

$$\mathbf{E}_a = \hat{\mathbf{a}}_z E_a \frac{e^{-jky}}{y}$$

where E_a is a constant.

112 FUNDAMENTAL PARAMETERS AND FIGURES-OF-MERIT OF ANTENNAS

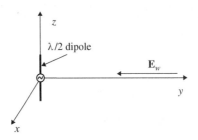

Assuming for this problem the receiving $l = \lambda/2$ dipole is *lossless* and it is perfectly-matched to the connecting transmission line, determine the:
(a) Polarization *of the impinging wave (linear, circular or elliptical). Why?*
(b) Sense of rotation *(CW or CCW), if any, of the impinging wave. Why?*
(c) Axial Ratio (AR) *of the impinging wave. Why?*
(d) Polarization Loss Factor (PLF) *between the impinging wave and the receiving $\lambda/2$ dipole antenna (dimensionless and in dB).*

2.40. A ground-based helical antenna is placed at the origin of a coordinate system and it is used as a receiving antenna. The normalized far-zone electric-field pattern of the helical antenna in the transmitting mode is represented in the direction θ_o, ϕ_o by

$$\mathbf{E}_a = E_0(j\hat{\mathbf{a}}_\theta + 2\hat{\mathbf{a}}_\phi)f_o(\theta_o, \phi_o)\frac{e^{-jkr}}{r}$$

The far-zone electric field transmitted by an antenna on a flying aircraft towards θ_o, ϕ_o, which is received by the ground-based helical antenna, is represented by

$$\mathbf{E}_w = E_1(2\hat{\mathbf{a}}_\theta + j\hat{\mathbf{a}}_\phi)f_1(\theta_o, \phi_o)\frac{e^{+jkr}}{r}$$

Determine the following:
(a) Polarization *(linear, circular, or elliptical)* of the helical antenna in the transmitting mode. *State also the sense of rotation, if any.*
(b) Polarization *(linear, circular, or elliptical)* of the incoming wave that impinges upon the helical antenna. *State also the sense of rotation, if any.*
(c) Polarization loss *(dimensionless and in dB)* due to match/mismatch of the polarizations of the antenna and incoming wave.

2.41. A circularly polarized wave, traveling in the positive z-direction, is incident upon a circularly polarized antenna. Find the polarization loss factor PLF (dimensionless and in dB) for right-hand (CW) and left-hand (CCW) wave and antenna.

2.42. The electric field radiated by a rectangular aperture, mounted on an infinite ground plane with z perpendicular to the aperture, is given by

$$\mathbf{E} = [\hat{\mathbf{a}}_\theta \cos\phi - \hat{\mathbf{a}}_\phi \sin\phi \cos\theta]f(r,\theta,\phi)$$

where $f(r,\theta,\phi)$ is a scalar function which describes the field variation of the antenna. Assuming that the receiving antenna is linearly polarized along the x-axis, find the polarization loss factor (PLF).

2.43. A circularly polarized wave, traveling in the positive z-direction, is received by an elliptically polarized antenna whose reception characteristics near the main lobe are given approximately by

$$\mathbf{E}_a \simeq [2\hat{\mathbf{a}}_x + j\hat{\mathbf{a}}_y]f(r,\theta,\phi)$$

Find the polarization loss factor PLF (*dimensionless* and *in dB*) when the incident wave is
(a) right-hand (CW) (b) left-hand (CCW)
circularly polarized. Repeat the problem when

$$\mathbf{E}_a \simeq [2\hat{\mathbf{a}}_x - j\hat{\mathbf{a}}_y]f(r,\theta,\phi)$$

In each case, what is the polarization of the antenna? How does it match with that of the wave?

2.44. A linearly polarized wave traveling in the negative z-direction has a tilt angle (τ) of 45°. It is incident upon an antenna whose polarization characteristics are given by

$$\hat{\rho}_a = \frac{4\hat{\mathbf{a}}_x + j\hat{\mathbf{a}}_y}{\sqrt{17}}$$

Find the polarization loss factor PLF (*dimensionless* and *in dB*).

2.45. An elliptically polarized wave traveling in the negative z-direction is received by a circularly polarized antenna whose main lobe is along the $\theta = 0$ direction. The unit vector describing the polarization of the incident wave is given by

$$\hat{\rho}_w = \frac{2\hat{\mathbf{a}}_x + j\hat{\mathbf{a}}_y}{\sqrt{5}}$$

Find the polarization loss factor PLF (*dimensionless* and *in dB*) when the wave that would be transmitted by the antenna is
(a) right-hand CP (b) left-hand CP

2.46. A CW circularly polarized uniform plane wave is traveling in the positive z-direction. Find the polarization loss factor PLF (*dimensionless* and *in dB*) assuming the receiving antenna (in its transmitting mode) is
(a) CW circularly polarized (b) CCW circularly polarized

2.47. The polarization of the field radiated by a helical antenna which is placed at the origin of a spherical coordinate system, which is used as a receiving antenna, is given by

$$\mathbf{E}_a = (2\hat{\mathbf{a}}_\theta + j4\hat{\mathbf{a}}_\phi)E_a\frac{e^{-jkr}}{r}$$

where E_a is a constant. The polarization of an incoming wave, which is received by the helical antenna, is given by

$$\mathbf{E}_w = (j4\hat{\mathbf{a}}_\theta + 2\hat{\mathbf{a}}_\phi)E_w\frac{e^{+jkr}}{r}$$

where E_w is a constant. Determine the:
(a) Polarization of the helical antenna (*linear, circular, elliptical*).
(b) The sense of rotation of the polarization of the helical antenna (CW or CCW).
(c) Axial Ratio of the polarization of the helical antenna (AR).
(d) Polarization of the incoming wave (linear, circular, elliptical).
(e) The sense of rotation of the incoming wave (CW or CCW).
(f) Axial Ratio of the incoming wave (AR).
(g) Polarization Loss factor (PLF) between the polarization of the helical antenna and that of the incoming wave (*dimensionless* and *in dB*).

2.48. A linearly polarized uniform plane wave traveling in the positive z-direction, with a power density of 10 milliwatts per square meter, is incident upon a CW circularly polarized antenna whose gain is 10 dB at 10 GHz. Find the
(a) maximum effective area of the antenna (in square meters)
(b) power (in watts) that will be delivered to a load attached directly to the terminals of the antenna.

2.49. A wave, whose electric field is given by

$$\mathbf{E}_w \simeq \hat{\mathbf{a}}_z E_w \frac{e^{+jky}}{y}$$

where E_w is a constant, is traveling along the $-y$ axis and impinges upon the *maximum intensity direction* of the $\lambda/2$ dipole, whose maximum electric field is given by

$$\mathbf{E}_a\Big|_{\max} = \hat{\mathbf{a}}_\theta E_a \sin^{1.5}\theta \frac{e^{-jkr}}{r}\Bigg|_{\substack{\max \\ \theta=90°, r=y \\ \hat{\mathbf{a}}_\theta=-\hat{\mathbf{a}}_z}} = -\hat{\mathbf{a}}_z E_a \frac{e^{-jky}}{y}$$

where E_a is a constant. Assuming the incoming wave has, at a frequency of 10 GHz, a power density of *100 mwatts/cm²*, determine at 10 GHz the:
(a) Polarization Loss Factor (PLF) (*dimensionless* and *in dB*)
(b) *Maximum power (in watts)* that can be delivered to a load connected to the $\lambda/2$ receiving antenna whose input impedance is

$$Z_{in} = 73 + j42.5$$

Its *total loss resistance* is 5 ohms and the antenna is connected to a transmission line with *characteristic impedance of 50 ohms*.

2.50. A linearly polarized plane wave traveling along the negative z-axis is incident upon an elliptically polarized antenna (either CW or CCW). The axial ratio of the antenna polarization ellipse is 2:1 and its major axis coincides with the principal x-axis. Find the polarization loss factor (PLF) assuming the incident wave is linearly polarized in the
(a) x-direction (b) y-direction

2.51. A wave traveling normally outward from the page (toward the reader) is the resultant of two elliptically polarized waves, one with components of $\mathbf{E}$ given by:

$$\mathscr{E}'_y = 3\cos\omega t; \quad \mathscr{E}'_x = 7\cos\left(\omega t + \frac{\pi}{2}\right)$$

and the other with components given by:

$$\mathscr{E}_y'' = 2\cos\omega t; \quad \mathscr{E}_x'' = 3\cos\left(\omega t - \frac{\pi}{2}\right)$$

(a) What is the axial ratio of the resultant wave?
(b) Does the resultant vector **E** rotate clockwise or counterclockwise?

2.52. A linearly polarized antenna lying in the *x-y* plane is used to determine the polarization axial ratio of incoming plane waves traveling in the negative *z*-direction. The polarization of the antenna is described by the unit vector

$$\hat{\boldsymbol{\rho}}_a = \hat{\mathbf{a}}_x \cos\psi + \hat{\mathbf{a}}_y \sin\psi$$

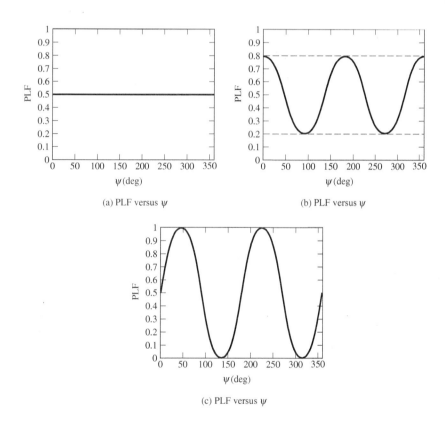

(a) PLF versus ψ

(b) PLF versus ψ

(c) PLF versus ψ

where ψ is an angle describing the orientation in the *x-y* plane of the receiving antenna. Above are the polarization loss factor (PLF) versus receiving antenna orientation curves obtained for three different incident plane waves. For each curve determine the axial ratio of the incident plane wave.

2.53. A $\lambda/2$ dipole, with a total loss resistance of 1 ohm, is connected to a generator whose internal impedance is $50 + j25$ ohms. Assuming that the peak voltage of the generator is 2 V and the impedance of the dipole, excluding the loss resistance, is $73 + j42.5$ ohms, find the power
(a) supplied by the source (real) (b) radiated by the antenna
(c) dissipated by the antenna

2.54. The antenna and generator of Problem 2.53 are connected via a 50-ohm $\lambda/2$-long lossless transmission line. Find the power
(a) supplied by the source (real)
(b) radiated by the antenna
(c) dissipated by the antenna

2.55. An antenna with a radiation resistance of 48 ohms, a loss resistance of 2 ohms, and a reactance of 50 ohms is connected to a generator with open-circuit voltage of 10 V and internal impedance of 50 ohms via a $\lambda/4$-long transmission line with characteristic impedance of 100 ohms.
(a) Draw the equivalent circuit
(b) Determine the power supplied by the generator
(c) Determine the power radiated by the antenna

2.56. A transmitter, with an internal impedance Z_0 (real), is connected to an antenna through a lossless transmission line of length l and characteristic impedance Z_0. Find a *simple* expression for the ratio between the antenna gain and its realized gain.

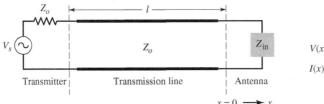

V_s = strength of voltage source
$Z_{in} = R_{in} + jX_{in}$ = input impedance of the antenna
$Z_0 = R_0$ = characteristic impedance of the line
$P_{accepted}$ = power accepted by the antenna $\{P_{accepted} = \text{Re}[V(0)I^*(0)]\}$
$P_{available}$ = power delivered to a matched load [i.e., $Z_{in} = Z_0^* = Z_0$]

2.57. The input reactance of an infinitesimal linear dipole of length $\lambda/60$ and radius $a = \lambda/200$ is given by

$$X_{in} \simeq -120 \frac{[\ln(l/2a) - 1]}{\tan(kl/2)}$$

Assuming the wire of the dipole is copper with a conductivity of 5.7×10^7 S/m, determine at $f = 1$ GHz the
(a) loss resistance
(b) radiation resistance
(c) radiation efficiency
(d) VSWR when the antenna is connected to a 50-ohm line

2.58. A dipole antenna consists of a circular wire of length l. Assuming the current distribution on the wire is cosinusoidal, i.e.,

$$I_z(z) = I_0 \cos\left(\frac{\pi}{l}z'\right) \quad -l/2 \leq z' \leq l/2$$

where I_0 is a constant, derive an expression for the loss resistance R_L, which is one-half of (2-90b).

2.59. The *E*-field pattern of an antenna, independent of ϕ, varies as follows:

$$E = \begin{cases} 1 & 0° \leq \theta \leq 45° \\ 0 & 45° < \theta \leq 90° \\ \frac{1}{2} & 90° < \theta \leq 180° \end{cases}$$

(a) What is the directivity of this antenna?
(b) What is the radiation resistance of the antenna at 200 m from it if the electric field is equal to 10 V/m (rms) for $\theta = 0°$ at that distance and the terminal current is 5 A (rms)?

2.60. The *approximate far-zone normalized electric field* radiated by a ground-based end-fire helical antenna used in communication systems for space-borne applications, positioned symmetrically at the origin along the z-axis, is given by

$$\mathbf{E}_a \simeq \begin{cases} \hat{\mathbf{a}}_\theta E_a \cos^3 \theta \frac{e^{-jkr}}{r}, & 0 \leq \theta \leq 90° \\ & 0 \leq \phi \leq 360° \\ 0 & \text{Elshewhere} \end{cases}$$

where E_a is a constant and r is the spherical radial distance measured from the origin of the coordinate system.
Determine the:
(a) *Half power beamwidth* (HPBW) (*in degrees*)
(b) *Exact maximum directivity* (*dimensionless* and *in dB*).
(c) *Approximate maximum directivity* (*dimensionless* and *in dB*) using the *most accurate* approximate formula for this problem.
Indicate which *most accurate* approximate formula you are using and why?
(d) *Maximum directivity* (*dimensionless* and *in dB*) using the computer program **Directivity**.
(e) *Maximum effective area* (A_{em})(*in* λ^2), based on the exact directivity.

2.61. The far-zone field radiated by a rectangular aperture mounted on a ground plane, with dimensions *a* and *b* and uniform aperture distribution, is given by (see Table 12.1)

$$E \approx \hat{\mathbf{a}}_\theta E_\theta + \hat{\mathbf{a}}_\phi E_\phi$$
$$E_\theta = C \sin\phi \frac{\sin X}{X} \frac{\sin Y}{Y} \quad X = \frac{ka}{2}\sin\theta\cos\phi; \quad 0 \leq \theta \leq 90°$$
$$E_\phi = C \cos\theta \cos\phi \frac{\sin X}{X} \frac{\sin Y}{Y} \quad Y = \frac{kb}{2}\sin\theta\sin\phi; \quad 0 \leq \phi \leq 360°$$

where *C* is a constant. For an aperture with $a = 3\lambda, b = 2\lambda$, determine the
(a) maximum partial directivities D_θ, D_ϕ (*dimensionless* and *in dB*) and
(b) total maximum directivity D_o (*dimensionless* and *in dB*). Compare with that computed using the equation in Table 12.1.
Use the computer program **Directivity** of this chapter.

2.62. Repeat Problem 2.61 when the aperture distribution is that of the dominant TE_{10} mode of a rectangular waveguide, or from Table 12.1

$$E \approx \hat{a}_\theta E_\theta + \hat{a}_\phi E_\phi$$

$$\left. \begin{array}{l} E_\theta = -\dfrac{\pi}{2} C \sin\phi \dfrac{\cos X}{(X)^2 - \left(\dfrac{\pi}{2}\right)^2} \dfrac{\sin Y}{Y} \\[2ex] E_\phi = -\dfrac{\pi}{2} C \cos\theta \cos\phi \dfrac{\cos X}{(X)^2 - \left(\dfrac{\pi}{2}\right)^2} \dfrac{\sin Y}{Y} \end{array} \right\} \quad \begin{array}{l} X = \dfrac{ka}{2} \sin\theta \cos\phi \\[2ex] Y = \dfrac{kb}{2} \sin\theta \sin\phi \end{array}$$

2.63. Repeat Problem 2.62 when the aperture dimensions are those of an X-band rectangular waveguide with $a = 2.286$ cm (0.9 in.), $b = 1.016$ cm (0.4 in.) and frequency of operation is 10 GHz.

2.64. Repeat Problem 2.61 for a circular aperture with a uniform distribution and whose far-zone fields are, from Table 12.2

$$E \approx \hat{a}_\theta E_\theta + \hat{a}_\phi E_\phi$$

$$\left. \begin{array}{l} E_\theta = jC_1 \sin\phi \dfrac{J_1(Z)}{Z} \\[2ex] E_\phi = jC_1 \cos\theta \cos\phi \dfrac{J_1(Z)}{Z} \end{array} \right\} \quad \begin{array}{l} Z = ka \sin\theta; \quad 0 \leq \theta \leq 90° \\[1ex] 0 \leq \phi \leq 360° \end{array}$$

where C_1 is a constant and $J_1(Z)$ is the Bessel function of the first kind. Assume $a = 1.5\lambda$.

2.65. Repeat Problem 2.64 when the aperture distribution is that of the dominant TE_{11} mode of a circular waveguide, or from Table 12.2

$$E \approx \hat{a}_\theta E_\theta + \hat{a}_\phi E_\phi$$

$$\left. \begin{array}{l} E_\theta = C_2 \sin\phi \dfrac{J_1(Z)}{Z} \\[2ex] E_\phi = C_2 \cos\theta \cos\phi \dfrac{J_z'(Z)}{(1) - \left(\dfrac{Z}{\chi_{11}'}\right)^2} \end{array} \right\} \quad \begin{array}{l} Z = ka \sin\theta; \quad 0 \leq \theta \leq 90° \\[1ex] J_z'(Z) = J_0(Z) \quad 0 \leq \phi \leq 360° \\[1ex] - J_1(Z)/Z; \end{array}$$

where C_2 is a constant, $J_1'(Z)$ is the derivative of $J_1(Z)$, $\chi_{11}' = 1.841$ is the first zero of $J_1'(Z)$, and $J_0(Z)$ is the Bessel function of the first find of order zero.

2.66. Repeat 2.65 when the radius of the aperture is $a = 1.143$ cm (0.45 in.) and the frequency of operation is 10 GHz.

2.67. The normalized *far-field total* electric field radiated by an antenna, consisting of an infinitesimal vertical dipole (oriented along the z-axis) plus a small circular loop (parallel to the xy-plane), *placed at the origin of a spherical coordinate system* is given by:

$$\underline{E}_a = (\hat{a}_\theta + j2\hat{a}_\phi) \sin\theta E_0 \dfrac{e^{-jkr}}{r}; \quad (0° \leq \theta \leq 180°, 0° \leq \phi \leq 360°)$$

Determine the:
(a) Polarization of the wave (linear, circular or elliptical). *Justify it. If circular*, state the sense of rotation. *If elliptical*, state the sense of rotation and the axial ratio (AR).
(b) partial *maximum* directivities $(D_0)_\theta$ and $(D_0)_\phi$ (*dimensionless* and *in dB*).
(c) Total maximum directivity D_o (*dimensionless* and *in dB*).
   Hint: For this problem, there are two different and distinct partial directivities.

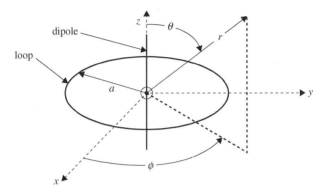

2.68. A 1-m long dipole antenna is driven by a 150 MHz source having a source resistance of 50 ohms and a voltage of 100 V. If the ohmic resistance of the antennas is given by $R_L = 0.625$ ohms, find the:
(a) Current going into the antenna (I_{ant});
(b) Power dissipated by the antenna
(c) Power radiated by the antenna;
(d) Radiation efficiency of the antenna

2.69. The field radiated by an infinitesimal dipole of very small length ($l \leq \lambda/50$), and of uniform current distribution I_o, is given by (4-26a) or

$$\mathbf{E} = \hat{\mathbf{a}}_\theta E_\theta \approx \hat{\mathbf{a}}_\theta j\eta \frac{kI_o l}{4\pi r} e^{-jkr} \sin\theta$$

Determine the
(a) vector effective length
(b) maximum value of the vector effective length. Specify the angle.
(c) ratio of the maximum effective length to the physical length l.

2.70. The field radiated by a half-wavelength dipole ($l = \lambda/2$), with a sinusoidal current distribution, is given by (4-84) or

$$\mathbf{E} = \hat{\mathbf{a}}_\theta E_\theta \approx \hat{\mathbf{a}}_\theta j\eta \frac{I_o}{2\pi r} e^{-jkr} \left[\frac{\cos\left(\frac{\pi}{2}\cos\theta\right)}{\sin\theta} \right]$$

where I_o is the maximum current. Determine the
(a) vector effective length
(b) maximum value of the vector effective length. Specify the angle.
(c) ratio of the maximum effective length to the physical length l.

2.71. A uniform plane wave, of 10^{-3} watts/cm^2 power density, is incident upon an infinitesimal dipole of length $l = \lambda/50$ and uniform current distribution, as shown in Figure 2.29(a). For a frequency of 10 GHz, determine the maximum open-circuited voltage at the terminals of the antenna. See Problem 2.69.

2.72. Repeat Problem 2.71 for a small dipole with triangular current distribution and length $l = \lambda/10$. See Example 2.14.

2.73. Repeat Problem 2.71 for a half-wavelength dipole ($\ell = \lambda/2$) with sinusoidal current distribution. See Problem 2.70.

2.74. Show that the effective length of a linear antenna can be written as

$$l_e = \sqrt{\frac{A_e |Z_t|^2}{\eta R_T}}$$

which for a lossless antenna and maximum power transfer reduces to

$$l_e = 2\sqrt{\frac{A_{em} R_r}{\eta}}$$

A_e and A_{em} represent, respectively, the effective and maximum effective apertures of the antenna while η is the intrinsic impedance of the medium.

2.75. An antenna has a maximum effective aperture of 2.147 m² at its operating frequency of 100 MHz. It has no conduction or dielectric losses. The input impedance of the antenna itself is 75 ohms, and it is connected to a 50-ohm transmission line. Find the directivity of the antenna system ("system" meaning includes any effects of connection to the transmission line). Assume no polarization losses.

2.76. A small circular parabolic reflector, often referred to as dish, is now being advertised as a TV antenna for direct broadcast. Assuming the diameter of the antenna is 1 meter, the frequency of operation is 3 GHz, and its aperture efficiency is 68%, determine the following:
   (a) Physical area of the reflector (*in m²*).
   (b) Maximum effective area of the antenna (*in m²*).
   (c) Maximum directivity (*dimensionless* and *in dB*).
   (d) Maximum power (*in watts*) that can be delivered to the TV if the power density of the wave incident upon the antenna is *10 μwatts/m²*. Assume *no losses* between the incident wave and the receiver (TV).

2.77. A uniform plane wave, with a power density of 10 mwatts/cm², is impinging upon a *half wavelength* dipole at an angle normal/perpendicular to the axis of the dipole. Determine the:
   (a) *Maximum effective area (in λ^2)* of the *lossless dipole element*. Assume the dipole has a directivity of 2.148 dB, it is *polarized matched* to the incident wave and it is *mismatched*, with reflection coefficieitn of 0.2. to the transmission line it is connected.
   (b) *Physical area (in λ^2)*. Assume the physical area of the dipole is equal to its lengthwise cross sectional area; the dipole *diameter is $\lambda/300$*.
   (c) *Aperture efficiency (in %)*.
   (d) *Maximum power* the dipole will interecept and deliver to a load. Assume a frequency of operation of *1 GHz*.

2.78. An incoming wave, with a uniform power density equal to 10^{-3} W/m² is incident normally upon a lossless horn antenna whose directivity is 20 dB. At a frequency of 10 GHz, determine the very maximum possible power that can be expected to be delivered to a receiver or a load connected to the horn antenna. There are no losses between the antenna and the receiver or load.

2.79. You are a communication/antenna engineer and you are asked to determine whether the signal received by a space borne communication system operating at 10 GHz will be of sufficient strength to be detected by the receiver. The power density incident upon the receiving antenna of the space borne system is 10×10^{-3} Watts/cm^2. *The polarization of the incident wave is right-hand circularly polarized while that of the receiving antenna is linearly polarized.* The *maximum directivity* of the receiving space borne antenna is 12 dB, its input impednace is 100 ohms, and its radiation efficiency is 75%. The characteristic impednace of the transmission line from the receiving space borne antenna to the space borne receiver has a characteistic impedance of 50 ohms. Determine the:
(a) maximum effective area (*in cm^2*) of the antenna *assuming no losses.*
(b) maximum effective area (*in cm^2*) of the antenna *taking into account all the stated losses. Identify each of the losses in %.*
(c) Maximum power (*in Watts*) delivered to the receiver *taking into account all of the stated losses.*

2.80. A linearly polarized aperture antenna, with a uniform field distribution over its area, is used as a receiving antenna. The antenna physical area over its aperture is 10 cm^2, and it is operating at 10 GHz. The antenna is illuminated with a circularly polarized plane wave whose incident power density is 10 mwatts/cm^2. Assuming the antenna element itself is lossless, determine its
(a) gain (*dimensionless* and *in dB*).
(b) maximum power (*in watts*) that can be delivered to a load connected to the antenna. Assume no other losses between the antenna and the load.

2.81. A uniform plane wave traveling along the negative z-axis, and whose normalized electric field is given by

$$\underline{E}_w = (j\hat{a}_x + 2\hat{a}_y)e^{+jkz}$$

it impinges upon an antenna whose polarization along the z-axis is represented by the normalized electric field of

$$\underline{E}_a = j\hat{a}_y e^{-jkz}$$

The power density of the wave which impinges upon the antenna is *10 mwatts/λ^2*. The antenna has a *maximum derectivity* of 2.15 dB along the z-direction.
Determine the:
(a) *Polarization* of the impinging/incident wave, including it *Axial Ratio* and *sense of rotation*, if any.
(b) *Polarization* of the antenna, including its *Axial Ratio* and *sense of rotatio*, if any.
(c) *Very maximum efective area* (*in λ^2*) of the *antenna assuming it has no losses of any kind.*
(d) *Maximum effective area* (*in λ^2*) *assuming the antenna is losseless*, but it possesses a reflection coefficeint of 0.5 to the transmission line to which it is connected. Include any other losses that should be accounted for.

(e) *Maximum power* the antenna will intercept and deliver to a load/receiver connected to its trasmission line, taking into account all the losses associated with this problem. Assume a frequency of operation of 1 GHz.

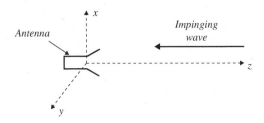

2.82. The *far-zone power density* radiated by a helical antenna can be approximated by

$$\mathbf{W}_{rad} = \mathbf{W}_{ave} \approx \hat{\mathbf{a}}_r C_o \frac{1}{r^2} \cos^4 \theta$$

The radiated power density is symmetrical with respect to ϕ, and *it exists only in the upper hemisphere* $(0 \leq \theta \leq \pi/2, 0 \leq \phi \leq 2\pi)$; C_o is a constant.
Determine the following:
(a) Power radiated by the antenna (*in watts*).
(b) Maximum directivity of the antenna (*dimensionless* and *in dB*)
(c) Direction (*in degrees*) along which the maximum directivity occurs.
(d) Maximum effective area (*in* m^2) at 1 GHz.
(e) Maximum power (*in watts*) received by the antenna, assuming no losses, at 1 GHz when the antenna is used as a receiver and the incident power density is *10 mwatts/m^2*.

2.83. The antenna used at a base station is a $\lambda/2$ dipole which has a *maximum directivity of 2.286 dB*. The power radiated by the $\lambda/2$ dipole is 10 watts. Assume an operating frequency of 1,900 MHz:
(a) Determine the *maximum power density* (*in watts/cm^2*) radiated by the $\lambda/2$ dipole at a distance of 1,000 meters.
(b) Assuming the receving antenna used in a mobile unit, such as an automobile, is a $\lambda/4$ monopole, with a *maximum directivity of 5.286 dB*, determine the maximum effective area (*in cm^2*) of the monopole antenna.
(c) If the receiving mobile unit, the automobile, is at a distance of 1,000 meters from the base station, what is the *maximum power* (*in watts*) that can be received by the antenna ($\lambda/4$ monopole), which is mounted on the top of the automobile, and delivered to a matched load/receiver. Assume the antennas are polarization matched, and that there are *no losses* of any kind in both antennas, including no matching/reflection losses.

2.84. For an X-band (8.2–12.4 GHz) rectangular horn, with aperture dimensions of 5.5 cm and 7.4 cm, find its maximum effective aperture (*in cm^2*) when its gain (over isotropic) is
(a) 14.8 dB at 8.2 GHz   (b) 16.5 dB at 10.3 GHz   (c) 18.0 dB at 12.4 GHz

2.85. A base station is installed near your neighborhood. One of the concerns of the residents living nearby is the exposure to electromagnetic radiation. The *input power inside the transmission line feeding the base station antenna is 100 Watts* while the *omnidirectional* radiation amplitude pattern of the base station antenna can be approximated by

$$U(\theta, \phi) = B_o \sin(\theta) \qquad 0 \leq \theta \leq 180°, 0 \leq \phi \leq 360°$$

where B_o is a constant. The characteristic impedance of the transmission line feeding the base station antenna is 75 ohms while the input impedance of the base station antenna is 100 ohms. The *radiation (conduction/dielectric) efficiency of the base station antenna is 50%*. Determine the:

(a) Reflection/mismatch efficiency of the antenna (*in %*)
(b) Total efficiency (*in %*) of the antenna
(c) Value of B_o. *Must do the integration in closed form and show the details.*
(d) Maximum exact directivity (*dimensionless* and *in dB*)
(e) Maximum *power density* (*in Watts/cm²*) at a distance of 1,000 meters. This may represent the distance from the base station to your house.

2.86. For Problem 2.61 compute the
(a) maximum effective area (in λ^2) using the computer program **Directivity** of this chapter. Compare with that computed using the equation in Table 12.1.
(b) aperture efficiencies of part (a). Are they smaller or larger than unity and why?

2.87. Repeat Problem 2.86 for Problem 2.62.

2.88. Repeat Problem 2.86 for Problem 2.63.

2.89. Repeat Problem 2.86 for Problem 2.64. Compare with those in Table 12.2.

2.90. Repeat Problem 2.86 for Problem 2.65. Compare with those in Table 12.2.

2.91. Repeat Problem 2.86 for Problem 2.66. Compare with those in Table 12.2.

2.92. A 30-dB, right-circularly polarized antenna in a radio link radiates (in the negative z-direction) 5 W of power at 2 GHz. The receiving antenna has an impedance mismatch at its terminals, which leads to a VSWR of 2. The receiving antenna is about 95% efficient and has a field pattern near the beam maximum (in the positive z-direction) given by $\mathbf{E}_r = (2\hat{\mathbf{a}}_x + j\hat{\mathbf{a}}_y)F_r(\theta, \phi)$. The distance between the two antennas is 4,000 km, and the receiving antenna is required to deliver 10^{-14} W to the receiver. Determine the maximum effective aperture of the receiving antenna.

2.93. The radiation intensity of an antenna can be approximated by

$$U(\theta, \phi) = \begin{cases} \cos^4(\theta) & 0° \leq \theta < 90° \\ 0 & 90° \leq \theta \leq 180° \end{cases} \text{ with } 0° \leq \phi \leq 360°$$

Determine the maximum effective aperture (*in* m^2) of the antenna if its frequency of operation is $f = 10$ GHz.

2.94. A communication satellite is in stationary (synchronous) orbit about the earth (assume altitude of 22,300 statute miles). Its transmitter generates 8.0 W. Assume the transmitting antenna is isotropic. Its signal is received by the 210-ft diameter tracking paraboloidal antenna on the earth at the NASA tracking station at Goldstone, California. Also assume no resistive losses in either antenna, perfect polarization match, and perfect impedance match at both antennas. At a frequency of 2 GHz, determine the:

(a) power density (*in watts/m²*) incident on the receiving antenna.
(b) power received by the ground-based antenna whose gain is 60 dB.

2.95. A lossless ($e_{cd} = 1$) antenna is operating at 100 MHz and its maximum effective aperture is 0.7162 m^2 at this frequency. The input impedance of this antenna is 75 ohms, and it is attached to a 50-ohm transmission line. Find the directivity (dimensionless) of this antenna if it is polarization-matched.

2.96. A resonant, lossless ($e_{cd} = 1.0$) half-wavelength dipole antenna, having a directivity of 2.156 dB, has an input impedance of 73 ohms and is connected to a lossless, 50 ohms transmission line. A wave, having the same polarization as the antenna, is incident upon the antenna with a power density of 5 W/m^2 at a frequency of 10 MHz. Find the received power available at the end of the transmission line.

2.97. Two X-band (8.2–12.4 GHz) rectangular horns, with aperture dimensions of 5.5 cm and 7.4 cm and each with a gain of 16.3 dB (over isotropic) at 10 GHz, are used as transmitting and receiving antennas. Assuming that the input power is 200 mW, the VSWR of each is 1.1, the conduction-dielectric efficiency is 100%, and the antennas are polarization-matched, find the maximum received power when the horns are separated in air by
(a) 5 m (b) 50 m (c) 500 m

2.98. Transmitting and receiving antennas operating at 1 GHz with gains (over isotropic) of 20 and 15 dB, respectively, are separated by a distance of 1 km. Find the maximum power delivered to the load when the input power is 150 W. Assume that the
(a) antennas are polarization-matched
(b) transmitting antenna is circularly polarized (either right- or left-hand) and the receiving antenna is linearly polarized.

2.99. Two lossless, polarization-matched antennas are aligned for maximum radiation between them, and are separated by a distance of 50λ. The antennas are matched to their transmission lines and have directivities of 20 dB. Assuming that the power at the input terminals of the transmitting antenna is 10 W, find the power at the terminals of the receiving antenna.

2.100. Repeat Problem 2.99 for two antennas with 30 dB directivities and separated by 100λ. The power at the input terminals is 20 W.

2.101. Transmitting and receiving antennas operating at 1 GHz with gains of 20 and 15 dB, respectively, are separated by a distance of 1 km. Find the power delivered to the load when the input power is 150 W. Assume the PLF = 1.

2.102. A series of microwave repeater links operating at 10 GHz are used to relay television signals into a valley that is surrounded by steep mountain ranges. Each repeater consists of a receiver, transmitter, antennas, and associated equipment. The transmitting and receiving antennas are identical horns, each having gain over isotropic of 15 dB. The repeaters are separated in distance by 10 km. For acceptable signal-to-noise ratio, the power received at each repeater must be greater than 10 nW. Loss due to polarization mismatch is not expected to exceed 3 dB. Assume matched loads and free-space propagation conditions. Determine the minimum transmitter power that should be used.

2.103. A one-way communication system, operating at 100 MHz, uses two identical $\lambda/2$ vertical, resonant, and lossless dipole antennas as transmitting and receiving elements separated by 10 km. In order for the signal to be detected by the receiver, the power level at the receiver terminals must be at least 1 µW. Each antenna is connected to the transmitter and receiver by a lossless 50-Ω transmission line. Assuming the antennas are polarization-matched and are aligned so that the maximum intensity of one is directed toward the maximum radiation intensity of the other, determine the minimum power that must be generated by the transmitter so that the signal will be detected by the receiver. Account for the proper losses from the transmitter to the receiver.

2.104. In a long-range microwave communication system operating at 9 GHz, the transmitting and receiving antennas are identical, and they are separated by 10,000 m. To meet the signal-to-noise ratio of the receiver, the received power must be at least 10 μW. Assuming the two antennas are aligned for maximum reception to each other, including being polarization-matched, what should the gains (in dB) of the transmitting and receiving antennas be when the input power to the transmitting antenna is 10 W?

2.105. A mobile wireless communication system operating at 2 GHz utilizes two antennas, one at the base station and the other at the mobile unit, which are separated by *16 kilometers*. The transmitting antenna, at the base station, is circularly-polarized while the receiving antenna, at the mobile station, is linearly polarized. The *maximum gain of the transmitting antenna is 20 dB* while the gain of the receiving antennas is unknown. The input power to the transmitting antenna is *100 watts* and the power received at the receiver, which is connected to the receiving antenna, is *5 nanowatts*. Assuming that the two antennas are aligned so that the maximum of one is directed toward the maximum of the other, *and also assuming no reflection/mismatch losses at the transmitter or the receiver*, what is the maximum gain of the receiving antenna (*dimensions* and *in dB*)?

2.106. A rectangular *X*-band horn, with aperture dimensions of 5.5 cm and 7.4 cm and a gain of 16.3 dB (over isotropic) at 10 GHz, is used to transmit and receive energy scattered from a perfectly conducting sphere of radius $a = 5\lambda$. Find the maximum scattered power delivered to the load when the distance between the horn and the sphere is
(a) 200λ (b) 500λ

Assume that the input power is 200 mW, and the radar cross section is equal to the geometrical cross section.

2.107. A radar antenna, used for both transmitting and receiving, has a gain of 150 (dimensionless) at its operating frequency of 5 GHz. It transmits 100 kW, and is aligned for maximum directional radiation and reception to a target 1 km away having a radar cross section of 3 m². The received signal matches the polarization of the transmitted signal. Find the received power.

2.108. In an experiment to determine the radar cross section of a Tomahawk cruise missile, a 100 W, 10 GHz signal was transmitted toward the target, and the received power was measured to be −160 dB. The same antenna, whose gain was 80 (*dimensionless*), was used for both transmitting and receiving. The polarizations of both signals were identical (PLF = 1), and the distance between the antenna and missile was 10^4 m. What is the radar cross section of the cruise missile?

2.109. Repeat Problem 2.108 for a radar system with 100 W, 3 GHz transmitted signal, −160 dB received signal, an antenna with a gain of 80 (*dimensionless*), and separation between the antenna and target of 10^4 m.

2.110. The maximum radar cross section of a resonant linear $\lambda/2$ dipole is approximately $0.86\lambda^2$. For a monostatic system (i.e., transmitter and receiver at the same location), find the received power (in W) if the transmitted power is 100 W, the distance of the dipole from the transmitting and receiving antennas is 100 m, the gain of the transmitting and receiving antennas is 15 dB each, and the frequency of operation is 3 GHz. Assume a polarization loss factor of −1 dB.

2.111. The effective antenna temperature of an antenna looking toward zenith is approximately 5 K. Assuming that the temperature of the transmission line (waveguide) is 72°F, find the

effective temperature at the receiver terminals when the attenuation of the transmission line is 4 dB/100 ft and its length is
(a) 2 ft   (b) 100 ft
Compare it to a receiver noise temperature of about 54 K.

2.112. Derive (2-140). Begin with an expression that assumes that the physical temperature and the attenuation of the transmission line are not constant.

CHAPTER 3

Radiation Integrals and Auxiliary Potential Functions

3.1 INTRODUCTION

In the analysis of radiation problems, the usual procedure is to specify the sources and then require the fields radiated by the sources. This is in contrast to the synthesis problem where the radiated fields are specified, and we are required to determine the sources.

It is a very common practice in the analysis procedure to introduce auxiliary functions, known as *vector potentials*, which will aid in the solution of the problems. The most common vector potential functions are the **A** (magnetic vector potential) and **F** (electric vector potential). Another pair is the Hertz potentials $\mathbf{\Pi}_e$ and $\mathbf{\Pi}_h$. *Although the electric and magnetic field intensities (**E** and **H**) represent physically measurable quantities, among most engineers the potentials are strictly mathematical tools.* The introduction of the potentials often simplifies the solution even though it may require determination of additional functions. While it is possible to determine the **E** and **H** fields directly from the source-current densities **J** and **M**, as shown in Figure 3.1, it is usually much simpler to find the auxiliary potential functions first and then determine the **E** and **H**. This two-step procedure is also shown in Figure 3.1.

The one-step procedure, through path 1, relates the **E** and **H** fields to **J** and **M** by integral relations. The two-step procedure, through path 2, relates the **A** and **F** (or $\mathbf{\Pi}_e$ and $\mathbf{\Pi}_h$) potentials to **J** and **M** by integral relations. The **E** and **H** are then determined simply by differentiating **A** and **F** (or $\mathbf{\Pi}_e$ and $\mathbf{\Pi}_h$). Although the two-step procedure requires both integration and differentiation, where path 1 requires only integration, the integrands in the two-step procedure are much simpler.

The most difficult operation in the two-step procedure is the integration to determine **A** and **F** (or $\mathbf{\Pi}_e$ and $\mathbf{\Pi}_h$). Once the vector potentials are known, then **E** and **H** can always be determined because any well-behaved function, no matter how complex, can always be differentiated.

The integration required to determine the potential functions is restricted over the bounds of the sources **J** and **M**. This will result in the **A** and **F** (or $\mathbf{\Pi}_e$ and $\mathbf{\Pi}_h$) to be functions of the observation point coordinates; the differentiation to determine **E** and **H** must be done in terms of the observation point coordinates. The integration in the one-step procedure also requires that its limits be determined by the bounds of the sources.

The vector Hertz potential $\mathbf{\Pi}_e$ is analogous to **A** and $\mathbf{\Pi}_h$ is analogous to **F**. The functional relation between them is a proportionality constant which is a function of the frequency and the constitutive parameters of the medium. In the solution of a problem, only one set, **A** and **F** or $\mathbf{\Pi}_e$ and $\mathbf{\Pi}_h$,

---

Antenna Theory: Analysis and Design, Fourth Edition. Constantine A. Balanis.
© 2016 John Wiley & Sons, Inc. Published 2016 by John Wiley & Sons, Inc.
Companion Website: www.wiley.com/go/antennatheory4e

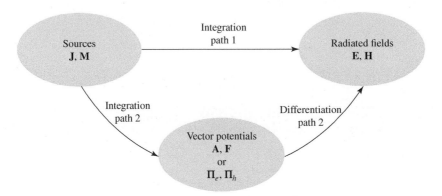

Figure 3.1 Block diagram for computing fields radiated by electric and magnetic sources.

is required. The author prefers the use of **A** and **F**, which will be used throughout the book. The derivation of the functional relations between **A** and $\mathbf{\Pi}_e$, and **F** and $\mathbf{\Pi}_h$ are assigned at the end of the chapter as problems. (Problems 3.1 and 3.2).

3.2 THE VECTOR POTENTIAL A FOR AN ELECTRIC CURRENT SOURCE J

The vector potential **A** is useful in solving for the EM field generated by a given harmonic electric current **J**. The magnetic flux **B** is always solenoidal; that is, $\nabla \cdot \mathbf{B} = 0$. Therefore, it can be represented as the curl of another vector because it obeys the vector identity

$$\nabla \cdot \nabla \times \mathbf{A} = 0 \quad (3\text{-}1)$$

where **A** is an arbitrary vector. Thus we define

$$\mathbf{B}_A = \mu \mathbf{H}_A = \nabla \times \mathbf{A} \quad (3\text{-}2)$$

or

$$\boxed{\mathbf{H}_A = \frac{1}{\mu} \nabla \times \mathbf{A}} \quad (3\text{-}2a)$$

where subscript A indicates the field due to the **A** potential. Substituting (3-2a) into Maxwell's curl equation

$$\nabla \times \mathbf{E}_A = -j\omega\mu \mathbf{H}_A \quad (3\text{-}3)$$

reduces it to

$$\nabla \times \mathbf{E}_A = -j\omega\mu \mathbf{H}_A = -j\omega \nabla \times \mathbf{A} \quad (3\text{-}4)$$

which can also be written as

$$\nabla \times [\mathbf{E}_A + j\omega \mathbf{A}] = 0 \quad (3\text{-}5)$$

From the vector identity

$$\nabla \times (-\nabla \phi_e) = 0 \tag{3-6}$$

and (3-5), it follows that

$$\mathbf{E}_A + j\omega \mathbf{A} = -\nabla \phi_e \tag{3-7}$$

or

$$\boxed{\mathbf{E}_A = -\nabla \phi_e - j\omega \mathbf{A}} \tag{3-7a}$$

The scalar function ϕ_e represents an arbitrary electric scalar potential which is a function of position. Taking the curl of both sides of (3-2) and using the vector identity

$$\nabla \times \nabla \times \mathbf{A} = \nabla(\nabla \cdot \mathbf{A}) - \nabla^2 \mathbf{A} \tag{3-8}$$

reduces it to

$$\nabla \times (\mu \mathbf{H}_A) = \nabla(\nabla \cdot \mathbf{A}) - \nabla^2 \mathbf{A} \tag{3-8a}$$

For a homogeneous medium, (3-8a) reduces to

$$\mu \nabla \times \mathbf{H}_A = \nabla(\nabla \cdot \mathbf{A}) - \nabla^2 \mathbf{A} \tag{3-9}$$

Equating Maxwell's equation

$$\boxed{\nabla \times \mathbf{H}_A = \mathbf{J} + j\omega \varepsilon \mathbf{E}_A} \tag{3-10}$$

to (3-9) leads to

$$\mu \mathbf{J} + j\omega \mu \varepsilon \mathbf{E}_A = \nabla(\nabla \cdot \mathbf{A}) - \nabla^2 \mathbf{A} \tag{3-11}$$

Substituting (3-7a) into (3-11) reduces it to

$$\begin{aligned}\nabla^2 \mathbf{A} + k^2 \mathbf{A} &= -\mu \mathbf{J} + \nabla(\nabla \cdot \mathbf{A}) + \nabla(j\omega\mu\varepsilon\phi_e) \\ &= -\mu \mathbf{J} + \nabla(\nabla \cdot \mathbf{A} + j\omega\mu\varepsilon\phi_e)\end{aligned} \tag{3-12}$$

where $k^2 = \omega^2 \mu \varepsilon$.

In (3-2), the curl of $\mathbf{A}$ was defined. Now we are at liberty to define the divergence of $\mathbf{A}$, which is independent of its curl. In order to simplify (3-12), let

$$\boxed{\nabla \cdot \mathbf{A} = -j\omega\varepsilon\mu\phi_e \Rightarrow \phi_e = -\frac{1}{j\omega\mu\varepsilon}\nabla \cdot \mathbf{A}} \tag{3-13}$$

which is known as the *Lorentz condition*. Substituting (3-13) into (3-12) leads to

$$\boxed{\nabla^2 \mathbf{A} + k^2 \mathbf{A} = -\mu \mathbf{J}} \tag{3-14}$$

130 RADIATION INTEGRALS AND AUXILIARY POTENTIAL FUNCTIONS

In addition, (3-7a) reduces to

$$\boxed{\mathbf{E}_A = -\nabla \phi_e - j\omega \mathbf{A} = -j\omega \mathbf{A} - j\frac{1}{\omega\mu\varepsilon}\nabla(\nabla \cdot \mathbf{A})} \quad (3\text{-}15)$$

Once $\mathbf{A}$ is known, $\mathbf{H}_A$ can be found from (3-2a) and $\mathbf{E}_A$ from (3-15). $\mathbf{E}_A$ can just as easily be found from Maxwell's equation (3-10) with $\mathbf{J} = 0$. It will be shown later how to find $\mathbf{A}$ in terms of the current density $\mathbf{J}$. It will be a solution to the inhomogeneous Helmholtz equation of (3-14).

3.3 THE VECTOR POTENTIAL F FOR A MAGNETIC CURRENT SOURCE M

Although magnetic currents appear to be physically unrealizable, equivalent magnetic currents arise when we use the volume or the surface equivalence theorems. The fields generated by a harmonic magnetic current in a homogeneous region, with $\mathbf{J} = 0$ but $\mathbf{M} \neq 0$, must satisfy $\nabla \cdot \mathbf{D} = 0$. Therefore, $\mathbf{E}_F$ can be expressed as the curl of the vector potential $\mathbf{F}$ by

$$\boxed{\mathbf{E}_F = -\frac{1}{\varepsilon}\nabla \times \mathbf{F}} \quad (3\text{-}16)$$

Substituting (3-16) into Maxwell's curl equation

$$\nabla \times \mathbf{H}_F = j\omega\varepsilon \mathbf{E}_F \quad (3\text{-}17)$$

reduces it to

$$\nabla \times (\mathbf{H}_F + j\omega \mathbf{F}) = 0 \quad (3\text{-}18)$$

From the vector identity of (3-6), it follows that

$$\boxed{\mathbf{H}_F = -\nabla \phi_m - j\omega \mathbf{F}} \quad (3\text{-}19)$$

where ϕ_m represents an arbitrary magnetic scalar potential which is a function of position. Taking the curl of (3-16)

$$\nabla \times \mathbf{E}_F = -\frac{1}{\varepsilon}\nabla \times \nabla \times \mathbf{F} = -\frac{1}{\varepsilon}[\nabla\nabla \cdot \mathbf{F} - \nabla^2 \mathbf{F}] \quad (3\text{-}20)$$

and equating it to Maxwell's equation

$$\boxed{\nabla \times \mathbf{E}_F = -\mathbf{M} - j\omega\mu \mathbf{H}_F} \quad (3\text{-}21)$$

leads to

$$\nabla^2 \mathbf{F} + j\omega\mu\varepsilon \mathbf{H}_F = \nabla\nabla \cdot \mathbf{F} - \varepsilon \mathbf{M} \quad (3\text{-}22)$$

Substituting (3-19) into (3-22) reduces it to

$$\nabla^2 \mathbf{F} + k^2 \mathbf{F} = -\varepsilon \mathbf{M} + \nabla(\nabla \cdot \mathbf{F}) + \nabla(j\omega\mu\varepsilon\phi_m) \quad (3\text{-}23)$$

By letting

$$\nabla \cdot \mathbf{F} = -j\omega\mu\varepsilon\phi_m \Rightarrow \phi_m = -\frac{1}{j\omega\mu\varepsilon}\nabla \cdot \mathbf{F} \quad (3\text{-}24)$$

reduces (3-23) to

$$\nabla^2 \mathbf{F} + k^2 \mathbf{F} = -\varepsilon\mathbf{M} \quad (3\text{-}25)$$

and (3-19) to

$$\mathbf{H}_F = -j\omega\mathbf{F} - \frac{j}{\omega\mu\varepsilon}\nabla(\nabla \cdot \mathbf{F}) \quad (3\text{-}26)$$

Once $\mathbf{F}$ is known, $\mathbf{E}_F$ can be found from (3-16) and $\mathbf{H}_F$ from (3-26) or (3-21) with $\mathbf{M} = 0$. It will be shown later how to find $\mathbf{F}$ once $\mathbf{M}$ is known. It will be a solution to the inhomogeneous Helmholtz equation of (3-25).

3.4 ELECTRIC AND MAGNETIC FIELDS FOR ELECTRIC (J) AND MAGNETIC (M) CURRENT SOURCES

In the previous two sections we have developed equations that can be used to find the electric and magnetic fields generated by an electric current source $\mathbf{J}$ and a magnetic current source $\mathbf{M}$. The procedure requires that the auxiliary potential functions $\mathbf{A}$ and $\mathbf{F}$ generated, respectively, by $\mathbf{J}$ and $\mathbf{M}$ are found first. In turn, the corresponding electric and magnetic fields are then determined ($\mathbf{E}_A, \mathbf{H}_A$ due to $\mathbf{A}$ and $\mathbf{E}_F, \mathbf{H}_F$ due to $\mathbf{F}$). The total fields are then obtained by the superposition of the individual fields due to $\mathbf{A}$ and $\mathbf{F}$ ($\mathbf{J}$ and $\mathbf{M}$).

In summary form, the procedure that can be used to find the fields is as follows:

Summary

1. Specify $\mathbf{J}$ and $\mathbf{M}$ (electric and magnetic current density sources).
2. a. Find $\mathbf{A}$ (due to $\mathbf{J}$) using

$$\mathbf{A} = \frac{\mu}{4\pi}\iiint_V \mathbf{J}\frac{e^{-jkR}}{R} dv' \quad (3\text{-}27)$$

which is a solution of the inhomogeneous vector wave equation of (3-14).

b. Find $\mathbf{F}$ (due to $\mathbf{M}$) using

$$\mathbf{F} = \frac{\varepsilon}{4\pi}\iiint_V \mathbf{M}\frac{e^{-jkR}}{R} dv' \quad (3\text{-}28)$$

which is a solution of the inhomogeneous vector wave equation of (3-25). In (3-27) and (3-28), $k^2 = \omega^2\mu\varepsilon$ and R is the distance from any point in the source to the observation point. In a latter section, we will demonstrate that (3-27) is a solution to (3-14) as (3-28) is to (3-25).

3. a. Find $\mathbf{H}_A$ using (3-2a) and $\mathbf{E}_A$ using (3-15). $\mathbf{E}_A$ can also be found using Maxwell's equation of (3-10) with $\mathbf{J} = 0$.
   b. Find $\mathbf{E}_F$ using (3-16) and $\mathbf{H}_F$ using (3-26). $\mathbf{H}_F$ can also be found using Maxwell's equation of (3-21) with $\mathbf{M} = 0$.

4. The total fields are then determined by

$$\mathbf{E} = \mathbf{E}_A + \mathbf{E}_F = -j\omega\mathbf{A} - j\frac{1}{\omega\mu\varepsilon}\nabla(\nabla \cdot \mathbf{A}) - \frac{1}{\varepsilon}\nabla \times \mathbf{F} \quad (3\text{-}29)$$

or

$$\mathbf{E} = \mathbf{E}_A + \mathbf{E}_F = \frac{1}{j\omega\varepsilon}\nabla \times \mathbf{H}_A - \frac{1}{\varepsilon}\nabla \times \mathbf{F} \quad (3\text{-}29a)$$

and

$$\mathbf{H} = \mathbf{H}_A + \mathbf{H}_F = \frac{1}{\mu}\nabla \times \mathbf{A} - j\omega\mathbf{F} - j\frac{1}{\omega\mu\varepsilon}\nabla(\nabla \cdot \mathbf{F}) \quad (3\text{-}30)$$

or

$$\mathbf{H} = \mathbf{H}_A + \mathbf{H}_F = \frac{1}{\mu}\nabla \times \mathbf{A} - \frac{1}{j\omega\mu}\nabla \times \mathbf{E}_F \quad (3\text{-}30a)$$

Whether (3-15) or (3-10) is used to find $\mathbf{E}_A$ and (3-26) or (3-21) to find $\mathbf{H}_F$ depends largely upon the problem. In many instances one may be more complex than the other or vice versa. In computing fields in the far-zone, it will be easier to use (3-15) for $\mathbf{E}_A$ and (3-26) for $\mathbf{H}_F$ because, as it will be shown, the second term in each expression becomes negligible in that region.

3.5 SOLUTION OF THE INHOMOGENEOUS VECTOR POTENTIAL WAVE EQUATION

In the previous section we indicated that the solution of the inhomogeneous vector wave equation of (3-14) is (3-27).

To derive it, let us assume that a source with current density J_z, which in the limit is an infinitesimal source, is placed at the origin of a x, y, z coordinate system, as shown in Figure 3.2(a). Since the current density is directed along the z-axis (J_z), only an A_z component will exist. Thus we can write (3-14) as

$$\nabla^2 A_z + k^2 A_z = -\mu J_z \quad (3\text{-}31)$$

SOLUTION OF THE INHOMOGENEOUS VECTOR POTENTIAL WAVE EQUATION   133

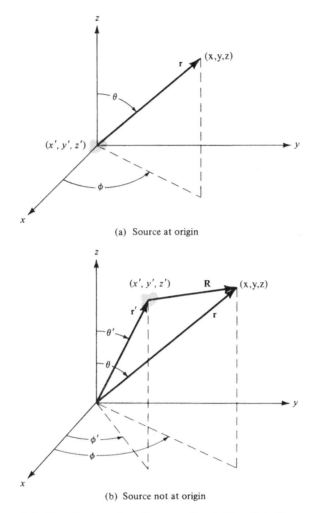

(a) Source at origin

(b) Source not at origin

Figure 3.2   Coordinate systems for computing fields radiated by sources.

At points removed from the source ($J_z = 0$), the wave equation reduces to

$$\nabla^2 A_z + k^2 A_z = 0 \tag{3-32}$$

Since in the limit the source is a point, it requires that A_z is not a function of direction (θ and ϕ); in a spherical coordinate system, $A_z = A_z(r)$ where r is the radial distance. Thus (3-32) can be written as

$$\nabla^2 A_z(r) + k^2 A_z(r) = \frac{1}{r^2}\frac{\partial}{\partial r}\left[r^2 \frac{\partial A_z(r)}{\partial r}\right] + k^2 A_z(r) = 0 \tag{3-33}$$

which when expanded reduces to

$$\frac{d^2 A_z(r)}{dr^2} + \frac{2}{r}\frac{dA_z(r)}{dr} + k^2 A_z(r) = 0 \tag{3-34}$$

The partial derivative has been replaced by the ordinary derivative since A_z is only a function of the radial coordinate.

The differential equation of (3-34) has two independent solutions

$$A_{z1} = C_1 \frac{e^{-jkr}}{r} \tag{3-35}$$

$$A_{z2} = C_2 \frac{e^{+jkr}}{r} \tag{3-36}$$

Equation (3-35) represents an outwardly (in the radial direction) traveling wave and (3-36) describes an inwardly traveling wave (assuming an $e^{j\omega t}$ time variation). For this problem, the source is placed at the origin with the radiated fields traveling in the outward radial direction. Therefore, we choose the solution of (3-35), or

$$A_z = A_{z1} = C_1 \frac{e^{-jkr}}{r} \tag{3-37}$$

In the static case ($\omega = 0, k = 0$), (3-37) simplifies to

$$A_z = \frac{C_1}{r} \tag{3-38}$$

which is a solution to the wave equation of (3-32), (3-33), or (3-34) when $k = 0$. Thus at points removed from the source, the time-varying and the static solutions of (3-37) and (3-38) differ only by the e^{-jkr} factor; or the time-varying solution of (3-37) can be obtained by multiplying the static solution of (3-38) by e^{-jkr}.

In the presence of the source ($J_z \neq 0$) and $k = 0$, the wave equation of (3-31) reduces to

$$\nabla^2 A_z = -\mu J_z \tag{3-39}$$

This equation is recognized to be Poisson's equation whose solution is widely documented. The most familiar equation with Poisson's form is that relating the scalar electric potential ϕ to the electric charge density ρ. This is given by

$$\nabla^2 \phi = -\frac{\rho}{\varepsilon} \tag{3-40}$$

whose solution is

$$\phi = \frac{1}{4\pi\varepsilon} \iiint_V \frac{\rho}{r} dv' \tag{3-41}$$

where r is the distance from any point on the charge density to the observation point. Since (3-39) is similar in form to (3-40), its solution is similar to (3-41), or

$$A_z = \frac{\mu}{4\pi} \iiint_V \frac{J_z}{r} dv' \tag{3-42}$$

Equation (3-42) represents the solution to (3-31) when $k = 0$ (static case). Using the comparative analogy between (3-37) and (3-38), the time-varying solution of (3-31) can be obtained by multiplying the static solution of (3-42) by e^{-jkr}. Thus

$$A_z = \frac{\mu}{4\pi} \iiint_V J_z \frac{e^{-jkr}}{r} \, dv' \tag{3-43}$$

which is a solution to (3-31).

If the current densities were in the x- and y-directions (J_x and J_y), the wave equation for each would reduce to

$$\nabla^2 A_x + k^2 A_x = -\mu J_x \tag{3-44}$$

$$\nabla^2 A_y + k^2 A_y = -\mu J_y \tag{3-45}$$

with corresponding solutions similar in form to (3-43), or

$$A_x = \frac{\mu}{4\pi} \iiint_V J_x \frac{e^{-jkr}}{r} \, dv' \tag{3-46}$$

$$A_y = \frac{\mu}{4\pi} \iiint_V J_y \frac{e^{-jkr}}{r} \, dv' \tag{3-47}$$

The solutions of (3-43), (3-46), and (3-47) allow us to write the solution to the vector wave equation of (3-14) as

$$\mathbf{A} = \frac{\mu}{4\pi} \iiint_V \mathbf{J} \frac{e^{-jkr}}{r} \, dv' \tag{3-48}$$

If the source is removed from the origin and placed at a position represented by the primed coordinates (x', y', z'), as shown in Figure 3.2(b), (3-48) can be written as

$$\mathbf{A}(x, y, z) = \frac{\mu}{4\pi} \iiint_V \mathbf{J}(x', y', z') \frac{e^{-jkR}}{R} \, dv' \tag{3-49}$$

where the primed coordinates represent the source, the unprimed the observation point, and R the distance from any point on the source to the observation point. In a similar fashion we can show that the solution of (3-25) is given by

$$\mathbf{F}(x, y, z) = \frac{\varepsilon}{4\pi} \iiint_V \mathbf{M}(x', y', z') \frac{e^{-jkR}}{R} \, dv' \tag{3-50}$$

136 RADIATION INTEGRALS AND AUXILIARY POTENTIAL FUNCTIONS

If **J** and **M** represent linear densities (m^{-1}), (3-49) and (3-50) reduce to surface integrals, or

$$\mathbf{A} = \frac{\mu}{4\pi} \iint_S \mathbf{J}_s(x',y',z') \frac{e^{-jkR}}{R}\, ds' \tag{3-51}$$

$$\mathbf{F} = \frac{\varepsilon}{4\pi} \iint_S \mathbf{M}_s(x',y',z') \frac{e^{-jkR}}{R}\, ds' \tag{3-52}$$

For electric and magnetic currents $\mathbf{I}_e$ and $\mathbf{I}_m$, (3-51) and (3-52) reduce to line integrals of the form

$$\mathbf{A} = \frac{\mu}{4\pi} \int_C \mathbf{I}_e(x',y',z') \frac{e^{-jkR}}{R}\, dl' \tag{3-53}$$

$$\mathbf{F} = \frac{\varepsilon}{4\pi} \int_C \mathbf{I}_m(x',y',z') \frac{e^{-jkR}}{R}\, dl' \tag{3-54}$$

3.6 FAR-FIELD RADIATION

The fields radiated by antennas of finite dimensions are spherical waves. For these radiators, a general solution to the vector wave equation of (3-14) in spherical components, each as a function of r, θ, ϕ, takes the general form of

$$\mathbf{A} = \hat{\mathbf{a}}_r A_r(r,\theta,\phi) + \hat{\mathbf{a}}_\theta A_\theta(r,\theta,\phi) + \hat{\mathbf{a}}_\phi A_\phi(r,\theta,\phi) \tag{3-55}$$

The amplitude variations of r in each component of (3-55) are of the form $1/r^n, n = 1, 2, \ldots$ [1], [2]. Neglecting higher order terms of $1/r^n (1/r^n = 0, n = 2, 3, \ldots)$ reduces (3-55) to

$$\mathbf{A} \simeq [\hat{\mathbf{a}}_r A'_r(\theta,\phi) + \hat{\mathbf{a}}_\theta A'_\theta(\theta,\phi) + \hat{\mathbf{a}}_\phi A'_\phi(\theta,\phi)] \frac{e^{-jkr}}{r}, \quad r \to \infty \tag{3-56}$$

The r variations are separable from those of θ and ϕ. This will be demonstrated in the chapters that follow by many examples.

Substituting (3-56) into (3-15) reduces it to

$$\mathbf{E} = \frac{1}{r}\{-j\omega e^{-jkr}[\hat{\mathbf{a}}_r(0) + \hat{\mathbf{a}}_\theta A'_\theta(\theta,\phi) + \hat{\mathbf{a}}_\phi A'_\phi(\theta,\phi)]\} + \frac{1}{r^2}\{\cdots\} + \cdots \tag{3-57}$$

The radial E-field component has no $1/r$ terms, because its contributions from the first and second terms of (3-15) cancel each other.

Similarly, by using (3-56), we can write (3-2a) as

$$\mathbf{H} = \frac{1}{r}\left\{j\frac{\omega}{\eta} e^{-jkr}[\hat{\mathbf{a}}_r(0) + \hat{\mathbf{a}}_\theta A'_\phi(\theta,\phi) - \hat{\mathbf{a}}_\phi A'_\theta(\theta,\phi)]\right\} + \frac{1}{r^2}\{\cdots\} + \cdots \tag{3-57a}$$

where $\eta = \sqrt{\mu/\varepsilon}$ is the intrinsic impedance of the medium.

Neglecting higher order terms of $1/r^n$, the radiated **E**- and **H**-fields have only θ and ϕ components. They can be expressed as

Far-Field Region

$$\left.\begin{array}{l} E_r \simeq 0 \\ E_\theta \simeq -j\omega A_\theta \\ E_\phi \simeq -j\omega A_\phi \end{array}\right\} \Rightarrow \boxed{\mathbf{E}_A \simeq -j\omega \mathbf{A}} \quad \text{(for the } \theta \text{ and } \phi \text{ components only since } E_r \simeq 0\text{)} \tag{3-58a}$$

$$\left.\begin{array}{l} H_r \simeq 0 \\ H_\theta \simeq +j\dfrac{\omega}{\eta}A_\phi = -\dfrac{E_\phi}{\eta} \\ H_\phi \simeq -j\dfrac{\omega}{\eta}A_\theta = +\dfrac{E_\theta}{\eta} \end{array}\right\} \Rightarrow \boxed{\mathbf{H}_A \simeq \dfrac{\hat{\mathbf{a}}_r}{\eta} \times \mathbf{E}_A = -j\dfrac{\omega}{\eta}\hat{\mathbf{a}}_r \times \mathbf{A}} \quad \text{(for the } \theta \text{ and } \phi \text{ components only since } H_r \simeq 0\text{)} \tag{3-58b}$$

Radial field components exist only for higher order terms of $1/r^n$.

In a similar manner, the far-zone fields due to a magnetic source **M** (potential **F**) can be written as

Far-Field Region

$$\left.\begin{array}{l} H_r \simeq 0 \\ H_\theta \simeq -j\omega F_\theta \\ H_\phi \simeq -j\omega F_\phi \end{array}\right\} \Rightarrow \boxed{\mathbf{H}_F \simeq -j\omega \mathbf{F}} \quad \text{(for the } \theta \text{ and } \phi \text{ components only since } H_r \simeq 0\text{)} \tag{3-59a}$$

$$\left.\begin{array}{l} E_r \simeq 0 \\ E_\theta \simeq -j\omega\eta F_\phi = \eta H_\phi \\ E_\phi \simeq +j\omega\eta F_\theta = -\eta H_\theta \end{array}\right\} \Rightarrow \boxed{\mathbf{E}_F = -\eta\hat{\mathbf{a}}_r \times \mathbf{H}_F = j\omega\eta\hat{\mathbf{a}}_r \times \mathbf{F}} \quad \text{(for the } \theta \text{ and } \phi \text{ components only since } E_r \simeq 0\text{)} \tag{3-59b}$$

Simply stated, *the corresponding far-zone **E**- and **H**-field components are orthogonal to each other and form TEM (to r) mode fields*. This is a very useful relation, and it will be adopted in the chapters that follow for the solution of the far-zone radiated fields. The far-zone (far-field) region for a radiator is defined in Figures 2.7 and 2.8. Its smallest radial distance is $2D^2/\lambda$ where D is the largest dimension of the radiator.

3.7 DUALITY THEOREM

When two equations that describe the behavior of two different variables are of the same mathematical form, their solutions will also be identical. The variables in the two equations that occupy identical positions are known as *dual* quantities and a solution of one can be formed by a systematic interchange of symbols to the other. This concept is known as the *duality theorem*.

Comparing Equations (3-2a), (3-3), (3-10), (3-14), and (3-15) to (3-16), (3-17), (3-21), (3-25), and (3-26), respectively, it is evident that they are to each other dual equations and their variables dual quantities. Thus knowing the solutions to one set (i.e., $\mathbf{J} \neq 0, \mathbf{M} = 0$), the solution to the other

TABLE 3.1 Dual Equations for Electric (J) and Magnetic (M) Current Sources

Electric Sources ($\mathbf{J} \neq 0, \mathbf{M} = 0$)	Magnetic Sources ($\mathbf{J} = 0, \mathbf{M} \neq 0$)
$\nabla \times \mathbf{E}_A = -j\omega\mu \mathbf{H}_A$	$\nabla \times \mathbf{H}_F = j\omega\varepsilon \mathbf{E}_F$
$\nabla \times \mathbf{H}_A = \mathbf{J} + j\omega\varepsilon \mathbf{E}_A$	$-\nabla \times \mathbf{E}_F = \mathbf{M} + j\omega\mu \mathbf{H}_F$
$\nabla^2 \mathbf{A} + k^2 \mathbf{A} = -\mu \mathbf{J}$	$\nabla^2 \mathbf{F} + k^2 \mathbf{F} = -\varepsilon \mathbf{M}$
$\mathbf{A} = \dfrac{\mu}{4\pi} \iiint\limits_V \mathbf{J} \dfrac{e^{-jkR}}{R} dv'$	$\mathbf{F} = \dfrac{\varepsilon}{4\pi} \iiint\limits_V \mathbf{M} \dfrac{e^{-jkR}}{R} dv'$
$\mathbf{H}_A = \dfrac{1}{\mu} \nabla \times \mathbf{A}$	$\mathbf{E}_F = -\dfrac{1}{\varepsilon} \nabla \times \mathbf{F}$
$\mathbf{E}_A = -j\omega \mathbf{A}$ $\quad - j\dfrac{1}{\omega\mu\varepsilon}\nabla(\nabla \cdot \mathbf{A})$	$\mathbf{H}_F = -j\omega \mathbf{F}$ $\quad - j\dfrac{1}{\omega\mu\varepsilon}\nabla(\nabla \cdot \mathbf{F})$

set ($\mathbf{J} = 0, \mathbf{M} \neq 0$) can be formed by a proper interchange of quantities. The dual equations and their dual quantities are listed, respectively in Tables 3.1 and 3.2 for electric and magnetic sources. Duality only serves as a guide to form mathematical solutions. It can be used in an abstract manner to explain the motion of magnetic charges giving rise to magnetic currents, when compared to their dual quantities of moving electric charges creating electric currents. It must, however, be emphasized that this is purely mathematical in nature since it is known, as of today, that there are no magnetic charges or currents in nature.

3.8 RECIPROCITY AND REACTION THEOREMS

We are all well familiar with the reciprocity theorem, as applied to circuits, which states that "in any network composed of linear, bilateral, lumped elements, if one places a constant **current** *(voltage)* generator between two **nodes** *(in any branch)* and places a **voltage** *(current)* meter between any other two **nodes** *(in any other branch)*, makes observation of the meter reading, then interchanges the locations of the source and the meter, the meter reading will be unchanged" [3]. We want now to discuss the reciprocity theorem as it applies to electromagnetic theory. This is done best by the use of Maxwell's equations.

TABLE 3.2 Dual Quantities for Electric (J) and Magnetic (M) Current Sources

Electric Sources ($\mathbf{J} \neq 0, \mathbf{M} = 0$)	Magnetic Sources ($\mathbf{J} = 0, \mathbf{M} \neq 0$)
$\mathbf{E}_A$	$\mathbf{H}_F$
$\mathbf{H}_A$	$-\mathbf{E}_F$
$\mathbf{J}$	$\mathbf{M}$
$\mathbf{A}$	$\mathbf{F}$
ε	μ
μ	ε
k	k
η	$1/\eta$
$1/\eta$	η

Let us assume that within a linear and isotropic medium, but not necessarily homogeneous, there exist two sets of sources $\mathbf{J}_1, \mathbf{M}_1$, and $\mathbf{J}_2, \mathbf{M}_2$ which are allowed to radiate simultaneously or individually inside the same medium at the same frequency and produce fields $\mathbf{E}_1, \mathbf{H}_1$ and $\mathbf{E}_2, \mathbf{H}_2$, respectively. It can be shown [1], [2] that the sources and fields satisfy

$$-\nabla \cdot (\mathbf{E}_1 \times \mathbf{H}_2 - \mathbf{E}_2 \times \mathbf{H}_1) = \mathbf{E}_1 \cdot \mathbf{J}_2 + \mathbf{H}_2 \cdot \mathbf{M}_1 - \mathbf{E}_2 \cdot \mathbf{J}_1 - \mathbf{H}_1 \cdot \mathbf{M}_2 \qquad (3\text{-}60)$$

which is called the *Lorentz Reciprocity Theorem* in differential form.

Taking a volume integral of both sides of (3-60) and using the divergence theorem on the left side, we can write it as

$$-\oiint_S (\mathbf{E}_1 \times \mathbf{H}_2 - \mathbf{E}_2 \times \mathbf{H}_1) \cdot d\mathbf{s}'$$

$$= \iiint_V (\mathbf{E}_1 \cdot \mathbf{J}_2 + \mathbf{H}_2 \cdot \mathbf{M}_1 - \mathbf{E}_2 \cdot \mathbf{J}_1 - \mathbf{H}_1 \cdot \mathbf{M}_2) \, dv' \qquad (3\text{-}61)$$

which is designated as the *Lorentz Reciprocity Theorem* in integral form.

For a source-free ($\mathbf{J}_1 = \mathbf{J}_2 = \mathbf{M}_1 = \mathbf{M}_2 = 0$) region, (3-60) and (3-61) reduce, respectively, to

$$\boxed{\nabla \cdot (\mathbf{E}_1 \times \mathbf{H}_2 - \mathbf{E}_2 \times \mathbf{H}_1) = 0} \qquad (3\text{-}62)$$

and

$$\boxed{\oiint_S (\mathbf{E}_1 \times \mathbf{H}_2 - \mathbf{E}_2 \times \mathbf{H}_1) \cdot d\mathbf{s}' = 0} \qquad (3\text{-}63)$$

Equations (3-62) and (3-63) are special cases of the Lorentz Reciprocity Theorem and must be satisfied in source-free regions.

As an example of where (3-62) and (3-63) may be applied and what they would represent, consider a section of a waveguide where two different modes exist with fields $\mathbf{E}_1, \mathbf{H}_1$ and $\mathbf{E}_2, \mathbf{H}_2$. For the expressions of the fields for the two modes to be valid, they must satisfy (3-62) and/or (3-63).

Another useful form of (3-61) is to consider that the fields ($\mathbf{E}_1, \mathbf{H}_1, \mathbf{E}_2, \mathbf{H}_2$) and the sources ($\mathbf{J}_1, \mathbf{M}_1, \mathbf{J}_2, \mathbf{M}_2$) are within a medium that is enclosed by a sphere of infinite radius. Assume that the sources are positioned within a finite region and that the fields are observed in the far field (ideally at infinity). Then the left side of (3-61) is equal to zero, or

$$\oiint_S (\mathbf{E}_1 \times \mathbf{H}_2 - \mathbf{E}_2 \times \mathbf{H}_1) \cdot d\mathbf{s}' = 0 \qquad (3\text{-}64)$$

which reduces (3-61) to

$$\iiint_V (\mathbf{E}_1 \cdot \mathbf{J}_2 + \mathbf{H}_2 \cdot \mathbf{M}_1 - \mathbf{E}_2 \cdot \mathbf{J}_1 - \mathbf{H}_1 \cdot \mathbf{M}_2) \, dv' = 0 \qquad (3\text{-}65)$$

Equation (3-65) can also be written as

$$\iiint_V (\mathbf{E}_1 \cdot \mathbf{J}_2 - \mathbf{H}_1 \cdot \mathbf{M}_2)\, dv' = \iiint_V (\mathbf{E}_2 \cdot \mathbf{J}_1 - \mathbf{H}_2 \cdot \mathbf{M}_1)\, dv' \qquad (3\text{-}66)$$

The reciprocity theorem, as expressed by (3-66), is the most useful form.

A close observation of (3-61) reveals that it does not, in general, represent relations of power because no conjugates appear. The same is true for the special cases represented by (3-63) and (3-66). Each of the integrals in (3-66) can be interpreted as a coupling between a set of fields and a set of sources, which produce another set of fields. This coupling has been defined as *Reaction* [4] and each of the integrals in (3-66) are denoted by

$$\langle 1, 2 \rangle = \iiint_V (\mathbf{E}_1 \cdot \mathbf{J}_2 - \mathbf{H}_1 \cdot \mathbf{M}_2)\, dv \qquad (3\text{-}67)$$

$$\langle 2, 1 \rangle = \iiint_V (\mathbf{E}_2 \cdot \mathbf{J}_1 - \mathbf{H}_2 \cdot \mathbf{M}_1)\, dv \qquad (3\text{-}68)$$

The relation $\langle 1, 2 \rangle$ of (3-67) relates the reaction (coupling) of fields $(\mathbf{E}_1, \mathbf{H}_1)$, which are produced by sources $\mathbf{J}_1, \mathbf{M}_1$ to sources $(\mathbf{J}_2, \mathbf{M}_2)$, which produce fields $\mathbf{E}_2, \mathbf{H}_2$; $\langle 2, 1 \rangle$ relates the reaction (coupling) of fields $(\mathbf{E}_2, \mathbf{H}_2)$ to sources $(\mathbf{J}_1, \mathbf{M}_1)$. For reciprocity to hold, it requires that the reaction (coupling) of one set of sources with the corresponding fields of another set of sources must be equal to the reaction (coupling) of the second set of sources with the corresponding fields of the first set of sources, and vice versa. In equation form, it is written as

$$\langle 1, 2 \rangle = \langle 2, 1 \rangle \qquad (3\text{-}69)$$

3.8.1 Reciprocity for Two Antennas

There are many applications of the reciprocity theorem. To demonstrate its potential, an antenna example will be considered. Two antennas, whose input impedances are Z_1 and Z_2, are separated by a linear and isotropic (but not necessarily homogeneous) medium, as shown in Figure 3.3. One antenna (#1) is used as a transmitter and the other (#2) as a receiver. The equivalent network of each antenna is given in Figure 3.4. The internal impedance of the generator Z_g is assumed to be the conjugate of the impedance of antenna #1 ($Z_g = Z_1^* = R_1 - jX_1$) while the load impedance Z_L is equal to the conjugate of the impedance of antenna #2 ($Z_L = Z_2^* = R_2 - jX_2$). These assumptions are made only for convenience.

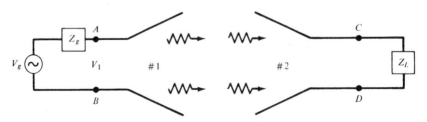

Figure 3.3 Transmitting and receiving antenna systems.

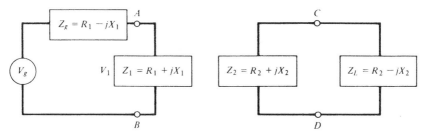

Figure 3.4 Two-antenna system with conjugate loads.

The power delivered by the generator to antenna #1 is given by (2-83) or

$$P_1 = \frac{1}{2}\mathrm{Re}[V_1 I_1^*] = \frac{1}{2}\mathrm{Re}\left[\left(\frac{V_g Z_1}{Z_1 + Z_g}\right)\frac{V_g^*}{(Z_1 + Z_g)^*}\right] = \frac{|V_g|^2}{8R_1} \quad (3\text{-}70)$$

If the transfer admittance of the combined network consisting of the generator impedance, antennas, and load impedance is Y_{21}, the current through the load is $V_g Y_{21}$ and the power delivered to the load is

$$P_2 = \frac{1}{2}\mathrm{Re}[Z_2(V_g Y_{21})(V_g Y_{21})^*] = \frac{1}{2}R_2|V_g|^2|Y_{21}|^2 \quad (3\text{-}71)$$

The ratio of (3-71) to (3-70) is

$$\frac{P_2}{P_1} = 4R_1 R_2 |Y_{21}|^2 \quad (3\text{-}72)$$

In a similar manner, we can show that when antenna #2 is transmitting and #1 is receiving, the power ratio of P_1/P_2 is given by

$$\frac{P_1}{P_2} = 4R_2 R_1 |Y_{12}|^2 \quad (3\text{-}73)$$

Under conditions of reciprocity ($Y_{12} = Y_{21}$), the power delivered in either direction is the same.

3.8.2 Reciprocity for Antenna Radiation Patterns

The radiation pattern is a very important antenna characteristic. Although it is usually most convenient and practical to measure the pattern in the receiving mode, it is identical, because of reciprocity, to that of the transmitting mode.

Reciprocity for antenna patterns is general provided the materials used for the antennas and feeds, and the media of wave propagation are linear. Nonlinear devices, such as diodes, can make the antenna system nonreciprocal. The antennas can be of any shape or size, and they do not have to be matched to their corresponding feed lines or loads provided there is a distinct single propagating mode at each port. The only other restriction for reciprocity to hold is for the antennas in the transmit and receive modes to be polarization matched, including the sense of rotation. This is necessary so that the antennas can transmit and receive the same field components, and thus total power. If the antenna that is used as a probe to measure the fields radiated by the antenna under test is not of the same polarization, then in some situations the transmit and receive patterns can still be the same. For example, if the transmit antenna is circularly polarized and the probe antenna is linearly polarized,

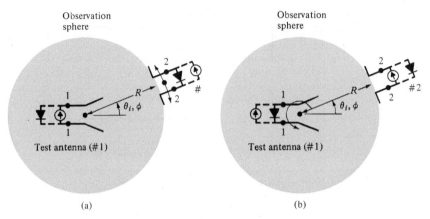

Figure 3.5 Antenna arrangement for pattern measurements and reciprocity theorem.

then if the linearly polarized probe antenna is used twice and it is oriented one time to measure the θ-component and the other the ϕ-component, then the sum of the two components can represent the pattern of the circularly polarized antenna in either the transmit or receive modes. During this procedure, the power level and sensitivities must be held constant.

To detail the procedure and foundation of pattern measurements and reciprocity, let us refer to Figures 3.5(a) and (b). The antenna under test is #1 while the probe antenna (#2) is oriented to transmit or receive maximum radiation. The voltages and currents V_1, I_1 at terminals 1–1 of antenna #1 and V_2, I_2 at terminals 2–2 of antenna #2 are related by

$$V_1 = Z_{11}I_1 + Z_{12}I_2$$
$$V_2 = Z_{21}I_1 + Z_{22}I_2$$
(3-74)

where

$$Z_{11} = \text{self-impedance of antenna \#1}$$
$$Z_{22} = \text{self-impedance of antenna \#2}$$
$$Z_{12}, Z_{21} = \text{mutual impedances between antennas \#1 and \#2}$$

If a current I_1 is applied at the terminals 1–1 and voltage V_2 (designated as V_{2oc}) is measured at the *open* ($I_2 = 0$) terminals of antenna #2, then an equal voltage V_{1oc} will be measured at the *open* ($I_1 = 0$) terminals of antenna #1 provided the current I_2 of antenna #2 is equal to I_1. In equation form, we can write

$$Z_{21} = \left.\frac{V_{2oc}}{I_1}\right|_{I_2=0}$$
(3-75a)

$$Z_{12} = \left.\frac{V_{1oc}}{I_2}\right|_{I_1=0}$$
(3-75b)

If the medium between the two antennas is linear, passive, isotropic, and the waves monochromatic, then because of reciprocity

$$Z_{21} = \left.\frac{V_{2oc}}{I_1}\right|_{I_2=0} = \left.\frac{V_{1oc}}{I_2}\right|_{I_1=0} = Z_{12}$$
(3-76)

If in addition $I_1 = I_2$, then

$$V_{2oc} = V_{1oc} \tag{3-77}$$

The above are valid for any position and any configuration of operation between the two antennas.

Reciprocity will now be reviewed for two modes of operation. In one mode, antenna #1 is held stationary while #2 is allowed to move on the surface of a constant radius sphere, as shown in Figure 3.5(a). In the other mode, antenna #2 is maintained stationary while #1 pivots about a point, as shown in Figure 3.5(b).

In the mode of Figure 3.5(a), antenna #1 can be used either as a transmitter or receiver. In the transmitting mode, while antenna #2 is moving on the constant radius sphere surface, the open terminal voltage V_{2oc} is measured. In the receiving mode, the open terminal voltage V_{1oc} is recorded. The three-dimensional plots of V_{2oc} and V_{1oc}, as a function of θ and ϕ, have been defined in Section 2.2 as *field patterns*. Since the three-dimensional graph of V_{2oc} is identical to that of V_{1oc} (due to reciprocity), the *transmitting* (V_{2oc}) *and receiving* (V_{1oc}) field patterns are also equal. The same conclusion can be arrived at if antenna #2 is allowed to remain stationary while #1 rotates, as shown in Figure 3.5(b).

The conditions of reciprocity hold whether antenna #1 is used as a transmitter and #2 as a receiver *or* antenna #2 as a transmitter and #1 as a receiver. In practice, the most convenient mode of operation is that of Figure 3.5(b) with the test antenna used as a receiver. Antenna #2 is usually placed in the far-field of the test antenna (#1), and vice versa, in order that its radiated fields are plane waves in the vicinity of #1.

The receiving mode of operation of Figure 3.5(b) for the test antenna is most widely used to measure antenna patterns because the transmitting equipment is, in most cases, bulky and heavy while the receiver is small and lightweight. In some cases, the receiver is nothing more than a simple diode detector. The transmitting equipment usually consists of sources and amplifiers. To make precise measurements, especially at microwave frequencies, it is necessary to have frequency and power stabilities. Therefore, the equipment must be placed on stable and vibration-free platforms. This can best be accomplished by allowing the transmitting equipment to be held stationary and the receiving equipment to rotate.

An excellent manuscript on test procedures for antenna measurements of amplitude, phase, impedance, polarization, gain, directivity, efficiency, and others has been published by IEEE [5]. A condensed summary of it is found in [6], and a review is presented in Chapter 17 of this text.

REFERENCES

1. R. F. Harrington, *Time-Harmonic Electromagnetic Fields*, McGraw-Hill, New York, 1961.
2. C. A. Balanis, *Advanced Engineering Electromagnetics*, Second edition, John Wiley & Sons, New York, 2012.
3. P. E. Mayes, personal communication.
4. V. H. Rumsey, "The Reaction Concept in Electromagnetic Theory," *Physical Review*, Series 2, Vol. 94, No. 6, June 15, 1954, pp. 1483–1491.
5. *IEEE Standard Test Procedures for Antennas*, IEEE Std 149–1979, IEEE, Inc., New York, 1979.
6. W. H. Kummer and E. S. Gillespie, "Antenna Measurements–1978," *Proc. IEEE*, Vol. 66, No. 4, April 1978, pp. 483–507.

PROBLEMS

3.1. If $\mathbf{H}_e = j\omega\varepsilon \, \nabla \times \mathbf{\Pi}_e$, where $\mathbf{\Pi}_e$ is the electric Hertzian potential, show that

(a) $\nabla^2 \mathbf{\Pi}_e + k^2 \mathbf{\Pi}_e = j\dfrac{1}{\omega\varepsilon}\mathbf{J}$ (b) $\mathbf{E}_e = k^2 \mathbf{\Pi}_e + \nabla(\nabla \cdot \mathbf{\Pi}_e)$

(c) $\mathbf{\Pi}_e = -j\dfrac{1}{\omega\mu\varepsilon}\mathbf{A}$

3.2. If $\mathbf{E}_h = -j\omega\mu\nabla \times \mathbf{\Pi}_h$, where $\mathbf{\Pi}_h$ is the magnetic Hertzian potential, show that
(a) $\nabla^2\mathbf{\Pi}_h + k^2\mathbf{\Pi}_h = j\dfrac{1}{\omega\mu}\mathbf{M}$ (b) $\mathbf{H}_h = k^2\mathbf{\Pi}_h + \nabla(\nabla\cdot\mathbf{\Pi}_h)$
(c) $\mathbf{\Pi}_h = -j\dfrac{1}{\omega\mu\varepsilon}\mathbf{F}$

3.3. Verify that (3-35) and (3-36) are solutions to (3-34).

3.4. Show that (3-42) is a solution to (3-39) and (3-43) is a solution to (3-31).

3.5. Verify (3-57) and (3-57a).

3.6. Derive (3-60) and (3-61).

CHAPTER 4

Linear Wire Antennas

4.1 INTRODUCTION

Wire antennas, linear or curved, are some of the oldest, simplest, cheapest, and in many cases the most versatile for many applications. It should not then come as a surprise to the reader that we begin our analysis of antennas by considering some of the oldest, simplest, and most basic configurations. Initially we will try to minimize the complexity of the antenna structure and geometry to keep the mathematical details to a minimum.

4.2 INFINITESIMAL DIPOLE

An infinitesimal linear wire ($l \ll \lambda$) is positioned symmetrically at the origin of the coordinate system and oriented along the z axis, as shown in Figure 4.1(a). Although infinitesimal dipoles are not very practical, they are used to represent capacitor-plate (also referred to as *top-hat-loaded*) antennas. In addition, they are utilized as building blocks of more complex geometries. The end plates are used to provide capacitive loading in order to maintain the current on the dipole nearly uniform. Since the end plates are assumed to be small, their radiation is usually negligible. The wire, in addition to being very small ($l \ll \lambda$), is very thin ($a \ll \lambda$). The spatial variation of the current is assumed to be constant and given by

$$\mathbf{I}(z') = \hat{\mathbf{a}}_z I_0 \qquad (4\text{-}1)$$

where I_0 = constant.

4.2.1 Radiated Fields

To find the fields radiated by the current element, the two-step procedure of Figure 3.1 is used. It will be required to determine first **A** and **F** and then find the **E** and **H**. The functional relation between **A** and the source **J** is given by (3-49), (3-51), or (3-53). Similar relations are available for **F** and **M**, as given by (3-50), (3-52), and (3-54).

Antenna Theory: Analysis and Design, Fourth Edition. Constantine A. Balanis.
© 2016 John Wiley & Sons, Inc. Published 2016 by John Wiley & Sons, Inc.
Companion Website: www.wiley.com/go/antennatheory4e

146   LINEAR WIRE ANTENNAS

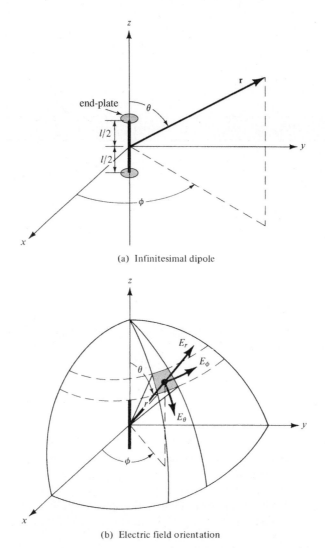

(a) Infinitesimal dipole

(b) Electric field orientation

Figure 4.1   Geometrical arrangement of an infinitesimal dipole and its associated electric-field components on a spherical surface.

Since the source only carries an electric current $\mathbf{I}_e$, $\mathbf{I}_m$ and the potential function $\mathbf{F}$ are zero. To find $\mathbf{A}$ we write

$$\mathbf{A}(x, y, z) = \frac{\mu}{4\pi} \int_C \mathbf{I}_e(x', y', z') \frac{e^{-jkR}}{R} \, dl' \qquad (4\text{-}2)$$

where (x, y, z) represent the observation point coordinates, (x', y', z') represent the coordinates of the source, R is the distance from any point on the source to the observation point, and path C is along the length of the source. For the problem of Figure 4.1

$$\mathbf{I}_e(x', y', z') = \hat{\mathbf{a}}_z I_0 \qquad (4\text{-}3a)$$

$$x' = y' = z' = 0 \text{ (infinitesimal dipole)} \qquad (4\text{-}3b)$$

$$R = \sqrt{(x-x')^2 + (y-y')^2 + (z-z')^2} = \sqrt{x^2 + y^2 + z^2}$$
$$= r = \text{constant} \tag{4-3c}$$
$$dl' = dz' \tag{4-3d}$$

so we can write (4-2) as

$$\mathbf{A}(x,y,z) = \hat{\mathbf{a}}_z \frac{\mu I_0}{4\pi r} e^{-jkr} \int_{-l/2}^{+l/2} dz' = \hat{\mathbf{a}}_z \frac{\mu I_0 l}{4\pi r} e^{-jkr} \tag{4-4}$$

The next step of the procedure is to find $\mathbf{H}_A$ using (3-2a) and then $\mathbf{E}_A$ using (3-15) or (3-10) with $\mathbf{J} = 0$. To do this, it is often much simpler to transform (4-4) from rectangular to spherical components and then use (3-2a) and (3-15) or (3-10) in spherical coordinates to find $\mathbf{H}$ and $\mathbf{E}$.

The transformation between rectangular and spherical components is given, in matrix form, by (VII-12a) (see Appendix VII)

$$\begin{bmatrix} A_r \\ A_\theta \\ A_\phi \end{bmatrix} = \begin{bmatrix} \sin\theta\cos\phi & \sin\theta\sin\phi & \cos\theta \\ \cos\theta\cos\phi & \cos\theta\sin\phi & -\sin\theta \\ -\sin\phi & \cos\phi & 0 \end{bmatrix} \begin{bmatrix} A_x \\ A_y \\ A_z \end{bmatrix} \tag{4-5}$$

For this problem, $A_x = A_y = 0$, so (4-5) using (4-4) reduces to

$$A_r = A_z \cos\theta = \frac{\mu I_0 l e^{-jkr}}{4\pi r} \cos\theta \tag{4-6a}$$

$$A_\theta = -A_z \sin\theta = -\frac{\mu I_0 l e^{-jkr}}{4\pi r} \sin\theta \tag{4-6b}$$

$$A_\phi = 0 \tag{4-6c}$$

Using the symmetry of the problem (no ϕ variations), (3-2a) can be expanded in spherical coordinates and written in simplified form as

$$\mathbf{H} = \hat{\mathbf{a}}_\phi \frac{1}{\mu r} \left[\frac{\partial}{\partial r}(rA_\theta) - \frac{\partial A_r}{\partial \theta} \right] \tag{4-7}$$

Substituting (4-6a)–(4-6c) into (4-7) reduces it to

$$H_r = H_\theta = 0 \tag{4-8a}$$

$$H_\phi = j\frac{kI_0 l \sin\theta}{4\pi r}\left[1 + \frac{1}{jkr}\right] e^{-jkr} \tag{4-8b}$$

The electric field $\mathbf{E}$ can now be found using (3-15) or (3-10) with $\mathbf{J} = 0$. That is,

$$\mathbf{E} = \mathbf{E}_A = -j\omega\mathbf{A} - j\frac{1}{\omega\mu\varepsilon}\nabla(\nabla\cdot\mathbf{A}) = \frac{1}{j\omega\varepsilon}\nabla\times\mathbf{H} \tag{4-9}$$

148 LINEAR WIRE ANTENNAS

Substituting (4-6a)–(4-6c) or (4-8a)–(4-8b) into (4-9) reduces it to

$$E_r = \eta \frac{I_0 l \cos\theta}{2\pi r^2} \left[1 + \frac{1}{jkr}\right] e^{-jkr} \tag{4-10a}$$

$$E_\theta = j\eta \frac{k I_0 l \sin\theta}{4\pi r} \left[1 + \frac{1}{jkr} - \frac{1}{(kr)^2}\right] e^{-jkr} \tag{4-10b}$$

$$E_\phi = 0 \tag{4-10c}$$

The **E**- and **H**-field components are valid everywhere, except on the source itself, and they are sketched in Figure 4.1(b) on the surface of a sphere of radius r. It is a straightforward exercise to verify Equations (4-10a)–(4-10c), and this is left as an exercise to the reader (Prob. 4.14).

4.2.2 Power Density and Radiation Resistance

The input impedance of an antenna, which consists of real and imaginary parts, was discussed in Section 2.13. For a lossless antenna, the real part of the input impedance was designated as radiation resistance. It is through the mechanism of the radiation resistance that power is transferred from the guided wave to the free-space wave. To find the input resistance for a lossless antenna, the Poynting vector is formed in terms of the **E**- and **H**-fields radiated by the antenna. By integrating the Poynting vector over a closed surface (usually a sphere of constant radius), the total power radiated by the source is found. The real part of it is related to the input resistance.

For the infinitesimal dipole, the complex Poynting vector can be written using (4-8a)–(4-8b) and (4-10a)–(4-10c) as

$$\mathbf{W} = \tfrac{1}{2}(\mathbf{E} \times \mathbf{H}^*) = \tfrac{1}{2}(\hat{\mathbf{a}}_r E_r + \hat{\mathbf{a}}_\theta E_\theta) \times (\hat{\mathbf{a}}_\phi H_\phi^*)$$
$$= \tfrac{1}{2}(\hat{\mathbf{a}}_r E_\theta H_\phi^* - \hat{\mathbf{a}}_\theta E_r H_\phi^*) \tag{4-11}$$

whose radial W_r and transverse W_θ components are given, respectively, by

$$W_r = \frac{\eta}{8} \left|\frac{I_0 l}{\lambda}\right|^2 \frac{\sin^2\theta}{r^2} \left[1 - j\frac{1}{(kr)^3}\right] \tag{4-12a}$$

$$W_\theta = j\eta \frac{k|I_0 l|^2 \cos\theta \sin\theta}{16\pi^2 r^3} \left[1 + \frac{1}{(kr)^2}\right] \tag{4-12b}$$

The complex power moving in the radial direction is obtained by integrating (4-11)–(4-12b) over a closed sphere of radius r. Thus it can be written as

$$P = \oiint_S \mathbf{W} \cdot d\mathbf{s} = \int_0^{2\pi} \int_0^\pi (\hat{\mathbf{a}}_r W_r + \hat{\mathbf{a}}_\theta W_\theta) \cdot \hat{\mathbf{a}}_r r^2 \sin\theta \, d\theta \, d\phi \tag{4-13}$$

which reduces to

$$P = \int_0^{2\pi} \int_0^\pi W_r r^2 \sin\theta \, d\theta \, d\phi = \eta \frac{\pi}{3} \left|\frac{I_0 l}{\lambda}\right|^2 \left[1 - j\frac{1}{(kr)^3}\right] \tag{4-14}$$

The transverse component W_θ of the power density does not contribute to the integral. Thus (4-14) does not represent the total complex power radiated by the antenna. Since W_θ, as given by (4-12b), is purely imaginary, it will not contribute to any real radiated power. However, it does contribute to the imaginary (reactive) power which along with the second term of (4-14) can be used to determine the total reactive power of the antenna. *The reactive power density, which is most dominant for small values of kr, has both radial and transverse components. It merely changes between outward and inward directions to form a standing wave at a rate of twice per cycle. It also moves in the transverse direction as suggested by (4-12b).*

Equation (4-13), which gives the real and imaginary power that is moving outwardly, can also be written as

$$P = \frac{1}{2} \iint_S \mathbf{E} \times \mathbf{H}^* \cdot d\mathbf{s} = \eta \left(\frac{\pi}{3}\right) \left|\frac{I_0 l}{\lambda}\right|^2 \left[1 - j\frac{1}{(kr)^3}\right]$$

$$= P_{rad} + j2\omega(\tilde{W}_m - \tilde{W}_e) \tag{4-15}$$

where

P = power (in radial direction)

P_{rad} = time-average power radiated

$\tilde{W}_m$ = time-average magnetic energy density (in radial direction)

$\tilde{W}_e$ = time-average electric energy density (in radial direction)

$2\omega(\tilde{W}_m - \tilde{W}_e)$ = time-average imaginary (reactive) power (in radial direction)

From (4-14)

$$P_{rad} = \eta \left(\frac{\pi}{3}\right) \left|\frac{I_0 l}{\lambda}\right|^2 \tag{4-16}$$

and

$$2\omega(\tilde{W}_m - \tilde{W}_e) = -\eta \left(\frac{\pi}{3}\right) \left|\frac{I_0 l}{\lambda}\right|^2 \frac{1}{(kr)^3} \tag{4-17}$$

It is clear from (4-17) that the radial electric energy must be larger than the radial magnetic energy. For large values of kr ($kr \gg 1$ or $r \gg \lambda$), the reactive power diminishes and vanishes when $kr = \infty$.

Since the antenna radiates its real power through the radiation resistance, for the infinitesimal dipole it is found by equating (4-16) to

$$\boxed{P_{rad} = \eta \left(\frac{\pi}{3}\right) \left|\frac{I_0 l}{\lambda}\right|^2 = \frac{1}{2}|I_0|^2 R_r} \tag{4-18}$$

where R_r is the radiation resistance. Equation (4-18) reduces to

$$\boxed{R_r = \eta \left(\frac{2\pi}{3}\right) \left(\frac{l}{\lambda}\right)^2 = 80\pi^2 \left(\frac{l}{\lambda}\right)^2} \tag{4-19}$$

for a free-space medium ($\eta \simeq 120\pi$). It should be pointed out that the radiation resistance of (4-19) represents the total radiation resistance since (4-12b) does not contribute to it.

For a wire antenna to be classified as an infinitesimal dipole, its overall length must be very small (usually $l \leq \lambda/50$).

Example 4.1

Find the radiation resistance of an infinitesimal dipole whose overall length is $l = \lambda/50$.
Solution: Using (4-19)

$$R_r = 80\pi^2 \left(\frac{l}{\lambda}\right)^2 = 80\pi^2 \left(\frac{1}{50}\right)^2 = 0.316 \text{ ohms}$$

Since the radiation resistance of an infinitesimal dipole is about 0.3 ohms, it will present a very large mismatch when connected to practical transmission lines, many of which have characteristic impedances of 50 or 75 ohms. The reflection efficiency (e_r) and hence the overall efficiency (e_0) will be very small.

The reactance of an infinitesimal dipole is capacitive. This can be illustrated by considering the dipole as a flared open-circuited transmission line, as discussed in Section 1.4. Since the input impedance of an open-circuited transmission line a distance $l/2$ from its open end is given by $Z_{in} = -jZ_c \cot(\beta l/2)$, where Z_c is its characteristic impedance, it will always be negative (capacitive) for $l \ll \lambda$.

4.2.3 Radian Distance and Radian Sphere

The **E**- and **H**-fields for the infinitesimal dipole, as represented by (4-8a)–(4-8b) and (4-10a)–(4-10c), are valid everywhere (except on the source itself). An inspection of these equations reveals the following:

a. At a distance $r = \lambda/2\pi$ (or $kr = 1$), which is referred to as the *radian distance*, the magnitude of the first and second terms within the brackets of (4-8b) and (4-10a) is the same. Also at the radian distance the magnitude of all three terms within the brackets of (4-10b) is identical; the only term that contributes to the total field is the second, because the first and third terms cancel each other. This is illustrated in Figure 4.2.
b. At distances less than the radian distance $r < \lambda/2\pi$ ($kr < 1$), the magnitude of the second term within the brackets of (4-8b) and (4-10a) is greater than the first term and begins to dominate as $r \ll \lambda/2\pi$. For (4-10b) and $r < \lambda/2\pi$, the magnitude of the third term within the brackets is greater than the magnitude of the first and second terms while the magnitude of the second term is greater than that of the first one; each of these terms begins to dominate as $r \ll \lambda/2\pi$. This is illustrated in Figure 4.2. The region $r < \lambda/2\pi$ ($kr < 1$) is referred to as the *near-field* region, and the energy in that region is basically imaginary (stored).
c. At distances greater than the radian distance $r > \lambda/2\pi$ ($kr > 1$), the first term within the brackets of (4-8b) and (4-10a) is greater than the magnitude of the second term and begins to dominate as $r \gg \lambda/2\pi$ ($kr \gg 1$). For (4-10b) and $r > \lambda/2\pi$, the first term within the brackets is greater than the magnitude of the second and third terms while the magnitude of the second term is greater than that of the third; each of these terms begins to dominate as $r \gg \lambda/2\pi$. This is illustrated in Figure 4.2. The region $r > \lambda/2\pi$ ($kr > 1$) is referred to as the *intermediate-field* region while that for $r \gg \lambda/2\pi$ ($kr \gg 1$) is referred to as the *far-field* region, and the energy in that region is basically real (radiated).

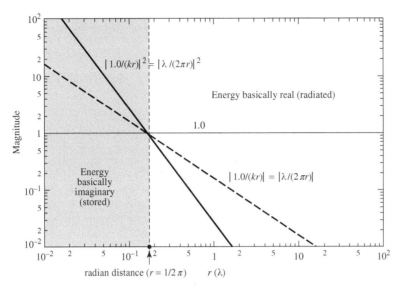

Figure 4.2 Magnitude variation, as a function of the radial distance, of the field terms radiated by an infinitesimal dipole.

d. The sphere with radius equal to the radian distance $(r = \lambda/2\pi)$ is referred as the *radian sphere*, and it defines the region within which the reactive power density is greater than the radiated power density [1]–[3]. For an antenna, the radian sphere represents the volume occupied mainly by the stored energy of the antenna's electric and magnetic fields. Outside the radian sphere the radiated power density is greater than the reactive power density and begins to dominate as $r \gg \lambda/2\pi$. Therefore the radian sphere can be used as a reference, and it defines the transition between stored energy pulsating primarily in the $\pm\theta$ direction [represented by (4-12b)] and energy radiating in the radial (r) direction [represented by the first term of (4-12a); the second term represents stored energy pulsating inwardly and outwardly in the radial (r) direction]. Similar behavior, where the power density near the antenna is primarily reactive and far away is primarily real, is exhibited by all antennas, although not exactly at the radian distance.

4.2.4 Near-Field ($kr \ll 1$) Region

An inspection of (4-8a)–(4-8b) and (4-10a)–(4-10c) reveals that for $kr \ll \lambda$ or $r \ll \lambda/2\pi$ they can be reduced in much simpler form and can be approximated by

$$E_r \simeq -j\eta \frac{I_0 l e^{-jkr}}{2\pi k r^3} \cos\theta \quad (4\text{-}20a)$$

$$E_\theta \simeq -j\eta \frac{I_0 l e^{-jkr}}{4\pi k r^3} \sin\theta \quad (4\text{-}20b)$$

$$E_\phi = H_r = H_\theta = 0 \quad (4\text{-}20c)$$

$$H_\phi \simeq \frac{I_0 l e^{-jkr}}{4\pi r^2} \sin\theta \quad (4\text{-}20d)$$

$kr \ll 1$

The **E**-field components, E_r and E_θ, are in time-phase but they are in time-phase quadrature with the **H**-field component H_ϕ; therefore there is no time-average power flow associated with them. This is demonstrated by forming the time-average power density as

$$\mathbf{W}_{av} = \tfrac{1}{2}\text{Re}[\mathbf{E} \times \mathbf{H}^*] = \tfrac{1}{2}\text{Re}[\hat{\mathbf{a}}_r E_\theta H^*_\phi - \hat{\mathbf{a}}_\theta E_r H^*_\phi] \qquad (4\text{-}21)$$

which by using (4-20a)–(4-20d) reduces to

$$\mathbf{W}_{av} = \frac{1}{2}\text{Re}\left[-\hat{\mathbf{a}}_r j \frac{\eta}{k}\left|\frac{I_0 l}{4\pi}\right|^2 \frac{\sin^2\theta}{r^5} + \hat{\mathbf{a}}_\theta j \frac{\eta}{k} \frac{|I_0 l|^2}{8\pi^2} \frac{\sin\theta\cos\theta}{r^5}\right] = 0 \qquad (4\text{-}22)$$

The condition of $kr \ll 1$ can be satisfied at moderate distances away from the antenna provided that the frequency of operation is very low. Equations (4-20a) and (4-20b) are similar to those of a static electric dipole and (4-20d) to that of a static current element. Thus we usually refer to (4-20a)–(4-20d) as the *quasistationary fields*.

4.2.5 Intermediate-Field ($kr > 1$) Region

As the values of kr begin to increase and become greater than unity, the terms that were dominant for $kr \ll 1$ become smaller and eventually vanish. For moderate values of kr the **E**-field components lose their in-phase condition and approach time-phase quadrature. Since their magnitude is not the same, in general, they form a rotating vector whose extremity traces an ellipse. This is analogous to the polarization problem except that the vector rotates in a plane parallel to the direction of propagation and is usually referred to as the *cross field*. At these intermediate values of kr, the E_θ and H_ϕ components approach time-phase, which is an indication of the formation of time-average power flow in the outward (radial) direction (radiation phenomenon).

As the values of kr become moderate ($kr > 1$), the field expressions can be approximated again but in a different form. In contrast to the region where $kr \ll 1$, the first term within the brackets in (4-8b) and (4-10a) becomes more dominant and the second term can be neglected. The same is true for (4-10b) where the second and third terms become less dominant than the first. Thus we can write for $kr > 1$

$$E_r \simeq \eta \frac{I_0 l e^{-jkr}}{2\pi r^2} \cos\theta \qquad (4\text{-}23\text{a})$$

$$E_\theta \simeq j\eta \frac{k I_0 l e^{-jkr}}{4\pi r} \sin\theta \qquad kr > 1 \qquad (4\text{-}23\text{b})$$

$$E_\phi = H_r = H_\theta = 0 \qquad (4\text{-}23\text{c})$$

$$H_\phi \simeq j \frac{k I_0 l e^{-jkr}}{4\pi r} \sin\theta \qquad (4\text{-}23\text{d})$$

The total electric field is given by

$$\mathbf{E} = \hat{\mathbf{a}}_r E_r + \hat{\mathbf{a}}_\theta E_\theta \qquad (4\text{-}24)$$

whose magnitude can be written as

$$|\mathbf{E}| = \sqrt{|E_r|^2 + |E_\theta|^2} \qquad (4\text{-}25)$$

4.2.6 Far-Field ($kr \gg 1$) Region

Since (4-23a)–(4-23d) are valid only for values of $kr > 1$ ($r > \lambda$), then E_r will be smaller than E_θ because E_r is inversely proportional to r^2 where E_θ is inversely proportional to r. In a region where $kr \gg 1$, (4-23a)–(4-23d) can be simplified and approximated by

$$E_\theta \simeq j\eta \frac{kI_0 l e^{-jkr}}{4\pi r} \sin\theta \quad \quad (4\text{-}26a)$$

$$E_r \simeq E_\phi = H_r = H_\theta = 0 \quad \Big\} \quad kr \gg 1 \quad (4\text{-}26b)$$

$$H_\phi \simeq j\frac{kI_0 l e^{-jkr}}{4\pi r} \sin\theta \quad \quad (4\text{-}26c)$$

The ratio of E_θ to H_ϕ is equal to

$$Z_w = \frac{E_\theta}{H_\phi} \simeq \eta \quad \quad (4\text{-}27)$$

where

Z_w = wave impedance

η = intrinsic impedance ($377 \simeq 120\pi$ ohms for free-space)

The **E**- and **H**-field components are perpendicular to each other, transverse to the radial direction of propagation, and the r variations are separable from those of θ and ϕ. The shape of the pattern is not a function of the radial distance r, and the fields form a Transverse ElectroMagnetic (TEM) wave whose wave impedance is equal to the intrinsic impedance of the medium. As it will become even more evident in later chapters, *this relationship is applicable in the far-field region of all antennas of finite dimensions.* Equations (4-26a)–(4-26c) can also be derived using the procedure outlined and relationships developed in Section 3.6. This is left as an exercise to the reader (Prob. 4.15).

Example 4.2

For an infinitesimal dipole determine and interpret the vector effective length [see Section 2.15, Figure 2.29(a)]. At what incidence angle does the open-circuit maximum voltage occurs at the output terminals of the dipole if the electric-field intensity of the incident wave is 10 mV/m? The length of the dipole is 10 cm.

Solution: Using (4-26a) and the effective length as defined by (2-92), we can write that

$$E_\theta = j\eta \frac{kI_0 l e^{-jkr}}{4\pi r} \sin\theta = -\hat{\mathbf{a}}_\theta j\eta \frac{kI_0 e^{-jkr}}{4\pi r} \cdot (-\hat{\mathbf{a}}_\theta l \sin\theta)$$

$$= -\hat{\mathbf{a}}_\theta j\eta \frac{kI_0 e^{-jkr}}{4\pi r} \cdot \boldsymbol{\ell}_e$$

Therefore, the effective length is

$$\boldsymbol{\ell}_e = -\hat{\mathbf{a}}_\theta l \sin\theta$$

154   LINEAR WIRE ANTENNAS

whose maximum value occurs when $\theta = 90°$, and it is equal to l. Therefore, to achieve maximum output the wave must be incident upon the dipole at a normal incidence angle ($\theta = 90°$).

The open-circuit maximum voltage is equal to

$$V_{oc}|_{max} = |\mathbf{E}^i \cdot \boldsymbol{\ell}_e|_{max} = |\hat{\mathbf{a}}_\theta 10 \times 10^{-3} \cdot (-\hat{\mathbf{a}}_\theta l \sin\theta)|_{max}$$
$$= 10 \times 10^{-3} l = 10^{-3} \text{ volts}$$

4.2.7 Directivity

The real power P_{rad} radiated by the dipole was found in Section 4.2.2, as given by (4-16). The same expression can be obtained by first forming the average power density, using (4-26a)–(4-26c). That is,

$$\mathbf{W}_{av} = \frac{1}{2}\text{Re}(\mathbf{E} \times \mathbf{H}^*) = \hat{\mathbf{a}}_r \frac{1}{2\eta}|E_\theta|^2 = \hat{\mathbf{a}}_r \frac{\eta}{2}\left|\frac{kI_0 l}{4\pi}\right|^2 \frac{\sin^2\theta}{r^2} \qquad (4\text{-}28)$$

Integrating (4-28) over a closed sphere of radius r reduces it to (4-16). This is left as an exercise to the reader (Prob. 4.15).

Associated with the average power density of (4-28) is a radiation intensity U which is given by

$$U = r^2 W_{av} = \frac{\eta}{2}\left(\frac{kI_0 l}{4\pi}\right)^2 \sin^2\theta = \frac{r^2}{2\eta}|E_\theta(r, \theta, \phi)|^2 \qquad (4\text{-}29)$$

and it conforms with (2-12a). The normalized pattern of (4-29) is shown in Figure 4.3. The maximum value occurs at $\theta = \pi/2$ and it is equal to

$$U_{max} = \frac{\eta}{2}\left(\frac{kI_0 l}{4\pi}\right)^2 \qquad (4\text{-}30)$$

Using (4-16) and (4-30), the directivity reduces to

$$D_0 = 4\pi \frac{U_{max}}{P_{rad}} = \frac{3}{2} \qquad (4\text{-}31)$$

and the maximum effective aperture to

$$A_{em} = \left(\frac{\lambda^2}{4\pi}\right) D_0 = \frac{3\lambda^2}{8\pi} \qquad (4\text{-}32)$$

The radiation resistance of the dipole can be obtained by the definition of (4-18). Since the radiated power obtained by integrating (4-28) over a closed sphere is the same as that of (4-16), the radiation resistance using it will also be the same as obtained previously and given by (4-19).

Integrating the complex Poynting vector over a closed sphere, as was done in (4-13), results in the power (real and imaginary) directed in the radial direction. Any transverse components of power density, as given by (4-12b), will not be captured by the integration even though they are part of the

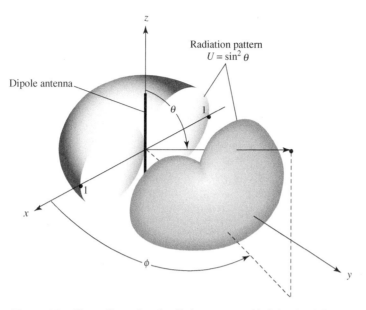

Figure 4.3 Three-dimensional radiation pattern of infinitesimal dipole.

overall power. *Because of this limitation, this method cannot be used to derive the input reactance of the antenna.*

The procedure that can be used to derive the far-zone electric and magnetic fields radiated by an antenna, along with some of the most important parameters/figures of merit that are used to describe the performance of an antenna, are summarized in Table 4.1.

4.3 SMALL DIPOLE

The creation of the current distribution on a thin wire was discussed in Section 1.4, and it was illustrated with some examples in Figure 1.16. The radiation properties of an infinitesimal dipole, which is usually taken to have a length $l \leq \lambda/50$, were discussed in the previous section. Its current distribution was assumed to be constant. Although a constant current distribution is not realizable (other than top-hat-loaded elements), it is a mathematical quantity that is used to represent actual current distributions of antennas that have been incremented into many small lengths.

A better approximation of the current distribution of wire antennas, whose lengths are usually $\lambda/50 < l \leq \lambda/10$, is the triangular variation of Figure 1.16(a). The sinusoidal variations of Figures 1.16(b)–(c) are more accurate representations of the current distribution of any length wire antenna.

The most convenient geometrical arrangement for the analysis of a dipole is usually to have it positioned symmetrically about the origin with its length directed along the z-axis, as shown in Figure 4.4(a). This is not necessary, but it is usually the most convenient. The current distribution of a small dipole ($\lambda/50 < l \leq \lambda/10$) is shown in Figure 4.4(b), and it is given by

$$\mathbf{I}_e(x', y', z') = \begin{cases} \hat{\mathbf{a}}_z I_0 \left(1 - \frac{2}{l} z'\right), & 0 \leq z' \leq l/2 \\ \hat{\mathbf{a}}_z I_0 \left(1 + \frac{2}{l} z'\right), & -l/2 \leq z' \leq 0 \end{cases} \quad (4\text{-}33)$$

where $I_0 = $ constant.

TABLE 4.1 Summary of Procedure to Determine the Far-Field Radiation Characteristics of an Antenna

1. Specify electric and/or magnetic current densities **J**, **M** [physical or equivalent (see Chapter 3, Figure 3.1)]
2. Determine vector potential components A_θ, A_ϕ and/or F_θ, F_ϕ using (3-46)–(3-54) in far field
3. Find far-zone **E** and **H** radiated fields (E_θ, E_ϕ; H_θ, H_ϕ) using (3-58a)–(3-58b)
4. Form either
   a.
   $$\mathbf{W}_{rad}(r,\theta,\phi) = \mathbf{W}_{av}(r,\theta,\phi) = \frac{1}{2}\text{Re}[\mathbf{E} \times \mathbf{H}^*]$$
   $$\simeq \frac{1}{2}\text{Re}\,[(\hat{\mathbf{a}}_\theta E_\theta + \hat{\mathbf{a}}_\phi E_\phi) \times (\hat{\mathbf{a}}_\theta H_\theta^* + \hat{\mathbf{a}}_\phi H_\phi^*)]$$
   $$\mathbf{W}_{rad}(r,\theta,\phi) = \hat{\mathbf{a}}_r \frac{1}{2}\left[\frac{|E_\theta|^2 + |E_\phi|^2}{\eta}\right] = \hat{\mathbf{a}}_r \frac{1}{r^2}|f(\theta,\phi)|^2$$

   or

   b. $U(\theta,\phi) = r^2\, W_{rad}(r,\theta,\phi) = |f(\theta,\phi)|^2$
5. Determine either

   a. $P_{rad} = \displaystyle\int_0^{2\pi}\int_0^\pi W_{rad}(r,\theta,\phi) r^2 \sin\theta\, d\theta\, d\phi$

   or

   b. $P_{rad} = \displaystyle\int_0^{2\pi}\int_0^\pi U(\theta,\phi)\sin\theta\, d\theta\, d\phi$
6. Find directivity using

   $$D(\theta,\phi) = \frac{U(\theta,\phi)}{U_0} = \frac{4\pi U(\theta,\phi)}{P_{rad}}$$

   $$D_0 = D_{max} = D(\theta,\phi)|_{max} = \frac{U(\theta,\phi)|_{max}}{U_0} = \frac{4\pi U(\theta,\phi)|_{max}}{P_{rad}}$$

7. Form *normalized* power amplitude pattern:

   $$P_n(\theta,\phi) = \frac{U(\theta,\phi)}{U_{max}}$$

8. Determine radiation and input resistance:

   $$R_r = \frac{2P_{rad}}{|I_0|^2}; \qquad R_{in} = \frac{R_r}{\sin^2\left(\frac{kl}{2}\right)}$$

9. Determine maximum effective area

   $$A_{em} = \frac{\lambda^2}{4\pi} D_0$$

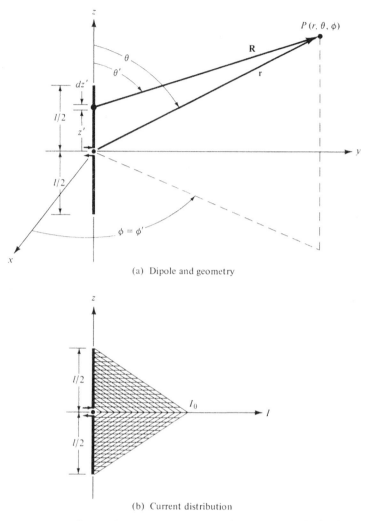

(a) Dipole and geometry

(b) Current distribution

Figure 4.4 Geometrical arrangement of dipole and current distribution.

Following the procedure established in the previous section, the vector potential of (4-2) can be written using (4-33) as

$$\mathbf{A}(x,y,z) = \frac{\mu}{4\pi} \left[\hat{\mathbf{a}}_z \int_{-l/2}^{0} I_0 \left(1 + \frac{2}{l}z'\right) \frac{e^{-jkR}}{R} \, dz' \right. \\ \left. + \hat{\mathbf{a}}_z \int_{0}^{l/2} I_0 \left(1 - \frac{2}{l}z'\right) \frac{e^{-jkR}}{R} \, dz' \right]$$

(4-34)

Because the overall length of the dipole is very small (usually $l \leq \lambda/10$), the values of R for different values of z' along the length of the wire ($-l/2 \leq z' \leq l/2$) are not much different from r. Thus R can be approximated by $R \simeq r$ throughout the integration path. The maximum phase error in (4-34) by allowing $R = r$ for $\lambda/50 < l \leq \lambda/10$, will be $kl/2 = \pi/10$ rad $= 18°$ for $l = \lambda/10$. Smaller values will occur for the other lengths. As it will be shown in the next section, this amount of phase error is usually considered negligible and has very little effect on the overall radiation characteristics.

Performing the integration, (4-34) reduces to

$$\mathbf{A} = \hat{\mathbf{a}}_z A_z = \hat{\mathbf{a}}_z \frac{1}{2} \left[\frac{\mu I_0 l e^{-jkr}}{4\pi r} \right] \quad (4\text{-}35)$$

which is one-half of that obtained in the previous section for the infinitesimal dipole and given by (4-4).

The potential function given by (4-35) becomes a more accurate approximation as $kr \to \infty$. This is also the region of most practical interest, and it has been designated as the *far-field* region. Since the potential function for the triangular distribution is one-half of the corresponding one for the constant (uniform) current distribution, the corresponding fields of the former are one-half of the latter. Thus we can write the **E**- and **H**-fields radiated by a small dipole as

$$E_\theta \simeq j\eta \frac{kI_0 l e^{-jkr}}{8\pi r} \sin\theta \quad (4\text{-}36a)$$

$$E_r \simeq E_\phi = H_r = H_\theta = 0 \quad\quad kr \gg 1 \quad (4\text{-}36b)$$

$$H_\phi \simeq j \frac{kI_0 l e^{-jkr}}{8\pi r} \sin\theta \quad (4\text{-}36c)$$

with the wave impedance equal, as before, to (4-27).

Since the directivity of an antenna is controlled by the relative shape of the field or power pattern, the directivity, and maximum effective area of this antenna are the same as the ones with the constant current distribution given by (4-31) and (4-32), respectively.

The radiation resistance of the antenna is strongly dependent upon the current distribution. Using the procedure established for the infinitesimal dipole, it can be shown that for the small dipole its radiated power is one-fourth ($\frac{1}{4}$) of (4-18). Thus the radiation resistance reduces to

$$R_r = \frac{2P_{\text{rad}}}{|I_0|^2} = 20\pi^2 \left(\frac{l}{\lambda}\right)^2 \quad (4\text{-}37)$$

which is also one-fourth ($\frac{1}{4}$) of that obtained for the infinitesimal dipole as given by (4-19). Their relative patterns (shapes) are the same and are shown in Figure 4.3.

4.4 REGION SEPARATION

Before we attempt to solve for the fields radiated by a finite dipole of any length, it would be very desirable to discuss the separation of the space surrounding an antenna into three regions; namely, the *reactive near-field, radiating near-field* (*Fresnel*) and the *far-field* (*Fraunhofer*) which were introduced briefly in Section 2.2.4. This is necessary because for a dipole antenna of any length and any current distribution, it will become increasingly difficult to solve for the fields everywhere. Approximations can be made, especially for the far-field (Fraunhofer) region, which is usually the one of most practical interest, to simplify the formulation to yield closed form solutions. The same approximations used to simplify the formulation of the fields radiated by a finite dipole are also used to formulate the fields radiated by most practical antennas. So it will be very important to introduce them properly and understand their implications upon the solution.

The difficulties in obtaining closed form solutions that are valid everywhere for any practical antenna stem from the inability to perform the integration of

$$\mathbf{A}(x, y, z) = \frac{\mu}{4\pi} \int_C \mathbf{I}_e(x', y', z') \frac{e^{-jkR}}{R} \, dl' \tag{4-38}$$

where

$$R = \sqrt{(x - x')^2 + (y - y')^2 + (z - z')^2} \tag{4-38a}$$

For a finite dipole with sinusoidal current distribution, the integral of (4-38) can be reduced to a closed form that is valid everywhere! This will be shown in Chapter 8. The length R is defined as the distance from any point on the source to the observation point. The integral of (4-38) was used to solve for the fields of infinitesimal and small dipoles in Sections 4.1 and 4.2. However in the first case (infinitesimal dipole) $R = r$ and in the second case (small dipole) R was approximated by $r(R \simeq r)$ because the length of the dipole was restricted to be $l \leq \lambda/10$. The major simplification of (4-38) will be in the approximation of R.

A very thin dipole of finite length l is symmetrically positioned about the origin with its length directed along the z-axis, as shown in Figure 4.5(a). Because the wire is assumed to be very thin ($x' = y' = 0$), we can write (4-38) as

$$R = \sqrt{(x - x')^2 + (y - y')^2 + (z - z')^2} = \sqrt{x^2 + y^2 + (z - z')^2} \tag{4-39}$$

which when expanded can be written as

$$R = \sqrt{(x^2 + y^2 + z^2) + (-2zz' + z'^2)} = \sqrt{r^2 + (-2rz' \cos\theta + z'^2)} \tag{4-40}$$

where

$$r^2 = x^2 + y^2 + z^2 \tag{4-40a}$$

$$z = r \cos\theta \tag{4-40b}$$

Using the binomial expansion, we can write (4-40) in a series as

$$R = r - z' \cos\theta + \frac{1}{r}\left(\frac{z'^2}{2} \sin^2\theta\right) + \frac{1}{r^2}\left(\frac{z'^3}{2} \cos\theta \sin^2\theta\right) + \cdots \tag{4-41}$$

whose higher order terms become less significant provided $r \gg z'$.

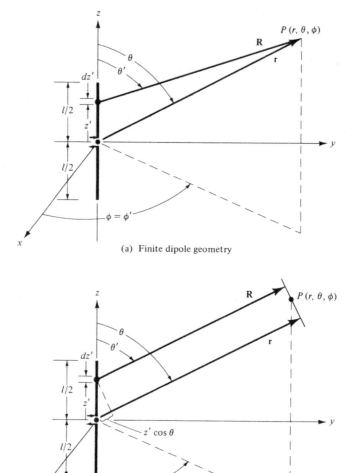

(a) Finite dipole geometry

(b) Geometrical arrangement for far-field approximations

Figure 4.5 Finite dipole geometry and far-field approximations.

4.4.1 Far-Field (Fraunhofer) Region

The most convenient simplification of (4-41), other than $R \simeq r$, will be to approximate it by its first two terms, or

$$R \simeq r - z' \cos\theta \tag{4-42}$$

The most significant neglected term of (4-41) is the third whose maximum value is

$$\frac{1}{r}\left(\frac{z'^2}{2}\sin^2\theta\right)_{max} = \frac{z'^2}{2r} \quad \text{when } \theta = \pi/2 \tag{4-43}$$

When (4-43) attains its maximum value, the fourth term of (4-41) vanishes because $\theta = \pi/2$. It can be shown that the higher order terms not shown in (4-41) also vanish. Therefore approximating (4-41) by (4-42) introduces a *maximum* error given by (4-43).

It has been shown by many investigators through numerous examples that for most practical antennas, *with overall lengths greater than a wavelength* ($l > \lambda$), a maximum total phase error of $\pi/8$ rad (22.5°) is not very detrimental in the analytical formulations. Using that as a criterion we can write, using (4-43), that the maximum phase error should always be

$$\frac{k(z')^2}{2r} \leq \frac{\pi}{8} \tag{4-44}$$

which for $-l/2 \leq z' \leq l/2$ reduces to

$$r \geq 2\left(\frac{l^2}{\lambda}\right) \tag{4-45}$$

Equation (4-45) simply states that to maintain the maximum phase error of an antenna equal to or less than $\pi/8$ rad (22.5°), the observation distance r must equal or be greater than $2l^2/\lambda$ where l is the largest* dimension of the antenna structure. The usual simplification for the far-field region is to approximate the R in the exponential (e^{-jkR}) of (4-38) by (4-42) and the R in the denominator of (4-38) by $R \simeq r$. These simplifications are designated as the far-field approximations and are usually denoted in the literature as

Far-field Approximations

$$\begin{aligned} R &\simeq r - z'\cos\theta \quad &\text{for phase terms} \\ R &\simeq r \quad &\text{for amplitude terms} \end{aligned} \tag{4-46}$$

provided r satisfies (4-45).

It may be advisable to illustrate the approximation (4-46) geometrically. For $R \simeq r - z'\cos\theta$, where θ is the angle measured from the z-axis, the radial vectors **R** and **r** must be parallel to each other, as shown in Figure 4.5(b). For any other antenna whose maximum dimension is D, the approximation of (4-46) is valid provided the observations are made at a distance

$$r \geq 2\frac{D^2}{\lambda} \tag{4-47}$$

For an aperture antenna the maximum dimension is taken to be its diagonal.

For most practical antennas, whose overall length is large compared to the wavelength, these are adequate approximations which have been shown by many investigators through numerous examples to give valid results in pattern predictions. Some discrepancies are evident in regions of low intensity (usually below −25 dB). This is illustrated in Figure 2.9 where the patterns of a paraboloidal antenna for $R = \infty$ and $R = 2D^2/\lambda$ differ at levels below −25 dB. Allowing R to have a value of $R = 4D^2/\lambda$ gives better results.

It would seem that the approximation of R in (4-46) for the amplitude is more severe than that for the phase. However a close observation reveals this is not the case. Since the observations are made at a distance where r is very large, any small error in the approximation of the denominator (amplitude) will not make much difference in the answer. However, because of the periodic nature of the phase (repeats every 2π rad), it can be a major fraction of a period. The best way to illustrate it will be to consider an example.

---

*Provided the overall length (l) of the antenna is large compared to the wavelength [see IEEE Standard Definitions of Terms for Antennas, IEEE Std (145-1983)].

Example 4.3

For an antenna with an overall length $l = 5\lambda$, the observations are made at $r = 60\lambda$. Find the errors in phase and amplitude using (4-46).

Solution: For $\theta = 90°, z' = 2.5\lambda$, and $r = 60\lambda$, (4-40) reduces to

$$R_1 = \lambda\sqrt{(60)^2 + (2.5)^2} = 60.052\lambda$$

and (4-46) to

$$R_2 = r = 60\lambda$$

Therefore the phase difference is

$$\Delta\phi = k\Delta R = \frac{2\pi}{\lambda}(R_1 - R_2) = 2\pi(0.052) = 0.327 \text{ rad} = 18.74°$$

which in an appreciable fraction ($\simeq \frac{1}{20}$) of a full period (360°).
The difference of the inverse values of R is

$$\frac{1}{R_2} - \frac{1}{R_1} = \frac{1}{\lambda}\left(\frac{1}{60} - \frac{1}{60.052}\right) = \frac{1.44 \times 10^{-5}}{\lambda}$$

which should always be a very small value in amplitude.

4.4.2 Radiating Near-Field (Fresnel) Region

If the observation point is chosen to be smaller than $r = 2l^2/\lambda$, the maximum phase error by the approximation of (4-46) is greater than $\pi/8$ rad (22.5°) which may be undesirable in many applications. If it is necessary to choose observation distances smaller than (4-45), another term (the third) in the series solution of (4-41) must be retained to maintain a maximum phase error of $\pi/8$ rad (22.5°). Doing this, the infinite series of (4-41) can be approximated by

$$R \simeq r - z'\cos\theta + \frac{1}{r}\left(\frac{z'^2}{2}\sin^2\theta\right) \quad (4\text{-}48)$$

The most significant term that we are neglecting from the infinite series of (4-41) is the fourth. To find the maximum phase error introduced by the omission of the next most significant term, the angle θ at which this occurs must be found. To do this, the neglected term is differentiated with respect to θ and the result is set equal to zero. Thus

$$\frac{\partial}{\partial\theta}\left[\frac{1}{r^2}\left(\frac{z'^3}{2}\cos\theta\sin^2\theta\right)\right] = \frac{z'^3}{2r^2}\sin\theta[-\sin^2\theta + 2\cos^2\theta] = 0 \quad (4\text{-}49)$$

The angle $\theta = 0$ is not chosen as a solution because for that value the fourth term is equal to zero. In other words, $\theta = 0$ gives the minimum error. The maximum error occurs when the second term of (4-49) vanishes; that is when

$$[-\sin^2\theta + 2\cos^2\theta]_{\theta=\theta_1} = 0 \quad (4\text{-}50)$$

or

$$\theta_1 = \tan^{-1}(\pm\sqrt{2}) \qquad (4\text{-}50a)$$

If the maximum phase error is allowed to be equal or less than $\pi/8$ rad, the distance r at which this occurs can be found from

$$\left.\frac{kz'^3}{2r^2}\cos\theta\sin^2\theta\right|_{\substack{z'=l/2 \\ \theta=\tan^{-1}\sqrt{2}}} = \frac{\pi}{\lambda}\frac{l^3}{8r^2}\left(\frac{1}{\sqrt{3}}\right)\left(\frac{2}{3}\right) = \frac{\pi}{12\sqrt{3}}\left(\frac{l^3}{\lambda r^2}\right) \leq \frac{\pi}{8} \qquad (4\text{-}51)$$

which reduces to

$$r^2 \geq \frac{2}{3\sqrt{3}}\left(\frac{l^3}{\lambda}\right) = 0.385\left(\frac{l^3}{\lambda}\right) \qquad (4\text{-}52)$$

or

$$r \geq 0.62\sqrt{l^3/\lambda} \qquad (4\text{-}52a)$$

A value of r greater than that of (4-52a) will lead to an error less than $\pi/8$ rad (22.5°). Thus the region where the first three terms of (4-41) are significant, and the omission of the fourth introduces a maximum phase error of $\pi/8$ rad (22.5°), is defined by

$$2l^2/\lambda > r \geq 0.62\sqrt{l^3/\lambda} \qquad (4\text{-}53)$$

where l is the length of the antenna. This region is designated as *radiating near-field* because the radiating power density is greater than the reactive power density and the field pattern (its shape) is a function of the radial distance r. This region is also called the *Fresnel region* because the field expressions in this region reduce to Fresnel integrals.

The discussion has centered around the finite length antenna of length l with the observation considered to be a point source. If the antenna is not a line source, l in (4-53) must represent the largest dimension of the antenna (which for an aperture is the diagonal). *Also if the transmitting antenna has maximum length l_t and the receiving antenna has maximum length l_r, then the sum of l_t and l_r must be used in place of l in (4-53).*

The boundaries for separating the far-field (Fraunhofer), the radiating near-field (Fresnel), and the reactive near-field regions are not very rigid. Other criteria have also been established [4] but the ones introduced here are the most "popular." Also the fields, as the boundaries from one region to the other are crossed, do not change abruptly but undergo a very gradual transition.

4.4.3 Reactive Near-Field Region

If the distance of observation is smaller than the inner boundary of the Fresnel region, this region is usually designated as *reactive near-field* with inner and outer boundaries defined by

$$0.62\sqrt{l^3/\lambda} > r > 0 \qquad (4\text{-}54)$$

where l is the length of the antenna. In this region the reactive power density predominates, as was demonstrated in Section 4.1 for the infinitesimal dipole.

In summary, the space surrounding an antenna is divided into three regions whose boundaries are determined by

$$\text{reactive near-field } [0.62\sqrt{D^3/\lambda} > r > 0] \quad (4\text{-}55a)$$

$$\text{radiating near-field (Fresnel) } [2D^2/\lambda > r \geq 0.62\sqrt{D^3/\lambda}] \quad (4\text{-}55b)$$

$$\text{far-field (Fraunhofer) } [\infty \geq r \geq 2D^2/\lambda] \quad (4\text{-}55c)$$

where D is the largest dimension of the antenna ($D = l$ for a wire antenna).

4.5 FINITE LENGTH DIPOLE

The techniques that were developed previously can also be used to analyze the radiation characteristics of a linear dipole of any length. To reduce the mathematical complexities, it will be assumed in this chapter that the dipole has a negligible diameter (ideally zero). This is a good approximation provided the diameter is considerably smaller than the operating wavelength. Finite radii dipoles will be analyzed in Chapters 8 and 9.

4.5.1 Current Distribution

For a very thin dipole (ideally zero diameter), the current distribution can be written, to a good approximation, as

$$\mathbf{I}_e(x' = 0, y' = 0, z') = \begin{cases} \hat{\mathbf{a}}_z I_0 \sin\left[k\left(\frac{l}{2} - z'\right)\right], & 0 \leq z' \leq l/2 \\ \hat{\mathbf{a}}_z I_0 \sin\left[k\left(\frac{l}{2} + z'\right)\right], & -l/2 \leq z' \leq 0 \end{cases} \quad (4\text{-}56)$$

This distribution assumes that the antenna is *center-fed and the current vanishes at the end points* ($z' = \pm l/2$). Experimentally it has been verified that the current in a center-fed wire antenna has sinusoidal form with nulls at the end points. For $l = \lambda/2$ and $\lambda/2 < l < \lambda$ the current distribution of (4-56) is shown plotted in Figures 1.16(b) and 1.12(c), respectively. The geometry of the antenna is that shown in Figure 4.5.

4.5.2 Radiated Fields: Element Factor, Space Factor, and Pattern Multiplication

For the current distribution of (4-56) it will be shown in Chapter 8 that closed form expressions for the **E**- and **H**-fields can be obtained which are valid in all regions (any observation point except on the source itself). In general, however, this is not the case. Usually we are limited to the far-field region, because of the mathematical complications provided in the integration of the vector potential **A** of (4-2). Since closed form solutions, which are valid everywhere, cannot be obtained for many antennas, the observations will be restricted to the far-field region. This will be done first in order to illustrate the procedure. In some cases, even in that region it may become impossible to obtain closed form solutions.

The finite dipole antenna of Figure 4.5 is subdivided into a number of infinitesimal dipoles of length $\Delta z'$. As the number of subdivisions is increased, each infinitesimal dipole approaches a length dz'. For an infinitesimal dipole of length dz' positioned along the z-axis at z', the electric and magnetic field components in the far field are given, using (4-26a)–(4-26c), as

$$dE_\theta \simeq j\eta \frac{kI_e(x',y',z')e^{-jkR}}{4\pi R} \sin\theta \, dz' \tag{4-57a}$$

$$dE_r \simeq dE_\phi = dH_r = dH_\theta = 0 \tag{4-57b}$$

$$dH_\phi \simeq j\frac{kI_e(x',y',z')e^{-jkR}}{4\pi R} \sin\theta \, dz' \tag{4-57c}$$

where R is given by (4-39) or (4-40).

Using the far-field approximations given by (4-46), (4-57a) can be written as

$$dE_\theta \simeq j\eta \frac{kI_e(x',y',z')e^{-jkr}}{4\pi r} \sin\theta \, e^{+jkz'\cos\theta} \, dz' \tag{4-58}$$

Summing the contributions from all the infinitesimal elements, the summation reduces, in the limit, to an integration. Thus

$$E_\theta = \int_{-l/2}^{+l/2} dE_\theta = j\eta \frac{ke^{-jkr}}{4\pi r} \sin\theta \left[\int_{-l/2}^{+l/2} I_e(x',y',z')e^{jkz'\cos\theta} \, dz' \right] \tag{4-58a}$$

The factor outside the brackets is designated as the *element factor* and that within the brackets as the *space factor*. For this antenna, the element factor is equal to the field of a unit length infinitesimal dipole located at a reference point (the origin). In general, the element factor depends on the type of current and its direction of flow while the space factor is a function of the current distribution along the source.

The total field of the antenna is equal to the product of the element and space factors. This is referred to as *pattern multiplication* for continuously distributed sources (see also Chapter 7), and it can be written as

$$\text{total field} = (\text{element factor}) \times (\text{space factor}) \tag{4-59}$$

The pattern multiplication for continuous sources is analogous to the pattern multiplication of (6-5) for discrete-element antennas (arrays).

For the current distribution of (4-56), (4-58a) can be written as

$$E_\theta \simeq j\eta \frac{kI_0 e^{-jkr}}{4\pi r} \sin\theta \left\{ \int_{-l/2}^{0} \sin\left[k\left(\frac{l}{2}+z'\right)\right] e^{+jkz'\cos\theta} \, dz' \right.$$
$$\left. + \int_{0}^{+l/2} \sin\left[k\left(\frac{l}{2}-z'\right)\right] e^{+jkz'\cos\theta} \, dz' \right\} \tag{4-60}$$

Each one of the integrals in (4-60) can be integrated using

$$\int e^{\alpha x} \sin(\beta x + \gamma) \, dx = \frac{e^{\alpha x}}{\alpha^2 + \beta^2} [\alpha \sin(\beta x + \gamma) - \beta \cos(\beta x + \gamma)] \tag{4-61}$$

166 LINEAR WIRE ANTENNAS

where

$$\alpha = \pm jk\cos\theta \tag{4-61a}$$

$$\beta = \pm k \tag{4-61b}$$

$$\gamma = kl/2 \tag{4-61c}$$

After some mathematical manipulations, (4-60) takes the form of

$$E_\theta \simeq j\eta \frac{I_0 e^{-jkr}}{2\pi r}\left[\frac{\cos\left(\frac{kl}{2}\cos\theta\right) - \cos\left(\frac{kl}{2}\right)}{\sin\theta}\right] \tag{4-62a}$$

In a similar manner, or by using the established relationship between the E_θ and H_ϕ in the far field as given by (3-58b) or (4-27), the total H_ϕ component can be written as

$$H_\phi \simeq \frac{E_\theta}{\eta} \simeq j\frac{I_0 e^{-jkr}}{2\pi r}\left[\frac{\cos\left(\frac{kl}{2}\cos\theta\right) - \cos\left(\frac{kl}{2}\right)}{\sin\theta}\right] \tag{4-62b}$$

4.5.3 Power Density, Radiation Intensity, and Radiation Resistance

For the dipole, the average Poynting vector can be written as

$$\mathbf{W}_{av} = \frac{1}{2}\text{Re}[\mathbf{E} \times \mathbf{H}^*] = \frac{1}{2}\text{Re}[\hat{\mathbf{a}}_\theta E_\theta \times \hat{\mathbf{a}}_\phi H_\phi^*] = \frac{1}{2}\text{Re}\left[\hat{\mathbf{a}}_\theta E_\theta \times \hat{\mathbf{a}}_\phi \frac{E_\theta^*}{\eta}\right]$$

$$\mathbf{W}_{av} = \hat{\mathbf{a}}_r W_{av} = \hat{\mathbf{a}}_r \frac{1}{2\eta}|E_\theta|^2 = \hat{\mathbf{a}}_r \eta \frac{|I_0|^2}{8\pi^2 r^2}\left[\frac{\cos\left(\frac{kl}{2}\cos\theta\right) - \cos\left(\frac{kl}{2}\right)}{\sin\theta}\right]^2 \tag{4-63}$$

and the radiation intensity as

$$U = r^2 W_{av} = \eta \frac{|I_0|^2}{8\pi^2}\left[\frac{\cos\left(\frac{kl}{2}\cos\theta\right) - \cos\left(\frac{kl}{2}\right)}{\sin\theta}\right]^2 \tag{4-64}$$

The normalized (to 0 dB) elevation power patterns, as given by (4-64) for $l = \lambda/4, \lambda/2, 3\lambda/4$, and λ are shown plotted in Figure 4.6. The current distribution of each is given by (4-56). The power patterns for an infinitesimal dipole $l \ll \lambda$ ($U \sim \sin^2\theta$) is also included for comparison. As the length of the antenna increases, the beam becomes narrower. Because of that, the directivity should also

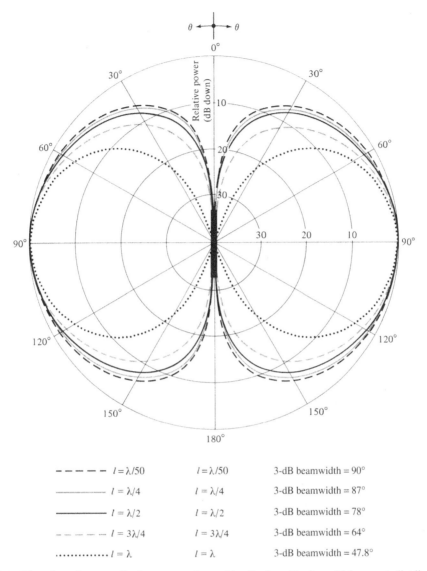

Figure 4.6 Elevation plane amplitude patterns for a thin dipole with sinusoidal current distribution ($l = \lambda/50, \lambda/4, \lambda/2, 3\lambda/4, \lambda$).

increase with length. It is found that the 3-dB beamwidth of each is equal to

$$
\begin{array}{ll}
l \ll \lambda & \text{3-dB beamwidth} = 90° \\
l = \lambda/4 & \text{3-dB beamwidth} = 87° \\
l = \lambda/2 & \text{3-dB beamwidth} = 78° \\
l = 3\lambda/4 & \text{3-dB beamwidth} = 64° \\
l = \lambda & \text{3-dB beamwidth} = 47.8°
\end{array}
\tag{4-65}
$$

As the length of the dipole increases beyond one wavelength ($l > \lambda$), the number of lobes begin to increase. The normalized power pattern for a dipole with $l = 1.25\lambda$ is shown in Figure 4.7. In

168 LINEAR WIRE ANTENNAS

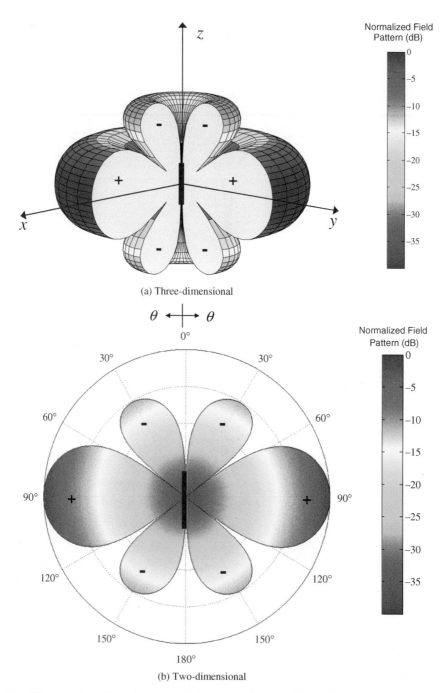

(a) Three-dimensional

(b) Two-dimensional

Figure 4.7 Three- and two-dimensional amplitude patterns for a thin dipole of $l = 1.25\lambda$ and sinusoidal current distribution.

Figure 4.7(a) the three-dimensional pattern in color is illustrated, while in Figure 4.7(b) the two-dimensional (elevation pattern) in color is depicted. For the three-dimensional illustration, a 90° angular section of the pattern has been omitted to illustrate the elevation plane directional pattern variations. The current distribution for the dipoles with $l = \lambda/4, \lambda/2, \lambda, 3\lambda/2$, and 2λ, as given by (4-56), is shown in Figure 4.8.

FINITE LENGTH DIPOLE **169**

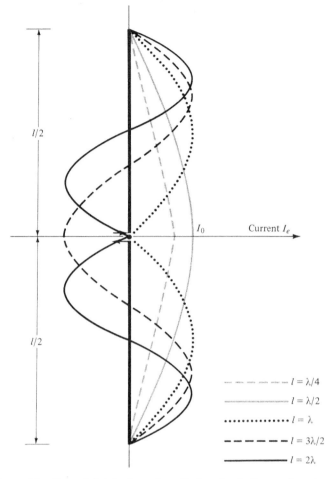

Figure 4.8 Current distributions along the length of a linear wire antenna.

To find the total power radiated, the average Poynting vector of (4-63) is integrated over a sphere of radius r. Thus

$$P_{\text{rad}} = \oiint_S \mathbf{W}_{\text{av}} \cdot d\mathbf{s} = \int_0^{2\pi} \int_0^{\pi} \hat{\mathbf{a}}_r W_{\text{av}} \cdot \hat{\mathbf{a}}_r r^2 \sin\theta \, d\theta \, d\phi$$

$$= \int_0^{2\pi} \int_0^{\pi} W_{\text{av}} r^2 \sin\theta \, d\theta \, d\phi \qquad (4\text{-}66)$$

Using (4-63), we can write (4-66) as

$$P_{\text{rad}} = \int_0^{2\pi} \int_0^{\pi} W_{\text{av}} r^2 \sin\theta \, d\theta \, d\phi$$

$$= \eta \frac{|I_0|^2}{4\pi} \int_0^{\pi} \frac{\left[\cos\left(\frac{kl}{2}\cos\theta\right) - \cos\left(\frac{kl}{2}\right)\right]^2}{\sin\theta} \, d\theta \qquad (4\text{-}67)$$

170 LINEAR WIRE ANTENNAS

After some extensive mathematical manipulations, it can be shown that (4-67) reduces to

$$P_{\text{rad}} = \eta \frac{|I_0|^2}{4\pi} \{C + \ln(kl) - C_i(kl) + \tfrac{1}{2}\sin(kl)[S_i(2kl) - 2S_i(kl)]$$
$$+ \tfrac{1}{2}\cos(kl)[C + \ln(kl/2) + C_i(2kl) - 2C_i(kl)]\} \tag{4-68}$$

where $C = 0.5772$ (Euler's constant) and $C_i(x)$ and $S_i(x)$ are the cosine and sine integrals (see Appendix III) given by

$$C_i(x) = -\int_x^\infty \frac{\cos y}{y}\, dy = \int_\infty^x \frac{\cos y}{y}\, dy \tag{4-68a}$$

$$S_i(x) = \int_0^x \frac{\sin y}{y}\, dy \tag{4-68b}$$

The derivation of (4-68) from (4-67) is assigned as a problem at the end of the chapter (Prob. 4.22). $C_i(x)$ is related to $C_{in}(x)$ by

$$C_{in}(x) = \ln(\gamma x) - C_i(x) = \ln(\gamma) + \ln(x) - C_i(x)$$
$$= 0.5772 + \ln(x) - C_i(x) \tag{4-69}$$

where

$$C_{in}(x) = \int_0^x \left(\frac{1 - \cos y}{y}\right) dy \tag{4-69a}$$

$C_i(x)$, $S_i(x)$ and $C_{in}(x)$ are tabulated in Appendix III.

The radiation resistance can be obtained using (4-18) and (4-68) and can be written as

$$R_r = \frac{2P_{\text{rad}}}{|I_0|^2} = \frac{\eta}{2\pi}\{C + \ln(kl) - C_i(kl)$$
$$+ \tfrac{1}{2}\sin(kl) \times [S_i(2kl) - 2S_i(kl)]$$
$$+ \tfrac{1}{2}\cos(kl) \times [C + \ln(kl/2) + C_i(2kl) - 2C_i(kl)]\} \tag{4-70}$$

Shown in Figure 4.9(a) is a plot of R_r as a function of l (in wavelengths) when the antenna is radiating into free-space ($\eta \simeq 120\pi$).

The imaginary part of the impedance cannot be derived using the same method as the real part because, as was explained in Section 4.2.2, the integration over a closed sphere in (4-13) does not capture the imaginary power contributed by the transverse component W_θ of the power density. Therefore, the EMF method is used in Chapter 8 as an alternative approach. Using the EMF method,

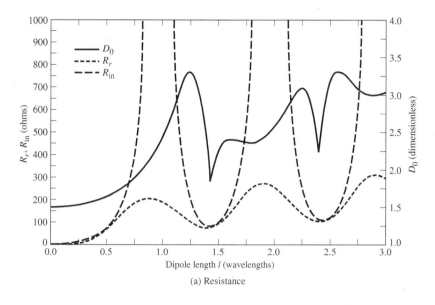

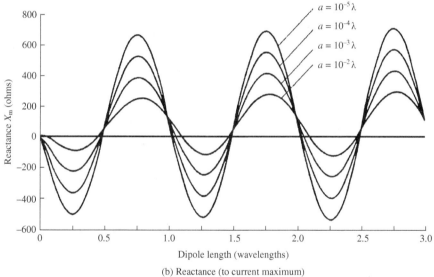

Figure 4.9 Radiation resistance and reactance, input resistance and directivity of a thin dipole with sinusoidal current distribution.

the imaginary part of the impedance, *relative to the current maximum*, is given by (8-57b) or

$$X_m = \frac{\eta}{4\pi} \left\{ 2S_i(kl) + \cos(kl)[2S_i(kl) - S_i(2kl)] \right.$$
$$\left. - \sin(kl) \left[2C_i(kl) - C_i(2kl) - C_i\left(\frac{2ka^2}{l}\right) \right] \right\} \quad (4\text{-}70\text{a})$$

An approximate form of (4-57b) for small dipoles is given by (8-59).

Ideally, the radius of the wire does not affect the input resistance, as is indicated by (4-70). However, in practice, it does, although the wire radius is not as significant as it is for the input reactance.

172 LINEAR WIRE ANTENNAS

To examine the effect the radius has on the values of the reactance, its values, as given by (4-70a), are plotted in Figure 4.9(b) for $a = 10^{-5}\lambda$, $10^{-4}\lambda$, $10^{-3}\lambda$, and $10^{-2}\lambda$. The overall length of the wire is taken to be $0 < l \leq 3\lambda$. The same ones are displayed in Figure 8.17, and they are derived in Chapter 8 based on the EMF method. It is apparent that the reactance can be reduced to zero provided that the overall length is slightly less than $n\lambda/2$, $n = 1, 3, \ldots$, or slightly greater than $n\lambda/2$, $n = 2, 4, \ldots$. This is often done, in practice, for the $l \simeq \lambda/2$ because the input resistance is close to 50 ohms, an almost ideal match for the widely used 50-ohm lines. How much smaller than $\lambda/2$ should it be reduced depends on the radius of the wire; the thicker the radius, the more there needs to be cut off. Typical dipole lengths for the first resonance range around $\lambda \simeq (0.46$–$0.48)\lambda$. For very small radii, the reactance for $l = \lambda/2$ equals 42.5 ohms.

4.5.4 Directivity

As was illustrated in Figure 4.6, the radiation pattern of a dipole becomes more directional as its length increases. When the overall length is greater than one wavelength, the number of lobes increases and the antenna loses its directional properties. The parameter that is used as a "figure of merit" for the directional properties of the antenna is the directivity which was defined in Section 2.6.

The directivity was defined mathematically by (2-22), or

$$D_0 = 4\pi \frac{F(\theta, \phi)|_{\max}}{\int_0^{2\pi} \int_0^{\pi} F(\theta, \phi) \sin\theta \, d\theta \, d\phi} \tag{4-71}$$

where $F(\theta, \phi)$ is related to the radiation intensity U by (2-19), or

$$U = B_0 F(\theta, \phi) \tag{4-72}$$

From (4-64), the dipole antenna of length l has

$$F(\theta, \phi) = F(\theta) = \left[\frac{\cos\left(\frac{kl}{2}\cos\theta\right) - \cos\left(\frac{kl}{2}\right)}{\sin\theta} \right]^2 \tag{4-73}$$

and

$$B_0 = \eta \frac{|I_0|^2}{8\pi^2} \tag{4-73a}$$

Because the pattern is not a function of ϕ, (4-71) reduces to

$$D_0 = \frac{2F(\theta)|_{\max}}{\int_0^{\pi} F(\theta) \sin\theta \, d\theta} \tag{4-74}$$

Equation (4-74) can be written, using (4-67), (4-68), and (4-73), as

$$D_0 = \frac{2F(\theta)|_{\max}}{Q} \tag{4-75}$$

where

$$Q = \{C + \ln(kl) - C_i(kl) + \tfrac{1}{2}\sin(kl)[S_i(2kl) - 2S_i(kl)]$$
$$+ \tfrac{1}{2}\cos(kl)[C + \ln(kl/2) + C_i(2kl) - 2C_i(kl)]\} \quad (4\text{-}75a)$$

The maximum value of $F(\theta)$ varies and depends upon the length of the dipole.

Values of the directivity, as given by (4-75) and (4-75a), have been obtained for $0 < l \leq 3\lambda$ and are shown plotted in Figure 4.9. The corresponding values of the maximum effective aperture are related to the directivity by

$$A_{em} = \frac{\lambda^2}{4\pi} D_0 \quad (4\text{-}76)$$

4.5.5 Input Resistance

In Section 2.13 the input impedance was defined as "the ratio of the voltage to current at a pair of terminals or the ratio of the appropriate components of the electric to magnetic fields at a point." The real part of the input impedance was defined as the input resistance which for a lossless antenna reduces to the radiation resistance, a result of the radiation of real power.

In Section 4.2.2, the radiation resistance of an infinitesimal dipole was derived using the definition of (4-18). The radiation resistance of a dipole of length l with sinusoidal current distribution, of the form given by (4-56), is expressed by (4-70). By this definition, the radiation resistance is referred to the maximum current which for some lengths ($l = \lambda/4, 3\lambda/4, \lambda$, etc.) does not occur at the input terminals of the antenna (see Figure 4.8). To refer the radiation resistance to the input terminals of the antenna, the antenna itself is first assumed to be lossless ($R_L = 0$). Then the power at the input terminals is equated to the power at the current maximum.

Referring to Figure 4.10, we can write

$$\frac{|I_{in}|^2}{2} R_{in} = \frac{|I_0|^2}{2} R_r \quad (4\text{-}77)$$

or

$$R_{in} = \left[\frac{I_0}{I_{in}}\right]^2 R_r \quad (4\text{-}77a)$$

where
R_{in} = radiation resistance at input (feed) terminals
R_r = radiation resistance at current maximum Eq. (4-70)
I_0 = current maximum
I_{in} = current at input terminals

For a dipole of length l, the current at the input terminals (I_{in}) is related to the current maximum (I_0) referring to Figure 4.10, by

$$I_{in} = I_0 \sin\left(\frac{kl}{2}\right) \quad (4\text{-}78)$$

174   LINEAR WIRE ANTENNAS

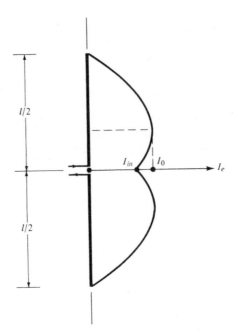

Figure 4.10   Current distribution of a linear wire antenna when current maximum does not occur at the input terminals.

Thus the input radiation resistance of (4-77a) can be written as

$$R_{in} = \frac{R_r}{\sin^2\left(\frac{kl}{2}\right)} \qquad (4\text{-}79)$$

Values of R_{in} for $0 < l \le 3\lambda$ are shown in Figure 4.9(a).

To compute the radiation resistance (in ohms), directivity (dimensionless and in dB), and input resistance (in ohms) for a dipole of length l, a MATLAB and FORTRAN computer program has been developed. The program is based on the definitions of each as given by (4-70), (4-71), and (4-79). The radiated power P_{rad} is computed by numerically integrating (over a closed sphere) the radiation intensity of (4-72)–(4-73a). The program, both in MATLAB and FORTRAN, is included in the publisher's website for this book. The length of the dipole (in wavelengths) must be inserted as an input.

When the overall length of the antenna is a multiple of λ (i.e., $l = n\lambda, n = 1, 2, 3, \ldots$), it is apparent from (4-56) and from Figure 4.8 that $I_{in} = 0$. That is,

$$I_{in} = I_0 \sin\left[k\left(\frac{l}{2} \pm z'\right)\right]\Bigg|_{\substack{z'=0 \\ l=n\lambda, n=0,1,2,\ldots}} = 0 \qquad (4\text{-}80)$$

which indicates that the radiation resistance at the input terminals, as given by (4-77a) or (4-79) is infinite. In practice this is not the case because the current distribution does not follow an exact sinusoidal distribution, especially at the feed point. It has, however, very high values (see Figure 4.11). Two of the primary factors which contribute to the nonsinusoidal current distribution on an actual wire antenna are the nonzero radius of the wire and finite gap spacing at the terminals.

The radiation resistance and input resistance, as predicted, respectively, by (4-70) and (4-79), are based on the ideal current distribution of (4-56) and do not account for the finite radius of the wire or the gap spacing at the feed. Although the radius of the wire does not strongly influence the resistances, the gap spacing at the feed does play a significant role especially when the current at and near the feed point is small.

4.5.6 Finite Feed Gap

To analytically account for a nonzero current at the feed point for antennas with a finite gap at the terminals, Schelkunoff and Friis [6] have changed the current of (4-56) by including a quadrature term in the distribution. The additional term is inserted to take into account the effects of radiation on the antenna current distribution. In other words, once the antenna is excited by the "ideal" current distribution of (4-56), electric and magnetic fields are generated which in turn disturb the "ideal" current distribution. This reaction is included by modifying (4-56) to

$$\mathbf{I}_e(x', y', z') = \begin{cases} \hat{\mathbf{a}}_z \left\{ I_0 \sin\left[k\left(\frac{l}{2} - z'\right)\right] + jpI_0 \left[\cos(kz') - \cos\left(\frac{k}{2}l\right)\right] \right\}, \\ \qquad 0 \leq z' \leq l/2 \\ \hat{\mathbf{a}}_z \left\{ I_0 \sin\left[k\left(\frac{l}{2} + z'\right)\right] + jpI_0 \left[\cos(kz') - \cos\left(\frac{k}{2}l\right)\right] \right\}, \\ \qquad -l/2 \leq z' \leq 0 \end{cases} \quad (4\text{-}81)$$

where p is a coefficient that is dependent upon the overall length of the antenna and the gap spacing at the terminals. The values of p become smaller as the radius of the wire and the gap decrease.

When $l = \lambda/2$,

$$\mathbf{I}_e(x', y', z') = \hat{\mathbf{a}}_z I_0 (1 + jp) \cos(kz') \quad 0 \leq |z'| \leq \lambda/4 \qquad (4\text{-}82)$$

and for $l = \lambda$

$$\mathbf{I}_e(x', y', z') = \begin{cases} \hat{\mathbf{a}}_z I_0 \{ \sin(kz') + jp[1 + \cos(kz')] \} & 0 \leq z' \leq \lambda/2 \\ \hat{\mathbf{a}}_z I_0 \{ -\sin(kz') + jp[1 + \cos(kz')] \} & -\lambda/2 \leq z' \leq 0 \end{cases} \qquad (4\text{-}83)$$

Thus for $l = \lambda/2$ the shape of the current is not changed while for $l = \lambda$ it is modified by the second term which is more dominant for small values of z'. The current distribution based on (4-83) is displayed in Figure 4.11.

The variations of the current distribution and impedances, especially of wire-type antennas, as a function of the radius of the wire and feed gap spacing can be easily taken into account by using advanced computational methods and numerical techniques, especially Integral Equations and Moment Method [7]–[12], which are introduced in Chapter 8.

To illustrate the point, the current distribution of an $l = \lambda/2$ and $l = \lambda$ dipole has been computed using an integral equation formulation with a moment method numerical solution, and it is shown in Figure 8.13(b) where it is compared with the ideal distribution of (4-56) and other available data. For the moment method solution, a gap at the feed has been inserted. As expected and illustrated in Figure 8.13(b), the current distribution for the $l = \lambda/2$ dipole based on (4-56) is not that different from that based on the moment method. This is also illustrated by (4-82). Therefore the input resistance based on these two methods will not be that different. However, for the $l = \lambda$ dipole, the current distribution based on (4-56) is quite different, especially at and near the feed point, compared to that based on the moment method, as shown in Figure 8.13(b). This is expected since the current distribution based on the ideal current distribution is zero at the feed point; for practical antennas it is very small. Therefore the gap at the feed plays an important role on the current distribution at and near the feed point. In turn, the values of the input resistance based on the two methods will be quite different, since there is a significant difference in the current between the two methods. This is discussed further in Chapter 8.

176 LINEAR WIRE ANTENNAS

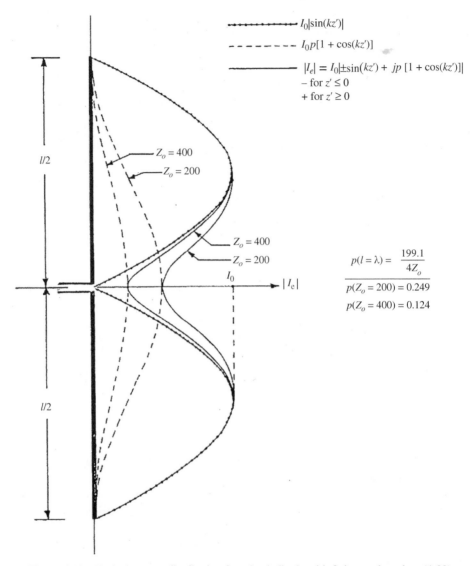

Figure 4.11 Typical current distribution for a $l = \lambda$ dipole with finite gap based on (4-83).

4.6 HALF-WAVELENGTH DIPOLE

One of the most commonly used antennas is the half-wavelength ($l = \lambda/2$) dipole. Because its radiation resistance is 73 ohms, which is very near the 50-ohm or 75-ohm characteristic impedances of some transmission lines, its matching to the line is simplified especially at resonance. Because of its wide acceptance in practice, we will examine in a little more detail its radiation characteristics.

The electric and magnetic field components of a half-wavelength dipole can be obtained from (4-62a) and (4-62b) by letting $l = \lambda/2$. Doing this, they reduce to

$$E_\theta \simeq j\eta \frac{I_0 e^{-jkr}}{2\pi r} \left[\frac{\cos\left(\frac{\pi}{2}\cos\theta\right)}{\sin\theta} \right] \qquad (4\text{-}84)$$

$$H_\phi \simeq j\frac{I_0 e^{-jkr}}{2\pi r} \left[\frac{\cos\left(\frac{\pi}{2}\cos\theta\right)}{\sin\theta}\right] \quad (4\text{-}85)$$

In turn, the time-average power density and radiation intensity can be written, respectively, as

$$W_{av} = \eta\frac{|I_0|^2}{8\pi^2 r^2}\left[\frac{\cos\left(\frac{\pi}{2}\cos\theta\right)}{\sin\theta}\right]^2 \simeq \eta\frac{|I_0|^2}{8\pi^2 r^2}\sin^3\theta \quad (4\text{-}86)$$

and

$$U = r^2 W_{av} = \eta\frac{|I_0|^2}{8\pi^2}\left[\frac{\cos\left(\frac{\pi}{2}\cos\theta\right)}{\sin\theta}\right]^2 \simeq \eta\frac{|I_0|^2}{8\pi^2}\sin^3\theta \quad (4\text{-}87)$$

whose two-dimensional pattern is shown plotted in Figure 4.6 while the three-dimensional pattern is depicted in Figure 4.12a. For the three-dimensional pattern of Figure 4.12a, a 90° angular sector has been removed to illustrate the figure-eight elevation plane pattern variations.

The radiation intensity of the $\lambda/2$ dipole can be approximated by a sine function with integer exponent of three, as represented in (4-87); that is, $U \sim \sin^3\theta$. Actually, noninteger exponent values, slightly less than three, match the exact pattern even better. This is indicated in Figure 4.12(b) where a sine function, with exponent values of 2.6 and 2.8, is used to plot the normalized pattern of (4-87) and to compare it to that of $\sin^3\theta$. It is apparent that a noninteger exponent of nearly 2.6 for the sine function is a better match to the exact pattern.

The total power radiated can be obtained as a special case of (4-67), or

$$P_{rad} = \eta\frac{|I_0|^2}{4\pi}\int_0^\pi \frac{\cos^2\left(\frac{\pi}{2}\cos\theta\right)}{\sin\theta}d\theta \quad (4\text{-}88)$$

which when integrated reduces, as a special case of (4-68), to

$$P_{rad} = \eta\frac{|I_0|^2}{8\pi}\int_0^{2\pi}\left(\frac{1-\cos y}{y}\right)dy = \eta\frac{|I_0|^2}{8\pi}C_{in}(2\pi) \quad (4\text{-}89)$$

By the definition of $C_{in}(x)$, as given by (4-69), $C_{in}(2\pi)$ is equal to

$$C_{in}(2\pi) = 0.5772 + \ln(2\pi) - C_i(2\pi) = 0.5772 + 1.838 - (-0.02) \simeq 2.435 \quad (4\text{-}90)$$

where $C_i(2\pi)$ is obtained from the tables in Appendix III.

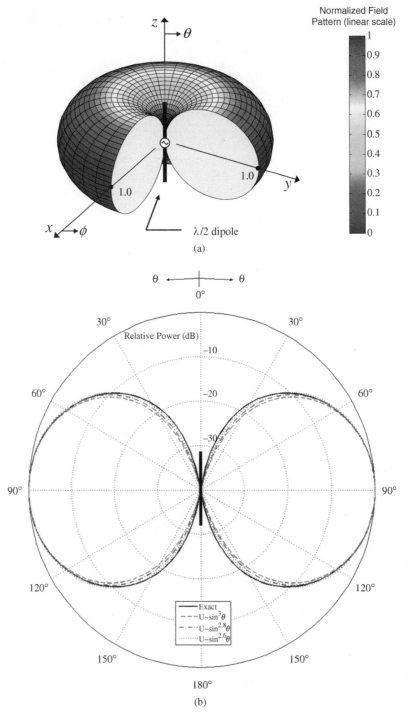

Figure 4.12 Three- and two-dimensional patterns of a $\lambda/2$ dipole (a) three-dimensional pattern of a $\lambda/2$ dipole. (b) comparison of two-dimensional patterns for a $\lambda/2$ dipole.

Using (4-87), (4-89), and (4-90), the maximum directivity of the half-wavelength dipole reduces to

$$D_0 = 4\pi \frac{U_{\max}}{P_{rad}} = 4\pi \frac{U|_{\theta=\pi/2}}{P_{rad}} = \frac{4}{C_{in}(2\pi)} = \frac{4}{2.435} \simeq 1.643 \qquad (4\text{-}91)$$

The corresponding maximum effective area is equal to

$$A_{em} = \frac{\lambda^2}{4\pi} D_0 = \frac{\lambda^2}{4\pi}(1.643) \simeq 0.13\lambda^2 \qquad (4\text{-}92)$$

and the radiation resistance, for a free-space medium ($\eta \simeq 120\pi$), is given by

$$R_r = \frac{2P_{rad}}{|I_0|^2} = \frac{\eta}{4\pi} C_{in}(2\pi) = 30(2.435) \simeq 73 \qquad (4\text{-}93)$$

The radiation resistance of (4-93) is also the radiation resistance at the input terminals (input resistance) since the current maximum for a dipole of $l = \lambda/2$ occurs at the input terminals (see Figure 4.8). As it will be shown in Chapter 8, the imaginary part (reactance) associated with the input impedance of a dipole is a function of its length (for $l = \lambda/2$, it is equal to $j42.5$). Thus the total input impedance for $l = \lambda/2$ is equal to

$$Z_{in} = 73 + j42.5 \qquad (4\text{-}93a)$$

To reduce the imaginary part of the input impedance to zero, the antenna is matched or reduced in length until the reactance vanishes. The latter is most commonly used in practice for half-wavelength dipoles.

Depending on the radius of the wire, the length of the dipole for first resonance is about $l = 0.47\lambda$ to 0.48λ; the thinner the wire, the closer the length is to 0.48λ. Thus, for thicker wires, a larger segment of the wire has to be removed from $\lambda/2$ to achieve resonance. The variations of the reactance as a function of the dipole length l, for different wire radii, are displayed in Figures 4.9(b) and 8.17. A summary of the dipole directivity, gain and realized gain are listed in Table 4.2.

4.7 LINEAR ELEMENTS NEAR OR ON INFINITE PERFECT ELECTRIC CONDUCTORS (PEC), PERFECT MAGNETIC CONDUCTORS (PMC) AND ELECTROMAGNETIC BAND-GAP (EBG) SURFACES

Thus far we have considered the radiation characteristics of antennas radiating into an unbounded medium. The presence of an obstacle, especially when it is near the radiating element, can significantly alter the overall radiation properties of the antenna system. In practice the most common obstacle that is always present, even in the absence of anything else, is the ground. Any energy from the radiating element directed toward the ground undergoes a reflection. The amount of reflected energy and its direction are controlled by the geometry and constitutive parameters of the ground.

In general, the ground is a lossy medium ($\sigma \neq 0$) whose effective conductivity increases with frequency. Therefore it should be expected to act as a very good conductor above a certain frequency, depending primarily upon its composition and moisture content. To simplify the analysis, it will first

TABLE 4.2 Summary of Dipole Directivity, Gain and Realized Gain (Resonant $X_A = 0; f = 100$ MHz; $\sigma = 5.7 \times 10^7$ S/m; $Z_c = 50; b = 3 \times 10^{-4}\lambda$)

	$l = \lambda/50$	$l = \lambda/10$	$l = \lambda/2$	$l = \lambda$
R_{hf}	0.0279	0.2792	0.698	1.3692
R_L	0.0279	0.1396	0.349	0.6981
R_r	0.3158	1.9739	73	199
R_{in}	0.3158	1.9739	73	∞
e_{cd}	0.9188	0.9339	0.9952	0.9965
	(−0.368 dB)	(−0.296 dB)	(−0.021 dB)	(−0.015 dB)
D_0	1.5	1.5	1.6409	2.411
	(1.761 dB)	(1.761 dB)	(2.151 dB)	(3.822 dB)
G_0	1.3782	1.4009	1.6331	2.4026
	(1.393 dB)	(1.464 dB)	(2.13 dB)	(3.807 dB)
Γ	−0.9863	−0.9189	0.18929	1
e_r	0.0271	0.1556	0.9642	0
	(−15.67 dB)	(−8.08 dB)	(−0.158 dB)	(−∞ dB)
G_{re0}	0.0374	0.2181	1.5746	0
	(−14.27 dB)	(−6.613 dB)	(1.972 dB)	(−∞ dB)

be assumed that the ground is a perfect electric conductor, flat, and infinite in extent. The effects of finite conductivity and earth curvature will be incorporated later. The same procedure can also be used to investigate the characteristics of any radiating element near any other infinite, flat, perfect electric conductor. Although infinite structures are not realistic, the developed procedures can be used to simulate very large (electrically) obstacles. The effects that finite dimensions have on the radiation properties of a radiating element can be conveniently accounted for by the use of the Geometrical Theory of Diffraction (Chapter 12, Section 12.10) and/or the Moment Method (Chapter 8, Section 8.4).

Magnetic conductors are nonphysical, meaning they do not exist in nature. However, in recent years, techniques have been developed to synthesize and fabricate materials which exhibit interesting, attractive, and exciting electromagnetic properties. These properties have captured the attention and imagination of leading engineers and scientists in academia, industry, and government. When such materials are further integrated with electromagnetic devices and interact with electromagnetic waves, they exhibit some unique and intriguing characteristics and phenomena. For example, they can be used to control, advance, and optimize the performance of antennas, microwave components and circuits, transmission lines, scatterers, and optical devices. A brief introduction to engineered synthesized magnetic surfaces, especially as used as ground planes, follows in Section 4.7.1.

4.7.1 Ground Planes: Electric and Magnetic

Surfaces that exhibit ideal electric conducting properties, and accordingly satisfy the electromagnetic boundary conditions such that the tangential components of the electric field vanish over their surface, are usually referred to as Perfect Electric Conductors (PEC). Such surfaces exist in nature and metals, with electric conductivities on the order of 10^7–10^8, are good approximations for most electrical characteristics, especially in their utilization as ground planes for antenna applications. The conductivity can be increased even further by applying superconductivity technology [7].

In comparison, materials that exhibit ideal magnetic conductivities such that the tangential components of the magnetic field vanish over their surface, although used previously as equivalents in electromagnetic boundary value problems, do not exist in nature and are nonphysical [13]. Yet, in recent years, technologies have been developed to synthesize and fabricate materials which exhibit interesting, attractive and exciting electromagnetic properties that have captured the attention and imagination of leading engineers and scientists from academia, industry and government. When such materials are integrated with electromagnetic devices and interact with electromagnetic waves, they exhibit some unique and intriguing characteristics and phenomena which can be used, for example, to control, advance, and optimize the performance of antennas, microwave components and circuits, transmission lines, scatterers, and optical devices. Examples as to how the amplitude patterns of a monopole and an aperture are influenced and controlled by such artificially synthesized surfaces are illustrated in [7].

In general, materials that do not exist in nature, but can be artificially synthesized, are referred to as *metamaterials* (beyond materials; *meta* in Greek for beyond/after) [15]. Such synthesized surfaces behave as nearly magnetic conductors only over a limited frequency range, and this limited range is often referred to as *band-gap*, although technologies are being pursued to advance and extend their frequency range [16]–[17]. There have been many other designations for such materials as well:

- AMC (artificial magnetic conductors)
- AIS (artificial impedance surfaces)
- EES (engineered electromagnetic surfaces)
- PBG (photonic band-gap)
- EBG (electromagnetic band-gap)
- HIS (high impedance surfaces)

There are too many different types of synthesized magnetic conductors to list them all here; a number of them are mentioned in [7], [13]–[23]. An extensive discussion and full list can be found in [7].

A surface can be synthesized to exhibit nearly magnetic properties by modifying its geometry and/or adding other layers, so that the surface waves and/or the phase of the reflection coefficient of the modified surface can be controlled. Although the magnitude of the reflection coefficient will also be affected, it is the phase that primarily has the largest impact, especially when the surface is used as a ground plane. While an ideal PMC surface introduces, through its image, a zero-phase shift in the reflected field, in contrast to a PEC, which presents a 180° phase shift, the reflection phase of an EBG surface can, in general, vary from −180° to +180°, which makes the EBG more versatile and unique [7].

One of the first and most widely utilized PMC surfaces is that shown in Figure 4.13. This surface consists of an array of periodic patches of different shapes, in this case, squares, placed above a very thin substrate (which can be air) and connected to the ground plane by posts through vias, if an actual substrate is utilized. The height of the substrate is usually less than a tenth of a wavelength ($h < \lambda/10$). The vias are necessary to suppress surface waves within the substrate. This surface structure is also referred to as EBG and PBG. It is a practical form of engineered textured surfaces or metamaterials. Because of the directional characteristics of EBG/PBG structures, integration of antenna elements with such structures can have some unique characteristics. A semi-empirical model of the mushroom EBG surface in Figure 4.13 was developed in [21], [22]; also reported in [7]. While EBG surfaces exhibit similar characteristics to PMCs when radiating elements are mounted on them, they have the additional ability to suppress surface waves of low-profile antenna designs, such as microstrip arrays. The surface waves introduced in microstrip arrays primarily travel within the substrate and are instrumental in developing coupling between the array elements. This can limit the beam scanning capabilities of the microstrip arrays; ultimately, surface waves and coupling may even lead to *scan blindness* (as discussed in Chapter 14, Section 14.8). When a plane wave is normally incident upon a surface, such as that of Figure 4.13 with a surface impedance Z_s, the +90° to −90° phase variation is also evident when the magnitude of the surface impedance exceeds the free-space intrinsic impedance η_0 [21, 22]. An EBG surface that does not include the vias does not suppress the surface waves, even though its reflection phase changes between +180° and −180°.

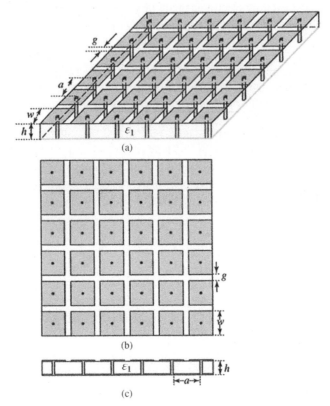

Figure 4.13 Geometry of PMC textured surface of square patches [7]. (a) Perspective view. (b) Top view. (c) Side view.

The performance of the mushroom PMC surface is verified based on the specified and obtained geometrical dimensions. The plane wave normal incidence reflection phase variations of S_{11} of the mushroom textured surface of square patches of Figure 4.13, between +90° and −90°, are shown in Figure 4.14 where they are compared with the results based on the design equations of Section 8.8.4 of [7]. Very good agreement is apparent between the two. The simulated data indicate a bandwidth of 3.9 GHz (f_l = 10.35 GHz and f_h = 14.25 GHz), compared to the specified one of 4 GHz (f_l = 10 GHz and f_h = 14 GHz), a center frequency of 12.15 GHz (compared to 12 GHz), and a fractional bandwidth of 0.321 (compared to 0.333). Overall, the performance indicates the design to be very favorable.

4.7.2 Image Theory

To analyze the performance of an antenna near an infinite plane conductor, virtual sources (images) will be introduced to account for the reflections. As the name implies, these are not real sources but imaginary ones, which when combined with the real sources, form an equivalent system. For analysis purposes only, the equivalent system gives the same radiated field on and above the conductor as the actual system itself. Below the conductor, the equivalent system does not give the correct field. However, in this region the field is zero and there is no need for the equivalent.

To begin the discussion, let us assume that a vertical electric dipole is placed a distance h above an infinite, flat, perfect electric conductor as shown in Figure 4.15(a). The arrow indicates the polarity of the source. Energy from the actual source is radiated in all directions in a manner determined by its unbounded medium directional properties. For an observation point P_1, there is a direct wave. In addition, a wave from the actual source radiated toward point R_1 of the interface undergoes a reflection. The direction is determined by the law of reflection ($\theta_1^i = \theta_1^r$) which assures that the energy

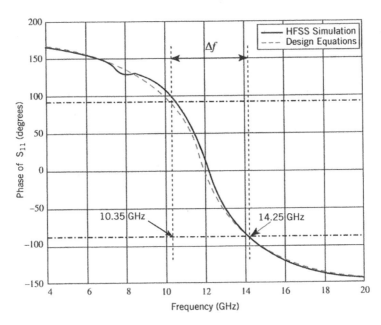

Figure 4.14 Phase of reflection coefficient S_{11} of PMC textured surface with square patches simulated using HFSS and design equations [7].

in homogeneous media travels in straight lines along the shortest paths. This wave will pass through the observation point P_1. By extending its actual path below the interface, it will seem to originate from a virtual source positioned a distance h below the boundary. For another observation point P_2 the point of reflection is R_2, but the virtual source is the same as before. The same is concluded for all other observation points above the interface.

The amount of reflection is generally determined by the respective constitutive parameters of the media below and above the interface. For a perfect electric conductor below the interface, the incident wave is completely reflected and the field below the boundary is zero. According to the boundary conditions, the tangential components of the electric field must vanish at all points along the interface. Thus for an incident electric field with vertical polarization shown by the arrows, the polarization of the reflected waves must be as indicated in the figure to satisfy the boundary conditions. To excite the polarization of the reflected waves, the virtual source must also be vertical and with a polarity in the same direction as that of the actual source (thus a reflection coefficient of +1).

Another orientation of the source will be to have the radiating element in a horizontal position, as shown in Figure 4.27. Following a procedure similar to that of the vertical dipole, the virtual source (image) is also placed a distance h below the interface but with a 180° polarity difference relative to the actual source (thus a reflection coefficient of −1).

In addition to electric sources, nonphysical equivalent "magnetic" sources and magnetic conductors have been introduced to aid in the analyses of electromagnetic boundary-value problems. Figure 4.16(a) displays the sources and their images for an electric plane conductor. The single arrow indicates an electric element and the double a magnetic one. The direction of the arrow identifies the polarity. Since many problems can be solved using duality, Figure 4.16(b) illustrates the sources and their images when the obstacle is an infinite, flat, perfect "magnetic" conductor.

4.7.3 Vertical Electric Dipole

The analysis procedure for vertical and horizontal electric and magnetic elements near infinite electric and magnetic plane conductors, using image theory, was illustrated graphically in the previous

184 LINEAR WIRE ANTENNAS

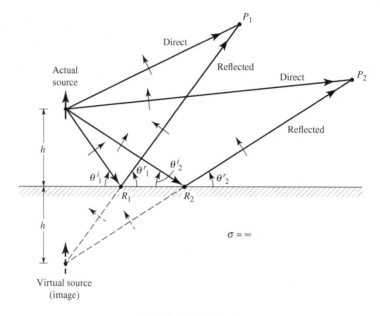

(a) Vertical electric dipole

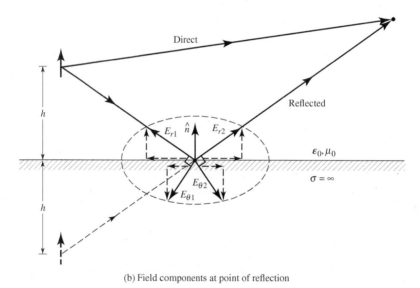

(b) Field components at point of reflection

Figure 4.15 Vertical electric dipole above an infinite, flat, perfect electric conductor.

section. Based on the graphical model of Figure 4.15, the mathematical expressions for the fields of a vertical linear element near a perfect electric conductor will now be developed. For simplicity, only far-field observations will be considered.

Referring to the geometry of Figure 4.15(a), the far-zone direct component of the electric field of the infinitesimal dipole of length l, constant current I_0, and observation point P is given according to (4-26a) by

$$E_\theta^d = j\eta \frac{kI_0 l e^{-jkr_1}}{4\pi r_1} \sin\theta_1 \qquad (4\text{-}94)$$

LINEAR ELEMENTS NEAR OR ON INFINITE PEC, PMC AND EBG SURFACES 185

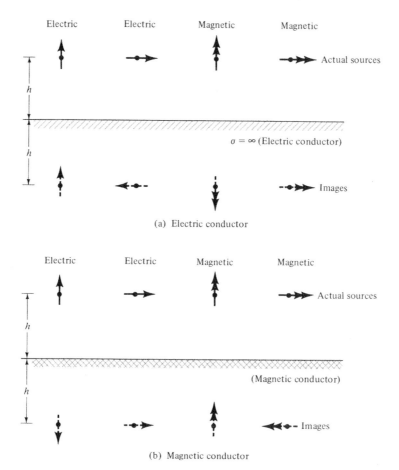

Figure 4.16 Electric and magnetic sources and their images near electric (PEC) and magnetic (PMC) conductors.

The reflected component can be accounted for by the introduction of the virtual source (image), as shown in Figure 4.14(a), and it can be written as

$$E_\theta^r = jR_v \eta \frac{kI_0 l e^{-jkr_2}}{4\pi r_2} \sin\theta_2 \tag{4-95}$$

or

$$E_\theta^r = j\eta \frac{kI_0 l e^{-jkr_2}}{4\pi r_2} \sin\theta_2 \tag{4-95a}$$

since the reflection coefficient R_v is equal to unity.

The total field above the interface ($z \geq 0$) is equal to the sum of the direct and reflected components as given by (4-94) and (4-95a). Since a field cannot exist inside a perfect electric conductor, it is equal to zero below the interface. To simplify the expression for the total electric field, it is referred to the origin of the coordinate system ($z = 0$).

In general, we can write that

$$r_1 = [r^2 + h^2 - 2rh\cos\theta]^{1/2} \quad (4\text{-}96a)$$

$$r_2 = [r^2 + h^2 - 2rh\cos(\pi - \theta)]^{1/2} \quad (4\text{-}96b)$$

For far-field observations ($r \gg h$), (4-96a) and (4-96b) reduce using the binomial expansion to

$$r_1 \simeq r - h\cos\theta \quad (4\text{-}97a)$$

$$r_2 \simeq r + h\cos\theta \quad (4\text{-}97b)$$

As shown in Figure 4.17(b), geometrically (4-97a) and (4-97b) represent parallel lines. Since the amplitude variations are not as critical

$$r_1 \simeq r_2 \simeq r \quad \text{for amplitude variations} \quad (4\text{-}98)$$

Using (4-97a)–(4-98), the sum of (4-94) and (4-95a) can be written as

$$\left. \begin{array}{l} E_\theta \simeq j\eta \dfrac{kI_0 l e^{-jkr}}{4\pi r} \sin\theta [2\cos(kh\cos\theta)] \quad z \geq 0 \\ E_\theta = 0 \quad\quad\quad\quad\quad\quad\quad\quad\quad\quad\quad\quad\quad\quad\quad z < 0 \end{array} \right\} \quad (4\text{-}99)$$

It is evident that the total electric field is equal to the product of the field of a single source positioned symmetrically about the origin and a factor [within the brackets in (4-99)] which is a function of the antenna height (h) and the observation angle (θ). This is referred to as *pattern multiplication* and the factor is known as the *array factor* [see also (6-5)]. This will be developed and discussed in more detail and for more complex configurations in Chapter 6.

The shape and amplitude of the field is not only controlled by the field of the single element but also by the positioning of the element relative to the ground. To examine the field variations as a function of the height h, the normalized (to 0 dB) power patterns for $h = 0, \lambda/8, \lambda/4, 3\lambda/8, \lambda/2$, and λ have been plotted in Figure 4.18. Because of symmetry, only half of each pattern is shown. For $h > \lambda/4$ more minor lobes, in addition to the major ones, are formed. As h attains values greater than λ, an even greater number of minor lobes is introduced. These are shown in Figure 4.19 for $h = 2\lambda$ and 5λ. The introduction of the additional lobes in Figure 4.19 is usually called *scalloping*. In general, the total number of lobes is equal to the integer that is closest to

$$\boxed{\text{number of lobes} \simeq \dfrac{2h}{\lambda} + 1} \quad (4\text{-}100)$$

Since the total field of the antenna system is different from that of a single element, the directivity and radiation resistance are also different. To derive expressions for them, we first find the total radiated power over the upper hemisphere of radius r using

$$P_{\text{rad}} = \oiint_S \mathbf{W}_{\text{av}} \cdot d\mathbf{s} = \dfrac{1}{2\eta} \int_0^{2\pi} \int_0^{\pi/2} |E_\theta|^2 r^2 \sin\theta \, d\theta \, d\phi$$

$$= \dfrac{\pi}{\eta} \int_0^{\pi/2} |E_\theta|^2 r^2 \sin\theta \, d\theta \quad (4\text{-}101)$$

LINEAR ELEMENTS NEAR OR ON INFINITE PEC, PMC AND EBG SURFACES **187**

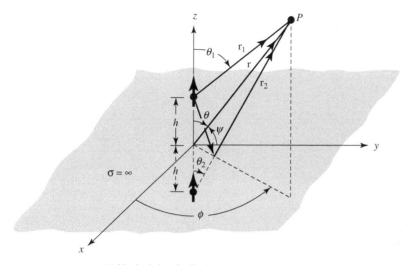

(a) Vertical electric dipole above ground plane

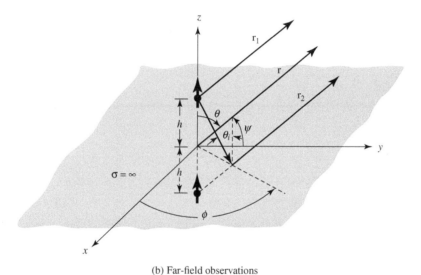

(b) Far-field observations

Figure 4.17 Vertical electric dipole above infinite perfect electric conductor.

which simplifies, with the aid of (4-99), to

$$P_{rad} = \pi\eta \left|\frac{I_0 l}{\lambda}\right|^2 \left[\frac{1}{3} - \frac{\cos(2kh)}{(2kh)^2} + \frac{\sin(2kh)}{(2kh)^3}\right] \quad (4\text{-}102)$$

As $kh \to \infty$ the radiated power, as given by (4-102), is equal to that of an isolated element. However, for $kh \to 0$, it can be shown by expanding the sine and cosine functions into series that the power is twice that of an isolated element. Using (4-99), the radiation intensity can be written as

$$U = r^2 W_{av} = r^2 \left(\frac{1}{2\eta}|E_\theta|^2\right) = \frac{\eta}{2}\left|\frac{I_0 l}{\lambda}\right|^2 \sin^2\theta \cos^2(kh\cos\theta) \quad (4\text{-}103)$$

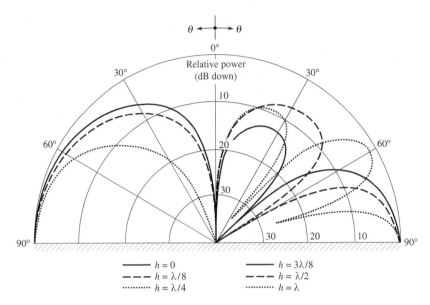

Figure 4.18 Elevation plane amplitude patterns of a vertical infinitesimal electric dipole for different heights above an infinite perfect electric conductor.

The maximum value of (4-103) occurs at $\theta = \pi/2$ and is given, excluding $kh \to \infty$, by

$$U_{max} = U|_{\theta=\pi/2} = \frac{\eta}{2}\left|\frac{I_0 l}{\lambda}\right|^2 \qquad (4\text{-}103a)$$

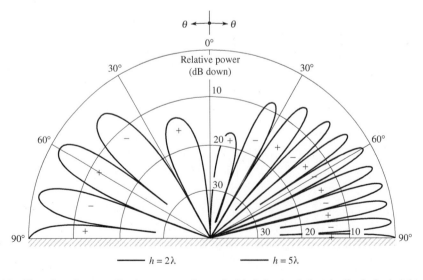

Figure 4.19 Elevation plane amplitude patterns of a vertical infinitesimal electric dipole for heights of 2λ and 5λ above an infinite perfect electric conductor.

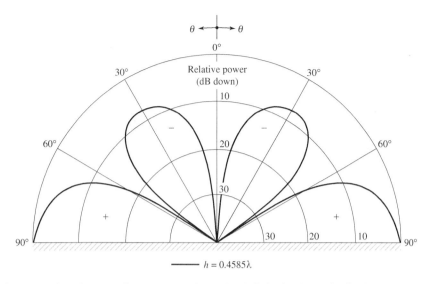

Figure 4.20 Elevation plane amplitude pattern of a vertical infinitesimal electric dipole at a height of 0.4585λ above an infinite perfect electric conductor.

which is four times greater than that of an isolated element. With (4-102) and (4-103a), the directivity can be written as

$$D_0 = \frac{4\pi U_{\max}}{P_{\text{rad}}} = \frac{2}{\left[\frac{1}{3} - \frac{\cos(2kh)}{(2kh)^2} + \frac{\sin(2kh)}{(2kh)^3}\right]} \quad (4\text{-}104)$$

whose value for $kh = 0$ is 3. The maximum value occurs when $kh = 2.881$ ($h = 0.4585\lambda$), and it is equal to 6.566 which is greater than four times that of an isolated element (1.5). The pattern for $h = 0.4585\lambda$ is shown plotted in Figure 4.20 while the directivity, as given by (4-104), is displayed in Figure 4.21 for $0 \leq h \leq 5\lambda$.

Using (4-102), the radiation resistance can be written as

$$R_r = \frac{2P_{\text{rad}}}{|I_0|^2} = 2\pi\eta \left(\frac{l}{\lambda}\right)^2 \left[\frac{1}{3} - \frac{\cos(2kh)}{(2kh)^2} + \frac{\sin(2kh)}{(2kh)^3}\right] \quad (4\text{-}105)$$

whose value for $kh \to \infty$ is the same and for $kh = 0$ is twice that of the isolated element as given by (4-19). When $kh = 0$, the value of R_r as given by (4-105) is only one-half the value of an $l' = 2l$ isolated element according to (4-19). The radiation resistance, as given by (4-105), is plotted in Figure 4.19 for $0 \leq h \leq 5\lambda$ when $l = \lambda/50$ and the element is radiating into free-space ($\eta \simeq 120\pi$). It can be compared to the value of $R_r = 0.316$ ohms for the isolated element of Example 4.1.

In practice, a wide use has been made of a quarter-wavelength monopole ($l = \lambda/4$) mounted above a ground plane, and fed by a coaxial line, as shown in Figure 4.22(a). For analysis purposes, a $\lambda/4$ image is introduced and it forms the $\lambda/2$ equivalent of Figure 4.22(b). It should be emphasized that the $\lambda/2$ equivalent of Figure 4.22(b) gives the correct field values for the actual system of Figure 4.22(a) only above the interface ($z \geq 0, 0 \leq \theta \leq \pi/2$). Thus, the far-zone electric and magnetic fields for the $\lambda/4$ monopole above the ground plane are given, respectively, by (4-84) and (4-85).

190 LINEAR WIRE ANTENNAS

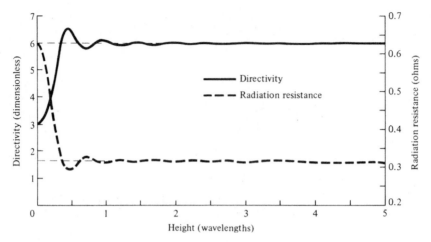

Figure 4.21 Directivity and radiation resistance of a vertical infinitesimal electric dipole as a function of its height above an infinite perfect electric conductor.

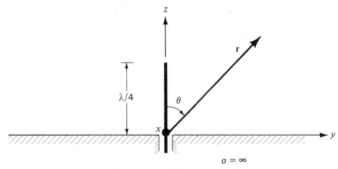

(a) λ/4 monopole on infinite electric conductor

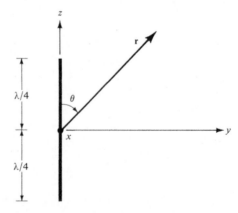

(b) Equivalent of λ/4 monopole on infinite electric conductor

Figure 4.22 Quarter-wavelength monopole on an infinite perfect electric conductor.

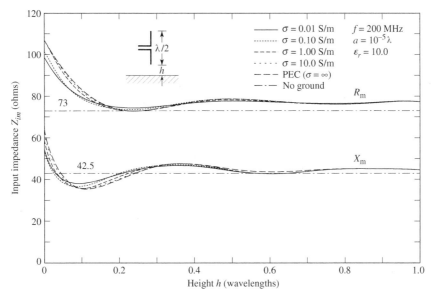

Figure 4.23 Input impedance of a vertical $\lambda/2$ dipole above a flat lossy electric conducting surface.

From the discussions of the resistance of an infinitesimal dipole above a ground plane for $kh = 0$, it follows that the input impedance of a $\lambda/4$ monopole above a ground plane is equal to one-half that of an isolated $\lambda/2$ dipole. Thus, referred to the current maximum, the input impedance Z_{im} is given by

$$Z_{im} \text{ (monopole)} = \tfrac{1}{2} Z_{im} \text{ (dipole)} = \tfrac{1}{2}[73 + j42.5] = 36.5 + j21.25 \qquad (4\text{-}106)$$

where $73 + j42.5$ is the input impedance (and also the impedance referred to the current maximum) of a $\lambda/2$ dipole as given by (4-93a).

The same procedure can be followed for any other length. The input impedance $Z_{im} = R_{im} + jX_{im}$ (referred to the current maximum) of a vertical $\lambda/2$ dipole placed near a flat lossy electric conductor, as a function of height above the ground plane, is plotted in Figure 4.23, for $0 \leq h \leq \lambda$. Conductivity values considered were 10^{-2}, 10^{-1}, 1, 10 S/m, and infinity (PEC). It is apparent that the conductivity does not strongly influence the impedance values. The conductivity values used are representative of dry to wet earth. It is observed that the values of the resistance and reactance approach, as the height increases, the corresponding ones of the isolated element (73 ohms for the resistance and 42.5 ohms for the reactance).

4.7.4 Approximate Formulas for Rapid Calculations and Design

Although the input resistance of a dipole of any length can be computed using (4-70) and (4-79), while that of the corresponding monopole using (4-106), very good answers can be obtained using simpler but approximate expressions. Defining G as

$$G = kl/2 \text{ for dipole} \qquad (4\text{-}107a)$$

$$G = kl \text{ for monopole} \qquad (4\text{-}107b)$$

where l is the total length of each respective element, it has been shown that the input resistance of the dipole and monopole can be computed approximately using [13]

$$0 < G < \pi/4$$
(maximum input resistance of dipole is less than 12.337 ohms)

$$R_{in} \text{ (dipole)} = 20G^2 \quad 0 < l < \lambda/4 \quad (4\text{-}108a)$$

$$R_{in} \text{ (monopole)} = 10G^2 \quad 0 < l < \lambda/8 \quad (4\text{-}108b)$$

$$\pi/4 \leq G < \pi/2$$
(maximum input resistance of dipole is less than 76.383 ohms)

$$R_{in} \text{ (dipole)} = 24.7G^{2.5} \quad \lambda/4 \leq l < \lambda/2 \quad (4\text{-}109a)$$

$$R_{in} \text{ (monopole)} = 12.35G^{2.5} \quad \lambda/8 \leq l < \lambda/4 \quad (4\text{-}109b)$$

$$\pi/2 \leq G < 2$$
(maximum input resistance of dipole is less than 200.53 ohms)

$$R_{in} \text{ (dipole)} = 11.14G^{4.17} \quad \lambda/2 \leq l < 0.6366\lambda \quad (4\text{-}110a)$$

$$R_{in} \text{ (monopole)} = 5.57G^{4.17} \quad \lambda/4 \leq l < 0.3183\lambda \quad (4\text{-}110b)$$

Besides being much simpler in form, these formulas are much more convenient in design (synthesis) problems where the input resistance is given and it is desired to determine the length of the element. These formulas can be verified by plotting the actual resistance versus length on a log–log scale and observe the slope of the line [24]. For example, the slope of the line for values of G up to about $\pi/4 \simeq 0.75$ is 2.

Example 4.4
Determine the length of a dipole whose input resistance is 50 ohms. Verify the answer.
Solution: Using (4-109a)

$$50 = 24.7G^{2.5}$$

or

$$G = 1.3259 = kl/2$$

Therefore

$$l = 0.422\lambda$$

Using (4-70) and (4-79) R_{in} for 0.422λ is 45.816 ohms, which closely agrees with the desired value of 50 ohms. To obtain 50 ohms using (4-70) and (4-79), $l = 0.4363\lambda$.

4.7.5 Mobile Communication Devices and Antennas for Mobile Communication Systems

The cellular era officially began in March 1982 when the FCC (Federal Communication Commission) gave communication carriers the official go-ahead to develop cellular technology. On March

6, 1983, Motorola officially unveiled the DynaTAC 8000X cellular telephone, which then weighted around 1.75 pounds (0.79 kg) and was nearly 13 inches (33 cm) long, and the race began. Since then, there has been an explosion in the advancement, miniaturization, and utilization of wireless communication devices, especially cell phones, smartphones, tablets and pads. While in 1998 there were worldwide about 200 million of cellular handset units sold, the figure grew to nearly 750 million units in 2006 and to nearly 1,000 million in 2013; over 1 billion in 2015. Many companies played key roles in this evolution; most prominent among them were Motorola, Qualcomm, Nokia, Ericsson, Apple, Samsung, LG, Huawei, and Lenovo. During this period, these devices provided vast services in the exchange of information, via emails, text messaging, news, stock quotes, weather, traveling maps, GPS, TV, search engines, just to name a few. Antenna technology, the "eyes and ears" of wireless communication systems, led this evolution, starting from the design and utilization of external radiating elements (primarily monopole and dipole type), such as in the DynaTAC, to embedded elements (primarily planar elements), such as microstrip, IFA, and PIFA employed in almost all of today's smartphones, tablets, pads, and other mobile units. In this chapter we introduce the basic radiation characteristics of dipoles and monopoles while those of planar elements, such as microstrips and PIFAs, are discussed in Chapter 14.

The dipole and monopole are two of the most widely used antennas for wireless mobile communication systems [25]–[29]. An array of dipole elements is extensively used as an antenna at the base station of a land mobile system while the monopole, because of its broadband characteristics and simple construction, is perhaps the most common antenna element for portable equipment, such as cellular telephones, cordless telephones, automobiles, trains, etc. The radiation efficiency and gain characteristics of both of these elements are strongly influenced by their electrical length which is related to the frequency of operation. In a handheld unit, such as a cellular telephone, the position of the monopole element on the unit influences the pattern while it does not strongly affect the input impedance and resonant frequency. In addition to its use in mobile communication systems, the quarter-wavelength monopole is very popular in many other applications. An alternative to the monopole for the handheld unit is the loop, which is discussed in Chapter 5. Other elements include the inverted F, planar inverted F antenna (PIFA), microstrip (patch), spiral, and others [25]–[29].

The variation of the input impedance, real and imaginary parts, of a vertical monopole antenna mounted on an experimental unit, simulating a cellular telephone, are shown in Figure 4.24(a,b) [28]–[29]. It is apparent that the first resonance, around 1,000 MHz, is of the *series type* with slowly varying values of impedance versus frequency, and of desirable magnitude, for practical implementation. For frequencies below the first resonance, the impedance is capacitive (imaginary part is negative), as is typical of linear elements of small lengths (see Figure 4.9); above the first resonance, the impedance is inductive (positive imaginary part). The second resonance, around 1,500 MHz, is of the *parallel type* (*antiresonance*) with large and rapid changes in the values of the impedance. These values and variation of impedance are usually undesirable for practical implementation. The order of the types of resonance (*series* vs. *parallel*) can be interchanged by choosing another element, such as a loop, as illustrated in Chapter 5, Section 5.8, Figure 5.20 [29]. The radiation amplitude patterns are those of a typical dipole with intensity in the lower hemisphere.

Examples of monopole type antennas used in cellular and cordless telephones, walkie-talkies, and CB radios are shown in Figure 4.25. The monopoles used in these units are either stationary, as it was in the first cell phone (Motorola DynaTAC) introduced in 1982–1984, or retractable/telescopic. The length of the retractable/telescopic monopole, such as the one used in the Motorola StarTAC and in others, is varied during operation to improve the radiation characteristics, such as the amplitude pattern and especially the input impedance. During nonusage, the element is usually retracted within the body of the device to prevent it from damage. Units that do not utilize a visible monopole type of antenna, especially in modern smart phones and similar devices like pads and tablets, some of which are shown in Figure 4.25, use embedded/internal type of antenna element. One such embedded/internal element that is often used is a Planar Inverted F Antenna (PIFA) [27], which will be discussed in Chapter 14; there are others. Many of the stationary monopoles are often covered with

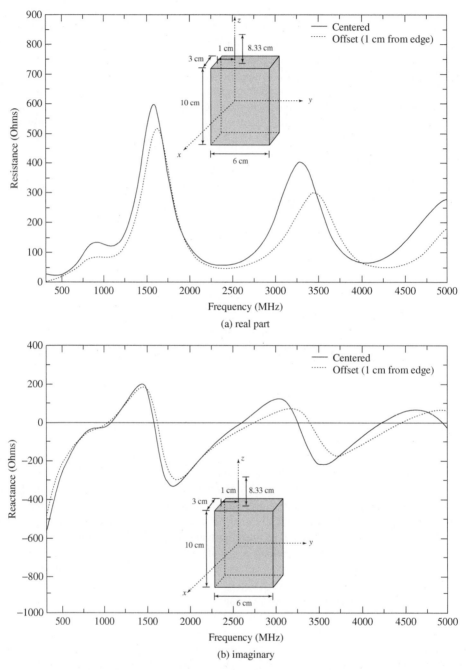

Figure 4.24 Input impedance, real and imaginary parts, of a vertical monopole mounted on an experimental cellular telephone device.

a dielectric cover. Within the cover, there is typically a straight wire. *However, another design that is often used is a helix antenna* (see Chapter 10, Section 10.3.1) *with a very small circumference and overall length so that the helix operates in the normal mode, whose relative pattern is exhibited in Figure 10.14(a) and which resembles that of a straight-wire monopole. The helix is used, in lieu of a straight wire, because it can be designed to have larger input resistance, which is more attractive for matching to typical feed lines, such as a coaxial line* (see Problem 10.20).

Figure 4.25 Examples of external and embedded/internal antennas used in commercial cellular and CB radios. (SOURCE: Reproduced with permissions from Motorola, Inc. © Motorola, Inc.; Nokia; Samsung © Samsung; Microsoft; HTC; Midland Radio Corporation © Midland Radio Corporation).

An antenna configuration that is widely used as a base-station antenna for mobile communication and is seen almost everywhere is shown in Figure 4.26. It is a triangular array configuration consisting of twelve dipoles, with four dipoles on each side of the triangle. Each four-element array, on each side of the triangle, is used to cover an angular sector of 120°, forming what is usually referred to as a sectoral array [see Section 16.3.1(B) and Figure 16.6(a)].

4.7.6 Horizontal Electric Dipole

Another dipole configuration is when the linear element is placed horizontally relative to the infinite electric ground plane, as shown in Figure 4.27. The analysis procedure of this is identical to the

196 LINEAR WIRE ANTENNAS

Figure 4.26 Triangular array of dipoles used as a sectoral base-station antenna for mobile communication.

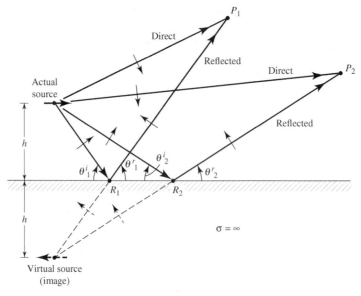

Figure 4.27 Horizontal electric dipole, and its associated image, above an infinite, flat, perfect electric conductor.

one of the vertical dipole. Introducing an image and assuming far-field observations, as shown in Figure 4.28(a,b), the direct component can be written as

$$E_\psi^d = j\eta \frac{kI_0 l e^{-jkr_1}}{4\pi r_1} \sin\psi \tag{4-111}$$

and the reflected one by

$$E_\psi^r = jR_h\eta \frac{kI_0 l e^{-jkr_2}}{4\pi r_2} \sin\psi \tag{4-111a}$$

or

$$E_\psi^r = -j\eta \frac{kI_0 l e^{-jkr_2}}{4\pi r_2} \sin\psi \tag{4-111b}$$

since the reflection coefficient is equal to $R_h = -1$.

To find the angle ψ, which is measured from the y-axis toward the observation point, we first form

$$\cos\psi = \hat{\mathbf{a}}_y \cdot \hat{\mathbf{a}}_r = \hat{\mathbf{a}}_y \cdot (\hat{\mathbf{a}}_x \sin\theta \cos\phi + \hat{\mathbf{a}}_y \sin\theta \sin\phi + \hat{\mathbf{a}}_z \cos\theta) = \sin\theta \sin\phi \tag{4-112}$$

from which we find

$$\sin\psi = \sqrt{1 - \cos^2\psi} = \sqrt{1 - \sin^2\theta \sin^2\phi} \tag{4-113}$$

Since for far-field observations

$$\left.\begin{array}{l} r_1 \simeq r - h\cos\theta \\ r_2 \simeq r + h\cos\theta \end{array}\right\} \quad \text{for phase variations} \tag{4-114a}$$

$$r_1 \simeq r_2 \simeq r \quad \text{for amplitude variations} \tag{4-114b}$$

the total field, which is valid only above the ground plane ($z \geq h; 0 \leq \theta \leq \pi/2, 0 \leq \phi \leq 2\pi$), can be written as

$$E_\psi = E_\psi^d + E_\psi^r = j\eta \frac{kI_0 l e^{-jkr}}{4\pi r} \sqrt{1 - \sin^2\theta \sin^2\phi}\, [2j\sin(kh\cos\theta)] \tag{4-115}$$

Equation (4-115) again consists of the product of the field of a single isolated element placed symmetrically at the origin and a factor (within the brackets) known as the *array factor*. This again is the pattern multiplication rule of (6-5), which is discussed in more detail in Chapter 6.

The general electric field expression E_ψ of (4-111), when the horizontal dipole is placed at the origin of the coordinate system of Figure 4.28(a), reduces in the principal planes (*xy, yz, xz*) to the

198 LINEAR WIRE ANTENNAS

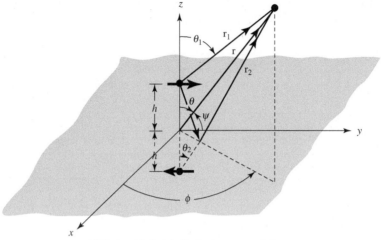

(a) Horizontal electric dipole above ground plane

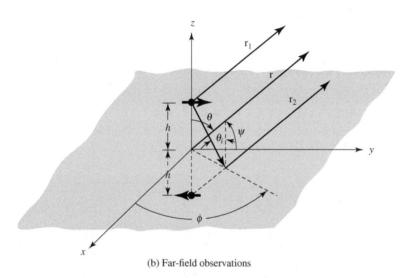

(b) Far-field observations

Figure 4.28 Horizontal electric dipole above an infinite perfect electric conductor.

spherical components E_θ and E_ϕ, as follows:

xy plane ($\theta = 90°$):

$$E_\psi = \begin{cases} E_\theta = 0 \\ E_\phi = -j\,\eta\dfrac{kI_o l e^{-jkr}}{4\pi r}\sqrt{1 - \sin^2\theta \sin^2\phi}\bigg|_{\theta=90°} = -j\dfrac{\omega\mu I_o l e^{-jkr}}{4\pi r}\cos\phi \end{cases} \quad (4\text{-}116a)$$

yz plane ($\phi = 90°$):

$$E_\psi = \begin{cases} E_\theta = -j\,\eta\dfrac{kI_o l e^{-jkr}}{4\pi r}\sqrt{1 - \sin^2\theta \sin^2\phi}\bigg|_{\phi=90°} = -j\dfrac{\omega\mu I_o l e^{-jkr}}{4\pi r}\cos\theta \\ E_\phi = 0 \end{cases} \quad (4\text{-}116b)$$

xz plane ($\phi = 0°$):

$$E_\psi = \begin{cases} E_\theta = 0 \\ E_\phi = -j\dfrac{\omega\mu I_0 l e^{-jkr}}{4\pi r} = \text{constant (isotropic)} \end{cases} \quad (4\text{-}116c)$$

These components match those of Example 4.5 that follows, which are derived based on the vector potential approach. Also, based on (4-116a)–(4-116c), it is easier to decide on the shape of the amplitude pattern and ascertain the polarization of the wave in the three principal planes.

Example 4.5

Using the vector potential **A** and the procedure outlined in Section 3.6 of Chapter 3, derive the far-zone spherical electric and magnetic field components of a horizontal infinitesimal dipole placed at the origin of the coordinate system of Figure 4.1.

Solution: Using (4-4), but for a horizontal infinitesimal dipole of uniform current directed along the y-axis, the corresponding vector potential can be written as

$$\mathbf{A} = \hat{\mathbf{a}}_y \frac{\mu I_0 l e^{-jkr}}{4\pi r}$$

with the corresponding spherical components, using the rectangular to spherical components transformation of (4-5), expressed as

$$A_\theta = A_y \cos\theta \sin\phi = \frac{\mu I_0 l e^{-jkr}}{4\pi r} \cos\theta \sin\phi$$

$$A_\phi = A_y \cos\phi = \frac{\mu I_0 l e^{-jkr}}{4\pi r} \cos\phi$$

Using (3-58a) and (3-58b), we can write the corresponding far-zone electric and magnetic field components as

$$E_\theta \cong -j\omega A_\theta = -j\frac{\omega\mu I_0 l e^{-jkr}}{4\pi r} \cos\theta \sin\phi$$

$$E_\phi \cong -j\omega A_\phi = -j\frac{\omega\mu I_0 l e^{-jkr}}{4\pi r} \cos\phi$$

$$H_\theta \cong -\frac{E_\phi}{\eta} = j\frac{\omega\mu I_0 l e^{-jkr}}{4\pi\eta r} \cos\phi$$

$$H_\phi \cong +\frac{E_\theta}{\eta} = -j\frac{\omega\mu I_0 l e^{-jkr}}{4\pi\eta r} \cos\theta \sin\phi$$

Although the electric-field components, and thus the magnetic field components, take a different analytical form than (4-111), the patterns are the same.

To examine the variations of the total field as a function of the element height above the ground plane, the two-dimensional elevation plane patterns (normalized to 0 dB) for $\phi = 90°$ (y-z plane) when $h = 0, \lambda/8, \lambda/4, 3\lambda/8, \lambda/2$, and λ are plotted in Figure 4.29. Since this antenna system is not symmetric with respect to the z axis, the azimuthal plane (x-y plane) pattern is not isotropic.

200 LINEAR WIRE ANTENNAS

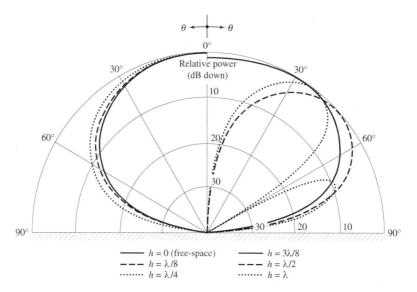

- —— $h = 0$ (free-space)
- - - - $h = \lambda/8$
- ······ $h = \lambda/4$
- —— $h = 3\lambda/8$
- - - - $h = \lambda/2$
- ······ $h = \lambda$

Figure 4.29 Elevation plane ($\phi = 90°$) amplitude patterns of a horizontal infinitesimal electric dipole for different heights above an infinite perfect electric conductor.

To obtain a better visualization of the radiation intensity in all directions above the interface, the three-dimensional pattern for $h = \lambda$ is shown plotted in Figure 4.30. The radial distance on the x-y plane represents the elevation angle θ from 0° to 90°, and the z-axis represents the normalized amplitude of the radiation field intensity from 0 to 1. The azimuthal angle ϕ ($0 \leq \phi \leq 2\pi$) is measured from the x- toward the y-axis on the x-y plane.

As the height increases beyond one wavelength ($h > \lambda$), a larger number of lobes is again formed. This is illustrated in Figure 4.31 for $h = 2\lambda$ and 5λ. The scalloping effect is evident here, as in Figure 4-19 for the vertical dipole. The total number of lobes is equal to the integer that most closely

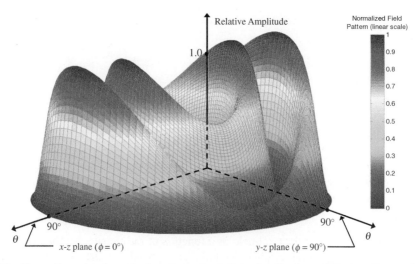

Figure 4.30 Three-dimensional amplitude pattern of an infinitesimal horizontal dipole a distance $h = \lambda$ above an infinite perfect electric conductor.

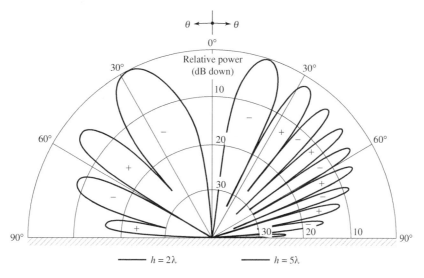

Figure 4.31 Elevation plane ($\phi = 90°$) amplitude patterns of a horizontal infinitesimal electric dipole for heights 2λ and 5λ above an infinite perfect electric conductor.

is equal to

$$\text{number of lobes} \simeq 2\left(\frac{h}{\lambda}\right) \qquad (4\text{-}117)$$

with unity being the smallest number.

Following a procedure similar to the one performed for the vertical dipole, the radiated power can be written as

$$P_{\text{rad}} = \eta \frac{\pi}{2} \left|\frac{I_0 l}{\lambda}\right|^2 \left[\frac{2}{3} - \frac{\sin(2kh)}{2kh} - \frac{\cos(2kh)}{(2kh)^2} + \frac{\sin(2kh)}{(2kh)^3}\right] \qquad (4\text{-}118)$$

and the radiation resistance as

$$R_r = \eta\pi \left(\frac{l}{\lambda}\right)^2 \left[\frac{2}{3} - \frac{\sin(2kh)}{2kh} - \frac{\cos(2kh)}{(2kh)^2} + \frac{\sin(2kh)}{(2kh)^3}\right] \qquad (4\text{-}119)$$

By expanding the sine and cosine functions into series, it can be shown that (4-119) reduces for small values of kh to

$$R_r \stackrel{kh \to 0}{=} \eta\pi \left(\frac{l}{\lambda}\right)^2 \left[\frac{2}{3} - \frac{2}{3} + \frac{8}{15}\left(\frac{2\pi h}{\lambda}\right)^2\right] = \eta \frac{32\pi^3}{15}\left(\frac{l}{\lambda}\right)^2 \left(\frac{h}{\lambda}\right)^2 \qquad (4\text{-}120)$$

For $kh \to \infty$, (4-119) reduces to that of an isolated element. The radiation resistance, as given by (4-119), is plotted in Figure 4.32 for $0 \leq h \leq 5\lambda$ when $l = \lambda/50$ and the antenna is radiating into free-space ($\eta \simeq 120\pi$).

The radiation intensity is given by

$$U \simeq \frac{r^2}{2\eta}|E_\psi|^2 = \frac{\eta}{2}\left|\frac{I_0 l}{\lambda}\right|^2 (1 - \sin^2\theta \sin^2\phi)\sin^2(kh\cos\theta) \qquad (4\text{-}121)$$

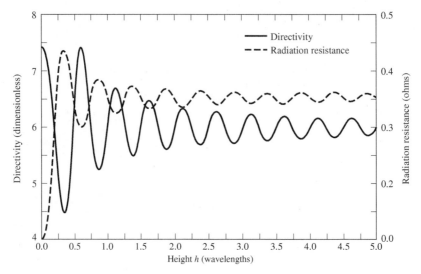

Figure 4.32 Radiation resistance and maximum directivity of a horizontal infinitesimal electric dipole as a function of its height above an infinite perfect electric conductor.

The maximum value of (4-121) depends on the value of kh (whether $kh \leq \pi/2, h \leq \lambda/4$ or $kh > \pi/2, h > \lambda/4$). It can be shown that the maximum of (4-121) is:

$$U_{max} = \begin{cases} \dfrac{\eta}{2}\left|\dfrac{I_0 l}{\lambda}\right|^2 \sin^2(kh) & \begin{array}{l} kh \leq \pi/2 \ (h \leq \lambda/4) \\ (\theta = 0°) \end{array} & (4\text{-}122a) \\[2ex] \dfrac{\eta}{2}\left|\dfrac{I_0 l}{\lambda}\right|^2 & \begin{array}{l} kh > \pi/2 \ (h > \lambda/4) \\ [\phi = 0° \text{ and } \sin(kh\cos\theta_{max}) = 1 \\ \text{ or } \theta_{max} = \cos^{-1}(\pi/2kh)] \end{array} & (4\text{-}122b) \end{cases}$$

Using (4-118) and (4-122a), (4-122b), the directivity can be written as

$$D_0 = \frac{4\pi U_{max}}{P_{rad}} = \begin{cases} \dfrac{4\sin^2(kh)}{R(kh)} & kh \leq \pi/2 \ (h \leq \lambda/4) & (4\text{-}123a) \\[2ex] \dfrac{4}{R(kh)} & kh > \pi/2 \ (h > \lambda/4) & (4\text{-}123b) \end{cases}$$

where

$$R(kh) = \left[\frac{2}{3} - \frac{\sin(2kh)}{2kh} - \frac{\cos(2kh)}{(2kh)^2} + \frac{\sin(2kh)}{(2kh)^3}\right] \quad (4\text{-}123c)$$

For small values of kh ($kh \to 0$), (4-123a) reduces to

$$D_0 \stackrel{kh \to 0}{=} \frac{4\sin^2(kh)}{\left[\frac{2}{3} - \frac{2}{3} + \frac{8}{15}(kh)^2\right]} = 7.5\left(\frac{\sin kh}{kh}\right)^2 \quad (4\text{-}124)$$

For $h = 0$ the element is shorted and it does not radiate. The directivity, as given by (4-123a)–(4-123b) is plotted for $0 \leq h \leq 5\lambda$ in Figure 4.32. It exhibits a maximum value of 7.5 for small values of h. Maximum values of slightly greater than 6 occur when $h \simeq (0.615 + n/2)\lambda, n = 1, 2, 3, \ldots$.

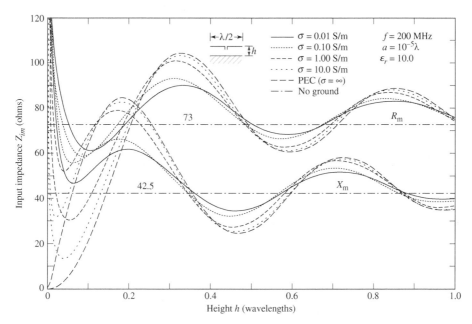

Figure 4.33 Input impedance of a horizontal $\lambda/2$ above a flat lossy electric conducting surface.

The input impedance $Z_{im} = R_{im} + jX_{im}$ (referred to the current maximum) of a horizontal $\lambda/2$ dipole above a flat lossy electric conductor is shown plotted in Figure 4.33 for $0 \leq h \leq \lambda$. Conductivities of $10^{-2}, 10^{-1}$, 1, 10 S/m, and infinity (PEC) were considered. It is apparent that the conductivity does have a more pronounced effect on the impedance values, compared to those of the vertical dipole shown in Figure 4.23. The conductivity values used are representative of those of the dry to wet earth. The values of the resistance and reactance approach, as the height increases, the corresponding values of the isolated element (73 ohms for the resistance and 42.5 ohms for the reactance).

4.8 GROUND EFFECTS

In the previous two sections the variations of the radiation characteristics (pattern, radiation resistance, directivity) of infinitesimal vertical and horizontal linear elements were examined when they were placed above plane perfect electric conductors. Although ideal electric conductors ($\sigma = \infty$) are not realizable, their effects can be used as guidelines for good conductors ($\sigma \gg \omega\varepsilon$, where ε is the permittivity of the medium).

One obstacle that is not an ideal conductor, and it is always present in any antenna system, is the ground (earth). In addition, the earth is not a plane surface. To simplify the analysis, however, the earth will initially be assumed to be flat. For pattern analysis, this is a very good engineering approximation provided the radius of the earth is large compared to the wavelength and the observation angles are greater than about $57.3/(ka)^{1/3}$ degrees from grazing (a is the earth radius) [30]. Usually these angles are greater than about $3°$.

In general, the characteristics of an antenna at low (LF) and medium (MF) frequencies are profoundly influenced by the lossy earth. This is particularly evident in the input resistance. When the antenna is located at a height that is small compared to the skin depth of the conducting earth, the input resistance may even be greater than its free-space values [30]. This leads to antennas with very low efficiencies. Improvements in the efficiency can be obtained by placing radial wires or metallic disks on the ground.

The analytical procedures that are introduced to examine the ground effects are based on the geometrical optics models of the previous sections. The image (virtual) source is again placed a distance h below the interface to account for the reflection. However, for each polarization nonunity reflection coefficients are introduced which, in general, will be a function of the angles of incidence and the constitutive parameters of the two media. Although plane wave reflection coefficients are used, even though spherical waves are radiated by the source, the error is small for conducting media [31]. The spherical nature of the wavefront begins to dominate the reflection phenomenon at grazing angles (i.e., as the point of reflection approaches the horizon) [32]. If the height (h) of the antenna above the interface is much less than the skin depth $\delta[\delta = \sqrt{2/(\omega\mu\sigma)}]$ of the ground, the image depth h below the interface should be increased [31] by a complex distance $\delta(1-j)$.

The geometrical optics formulations are valid provided the sources are located inside the lossless medium. When the sources are placed within the ground, the formulations should include possible surface-wave contributions. Exact boundary-value solutions, based on Sommerfeld integral formulations, are available [30]. However they are too complex to be included in an introductory chapter.

4.8.1 Vertical Electric Dipole

The field radiated by an electric infinitesimal dipole when placed above the ground can be obtained by referring to the geometry of Figures 4.17(a) and (b). Assuming the earth is flat and the observations are made in the far field, the direct component of the field is given by (4-94) and the reflected component by (4-95a) where the reflection coefficient R_v is given by

$$R_v = \frac{\eta_0 \cos\theta_i - \eta_1 \cos\theta_t}{\eta_0 \cos\theta_i + \eta_1 \cos\theta_t} = -R_\parallel \tag{4-125}$$

where $R_\parallel$ is the reflection coefficient for parallel polarization [7] and

$\eta_0 = \sqrt{\dfrac{\mu_0}{\varepsilon_0}}$ = intrinsic impedance of free-space (air)

$\eta_1 = \sqrt{\dfrac{j\omega\mu_1}{\sigma_1 + j\omega\varepsilon_1}}$ = intrinsic impedance of the ground

θ_i = angle of incidence (relative to the normal)

θ_t = angle of refraction (relative to the normal)

The angles θ_i and θ_t are related by Snell's law of refraction

$$\gamma_0 \sin\theta_i = \gamma_1 \sin\theta_t \tag{4-126}$$

where

$\gamma_0 = jk_0$ = propagation constant for free-space (air)

k_0 = phase constant for free-space (air)

$\gamma_1 = (\alpha_1 + jk_1)$ = propagation constant for the ground

α_1 = attenuation constant for the ground

k_1 = phase constant for the ground

Using the far-field approximations of (4-97a)–(4-98), the total electric field above the ground ($z \geq 0$) can be written as

$$E_\theta = j\eta \frac{kI_0 l e^{-jkr}}{4\pi r} \sin\theta [e^{jkh\cos\theta} + R_v e^{-jkh\cos\theta}] \quad z \geq 0 \tag{4-127}$$

where R_v is given by (4-125).

The permittivity and conductivity of the earth are strong functions of the ground's geological constituents, especially its moisture. Typical values for the relative permittivity ε_r (dielectric constant) are in the range of 5–100 and for the conductivity σ in the range of $10^{-4} - 10$ S/m.

Plots of the magnitude and phase of the reflection coefficient R_v, as a function of the incidence angle θ_i at a frequency of $f = 1$ GHz, are shown in Figure 4.34(a, b), respectively. As is apparent in Figure 4.34(a), for this polarization the reflection coefficient vanishes at the angles called *Brewster angles* [7]. At these angles the phase undergoes an 180° phase jump, as illustrated in Figure 4.34(b).

Normalized (to 0 dB) patterns of an infinitesimal dipole placed above the ground with height $h = \lambda/4$, above a flat interface, are shown plotted in Figure 4.35. In the presence of the ground, the radiation toward the vertical direction ($80° > \theta > 0°$) is more intense than that for the perfect electric conductor, but it vanishes for grazing angles ($\theta = 90°$). The null field toward the horizon ($\theta = 90°$) is formed because the reflection coefficient R_v approaches -1 as $\theta_i \to 90°$. Thus the ground effects on the pattern of a vertically polarized antenna are significantly different from those of a perfect conductor.

Significant changes also occur in the impedance. Because the formulation for the impedance is much more complex [30], it will not be presented here. Graphical illustrations for the impedance change of a vertical dipole placed a height h above a homogeneous lossy half-space, as compared to those in free-space, can be found in [33].

4.8.2 Horizontal Electric Dipole

The analytical formulation of the horizontal dipole above the ground can also be obtained in a similar manner as for the vertical electric dipole. Referring to Figure 4.28(a) and (b), the direct component is given by (4-111) and the reflected by (4-111a) where the reflection coefficient R_h is given by

$$R_h = \begin{cases} R_\perp & \text{for } \phi = 0°, 180° \text{ plane} \\ R_\parallel & \text{for } \phi = 90°, 270° \text{ plane} \end{cases} \quad (4\text{-}128)$$

where $R_\parallel$ is the reflection coefficient for parallel polarization, as given by (4-125), and $R_\perp$ is the reflection coefficient for perpendicular polarization given by [7]

$$R_\perp = \frac{\eta_1 \cos\theta_i - \eta_0 \cos\theta_t}{\eta_1 \cos\theta_i + \eta_0 \cos\theta_t} \quad (4\text{-}128\text{a})$$

The angles θ_i and θ_t are again related by Snell's law of refraction as given by (4-126).

Using the far-field approximations of (4-114a) and (4-114b), the total field above the ground ($z \geq h$) can be written as

$$E_\psi = j\eta \frac{kI_0 e^{-jkr}}{4\pi r} \sqrt{1 - \sin^2\theta \sin^2\phi} \, [e^{jkh\cos\theta} + R_h e^{-jkh\cos\theta}], \quad z \geq h \quad (4\text{-}129)$$

where R_h is given by (4-128).

To give an insight why $R_\perp$ is used in the $\phi = 0°, 180°$ plane and why $R_\parallel$ is used in the $\phi = 90°, 270°$ plane, the reader is referred to electric field of (4-115), which is decomposed into the far-zone spherical electric field components E_θ and E_ϕ represented by (4-116a), (4-116b) and (4-116c) in the principal planes. From these expresssions, the polarization of the electric field radiated by the horizontal dipole, and that reflected by the PEC ground plane, is more apparent; thus the use of the appropriate reflection coeffients $R_\perp$ and $R_\parallel$ in the respective principal planes.

Plots of the magnitude and phase of the reflection coefficient R_h, as a function of the incidence angle θ_i on the $xz(\phi = 0°)$ plane, are shown in Figure 4.36(a, b), respectively. It is apparent from Figure 4.36(a) that for this polarization the reflection coefficient does not vanish, but rather monotonically increases as the angle increases [7]. Similarly, the corresponding phase remains nearly constant and at 180°. The corresponding normalized amplitude patterns in this plane ($\phi = 0°$) of a horizontal electric dipole placed at a height $h = \lambda/4$ above an interface are shown in Figure 4.37(a),

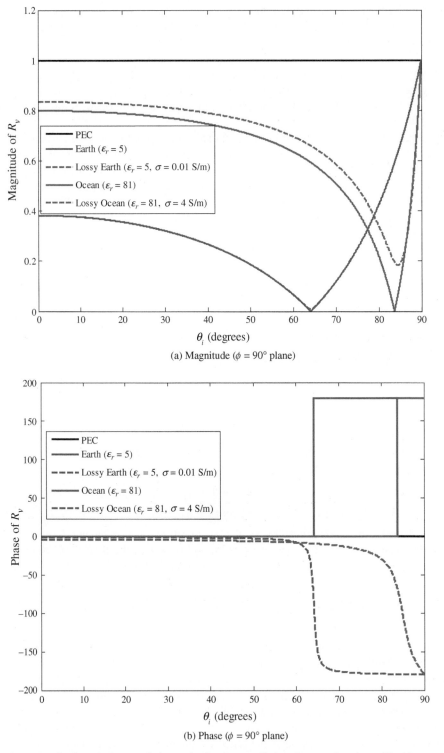

Figure 4.34 Magnitude and phase variations of reflection coefficient R_v, as a function of incidence angle θ_i, for $f = 1$ GHz.

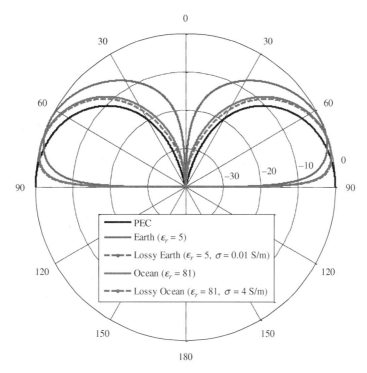

Figure 4.35 Elevation plane ($\phi = 0°$, $90°$ planes) amplitude patterns of an infinitesimal vertical electric dipole above an interface. ($h = \lambda/4, f = 1$ GHz).

whose shape is basically the same for the cases examined. This result is attributed to monotonic and small variations of the amplitude and basically constant phase, for all the cases examined, of the corresponding reflection coefficient R_h as a function of the angle of incidence, as displayed in Figure 4.36(a, b).

For the $yz(\phi = 90°)$ plane, the magnitude and phase variations of the reflection coefficient R_h, as a function of the incidence angle θ_i, are the same as those for the vertical polarization coefficient R_v, as are displayed in Figure 4.34(a, b). For this plane ($\phi = 90°$), the corresponding normalized amplitude pattern of a horizontal infinitesimal electric dipole placed at a height $h = \lambda/4$ above the interface are shown in Figure 4.37(b), whose shape is basically the same for the cases examined. This can be attributed to the small amplitude and phase variation of the reflection coefficient R_h/R_v in this plane for incidence angle up to the Brewster angle, as displayed in Figure 4.34(a, b), and it is due to the small field intensity of the horizontal dipole near and greater than the Brewster angle and as the observation angle approached $90°$; the field intensity vanishes at $\theta = 90°$.

4.8.3 PEC, PMC and EBG Surfaces

In general, PEC, PMC, and EBG surfaces individually possess attractive characteristics, but these surfaces also exhibit shortcomings when electromagnetic radiating elements are mounted on them, especially when the designs are judged based on aerodynamic, stealth, and conformal criteria. For example, when an electric element is mounted vertically on a PEC surface, radiation and system efficiency get reinforced; however, its low-profile geometry is undesirable for aerodynamic, stealth, and conformal designs. Yet, when the same electric radiating element is placed horizontally on a PEC surface, its radiation efficiency suffers because, at $h = 0$, the image possesses a $180°$ phase shift and its radiation cancels the radiation of the actual electric element. While the height stays small electrically, the radiation efficiency is low. For the radiation to attain maximum efficiency at a direction normal to the surface, the horizontal element must be placed at a height $\lambda/4$ above

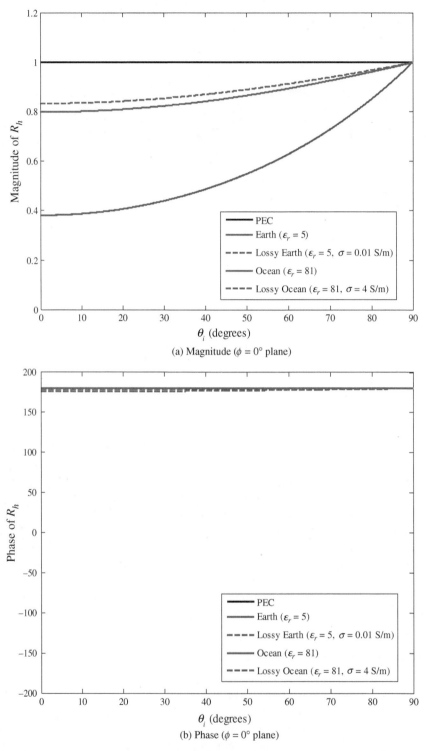

Figure 4.36 Magnitude and phase variations of the reflection coefficient R_h, as a function of incidence angle θ_i, for $f = 1$ GHz.

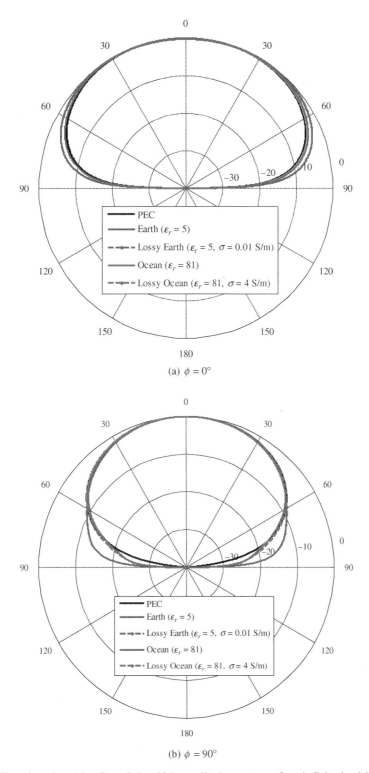

Figure 4.37 Elevation plane ($\phi = 0°$ and $\phi = 90°$) amplitude patterns of an infinitesimal horizontal electric dipole above an interface ($h = \lambda/4$, $f = 1$ GHz).

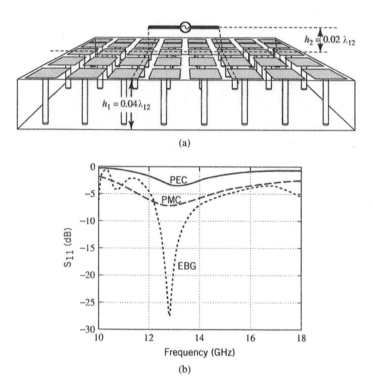

Figure 4.38 Geometry and S_{11} of horizontal dipole above PEC, PMC and EBG surfaces. (a) Geometry ([7] Reprinted with permission from John Wiley & Sons, Inc.) (b) Reflection coefficient (SOURCE: [23] © 2003 IEEE).

the surface. Such an arrangement is especially not desirable on space-borne platforms because of aerodynamic considerations. Furthermore, for stealth type of targets, configurations that are visible to radar can create a large radar cross section (RCS) signature, which is why low-profile designs are desirable for aerodynamic, stealth, and conformal applications. When the same electric radiating element is placed horizontally on a PMC surface, its image then has a low profile and 0° phase, which reinforces the radiation of the actual electric element. The characteristics of vertical and horizontal electric elements placed vertically and horizontally on PEC and PMC surfaces are based on image theory, and they are visually contrasted in Figure 4.16.

Whether a PEC, PMC, or EBG surface outperforms the others as a ground plane depends on the application. This is best illustrated by a basic example. In Figure 4.38, a $0.4\lambda_{12}$ dipole (λ_{12} is the free-space wavelength at $f = 12$ GHz) is placed horizontally above PEC, PMC, and EBG surfaces. The EBG surface has a height of $0.04\lambda_{12}$. The dipole is placed at a height h of $0.06\lambda_{12}(h = 0.06\lambda_{12})$ above a $\lambda_{12} \times \lambda_{12}$ PEC, PMC square surface, which in turn means that the dipole is placed at a height of $0.02\lambda_{12}$ above the EBG surface. The S_{11} of this system (based on a 50-ohm line impedance) was simulated, using the FDTD method, over a frequency range of 10–18 GHz [23], and the results are shown in Figure 4.38(b). From these results, it is clear that the EBG surface (which has a reflection phase variation from +180° to −180°; see an example in Figure 4.14) exhibits a best return loss of −27 dB while the PMC (which has a reflection phase of 0°) has a best return loss of −7.2 dB and the PEC (which has a reflection phase of 180°) has a best return loss of only −3.5 dB. For the PMC surface, the return loss is influenced by the mutual coupling, due to the close proximity between the main element and its in-phase image, whereas for the PEC the return loss is influenced by the 180° phase reversal, which severely impacts the radiation efficiency. In this example, the EBG surface, because of its +180° to −180° phase variation over the frequency band-gap of the EBG design, outperforms the PEC and PMC and serves as a good ground plane. The phase characteristics of the PEC and PMC surfaces are constant (out-of-phase and in-phase, respectively) over the entire frequency range.

4.8.4 Earth Curvature

Antenna pattern measurements on aircraft can be made using either scale models or full scale in-flight. Scale model measurements usually are made indoors using electromagnetic anechoic chambers, as described in Chapter 17. The indoor facilities provide a controlled environment, and all-weather capability, security, and minimize electromagnetic interference. However, scale model measurements may not always simulate real-life outdoor conditions, such as the reflecting surface of seawater. Therefore full-scale model measurements may be necessary. For in-flight measurements, reflecting surfaces, such as earth and seawater, introduce reflections, which usually interfere with the direct signal. These unwanted signals are usually referred to as *multipath*. Therefore the total measured signal in an outdoor system configuration is the combination of the direct signal and that due to multipath, and usually it cannot be easily separated in its parts using measuring techniques. Since the desired signal is that due to the direct path, it is necessary to subtract from the total response the contributions due to multipath. This can be accomplished by developing analytical models to predict the contributions due to multipath, which can then be subtracted from the total signal in order to be left with the desired direct path signal. In this section we will briefly describe techniques that have been used to accomplish this [34], [35].

The analytical formulations of Sections 4.8.1 and 4.8.2 for the patterns of vertical and horizontal dipoles assume that the earth is flat. This is a good approximation provided the curvature of the earth is large compared to the wavelength and the angle of observation is greater than about 3° from grazing [or more accurately greater than about $57.3/(ka)^{1/3}$ degrees, where a is the radius of the earth] from grazing [36]. The curvature of the earth has a tendency to spread out (weaken, diffuse, diverge) the reflected energy more than a corresponding flat surface. The spreading of the reflected energy from a curved surface as compared to that from a flat surface is taken into account by introducing a divergence factor D [32], [34], [35], defined as

$$D = \text{divergence factor} = \frac{\text{reflected field from curved surface}}{\text{reflected field from flat surface}} \quad (4\text{-}130)$$

The formula for D can be derived using purely geometrical considerations. It is accomplished by comparing the ray energy density in a small cone reflected from a sphere near the principal point of reflection with the energy density the rays (within the same cone) would have if they were reflected from a plane surface. Based on the geometrical optics energy conservation law for a bundle of rays within a cone, the reflected rays within the cone will subtend a circle on a perpendicular plane for reflections from a flat surface, as shown in Figure 4.39(a). However, according to the geometry of Figure 4.39(b), it will subtend an ellipse for a spherical reflecting surface. Therefore the divergence factor of (4-130) can also be defined as

$$D = \frac{E_s^r}{E_f^r} = \left[\frac{\text{area contained in circle}}{\text{area contained in ellipse}}\right]^{1/2} \quad (4\text{-}131)$$

where

E_s^r = reflected field from spherical surface

E_f^r = reflected field from flat surface

Using the geometry of Figure 4.40, the divergence factor can be written as [7] and [35]

$$D = \frac{\sqrt{\dfrac{\rho_1^r \rho_2^r}{(\rho_1^r + s)(\rho_2^r + s)}}}{\dfrac{s'}{s' + s}} \quad (4\text{-}132)$$

212   LINEAR WIRE ANTENNAS

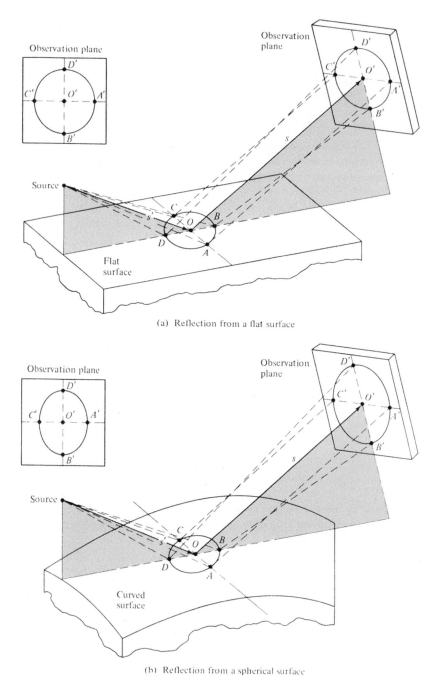

(a) Reflection from a flat surface

(b) Reflection from a spherical surface

Figure 4.39   Reflection from flat and spherical surfaces.

where ρ_1^r and ρ_2^r are the principal radii of curvature of the reflected wavefront at the point of reflection and are given, according to the geometry of Figure 4.40, by

$$\frac{1}{\rho_1^r} = \frac{1}{s'} + \frac{1}{\rho \sin \psi} + \sqrt{\frac{1}{(\rho \sin \psi)^2} - \frac{4}{a^2}} \qquad (4\text{-}132a)$$

GROUND EFFECTS 213

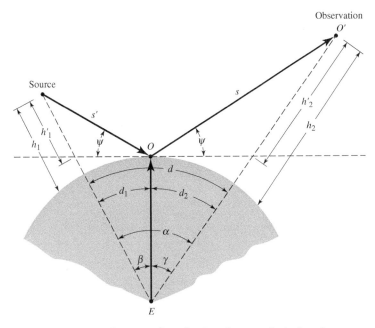

Figure 4.40 Geometry for reflections from a spherical surface.

$$\frac{1}{\rho_2^r} = \frac{1}{s'} + \frac{1}{\rho \sin \psi} - \sqrt{\frac{1}{(\rho \sin \psi)^2} - \frac{4}{a^2}} \tag{4-132b}$$

$$\rho = \frac{a}{1 + \sin^2 \psi} \tag{4-132c}$$

A simplified form of the divergence factor is that of [37]

$$D \cong \left[1 + \frac{2s's}{a(s' + s)\sin \psi}\right]^{-1/2} \cdot \left[1 + \frac{2s's}{a(s' + s)}\right]^{-1/2} \tag{4-133}$$

Both (4-132) and (4-133) take into account the earth curvature in two orthogonal planes.

Assuming that the divergence of rays in the azimuthal plane (plane vertical to the page) is negligible (two-dimensional case), the divergence factor can be written as

$$D \simeq \left[1 + 2\frac{ss'}{ad \tan \psi}\right]^{-1/2} \tag{4-134}$$

where ψ is the grazing angle. Thus the divergence factor of (4-134) takes into account energy spreading primarily in the elevation plane. According to Figure 4.40

h_1' = height of the source above the earth (with respect to the tangent at the point of reflection)

h_2' = height of the observation point above the earth (with respect to the tangent at the point of reflection)

d = range (along the surface of the earth) between the source and the observation point

a = radius of the earth (3,959 mi). Usually a $\frac{4}{3}$ radius ($\simeq$ 5,280 mi) is used.
ψ = reflection angle (with respect to the tangent at the point of reflection).
d_1 = distance (along the surface of the earth) from the source to the reflection point
d_2 = distance (along the surface of the earth) from the observation point to the reflection point

The divergence factor D can be included in the formulation of the fields radiated by a vertical or a horizontal dipole, in the presence of the earth, by modifying (4-127) and (4-129) and writing them, respectively, as

$$E_\theta = j\eta \frac{kI_0 le^{-jkr}}{4\pi r} \sin\theta [e^{jkh\cos\theta} + DR_v e^{-jkh\cos\theta}] \qquad (4\text{-}135a)$$

$$E_\psi = j\eta \frac{kI_0 le^{-jkr}}{4\pi r} \sqrt{1 - \sin^2\theta \sin^2\phi} [e^{jkh\cos\theta} + DR_h e^{-jkh\cos\theta}] \qquad (4\text{-}135b)$$

While the previous formulations are valid for smooth surfaces, they can still be used with rough surfaces, provided the surface geometry satisfies the Rayleigh criterion [32] and [37]

$$h_m < \frac{\lambda}{8\sin\psi} \qquad (4\text{-}136)$$

where h_m is the maximum height of the surface roughness. Since the dividing line between a smooth and a rough surface is not that well defined, (4-136) should only be used as a guideline.

The *coherent* contributions due to scattering by a surface with Gaussian rough surface statistics can be approximately accounted for by modifying the vertical and horizontal polarization smooth surface reflection coefficients of (4-125) and (4-128) and express them as

$$R^s_{v,h} = R^0_{v,h} e^{-2(k_0 h_0 \cos\theta_i)^2} \qquad (4\text{-}137)$$

where
$R^s_{v,h}$ = reflection coefficient of a rough surface for either vertical or horizontal polarization
$R^0_{v,h}$ = reflection coefficient of a smooth surface for either vertical (4-125) or horizontal (4-128) polarization
h_0^2 = mean-square roughness height

A *slightly rough surface* is defined as one whose rms height is much smaller than the wavelength, while a *very rough surface* is defined as one whose rms height is much greater than the wavelength.

Plots of the divergence factor as a function of the grazing angle ψ (or as a function of the observation point h'_2) for different source heights are shown in Figure 4.41. It is observed that the divergence factor is somewhat different and smaller than unity for small grazing angles, and it approaches unity as the grazing angle becomes larger. The variations of D displayed in Figure 4.41 are typical but not unique. For different positions of the source and observation point, the variations will be somewhat different. More detailed information on the variation of the divergence factor and its effect on the overall field pattern is available [35].

The most difficult task usually involves the determination of the reflection point from a knowledge of the heights of the source and observation points, and the range d between them. Procedures to do this have been developed [32], [34]–[38].

Using the analytical model developed here, computations were performed to see how well the predictions compared with measurements. For the computations it was assumed that the reflecting

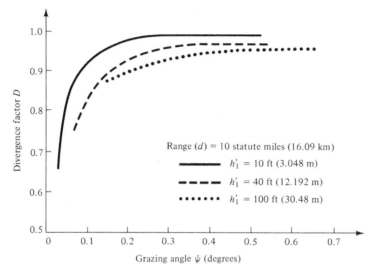

Figure 4.41 Divergence factor for a 4/3 radius earth (a_e = 5,280 mi = 8,497.3 km) as a function of grazing angle ψ.

surface is seawater possessing a dielectric constant of 81 and a conductivity of 4.64 S/m [34], [35]. To account for atmospheric refraction, a 4/3 earth was assumed [32], [34], [39] so the atmosphere above the earth can be considered homogeneous with propagation occurring along straight lines.

For computations using the earth as the reflecting surface, all three divergence factors of (4-132)–(4-134) gave the same results. However, for nonspherical reflecting surfaces and for those

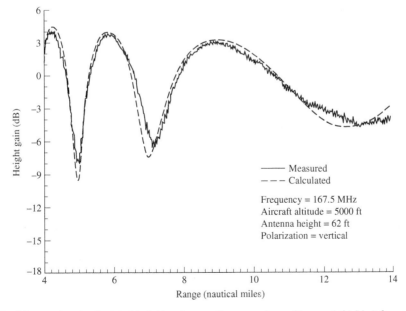

Figure 4.42 Measured and calculated height gain over the ocean (ε_r = 81, σ = 4.64 S/m) for vertical polarization.

with smaller radii of curvature, the divergence factor of (4-132) is slightly superior followed by (4-133) and then by (4-134). In Figure 4.42 we display and compare the predicted and measured *height gain* versus range d ($4 < d < 14$ nautical miles) for a vertical-vertical polarization system configuration at a frequency of 167.5 MHz. The *height gain* is defined as the ratio of the total field in the presence of the earth divided by the total field in the absence of the earth. A good agreement is noted between the two. The peaks and nulls are formed by constructive and destructive interferences between the direct and reflected components. If the reflecting surface were perfectly conducting, the maximum height gain would be 2 (6 dB). Because the modeled reflecting surface of Figure 4.42 was seawater with a dielectric constant of 81 and a conductivity of 4.64 S/m, the maximum height gain is less than 6 dB. The measurements were taken by aircraft and facilities of the Naval Air Warfare Center, Patuxent River, MD. Additional measurements were made but are not included here; they can be found in [40] and [41].

A summary of the pertinent parameters, and associated formulas and equation numbers for this chapter are listed in Table 4.3.

4.9 COMPUTER CODES

There are many computer codes that have been developed to analyze wire-type linear antennas, such as the dipole, and they are too numerous to mention here. One simple program to characterize the radiation characteristics of a dipole, designated as **Dipole** (both in FORTRAN and MATLAB), is included in the publisher's website for this book. Another much more advanced program, designated as the **Numerical Electromagnetics Code (NEC)**, is a user-oriented software developed by Lawrence Livermore National Laboratory [42]. It is a Method of Moments (MoM) code for analyzing the interaction of electromagnetic waves with arbitrary structures consisting of conducting wires and surfaces. In the 1970s and 1980s the NEC was the most widely distributed and used electromagnetics code. Included with the distribution are graphics programs for generating plots of the structure, antenna patterns, and impedance. There are also other commercial software that are based on the NEC. A compact version of the NEC is the **MININEC (Mini-Numerical Electromagnetics Code)** [42]–[44]. The MININEC is more convenient for the analysis of wire-type antennas.

4.10 MULTIMEDIA

In the publisher's website for this book, the following multimedia resources are included for the review, understanding, and visualization of the material of this chapter.

a. **Java**-based **interactive questionnaire**, with answers.
b. **Java**-based **applet** for computing and displaying the radiation characteristics of a dipole.
c. **Java**-based **visualization/animation** for displaying the radiation characteristics of a dipole of different lengths.
d. **Matlab** and **Fortran** computer program, designated **Dipole**, for computing the radiation characteristics of a dipole. The description of the program is found in the corresponding READ ME file in the publisher's website for this book.
e. **Matlab** computer program, designated **Ground_Reflections**, to compute the amplitude and phase variations of the reflection coefficients and the corresponding amplitude pattern of vertical and horizontal dipoles placed at a height h, above a planar interface.
f. **Power Point (PPT)** viewgraphs, in multicolor.

TABLE 4.3 Summary of Important Parameters and Associated Formulas and Equation Numbers for a Dipole in the Far Field

Parameter	Formula	Equation Number
	Infinitesimal Dipole *($l \leq \lambda/50$)*	
Normalized power pattern	$U = (E_{\theta n})^2 = C_0 \sin^2\theta$	(4-29)
Radiation resistance R_r	$R_r = \eta \left(\dfrac{2\pi}{3}\right)\left(\dfrac{l}{\lambda}\right)^2 = 80\pi^2 \left(\dfrac{l}{\lambda}\right)^2$	(4-19)
Input resistance R_{in}	$R_{in} = R_r = \eta \left(\dfrac{2\pi}{3}\right)\left(\dfrac{l}{\lambda}\right)^2 = 80\pi^2 \left(\dfrac{l}{\lambda}\right)^2$	(4-19)
Wave impedance Z_w	$Z_w = \dfrac{E_\theta}{H_\phi} \simeq \eta = 377$ ohms	
Directivity D_0	$D_0 = \dfrac{3}{2} = 1.761$ dB	(4-31)
Maximum effective area A_{em}	$A_{em} = \dfrac{3\lambda^2}{8\pi}$	(4-32)
Vector effective length ℓ_e	$\ell_e = -\hat{\mathbf{a}}_\theta l \sin\theta$	(2-92)
	$\|\ell_e\|_{max} = \lambda$	Example 4.2
Half-power beamwidth	HPBW = $90°$	(4-65)
Loss resistance R_L	$R_L = \dfrac{l}{P}\sqrt{\dfrac{\omega\mu_0}{2\sigma}} = \dfrac{l}{2\pi b}\sqrt{\dfrac{\omega\mu_0}{2\sigma}}$	(2-90b)
	Small Dipole *($\lambda/50 < l \leq \lambda/10$)*	
Normalized power pattern	$U = (E_{\theta n})^2 = C_1 \sin^2\theta$	(4-36a)
Radiation resistance R_r	$R_r = 20\pi^2 \left(\dfrac{l}{\lambda}\right)^2$	(4-37)
Input resistance R_{in}	$R_{in} = R_r = 20\pi^2 \left(\dfrac{l}{\lambda}\right)^2$	(4-37)
Wave impedance Z_w	$Z_w = \dfrac{E_\theta}{H_\phi} \simeq \eta = 377$ ohms	(4-36a), (4-36c)
Directivity D_0	$D_0 = \dfrac{3}{2} = 1.761$ dB	
Maximum effective area A_{em}	$A_{em} = \dfrac{3\lambda^2}{8\pi}$	
Vector effective length ℓ_e	$\ell_e = -\hat{\mathbf{a}}_\theta \dfrac{l}{2} \sin\theta$	(2-92)
	$\|\ell_e\|_{max} = \dfrac{l}{2}$	(4-36a)
Half-power beamwidth	HPBW = $90°$	(4-65)
	Half Wavelength Dipole *($l = \lambda/2$)*	
Normalized power pattern	$U = (E_{\theta n})^2 = C_2 \left[\dfrac{\cos\left(\dfrac{\pi}{2}\cos\theta\right)}{\sin\theta}\right]^2 \simeq C_2 \sin^3\theta$	(4-87)
Radiation resistance R_r	$R_r = \dfrac{\eta}{4\pi} C_{in}(2\pi) \simeq 73$ ohms	(4-93)

(continued overleaf)

218 LINEAR WIRE ANTENNAS

TABLE 4.3 (*continued*)

Parameter	Formula	Equation Number
Input resistance R_{in}	$R_{in} = R_r = \dfrac{\eta}{4\pi} C_{in}(2\pi) \simeq 73$ ohms	(4-79), (4-93)
Input impedance Z_{in}	$Z_{in} = 73 + j42.5$	(4-93a)
Wave impedance Z_w	$Z_w = \dfrac{E_\theta}{H_\phi} \simeq \eta = 377$ ohms	
Directivity D_0	$D_0 = \dfrac{4}{C_{in}(2\pi)} \simeq 1.643 = 2.156$ dB	(4-91)
Vector effective length ℓ_e	$\ell_e = -\hat{\mathbf{a}}_\theta \dfrac{\lambda}{\pi} \dfrac{\cos\left(\dfrac{\pi}{2}\cos\theta\right)}{\sin\theta}$	(2-91)
	$|\ell_e|_{max} = \dfrac{\lambda}{\pi} = 0.3183\lambda$	(4-84)
Half-power beamwidth	HPBW = 78°	(4-65)
Loss resistance R_L	$R_L = \dfrac{l}{2P}\sqrt{\dfrac{\omega\mu_0}{2\sigma}} = \dfrac{l}{4\pi b}\sqrt{\dfrac{\omega\mu_0}{2\sigma}}$	Example (2-13)

Quarter-Wavelength Monopole
($l = \lambda/4$)

Parameter	Formula	Equation Number
Normalized power pattern	$U = (E_{\theta n})^2 = C_2 \left[\dfrac{\cos\left(\dfrac{\pi}{2}\cos\theta\right)}{\sin\theta}\right]^2 \simeq C_2 \sin^3\theta$	(4-87)
Radiation resistance R_r	$R_r = \dfrac{\eta}{8\pi} C_{in}(2\pi) \simeq 36.5$ ohms	(4-106)
Input resistance R_{in}	$R_{in} = R_r = \dfrac{\eta}{8\pi} C_{in}(2\pi) \simeq 36.5$ ohms	(4-106)
Input impedance Z_{in}	$Z_{in} = 36.5 + j21.25$	(4-106)
Wave impedance Z_w	$Z_w = \dfrac{E_\theta}{H_\phi} \simeq \eta = 377$ ohms	
Directivity D_0	$D_0 = 3.286 = 5.167$ dB	
Vector effective length ℓ_e	$\ell_e = -\hat{\mathbf{a}}_\theta \dfrac{\lambda}{\pi}\cos\left(\dfrac{\pi}{2}\cos\theta\right)$	(2-91)
	$|\ell_e|_{max} = \dfrac{\lambda}{\pi} = 0.3183\lambda$	(4-84)

REFERENCES

1. W. A. Wheeler, "The Spherical Coil as an Inductor, Shield, or Antenna," *Proc. IRE*, Vol. 46, pp. 1595–1602, September 1958 (correction, Vol. 48, p. 328, March 1960).
2. W. A. Wheeler, "The Radiansphere Around a Small Antenna," *Proc. IRE*, Vol. 47, pp. 1325–1331, August 1959.
3. W. A. Wheeler, "Small Antennas," *IEEE Trans. Antennas Propagat.*, Vol. AP-23, No. 4, pp. 462–469, July 1975.
4. C. H. Walter, *Traveling Wave Antennas*, McGraw-Hill, 1965, pp. 32–44.
5. W. R. Scott, Jr., "A General Program for Plotting Three-Dimensional Antenna Patterns," *IEEE Antennas Propagat. Soc. Newsletter*, pp. 6–11, December 1989.

6. S. K. Schelkunoff and H. T. Friis, *Antennas: Theory and Practice*, Wiley, New York, 1952, pp. 229–244, 351–353.
7. C. A. Balanis, *Advanced Engineering Electromagnetics*, Second Edition, John Wiley & Sons, Inc., New York, 2012.
8. R. F. Harrington, "Matrix Methods for Field Problems," *Proc. IEEE*, Vol. 55, No. 2, pp. 136–149, February 1967.
9. R. F. Harrington, *Field Computation by Moment Methods*, Macmillan, New York, 1968.
10. R. Mittra (Ed.), *Computer Techniques for Electromagnetics*, Pergamon, New York, 1973.
11. J. Moore and P. Pizer (Eds.), *Moment Methods in Electromagnetics: Techniques and Applications*, Research Studies Press, Letchworth, UK, 1984.
12. J. J. Wang, *Generalized Moment Methods in Electromagnetics*, John Wiley & Sons, Inc., New York, 1991.
13. *IEEE Trans. Antennas Propagation*, Special Issue on *Artificial Magnetic Conductors, Soft/Hard Surfaces, and Other Complex Surfaces*, vol. 53, no. 1, Jan. 2005.
14. V. G. Veselago, "The Electromagnetics of Substances with Simultaneous Negative Values of ϵ and μ," *Sov. Phys.-Usp.*, vol. 47, pp. 509–514, Jan.–Feb. 1968.
15. P. M. Valanju, R. M. Walser, and A. P. Valanju, "Wave Refraction in Negative-Index Media; Always Positive and very Inhomogeneous," *Phys. Rev. Lett.*, vol. 88, no. 18, 187401:1–4, May 2002.
16. N. Engheta and R. W. Ziolkowski (editors), *Metamaterials: Physics and Engineering Explorations*, N. Engheta, R. W. Ziolkowski (editors), IEEE Press, Wiley Inter-Science, New York, 2006.
17. G. V. Eleftheriades and K. G. Balmain (editors), *Negative-Refraction Metamaterials: Fundamental Principles and Applications*, John Wiley & Sons, New York, 2005.
18. C. Caloz and T. Itoh, *Electromagnetic Metamaterials: Transmission Line Theory and Microwave Applications*, John Wiley & Sons, New York, 2006.
19. R. Marques, F. Martin, and M. Sorolla, *Metamaterials with Negative Parameters: Theory, Design and Microwave Applications*, John Wiley & Sons, New York, 2008.
20. E. Yablonovitch, "Photonic Band-Gap Structures," *J. Opt. Soc. Amer. B*, Vol. 10, No. 2, pp. 283–294, Feb. 1993.
21. D. Sievenpiper, High-Impedance Electromagnetic Surfaces, Ph.D. dissertation, Department of Electrical Engineering, UCLA, 1999.
22. D. Sievenpiper, "Artificial Impedance Surfaces," Chapter 15, in *Modern Antenna Handbook*, C. A. Balanis (editor), John Wiley & Sons, pp. 737–777, 2008.
23. F. Yang and Y. Rahmat-Samii, "Reflection Phase Characterization of the EBG Ground Plane for Low Profile Wire Antenna Applications," *IEEE Trans. Antenna Propagat.*, Vol. 51, No. 10, pp. 2691–2703, October 2003.
24. R. F. Schwartz, "Input Impedance of a Dipole or Monopole," *Microwave J.*, Vol. 15, No. 12, p. 22, December 1972.
25. K. Fujimoto and J. R. James, *Mobile Antenna Systems Handbook*, Artech House, Norwood, MA, 1994.
26. M. A. Jensen and Y. Rahmat-Samii, "Performance Analysis of Antennas for Hand-Held Transceivers Using FDTD," *IEEE Trans. Antennas Propagat.*, Vol. 42, No. 8, pp. 1106–1113, August 1994.
27. M. A. Jensen and Y. Rahmat-Samii, "EM Interaction of Handset Antennas and a Human in Personal Communications," *Proc. IEEE*, Vol. 83, No. 1, pp. 7–17, January 1995.
28. K. D. Katsibas, "Analysis and Design of Mobile Antennas for Handheld Units," Master's Thesis, Arizona State University, Tempe, AZ, August 1996.
29. K. D. Katsibas, C. A. Balanis, P. A. Tirkas, and C. R. Birtcher, "Folded Loop Antenna for Mobile Handheld Units," *IEEE Trans. Antennas Propagat.*, Vol. 46, No. 2, pp. 260–266, February 1998.
30. R. E. Collin and F. J. Zucker (Eds.), *Antenna Theory Part 2*, Chapters 23 and 24 (by J. R. Wait), McGraw-Hill, New York, 1969.
31. P. R. Bannister, "Image Theory Results for the Mutual Impedance of Crossing Earth Return Circuits," *IEEE Trans. Electromagn. Compat.*, Vol. 15, No. 4, pp. 158–160, 1973.

32. D. E. Kerr, *Propagation of Short Radio Waves*, MIT Radiation Laboratory Series, McGraw-Hill, New York, 1951, Vol. 13, pp. 98–109, 112–122, 396–444.
33. L. E. Vogler and J. L. Noble, "Curves of Input Impedance Change due to Ground for Dipole Antennas," U.S. National Bureau of Standards, Monograph 72, January 31, 1964.
34. H. R. Reed and C. M. Russell, *Ultra High Frequency Propagation*, Boston Technical Publishers, Inc., Lexington, Mass., 1964, Chapter 4, pp. 102–116.
35. C. A. Balanis, R. Hartenstein, and D. DeCarlo, "Multipath Interference for In-Flight Antenna Measurements," *IEEE Trans. Antennas Propagat.*, Vol. AP-32, No. 1, pp. 100–104, January 1984.
36. J. R. Wait and A. M. Conda, "Pattern of an Antenna on a Curved Lossy Surface," *IRE Trans. Antennas Propagat.*, Vol. AP-6, No. 4, pp. 348–359, October 1958.
37. P. Bechmann and A. Spizzichino, *The Scattering of Electromagnetic Waves from Rough Surfaces*, Macmillan, New York, 1963.
38. G. May, "Determining the Point of Reflection on MW Radio Links," *Microwave J.*, Vol. 20, No. 9, pp. 74, 76, September 1977.
39. D. T. Paris and F. K. Hurd, *Basic Electromagnetic Theory*, McGraw-Hill Book Co., pp. 385–386, 1969.
40. C. A. Balanis, "Multipath Interference in Airborne Antenna Measurements," Final Report, Prepared for Naval Air Station, Patuxent River, MD, May 28, 1982.
41. D. DeCarlo, "Automation of In-Flight Antenna Measurements," MSEE Problem Report, Dept. of Electrical Engineering, West Virginia University, July 1980.
42. G. J. Burke and A. J. Poggio, "Numerical Electromagnetics Code (NEC)-Method of Moments," Technical Document 116, Naval Ocean Systems Center, San Diego, CA, January 1981.
43. A. J. Julian, J. M. Logan, and J. W. Rockway, "MININEC: A Mini-Numerical Electromagnetics Code," Technical Document 516, Naval Ocean Systems Center, San Diego, CA, September 6, 1982.
44. J. Rockway, J. Logan, D. Tam, and S. Li, *The MININEC SYSTEM: Microcomputer Analysis of Wire Antennas*, Artech House, Inc., Norwood, MA, 1988.

PROBLEMS

4.1. A horizontal infinitesimal electric dipole of constant current I_0 is placed symmetrically about the origin and directed along the x-axis. Derive the
  (a) far-zone fields radiated by the dipole
  (b) directivity of the antenna

4.2. Repeat Problem 4.1 for a horizontal infinitesimal electric dipole directed along the y-axis.

4.3. Repeat Problem 4.1 using the procedure of Example 4.5.

4.4. For Example 4.5,
  (a) formulate an expression for the directivity.
  (b) determine the radiated power.
  (c) determine the maximum directivity by integrating the radiated power. Compare with that of Problem 4.2 or any other infinitesimal dipole.
  (d) determine the maximum directivity using the computer program **Dipole**; compare with that of part (c).

4.5. For Problem 4.1 determine the polarization of the radiated far-zone electric fields (E_θ, E_ϕ) and normalized amplitude pattern in the following planes:
  (a) $\phi = 0°$ (b) $\phi = 90°$ (c) $\theta = 90°$

4.6. Repeat Problem 4.5 for the horizontal infinitesimal electric dipole of Problem 4.2, which is directed along the *y*-axis.

4.7. For Problem 4.3, determine the polarization of the radiated far-zone fields (E_θ, E_ϕ) in the following planes:

(a) $\phi = 0°$ (b) $\phi = 90°$ (c) $\theta = 90°$

Compare with those of Problem 4.5.

4.8. For Example 4.5, determine the polarization of the radiated far-zone fields (E_θ, E_ϕ) in the following planes:

(a) $\phi = 0°$ (b) $\phi = 90°$ (c) $\theta = 90°$

Compare with those of Problem 4.6.

4.9. An infinitesimal magnetic dipole of constant current I_m and length l is symmetrically placed about the origin along the *z*-axis. Find the

(a) spherical **E**- and **H**-field components radiated by the dipole in all space

(b) directivity of the antenna

4.10. For the infinitesimal magnetic dipole of Problem 4.9, find the far-zone fields when the element is placed along the

(a) *x*-axis, (b) *y*-axis

4.11. An infinitesimal electric dipole is centered at the origin and lies on the *x-y* plane along a line which is at an angle of 45° with respect to the *x*-axis. Find the far-zone electric and magnetic fields radiated. The answer should be a function of spherical coordinates.

4.12. Repeat Problem 4.11 for an infinitesimal magnetic dipole.

4.13. An infinitesimal electric dipole of length l and constant current I_o is placed symmetrically about the origin and it is tilted at an angle of 45° *on the yz-plane. Using the vector potential approach,* determine for the infinitesimal dipole the:

(a) Far-zone electric and magnetic fields $(E_r, E_\theta, E_\phi, H_r, H_\theta, H_\phi)$ in terms of the spherical coordinates r, θ, ϕ. For example, $E_\theta(r, \theta, \phi)$. The same for the other components.

(b) Directivity (*dimensionless* and *in dB*).

(c) Polarization of the radiated fields (*linear circular or elliptical*).

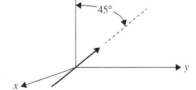

4.14. Derive (4-10a)–(4-10c) using (4-8a)–(4-9).

4.15. Derive the radiated power of (4-16) by forming the average power density, using (4-26a)–(4-26c), and integrating it over a sphere of radius *r*.

4.16. Derive the far-zone fields of an infinitesimal electric dipole, of length l and constant current I_0, using (4-4) and the procedure outlined in Section 3.6. Compare the results with (4-26a)–(4-26c).

4.17. Derive the fifth term of (4-41).

4.18. For an antenna with a maximum linear dimension of *D*, find the inner and outer boundaries of the Fresnel region so that the maximum phase error does not exceed

(a) $\pi/16$ rad (b) $\pi/4$ rad (c) 18° (d) 15°

4.19. The boundaries of the far-field (Fraunhofer) and Fresnel regions were selected based on a maximum phase error of 22.5°, which occur, respectively, at directions of 90° and 54.74° from the axis along the largest dimension of the antenna. For an antenna of maximum length of 5λ, what do these maximum phase errors reduce to at an angle of 30° from the axis along the length of the antenna? Assume that the phase error in each case is totally contributed by the respective first higher order term that is being neglected in the infinite series expansion of the distance from the source to the observation point.

4.20. The current distribution on a terminated and matched long linear (traveling wave) antenna of length l, positioned along the z-axis and fed at its one end, is given by

$$\mathbf{I} = \hat{\mathbf{a}}_z I_0 e^{-jkz'}, \quad 0 \le z' \le l$$

where I_0 is a constant. Derive expressions for the
(a) far-zone spherical electric and magnetic field components
(b) radiation power density

4.21. A line source of infinite length and constant current I_0 is positioned along the z-axis. Find the
(a) vector potential $\mathbf{A}$
(b) cylindrical $\mathbf{E}$- and $\mathbf{H}$-field components radiated

Hint: $\displaystyle\int_{-\infty}^{+\infty} \frac{e^{-j\beta\sqrt{b^2+t^2}}}{\sqrt{b^2+t^2}}\, dt = -j\pi H_0^{(2)}(\beta b)$

where $H_0^{(2)}(\alpha x)$ is the Hankel function of the second kind of order zero.

4.22. Show that (4-67) reduces to (4-68) and (4-88) to (4-89).

4.23. A thin linear dipole of length l is placed symmetrically about the z-axis. Find the far-zone spherical electric and magnetic components radiated by the dipole whose current distribution can be approximated by

(a) $I_z(z') = \begin{cases} I_0\left(1 + \frac{2}{l}z'\right), & -l/2 \le z' \le 0 \\ I_0\left(1 - \frac{2}{l}z'\right), & 0 \le z' \le l/2 \end{cases}$

(b) $I_z(z') = I_0 \cos\left(\frac{\pi}{l}z'\right), \quad -l/2 \le z' \le l/2$

(c) $I_z(z') = I_0 \cos^2\left(\frac{\pi}{l}z'\right), \quad -l/2 \le z' \le l/2$

4.24. A center-fed electric dipole of length l is attached to a balanced lossless transmission line whose characteristic impedance is 50 ohms. Assuming the dipole is resonant at the given length, find the input VSWR when

(a) $l = \lambda/4$ (b) $l = \lambda/2$ (c) $l = 3\lambda/4$ (d) $l = \lambda$

4.25. Use the equations in the book or the computer program of this chapter. Find the radiation efficiency of resonant linear electric dipoles of length

(a) $l = \lambda/50$ (b) $l = \lambda/4$ (c) $l = \lambda/2$ (d) $l = \lambda$

Assume that each dipole is made out of copper [$\sigma = 5.7 \times 10^7$ S/m], has a radius of $10^{-4}\lambda$, and is operating at $f = 10$ MHz. Use the computer program **Dipole** of this chapter to find the radiation resistances.

4.26. Write the far-zone electric and magnetic fields radiated by a magnetic dipole of $l = \lambda/2$ aligned with the z-axis. Assume a sinusoidal magnetic current with maximum value I_{m0}.

4.27. A resonant center-fed dipole is connected to a 50-ohm line. It is desired to maintain the input VSWR = 2.
  (a) What should the largest input resistance of the dipole be to maintain the VSWR = 2?
  (b) What should the length (in wavelengths) of the dipole be to meet the specification?
  (c) What is the radiation resistance of the dipole?

4.28. The radiation field of a particular antenna is given by:

$$\mathbf{E} = \hat{\mathbf{a}}_\theta j\omega\mu k \sin\theta \frac{I_0 A_1 e^{-jkr}}{4\pi r} + \hat{\mathbf{a}}_\phi \omega\mu \sin\theta \frac{I_0 A_2 e^{-jkr}}{2\pi r}$$

The values A_1 and A_2 depend on the antenna geometry. Obtain an expression for the radiation resistance. What is the polarization of the antenna?

4.29. The *approximate far zone electric field* radiated by a *very thin wire* linear dipole of length l, positioned symmetrically along the z-axis, is given by

$$E_\theta \simeq C_o \sin^{1.5}\theta \frac{e^{-jkr}}{r}$$

where C_o is a constant. Determine the:
  (a) Exact directivity (*dimensionless* and *in dB*).
  (b) Approximate directivity (*dimensionless* and *in dB*) using an approximate but appropriate formula (*state the formula you are using*).
  (c) Length of the dipole (*in wavelengths*).
  (d) Input impedance of the dipole. Assume the wire radius a is very small ($a \ll \lambda$).

4.30. For a $\lambda/2$ dipole placed symmetrical along the z-axis, determine the
  (a) vector effective height
  (b) maximum value (magnitude) of the vector effective height
  (c) ratio (in percent) of the maximum value (magnitude) of the vector effective height to its total length
  (d) maximum open-circuit output voltage when a uniform plane wave with an electric field of

$$\mathbf{E}^i|_{\theta=90°} = -\hat{\mathbf{a}}_\theta 10^{-3} \text{ volts/wavelength}$$

impinges at broadside incidence on the dipole.

4.31. A base-station cellular communication system utilizes arrays of $\lambda/2$ dipoles as transmitting and receiving antennas. Assuming that each element is *lossless* and that the *input power* to each of the $\lambda/2$ dipoles is 1 watt, determine at *1,900 MHz* and a distance of *5 km* the maximum
  (a) radiation intensity. *Specify also the units.*
  (b) radiation density (*in watts/m²*)

for each $\lambda/2$ dipole. This determines the safe level for human exposure to EM radiation.

4.32. A $\lambda/2$ dipole situated with its center at the origin radiates a time-averaged power of 600 W at a frequency of 300 MHz. A second $\lambda/2$ dipole is placed with its center at a point $P(r, \theta, \phi)$, where $r = 200$ m, $\theta = 90°$, $\phi = 40°$. It is oriented so that its axis is parallel to that of the transmitting antenna. What is the available power at the terminals of the second (receiving) dipole?

4.33. A half-wave dipole is radiating into free-space. The coordinate system is defined so that the origin is at the center of the dipole and the z-axis is aligned with the dipole. Input power to the dipole is 100 W. Assuming an overall efficiency of 50%, find the power density (in W/m²) at $r = 500$ m, $\theta = 60°$, $\phi = 0°$.

4.34. A small dipole of length $l = \lambda/20$ and of wire radius $a = \lambda/400$ is fed symmetrically, and it is used as a communications antenna at the lower end of the VHF band ($f = 30$ MHz). *The antenna is made of perfect electric conductor (PEC).* The *input reactance* of the dipole is given by

$$X_{in} = -j120\frac{[\ln(l/2a) - 1]}{\tan\left(\frac{\pi l}{\lambda}\right)}$$

Determine the following:
(a) Input impedance of the antenna. *State whether it is inductive or capacitive.*
(b) Radiation efficiency (*in percent*).
(c) Capacitor (*in farads*) *or* inductor (*in henries*) that must be connected *in series* with the dipole at the feed in order to resonate the element. *Specify which element is used and its value.*

4.35. A half-wavelength ($l = \lambda/2$) dipole is connected to a transmission line with a characteristic impedance of 75 ohms. Determine the following:
(a) Reflection coefficient. Magnitude and phase (*in degrees*).
(b) VSWR.

It is now desired to resonate the dipole using, *in series*, an inductor or capacitor. At a frequency of 100 MHz, determine:
(c) What kind of an element, inductor or capacitor, is needed to resonate the dipole?
(d) What is the inductance or capacitance?
(e) The new VSWR of the resonant dipole.

4.36. A $\lambda/2$ dipole is connected to a 50-ohm lossless transmission line. It is desired to resonate the element at 300 MHz by placing an *inductor or capacitor* in parallel/shunt at its feed points.
(a) What is the reflection coefficient and VSWR of the dipole before the insertion of the parallel/shunt element?
(b) What kind of an element is needed, inductor or capacitor, and what is its value in order to resonate the dipole?
(c) What is the new reflection coefficient and VSWR inside the transmission line after the insertion of the parallel/shunt element?

4.37. A $\lambda/2$ dipole is used as a radiating element while it is connected to a 50-ohm lossless transmission line. It is desired to resonate the element at *1.9 GHz* by placing *in series capacitor(s) or inductor(s)* (whichever are appropriate) at its input terminals. Determine the following:

(a) VSWR inside the transmission line *before the dipole is resonated [before the capacitor(s) or inductor(s) are placed in series]*.
(b) Total single capacitance C_T (in farads) or inductance L_T (in henries) that must be placed *in series* with the element at its input terminals in order to resonate it. (*See diagram a*).
(c) Individual two capacitances C_o (in farads) or inductances L_o (in henries) that must be placed *in series* with the element at its input terminals in order to resonate it. We need to use two capacitors or two inductors to keep the system balanced by placing in series one with each arm of the dipole (*see diagram b*).
(d) VSWR after the element is resonated with capacitor(s) or inductor(s).

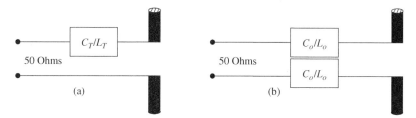

4.38. The input impedance of a $\lambda/2$ dipole, assuming the input (feed) terminals are at the center of the dipole, is equal to $73 + j42.5$. Assuming the dipole is lossless, find the
(a) input impedance (real and imaginary parts) assuming the input (feed) terminals have been shifted to a point on the dipole which is $\lambda/8$ from either end point of the dipole length
(b) capacitive or inductive reactance that must be placed across the new input terminals of part (a) so that the dipole is self-resonant
(c) VSWR at the new input terminals when the self-resonant dipole of part (b) is connected to a "twin-lead" 300-ohm line

4.39. A linear half-wavelength dipole is operating at a frequency of 1 GHz; determine the capacitance *or* inductance that must be placed *across* (in parallel) the input terminals of the dipole so that the antenna becomes resonant (make the total input impedance real). What is then the VSWR of the resonant half-wavelength dipole when it is connected to a 50-ohm line?

4.40. A folded dipole (whose length is $l = \lambda/2$ and spacing s between the two parallel length is much smaller than λ, $s \ll \lambda$), acts as an impedance transformer with a turns ratio of 2:1; i.e., its impedance is 4 times greater than that of a regular $\lambda/2$ dipole. Such a folded dipole was a basic feed element of a Yagi-Uda antenna, which was popular TV antenna prior to the cable, especially for 2-3 TV channels. Assuming the wire radius of the folded dipole is very small compared to the wavelength ($a \ll \lambda$):
(a) Write an expression for the input impedance of the folded dipole.
(b) If we want to resonate the folded dipole, what kind of an element (*inductor or capacitor*) shall we place in parallel across the input terminals of the folded dipole; i.e., which of the two is appropriate to accomplish the task?
(c) At a frequency of 100 MHz, what is the value of the inductor or capacitor, whichever is appropriate to resonate the folded dipole of Part b?

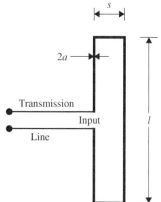

226 LINEAR WIRE ANTENNAS

(d) After the folded dipole has been resonated, what is its:
- Input impedance?
- Input reflection coefficient (assume a twin-lead transmission line connected to it with a characteristic impedance of 300 ohms)?
- VSWR?

4.41. The field radiated by an infinitesimal electric dipole, placed along the z-axis a distance s along the x-axis, is incident upon a waveguide aperture antenna of dimensions a and b, mounted on an infinite ground plane, as shown in the figure. The normalized electric field radiated by the aperture in the E-plane (x-z plane; $\phi = 0°$) is given by

$$\mathbf{E} = -\hat{\mathbf{a}}_\theta j \frac{\omega\mu b I_0 e^{-jkr}}{4\pi r} \frac{\sin\left(\frac{kb}{2}\cos\theta\right)}{\frac{kb}{2}\cos\theta}$$

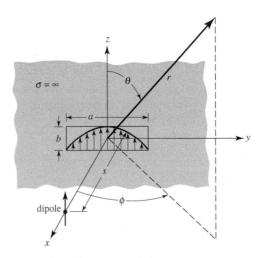

Assuming the dipole and aperture antennas are in the far field of each other, determine the polarization loss (in dB) between the two antennas.

4.42. We are given the following information about antenna A:

(a) When A is transmitting, its radiated far-field expression for the E field is given by:

$$\mathbf{E}_a(z) = E_0 \frac{e^{-jkz}}{4\pi z}\left(\frac{\hat{\mathbf{a}}_x + j\hat{\mathbf{a}}_y}{\sqrt{2}}\right) \text{ V/m}$$

(b) When A is receiving an incident plane wave given by:

$$\mathbf{E}_1(z) = \hat{\mathbf{a}}_y e^{jkz} \text{ V/m}$$

its open-circuit voltage is $V_1 = 4e^{j20°}$ V.

If we use the same antenna to receive a second incident plane given by:

$$\mathbf{E}_2(z) = 10(2\hat{\mathbf{a}}_x + \hat{\mathbf{a}}_y E^{j30°})e^{jkz} \text{ V/m}$$

find its received open-circuit voltage V_2.

4.43. A 3-cm long dipole carries a phasor current $I_0 = 10e^{j60}$ A. Assuming that $\lambda = 5$ cm, determine the E- and H-fields at 10 cm away from the dipole and at $\theta = 45°$.

4.44. The radiation resistance of a thin, lossless linear electric dipole of length $l = 0.6\lambda$ is 120 ohms. What is the input resistance?

4.45. A lossless, resonant, center-fed $3\lambda/4$ linear dipole, radiating in free-space is attached to a balanced, lossless transmission line whose characteristic impedance is 300 ohms. Assuming $a = 0.03\lambda$, calculate the:
 (a) radiation resistance (referred to the current maximum)
 (b) input impedance (referred to the input terminals)
 (c) VSWR on the transmission line

For parts (a) and (b) use the computer program **Dipole** at the end of the chapter.

4.46. Repeat Problem 4.45 for a center-fed $5\lambda/8$ dipole.

4.47. A dipole antenna, with a triangular current distribution, is used for communication with submarines at a frequency of 150 kHz. The overall length of the dipole is 200 m, and its radius is 1 m. Assume a loss resistance of 2 ohms in series with the radiation resistance of the antenna.
 (a) Evaluate the input impedance of the antenna including the loss resistance. The input reactance can be approximated by

$$X_{in} = -120\frac{[\ln(l/2a) - 1]}{\tan(\pi l/\lambda)}$$

 (b) Evaluate the radiation efficiency of the antenna.
 (c) Evaluate the radiation power factor of the antenna.
 (d) Design a conjugate-matching network to provide a perfect match between the antenna and a 50-ohm transmission line. Give the value of the series reactance X and the turns ratio n of the ideal transformer.
 (e) Assuming a conjugate match, evaluate the instantaneous 2:1 VSWR bandwidth of the antenna.

4.48. A uniform plane wave traveling along the negative z-axis given by

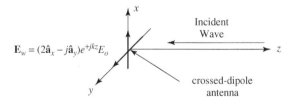

impinges upon an crossed-dipole antenna consisting of *two identical dipoles*, one directed along the x-axis and the other directed along the y-axis, *both fed with the same amplitude*. The *y-directed dipole is fed with a 90° phase lead* compared to the x-directed dipole.
 (a) Write an expression for the polarization unit vector of the incident wave.
 (b) Write an expression for the polarization unit vector of the receiving antenna *along the $+z$-axis*.
 (c) For the incident wave, state the following:
  1. Polarization (linear, circular, elliptical) and axial ratio.
  2. Rotation of the polarization vector (CW, CCW).

(d) For the receiving antenna, state the following:
   1. Polarization (linear, circular, elliptical) and axial ratio.
   2. Rotation of the polarization vector (CW, CCW).
(e) Determine the polarization loss factor (*dimensionless* and *in dB*) between the incident wave and the receiving antenna.

4.49. A half-wavelength ($l = \lambda/2$) dipole, positioned symmetrically about the origin along the z-axis, is used as a receiving antenna. A 300 MHz uniform plane wave, traveling along the x-axis in the negative x direction, impinges upon the $\lambda/2$ dipole. The incident plane wave has a power density of 2μ watts/m^2, and its electric field is given by

$$\mathbf{E}^i_w = (3\hat{\mathbf{a}}_z + j\hat{\mathbf{a}}_y)E_0 e^{+jkx}$$

where E_0 is a constant. Determine the following:
(a) Polarization of the incident wave (*including its axial ratio and sense of rotation, if applicable*).
(b) Polarization of the antenna toward the x-axis (*including its axial ratio and sense of direction, if applicable*).
(c) Polarization losses (*in dB*) between the antenna and the incoming wave (assume far-zone fields for the antenna).
(d) Maximum power (*in watts*) that can be delivered to a matched load connected to the $\lambda/2$ dipole (assume no other losses).

4.50. Derive (4-102) using (4-99).

4.51. Determine the smallest height that an infinitesimal vertical electric dipole of $l = \lambda/50$ must be placed above an electric ground plane so that its pattern has only one null (aside from the null toward the vertical), and it occurs at 30° from the vertical. For that height, find the directivity and radiation resistance.

4.52. A $\lambda/50$ linear dipole is placed vertically at a height $h = 2\lambda$ above an infinite electric ground plane. Determine the angles (in degrees) where all the nulls of its pattern occur.

4.53. A linear infinitesimal dipole of length l and constant current is placed vertically a distance h above an infinite electric ground plane. Find the first five smallest heights (in ascending order) so that a null is formed (for each height) in the far-field pattern at an angle of 60° from the vertical.

4.54. A vertical infinitesimal linear dipole is placed a distance $h = 3\lambda/2$ above an infinite perfectly conducting flat ground plane. Determine the
(a) angle(s) (*in degrees* from the vertical) where the *array factor* of the system will achieve its *maximum* value
(b) angle (*in degrees* from the vertical) where the maximum of the *total field* will occur
(c) relative (compared to its maximum) field strength (*in dB*) of the total field at the angles where the array factor of the system achieves its maximum value (as obtained in part a).

4.55. An infinitesimal dipole of length l is placed a distance s from an air-conductor interface and at an angle of $\theta = 60°$ from the vertical axis, as shown in the figure. Determine the location and direction of the image source which can be used to account for reflections. Be very clear in indicating the location and direction of the image. Your answer can be in the form of a very clear sketch.

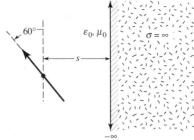

4.56. An *infinitesimal magnetic dipole* of length l, directed along the z-axis, is placed at a height h above a Perfect Electric Conductor (PEC).
  (a) Write (you do not have to derive it, as long as it is correct) an expression for the *normalized Array Factor* of the *equivalent system*.
  (b) For a height $h = 0.5\lambda$, find all the nulls in terms of the angle theta ($0° \leq \theta \leq 180°$) (*in degrees*).
  (c) If it is desired to place *a null* at $\theta = 60°$, find the two smallest heights, (other than $h = 0$) h (*in* λ) that will accomplish this.

4.57. It is desired to design an antenna system, which utilizes a vertical infinitesimal dipole of length l placed a height h above a flat, perfect electric conductor of infinite extent. The design specifications require that the pattern of the array factor of the source and its image has only one maximum, and that maximum is pointed at an angle of 60° from the vertical. Determine (in wavelengths) the height of the source to achieve this desired design specification.

4.58. A very short ($l \leq \lambda/50$) vertical electric dipole is mounted on a pole a height h above the ground, which is assumed to be flat, perfectly conducting, and of infinite extent. The dipole is used as a transmitting antenna in a VHF ($f = 50$ MHz) ground-to-air communication system. In order for the communication system transmitting antenna signal not to interfere with a nearby radio station, it is necessary to place a null in the vertical dipole system pattern at an angle of 80° from the vertical. What should the shortest height (in meters) of the dipole be to achieve the desired specifications?

4.59. A half-wavelength dipole is placed vertically on an infinite electric ground plane. Assuming that the dipole is fed at its base, find the
  (a) radiation impedance (referred to the current maximum)
  (b) input impedance (referred to the input terminals)
  (c) VSWR when the antenna is connected to a lossless 50-ohm transmission line.

4.60. A *lossless half-wavelength* ($l = \lambda/2$) dipole operating at 1 GHz, with an ideal sinusoidal current distribution, is *placed horizontally a height h above a flat, smooth and infinite in extent perfect electric conductor* (PEC).
  (a) Write an expression for the *normalized* array factor. Assume angle θ is measured from the vertical to the ground plane.
  (b) For a height $h = 1.5\lambda$, determine *all* the *physical* angles θ ($0° \leq \theta \leq 90°$) where the array factor achieves its maximum value.
  (c) For the same height $h = 1.5\lambda$, find all the physical angles θ ($0° \leq \theta \leq 90°$) where the array factor has null(s).

4.61. A *resonant* vertical $\lambda/8$ monopole, mounted on an infinite flat Perfect Electric Conductor (PEC), is connected to a lossless transmission line. It is desired to maintain the maximum reflection coefficient inside the transmission line to 0.2. Determine the:
  (a) Total far-zone electric field radiated by the $\lambda/8$ monopole on and above the PEC.
  (b) Input resistance of the monopole.
  (c) The desired characteristic impedance of the transmission line to maintain the maximum reflection coefficient to 0.2.

4.62. A vertical $\lambda/2$ *dipole* is the radiating element in a circular array used for over-the-horizon communication system operating at *1 GHz*. The circular array (*center of the dipoles*) is placed at a *height h* above the ground that is assumed to be flat, perfect electric conducting, and infinite in extent.

(a) In order for the array not to be interfered with by another communication system that is operating in the same frequency, it is desired to place *only one null* in the *elevation pattern of the array factor* of a single vertical $\lambda/2$ dipole at an angle of $\theta = 30°$ from zenith (axis of the dipole). Determine the *smallest nonzero height h* (*in meters*) above the ground at which the center of the dipole must be placed to accomplish this.

(b) If the height (*at its center*) of the vertical dipole is *0.3 m* above ground, determine *all the angles θ* from zenith (*in degrees*) where *all* the
  1. null(s) of the *array factor* of a single dipole in the elevation plane will be directed toward.
  2. main maximum (maxima) of the *array factor* of a single dipole in the elevation plane will be directed toward.

4.63. A vertical $\lambda/2$ dipole antenna is used as a ground-to-air, over-the-horizon communication antenna at the VHF band ($f = 200\ MHz$). The antenna is elevated at a height h (*measured from its center/feed point*) above ground (*assume the ground is flat, smooth, and perfect electric conductor extending to infinity*). In order to avoid interference with other simultaneously operating communication systems, it is desired to place a null in the far-field amplitude pattern of the antenna system at an angle of *60° from the vertical*.

Determine the *three smallest physical/nontrivial heights* (*in meters at 200 MHz*) above the ground at which the antenna can be placed to meet the desired pattern specifications.

4.64. A ground-based, *resonant, lossless* linear *vertical* half-wavelength dipole (of length $l = \lambda/2$), elevated at a height h about PEC (perfect electric conducting) ground plane, is used as the antenna for a ground-based communication system. It is expected that some interferers/threats to the ground-based communication system will appear at a height 1,000 meters and at a horizontal distance of 1,000 meters from the ground-based antenna, as shown in the figure below.

To eliminate the presence of the interferers/threats to the operation of the ground-based communication system, we want to choose the height h of the ground-based system above the PEC ground so that we place an ideal null in the angular direction θ of the interferers as measured from the reference of the coordinate system.

Determine, at a frequency of 300 MHz:

(a) The *normalized AF* (array factor) of the equivalent antenna system that is valid in all space on and above the PEC ground plane.

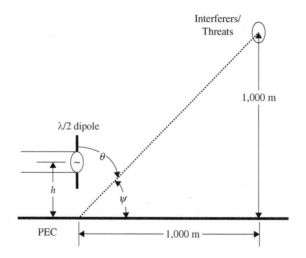

(b) *The two smallest heights of h* (in meters), *from the smallest to largest,* that we can place the ground-based antenna and achieve the goal; i.e., to place a null in the θ direction and eliminate the presence of the interferers/threats to the operation of the ground-based system. Assume far-field observations.

4.65. A base-station cellular communication systems lossless antenna, which is placed in a residential area of a city, has a maximum gain of *16 dB* (above isotropic) *toward the residential area* at *1,900 MHz*. Assuming the *input* power to the antenna is 8 watts, what is the

(a) maximum radiated power density (*in watts/cm²*) at a distance of 100 m (*line of sight*) from the base station to the residential area? This will determine the safe level for human exposure to electromagnetic radiation.

(b) power (*in watts*) received at that point of the residential area by a cellular telephone whose antenna is a *lossless* $\lambda/4$ vertical monopole and whose maximum value of the amplitude pattern is directed toward the maximum incident power density. *Assume the $\lambda/4$ monopole is mounted on an infinite ground plane.*

4.66. A vertical $\lambda/4$ monopole is used as the antenna on a cellular telephone operating at 1.9 GHz. Even though the monopole is mounted on a box-type cellular telephone, for simplicity purposes, assume here that it is mounted on a perfectly electric conducting (PEC) ground plane. Assuming an incident maximum power density of 10^{-6} watts/m², *state or determine*, for the monopole's omnidirectional pattern, the

(a) maximum directivity (*dimensionless and in dB*). *You must state the rationale or method you are using to find the directivity.*

(b) maximum power that can be delivered to the cellular telephone receiver. *Assume no losses.*

4.67. A vertical, infinitesimal in length ($l = \lambda/50$), monopole is placed on top of a police car, and it is used as an antenna for an emergency radio receiver system operating at *10 MHz*. Consider that the top of the police car to be an *infinite and planar PEC*. The sensitivity (*minimum power*) of the system receiver, to be able to detect an incoming signal, is *10 μwatts*. Assume the incoming signal is *circularly polarized* and it incident from a horizontal direction (grazing angle; $\theta = 90°$), what is the:

(a) Maximum directivity of the monopole in the presence of the PEC (*dimensionless* and *in dB*)?

(b) Minimum required power density (*in watts/cm²*) of the incoming signal to be detected by the radio receiver?

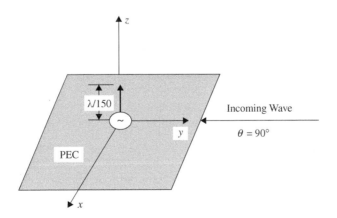

4.68. A homeowner uses a CB antenna mounted on the top of his house. Let us assume that the operating frequency is 900 MHz and the radiated power is *1,000 watts*. In order not to be exposed to a long-term microwave radiation, there have been some standards, although controversial, developed that set the maximum safe power density that humans can be exposed to and not be subject to any harmful effects. Let us assume that the maximum safe power density of long-term human RF exposure is 10^{-3} *watts/cm²* or *10 watts/m²*. Assuming no losses, determine the *shortest distance (in meters)* from the CB antenna you must be in order not to exceed the safe level of power density exposure. Assume that the CB antenna is radiating into free-space and it is

(a) an isotropic radiator.

(b) a $\lambda/4$ monopole mounted on an infinite PEC and radiating towards its maximum.

4.69. Derive (4-118) using (4-115).

4.70. An infinitesimal horizontal electric dipole of length $l = \lambda/50$ is placed parallel to the y-axis a height h above an infinite electric ground plane.

(a) Find the smallest height h (excluding $h = 0$) that the antenna must be elevated so that a null in the $\phi = 90°$ plane will be formed at an angle of $\theta = 45°$ from the vertical axis.

(b) For the height of part (a), determine the (1) radiation resistance and (2) directivity (for $\theta = 0°$) of the antenna system.

4.71. A horizontal $\lambda/50$ infinitesimal dipole of constant current and length l is placed parallel to the y-axis a distance $h = 0.707\lambda$ above an infinite electric ground plane. Find *all* the nulls formed by the antenna system in the $\phi = 90°$ plane.

4.72. An infinitesimal electric dipole of length $l = \lambda/50$ is placed horizontally at a height of $h = 2\lambda$ above a flat, smooth, perfect electric conducting plane which extends to infinity. It is desired to measure its far-field radiation characteristics (e.g. amplitude pattern, phase pattern, polarization pattern, etc.). The system is operating at 300 MHz. What should the minimum radius *(in meters)* of the circle be where the measurements should be carried out? The radius should be measured from the origin of the coordinate system, which is taken at the interface between the actual source and image.

4.73. An infinitesimal magnetic dipole is placed vertically a distance h above an infinite, perfectly conducting electric ground plane. Derive the far-zone fields radiated by the element above the ground plane.

4.74. Repeat Problem 4.73 for an electric dipole above an infinite, perfectly conducting magnetic ground plane.

4.75. A $\lambda/50$ infinitesimal linear electric dipole operating at 500 MHz is placed horizontally a height h above a simulated flat, smooth and infinite in extent *perfect magnetic conductor* (PMC). Determine/write the:

(a) *Array factor* for the system that can be used to determine the far-zone field on and above the PMC. *Justify as to why you chose that array factor.*

(b) *Smallest* height h *(in λ and in cm)* that the $\lambda/50$ electric dipole must be placed so that the total far-zone amplitude pattern has a null at an angle of $\theta = 60°$ from the normal/vertical to the PMC interface.

4.76. Repeat Problem 4.73 for a magnetic dipole above an infinite, perfectly conducting magnetic ground plane.

4.77. An infinitesimal vertical electric dipole is placed at height h above an infinite PMC (perfect magnetic conductor) ground plane.

(a) Find the smallest height h (excluding $h = 0$) to which the antenna must be elevated so that a null is formed at an angle $\theta = 60°$ from the vertical axis

(b) For the value of h found in part (a), determine
1. the directive gain of the antenna in the $\theta = 45°$ direction
2. the radiation resistance of the antenna normalized to the intrinsic impedance of the medium above the ground plane

Assume that the length of the antenna is $l = \lambda/100$.

4.78. A vertical $\lambda/2$ dipole, operating at 1 GHz, is placed a distance of 5 m (with respect to the tangent at the point of reflections) above the earth. Find the total field at a point 20 km from the source ($d = 20 \times 10^3$ m), at a height of 1,000 m (with respect to the tangent) above the ground. Use a 4/3 radius earth and assume that the electrical parameters of the earth are $\varepsilon_r = 5, \sigma = 10^{-2}$ S/m.

4.79. A ground-based, *resonant, lossless* linear *vertical* half-wavelength dipole (of length $l = \lambda/2$), is used to communicate with a space-borne *lossless, resonant,* linear wave-wavelength dipole (of length $l = \lambda/2$). *Both dipoles are oriented along the z axis.* While one dipole is assumed to be at ground level, the other is elevated at a height of 1,000 meter; the two dipoles are separated horizontally by 1,000 meters, as shown below.

Assuming the input power (in the 50-ohm transmission line) feeding the dipole at the ground level is 100 mwatts, determine, at a frequency of 3 GHz, the power (*in watts*) received in the 50-ohm transmission line which is connected to the space borne dipole.

Assume both dipoles are radiating in an unbounded free space and each is in the far-field of the other.

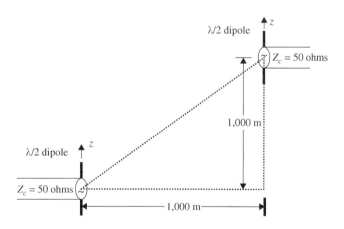

4.80. Two astronauts equipped with handheld radios land on different parts of a large asteroid. The radios are identical and transmit 5 W average power at 300 MHz. Assume the asteroid is a smooth sphere with physical radius of 1,000 km, has no atmosphere, and consists of a lossless dielectric material with relative permittivity $\varepsilon_r = 9$. Assume that the radios' antennas can be modeled as vertical infinitesimal electric dipoles. Determine the signal power (in microwatts) received by each radio from the other, if the astronauts are separated by a range (distance along the asteroid's surface) of 2 km, and hold their radios vertically at heights of 1.5 m above the asteroid's surface.

Additional Information Required to Answer this Question: Prior to landing on the asteroid the astronauts calibrated their radios. Separating themselves in outer space by 10 km, the astronauts found the received signal power at each radio from the other was 10 microwatts, when both antennas were oriented in the same direction.

4.81. A satellite S transmits an electromagnetic wave, at 10 GHz, via its transmitting antenna. The characteristics of the satellite-based transmitter are:
   (a) The power radiated from the satellite antenna is 10 W.
   (b) The distance between the satellite antenna and a point A on the earth's surface is 3.7×10^7 m, and
   (c) The satellite transmitting antenna directivity in the direction SA is 50 dB. Ignoring ground effects,
      1. Determine the magnitude of the E-field at A.
      2. If the receiver at point A is a $\lambda/2$ dipole, what would be the voltage reading at the terminals of the antenna?

4.82. Derive (4-132) based on geometrical optics as presented in section 13.2 of [7].

CHAPTER 5

Loop Antennas

5.1 INTRODUCTION

Another simple, inexpensive, and very versatile antenna type is the loop antenna. Loop antennas take many different forms such as a rectangle, square, triangle, ellipse, circle, and many other configurations. Because of the simplicity in analysis and construction, the circular loop is the most popular and has received the widest attention. It will be shown that a small loop (circular or square) is equivalent to an infinitesimal magnetic dipole whose axis is perpendicular to the plane of the loop. That is, the fields radiated by an electrically small circular or square loop are of the same mathematical form as those radiated by an infinitesimal magnetic dipole.

Loop antennas are usually classified into two categories, electrically small and electrically large. Electrically small antennas are those whose overall length (circumference) is usually less than about one-tenth of a wavelength ($C < \lambda/10$). However, electrically large loops are those whose circumference is about a free-space wavelength ($C \sim \lambda$). Most of the applications of loop antennas are in the HF (3–30 MHz), VHF (30–300 MHz), and UHF (300–3,000 MHz) bands. When used as field probes, they find applications even in the microwave frequency range.

Loop antennas with electrically small circumferences or perimeters have small radiation resistances that are usually smaller than their loss resistances. Thus they are very poor radiators, and they are seldom employed for transmission in radio communication. When they are used in any such application, it is usually in the receiving mode, such as in portable radios and pagers, where antenna efficiency is not as important as the signal-to-noise ratio. They are also used as probes for field measurements and as directional antennas for radiowave navigation. The field pattern of electrically small antennas of any shape (circular, elliptical, rectangular, square, etc.) is similar to that of an infinitesimal dipole with a null perpendicular to the plane of the loop and with its maximum along the plane of the loop. As the overall length of the loop increases and its circumference approaches one free-space wavelength, the maximum of the pattern shifts from the plane of the loop to the axis of the loop which is perpendicular to its plane.

The radiation resistance of the loop can be increased, and made comparable to the characteristic impedance of practical transmission lines, by increasing (electrically) its perimeter and/or the number of turns. Another way to increase the radiation resistance of the loop is to insert, within its circumference or perimeter, a ferrite core of very high permeability which will raise the magnetic field intensity and hence the radiation resistance. This forms the so-called ferrite loop.

Electrically large loops are used primarily in directional arrays, such as in helical antennas (see Section 10.3.1), Yagi-Uda arrays (see Section 10.3.3), quad arrays (see Section 10.3.4), and so on.

236 LOOP ANTENNAS

(a) Single element   (b) Array of eight elements

Figure 5.1 Commercial loop antenna as a single vertical element and in the form of an eight-element linear array. (Courtesy: TCI, A Dielectric Company).

For these and other similar applications, the maximum radiation is directed toward the axis of the loop forming an end-fire antenna. To achieve such directional pattern characteristics, the circumference (perimeter) of the loop should be about one free-space wavelength. The proper phasing between turns enhances the overall directional properties.

Loop antennas can be used as single elements, as shown in Figure 5.1(a), whose plane of its area is perpendicular to the ground. The relative orientation of the loop can be in other directions, including its plane being parallel relative to the ground. Thus, its mounting orientation will determine its radiation characteristics relative to the ground. Loops are also used in arrays of various forms. The particular array configuration will determine its overall pattern and radiation characteristics. One form of arraying is shown in Figure 5.1(b), where eight loops of Figure 5.1(a) are placed to form a linear array of eight vertical elements.

5.2 SMALL CIRCULAR LOOP

The most convenient geometrical arrangement for the field analysis of a loop antenna is to position the antenna symmetrically on the x-y plane, at $z = 0$, as shown in Figure 5.2(a). The wire is assumed to be very thin and the current spatial distribution is given by

$$I_\phi = I_0 \tag{5-1}$$

where I_0 is a constant. Although this type of current distribution is accurate only for a loop antenna with a very small circumference, a more complex distribution makes the mathematical formulation quite cumbersome.

5.2.1 Radiated Fields

To find the fields radiated by the loop, the same procedure is followed as for the linear dipole. The potential function **A** given by (3-53) as

$$\mathbf{A}(x, y, z) = \frac{\mu}{4\pi} \int_C \mathbf{I}_e(x', y', z') \frac{e^{-jkR}}{R} \, dl' \tag{5-2}$$

SMALL CIRCULAR LOOP   237

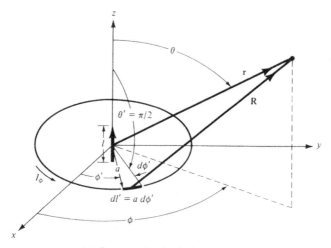

(a) Geometry for circular loop

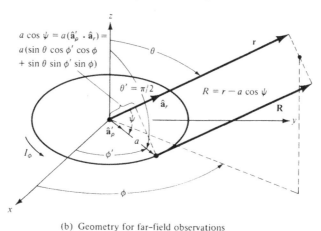

(b) Geometry for far-field observations

Figure 5.2 Geometrical arrangement for loop antenna analysis.

is first evaluated. Referring to Figure 5.2(a), R is the distance from any point on the loop to the observation point and dl' is an infinitesimal section of the loop antenna. In general, the current spatial distribution $\mathbf{I}_e(x', y', z')$ can be written as

$$\mathbf{I}_e(x', y', z') = \hat{\mathbf{a}}_x I_x(x', y', z') + \hat{\mathbf{a}}_y I_y(x', y', z') + \hat{\mathbf{a}}_z I_z(x', y', z') \tag{5-3}$$

whose form is more convenient for linear geometries. For the circular-loop antenna of Figure 5.2(a), whose current is directed along a circular path, it would be more convenient to write the rectangular current components of (5-3) in terms of the cylindrical components using the transformation (see Appendix VII)

$$\begin{bmatrix} I_x \\ I_y \\ I_z \end{bmatrix} = \begin{bmatrix} \cos\phi' & -\sin\phi' & 0 \\ \sin\phi' & \cos\phi' & 0 \\ 0 & 0 & 1 \end{bmatrix} \begin{bmatrix} I_\rho \\ I_\phi \\ I_z \end{bmatrix} \tag{5-4}$$

238 LOOP ANTENNAS

which when expanded can be written as

$$\left.\begin{array}{l} I_x = I_\rho \cos\phi' - I_\phi \sin\phi' \\ I_y = I_\rho \sin\phi' + I_\phi \cos\phi' \\ I_z = I_z \end{array}\right\} \quad (5\text{-}5)$$

Since the radiated fields are usually determined in spherical components, the rectangular unit vectors of (5-3) are transformed to spherical unit vectors using the transformation matrix given by (4-5). That is,

$$\left.\begin{array}{l} \hat{\mathbf{a}}_x = \hat{\mathbf{a}}_r \sin\theta\cos\phi + \hat{\mathbf{a}}_\theta \cos\theta\cos\phi - \hat{\mathbf{a}}_\phi \sin\phi \\ \hat{\mathbf{a}}_y = \hat{\mathbf{a}}_r \sin\theta\sin\phi + \hat{\mathbf{a}}_\theta \cos\theta\sin\phi + \hat{\mathbf{a}}_\phi \cos\phi \\ \hat{\mathbf{a}}_z = \hat{\mathbf{a}}_r \cos\theta \quad\quad - \hat{\mathbf{a}}_\theta \sin\theta \end{array}\right\} \quad (5\text{-}6)$$

Substituting (5-5) and (5-6) in (5-3) reduces it to

$$\begin{aligned} \mathbf{I}_e = &\ \hat{\mathbf{a}}_r [I_\rho \sin\theta\cos(\phi-\phi') + I_\phi \sin\theta\sin(\phi-\phi') + I_z \cos\theta] \\ &+ \hat{\mathbf{a}}_\theta [I_\rho \cos\theta\cos(\phi-\phi') + I_\phi \cos\theta\sin(\phi-\phi') - I_z \sin\theta] \\ &+ \hat{\mathbf{a}}_\phi [-I_\rho \sin(\phi-\phi') + I_\phi \cos(\phi-\phi')] \end{aligned} \quad (5\text{-}7)$$

allowing for I_ρ, I_ϕ and I_z current components.

It should be emphasized that the source coordinates are designated as primed (ρ',ϕ',z') and the observation coordinates as unprimed (r,θ,ϕ). For a very thin wire radius circular loop, the current is flowing in the ϕ direction (I_ϕ) so that (5-7) reduces to

$$\mathbf{I}_e = \hat{\mathbf{a}}_r I_\phi \sin\theta\sin(\phi-\phi') + \hat{\mathbf{a}}_\theta I_\phi \cos\theta\sin(\phi-\phi') + \hat{\mathbf{a}}_\phi I_\phi \cos(\phi-\phi') \quad (5\text{-}8)$$

The distance R, from any point on the loop to the observation point, can be written as

$$R = \sqrt{(x-x')^2 + (y-y')^2 + (z-z')^2} \quad (5\text{-}9)$$

Since

$$\begin{aligned} x &= r\sin\theta\cos\phi \\ y &= r\sin\theta\sin\phi \\ z &= r\cos\theta \\ x^2 + y^2 + z^2 &= r^2 \\ x' &= a\cos\phi' \\ y' &= a\sin\phi' \\ z' &= 0 \\ x'^2 + y'^2 + z'^2 &= a^2 \end{aligned} \quad (5\text{-}10)$$

(5-9) reduces to

$$R = \sqrt{r^2 + a^2 - 2ar\sin\theta\cos(\phi - \phi')} \tag{5-11}$$

By referring to Figure 5.2(a), the differential element length is given by

$$dl' = a\,d\phi' \tag{5-12}$$

Using (5-8), (5-11), and (5-12), the ϕ-component of (5-2) can be written as

$$A_\phi = \frac{a\mu}{4\pi} \int_0^{2\pi} I_\phi \cos(\phi - \phi') \frac{e^{-jk\sqrt{r^2+a^2-2ar\sin\theta\cos(\phi-\phi')}}}{\sqrt{r^2 + a^2 - 2ar\sin\theta\cos(\phi - \phi')}} \, d\phi' \tag{5-13}$$

Since the spatial current I_ϕ as given by (5-1) is constant, the field radiated by the loop will not be a function of the observation angle ϕ. Thus any observation angle ϕ can be chosen; for simplicity $\phi = 0$. Therefore (5-13) reduces to

$$A_\phi = \frac{a\mu I_0}{4\pi} \int_0^{2\pi} \cos\phi' \frac{e^{-jk\sqrt{r^2+a^2-2ar\sin\theta\cos\phi'}}}{\sqrt{r^2 + a^2 - 2ar\sin\theta\cos\phi'}} \, d\phi' \tag{5-14}$$

The integration of (5-14), for very thin circular loop of any radius, can be carried out and is represented by a complex infinite series whose real part contains complete elliptic integrals of the first and second kind while the imaginary part consists of elementary functions [1]. This treatment is only valid provided the observation distance is greater than the radius of the loop ($r > a$). Another very detailed and systematic treatment is that of [2], [3] which is valid for any observation distance ($r < a, r > a$) except when the observation point is on the loop itself ($r = a, \theta = \pi/2$). The development in [2], [3] has been applied to circular loops whose current distribution is uniform, cosinusoidal, and Fourier cosine series. Asymptotic expansions have been presented in [2], [3] to find simplified and approximate forms for far-field observations.

Both treatments, [1]–[3], are too complex to be presented here. The reader is referred to the literature. In this chapter a method will be presented that approximates the integration of (5-14). For small loops, the function

$$f = \frac{e^{-jk\sqrt{r^2+a^2-2ar\sin\theta\cos\phi'}}}{\sqrt{r^2 + a^2 - 2ar\sin\theta\cos\phi'}} \tag{5-15}$$

which is part of the integrand of (5-14), can be expanded in a Maclaurin series in a using

$$f = f(0) + f'(0)a + \frac{1}{2!}f''(0)a^2 + \cdots + \frac{1}{(n-1)!}f^{(n-1)}(0)a^{n-1} + \cdots \tag{5-15a}$$

where $f'(0) = \partial f/\partial a|_{a=0}, f''(0) = \partial^2 f/\partial a^2|_{a=0}$, and so forth. Taking into account only the first two terms of (5-15a), or

$$f(0) = \frac{e^{-jkr}}{r} \tag{5-15b}$$

$$f'(0) = \left(\frac{jk}{r} + \frac{1}{r^2}\right) e^{-jkr} \sin\theta \cos\phi' \tag{5-15c}$$

$$f \simeq \left[\frac{1}{r} + a\left(\frac{jk}{r} + \frac{1}{r^2}\right) \sin\theta \cos\phi'\right] e^{-jkr} \tag{5-15d}$$

reduces (5-14) to

$$A_\phi \simeq \frac{a\mu I_0}{4\pi} \int_0^{2\pi} \cos\phi' \left[\frac{1}{r} + a\left(\frac{jk}{r} + \frac{1}{r^2}\right) \sin\theta \cos\phi'\right] e^{-jkr} \, d\phi'$$

$$A_\phi \simeq \frac{a^2 \mu I_0}{4} e^{-jkr} \left(\frac{jk}{r} + \frac{1}{r^2}\right) \sin\theta \tag{5-16}$$

In a similar manner, the r- and θ-components of (5-2) can be written as

$$A_r \simeq \frac{a\mu I_0}{4\pi} \sin\theta \int_0^{2\pi} \sin\phi' \left[\frac{1}{r} + a\left(\frac{jk}{r} + \frac{1}{r^2}\right) \sin\theta \cos\phi'\right] e^{-jkr} \, d\phi' \tag{5-16a}$$

$$A_\theta \simeq -\frac{a\mu I_0}{4\pi} \cos\theta \int_0^{2\pi} \sin\phi' \left[\frac{1}{r} + a\left(\frac{jk}{r} + \frac{1}{r^2}\right) \sin\theta \cos\phi'\right] e^{-jkr} \, d\phi' \tag{5-16b}$$

which when integrated reduce to zero. Thus

$$\mathbf{A} \simeq \hat{\mathbf{a}}_\phi A_\phi = \hat{\mathbf{a}}_\phi \frac{a^2 \mu I_0}{4} e^{-jkr} \left[\frac{jk}{r} + \frac{1}{r^2}\right] \sin\theta$$

$$= \hat{\mathbf{a}}_\phi j \frac{k\mu a^2 I_0 \sin\theta}{4r} \left[1 + \frac{1}{jkr}\right] e^{-jkr} \tag{5-17}$$

Substituting (5-17) into (3-2a) reduces the magnetic field components to

$$H_r = j\frac{ka^2 I_0 \cos\theta}{2r^2} \left[1 + \frac{1}{jkr}\right] e^{-jkr} \tag{5-18a}$$

$$H_\theta = -\frac{(ka)^2 I_0 \sin\theta}{4r} \left[1 + \frac{1}{jkr} - \frac{1}{(kr)^2}\right] e^{-jkr} \tag{5-18b}$$

$$H_\phi = 0 \tag{5-18c}$$

Using (3-15) or (3-10) with $\mathbf{J} = 0$, the corresponding electric-field components can be written as

$$E_r = E_\theta = 0 \tag{5-19a}$$

$$E_\phi = \eta \frac{(ka)^2 I_0 \sin\theta}{4r} \left[1 + \frac{1}{jkr}\right] e^{-jkr} \tag{5-19b}$$

5.2.2 Small Loop and Infinitesimal Magnetic Dipole

A comparison of (5-18a)–(5-19b) with those of the infinitesimal magnetic dipole indicates that they have similar forms. In fact, the electric and magnetic field components of an infinitesimal magnetic dipole of length l and constant "magnetic" spatial current I_m are given by

$$E_r = E_\theta = H_\phi = 0 \tag{5-20a}$$

$$E_\phi = -j\frac{kI_m l \sin\theta}{4\pi r} \left[1 + \frac{1}{jkr}\right] e^{-jkr} \tag{5-20b}$$

$$H_r = \frac{I_m l \cos\theta}{2\pi\eta r^2} \left[1 + \frac{1}{jkr}\right] e^{-jkr} \tag{5-20c}$$

$$H_\theta = j\frac{kI_m l \sin\theta}{4\pi\eta r} \left[1 + \frac{1}{jkr} - \frac{1}{(kr)^2}\right] e^{-jkr} \tag{5-20d}$$

These can be obtained, using duality, from the fields of an infinitesimal electric dipole, (4-8a)–(4-10c). When (5-20a)–(5-20d) are compared with (5-18a)–(5-19b), they indicate that *a magnetic dipole of magnetic moment $I_m l$ is equivalent to a small electric loop of radius a and constant electric current I_0 provided that*

$$I_m l = jS\omega\mu I_0 \tag{5-21}$$

where $S = \pi a^2$ (area of the loop). Thus, for analysis purposes, the small electric loop can be replaced by a small linear magnetic dipole of constant current. The geometrical equivalence is illustrated in Figure 5.2(a) where the magnetic dipole is directed along the z-axis which is also perpendicular to the plane of the loop.

5.2.3 Power Density and Radiation Resistance

The fields radiated by a small loop, as given by (5-18a)–(5-19b), are valid everywhere except at the origin. As was discussed in Section 4.1 for the infinitesimal dipole, the power in the region very close to the antenna (near field, $kr \ll 1$) is predominantly reactive and in the far field ($kr \gg 1$) is predominantly real. To illustrate this for the loop, the complex power density

$$\mathbf{W} = \tfrac{1}{2}(\mathbf{E} \times \mathbf{H}^*) = \tfrac{1}{2}[(\hat{\mathbf{a}}_\phi E_\phi) \times (\hat{\mathbf{a}}_r H_r^* + \hat{\mathbf{a}}_\theta H_\theta^*)]$$

$$= \tfrac{1}{2}(-\hat{\mathbf{a}}_r E_\phi H_\theta^* + \hat{\mathbf{a}}_\theta E_\phi H_r^*) \tag{5-22}$$

is first formed. When (5-22) is integrated over a closed sphere, only its radial component given by

$$W_r = \eta \frac{(ka)^4}{32} |I_0|^2 \frac{\sin^2\theta}{r^2} \left[1 + j\frac{1}{(kr)^3}\right] \qquad (5\text{-}22\text{a})$$

contributes to the complex power P_r. Thus

$$P_r = \oiint_S \mathbf{W} \cdot d\mathbf{s} = \eta \frac{(ka)^4}{32} |I_0|^2 \int_0^{2\pi}\int_0^\pi \left[1 + j\frac{1}{(kr)^3}\right] \sin^3\theta \, d\theta \, d\phi \qquad (5\text{-}23)$$

which reduces to

$$P_r = \eta\left(\frac{\pi}{12}\right)(ka)^4 |I_0|^2 \left[1 + j\frac{1}{(kr)^3}\right] \qquad (5\text{-}23\text{a})$$

and whose real part is equal to

$$P_{\text{rad}} = \eta\left(\frac{\pi}{12}\right)(ka)^4 |I_0|^2 \qquad (5\text{-}23\text{b})$$

For small values of kr ($kr \ll 1$), the second term within the brackets of (5-23a) is dominant which makes the power mainly reactive. In the far field ($kr \gg 1$), the second term within the brackets diminishes, which makes the power real. *A comparison between (5-23a) with (4-14) indicates a difference in sign between the terms within the brackets. Whereas for the infinitesimal dipole the radial power density in the near field is capacitive, for the small loop it is inductive.* This is illustrated in Figure 4.21 for the dipole and in Figures 5.13 and 5.20 for the loop.

The radiation resistance of the loop is found by equating (5-23b) to $|I_0|^2 R_r/2$. Doing this, the radiation resistance can be written as

$$R_r = \eta\left(\frac{\pi}{6}\right)(k^2 a^2)^2 = \eta\frac{2\pi}{3}\left(\frac{kS}{\lambda}\right)^2 = 20\pi^2\left(\frac{C}{\lambda}\right)^4 \simeq 31{,}171\left(\frac{S^2}{\lambda^4}\right) \qquad (5\text{-}24)$$

where $S = \pi a^2$ is the area and $C = 2\pi a$ is the circumference of the loop. The last form of (5-24) holds for loops of other configurations, such as rectangular, elliptical, etc. (See Problem 5.30).

The radiation resistance as given by (5-24) is only for a single-turn loop. If the loop antenna has N turns wound so that the magnetic field passes through all the loops, the radiation resistance is equal to that of single turn multiplied by N^2. That is,

$$R_r = \eta\left(\frac{2\pi}{3}\right)\left(\frac{kS}{\lambda}\right)^2 N^2 = 20\pi^2\left(\frac{C}{\lambda}\right)^4 N^2 \simeq 31{,}171\, N^2 \left(\frac{S^2}{\lambda^4}\right) \qquad (5\text{-}24\text{a})$$

Even though the radiation resistance of a single-turn loop may be small, the overall value can be increased by including many turns. This is a very desirable and practical mechanism that is not available for the infinitesimal dipole.

Example 5.1

Find the radiation resistance of a single-turn and an eight-turn small circular loop. The radius of the loop is $\lambda/25$ and the medium is free-space.

Solution:

$$S = \pi a^2 = \pi \left(\frac{\lambda}{25}\right)^2 = \frac{\pi \lambda^2}{625}$$

$$R_r \text{ (single turn)} = 120\pi \left(\frac{2\pi}{3}\right)\left(\frac{2\pi^2}{625}\right)^2 = 0.788 \text{ ohms}$$

$$R_r \text{ (8 turns)} = 0.788(8)^2 = 50.43 \text{ ohms}$$

The radiation and loss resistances of an antenna determine the radiation efficiency, as defined by (2-90). The loss resistance of a single-turn small loop is, in general, much larger than its radiation resistance; thus the corresponding radiation efficiencies are very low and depend on the loss resistance. To increase the radiation efficiency, multiturn loops are often employed. However, because the current distribution in a multiturn loop is quite complex, great confidence has not yet been placed in analytical methods for determining the radiation efficiency. Therefore greater reliance has been placed on experimental procedures. Two experimental techniques that can be used to measure the radiation efficiency of a small multiturn loop are those that are usually referred to as the *Wheeler method* and the *Q method* [4].

Usually it is assumed that the loss resistance of a small loop is the same as that of a straight wire whose length is equal to the circumference of the loop, and it is computed using (2-90b). Although this assumption is adequate for single-turn loops, it is not valid for multiturn loops. In a multiturn loop, the current is not uniformly distributed around the wire but depends on the skin and proximity effects [5]. In fact, for close spacings between turns, the contribution to the loss resistance due to the proximity effect can be larger than that due to the skin effect.

The total ohmic resistance for an N-turn circular-loop antenna with loop radius a, wire radius b, and loop separation $2c$, shown in Figure 5.3(a) is given by [6]

$$R_{\text{ohmic}} = \frac{Na}{b} R_s \left(\frac{R_p}{R_0} + 1\right) \qquad (5\text{-}25)$$

where

$$R_s = \sqrt{\frac{\omega \mu_0}{2\sigma}} = \text{surface impedance of conductor}$$

R_p = ohmic resistance per unit length due to proximity effect

$$R_0 = \frac{NR_s}{2\pi b} = \text{ohmic skin effect resistance per unit length (ohms/m)}$$

The ratio of R_p/R_0 has been computed [6] as a function of the spacing c/b for loops with $2 \leq N \leq 8$ and it is shown plotted in Figure 5.3(b). It is evident that for close spacing the ohmic resistance is twice as large as that in the absence of the proximity effect ($R_p/R_0 = 0$).

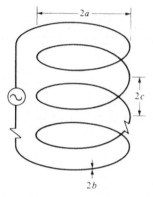

(a) *N*-turn circular loop

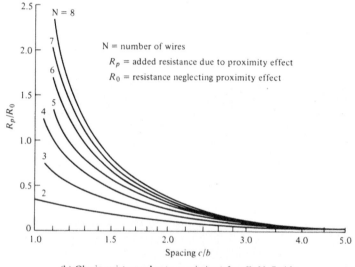

(b) Ohmic resistance due to proximity (after G. N. Smith)

Figure 5.3 *N*-turn circular loop and ohmic resistance due to proximity effect. (SOURCE: G. S. Smith, "Radiation Efficiency of Electrically Small Multiturn Loop Antennas," *IEEE Trans. Antennas Propagat.*, Vol. AP-20, No. 5, September, pp. 656–657. 1972 © 1972 IEEE).

Example 5.2

Find the radiation efficiency of a single-turn and an eight-turn small circular loop at $f = 100$ MHz. The radius of the loop is $\lambda/25$, the radius of the wire is $10^{-4}\lambda$, and the turns are spaced $4 \times 10^{-4}\lambda$ apart. Assume the wire is copper with a conductivity of 5.7×10^7 (S/m) and the antenna is radiating into free-space.

Solution: From Example 5.1

$$R_r \text{ (single turn)} = 0.788 \text{ ohms}$$

$$R_r \text{ (8 turns)} = 50.43 \text{ ohms}$$

The loss resistance for a single turn is given, according to (2-90b), by

$$R_L = R_{hf} = \frac{a}{b}\sqrt{\frac{\omega\mu_0}{2\sigma}} = \frac{1}{25(10^{-4})}\sqrt{\frac{\pi(10^8)(4\pi \times 10^{-7})}{5.7 \times 10^7}}$$

$$= 1.053 \text{ ohms}$$

and the radiation efficiency, according to (2-90), by

$$e_{cd} = \frac{0.788}{0.788 + 1.053} = 0.428 = 42.8\%$$

From Figure 5.3(b)

$$\frac{R_p}{R_0} = 0.38$$

and from (5-25)

$$R_L = R_{\text{ohmic}} = \frac{8}{25(10^{-4})} \sqrt{\frac{\pi(10^8)(4\pi \times 10^{-7})}{5.7 \times 10^7}} (1.38) = 11.62$$

Thus

$$e_{cd} = \frac{50.43}{50.43 + 11.62} = 0.813 = 81.3\%$$

5.2.4 Near-Field ($kr \ll 1$) Region

The expressions for the fields, as given by (5-18a)–(5-19b), can be simplified if the observations are made in the near field ($kr \ll 1$). As for the infinitesimal dipole, the predominant term in each expression for the field in the near-zone region is the last one within the parentheses of (5-18a)–(5-19b). Thus for $kr \ll 1$

$$H_r \simeq \frac{a^2 I_0 e^{-jkr}}{2r^3} \cos\theta \quad (5\text{-}26\text{a})$$

$$H_\theta \simeq \frac{a^2 I_0 e^{-jkr}}{4r^3} \sin\theta \quad kr \ll 1 \quad (5\text{-}26\text{b})$$

$$H_\phi = E_r = E_\theta = 0 \quad (5\text{-}26\text{c})$$

$$E_\phi \simeq -j\frac{a^2 k I_0 e^{-jkr}}{4r^2} \sin\theta \quad (5\text{-}26\text{d})$$

The two **H**-field components are in time-phase. However, they are in time quadrature with those of the electric field. This indicates that the average power (real power) is zero, as is for the infinitesimal electric dipole. The condition of $kr \ll 1$ can be satisfied at moderate distances away from the antenna provided the frequency of operation is very low. The fields of (5-26a)–(5-26d) are usually referred to as *quasi-stationary*.

5.2.5 Far-Field ($kr \gg 1$) Region

The other space of interest where the fields can be approximated is the far-field ($kr \gg 1$) region. In contrast to the near field, the dominant term in (5-18a)–(5-19b) for $kr \gg 1$ is the first one within the parentheses. Since for $kr > 1$ the H_r component will be inversely proportional to r^2 whereas H_θ

will be inversely proportional to r. For large values of $kr(kr \gg 1)$, the H_r component will be small compared to H_θ. Thus it can be assumed that it is approximately equal to zero. Therefore for $kr \gg 1$,

$$H_\theta \simeq -\frac{k^2 a^2 I_0 e^{-jkr}}{4r} \sin\theta = -\frac{\pi S I_0 e^{-jkr}}{\lambda^2 r} \sin\theta \quad \Bigg\} \; kr \gg 1 \quad (5\text{-}27\text{a})$$

$$E_\phi \simeq \eta \frac{k^2 a^2 I_0 e^{-jkr}}{4r} \sin\theta = \eta \frac{\pi S I_0 e^{-jkr}}{\lambda^2 r} \sin\theta \quad (5\text{-}27\text{b})$$

$$H_r \simeq H_\phi = E_r = E_\theta = 0 \quad (5\text{-}27\text{c})$$

where $S = \pi a^2$ is the geometrical area of the loop.

Forming the ratio of $-E_\phi/H_\theta$, the wave impedance can be written as

$$Z_w = -\frac{E_\phi}{H_\theta} \simeq \eta \quad (5\text{-}28)$$

where

Z_w = wave impedance

η = intrinsic impedance

As for the infinitesimal dipole, the **E**- and **H**-field components of the loop in the far-field ($kr \gg 1$) region are perpendicular to each other and transverse to the direction of propagation. They form a *T*ransverse *E*lectro *M*agnetic (TEM) field whose wave impedance is equal to the intrinsic impedance of the medium. Equations (5-27a)–(5-27c) can also be derived using the procedure outlined and relationships developed in Section 3.6. This is left as an exercise to the reader (Problem 5.9).

5.2.6 Radiation Intensity and Directivity

The real power P_{rad} radiated by the loop was found in Section 5.2.3 and is given by (5-23b). The same expression can be obtained by forming the average power density, using (5-27a)–(5-27c), and integrating it over a closed sphere of radius r. This is left as an exercise to the reader (Problem 5.8). Associated with the radiated power P_{rad} is an average power density $\mathbf{W}_{\text{av}}$. It has only a radial component W_r which is related to the radiation intensity U by

$$U = r^2 W_r = \frac{\eta}{2}\left(\frac{k^2 a^2}{4}\right)^2 |I_0|^2 \sin^2\theta = \frac{r^2}{2\eta}|E_\phi(r,\theta,\phi)|^2 \quad (5\text{-}29)$$

and it conforms to (2-12a). The normalized pattern of the loop, as given by (5-29), is identical to that of the infinitesimal dipole shown in Figure 4.3. The maximum value occurs at $\theta = \pi/2$, and it is given by

$$U_{\max} = U|_{\theta=\pi/2} = \frac{\eta}{2}\left(\frac{k^2 a^2}{4}\right)^2 |I_0|^2 \quad (5\text{-}30)$$

Using (5-30) and (5-23b), the directivity of the loop can be written as

$$D_0 = 4\pi \frac{U_{\max}}{P_{\text{rad}}} = \frac{3}{2} \quad (5\text{-}31)$$

and its maximum effective area as

$$A_{em} = \left(\frac{\lambda^2}{4\pi}\right) D_0 = \frac{3\lambda^2}{8\pi} \qquad (5\text{-}32)$$

It is observed that the directivity, and as a result the maximum effective area, of a small loop is the same as that of an infinitesimal electric dipole. This should be expected since their patterns are identical.

The far-field expressions for a small loop, as given by (5-27a)–(5-27c), will be obtained by another procedure in the next section. In that section a loop of any radius but of constant current will be analyzed. The small loop far-field expressions will then be obtained as a special case of that problem.

Example 5.3

The radius of a small loop of constant current is $\lambda/25$. Find the physical area of the loop and compare it with its maximum effective aperture.

Solution:

$$S \text{ (physical)} = \pi a^2 = \pi \left(\frac{\lambda}{25}\right)^2 = \frac{\pi \lambda^2}{625} = 5.03 \times 10^{-3}\lambda^2$$

$$A_{em} = \frac{3\lambda^2}{8\pi} = 0.119\lambda^2$$

$$\frac{A_{em}}{S} = \frac{0.119\lambda^2}{5.03 \times 10^{-3}\lambda^2} = 23.66$$

Electrically the loop is about 24 times larger than its physical size, which should not be surprising. To be effective, a small loop must be larger electrically than its physical size.

5.2.7 Equivalent Circuit

A small loop is primarily inductive, and it can be represented by a lumped element equivalent circuit similar to those of Figure 2.28.

A. Transmitting Mode

The equivalent circuit for its input impedance when the loop is used as a transmitting antenna is that shown in Figure 5.4. This is similar to the equivalent circuit of Figure 2.28(b). Therefore its input impedance Z_{in} is represented by

$$Z_{in} = R_{in} + jX_{in} = (R_r + R_L) + j(X_A + X_i) \qquad (5\text{-}33)$$

where
R_r = radiation resistance as given by (5-24)
R_L = loss resistance of loop conductor
X_A = external inductive reactance of loop antenna = ωL_A
X_i = internal high-frequency reactance of loop conductor = ωL_i

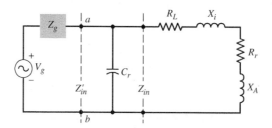

Figure 5.4 Equivalent circuit of loop antenna in transmitting mode.

In Figure 5.4 the capacitor C_r is used in parallel to (5-33) to resonate the antenna; it can also be used to represent distributed stray capacitances. In order to determine the capacitance of C_r at resonance, it is easier to represent (5-33) by its equivalent admittance Y_{in} of

$$Y_{in} = G_{in} + jB_{in} = \frac{1}{Z_{in}} = \frac{1}{R_{in} + jX_{in}} \tag{5-34}$$

where

$$G_{in} = \frac{R_{in}}{R_{in}^2 + X_{in}^2} \tag{5-34a}$$

$$B_{in} = -\frac{X_{in}}{R_{in}^2 + X_{in}^2} \tag{5-34b}$$

At resonance, the susceptance B_r of the capacitor C_r must be chosen to eliminate the imaginary part B_{in} of (5-34) given by (5-34b). This is accomplished by choosing C_r according to

$$C_r = \frac{B_r}{2\pi f} = -\frac{B_{in}}{2\pi f} = \frac{1}{2\pi f} \frac{X_{in}}{R_{in}^2 + X_{in}^2} \tag{5-35}$$

Under resonance the input impedance Z'_{in} is then equal to

$$Z'_{in} = R'_{in} = \frac{1}{G_{in}} = \frac{R_{in}^2 + X_{in}^2}{R_{in}} = R_{in} + \frac{X_{in}^2}{R_{in}} \tag{5-36}$$

The loss resistance R_L of the loop conductor can be computed using techniques illustrated in Example 5.2. The inductive reactance X_A of the loop is computed using the inductance L_A [7] of:

Circular loop of radius a and wire radius b:

$$L_A = \mu_0 a \left[\ln\left(\frac{8a}{b}\right) - 2 \right] \tag{5-37a}$$

Square loop with sides a and wire radius b:

$$L_A = 2\mu_0 \frac{a}{\pi} \left[\ln\left(\frac{a}{b}\right) - 0.774 \right] \tag{5-37b}$$

SMALL CIRCULAR LOOP   249

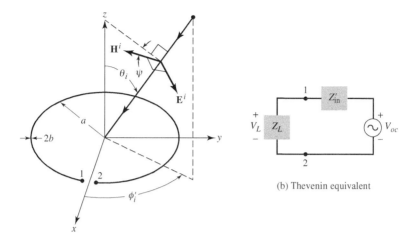

(a) Plane wave incident on a receiving loop (G.S. Smith, "Loop Antennas,"
Copyright © 1984, McGraw-Hill, Inc. Permission by McGraw-Hill, Inc.)

(b) Thevenin equivalent

Figure 5.5   Loop antenna and its equivalent in receiving mode.

The internal reactance of the loop conductor X_i can be found using the internal inductance L_i of the loop which for a single turn can be approximated by

$$L_i = \frac{l}{\omega P}\sqrt{\frac{\omega\mu_0}{2\sigma}} = \frac{a}{\omega b}\sqrt{\frac{\omega\mu_0}{2\sigma}} \quad \text{Circular loop} \tag{5-38a}$$

$$L_i = \frac{l}{\omega P}\sqrt{\frac{\omega\mu_0}{2\sigma}} = \frac{2a}{\omega\pi b}\sqrt{\frac{\omega\mu_0}{2\sigma}} \quad \text{Square loop} \tag{5-38b}$$

where l is the length and P is the perimeter (circumference) of the cross section of the wire of the loop.

B. Receiving Mode
The loop antenna is often used as a receiving antenna or as a probe to measure magnetic flux density. Therefore when a plane wave impinges upon it, as shown in Figure 5.5(a), an open-circuit voltage develops across its terminals. This open-circuit voltage is related according to (2-93) to its vector effective length and incident electric field. This open-circuit voltage is proportional to the incident magnetic flux density B_z^i, which is normal to the plane of the loop. Assuming the incident field is uniform over the plane of the loop, the open-circuit voltage for a single-turn loop can be written as [8]

$$V_{oc} = j\omega\pi a^2 B_z^i \tag{5-39}$$

Defining in Figure 5.5(a) the plane of incidence as the plane formed by the z axis and radial vector, then the open-circuit voltage of (5-39) can be related to the magnitude of the incident magnetic and electric fields by

$$V_{oc} = j\omega\pi a^2 \mu_0 H^i \cos\psi_i \sin\theta_i = jk_0\pi a^2 E^i \cos\psi_i \sin\theta_i \tag{5-39a}$$

where ψ_i is the angle between the direction of the magnetic field of the incident plane wave and the plane of incidence, as shown in Figure 5.5(a).

250 LOOP ANTENNAS

Since the open-circuit voltage is also related to the vector effective length by (2-93), then the effective length for a single-turn loop can be written as

$$\boldsymbol{\ell}_e = \hat{\mathbf{a}}_\phi \ell_e = \hat{\mathbf{a}}_\phi j k_0 \pi a^2 \cos\psi_i \sin\theta_i = \hat{\mathbf{a}}_\phi j k_0 S \cos\psi_i \sin\theta_i \qquad (5\text{-}40)$$

where S is the area of the loop. The factor $\cos\psi_i \sin\theta_i$ is introduced because the open-circuit voltage is proportional to the magnetic flux density component B_z^i which is normal to the plane of the loop.

When a load impedance Z_L is connected to the output terminals of the loop as shown in Figure 5.5(b), the voltage V_L across the load impedance Z_L is related to the input impedance Z'_{in} of Figure 5.5(b) and the open-circuit voltage of (5-39a) by

$$V_L = V_{oc} \frac{Z_L}{Z'_{in} + Z_L} \qquad (5\text{-}41)$$

5.3 CIRCULAR LOOP OF CONSTANT CURRENT

Let us now reconsider the loop antenna of Figure 5.2(a) but with a radius that may not necessarily be small. The current in the loop will again be assumed to be constant, as given by (5-1). For this current distribution, the vector potential is given by (5-14). The integration in (5-14) is quite complex, as is indicated right after (5-14). However, if the observation are restricted in the far-field ($r \gg a$) region, the small radius approximation is not needed to simplify the integration of (5-14).

Although the uniform current distribution along the perimeter of the loop is only valid provided the circumference is less than about 0.2λ (radius less than about 0.032λ), the procedure developed here for a constant current can be followed to find the far-zone fields of any size loop with not necessarily uniform current.

5.3.1 Radiated Fields

To find the fields in the far-field region, the distance R can be approximated by

$$R = \sqrt{r^2 + a^2 - 2ar\sin\theta\cos\phi'} \simeq \sqrt{r^2 - 2ar\sin\theta\cos\phi'} \quad \text{for } r \gg a \qquad (5\text{-}42)$$

which can be reduced, using the binomial expansion, to

$$\left.\begin{array}{c} R \simeq r\sqrt{1 - \dfrac{2a}{r}\sin\theta\cos\phi'} = r - a\sin\theta\cos\phi' = r - a\cos\psi_0 \\ \text{for phase terms} \\ R \simeq r \qquad \text{for amplitude terms} \end{array}\right\} \qquad (5\text{-}43)$$

since

$$\cos\psi_0 = \hat{\mathbf{a}}'_\rho \cdot \hat{\mathbf{a}}_r|_{\phi=0} = (\hat{\mathbf{a}}_x \cos\phi' + \hat{\mathbf{a}}_y \sin\phi')$$
$$\cdot (\hat{\mathbf{a}}_x \sin\theta\cos\phi + \hat{\mathbf{a}}_y \sin\theta\sin\phi + \hat{\mathbf{a}}_z \cos\theta)|_{\phi=0}$$
$$= \sin\theta\cos\phi' \qquad (5\text{-}43a)$$

CIRCULAR LOOP OF CONSTANT CURRENT   251

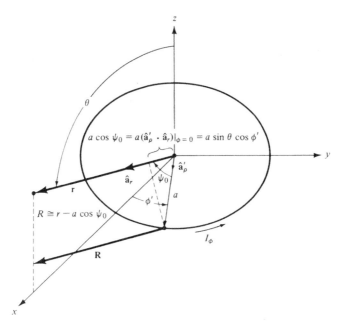

Figure 5.6   Geometry for far-field analysis of a loop antenna.

The geometrical relation between R and r, for any observation angle ϕ in the far-field region, is shown in Figure 5.2(b). For observations at $\phi = 0$, it simplifies to that given by (5-43) and shown in Figure 5.6. Thus (5-14) can be simplified to

$$A_\phi \simeq \frac{a\mu I_0 e^{-jkr}}{4\pi r} \int_0^{2\pi} \cos\phi' e^{+jka\sin\theta\cos\phi'}\, d\phi' \tag{5-44}$$

and it can be separated into two terms as

$$A_\phi \simeq \frac{a\mu I_0 e^{-jkr}}{4\pi r} \left[\int_0^{\pi} \cos\phi' e^{+jka\sin\theta\cos\phi'}\, d\phi' + \int_{\pi}^{2\pi} \cos\phi' e^{+jka\sin\theta\cos\phi'}\, d\phi' \right] \tag{5-45}$$

The second term within the brackets can be rewritten by making a change of variable of the form

$$\phi' = \phi'' + \pi \tag{5-46}$$

Thus (5-45) can also be written as

$$A_\phi \simeq \frac{a\mu I_0 e^{-jkr}}{4\pi r} \left[\int_0^{\pi} \cos\phi' e^{+jka\sin\theta\cos\phi'}\, d\phi' - \int_0^{\pi} \cos\phi'' e^{-jka\sin\theta\cos\phi''}\, d\phi'' \right] \tag{5-47}$$

Each of the integrals in (5-47) can be integrated by the formula (see Appendix V)

$$\pi j^n J_n(z) = \int_0^{\pi} \cos(n\phi) e^{+jz\cos\phi}\, d\phi \tag{5-48}$$

where $J_n(z)$ is the Bessel function of the first kind of order n. Using (5-48) reduces (5-47) to

$$A_\phi \simeq \frac{a\mu I_0 e^{-jkr}}{4\pi r}[\pi j J_1(ka\sin\theta) - \pi j J_1(-ka\sin\theta)] \tag{5-49}$$

The Bessel function of the first kind and order n is defined (see Appendix V) by the infinite series

$$J_n(z) = \sum_{m=0}^{\infty} \frac{(-1)^m (z/2)^{n+2m}}{m!(m+n)!} \tag{5-50}$$

By a simple substitution into (5-50), it can be shown that

$$J_n(-z) = (-1)^n J_n(z) \tag{5-51}$$

which for $n = 1$ is equal to

$$J_1(-z) = -J_1(z) \tag{5-52}$$

Using (5-52) we can write (5-49) as

$$A_\phi \simeq j\frac{a\mu I_0 e^{-jkr}}{2r} J_1(ka\sin\theta) \tag{5-53}$$

The next step is to find the **E-** and **H-**fields associated with the vector potential of (5-53). Since (5-53) is only valid for far-field observations, the procedure outlined in Section 3.6 can be used. The vector potential **A**, as given by (5-53), is of the form suggested by (3-56). That is, the r variations are separable from those of θ and ϕ. Therefore, according to (3-58a)–(3-58b) and (5-53)

$$E_r \simeq E_\theta = 0 \tag{5-54a}$$

$$E_\phi \simeq \frac{ak\eta I_0 e^{-jkr}}{2r} J_1(ka\sin\theta) \tag{5-54b}$$

$$H_r \simeq H_\phi = 0 \tag{5-54c}$$

$$H_\theta \simeq -\frac{E_\phi}{\eta} = -\frac{akI_0 e^{-jkr}}{2r} J_1(ka\sin\theta) \tag{5-54d}$$

5.3.2 Power Density, Radiation Intensity, Radiation Resistance, and Directivity

The next objective for this problem will be to find the power density, radiation intensity, radiation resistance, and directivity. To do this, the time-average power density is formed. That is,

$$\mathbf{W}_{av} = \frac{1}{2}\text{Re}[\mathbf{E}\times\mathbf{H}^*] = \frac{1}{2}\text{Re}[\hat{\mathbf{a}}_\phi E_\phi \times \hat{\mathbf{a}}_\theta H_\theta^*] = \hat{\mathbf{a}}_r \frac{1}{2\eta}|E_\phi|^2 \tag{5-55}$$

which can be written using (5-54b) as

$$\mathbf{W}_{av} = \hat{\mathbf{a}}_r W_r = \hat{\mathbf{a}}_r \frac{(a\omega\mu)^2 |I_0|^2}{8\eta r^2} J_1^2(ka\sin\theta) \tag{5-56}$$

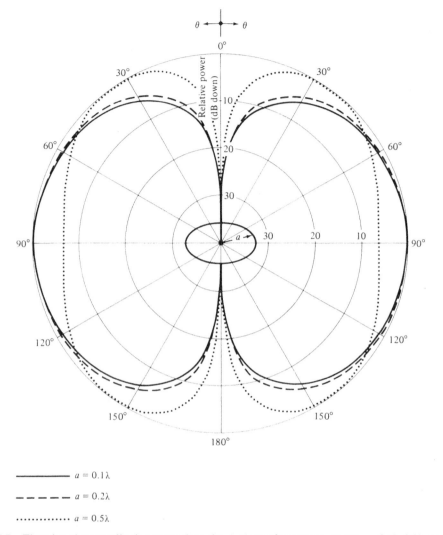

Figure 5.7 Elevation plane amplitude patterns for a circular loop of constant current ($a = 0.1\lambda, 0.2\lambda$, and 0.5λ).

with the radiation intensity given by

$$U = r^2 W_r = \frac{(a\omega\mu)^2 |I_0|^2}{8\eta} J_1^2(ka \sin\theta) \qquad (5\text{-}57)$$

The 2-D radiation patterns for $a = \lambda/10, \lambda/5$, and $\lambda/2$, based on a uniform current distribution, are shown in Figure 5.7. These patterns indicate that the field radiated by the loop along its axis ($\theta = 0°$) is zero. Also the shape of these patterns is similar to that of a linear dipole with $l \leq \lambda$ (a figure-eight shape). As the radius a increases beyond 0.5λ, the field intensity along the plane of the loop ($\theta = 90°$) diminishes and eventually it forms a null when $a \simeq 0.61\lambda$. This is left as an exercise to the reader for verification (Prob. 5.21). Beyond $a = 0.61\lambda$, the radiation along the plane of the loop begins to intensify and the pattern attains a multilobe form.

Three-dimensional patterns for loop circumferences of $C = 0.1\lambda$ and 5λ, assuming a uniform current distribution, are shown in Figure 5.8. It is apparent that for the 0.1λ circumference the pattern

254 LOOP ANTENNAS

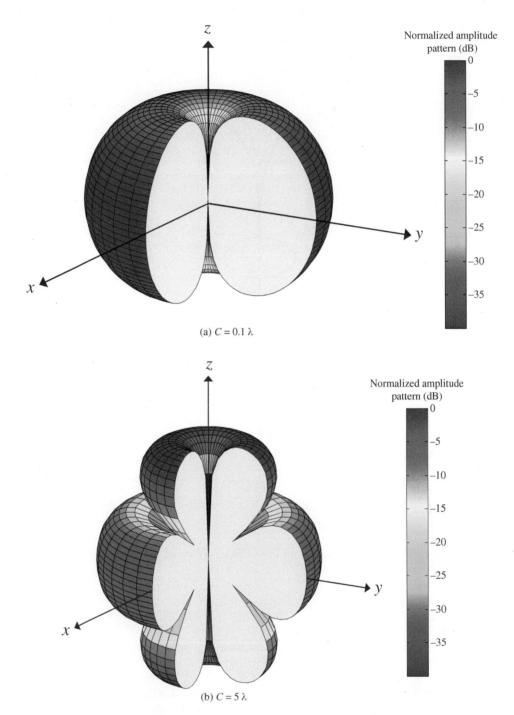

(a) $C = 0.1\lambda$

(b) $C = 5\lambda$

Figure 5.8 Three-dimensional amplitude patterns of a circular loop with constant current distribution.

is basically that of figure eight ($\sin\theta$), while for the 5λ loop it exhibits multiple lobes. The multiple lobes in a large loop begin to form when the circumference exceeds about 3.83λ (radius exceeds about 0.61λ); see Problem 5.21.

The patterns represented by (5-57) (some of them are illustrated in Figure 5.7) assume that the current distribution, no matter what the loop size, is constant. This is not a valid assumption if the loop circumference $C(C = 2\pi a)$ exceeds about 0.2λ (i.e., $a > 0.032\lambda$) [9]. For radii much greater than about 0.032λ, the current variation along the circumference of the loop begins to attain a distribution that is best represented by a Fourier series [9]. Although a most common assumption is that the current distribution is nearly cosinusoidal, it is not physical and satisfactory particularly near the driving point of the antenna.

A uniform and nonuniform in-phase current distribution can be attained on a loop antenna even if the radius is large. To accomplish this, the loop is subdivided into sections, with each section/arc of the loop fed with a different feed line; all feed lines are typically fed from a common feed source. Such an arrangement, although more complex, can approximate either uniform or nonuniform in-phase current distribution.

It has been shown [10] that when the circumference of the loop is about one wavelength ($C \simeq \lambda$), its maximum radiation based on a nonuniform current distribution is along its axis ($\theta = 0°, 180°$) which is perpendicular to the plane of the loop. This will also be discussed in Section 5.4 that follows. Feature of the loop antenna has been utilized to design Yagi-Uda arrays whose basic elements (feed, directors, and reflectors) are circular loops [11]–[14]. Because of its many applications, the one-wavelength circumference circular-loop antenna is considered as fundamental as a half-wavelength dipole.

The radiated power can be written using (5-56) as

$$P_{rad} = \iint_S \mathbf{W}_{av} \cdot d\mathbf{s} = \frac{\pi(a\omega\mu)^2 |I_0|^2}{4\eta} \int_0^\pi J_1^2(ka\sin\theta)\sin\theta\, d\theta \tag{5-58}$$

The integral in (5-58) can be rewritten [15] as

$$\int_0^\pi J_1^2(ka\sin\theta)\sin\theta\, d\theta = \frac{1}{ka}\int_0^{2ka} J_2(x)\, dx \tag{5-59}$$

The evaluation of the integral of (5-59) has been the subject of papers [16]–[19]. In these references, *along with some additional corrections*, the integral of (5-59)

$$Q_{11}^{(1)}(ka) = \frac{1}{2}\int_0^\pi J_1^2(ka\sin\theta)\sin\theta\, d\theta = \frac{1}{2ka}\int_0^{2ka} J_2(x)\, dx \tag{5-59a}$$

can be represented by a series of Bessel functions [20]

$$Q_{11}^{(1)}(ka) = \frac{1}{ka}\sum_{m=0}^\infty J_{2m+3}(2ka) \tag{5-59b}$$

where $J_m(x)$ is the Bessel function of the first kind, mth order. This is a highly convergent series (typically no more than $2ka$ terms are necessary), and its numerical evaluation is very efficient. Approximations to (5-59) can be made depending upon the values of the upper limit (large or small radii of the loop).

A. Large Loop Approximation ($a \geq \lambda/2$)

To evaluate (5-59), the first approximation will be to assume that the radius of the loop is large ($a \geq \lambda/2$). For that case, the integral in (5-59) can be approximated by

$$\int_0^\pi J_1^2(ka\sin\theta)\sin\theta \, d\theta = \frac{1}{ka}\int_0^{2ka} J_2(x)\, dx \simeq \frac{1}{ka} \tag{5-60}$$

and (5-58) by

$$P_{rad} \simeq \frac{\pi(a\omega\mu)^2|I_0|^2}{4\eta(ka)} \tag{5-61}$$

The maximum radiation intensity occurs when $ka\sin\theta = 1.84$ so that

$$U|_{max} = \frac{(a\omega\mu)^2|I_0|^2}{8\eta} J_1^2(ka\sin\theta)|_{ka\sin\theta=1.84} = \frac{(a\omega\mu)^2|I_0|^2}{8\eta}(0.582)^2 \tag{5-62}$$

Thus

$$R_r = \frac{2P_{rad}}{|I_0|^2} = \frac{2\pi(a\omega\mu)^2}{4\eta(ka)} = \eta\left(\frac{\pi}{2}\right)ka = 60\pi^2(ka) = 60\pi^2\left(\frac{C}{\lambda}\right) \tag{5-63a}$$

$$D_0 = 4\pi\frac{U_{max}}{P_{rad}} = 4\pi\frac{ka(0.582)^2}{2\pi} = 2ka(0.582)^2 = 0.677\left(\frac{C}{\lambda}\right) \tag{5-63b}$$

$$A_{em} = \frac{\lambda^2}{4\pi}D_0 = \frac{\lambda^2}{4\pi}\left[0.677\left(\frac{C}{\lambda}\right)\right] = 5.39\times10^{-2}\lambda C \tag{5-63c}$$

where C(circumference) $= 2\pi a$ and $\eta \simeq 120\pi$.

B. Intermediate Loop Approximation ($\lambda/6\pi \leq a < \lambda/2$)

If the radius of the loop is $\lambda/(6\pi) = 0.053\lambda \leq a < \lambda/2$, the integral of (5-59) for $Q_{11}^{(1)}(ka)$ is approximated by (5-59a) and (5-59b), and the radiation resistance and directivity can be expressed, respectively, as

$$R_r = \frac{2P_{rad}}{|I_0|^2} = \eta\pi(ka)^2 Q_{11}^{(1)}(ka) \tag{5-64a}$$

$$D_0 = \frac{4\pi U_{max}}{P_{rad}} = \frac{F_m(ka)}{Q_{11}^{(1)}(ka)} \tag{5-64b}$$

where

$$F_m(ka) = J_1^2(ka\sin\theta)|_{max} = \begin{cases} J_1^2(1.840) = (0.582)^2 = 0.339 & \\ \quad ka > 1.840 \; (a > 0.293\lambda) & (5\text{-}64c) \\ J_1^2(ka) & \\ \quad ka < 1.840 \; (a < 0.293\lambda) & (5\text{-}64d) \end{cases}$$

C. Small Loop Approximation ($a < \lambda/6\pi$)

If the radius of the loop is small ($a < \lambda/6\pi$), the expressions for the fields as given by (5-54a)–(5-54d) can be simplified. To do this, the Bessel function $J_1(ka\sin\theta)$ is expanded, by the definition of (5-50), in an infinite series of the form (see Appendix V)

$$J_1(ka\sin\theta) = \tfrac{1}{2}(ka\sin\theta) - \tfrac{1}{16}(ka\sin\theta)^3 + \cdots \tag{5-65}$$

For small values of ka ($ka < \tfrac{1}{3}$), (5-65) can be approximated by its first term, or

$$J_1(ka\sin\theta) \simeq \frac{ka\sin\theta}{2} \tag{5-65a}$$

Thus (5-54a)–(5-54d) can be written as

$$E_r \simeq E_\theta = 0 \tag{5-66a}$$

$$E_\phi \simeq \frac{a^2\omega\mu k I_0 e^{-jkr}}{4r}\sin\theta = \eta\frac{a^2 k^2 I_0 e^{-jkr}}{4r}\sin\theta \tag{5-66b}$$

$$H_r \simeq H_\phi = 0 \tag{5-66c}$$

$$H_\theta \simeq -\frac{a^2\omega\mu k I_0 e^{-jkr}}{4\eta r}\sin\theta = -\frac{a^2 k^2 I_0 e^{-jkr}}{4r}\sin\theta \tag{5-66d}$$

$$a < \lambda/6\pi$$

which are identical to those of (5-27a)–(5-27c). Thus the expressions for the radiation resistance, radiation intensity, directivity, and maximum effective aperture are those given by (5-24), (5-29), (5-31), and (5-32).

To demonstrate the variation of the radiation resistance as a function of the radius a of the loop, it is plotted in Figure 5.9 for $\lambda/100 \le a \le \lambda/30$ using (5-24), based on the approximation of (5-65a).

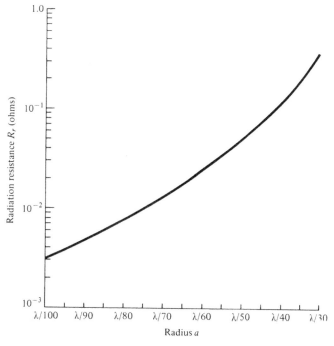

Figure 5.9 Radiation resistance for a constant current circular-loop antenna based on the approximation of (5-65a).

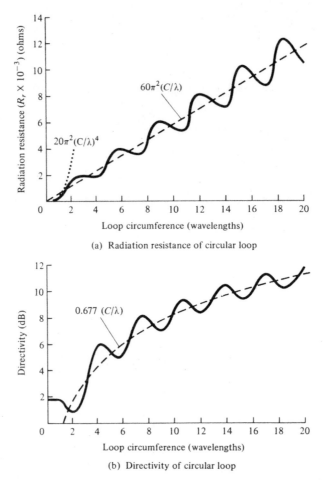

Figure 5.10 Radiation resistance and directivity for circular loop of constant current. (SOURCE: E. A. Wolff, *Antenna Analysis*, Wiley, New York, 1966).

It is evident that the values are extremely low (less than 1 ohm), and they are usually smaller than the loss resistances of the wires. These radiation resistances also lead to large mismatch losses when connected to practical transmission lines of 50 or 75 ohms. To increase the radiation resistance, it would require multiple turns as suggested by (5-24a). This, however, also increases the loss resistance which contributes to the inefficiency of the antenna. A plot of the radiation resistance for $0 < ka = C/\lambda < 20$, based on the evaluation of (5-59) by numerical techniques, is shown in Figure 5.10. The dashed line represents (5-63a) based on the large loop approximation of (5-60) and the dotted ($\cdots\cdots$) represents (5-24) based on the small loop approximation of (5-65a).

In addition to the real part of the input impedance, there is also an imaginary component which would increase the mismatch losses, even if the real part is equal to the characteristic impedance of the lossless transmission line. However, the imaginary component can always, in principle at least, be eliminated by connecting a reactive element (capacitive or inductive) across the terminals of the loop, as shown in Figure 5.4, to make the antenna a resonant circuit.

To facilitate the computations for the directivity and radiation resistance of a circular loop with a constant current distribution, a MATLAB and FORTRAN computer program has been developed. The program utilizes (5-62) and (5-58) to compute the directivity [(5-58) is integrated numerically]. The program requires as an input the radius of the loop (in wavelengths). A Bessel function subroutine is contained within the FORTRAN program. A listing of the program is included in the CD attached with the book.

5.4 CIRCULAR LOOP WITH NONUNIFORM CURRENT

The analysis in the previous sections was based on a uniform current, which would be a valid approximation when the radius of the loop is small electrically (usually about $a < 0.032\lambda$). As the dimensions of the loop increase, the current variations along the circumference of the loop must be taken into account.

A common assumption is a cosinusoidal variation for the current distribution [21], [22]. This, however, is not a physical and satisfactory representation, particularly near the driving point [9]. A better distribution would be to represent the current by a Fourier series, based on a delta gap voltage V across an infinitesimal gap at $\phi' = 0$ on the loop, and represented by [23]–[26]:

$$I(\phi') = \sum_{n=-\infty}^{\infty} I_n e^{jn\phi'} = \sum_{n=0}^{\infty} I_n \cos(n\phi') = \frac{V\delta(\phi')}{j\pi\eta_0}\left[\frac{1}{c_0} + 2\sum_{n=1}^{\infty}\frac{\cos(n\phi')}{c_n}\right] \quad (5\text{-}67)$$

where $\delta(\phi') =$ delta function, $\eta_0 = 377$ ohms, and ϕ' is measured from the feed point of the loop along the circumference, as shown in the inset of Figure 5.11.

A complete analysis of the fields radiated by a loop with nonuniform current distribution of (5-67) is somewhat complex and quite lengthy [2], [3]. Instead of attempting to include the analytical formulations, which are advanced but well documented in the cited references, a summary will be presented along with number of graphical illustrations of numerical and experimental data. These curves can be used in facilitating designs.

Based on the current distribution of (5-67), it is shown in [2] that the far-zone electric fields radiated by the loop are represented by

$$E_\theta \approx -\frac{V\cot\theta}{2\pi}\frac{e^{-jkr}}{r}\sum_{n=1}^{\infty} n\frac{(j)^n}{c_n}\sin(n\phi)J_n(ka\sin\theta) \quad (5\text{-}68a)$$

$$E_\phi \approx -\frac{Vka}{2\pi}\frac{e^{-jkr}}{r}\sum_{n=0}^{\infty}\frac{(j)^n}{c_n}\cos(n\phi)J'_n(ka\sin\theta) \quad (5\text{-}68b)$$

where c_n is evaluated using (5-70a) of page 262.

To illustrate that the current distribution of a wire loop antenna is not uniform unless its radius is very small, the magnitude and phase of it have been plotted in Figure 5.11 as a function of ϕ' (in degrees). The loop circumference C is $ka = C/\lambda = 0.05, 0.1, 0.2, 0.3,$ and 0.4 and the wire radius was chosen so that $\Omega = 2\ln(2\pi a/b) = 10$. It is apparent that for $ka = 0.2$ the current is nearly uniform. For $ka = 0.3$ the variations are slightly greater and become even larger as ka increases. On the basis of these results, loops much larger than $ka = 0.2$ (radius much greater than 0.032λ) cannot be considered small.

The maximum of the far-field radiation pattern shifts from $\theta = 90°$ [x-y plane; Figure 5.8(a) for small loop, $C = 0.1\lambda$, with uniform current] to $\theta = 0°, 180°$ (for large loops with nonuniform current). Evidence that this shift occurs was computed using 3-D and 2-D patterns (elevation plane for $\phi = 0°, 45°,$ and $90°$), based on (5-68a) and (5-68b), for a circular loop of $C = \lambda$. These results are displayed in Figure 5.12. As is apparent from the 3-D and 2-D patterns of Figure 5.12, the maximum is along $\theta = 0°$ and $180°$. The three 2-D elevation plane patterns of Figure 5.12(b) are not identical, as they should not be, because the current distribution is not uniform along its circumference; they

260 LOOP ANTENNAS

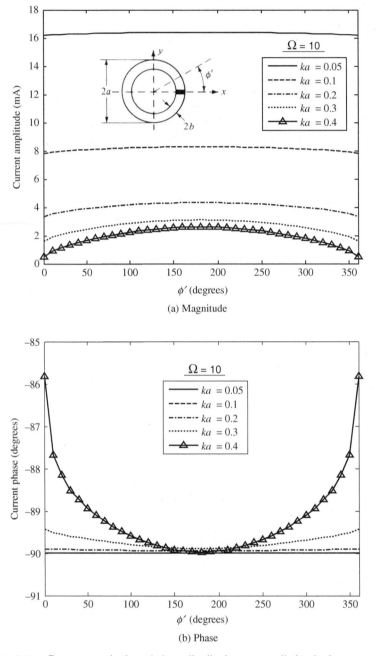

Figure 5.11 Current magnitude and phase distributions on small circular loop antennas.

are identical if the current distribution is uniform for loops of small radii (radius less than about 0.032λ).

As was indicated above, the maximum of the pattern for a loop antenna shifts from the plane of the loop ($\theta = 90°$) to its axis ($\theta = 0°, 180°$) as the circumference of the loop approaches one wavelength, as the current changes from uniform to nonuniform. Based on the nonuniform current distribution of (5-67), the directivity of the loop along $\theta = 0°$ has been computed, and it is plotted in Figure 5.13 versus the circumference of the loop in wavelengths. The maximum directivity is about 4.63 dB,

CIRCULAR LOOP WITH NONUNIFORM CURRENT 261

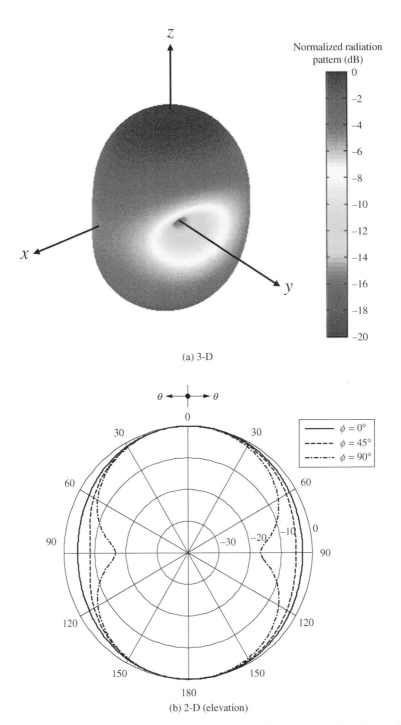

Figure 5.12 Far-field normalized three- and two-dimensional amplitude patterns for a loop with $C = \lambda$ and $\Omega = 10$.

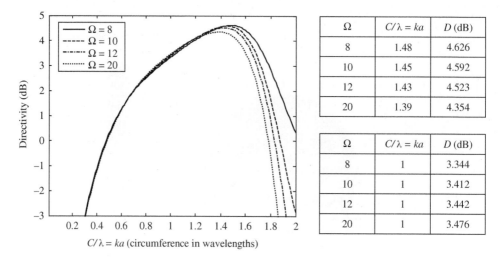

Figure 5.13 Directivity of circular-loop antenna for $\theta = 0, \pi$ versus electrical size (C/λ).

and it occurs when the circumference is about 1.48λ. For a one-wavelength circumference, which is usually the optimum design for a helical antenna, the directivity is about 3.476 dB for $\Omega = 20$. It is also apparent that the directivity is basically independent of the radius of the wire, as long as the circumference is equal or less than about 1.3 wavelengths; there are differences in directivity as a function of the wire radius for greater circumferences.

While the maximum directivity toward $\theta = 0°, 180°$ is displayed in Figure 5.13, the *overall maximum directivity* may not occur along $\theta = 0°, 180°$ for all radii of the loop. For example, for small radii loops the maximum directivity occurs along $\theta = 90°$, and it is equal to 1.5 (1.76 dB). To find the *overall maximum directivity*, a search procedure was employed over all observation angles ($0° \leq \theta \leq 180°, 0° \leq \phi \leq 360°$) for $\Omega = 8 - 12$, and the results are displayed in Figure 5.14. As can be seen, the maximum overall directivity of $\Omega = 8 - 12$ occurs for $\Omega = 8, C = 1.6\lambda$, it is equal to 4.88 dB, which is about 0.25 dB greater than the corresponding one along $\theta = 0°, 180°$ of Figure 5.13.

The input impedance Z_{in} at $\phi' = 0$ is given by

$$Z_{in} = \frac{j\pi\eta_0}{\frac{1}{c_0} + 2\sum_{n=1}^{\infty}\frac{1}{C_n}} = \frac{1}{\frac{1}{Z_0} + \sum_{n=1}^{\infty}\frac{1}{Z_n}} \tag{5-69}$$

where $Z_0 = j\pi\eta_0 c_0$ and $Z_n = j\pi\eta_0 (C_n/2)$. In both (5.67) and (5-69), the maximum number of terms needed for the infinite summation to converge is the maximum value of either 5 or $3ka$; that is, the max number is max[5, 3ka] [27].

When the loop is a perfect electric conductor and the wire radius b is much smaller than the radius a of the loop ($b \ll a$), the coefficients are represented by [25], [26]:

$$c_n = c_{-n} = ka\left(\frac{N_{n+1} + N_{n-1}}{2}\right) - \frac{n^2}{ka}N_n \tag{5-70a}$$

$$N_n = N_{-n} = \frac{1}{\pi}\left[K_0\left(\frac{nb}{a}\right)I_0\left(\frac{nb}{a}\right) + C_n\right] - \frac{1}{2}\int_0^{2ka}[\Omega_{2n}(x) + jJ_{2n}(x)] \quad \text{for } n > 1 \tag{5-70b}$$

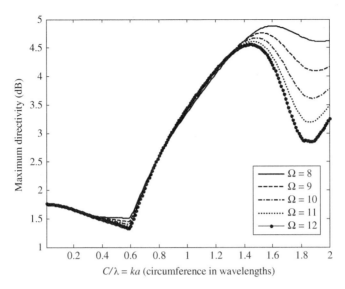

Figure 5.14 Circular loop maximum directivity as a function of circumference.

$C_n = \ln(4n) + \gamma - 2 \sum_{p=0}^{p-1} 1/(2p+1)$ and $\gamma = 0.5772$ (Euler's constant). The zero order term for N_n reduces to

$$N_0 = \frac{1}{\pi} \ln\left(\frac{8a}{b}\right) - \frac{1}{2}\left[\int_0^{2ka} [\Omega_0(x) + jJ_0(x)]dx\right] \quad (5\text{-}71)$$

where

$\Omega_n(x)$: Lommel–Weber function
$J_n(x)$: Bessel function of the first kind
$I_0(x) = 1 + O(x^2) \approx 1$: modified Bessel function of the first kind
$K_0(x) = -(\ln(x/2) + \gamma)I_0(x) + O(x^2) \approx -\ln(x/2) - \gamma$: modified Bessel function of the second kind

The preceding equations are commonly used to find the input impedance of a circular loop antenna. Recently, an RCL representation was proposed to solve this problem using an equivalent circuit model [28], [29]. One of the objectives of this new method is to demonstrate that each natural mode of the loop can be represented as a series of resonant circuits; thus, the overall performance of the circular loop is obtained by combining the modal impedances in parallel. In using this analogy, it is easier to understand intuitively the impedance of a circular loop antenna.

Computed impedances from (5-67), based on the Fourier series representation of the current, are shown plotted in Figure 5.15. The input resistance and reactance are plotted as a function of the circumference C (in wavelengths) for $0 \leq ka = C/\lambda \leq 2.5$. The diameter of the wire was chosen so that $\Omega = 2\ln(2\pi a/b) = 8, 9, 10, 11,$ and 12. It is apparent that the first antiresonance occurs when the circumference of the loop is about $\lambda/2$, and it is extremely sharp. It is also noted that as the loop wire increases in thickness, there is a rapid disappearance of the resonances. As a matter of fact, for $\Omega < 9$ there is only one antiresonance point. These curves (for $C > \lambda$) are similar, both qualitatively and quantitatively, to those of a linear dipole. The major difference is that the loop is more capacitive (by about 130 ohms) than a dipole. This shift in reactance allows the dipole to have several resonances and antiresonances while moderately thick loops ($\Omega < 9$) have only one antiresonance. *Also small loops are primarily inductive while small dipoles are primarily capacitive*. The resistance curves for the loop and the dipole are very similar.

264   LOOP ANTENNAS

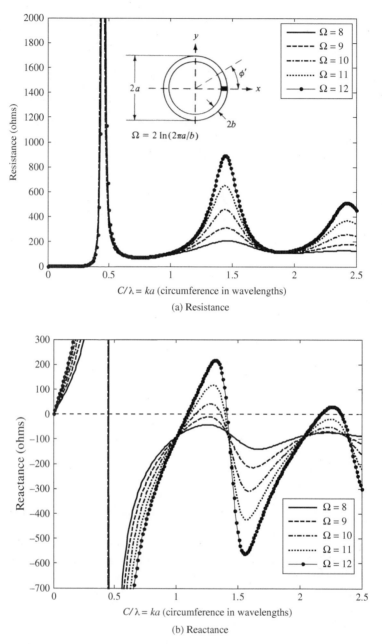

Figure 5.15 Input impedance of circular-loop antennas.

To verify the analytical formulations and the numerical computations, loop antennas were built and measurements of impedance were made [9]. The measurements were conducted using a half-loop over an image plane, and it was driven by a two-wire line. An excellent agreement between theory and experiment was indicated everywhere except near resonances where computed conductance curves were slightly higher than those measured. This is expected since ohmic losses were not taken into account in the analytical formulation. It was also noted that the measured susceptance curve was slightly displaced vertically by a constant value. This can be attributed to the "end effect" of the experimental feeding line and the "slice generator" used in the analytical modeling of the feed. For a dipole, the correction to the analytical model is usually a negative capacitance in shunt with the antenna [30]. A similar correction for the loop would result in a better agreement between the computed and measured susceptances. Computations for a half-loop above a ground plane were also performed by J. E. Jones [31] using the Moment Method.

The radiation resistance and maximum directivity of a loop antenna with a cosinusoidal current distribution $I_\phi(\phi) = I_0 \cos\phi$ was derived in [2],[16] and evaluated by integrating the far-zone fields. Doing this, the values are plotted, respectively, in Figures 5.16(a,b) where they are compared with those based on a uniform and nonuniform current distribution.

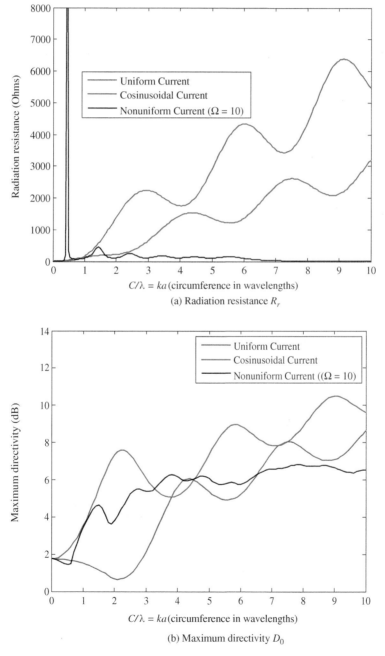

Figure 5.16 Radiation resistance (R_r) and maximum directivity (D_0) of a circular loop with uniform, cosinusoidal and nonuniform current distributions.

A general Matlab computer program **Circular_Loop_Nonuniform** has been developed to compute the following current distributions for *uniform, cosinusoidal*, and *Fourier series*:

- Current distribution, based on (5-67)
- Input impedance, based on (5-69)
- Far-field amplitude radiation pattern, based on (5-68a), (5-68b)
- Directivity pattern (in dB)
- Maximum directivity (dimensionless and in dB)

The *uniform* ($n = 0$) and *cosinusoidal* ($n = 1$) current distributions are treated as special cases of the Fourier series distribution. The Matlab program was advanced by the Matlab programs written by A. F. McKinley and associates, based on [28], [29], and made available to this author. The programs by A. F. McKinley are more general while the one in this book is included to primarily aid the reader in computing the parameters listed above.

5.4.1 Arrays

In addition to being used as single elements and in arrays, as shown in Figure 5.1(a,b), there are some other classic arrays of loop configurations. Two of the most popular arrays of loop antennas are the helical antenna and the Yagi-Uda array. The loop is also widely used to form a solenoid which in conjunction with a ferrite cylindrical rod within its circumference is used as a receiving antenna and as a tuning element, especially in transistor radios. This is discussed in Section 5.7.

The helical antenna, which is discussed in more detail in Section 10.3.1, is a wire antenna, which is wound in the form of a helix, as shown in Figure 10.13. It is shown that it can be modeled approximately by a series of loops and vertical dipoles, as shown in Figure 10.15. The helical antenna possesses in general elliptical polarization, but it can be designed to achieve nearly circular polarization. There are two primary modes of operation for a helix, the normal mode and the axial mode. The helix operates in its normal mode when its overall length is small compared to the wavelength, and it has a pattern with a null along its axis and the maximum along the plane of the loop. This pattern (figure-eight type in the elevation plane) is similar to that of a dipole or a small loop. *A helical antenna operating in the normal mode is sometimes used as a monopole antenna for mobile cell and cordless telephones, and it is usually covered with a plastic cover. This helix monopole is used because its input impedance is larger than that of a regular monopole and more attractive for matching to typical transmission lines used as feed lines, such as a coaxial line* (see Problem 10.18).

The helix operates in the axial mode when the circumference of the loop is between $3/4\lambda < C < 4/3\lambda$ with an optimum design when the circumference is nearly one wavelength. When the circumference of the loop approaches one wavelength, the maximum of the pattern is along its axis. In addition, the phasing among the turns is such that overall the helix forms an end-fire antenna with attractive impedance and polarization characteristics (see Example 10.1). In general, the helix is a popular communication antenna in the VHF and UHF bands.

The Yagi-Uda antenna is primarily an array of linear dipoles with one element serving as the feed while the others act as parasitic. However this arrangement has been extended to include arrays of loop antennas, as shown in Figure 10.30. As for the helical antenna, in order for this array to perform as an end-fire array, the circumference of each of the elements is near one wavelength. More details can be found in Section 10.3.4 and especially in [11]–[14]. A special case is the quad antenna which is very popular amongst ham radio operators. It consists of two square loops, one serving as the excitation while the other is acting as a reflector; there are no directors. The overall perimeter of each loop is one wavelength.

5.4.2 Design Procedure

The design of small loops is based on the equations for the radiation resistance (5-24), (5-24a), directivity (5-31), maximum effective aperture (5-32), resonance capacitance (5-35), resonance input impedance (5-36) and inductance (5-37a), (5-37b). In order to resonate the element, the capacitor C_r of Figure 5.4 is chosen based on (5-35) so as to cancel out the imaginary part of the input impedance Z_{in}.

For large loops with a nonuniform current distribution, the design is accomplished using the curves of Figure 5.13 for the axial directivity and those of Figure 5.15 for the impedance. To resonate the loop, usually a capacitor in parallel or an inductor in series is added, depending on the radius of the loop and that of the wire.

Example 5.4

Design a resonant loop antenna to operate at 100 MHz so that the pattern maximum is along the axis of the loop. Determine the radius of the loop and that of the wire (in meters), the axial directivity (in dB), and the parallel lumped element (capacitor in parallel or inductor in series) that must be used in order to resonate the antenna.

Solution: In order for the pattern maximum to be along the axis of the loop, the circumference of the loop must be large compared to the wavelength. Therefore the current distribution will be nonuniform. To accomplish this, Figure 5.15 should be used. There is not only one unique design which meets the specifications, but there are many designs that can accomplish the goal.

One design is to select a circumference where the loop is self resonant, and there is no need for a resonant capacitor. For example, referring to Figure 5.15(b) and choosing an $\Omega = 12$, the circumference of the loop is nearly 1.089λ. Since the free-space wavelength at 100 MHz is 3 meters, then the circumference is

$$\text{circumference} \simeq 1.089(3) = 3.267 \text{ meters}$$

while the radius of the loop is

$$\text{radius} = a = \frac{3.267}{2\pi} = 0.52 \text{ meters}$$

The radius of the wire is obtained using

$$\Omega = 12 = 2 \ln\left(\frac{2\pi a}{b}\right)$$

or

$$\frac{a}{b} = 64.2077$$

Therefore the radius of the wire is

$$b = \frac{a}{64.2077} = \frac{0.52}{64.2077} = 0.8099 \text{ cm} = 8.099 \times 10^{-3} \text{ meters}$$

Using Figure 5.13, the axial directivity for this design is approximately 3.7 dB. Using Figure 5.15(a), the input impedance is approximately

$$Z_{in} = Z'_{in} \simeq 149 \text{ ohms}$$

268   LOOP ANTENNAS

Since the antenna chosen is self resonant, there is no need for a lumped element to resonate the radiator.

Another design will be to use another circumference where the loop is not self resonant. This will necessitate the use of a capacitor C_r to resonate the antenna. This is left as an end of the chapter exercise.

5.5   GROUND AND EARTH CURVATURE EFFECTS FOR CIRCULAR LOOPS

The presence of a lossy medium can drastically alter the performance of a circular loop. The parameters mostly affected are the pattern, directivity, input impedance, and antenna efficiency. The amount of energy dissipated as heat by the lossy medium directly affects the antenna efficiency. As for the linear elements, geometrical optics techniques can be used to analyze the radiation characteristics of loops in the presence of conducting surfaces. The reflections are taken into account by introducing appropriate image (virtual) sources. Divergence factors are introduced to take into account the effects of the ground curvature. Because the techniques are identical to the formulations of Section 4.8, they will not be repeated here. The reader is directed to that section for the details. It should be pointed out, however, that a horizontal loop has horizontal polarization in contrast to the vertical polarization of a vertical electric dipole. Exact boundary-value solutions, based on Sommerfeld integral formulations, are available [31]. However they are too complex to be included in an introductory chapter.

By placing the loop above a reflector, the pattern is made unidirectional and the directivity is increased. To simplify the problem, initially the variations of the axial directivity ($\theta = 0°$) of a circular loop with a circumference of one wavelength ($ka = 1$) when placed horizontally a height h above an infinite in extent perfect electric conductor are examined as a function of the height above the ground plane. These were obtained using image theory and the array factor of two loops, and they are shown for $10 < \Omega < 20$ in Figure 5.17 [8], [32]. Since only one curve is shown for $10 < \Omega < 20$, it is evident that the directivity variations as a function of the height are not strongly dependent on the radius of the wire of the loop. It is also apparent that for $0.05\lambda < h < 0.2\lambda$ and $0.65\lambda < h < 0.75\lambda$

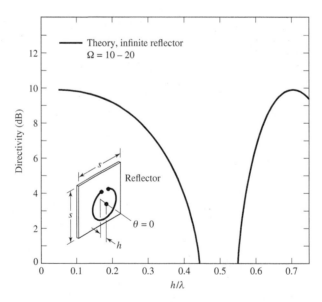

Figure 5.17   Directivity of circular-loop antenna, $C/\lambda = ka = 1$, for $\theta = 0$ versus distance from reflector h/λ. Theoretical curve is for infinite planar reflector. (SOURCE: G. S. Smith, "Loop Antennas," Chapter 5 of *Antenna Engineering Handbook*, 1984, © 1984 McGraw-Hill, Inc. Permission by McGraw-Hill, Inc).

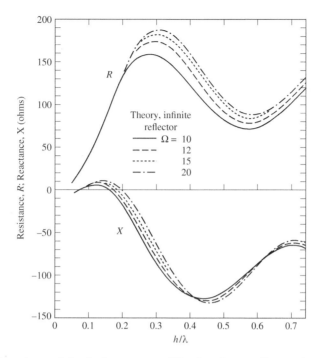

Figure 5.18 Input impedance of circular-loop antenna $C/\lambda = ka = 1$ versus distance from reflector h/λ. Theoretical curves are for infinite planar reflector; measured points are for square reflector. (SOURCE: G. S. Smith, "Loop Antennas," Chapter 5 of *Antenna Engineering Handbook*, 1984, © 1984, McGraw-Hill, Inc. Permission by McGraw-Hill, Inc).

the directivity is about 9-10 dB. For the same size loop, the corresponding variations of the impedance as a function of the height are shown in Figure 5.18 [8], [32]. While the directivity variations are not strongly influenced by the radius of the wire, the variations of the impedance do show a dependence on the radius of the wire of the loop for $10 < \Omega < 20$.

A qualitative criterion that can be used to judge the antenna performance is the ratio of the radiation resistance in free-space to that in the presence of the homogeneous lossy medium [33]. This is a straightforward but very tedious approach. A much simpler method [34] is to find directly the self-impedance changes (real and imaginary) that result from the presence of the conducting medium.

Since a small horizontal circular loop is equivalent to a small vertical magnetic dipole (see Section 5.2.2), computations [35] were carried out for a vertical magnetic dipole placed a height h above a homogeneous lossy half-space. The changes in the self-impedance, normalized with respect to the free-space radiation resistance R_0 given by (5-24), are found in [35]. Significant changes, compared to those of a perfect conductor, are introduced by the presence of the ground.

The effects that a stratified lossy half-space have on the characteristics of a horizontal small circular loop have also been investigated and documented [36]. It was found that when a resonant loop is close to the interface, the changes in the input admittance as a function of the antenna height and the electrical properties of the lossy medium were very pronounced. This suggests that a resonant loop can be used effectively to sense and to determine the electrical properties of an unknown geological structure.

5.6 POLYGONAL LOOP ANTENNAS

The most attractive polygonal loop antennas are the square, rectangular, triangular, and rhombic. These antennas can be used for practical applications such as for aircraft, missiles, and communications systems. However, because of their more complex structure, theoretical analyses seem to be unsuccessful [37]. Thus the application of these antennas has received much less attention. However design curves, computed using the Moment Method, do exist [38] and can be used to design polygonal loop antennas for practical applications. Usually the circular loop has been used in the UHF range because of its higher directivity while triangular and square loops have been applied in the HF and UHF bands because of advantages in their mechanical construction. Broadband impedance characteristics can be obtained from the different polygonal loops.

The input impedance ($Z = R + jX$) variations, for the following four configurations are found in [38]:

- Top-driven triangular
- Base-driven triangular
- Rectangular
- Rhombic

If the appropriate shape and feed point are chosen, a polygonal loop will have broadband impedance characteristics and be matched to the 50-ohm lines. From the four configurations listed above, the two most attractive configurations are the triangular loop (isosceles triangle) and the rectangular loop with a height/length $= 0.5$ ratio.

5.7 FERRITE LOOP

Because the loss resistance is comparable to the radiation resistance, electrically small loops are very poor radiators and are seldom used in the transmitting mode. However, they are often used for receiving signals, such as in radios and pagers, where the signal-to-noise ratio is much more important than the efficiency.

5.7.1 Radiation Resistance

The radiation resistance, and in turn the antenna efficiency, can be raised by increasing the circumference of the loop. Another way to increase the radiation resistance, without increasing the electrical dimensions of the antenna, would be to insert within its circumference a ferrite core that has a tendency to increase the magnetic flux, the magnetic field, the open-circuit voltage, and in turn the radiation resistance of the loop [39], [40]. This is the so-called *ferrite loop* and the ferrite material can be a rod of very few inches in length. The radiation resistance of the ferrite loop is given by

$$\frac{R_f}{R_r} = \left(\frac{\mu_{ce}}{\mu_0}\right)^2 = \mu_{cer}^2 \qquad (5\text{-}72)$$

where
R_f = radiation resistance of ferrite loop
R_r = radiation resistance of air core loop
μ_{ce} = effective permeability of ferrite core
μ_0 = permeability of free-space
μ_{cer} = relative effective permeability of ferrite core

Using (5-24), the radiation resistance of (5-72) for a single-turn small ferrite loop can be written as

$$\boxed{R_f = 20\pi^2 \left(\frac{C}{\lambda}\right)^4 \left(\frac{\mu_{ce}}{\mu_0}\right)^2 = 20\pi^2 \left(\frac{C}{\lambda}\right)^4 \mu_{cer}^2} \qquad (5\text{-}73)$$

and for an *N*-turn loop, using (5-24a), as

$$R_f = 20\pi^2 \left(\frac{C}{\lambda}\right)^4 \left(\frac{\mu_{ce}}{\mu_0}\right)^2 N^2 = 20\pi^2 \left(\frac{C}{\lambda}\right)^4 \mu_{cer}^2 N^2 \qquad (5\text{-}74)$$

The relative effective permeability of the ferrite core μ_{cer} is related to the relative intrinsic permeability of the unbounded ferrite material $\mu_{fr}(\mu_{fr} = \mu_f/\mu_0)$ by

$$\mu_{cer} = \frac{\mu_{ce}}{\mu_0} = \frac{\mu_{fr}}{1 + D(\mu_{fr} - 1)} \qquad (5\text{-}75)$$

where D is the demagnetization factor which has been found experimentally for different core geometries, as shown in Figure 5.19. For most ferrite material, the relative intrinsic permeability μ_{fr} is very large ($\mu_{fr} \gg 1$) so that the relative effective permeability of the ferrite core μ_{cer} is approximately inversely proportional to the demagnetization factor, or $\mu_{cer} \sim 1/D = D^{-1}$. In general, the demagnetization factor is a function of the geometry of the ferrite core. For example, the demagnetization factor for a sphere is $D = \frac{1}{3}$ while that for an ellipsoid of length $2l$ and radius a, such that $l \gg a$, is

$$D = \left(\frac{a}{l}\right)^2 \left[\ln\left(\frac{2l}{a}\right) - 1\right], \qquad l \gg a \qquad (5\text{-}75a)$$

5.7.2 Ferrite-Loaded Receiving Loop

Because of their smallness, ferrite loop antennas of few turns wound around a small ferrite rod are used as antennas, especially in the older generation pocket transistor radios. The antenna is usually connected in parallel with the RF amplifier tuning capacitance and, in addition to acting as an antenna, it furnishes the necessary inductance to form a tuned circuit. Because the inductance is

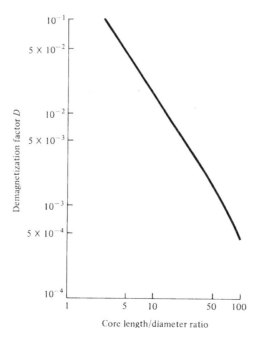

Figure 5.19 Demagnetization factor as a function of core length/diameter ratio. (SOURCE: E. A. Wolff, *Antenna Analysis*, Wiley, New York, 1966).

obtained with only few turns, the loss resistance is kept small. Thus the Q is usually very high, and it results in high selectivity and greater induced voltage.

The equivalent circuit for a ferrite-loaded loop antenna is similar to that of Figure 5.4 except that a loss resistance R_M, in addition to R_L, is needed to account for the power losses in the ferrite core. Expressions for the loss resistance R_M and inductance L_A for the ferrite-loaded loop of N turns can be found in [7] and depend on some empirical factors which are determined from an average of experimental results. The inductance L_i is the same as that of the unloaded loop.

5.8 MOBILE COMMUNICATION SYSTEMS APPLICATIONS

As was indicated in Section 4.7.4 of Chapter 4, the monopole was one of the most widely used elements for handheld units of mobile communication systems. An alternative to the monopole is the loop, [41]–[46], which has been often used in pagers but has found very few applications in handheld transceivers. This is probably due to loop's high resistance and inductive reactance which are more difficult to match to standard feed lines. The fact that loop antennas are more immune to noise makes them more attractive for an interfering and fading environment, like that of mobile communication systems. In addition, loop antennas become more viable candidates for wireless communication systems which utilize devices operating at higher frequency bands, particularly in designs where balanced amplifiers must interface with the antenna. Relative to top side of the handheld unit, such as the telephone, the loop can be placed either horizontally [42] or vertically [44]–[46]. Either configuration presents attractive radiation characteristics for land-based mobile systems.

The radiation characteristics, normalized pattern and input impedance, of a monopole and vertical loop mounted on an experimental mobile handheld device were examined in [44]–[46]. The loop was in the form of a folded configuration mounted vertically on the handheld conducting device with its one end either grounded or ungrounded to the device. The predicted and measured input impedance of the folded loop, when its terminating end was grounded to the box, are displayed in Figure 5.20(a,b). It is evident that the first resonance, around 900 MHz, of the folded loop is of the *parallel type* (*antiresonance*) with a very high, and rapidly changing versus frequency, resistance, and reactance. These values and variations of impedance are usually undesirable for practical implementation. For frequencies below the first resonance, the impedance is inductive (imaginary part is positive), as is typical of small loop antennas (see Figure 5.15); above the first resonance, the impedance is capacitive (negative imaginary part). The second resonance, around 2,100 MHz, is of the *series type* with slowly varying values of impedance, and of desirable magnitude, for practical implementation. The resonance forms (*parallel* vs. *series*) can be interchanged if the terminating end of the folded loop is ungrounded with the element then operating as an L monopole [44]–[46] and exhibiting the same resonance behavior as that of a monopole mounted on the device (see Chapter 4, Section 4.7.5, Figure 4.24). Even though the radiating element is a loop whose plane is vertical to the box, the amplitude pattern, in both cases (loop and L), is similar and nearly omnidirectional as that of the monopole of Figure 4.24 because the PEC box is also part of the radiating system.

A summary of the pertinent parameters and associated formulas and equation numbers for this chapter are listed in Table 5.1.

5.9 MULTIMEDIA

In the publisher's website for this book, the following multimedia resources are included for the review, understanding, and visualization of the material of this chapter:

a. **Java**-based **interactive questionnaire**, with answers.
b. **Java**-based **applet** for computing and displaying the radiation characteristics of a loop.
c. **Java**-based animation of loop amplitude pattern.

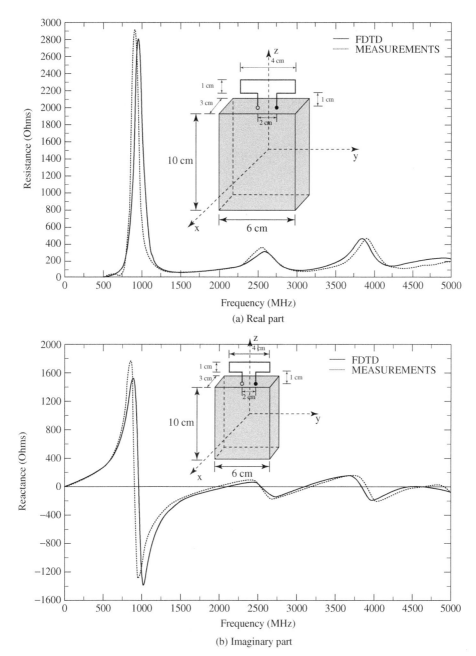

Figure 5.20 Input impedance, real and imaginary parts of a wire folded loop mounted vertically on a conducting mobile hand-held unit (SOURCE: K. D. Katsibas, et al., "Folded Loop Antenna for Mobile Hand-Held Units," *IEEE Transactions Antennas Propagat.*, Vol. 46, No. 2, February 1998, pp. 260–266. © 1998 IEEE).

d. **Matlab** computer program, designated **Circular_Loop_Uniform**, for computing the radiation characteristics of a loop. A description of the program is found in the READ ME file of the corresponding program in the publisher's website for this book.

e. The Matlab computer program **Circular_Loop_Nonuniform** can be used to compute the radiation characteristics (current distribution, input impedance, amplitude pattern, and directivity pattern) of a circular loop with uniform cosinusoidal and Fourier series current distributions.

f. **Power Point (PPT)** viewgraphs, in multicolor.

TABLE 5.1 Summary of Important Parameters, and Associated Formulas and Equation Numbers for Loop in Far Field

Parameter	Formula	Equation Number
	Small Circular Loop $(a < \lambda/6\pi, C < \lambda/3)$ **(Uniform Current)**	
Normalized power pattern	$U = |E_{\phi n}|^2 = C_0 \sin^2\theta$	(5-27b)
Wave impedance Z_w	$Z_w = -\dfrac{E_\phi}{H_\theta} \simeq \eta = 377$ Ohms	(5-28)
Directivity D_0	$D_0 = \dfrac{3}{2} = 1.761$ dB	(5-31)
Maximum effective area A_{em}	$A_{em} = \dfrac{3\lambda^2}{8\pi}$	(5-32)
Radiation resistance R_r (one turn)	$R_r = 20\pi^2 \left(\dfrac{C}{\lambda}\right)^4$	(5-24)
Radiation resistance R_r (N turns)	$R_r = 20\pi^2 \left(\dfrac{C}{\lambda}\right)^4 N^2$	(5-24a)
Input resistance R_{in}	$R_{in} = R_r = 20\pi^2 \left(\dfrac{C}{\lambda}\right)^4$	(5-24)
Loss resistance R_L (one turn)	$R_L = \dfrac{l}{P}\sqrt{\dfrac{\omega\mu_0}{2\sigma}} = \dfrac{C}{2\pi b}\sqrt{\dfrac{\omega\mu_0}{2\sigma}}$	(2-90b)
Loss resistance R_L (N turns)	$R_L = \dfrac{Na}{b}R_s\left(\dfrac{R_p}{R_0} + 1\right)$	(5-25)
Circular loop external inductance L_A	$L_A = \mu_0 a\left[\ln\left(\dfrac{8a}{b}\right) - 2\right]$	(5-37a)
Square loop external inductance L_A	$L_A = 2\mu_0 \dfrac{a}{\pi}\left[\ln\left(\dfrac{a}{b}\right) - 0.774\right]$	(5-37b)
Circular loop internal inductance L_i	$L_i = \dfrac{a}{\omega b}\sqrt{\dfrac{\omega\mu_0}{2\sigma}}$	(5-38a)
Square loop internal inductance L_i	$L_A = \dfrac{2a}{\omega\pi b}\sqrt{\dfrac{\omega\mu_0}{2\sigma}}$	(5-38b)
Vector effective length ℓ_e	$\ell_e = \hat{\mathbf{a}}_\phi jk_0\pi a^2 \cos\psi_i \sin\theta_i$	(5-40)
Half-power beamwidth	HPBW = 90°	(4-65)
	Large Circular Loop $(a \geq \lambda/2, C \geq 3.14\lambda)$ **(Uniform Current)**	
Normalized power pattern	$U = |E_{\phi n}|^2 = C_1 J_1^2(ka\sin\theta)$	(5-57)
Wave impedance Z_w	$Z_w = -\dfrac{E_\phi}{H_\theta} \simeq \eta = 377$ Ohms	(5-28)
Directivity D_0 $(a > \lambda/2)$	$D_0 = 0.677\left(\dfrac{C}{\lambda}\right)$	(5-63b)
Maximum effective area A_{em} $(a > \lambda/2)$	$A_{em} = \dfrac{\lambda^2}{4\pi}\left[0.677\left(\dfrac{C}{\lambda}\right)\right]$	(5-63c)
Radiation resistance $(a > \lambda/2)$, (one turn)	$R_r = 60\pi^2\left(\dfrac{C}{\lambda}\right)$	(5-63a)
Input resistance $(a > \lambda/2)$, (one turn)	$R_{in} = R_r = 60\pi^2\left(\dfrac{C}{\lambda}\right)$	(5-63a)
Loss resistance R_L (one turn)	$R_L = \dfrac{l}{P}\sqrt{\dfrac{\omega\mu_0}{2\sigma}} = \dfrac{C}{2\pi b}\sqrt{\dfrac{\omega\mu_0}{2\sigma}}$	(2-90b)

TABLE 5.1 (*continued*)

Parameter	Formula	Equation Number
Loss resistance R_L (N turns)	$R_L = \dfrac{Na}{b} R_s \left(\dfrac{R_p}{R_0} + 1 \right)$	(5-25)
External inductance L_A	$L_A = \mu_0 a \left[\ln\left(\dfrac{8a}{b}\right) - 2 \right]$	(5-37a)
Internal inductance L_i	$L_i = \dfrac{a}{\omega b} \sqrt{\dfrac{\omega \mu_0}{2\sigma}}$	(5-38a)
Vector effective length $\boldsymbol{\ell}_e$	$\boldsymbol{\ell}_e = \hat{\mathbf{a}}_\phi\, jk_0 \pi a^2 \cos\psi_i \sin\theta_i$	(5-40)
	Ferrite Circular Loop ($a < \lambda/6\pi, C < \lambda/3$) (uniform current)	
Radiation resistance R_f (one turn)	$R_f = 20\pi^2 \left(\dfrac{C}{\lambda}\right)^4 \mu_{cer}^2$	(5-73)
	$\mu_{cer} = \dfrac{\mu_{fr}}{1 + D(\mu_{fr} - 1)}$	(5-75)
Radiation resistance R_f (N turns)	$R_f = 20\pi^2 \left(\dfrac{C}{\lambda}\right)^4 \mu_{cer}^2 N^2$	(5-74)
Demagnetizing factor D	Ellipsoid: $D = \left(\dfrac{a}{l}\right)^2 \left[\ln\left(\dfrac{2l}{a}\right) - 1\right]$ $l \gg a$ Sphere: $D = \dfrac{1}{3}$	(5-75a)

REFERENCES

1. P. L. Overfelt, "Near Fields of the Constant Current Thin Circular Loop Antenna of Arbitrary Radius," *IEEE Trans. Antennas Propagat.*, Vol. 44, No. 2, February 1996, pp. 166–171.
2. D. H. Werner, "An Exact Integration Procedure for Vector Potentials of Thin Circular Loop Antennas," *IEEE Trans. Antennas Propagat.*, Vol. 44, No. 2, February 1996, pp. 157–165.
3. D. H. Werner, "Lommel Expansions in Electromagnetics," Chapter in *Frontiers in Electromagnetics*, (D. H. Werner and R. Mittra, eds.), IEEE Press/John Wiley, New York, 2000.
4. E. H. Newman, P. Bohley, and C. H. Walter, "Two Methods for Measurement of Antenna Efficiency," *IEEE Trans. Antennas Propagat.*, Vol. AP-23, No. 4, July 1975, pp. 457–461.
5. G. S. Smith, "Radiation Efficiency of Electrically Small Multiturn Loop Antennas," *IEEE Trans. Antennas Propagat.*, Vol. AP-20, No. 5, September 1972, pp. 656–657.
6. G. S. Smith, "The Proximity Effect in Systems of Parallel Conductors," *J. Appl. Phys.*, Vol. 43, No. 5, May 1972, pp. 2196–2203.
7. J. D. Kraus, *Electromagnetics*, 4th ed., McGraw-Hill Book Co., New York, 1992.
8. G. S. Smith, "Loop Antennas," Chapter 5 in *Antenna Engineering Handbook*, 2nd ed., McGraw-Hill Book Co., New York, 1984.
9. J. E. Storer, "Impedance of Thin-Wire Loop Antennas," *AIEE Trans., (Part I. Communication and Electronics)*, Vol. 75, Nov. 1956, pp. 606–619.
10. S. Adachi and Y. Mushiake, "Studies of Large Circular Loop Antenna," *Sci. Rep. Research Institute of Tohoku University (RITU), B*, Vol. 9, No. 2, 1957, pp. 79–103.
11. S. Ito, N. Inagaki, and T. Sekiguchi, "An Investigation of the Array of Circular-Loop Antennas," *IEEE Trans. Antennas Propagat.*, Vol. AP-19, No. 4, July 1971, pp. 469–476.
12. A. Shoamanesh and L. Shafai, "Properties of Coaxial Yagi Loop Arrays," *IEEE Trans. Antennas Propagat.*, Vol. AP-26, No. 4, July 1978, pp. 547–550.

13. A. Shoamanesh and L. Shafai, "Design Data for Coaxial Yagi Array of Circular Loops," *IEEE Trans. Antennas Propagat.*, Vol. AP-27, September 1979, pp. 711–713.
14. D. DeMaw (ed.), *The Radio Amateur's Handbook*, American Radio Relay League, 56th ed., 1979, pp. 20–18.
15. G. N. Watson, *A Treatise on the Theory of Bessel Functions*, Cambridge University Press, London, 1922.
16. S. V. Savov, "An Efficient Solution of a Class of Integrals Arising in Antenna Theory," *IEEE Antennas Propagat. Mag.*, Vol. 44, No. 5, October 2002, pp. 98–101.
17. J. D. Mahony, "Circular Microstrip-Patch Directivity Revisited: An Easily Computable Exact Expression," *IEEE Antennas Propagat. Mag.*, Vol. 45, No. 1, February 2003, pp. 120–122.
18. J. D. Mahony, "A Comment on Q-Type Integrals and Their Use in Expressions for Radiated Power," *IEEE Antennas Propagat. Mag.*, Vol. 45, No. 3, June 2003, pp. 127–138.
19. S. V. Savov, "A Comment on the Radiation Resistance," *IEEE Antennas Propagat. Mag.*, Vol. 45, No. 3, June 2003, p. 129.
20. I. Gradshteyn and I. Ryzhik, *Tables of Integrals, Series and Products*, Academic Press, New York, 1965.
21. J. E. Lindsay, Jr., "A Circular Loop Antenna with Non-Uniform Current Distribution," *IRE Trans. Antennas Propagat.*, Vol. AP-8, No. 4, July 1960, pp. 438–441.
22. E. A. Wolff, *Antenna Analysis*, Wiley, New York, 1966.
23. H. C. Pocklington, "Electrical Oscillations in Wire," Cambridge Philosophical Society Proceedings, London, England, Vol. 9, 1897, p. 324.
24. E. Hallén, "Theoretical Investigations into the Transmitting and Receiving Qualities of Antennae," *Nova Acta Regiae Soc. Sci. Upsaliensis*, Ser. IV, No. 4, pp. 1–44, 1938.
25. J. E. Storer, "Impedance of Thin-Wire Loop Antennas," AIEE Trans., (*Part I. Communication and Electronics*), Vol. 75, pp. 606–619, Nov. 1956.
26. T. T. Wu, "Theory of Thin Circular Antenna," *J. Math Phys.*, Vol. 3, pp. 1301–1304, Nov.–Dec. 1962.
27. L. D. Licking and D. E. Merewether, "An Analysis of Thin-Wire Circular Loop Antennas of Arbitrary Size," Report No. SC-RR-70-433, Sandia National Lab., Albuquerque, NM, Aug. 1970.
28. A. F. McKinley, T. P. White, I. S. Maksymov, and K. R. Catchpole, "The Analytical Basis for the Resonances and Anti-Resonances of Loop Antennas and Meta-Material Ring Resonators," *Journal of Applied Physics*, Vol. 112, No. 9, pp. 094911-094911-9, Nov. 2012.
29. A. F. McKinley, T. P. White, K. R. Catchpole, "Theory of the Circular Loop Antenna in the Terahertz, Infrared, and Optical Regions," *Journal of Applied Physics*, Vol. 114, No. 4, pp. 044317-044317-10, 2013.
30. R. King, "Theory of Antennas Driven from Two-Wire Line," *J. Appl. Phys.*, Vol. 20, 1949, p. 832.
31. D. G. Fink (ed.), *Electronics Engineers' Handbook*, Section 18, "Antennas" (by W. F. Croswell), McGraw-Hill, New York, pp. 18–22.
32. K. Iizuka, R. W. P. King, and C. W. Harrison, Jr., "Self- and Mutual Admittances of Two Identical Circular Loop Antennas in a Conducting Medium and in Air," *IEEE Trans. Antennas Propagat.*, Vol. AP-14, No. 4, July 1966, pp. 440–450.
33. R. E. Collin and F. J. Zucher (eds.), *Antenna Theory Part 2*, Chapter 23 (by J. R. Wait), McGraw-Hill, New York, 1969.
34. J. R. Wait, "Possible Influence of the Ionosphere on the Impedance of a Ground-Based Antenna," *J. Res. Natl. Bur. Std.* (U.S.), Vol. 66D, September–October 1962, pp. 563–569.
35. L. E. Vogler and J. L. Noble, "Curves of Input Impedance Change Due to Ground for Dipole Antennas," U.S. National Bureau of Standards, Monograph 72, January 31, 1964.
36. D. C. Chang, "Characteristics of a Horizontal Circular Loop Antenna over a Multilayered, Dissipative Half-Space," *IEEE Trans. Antennas Propagat.*, Vol. AP-21, No. 6, November 1973, pp. 871–874.
37. R. W. P. King, "Theory of the Center-Driven Square Loop Antenna," *IRE Trans. Antennas Propagat.*, Vol. AP-4, No. 4, July 1956, p. 393.
38. T. Tsukiji and S. Tou, "On Polygonal Loop Antennas," *IEEE Trans. Antennas Propagat.*, Vol. AP-28, No. 4, July 1980, pp. 571–575.
39. M. A. Islam, "A Theoretical Treatment of Low-Frequency Loop Antennas with Permeable Cores," *IEEE Trans. Antennas Propagat.*, Vol. AP-11, No. 2, March 1963, pp. 162–169.

40. V. H. Rumsey and W. L. Weeks, "Electrically Small Ferrite Loaded Loop Antennas," *IRE Convention Rec.*, Vol. 4, Part 1, 1956, pp. 165–170.
41. K. Fujimoto and J. R. James, *Mobile Antenna Systems Handbook*, Artech House, Norwood, MA, 1994.
42. M. A. Jensen and Y. Rahmat-Samii, "Performance Analysis of Antennas for Hand-Held Transceivers Using FDTD," *IEEE Trans. Antennas Propagat.*, Vol. 42, No. 8, August 1994, pp. 1106–1113.
43. M. A. Jensen and Y. Rahmat-Samii, "EM Interaction of Handset Antennas and a Human in Personal Communications," *Proc. IEEE*, Vol. 83, No. 1, January 1995, pp. 7–17.
44. K. D. Katsibas, C. A. Balanis, P. A. Tirkas, and C. R. Birtcher, "Folded Loop Antenna for Mobile Communication Systems," 1996 IEEE Antennas and Propagation Society International Symposium, Baltimore, MD, July 21–26, 1996, pp. 1582–1585.
45. C. A. Balanis, K. D. Katsibas, P. A. Tirkas, and C. R. Birtcher, "Loop Antenna for Mobile and Personal Communication Systems," IEEE International Vehicular Technology Conference (IEEE VTC '97), Phoenix, AZ, May 5–7, 1997.
46. K. D. Katsibas, C. A. Balanis, P. A. Tirkas, and C. R. Birtcher, "Folded Loop Antenna for Mobile Handheld Units," *IEEE Trans. Antennas Propagat.*, Vol. 46, No. 2, February 1998, pp. 260–266.

PROBLEMS

5.1. Derive
   (a) (5-18a)–(5-18c) using (5-17) and (3-2a)
   (b) (5-19a)–(5-19b) using (5-18a)–(5-18c)

5.2. Write the fields of an infinitesimal linear magnetic dipole of constant current I_m, length l, and positioned along the z-axis. Use the fields of an infinitesimal electric dipole, (4-8a)–(4-10c), and apply the principle of duality. Compare with (5-20a)–(5-20d).

5.3. A circular loop, *of loop radius $\lambda/30$ and wire radius $\lambda/1000$*, is used as a transmitting/receiving antenna in a back-pack radio communication system at *10 MHz*. The wire of the loop is made of copper with a conductivity of 5.7×10^7 *S/m*. Assuming the antenna is *radiating in free space*, determine the
   (a) radiation resistance of the loop;
   (b) loss resistance of the loop (*assume that its value is the same as if the wire were straight*);
   (c) input resistance;   (d) input impedance;   (e) radiation efficiency.

5.4. A small circular loop with a uniform current distribution, and with its classical omnidirectional pattern, is used as a receiving antenna. Determine the maximum directivity (*dimensionless* and *in dB*) using:
   (a) Exact method.
   (b) An approximate method appropriate for this pattern. Specify the method used.
   (c) Another approximate method appropriate for this pattern. Specify the method used.
   Hint: For the approximate methods, the word omnidirectional is a clue.

5.5. A *N-turn* resonant circular loop *with a uniform current distribution* and with a *circumference of $\lambda/4$*, is fed by a lossless balanced twin-lead transmission line with a characteristic impedance of 300 ohms. *Neglecting proximity effects*, determine the
   (a) *closest integer number* of turns so that the input impedance is nearly 300 ohms;
   (b) input impedance of the antenna;   (c) reflection coefficient;
   (d) VSWR inside the transmission line.

5.6. A small circular loop with circumference $C < \lambda/20$ is used as a receiving antenna. A uniform plane wave traveling along the x-axis and toward the positive (+) x direction (as shown in the figure), whose electric field is given by

$$\mathbf{E}_w^i = (\hat{\mathbf{a}}_y + 2\hat{\mathbf{a}}_z)e^{-jkx}$$

is incident upon the antenna. Determine the
(a) polarization of the incident wave. Justify your answer.
(b) axial ratio of the polarization ellipse of the incident wave.
(c) polarization of the loop antenna toward the x-axis.
(d) polarization loss factor (*dimensionless* and *in dB*).
(e) maximum power at *1 GHz* that can be delivered to a load connected to the antenna, if the power density of the above incident wave is 5 mwatts/cm^2. Assume no other losses.

Hint: $\hat{\mathbf{a}}_\phi = -\hat{\mathbf{a}}_x \sin\phi + \hat{\mathbf{a}}_y \cos\phi$

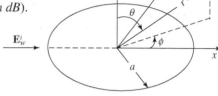

5.7. A combination of a horizontal small loop of *uniform current* and a vertical infinitesimal dipole, as shown in the figure below, *are used as one antenna*. The currents in the two elements are adjusted so that the *magnitudes* of the corresponding *far-zone* electric field components radiated by each element are equal. For the entire antenna system, loop plus dipole:

(a) Write an expression for the *normalized electric filed* radiated by the combination of the two elements.
(b) State the *polarization* of the entire antenna system (linear, circular, elliptical).
(c) Determine the *polarization loss factor (dimensionless and in dB) if a linearly polarized wave*, coming from any direction, is incident upon this antenna which is used as a receiving antenna.

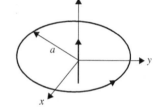

5.8. Find the radiation efficiency of a single-turn and a four-turn circular loop each of radius $\lambda/(10\pi)$ and operating at 10 MHz. The radius of the wire is $10^{-3}\lambda$ and the turns are spaced $3 \times 10^{-3}\lambda$ apart. Assume the wire is copper with a conductivity of 5.7×10^7 S/m, and the antenna is radiating into free-space.

5.9. Find the power radiated by a small loop by forming the average power density, using (5-27a)–(5-27c), and integrating over a sphere of radius r. Compare the answer with (5-23b).

5.10. For a small loop of constant current, derive its far-zone fields using (5-17) and the procedure outlined and relationships developed in Section 3.6. Compare the answers with (5-27a)–(5-27c).

5.11. A single-turn resonant circular loop with a $\lambda/8\pi$ radius is made of copper wire with a wire radius of $10^{-4}\lambda/2\pi$ and conductivity of 5.7×10^7 S/m. For a frequency of 100 MHz, determine, *assuming uniform current*, the
(a) radiation efficiency (assume the wire is straight);
(b) *maximum gain* of the antenna (*dimensionless* and *in dB*).

5.12. A horizontal, one-turn, loop antenna with a circumference of $C = \pi\lambda$ is radiating in free space, and it used as a ground-based receiving antenna for an over-the-horizon communication system. *Assuming the current distribution is uniform*, determine the
(a) *Maximum* directivity of the antenna (*dimensionless and in dB*).
(b) *Loss resistance* of the wire of the loop. Assume the wire is straight, has a radius of $10^{-4}\lambda$, a conductivity of 10^7 S/m, and the loop is operating at 100 MHz.

(c) *Radiation efficiency* of the loop (*in %*).
(d) *Maximum gain* (*dimensionless and in dB*) of the loop.

5.13. Design a lossless resonant circular loop operating at 10 MHz so that its single-turn radiation resistance is 0.73 ohms. The resonant loop is to be connected to a matched load through a balanced "twin-lead" 300-ohm transmission line.
   (a) Determine the radius of the loop (in meters and wavelengths).
   (b) To minimize the matching reflections between the resonant loop and the 300-ohm transmission line, determine the closest number of integer turns the loop must have.
   (c) For the loop of part b, determine the maximum power that can be expected to be delivered to a receiver matched load if the incident wave is polarization matched to the lossless resonant loop. The power density of the incident wave is 10^{-6} watts/m².

5.14. A resonant six-turn loop of *closely spaced turns* is operating at 50 MHz. The radius of the loop is $\lambda/30$, and the loop is connected to a 50-ohm transmission line. The radius of the wire is $\lambda/300$, its conductivity is $\sigma = 5.7 \times 10^7$ S/m, and the spacing between the turns is $\lambda/100$. Determine the
   (a) directivity of the antenna (in dB)
   (b) radiation efficiency taking into account the proximity effects of the turns
   (c) reflection efficiency   (d) gain of the antenna (in dB)

5.15. A horizontal, lossless, one-turn circular loop of circumference $C = \lambda$, with a nonuniform current distribution, is radiating in free space. The far-field pattern of the antenna can be approximated by

$$E_\phi \simeq C_0 \cos^2 \theta \, \frac{e^{-jkr}}{r} \left. \begin{matrix} 0° \leq \theta \leq 90° \\ 0° \leq \phi \leq 360° \end{matrix} \right\}$$

where C_0 is a constant and θ is measured from the normal to the plane/area of the loop. Determine the
   (a) Maximum exact directivity (*dimensionless* and *in dB*) of the antenna.
   (b) *Approximate* input impedance of the loop.
   (c) Input *reflection coefficient* when the antenna is connected to a balanced "twin-lead" transmission line with a characteristic impedance of 300 ohms.
   (d) *Maximum gain* of the loop (*dimensionless* and *in dB*).
   (e) *Maximum realized gain* of the loop (*dimensionless* and *in dB*).

5.16. Find the radiation efficiency (in percent) of an eight-turn circular-loop antenna operating at 30 MHz. The radius of each turn is $a = 15$ cm, the radius of the wire is $b = 1$ mm, and the spacing between turns is $2c = 3.6$ mm. Assume the wire is copper ($\sigma = 5.7 \times 10^7$ S/m), and the antenna is radiating into free-space. Account for the *proximity effect*.

5.17. A very small circular loop of radius $a(a < \lambda/6\pi)$ and constant current I_0 is symmetrically placed about the origin at $x = 0$ and with the plane of its area parallel to the y-z plane. Find the
   (a) spherical **E**- and **H**-field components radiated by the loop in the far zone
   (b) directivity of the antenna

5.18. Repeat Problem 5.17 when the plane of the loop is parallel to the x-z plane at $y = 0$.

5.19. Using the computer program of this chapter, compute the radiation resistance and the directivity of a circular loop of constant current with a radius of
   (a) $a = \lambda/50$ (b) $a = \lambda/10$ (c) $a = \lambda/4$ (d) $a = \lambda/2$

5.20. A constant current circular loop of radius $a = 5\lambda/4$ is placed on the x-y plane. Find the *two* smallest angles (excluding $\theta = 0°$) where a null is formed in the far-field pattern.

5.21. Design a circular loop of constant current such that its field intensity vanishes only at $\theta = 0° (\theta = 180°)$ and $90°$. Find its
  (a) radius   (b) radiation resistance   (c) directivity

5.22. Design a constant current circular loop so that its first minimum, aside from $\theta = 0°$, in its far-field pattern is at $30°$ from a normal to the plane of the loop. Find the
  (a) smallest radius of the antenna (in wavelengths)
  (b) relative (to the maximum) radiation intensity (in dB) in the plane of the loop

5.23. Design a constant current circular loop so that its pattern has a null in the plane of the loop, and two nulls above and two nulls below the plane of the loop. Find the
  (a) radius of the loop   (b) angles where the nulls occur

5.24. A constant current circular loop is placed on the x-y plane. Find the far-field position, relative to that of the loop, that a linearly polarized probe antenna must have so that the polarization loss factor (PLF) is maximized.

5.25. A very small $(a \ll \lambda)$ circular loop of constant current is placed a distance h above an infinite electric ground plane. Assuming z is perpendicular to the ground plane, find the total far-zone field radiated by the loop when its plane is parallel to the
  (a) x-z plane   (b) y-z plane

5.26. A very small loop antenna $(a \ll \lambda/30)$ of constant current is placed a height h above a flat, perfectly conducting ground plane of infinite extent. The area plane of the loop is parallel to the interface (x-y plane). For far-field observations
  (a) find the total electric field radiated by the loop in the presence of the ground plane
  (b) all the angles (in degrees) from the vertical to the interface where the total field will vanish when the height is λ
  (c) the smallest nonzero height (in λ) such that the total far-zone field exhibits a null at an angle of $60°$ from the vertical

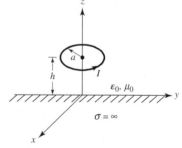

5.27. The antenna of the VOR (VHF Omni Range) airport guidance system consists of a small radius a circular loop ($a \ll \lambda$, so that its current is uniform). The circular loop is placed on a plane horizontal and at a height h above an ideal planar and infinite in extent synthesized PMC (perfect magnetic conductor) ground plane. To make sure the antenna remains operational at all times, the antenna is placed at a height $h = 0.75\lambda$, above the PMC ground plane.
  Assuming far-field observations, determine
  (a) The *normalized array factor* of the equivalent antenna system that is valid in all space on and above the PMC ground plane.
  (b) All the angles θ (in degrees) that th AF (array factor) of the equivalent system will achieve its maximum radiation and allow safe operation of the VOR navigation system.

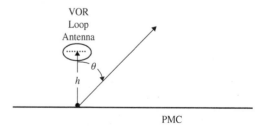

5.28. A very small circular loop, of radius a and constant current I_0, is placed a height h above an infinite and flat Perfect Magnetic Conductor (PMC). The area of the loop is parallel to the PMC, which is on the xy-plane; the z-axis is perpendicular to the PMC interface. Determine the

(a) Total far-zone electric field radiated by the loop in the presence of the PMC.

(b) Smallest height h (*in wavelengths*) so that the total field pattern possesses simultaneously nulls only at $\theta = 0°$ and $30°$.

5.29. A small circular loop, with its area parallel to the x-z plane, is placed a height h above an infinite flat perfectly electric conducting ground plane. Determine

(a) the array factor for the equivalent problem which allows you to find the total field on and above the ground plane

(b) angle(s) θ (in degrees) where the array factor will vanish when the loop is placed at a height $\lambda/2$ above the ground plane

5.30. A small circular electric loop, *of uniform current I_0*, is placed horizontally/parallel at zero height ($h = 0$) above a *Perfect Magnetic Conductor* (PMC), and it is used as a receiving antenna at a frequency of *100 MHz*. The circumference of the loop is $C = \lambda/20$. Assuming the wire radius is very small ($b \ll \lambda$), determine the:

(a) *Maximum* directivity (*dimensionless* and *in dB*). *Justify your answer.*

(b) *Maximum* effective area (*in cm²*).

(c) *Maximum* power (*in watts*) that can be delivered to a *matched load*, connected to the loop, when a *circularly polarized wave*, with an *power density of 10^{-4} watts/cm²*, is incident (*in the direction of maximum directivity*) upon the loop. Assume no other losses.

5.31. A small circular loop with its area parallel to the x-z plane is placed at a height h above an infinite perfectly conducting ground plane, as shown in the figure for Problem 5.29. Determine the

(a) array factor for the equivalent problem which will allow you to find the total field *on and above* the ground plane.

(b) two *smallest* heights h (*in λ*) greater than $h = 0$ (i.e., $h > 0$) that will form a maximum on the magnitude of the array factor toward $\theta = 0°$.

5.32. The emergency radio police system *of Problem 1* ($f = 10$ MHz) now uses a *very small circular* loop of *constant current distribution*. The circular loop is placed horizontally, as shown below, a height h above the top of the police car. Consider that the top of the police car to be an *infinite and planar artificial PMC* surface. The sensitivity (*minimum power*) of the system receiver, to be able to detect an incoming signal, is 10 μwatts. Assuming the incoming signal is *circularly polarized* and it incident from a horizontal direction (grazing angle; $\theta = 90°$), what is the

(a) *Smallest obvious height h* (in λ) of the loop *above the PMC* to maximize the directivity?

(b) What is this *maximum directivity, using the smallest height h from part a, (dimensionless and in dB)*?

(c) *Minimum power density* (in watts/cm²) of the incoming signal to be detected by the radio receiver, *using the smallest height h from part a*?

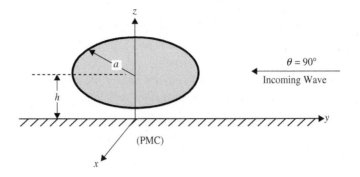

5.33. For the loop of Problem 5.25(a), find the smallest height h so that a null is formed in the y-z plane at an angle of 45° above the ground plane.

5.34. A circular loop with a radius of $a = \lambda/20\pi$ is placed *vertically* a height h above a PEC ground plane, as shown in the figure (the yz-plane of its area is perpendicular to the ground plane). The height h is measured from the center of the loop. For this configuration, determine the

(a) *Normalized array factor.* Indicate how you obtained it. *You do not need to derive it as long as you explain the rationale.*

(b) Smallest height (*in wavelengths*) that the loop must be placed above the ground plane to introduce the first null in the array factor at an angle of $\theta = 60$ degrees from the vertical direction.

5.35. The transmitting and receiving antennas of a wireless communication system consist, respectively, of a small horizontal circular loop (with radius $a \ll \lambda$, so that its current is uniform) and an ideal infinitesimal electric dipole ($l \ll \lambda/50$). The two antennas are at the same level and separated by a distance d so that one is in the far-field of the other. Assuming both radiate in an unbounded infinite free-space medium:

Determine the PLF (polarization loss factor, *dimensionless and in dB*) of the two-antenna communication system for two different dipole orientations: i.e., when the linear dipole is oriented along the

(a) z direction   (b) y direction

In both cases, the plane (area) of the loop lies on the horizontal plane (parallel to the xy-plane); i.e., the loop does NOT change orientation; stays the same for both dipole orientations.

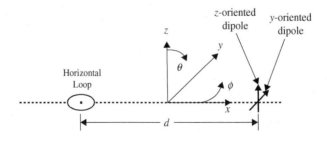

5.36. A small single-turn circular loop of radius $a = 0.05\lambda$ is operating at 300 MHz. Assuming the radius of the wire is $10^{-4}\lambda$, determine the

(a) loss resistance   (b) radiation resistance   (c) loop inductance

Show that the loop inductive reactance is much greater than the loss resistance and radiation resistance indicating that a small loop acts primarily as an inductor.

5.37. Determine the radiation resistance of a single-turn small loop, assuming the geometrical shape of the loop is

(a) rectangular with dimensions a and b $(a, b \ll \lambda)$
(b) elliptical with major axis a and minor axis b $(a, b \ll \lambda)$

5.38. A one-turn small circular loop is used as a radiating element for a VHF ($f = 100$ MHz) communications system. The circumference of the loop is $C = \lambda/20$ while the radius of the wire is $\lambda/400$. Determine, using a wire conductivity of $\sigma = 5.7 \times 10^7$ S/m, the

(a) input resistance of the wire for a single turn.
(b) input reactance of the loop. *Is it inductive or capacitive? Be specific.*
(c) inductance (*in henries*) or capacitance (*in farads*) that can be placed *in series* with the loop at the feed to resonate the antenna at $f = 100$ MHz; choose the element that will accomplish the desired objective.

5.39. Show that for the rectangular loop the radiation resistance is represented by

$$R_r = 31{,}171 \left(\frac{a^2 b^2}{\lambda^4}\right)$$

while for the elliptical loop is represented by

$$R_r = 31{,}171 \left(\frac{\pi^2 a^2 b^2}{16\lambda^4}\right)$$

5.40. Assuming the direction of the magnetic field of the incident plane wave coincides with the plane of incidence, derive the effective length of a small circular loop of radius a based on the definition of (2-92). Show that its effective length is $(S = \pi a^2)$

$$\boldsymbol{\ell}_e = \hat{\mathbf{a}}_\phi \, jkS \sin(\theta)$$

5.41. A circular loop of nonconstant current distribution, with circumference of 1.4λ, is attached to a 300-ohm line. Assuming the radius of the wire is $1.555 \times 10^{-2}\lambda$, find the

(a) input impedance of the loop   (b) VSWR of the system
(c) inductance or capacitance that must be placed across the feed points so that the loop becomes resonant at $f = 100$ MHz.

5.42. A very popular antenna for amateur radio operators is a square loop antenna (referred to as *quad antenna*) whose circumference is one wavelength. Assuming the radiation characteristics of the square loop are well represented by those of a circular loop:

(a) What is the input impedance (real and imaginary parts) of the antenna?
(b) What element (inductor or capacitor), and of what value, must be placed in series with the loop at the feed point to resonate the radiating element at a frequency of 1 GHz?
(c) What is the input VSWR, having the inductor or capacitor in place, if the loop is connected to a 78-ohm coaxial cable?

5.43. A circular loop of *nonuniform current*, circumference $C = \lambda$, and wire radius $b = 2.47875 \times 10^{-3}\lambda$, is used for end-fire (over-the-head; toward zenith; $\theta = 0°$) communication. The loop is connected to a 75-ohm transmission line. Determine the

(a) Approximate input impedance (real and imaginary parts). To get total credit, state how or where you got the answer. *You do not necessarily have to compute it. Equations (5-37a)–(5-38) are valid only for uniform current distribution.*

(b) Is the input impedance capacitive or inductive?

(c) What kind of a lumped element, *capacitor or inductor*, must be placed in *parallel* to resonate the loop?

(d) At a frequency of 500 MHz, what is the *capacitance or inductance* of the parallel lumped element?

(e) What is the new input impedance of the resonated loop (*in the presence of the parallel capacitor or inductor*)?

(f) What is the input VSWR of the resonated loop (*in the presence of the parallel capacitor or inductor*)?

5.44. Design circular loops of wire radius b, which resonate at the first resonance. Find

(a) four values of a/b where the first resonance occurs (a is the radius of the loop)

(b) the circumference of the loops and the corresponding radii of the wires for the antennas of part (a).

5.45. Using (5-54b) the asymptotic form of (5-65a) for small argument, show that the radiation resistance R_r for a small radius ($a \ll \lambda$) loop of uniform current is given by

$$R_r = 20\pi^2(ka)^4 = 20\pi^2 \left(\frac{C}{\lambda}\right)^4$$

5.46. Consider a circular loop of wire of radius a on the *x-y* plane and centered about the origin. Assume the current on the loop is given by

$$I_\phi(\phi') = I_0 \cos(\phi')$$

(a) Show that the far-zone electric field of the loop is given by

$$E_\theta = \frac{j\eta ka}{2} I_0 \frac{e^{-jkr}}{r} \frac{J_1(ka\sin\theta)}{ka\sin\theta} \cos\theta \sin\phi$$

$$E_\phi = \frac{j\eta ka}{2} I_0 \frac{e^{-jkr}}{r} J_1'(ka\sin\theta) \cos\phi$$

$$J_1'(x) = \frac{dJ_1(x)}{dx}$$

(b) Evaluate the radiation intensity $U(\theta, \phi)$ in the direction $\theta = 0$ and $\phi = \frac{\pi}{2}$ as a function of ka.

CHAPTER 6

Arrays: Linear, Planar, and Circular

6.1 INTRODUCTION

In the previous chapter, the radiation characteristics of single-element antennas were discussed and analyzed. Usually the radiation pattern of a single element is relatively wide, and each element provides low values of directivity (gain). In many applications it is necessary to design antennas with very directive characteristics (very high gains) to meet the demands of long distance communication. This can only be accomplished by increasing the electrical size of the antenna.

Enlarging the dimensions of single elements often leads to more directive characteristics. Another way to enlarge the dimensions of the antenna, without necessarily increasing the size of the individual elements, is to form an assembly of radiating elements in an electrical and geometrical configuration. This new antenna, formed by multielements, is referred to as an *array*. In most cases, the elements of an array are identical. This is not necessary, but it is often convenient, simpler, and more practical. The individual elements of an array may be of any form (wires, apertures, etc.).

The total field of the array is determined by the vector addition of the fields radiated by the individual elements. This assumes that the current in each element is the same as that of the isolated element (neglecting coupling). This is usually not the case and depends on the separation between the elements. To provide very directive patterns, it is necessary that the fields from the elements of the array interfere constructively (add) in the desired directions and interfere destructively (cancel each other) in the remaining space. Ideally this can be accomplished, but practically it is only approached. In an array of identical elements, there are at least five controls that can be used to shape the overall pattern of the antenna. These are:

1. the geometrical configuration of the overall array (linear, circular, rectangular, spherical, etc.)
2. the relative displacement between the elements
3. the excitation amplitude of the individual elements
4. the excitation phase of the individual elements
5. the relative pattern of the individual elements

The influence that each one of the above has on the overall radiation characteristics will be the subject of this chapter. In many cases the techniques will be illustrated with examples.

Antenna Theory: Analysis and Design, Fourth Edition. Constantine A. Balanis.
© 2016 John Wiley & Sons, Inc. Published 2016 by John Wiley & Sons, Inc.
Companion Website: www.wiley.com/go/antennatheory4e

There are a plethora of antenna arrays used for personal, commercial, and military applications utilizing different elements including dipoles, loops, apertures, microstrips, horns, reflectors, and so on. Arrays of dipoles are shown in Figures 4.26, 10.19, and 11.15. The one in Figure 4.26 is an array that is widely used as a base-station antenna for mobile communication. It is a triangular array consisting of twelve dipoles, with four dipoles on each side of the triangle. Each four-element array, on each side of the triangle, is basically used to cover an angular sector of 120° forming what is usually referred to as a *sectoral array*. The one in Figure 10.19 is a classic array of dipoles, referred to as the *Yagi-Uda* array, and it is primarily used for TV and amateur radio applications. The array of Figure 11.15 is also an array of dipoles, which is referred to as the *log-periodic* antenna, which is primarily used for TV reception and has wider bandwidth than the Yagi-Uda array but slightly smaller directivity. An array of loops is shown in Figure 5.1 and one utilizing microstrips as elements is displayed in Figure 14.35. An advanced array design of slots, used in the AWACS, is shown in Figure 6.29.

The simplest and one of the most practical arrays is formed by placing the elements along a line. To simplify the presentation and give a better physical interpretation of the techniques, a two-element array will first be considered. The analysis of an N-element array will then follow. Two-dimensional analysis will be the subject at first. In latter sections, three-dimensional techniques will be introduced.

6.2 TWO-ELEMENT ARRAY

Let us assume that the antenna under investigation is an array of two infinitesimal horizontal dipoles positioned along the z-axis, as shown in Figure 6.1(a). The total field radiated by the two elements, *assuming no coupling between the elements*, is equal to the sum of the two and in the y-z plane it is given by

$$\mathbf{E}_t = \mathbf{E}_1 + \mathbf{E}_2 = \hat{\mathbf{a}}_\theta j\eta \frac{kI_0 l}{4\pi} \left\{ \frac{e^{-j[kr_1-(\beta/2)]}}{r_1} \cos\theta_1 + \frac{e^{-j[kr_2+(\beta/2)]}}{r_2} \cos\theta_2 \right\} \quad (6\text{-}1)$$

where β is the difference in phase excitation between the elements. The magnitude excitation of the radiators is identical. Assuming far-field observations and referring to Figure 6.1(b),

$$\theta_1 \simeq \theta_2 \simeq \theta \quad (6\text{-}2\text{a})$$

$$\left.\begin{array}{l} r_1 \simeq r - \dfrac{d}{2}\cos\theta \\[4pt] r_2 \simeq r + \dfrac{d}{2}\cos\theta \end{array}\right\} \text{ for phase variations} \quad (6\text{-}2\text{b})$$

$$r_1 \simeq r_2 \simeq r \quad \text{for amplitude variations} \quad (6\text{-}2\text{c})$$

Equation 6-1 reduces to

$$\mathbf{E}_t = \hat{\mathbf{a}}_\theta j\eta \frac{kI_0 l e^{-jkr}}{4\pi r} \cos\theta [e^{+j(kd\cos\theta+\beta)/2} + e^{-j(kd\cos\theta+\beta)/2}]$$

$$\mathbf{E}_t = \hat{\mathbf{a}}_\theta j\eta \frac{kI_0 l e^{-jkr}}{4\pi r} \cos\theta \left\{ 2\cos\left[\frac{1}{2}(kd\cos\theta + \beta)\right] \right\} \quad (6\text{-}3)$$

It is apparent from (6-3) that the total field of the array is equal to the field of a single element positioned at the origin multiplied by a factor which is widely referred to as the *array factor*. Thus

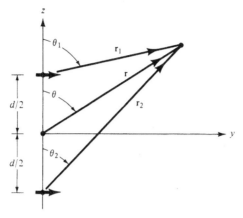

(a) Two infinitesimal dipoles

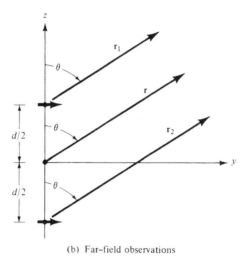

(b) Far-field observations

Figure 6.1 Geometry of a two-element array positioned along the z-axis.

for the two-element array of constant amplitude, the array factor is given by

$$\text{AF} = 2\cos[\tfrac{1}{2}(kd\cos\theta + \beta)] \tag{6-4}$$

which in normalized form can be written as

$$(\text{AF})_n = \cos[\tfrac{1}{2}(kd\cos\theta + \beta)] \tag{6-4a}$$

The array factor is a function of the geometry of the array and the excitation phase. By varying the separation d and/or the phase β between the elements, the characteristics of the array factor and of the total field of the array can be controlled.

It has been illustrated that the far-zone field of a uniform two-element array of identical elements is equal to the *product of the field of a single element, at a selected reference point (usually the origin), and the array factor of that array*. That is,

$$\boxed{\mathbf{E}(\text{total}) = [\mathbf{E}(\text{single element at reference point})] \times [\text{array factor}]} \tag{6-5}$$

This is referred to as *pattern multiplication* for arrays of identical elements, and it is analogous to the pattern multiplication of (4-59) for continuous sources. Although it has been illustrated only for an array of two elements, each of identical magnitude, it is also valid for arrays with any number of identical elements which do not necessarily have identical magnitudes, phases, and/or spacings between them. This will be demonstrated in this chapter by a number of different arrays.

Each array has its own array factor. The array factor, in general, is a function of the number of elements, their geometrical arrangement, their relative magnitudes, their relative phases, and their spacings. The array factor will be of simpler form if the elements have identical amplitudes, phases, and spacings. Since the array factor does not depend on the directional characteristics of the radiating elements themselves, it can be formulated by replacing the actual elements with isotropic (point) sources. Once the array factor has been derived using the point-source array, the total field of the actual array is obtained by the use of (6-5). Each point-source is assumed to have the amplitude, phase, and location of the corresponding element it is replacing.

In order to synthesize the total pattern of an array, the designer is not only required to select the proper radiating elements but the geometry (positioning) and excitation of the individual elements. To illustrate the principles, let us consider some examples.

Example 6.1

Given the array of Figures 6.1(a) and (b), find the nulls of the total field when $d = \lambda/4$ and

a. $\beta = 0$
b. $\beta = +\dfrac{\pi}{2}$
c. $\beta = -\dfrac{\pi}{2}$

Solution:

a. $\beta = 0$

The normalized field is given by

$$E_{tn} = \cos\theta \cos\left(\frac{\pi}{4}\cos\theta\right)$$

The nulls are obtained by setting the total field equal to zero, or

$$E_{tn} = \cos\theta \cos\left(\frac{\pi}{4}\cos\theta\right)\Big|_{\theta=\theta_n} = 0$$

Thus

$$\cos\theta_n = 0 \Rightarrow \theta_n = 90°$$

and

$$\cos\left(\frac{\pi}{4}\cos\theta_n\right) = 0 \Rightarrow \frac{\pi}{4}\cos\theta_n = \frac{\pi}{2}, -\frac{\pi}{2} \Rightarrow \theta_n = \text{does not exist}$$

The only null occurs at $\theta = 90°$ and is due to the pattern of the individual elements. The array factor does not contribute any additional nulls because there is not enough separation between the elements to introduce a phase difference of 180° between the elements, for any observation angle.

b. $\beta = +\dfrac{\pi}{2}$

The normalized field is given by

$$E_{tn} = \cos\theta \cos\left[\dfrac{\pi}{4}(\cos\theta + 1)\right]$$

The nulls are found from

$$E_{tn} = \cos\theta \cos\left[\dfrac{\pi}{4}(\cos\theta + 1)\right]\Big|_{\theta=\theta_n} = 0$$

Thus

$$\cos\theta_n = 0 \Rightarrow \theta_n = 90°$$

and

$$\cos\left[\dfrac{\pi}{4}(\cos\theta + 1)\right]\Big|_{\theta=\theta_n} = 0 \Rightarrow \dfrac{\pi}{4}(\cos\theta_n + 1) = \dfrac{\pi}{2} \Rightarrow \theta_n = 0°$$

and

$$\Rightarrow \dfrac{\pi}{4}(\cos\theta_n + 1) = -\dfrac{\pi}{2} \Rightarrow \theta_n = \text{does not exist}$$

The nulls of the array occur at $\theta = 90°$ and $0°$. The null at $0°$ is introduced by the arrangement of the elements (array factor). This can also be shown by physical reasoning, as shown in Figure 6.2(a). The element in the negative z-axis has an initial phase lag of $90°$ relative to the other element. As the wave from that element travels toward the positive z-axis ($\theta = 0°$ direction), it undergoes an additional $90°$ phase retardation when it arrives at the other element on the positive z-axis. Thus there is a total of $180°$ phase difference between the waves of the two elements when travel is toward the positive z-axis ($\theta = 0°$). The waves of the two elements are in phase when they travel in the negative z-axis ($\theta = 180°$), as shown in Figure 6.2(b).

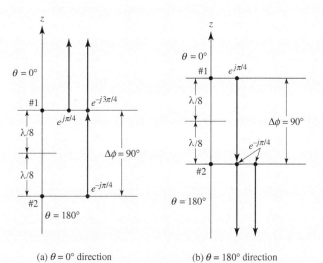

(a) $\theta = 0°$ direction (b) $\theta = 180°$ direction

Figure 6.2 Phase accumulation for two-element array for null formation toward $\theta = 0°$ and $180°$.

c. $\beta = -\dfrac{\pi}{2}$

The normalized field is given by

$$E_{tn} = \cos\theta \cos\left[\dfrac{\pi}{4}(\cos\theta - 1)\right]$$

and the nulls by

$$E_{tn} = \cos\theta \cos\left[\dfrac{\pi}{4}(\cos\theta - 1)\right]\bigg|_{\theta=\theta_n} = 0$$

Thus

$$\cos\theta_n = 0 \Rightarrow \theta_n = 90°$$

and

$$\cos\left[\dfrac{\pi}{4}(\cos\theta_n - 1)\right] = 0 \Rightarrow \dfrac{\pi}{4}(\cos\theta_n - 1) = \dfrac{\pi}{2} \Rightarrow \theta_n = \text{does not exist}$$

and

$$\Rightarrow \dfrac{\pi}{4}(\cos\theta_n - 1) = -\dfrac{\pi}{2} \Rightarrow \theta_n = 180°$$

The nulls occur at 90° and 180°. The element at the positive z-axis has a phase lag of 90° relative to the other, and the phase difference is 180° when travel is restricted toward the negative z-axis. There is no phase difference when the waves travel toward the positive z-axis. A diagram similar to that of Figure 6.2 can be used to illustrate this case.

To better illustrate the pattern multiplication rule, the normalized patterns of the single element, the array factor, and the total array for each of the above array examples are shown in Figures 6.3, 6.4(a), and 6.4(b). In each figure, the total pattern of the array is obtained by multiplying the pattern of the single element by that of the array factor. *In each case, the pattern is normalized to its own maximum.* Since the array factor for the example of Figure 6.3 is nearly isotropic (within 3 dB), the element pattern and the total pattern are almost identical in shape. The largest magnitude difference between the two is about 3 dB, and for each case it occurs toward the direction along which the phases of the two elements are in phase quadrature (90° out of phase). For Figure 6.3 this occurs along $\theta = 0°$ while for Figures 6.4(a,b) this occurs along $\theta = 90°$. Because the array factor for Figure 6.4(a) is of cardioid form, its corresponding element and total patterns are considerably different. In the total pattern, the null at $\theta = 90°$ is due to the element pattern while that toward $\theta = 0°$ is due to the array factor. Similar results are displayed in Figure 6.4(b).

Example 6.2

Consider an array of two identical infinitesimal dipoles oriented as shown in Figures 6.1(a) and (b). For a separation d and phase excitation difference β between the elements, find the angles of observation where the nulls of the array occur. The magnitude excitation of the elements is the same.

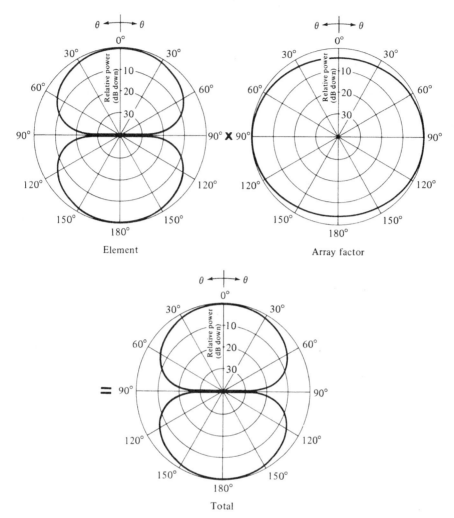

Figure 6.3 Element, array factor, and total field patterns of a two-element array of infinitesimal horizontal dipoles with identical phase excitation ($\beta = 0°, d = \lambda/4$).

Solution: The normalized total field of the array is given by (6-3) as

$$E_{tn} = \cos\theta \cos[\tfrac{1}{2}(kd\cos\theta + \beta)]$$

To find the nulls, the field is set equal to zero, or

$$E_{tn} = \cos\theta \cos[\tfrac{1}{2}(kd\cos\theta + \beta)]|_{\theta=\theta_n} = 0$$

Thus

$$\cos\theta_n = 0 \Rightarrow \theta_n = 90°$$

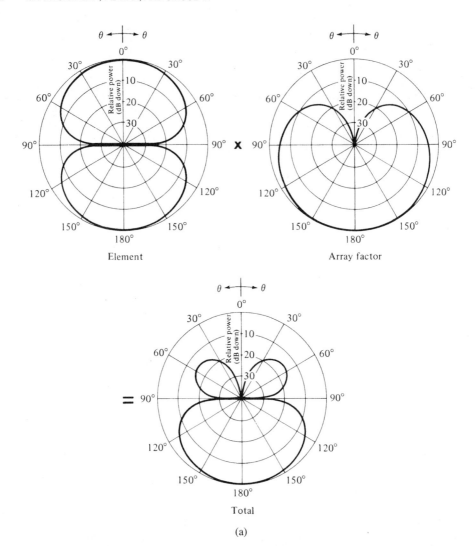

Figure 6.4 Pattern multiplication of element, array factor, and total array patterns of a two-element array of infinitesimal horizontal dipoles with (a) $\beta = +90°$, $d = \lambda/4$. (*continued*)

and

$$\cos\left[\frac{1}{2}(kd\cos\theta_n + \beta)\right] = 0 \Rightarrow \frac{1}{2}(kd\cos\theta_n + \beta) = \pm\left(\frac{2n+1}{2}\right)\pi$$

$$\Rightarrow \theta_n = \cos^{-1}\left(\frac{\lambda}{2\pi d}[-\beta \pm (2n+1)\pi]\right),$$

$$n = 0, 1, 2, \ldots$$

The null at $\theta = 90°$ is attributed to the pattern of the individual elements of the array while the remaining ones are due to the formation of the array. For no phase difference between the elements ($\beta = 0$), the separation d must be equal or greater than half a wavelength ($d \geq \lambda/2$) in order for at least one null, due to the array, to occur.

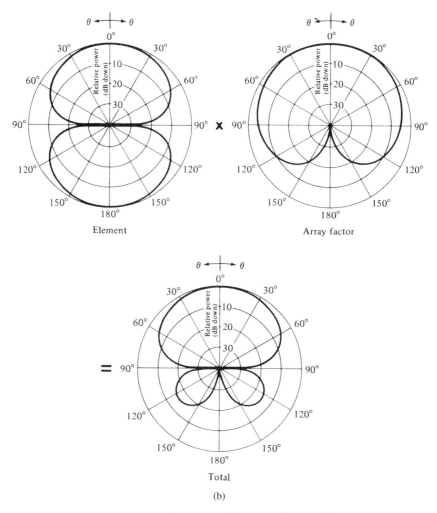

Figure 6.4 *(Continued)* (b) $\beta = -90°$, $d = \lambda/4$.

6.3 N-ELEMENT LINEAR ARRAY: UNIFORM AMPLITUDE AND SPACING

Now that the arraying of elements has been introduced and it was illustrated by the two-element array, let us generalize the method to include N elements. Referring to the geometry of Figure 6.5(a), let us assume that all the elements have identical amplitudes but each succeeding element has a β progressive phase lead current excitation relative to the preceding one (β represents the phase by which the current in each element leads the current of the preceding element). *An array of identical elements all of identical magnitude and each with a progressive phase is referred to as a uniform array.* The array factor can be obtained by considering the elements to be point sources. If the actual elements are not isotropic sources, the total field can be formed by multiplying the array factor of the isotropic sources by the field of a single element. This is the pattern multiplication rule of (6-5), and it applies only for arrays of identical elements. The array factor is given by

$$AF = 1 + e^{+j(kd\cos\theta+\beta)} + e^{+j2(kd\cos\theta+\beta)} + \cdots + e^{j(N-1)(kd\cos\theta+\beta)}$$

$$AF = \sum_{n=1}^{N} e^{j(n-1)(kd\cos\theta+\beta)} \tag{6-6}$$

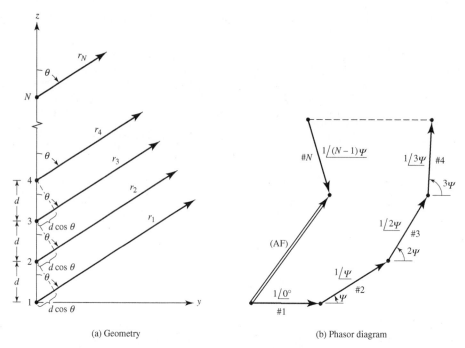

Figure 6.5 Far-field geometry and phasor diagram of N-element array of isotropic sources positioned along the z-axis.

which can be written as

$$\text{AF} = \sum_{n=1}^{N} e^{j(n-1)\psi} \tag{6-7}$$

$$\text{where} \quad \psi = kd\cos\theta + \beta \tag{6-7a}$$

Since the total array factor for the uniform array is a summation of exponentials, it can be represented by the vector sum of N phasors each of unit amplitude and progressive phase ψ relative to the previous one. Graphically this is illustrated by the phasor diagram in Figure 6.5(b). It is apparent from the phasor diagram that the amplitude and phase of the AF can be controlled in uniform arrays by properly selecting the relative phase ψ between the elements; in nonuniform arrays, the amplitude as well as the phase can be used to control the formation and distribution of the total array factor.

The array factor of (6-7) can also be expressed in an alternate, compact and closed form whose functions and their distributions are more recognizable. This is accomplished as follows.

Multiplying both sides of (6-7) by $e^{j\psi}$, it can be written as

$$(\text{AF})e^{j\psi} = e^{j\psi} + e^{j2\psi} + e^{j3\psi} + \cdots + e^{j(N-1)\psi} + e^{jN\psi} \tag{6-8}$$

Subtracting (6-7) from (6-8) reduces to

$$\text{AF}(e^{j\psi} - 1) = (-1 + e^{jN\psi}) \tag{6-9}$$

which can also be written as

$$\text{AF} = \left[\frac{e^{jN\psi}-1}{e^{j\psi}-1}\right] = e^{j[(N-1)/2]\psi}\left[\frac{e^{j(N/2)\psi}-e^{-j(N/2)\psi}}{e^{j(1/2)\psi}-e^{-j(1/2)\psi}}\right]$$

$$= e^{j[(N-1)/2]\psi}\left[\frac{\sin\left(\frac{N}{2}\psi\right)}{\sin\left(\frac{1}{2}\psi\right)}\right] \qquad (6\text{-}10)$$

If the reference point is the physical center of the array, the array factor of (6-10) reduces to

$$\text{AF} = \left[\frac{\sin\left(\frac{N}{2}\psi\right)}{\sin\left(\frac{1}{2}\psi\right)}\right] \qquad (6\text{-}10\text{a})$$

For small values of ψ, the above expression can be approximated by

$$\text{AF} \simeq \left[\frac{\sin\left(\frac{N}{2}\psi\right)}{\frac{\psi}{2}}\right] \qquad (6\text{-}10\text{b})$$

The maximum value of (6-10a) or (6-10b) is equal to N. To normalize the array factors so that the maximum value of each is equal to unity, (6-10a) and (6-10b) are written in normalized form as (see Appendix II)

$$(\text{AF})_n = \frac{1}{N}\left[\frac{\sin\left(\frac{N}{2}\psi\right)}{\sin\left(\frac{1}{2}\psi\right)}\right] \qquad (6\text{-}10\text{c})$$

and (see Appendix I)

$$(\text{AF})_n \simeq \left[\frac{\sin\left(\frac{N}{2}\psi\right)}{\frac{N}{2}\psi}\right] \qquad (6\text{-}10\text{d})$$

To find the nulls of the array, (6-10c) or (6-10d) is set equal to zero. That is,

$$\sin\left(\frac{N}{2}\psi\right) = 0 \Rightarrow \frac{N}{2}\psi|_{\theta=\theta_n} = \pm n\pi \Rightarrow \theta_n = \cos^{-1}\left[\frac{\lambda}{2\pi d}\left(-\beta \pm \frac{2n}{N}\pi\right)\right]$$
$$n = 1, 2, 3, \ldots \qquad (6\text{-}11)$$
$$n \neq N, 2N, 3N, \ldots \text{ with (6-10c)}$$

For $n = N, 2N, 3N, \ldots$, (6-10c) attains its maximum values because it reduces to a sin(0)/0 form. The values of n determine the order of the nulls (first, second, etc.). For a zero to exist, the argument of

296 ARRAYS: LINEAR, PLANAR, AND CIRCULAR

the arccosine cannot exceed unity. Thus the number of nulls that can exist will be a function of the element separation d and the phase excitation difference β.

The maximum values of (6-10c) occur when

$$\frac{\psi}{2} = \frac{1}{2}(kd\cos\theta + \beta)|_{\theta=\theta_m} = \pm m\pi \Rightarrow \theta_m = \cos^{-1}\left[\frac{\lambda}{2\pi d}(-\beta \pm 2m\pi)\right]$$
$$m = 0, 1, 2, \ldots \quad (6\text{-}12)$$

The array factor of (6-10d) has only one maximum and occurs when $m = 0$ in (6-12). That is,

$$\theta_m = \cos^{-1}\left(\frac{\lambda\beta}{2\pi d}\right) \quad (6\text{-}13)$$

which is the observation angle that makes $\psi = 0$.

The 3-dB point for the array factor of (6-10d) occurs when (see Appendix I)

$$\frac{N}{2}\psi = \frac{N}{2}(kd\cos\theta + \beta)|_{\theta=\theta_h} = \pm 1.391$$

$$\Rightarrow \theta_h = \cos^{-1}\left[\frac{\lambda}{2\pi d}\left(-\beta \pm \frac{2.782}{N}\right)\right] \quad (6\text{-}14)$$

which can also be written as

$$\theta_h = \frac{\pi}{2} - \sin^{-1}\left[\frac{\lambda}{2\pi d}\left(-\beta \pm \frac{2.782}{N}\right)\right] \quad (6\text{-}14\text{a})$$

For large values of $d(d \gg \lambda)$, it reduces to

$$\theta_h \simeq \left[\frac{\pi}{2} - \frac{\lambda}{2\pi d}\left(-\beta \pm \frac{2.782}{N}\right)\right] \quad (6\text{-}14\text{b})$$

The half-power beamwidth Θ_h can be found once the angles of the first maximum (θ_m) and the half-power point (θ_h) are determined. For a symmetrical pattern

$$\Theta_h = 2|\theta_m - \theta_h| \quad (6\text{-}14\text{c})$$

For the array factor of (6-10d), there are secondary maxima (maxima of minor lobes) which occur *approximately* when the numerator of (6-10d) attains its maximum value. That is,

$$\sin\left(\frac{N}{2}\psi\right) = \sin\left[\frac{N}{2}(kd\cos\theta + \beta)\right]\bigg|_{\theta=\theta_s} \simeq \pm 1 \Rightarrow \frac{N}{2}(kd\cos\theta + \beta)|_{\theta=\theta_s}$$

$$\simeq \pm\left(\frac{2s+1}{2}\right)\pi \Rightarrow \theta_s \simeq \cos^{-1}\left\{\frac{\lambda}{2\pi d}\left[-\beta \pm \left(\frac{2s+1}{N}\right)\pi\right]\right\},$$
$$s = 1, 2, 3, \ldots \quad (6\text{-}15)$$

which can also be written as

$$\theta_s \simeq \frac{\pi}{2} - \sin^{-1}\left\{\frac{\lambda}{2\pi d}\left[-\beta \pm \left(\frac{2s+1}{N}\right)\pi\right]\right\}, \quad s = 1, 2, 3, \ldots \quad (6\text{-}15\text{a})$$

For large values of $d(d \gg \lambda)$, it reduces to

$$\theta_s \simeq \frac{\pi}{2} - \frac{\lambda}{2\pi d}\left[-\beta \pm \left(\frac{2s+1}{N}\right)\pi\right], \quad s = 1, 2, 3, \ldots \quad (6\text{-}15b)$$

The maximum of the first minor lobe of (6-10c) occurs *approximately* when (see Appendix I)

$$\frac{N}{2}\psi = \frac{N}{2}(kd\cos\theta + \beta)|_{\theta=\theta_s} \simeq \pm\left(\frac{3\pi}{2}\right) \quad (6\text{-}16)$$

or when

$$\theta_s = \cos^{-1}\left\{\frac{\lambda}{2\pi d}\left[-\beta \pm \frac{3\pi}{N}\right]\right\} \quad (6\text{-}16a)$$

At that point, the magnitude of (6-10d) reduces to

$$(AF)_n \simeq \left[\frac{\sin\left(\frac{N}{2}\psi\right)}{\frac{N}{2}\psi}\right]_{\substack{\theta=\theta_s \\ s=1}} = \frac{2}{3\pi} = 0.212 \quad (6\text{-}17)$$

which in dB is equal to

$$(AF)_n = 20\log_{10}\left(\frac{2}{3\pi}\right) = -13.46 \text{ dB} \quad (6\text{-}17a)$$

Thus the maximum of the first minor lobe of the array factor of (6-10d) is 13.46 dB down from the maximum at the major lobe. More accurate expressions for the angle, beamwidth, and magnitude of first minor lobe of the array factor of (6-10d) can be obtained. These will be discussed in Chapter 12.

6.3.1 Broadside Array

In many applications it is desirable to have the maximum radiation of an array directed normal to the axis of the array [broadside; $\theta_0 = 90°$ of Figure 6.5(a)]. To optimize the design, the maxima of the single element and of the array factor should both be directed toward $\theta_0 = 90°$. The requirements of the single elements can be accomplished by the judicious choice of the radiators, and those of the array factor by the proper separation and excitation of the individual radiators. In this section, the requirements that allow the array factor to "radiate" efficiently broadside will be developed.

Referring to (6-10c) or (6-10d), the first maximum of the array factor occurs when

$$\psi = kd\cos\theta + \beta = 0 \quad (6\text{-}18)$$

Since it is desired to have the first maximum directed toward $\theta_0 = 90°$, then

$$\boxed{\psi = kd\cos\theta + \beta|_{\theta=90°} = \beta = 0} \quad (6\text{-}18a)$$

Thus to have the maximum of the array factor of a uniform linear array directed broadside to the axis of the array, it is necessary that all the elements have the same phase excitation (in addition to the same amplitude excitation). The separation between the elements can be of any value. To ensure that there are no principal maxima in other directions, which are referred to as *grating lobes*, the separation between the elements should not be equal to multiples of a wavelength ($d \neq n\lambda, n = 1, 2, 3 \ldots$) when $\beta = 0$. If $d = n\lambda, n = 1, 2, 3, \ldots$ and $\beta = 0$, then

$$\psi = kd \cos\theta + \beta \Big|_{\substack{d=n\lambda \\ \beta=0 \\ n=1,2,3,\ldots}} = 2\pi n \cos\theta \Big|_{\theta=0°,180°} = \pm 2n\pi \quad (6\text{-}19)$$

This value of ψ when substituted in (6-10c) makes the array factor attain its maximum value. Thus for a uniform array with $\beta = 0$ and $d = n\lambda$, in addition to having the maxima of the array factor directed broadside ($\theta_0 = 90°$) to the axis of the array, there are additional maxima directed along the axis ($\theta_0 = 0°, 180°$) of the array (end-fire radiation).

One of the objectives in many designs is to avoid multiple maxima, in addition to the main maximum, which are referred to as *grating lobes*. Often it may be required to select the largest spacing between the elements but with no grating lobes. *To avoid any grating lobe, the largest spacing between the elements should be less than one wavelength* ($d_{\max} < \lambda$).

To illustrate the method, the three-dimensional array factor of a 10-element ($N = 10$) uniform array with $\beta = 0$ and $d = \lambda/4$ is shown plotted in Figure 6.6(a). A 90° angular sector has been

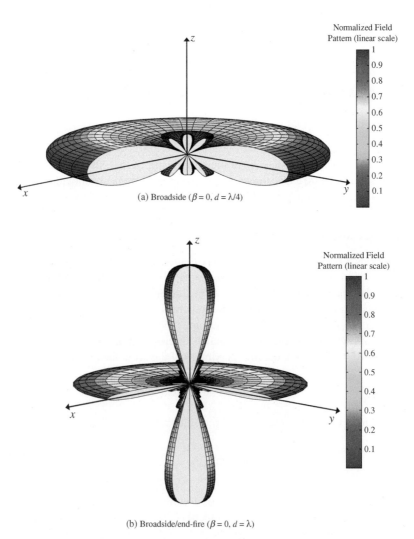

(a) Broadside ($\beta = 0, d = \lambda/4$)

(b) Broadside/end-fire ($\beta = 0, d = \lambda$)

Figure 6.6 Three-dimensional amplitude patterns for broadside, and broadside/end-fire arrays ($N = 10$).

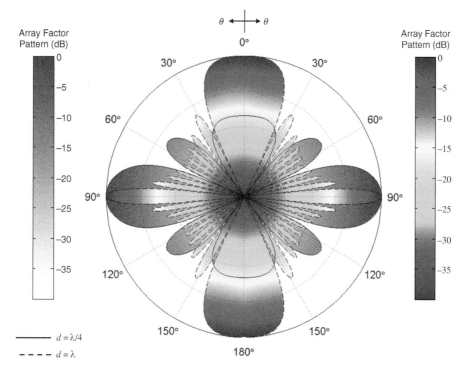

Figure 6.7 Array factor patterns of a 10-element uniform amplitude broadside array ($N = 10, \beta = 0$).

removed for better view of the pattern distribution in the elevation plane. The only maximum occurs at broadside ($\theta_0 = 90°$). To form a comparison, the three-dimensional pattern of the same array but with $d = \lambda$ is also plotted in Figure 6.6(b). For this pattern, in addition to the maximum at $\theta_0 = 90°$, there are additional maxima directed toward $\theta_0 = 0°, 180°$. The corresponding two-dimensional patterns of Figures 6.6(a,b) are shown in Figure 6.7.

If the spacing between the elements is chosen between $\lambda < d < 2\lambda$, then the maximum of Figure 6.6 toward $\theta_0 = 0°$ shifts toward the angular region $0° < \theta_0 < 90°$ while the maximum toward $\theta_0 = 180°$ shifts toward $90° < \theta_0 < 180°$. When $d = 2\lambda$, there are maxima toward $0°, 60°, 90°, 120°$ and $180°$.

In Tables 6.1 and 6.2 the expressions for the nulls, maxima, half-power points, minor lobe maxima, and beamwidths for broadside arrays have been listed. They are derived from (6-10c)–(6-16a).

6.3.2 Ordinary End-Fire Array

Instead of having the maximum radiation broadside to the axis of the array, it may be desirable to direct it along the axis of the array (end-fire). As a matter of fact, it may be necessary that it radiates toward only one direction (either $\theta_0 = 0°$ or $180°$ of Figure 6.5).

To direct the first maximum toward $\theta_0 = 0°$,

$$\psi = kd \cos\theta + \beta|_{\theta=0°} = kd + \beta = 0 \Rightarrow \beta = -kd \qquad (6\text{-}20a)$$

If the first maximum is desired toward $\theta_0 = 180°$, then

$$\psi = kd \cos\theta + \beta|_{\theta=180°} = -kd + \beta = 0 \Rightarrow \beta = kd \qquad (6\text{-}20b)$$

TABLE 6.1 Nulls, Maxima, Half-Power Points, and Minor Lobe Maxima for Uniform Amplitude Broadside Arrays

NULLS	$\theta_n = \cos^{-1}\left(\pm \dfrac{n}{N}\dfrac{\lambda}{d}\right)$ $n = 1, 2, 3, \ldots$ $n \neq N, 2N, 3N, \ldots$
MAXIMA	$\theta_m = \cos^{-1}\left(\pm \dfrac{m\lambda}{d}\right)$ $m = 0, 1, 2, \ldots$
HALF-POWER POINTS	$\theta_h \simeq \cos^{-1}\left(\pm \dfrac{1.391\lambda}{\pi N d}\right)$ $\pi d/\lambda \ll 1$
MINOR LOBE MAXIMA	$\theta_s \simeq \cos^{-1}\left[\pm \dfrac{\lambda}{2d}\left(\dfrac{2s+1}{N}\right)\right]$ $s = 1, 2, 3, \ldots$ $\pi d/\lambda \ll 1$

Thus end-fire radiation is accomplished when $\beta = -kd$ (for $\theta_0 = 0°$) or $\beta = kd$ (for $\theta_0 = 180°$).

If the element separation is $d = \lambda/2$, end-fire radiation exists simultaneously in both directions ($\theta_0 = 0°$ and $\theta_0 = 180°$). If the element spacing is a multiple of a wavelength ($d = n\lambda$, $n = 1, 2, 3, \ldots$), then in addition to having end-fire radiation in both directions, there also exist maxima in the broadside directions. Thus for $d = n\lambda, n = 1, 2, 3, \ldots$ there exist four maxima; two in the broadside directions and two along the axis of the array. *To have only one end-fire maximum and to avoid any grating lobes, the maximum spacing between the elements should be less than* $d_{\max} < \lambda/2$.

The three-dimensional radiation patterns of a 10-element ($N = 10$) array with $d = \lambda/4$, $\beta = +kd$ are plotted in Figure 6.8. When $\beta = -kd$, the maximum is directed along $\theta_0 = 0°$ and the three-dimensional pattern is shown in Figure 6.8(a). However, when $\beta = +kd$, the maximum is oriented toward $\theta_0 = 180°$, and the three-dimensional pattern is shown in Figure 6.8(b). The two-dimensional patterns of Figures 6.8(a,b) are shown in Figure 6.9. To form a comparison, the array factor of the same array ($N = 10$) but with $d = \lambda$ and $\beta = -kd$ has been calculated. Its pattern is identical to that of a broadside array with $N = 10, d = \lambda$, and it is shown plotted in Figure 6.7. It is seen that there are four maxima; two broadside and two along the axis of the array.

The expressions for the nulls, maxima, half-power points, minor lobe maxima, and beamwidths, as applied to ordinary end-fire arrays, are listed in Tables 6.3 and 6.4.

TABLE 6.2 Beamwidths for Uniform Amplitude Broadside Arrays

FIRST-NULL BEAMWIDTH (FNBW)	$\Theta_n = 2\left[\dfrac{\pi}{2} - \cos^{-1}\left(\dfrac{\lambda}{Nd}\right)\right]$
HALF-POWER BEAMWIDTH (HPBW)	$\Theta_h \simeq 2\left[\dfrac{\pi}{2} - \cos^{-1}\left(\dfrac{1.391\lambda}{\pi Nd}\right)\right]$ $\pi d/\lambda \ll 1$
FIRST SIDE LOBE BEAMWIDTH (FSLBW)	$\Theta_s \simeq 2\left[\dfrac{\pi}{2} - \cos^{-1}\left(\dfrac{3\lambda}{2dN}\right)\right]$ $\pi d/\lambda \ll 1$

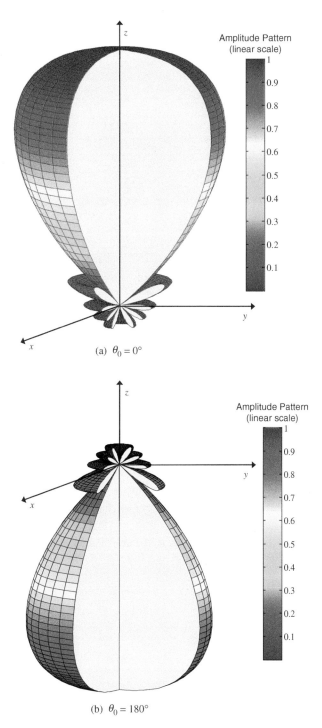

Figure 6.8 Three-dimensional amplitude patterns for end-fire arrays toward $\theta_0 = 0°$ and $180°$ ($N = 10$, $d = \lambda/4$).

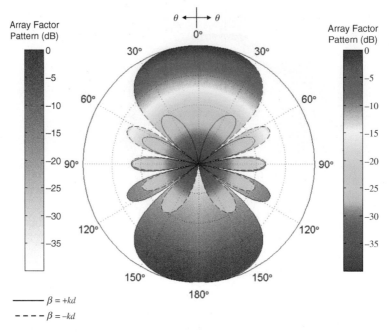

Figure 6.9 Array factor patterns of a 10-element uniform amplitude end-fire array ($N = 10, d = \lambda/4$).

6.3.3 Phased (Scanning) Array

In the previous two sections it was shown how to direct the major radiation from an array, by controlling the phase excitation between the elements, in directions normal (broadside) and along the axis (end fire) of the array. It is then logical to assume that the maximum radiation can be oriented in any direction to form a scanning array. The procedure is similar to that of the previous two sections.

Let us assume that the maximum radiation of the array is required to be oriented at an angle $\theta_0 (0° \leq \theta_0 \leq 180°)$. To accomplish this, the phase excitation β between the elements must be

TABLE 6.3 Nulls, Maxima, Half-Power Points, and Minor Lobe Maxima for Uniform Amplitude Ordinary End-Fire Arrays (For $\theta_0 = 0°$)*

NULLS	$\theta_n = \cos^{-1}\left(1 - \dfrac{n\lambda}{Nd}\right)$ $n = 1, 2, 3, \ldots$ $n \neq N, 2N, 3N, \ldots$
MAXIMA	$\theta_m = \cos^{-1}\left(1 - \dfrac{m\lambda}{d}\right)$ $m = 0, 1, 2, \ldots$
HALF-POWER POINTS	$\theta_h \simeq \cos^{-1}\left(1 - \dfrac{1.391\lambda}{\pi dN}\right)$ $\pi d/\lambda \ll 1$
MINOR LOBE MAXIMA	$\theta_s \simeq \cos^{-1}\left[1 - \dfrac{(2s+1)\lambda}{2Nd}\right]$ $s = 1, 2, 3, \ldots$ $\pi d/\lambda \ll 1$

* For $\theta_0 = 180°$: $\theta_j(\theta_0 = 180°) = 180° - \theta_j(\theta_0 = 0°)$, $j = n, m, h, s$

TABLE 6.4 Beamwidths for Uniform Amplitude Ordinary End-Fire Arrays

FIRST-NULL BEAMWIDTH (FNBW) $\Theta_n = 2\cos^{-1}\left(1 - \dfrac{\lambda}{Nd}\right)$

HALF-POWER BEAMWIDTH (HPBW) $\Theta_h \simeq 2\cos^{-1}\left(1 - \dfrac{1.391\lambda}{\pi dN}\right)$
$\pi d/\lambda \ll 1$

FIRST SIDE LOBE BEAMWIDTH (FSLBW) $\Theta_s \simeq 2\cos^{-1}\left(1 - \dfrac{3\lambda}{2Nd}\right)$
$\pi d/\lambda \ll 1$

adjusted so that

$$\psi = kd\cos\theta + \beta\big|_{\theta=\theta_0} = kd\cos\theta_0 + \beta = 0 \Rightarrow \beta = -kd\cos\theta_0 \tag{6-21}$$

Thus by controlling the progressive phase difference between the elements, the maximum radiation can be squinted in any desired direction to form a scanning array. This is the basic principle of electronic scanning phased array operation. Since in phased array technology the scanning must be continuous, the system should be capable of continuously varying the progressive phase between the elements. In practice, this is accomplished electronically by the use of ferrite or diode phase shifters. For ferrite phase shifters, the phase shift is controlled by the magnetic field within the ferrite, which in turn is controlled by the amount of current flowing through the wires wrapped around the phase shifter.

For diode phase shifter using balanced, hybrid-coupled varactors, the actual phase shift is controlled either by varying the analog bias dc voltage (typically 0–30 volts) or by a digital command through a digital-to-analog (D/A) converter [1]–[3].

Shown in Figure 6.10 is an *incremental switched-line* PIN-diode phase shifter [2]–[3]. This design is simple, straightforward, lightweight, and high speed. The lines of lengths l_1 and l_2 are switched on and off by controlling the bias of the PIN diodes, using two single-pole double-throw switches, as illustrated in Figure 6.10. The differential phase shift, provided by switching on and off the two paths, is given by

$$\Delta\phi = k(l_2 - l_1) \tag{6-21a}$$

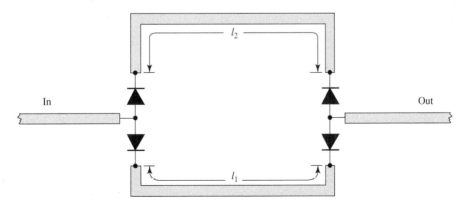

Figure 6.10 Incremental switched-line phase shifter using PIN diodes. (SOURCE: D.M. Pozar, *Microwave Engineering*, John Wiley & Sons, Inc. 2004).

By properly choosing l_1 and l_2, and the operating frequency, the differential phase shift (in degrees) provided by each incremental line phase shifter can be as small as desired, and it determines the resolution of the phase shifter. The design of an entire phase shifter typically utilizes several such incremental phase shifters to cover the entire range $(0-180°)$ of phase. However, the *switched-line* phase shifter, as well as many other ones, are usually designed for binary phase shifts of $\Delta\phi = 180°, 90°, 45°, 22.5°$, etc. [3]. There are other designs of PIN-diode phase shifters, including those that utilize open-circuited stubs and reactive elements [2]. The basic designs of a phase shifter utilizing PIN diodes are typically classified into three categories: *switched line*, *loaded line*, and *reflection type* [3]. The *loaded-line* phase shifter can be used for phase shifts generally 45° or smaller. Phase shifters that utilize PIN diodes are not ideal switches since the PIN diodes usually possess finite series resistance and reactance that can contribute significant insertion loss if several of them are used. These phase shifters can also be used as time-delay devices.

To demonstrate the principle of scanning, the three-dimensional radiation pattern of a 10-element array, with a separation of $\lambda/4$ between the elements and with the maximum squinted in the $\theta_0 = 60°$ direction, is plotted in Figure 6.11(a). The corresponding two-dimensional pattern is shown in Figure 6.11(b).

The half-power beamwidth of the scanning array is obtained using (6-14) with $\beta = -kd \cos\theta_0$. Using the minus sign in the argument of the inverse cosine function in (6-14) to represent one angle of the half-power beamwidth and the plus sign to represent the other angle, then the total beamwidth is the difference between these two angles and can be written as

$$\Theta_h = \cos^{-1}\left[\frac{\lambda}{2\pi d}\left(kd\cos\theta_0 - \frac{2.782}{N}\right)\right] - \cos^{-1}\left[\frac{\lambda}{2\pi d}\left(kd\cos\theta_0 + \frac{2.782}{N}\right)\right]$$
$$= \cos^{-1}\left(\cos\theta_0 - \frac{2.782}{Nkd}\right) - \cos^{-1}\left(\cos\theta_0 + \frac{2.782}{Nkd}\right) \quad (6\text{-}22)$$

Since $N = (L+d)/d$, (6-22) reduces to [4]

$$\boxed{\begin{aligned}\Theta_h = &\cos^{-1}\left[\cos\theta_0 - 0.443\frac{\lambda}{(L+d)}\right] \\ &- \cos^{-1}\left[\cos\theta_0 + 0.443\frac{\lambda}{(L+d)}\right]\end{aligned}} \quad (6\text{-}22a)$$

where L is the length of the array. Equation (6-22a) can also be used to compute the half-power beamwidth of a broadside array. However, it is not valid for an end-fire array. A plot of the half-power beamwidth (in degrees) as a function of the array length is shown in Figure 6.12. These curves are valid for broadside, ordinary end-fire, and scanning uniform arrays (constant magnitude but with progressive phase shift). In a later section it will be shown that the curves of Figure 6.12 can be used, in conjunction with a beam broadening factor [4], to compute the directivity of nonuniform amplitude arrays.

6.3.4 Hansen-Woodyard End-Fire Array

The conditions for an ordinary end-fire array were discussed in Section 6.3.2. It was concluded that the maximum radiation can be directed along the axis of the uniform array by allowing the progressive phase shift β between elements to be equal to (6-20a) for $\theta_0 = 0°$ and (6-20b) for $\theta_0 = 180°$.

To enhance the directivity of an end-fire array without destroying any of the other characteristics, Hansen and Woodyard [5] in 1938 proposed that the required phase shift between *closely spaced*

N-ELEMENT LINEAR ARRAY: UNIFORM AMPLITUDE AND SPACING 305

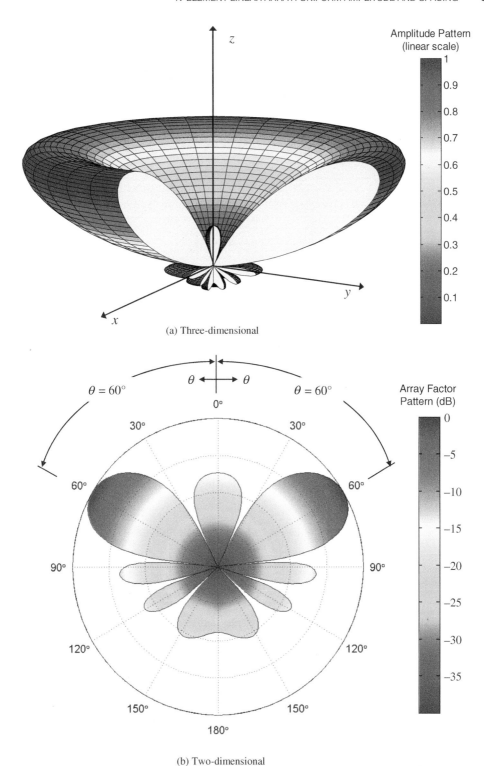

(a) Three-dimensional

(b) Two-dimensional

Figure 6.11 Three- and two-dimensional array factor patterns of a 10-element uniform amplitude scanning array ($N = 10, \beta = -kd\cos\theta_0, \theta_0 = 60°, d = \lambda/4$).

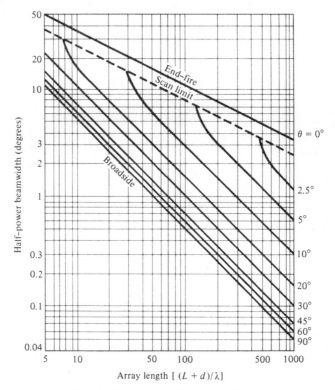

Figure 6.12 Half-power beamwidth for broadside, ordinary end-fire, and scanning uniform linear arrays. (SOURCE: R. S. Elliott, "Beamwidth and Directivity of Large Scanning Arrays," First of Two Parts, *The Microwave Journal*, December 1963).

elements of a very long array[†] *should be*

$$\beta = -\left(kd + \frac{2.94}{N}\right) \simeq -\left(kd + \frac{\pi}{N}\right) \Rightarrow \text{for maximum in } \theta_0 = 0° \quad (6\text{-}23a)$$

$$\beta = +\left(kd + \frac{2.94}{N}\right) \simeq +\left(kd + \frac{\pi}{N}\right) \Rightarrow \text{for maximum in } \theta_0 = 180° \quad (6\text{-}23b)$$

These requirements are known today as the *Hansen-Woodyard conditions for end-fire radiation*. They lead to a *larger* directivity than the conditions given by (6-20a) and (6-20b). It should be pointed out, however, that *these conditions do not necessarily yield the maximum possible directivity*. In fact, the maximum may not even occur at $\theta_0 = 0°$ or $180°$, its value found using (6-10c) or (6-10d) may not be unity, and the side lobe level may not be -13.46 dB. Both of them, maxima and side lobe levels, depend on the number of array elements, as will be illustrated.

To realize the increase in directivity as a result of the Hansen-Woodyard conditions, it is necessary that, in addition to the conditions of (6-23a) and (6-23b), $|\psi|$ assumes values of

For maximum radiation along $\theta_0 = 0°$

$$|\psi| = |kd\cos\theta + \beta|_{\theta=0°} = \frac{\pi}{N} \quad \text{and} \quad |\psi| = |kd\cos\theta + \beta|_{\theta=180°} \simeq \pi \quad (6\text{-}24a)$$

[†]In principle, the Hansen-Woodyard condition was derived for an infinitely long antenna with continuous distribution. It thus gives good results for very long, finite length discrete arrays with closely spaced elements.

For maximum radiation along $\theta_0 = 180°$

$$|\psi| = |kd\cos\theta + \beta|_{\theta=180°} = \frac{\pi}{N} \quad \text{and} \quad |\psi| = |kd\cos\theta + \beta|_{\theta=0°} \simeq \pi \qquad (6\text{-}24\text{b})$$

The condition of $|\psi| = \pi/N$ in (6-24a) or (6-24b) is realized by the use of (6-23a) or (6-23b), respectively. Care must be exercised in meeting the requirement of $|\psi| \simeq \pi$ for each array. For an array of N elements, the condition of $|\psi| \simeq \pi$ is satisfied by using (6-23a) for $\theta = 0°$, (6-23b) for $\theta = 180°$, and choosing for each a spacing of

$$d = \left(\frac{N-1}{N}\right)\frac{\lambda}{4} \qquad (6\text{-}25)$$

If the number of elements is large, (6-25) can be approximated by

$$d \simeq \frac{\lambda}{4} \qquad (6\text{-}25\text{a})$$

Thus for a large uniform array, the Hansen-Woodyard condition can only yield an improved directivity provided the spacing between the elements is approximately $\lambda/4$.

This is also illustrated in Figure 6.13 where the 3-D field patterns of the *ordinary* and the *Hansen-Woodyard* end-fire designs, for $N = 10$ and $d = \lambda/4$, are placed next to each other. It is apparent that the major lobe of the ordinary end-fire is wider (HPBW = 74°) than that of the Hansen-Woodyard (HPBW = 37°); thus, higher directivity for the Hansen-Woodyard. However, the side lobe of the ordinary end-fire is lower (about -13.5 dB) compared to that of the Hansen-Woodyard, which is about -8.9 dB. *The lower side lobe by the ordinary end-fire is not sufficient to offset the benefit from the narrower beamwidth of the Hansen-Woodyard that leads to the higher directivity.* A comparison between the ordinary and Hansen-Woodyard end-fire array patterns is also illustrated in Figure 10.16 for the design of a helical antenna.

To make the comparisons more meaningful, the directivities for each of the patterns of Figures 6.13 have been calculated, using numerical integration, and it is found that they are equal to 11 and 19, respectively. Thus the Hansen-Woodyard conditions realize a 73% increase in directivity for this case.

As will be shown in Section 6.4 and listed in Table 6.8, the directivity of a Hansen-Woodyard end-fire array is always approximately 1.805 times (or 2.56 dB) greater than the directivity of an ordinary end-fire array. The increase in directivity of the pattern in Figure 6.13 for the Hansen-Woodyard design is at the expense of an increase of about 4 dB in side lobe level. Therefore in the design of an array, there is a trade-off between directivity (or half-power beamwidth) and side lobe level.

To show that (6-23a) and (6-23b) do *not* lead to improved directivities over those of (6-20a) and (6-20b) if (6-24a) and (6-24b) are not satisfied, the pattern for the same array ($N = 10$) with $d = \lambda/4(\beta = -3\pi/5)$ and $d = \lambda/2(\beta = -11\pi/10)$ are plotted in Figure 6.14. Even though the $d = \lambda/2$ pattern exhibits a very narrow lobe in the $\theta_0 = 0°$ direction, its back lobes are larger than its main lobe. The $d = \lambda/2$ pattern fails to realize a larger directivity because the necessary $|\psi|_{\theta=180°} \simeq \pi$ condition of (6-24a) is not satisfied. That is,

$$|\psi| = |(kd\cos\theta + \beta)|\begin{subarray}{l}\theta=180°\\ \beta=-(kd+\pi/N)\end{subarray} = |-(2kd+\pi/N)|\begin{subarray}{l}d=\lambda/2\\ N=10\end{subarray} = 2.1\pi \qquad (6\text{-}26)$$

which is not equal to π as required by (6-24a). Similar results occur for spacings other than those specified by (6-25) or (6-25a).

To better understand and appreciate the Hansen-Woodyard conditions, a succinct derivation of (6-23a) will be outlined. The procedure is identical to that reported by Hansen and Woodyard in their classic paper [5].

308 ARRAYS: LINEAR, PLANAR, AND CIRCULAR

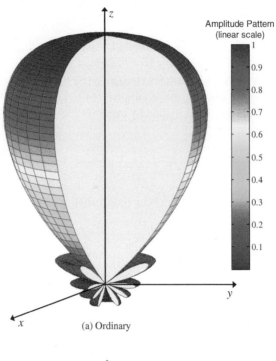

(a) Ordinary

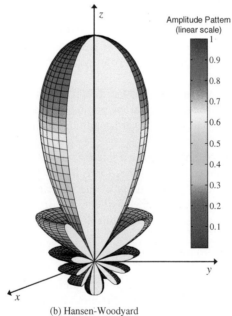

(b) Hansen-Woodyard

Figure 6.13 Three-dimensional patterns for ordinary and Hansen-Woodyard end-fire designs ($N = 10$, $d = \lambda/4$).

The array factor of an N-element array is given by (6-10c) as

$$(\text{AF})_n = \frac{1}{N} \left\{ \frac{\sin\left[\frac{N}{2}(kd\cos\theta + \beta)\right]}{\sin\left[\frac{1}{2}(kd\cos\theta + \beta)\right]} \right\} \tag{6-27}$$

N-ELEMENT LINEAR ARRAY: UNIFORM AMPLITUDE AND SPACING   **309**

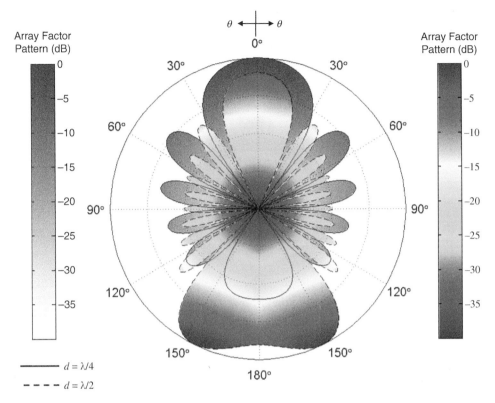

Figure 6.14 Array factor patterns of a 10-element uniform amplitude Hansen-Woodyard end-fire array $[N = 10, \beta = -(kd + \pi/N)]$.

and approximated, for small values of ψ ($\psi = kd \cos \theta + \beta$), by (6-10d) or

$$(\text{AF})_n \simeq \frac{\sin\left[\frac{N}{2}(kd \cos \theta + \beta)\right]}{\left[\frac{N}{2}(kd \cos \theta + \beta)\right]} \qquad (6\text{-}27a)$$

If the progressive phase shift between the elements is equal to

$$\beta = -pd \qquad (6\text{-}28)$$

where p is a constant, (6-27a) can be written as

$$(\text{AF})_n = \left\{\frac{\sin[q(k \cos \theta - p)]}{q(k \cos \theta - p)}\right\} = \left[\frac{\sin(Z)}{Z}\right] \qquad (6\text{-}29)$$

where

$$q = \frac{Nd}{2} \qquad (6\text{-}29a)$$

$$Z = q(k \cos \theta - p) \qquad (6\text{-}29b)$$

The radiation intensity can be written as

$$U(\theta) = [(\text{AF})_n]^2 = \left[\frac{\sin(Z)}{Z}\right]^2 \qquad (6\text{-}30)$$

whose value at $\theta = 0°$ is equal to

$$U(\theta)|_{\theta=0°} = \left\{\frac{\sin[q(k\cos\theta - p)]}{q(k\cos\theta - p)}\right\}^2\bigg|_{\theta=0°} = \left\{\frac{\sin[q(k-p)]}{q(k-p)}\right\}^2 \tag{6-30a}$$

Dividing (6-30) by (6-30a), so that the value of the array factor is equal to unity at $\theta = 0°$, leads to

$$U(\theta)_n = \left\{\frac{q(k-p)}{\sin[q(k-p)]}\frac{\sin[q(k\cos\theta - p)]}{[q(k\cos\theta - p)]}\right\}^2 = \left[\frac{v}{\sin(v)}\frac{\sin(Z)}{Z}\right]^2 \tag{6-31}$$

where

$$v = q(k-p) \tag{6-31a}$$

$$Z = q(k\cos\theta - p) \tag{6-31b}$$

The directivity of the array factor can be evaluated using

$$D_0 = \frac{4\pi U_{max}}{P_{rad}} = \frac{U_{max}}{U_0} \tag{6-32}$$

where U_0 is the average radiation intensity and it is given by

$$U_0 = \frac{P_{rad}}{4\pi} = \frac{1}{4\pi}\int_0^{2\pi}\int_0^{\pi} U(\theta)\sin\theta\, d\theta\, d\phi$$

$$= \frac{1}{2}\left[\frac{v}{\sin(v)}\right]^2 \int_0^{\pi}\left[\frac{\sin(Z)}{Z}\right]^2 \sin\theta\, d\theta \tag{6-33}$$

By using (6-31a) and (6-31b), (6-33) can be written as

$$U_0 = \frac{1}{2}\left[\frac{q(k-p)}{\sin[q(k-p)]}\right]^2 \int_0^{\pi}\left[\frac{\sin[q(k\cos\theta - p)]}{q(k\cos\theta - p)}\right]^2 \sin\theta\, d\theta \tag{6-33a}$$

To maximize the directivity, as given by (6-32), (6-33a) must be minimized. Performing the integration, (6-33a) reduces to

$$U_0 = \frac{1}{2kq}\left[\frac{v}{\sin(v)}\right]^2\left[\frac{\pi}{2} + \frac{[\cos(2v) - 1]}{2v} + S_i(2v)\right] = \frac{1}{2kq}g(v) \tag{6-34}$$

where

$$v = q(k-p) \tag{6-34a}$$

$$S_i(z) = \int_0^z \frac{\sin t}{t}\, dt \tag{6-34b}$$

$$g(v) = \left[\frac{v}{\sin(v)}\right]^2\left[\frac{\pi}{2} + \frac{[\cos(2v) - 1]}{2v} + S_i(2v)\right] \tag{6-34c}$$

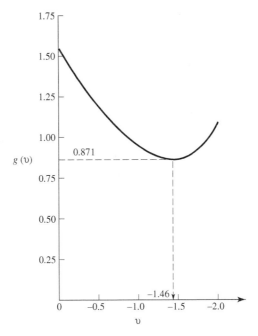

Figure 6.15 Variation of $g(v)$ (see Eq. 6-34c) as a function of v.

The function $g(v)$ is plotted in Figure 6.15 and its minimum value occurs when

$$v = q(k-p) = \frac{Nd}{2}(k-p) = -1.46 \tag{6-35}$$

Thus

$$\beta = -pd = -\left(kd + \frac{2.92}{N}\right) \tag{6-36}$$

which is the condition for end-fire radiation with improved directivity (Hansen-Woodyard condition) along $\theta_0 = 0°$, as given by (6-23a). Similar procedures can be followed to establish (6-23b).

Ordinarily, (6-36) is approximated by

$$\beta = -\left(kd + \frac{2.92}{N}\right) \simeq -\left(kd + \frac{\pi}{N}\right) \tag{6-36a}$$

with not too much relaxation in the condition since the curve of Figure 6.15 is broad around the minimum point $v = -1.46$. Its value at $v = -1.57$ is almost the same as the minimum at $v = -1.46$.

The expressions for the nulls, maxima, half-power points, minor lobe maxima, and beamwidths are listed in Tables 6.5 and 6.6.

For the broadside, end-fire and scanning linear designs, there is a maximum spacing d_{max} that should not be exceeded to maintain in the amplitude pattern either one or two maxima. A second maximum of the array factor, if it exists, will begin to appear at $\theta = 0°$ as the separation between the elements increases and the ψ of (6-21) approaches 2π. This is indicated in Figure 6.7 for the broadside array with $d = \lambda$ when a second maximum appears in both of the end-fire directions ($\theta_o = 0°$ and $180°$). The maximum spacing separation d_{max} is intended to keep a second lobe from appearing, whose amplitude would equal the amplitude of the main lobe (referred to as *grating lobe*). This

TABLE 6.5 Nulls, Maxima, Half-Power Points, and Minor Lobe Maxima for Uniform Amplitude Hansen-Woodyard End-Fire Arrays (For $\theta_0 = 0°$) *

NULLS	$\theta_n = \cos^{-1}\left[1 + (1-2n)\dfrac{\lambda}{2dN}\right]$ $n = 1, 2, 3, \ldots$ $n \neq N, 2N, 3N, \ldots$
MAXIMA	$\theta_m = \cos^{-1}\left\{1 + [1-(2m+1)]\dfrac{\lambda}{2Nd}\right\}$ $m = 1, 2, 3, \ldots$ $\pi d/\lambda \ll 1$
HALF-POWER POINTS	$\theta_h = \cos^{-1}\left(1 - 0.1398\dfrac{\lambda}{Nd}\right)$ $\pi d/\lambda \ll 1$ N large
MINOR LOBE MAXIMA	$\theta_s = \cos^{-1}\left(1 - \dfrac{s\lambda}{Nd}\right)$ $s = 1, 2, 3, \ldots$ $\pi d/\lambda \ll 1$

* For $\theta_0 = 180°$: $\theta_j(\theta_0 = 180°) = 180° - \theta_j(\theta_0 = 0°)$, $j = n, m, h, s$

TABLE 6.6 Beamwidths for Uniform Amplitude Hansen-Woodyard End-Fire Arrays

FIRST-NULL BEAMWIDTH (FNBW)	$\Theta_n = 2\cos^{-1}\left(1 - \dfrac{\lambda}{2dN}\right)$
HALF-POWER BEAMWIDTH (HPBW)	$\Theta_h = 2\cos^{-1}\left(1 - 0.1398\dfrac{\lambda}{Nd}\right)$ $\pi d/\lambda \ll 1$ N large
FIRST SIDE LOBE BEAMWIDTH (FSLBW)	$\Theta_s = 2\cos^{-1}\left(1 - \dfrac{\lambda}{Nd}\right)$ $\pi d/\lambda \ll 1$

separation can be obtained by equating ψ of (6-21), with $\theta = 0°$, to less than 2π, or

$$\psi = kd(\cos\theta + |\cos\theta_o|)\Big|_{\substack{\theta=0 \\ d=d_{\max}}} < 2\pi \Rightarrow d_{\max} < \frac{\lambda}{(1 + |\cos\theta_o|)} \qquad (6\text{-}37)$$

where θ_o is the scan angle of the main lobe, which allows scanning toward the broadside ($\theta_o = 90°$), in both end-fire directions ($\theta_o = 0°, 180°$) and all other angles ($0° \leq \theta_o \leq 180°$). *Table 6.7 lists the maximum element spacing $d_{\max}$ for the various linear and planar arrays, uniform and nonuniform, in order to maintain either one or two amplitude maxima.* The $d_{\max}$ for broadside, end-fire and scanning arrays are the same as those obtained from (6-37).

6.4 N-ELEMENT LINEAR ARRAY: DIRECTIVITY

The criteria that must be met to achieve broadside and end-fire radiation by a uniform linear array of N elements were discussed in the previous section. It would be instructive to investigate the directivity of each of the arrays, since it represents a figure of merit on the operation of the system.

N-ELEMENT LINEAR ARRAY: DIRECTIVITY

TABLE 6.7 Maximum Element Spacing d_{max} to Maintain Either One or Two Amplitude Maxima of a Linear Array

Array	Distribution	Type	Direction of Maximum	Element Spacing
Linear	Uniform	Broadside	$\theta_0 = 90°$ only	$d_{max} < \lambda$
			$\theta_0 = 0°, 90°, 180°$ simultaneously	$d = \lambda$
Linear	Uniform	Ordinary end-fire	$\theta_0 = 0°$ only	$d_{max} < \lambda/2$
			$\theta_0 = 180°$ only	$d_{max} < \lambda/2$
			$\theta_0 = 0°, 90°, 180°$ simultaneously	$d = \lambda$
Linear	Uniform	Hansen-Woodyard end-fire	$\theta_0 = 0°$ only	$d \simeq \lambda/4$
			$\theta_0 = 180°$ only	$d \simeq \lambda/4$
Linear	Uniform	Scanning	$\theta_0 = \theta_{max}$ $0 < \theta_0 < 180°$	$d_{max} < \lambda$
Linear	Nonuniform	Binomial	$\theta_0 = 90°$ only	$d_{max} < \lambda$
			$\theta_0 = 0°, 90°, 180°$ simultaneously	$d = \lambda$
Linear	Nonuniform	Dolph-Tschebyscheff	$\theta_0 = 90°$ only	$d_{max} \leq \frac{\lambda}{\pi} \cos^{-1}\left(-\frac{1}{z_o}\right)$
			$\theta_0 = 0°, 90°, 180°$ simultaneously	$d = \lambda$
Planar	Uniform	Planar	$\theta_0 = 0°$ only	$d_{max} < \lambda$
			$\theta_0 = 0°, 90°$ and $180°$; $\phi_0 = 0°, 90°, 180°, 270°$ simultaneously	$d = \lambda$

6.4.1 Broadside Array

As a result of the criteria for broadside radiation given by (6-18a), the array factor for this form of the array reduces to

$$(AF)_n = \frac{1}{N} \left[\frac{\sin\left(\frac{N}{2} kd \cos\theta\right)}{\sin\left(\frac{1}{2} kd \cos\theta\right)} \right] \quad (6\text{-}38)$$

which for a small spacing between the elements ($d \ll \lambda$) can be approximated by

$$(AF)_n \simeq \left[\frac{\sin\left(\frac{N}{2} kd \cos\theta\right)}{\left(\frac{N}{2} kd \cos\theta\right)} \right] \quad (6\text{-}38a)$$

The radiation intensity can be written as

$$U(\theta) = [(AF)_n]^2 = \left[\frac{\sin\left(\frac{N}{2}kd\cos\theta\right)}{\frac{N}{2}kd\cos\theta}\right]^2 = \left[\frac{\sin(Z)}{Z}\right]^2 \quad (6\text{-}39)$$

$$Z = \frac{N}{2}kd\cos\theta \quad (6\text{-}39a)$$

The directivity can be obtained using (6-32) where U_{max} of (6-39) is equal to unity ($U_{max} = 1$) and it occurs at $\theta = 90°$. The average value U_0 of the intensity reduces to

$$U_0 = \frac{1}{4\pi}P_{rad} = \frac{1}{2}\int_0^\pi \left[\frac{\sin(Z)}{Z}\right]^2 \sin\theta \, d\theta$$

$$= \frac{1}{2}\int_0^\pi \left[\frac{\sin\left(\frac{N}{2}kd\cos\theta\right)}{\frac{N}{2}kd\cos\theta}\right]^2 \sin\theta \, d\theta \quad (6\text{-}40)$$

By making a change of variable, that is,

$$Z = \frac{N}{2}kd\cos\theta \quad (6\text{-}40a)$$

$$dZ = -\frac{N}{2}kd\sin\theta \, d\theta \quad (6\text{-}40b)$$

(6-40) can be written as

$$U_0 = -\frac{1}{Nkd}\int_{+Nkd/2}^{-Nkd/2}\left[\frac{\sin Z}{Z}\right]^2 dZ = \frac{1}{Nkd}\int_{-Nkd/2}^{+Nkd/2}\left[\frac{\sin Z}{Z}\right]^2 dZ \quad (6\text{-}41)$$

For a large array ($Nkd/2 \to$ large), (6-41) can be approximated by extending the limits to infinity. That is,

$$U_0 = \frac{1}{Nkd}\int_{-Nkd/2}^{+Nkd/2}\left[\frac{\sin Z}{Z}\right]^2 dZ \simeq \frac{1}{Nkd}\int_{-\infty}^{+\infty}\left[\frac{\sin Z}{Z}\right]^2 dZ \quad (6\text{-}41a)$$

Since

$$\int_{-\infty}^{+\infty}\left[\frac{\sin(Z)}{Z}\right]^2 dZ = \pi \quad (6\text{-}41b)$$

(6-41a) reduces to

$$U_0 \simeq \frac{\pi}{Nkd} \quad (6\text{-}41c)$$

The directivity of (6-32) can now be written as

$$D_0 = \frac{U_{max}}{U_0} \simeq \frac{Nkd}{\pi} = 2N\left(\frac{d}{\lambda}\right) \quad (6\text{-}42)$$

Using

$$L = (N-1)d \tag{6-43}$$

where L is the overall length of the array, (6-42) can be expressed as

$$D_0 \simeq 2N\left(\frac{d}{\lambda}\right) \simeq 2\left(1+\frac{L}{d}\right)\left(\frac{d}{\lambda}\right) \tag{6-44}$$

which for a large array ($L \gg d$) reduces to

$$D_0 \simeq 2N\left(\frac{d}{\lambda}\right) = 2\left(1+\frac{L}{d}\right)\left(\frac{d}{\lambda}\right) \stackrel{L \gg d}{\simeq} 2\left(\frac{L}{\lambda}\right) \tag{6-44a}$$

Example 6.3
Given a linear, broadside, uniform array of 10 isotropic elements ($N = 10$) with a separation of $\lambda/4$ ($d = \lambda/4$) between the elements, find the directivity of the array.
Solution: Using (6-44a)

$$D_0 \simeq 2N\left(\frac{d}{\lambda}\right) = 5 \text{ (dimensionless)} = 10\log_{10}(5) = 6.99 \text{ dB}$$

6.4.2 Ordinary End-Fire Array

For an end-fire array, with the maximum radiation in the $\theta_0 = 0°$ direction, the array factor is given by

$$(AF)_n = \left[\frac{\sin\left[\frac{N}{2}kd(\cos\theta - 1)\right]}{N\sin\left[\frac{1}{2}kd(\cos\theta - 1)\right]}\right] \tag{6-45}$$

which, for a small spacing between the elements ($d \ll \lambda$), can be approximated by

$$(AF)_n \simeq \left[\frac{\sin\left[\frac{N}{2}kd(\cos\theta - 1)\right]}{\left[\frac{N}{2}kd(\cos\theta - 1)\right]}\right] \tag{6-45a}$$

The corresponding radiation intensity can be written as

$$U(\theta) = [(AF)_n]^2 = \left[\frac{\sin\left[\frac{N}{2}kd(\cos\theta - 1)\right]}{\frac{N}{2}kd(\cos\theta - 1)}\right]^2 = \left[\frac{\sin(Z)}{Z}\right]^2 \tag{6-46}$$

$$Z = \frac{N}{2}kd(\cos\theta - 1) \tag{6-46a}$$

whose maximum value is unity ($U_{max} = 1$) and it occurs at $\theta = 0°$. The average value of the radiation intensity is given by

$$U_0 = \frac{1}{4\pi} \int_0^{2\pi} \int_0^{\pi} \left[\frac{\sin\left[\frac{N}{2} kd(\cos\theta - 1)\right]}{\frac{N}{2} kd(\cos\theta - 1)} \right]^2 \sin\theta \, d\theta \, d\phi$$

$$= \frac{1}{2} \int_0^{\pi} \left[\frac{\sin\left[\frac{N}{2} kd(\cos\theta - 1)\right]}{\frac{N}{2} kd(\cos\theta - 1)} \right]^2 \sin\theta \, d\theta \qquad (6\text{-}47)$$

By letting

$$Z = \frac{N}{2} kd(\cos\theta - 1) \qquad (6\text{-}47a)$$

$$dZ = -\frac{N}{2} kd \sin\theta \, d\theta \qquad (6\text{-}47b)$$

(6-47) can be written as

$$U_0 = -\frac{1}{Nkd} \int_0^{-Nkd} \left[\frac{\sin(Z)}{Z}\right]^2 dZ = \frac{1}{Nkd} \int_0^{Nkd} \left[\frac{\sin(Z)}{Z}\right]^2 dZ \qquad (6\text{-}48)$$

For a large array ($Nkd \to$ large), (6-48) can be approximated by extending the limits to infinity. That is,

$$U_0 = \frac{1}{Nkd} \int_0^{Nkd} \left[\frac{\sin(Z)}{Z}\right]^2 dZ \simeq \frac{1}{Nkd} \int_0^{\infty} \left[\frac{\sin(Z)}{Z}\right]^2 dZ \qquad (6\text{-}48a)$$

Using (6-41b) reduces (6-48a) to

$$U_0 \simeq \frac{\pi}{2Nkd} \qquad (6\text{-}48b)$$

and the directivity to

$$D_0 = \frac{U_{max}}{U_0} \simeq \frac{2Nkd}{\pi} = 4N\left(\frac{d}{\lambda}\right) \qquad (6\text{-}49)$$

Another form of (6-49), using (6-43), is

$$D_0 \simeq 4N\left(\frac{d}{\lambda}\right) = 4\left(1 + \frac{L}{d}\right)\left(\frac{d}{\lambda}\right) \qquad (6\text{-}49a)$$

which for a large array ($L \gg d$) reduces to

$$D_0 \simeq 4N\left(\frac{d}{\lambda}\right) = 4\left(1 + \frac{L}{d}\right)\left(\frac{d}{\lambda}\right) \stackrel{L \gg d}{\simeq} 4\left(\frac{L}{\lambda}\right) \qquad (6\text{-}49b)$$

It should be noted that the directivity of the end-fire array, as given by (6-49)–(6-49b), is twice that for the broadside array as given by (6-42)–(6-44a).

Example 6.4
Given a linear, end-fire, uniform array of 10 elements ($N = 10$) with a separation of $\lambda/4$ ($d = \lambda/4$) between the elements, find the directivity of the array factor. This array is identical to the broadside array of Example 6.3.

Solution: Using (6-49)

$$D_0 \simeq 4N\left(\frac{d}{\lambda}\right) = 10 \text{ (dimensionless)} = 10\log_{10}(10) = 10 \text{ dB}$$

This value for the directivity ($D_0 = 10$) is approximate, based on the validity of (6-48a). However, it compares very favorably with the value of $D_0 = 10.05$ obtained by numerically integrating (6-45) using the **Directivity** computer program of Chapter 2.

6.4.3 Hansen-Woodyard End-Fire Array

For an end-fire array with improved directivity (Hansen-Woodyard designs) and maximum radiation in the $\theta_0 = 0°$ direction, the radiation intensity (for small spacing between the elements, $d \ll \lambda$) is given by (6-31)–(6-31b). The maximum radiation intensity is unity ($U_{\max} = 1$), and the average radiation intensity is given by (6-34) where q and v are defined, respectively, by (6-29a) and (6-34a). Using (6-29a), (6-34a), (6-35), and (6-37), the radiation intensity of (6-34) reduces to

$$U_0 = \frac{1}{Nkd}\left(\frac{\pi}{2}\right)^2\left[\frac{\pi}{2} + \frac{2}{\pi} - 1.8515\right] = \frac{0.871}{Nkd} \tag{6-50}$$

which can also be written as

$$U_0 = \frac{0.871}{Nkd} = \frac{1.742}{2Nkd} = 0.554\left(\frac{\pi}{2Nkd}\right) \tag{6-50a}$$

The average value of the radiation intensity, as given by (6-50a), is 0.554 times that for the ordinary end-fire array of (6-48b). Thus the directivity can be expressed, using (6-50a), as

$$D_0 = \frac{U_{\max}}{U_0} = \frac{1}{0.554}\left[\frac{2Nkd}{\pi}\right] = 1.805\left[4N\left(\frac{d}{\lambda}\right)\right] \tag{6-51}$$

which is 1.805 times that of the ordinary end-fire array as given by (6-49). Using (6-43), (6-51) can also be written as

$$D_0 = 1.805\left[4N\left(\frac{d}{\lambda}\right)\right] = 1.805\left[4\left(1 + \frac{L}{d}\right)\frac{d}{\lambda}\right] \tag{6-51a}$$

which for a large array ($L \gg d$) reduces to

$$D_0 = 1.805\left[4N\left(\frac{d}{\lambda}\right)\right] = 1.805\left[4\left(1 + \frac{L}{d}\right)\left(\frac{d}{\lambda}\right)\right]$$

$$\simeq 1.805\left[4\left(\frac{L}{\lambda}\right)\right] \tag{6-51b}$$

TABLE 6.8 Directivities for Broadside and End-Fire Arrays

Array	Directivity	
BROADSIDE	$D_0 = 2N\left(\dfrac{d}{\lambda}\right) = 2\left(1+\dfrac{L}{d}\right)\dfrac{d}{\lambda} \simeq 2\left(\dfrac{L}{\lambda}\right)$ $N\pi d/\lambda \to \infty, L \gg d$	
END-FIRE (ORDINARY)	$D_0 = 4N\left(\dfrac{d}{\lambda}\right) = 4\left(1+\dfrac{L}{d}\right)\dfrac{d}{\lambda} \simeq 4\left(\dfrac{L}{\lambda}\right)$ $2N\pi d/\lambda \to \infty, L \gg d$	Only one maximum ($\theta_0 = 0°$ or $180°$)
	$D_0 = 2N\left(\dfrac{d}{\lambda}\right) = 2\left(1+\dfrac{L}{d}\right)\dfrac{d}{\lambda} \simeq 2\left(\dfrac{L}{\lambda}\right)$	Two maxima ($\theta_0 = 0°$ and $180°$)
END-FIRE (HANSEN-WOODYARD)	$D_0 = 1.805\left[4N\left(\dfrac{d}{\lambda}\right)\right] = 1.805\left[4\left(1+\dfrac{L}{d}\right)\dfrac{d}{\lambda}\right] = 1.805\left[4\left(\dfrac{L}{\lambda}\right)\right]$ $2N\pi d/\lambda \to \infty, L \gg d$	

Example 6.5

Given a linear, end-fire (with improved directivity) Hansen-Woodyard, uniform array of 10 elements ($N = 10$) with a separation of $\lambda/4$ ($d = \lambda/4$) between the elements, find the directivity of the array factor. This array is identical to that of Examples 6.3 (broadside) and 6.4 (ordinary end-fire), and it is used for comparison.

Solution: Using (6-51b)

$$D_0 = 1.805\left[4N\left(\dfrac{d}{\lambda}\right)\right] = 18.05 \text{ (dimensionless)} = 10\log_{10}(18.05) = 12.56 \text{ dB}$$

The value of this directivity ($D_0 = 18.05$) is 1.805 times greater than that of Example 6.4 (ordinary end-fire) and 3.61 times greater than that found in Example 6.3 (broadside).

Table 6.8 lists the directivities for broadside, ordinary end fire, and Hansen-Woodyard arrays.

6.5 DESIGN PROCEDURE

In the design of any antenna system, the most important design parameters are usually the number of elements, spacing between the elements, excitation (amplitude and phase), half-power beamwidth, directivity, and side lobe level. In a design procedure some of these parameters are specified and the others are then determined.

The parameters that are specified and those that are determined vary among designs. For a uniform array, other than for the Hansen-Woodyard end-fire, the side lobe is always approximately -13.5 dB. For the Hansen-Woodyard end-fire array the side lobe level is somewhat compromised above the -13.5 dB in order to gain about 1.805 (or 2.56 dB) in directivity. The order in which the other parameters are specified and determined varies among designs. For each of the uniform linear arrays that have been discussed, equations and some graphs have been presented which can be used to determine the half-power beamwidth and directivity, once the number of elements and spacing (or the total length of the array) are specified. In fact, some of the equations have been boxed or listed in tables. This may be considered more of an analysis procedure. The other approach is to specify the half-power beamwidth or directivity and to determine most of the other parameters. This can be viewed more as a design approach, and can be accomplished to a large extent with equations or graphs that have been presented. More exact values can be obtained, if necessary, using iterative or numerical methods.

Example 6.6

Design a uniform linear scanning array whose maximum of the array factor is 30° from the axis of the array ($\theta_0 = 30°$). The desired half-power beamwidth is 2° while the spacing between the elements is $\lambda/4$. Determine the excitation of the elements (amplitude and phase), length of the array (in wavelengths), number of elements, and directivity (in dB).

Solution: Since the desired design is a uniform linear scanning array, the amplitude excitation is uniform. However, the progressive phase between the elements is, using (6-21)

$$\beta = -kd\cos\theta_0 = -\frac{2\pi}{\lambda}\left(\frac{\lambda}{4}\right)\cos(30°) = -1.36 \text{ radians} = -77.94°$$

The length of the array is obtained using an iterative procedure of (6-22) or its graphical solution of Figure 6.12. Using the graph of Figure 6.12 for a scan angle of 30° and 2° half-power beamwidth, the approximate length plus one spacing $(L + d)$ of the array is 50λ. For the 50λ length plus one spacing dimension from Figure 6.12 and 30° scan angle, (6-22) leads to a half-power beamwidth of 2.03°, which is very close to the desired value of 2°. Therefore, the length of the array for a spacing of $\lambda/4$ is 49.75λ.

Since the length of the array is 49.75λ and the spacing between the elements is $\lambda/4$, the total number of elements is

$$N = \frac{L}{d} + 1 = \left(\frac{L+d}{d}\right) = \frac{50}{1/4} = 200$$

The directivity of the array is obtained using the radiation intensity and the computer program **Directivity** of Chapter 2, and it is equal to 100.72 or 20.03 dB.

6.6 N-ELEMENT LINEAR ARRAY: THREE-DIMENSIONAL CHARACTERISTICS

Up to now, the two-dimensional array factor of an N-element linear array has been investigated. Although in practice only two-dimensional patterns can be measured, a collection of them can be used to reconstruct the three-dimensional characteristics of an array. It would then be instructive to examine the three-dimensional patterns of an array of elements. Emphasis will be placed on the array factor.

6.6.1 N-Elements Along Z-Axis

A linear array of N isotropic elements are positioned along the z-axis and are separated by a distance d, as shown in Figure 6.5(a). The amplitude excitation of each element is a_n and there exists a progressive phase excitation β between the elements. For far-field observations, the array factor can be written according to (6-6) as

$$\text{AF} = \sum_{n=1}^{N} a_n e^{j(n-1)(kd\cos\gamma + \beta)} = \sum_{n=1}^{N} a_n e^{j(n-1)\psi} \quad (6\text{-}52)$$

$$\psi = kd\cos\gamma + \beta \quad (6\text{-}52a)$$

where the a_n's are the amplitude excitation coefficients and γ is the angle between the axis of the array (z-axis) and the radial vector from the origin to the observation point.

In general, the angle γ can be obtained from the dot product of a unit vector along the axis of the array with a unit vector directed toward the observation point. For the geometry of Figure 6.5(a)

$$\cos\gamma = \hat{a}_z \cdot \hat{a}_r = \hat{a}_z \cdot (\hat{a}_x \sin\theta \cos\phi + \hat{a}_y \sin\theta \sin\phi + \hat{a}_z \cos\theta) = \cos\theta \Rightarrow \gamma = \theta \quad (6\text{-}53)$$

Thus (6-52) along with (6-53) is identical to (6-6), because the system of Figure 6.5(a) possesses a symmetry around the z-axis (no ϕ variations). This is not the case when the elements are placed along any of the other axes, as will be shown next.

6.6.2 N-Elements Along X- or Y-Axis

To demonstrate the simplicity that a judicious coordinate system and geometry can provide in the solution of a problem, let us consider an array of N isotropic elements along the x-axis, as shown in Figure 6.16. The far-zone array factor for this array is identical in form to that of Figure 6.5(a) except for the phase factor ψ. For this geometry

$$\cos\gamma = \hat{a}_x \cdot \hat{a}_r = \hat{a}_x \cdot (\hat{a}_x \sin\theta \cos\phi + \hat{a}_y \sin\theta \sin\phi + \hat{a}_z \cos\theta) = \sin\theta \cos\phi \quad (6\text{-}54)$$

$$\cos\gamma = \sin\theta \cos\phi \Rightarrow \gamma = \cos^{-1}(\sin\theta \cos\phi) \quad (6\text{-}54a)$$

The array factor of this array is also given by (6-52) but with γ defined by (6-54a). For this system, the array factor is a function of both angles (θ and ϕ).

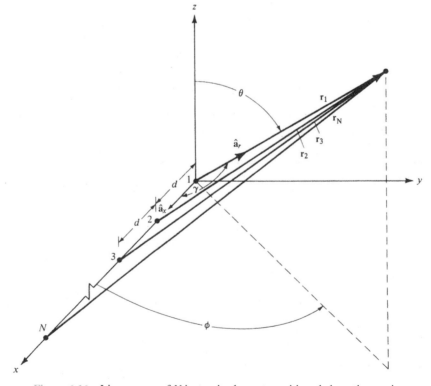

Figure 6.16 Linear array of N isotropic elements positioned along the x-axis.

In a similar manner, the array factor for N isotropic elements placed along the y-axis is that of (6-52) but with γ defined by

$$\cos\gamma = \hat{\mathbf{a}}_y \cdot \hat{\mathbf{a}}_r = \sin\theta\sin\phi \Rightarrow \gamma = \cos^{-1}(\sin\theta\sin\phi) \tag{6-55}$$

Physically placing the elements along the z-, x-, or y-axis does not change the characteristics of the array. Numerically they yield identical patterns even though their mathematical forms are different.

Example 6.7

Two half-wavelength dipole ($l = \lambda/2$) are positioned along the x-axis and are separated by a distance d, as shown in Figure 6.17. The lengths of the dipoles are parallel to the z-axis. Find the total field of the array. Assume uniform amplitude excitation and a progressive phase difference of β.

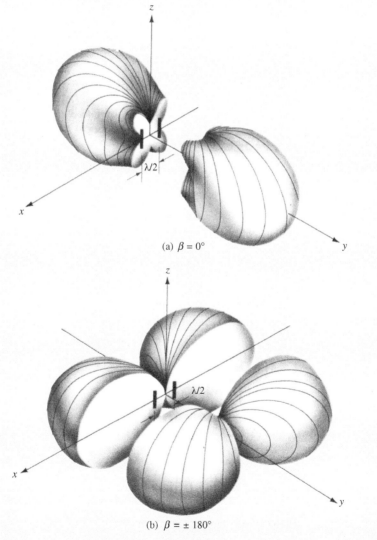

(a) $\beta = 0°$

(b) $\beta = \pm 180°$

Figure 6.17 Three-dimensional patterns for two $\lambda/2$ dipoles spaced $\lambda/2$. (SOURCE: P. Lorrain and D. R. Corson, *Electromagnetic Fields and Waves*, 2nd ed., W. H. Freeman and Co., Copyright © 1970).

Solution: The field pattern of a single element placed at the origin is given by (4-84) as

$$E_\theta = j\eta \frac{I_0 e^{-jkr}}{2\pi r} \left[\frac{\cos\left(\frac{\pi}{2} \cos\theta\right)}{\sin\theta} \right]$$

Using (6-52), (6-54a), and (6-10c), the array factor can be written as

$$(AF)_n = \frac{\sin(kd \sin\theta \cos\phi + \beta)}{2 \sin\left[\frac{1}{2}(kd \sin\theta \cos\phi + \beta)\right]}$$

The total field of the array is then given, using the pattern multiplication rule of (6-5), by

$$E_{\theta t} = E_\theta \cdot (AF)_n = j\eta \frac{I_0 e^{-jkr}}{2\pi r} \frac{\cos\left(\frac{\pi}{2}\cos\theta\right)}{\sin\theta} \left[\frac{\sin(kd \sin\theta \cos\phi + \beta)}{2\sin\left[\frac{1}{2}(kd\sin\theta\cos\phi + \beta)\right]} \right]$$

To illustrate the techniques, the three-dimensional patterns of the two-element array of Example 6.7 have been sketched in Figures 6.17(a) and (b). For both, the element separation is $\lambda/2(d = \lambda/2)$. For the pattern of Figure 6.17(a), the phase excitation between the elements is identical ($\beta = 0$). In addition to the nulls in the $\theta = 0°$ direction, provided by the individual elements of the array, there are additional nulls along the x-axis ($\theta = \pi/2, \phi = 0$ and $\phi = \pi$) provided by the formation of the array. The 180° phase difference required to form the nulls along the x-axis is a result of the separation of the elements [$kd = (2\pi/\lambda)(\lambda/2) = \pi$].

To form a comparison, the three-dimensional pattern of the same array but with a 180° phase excitation ($\beta = 180°$) between the elements is sketched in Figure 6.17(b). The overall pattern of this array is quite different from that shown in Figure 6.17(a). In addition to the nulls along the z-axis ($\theta = 0°$) provided by the individual elements, there are nulls along the y-axis formed by the 180° excitation phase difference.

6.7 RECTANGULAR-TO-POLAR GRAPHICAL SOLUTION

In antenna theory, many solutions are of the form

$$f(\zeta) = f(C \cos\gamma + \delta) \tag{6-56}$$

where C and δ are constants and γ is a variable. For example, the approximate array factor of an N-element, uniform amplitude linear array [Equation (6-10d)] is that of a $\sin(\zeta)/\zeta$ form with

$$\zeta = C\cos\gamma + \delta = \frac{N}{2}\psi = \frac{N}{2}(kd\cos\theta + \beta) \tag{6-57}$$

where

$$C = \frac{N}{2}kd \tag{6-57a}$$

$$\delta = \frac{N}{2}\beta \tag{6-57b}$$

Usually the $f(\zeta)$ function can be sketched as a function of ζ in rectilinear coordinates. Since ζ in (6-57) has no physical analog, in many instances it is desired that a graphical representation of $|f(\zeta)|$ be obtained as a function of the physically observable angle θ. This can be constructed graphically from the rectilinear graph, and it forms a polar plot.

The procedure that must be followed in the construction of the polar graph is as follows:

1. Plot, using rectilinear coordinates, the function $|f(\zeta)|$.
2. a. Draw a circle with radius C and with its center on the abscissa at $\zeta = \delta$.
   b. Draw vertical lines to the abscissa so that they will intersect the circle.
   c. From the center of the circle, draw radial lines through the points on the circle intersected by the vertical lines.
   d. Along the radial lines, mark off corresponding magnitudes from the linear plot.
   e. Connect all points to form a continuous graph.

To better illustrate the procedure, the polar graph of the function

$$f(\zeta) = \frac{\sin\left(\frac{N}{2}\psi\right)}{N \sin\left(\frac{\psi}{2}\right)}, \quad \zeta = \frac{5\pi}{2}\cos\theta - \frac{5\pi}{4} \tag{6-58}$$

has been constructed in Figure 6.18. The function $f(\zeta)$ of (6-58) represents the array factor of a 10-element ($N = 10$) uniform linear array with a spacing of $\lambda/4(d = \lambda/4)$ and progressive phase shift of $-\pi/4(\beta = -\pi/4)$ between the elements. The constructed graph can be compared with its exact form shown in Figure 6.11.

From the construction of Figure 6.18, it is evident that the angle at which the maximum is directed is controlled by the radius of the circle C and the variable δ. For the array factor of Figure 6.18, the radius C is a function of the number of elements (N) and the spacing between the elements (d). In turn, δ is a function of the number of elements (N) and the progressive phase shift between the elements (β). Making $\delta = 0$ directs the maximum toward $\theta = 90°$ (broadside array). The part of the linear graph that is used to construct the polar plot is determined by the radius of the circle and the relative position of its center along the abscissa. The usable part of the linear graph is referred to as the *visible* region and the remaining part as the *invisible* region. Only the *visible* region of the linear graph is related to the physically observable angle θ (hence its name).

6.8 N-ELEMENT LINEAR ARRAY: UNIFORM SPACING, NONUNIFORM AMPLITUDE

The theory to analyze linear arrays with uniform spacing, uniform amplitude, and a progressive phase between the elements was introduced in the previous sections of this chapter. A number of numerical and graphical solutions were used to illustrate some of the principles. In this section, broadside arrays with uniform spacing but nonuniform amplitude distribution will be considered. Most of the discussion will be directed toward binomial [6] and Dolph-Tschebyscheff [7] broadside arrays (also spelled Tchebyscheff or Chebyshev).

Of the three distributions (uniform, binomial, and Tschebyscheff), a uniform amplitude array yields the smallest half-power beamwidth. It is followed, in order, by the Dolph-Tschebyscheff and binomial arrays. In contrast, binomial arrays usually possess the smallest side lobes followed, in order, by the Dolph-Tschebyscheff and uniform arrays. As a matter of fact, binomial arrays with

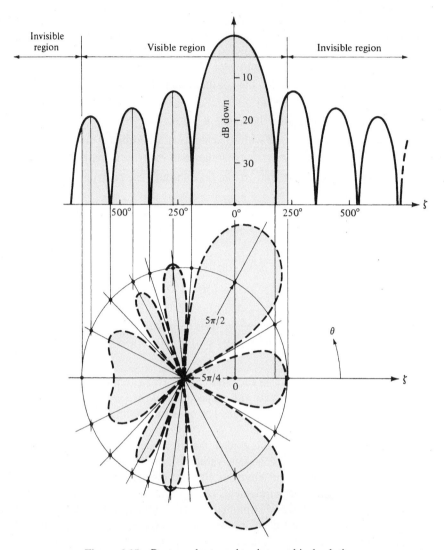

Figure 6.18 Rectangular-to-polar plot graphical solution.

element spacing equal or less than $\lambda/2$ have no side lobes. It is apparent that the designer must compromise between side lobe level and beamwidth.

A criterion that can be used to judge the relative beamwidth and side lobe level of one design to another is the amplitude distribution (tapering) along the source. *It has been shown analytically that for a given side lobe level the Dolph-Tschebyscheff array produces the smallest beamwidth between the first nulls. Conversely, for a given beamwidth between the first nulls, the Dolph-Tschebyscheff design leads to the smallest possible side lobe level.*

Uniform arrays usually possess the largest directivity. However, superdirective (or super gain as most people refer to them) antennas possess directivities higher than those of a uniform array [8]. Although a certain amount of superdirectivity is practically possible, superdirective arrays usually require very large currents with opposite phases between adjacent elements. Thus the net total current and efficiency of each array are very small compared to the corresponding values of an individual element.

N-ELEMENT LINEAR ARRAY: UNIFORM SPACING, NONUNIFORM AMPLITUDE

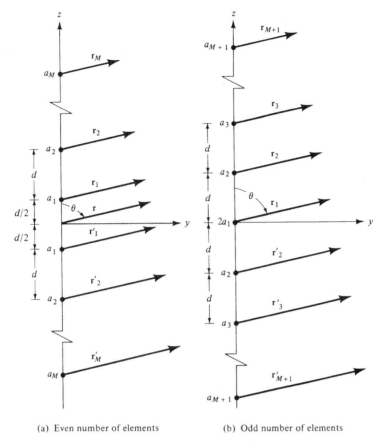

(a) Even number of elements   (b) Odd number of elements

Figure 6.19 Nonuniform amplitude arrays of even and odd number of elements.

Before introducing design methods for specific nonuniform amplitude distributions, let us first derive the array factor.

6.8.1 Array Factor

An array of an even number of isotropic elements $2M$ (where M is an integer) is positioned symmetrically along the z-axis, as shown in Figure 6.19(a). The separation between the elements is d, and M elements are placed on each side of the origin. Assuming that the amplitude excitation is symmetrical about the origin, the array factor for a nonuniform amplitude broadside array can be written as

$$(AF)_{2M} = a_1 e^{+j(1/2)kd\cos\theta} + a_2 e^{+j(3/2)kd\cos\theta} + \cdots$$
$$+ a_M e^{+j[(2M-1)/2]kd\cos\theta}$$
$$+ a_1 e^{-j(1/2)kd\cos\theta} + a_2 e^{-j(3/2)kd\cos\theta} + \cdots$$
$$+ a_M e^{-j[(2M-1)/2]kd\cos\theta}$$

$$(AF)_{2M} = 2\sum_{n=1}^{M} a_n \cos\left[\frac{(2n-1)}{2}kd\cos\theta\right] \qquad (6\text{-}59)$$

which in normalized form reduces to

$$(AF)_{2M} = \sum_{n=1}^{M} a_n \cos\left[\frac{(2n-1)}{2} kd \cos\theta\right] \quad (6\text{-}59a)$$

where a_n's are the excitation coefficients of the array elements.

If the total number of isotropic elements of the array is odd $2M + 1$ (where M is an integer), as shown in Figure 6.19(b), the array factor can be written as

$$(AF)_{2M+1} = 2a_1 + a_2 e^{+jkd\cos\theta} + a_3 e^{j2kd\cos\theta} + \cdots + a_{M+1} e^{jMkd\cos\theta}$$
$$+ a_2 e^{-jkd\cos\theta} + a_3 e^{-j2kd\cos\theta} + \cdots + a_{M+1} e^{-jMkd\cos\theta}$$

$$(AF)_{2M+1} = 2 \sum_{n=1}^{M+1} a_n \cos[(n-1)kd\cos\theta] \quad (6\text{-}60)$$

which in normalized form reduces to

$$(AF)_{2M+1} = \sum_{n=1}^{M+1} a_n \cos[(n-1)kd\cos\theta] \quad (6\text{-}60a)$$

The amplitude excitation of the center element is $2a_1$.

Equations (6-59a) and (6-60a) can be written in normalized form as

$$(AF)_{2M}(\text{even}) = \sum_{n=1}^{M} a_n \cos[(2n-1)u] \quad (6\text{-}61a)$$

$$(AF)_{2M+1}(\text{odd}) = \sum_{n=1}^{M+1} a_n \cos[2(n-1)u] \quad (6\text{-}61b)$$

where

$$u = \frac{\pi d}{\lambda} \cos\theta \quad (6\text{-}61c)$$

The next step will be to determine the values of the excitation coefficients (a_n's).

6.8.2 Binomial Array

The array factor for the binomial array is represented by (6-61a)–(6-61c) where the a_n's are the excitation coefficients which will now be derived.

A. Excitation Coefficients

To determine the excitation coefficients of a binomial array, J. S. Stone [6] suggested that the function $(1 + x)^{m-1}$ be written in a series, using the binomial expansion, as

$$(1+x)^{m-1} = 1 + (m-1)x + \frac{(m-1)(m-2)}{2!}x^2$$
$$+ \frac{(m-1)(m-2)(m-3)}{3!}x^3 + \cdots \quad (6\text{-}62)$$

The positive coefficients of the series expansion for different values of m are

$m = 1$										1									
$m = 2$									1		1								
$m = 3$									1	2	1								
$m = 4$								1	3		3	1							
$m = 5$							1	4		6		4	1						
$m = 6$						1		5	10		10	5		1					
$m = 7$					1		6	15		20		15	6		1				
$m = 8$				1		7	21		35		35		21	7		1			
$m = 9$			1		8	28		56		70		56		28	8		1		
$m = 10$		1		9	36		84		126		126		84		36	9		1	

(6-63)

The above represents Pascal's triangle. If the values of m are used to represent the number of elements of the array, then the coefficients of the expansion represent the relative amplitudes of the elements. Since the coefficients are determined from a binomial series expansion, the array is known as a *binomial array*.

Referring to (6-61a), (6-61b), and (6-63), the amplitude coefficients for the following arrays are:

1. Two elements ($2M = 2$)

   $a_1 = 1$

2. Three elements ($2M + 1 = 3$)

   $2a_1 = 2 \Rightarrow a_1 = 1$

   $a_2 = 1$

3. Four elements ($2M = 4$)

   $a_1 = 3$

   $a_2 = 1$

4. Five elements ($2M + 1 = 5$)

   $2a_1 = 6 \Rightarrow a_1 = 3$

   $a_2 = 4$

   $a_3 = 1$

The coefficients for other arrays can be determined in a similar manner.

B. Design Procedure

One of the objectives of any method is its use in a design. For the binomial method, as for any other nonuniform array method, one of the requirements is the amplitude excitation coefficients for a given number of elements. This can be accomplished using either (6-62) or the Pascal triangle of (6-63) or extensions of it. Other figures of merit are the directivity, half-power beamwidth and side lobe level. It already has been stated that *binomial arrays do not exhibit any minor lobes provided the spacing between the elements is equal or less than one-half of a wavelength*. Unfortunately,

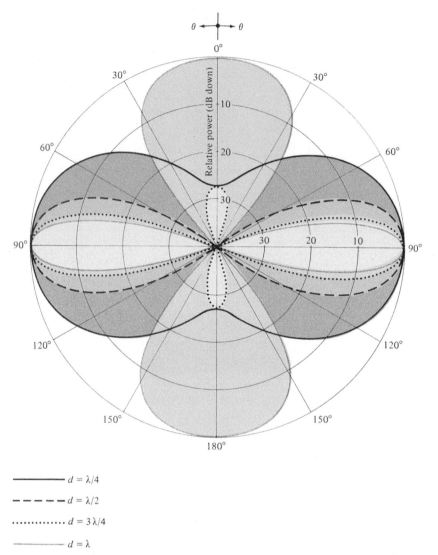

- ———— $d = \lambda/4$
- — — — — $d = \lambda/2$
- ·············· $d = 3\lambda/4$
- ———————— $d = \lambda$

Figure 6.20 Array factor power patterns for a 10-element broadside binomial array with $N = 10$ and $d = \lambda/4, \lambda/2, 3\lambda/4,$ and λ.

closed-form expressions for the directivity and half-power beamwidth for binomial arrays of any spacing between the elements are not available. However, because the design using a $\lambda/2$ spacing leads to a pattern with no minor lobes, approximate closed-form expressions for the half-power beamwidth and maximum directivity for the $d = \lambda/2$ spacing only have been derived [9] in terms of the numbers of elements or the length of the array, and they are given, respectively, by

$$\text{HPBW}(d = \lambda/2) \simeq \frac{1.06}{\sqrt{N-1}} = \frac{1.06}{\sqrt{2L/\lambda}} = \frac{0.75}{\sqrt{L/\lambda}} \tag{6-64}$$

$$D_0 = \frac{2}{\int_0^\pi \left[\cos\left(\frac{\pi}{2}\cos\theta\right)\right]^{2(N-1)} \sin\theta \, d\theta} \tag{6-65}$$

$$D_0 = \frac{(2N-2)(2N-4)\cdots 2}{(2N-3)(2N-5)\cdots 1} \tag{6-65a}$$

$$D_0 \simeq 1.77\sqrt{N} = 1.77\sqrt{1+2L/\lambda} \tag{6-65b}$$

These expressions can be used effectively to design binomial arrays with a desired half-power beamwidth or directivity. The value of the directivity as obtained using (6-65) to (6-65b) can be compared with the value using the array factor and the computer program **Directivity** of Chapter 2.

To illustrate the method, the patterns of a 10-element binomial array ($2M = 10$) with spacings between the elements of $\lambda/4, \lambda/2, 3\lambda/4$, and λ, respectively, have been plotted in Figure 6.20. The patterns are plotted using (6-61a) and (6-61c) with the coefficients of $a_1 = 126, a_2 = 84, a_3 = 36, a_4 = 9$, and $a_5 = 1$. It is observed that there are no minor lobes for the arrays with spacings of $\lambda/4$ and $\lambda/2$ between the elements. While binomial arrays have very low level minor lobes, they exhibit larger beamwidths (compared to uniform and Dolph-Tschebyscheff designs). A major practical disadvantage of binomial arrays is the wide variations between the amplitudes of the different elements of an array, especially for an array with a large number of elements. This leads to very low efficiencies for the feed network, and it makes the method not very desirable in practice. For example, the relative amplitude coefficient of the end elements of a 10-element array is 1 while that of the center element is 126. Practically, it would be difficult to obtain and maintain such large amplitude variations among the elements. They would also lead to very inefficient antenna systems. Because the magnitude distribution is monotonically decreasing from the center toward the edges and the magnitude of the extreme elements is negligible compared to those toward the center, a very low side lobe level is expected.

Table 6.7 lists the maximum element spacing d_{max} for the various linear and planar arrays, including binomial arrays, in order to maintain either one or two amplitude maxima.

Example 6.8

For a 10-element binomial array with a spacing of $\lambda/2$ between the elements, whose amplitude pattern is displayed in Figure 6.20, determine the half-power beamwidth (in degrees) and the maximum directivity (in dB). Compare the answers with other available data.

Solution: Using (6-64), the half-power beamwidth is equal to

$$\text{HPBW} \simeq \frac{1.06}{\sqrt{10-1}} = \frac{1.06}{3} = 0.353 \text{ radians} = 20.23°$$

The value obtained using the array factor, whose pattern is shown in Figure 6.20, is 20.5° which compares well with the approximate value.

Using (6-65a), the value of the directivity is equal for $N = 10$

$$D_0 = 5.392 = 7.32 \text{ dB}$$

while the value obtained using (6-65b) is

$$D_0 = 1.77\sqrt{10} = 5.597 = 7.48 \text{ dB}$$

The value obtained using the array factor and the computer program **Directivity** is $D_0 = 5.392$ (dimensionless) $= 7.32$ dB. These values compare favorably with each other.

6.8.3 Dolph-Tschebyscheff Array: Broadside

Another array, with many practical applications, is the *Dolph-Tschebyscheff array*. The method was originally introduced by Dolph [7] and investigated afterward by others [10]–[13]. It is primarily a compromise between uniform and binomial arrays. Its excitation coefficients are related to Tschebyscheff polynomials. A Dolph-Tschebyscheff array with no side lobes (or side lobes of $-\infty$ dB) reduces to the binomial design. The excitation coefficients for this case, as obtained by both methods, would be identical.

A. Array Factor

Referring to (6-61a) and (6-61b), the array factor of an array of even or odd number of elements with symmetric amplitude excitation is nothing more than a summation of M or $M + 1$ cosine terms. The largest harmonic of the cosine terms is one less than the total number of elements of the array. Each cosine term, whose argument is an integer times a fundamental frequency, can be rewritten as a series of cosine functions with the fundamental frequency as the argument. That is,

$$
\begin{aligned}
m &= 0 \quad \cos(mu) = 1 \\
m &= 1 \quad \cos(mu) = \cos u \\
m &= 2 \quad \cos(mu) = \cos(2u) = 2\cos^2 u - 1 \\
m &= 3 \quad \cos(mu) = \cos(3u) = 4\cos^3 u - 3\cos u \\
m &= 4 \quad \cos(mu) = \cos(4u) = 8\cos^4 u - 8\cos^2 u + 1 \\
m &= 5 \quad \cos(mu) = \cos(5u) = 16\cos^5 u - 20\cos^3 u + 5\cos u \\
m &= 6 \quad \cos(mu) = \cos(6u) = 32\cos^6 u - 48\cos^4 u + 18\cos^2 u - 1 \\
m &= 7 \quad \cos(mu) = \cos(7u) = 64\cos^7 u - 112\cos^5 u + 56\cos^3 u - 7\cos u \\
m &= 8 \quad \cos(mu) = \cos(8u) = 128\cos^8 u - 256\cos^6 u + 160\cos^4 u - 32\cos^2 u + 1 \\
m &= 9 \quad \cos(mu) = \cos(9u) = 256\cos^9 u - 576\cos^7 u + 432\cos^5 u - 120\cos^3 u + 9\cos u
\end{aligned}
\tag{6-66}
$$

The above are obtained by the use of Euler's formula

$$[e^{ju}]^m = (\cos u + j\sin u)^m = e^{jmu} = \cos(mu) + j\sin(mu) \tag{6-67}$$

and the trigonometric identity $\sin^2 u = 1 - \cos^2 u$.

If we let

$$z = \cos u \tag{6-68}$$

(6-66) can be written as

$$
\begin{aligned}
m &= 0 \quad \cos(mu) = 1 = T_0(z) \\
m &= 1 \quad \cos(mu) = z = T_1(z) \\
m &= 2 \quad \cos(mu) = 2z^2 - 1 = T_2(z) \\
m &= 3 \quad \cos(mu) = 4z^3 - 3z = T_3(z) \\
m &= 4 \quad \cos(mu) = 8z^4 - 8z^2 + 1 = T_4(z) \\
m &= 5 \quad \cos(mu) = 16z^5 - 20z^3 + 5z = T_5(z) \\
m &= 6 \quad \cos(mu) = 32z^6 - 48z^4 + 18z^2 - 1 = T_6(z) \\
m &= 7 \quad \cos(mu) = 64z^7 - 112z^5 + 56z^3 - 7z = T_7(z) \\
m &= 8 \quad \cos(mu) = 128z^8 - 256z^6 + 160z^4 - 32z^2 + 1 = T_8(z) \\
m &= 9 \quad \cos(mu) = 256z^9 - 576z^7 + 432z^5 - 120z^3 + 9z = T_9(z)
\end{aligned}
\tag{6-69}
$$

and each is related to a Tschebyscheff (Chebyshev) polynomial $T_m(z)$. These relations between the cosine functions and the Tschebyscheff polynomials are valid only in the $-1 \leq z \leq +1$ range. Because $|\cos(mu)| \leq 1$, each Tschebyscheff polynomial is $|T_m(z)| \leq 1$ for $-1 \leq z \leq +1$. For $|z| > 1$, the Tschebyscheff polynomials are related to the hyperbolic cosine functions.

The recursion formula for Tschebyscheff polynomials is

$$T_m(z) = 2zT_{m-1}(z) - T_{m-2}(z) \tag{6-70}$$

It can be used to find one Tschebyscheff polynomial if the polynomials of the previous two orders are known. Each polynomial can also be computed using

$$T_m(z) = \cos[m\cos^{-1}(z)] \qquad -1 \leq z \leq +1 \tag{6-71a}$$

$$T_m(z) = \cosh[m\cosh^{-1}(z)]^\dagger \qquad z < -1, z > +1 \tag{6-71b}$$

In Figure 6.21 the first six Tschebyscheff polynomials have been plotted. The following properties of the polynomials are of interest:

1. All polynomials, of any order, pass through the point (1, 1).
2. Within the range $-1 \leq z \leq 1$, the polynomials have values within -1 to $+1$.
3. All roots occur within $-1 \leq z \leq 1$, and all maxima and minima have values of $+1$ and -1, respectively.

Since the array factor of an even or odd number of elements is a summation of cosine terms whose form is the same as the Tschebyscheff polynomials, the unknown coefficients of the array factor can be determined by equating the series representing the cosine terms of the array factor to the appropriate Tschebyscheff polynomial. *The order of the polynomial should be one less than the total number of elements of the array.*

The design procedure will be outlined first, and it will be illustrated with an example. In outlining the procedure, it will be assumed that the number of elements, spacing between the elements, and ratio of major-to-minor lobe intensity (R_0) are known. The requirements will be to determine the excitation coefficients and the array factor of a Dolph-Tschebyscheff array.

B. Array Design

Statement. Design a broadside Dolph-Tschebyscheff array of $2M$ or $2M + 1$ elements with spacing d between the elements. The side lobes are R_0 dB below the maximum of the major lobe. Find the excitation coefficients and form the array factor.

Procedure

a. Select the appropriate array factor as given by (6-61a) or (6-61b).
b. Expand the array factor. Replace each $\cos(mu)$ function ($m = 0, 1, 2, 3, \ldots$) by its appropriate series expansion found in (6-66).
c. Determine the point $z = z_0$ such that $T_m(z_0) = R_0$ (voltage ratio). *The order m of the Tschebyscheff polynomial is always one less than the total number of elements.* The design procedure requires that the Tschebyscheff polynomial in the $-1 \leq z \leq z_1$, where z_1 is the null

---

$^\dagger x = \cosh^{-1}(y) = \ln[y \pm (y^2 - 1)^{1/2}]$

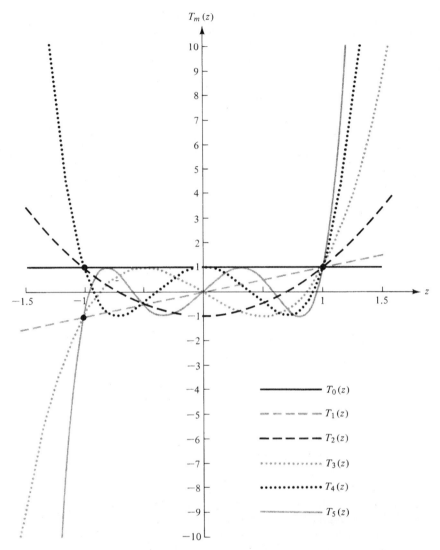

Figure 6.21 Tschebyscheff polynomials of orders zero through five.

nearest to $z = +1$, be used to represent the minor lobes of the array. The major lobe of the pattern is formed from the remaining part of the polynomial up to point $z_0 (z_1 < z \leq z_0)$.

d. Substitute

$$\cos(u) = \frac{z}{z_0} \qquad (6\text{-}72)$$

in the array factor of step b. The $\cos(u)$ is replaced by z/z_0, *and not by z*, so that (6-72) would be valid for $|z| \leq |z_0|$. At $|z| = |z_0|$, (6-72) attains its maximum value of unity.

e. Equate the array factor from step b, after substitution of (6-72), to a $T_m(z)$ from (6-69). The $T_m(z)$ chosen should be of order m where m is an integer equal to one less than the total number of elements of the designed array. This will allow the determination of the excitation coefficients a_n's.

f. Write the array factor of (6-61a) or (6-61b) using the coefficients found in step e.

Example 6.9

Design a broadside Dolph-Tschebyscheff array of 10 elements with spacing d between the elements and with a major-to-minor lobe ratio of 26 dB. Find the excitation coefficients and form the array factor.

Solution:

1. The array factor is given by (6-61a) and (6-61c). That is,

$$(AF)_{2M} = \sum_{n=1}^{M=5} a_n \cos[(2n-1)u]$$

$$u = \frac{\pi d}{\lambda} \cos\theta$$

2. When expanded, the array factor can be written as

$$(AF)_{10} = a_1 \cos(u) + a_2 \cos(3u) \\ + a_3 \cos(5u) + a_4 \cos(7u) + a_5 \cos(9u)$$

Replace $\cos(u)$, $\cos(3u)$, $\cos(5u)$, $\cos(7u)$, and $\cos(9u)$ by their series expansions found in (6-66).

3. R_0 (dB) $= 26 = 20\log_{10}(R_0)$ or R_0 (voltage ratio) $= 20$. Determine z_0 by equating R_0 to $T_9(z_0)$. Thus

$$R_0 = 20 = T_9(z_0) = \cosh[9\cosh^{-1}(z_0)]$$

or

$$z_0 = \cosh[\tfrac{1}{9}\cosh^{-1}(20)] = 1.0851$$

Another equation which can, in general, be used to find z_0 and does not require hyperbolic functions is [10]

$$z_0 = \frac{1}{2}\left[\left(R_0 + \sqrt{R_0^2-1}\right)^{1/P} + \left(R_0 - \sqrt{R_0^2-1}\right)^{1/P}\right] \quad (6\text{-}73)$$

where P is an integer equal to one less than the number of array elements (in this case $P = 9$). $R_0 = H_0/H_1$ and z_0 are identified in Figure 6.22.

4. Substitute

$$\cos(u) = \frac{z}{z_0} = \frac{z}{1.0851}$$

in the array factor found in step 2.

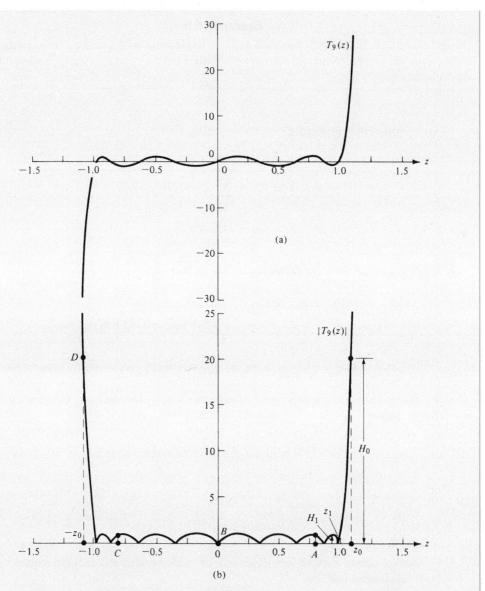

Figure 6.22 Tschebyscheff polynomial of order nine (a) amplitude (b) magnitude.

5. Equate the array factor of step 2, after the substitution from step 4, to $T_9(z)$. The polynomial $T_9(z)$ is shown plotted in Figure 6.22. Thus

$$
\begin{aligned}
(AF)_{10} = \ & z[(a_1 - 3a_2 + 5a_3 - 7a_4 + 9a_5)/z_0] \\
& + z^3[(4a_2 - 20a_3 + 56a_4 - 120a_5)/z_0^3] \\
& + \quad z^5[(16a_3 - 112a_4 + 432a_5)/z_0^5] \\
& + \qquad\qquad z^7[(64a_4 - 576a_5)/z_0^7] \\
& + \qquad\qquad\qquad z^9[(256a_5)/z_0^9] \\
= \ & 9z - 120z^3 + 432z^5 - 576z^7 + 256z^9
\end{aligned}
$$

Matching similar terms allows the determination of the a_n's. That is,

$$256a_5/z_0^9 = 256 \quad\Rightarrow a_5 = 2.0860$$
$$(64a_4 - 576a_5)/z_0^7 = -576 \quad\Rightarrow a_4 = 2.8308$$
$$(16a_3 - 112a_4 + 432a_5)/z_0^5 = 432 \quad\Rightarrow a_3 = 4.1184$$
$$(4a_2 - 20a_3 + 56a_4 - 120a_5)/z_0^3 = -120 \quad\Rightarrow a_2 = 5.2073$$
$$(a_1 - 3a_2 + 5a_3 - 7a_4 + 9a_5)/z_0 = 9 \quad\Rightarrow a_1 = 5.8377$$

In normalized form, the a_n coefficients can be written as

$$\begin{array}{ll} a_5 = 1 & a_5 = 0.357 \\ a_4 = 1.357 & a_4 = 0.485 \\ a_3 = 1.974 \quad \text{or} & a_3 = 0.706 \\ a_2 = 2.496 & a_2 = 0.890 \\ a_1 = 2.798 & a_1 = 1 \end{array}$$

The first (left) set is normalized with respect to the amplitude of the elements at the edge while the other (right) is normalized with respect to the amplitude of the center element.

6. Using the first (left) set of normalized coefficients, the array factor can be written as

$$(AF)_{10} = 2.798\cos(u) + 2.496\cos(3u) + 1.974\cos(5u)$$
$$+ 1.357\cos(7u) + \cos(9u)$$

where $u = [(\pi d/\lambda)\cos\theta]$.

The array factor patterns of Example 6.9 for $d = \lambda/4$ and $\lambda/2$ are shown plotted in Figure 6.23. Since the spacing is less than $\lambda(d < \lambda)$, maxima exist only at broadside ($\theta_0 = 90°$). However when the spacing is equal to $\lambda(d = \lambda)$, two more maxima appear (one toward $\theta_0 = 0°$ and the other toward $\theta_0 = 180°$). For $d = \lambda$ the array has four maxima, and it acts as an *end-fire* as well as a *broadside* array.

Table 6.7 lists the maximum element spacing $d_{\max}$ for the various linear and planar arrays, including Dolph-Tschebyscheff arrays, in order to maintain either one or two amplitude maxima.

To better illustrate how the pattern of a Dolph-Tschebyscheff array is formed from the Tschebyscheff polynomial, let us again consider the 10-element array whose corresponding Tschebyscheff polynomial is of order 9 and is shown plotted in Figure 6.22. The abscissa of Figure 6.22, in terms of the spacing between the elements (d) and the angle θ, is given by (6-72) or

$$z = z_0\cos u = z_0\cos\left(\frac{\pi d}{\lambda}\cos\theta\right) = 1.0851\cos\left(\frac{\pi d}{\lambda}\cos\theta\right) \qquad (6\text{-}74)$$

For $d = \lambda/4, \lambda/2, 3\lambda/4$, and λ the values of z for angles from $\theta = 0°$ to $90°$ to $180°$ are shown tabulated in Table 6.9. Referring to Table 6.9 and Figure 6.22, it is interesting to discuss the pattern formation for the different spacings.

336 ARRAYS: LINEAR, PLANAR, AND CIRCULAR

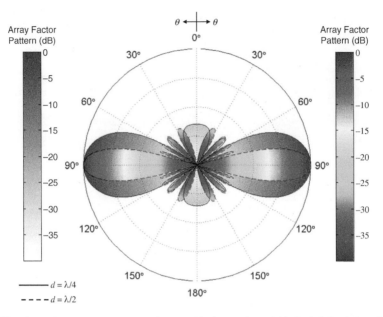

Figure 6.23 Array factor power pattern of a $N = 10$ element broadside Dolph-Tschebyscheff array.

1. $d = \lambda/4, N = 10, R_0 = 20$

   At $\theta = 0°$ the value of z is equal to 0.7673 (point A). As θ attains larger values, z increases until it reaches its maximum value of 1.0851 for $\theta = 90°$. Beyond 90°, z begins to decrease and reaches its original value of 0.7673 for $\theta = 180°$. Thus for $d = \lambda/4$, only the Tschebyscheff polynomial between the values $0.7673 \leq z \leq 1.0851 (A \leq z \leq z_0)$ is used to form the pattern of the array factor.

TABLE 6.9 Values of the Abscissa z as a Function of θ for a 10-Element Dolph-Tschebyscheff Array with $R_0 = 20$

θ	$d = \lambda/4$ z (Eq. 6-74)	$d = \lambda/2$ z (Eq. 6-74)	$d = 3\lambda/4$ z (Eq. 6-74)	$d = \lambda$ z (Eq. 6-74)
0°	0.7673	0.0	−0.7673	−1.0851
10°	0.7764	0.0259	−0.7394	−1.0839
20°	0.8028	0.1026	−0.6509	−1.0657
30°	0.8436	0.2267	−0.4912	−0.9904
40°	0.8945	0.3899	−0.2518	−0.8049
50°	0.9497	0.5774	0.0610	−0.4706
60°	1.0025	0.7673	0.4153	0.0
70°	1.0462	0.9323	0.7514	0.5167
80°	1.0750	1.0450	0.9956	0.9276
90°	1.0851	1.0851	1.0851	1.0851
100°	1.0750	1.0450	0.9956	0.9276
110°	1.0462	0.9323	0.7514	0.5167
120°	1.0025	0.7673	0.4153	0.0
130°	0.9497	0.5774	0.0610	−0.4706
140°	0.8945	0.3899	−0.2518	−0.8049
150°	0.8436	0.2267	−0.4912	−0.9904
160°	0.8028	0.1026	−0.6509	−1.0657
170°	0.7764	0.0259	−0.7394	−1.0839
180°	0.7673	0.0	−0.7673	−1.0851

2. $d = \lambda/2, N = 10, R_0 = 20$

   At $\theta = 0°$ the value of z is equal to 0 (point B). As θ becomes larger, z increases until it reaches its maximum value of 1.0851 for $\theta = 90°$. Beyond 90°, z decreases and comes back to the original point for $\theta = 180°$. For $d = \lambda/2$, a larger part of the Tschebyscheff polynomial is used ($0 \leq z \leq 1.0851; B \leq z \leq z_0$).

3. $d = 3\lambda/4, N = 10, R_0 = 20$

   For this spacing, the value of z for $\theta = 0°$ is -0.7673 (point C), and it increases as θ becomes larger. It attains its maximum value of 1.0851 at $\theta = 90°$. Beyond 90°, it traces back to its original value ($-0.7673 \leq z \leq z_0; C \leq z \leq z_0$).

4. $d = \lambda, N = 10, R_0 = 20$

As the spacing increases, a larger portion of the Tschebyscheff polynomial is used to form the pattern of the array factor. When $d = \lambda$, the value of z for $\theta = 0°$ is equal to -1.0851 (point D) which in magnitude is equal to the maximum value of z. As θ attains values larger than 0°, z increases until it reaches its maximum value of 1.0851 for $\theta = 90°$. At that point the polynomial (and thus the array factor) again reaches its maximum value. Beyond $\theta = 90°$, z and in turn the polynomial and array factor retrace their values ($-1.0851 \leq z \leq +1.0851; D \leq z \leq z_0$). For $d = \lambda$ there are four maxima, and a *broadside* and an *end-fire* array have been formed simultaneously.

It is often desired in some Dolph-Tschebyscheff designs to take advantage of the largest possible spacing between the elements while maintaining the same level of all minor lobes, including the one toward $\theta = 0°$ and 180°. In general, as well as in Example 6.8 (Figure 6.20), the only minor lobe that can exceed the level of the others, when the spacing exceeds a certain maximum spacing between the elements, is the one toward end fire ($\theta = 0°$ or 180° or $z = -1$ in Figure 6.21 or Figure 6.22). The maximum spacing which can be used while meeting the requirements is obtained using (6-72) or

$$z = z_0 \cos(u) = z_0 \cos\left(\frac{\pi d}{\lambda} \cos \theta\right) \tag{6-75}$$

The requirement not to introduce a minor lobe with a level exceeding the others is accomplished by utilizing the Tschebyscheff polynomial up to, but not going beyond $z = -1$. Therefore, for $\theta = 0°$ or 180°

$$-1 \geq z_0 \cos\left(\frac{\pi d_{\max}}{\lambda}\right) \tag{6-76}$$

or

$$\boxed{d_{\max} \leq \frac{\lambda}{\pi} \cos^{-1}\left(-\frac{1}{z_0}\right)} \tag{6-76a}$$

Equation (6-76a) provides the maximum spacing $d_{\max}$. With this separation, for a given broadside Tschebyscheff array with fixed number of N elements and specified side lobe level, the array factor maintains the same side lobe level for all of its minor lobes. There is another element separation d_{opt}, referred to as *optimum* separation, which, for a broadside Tschebyscheff array with fixed number of elements and side lobe level, leads to smallest possible HPBW, and it is

$$d_{\text{opt}} = \lambda \left[1 - \frac{\cos^{-1}(1/\gamma)}{\pi}\right] \tag{6-76b}$$

given by [14]

$$\gamma = \cosh\left[\left(\frac{1}{N-1}\right)\ln\left(R + \sqrt{R^2 - 1}\right)\right] \quad (6\text{-}76c)$$

where R represents the side lobe level (as a voltage ratio).

Design curves for $d_{max} = d_{opt}$ (from -10 dB to -100 dB), for a broadside Dolph-Tschebyscheff array with elements $N = 10, 20, 30$, and 40, are displayed in Figure 6.24(b). As expected, the maximum/optimum element separation, for each array with fixed number of elements, it gets smaller as the side lobe level decreases. Also the design curves for $d_{max} = d_{opt}$ as a function of the number of elements, $N = 3-20$ for side lobe levels $-30, -40, -50$, and -60 dB, are displayed in Figure 6.24(b). As expected, for each of the side lobe levels, the maximum/optimum element separation increases as the number of elements increase.

The excitation coefficients of a Dolph-Tschebyscheff array can be derived using various documented techniques [11]–[13] and others. One method, whose results are suitable for computer calculations, is that by Barbiere [11]. The coefficients using this method can be obtained using

$$a_n = \begin{cases} \displaystyle\sum_{q=n}^{M}(-1)^{M-q}(z_0)^{2q-1}\frac{(q+M-2)!(2M-1)}{(q-n)!(q+n-1)!(M-q)!} & (6\text{-}77a) \\ \qquad\qquad\text{for even } 2M \text{ elements} \\ \qquad\qquad n = 1, 2, \ldots M \\[1em] \displaystyle\sum_{q=n}^{M+1}(-1)^{M-q+1}(z_0)^{2(q-1)}\frac{(q+M-2)!(2M)}{\varepsilon_n(q-n)!(q+n-2)!(M-q+1)!} & (6\text{-}77b) \\ \qquad\qquad\text{for odd } 2M+1 \text{ elements} \\ \qquad\qquad n = 1, 2, \ldots M+1 \end{cases}$$

$$\text{where } \varepsilon_n = \begin{cases} 2 & n = 1 \\ 1 & n \neq 1 \end{cases}$$

C. Beamwidth and Directivity

For large Dolph-Tschebyscheff arrays scanned not too close to end-fire and with side lobes in the range from -20 to -60 dB, the half-power beamwidth and directivity can be found by introducing a beam broadening factor given approximately by [4]

$$\boxed{f = 1 + 0.636\left\{\frac{2}{R_0}\cosh\left[\sqrt{(\cosh^{-1} R_0)^2 - \pi^2}\right]\right\}^2} \quad (6\text{-}78)$$

where R_0 is the major-to-side lobe voltage ratio. The beam broadening factor is plotted in Figure 6.25(a) as a function of side lobe level (in dB).

The half-power beamwidth of a Dolph-Tschebyscheff array can be determined by

1. calculating the beamwidth of a uniform array (of the same number of elements and spacing) using (6-22a) or reading it off Figure 6.12
2. multiplying the beamwidth of part (1) by the appropriate beam broadening factor f computed using (6-78) or reading it off Figure 6.25(a)

The same procedure can be used to determine the beamwidth of arrays with a cosine-on-a-pedestal distribution [4].

N-ELEMENT LINEAR ARRAY: UNIFORM SPACING, NONUNIFORM AMPLITUDE   **339**

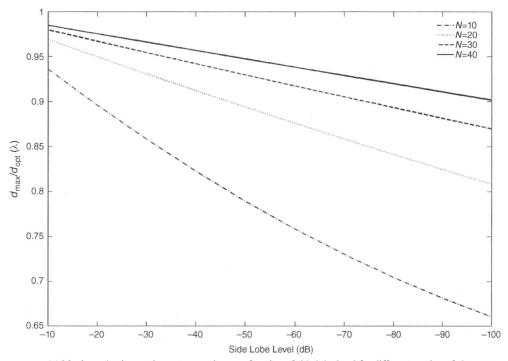

(a) Maximum/optimum element separation as a function of side lobe level for different number of elements

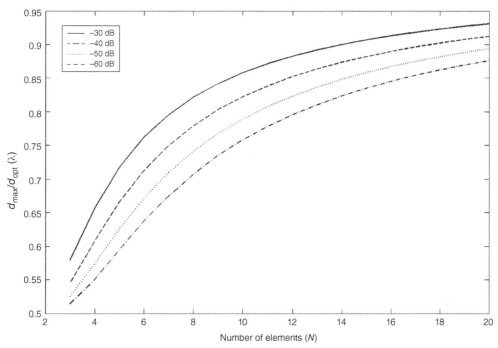

(b) Maximum/optimum element separation as a function of number of elements for different side lobe levels

Figure 6.24 Maximum/optimum element seperation for broadside Dolph-Tschebyscheff array as a function of side lobe level and number of elements.

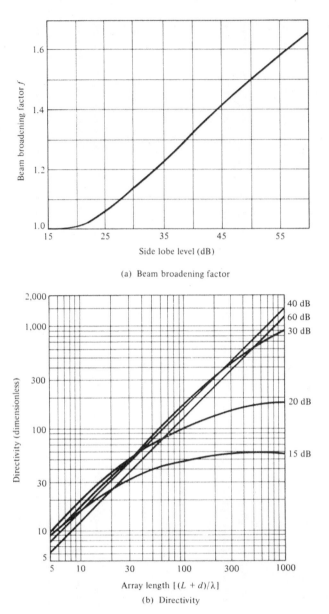

Figure 6.25 Beam broadening factor and directivity of Tschebyscheff arrays. (SOURCE: R. S. Elliott, "Beamwidth and Directivity of Large Scanning Arrays," First of Two Parts, *The Microwave Journal*, December 1963).

The beam broadening factor f can also be used to determine the directivity of large Dolph-Tschebyscheff arrays, scanned near broadside, with side lobes in the -20 to -60 dB range [4]. That is,

$$D_0 = \frac{2R_0^2}{1 + (R_0^2 - 1)f \dfrac{\lambda}{(L+d)}} \tag{6-79}$$

which is shown plotted in Figure 6.25(b) as a function of $L + d$ (in wavelengths).

From the data in Figure 6.25(b) it can be concluded that:

1. The directivity of a Dolph-Tschebyscheff array, with a given side lobe level, increases as the array size or number of elements increases.
2. For a given array length, or a given number of elements in the array, the directivity does not necessarily increase as the side lobe level decreases. As a matter of fact, a −15 dB side lobe array has smaller directivity than a −20 dB side lobe array (see Figure 6.27). This may not be the case for all other side lobe levels.

The beamwidth and the directivity of an array depend linearly, but not necessarily at the same rate, on the overall length or total number of elements of the array. Therefore, the beamwidth and directivity must be related to each other. For a uniform broadside array this relation is [4]

$$D_0 = \frac{101.5}{\Theta_d} \quad (6\text{-}80)$$

where Θ_d is the 3-dB beamwidth (in degrees). The above relation can be used as a good approximation between beamwidth and directivity for most linear broadside arrays with practical distributions (including the Dolph-Tschebyscheff array). Equation (6-80) states that for a linear broadside array the product of the 3-dB beamwidth and the directivity is approximately equal to 100. This is analogous to the product of the gain and bandwidth for electronic amplifiers.

D. Design

The design of a Dolph-Tschebyscheff array is very similar to those of other methods. Usually a certain number of parameters is specified, and the remaining are obtained following a certain procedure. In this section we will outline an alternate method that can be used, in addition to the one outlined and followed in Example 6.9, to design a Dolph-Tschebyscheff array. This method leads to the excitation coefficients more directly.

Specify

  a. The side lobe level (in dB).
  b. The number of elements.

Design Procedure

  a. Transform the side lobe level from decibels to a voltage ratio using

$$R_0(\text{Voltage Ratio}) = [R_0(VR)] = 10^{R_0(\text{dB})/20} \quad (6\text{-}81)$$

  b. Calculate P, which also represents the order of the Tschebyscheff polynomial, using

$$P = \text{number of elements} - 1$$

  c. Determine z_0 using (6-73) or

$$z_0 = \cosh\left[\frac{1}{P}\cosh^{-1}[R_0(VR)]\right] \quad (6\text{-}82)$$

  d. Calculate the excitation coefficients using (6-77a) or (6-77b).
  e. Determine the beam broadening factor using (6-78).

f. Calculate using (6-22a) the half-power beamwidth of a uniform array with the same number of elements and spacing between them.
g. Find the half-power beamwidth of the Tschebyscheff array by multiplying the half-power beamwidth of the uniform array by the beam broadening factor.
h. The maximum spacing between the elements should not exceed that of (6-76a).
i. Determine the directivity using (6-79).
j. The number of minor lobes for the three-dimensional pattern on either side of the main maximum ($0° \leq \theta \leq 90°$), using the maximum permissible spacing, is equal to $N - 1$.
k. Calculate the array factor using (6-61a) or (6-61b).

This procedure leads to the same results as any other.

Example 6.10

Calculate the half-power beamwidth and the directivity for the Dolph-Tschebyscheff array of Example 6.9 for a spacing of $\lambda/2$ between the elements.
Solution: For Example 6.9,

$$R_0 = 26 \text{ dB} \Rightarrow R_0 = 20 \quad \text{(voltage ratio)}$$

Using (6-78) or Figure 6.24(a), the beam broadening factor f is equal to

$$f = 1.079$$

According to (6-22a) or Figure 6.12, the beamwidth of a uniform broadside array with $L + d = 5\lambda$ is equal to

$$\Theta_h = 10.17°$$

Thus the beamwidth of a Dolph-Tschebyscheff array is equal to

$$\Theta_h = 10.17°f = 10.17°(1.079) = 10.97°$$

The directivity can be obtained using (6-79), and it is equal to

$$D_0 = \frac{2(20)^2}{1 + [(20)^2 - 1]\frac{1.079}{5}} = 9.18 \text{ (dimensionless)} = 9.63 \text{ dB}$$

which closely agrees with the results of Figure 6.24(b).

In designing nonuniform arrays, the amplitude distribution between the elements is used to control the side lobe level. Shown in Figure 6.26 are the excitation amplitude distributions of Dolph-Tschebyscheff arrays each with $N = 10$ elements, uniform element spacing of $d = \lambda/4$, and different side lobe levels. It is observed that as the side lobe level increases the distribution from the center element(s) toward those at the edges is smoother and monotonically decreases for all levels except that of -20 dB. For this particular design ($N = 10, d = 0.25\lambda$), the smallest side lobe level, which still maintains a monotonic amplitude distribution from the center toward the edges, is about -21.05 dB, which is also displayed in Figure 6.26. Smaller side lobe levels than -20 dB will lead to even more abrupt amplitude distribution at the edges.

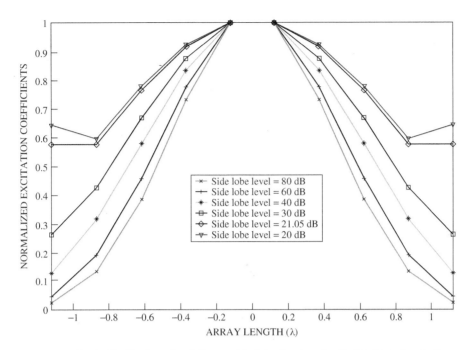

Figure 6.26 Amplitude distribution, for different side lobe levels, of a Dolph-Tschebyscheff array with $N = 10, d = \lambda/4$.

In designing nonuniform arrays, there is a compromise between side lobe level and half-power beamwidth/directivity. While the side lobe level decreases, the half-power beamwidth (HPBW) decreases and the directivity usually increases. This is demonstrated in Figure 6.27 for a 10-element Dolph-Tschebyscheff linear array with a uniform spacing of $\lambda/4$ between the elements. Similar trends can be expected for other designs.

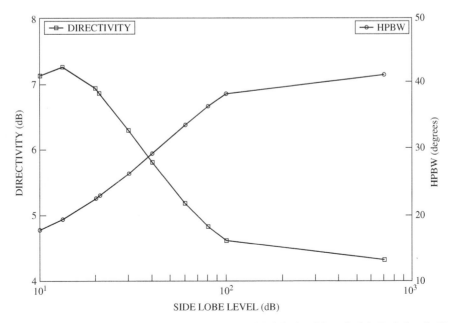

Figure 6.27 Directivity and half-power beamwidth versus side lobe level for a Dolph-Tschebyscheff array of $N = 10, d = \lambda/4$.

6.8.4 Tschebysheff Design: Scanning

In Section 6.8.3, the formulation for the broadside ($\theta_0 = 90°$) array factor design was developed. The Tschebysheff design can be extended to allow for scanning the main beam in other directions ($0° \leq \theta_0 \leq 180°$) while maintaining all the minor lobes at the same level. The approach is to utilize (6-21), which allows the phase excitation for scanning the main beam toward any angle θ_0 in the angular range of $0° \leq \theta_0 \leq 180°$. By doing this, we can rewrite (6-21) as

$$\psi = kd(\cos\theta - \cos\theta_0) \tag{6-83}$$

while (6-75) can be expressed as

$$z = z_0 \cos[kd(\cos\theta - \cos\theta_0)] = z_0 \cos\left[\frac{\pi d}{\lambda}(\cos\theta - \cos\theta_0)\right] \tag{6-84}$$

The maximum spacing $d_{\max}$, while allowing all minor lobes at the same level, can be derived in a manner similar to (6-75)–(6-76a) but utilizing (6-84) in this case. As before, the requirement not to introduce a minor lobe with a level exceeding the others is accomplished by utilizing the Tschebysheff polynomial up to, but not beyond $z = -1$ in Figure 6.21. Doing this, (6-76) can be written for $\theta = 0°$ or $180°$, while allowing $0° \leq \theta_0 \leq 180°$, as

$$-1 \geq z_0 \cos\left[\frac{\pi d_{\max}}{\lambda}(1 - \cos\theta_0)\right] \tag{6-85}$$

When solved for $d_{\max}$, equation (6-85) reduces to

$$d_{\max} \leq \frac{\lambda}{\pi(1 - \cos\theta_0)} \cos^{-1}\left(-\frac{1}{z_0}\right) \tag{6-85a}$$

To illustrate the design procedure, an example is used.

Example 6.11

Design a 10-element ($N = 10$) scanning Tschebysheff array that meets the following specifications:

- The maximum of the main beam directed toward $\theta = 60°$ ($\theta_0 = 60°$).
- All minor lobes are maintained at the same level of -26 dB over the entire angular range of $0° \leq \theta \leq 180°$.
- Use the maximum allowable spacing $d_{\max}$ between the elements.
- Plot the pattern and compare it with that of the corresponding design but with the maximum of the main beam directed toward $\theta = 90°$ (broadside design; $\theta_0 = 90°$).

Solution:

- Using (6-82), $z_0 = 1.0851$, which is the same as for the broadside design of Example 6.9.
- Using (6-85a), $d_{\max} = 0.5821\lambda$ while for the broadside design of Example 6.9 it is $d_{\max} = 0.8731\lambda$.
- The amplitude excitation coefficients are the same as those of Example 6.9.

- The normalized amplitude pattern for the scanning design displayed in Figure 6.28, where it is compared with the pattern of the corresponding broadside design. As expected, the pattern for the scanned design is asymmetrical about the $\theta = 90°$ direction while that of the broadside design is symmetrical. Also, it is evident that both designs maintain the -26 dB side lobe level over the entire $0° \leq \theta \leq 180°$ angular range.

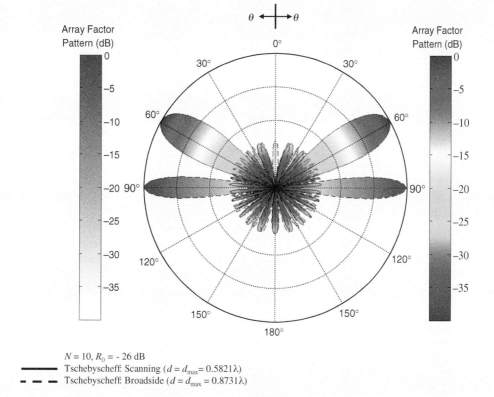

$N = 10$, $R_0 = -26$ dB
Tschebyscheff: Scanning ($d = d_{max} = 0.5821\lambda$)
— — Tschebyscheff: Broadside ($d = d_{max} = 0.8731\lambda$)

Figure 6.28 Normalized amplitude patterns of 10-element Tschebyscheff design for scanning and broadside arrays.

An interactive MATLAB and FORTRAN computer program entitled **Arrays** has been developed, and it performs the analysis for uniform and nonuniform linear arrays, and uniform planar arrays. The MATLAB version of the program also analyzes uniform circular arrays. The description of the program is provided in the corresponding READ ME file in the publisher's website for this book.

6.9 SUPERDIRECTIVITY

Antennas whose directivities are much larger than the directivity of a reference antenna of the same size are known as superdirective antennas. Thus a superdirective array is one whose directivity is larger than that of a reference array (usually a uniform array of the same length). In an array, superdirectivity is accomplished by inserting more elements within a fixed length (decreasing the spacing). Doing this leads eventually to very large magnitudes and rapid changes of phase in the excitation coefficients of the elements of the array. Thus adjacent elements have very large and oppositely directed currents. This necessitates a very precise adjustment of their values. Associated with this are increases in reactive power (relative to the radiated power) and the Q of the array.

6.9.1 Efficiency and Directivity

Because of the very large currents in the elements of superdirective arrays, the ohmic losses increases and the antenna efficiency decreases very sharply. Although practically the ohmic losses can be reduced by the use of superconductive materials, there is no easy solution for the precise adjustment of the amplitudes and phases of the array elements. High radiation efficiency superdirective arrays can be designed utilizing array functions that are insensitive to changes in element values [15].

In practice, superdirective arrays are usually referred to as *super gain*. However, super gain is a misnomer because such antennas have actual overall gains (because of very low efficiencies) less than uniform arrays of the same length. Although significant superdirectivity is very difficult and usually very impractical, a moderate amount can be accomplished. Superdirective antennas are very intriguing, and they have received much attention in the literature.

The length of the array is usually the limiting factor to the directivity of an array. Schelkunoff [16] pointed out that theoretically very high directivities can be obtained from linear end-fire arrays. Bowkamp and de Bruijn [17], however, concluded that theoretically there is no limit in the directivity of a linear antenna. More specifically, Riblet [10] showed that Dolph-Tschebyscheff arrays with element spacing less than $\lambda/2$ can yield any desired directivity. A numerical example of a Dolph-Tschebyscheff array of nine elements, $\lambda/32$ spacing between the elements (total length of $\lambda/4$), and a 1/19.5 (-25.8 dB) side lobe level was carried out by Yaru [8]. It was found that to produce a directivity of 8.5 times greater than that of a single element, the currents on the individual elements must be on the order of 14×10^6 amperes and their values adjusted to an accuracy of better than one part in 10^{11}. The maximum radiation intensity produced by such an array is equivalent to that of a single element with a current of only 19.5×10^{-3} amperes. If the elements of such an array are 1-cm diameter, of copper, $\lambda/2$ dipoles operating at 10 MHz, the efficiency of the array is less than $10^{-14}\%$.

6.9.2 Designs with Constraints

To make the designs more practical, applications that warrant some superdirectivity should incorporate constraints. One constraint is based on the sensitivity factor, and it was utilized for the design of superdirective arrays [18]. The sensitivity factor (designated as K) is an important parameter which is related to the electrical and mechanical tolerances of an antenna, and it can be used to describe its performance (especially its practical implementation). For an N-element array, such as that shown in Figure 6.5(a), it can be written as [18]

$$K = \frac{\sum_{n=1}^{N} |a_n|^2}{\left| \sum_{n=1}^{N} a_n e^{-jkr_n} \right|^2} \tag{6-86}$$

where a_n is the current excitation of the nth element, and r'_n is the distance from the nth element to the far-field observation point (*in the direction of maximum radiation*).

In practice, the excitation coefficients and the positioning of the elements, which result in a desired pattern, cannot be achieved as specified. A certain amount of error, both electrical and mechanical, will always be present. Therefore the desired pattern will not be realized exactly, as required. However, if the design is accomplished based on specified constraints, the realized pattern will approximate the desired one within a specified deviation.

To derive design constraints, the realized current excitation coefficients c_n's are related to the desired ones a_n's by

$$c_n = a_n + \alpha_n a_n = a_n(1 + \alpha_n) \tag{6-86a}$$

where $\alpha_n a_n$ represents the error in the nth excitation coefficient. The mean square value of α_n is denoted by

$$\varepsilon^2 = \langle |\alpha_n| \rangle^2 \tag{6-86b}$$

To take into account the error associated with the positioning of the elements, we introduce

$$\delta^2 = \frac{(k\sigma)^2}{3} \tag{6-86c}$$

where σ is the root-mean-square value of the element position error. Combining (6-86b) and (6-86c) reduces to

$$\Delta^2 = \delta^2 + \varepsilon^2 \tag{6-86d}$$

where Δ is a measure of the combined electrical and mechanical errors.

For uncorrelated errors [18]

$$K\Delta^2 = \frac{\text{average radiation intensity of realized pattern}}{\text{maximum radiation intensity of desired pattern}}$$

If the realized pattern is to be very close to the desired one, then

$$K\Delta^2 \ll 1 \Rightarrow \Delta \ll \frac{1}{\sqrt{K}} \tag{6-86e}$$

Equation (6-86e) can be rewritten, by introducing a safety factor S, as

$$\Delta = \frac{1}{\sqrt{SK}} \tag{6-86f}$$

S is chosen large enough so that (6-86e) is satisfied. When Δ is multiplied by 100, 100Δ represents the percent tolerance for combined electrical and mechanical errors.

The choice of the value of S depends largely on the required accuracy between the desired and realized patterns. For example, if the focus is primarily on the realization of the main beam, a value of $S = 10$ will probably be satisfactory. For side lobes of 20 dB down, S should be about 1,000. In general, an approximate value of S should be chosen according to

$$S \simeq 10 \times 10^{b/10} \tag{6-86g}$$

where b represents the pattern level (in dB down) whose shape is to be accurately realized.

The above method can be used to design, with the safety factor K constrained to a certain value, arrays with maximum directivity. Usually one first plots, for each selected excitation distribution and positioning of the elements, the directivity D of the array under investigation versus the corresponding sensitivity factor K (using 6-86) of the same array. The design usually begins with the excitation and positioning of a uniform array (i.e., uniform amplitudes, a progressive phase, and equally spaced elements). The directivity associated with it is designated as D_0 while the corresponding sensitivity factor, computed using (6-86), is equal to $K_0 = 1/N$.

As the design deviates from that of the uniform array and becomes superdirective, the values of the directivity increase monotonically with increases in K. Eventually a maximum directivity is attained (designated as D_{max}), and it corresponds to a $K = K_{max}$; beyond that point ($K > K_{max}$), the directivity decreases monotonically. The antenna designer should then select the design for which $D_0 < D < D_{max}$ and $K_0 = 1/N < K < K_{max}$.

The value of D is chosen subject to the constraint that K is a certain number whose corresponding tolerance error Δ of (6-86f), for the desired safety factor S, can be achieved practically. Tolerance errors of less than about 0.3 percent are usually not achievable in practice. In general, the designer must trade-off between directivity and sensitivity factor; larger D's (provided $D \leq D_{max}$) result in larger K's ($K \leq K_{max}$), and vice versa.

A number of constrained designs can be found in [18]. For example, an array of cylindrical monopoles above an infinite and perfectly conducting ground plane was designed for optimum directivity at $f = 30$ MHz, with a constraint on the sensitivity factor. The spacing d between the elements was maintained uniform.

For a four-element array, it was found that for $d = 0.3\lambda$ the maximum directivity was 14.5 dB and occurred at a sensitivity factor of $K = 1$. However for $d = 0.1\lambda$ the maximum directivity was up to 15.8 dB, with the corresponding sensitivity factor up to about 10^3. At $K_0 = 1/N = 1/4$, the directivities for $d = 0.3\lambda$ and 0.1λ were about 11.3 and 8 dB, respectively. When the sensitivity factor was maintained constant and equal to $K = 1$, the directivity for $d = 0.3\lambda$ was 14.5 dB and only 11.6 dB for $d = 0.1\lambda$. It should be noted that the directivity of a single monopole above an infinite ground plane is twice that of the corresponding dipole in free-space and equal to about 3.25 (or about 5.1 dB).

6.10 PLANAR ARRAY

In addition to placing elements along a line (to form a linear array), individual radiators can be positioned along a rectangular grid to form a rectangular or planar array. Planar arrays provide additional variables which can be used to control and shape the pattern of the array. Planar arrays are more versatile and can provide more symmetrical patterns with lower side lobes. In addition, they can be used to scan the main beam of the antenna toward any point in space. Applications include tracking radar, search radar, remote sensing, communications, and many others.

A planar array of slots, used in the Airborne Warning and Control System (AWACS), is shown in Figure 6.29. It utilizes rectangular waveguide sticks placed vertically, with slots on the narrow wall of the waveguides. The system has 360° view of the area, and at operating altitudes can detect targets hundreds of kilometers away. It is usually mounted at a height above the fuselage of an aircraft.

6.10.1 Array Factor

To derive the array factor for a planar array, let us refer to Figure 6.30. If M elements are initially placed along the x-axis, as shown in Figure 6.30(a), the array factor of it can be written according to (6-52) and (6-54) as

$$\text{AF} = \sum_{m=1}^{M} I_{m1} e^{j(m-1)(kd_x \sin\theta \cos\phi + \beta_x)} \qquad (6\text{-}87)$$

where I_{m1} is the excitation coefficient of each element. The spacing and progressive phase shift between the elements along the x-axis are represented, respectively, by d_x and β_x. If N such arrays are placed next to each other in the y-direction, a distance d_y apart and with a progressive phase β_y,

Figure 6.29 AWACS antenna array of waveguide slots. (PHOTOGRAPH COURTESY: Northrop Grumman Corporation).

a rectangular array will be formed as shown in Figure 6.30(b). The array factor for the entire planar array can be written as

$$\text{AF} = \sum_{n=1}^{N} I_{1n} \left[\sum_{m=1}^{M} I_{m1} e^{j(m-1)(kd_x \sin\theta \cos\phi + \beta_x)} \right] e^{j(n-1)(kd_y \sin\theta \sin\phi + \beta_y)} \tag{6-87a}$$

or

$$\text{AF} = S_{xm} S_{yn} \tag{6-88}$$

where

$$S_{xm} = \sum_{m=1}^{M} I_{m1} e^{j(m-1)(kd_x \sin\theta \cos\phi + \beta_x)} \tag{6-88a}$$

$$S_{yn} = \sum_{n=1}^{N} I_{1n} e^{j(n-1)(kd_y \sin\theta \sin\phi + \beta_y)} \tag{6-88b}$$

Equation (6-88) indicates that the pattern of a rectangular array is the product of the array factors of the arrays in the x- and y-directions.

350 ARRAYS: LINEAR, PLANAR, AND CIRCULAR

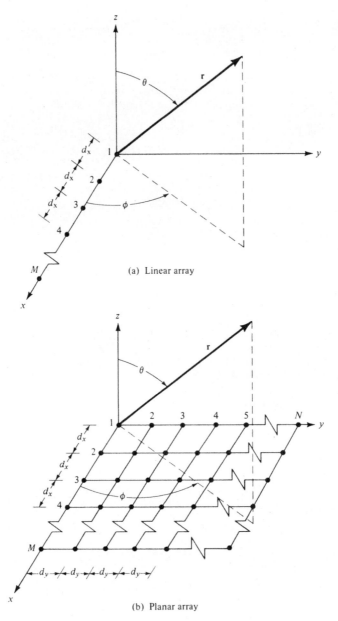

Figure 6.30 Linear and planar array geometries.

If the amplitude excitation coefficients of the elements of the array in the y-direction are proportional to those along the x, the amplitude of the (m, n)th element can be written as

$$I_{mn} = I_{m1} I_{1n} \qquad (6\text{-}89)$$

If in addition the amplitude excitation of the entire array is uniform ($I_{mn} = I_0$), (6-87a) can be expressed as

$$\mathrm{AF} = I_0 \sum_{m=1}^{M} e^{j(m-1)(kd_x \sin\theta \cos\phi + \beta_x)} \sum_{n=1}^{N} e^{j(n-1)(kd_y \sin\theta \sin\phi + \beta_y)} \qquad (6\text{-}90)$$

According to (6-6), (6-10), and (6-10c), the normalized form of (6-90) can also be written as

$$\text{AF}_n(\theta, \phi) = \left\{ \frac{1}{M} \frac{\sin\left(\frac{M}{2}\psi_x\right)}{\sin\left(\frac{\psi_x}{2}\right)} \right\} \left\{ \frac{1}{N} \frac{\sin\left(\frac{N}{2}\psi_y\right)}{\sin\left(\frac{\psi_y}{2}\right)} \right\} \tag{6-91}$$

where

$$\psi_x = kd_x \sin\theta \cos\phi + \beta_x \tag{6-91a}$$
$$\psi_y = kd_y \sin\theta \sin\phi + \beta_y \tag{6-91b}$$

When the spacing between the elements is equal or greater than $\lambda/2$, multiple maxima of equal magnitude can be formed. The principal maximum is referred to as the *major lobe* and the remaining as the *grating lobes*. A *grating lobe* is defined as "a lobe, other than the main lobe, produced by an array antenna when the inter element spacing is sufficiently large to permit the in-phase addition of radiated fields in more than one direction." To form or avoid grating lobes in a rectangular array, the same principles must be satisfied as for a linear array. To avoid grating lobes in the x-z and y-z planes, the spacing between the elements in the x- and y-directions, respectively, must be less than $\lambda/2$ ($d_x < \lambda/2$ and $d_y < \lambda/2$).

Table 6.7 lists the maximum element spacing $d_{\max}$ (for either d_x, d_y or both) for the various uniform and nonuniform arrays, including planar arrays, in order to maintain either one or two amplitude maxima.

For a rectangular array, the major lobe and grating lobes of S_{xm} and S_{yn} in (6-88a) and (6-88b) are located at

$$kd_x \sin\theta \cos\phi + \beta_x = \pm 2m\pi \quad m = 0, 1, 2, \ldots \tag{6-92a}$$
$$kd_y \sin\theta \sin\phi + \beta_y = \pm 2n\pi \quad n = 0, 1, 2, \ldots \tag{6-92b}$$

The phases β_x and β_y are independent of each other, and they can be adjusted so that the main beam of S_{xm} is not the same as that of S_{yn}. However, in most practical applications it is required that the conical main beams of S_{xm} and S_{yn} intersect and their maxima be directed toward the same direction. If it is desired to have only one main beam that is directed along $\theta = \theta_0$ and $\phi = \phi_0$, the progressive phase shift between the elements in the x- and y-directions must be equal to

$$\beta_x = -kd_x \sin\theta_0 \cos\phi_0 \tag{6-93a}$$
$$\beta_y = -kd_y \sin\theta_0 \sin\phi_0 \tag{6-93b}$$

When solved simultaneously, (6-93a) and (6-93b) can also be expressed as

$$\tan\phi_0 = \frac{\beta_y d_x}{\beta_x d_y} \tag{6-94a}$$

$$\sin^2\theta_0 = \left(\frac{\beta_x}{kd_x}\right)^2 + \left(\frac{\beta_y}{kd_y}\right)^2 \tag{6-94b}$$

The principal maximum ($m = n = 0$) and the grating lobes can be located by

$$kd_x(\sin\theta\cos\phi - \sin\theta_0\cos\phi_0) = \pm 2m\pi, \quad m = 0, 1, 2, \ldots \quad \text{(6-95a)}$$

$$kd_y(\sin\theta\sin\phi - \sin\theta_0\sin\phi_0) = \pm 2n\pi, \quad n = 0, 1, 2, \ldots \quad \text{(6-95b)}$$

or

$$\sin\theta\cos\phi - \sin\theta_0\cos\phi_0 = \pm\frac{m\lambda}{d_x}, \quad m = 0, 1, 2, \ldots \quad \text{(6-96a)}$$

$$\sin\theta\sin\phi - \sin\theta_0\sin\phi_0 = \pm\frac{n\lambda}{d_y}, \quad n = 0, 1, 2, \ldots \quad \text{(6-96b)}$$

which, when solved simultaneously, reduce to

$$\boxed{\phi = \tan^{-1}\left[\frac{\sin\theta_0\sin\phi_0 \pm n\lambda/d_y}{\sin\theta_0\cos\phi_0 \pm m\lambda/d_x}\right]} \quad \text{(6-97a)}$$

and

$$\boxed{\theta = \sin^{-1}\left[\frac{\sin\theta_0\cos\phi_0 \pm m\lambda/d_x}{\cos\phi}\right] = \sin^{-1}\left[\frac{\sin\theta_0\sin\phi_0 \pm n\lambda/d_y}{\sin\phi}\right]} \quad \text{(6-97b)}$$

In order for a true grating lobe to occur, both forms of (6-97b) must be satisfied simultaneously (i.e., lead to the same θ value).

To demonstrate the principles of planar array theory, the three-dimensional pattern of a 5×5 element array of uniform amplitude, $\beta_x = \beta_y = 0$, and $d_x = d_y = \lambda/4$, is shown in Figure 6.31. The maximum is oriented along $\theta_0 = 0°$ and only the pattern above the x-y plane is shown. An identical pattern is formed in the lower hemisphere which can be diminished by the use of a ground plane.

To examine the pattern variation as a function of the element spacing, the three-dimensional pattern of the same 5×5 element array of isotropic sources with $d_x = d_y = \lambda/2$ and $\beta_x = \beta_y = 0$ is displayed in Figure 6.32. As contrasted with Figure 6.31, the pattern of Figure 6.32 exhibits complete minor lobes in all planes. Figure 6.33 displays the corresponding two-dimensional elevation patterns with cuts at $\phi = 0°$ (x-z plane), $\phi = 90°$ (y-z plane), and $\phi = 45°$. The two principal patterns ($\phi = 0°$ and $\phi = 90°$) are identical. The patterns of Figures 6.31 and 6.32 display a fourfold symmetry.

As discussed previously, arrays possess wide versatility in their radiation characteristics. The most common characteristic of an array is its scanning mechanism. To illustrate that, the three-dimensional pattern of the same 5×5 element array, with its maximum oriented along the $\theta_0 = 30°$, $\phi_0 = 45°$, is plotted in Figure 6.34. In Figure 6.34(a) is plotted in 'cylindrical' format while in Figure 6.34(b) is plotted in 'spherical' format. The element spacing is $d_x = d_y = \lambda/2$. The maximum is found in the first quadrant of the upper hemisphere. The two-dimensional patterns are shown in Figure 6.35, and they exhibit only a twofold symmetry. The principal-plane pattern ($\phi = 0°$ or $\phi = 90°$) is normalized relative to the maximum which occurs at $\theta_0 = 30°$, $\phi_0 = 45°$. Its maximum along the principal planes ($\phi = 0°$ or $\phi = 90°$) occurs when $\theta = 21°$ and it is 17.37 dB down from the maximum at $\theta_0 = 30°$, $\phi_0 = 45°$.

To illustrate the formation of the grating lobes, when the spacing between the elements is large, the three-dimensional pattern of the 5×5 element array with $d_x = d_y = \lambda$ and $\beta_x = \beta_y = 0$ are displayed in Figure 6.36. Its corresponding two-dimensional elevation patterns at $\phi = 0°(\phi = 90°)$ and

PLANAR ARRAY 353

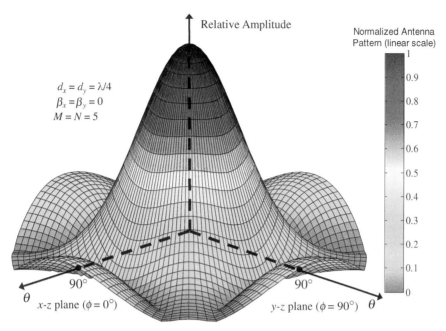

Figure 6.31 Three-dimensional antenna pattern of a planar array of isotropic elements with a spacing of $d_x = d_y = \lambda/4$, and equal amplitude and phase excitations.

$\phi = 45°$ are exhibited in Figure 6.37. Besides the maxima along $\theta = 0°$ and $\theta = 180°$, additional maxima with equal intensity, referred to as *grating lobes*, appear along the principal planes (x-z and y-z planes) when $\theta = 90°$. Further increase of the spacing to $d_x = d_y = 2\lambda$ would result in additional grating lobes.

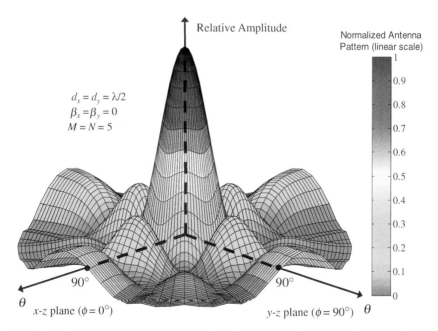

Figure 6.32 Three-dimensional antenna pattern of a planar array of isotropic elements with a spacing of $d_x = d_y = \lambda/2$, and equal amplitude and phase excitations.

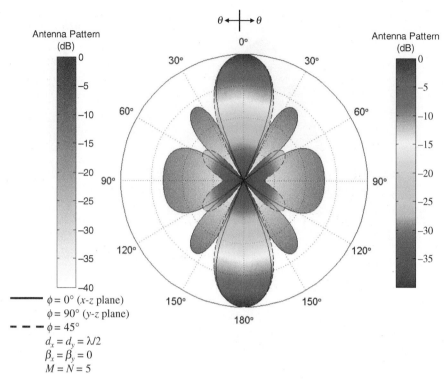

Figure 6.33 Two-dimensional antenna patterns of a planar array of isotropic elements with a spacing of $d_x = d_y = \lambda/2$, and equal amplitude and phase excitations.

The array factor of the planar array has been derived assuming that each element is an isotropic source. If the antenna is an array of *identical* elements, the total field can be obtained by applying the pattern multiplication rule of (6-5) in a manner similar as for the linear array.

When only the central element of a large planar array is excited and the others are passively terminated, it has been observed experimentally that additional nulls in the pattern of the element are developed which are not accounted for by theory which does not include coupling. The nulls were observed to become deeper and narrower [19] as the number of elements surrounding the excited element increased and approached a large array. These effects became more noticeable for arrays of open waveguides. It has been demonstrated [20] that dips at angles interior to grating lobes are formed by coupling through surface wave propagation. The coupling decays very slowly with distance, so that even distant elements from the driven elements experience substantial parasitic excitation. The angles where these large variations occur can be placed outside scan angles of interest by choosing smaller element spacing than would be used in the absence of such coupling. Because of the complexity of the problem, it will not be pursued here any further but the interested reader is referred to the published literature.

6.10.2 Beamwidth

The task of finding the beamwidth of nonuniform amplitude planar arrays is quite formidable. Instead, a very simple procedure will be outlined which can be used to compute these parameters for large arrays whose maximum is not scanned too far off broadside. The method [20] utilizes results of a uniform linear array and the beam broadening factor of the amplitude distribution.

PLANAR ARRAY 355

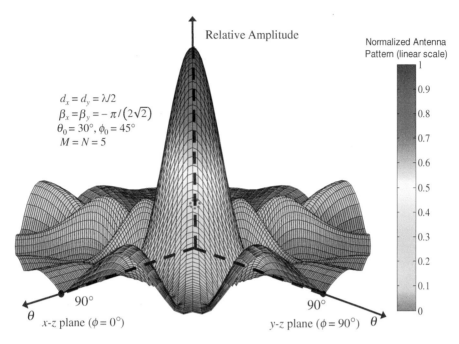

(a) in 'cylindrical' format

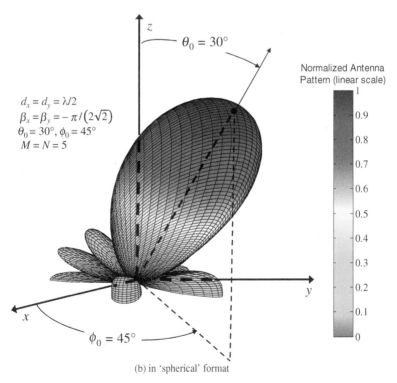

(b) in 'spherical' format

Figure 6.34 Three-dimensional antenna patterns of a planar array of isotropic elements with a spacing of $d_x = d_y = \lambda/2$, equal amplitude, and progressive phase excitation.

356 ARRAYS: LINEAR, PLANAR, AND CIRCULAR

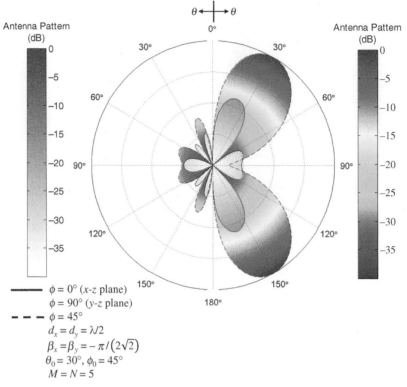

Figure 6.35 Two-dimensional antenna patterns of a planar array of isotropic elements with a spacing of $d_x = d_y = \lambda/2$, equal amplitude, and progressive phase excitation.

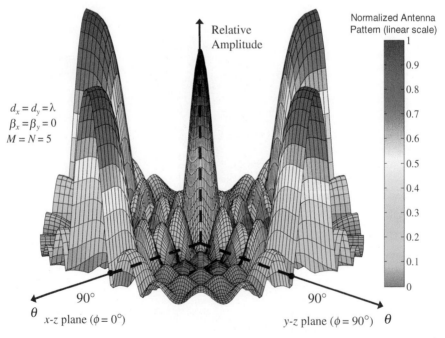

Figure 6.36 Three-dimensional antenna pattern of a planar array of isotropic elements with a spacing of $d_x = d_y = \lambda$, and equal amplitude and phase excitations.

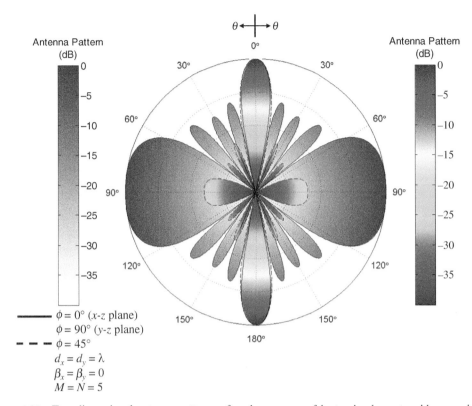

Figure 6.37 Two-dimensional antenna patterns of a planar array of isotropic elements with a spacing of $d_x = d_y = \lambda$, and equal amplitude and phase excitations.

The maximum of the conical main beam of the array is assumed to be directed toward θ_0, ϕ_0 as shown in Figure 6.38. To define a beamwidth, two planes are chosen. One is the elevation plane defined by the angle $\phi = \phi_0$ and the other is a plane that is perpendicular to it. The corresponding half-power beamwidth of each is designated, respectively, by Θ_h and Ψ_h. For example, if the array maximum is pointing along $\theta_0 = \pi/2$ and $\phi_0 = \pi/2$, Θ_h represents the beamwidth in the y-z plane and Ψ_h, the beamwidth in the x-y plane.

For a large array, with its maximum near broadside, the elevation plane half-power beamwidth Θ_h is given approximately by [21]

$$\Theta_h = \sqrt{\frac{1}{\cos^2\theta_0[\Theta_{x0}^{-2}\cos^2\phi_0 + \Theta_{y0}^{-2}\sin^2\phi_0]}} \qquad (6\text{-}98)$$

where Θ_{x0} represents the half-power beamwidth of a *broadside* linear array of M elements. Similarly, Θ_{y0} represents the half-power beamwidth of a *broadside* array of N elements.

The values of Θ_{x0} and Θ_{y0} can be obtained by using previous results. For a uniform distribution, for example, the values of Θ_{x0} and Θ_{y0} can be obtained by using, respectively, the lengths $(L_x + d_x)/\lambda$ and $(L_y + d_y)/\lambda$ and reading the values from the broadside curve of Figure 6.12. For a Tschebyscheff distribution, the values of Θ_{x0} and Θ_{y0} are obtained by multiplying each uniform distribution value by the beam broadening factor of (6-78) or Figure 6.25(a). The same concept can be

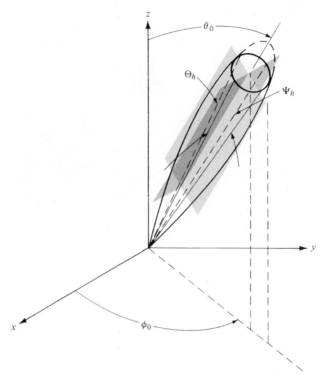

Figure 6.38 Half-power beamwidths for a conical main beam oriented toward $\theta = \theta_0$, $\phi = \phi_0$. (SOURCE: R. S. Elliott, "Beamwidth and Directivity of Large Scanning Arrays," Last of Two Parts, *The Microwave Journal*, January 1964.).

used to obtain the beamwidth of other distributions as long as their corresponding beam broadening factors are available.

For a square array ($M = N, \Theta_{x0} = \Theta_{y0}$), (6-95) reduces to

$$\Theta_h = \Theta_{x0} \sec \theta_0 = \Theta_{y0} \sec \theta_0 \tag{6-98a}$$

Equation (6-98a) indicates that for $\theta_0 > 0$ the beamwidth increases proportionally to $\sec \theta_0 = 1/\cos \theta_0$. The broadening of the beamwidth by $\sec \theta_0$, as θ_0 increases, is consistent with the reduction by $\cos \theta_0$ of the projected area of the array in the pointing direction.

The half-power beamwidth Ψ_h, in the plane that is perpendicular to the $\phi = \phi_0$ elevation, is given by [20]

$$\boxed{\Psi_h = \sqrt{\frac{1}{\Theta_{x0}^{-2} \sin^2 \phi_0 + \Theta_{y0}^{-2} \cos^2 \phi_0}}} \tag{6-99}$$

and it does not depend on θ_0. For a square array, (6-99) reduces to

$$\Psi_h = \Theta_{x0} = \Theta_{y0} \tag{6-99a}$$

The values of Θ_{x0} and Θ_{y0} are the same as in (6-98) and (6-98a).

For a planar array, it is useful to define a beam solid angle Ω_A by

$$\Omega_A = \Theta_h \Psi_h \tag{6-100}$$

as it was done in (2-23), (2-24), and (2-26a). Using (6-98) and (6-99), (6-100) can be expressed as

$$\Omega_A = \frac{\Theta_{x0} \Theta_{y0} \sec \theta_0}{\left[\sin^2 \phi_0 + \frac{\Theta_{y0}^2}{\Theta_{x0}^2} \cos^2 \phi_0\right]^{1/2} \left[\sin^2 \phi_0 + \frac{\Theta_{x0}^2}{\Theta_{y0}^2} \cos^2 \phi_0\right]^{1/2}} \tag{6-101}$$

6.10.3 Directivity

The directivity of the array factor AF(θ, ϕ) whose major beam is pointing in the $\theta = \theta_0$ and $\phi = \phi_0$ direction, can be obtained by employing the definition of (2-22) and writing it as

$$D_0 = \frac{4\pi [AF(\theta_0, \phi_0)][AF(\theta_0, \phi_0)]^*|_{max}}{\int_0^{2\pi} \int_0^{\pi} [AF(\theta, \phi)][AF(\theta, \phi)]^* \sin\theta \, d\theta \, d\phi} \tag{6-102}$$

A novel method has been introduced [22] for integrating the terms of the directivity expression for isotropic and conical patterns.

As in the case of the beamwidth, the task of evaluating (6-102) for nonuniform amplitude distribution is formidable. Instead, a very simple procedure will be outlined to compute the directivity of a planar array using data from linear arrays.

It should be pointed out that the directivity of an array with bidirectional characteristics (two-sided pattern in free space) would be half the directivity of the same array with unidirectional (one-sided pattern) elements (e.g., dipoles over ground plane).

For large planar arrays, which are nearly broadside, the directivity reduces to [21]

$$D_0 = \pi \cos \theta_0 D_x D_y \tag{6-103}$$

where D_x and D_y are the directivities of broadside linear arrays each, respectively, of length and number of elements L_x, M and L_y, N. The factor $\cos \theta_0$ accounts for the decrease of the directivity because of the decrease of the projected area of the array. Each of the values, D_x and D_y, can be obtained by using (6-79) with the appropriate beam broadening factor f. For Tschebyscheff arrays, D_x and D_y can be obtained using (6-78) or Figure 6-25(a) and (6-79). Alternatively, they can be obtained using the graphical data of Figure 6.25(b).

For most practical amplitude distributions, the directivity of (6-103) is related to the beam solid angle of the same array by

$$D_0 \simeq \frac{\pi^2}{\Omega_A (\text{rads}^2)} = \frac{32,400}{\Omega_A (\text{degrees}^2)} \tag{6-104}$$

where Ω_A is expressed in square radians or square degrees. Equation (6-104) should be compared with (2-26) or (2-27) given by Kraus.

Example 6.12

Compute the half-power beamwidths, beam solid angle, and directivity of a planar square array of 100 isotropic elements (10 × 10). Assume a Tschebyscheff distribution, $\lambda/2$ spacing between the elements, -26 dB side lobe level, and the maximum oriented along $\theta_0 = 30°, \phi_0 = 45°$.

Solution: Since in the x- and y-directions

$$L_x + d_x = L_y + d_y = 5\lambda$$

and each is equal to $L + d$ of Example 6.10, then

$$\Theta_{x0} = \Theta_{y0} = 10.97°$$

According to (6-98a)

$$\Theta_h = \Theta_{x0} \sec \theta_0 = 10.97° \sec(30°) = 12.67°$$

and (6-99a)

$$\Psi_h = \Theta_{x0} = 10.97°$$

and (6-100)

$$\Omega_A = \Theta_h \Psi_h = 12.67(10.97) = 138.96 \text{ (degrees}^2)$$

The directivity can be obtained using (6-103). Since the array is square, $D_x = D_y$, each one is equal to the directivity of Example 6.10. Thus

$$D_0 = \pi \cos(30°)(9.18)(9.18) = 229.28 \text{ (dimensionless)} = 23.60 \text{ dB}$$

Using (6-104)

$$D_0 \simeq \frac{32{,}400}{\Omega_A(\text{degrees}^2)} = \frac{32{,}400}{138.96} = 233.16 \text{ (dimensionless)} = 23.67 \text{ dB}$$

Obviously we have an excellent agreement.

An interactive MATLAB and FORTRAN computer program entitled **Arrays** has been developed, and it performs the analysis for uniform and nonuniform linear arrays, and uniform planar arrays. The MATLAB version of the program also analyzes uniform circular arrays. The description of the program is provided in the corresponding READ ME file.

6.11 DESIGN CONSIDERATIONS

Antenna arrays can be designed to control their radiation characteristics by properly selecting the phase and/or amplitude distribution between the elements. It has already been shown that a control of the phase can significantly alter the radiation pattern of an array. In fact, the principle of scanning arrays, where the maximum of the array pattern can be pointed in different directions, is based primarily on control of the phase excitation of the elements. In addition, it has been shown that a proper amplitude excitation taper between the elements can be used to control the beamwidth and side lobe level. Typically the level of the minor lobes can be controlled by tapering the distribution across the array; the smoother the taper from the center of the array toward the edges, the lower the side lobe level and the larger the half-power beamwidth, and conversely. Therefore a very smooth taper, such as that represented by a binomial distribution or others, would result in very low side lobe but larger half-power beamwidth. In contrast, an abrupt distribution, such as that of uniform illumination, exhibits the smaller half-power beamwidth but the highest side lobe level (about -13.5 dB). Therefore, if it is desired to achieve simultaneously both a very low side lobe level, as well as a small half-power beamwidth, a compromise design has to be selected. The Dolph-Tschebyscheff design of Section 6.8.3 is one such distribution. There are other designs that can be used effectively to achieve a good compromise between side lobe level and beamwidth. Two such examples are the Taylor Line-Source (Tschebyscheff-Error) and the Taylor Line-Source (One-Parameter). These are discussed in detail in Sections 7.6 and 7.7 of Chapter 7, respectively. Both of these are very similar to the Dolph-Tschebyscheff, with primarily the following exceptions.

For the Taylor Tschebyscheff-Error design, the number of minor lobes with the same level can be controlled as part of the design; the level of the remaining one is monotonically decreasing. This is in contrast to the Dolph-Tschebyscheff where all the minor lobes are of the same level. Therefore, given the same side lobe level, the half-power beamwidth of the Taylor Tschebyscheff-Error is slightly greater than that of the Dolph-Tschebyscheff. For the Taylor One-Parameter design, the level of the first minor lobe (closest to the major lobe) is controlled as part of the design; the level of the remaining ones are monotonically decreasing. Therefore, given the same side lobe level, the half-power beamwidth of the Taylor One-Parameter is slightly greater than that of the Taylor Tschebyscheff-Error, which in turn is slightly greater than that of the Dolph-Tschebyscheff design. More details of these two methods, and other ones, can be found in Chapter 7. However there are some other characteristics that can be used to design arrays.

Uniform arrays are usually preferred in design of direct-radiating active-planar arrays with a large number of elements [23]. One design consideration in satellite antennas is the beamwidth which can be used to determine the "footprint" area of the coverage. It is important to relate the beamwidth to the size of the antenna. In addition, it is also important to maximize the directivity of the antenna within the angular sector defined by the beamwidth, especially at the edge-of-the-coverage (EOC) [23]. For engineering design purposes, closed-form expressions would be desirable.

To relate the half-power beamwidth, or any other beamwidth, to the length of the array in closed form, it is easier to represent the uniform array with a large number of elements as an aperture. The normalized array factor for a rectangular array is that of (6-91). For broadside radiation ($\theta_0 = 0°$) and small spacings between the elements ($d_x \ll \lambda$ and $d_y \ll \lambda$), (6-91) can be used to approximate the pattern of a uniform illuminated aperture. In one principal plane (i.e., x-z plane; $\phi = 0°$) of Figure 6.30, (6-101) reduces for small element spacing and large number of elements to

$$(AF)_n(\theta, \phi = 0) = \frac{1}{M} \frac{\sin\left(\frac{Mkd_x}{2}\sin\theta\right)}{\sin\left(\frac{kd_x}{2}\sin\theta\right)} \simeq \frac{\sin\left(\frac{Mkd_x}{2}\sin\theta\right)}{\frac{Mkd_x}{2}\sin\theta} \simeq \frac{\sin\left(\frac{kL_x}{2}\sin\theta\right)}{\frac{kL_x}{2}\sin\theta} \quad (6\text{-}105)$$

where L_x is the length of the array in the x direction. The array factor of (6-105) can be used to represent the field in a principal plane of a uniform aperture (see Sections 12.5.1, 12.5.2 and Table 12.1). Since the maximum effective area of a uniform array is equal to its physical area $A_{em} = A_p$ [see (12-37)], the maximum directivity is equal to

$$D_0 = \frac{4\pi}{\lambda^2} A_{em} = \frac{4\pi}{\lambda^2} A_p = \frac{4\pi}{\lambda^2} L_x L_y \quad (6\text{-}106)$$

Therefore the normalized power pattern in the xz-plane, multiplied by the maximum directivity, can be written as the product of (6-105) and (6-106), and it can be expressed as

$$P(\theta, \phi = 0) = \left(\frac{4\pi L_x L_y}{\lambda^2}\right) \left[\frac{\sin\left(\frac{kL_x}{2}\sin\theta\right)}{\frac{kL_x}{2}\sin\theta}\right]^2 \tag{6-107}$$

The maximum of (6-107) occurs when $\theta = 0°$. However, for any other angle $\theta = \theta_c$, the maximum of the pattern occurs when

$$\sin\left(\frac{kL_x}{2}\sin\theta_c\right) = 1 \tag{6-108}$$

or

$$\boxed{L_x = \frac{\pi}{k\sin\theta_c} = \frac{\lambda}{2\sin\theta_c}} \tag{6-108a}$$

Therefore to maximize the directivity at the edge $\theta = \theta_c$ of a given angular sector $0° \leq \theta \leq \theta_c$, the optimum aperture dimension must be chosen according to (6-108a). Doing otherwise leads to a decrease in directivity at the edge-of-the-coverage.

For a square aperture ($L_y = L_x$) the maximum value of the normalized power pattern of (6-107) occurs when $\theta = 0°$, and it is equal to

$$P(\theta = 0°)|_{max} = 4\pi\left(\frac{L_x}{\lambda}\right)^2 \tag{6-109}$$

while that at the edge of the covering, using the optimum dimension, is

$$P(\theta = \theta_c) = 4\pi\left(\frac{L_x}{\lambda}\right)^2\left(\frac{2}{\pi}\right)^2 \tag{6-110}$$

Therefore the value of the directivity at the edge of the desired coverage ($\theta = \theta_c$), relative to its maximum value at $\theta = 0°$, is

$$\frac{P(\theta = \theta_c)}{P(\theta = 0°)} = \left(\frac{2}{\pi}\right)^2 = 0.4053 \text{ (dimensionless)} = -3.92 \text{ dB} \tag{6-111}$$

Thus the variation of the directivity within the desired coverage ($0° \leq \theta \leq \theta_c$) is less than 4 dB.

If, for example, the length of the array for a maximum half-power beamwidth coverage is changed from the optimum or chosen to be optimized at another angle, then the directivity at the edge of the half-power beamwidth is reduced from the optimum.

Similar expressions have been derived for circular apertures with uniform, parabolic and parabolic with -10 dB cosine-on-a-pedestal distributions [23], and they can be found in Chapter 12, Section 12.7.

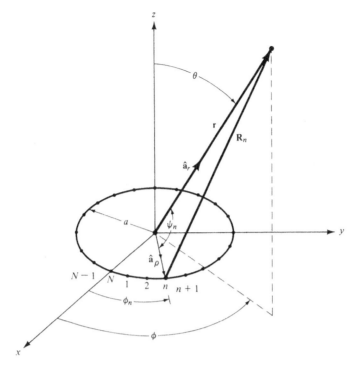

Figure 6.39 Geometry of an N-element circular array.

6.12 CIRCULAR ARRAY

The circular array, in which the elements are placed in a circular ring, is an array configuration of very practical interest. Over the years, applications span radio direction finding, air and space navigation, underground propagation, radar, sonar, and many other systems. More recently, circular arrays have been proposed for wireless communication, and in particular for smart antennas [24]. For more details see Section 16.12.

6.12.1 Array Factor

Referring to Figure 6.39, let us assume that N isotropic elements are equally spaced on the x-y plane along a circular ring of the radius a. The normalized field of the array can be written as

$$E_n(r, \theta, \phi) = \sum_{n=1}^{N} a_n \frac{e^{-jkR_n}}{R_n} \tag{6-112}$$

where R_n is the distance from the nth element to the observation point. In general

$$R_n = (r^2 + a^2 - 2ar \cos \psi)^{1/2} \tag{6-112a}$$

which for $r \gg a$ reduces to

$$R_n \simeq r - a \cos \psi_n = r - a(\hat{\mathbf{a}}_\rho \cdot \hat{\mathbf{a}}_r) = r - a \sin \theta \cos(\phi - \phi_n) \tag{6-112b}$$

where

$$\hat{a}_\rho \cdot \hat{a}_r = (\hat{a}_x \cos\phi_n + \hat{a}_y \sin\phi_n) \cdot (\hat{a}_x \sin\theta\cos\phi + \hat{a}_y \sin\theta\sin\phi + \hat{a}_z \cos\theta)$$
$$= \sin\theta\cos(\phi - \phi_n) \qquad (6\text{-}112c)$$

Thus (6-112) reduces, assuming that for amplitude variations $R_n \simeq r$, to

$$E_n(r, \theta, \phi) = \frac{e^{-jkr}}{r} \sum_{n=1}^{N} a_n e^{+jka\sin\theta\cos(\phi-\phi_n)} \qquad (6\text{-}113)$$

where

a_n = excitation coefficients (amplitude and phase) of nth element

$\phi_n = 2\pi\left(\dfrac{n}{N}\right)$ = angular position of nth element on x-y plane

In general, the excitation coefficient of the nth element can be written as

$$a_n = I_n e^{j\alpha_n} \qquad (6\text{-}114)$$

where

I_n = amplitude excitation of the nth element

α_n = phase excitation (relative to the array center) of the nth element

With (6-114), (6-113) can be expressed as

$$E_n(r, \theta, \phi) = \frac{e^{-jkr}}{r} [\text{AF}(\theta, \phi)] \qquad (6\text{-}115)$$

where

$$\text{AF}(\theta, \phi) = \sum_{n=1}^{N} I_n e^{j[ka\sin\theta\cos(\phi-\phi_n)+\alpha_n]} \qquad (6\text{-}115a)$$

Equation (6-115a) represents the array factor of a circular array of N equally spaced elements. To direct the peak of the main beam in the (θ_0, ϕ_0) direction, the phase excitation of the nth element can be chosen to be

$$\alpha_n = -ka\sin\theta_0 \cos(\phi_0 - \phi_n) \qquad (6\text{-}116)$$

Thus the array factor of (6-115a) can be written as

$$\text{AF}(\theta, \phi) = \sum_{n=1}^{N} I_n e^{jka[\sin\theta\cos(\phi-\phi_n)-\sin\theta_0\cos(\phi_0-\phi_n)]}$$

$$= \sum_{n=1}^{N} I_n e^{jka[\cos\psi - \cos\psi_0)} \qquad (6\text{-}117)$$

To reduce (6-117) to a simpler form, we define ρ_0 as

$$\rho_0 = a[(\sin\theta\cos\phi - \sin\theta_0\cos\phi_0)^2 + (\sin\theta\sin\phi - \sin\theta_0\sin\phi_0)^2]^{1/2} \tag{6-118}$$

Thus the exponential in (6-117) takes the form of

$$ka(\cos\psi - \cos\psi_0)$$
$$= \frac{k\rho_0[\sin\theta\cos(\phi - \phi_n) - \sin\theta_0\cos(\phi_0 - \phi_n)]}{[(\sin\theta\cos\phi - \sin\theta_0\cos\phi_0)^2 + (\sin\theta\sin\phi - \sin\theta_0\sin\phi_0)^2]^{1/2}} \tag{6-119}$$

which when expanded reduces to

$$ka(\cos\psi - \cos\psi_0)$$
$$= k\rho_0 \left\{ \frac{\cos\phi_n(\sin\theta\cos\phi - \sin\theta_0\cos\phi_0) + \sin\phi_n(\sin\theta\sin\phi - \sin\theta_0\sin\phi_0)}{[(\sin\theta\cos\phi - \sin\theta_0\cos\phi_0)^2 + (\sin\theta\sin\phi - \sin\theta_0\sin\phi_0)^2]^{1/2}} \right\} \tag{6-119a}$$

Defining

$$\cos\xi = \frac{\sin\theta\cos\phi - \sin\theta_0\cos\phi_0}{[(\sin\theta\cos\phi - \sin\theta_0\cos\phi_0)^2 + (\sin\theta\sin\phi - \sin\theta_0\sin\phi_0)^2]^{1/2}} \tag{6-120}$$

then

$$\sin\xi = [1 - \cos^2\xi]^{1/2}$$
$$= \frac{\sin\theta\sin\phi - \sin\theta_0\sin\phi_0}{[(\sin\theta\cos\phi - \sin\theta_0\cos\phi_0)^2 + (\sin\theta\sin\phi - \sin\theta_0\sin\phi_0)^2]^{1/2}} \tag{6-121}$$

Thus (6-119a) and (6-117) can be rewritten, respectively, as

$$ka(\cos\psi - \cos\psi_0) = k\rho_0(\cos\phi_n\cos\xi + \sin\phi_n\sin\xi) = k\rho_0\cos(\phi_n - \xi) \tag{6-122}$$

$$\mathrm{AF}(\theta,\phi) = \sum_{n=1}^{N} I_n e^{jka(\cos\psi - \cos\psi_0)} = \sum_{n=1}^{N} I_n e^{jk\rho_0\cos(\phi_n - \xi)} \tag{6-123}$$

where

$$\xi = \tan^{-1}\left[\frac{\sin\theta\sin\phi - \sin\theta_0\sin\phi_0}{\sin\theta\cos\phi - \sin\theta_0\cos\phi_0}\right] \tag{6-123a}$$

and ρ_0 is defined by (6-118).

Equations (6-123), (6-118), and (6-123a) can be used to calculate the array factor once N, I_n, a, θ_0, and ϕ_0 are specified. This is usually very time consuming, even for moderately large values of N. The three-dimensional pattern of the array factor for a 10-element uniform circular array of $ka = 10$ is shown in Figure 6.40. The corresponding two-dimensional principal-plane patterns are displayed in Figure 6.41. As the radius of the array becomes very large, the directivity of a uniform circular array approaches the value of N, where N is equal to the number of elements. An excellent discussion on circular arrays can be found in [25].

366 ARRAYS: LINEAR, PLANAR, AND CIRCULAR

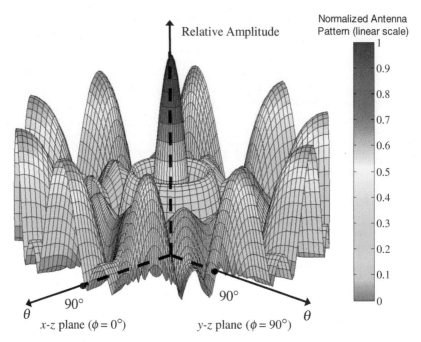

Figure 6.40 Three-dimensional amplitude pattern of the array factor for a uniform circular array of $N = 10$ elements ($C/\lambda = ka = 10$).

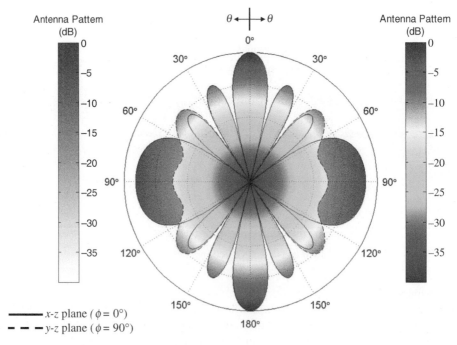

Figure 6.41 Principal-plane amplitude patterns of the array factor for a uniform circular array of $N = 10$ elements ($C/\lambda = ka = 10$).

For a uniform amplitude excitation of each element ($I_n = I_0$), (6-123) can be written as

$$\text{AF}(\theta, \phi) = NI_0 \sum_{m=-\infty}^{+\infty} J_{mN}(k\rho_0) e^{jmN(\pi/2-\xi)} \qquad (6\text{-}124)$$

where $J_p(x)$ is the Bessel function of the first kind (see Appendix V). The part of the array factor associated with the zero order Bessel function $J_0(k\rho_0)$ is called the *principal term* and the remaining terms are noted as the *residuals*. For a circular array with a large number of elements, the term $J_0(k\rho_0)$ alone can be used to approximate the two-dimensional principal-plane patterns. The remaining terms in (6-124) contribute negligibly because Bessel functions of larger orders are very small.

The MATLAB computer program **Arrays**, which is used to compute radiation characteristics of planar and circular arrays, does compute the radiation patterns of a circular array based on (6-123) and (6-124). The one based on (6-124) computes two patterns; one based on the principal term and the other based on the principal term plus two residual terms.

6.13 MULTIMEDIA

In the publisher's website for this book, the following multimedia resources are included for the review, understanding, and visualization of the material of this chapter:

a. **Java**-based **interactive questionnaire**, with answers.
b. **Java**-based **applet** for computing and displaying the radiation characteristics, directivity, and pattern of uniform and nonuniform linear arrays.
c. **Java**-based pattern **animation** of uniform and nonuniform linear arrays.
d. **Matlab** and **Fortran** computer program, designated as **Arrays**, for computing the radiation characteristics linear, planar, and circular arrays.
e. **Power Point (PPT)** viewgraphs, in multicolor.

REFERENCES

1. G. Aspley, L. Coltum and M. Rabinowitz, "Quickly Devise a Fast Diode Phase Shifter," *Microwaves*, May 1979, pp. 67–68.
2. R. E. Collin, *Foundations for Microwave Engineering* (2nd edition), McGraw-Hill, Inc., New York, 1992.
3. D. M. Pozar, *Microwave Engineering*, 3rd edition, John Wiley & Sons, Inc., New York, 2004.
4. R. S. Elliott, "Beamwidth and Directivity of Large Scanning Arrays," First of Two Parts, *Microwave Journal*, December 1963, pp. 53–60.
5. W. W. Hansen and J. R. Woodyard, "A New Principle in Directional Antenna Design," *Proc. IRE*, Vol. 26, No. 3, March 1938, pp. 333–345.
6. J. S. Stone, United States Patents No. 1,643,323 and No. 1,715,433.
7. C. L. Dolph, "A Current Distribution for Broadside Arrays Which Optimizes the Relationship Between Beamwidth and Side-Lobe Level," *Proc. IRE and Waves and Electrons*, June 1946.
8. N. Yaru, "A Note on Super-Gain Arrays," *Proc. IRE*, Vol. 39, September 1951, pp. 1081–1085.
9. L. J. Ricardi, "Radiation Properties of the Binomial Array," *Microwave J.*, Vol. 15, No. 12, December 1972, pp. 20–21.
10. H. J. Riblet, Discussion on "A Current Distribution for Broadside Arrays Which Optimizes the Relationship Between Beamwidth and Side-Lobe Level," *Proc. IRE*, May 1947, pp. 489–492.
11. D. Barbiere, "A Method for Calculating the Current Distribution of Tschebyscheff Arrays," *Proc. IRE*, January 1952, pp. 78–82.

12. R. J. Stegen, "Excitation Coefficients and Beamwidths of Tschebyscheff Arrays," *Proc. IRE*, November 1953, pp. 1671–1674.
13. C. J. Drane, Jr., "Useful Approximations for the Directivity and Beamwidth of Large Scanning Dolph-Chebyshev Arrays," *Proc. IEEE*, November 1968, pp. 1779–1787.
14. A. Safaai-Jazi, "A New Formulation for the Design of Chebyshev Arrays," *IEEE Trans. Antennas Propagat.*, Vol. 42, No. 3, March 1994, pp. 439–443.
15. M. M. Dawoud and A. P. Anderson, "Design of Superdirective Arrays with High Radiation Efficiency." *IEEE Trans. Antennas Propagat.*, Vol. AP-26. No. 6, January 1978, pp. 819–823.
16. S. A. Schelkunoff, "A Mathematical Theory of Linear Arrays," *Bell System Tech. Journal*, Vol. 22, January 1943, pp. 80–87.
17. C. J. Bowkamp and N. G. de Bruijn, "The Problem of Optimum Antenna Current Distribution," *Phillips Res. Rept.*, Vol. 1, January 1946, pp. 135–158.
18. E. H. Newman, J. H. Richmond, and C. H. Walter, "Superdirective Receiving Arrays," *IEEE Trans. Antennas Propagat.*, Vol. AP-26, No. 5, September 1978, pp. 629–635.
19. J. L. Allen, "On Surface-Wave Coupling Between Elements of Large Arrays," *IEEE Trans. Antennas Propagat.*, Vol. AP-13, No. 4, July 1965, pp. 638–639.
20. R. H. T. Bates, "Mode Theory Approach to Arrays," *IEEE Trans. Antennas Propagat.*, Vol. AP-13, No. 2, March 1965, pp. 321–322.
21. R. S. Elliott, "Beamwidth and Directivity of Large Scanning Arrays," Last of Two Parts, *Microwave Journal*, January 1964, pp. 74–82.
22. B. J. Forman, "A Novel Directivity Expression for Planar Antenna Arrays," *Radio Science*, Vol. 5, No. 7, July 1970, pp. 1077–1083.
23. K. Praba, "Optimal Aperture for Maximum Edge-of-Coverage (EOC) Directivity," *IEEE Antennas Propagat. Magazine*, Vol. 36, No. 3, June 1994, pp. 72–74.
24. P. Ioannides and C. A. Balanis, "Uniform Circular Arrays for Smart Antennas," IEEE Trans. Antennas and Propagat. Society International Symposium, Monterey, CA, June 20–25, 2004, Vol. 3, pp. 2796–2799.
25. M. T. Ma, *Theory and Application of Antenna Arrays*, Wiley, 1974, Chapter 3, pp. 191–202.

PROBLEMS

6.1. Three isotropic sources, with spacing d between them, are placed along the z-axis. The excitation coefficient of each outside element is unity while that of the center element is 2. For a spacing of $d = \lambda/4$ between the elements, find the
   (a) array factor
   (b) angles (in degrees) where the nulls of the pattern occur ($0° \leq \theta \leq 180°$)
   (c) angles (in degrees) where the maxima of the pattern occur ($0° \leq \theta \leq 180°$)
   (d) directivity using the computer program **Directivity** of Chapter 2.

6.2. Two very short dipoles ("infinitesimal") of equal length are equidistant from the origin with their centers lying on the y-axis, and oriented parallel to the z-axis. They are excited with currents of equal amplitude. The current in dipole 1 (at $y = -d/2$) leads the current in dipole 2 (at $y = +d/2$) by 90° in phase. The spacing between dipoles is one quarter wavelength. To simplify the notation, let E_0 equal the maximum magnitude of the far field at distance r due to either source alone.
   (a) Derive expressions for the following six principal-plane patterns:
   1. $|E_\theta(\theta)|$ for $\phi = 0°$
   2. $|E_\theta(\theta)|$ for $\phi = 90°$
   3. $|E_\theta(\phi)|$ for $\theta = 90°$
   4. $|E_\phi(\theta)|$ for $\phi = 0°$
   5. $|E_\phi(\theta)|$ for $\phi = 90°$
   6. $|E_\phi(\phi)|$ for $\theta = 90°$
   (b) Sketch the six field patterns.

6.3. A three-element array of isotropic sources has the phase and magnitude relationships shown. The spacing between the elements is $d = \lambda/2$.
   (a) Find the array factor.   (b) Find all the nulls.

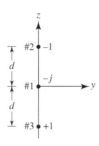

6.4. Repeat Problem 6.3 when the excitation coefficients for elements #1, #2 and #3 are, respectively, $+1, +j$ and $-j$.

6.5. Four isotropic sources are placed along the z-axis as shown. Assuming that the amplitudes of elements #1 and #2 are $+1$ and the amplitudes of elements #3 and #4 are -1 (or 180 degrees out of phase with #1 and #2), find
   (a) the array factor *in simplified form*   (b) all the nulls when $d = \lambda/2$

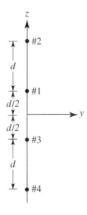

6.6. A *uniform* linear *broadside* array of *4 elements* are placed along the z-axis each a distance d apart.
   (a) Write the normalized array factor in *simplified form*.
   (b) For a separation of $d = 3\lambda/8$ between the elements, determine the:
      1. Approximate half-power beamwidth (*in degrees*).
      2. Approximate directivity (*dimensionless* and *in dB*).

6.7. Three isotropic elements of equal excitation phase are placed along the y-axis, as shown in the figure. If the relative amplitude of #1 is $+2$ and of #2 and #3 is $+1$, find a *simplified* expression for the three-dimensional unnormalized array factor.

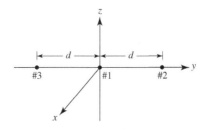

6.8. Design a *uniform broadside* linear array of N elements placed along the z-axis with a uniform spacing $d = \lambda/10$ between the elements. Determine the *closest integer number* of elements so that in the *elevation plane* the
   (a) Half-power beamwidth of the array factor is approximately $60°$.
   (b) First-null beamwidth of the array factor is $60°$.

6.9. A uniform array of 3 elements is designed so that its maximum is directed toward broadside. The spacing between the elements is $\lambda/2$. For the array factor of the antenna, determine
   (a) all the angles (*in degrees*) where the nulls will occur.
   (b) all the angles (*in degrees*) where all the maxima will occur.
   (c) the half-power beamwidth (*in degrees*).
   (d) directivity (*dimensionless* and *in dB*).
   (e) the *relative* value (*in dB*) of the magnitude of the array factor toward end-fire ($\theta_0 = 0°$) compared to that toward broadside ($\theta_0 = 90°$).

6.10. Design a two-element uniform array of isotropic sources, positioned along the z-axis a distance $\lambda/4$ apart, so that its only maximum occurs along $\theta_0 = 0°$. Assuming ordinary end-fire conditions, find the
   (a) relative phase excitation of each element   (b) array factor of the array
   (c) directivity using the computer program **Directivity** of Chapter 2. Compare it with Kraus' approximate formula.

6.11. Repeat the design of Problem 6.10 so that its only maximum occurs along $\theta = 180°$.

6.12. Design a four-element ordinary end-fire array with the elements placed along the z-axis a distance d apart. For a spacing of $d = \lambda/2$ between the elements find the
   (a) progressive phase excitation between the elements to accomplish this
   (b) angles (*in degrees*) where the nulls of the array factor occur
   (c) angles (*in degrees*) where the maximum of the array factor occur
   (d) beamwidth (*in degrees*) between the first nulls of the array factor
   (e) directivity (*in dB*) of the array factor. Verify using the computer program **Directivity** of Chapter 2.

6.13. Design an ordinary end-fire uniform linear array with only one maximum so that its directivity is 20 dB (above isotropic). The spacing between the elements is $\lambda/4$, and its length is much greater than the spacing. Determine the
   (a) number of elements
   (b) overall length of the array (in wavelengths)
   (c) approximate half-power beamwidth (in degrees)
   (d) amplitude level (compared to the maximum of the major lobe) of the first minor lobe (in dB)
   (e) progressive phase shift between the elements (in degrees).

6.14. Design a *uniform ordinary end-fire* linear array of 8 elements placed along the z-axis so that the *maximum amplitude* of the array factor is oriented in different directions. Determine the *range* (*in* λ) of the *spacing d* between the elements when the main maximum of the array factor is directed toward
   (a) $\theta_0 = 0°$ only;
   (b) $\theta_0 = 180°$ only;
   (c) $\theta_0 = 0°$ and $180°$ only;
   (d) $\theta_0 = 0°, 90°$, and $180°$ only.

6.15. It is desired to design a *uniform ordinary end-fire* array of 6 elements with a maximum toward $\theta_0 = 0°$ and $\theta_0 = 180°$, simultaneously. Determine the
  (a) smallest separation between the elements (*in* λ).
  (b) excitation progressive phase shift (*in degrees*) that should be used
  (c) *approximate* directivity of the array (*dimensionless* and in *dB*)
  (d) *relative* value (*in dB*) of the magnitude of the array factor toward broadside ($\theta_0 = 90°$) compared to that toward the maximum ($\theta_0 = 0°$ or $180°$).

6.16. Redesign the end-fire uniform array of Problem 6.13 in order to increase its directivity while maintaining the same, as in Problem 6.13, uniformity, number of elements, spacing between them, and end-fire radiation.
  (a) What different from the design of Problem 6.13 are you going to do to achieve this? Be very specific, and give values.
  (b) By how many decibels (maximum) can you increase the directivity, compared to the design of Problem 6.13?
  (c) Are you expecting the half-power beamwidth to increase or decrease? Why increase or decrease and by how much?
  (d) What antenna figure of merit will be degraded by this design? Be very specific in naming it, and why is it degraded?

6.17. Ten isotropic elements are placed along the z-axis. Design a Hansen-Woodyard end-fire array with the maximum directed toward $\theta_0 = 180°$. Find the:
  (a) desired spacing   (b) progressive phase shift β (in radians)
  (c) location of all the nulls (in degrees)   (d) first-null beamwidth (in degrees)
  (e) directivity; verify using the computer program **Directivity** of Chapter 2.

6.18. Design *a uniform ordinary* end-fire array of 6 elements placed along the z-axis and with the maximum of the array factor *directed only* along $\theta_0 = 0°$ (*end-fire only in one direction*). Determine the
  (a) maximum spacing (*in* λ) that can be used between the elements.
  (b) maximum directivity (*in dB*) of the array factor using the maximum allowable spacing.
  If the array was designed to be a *Hansen-Woodyard* end-fire array of the same number of elements, what would the following parameters be for the new array?
  (c) directivity (*in dB*).   (d) Spacing (*in* λ) between the elements.

6.19. Design a uniform linear array of elements placed long the z-axis with a *uniform spacing of* 0.2λ between the elements. It is desired that the array factor has *end-fire radiation along the* $\theta = 0°$ direction only and it achieves its maximum possible directivity as we presently know. Determine the:
  (a) Linear array design that will achieve the desired specifications. State the name of the design.
  (b) Number of elements required for the design.
  (c) Half-power beamwidth (*in degrees*) of the array factor.
  (d) Directivity (*dimensionless* and *in dB*) of the array factor.
  (e) Approximate side lobe level (*in dB*) of the array factor for this design.

6.20. It is desired to design a linear *uniform end-fire* array *that will maximize its directivity* along the $\theta = 0°$ direction only. The array elements are all placed along the z-axis with a uniform spacing d between them. The desired *maximum* directivity is *9.5545 dB*. Determine the:

(a) Array design; state its name. (b) Number of elements.
(c) *Exact* spacing between the elements (in λ).
(d) *Exact* progressive phase difference between the elements (*in degrees*).
(e) *Relative exact* phase excitation of each of the elements (*in degrees*).
Take element #1 as a reference (0 degrees).

6.21. An array of 10 isotropic elements are placed along the z-axis a distance d apart. Assuming uniform distribution, find the progressive phase (in degrees), half-power beamwidth (in degrees), first-null beamwidth (in degrees), first side lobe level maximum beamwidth (in degrees), relative side lobe level maximum (in dB), and directivity (in dB) (using equations and the computer program **Directivity** of Chapter 2, and compare) for
(a) broadside   (b) ordinary end-fire
(c) Hansen-Woodyard end-fire
arrays when the spacing between the elements is $d = \lambda/4$.

6.22. Find the beamwidth and directivity of a 10-element uniform scanning array of isotropic sources placed along the z-axis. The spacing between the elements is $\lambda/4$ and the maximum is directed at 45° from its axis.

6.23. It is desired to design a *linear adaptive array of 6 isotropic elements* placed *along the z-axis* a distance $\lambda/2$ apart. The linear adaptive array is to be able to receive a signal-of-interest (SOI), (i.e., maximum) from a direction of $\theta = 30°$ form the axis (*z-axis*) of the array. *No other specifications are placed upon the design.* Choose the *simplest type of a linear array design* that will accomplish this.
(a) State the array design; i.e., *give its name*.
(b) What should be the normalized amplitude excitation of each element?
(c) What should be the progressive phase difference (*in degrees*) between the elements.

6.24. It is desired to design a *linear adaptive array of 6 isotropic elements* placed *along the z-axis* a distance $\lambda/2$ apart. The linear adaptive array is to be able to simultaneously (all the SOI and SNOIs at the same time):
- Receive a signal-of-interest (SOI), (i.e., maximum) from a direction of $\theta_m = 90°$ (perpendicular to the z axiz)
- SNOIs (interferers; nulls) from

$$\theta_n = \begin{cases} 70.529° \text{ and } 109.471° \\ 48.189° \text{ and } 131.811° \\ 0° \text{ and } 180° \end{cases} \quad (0° \leq \theta \leq 180°)$$

Choose the *simplest type of a linear array design* that will accomplish placing the SOI and all the SNOIs all at the same time.
(a) State the array design; i.e., *give its name*.
(b) What should be the normalized amplitude excitation of each element?
(c) Verify (compute) all the angles of θ (in degrees) of the SOI and the SNOIs.
(d) What is the progressive phase difference (*in degrees*) between the elements.

6.25. Show that in order for a uniform array of N elements not to have any minor lobes, the spacing and the progressive phase shift between the elements must be
(a) $d = \lambda/N, \beta = 0$ for a broadside array.
(b) $d = \lambda/(2N), \beta = \pm kd$ for an ordinary end-fire array.

6.26. A uniform array of 20 isotropic elements is placed along the z-axis a distance $\lambda/4$ apart with a progressive phase shift of β rad. Calculate β (give the answer in radians) for the following array designs:
(a) broadside
(b) end-fire with maximum at $\theta_0 = 0°$
(c) end-fire with maximum at $\theta_0 = 180°$
(d) phased array with maximum aimed at $\theta_0 = 30°$
(e) Hansen-Woodyard with maximum at $\theta_0 = 0°$
(f) Hansen-Woodyard with maximum at $\theta_0 = 180°$

6.27. Design a 19-element uniform linear scanning array with a spacing of $\lambda/4$ between the elements.
(a) What is the progressive phase excitation between the elements so that the maximum of the array factor is 30° from the line where the elements are placed?
(b) What is the half-power beamwidth (in degrees) of the array factor of part a?
(c) What is the value (in dB) of the maximum of the first minor lobe?
Verify using the computer program **Arrays** of this chapter.

6.28. For a uniform broadside linear array of 10 isotropic elements, determine the approximate directivity (in dB) when the spacing between the elements is
(a) $\lambda/4$ (b) $\lambda/2$ (c) $3\lambda/4$ (d) λ
Compare the values with those obtained using the computer program **Arrays**.

6.29. The maximum distance d between the elements in a linear scanning array to suppress grating lobes is

$$d_{max} = \frac{\lambda}{1 + |\cos(\theta_0)|}$$

where θ_0 is the direction of the pattern maximum. What is the maximum distance between the elements, without introducing grating lobes, when the array is designed to scan to maximum angles of
(a) $\theta_0 = 30°$ (b) $\theta_0 = 45°$ (c) $\theta_0 = 60°$

6.30. An array of 4 isotropic sources is formed by placing one at the origin, and one along the x-, y-, and z-axes a distance d from the origin. Find the array factor for all space. The excitation coefficient of each element is identical.

6.31. The normalized array factor of a linear array of discrete elements placed along the z-axis can be approximated by

$$(AF)_n \approx \cos\theta \begin{cases} 0 \le \theta \le 90° \\ 0 \le \phi \le 360° \end{cases}$$

Assume now that the same physical linear array, with the same number of elements, is placed along the y-axis, *and it is radiating in the* $0 \le \theta \le 180°, 0 \le \phi \le 180°$ *angular space. For the array with its elements along the y-axis*
(a) Write the new approximate array factor.
(b) Find the half-power beamwidth (*in degrees*) in the two principal planes.

1. xy–plane 2. yz–plane
(c) Exact directivity (*dimensionless* and *in dB*) based on the approximate expression for the array factor.

6.32. Repeat Problem 6.31 for an array factor of

$$(AF)_n \approx \cos^2\theta \begin{cases} 0 \leq \theta \leq 90° \\ 0 \leq \phi \leq 360° \end{cases}$$

6.33. Design a linear array of isotropic elements placed along the z-axis such that the nulls of the array factor occur at $\theta = 0°$ and $\theta = 45°$. Assume that the elements are spaced a distance of $\lambda/4$ apart and that $\beta = 0°$.
(a) Sketch and label the visible region on the unit circle
(b) Find the required number of elements
(c) Determine their excitation coefficients

6.34. Design a linear array of isotropic elements placed along the z-axis such that the zeros of the array factor occur at $\theta = 10°, 70°$, and $110°$. Assume that the elements are spaced a distance of $\lambda/4$ apart and that $\beta = 45°$.
(a) Sketch and label the visible region on the unit circle
(b) Find the required number of elements
(c) Determine their excitation coefficients

6.35. Repeat Problem 6.34 so that the nulls occur at $\theta = 0°, 50°$ and $100°$. Assume a spacing of $\lambda/5$ and $\beta = 0°$ between the elements.

6.36. Design a three-element binomial array of isotropic elements positioned along the z-axis a distance d apart. Find the
(a) normalized excitation coefficients   (b) array factor
(c) nulls of the array factor for $d = \lambda$ (d) maxima of the array factor for $d = \lambda$

6.37. Show that a three-element binomial array with a spacing of $d \leq \lambda/2$ between the elements does not have a side lobe.

6.38. Four isotropic sources are placed symmetrically along the z-axis a distance d apart. Design a binomial array. Find the
(a) normalized excitation coefficients   (b) array factor
(c) angles (in degrees) where the array factor nulls occur when $d = 3\lambda/4$

6.39. Five isotropic sources are placed symmetrically along the z-axis, each separated from its neighbor by an electrical distance $kd = 5\pi/4$. For a binomial array, find the
(a) excitation coefficients   (b) array factor
(c) normalized power pattern
(d) angles (in degrees) where the nulls (if any) occur
Verify parts of the problem using the computer program **Arrays**.

6.40. Design a four-element binomial array of $\lambda/2$ dipoles, placed symmetrically along the x-axis a distance d apart. The length of each dipole is parallel to the z-axis.
(a) Find the normalized excitation coefficients.
(b) Write the array factor for all space.
(c) Write expressions for the E-fields for all space.

6.41. Repeat the design of Problem 6.40 when the $\lambda/2$ dipoles are placed along the y-axis.

6.42. Design a broadside binomial array of six elements placed along the z-axis separated by a distance $d = \lambda/2$.
  (a) Find the amplitude excitation coefficients (a_n's).
  (b) What is the progressive phase excitation between the elements?
  (c) Write the array factor.
  (d) Now assume that the elements are $\lambda/4$ dipoles oriented in the z-direction. Write the expression for the electric field *vector* in the far field.

  Verify parts of the problem using the computer program **Arrays**.

6.43. Repeat Problem 6.42 for an array of seven elements.

6.44. Five isotropic elements, with spacing d between them, are placed along the z-axis. For a binomial amplitude distribution,
  (a) write the array factor in its most simplified form
  (b) compute the directivity (in dB) and compare using the computer program **Arrays** of this chapter ($d = \lambda/2$)
  (c) find the nulls of the array when $d = \lambda (0° \leq \theta \leq 180°)$

6.45. A typical base station that you see as you travel around the city consists of an equilateral/triangular array of dipoles. Assume that each side of the equilateral triangle consists of three dipoles. Let us assume that each of the dipoles, at a frequency of 1.9 GHz, is $\lambda/2$ in length. The dipoles are placed along the y-axis, are separated by a distance of $\lambda/2$, and are pointing along the z-axis. The center element is placed at the origin and the other two are placed one on each side of the center element. Assuming that the elements are fed in phase and are designed for a broadside binomial amplitude distribution:

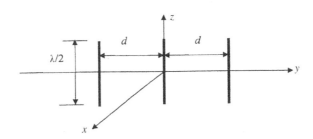

  (a) Determine the *total* normalized amplitude excitation coefficient of each element.
  (b) Write an expression for the normalized array factor.
  (c) Determine the maximum directivity (*dimensionless* and *in dB*) of the *array factor* when $d = \lambda/2$.
  (d) State the directivity (*dimensionless* and *in dB*) of each individual element.
  (e) Making an *educated guess*, what would you expect the *very maximum directivity (dimensionless* and *in dB)* of the entire 3-element array, *which takes into account the element pattern and the array factor*, could not exceed?

6.46. A *nonuniform* linear array has *3 elements* placed symmetrically along the z-axis and spaced $d = \lambda/4$ apart, and *all are fed with the same phase*. However, the total amplitude excitation coefficients of the elements are as follows:
  • 2 for the center element   • Unity for each of the edge elements

For the array factor of the array, determine the:
(a) Angle θ (*in degrees*) where the maximum of the main lobe occurs.
(b) Angles θ (*in degrees*) where the 2 *half-power points* of the main lobe occur.
(c) Half-power beamwidth (*in degrees*) of the main lobe.
(d) Approximate maximum directivity (*dimensionless* and *in dB*).

6.47. Design a *five-element binomial array* with elements placed *along the z-axis*.
(a) Derive the excitation coefficients.
(b) Write a *simplified expression* for the array factor.
(c) For a spacing of $d = 3\lambda/4$, determine all the angles θ (*in degrees*) where the array factor possesses nulls.
(d) For a spacing of $d = 3\lambda/4$, determine all the angles θ (*in degrees*) where the array factor possesses *main* maxima.

6.48. Design a five-element binomial array with the elements placed along the z-axis. It is desired that the amplitude pattern of the array factor has nulls *only* at $\theta = 0°$ and $180°$, *one major lobe* with the maximum at $\theta_0 = 90°$, and *no minor lobes*. To meet the requirements of this array, determine the:
(a) Spacing between the elements (*in λ*).
(b) Total amplitude excitation coefficient of each element.
(c) Directivity (*dimensionless* and *in dB*).
(d) Half-power beamwidth (*in degrees*).
(e) Verify the design using the computer program **Arrays**.

6.49. It is desired to design a binomial array with a uniform spacing between the elements of $\lambda/2$ placed along the z-axis, and with an elevation half-power beamwidth for its array factor of 15.18 degrees. To accomplish this, determine the:
(a) Number of elements.
(b) Directivity (*dimensionless* and *in dB*).
(c) Sidelobe level of the array factor (in *dB*).

6.50. It is to design a *linear nonuniform broadside* array whose directivity is 6 dB and whose pattern has nulls only at $\theta = 0°$ and $180°$. Assume the elements are placed along the z-axis.
(a) From the nonuniform designs you have been exposed to, select one that will accomplish this. *State its name or distribution.*
(b) Determine the closest integer number of elements.
(c) State the total amplitude excitation of each element.
(d) What is the spacing between the elements (*in λ*).
(e) Determine the half-power beamwidth (*in degrees*).

6.51. It is desired to design two separate *broadside square planar arrays* (with 9x9 elements; a total of 81 elements) with the elements placed along the *xy*-plane; the spacing between the elements in both directions is $\lambda/2$. One of the designs is a uniform design and the other one is a binomial design. *Assume that both arrays only radiate in the upper hemisphere (above the xy-plane).*
(a) For the *Uniform* planar array design: Determine, for the *entire planar array*, the:
 • Half-power beamwidth in the *elevation plane* (*in degrees*).
 • Half-power beamwidth in a *plane perpendicular to the elevation plane* (*in degrees*).
 • *Maximum directivity* (*dimensionless* and *in dB*).

(b) For the *Binomial* planar array design: Determine, for the *entire planar array*, the:
  • Half-power beamwidth in the *elevation plane (in degrees)*.
  • Half-power beamwidth in a *plane perpendicular to the elevation plane (in degrees)*.
  • *Maximum directivity (dimensionless* and *in dB)*.
(c) Is the half-power beamwidth of the binomial array design, compared to that of the uniform design:
  • Smaller or larger? Why? Justify the answer. Is this what you expected?
(d) Is the maximum directivity of the binomial array design, compared to that of the uniform design:
  • Smaller or larger? Why? Justify the answer. Is this what you expected?

6.52. Design a nonuniform binomial broadside linear array of N elements, with a uniform spacing d between the elements, which is desired to *have no minor lobes*.
  (a) Determine the *largest spacing d (in λ)*.
  (b) Find the *closest integer number* of elements so that the half-power beamwidth is 18°.
  (c) Compare the directivities (in dB) of the *uniform* and *binomial* broadside arrays with the same number of elements and spacing between them. *Which is smaller and by how many dB?*

6.53. It is desired to design a broadside uniform linear array with the elements placed along the x-axis. The design requires a total length (edge-to-edge) of the array is 4λ and the spacing between the elements must be $\lambda/2$. Determine the:
  (a) Number of elements of the array.
  (b) Half-power beamwidth (*in degrees*) for the uniform broadside array.
  (c) Maximum directivity (*dimensionless* and *in dB*) for the uniform broadside array.
  (d) Side lobe level (*in dB*) of the uniform broadside array.
  (e) Total excitation coefficients if the linear array is a binomial broadside design (normalize the coefficients so that those of the edge elements are unity).
  (f) Half-power beamwidth (*in degrees*) if the linear array is a binomial broadside design.
  (g) Maximum directivity (*dimensionless* and *in dB*) if the array is a binomial broadside design.
  (h) Sidelobe level (*in dB*) if the linear array is a binomial broadside design.

6.54. Design a *nonuniform broadside binomial* array of *three* elements that its pattern has one major lobe and no minor lobes in $0° \leq \theta \leq 180°$. It is also required that the pattern exhibits nulls toward $\theta = 0°$ and $\theta = 180°$. Determine the:
  (a) Normalized total amplitude coefficient of each of the three elements (*the ones at the edges to be of unity amplitude*).
  (b) Half-power beamwidth (*in degrees*).
  (c) Directivity (*dimensionless* and *in dB*).

6.55. The normalized array factor of a nonuniform linear broadside array of N elements, with a uniform spacing of d between them, is given by

$$\text{AF}(\theta) = 2\cos^2\left(\frac{kd}{2}\cos\theta\right) = 2\cos^2(u), \quad u = \frac{\pi d}{\lambda}\cos\theta$$

  (a) Determine the number of elements of the array with this specific array factor.
  (b) What is the normalized (*relative to the ones at the edge*) *total* excitation coefficient of each of the elements?

(c) What is the *specific name* of this nonuniform classic broadside array design?
(d) For $d = 3\lambda/4$, determine all the angles ($0° \leq \theta \leq 180°$) where the array factor exhibits nulls.

6.56. The array factor of a *nonuniform* linear array, with the elements placed along the z-axis and with a uniform spacing d among the elements, is given by

$$\text{AF} = 1 + \cos\left(2\pi\frac{d}{\lambda}\cos\theta\right)$$

Determine the:
(a) *Total* number of elements of the array.
(b) *Total* excitation coefficient of each element. Identify the *total value* of the coefficient with the appropriate element.
(c) All the nulls (*in degrees*; $0° \leq \theta \leq 180°$) of the array factor when the spacing between the elements is $d = 3\lambda/4$.
(d) All the *major* maxima (*in degrees*; $0° \leq \theta \leq 180°$) of the array factor when the spacing between the elements is $d = 3\lambda/4$.

6.57. A 4-element array of isotropic elements are placed along the z-axis, symmetrically about the origin. The separation between all of the adjoining elements is d. The *normalized* amplitude excitation of the *two inner most elements is 3* while that of the *two outer most elements (at each of the two outer edges of the array) is unity*. All of the elements are excited by the same phase ($\beta = 0$).
(a) Write the array factor *in simplified form*. You may want to refer to (6-66).
(b) Assuming a spacing of $d = \lambda/2$ between the elements, determine *analytically*:
- All the nulls ($0° \leq \theta \leq 180°$) *in degrees*.
- All the main maxima ($0° \leq \theta \leq 180°$) *in degrees*.
- Half-power beamwidth (*in degrees*).
- Directivity (*dimensionless* and *in dB*).

6.58. Design a *four-element nonuniform linear broadside array* with a uniform spacing between the elements. It is desired that the array factor pattern *has no minor lobes* and the *maximum permissible spacing* between the elements is selected. *The selected spacing should be greater than $\lambda/4$*.
(a) Select the appropriate *nonuniform* array design. State its name.
(b) Determine the maximum permissible spacing (*in λ*) between the elements.
(c) Determine the normalized amplitude excitation coefficients.
(d) Write the array factor *assuming the elements are placed along the x-axis*.
(e) Write the array factor *assuming the elements are placed along the y-axis*.
(f) Half-power beamwidth (*in degrees*) of the array factor.
(g) Directivity (*dimensionless* and *in dB*) of the array factor.

6.59. The normalized Array Factor of a *broadside* array is given by

$$(\text{AF})_n = 3\cos\left(\frac{\pi d}{\lambda}\sin\theta\cos\phi\right) + \cos\left(3\frac{\pi d}{\lambda}\sin\theta\cos\phi\right)$$

which can also be written, using trigonometric identities, as

$$(\text{AF})_n = 4\left[\cos\left(\frac{\pi d}{\lambda}\sin\theta\cos\phi\right)\right]^3 = 4\cos^3\left(\frac{\pi d}{\lambda}\sin\theta\cos\phi\right)$$

where d is the spacing between the elements, and θ and ϕ are the standard spherical coordinate angles.

(a) What type of an array we have (linear, square, rectangular, circular, other)?

(b) Along what axis (x, y or z) are the elements positioned?

(c) What is the amplitude distribution of the elements along the respective axis (uniform, binomial, Tschebysheff, cosine, cosine squared, other)?

(d) How many array elements are there along the respective axis?

(e) Assuming $d = \lambda/2$, determine the:
- *Approximate* half-power beam width (*in degrees*) of the array, assuming isotropic elements.
- *Exact* directivity (*dimensionless* and *in dB*) of the array, assuming isotropic elements (*dimensionless* and *in dB*).
- Approximate directivity (*dimensionless* and *in dB*) of the array, *using another/alternate formula/expression which uses the half-power beamwidth information. Stale which one you are using and why*. Be specific.
- Silelobe level (*in dB*).

6.60. Show that the:

(a) Maximum spacing d_{max} of (6-76a) reduces to the optimum spacing d_{opt} of (6-76b).

(b) Optimum spacing d_{opt} of (6-76b) reduces to the maximum spacing d_{max} of (6-76a).

6.61. Repeat the design of Problem 6.36 for a Dolph-Tschebyscheff array with a side lobe level of -20 dB.

6.62. Design a three-element, -40 dB side lobe level Dolph-Tschebyscheff array of isotropic elements placed symmetrically along the z-axis. Find the

(a) amplitude excitation coefficients (b) array factor

(c) angles where the nulls occur for $d = 3\lambda/4 (0° \leq \theta \leq 180°)$

(d) directivity for $d = 3\lambda/4$ (e) half-power beamwidth for $d = 3\lambda/4$

6.63. Design a four-element, -40 dB side lobe level Dolph-Tschebyscheff array of isotropic elements placed symmetrically about the z-axis. Find the

(a) amplitude excitation coefficients (b) array factor

(c) angles where the nulls occur for $d = 3\lambda/4$.

Are all of the minor lobes of the same level? Why not? What needs to be changed to make them of the same level?

Verify parts of the problem using the computer program **Arrays**.

6.64. Repeat the design of Problem 6.63 for a five-element, -20 dB Dolph-Tschebyscheff array.

6.65. Repeat the design of Problem 6.63 for a six-element, -20 dB Dolph-Tschebyscheff array.

6.66. Repeat the design of Problem 6.40 for a Dolph-Tschebyscheff distribution of -40 dB side lobe level and $\lambda/4$ spacing between the elements. In addition, find the

(a) directivity of the entire array

(b) half-power beamwidths of the entire array in the x-y and y-z planes

6.67. Repeat the design of Problem 6.41 for a Dolph-Tschebyscheff distribution of -40 dB side lobe level and $\lambda/4$ spacing between the elements. In addition, find the

(a) directivity of the entire array

(b) half-power beamwidths of the entire array in the x-y and y-z planes

6.68. Design a five-element, −40 dB side lobe level Dolph-Tschebyscheff array of isotropic elements. The elements are placed along the x-axis with a spacing of $\lambda/4$ between them. Determine the:
(a) normalized amplitude coefficients
(b) array factor
(c) directivity
(d) half-power beamwidth

6.69. The total length of a discrete-element array is 4λ. For a −30 dB side lobe level Dolph-Tschebyscheff design and a spacing of $\lambda/2$ between the elements along the z-axis, find the:
(a) number of elements
(b) excitation coefficients
(c) directivity
(d) half-power beamwidth

6.70. Design a broadside three-element, −26 dB side lobe level Dolph-Tschebyscheff array of isotopic sources placed along the z-axis. For this design, find the:
(a) normalized excitation coefficients
(b) array factor
(c) nulls of the array factor when $d = \lambda/2$ (in degrees)
(d) maxima of the array factor when $d = \lambda/2$ (in degrees)
(e) HPBW beamwidth (in degrees) of the array factor when $d = \lambda/2$
(f) directivity (in dB) of the array factor when $d = \lambda/2$
(g) verify the design using the computer program **Arrays**.

6.71. Design a broadside uniform array, with its elements placed along the z axis, in order the directivity of the array factor is 33 dB (above isotropic). Assuming the spacing between the elements is $\lambda/16$, and it is very small compared to the overall length of the array, determine the:
(a) closest number of integer elements to achieve this.
(b) overall length of the array (in wavelengths).
(c) half-power beamwidth (in degrees).
(d) amplitude level (in dB) of the maximum of the first minor lobe compared to the maximum of the major lobe.

6.72. The design of Problem 6.71 needs to be changed to a nonuniform Dolph-Tschebyscheff in order to lower the side lobe amplitude level to −30 dB, while maintaining the same number of elements and spacing. For the new nonuniform design, what is the:
(a) half-power beamwidth (in degrees).
(b) directivity (in dB).

6.73. Design a Dolph-Tschebyscheff linear array of 6 elements with uniform spacing between them. The array factor must meet the following specifications:
1. −40 dB side lobe level.
2. Minor lobes from $0° \leq \theta \leq 90°$ all of the same level.
3. Largest allowable spacing between the elements (in wavelengths) and still meet above specifications.

Determine:
(a) excitation coefficients, normalized so that the ones of the edge elements is unity.
(b) maximum allowable spacing (in wavelengths) between the elements and still meet specifications.
(c) plot (in 1° increments) the normalized (max = 0 dB) array factor (in dB). Check to see that the array factor meets the specifications. If not, find out what is wrong with it.

Verify parts of the problem using the computer program **Arrays**.

6.74. Design the array factor of a three-element Dolph-Tschebyscheff broadside array with a side lobe level of −40 dB. Determine the

(a) *normalized* amplitude coefficients.
(b) *maximum allowable* spacing (*in* λ) to maintain the same sidelobe level for all minor lobes.
(c) approximate half-power beamwidth (*in degrees*) *using the spacing from part b*.
(d) approximate directivity (*in dB*) *using the spacing from part b*.

6.75. Design a *Dolph-Tschebyscheff broadside* array of 5 elements with a −30 dB sidelobe level.

(a) Determine the normalized amplitude excitation coefficients. *Make the ones at the edges of the array unity*.
(b) Determine the maximum spacing between the elements (*in* λ) *so that all sidelobes are maintained at the same level of* −30 dB.
(c) For the spacing of Part (b), determine the half-power beamwidth (*in degrees*). *Compare it to that of a uniform array of the same number of elements and spacing*.
(d) For the spacing of Part b, determine the directivity (*dimensionless and in dB*).

6.76. It is desired to design a Dolph-Tschebyscheff nonuniform linear broadside array. The desired array should have 20 elements with a uniform spacing between them. The required sidelobe level −40 dB down from the maximum. Determine the:

(a) Maximum uniform spacing that can be used between the elements and still maintain a constant sidelobe level of −40 dB for all minor lobes.
(b) Half-power beamwidth (in *degrees*) of a *uniform* broadside linear array of the same number of elements and spacing as the Dolph-Tschebyscheff array. Assume $d = \lambda/2$.
(c) Half-power beamwidth (in *degrees*) of the Dolph-Tschebyscheff array with $d = \lambda/2$.
(d) Directivity of the Dolph-Tschebyscheff array of $d = \lambda/2$ (*dimensionless and in dB*).
(e) Directivity of the uniform broadside array of $d = \lambda/2$ (*dimensionless and in dB*).

6.77. A *nonuniform* Dolph-Tschebyscheff linear *broadside* array of 4 *elements* are placed along the z-axis each a distance d apart. For a separation of $d = 3\lambda/8$ between the elements and a −27.959 dB sidelobe level, determine *analytically (not graphically)* the:

(a) Maximum element separation d_{max} (*in* λ) that can be used to maintain the constant −27.959 dB sidelobe level?
(b) Approximate half-power beamwidth (*in degrees*) using $d = 3\lambda/8$.
(c) Approximate directivity (*dimensionless and in dB*) using $d = 3\lambda/8$.

6.78. Design a broadside linear Dolph-Tschebyscheff array with the elements placed along the z-axis so that its array factor pattern, *using the largest possible spacing between the elements while still maintaining the same sidelobe level*, has 9 minor lobes on each side of the three-dimensional pattern. The desired sidelobe level is −60 dB. To accomplish this, determine the:

(a) Order of the Tschebyscheff polynomial.
(b) Number of elements.
(c) Maximum spacing between the elements (in λ).

6.79. It is desired to design a *broadside Tschebyscheff* linear array of $N = 10$ elements, placed along the z-axis, with a uniform spacing of $d = \lambda/2$ between the elements and with a uniform sidelobe level of −26 dB from the main maximum. Determine the:

(a) Progressive phase (*in degrees*) excitation between the elements.
(b) Number of *complete minor lobes* in the elevation plane *between* $0° \leq \theta \leq 90°$.
(c) *Maximum* spacing between the elements to *maintain the same sidelobe level over all the minor lobes*.

6.80. It is desired to design a -25 dB broadside Dolph-Tschebyscheff array of 6 elements with a spacing of $d = \lambda/4$ between the elements. Determine the:
(a) HPBW (*in degrees*).
(b) Maximum directivity (*dimensionless* and *in dB*).

6.81. A Dolph-Tschebyscheff broadside array of 6 elements, 50-dB sidelobe level, and with a spacing of $\lambda/2$ between the elements is designed to operate at 9 GHz. For the array factor of the antenna, determine the:
(a) Half-power beamwidth (*in degrees*). How much narrower or wider (*in degrees*) is this half-power beamwidth compared to that of a uniform array of the same number of elements and spacing? *Justify your answer. Do you expect it? Why?*
(b) Directivity of array factor (*in dB*). How much smaller or larger (*in dB*) is this directivity (*in dB*) compared to that of a uniform array of the same number of elements and spacing? *Justify you answer. Do you expect it? Why?*

6.82. Dolph-Tschebyscheff designs are practical because you can perform the design to meet the required specifications (like the selection of the number of elements, spacing between the elements, and to maintain all the minor lobes at a required uniform/constant level). Such a design leads to amplitude coefficients, which for modest sidelobe levels usually do not vary more than 4:1, and can be achieved with efficient feed designs. This is usually not the case for binomial designs where the amplitude excitation coefficients usually vary more than 100:1, especially when a large number of elements (equal or greater than 10) are required. To demonstrate this, design two separate *broadside linear arrays*, with each design of 9 elements placed along the z-axis. The spacing between the elements is $\lambda/2$ and a sidelobe level of -30 dB for the Dolph-Tschebyscheff design.
(a) For the *binomial broadside linear array* design: Determine the:
- Sidelobe level (*in dB*).
- Half-power beamwidth (*in degrees*).
- Maximum directivity (*dimensionless* and *in dB*).
(b) For the *Dolph-Tschebyscheff broadside linear array* design: Determine the:
- Half-power beamwidth (*in degrees*).
- Maximum directivity (*dimensionless* and *in dB*).
(c) Is the half-power beamwidth of the binomial broadside linear array design, compared to that of the Dolph-Tschebysceff broadside linear array design:
- Smaller or larger? Why? Justify the answer. Is this what you expected?
(d) Is the maximum directivity of the binomial broadside linear array design, compared to that of the Dolph-Tschebyscheff broadside linea design:
- Smaller or larger? Why? Justify the answer. Is this what you expected?

6.83. In the design of antenna arrays, *with a spacing of $d \leq \lambda/2$*, there is a choice between *uniform, binomial, cosine-squared*, and *Dolph-Tschebyscheff* (of -25 dB sidelobe level) distributions. If it is desired to:
(a) Select the design distributions with the *smallest half-power beamwidths*, place the antennas in order of *smaller-to-larger half-power beamwidths*.
(b) Select the design distributions with the *lowest sidelobe level*, place the antennas in order of *lower-to-higher sidelobe level*.

6.84. It is desired to design a planar *square* array with uniform illumination so that its *approximate* half-power beamwidth is $1°$, when the main beam maximum, is pointed in some direction θ_0. Determine the total dimension (*in λ*) on each side of the square array when its maximum is directed toward (*z-axis is perpendicular to the plane of the array*):

(a) Broadside ($\theta_0 = 0°$);   (b) $\theta_0 = 60°$ from broadside.

Treat the planar array in each plane as a source with a continuous distribution (like an aperture), and assume it is large in terms of a wavelength.

6.85. In high-performance radar arrays low-sidelobes are very desirable. In a particular application it is desired to design a broadside linear array which maintains all the sidelobes at the same level of −30 dB. The number of elements must be 3 and the spacing between them must be $\lambda/4$.
(a) State the design that will meet the specifications.
(b) What are the amplitude excitations of the elements?
(c) What is the half-power beamwidth (in degrees) of the main lobe?
(d) What is the directivity (in dB) of the array?

6.86. Design a nonuniform amplitude broadside linear array of 5 elements. The total length of the array is 2λ. To meet the sidelobe and half-power beamwidth specifications, the amplitude excitations of the elements must be that of a cosine-on-a-pedestal distribution represented by

$$\text{Amplitude distribution} = 1 + \cos(\pi x_n/L)$$

where x_n is the position of the nth element (in terms of L) measured from the center of the array. Determine the amplitude excitation coefficients a_n's of the five elements. Assume uniform spacing between the elements and the end elements are located at the edges of the array length.

6.87. It is desired to design a uniform square scanning array whose elevation half-power beamwidth is 2°. Determine the minimum dimensions of the array when the scan maximum angle is
(a) $\theta_0 = 30°$ (b) $\theta_0 = 45°$ (c) $\theta_0 = 60°$

6.88. Determine the azimuthal and elevation angles of the grating lobes for a 10×10 element uniform planar array when the spacing between the elements is λ. The maximum of the main beam is directed toward $\theta_0 = 60°, \phi_0 = 90°$ and the array is located on the x-y plane.

6.89. Four isotropic elements are placed on the *xy*-plane, *and all four are excited with the same amplitude; uniform array*. Two of the elements are placed along the *x*-axis, symmetrically about the *z*-axis, with a total separation of d_x between them, and a phase excitation of $+\beta_x$ between them. The other two elements are placed along the *y*-axis, symmetrically about the *z*-axis, with a total separation of d_y between them, and a phase excitation of β_y between them. *For far-field observations,*
(a) Write (you do not have to derive it) the normalized array factor, *in simplified form (sine and/or cosines only)*, for only the two elements along the *x*-axis.
(b) Write (you do not have to derive it) the normalized array factor, *in simplified form (sine and/or cosines only)*, for only the two elements along the *y*-axis.
(c) Write (you do not have to derive it) the normalized array factor, *in simplified form (sine and/or cosines only)*, of the four elements along the *x*- and *y*- axes. *You can use superposition.*
(d) *Without having to derive anything but based on the fundamental phasing principles that we covered in class and discussed in the book, what should the phases β_x and β_y be (in degrees) so that there is (assuming $d_x = d_y = \lambda/4$):*
 • A maximum along the +z direction; i.e., perpendicular to the xy-plane.
 • A minimum along the +z direction; i.e., perpendicular to the xy-plane.

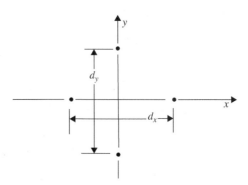

6.90. Design a 10×8 (10 in the x-direction and 8 in the y) element uniform planar array so that the main maximum is oriented along $\theta_0 = 10°$, $\phi_0 = 90°$. For a spacing of $d_x = d_y = \lambda/8$ between the elements, find the
  (a) progressive phase shift between the elements in the x and y directions
  (b) directivity of the array
  (c) half-power beamwidths (in two perpendicular planes) of the array.
  Verify the design using the computer program **Arrays** of this chapter.

6.91. It is desired to design a *nonuniform binomial planar array of 4x4 elements placed along the x-y plane* with a uniform spacing of $\lambda/2$ between the elements along both the x- and y-directions. The z-axis is perpendicular to the array. During the operation, the array factor is scanned with its maximum directed along the $\theta_0 = 30°$, $\phi_0 = 45°$ direction. Determine the:
  (a) Half-power beamwidths (*in degrees*) of the array amplitude pattern *in two mutually perpendicular planes that both pass through the maximum of the array factor*.
  (b) Exact directivity (*dimensionless* and *in dB*) of the array factor.
  (c) Approximate directivity (*dimensionless* and *in dB*) of the array factor. *To get credit, select the most appropriate formula to compute the approximate directivity. State which one you are using.*

6.92. The main beam maximum of a 10×10 planar array of isotropic elements (100 elements) is directed toward $\theta_0 = 10°$ and $\phi_0 = 45°$. Find the directivity, beamwidths (in two perpendicular planes), and beam solid angle for a Tschebyscheff distribution design with side lobes of -26 dB. The array is placed on the x-y plane and the elements are equally spaced with $d = \lambda/4$. It should be noted that an array with bidirectional (two-sided pattern) elements would have a directivity which would be half of that of the same array but with unidirectional (one-sided pattern) elements. Verify the design using the computer program **Arrays** of this chapter.

6.93. Repeat Problem 6.90 for a Tschebyscheff distribution array of -30 dB side lobes.

6.94. In the design of uniform linear arrays, the maximum usually occurs at $\theta = \theta_0$ at the design frequency $f = f_0$, which has been used to determine the progressive phase between the elements. As the frequency shifts from the designed center frequency f_0 to f_h, the maximum amplitude of the array factor at $f = f_h$ is 0.707 the normalized maximum amplitude of unity at $f = f_0$. The frequency f_h is referred to as the half-power frequency, and it is used to determine the frequency bandwidth over which the pattern maximum varies over an amplitude of 3 dB. Using the array factor of a linear uniform array, derive an expression for the 3-dB frequency bandwidth in terms of the length L of the array and the scan angle θ_0.

CHAPTER 7

Antenna Synthesis and Continuous Sources

7.1 INTRODUCTION

Thus far in the book we have concentrated primarily on the analysis and design of antennas. In the analysis problem an antenna model is chosen, and it is analyzed for its radiation characteristics (pattern, directivity, impedance, beamwidth, efficiency, polarization, and bandwidth). This is usually accomplished by initially specifying the current distribution of the antenna, and then analyzing it using standard procedures. If the antenna current is not known, it can usually be determined from integral equation formulations. Numerical techniques, such as the Moment Method of Chapter 8, can be used to numerically solve the integral equations.

In practice, it is often necessary to design an antenna system that will yield desired radiation characteristics. For example, a very common request is to design an antenna whose far-field pattern possesses nulls in certain directions. Other common requests are for the pattern to exhibit a desired distribution, narrow beamwidth and low sidelobes, decaying minor lobes, and so forth. The task, in general, is to find not only the antenna configuration but also its geometrical dimensions and excitation distribution. The designed system should yield, either exactly or approximately, an acceptable radiation pattern, and it should satisfy other system constraints. This method of design is usually referred to as *synthesis*. Although synthesis, in its broadest definition, usually refers to antenna pattern synthesis, it is often used interchangeably with design. Since design methods have been outlined and illustrated previously, as in Chapter 6, in this chapter we want to introduce and illustrate antenna pattern synthesis methods.

Antenna pattern synthesis usually requires that first an approximate analytical model is chosen to represent, either exactly or approximately, the desired pattern. The second step is to match the analytical model to a physical antenna model. Generally speaking, antenna pattern synthesis can be classified into three categories. One group requires that the antenna patterns possess nulls in desired directions. The method introduced by Schelkunoff [1] can be used to accomplish this; it will be discussed in Section 7.3. Another category requires that the patterns exhibit a desired distribution in the entire visible region. This is referred to as *beam shaping*, and it can be accomplished using the Fourier transform [2] and the Woodward-Lawson [3], [4] methods. They will be discussed and illustrated in Sections 7.4 and 7.5, respectively. A third group includes techniques that produce patterns with narrow beams and low sidelobes. Some methods that accomplish this have already been discussed; namely the binomial method (Section 6.8.2) and the Dolph-Tschebyscheff method (also spelled Tchebyscheff or Chebyshev) of Section 6.8.3. Other techniques that belong to this family

---

Antenna Theory: Analysis and Design, Fourth Edition. Constantine A. Balanis.
© 2016 John Wiley & Sons, Inc. Published 2016 by John Wiley & Sons, Inc.
Companion Website: www.wiley.com/go/antennatheory4e

are the Taylor line-source (Tschebyscheff-error) [5] and the Taylor line-source (one parameter) [6]. They will be outlined and illustrated in Sections 7.6 and 7.7, respectively.

The synthesis methods will be utilized to design line-sources and linear arrays whose space factors [as defined by (4-58a)] and array factors [as defined by (6-52)] will yield desired far-field radiation patterns. The total pattern is formed by multiplying the space factor (or array factor) by the element factor (or element pattern) as dictated by (4-59) [or (6-5)]. For very narrow beam patterns, the total pattern is nearly the same as the space factor or array factor. This is demonstrated by the dipole antenna ($l \ll \lambda$) of Figure 4.3 whose element factor, as given by (4-58a), is $\sin\theta$; for values of θ near $90°(\theta \simeq 90°), \sin\theta \simeq 1$.

The synthesis techniques will be followed with a brief discussion of some very popular line-source distributions (triangular, cosine, cosine-squared) and continuous aperture distributions (rectangular and circular).

7.2 CONTINUOUS SOURCES

Very long (in terms of a wavelength) arrays of discrete elements usually are more difficult to implement, more costly, and have narrower bandwidths. For such applications, antennas with continuous distributions would be convenient to use. A very long wire and a large reflector represent, respectively, antennas with continuous line and aperture distributions. Continuous distribution antennas usually have larger sidelobes, are more difficult to scan, and in general, they are not as versatile as arrays of discrete elements. The characteristics of continuously distributed sources can be approximated by discrete-element arrays, and vice-versa, and their development follows and parallels that of discrete-element arrays.

7.2.1 Line-Source

Continuous line-source distributions are functions of only one coordinate, and they can be used to approximate linear arrays of discrete elements and vice-versa.

The array factor of a discrete-element array, placed along the z-axis, is given by (6-52) and (6-52a). As the number of elements increases in a fixed-length array, the source approaches a continuous distribution. In the limit, the array factor summation reduces to an integral. For a continuous distribution, the factor that corresponds to the array factor is known as the *space factor*. For a line-source distribution of length l placed symmetrically along the z-axis as shown in Figure 7.1(a), the space factor (SF) is given by

$$\text{SF}(\theta) = \int_{-l/2}^{+l/2} I_n(z') e^{j[kz'\cos\theta + \phi_n(z')]} \, dz' \quad (7\text{-}1)$$

where $I_n(z')$ and $\phi_n(z')$ represent, respectively, the amplitude and phase distributions along the source. For a uniform phase distribution $\phi_n(z') = 0$.

Equation (7-1) is a finite one-dimensional Fourier transform relating the far-field pattern of the source to its excitation distribution. Two-dimensional Fourier transforms are used to represent the space factors for two-dimensional source distributions. These relations are results of the angular spectrum concept for plane waves, introduced first by Booker and Clemmow [2], and it relates the angular spectrum of a wave to the excitation distribution of the source.

For a continuous source distribution, the total field is given by the product of the *element* and *space* factors as defined in (4-59). This is analogous to the pattern multiplication of (6-5) for arrays. *The type of current and its direction of flow on a source determine the element factor.* For a finite length linear dipole, for example, the total field of (4-58a) is obtained by summing the contributions

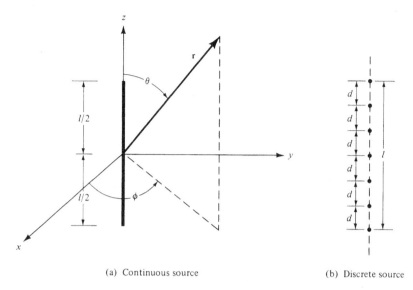

(a) Continuous source (b) Discrete source

Figure 7.1 Continuous and discrete linear sources.

of small infinitesimal elements which are used to represent the entire dipole. In the limit, as the infinitesimal lengths become very small, the summation reduces to an integration. In (4-58a), the factor outside the brackets is the element factor and the one within the brackets is the space factor and corresponds to (7-1).

7.2.2 Discretization of Continuous Sources

The radiation characteristics of continuous sources can be approximated by discrete-element arrays, and vice-versa. This is illustrated in Figure 7.1(b) whereby discrete elements, with a spacing d between them, are placed along the length of the continuous source. Smaller spacings between the elements yield better approximations, and they can even capture the fine details of the continuous distribution radiation characteristics. For example, the continuous line-source distribution $I_n(z')$ of (7-1) can be approximated by a discrete-element array whose element excitation coefficients, at the specified element positions within $-l/2 \leq z' \leq l/2$, are determined by the sampling of $I_n(z')e^{j\phi_n(z')}$. The radiation pattern of the digitized discrete-element array will approximate the pattern of the continuous source.

The technique can be used for the discretization of any continuous distribution. The accuracy increases as the element spacing decreases; in the limit, the two patterns will be identical. For large element spacing, the patterns of the two antennas will not match well. To avoid this, another method known as *root-matching* can be used [7]. Instead of sampling the continuous current distribution to determine the element excitation coefficients, the root-matching method requires that the nulls of the continuous distribution pattern also appear in the initial pattern of the discrete-element array. If the synthesized pattern using this method still does not yield (within an acceptable accuracy) the desired pattern, a perturbation technique [7] can then be applied to the distribution of the discrete-element array to improve its accuracy.

7.3 SCHELKUNOFF POLYNOMIAL METHOD

A method that is conducive to the synthesis of arrays whose patterns possess nulls in desired directions is that introduced by Schelkunoff [1]. To complete the design, this method requires information

388 ANTENNA SYNTHESIS AND CONTINUOUS SOURCES

on the number of nulls and their locations. The number of elements and their excitation coefficients are then derived. The analytical formulation of the technique follows.

Referring to Figure 6.5(a), the array factor for an N-element, equally spaced, nonuniform amplitude, and progressive phase excitation is given by (6-52) as

$$\text{AF} = \sum_{n=1}^{N} a_n e^{j(n-1)(kd\cos\theta+\beta)} = \sum_{n=1}^{N} a_n e^{j(n-1)\psi} \qquad (7\text{-}2)$$

where a_n accounts for the nonuniform amplitude excitation of each element. The spacing between the elements is d and β is the progressive phase shift.

Letting

$$z = x + jy = e^{j\psi} = e^{j(kd\cos\theta+\beta)} \qquad (7\text{-}3)$$

we can rewrite (7-2) as

$$\text{AF} = \sum_{n=1}^{N} a_n z^{n-1} = a_1 + a_2 z + a_3 z^2 + \cdots + a_N z^{N-1} \qquad (7\text{-}4)$$

which is a polynomial of degree $(N-1)$. From the mathematics of complex variables and algebra, any polynomial of degree $(N-1)$ has $(N-1)$ roots and can be expressed as a product of $(N-1)$ linear terms. Thus we can write (7-4) as

$$\boxed{\text{AF} = a_n(z-z_1)(z-z_2)(z-z_3)\cdots(z-z_{N-1})} \qquad (7\text{-}5)$$

where $z_1, z_2, z_3, \ldots, z_{N-1}$ are the roots, which may be complex, of the polynomial. The magnitude of (7-5) can be expressed as

$$\boxed{|\text{AF}| = |a_n||z-z_1||z-z_2||z-z_3|\cdots|z-z_{N-1}|} \qquad (7\text{-}6)$$

Some very interesting observations can be drawn from (7-6) which can be used judiciously for the analysis and synthesis of arrays. Before tackling that phase of the problem, let us first return and examine the properties of (7-3).

The complex variable z of (7-3) can be written in another form as

$$z = |z|e^{j\psi} = |z|\angle\psi = 1\angle\psi \qquad (7\text{-}7)$$

$$\psi = kd\cos\theta + \beta = \frac{2\pi}{\lambda}d\cos\theta + \beta \qquad (7\text{-}7\text{a})$$

It is clear that for any value of $d, \theta,$ or β the magnitude of z lies always on a *unit* circle; however its phase depends upon $d, \theta,$ and β. For $\beta = 0$, we have plotted in Figures 7-2(a)–(d) the value of z, magnitude and phase, as θ takes values of 0 to π rad. It is observed that for $d = \lambda/8$ the values of z, for all the physically observable angles of θ, only exist over the part of the circle shown in Figure 7.2(a). Any values of z outside that arc are not realizable by any physical observation angle θ for the spacing $d = \lambda/8$. We refer to the realizable part of the circle as the *visible region* and the remaining as the *invisible region*. In Figure 7.2(a) we also observe the path of the z values as θ changes from 0° to 180°.

SCHELKUNOFF POLYNOMIAL METHOD 389

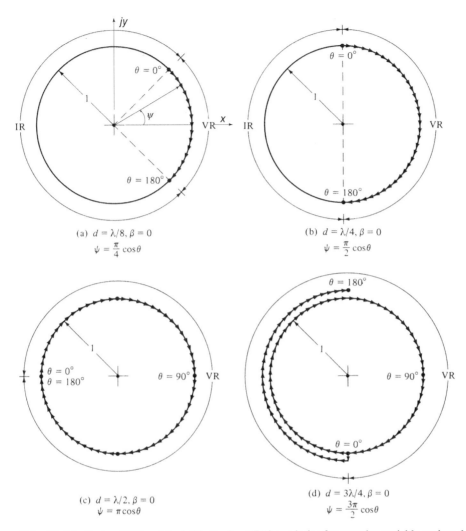

Figure 7.2 Visible Region (VR) and Invisible Region (IR) boundaries for complex variable z when $\beta = 0$.

In Figures 7.2(b)–(d) we have plotted the values of z when the spacing between the elements is $\lambda/4, \lambda/2$, and $3\lambda/4$. It is obvious that the visible region can be extended by increasing the spacing between the elements. It requires a spacing of at least $\lambda/2$ to encompass, at least once, the entire circle. Any spacing greater than $\lambda/2$ leads to multiple values for z. In Figure 7.2(d) we have double values for z for half of the circle when $d = 3\lambda/4$.

To demonstrate the versatility of the arrays, in Figures 7.3(a)–(d) we have plotted the values of z for the same spacings as in Figure 7.2(a)–(d) but with a $\beta = \pi/4$. A comparison between the corresponding figures indicates that the overall visible region for each spacing has not changed but its relative position on the unit circle has rotated counterclockwise by an amount equal to β.

We can conclude then that the overall extent of the visible region can be controlled by the spacing between the elements and its relative position on the unit circle by the progressive phase excitation of the elements. These two can be used effectively in the design of the array factors.

Now let us return to (7-6). The magnitude of the array factor, its form as shown in (7-6), has a geometrical interpretation. For a given value of z in the visible region of the unit circle, corresponding

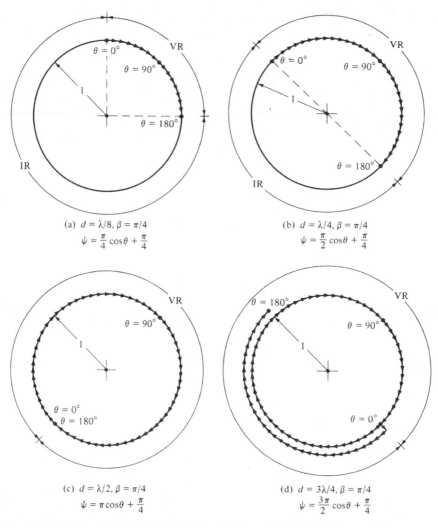

Figure 7.3 Visible Region (VR) and Invisible Region (IR) boundaries for complex variable z when $\beta = \pi/4$.

to a value of θ as determined by (7-3), |AF| is proportional to the product of the distances between z and $z_1, z_2, z_3, \ldots, z_{N-1}$, the roots of AF. In addition, apart from a constant, the phase of AF is equal to the sum of the phases between z and each of the zeros (roots). This is best demonstrated geometrically in Figure 7.4(a). If all the roots $z_1, z_2, z_3, \ldots, z_{N-1}$ are located in the visible region of the unit circle, then each one corresponds to a null in the pattern of |AF| because as θ changes z changes and eventually passes through each of the z_n's. When it does, the length between z and that z_n is zero and (7-6) vanishes. When all the zeros (roots) are not in the visible region of the unit circle, but some lie outside it and/or any other point not on the unit circle, then only those zeros on the visible region will contribute to the nulls of the pattern. This is shown geometrically in Figure 7.4(b). If no zeros exist in the visible region of the unit circle, then that particular array factor has no nulls for any value of θ. However, if a given zero lies on the unit circle but not in its visible region, that zero can be included in the pattern by changing the phase excitation β so that the visible region is rotated until it encompasses that root. Doing this, and not changing d, may exclude some other zero(s).

To demonstrate all the principles, we will consider an example along with some computations.

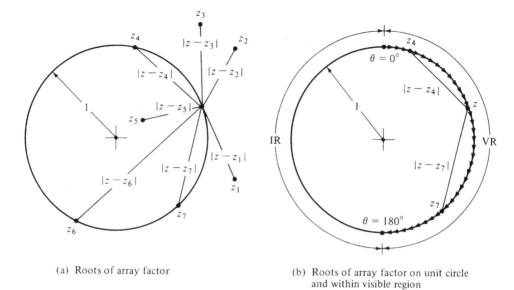

(a) Roots of array factor

(b) Roots of array factor on unit circle and within visible region

Figure 7.4 Array factor roots within and outside unit circle, and visible and invisible regions.

Example 7.1

Design a linear array with a spacing between the elements of $d = \lambda/4$ such that it has zeros at $\theta = 0°, 90°$, and $180°$. Determine the number of elements, their excitation, and plot the derived pattern. Use Schelkunoff's method.

Solution: For a spacing of $\lambda/4$ between the elements and a phase shift $\beta = 0°$, the visible region is shown in Figure 7.2(b). If the desired zeros of the array factor must occur at $\theta = 0°, 90°$, and $180°$, then these correspond to $z = j, 1, -j$ on the unit circle. Thus a normalized form of the array factor is given by

$$AF = (z - z_1)(z - z_2)(z - z_3) = (z - j)(z - 1)(z + j)$$
$$AF = z^3 - z^2 + z - 1$$

Referring to (7-4), the above array factor and the desired radiation characteristics can be realized when there are four elements and their excitation coefficients are equal to

$$a_1 = -1$$
$$a_2 = +1$$
$$a_3 = -1$$
$$a_4 = +1$$

To illustrate the method, we plotted in Figure 7.5 the pattern of that array; it clearly meets the desired specifications. Because of the symmetry of the array, the pattern of the left hemisphere is identical to that of the right.

392   ANTENNA SYNTHESIS AND CONTINUOUS SOURCES

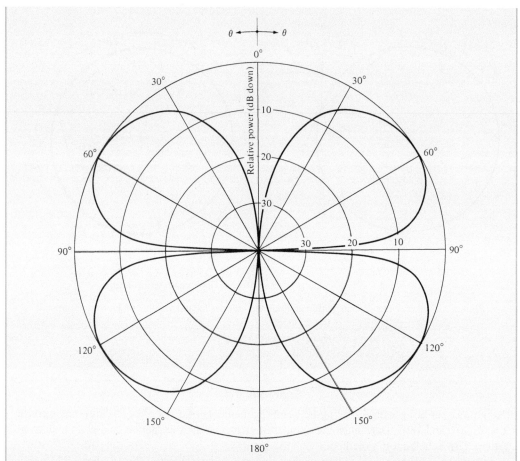

Figure 7.5 Amplitude radiation pattern of a four-element array of isotropic sources with a spacing of $\lambda/4$ between them, zero degrees progressive phase shift, and zeros at $\theta = 0°, 90°$, and $180°$.

7.4 FOURIER TRANSFORM METHOD

This method can be used to determine, given a complete description of the desired pattern, the excitation distribution of a continuous or a discrete source antenna system. The derived excitation will yield, either exactly or approximately, the desired antenna pattern. The pattern synthesis using this method is referred to as *beam shaping*.

7.4.1 Line-Source

For a continuous line-source distribution of length l, as shown in Figure 7.1, the normalized space factor of (7-1) can be written as

$$\text{SF}(\theta) = \int_{-l/2}^{l/2} I(z')e^{j(k\cos\theta - k_z)z'}\,dz' = \int_{-l/2}^{l/2} I(z')e^{j\xi z'}\,dz' \qquad (7\text{-}8)$$

$$\xi = k\cos\theta - k_z \Rightarrow \theta = \cos^{-1}\left(\frac{\xi + k_z}{k}\right) \qquad (7\text{-}8a)$$

where k_z is the excitation phase constant of the source. For a normalized uniform current distribution of the form $I(z') = I_0/l$, (7-8) reduces to

$$\text{SF}(\theta) = I_0 \frac{\sin\left[\frac{kl}{2}\left(\cos\theta - \frac{k_z}{k}\right)\right]}{\frac{kl}{2}\left(\cos\theta - \frac{k_z}{k}\right)} \tag{7-9}$$

The observation angle θ of (7-9) will have real values (visible region) provided that $-(k + k_z) \leq \xi \leq (k - k_z)$ as obtained from (7-8a).

Since the current distribution of (7-8) extends only over $-l/2 \leq z' \leq l/2$ (and it is zero outside it), the limits can be extended to infinity and (7-8) can be written as

$$\text{SF}(\theta) = \text{SF}(\xi) = \int_{-\infty}^{+\infty} I(z')e^{j\xi z'}\, dz' \tag{7-10a}$$

The form of (7-10a) is a Fourier transform, and it relates the excitation distribution $I(z')$ of a continuous source to its far-field space factor $\text{SF}(\theta)$. The transform pair of (7-10a) is given by

$$I(z') = \frac{1}{2\pi}\int_{-\infty}^{+\infty} \text{SF}(\xi)e^{-jz'\xi}\, d\xi = \frac{1}{2\pi}\int_{-\infty}^{+\infty} \text{SF}(\theta)e^{-jz'\xi}\, d\xi \tag{7-10b}$$

Whether (7-10a) represents the direct transform and (7-10b) the inverse transform, or vice-versa, does not matter here. The most important thing is that the excitation distribution and the far-field space factor are related by Fourier transforms.

Equation (7-10b) indicates that if $\text{SF}(\theta)$ represents the desired pattern, the excitation distribution $I(z')$ that will yield the exact desired pattern must in general exist for all values of $z'(-\infty \leq z' \leq \infty)$. Since physically only sources of finite dimensions are realizable, the excitation distribution of (7-10b) is usually truncated at $z' = \pm l/2$ (beyond $z' = \pm l/2$ it is set to zero). Thus the approximate source distribution is given by

$$I_a(z') \simeq \begin{cases} I(z') = \frac{1}{2\pi}\int_{-\infty}^{+\infty} \text{SF}(\xi)e^{-jz'\xi}\, d\xi & -l/2 \leq z' \leq l/2 \\ 0 & \text{elsewhere} \end{cases} \tag{7-11}$$

and it yields an approximate pattern $\text{SF}(\theta)_a$. The approximate pattern is used to represent, within certain error, the desired pattern $\text{SF}(\theta)_d$. Thus

$$\text{SF}(\theta)_d \simeq \text{SF}(\theta)_a = \int_{-l/2}^{l/2} I_a(z')e^{j\xi z'}\, dz' \tag{7-12}$$

It can be shown that, over all values of ξ, the synthesized approximate pattern $\text{SF}(\theta)_a$ yields the least-mean-square error or deviation from the desired pattern $\text{SF}(\theta)_d$. However that criterion is not satisfied when the values of ξ are restricted only in the visible region [8], [9].

To illustrate the principles of this design method, an example is taken.

Example 7.2

Determine the current distribution and the approximate radiation pattern of a line-source placed along the z-axis whose desired radiation pattern is symmetrical about $\theta = \pi/2$, and it is given by

$$\text{SF}(\theta) = \begin{cases} 1 & \pi/4 \leq \theta \leq 3\pi/4 \\ 0 & \text{elsewhere} \end{cases}$$

This is referred to as a sectoral pattern, and it is widely used in radar search and communication applications.

Solution: Since the pattern is symmetrical, $k_z = 0$. The values of ξ, as determined by (7-8a), are given by $k/\sqrt{2} \geq \xi \geq -k/\sqrt{2}$. In turn, the current distribution is given by (7-10b) or

$$I(z') = \frac{1}{2\pi} \int_{-\infty}^{+\infty} \text{SF}(\xi) e^{-jz'\xi} \, d\xi$$

$$= \frac{1}{2\pi} \int_{-k/\sqrt{2}}^{k/\sqrt{2}} e^{-jz'\xi} \, d\xi = \frac{k}{\pi\sqrt{2}} \left[\frac{\sin\left(\frac{kz'}{\sqrt{2}}\right)}{\frac{kz'}{\sqrt{2}}} \right]$$

and it exists over all values of $z'(-\infty \leq z' \leq \infty)$. Over the extent of the line-source, the current distribution is approximated by

$$I_a(z') \simeq I(z'), \quad -l/2 \leq z' \leq l/2$$

If the derived current distribution $I(z')$ is used in conjunction with (7-10a) and it is assumed to exist over all values of z', the exact and desired sectoral pattern will result. If however it is truncated at $z' = \pm l/2$ (and assumed to be zero outside), then the desired pattern is approximated by (7-12) or

$$\text{SF}(\theta)_d \simeq \text{SF}(\theta)_a = \int_{-l/2}^{l/2} I_a(z') e^{j\xi z'} \, dz'$$

$$= \frac{1}{\pi} \left\{ S_i \left[\frac{l}{\lambda} \pi \left(\cos\theta + \frac{1}{\sqrt{2}} \right) \right] - S_i \left[\frac{l}{\lambda} \pi \left(\cos\theta - \frac{1}{\sqrt{2}} \right) \right] \right\}$$

where $S_i(x)$ is the sine integral of (4-68b).

The approximate current distribution (normalized so that its maximum is unity) is plotted in Figure 7.6(a) for $l = 5\lambda$ and $l = 10\lambda$. The corresponding approximate normalized patterns are shown in Figure 7.6(b) where they are compared with the desired pattern. A very good reconstruction is indicated. The longer line-source ($l = 10\lambda$) provides a better realization. The sidelobes are about 0.102 (-19.83 dB) for $l = 5\lambda$ and 0.081 (-21.83 dB) for $l = 10\lambda$ (relative to the pattern at $\theta = 90°$).

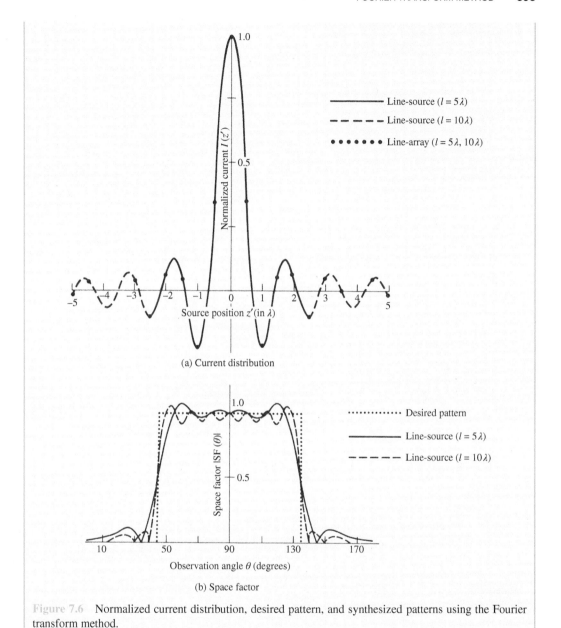

Figure 7.6 Normalized current distribution, desired pattern, and synthesized patterns using the Fourier transform method.

7.4.2 Linear Array

The array factor of an N-element linear array of equally spaced elements and nonuniform excitation is given by (7-2). If the reference point is taken at the physical center of the array, the array factor can also be written as

Odd Number of Elements $(N = 2M + 1)$

$$\mathrm{AF}(\theta) = \mathrm{AF}(\psi) = \sum_{m=-M}^{M} a_m e^{jm\psi} \qquad (7\text{-}13\mathrm{a})$$

Even Number of Elements ($N = 2M$)

$$\text{AF}(\theta) = \text{AF}(\psi) = \sum_{m=-M}^{-1} a_m e^{j[(2m+1)/2]\psi} + \sum_{m=1}^{M} a_m e^{j[(2m-1)/2]\psi} \quad (7\text{-}13\text{b})$$

where

$$\psi = kd\cos\theta + \beta \quad (7\text{-}13\text{c})$$

For an odd number of elements ($N = 2M + 1$), the elements are placed at

$$z'_m = md, \quad m = 0, \pm 1, \pm 2, \ldots, \pm M \quad (7\text{-}13\text{d})$$

and for an even number ($N = 2M$) at

$$z'_m = \begin{cases} \dfrac{2m-1}{2}d, & 1 \le m \le M \\ \dfrac{2m+1}{2}d, & -M \le m \le -1 \end{cases} \quad (7\text{-}13\text{e})$$

An odd number of elements must be utilized to synthesize a desired pattern whose average value, over all angles, is not equal to zero. The $m = 0$ term of (7-13a) is analogous to the d.c. term in a Fourier series expansion of functions whose average value is not zero.

In general, the array factor of an antenna is a periodic function of ψ, and it must repeat for every 2π radians. In order for the array factor to satisfy the periodicity requirements for real values of θ (visible region), then $2kd = 2\pi$ or $d = \lambda/2$. The periodicity and visible region requirement of $d = \lambda/2$ can be relaxed; in fact, it can be made $d < \lambda/2$. However, the array factor $\text{AF}(\psi)$ must be made pseudoperiodic by using fill-in functions, as is customarily done in Fourier series analysis. Such a construction leads to nonunique solutions, because each new fill-in function will result in a different solution. In addition, spacings smaller than $\lambda/2$ lead to superdirective arrays that are undesirable and impractical. If $d > \lambda/2$, the derived patterns exhibit undesired grating lobes; in addition, they must be restricted to satisfy the periodicity requirements.

If $\text{AF}(\psi)$ represents the desired array factor, the excitation coefficients of the array can be obtained by the Fourier formula of

Odd Number of Elements ($N = 2M + 1$)

$$a_m = \frac{1}{T}\int_{-T/2}^{T/2} \text{AF}(\psi)e^{-jm\psi}\,d\psi = \frac{1}{2\pi}\int_{-\pi}^{\pi} \text{AF}(\psi)e^{-jm\psi}\,d\psi \quad -M \le m \le M$$

$$(7\text{-}14\text{a})$$

Even Number of Elements ($N = 2M$)

$$a_m = \begin{cases} \dfrac{1}{T}\displaystyle\int_{-T/2}^{T/2} \text{AF}(\psi)e^{-j[(2m+1)/2]\psi}\,d\psi \\[4pt] = \dfrac{1}{2\pi}\displaystyle\int_{-\pi}^{\pi} \text{AF}(\psi)e^{-j[(2m+1)/2]\psi}\,d\psi & -M \le m \le -1 \quad (7\text{-}14\text{b}) \\[10pt] \dfrac{1}{T}\displaystyle\int_{-T/2}^{T/2} \text{AF}(\psi)e^{-j[(2m-1)/2]\psi}\,d\psi \\[4pt] = \dfrac{1}{2\pi}\displaystyle\int_{-\pi}^{\pi} \text{AF}(\psi)e^{-j[(2m-1)/2]\psi}\,d\psi & 1 \le m \le M \quad (7\text{-}14\text{c}) \end{cases}$$

Simplifications in the forms of (7-13a)–(7-13b) and (7-14a)–(7-14c) can be obtained when the excitations are symmetrical about the physical center of the array.

Example 7.3

Determine the excitation coefficients and the resultant pattern for a broadside discrete-element array whose array factor closely approximates the desired symmetrical sectoral pattern of Example 7.2. Use 11 elements with a spacing of $d = \lambda/2$ between them. Repeat the design for 21 elements.

Solution: Since the array is broadside, the progressive phase shift between the elements as required by (6-18a) is zero ($\beta = 0$). Since the pattern is nonzero only for $\pi/4 \le \theta \le 3\pi/4$, the corresponding values of ψ are obtained from (7-13c) or $\pi/\sqrt{2} \ge \psi \ge -\pi/\sqrt{2}$. The excitation coefficients are obtained from (7-14a) or

$$a_m = \frac{1}{2\pi}\int_{-\pi/\sqrt{2}}^{\pi/\sqrt{2}} e^{-jm\psi}\,d\psi = \frac{1}{\sqrt{2}}\left[\frac{\sin\!\left(\dfrac{m\pi}{\sqrt{2}}\right)}{\dfrac{m\pi}{\sqrt{2}}}\right]$$

and they are symmetrical about the physical center of the array $[a_m(-z'_m) = a_m(z'_m)]$. The corresponding array factor is given by (7-13a).

The normalized excitation coefficients are

$$\begin{aligned}
a_0 &= 1.0000 & a_{\pm 4} &= 0.0578 & a_{\pm 8} &= -0.0496 \\
a_{\pm 1} &= 0.3582 & a_{\pm 5} &= -0.0895 & a_{\pm 9} &= 0.0455 \\
a_{\pm 2} &= -0.2170 & a_{\pm 6} &= 0.0518 & a_{\pm 10} &= -0.0100 \\
a_{\pm 3} &= 0.0558 & a_{\pm 7} &= 0.0101 &&
\end{aligned}$$

They are displayed graphically by a dot (•) in Figure 7.6(a) where they are compared with the continuous current distribution of Example 7.2. It is apparent that at the element positions, the line-source and linear array excitation values are identical. This is expected since the two antennas are of the same length (for $N = 11, d = \lambda/2 \Rightarrow l = 5\lambda$ and for $N = 21, d = \lambda/2 \Rightarrow l = 10\lambda$).

398 ANTENNA SYNTHESIS AND CONTINUOUS SOURCES

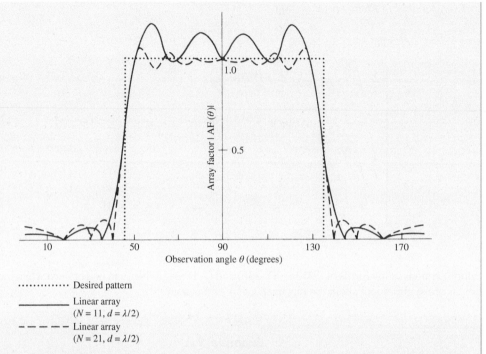

Figure 7.7 Desired array factor and synthesized normalized patterns for linear array of 11 and 21 elements using the Fourier transform method.

The corresponding normalized array factors are displayed in Figure 7.7. As it should be expected, the larger array ($N = 21, d = \lambda/2$) provides a better reconstruction of the desired pattern. The sidelobe levels, relative to the value of the pattern at $\theta = 90°$, are 0.061 (-24.29 dB) for $N = 11$ and 0.108 (-19.33 dB) for $N = 21$.

Discrete-element linear arrays only approximate continuous line-sources. Therefore, their patterns shown in Figure 7.7 do not approximate as well the desired pattern as the corresponding patterns of the line-source distributions shown in Figure 7.6(b).

Whenever the desired pattern contains discontinuities or its values in a given region change very rapidly, the reconstruction pattern will exhibit oscillatory overshoots which are referred to as *Gibbs' phenomena*. Since the desired sectoral patterns of Examples 7.2 and 7.3 are discontinuous at $\theta = \pi/4$ and $3\pi/4$, the reconstructed patterns displayed in Figures 7.6(b) and 7.7 exhibit these oscillatory overshoots.

7.5 WOODWARD-LAWSON METHOD

A very popular antenna pattern synthesis method used for beam shaping was introduced by Woodward and Lawson [3], [4], [10]. The synthesis is accomplished by sampling the desired pattern at various discrete locations. Associated with each pattern sample is a harmonic current of uniform amplitude distribution and uniform progressive phase, whose corresponding field is referred to as a *composing function*. For a line-source, each composing function is of an $b_m \sin(\psi_m)/\psi_m$ form whereas for a linear array it takes an $b_m \sin(N\phi_m)/N \sin(\phi_m)$ form. The excitation coefficient b_m of each harmonic current is such that its field strength is equal to the amplitude of the desired pattern at its corresponding sampled point. The total excitation of the source is comprised of a finite summation of space harmonics. The corresponding synthesized pattern is represented by a finite summation of composing functions with each term representing the field of a current harmonic with uniform amplitude distribution and uniform progressive phase.

The formation of the overall pattern using the Woodward-Lawson method is accomplished as follows. The first composing function produces a pattern whose main beam placement is determined by the value of its uniform progressive phase while its innermost sidelobe level is about -13.5 dB; the level of the remaining sidelobes monotonically decreases. The second composing function has also a similar pattern except that its uniform progressive phase is adjusted so that its main lobe maximum coincides with the innermost null of the first composing function. This results in the filling-in of the innermost null of the pattern of the first composing function; the amount of filling-in is controlled by the amplitude excitation of the second composing function. Similarly, the uniform progressive phase of the third composing function is adjusted so that the maximum of its main lobe occurs at the second innermost null of the first composing function; it also results in filling-in of the second innermost null of the first composing function. This process continues with the remaining finite number of composing functions.

The Woodward-Lawson method is simple, elegant, and provides insight into the process of pattern synthesis. However, because the pattern of each composing function perturbs the entire pattern to be synthesized, it lacks local control over the sidelobe level in the unshaped region of the entire pattern. In 1988 and 1989 a spirited and welcomed dialogue developed concerning the Woodward-Lawson method [11]–[14]. The dialogue centered whether the Woodward-Lawson method should be taught and even appear in textbooks, and whether it should be replaced by an alternate method [15] which overcomes some of the shortcomings of the Woodward-Lawson method. The alternate method of [15] is more of a numerical and iterative extension of Schelkunoff's polynomial method which may be of greater practical value because it provides superior beamshape and pattern control. One of the distinctions of the two methods is that the Woodward-Lawson method deals with the synthesis of *field patterns* while that of [15] deals with the synthesis of *power patterns*.

The analytical formulation of this method is similar to the Shannon sampling theorem used in communications which states that "if a function $g(t)$ is band-limited, with its highest frequency being f_h, the function $g(t)$ can be reconstructed using samples taken at a frequency f_s. To faithfully reproduce the original function $g(t)$, the sampling frequency f_s should be at least twice the highest frequency f_h ($f_s \geq 2f_h$) or the function should be sampled at points separated by no more than $\Delta t \leq 1/f_s \geq 1/(2f_h) = T_h/2$ where T_h is the period of the highest frequency f_h." In a similar manner, the radiation pattern of an antenna can be synthesized by sampling functions whose samples are separated by λ/l rad, where l is the length of the source [9], [10].

7.5.1 Line-Source

Let the current distribution of a continuous source be represented, within $-l/2 \leq z' \leq l/2$, by a finite summation of normalized sources each of constant amplitude and linear phase of the form

$$i_m(z') = \frac{b_m}{l} e^{-jkz' \cos \theta_m} \qquad -l/2 \leq z' \leq l/2 \qquad (7\text{-}15)$$

As it will be shown later, θ_m represents the angles where the desired pattern is sampled. The total current $I(z')$ is given by a finite summation of $2M$ (even samples) or $2M + 1$ (odd samples) current sources each of the form of (7-15). Thus

$$I(z') = \frac{1}{l} \sum_{m=-M}^{M} b_m e^{-jkz'\cos\theta_m} \tag{7-16}$$

where

$$m = \pm 1, \pm 2, \ldots, \pm M \quad \text{(for } 2M \text{ even number of samples)} \tag{7-16a}$$

$$m = 0, \pm 1, \pm 2, \ldots, \pm M \quad \text{(for } 2M+1 \text{ odd number of samples)} \tag{7-16b}$$

For simplicity use odd number of samples.

Associated with each current source of (7-15) is a corresponding field pattern of the form given by (7-9) or

$$s_m(\theta) = b_m \left\{ \frac{\sin\left[\frac{kl}{2}(\cos\theta - \cos\theta_m)\right]}{\frac{kl}{2}(\cos\theta - \cos\theta_m)} \right\} \tag{7-17}$$

whose maximum occurs when $\theta = \theta_m$. The total pattern is obtained by summing $2M$ (even samples) or $2M+1$ (odd samples) terms each of the form given by (7-17). Thus

$$\text{SF}(\theta) = \sum_{m=-M}^{M} b_m \left\{ \frac{\sin\left[\frac{kl}{2}(\cos\theta - \cos\theta_m)\right]}{\frac{kl}{2}(\cos\theta - \cos\theta_m)} \right\} \tag{7-18}$$

The maximum of each individual term in (7-18) occurs when $\theta = \theta_m$, and it is equal to $\text{SF}(\theta = \theta_m)$. In addition, when one term in (7-18) attains its maximum value at its sample at $\theta = \theta_m$, all other terms of (7-18) which are associated with the other samples are zero at $\theta = \theta_m$. In other words, all sampling terms (composing functions) of (7-18) are zero at all sampling points other than at their own. Thus at each sampling point the total field is equal to that of the sample. This is one of the most appealing properties of this method. If the desired space factor is sampled at $\theta = \theta_m$, the excitation coefficients b_m can be made equal to its value at the sample points θ_m. Thus

$$b_m = \text{SF}(\theta = \theta_m)_d \tag{7-19}$$

The reconstructed pattern is then given by (7-18), and it approximates closely the desired pattern.

In order for the synthesized pattern to satisfy the periodicity requirements of 2π for real values of θ (visible region) and to faithfully reconstruct the desired pattern, each sample should be separated by

$$kz'\Delta|_{|z'|=l} = 2\pi \Rightarrow \Delta = \frac{\lambda}{l} \tag{7-19a}$$

The location of each sample is given by

$$\cos\theta_m = m\Delta = m\left(\frac{\lambda}{l}\right), \quad m = 0, \pm 1, \pm 2, \ldots \quad \text{for } \textit{odd} \text{ samples} \tag{7-19b}$$

$$\cos\theta_m = \begin{cases} \dfrac{(2m-1)}{2}\Delta = \dfrac{(2m-1)}{2}\left(\dfrac{\lambda}{l}\right), \\ \qquad m = +1, +2, \ldots \quad \text{for } \textit{even} \text{ samples} \\ \dfrac{(2m+1)}{2}\Delta = \dfrac{(2m+1)}{2}\left(\dfrac{\lambda}{l}\right), \\ \qquad m = -1, -2, \ldots \quad \text{for } \textit{even} \text{ samples} \end{cases} \tag{7-19c}$$

Therefore, M should be the closest integer to $M = l/\lambda$.

As long as the location of each sample is determined by (7-19b or 7-19c), the pattern value at the sample points is determined solely by that of one sample and it is not correlated to the field of the other samples.

Example 7.4

Repeat the design of Example 7.2 for $l = 5\lambda$ using odd samples and the Woodward-Lawson line-source synthesis method.

Solution: Since $l = 5\lambda$, $M = 5$ and the sampling separation is 0.2. The total number of sampling points is 11. The angles where the sampling is performed are given, according to (7-19b), by

$$\theta_m = \cos^{-1}\left(m\frac{\lambda}{l}\right) = \cos^{-1}(0.2m), \quad m = 0, \pm 1, \ldots, \pm 5$$

The angles and the excitation coefficients at the sample points are listed below.

m	θ_m	$b_m = \text{SF}(\theta_m)_d$	m	θ_m	$b_m = \text{SF}(\theta_m)_d$
0	90°	1			
1	78.46°	1	−1	101.54°	1
2	66.42°	1	−2	113.58°	1
3	53.13°	1	−3	126.87°	1
4	36.87°	0	−4	143.13°	0
5	0°	0	−5	180°	0

The computed pattern is shown in Figure 7.8(a) where it is compared with the desired pattern. A good reconstruction is indicated. The sidelobe level, relative to the value of the pattern at $\theta = 90°$, is 0.160 (−15.92 dB).

To demonstrate the synthesis of the pattern using the sampling concept, we have plotted in Figure 7.8(b) all seven nonzero composing functions $s_m(\theta)$ used for the reconstruction of the $l = 5\lambda$ line-source pattern of Figure 7.8(a). Each nonzero $s_m(\theta)$ composing function was computed using (7-17) for $m = 0, \pm 1, \pm 2, \pm 3$. It is evident that at each sampling point all the composing functions are zero, except the one that represents that sample. Thus the value of the desired pattern at each sampling point is determined solely by the maximum value of a single composing function. The angles where the composing functions attain their maximum values are listed in the previous table.

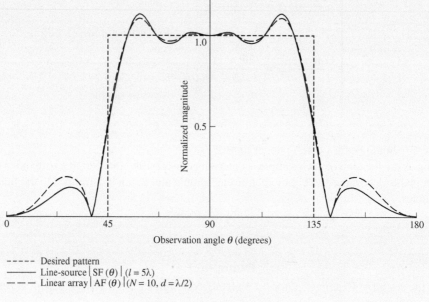

(a) Normalized amplitude patterns

----- Desired pattern
——— Line-source $|\text{SF}(\theta)|$ $(l = 5\lambda)$
— — — Linear array $|\text{AF}(\theta)|$ $(N = 10, d = \lambda/2)$

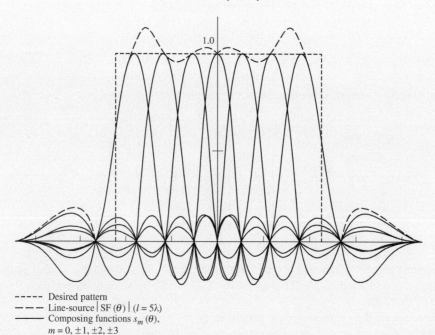

----- Desired pattern
— — — Line-source $|\text{SF}(\theta)|$ $(l = 5\lambda)$
——— Composing functions $s_m(\theta)$, $m = 0, \pm 1, \pm 2, \pm 3$

(b) Composing functions for line-source ($l = 5\lambda$)

Figure 7.8 Desired and synthesized patterns, and composing functions for Woodward-Lawson designs.

7.5.2 Linear Array

The Woodward-Lawson method can also be implemented to synthesize discrete linear arrays. The technique is similar to the Woodward-Lawson method for line-sources except that the pattern of each sample, as given by (7-17), is replaced by the array factor of a uniform array as given by (6-10c). The pattern of each sample can be written as

$$f_m(\theta) = b_m \frac{\sin\left[\frac{N}{2}kd(\cos\theta - \cos\theta_m)\right]}{N\sin\left[\frac{1}{2}kd(\cos\theta - \cos\theta_m)\right]} \quad (7\text{-}20)$$

$l = Nd$ assumes the array is equal to the length of the line-source (*for this design only, the length l of the line includes a distance $d/2$ beyond each end element*). The total array factor can be written as a superposition of $2M + 1$ sampling terms (as was done for the line-source) each of the form of (7-20). Thus

$$\text{AF}(\theta) = \sum_{m=-M}^{M} b_m \frac{\sin\left[\frac{N}{2}kd(\cos\theta - \cos\theta_m)\right]}{N\sin\left[\frac{1}{2}kd(\cos\theta - \cos\theta_m)\right]} \quad (7\text{-}21)$$

As for the line-sources, the excitation coefficients of the array elements at the sample points are equal to the value of the desired array factor at the sample points. That is,

$$b_m = \text{AF}(\theta = \theta_m)_d \quad (7\text{-}22)$$

The sample points are taken at

$$\cos\theta_m = m\Delta = m\left(\frac{\lambda}{l}\right), \, m = 0, \pm 1, \pm 2, \ldots \quad \text{for } \textit{odd} \text{ samples} \quad (7\text{-}23\text{a})$$

$$\cos\theta_m = \begin{cases} \dfrac{(2m-1)}{2}\Delta = \dfrac{(2m-1)}{2}\left(\dfrac{\lambda}{Nd}\right), \\ \qquad m = +1, +2, \ldots \quad \text{for } \textit{even} \text{ samples} \\ \dfrac{(2m+1)}{2}\Delta = \dfrac{(2m+1)}{2}\left(\dfrac{\lambda}{Nd}\right), \\ \qquad m = -1, -2, \ldots \quad \text{for } \textit{even} \text{ samples} \end{cases} \quad (7\text{-}23\text{b})$$

The normalized excitation coefficient of each array element, required to give the desired pattern, is given by

$$a_n(z') = \frac{1}{N}\sum_{m=-M}^{M} b_m e^{-jkz_n'\cos\theta_m} \quad (7\text{-}24)$$

where z_n' indicates the position of the nth element (element in question) symmetrically placed about the geometrical center of the array.

> **Example 7.5**
>
> Repeat the design of Example 7.4 for a linear array of 10 elements using the Woodward-Lawson method with odd samples and an element spacing of $d = \lambda/2$.
>
> *Solution*: According to (7-19), (7-19b), (7-22) and (7-23a), the excitation coefficients of the array at the sampling points are the same as those of the line-source. Using the values of b_m as listed in Example 7.4, the computed array factor pattern using (7-21) is shown in Figure 7.8(a). A good synthesis of the desired pattern is displayed. The sidelobe level, relative to the pattern value at $\theta = 90°$, is 0.221 (−13.1 dB). The agreement between the line-source and the linear array Woodward-Lawson designs are also good.
>
> The normalized pattern of the symmetrical discrete array can also be generated using the array factor of (6-61a) or (6-61b), where the normalized excitation coefficients a_n's of the array elements are obtained using (7-24). For this example, the excitation coefficients of the 10-element array, along with their symmetrical position, are listed below. To achieve the normalized amplitude pattern of unity at $\theta = 90°$ in Figure 7.8(a), the array factor of (6-61a) must be multiplied by $1/\Sigma a_n = 1/0.4482 = 2.2312$.
>
> | Element Number n | Element Position z'_n | Excitation Coefficient a_n |
> |---|---|---|
> | ±1 | ±0.25λ | 0.5696 |
> | ±2 | ±0.75λ | −0.0345 |
> | ±3 | ±1.25λ | −0.1000 |
> | ±4 | ±1.75λ | 0.1109 |
> | ±5 | ±2.25λ | −0.0460 |
>
> In general, the Fourier transform synthesis method yields reconstructed patterns whose mean-square error (or deviation) from the desired pattern is a minimum. However, the Woodward-Lawson synthesis method reconstructs patterns whose values at the sampled points are identical to the ones of the desired pattern; it does not have any control of the pattern between the sample points, and it does not yield a pattern with least-mean-square deviation.
>
> Ruze [9] points out that the least-mean-square error design is not necessarily the best. The particular application will dictate the preference between the two. However, the Fourier transform method is best suited for reconstruction of desired patterns which are analytically simple and which allow the integrations to be performed in closed form. Today, with the advancements in high-speed computers, this is not a major restriction since the integration can be performed (with high efficiency) numerically. In contrast, the Woodward-Lawson method is more flexible, and it can be used to synthesize any desired pattern. In fact, it can even be used to reconstruct patterns which, because of their complicated nature, cannot be expressed analytically. Measured patterns, either of analog or digital form, can also be synthesized using the Woodward-Lawson method.

7.6 TAYLOR LINE-SOURCE (TSCHEBYSCHEFF-ERROR)

In Chapter 6 we discussed the classic Dolph-Tschebyscheff array design which yields, for a given sidelobe level, the smallest possible first-null beamwidth (or the smallest possible sidelobe level for a given first-null beamwidth). Another classic design that is closely related to it, but is more applicable for continuous distributions, is that by Taylor [5] (this method is different from that by Taylor [6] which will be discussed in the next section).

The Taylor design [5] yields a pattern that is an optimum compromise between beamwidth and sidelobe level. In an ideal design, the minor lobes are maintained at an equal and specific level. Since the minor lobes are of equal ripple and extend to infinity, this implies an infinite power. More realistically, however, the technique as introduced by Taylor leads to a pattern whose first few minor lobes (closest to the main lobe) are maintained at an equal and specified level; the remaining lobes decay monotonically. Practically, even the level of the closest minor lobes exhibits a slight monotonic decay. This decay is a function of the space u over which these minor lobes are required to be maintained at an equal level. As this space increases, the rate of decay of the closest minor lobes decreases. For a very large space of u (over which the closest minor lobes are required to have an equal ripple), the rate of decay is negligible. It should be pointed out, however, that the other method by Taylor [6] (of Section 7.7) yields minor lobes, all of which decay monotonically.

The details of the analytical formulation are somewhat complex (for the average reader) and lengthy, and they will not be included here. The interested reader is referred to the literature [5], [16]. Instead, a succinct outline of the salient points of the method and of the design procedure will be included. The design is for far-field patterns, and it is based on the formulation of (7-1).

Ideally the normalized space factor that yields a pattern with equal-ripple minor lobes is given by

$$\mathrm{SF}(\theta) = \frac{\cosh[\sqrt{(\pi A)^2 - u^2}]}{\cosh(\pi A)} \quad (7\text{-}25)$$

$$u = \pi \frac{l}{\lambda} \cos\theta \quad (7\text{-}25\mathrm{a})$$

whose maximum value occurs when $u = 0$. The constant A is related to the maximum desired sidelobe level R_0 by

$$\cosh(\pi A) = R_0 \text{ (voltage ratio)} \quad (7\text{-}26)$$

The space factor of (7-25) can be derived from the Dolph-Tschebyscheff array formulation of Section 6.8.3, if the number of elements of the array are allowed to become infinite.

Since (7-25) is ideal and cannot be realized physically, Taylor [5] suggested that it be approximated (within a certain error) by a space factor comprised of a product of factors whose roots are the zeros of the pattern. Because of its approximation to the ideal Tschebyscheff design, it is also referred to as *Tschebyscheff-error*. The Taylor space factor is given by

$$\mathrm{SF}(u, A, \bar{n}) = \frac{\sin(u)}{u} \frac{\prod_{n=1}^{\bar{n}-1}\left[1 - \left(\frac{u}{u_n}\right)^2\right]}{\prod_{n=1}^{\bar{n}-1}\left[1 - \left(\frac{u}{n\pi}\right)^2\right]} \quad (7\text{-}27)$$

$$u = \pi v = \pi \frac{l}{\lambda} \cos\theta \quad (7\text{-}27\mathrm{a})$$

$$u_n = \pi v_n = \pi \frac{l}{\lambda} \cos\theta_n \quad (7\text{-}27\mathrm{b})$$

where θ_n represents the locations of the nulls. The parameter $\bar{n}$ is a constant chosen by the designer so that the minor lobes for $|v| = |u/\pi| \leq \bar{n}$ are maintained at a nearly constant voltage level of $1/R_0$ while for $|v| = |u/\pi| > \bar{n}$ the envelope, through the maxima of the remaining minor lobes, decays at a rate of $1/v = \pi/u$. In addition, the nulls of the pattern for $|v| \geq \bar{n}$ occur at integer values of v.

In general, there are $\bar{n} - 1$ inner nulls for $|v| < \bar{n}$ and an infinite number of outer nulls for $|v| \geq \bar{n}$. To provide a smooth transition between the inner and the outer nulls (at the expense of slight beam broadening), Taylor introduced a parameter σ. It is usually referred to as the *scaling factor*, and it spaces the inner nulls so that they blend smoothly with the outer ones. In addition, it is the factor by which the beamwidth of the Taylor design is greater than that of the Dolph-Tschebyscheff, and it is given by

$$\sigma = \frac{\bar{n}}{\sqrt{A^2 + \left(\bar{n} - \frac{1}{2}\right)^2}} \tag{7-28}$$

The location of the nulls are obtained using

$$u_n = \pi v_n = \pi \frac{l}{\lambda} \cos \theta_n = \begin{cases} \pm \pi \sigma \sqrt{A^2 + \left(n - \frac{1}{2}\right)^2} & 1 \leq n < \bar{n} \\ \pm n\pi & \bar{n} \leq n \leq \infty \end{cases} \tag{7-29}$$

The normalized line-source distribution, which yields the desired pattern, is given by

$$I(z') = \frac{\lambda}{l}\left[1 + 2\sum_{p=1}^{\bar{n}-1} \text{SF}(p, A, \bar{n}) \cos\left(2\pi p \frac{z'}{l}\right)\right] \tag{7-30}$$

The coefficients $\text{SF}(p, A, \bar{n})$ represent samples of the Taylor pattern, and they can be obtained from (7-27) with $u = \pi p$. They can also be found using

$$\text{SF}(p, A, \bar{n}) = \begin{cases} \dfrac{[(\bar{n} - 1)!]^2}{(\bar{n} - 1 + p)!(\bar{n} - 1 - p)!} \prod_{m=1}^{\bar{n}-1}\left[1 - \left(\dfrac{\pi p}{u_m}\right)^2\right] & |p| < \bar{n} \\ 0 & |p| \geq \bar{n} \end{cases} \tag{7-30a}$$

with $\text{SF}(-p, A, \bar{n}) = \text{SF}(p, A, \bar{n})$.

The half-power beamwidth is given approximately by [8]

$$\Theta_0 \simeq 2\sin^{-1}\left\{\frac{\lambda\sigma}{\pi l}\left[(\cosh^{-1} R_0)^2 - \left(\cosh^{-1} \frac{R_0}{\sqrt{2}}\right)^2\right]^{1/2}\right\} \tag{7-31}$$

7.6.1 Design Procedure

To initiate a Taylor design, you must

1. specify the normalized maximum tolerable sidelobe level $1/R_0$ of the pattern.

2. choose a positive integer value for $\bar{n}$ such that for $|v| = |(l/\lambda)\cos\theta| \leq \bar{n}$ the normalized level of the minor lobes is nearly constant at $1/R_0$. For $|v| > \bar{n}$, the minor lobes decrease monotonically. In addition, for $|v| < \bar{n}$ there exist $(\bar{n} - 1)$ nulls. The position of all the nulls is found using (7-29). Small values of $\bar{n}$ yield source distributions which are maximum at the center and monotonically decrease toward the edges. In contrast, large values of $\bar{n}$ result in sources which are peaked simultaneously at the center and at the edges, and they yield sharper main beams. Therefore, very small and very large values of $\bar{n}$ should be avoided. Typically, the value of $\bar{n}$ should be at least 3 and at least 6 for designs with sidelobes of -25 and -40 dB, respectively.

To complete the design, you do the following:

1. Determine A using (7-26), σ using (7-28), and the nulls using (7-29).
2. Compute the space factor using (7-27), the source distribution using (7-30) and (7-30a), and the half-power beamwidth using (7-31).

Example 7.6

Design a -20 dB Taylor, Tschebyscheff-error, distribution line-source with $\bar{n} = 5$. Plot the pattern and the current distribution for $l = 7\lambda (-7 \leq v = u/\pi \leq 7)$.

Solution: For a -20 dB sidelobe level

$$R_0 \text{ (voltage ratio)} = 10$$

Using (7-26)

$$A = \frac{1}{\pi} \cosh^{-1}(10) = 0.95277$$

and by (7-28)

$$\sigma = \frac{5}{\sqrt{(0.95277)^2 + (5 - 0.5)^2}} = 1.0871$$

The nulls are given by (7-29) or

$$v_n = u_n/\pi = \pm 1.17, \pm 1.932, \pm 2.91, \pm 3.943, \pm 5.00, \pm 6.00, \pm 7.00, \ldots$$

The corresponding null angles for $l = 7\lambda$ are

$$\theta_n = 80.38°(99.62°), 73.98°(106.02°), 65.45°(114.55°),$$
$$55.71°(124.29°), 44.41°(135.59°), \text{ and } 31.00°(149.00°)$$

The half-power beamwidth for $l = 7\lambda$ is found using (7-31), or

$$\Theta_0 \simeq 7.95°$$

The source distribution, as computed using (7-30) and (7-30a), is displayed in Figure 7.9(a). The corresponding radiation pattern for $-7 \leq v = u/\pi \leq 7$ ($0° \leq \theta \leq 180°$ for $l = 7\lambda$) is shown in Figure 7.9(b).

All the computed parameters compare well with results reported in [5] and [16].

408 ANTENNA SYNTHESIS AND CONTINUOUS SOURCES

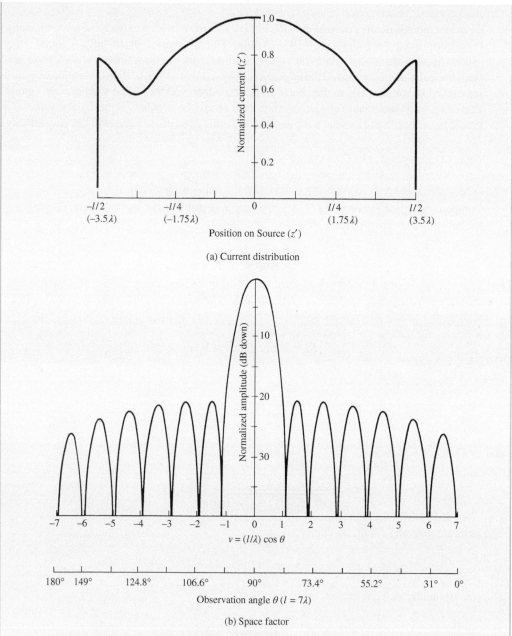

Figure 7.9 Normalized current distribution and far-field space factor pattern for a −20 dB sidelobe and $\bar{n} = 5$ Taylor (Tschebyscheff-error) line-source of $l = 7\lambda$.

7.7 TAYLOR LINE-SOURCE (ONE-PARAMETER)

The Dolph-Tschebyscheff array design of Section 6.8.3 yields minor lobes of equal intensity while the Taylor (Tschebyscheff-error) produces a pattern whose inner minor lobes are maintained at a constant level and the remaining ones decrease monotonically. For some applications, such as radar and low-noise systems, it is desirable to sacrifice some beamwidth and low inner minor lobes to have all the minor lobes decay as the angle increases on either side of the main beam. In radar applications this is preferable because interfering or spurious signals would be reduced further when they try to enter through the decaying minor lobes. Thus any significant contributions from interfering signals would be through the pattern in the vicinity of the major lobe. Since in practice it is easier to maintain pattern symmetry around the main lobe, it is also possible to recognize that such signals are false targets. In low-noise applications, it is also desirable to have minor lobes that decay away from the main beam in order to diminish the radiation accepted through them from the relatively "hot" ground.

A continuous line-source distribution that yields decaying minor lobes and, in addition, controls the amplitude of the sidelobe is that introduced by Taylor [6] in an unpublished classic memorandum. It is referred to as the *Taylor (one-parameter)* design and its source distribution is given by

$$I_n(z') = \begin{cases} J_0\left[j\pi B\sqrt{1 - \left(\frac{2z'}{l}\right)^2}\right] & -l/2 \leq z' \leq +l/2 \\ 0 & \text{elsewhere} \end{cases} \quad (7\text{-}32)$$

where J_0 is the Bessel function of the first kind of order zero, l is the total length of the continuous source [see Figure 7.1(a)], and B is a constant to be determined from the specified sidelobe level.

The space factor associated with (7-32) can be obtained by using (7-1). After some intricate mathematical manipulations, utilizing Gegenbauer's finite integral and Gegenbauer polynomials [17], the space factor for a Taylor amplitude distribution line-source with uniform phase $[\phi_n(z') = \phi_0 = 0]$ can be written as

$$SF(\theta) = \begin{cases} l\dfrac{\sinh[\sqrt{(\pi B)^2 - u^2}]}{\sqrt{(\pi B)^2 - u^2}}, & u^2 < (\pi B)^2 \\[1em] l\dfrac{\sin[\sqrt{u^2 - (\pi B)^2}]}{\sqrt{u^2 - (\pi B)^2}}, & u^2 > (\pi B)^2 \end{cases} \quad (7\text{-}33)$$

where

$$u = \pi\frac{l}{\lambda}\cos\theta \quad (7\text{-}33\text{a})$$

B = constant determined from sidelobe level

l = line-source dimension

The derivation of (7-33) is assigned as an exercise to the reader (Problem 7.28). When $(\pi B)^2 > u^2$, (7-33) represents the region near the main lobe. The minor lobes are represented by $(\pi B)^2 < u^2$ in (7-33). Either form of (7-33) can be obtained from the other by knowing that (see Appendix VI)

$$\begin{aligned} \sin(jx) &= j\sinh(x) \\ \sinh(jx) &= j\sin(x) \end{aligned} \quad (7\text{-}34)$$

ANTENNA SYNTHESIS AND CONTINUOUS SOURCES

When $u = 0$ ($\theta = \pi/2$ and maximum radiation), the normalized pattern height is equal to

$$(\text{SF})_{\max} = \frac{\sinh(\pi B)}{\pi B} = H_0 \tag{7-35}$$

For $u^2 \gg (\pi B)^2$, the normalized form of (7-33) reduces to

$$\text{SF}(\theta) = \frac{\sin[\sqrt{u^2 - (\pi B)^2}]}{\sqrt{u^2 - (\pi B)^2}} \simeq \frac{\sin(u)}{u} \quad u \gg \pi B \tag{7-36}$$

and it is identical to the pattern of a uniform distribution. The maximum height H_1 of the sidelobe of (7-36) is $H_1 = 0.217233$ (or 13.2 dB down from the maximum), and it occurs when (see Appendix I)

$$[u^2 - (\pi B)^2]^{1/2} \simeq u = 4.494 \tag{7-37}$$

Using (7-35), the maximum voltage height of the sidelobe (relative to the maximum H_0 of the major lobe) is equal to

$$\frac{H_1}{H_0} = \frac{1}{R_0} = \frac{0.217233}{\sinh(\pi B)/(\pi B)} \tag{7-38}$$

or

$$\boxed{R_0 = \frac{1}{0.217233}\frac{\sinh(\pi B)}{\pi B} = 4.603\frac{\sinh(\pi B)}{\pi B}} \tag{7-38a}$$

Equation (7-38a) can be used to find the constant B when the intensity ratio R_0 of the major-to-the-sidelobe is specified. Values of B for typical sidelobe levels are

Sidelobe Level (dB)	−10	−15	−20	−25	−30	−35	−40
B	j0.4597	0.3558	0.7386	1.0229	1.2761	1.5136	1.7415

The disadvantage of designing an array with decaying minor lobes as compared to a design with equal minor lobe level (Dolph-Tschebyscheff), is that it yields about 12 to 15% greater half-power beamwidth. However such a loss in beamwidth is a small penalty to pay when the extreme minor lobes decrease as $1/u$.

To illustrate the principles, let us consider an example.

Example 7.7

Given a continuous line-source, whose total length is 4λ, design a Taylor, one-parameter, distribution array whose sidelobe is 30 dB down from the maximum of the major lobe.

a. Find the constant B.
b. Plot the pattern (in dB) of the continuous line-source distribution.
c. For a spacing of $\lambda/4$ between the elements, find the number of discrete isotropic elements needed to approximate the continuous source. Assume that the two extreme elements are placed at the edges of the continuous line source.

d. Find the normalized coefficients of the discrete array of part (c).
e. Write the array factor of the discrete array of parts (c) and (d).
f. Plot the array factor (in dB) of the discrete array of part (e).
g. For a corresponding Dolph-Tschebyscheff array, find the normalized coefficients of the discrete elements.
h. Compare the patterns of the Taylor continuous line-source distribution and discretized array, and the corresponding Dolph-Tschebyscheff discrete-element array.

Solution: For a -30 dB maximum sidelobe, the voltage ratio of the major-to-the-sidelobe level is equal to

$$30 = 20 \log_{10}(R_0) \Rightarrow R_0 = 31.62$$

a. The constant B is obtained using (7-38a) or

$$R_0 = 31.62 = 4.603 \frac{\sinh(\pi B)}{\pi B} \Rightarrow B = 1.2761$$

b. The normalized space factor pattern is obtained using (7-33), and it is shown plotted in Figure 7.10.
c. For $d = \lambda/4$ and with elements at the extremes, the number of elements is 17.

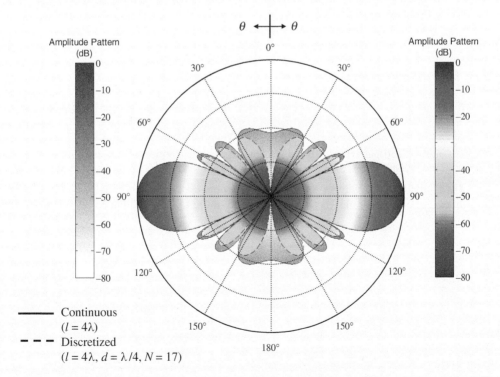

Figure 7.10 Far-field amplitude patterns of continuous and discretized Taylor (one-parameter) distributions.

d. The coefficients are obtained using (7-32). Since we have an odd number of elements, their positioning and excitation coefficients are those shown in Figure 6.19(b). Thus the total excitation coefficient of the center element is

$$2a_1 = I_n(z')|_{z'=0} = J_0(j4.009) = 11.400 \Rightarrow a_1 = 5.70$$

The coefficients of the elements on either side of the center element are identical (because of symmetry), and they are obtained from

$$a_2 = I(z')|_{z'=\pm\lambda/4} = J_0(j3.977) = 11.106$$

The coefficients of the other elements are obtained in a similar manner, and they are given by

$$a_3 = 10.192$$
$$a_4 = 8.889$$
$$a_5 = 7.195$$
$$a_6 = 5.426$$
$$a_7 = 3.694$$
$$a_8 = 2.202$$
$$a_9 = 1.000$$

e. The array factor is given by (6-61b) and (6-61c), or

$$(AF)_{17} = \sum_{n=1}^{9} a_n \cos[2(n-1)u]$$

$$u = \pi \frac{d}{\lambda} \cos\theta = \frac{\pi}{4} \cos\theta$$

where the coefficients (a_n's) are those found in part (d).

f. The normalized pattern (in dB) of the discretized distribution (discrete-element array) is shown in Figure 7.10.

g. The normalized coefficients of a 17-element Dolph-Tschebyscheff array, with -30 dB sidelobes, are obtained using the method outlined in the Design Section of Section 6.8.3 and are given by

Unnormalized	Normalized
$a_1 = 2.858$	$a_{1n} = 1.680$
$a_2 = 5.597$	$a_{2n} = 3.290$
$a_3 = 5.249$	$a_{3n} = 3.086$
$a_4 = 4.706$	$a_{4n} = 2.767$
$a_5 = 4.022$	$a_{5n} = 2.364$
$a_6 = 3.258$	$a_{6n} = 1.915$
$a_7 = 2.481$	$a_{7n} = 1.459$
$a_8 = 1.750$	$a_{8n} = 1.029$
$a_9 = 1.701$	$a_{9n} = 1.000$

As with the discretized Taylor distribution array, the coefficients are symmetrical, and the form of the array factor is that given in part (e).

h. The normalized pattern (in dB) is plotted in Figure 7.11 where it is compared with that of the discretized Taylor distribution. From the patterns in Figures 7.10 and 7.11, it can be concluded that

1. the main lobe of the continuous line-source Taylor design is well approximated by the discretized distribution with a $\lambda/4$ spacing between the elements. Even the minor lobes are well represented, and a better approximation can be obtained with more elements and smaller spacing between them.
2. the Taylor distribution array pattern has a wider main lobe than the corresponding Dolph-Tschebyscheff, but it displays decreasing minor lobes away from the main beam.

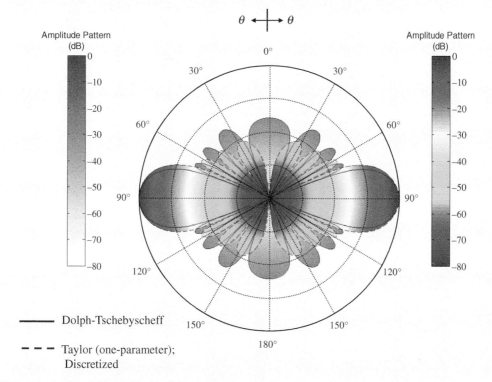

Figure 7.11 Far-field amplitude patterns of Taylor (discretized) and Dolph-Tschebyscheff distributions ($l = 4\lambda, d = \lambda/4, N = 17$).

A larger spacing between the elements does not approximate the continuous distribution as accurately. The design of Taylor and Dolph-Tschebyscheff arrays for $l = 4\lambda$ and $d = \lambda/2(N = 9)$ is assigned as a problem at the end of the chapter (Problem 7.38).

To qualitatively assess the performance between uniform, binomial, Dolph-Tschebyscheff, and Taylor (one-parameter) array designs, the amplitude distribution of each has been plotted in Figure 7.12(a). It is assumed that $l = 4\lambda, d = \lambda/4, N = 17$, and the maximum sidelobe is 30 dB down. The coefficients are normalized with respect to the amplitude of the corresponding element at the center of that array.

414 ANTENNA SYNTHESIS AND CONTINUOUS SOURCES

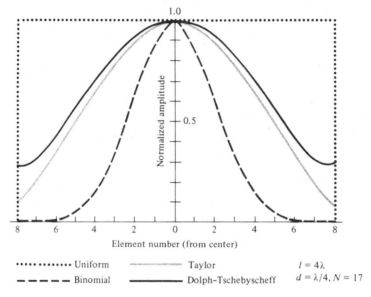

(a) Amplitude distribution of uniform, binomial, Taylor, and Dolph-Tschebyscheff discrete-element arrays

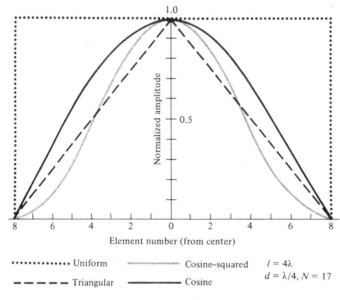

(b) Amplitude distribution of uniform, triangular, cosine, and cosine squared discrete-element arrays

Figure 7.12 Amplitude distribution of nonuniform amplitude linear arrays.

The binomial design possesses the smoothest amplitude distribution (between 1 and 0) from the center to the edges (the amplitude toward the edges is vanishingly small). Because of this characteristic, the binomial array displays the smallest sidelobes followed, in order, by the Taylor, Tschebyscheff, and the uniform arrays. In contrast, the uniform array possesses the smallest half-power beamwidth followed, in order, by the Tschebyscheff, Taylor, and binomial arrays. *As a rule of thumb, the array with the smoothest amplitude distribution (from the center to the edges) has the smallest sidelobes and the larger half-power beamwidths.* The best design is a trade-off between sidelobe level and beamwidth.

TRIANGULAR, COSINE, AND COSINE-SQUARED AMPLITUDE DISTRIBUTIONS

TABLE 7.1 Radiation Characteristics for Line-Sources and Linear Arrays with Uniform, Triangular, Cosine, and Cosine-Squared Distributions

Distribution	Uniform	Triangular	Cosine	Cosine-Squared
Distribution I_n (analytical)	I_0	$I_1\left(1 - \frac{2}{l}\lvert z'\rvert\right)$	$I_2 \cos\left(\frac{\pi}{l}z'\right)$	$I_3 \cos^2\left(\frac{\pi}{l}z'\right)$
Distribution (graphical)				
Space factor (SF) $u = \left(\frac{\pi l}{\lambda}\right)\cos\theta$	$I_0 l \frac{\sin(u)}{u}$	$I_1 \frac{l}{2}\left[\frac{\sin\left(\frac{u}{2}\right)}{\frac{u}{2}}\right]^2$	$I_2 l \frac{\pi}{2} \frac{\cos(u)}{(\pi/2)^2 - u^2}$	$I_3 \frac{l}{2} \frac{\sin(u)}{u}\left[\frac{\pi^2}{\pi^2 - u^2}\right]$
Space factor |SF|				
Half-power beamwidth (degrees) $l \gg \lambda$	$\frac{50.6}{(l/\lambda)}$	$\frac{73.4}{(l/\lambda)}$	$\frac{68.8}{(l/\lambda)}$	$\frac{83.2}{(l/\lambda)}$
First-null beamwidth (degrees) $l \gg \lambda$	$\frac{114.6}{(l/\lambda)}$	$\frac{229.2}{(l/\lambda)}$	$\frac{171.9}{(l/\lambda)}$	$\frac{229.2}{(l/\lambda)}$
First sidelobe max. (to main max.) (dB)	-13.2	-26.4	-23.2	-31.5
Directivity factor (l large)	$2\left(\frac{l}{\lambda}\right)$	$0.75\left[2\left(\frac{l}{\lambda}\right)\right]$	$0.810\left[2\left(\frac{l}{\lambda}\right)\right]$	$0.667\left[2\left(\frac{l}{\lambda}\right)\right]$

A MATLAB computer program entitled **Synthesis** has been developed, and it performs synthesis using the Schelkunoff, Fourier transform, Woodward-Lawson, Taylor (Tschebyscheff-error) and Taylor (one-parameter) methods. The program is included in the publisher's website for this the book. The description of the program is provided in the corresponding READ ME file.

7.8 TRIANGULAR, COSINE, AND COSINE-SQUARED AMPLITUDE DISTRIBUTIONS

Some other very common and simple line-source amplitude distributions are those of the triangular, cosine, cosine-squared, cosine on-a-pedestal, cosine-squared on-a-pedestal, Gaussian, inverse taper, and edge. Instead of including many details, the pattern, half-power beamwidth, first-null beamwidth, magnitude of sidelobes, and directivity for uniform, triangular, cosine, and cosine-squared amplitude distributions (with constant phase) are summarized in Table 7.1 [18], [19].

The normalized coefficients for a uniform, triangular, cosine, and cosine-squared arrays of $l = 4\lambda$, $d = \lambda/4$, $N = 17$ are shown plotted in Figure 7.12(b). The array with the smallest sidelobes and the larger half-power beamwidth is the cosine-squared, because it possesses the smoothest distribution. It is followed, in order, by the triangular, cosine, and uniform distributions. This is verified by examining the characteristics in Table 7.1.

Cosine on-a-pedestal distribution is obtained by the superposition of the uniform and the cosine distributions. Thus it can be represented by

$$I_n(z') = \begin{cases} I_0 + I_2 \cos\left(\frac{\pi}{l}z'\right), & -l/2 \leq z' \leq l/2 \\ 0 & \text{elsewhere} \end{cases} \quad (7\text{-}39)$$

where I_0 and I_2 are constants. The space factor pattern of such a distribution is obtained by the addition of the patterns of the uniform and the cosine distributions found in Table 7.1. That is,

$$\text{SF}(\theta) = I_0 l \frac{\sin(u)}{u} + I_2 \frac{\pi l}{2} \frac{\cos u}{(\pi/2)^2 - u^2} \qquad (7\text{-}40)$$

A similar procedure is used to represent and analyze a cosine-squared on-a-pedestal distribution.

7.9 LINE-SOURCE PHASE DISTRIBUTIONS

The amplitude distributions of the previous section were assumed to have uniform phase variations throughout the physical extent of the source. Practical radiators (such as reflectors, lenses, horns, etc.) have nonuniform phase fronts caused by one or more of the following:

1. displacement of the reflector feed from the focus
2. distortion of the reflector or lens surface
3. feeds whose wave fronts are not ideally cylindrical or spherical (as they are usually presumed to be)
4. physical geometry of the radiator

These are usually referred to *phase errors*, and they are more evident in radiators with tilted beams.

To simplify the analytical formulations, most of the phase fronts are represented with linear, quadratic, or cubic distributions. Each of the phase distributions can be associated with each of the amplitude distributions. In (7-1), the phase distribution of the source is represented by $\phi_n(z')$. For linear, quadratic, and cubic phase variations, $\phi_n(z')$ takes the form of

$$\text{linear:} \quad \phi_1(z') = \beta_1 \frac{2}{l} z', \qquad -l/2 \leq z' \leq l/2 \qquad (7\text{-}41a)$$

$$\text{quadratic:} \quad \phi_2(z') = \beta_2 \left(\frac{2}{l}\right)^2 z'^2, \qquad -l/2 \leq z' \leq l/2 \qquad (7\text{-}41b)$$

$$\text{cubic:} \quad \phi_3(z') = \beta_3 \left(\frac{2}{l}\right)^3 z'^3, \qquad -l/2 \leq z' \leq l/2 \qquad (7\text{-}41c)$$

and it is shown plotted in Figure 7.13. The quadratic distribution is used to represent the phase variations at the aperture of a horn and of defocused (along the symmetry axis) reflector and lens antennas.

The space factor patterns corresponding to the phase distributions of (7-41a)–(7-41c) can be obtained by using (7-1). Because the analytical formulations become lengthy and complex, especially for the quadratic and cubic distributions, they will not be included here. Instead, a general guideline of their effects will be summarized [18], [19].

Linear phase distributions have a tendency to tilt the main beam of an antenna by an angle θ_0 and to form an asymmetrical pattern. The pattern of this distribution can be obtained by replacing the u (for uniform phase) in Table 7.1 by $(u - \theta_0)$. In general, the half-power beamwidth of the tilted pattern is increased by $1/\cos \theta_0$ while the directivity is decreased by $\cos \theta_0$. This becomes more apparent by realizing that the projected length of the line-source toward the maximum is reduced by $\cos \theta_0$. Thus the effective length of the source is reduced.

Quadratic phase errors lead primarily to a reduction of directivity, and an increase in sidelobe level on either side of the main lobe. The symmetry of the original pattern is maintained. In addition, for moderate phase variations, ideal nulls in the patterns disappear. Thus the minor lobes blend into each other and into the main beam, and they represent shoulders of the main beam instead of appearing as

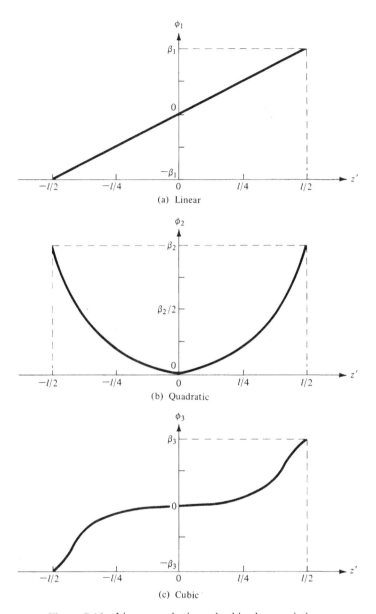

Figure 7.13 Linear, quadratic, and cubic phase variations.

separate lobes. Analytical formulations for quadratic phase distributions are introduced in Chapter 13 on horn antennas.

Cubic phase distributions introduce not only a tilt in the beam but also decrease the directivity. The newly formed patterns are asymmetrical. The minor lobes on one side are increased in magnitude and those on the other side are reduced in intensity.

7.10 CONTINUOUS APERTURE SOURCES

Space factors for aperture (two-dimensional) sources can be introduced in a similar manner as in Section 7.2.1 for line-sources.

7.10.1 Rectangular Aperture

Referring to the geometry of Figure 6.28(b), the space factor for a two-dimensional rectangular distribution along the *x-y* plane is given by

$$\text{SF} = \int_{-l_y/2}^{l_y/2} \int_{-l_x/2}^{l_x/2} A_n(x',y') e^{j[kx' \sin\theta \cos\phi + ky' \sin\theta \sin\phi + \phi_n(x',y')]} \, dx' \, dy' \qquad (7\text{-}42)$$

where l_x and l_y are, respectively, the linear dimensions of the rectangular aperture along the *x* and *y* axes. $A_n(x',y')$ and $\phi_n(x',y')$ represent, respectively, the amplitude and phase distributions on the aperture.

For many practical antennas (such as waveguides, horns, etc.) the aperture distribution (amplitude and phase) is separable. That is,

$$A_n(x',y') = I_x(x') I_y(y') \qquad (7\text{-}42a)$$

$$\phi_n(x',y') = \phi_x(x') + \phi_y(y') \qquad (7\text{-}42b)$$

so that (7-42) can be written as

$$\text{SF} = S_x S_y \qquad (7\text{-}43)$$

where

$$S_x = \int_{-l_x/2}^{l_x/2} I_x(x') e^{j[kx' \sin\theta \cos\phi + \phi_x(x')]} \, dx' \qquad (7\text{-}43a)$$

$$S_y = \int_{-l_y/2}^{l_y/2} I_y(y') e^{j[ky' \sin\theta \sin\phi + \phi_y(y')]} \, dy' \qquad (7\text{-}43b)$$

which is analogous to the array factor of (6-85)–(6-85b) for discrete-element arrays.

The evaluation of (7-42) can be accomplished either analytically or graphically. If the distribution is separable, as in (7-42a) and (7-42b), the evaluation can be performed using the results of a line-source distribution.

The total field of the aperture antenna is equal to the product of the element and space factors. As for the line-sources, the element factor for apertures depends on the type of equivalent current density and its orientation.

7.10.2 Circular Aperture

The space factor for a circular aperture can be obtained in a similar manner as for the rectangular distribution. Referring to the geometry of Figure 6.37, the space factor for a circular aperture with radius *a* can be written as

$$\text{SF}(\theta,\phi) = \int_0^{2\pi} \int_0^a A_n(\rho',\phi') e^{j[k\rho' \sin\theta \cos(\phi-\phi') + \zeta_n(\rho',\phi')]} \rho' \, d\rho' \, d\phi' \qquad (7\text{-}44)$$

CONTINUOUS APERTURE SOURCES

TABLE 7.2 Radiation Characteristics for Circular Apertures and Circular Planar Arrays with Circular Symmetry and Tapered Distribution

Distribution	Uniform	Radial Taper	Radial Taper Squared
Distribution (analytical)	$I_0\left[1-\left(\frac{\rho'}{a}\right)^2\right]^0$	$I_1\left[1-\left(\frac{\rho'}{a}\right)^2\right]^1$	$I_2\left[1-\left(\frac{\rho'}{a}\right)^2\right]^2$
Distribution (graphical)			
Space factor (SF) $u=\left(2\pi\frac{a}{\lambda}\right)\sin\theta$	$I_0 2\pi a^2 \frac{J_1(u)}{u}$	$I_1 4\pi a^2 \frac{J_2(u)}{u}$	$I_2 16\pi a^2 \frac{J_3(u)}{u}$
Half-power beamwidth (degrees) $a \gg \lambda$	$\frac{29.2}{(a/\lambda)}$	$\frac{36.4}{(a/\lambda)}$	$\frac{42.1}{(a/\lambda)}$
First-null beamwidth (degrees) $a \gg \lambda$	$\frac{69.9}{(a/\lambda)}$	$\frac{93.4}{(a/\lambda)}$	$\frac{116.3}{(a/\lambda)}$
First sidelobe max. (to main max.)	-17.6 dB	-24.6 dB	-30.6 dB
Directivity factor	$\left(\frac{2\pi a}{\lambda}\right)^2$	$0.75\left(\frac{2\pi a}{\lambda}\right)^2$	$0.56\left(\frac{2\pi a}{\lambda}\right)^2$

where ρ' is the radial distance ($0 \le \rho' \le a$), ϕ' is the azimuthal angle over the aperture ($0 \le \phi' \le 2\pi$ for $0 \le \rho' \le a$), and $A_n(\rho', \phi')$ and $\zeta_n(\rho', \phi')$ represent, respectively, the amplitude and phase distributions over the aperture. Equation (7-44) is analogous to the array factor of (6-112a) for discrete elements.

If the aperture distribution has uniform phase [$\zeta_n(\rho', \phi') = \zeta_0 = 0$] and azimuthal amplitude symmetry [$A_n(\rho', \phi') = A_n(\rho')$], (7-44) reduces, by using (5-48), to

$$\text{SF}(\theta) = 2\pi \int_0^a A_n(\rho') J_0(k\rho' \sin\theta) \rho' \, d\rho' \qquad (7\text{-}45)$$

where $J_0(x)$ is the Bessel function of the first kind and of order zero.

Many practical antennas, such as a parabolic reflector, have distributions that taper toward the edges of the apertures. These distributions can be approximated reasonably well by functions of the form

$$A_n(\rho') = \begin{cases} \left[1-\left(\frac{\rho'}{a}\right)^2\right]^n & 0 \le \rho' \le a, \quad n = 0, 1, 2, 3, \ldots \\ 0 & \text{elsewhere} \end{cases} \qquad (7\text{-}46)$$

For $n = 0$, (7-46) reduces to a uniform distribution.

The radiation characteristics of circular apertures or planar circular arrays with distributions (7-46) with $n = 0, 1, 2$ are shown tabulated in Table 7.2 [19]. It is apparent, as before, that distributions with lower taper toward the edges (larger values of n) have smaller sidelobes but larger beamwidths. In design, a compromise between sidelobe level and beamwidth is necessary.

7.11 MULTIMEDIA

In the publisher's website for this book, the following multimedia resources are included for the review, understanding, and visualization of the material of this chapter:

a. **Java**-based **interactive questionnaire**, with answers.
b. **Java**-based **applet** for computing and displaying the synthesis characteristics of Schelkunoff, Woodward-Lawson and Tschebyscheff-error designs.
c. **Matlab** computer program, designated **Synthesis**, for computing and displaying the radiation characteristics of
   - Schelkunoff
   - Fourier transform (line-source and linear array)
   - Woodward-Lawson (line-source and linear array)
   - Taylor (Tschebyscheff-error and One-parameter) synthesis designs.
d. **Power Point (PPT)** viewgraphs, in multicolor.

REFERENCES

1. S. A. Schelkunoff, "A Mathematical Theory of Linear Arrays," *Bell System Technical Journal*, Vol. 22, pp. 80–107, 1943.
2. H. G. Booker and P. C. Clemmow, "The Concept of an Angular Spectrum of Plane Waves, and Its Relation to That of Polar Diagram and Aperture Distribution," *Proc. IEE* (London), Paper No. 922, Radio Section, Vol. 97, pt. III, pp. 11–17, January 1950.
3. P. M. Woodward, "A Method for Calculating the Field over a Plane Aperture Required to Produce a Given Polar Diagram," *J. IEE*, Vol. 93, pt. IIIA, pp. 1554–1558, 1946.
4. P. M. Woodward and J. D. Lawson, "The Theoretical Precision with Which an Arbitrary Radiation-Pattern May be Obtained from a Source of a Finite Size," *J. IEE*, Vol. 95, pt. III, No. 37, pp. 363–370, September 1948.
5. T. T. Taylor, "Design of Line-Source Antennas for Narrow Beamwidth and Low Sidelobes," *IRE Trans. Antennas Propagat.*, Vol. AP-3, No. 1, pp. 16–28, January 1955.
6. T. T. Taylor, "One Parameter Family of Line-Sources Producing Modified $\text{Sin}(\pi u)/\pi u$ Patterns," Hughes Aircraft Co. Tech. Mem. 324, Culver City, Calif., Contract AF 19(604)-262-F-14, September 4, 1953.
7. R. S. Elliott, "On Discretizing Continuous Aperture Distributions," *IEEE Trans. Antennas Propagat.*, Vol. AP-25, No. 5, pp. 617–621, September 1977.
8. R. C. Hansen (ed.), *Microwave Scanning Antennas*, Vol. I, Academic Press, New York, 1964, p. 56.
9. J. Ruze, "Physical Limitations on Antennas," MIT Research Lab., Electronics Tech. Rept. 248, October 30, 1952.
10. M. I. Skolnik, *Introduction to Radar Systems*, McGraw-Hill, New York, 1962, pp. 320–330.
11. R. S. Elliott, "Criticisms of the Woodward-Lawson Method," *IEEE Antennas Propagation Society Newsletter*, Vol. 30, p. 43, June 1988.
12. H. Steyskal, "The Woodward-Lawson Method: A Second Opinion," *IEEE Antennas Propagation Society Newsletter*, Vol. 30, p. 48, October 1988.
13. R. S. Elliott, "More on the Woodward-Lawson Method," *IEEE Antennas Propagation Society Newsletter*, Vol. 30, pp. 28–29, December 1988.
14. H. Steyskal, "The Woodward-Lawson Method-To Bury or Not to Bury," *IEEE Antennas Propagation Society Newsletter*, Vol. 31, pp. 35–36, February 1989.
15. H. J. Orchard, R. S. Elliott, and G. J. Stern. "Optimizing the Synthesis of Shaped Beam Antenna Patterns," *IEE Proceedings*, Part H, pp. 63–68, 1985.

16. R. S. Elliott, "Design of Line-Source Antennas for Narrow Beamwidth and Asymmetric Low Sidelobes," *IEEE Trans. Antennas Propagat.*, Vol. AP-23, No. 1, pp. 100–107, January 1975.
17. G. N. Watson, *A Treatise on the Theory of Bessel Functions*, 2nd. Ed., Cambridge University Press, London, pp. 50 and 379, 1966.
18. S. Silver (ed.), *Microwave Antenna Theory and Design*, MIT Radiation Laboratory Series, Vol. 12, McGraw-Hill, New York, 1965, Chapter 6, pp. 169–199.
19. R. C. Johnson and H. Jasik (eds.), *Antenna Engineering Handbook*, 2nd. Ed., McGraw-Hill, New York, 1984, pp. 2–14 to 2-41.

PROBLEMS

7.1. A three-element array is placed along the z-axis. Assuming the spacing between the elements is $d = \lambda/4$ and the relative amplitude excitation is equal to $a_1 = 1, a_2 = 2, a_3 = 1$,
   (a) find the angles where the array factor vanishes when $\beta = 0, \pi/2, \pi$, and $3\pi/2$
   (b) plot the relative pattern for each array factor
   Use Schelkunoff's method.

7.2. Design a linear array of isotropic elements placed along the z-axis such that the zeros of the array factor occur at $\theta = 0°, 60°$, and $120°$. Assume that the elements are spaced $\lambda/4$ apart and that the progressive phase shift between them is $0°$.
   (a) Find the required number of elements.
   (b) Determine their excitation coefficients.
   (c) Write the array factor.
   (d) Plot the array factor pattern to verify the validity of the design.
   Verify using the computer program **Synthesis**.

7.3. To minimize interference between the operational system, whose antenna is a linear array with elements placed along the z-axis, and other undesired sources of radiation, it is required that nulls be placed at elevation angles of $\theta = 0°, 60°, 120°$, and $180°$. The elements will be separated with a uniform spacing of $\lambda/4$. Choose a synthesis method that will allow you to design such an array that will meet the requirements of the amplitude pattern of the array factor. To meet the requirements:
   (a) Specify the synthesis method you will use.
   (b) Determine the number of elements.
   (c) Find the excitation coefficients.

7.4. It is desired to synthesize a discrete array of vertical infinitesimal dipoles placed along the z-axis with a spacing of $d = \lambda/2$ between the adjacent elements. It is desired for the array factor to have nulls along $\theta = 60°, 90°$, and $120°$. *Assume there is no initial progressive phase excitation between the elements.* To achieve this, determine:
   (a) number of elements.
   (b) excitation coefficients.
   (c) angles (*in degrees*) of all the nulls of the entire array (*including of the actual elements*).

7.5. It is desired to synthesize a linear array of elements with spacing $d = 3\lambda/8$. It is important that the array factor (AF) exhibits nulls along $\theta = 0, 90$, and 180 *degrees*. Assume there is no initial progressive phase excitation between the elements (i.e., $\beta = 0$). To achieve this design, determine:
   (a) The number of elements

(b) The excitation coefficients (amplitude and phase)

If the design allows the progressive phase shift (β) to change, while maintaining the spacing constant ($d = 3\lambda/8$),

(c) What would it be the range of possible values for the progressive phase shift to cause the null at $\theta = 90$ *degrees* disappear (to place its corresponding root outside the visible region)?

7.6. The z-plane array factor of an array of isotropic elements placed along the z-axis is given by

$$\text{AF} = z(z^4 - 1)$$

Determine the

(a) number of elements of the array. If there are any elements with zero excitation coefficients (null elements), so indicate
(b) position of each element (including that of null elements) along the z axis
(c) magnitude and phase (in degrees) excitation of each element
(d) angles where the pattern vanishes when the total array length (including null elements) is 2λ

Verify using the computer program **Synthesis**.

7.7. Repeat Problem 7.6 when

$$\text{AF} = z(z^3 - 1)$$

7.8. The z-plane array factor of an array of isotropic elements placed along the z-axis is given by (assume $\beta = 0$)

$$\text{AF}(z) = (z+1)^3$$

Determine the

(a) Number of elements of the discrete array to have such an array factor.
(b) *Normalized* excitation coefficients of each of the elements of the array (*the ones at the edges to be unity*).
(c) *Classical name* of the array design with these excitation coefficients.
(d) Angles in theta (θ *in degrees*) of *all the nulls* of the array factor when the spacing d between the elements is $d = \lambda/2$.
(e) Half-power beamwidth (*in degrees*) of the array factor when $d = \lambda/2$.
(f) Maximum directivity (*dimensionless* and *in dB*) of the array factor when $d = \lambda/2$.

7.9. The Array Factor in the complex z-plane ($z = x + jy$) of a linear array, with its elements placed along the z-axis, is given by:

$$\text{AF}(z) = (z^2 + 1)(z - 1)$$

Assuming a phase excitation of $\beta = 45°$ *and a spacing of* $d = \lambda/8$ between the elements, determine *analytically*, the:

(a) *Actual number of elements* that *contribute in the visible region* ($0 \leq \theta \leq 180°$).
(b) Excitation coefficients of these actual number of elements.
(c) ALL the angles θ (in degrees) in the visible region ($0 \leq \theta \leq 180°$) where the nulls will occur.

7.10. The z-plane array factor of an array of isotropic elements placed along the z-axis is given by (assume $\beta = 0$)

$$\text{AF}(z) = (z+1)^4$$

Determine the
(a) Number of elements of the discrete array to have such an array factor.
(b) *Normalized* excitation coefficients of each of the elements of the array (*the ones at the edges to be unity*).
(c) *Classical name* of the array design with these excitation coefficients.
(d) Angles in theta (θ *in degrees*) of *all the nulls* of the array factor when the spacing d between the elements is
   1. $d = \lambda_0/4$ 2. $d = \lambda_0/2$
(e) Half-power beamwidth (*in degrees*) of the array factor when $d = \lambda_0/2$.
(f) Maximum directivity (*dimensionless* and *in dB*) of the array factor when $d = \lambda_0/2$.

7.11. The desired array factor in complex form (z-plane) of an array, with the elements along the z-axis, is given by

$$\text{AF}(z) = (z^4 - \sqrt{2}z^3 + 2z^2 - \sqrt{2}z + 1) = (z^2+1)(z^2 - \sqrt{2}z + 1)$$

$$= (z^2+1)\left[\left(z^2 - \sqrt{2}z + \frac{1}{2}\right) + \frac{1}{2}\right] = (z^2+1)\left[\left(z - \frac{1}{\sqrt{2}}\right)^2 + \frac{1}{2}\right]$$

where $z = x + jy$ in the complex z-plane.
(a) Determine the *number of elements* that you will need to realize this array factor.
(b) Determine *all the roots* of this array factor *on the unity circle* of the complex plane.
(c) For a spacing of $d = \lambda/4$ and zero initial phase ($\beta = 0$), determine *all the angles* θ (*in degrees*) where *this pattern possesses nulls*.

7.12. The z-plane ($z = x + jy$) array factor of a linear array of elements placed along the z-axis, with a uniform spacing d between them and with $\beta = 0$, is given by

$$\text{AF} = z(z^2 + 1)$$

Determine, analytically, the
(a) number of elements of the array;
(b) excitation coefficients;
(c) *all* the roots of the array factor in the *visible region only* ($0 \leq \theta \leq 180°$) when $d = \lambda/4$;
(d) *all* the nulls of the array factor (*in degrees*) in the *visible region only* ($0 \leq \theta \leq 180°$) when $d = \lambda/4$.

Verify using the computer program **Synthesis**.

7.13. The array factor, in the complex z-plane ($z = x + jy$), of a linear array with elements placed along the z-axis is given by

$$\text{AF} = z^4 - 1$$

For a uniform spacing of $d = \lambda/4$ between the elements and assuming a phase excitation of $\beta = 0$, determine, *analytically*:

(a) The number of elements of the array, including those that may have null excitation.
(b) The excitation coefficient of each element, including any if they are zero.
(c) *All* the roots of the array factor.
(d) *All* the roots of the array factor in the *visible region*.
(e) *All* the nulls of the array factor (*in degrees*) only in the visible region ($0° \leq \theta \leq 180°$).

7.14. The roots of the array factor of a linear array in the z-plane ($z = x + jy$) array, with the elements placed along the z-axis and with a uniform spacing of $d = \lambda/2$ between them, are given by

$$z_1 = +1, z_2 = -1, z_3 = +j, z_4 = -j$$

Assuming a phase excitation of $\beta = 0$, determine, *analytically*, the:
(a) Array factor.
(b) Number of elements of the array.
(c) Excitation coefficients of each element. Indicate any that may be zero/null.
(d) *All* the nulls of the array factor (*in degrees*) only in the visible ($0 \leq \theta \leq 180°$).

7.15. It is desired to design a linear array, with the elements along the z-axis, with a spacing of $d = \lambda/2$ and a progressive phase excitation of $\beta = 0°$ between the elements. It is also necessary that the array exhibits nulls along $\theta = 60°, 90°$ and $120°$.

(a) Select an array design that will meet these specifications (*give its name*).
(b) Determine the array factor (in expanded polynomial form) expressed in the complex z-plane.
(c) How many elements are necessary to meet the desired specifications?
(d) State the amplitude excitation coefficients for each of the array elements.

7.16. Repeat Example 7.2 when

$$SF(\theta) = \begin{cases} 1 & 40° \leq \theta \leq 140° \\ 0 & \text{elsewhere} \end{cases}$$

Verify using the computer program **Synthesis**.

7.17. Repeat the Fourier transform design of Example 7.2 for a line-source along the z-axis whose sectoral pattern is given by

$$SF(\theta) = \begin{cases} 1 & 60° \leq \theta \leq 120° \\ 0 & \text{elsewhere} \end{cases}$$

Use $l = 5\lambda$ and 10λ. Compare the reconstructed patterns with the desired one.
Verify using the computer program **Synthesis**.

7.18. Repeat the Fourier transform design of Problem 7.17 for a linear array with a spacing of $d = \lambda/2$ between the elements and
(a) $N = 11$ elements
(b) $N = 21$ elements

7.19. Repeat the design of Problem 7.17 using the Woodward-Lawson method for line sources.

7.20. Repeat the design of Problem 7.18 using the Woodward-Lawson method for linear arrays for $N = 10, 20$.

7.21. Design, using the Woodward-Lawson method, a line-source of $l = 5\lambda$ whose space factor pattern is given by

$$\text{SF}(\theta) = \sin^3(\theta) \qquad 0° \leq \theta \leq 180°$$

Determine the current distribution and compare the reconstructed pattern with the desired pattern. Verify using the computer program **Synthesis**.

7.22. The desired Space Factor SF of a *line-source* of $l = 3\lambda$ is the sectoral pattern of

$$\text{SF} = \begin{cases} \dfrac{1}{30}\theta - 2, & 60° \leq \theta \leq 90° \\ -\dfrac{1}{30}\theta + 4, & 90° \leq \theta \leq 120° \\ 0 & \text{elsewhere} \end{cases} \quad (\theta \text{ is in degrees})$$

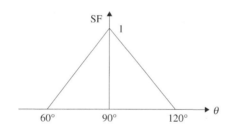

It is required to synthesize the desired pattern using the *Woodward–Lawson method with odd samples*. To accomplish this, determine *analytically* when $l = 3\lambda$, the
(a) number of samples;
(b) *all* the angles (*in degrees*) where the SF is sampled;
(c) *all* the excitation coefficients b_m at the sampling points.
Verify using the computer program **Synthesis**.

7.23. The space factor of a linear array of discrete elements placed along the z-axis, with a phase of $\beta = 0$ between the elements, is given by the polynomial

$$\text{SF} = (z^2 + 1)(z + 1)$$

Determine the:
(a) Number of elements necessary.
(b) Corresponding excitation coefficients of the elements.
(c) All the nulls in θ (*in degrees*) ($0° \leq \theta \leq 180°$) when the spacing between the elements is $d = \lambda/4$.
(d) All the nulls in θ (*in degrees*) ($0° \leq \theta \leq 180°$) when the spacing between the elements is $d = \lambda/2$.

7.24. Repeat Problem 7.22 using the *Woodward-Lawson method with even samples*.

7.25. Using the *Woodward-Lawson method*, design a *line source* with its length along the z-axis, whose Space Factor is represented by the normalized triangular distribution shown below. Assuming the length of the line source is $l = 4\lambda$ and the observation angular region is $0° \leq \theta \leq 180°$:

(a) Write expressions, in terms of the observation angle θ (*in degrees*), that represent the Space Factor that is displayed graphically below.

(b) Assuming *odd samples*, determine the angles θ (in degrees) where the Space Factor should be sampled.

(c) Determine the normalized amplitude of the Space Factor *at the sampled angles*.

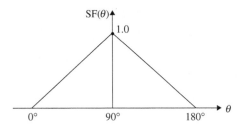

7.26. Repeat the design of Problem 7.21 for a linear array of $N = 10$ elements with a spacing of $d = \lambda/2$ between them.

7.27. It is desired to synthesize the space factor of a line source, of length $l = 3\lambda$, whose normalized ideal space factor is

$$SF = \sin^2(\theta)$$

Using the *Woodward-Lawson method* and assuming *even samples*, determine analytically the:

(a) *Total* number of samples.

(b) *All* the angles (*in degrees*) where the SF should be sampled.

(c) *All* the excitation coefficients b_m of the corresponding composing functions at the sampling points. If there are any excitation coefficients which are null (zero), state them.

7.28. In target-search, grounding-mapping radars, and in airport beacons it is desirable to have the echo power received from a target, of constant cross section, to be independent of its range R.

Generally, the far-zone field radiated by an antenna is given by

$$|E(R, \theta, \phi)| = C_0 \frac{|F(\theta, \phi)|}{R}$$

where C_0 is a constant. According to the geometry of the figure

$$R = h/\sin(\theta) = h \csc(\theta)$$

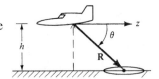

For a constant value of ϕ, the radiated field expression reduces to

$$|E(R, \theta, \phi = \phi_0)| = C_0 \frac{|F(\theta, \phi = \phi_0)|}{R} = C_1 \frac{|f(\theta)|}{R}$$

A constant value of field strength can be maintained provided the radar is flying at a constant altitude h and the far-field antenna pattern is equal to

$$f(\theta) = C_2 \csc(\theta)$$

This is referred to as a cosecant pattern, and it is used to compensate for the range variations. For very narrow beam antennas, the total pattern is approximately equal to the space or array factor. Design a line-source, using the Woodward-Lawson method, whose space factor is given by

$$SF(\theta) = \begin{cases} 0.342 \csc(\theta), & 20° \le \theta \le 60° \\ 0 & \text{elsewhere} \end{cases}$$

Plot the synthesized pattern for $l = 20\lambda$, and compare it with the desired pattern. Verify using the computer program **Synthesis**.

7.29. Repeat the design of Problem 7.28 for a linear array of $N = 41$ elements with a spacing of $d = \lambda/2$ between them.

7.30. For some radar search applications, it is more desirable to have an antenna which has a square beam for $0 \le \theta \le \theta_0$, a cosecant pattern for $\theta_0 \le \theta \le \theta_m$, and it is zero elsewhere. Design a line-source, using the Woodward-Lawson method, with a space factor of

$$SF(\theta) = \begin{cases} 1 & 15° \le \theta < 20° \\ 0.342 \csc(\theta) & 20° \le \theta \le 60° \\ 0 & \text{elsewhere} \end{cases}$$

Plot the reconstructed pattern for $l = 20\lambda$, and compare it with the desired pattern.

7.31. To maximize the aperture efficiency, and thus the directivity, of a paraboloidal (parabola of revolution) reflector, the desired normalized space factor power pattern $g(\theta)$ of the feed antenna is given by

$$SF(\theta) = g(\theta) = \sec^4\left(\frac{90° - \theta}{2}\right) \qquad 60° \le \theta \le 120°$$

It is desired to synthesize the desired antenna feed pattern using the *Woodward-Lawson method with a line source of total length* 3λ. Assuming *odd samples*, determine analytically the:

(a) Number of samples.
(b) *All* the angles (*in degrees*) where the SF should be sampled.
(c) *All* the normalized excitation coefficients b_m of the corresponding composing functions at the sampling points. If there are any excitation coefficients which are null (zero), so indicate.

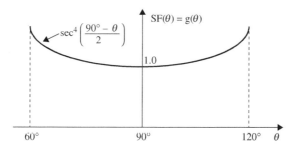

7.32. Repeat the design of Problem 7.30, using the Woodward-Lawson method, for a linear array of 41 elements with a spacing of $d = \lambda/2$ between them.

7.33. Repeat Problem 7.22 for a linear array of $l = 3\lambda$ with a spacing of $d = 0.5\lambda$ between the elements. Replacing the space factor (SF) of Problem 7.22 with an array factor (AF) and using the *Woodward-Lawson method with odd samples*, determine analytically the
(a) number of elements;   (b) number of samples;
(c) angles (*in degrees*) where the AF is sampled;
(d) *all* the excitation coefficients b_m at the sampling points.
(e) *all* the normalized excitation coefficients a_n of the elements.
Verify using the computer program **Synthesis**.

7.34. Repeat Problem 7.33 using the *Woodward-Lawson method with even samples*.

7.35. Design a Taylor (Tschebyscheff-error) line-source with a
(a) −25 dB sidelobe level and $\bar{n} = 5$
(b) −20 dB sidelobe level and $\bar{n} = 10$
For each, find the half-power beamwidth and plot the normalized current distribution and the reconstructed pattern when $l = 10\lambda$. Verify using the computer program **Synthesis**.

7.36. Derive (7-33) using (7-1), (7-32), and Gegenbauer's finite integral and polynomials.

7.37. Design a *Taylor (Tschebyscheff Error)* line source, with its length placed along the z-axis, so that the sidelobe level of the pattern for the 2 inner nulls is maintained ideally at −40 dB. Determine the:
(a) Constant A so that the sidelobes of the line source are maintained at the −40 dB level.
(b) Scaling factor so that the inner most nulls blend with the outer most.
(c) Angles θ (in degrees) ($0° \leq \theta \leq 180°$) where ALL the inner and outer most nulls occur. Assume the length l of the line source is $l = 3\lambda$.

7.38. Repeat the design of Example 7.7 for an array with $l = 4\lambda, d = \lambda/2, N = 9$.

7.39. Using a spacing of $d = \lambda/4$ between the elements, determine for a Taylor Line-Source (Tschebyscheff-error) of −30 dB, for $\bar{n} = 4$ and an *equivalent length* equal to that of *20 discrete elements*:
(a) The half-power beamwidth (*in degrees*).
(b) State whether the half-power beamwidth in *Part a* is larger or smaller than that of an equivalent design for a Dolph-Tschebyscheff array with the same sidelobe level. *Explain as to why one is larger or smaller than the other.*
(c) The number of complete *innermost minor lobes* that would have *approximately equal* sidelobe level.

7.40. It is desired to synthesize a Taylor (*Tschebyscheff-Error*) line-source continuous distribution source with a −26 dB side lobe level. The total length of the array is 5λ, and it is desired to maintain −26 dB side lobe level for $\bar{n} = 3$.
Determine the half-power beamwidth (*in degrees*) of the:
(a) Taylor (Tschebyscheff-Error) design.
(b) Corresponding Dolph-Tschebyscheff array design with the same side lobe level.

7.41. Design a broadside five-element, −40 dB sidelobe level Taylor (one-parameter) distribution array of isotropic sources. The elements are placed along the x-axis with a spacing of $\lambda/4$ between them. Determine the
(a) normalized excitation coefficients (amplitude and phase) of each element
(b) array factor
Verify using the computer program **Synthesis**.

7.42. Given a continuous line-source, whose total length is 4λ, design a *symmetrical Taylor One-Parameter* distribution array whose sidelobe is 35 dB down from the maximum of the major lobe.
   (a) Find the constant B.
   (b) For a spacing of λ between the elements, find the number of discrete elements needed to approximate the continuous source. *Assume that the two extreme elements are placed at the edges of the continuous line source.*
   (c) Find the total normalized coefficients (edge elements to be unity) of the discrete array of part (c). Identify the position of the corresponding elements.
   (d) Write the array factor of the discrete array of parts (c) and (d).
   Verify using the computer program **Synthesis**.

7.43. Derive the space factors for uniform, triangular, cosine, and cosine-squared line-source continuous distributions. Compare with the results in Table 7.1.

7.44. Compute the half-power beamwidth, first-null beamwidth, first sidelobe level (in dB), and directivity of a linear array of closely spaced elements with overall length of 4λ when its amplitude distribution is
   (a) uniform   (b) triangular
   (c) cosine   (d) cosine-squared

7.45. Design a continuous line-source of length $l = 5\lambda$ whose half-power beamwidth of the space factor is $16.64°$. Determine the:
   (a) Amplitude tapering over the length of the line source that will lead to the desired half-power beamwidth; i.e., state the desired amplitude distribution.
   (b) Corresponding side lobe level (*in dB*) of the space factor.
   (c) Corresponding directivity (*dimensionless and in dB*) of the space factor.

7.46. A line source of total length l, with a *continuous triangular current distribution*, is center-fed to achieve the desired excitation. Assuming the total length is $l = 3\lambda$, *calculate* (you do not have to derive them) the:
   (a) Half-power beamwidth (*in degrees*).
   (b) Side lobe level (*in dB*)
   (c) Directivity (*dimensionless and in dB*)
   (d) Maximum effective area (*in λ^2*). Assume no losses.
   (e) Aperture efficiency (*in %*) when the diameter of the wire representing the line source is $\lambda/300$. Assume that the physical area of the line source is the cross section of the wire along its length.
   (f) Maximum power delivered to a load connected to the line source when the *incident wave is circularly polarized* with an *incident power density of 10^{-3} Watts/cm^2*. Assume the operating frequency is 1 GHz.

7.47. A nonuniform linear array with *a triangular symmetrical distribution* consists of seven discrete elements placed $\lambda/2$ *apart*. The *total length (edge-to-edge with elements at the edges)* of the array is $l = 3\lambda$. Determine the following:
   (a) *Normalized amplitude* excitation coefficients (*maximum is unity*).
   (b) Half-power beamwidth (*in degrees*).
   (c) Maximum directivity (*dimensionless and in dB*) for the triangular distribution.
   (d) Maximum directivity (*dimensionless and in dB*) *if the distribution were uniform*. How does it compare with that of Part c? Is it smaller or larger, and *by how many dB?*

(e) Maximum effective length (*assuming that the distribution of the discrete array is the same as that of a continuous line-source*) compared to physical length l:
  1. If the distribution *is triangular*.
  2. If the distribution *is uniform*.

7.48. Synthesize a 7-element nonuniform array *with symmetrical amplitude excitation and with uniform spacing between the elements*. The desired amplitude distribution across the elements (*relative to the center of the array*) is $\cos^2$ [i.e., $\cos^2(\pi x'/L)$] where L is the total length of the array (*the end elements are at the edges of the array length*) and x' is the position of each element *relative to the center* of the array. *The overall length L of the array is 3λ*. The end elements are at the edges of the array length. Determine the:
  (a) Spacing between the elements (*in λ*).
  (b) Normalized *total* amplitude coefficients of each of the elements (*normalize them so that the amplitude of the center element is unity*). Identify each of the element position and its corresponding normalized amplitude coefficient.
  (c) Approximate half-power beamwidth (*in degrees*).
  (d) Approximate maximum directivity (*in dB*).

7.49. Derive the space factors for the uniform radial taper, and radial taper-squared circular aperture continuous distributions. Compare with the results in Table 7.2.

7.50. Design a *nonuniform* line source, placed along the z-axis, *of total length of 4λ*. It is desired that the *major lobe is at broadside ($\theta = 90°$)* and the *side lobe of the first minor lobe of the space factor is approximately -23 dB* from the maximum of the major lobe.
Choose the easiest design which will satisfy the requirement of -23 dB side lobe level of the first minor lobe.
  (a) Determine the nonuniform design/distribution. *State its name*.
  (b) If the continuous line source is approximated by a discrete array of 9 elements (end-to-end), determine the spacing (*in λ*) between the elements.
  (c) *For part b*, determine the *normalized* amplitude excitation coefficients of the elements (*normalized to the value of the element at the center; make the total amplitude excitation of the center element equal to unity*).
  (d) Determine the half-power beamwidth (*in degrees*).
  (e) Compute the directivity (*dimensionless and in dB*).

7.51. Compute the half-power beamwidth, first-null beamwidth, first sidelobe level (in dB), and gain factor of a circular planar array of closely spaced elements, with radius of 2λ when its amplitude distribution is
  (a) uniform   (b) radial taper   (c) radial taper-squared.

CHAPTER 8

Integral Equations, Moment Method, and Self and Mutual Impedances

8.1 INTRODUCTION

In Chapter 2 it was shown, by the Thévenin and Norton equivalent circuits of Figures 2.27 and 2.28, that an antenna can be represented by an equivalent impedance $Z_A[Z_A = (R_r + R_L) + jX_A]$. The equivalent impedance is attached across two terminals (terminals $a - b$ in Figures 2.27 and 2.28) which are used to connect the antenna to a generator, receiver, or transmission line. In general, this impedance is called the *driving-point* impedance. However, when the antenna is radiating in an unbounded medium, in the absence of any other interfering elements or objects, the driving-point impedance is the same as the *self-impedance* of the antenna. In practice, however, there is always the ground whose presence must be taken into account in determining the antenna driving-point impedance. The self- and driving-point impedances each have, in general, a real and an imaginary part. The real part is designated as the resistance and the imaginary part is called the reactance.

The impedance of an antenna depends on many factors including its frequency of operation, its geometry, its method of excitation, and its proximity to the surrounding objects. Because of their complex geometries, only a limited number of practical antennas have been investigated analytically. For many others, the input impedance has been determined experimentally.

The impedance of an antenna at a point is defined as the ratio of the electric to the magnetic fields at that point; alternatively, at a pair of terminals, it is defined as the ratio of the voltage to the current across those terminals. There are many methods that can be used to calculate the impedance of an antenna [1]. Generally, these can be classified into three categories: (1) the boundary-value method, (2) the transmission-line method, and (3) the Poynting vector method. Extensive and brief discussions and comparisons of these methods have been reported [1], [2].

The boundary-value approach is the most basic, and it treats the antenna as a boundary-value problem. The solution to this is obtained by enforcing the boundary conditions (usually that the tangential electric-field components vanish at the conducting surface). In turn, the current distribution and finally the impedance (ratio of applied emf to current) are determined, with no assumptions as to their distribution, as solutions to the problem. The principal disadvantage of this method is that it has limited applications. It can only be applied and solved exactly on simplified geometrical shapes where the scalar wave equation is separable.

The transmission-line method, which has been used extensively by Schelkunoff [3], treats the antenna as a transmission line, and it is most convenient for the biconical antenna. Since it utilizes

---

Antenna Theory: Analysis and Design, Fourth Edition. Constantine A. Balanis.
© 2016 John Wiley & Sons, Inc. Published 2016 by John Wiley & Sons, Inc.
Companion Website: www.wiley.com/go/antennatheory4e

tangential electric-field boundary conditions for its solution, this technique may also be classified as a boundary-value method.

The basic approach to the Poynting vector method is to integrate the Poynting vector (power density) over a closed surface. The closed surface chosen is usually either a sphere of a very large radius r ($r \geq 2D^2/\lambda$ where D is the largest dimension of the antenna) or a surface that coincides with the surface of the antenna. The large sphere closed surface method has been introduced in Chapters 4 and 5, but it lends itself to calculations only of the real part of the antenna impedance (radiation resistance). The method that utilizes the antenna surface has been designated as the induced emf method, and it has been utilized [4]-[6] for the calculation of antenna impedances.

The impedance of an antenna can also be found using an integral equation with a numerical technique solution, which is widely referred to as the *Integral Equation Method of Moments* [7]-[14]. This method, which in the late 1960s was extended to include electromagnetic problems, is analytically simple, it is versatile, but it requires large amounts of computation. The limitation of this technique is usually the speed and storage capacity of the computer.

In this chapter the integral equation method, with a Moment Method numerical solution, will be introduced and used first to find the self- and driving-point impedances, and mutual impedance of wire type of antennas. This method casts the solution for the induced current in the form of an integral (hence its name) where the unknown induced current density is part of the integrand. Numerical techniques, such as the *Moment Method* [7]-[14], can then be used to solve the current density. In particular two classical integral equations for linear elements, *Pocklington's and Hallén's Integral Equations*, will be introduced. This approach is very general, and it can be used with today's modern computational methods and equipment to compute the characteristics of complex configurations of antenna elements, including skewed arrangements. For special cases, closed-form expressions for the self, driving-point, and mutual impedances will be presented using the induced emf method. This method is limited to classical geometries, such as straight wires and arrays of collinear and parallel straight wires.

8.2 INTEGRAL EQUATION METHOD

The objective of the Integral Equation (IE) method for radiation or scattering is to cast the solution for the unknown current density, which is induced on the surface of the radiator/scatterer, in the form of an integral equation where the unknown induced current density is part of the integrand. The integral equation is then solved for the unknown induced current density using numerical techniques such as the *Moment Method* (MM). To demonstrate this technique, we will initially consider some specific problems. For introduction, we will start with an electrostatics problem and follow it with time-harmonic problems.

8.2.1 Electrostatic Charge Distribution

In electrostatics, the problem of finding the potential that is due to a given charge distribution is often considered. In physical situations, however, it is seldom possible to specify a charge distribution. Whereas we may connect a conducting body to a voltage source, and thus specify the potential throughout the body, the distribution of charge is obvious only for a few rotationally symmetric geometries. In this section we will consider an integral equation approach to solve for the electric charge distribution once the electric potential is specified. Some of the material here and in other sections is drawn from [15], [16].

From statics we know that a linear electric charge distribution $\rho(r')$ creates a scalar electric potential, $V(r)$, according to [17]

$$V(r) = \frac{1}{4\pi\varepsilon_0} \int_{\substack{\text{source} \\ \text{(charge)}}} \frac{\rho(r')}{R} dl' \qquad (8\text{-}1)$$

INTEGRAL EQUATION METHOD   **433**

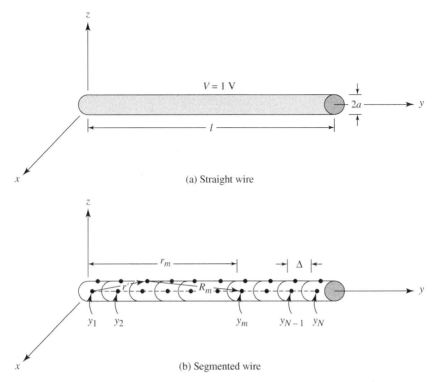

Figure 8.1   Straight wire of constant potential and its segmentation.

where $r'(x', y', z')$ denotes the source coordinates, $r(x, y, z)$ denotes the observation coordinates, dl' is the path of integration, and R is the distance from any one point on the source to the observation point, which is generally represented by

$$R(r, r') = |\mathbf{r} - \mathbf{r}'| = \sqrt{(x - x')^2 + (y - y')^2 + (z - z')^2} \qquad (8\text{-}1a)$$

We see that (8-1) may be used to calculate the potentials that are due to any known line charge density. However, the charge distribution on most configurations of practical interest, i.e., complex geometries, is not usually known, even when the potential on the source is given. It is the nontrivial problem of determining the charge distribution, for a specified potential, that is to be solved here using an integral equation-numerical solution approach.

A. Finite Straight Wire

Consider a straight wire of length l and radius a, placed along the y axis, as shown in Figure 8-1(a). The wire is given a normalized constant electric potential of 1 V.

Note that (8-1) is valid everywhere, including on the wire itself ($V_{\text{wire}} = 1$ V). Thus, choosing the observation along the wire axis ($x = z = 0$) and representing the charge density on the surface of the wire, (8-1) can be expressed as

$$1 = \frac{1}{4\pi\varepsilon_0} \int_0^l \frac{\rho(y')}{R(y, y')} dy', \quad 0 \le y \le l \qquad (8\text{-}2)$$

where

$$R(y, y') = R(r, r')|_{x=z=0} = \sqrt{(y-y')^2 + [(x')^2 + (z')^2]}$$
$$= \sqrt{(y-y')^2 + a^2} \tag{8-2a}$$

The observation point is chosen along the wire axis and the charge density is represented along the surface of the wire to avoid $R(y, y') = 0$, which would introduce a singularity in the integrand of (8-2).

It is necessary to solve (8-2) for the unknown $\rho(y')$ (an inversion problem). Equation (8-2) is an integral equation that can be used to find the charge density $\rho(y')$ based on the 1 V potential. The solution may be reached numerically by reducing (8-2) to a series of linear algebraic equations that may be solved by conventional matrix equation techniques. To facilitate this, let us approximate the unknown charge distribution $\rho(y')$ by an expansion of N known terms with constant, but unknown, coefficients, that is,

$$\rho(y') = \sum_{n=1}^{N} a_n g_n(y') \tag{8-3}$$

Thus, (8-2) may be written, using (8-3), as

$$4\pi\varepsilon_0 = \int_0^l \frac{1}{R(y, y')} \left[\sum_{n=1}^{N} a_n g_n(y') \right] dy' \tag{8-4}$$

Because (8-4) is a nonsingular integral, its integration and summation can be interchanged, and it can be written as

$$4\pi\varepsilon_0 = \sum_{n=1}^{N} a_n \int_0^l \frac{g_n(y')}{\sqrt{(y-y')^2 + a^2}} dy' \tag{8-4a}$$

The wire is now divided into N uniform segments, each of length $\Delta = l/N$, as illustrated in Figure 8.1(b). The $g_n(y')$ functions in the expansion (8-3) are chosen for their ability to accurately model the unknown quantity, while minimizing computation. They are often referred to as *basis* (or expansion) functions, and they will be discussed further in Section 8.4.1. To avoid complexity in this solution, subdomain piecewise constant (or "pulse") functions will be used. These functions, shown in Figure 8.8, are defined to be of a constant value over one segment and zero elsewhere, or

$$g_n(y') = \begin{cases} 0 & y' < (n-1)\Delta \\ 1 & (n-1)\Delta \leq y' \leq n\Delta \\ 0 & n\Delta < y' \end{cases} \tag{8-5}$$

Many other basis functions are possible, some of which will be introduced later in Section 8.4.1.

Replacing y in (8-4) by a fixed point such as y_m, results in an integrand that is solely a function of y', so the integral may be evaluated. Obviously, (8-4) leads to one equation with N unknowns a_n written as

$$4\pi\varepsilon_0 = a_1 \int_0^{\Delta} \frac{g_1(y')}{R(y_m,y')} dy' + a_2 \int_{\Delta}^{2\Delta} \frac{g_2(y')}{R(y_m,y')} dy' + \cdots \\ + a_n \int_{(n-1)\Delta}^{n\Delta} \frac{g_n(y')}{R(y_m,y')} dy' + \cdots + a_N \int_{(N-1)\Delta}^{l} \frac{g_N(y')}{R(y_m,y')} dy' \quad (8\text{-}6)$$

In order to obtain a solution for these N amplitude constants, N linearly independent equations are necessary. These equations may be produced by choosing N observation points y_m each at the center of each Δ length element as shown in Figure 8.1(b). This results in one equation of the form of (8-6) corresponding to each observation point. For N such points, we can reduce (8-6) to

$$4\pi\varepsilon_0 = a_1 \int_0^{\Delta} \frac{g_1(y')}{R(y_1,y')} dy' + \cdots + a_N \int_{(N-1)\Delta}^{l} \frac{g_N(y')}{R(y_1,y')} dy' \\ \vdots \\ 4\pi\varepsilon_0 = a_1 \int_0^{\Delta} \frac{g_1(y')}{R(y_N,y')} dy' + \cdots + a_N \int_{(N-1)\Delta}^{l} \frac{g_N(y')}{R(y_N,y')} dy' \quad (8\text{-}6a)$$

We may write (8-6a) more concisely using matrix notation as

$$[V_m] = [Z_{mn}][I_n] \quad (8\text{-}7)$$

where each Z_{mn} term is equal to

$$Z_{mn} = \int_0^l \frac{g_n(y')}{\sqrt{(y_m - y')^2 + a^2}} dy' \\ = \int_{(n-1)\Delta}^{n\Delta} \frac{1}{\sqrt{(y_m - y')^2 + a^2}} dy' \quad (8\text{-}7a)$$

and

$$[I_n] = [a_n] \quad (8\text{-}7b)$$

$$[V_m] = [4\pi\varepsilon_0]. \quad (8\text{-}7c)$$

The V_m column matrix has all terms equal to $4\pi\varepsilon_0$, and the $I_n = a_n$ values are the unknown charge distribution coefficients. Solving (8-7) for $[I_n]$ gives

$$[I_n] = [a_n] = [Z_{mn}]^{-1}[V_m] \quad (8\text{-}8)$$

Either (8-7) or (8-8) may readily be solved on a digital computer by using any of a number of matrix inversion or equation solving routines. Whereas the integral involved here may be evaluated in closed form by making appropriate approximations, this is not usually possible with more complicated problems. Efficient numerical integral computer subroutines are commonly available in easy-to-use forms.

One closed-form evaluation of (8-7a) is to reduce the integral and represent it by

$$Z_{mn} = \begin{cases} 2\ln\left(\dfrac{\dfrac{\Delta}{2}+\sqrt{a^2+\left(\dfrac{\Delta}{2}\right)^2}}{a}\right) & m = n \quad (\text{8-9a}) \\[2ex] \ln\left\{\dfrac{d_{mn}^+ + [(d_{mn}^+)^2 + a^2]^{1/2}}{d_{mn}^- + [(d_{mn}^-)^2 + a^2]^{1/2}}\right\} & m \neq n \text{ but } |m-n| \leq 2 \quad (\text{8-9b}) \\[2ex] \ln\left(\dfrac{d_{mn}^+}{d_{mn}^-}\right) & |m-n| > 2 \quad (\text{8-9c}) \end{cases}$$

where

$$d_{mn}^+ = l_m + \frac{\Delta}{2} \quad (\text{8-9d})$$

$$d_{mn}^- = l_m - \frac{\Delta}{2} \quad (\text{8-9e})$$

l_m is the distance between the mth matching point and the center of the nth source point.

In summary, the solution of (8-2) for the charge distribution on a wire has been accomplished by approximating the unknown with some basis functions, dividing the wire into segments, and then sequentially enforcing (8-2) at the center of each segment to form a set of linear equations.

Even for the relatively simple straight wire geometry we have discussed, the exact form of the charge distribution is not intuitively apparent. To illustrate the principles of the numerical solution, an example is now presented.

Example 8.1

A 1-m long straight wire of radius $a = 0.001$ m is maintained at a constant potential of 1 V. Determine the linear charge distribution on the wire by dividing the length into 5 and 20 uniform segments. Assume subdomain pulse basis functions.

Solution:

1. $N = 5$. When the 1-m long wire is divided into five uniform segments each of length $\Delta = 0.2$ m, (8-7) reduces to

$$\begin{bmatrix} 10.60 & 1.10 & 0.51 & 0.34 & 0.25 \\ 1.10 & 10.60 & 1.10 & 0.51 & 0.34 \\ 0.51 & 1.10 & 10.60 & 1.10 & 0.51 \\ 0.34 & 0.51 & 1.10 & 10.60 & 1.10 \\ 0.25 & 0.34 & 0.51 & 1.10 & 10.60 \end{bmatrix} \begin{bmatrix} a_1 \\ a_2 \\ a_3 \\ a_4 \\ a_5 \end{bmatrix} = \begin{bmatrix} 1.11 \times 10^{-10} \\ 1.11 \times 10^{-10} \\ \vdots \\ 1.11 \times 10^{-10} \end{bmatrix}$$

Inverting this matrix leads to the amplitude coefficients and subsequent charge distribution of

$$a_1 = 8.81 pC/m$$
$$a_2 = 8.09 pC/m$$
$$a_3 = 7.97 pC/m$$
$$a_4 = 8.09 pC/m$$
$$a_5 = 8.81 pC/m$$

The charge distribution is shown in Figure 8.2(a).

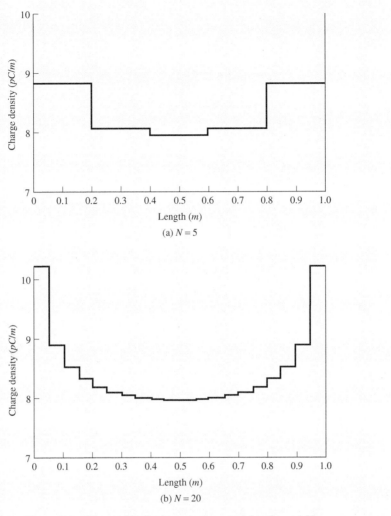

Figure 8.2 Charge distribution on a 1-m straight wire at 1 V.

2. $N = 20$. Increasing the number of segments to 20 results in a much smoother distribution, as shown plotted in Figure 8.2(b). As more segments are used, a better approximation of the actual charge distribution is attained, which has smaller discontinuities over the length of the wire.

B. Bent Wire

In order to illustrate the solution of a more complex structure, let us analyze a body composed of two noncollinear straight wires; that is, a bent wire. If a straight wire is bent, the charge distribution will be altered, although the solution to find it will differ only slightly from the straight wire case. We will assume a bend of angle α, which remains on the yz-plane, as shown in Figure 8.3.

For the first segment l_1 of the wire, the distance R can be represented by (8-2a). However, for the second segment l_2 we can express the distance as

$$R = \sqrt{(y - y')^2 + (z - z')^2} \tag{8-10}$$

438   INTEGRAL EQUATIONS, MOMENT METHOD, AND SELF AND MUTUAL IMPEDANCES

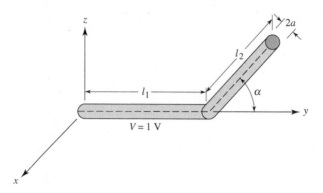

Figure 8.3   Geometry for bent wire.

Also because of the bend, the integral in (8-7a) must be separated into two parts of

$$Z_{mn} = \int_0^{l_1} \frac{\rho_n(l'_1)}{R}\,dl'_1 + \int_0^{l_2} \frac{\rho_n(l'_2)}{R}\,dl'_2 \tag{8-11}$$

where l_1 and l_2 are measured along the corresponding straight sections from their left ends.

Example 8.2

Repeat Example 8.1 assuming that the wire has been bent 90° at its midpoint. Subdivide the wire into 20 uniform segments.

Solution: The charge distribution for this case, calculated using (8-10) and (8-11), is plotted in Figure 8.4 for $N = 20$ segments. Note that the charge is relatively more concentrated near the ends of this structure than was the case for a straight wire of Figure 8.2(b). Further, the overall density, and thus capacitance, on the structure has decreased.

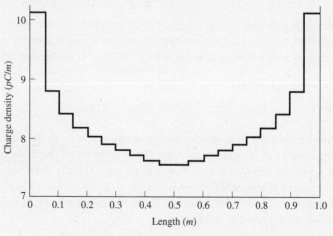

Figure 8.4   Charge distribution on a 1-m bent wire ($\alpha = 90°, N = 20$).

Arbitrary wire configurations, including numerous ends and even curved sections, may be analyzed by the methods already outlined here. As with the simple bent wire, the only alterations generally necessary are those required to describe the geometry analytically.

8.2.2 Integral Equation

Equation (8-2), for the 1 V potential on a wire of length l, is an integral equation, which can be used to solve for the charge distribution. Numerically this is accomplished using a method, which is usually referred to as *Moment Method* or *Method of Moments* [7]–[14]. To solve (8-2) numerically the unknown charge density $\rho(y')$ is represented by N terms, as given by (8-3). In (8-3) $g_n(y')$ are a set of N known functions, usually referred to as *basis* or *expansion* functions, while a_n represents a set of N constant, but unknown, coefficients. The basis or expansion functions are chosen to best represent the unknown charge distribution.

Equation (8-2) is valid at every point on the wire. By enforcing (8-2) at N discrete but different points on the wire, the integral equation of (8-2) is reduced to a set of N linearly independent algebraic equations, as given by (8-6a). This set is generalized by (8-7)–(8-7c), which is solved for the unknown coefficients a_n by (8-8) using matrix inversion techniques. Since the system of N linear equations each with N unknowns, as given by (8-6a)–(8-8), was derived by applying the boundary condition (constant 1 V potential) at N discrete points on the wire, the technique is referred to as *point-matching* (or *collocation*) method [7], [8]. Thus, by finding the elements of the [V] and [Z], and then the inverse $[Z]^{-1}$ matrices, we can then determine the coefficients a_n of the [I] matrix using (8-8). This in turn allows us to approximate the charge distribution $\rho(y')$ using (8-3). This was demonstrated by Examples 8.1 and 8.2 for the straight and bent wires, respectively.

In general, there are many forms of integral equations. For time-harmonic electromagnetics, two of the most popular integral equations are the *electric-field integral equation* (EFIE) and the *magnetic field integral equation* (MFIE) [14]. The EFIE enforces the boundary condition on the tangential electric field while the MFIE enforces the boundary condition on the tangential components of the magnetic field. The EFIE is valid for both closed or open surfaces while the MFIE is valid for closed surfaces. These integral equations can be used for both radiation and scattering problems. Two- and three-dimensional EFIE and MFIE equations for TE and TM polarizations are derived and demonstrated in [14]. For radiation problems, especially wire antennas, two popular EFIEs are the *Pocklington Integral Equation* and the *Hallén Integral Equation*. Both of these will be discussed and demonstrated in the section that follows.

8.3 FINITE DIAMETER WIRES

In this section we want to derive and apply two classic three-dimensional integral equations, referred to as *Pocklington's integrodifferential equation* and *Hallén's integral equation* [18]–[26], that can be used most conveniently to find the current distribution on conducting wires. Hallén's equation is usually restricted to the use of a *delta-gap* voltage source model at the feed of a wire antenna. Pocklington's equation, however, is more general and it is adaptable to many types of feed sources (through alteration of its excitation function or excitation matrix), including a magnetic frill [27]. In addition, Hallén's equation requires the inversion of an $N + 1$ order matrix (where N is the number of divisions of the wire) while Pocklington's equation requires the inversion of an N order matrix.

For very thin wires, the current distribution is usually assumed to be of sinusoidal form as given by (4-56). For finite diameter wires (usually diameters d of $d > 0.05\lambda$), the sinusoidal current distribution is representative but not accurate. To find a more accurate current distribution on a cylindrical wire, an integral equation is usually derived and solved. Previously, solutions to the integral equation were obtained using iterative methods [20]; presently, it is most convenient to use moment method techniques [7]–[9].

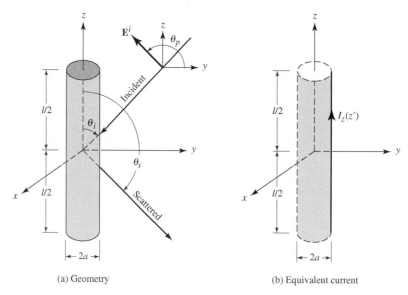

Figure 8.5 Uniform plane wave obliquely incident on a conducting wire.

If we know the voltage at the feed terminals of a wire antenna and find the current distribution, the input impedance and radiation pattern can then be obtained. Similarly, if a wave impinges upon the surface of a wire scatterer, it induces a current density that in turn is used to find the scattered field. Whereas the linear wire is simple, most of the information presented here can be readily extended to more complicated structures.

8.3.1 Pocklington's Integral Equation

To derive Pocklington's integral equation, refer to Figure 8.5. Although this derivation is general, it can be used either when the wire is a scatterer or an antenna. Let us assume that an incident wave impinges on the surface of a conducting wire, as shown in Figure 8.5(a), and it is referred to as the *incident electric field* $\mathbf{E}^i(\mathbf{r})$. When the wire is an antenna, the incident field is produced by the feed source at the gap, as shown in Figure 8.7. Part of the incident field impinges on the wire and induces on its surface a linear current density $\mathbf{J}_s$ (amperes per meter). The induced current density $\mathbf{J}_s$ reradiates and produces an electric field that is referred to as the *scattered electric field* $\mathbf{E}^s(\mathbf{r})$. Therefore, at any point in space the total electric field $\mathbf{E}^t(\mathbf{r})$ is the sum of the incident and scattered fields, or

$$\mathbf{E}^t(\mathbf{r}) = \mathbf{E}^i(\mathbf{r}) + \mathbf{E}^s(\mathbf{r}) \tag{8-12}$$

where

$\mathbf{E}^t(\mathbf{r})$ = total electric field

$\mathbf{E}^i(\mathbf{r})$ = incident electric field

$\mathbf{E}^s(\mathbf{r})$ = scattered electric field

When the observation point is moved to the surface of the wire ($\mathbf{r} = \mathbf{r}_s$) and the wire is perfectly conducting, the total tangential electric field vanishes. In cylindrical coordinates, the electric field radiated by the dipole has a radial component (E_ρ) and a tangential component (E_z). These are represented by (8-52a) and (8-52b). Therefore on the surface of the wire the tangential component of (8-12) reduces to

$$E_z^t(\mathbf{r} = \mathbf{r}_s) = E_z^i(\mathbf{r} = \mathbf{r}_s) + E_z^s(\mathbf{r} = \mathbf{r}_s) = 0 \tag{8-13}$$

or

$$E_z^s(\mathbf{r} = \mathbf{r}_s) = -E_z^i(\mathbf{r} = \mathbf{r}_s) \tag{8-13a}$$

In general, the scattered electric field generated by the induced current density $\mathbf{J}_s$ is given by (3-15), or

$$\mathbf{E}^s(\mathbf{r}) = -j\omega\mathbf{A} - j\frac{1}{\omega\mu\varepsilon}\nabla(\nabla\cdot\mathbf{A})$$
$$= -j\frac{1}{\omega\mu\varepsilon}[k^2\mathbf{A} + \nabla(\nabla\cdot\mathbf{A})] \tag{8-14}$$

However, for observations at the wire surface only the z component of (8-14) is needed, and we can write it as

$$E_z^s(r) = -j\frac{1}{\omega\mu\varepsilon}\left(k^2 A_z + \frac{\partial^2 A_z}{\partial z^2}\right) \tag{8-15}$$

According to (3-51) and neglecting edge effects

$$A_z = \frac{\mu}{4\pi}\iint_S J_z \frac{e^{-jkR}}{R}\,ds' = \frac{\mu}{4\pi}\int_{-l/2}^{+l/2}\int_0^{2\pi} J_z \frac{e^{-jkR}}{R}a\,d\phi'\,dz' \tag{8-16}$$

If the wire is very thin, the current density J_z is not a function of the azimuthal angle ϕ, and we can write it as

$$2\pi a J_z = I_z(z') \Rightarrow J_z = \frac{1}{2\pi a}I_z(z') \tag{8-17}$$

where $I_z(z')$ is assumed to be an equivalent filament line-source current located a radial distance $\rho = a$ from the z axis, as shown in Figure 8.6(a). Thus (8-16) reduces to

$$A_z = \frac{\mu}{4\pi}\int_{-l/2}^{+l/2}\left[\frac{1}{2\pi a}\int_0^{2\pi} I_z(z')\frac{e^{-jkR}}{R}a\,d\phi'\right]dz' \tag{8-18}$$

$$R = \sqrt{(x-x')^2 + (y-y')^2 + (z-z')^2}$$
$$= \sqrt{\rho^2 + a^2 - 2\rho a\cos(\phi-\phi') + (z-z')^2} \tag{8-18a}$$

where ρ is the radial distance to the observation point and a is the radius.

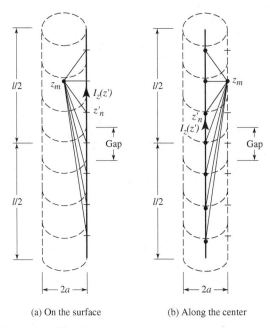

(a) On the surface   (b) Along the center

Figure 8.6 Dipole segmentation and its equivalent current.

Because of the symmetry of the scatterer, the observations are not a function of ϕ. For simplicity, let us then choose $\phi = 0$. For observations on the surface $\rho = a$ of the scatterer (8-18) and (8-18a) reduce to

$$A_z(\rho = a) = \mu \int_{-l/2}^{+l/2} I_z(z') \left(\frac{1}{2\pi} \int_0^{2\pi} \frac{e^{-jkR}}{4\pi R} d\phi' \right) dz'$$

$$= \mu \int_{-l/2}^{+l/2} I_z(z') G(z, z') \, dz' \qquad (8\text{-}19)$$

$$G(z, z') = \frac{1}{2\pi} \int_0^{2\pi} \frac{e^{-jkR}}{4\pi R} d\phi' \qquad (8\text{-}19a)$$

$$R(\rho = a) = \sqrt{4a^2 \sin^2\left(\frac{\phi'}{2}\right) + (z - z')^2} \qquad (8\text{-}19b)$$

Thus for observations on the surface $\rho = a$ of the scatterer, the z component of the scattered electric field can be expressed as

$$E_z^s(\rho = a) = -j\frac{1}{\omega\varepsilon}\left(k^2 + \frac{d^2}{dz^2}\right) \int_{-l/2}^{+l/2} I_z(z') G(z, z') \, dz' \qquad (8\text{-}20)$$

which by using (8-13a) reduces to

$$-j\frac{1}{\omega\varepsilon}\left(\frac{d^2}{dz^2} + k^2\right) \int_{-l/2}^{+l/2} I_z(z') G(z, z') \, dz' = -E_z^i(\rho = a) \qquad (8\text{-}21)$$

or

$$\left(\frac{d^2}{dz^2} + k^2\right) \int_{-l/2}^{+l/2} I_z(z')G(z,z')\,dz' = -j\omega\varepsilon E_z^i(\rho = a) \tag{8-21a}$$

Interchanging integration with differentiation, we can rewrite (8-21a) as

$$\boxed{\int_{-l/2}^{+l/2} I_z(z')\left[\left(\frac{\partial^2}{\partial z^2} + k^2\right)G(z,z')\right]dz' = -j\omega\varepsilon E_z^i(\rho = a)} \tag{8-22}$$

where $G(z, z')$ is given by (8-19a).

Equation (8-22) is referred to as *Pocklington's integral equation* [1], and it can be used to determine the equivalent filamentary line-source current of the wire, and thus current density on the wire, by knowing the incident field on the surface of the wire.

If we assume that the wire is very thin ($a \ll \lambda$), (8-19a) reduces to

$$G(z,z') = G(R) = \frac{e^{-jkR}}{4\pi R} \tag{8-23}$$

Substituting (8-23) into (8-22) reduces it to

$$\int_{-l/2}^{+l/2} I_z(z')\left[\left(\frac{\partial}{\partial z^2} + k^2\right)\frac{e^{-jkR}}{4\pi R}\right]dz'$$

$$= \int_{-l/2}^{+l/2} I_z(z')\frac{1}{4\pi}\left\{\frac{\partial}{\partial z}\left[\frac{\partial}{\partial z}\left(\frac{e^{-jkR}}{R}\right)\right] + k^2\frac{e^{-jkR}}{R}\right\}dz' = -j\omega\varepsilon E_z^i(\rho = a) \tag{8-24a}$$

We apply the chain rule to (8-24a) and that $R = \sqrt{a^2 + (z-z')^2}$ in order to rewrite (8-24a) as

$$\int_{-l/2}^{+l/2} I_z(z')\frac{1}{4\pi}\left\{\frac{\partial}{\partial z}\left[\frac{\partial}{\partial R}\left(\frac{e^{-jkR}}{R}\right)\frac{\partial}{\partial z}\left(\sqrt{a^2+(z-z')^2}\right)\right] + k^2\frac{e^{-jkR}}{R}\right\}dz' = -j\omega\varepsilon E_z^i(\rho = a)$$
$$\tag{8-24b}$$

After taking partial derivatives within the brackets of the integrand, we can rewrite the first part of the integrand as

$$\frac{\partial}{\partial z}\left[\frac{\partial}{\partial R}\left(\frac{e^{-jkR}}{R}\right)\frac{\partial}{\partial z}\left(\sqrt{a^2+(z-z')^2}\right)\right] = \frac{\partial}{\partial z}\left[-e^{-jkR}(z-z')\left(\frac{1+jkR}{R^3}\right)\right]$$

$$= \frac{e^{-jkR}}{R^5}\left\{-R^2(1+jkR) - [jkR(z-z')^2 - (1+jkR)(3+jkR)(z-z')^2]\right\} \tag{8-24c}$$

We use (8-24c) and $(z - z')^2 = R^2 - a^2$ to express (8-24a) in a more compact and convenient form as [22]

$$\boxed{\int_{-l/2}^{+l/2} I_z(z')\frac{e^{-jkR}}{4\pi R^5}[(1+jkR)(2R^2 - 3a^2) + (kaR)^2]\,dz' = -j\omega\varepsilon E_z^i(\rho = a)} \tag{8-25}$$

where for observations along the center of the wire ($\rho = 0$)

$$R = \sqrt{a^2 + (z - z')^2} \tag{8-25a}$$

In (8-22) or (8-25), $I_z(z')$ represents the equivalent filamentary line-source current located on the surface of the wire, as shown in Figure 8.5(b), and it is obtained by knowing the incident electric field on the surface of the wire. By point-matching techniques, this is solved by matching the boundary conditions at discrete points on the surface of the wire. Often it is easier to choose the matching points to be at the interior of the wire, especially along the axis as shown in Figure 8.6(a), where $I_z(z')$ is located on the surface of the wire. By reciprocity, the configuration of Figure 8.6(a) is analogous to that of Figure 8.6(b) where the equivalent filamentary line-source current is assumed to be located along the center axis of the wire and the matching points are selected on the surface of the wire. Either of the two configurations can be used to determine the equivalent filamentary line-source current $I_z(z')$; the choice is left to the individual.

8.3.2 Hallén's Integral Equation

Referring again to Figure 8.5(a), let us assume that the length of the cylinder is much larger than its radius ($l \gg a$) and its radius is much smaller than the wavelength ($a \ll \lambda$) so that the effects of the end faces of the cylinder can be neglected. Therefore the boundary conditions for a wire with infinite conductivity are those of vanishing *total tangential* **E** *fields on the surface of the cylinder and vanishing current at the ends of the cylinder* $[I_z(z' = \pm l/2) = 0]$.

Since only an electric current density flows on the cylinder and it is directed along the z axis ($\mathbf{J} = \hat{\mathbf{a}}_z J_z$), then according to (3-14) and (3-51) $\mathbf{A} = \hat{\mathbf{a}}_z A_z(z')$, which for small radii is assumed to be only a function of z'. Thus (3-15) reduces to

$$E_z^t = -j\omega A_z - j\frac{1}{\omega\mu\varepsilon}\frac{\partial^2 A_z}{\partial z^2} = -j\frac{1}{\omega\mu\varepsilon}\left[\frac{d^2 A_z}{dz^2} + \omega^2\mu\varepsilon A_z\right] \tag{8-26}$$

Since the total tangential electric field E_z^t vanishes on the surface of the PEC cylinder, (8-26) reduces to

$$\frac{d^2 A_z}{dz^2} + k^2 A_z = 0 \tag{8-26a}$$

The differential equation of (8-26a) is of the homogeneous form, and its solution for A_z is

$$A_z = -j\sqrt{\mu\varepsilon}\left[B_1 \cos(kz) + C_1 \sin(kz)\right] \tag{8-27}$$

Because the current density on the cylindrical wire is symmetrical, meaning $J_z(+z) = J_z(-z)$, the potential A_z is also symmetrical, $A_z(+z) = A_z(-z)$. Hence (8-27) reduces to

$$A_z = -j\sqrt{\mu\varepsilon}\left[B_1 \cos(kz) + C_1 \sin(k|z|)\right] \tag{8-27a}$$

where B_1 and C_1 are constants. To determine C_1, do the following:

- Differentiate A_z of (8-27a) with respect to z and let $z \to 0 \Rightarrow \lim_{z \to 0}\left(\dfrac{\partial A_z}{\partial z}\right) = -j\sqrt{\mu\varepsilon}kC_1$

- Use $\nabla \cdot \mathbf{A} = -j\omega\mu\varepsilon V \Rightarrow \dfrac{\partial A_z}{\partial z} = -j\omega\mu\varepsilon V$

- Set $\dfrac{\partial A_z(+z)}{\partial z} = -j\omega\mu\varepsilon V(+z)$ and $\dfrac{\partial A_z(-z)}{\partial z} = -j\omega\mu\varepsilon V(-z)$

- Since $V(+z) = V(-z) \Rightarrow V_i = 2\lim_{z \to 0}[V(+z)] = j\dfrac{2}{\omega\mu\varepsilon}\lim_{z \to 0}\left[\dfrac{\partial A_z(+z)}{\partial z}\right]$

- Thus, $V_i = j\dfrac{2}{\omega\mu\varepsilon}(-j\sqrt{\mu\varepsilon}kC_1) = 2C_1 \Rightarrow \boxed{C_1 = \dfrac{V_i}{2}}$ (8-27b)

For a current-carrying wire, its potential is also given by (3-53). Equating (8-27a) to (3-53) leads to

$$\boxed{\int_{-l/2}^{+l/2} I_z(z')\dfrac{e^{-jkR}}{4\pi R}\,dz' = -j\sqrt{\dfrac{\varepsilon}{\mu}}[B_1\cos(kz) + C_1\sin(k|z|)]} \qquad (8\text{-}28)$$

The constant B_1 is determined from the boundary condition that requires the current to vanish at the end points of the wire.

Equation (8-28) is referred to as *Hallén's integral equation* for a perfectly conducting wire. It was derived by solving the differential equation (3-15) or (8-26a) with the enforcement of the appropriate boundary conditions.

8.3.3 Source Modeling

Let us assume that the wire of Figure 8.5 is symmetrically fed by a voltage source, as shown in Figure 8.7(a), and the element is acting as a dipole antenna. To use, for example, Pocklington's integrodifferential equation (8-22) or (8-25) we need to know how to express $E_z^i(\rho = a)$. Traditionally there have been two methods used to model the excitation to represent $E_z^i(\rho = a, 0 \leq \phi \leq 2\pi, -l/2 \leq z \leq +l/2)$ at all points on the surface of the dipole: One is referred to as the *delta-gap* excitation and the other as the *equivalent magnetic ring current* (better known as *magnetic-frill generator*) [27].

A. Delta Gap
The delta-gap source modeling is the simplest and most widely used of the two, but it is also the least accurate, especially for impedances. Usually it is most accurate for smaller width gaps. Using the delta gap, it is assumed that the excitation voltage at the feed terminals is of a constant V_i value and zero elsewhere. Therefore the incident electric field $E_z^i(\rho = a, 0 \leq \phi \leq 2\pi, -l/2 \leq z \leq +l/2)$ is also a constant (V_s/Δ where Δ is the gap width) over the feed gap and zero elsewhere; hence the name delta gap. For the delta-gap model, the feed gap Δ of Figure 8.7(a) is replaced by a narrow band of strips of equivalent magnetic current density of

$$\mathbf{M}_g = -\hat{\mathbf{n}} \times \mathbf{E}^i = -\hat{\mathbf{a}}_\rho \times \hat{\mathbf{a}}_z\dfrac{V_s}{\Delta} = \hat{\mathbf{a}}_\phi\dfrac{V_s}{\Delta} \qquad -\dfrac{\Delta}{2} \leq z' \leq \dfrac{\Delta}{2} \qquad (8\text{-}29)$$

The magnetic current density $\mathbf{M}_g$ is sketched in Figure 8.7(a).

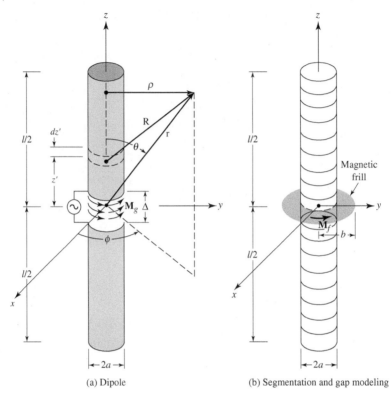

Figure 8.7 Cylindrical dipole, its segmentation, and gap modeling.

B. Magnetic-Frill Generator

The magnetic-frill generator was introduced to calculate the near- as well as the far-zone fields from coaxial apertures [27]. To use this model, the feed gap is replaced with a circumferentially directed magnetic current density that exists over an annular aperture with inner radius a, which is usually chosen to be the radius of the wire, and an outer radius b, as shown in Figure 8.7(b). Since the dipole is usually fed by transmission lines, the outer radius b of the equivalent annular aperture of the magnetic-frill generator is found using the expression for the characteristic impedance of the transmission line.

Over the annular aperture of the magnetic-frill generator, the electric field is represented by the TEM mode field distribution of a coaxial transmission line given by

$$\mathbf{E}_f = \hat{\mathbf{a}}_\rho \frac{V_s}{2\rho' \ln(b/a)} \qquad a \le \rho' \le b \tag{8-30}$$

where V_s is the voltage supplied by the source. *The 1/2 factor is used because it is assumed that the source impedance is matched to the input impedance of the antenna. The 1/2 should be replaced by unity if the voltage V_i present at the input connection to the antenna is used, instead of the voltage V_s supplied by the source.*

Therefore the corresponding equivalent magnetic current density $\mathbf{M}_f$ for the magnetic-frill generator used to represent the aperture is equal to

$$\mathbf{M}_f = -2\hat{\mathbf{n}} \times \mathbf{E}_f = -2\hat{\mathbf{a}}_z \times \hat{\mathbf{a}}_\rho E_\rho = -\hat{\mathbf{a}}_\phi \frac{V_s}{\rho' \ln(b/a)} \qquad a \le \rho' \le b \tag{8-31}$$

The fields generated by the magnetic-frill generator of (8-31) on the surface of the wire are found by using [27]

$$E_z^i\left(\rho = a, 0 \le \phi \le 2\pi, -\frac{l}{2} \le z \le \frac{l}{2}\right)$$
$$\simeq -V_s \left(\frac{k(b^2-a^2)e^{-jkR_0}}{8\ln(b/a)R_0^2} \left\{2\left[\frac{1}{kR_0} + j\left(1 - \frac{b^2-a^2}{2R_0^2}\right)\right]\right.\right.$$
$$+ \frac{a^2}{R_0}\left[\left(\frac{1}{kR_0} + j\left(1 - \frac{(b^2+a^2)}{2R_0^2}\right)\right)\left(-jk - \frac{2}{R_0}\right)\right.$$
$$\left.\left.+ \left(-\frac{1}{kR_0^2} + j\frac{b^2+a^2}{R_0^3}\right)\right]\right\}\right) \quad (8\text{-}32)$$

where

$$R_0 = \sqrt{z^2 + a^2} \quad (8\text{-}32\text{a})$$

The fields generated on the surface of the wire computed using (8-32) can be approximated by those found along the axis ($\rho = 0$). Doing this leads to a simpler expression of the form [27]

$$E_z^i\left(\rho = 0, -\frac{l}{2} \le z \le \frac{l}{2}\right) = -\frac{V_s}{2\ln(b/a)}\left[\frac{e^{-jkR_1}}{R_1} - \frac{e^{-jkR_2}}{R_2}\right] \quad (8\text{-}33)$$

where

$$R_1 = \sqrt{z^2 + a^2} \quad (8\text{-}33\text{a})$$
$$R_2 = \sqrt{z^2 + b^2} \quad (8\text{-}33\text{b})$$

To compare the results using the two-source modelings (delta-gap and magnetic-frill generator), an example is performed.

Example 8.3

For a center-fed linear dipole of $l = 0.47\lambda$ and $a = 0.005\lambda$, determine the induced voltage along the length of the dipole based on the incident electric field of the magnetic frill of (8-33). Subdivide the wire into 21 segments ($N = 21$). Compare the induced voltage distribution based on the magnetic frill to that of the delta gap. Assume a 50-ohm characteristic impedance with free-space between the conductors for the annular feed.

Solution: Since the characteristic impedance of the annular aperture is 50 ohms, then

$$Z_c = \sqrt{\frac{\mu_0}{\varepsilon_0}}\frac{\ln(b/a)}{2\pi} = 50 \Rightarrow \frac{b}{a} = 2.3$$

Subdividing the total length ($l = 0.47\lambda$) of the dipole to 21 segments makes

$$\Delta = \frac{0.47\lambda}{21} = 0.0224\lambda$$

Using (8-33) to compute E_z^i, the corresponding induced voltages, obtained by multiplying the value of $-E_z^i$ at each segment by the length of the segment are listed in Table 8.1, where they are compared with those of the delta gap. In Table 8.1, $n = 1$ represents the outermost segment and $n = 11$ represents the center segment. Because of the symmetry, only values for the center segment and half of the other segments are shown. Although the two distributions are not identical, the magnetic-frill distribution voltages decay quite rapidly away from the center segment and they very quickly reach almost vanishing values.

TABLE 8.1 Unnormalized and Normalized Dipole Induced Voltage[†] Differences for Delta-Gap and Magnetic-Frill Generator ($l = 0.47\lambda, a = 0.005\lambda, N = 21$)

Segment Number n	Delta-Gap Voltage		Magnetic-Frill Generator Voltage	
	Unnormalized	Normalized	Unnormalized	Normalized
1	0	0	1.11×10^{-4} $-26.03°$	7.30×10^{-5} $-26.03°$
2	0	0	1.42×10^{-4} $-20.87°$	9.34×10^{-5} $-20.87°$
3	0	0	1.89×10^{-4} $-16.13°$	1.24×10^{-4} $-16.13°$
4	0	0	2.62×10^{-4} $-11.90°$	1.72×10^{-4} $-11.90°$
5	0	0	3.88×10^{-4} $-8.23°$	2.55×10^{-4} $-8.23°$
6	0	0	6.23×10^{-4} $-5.22°$	4.10×10^{-4} $-5.22°$
7	0	0	1.14×10^{-3} $-2.91°$	7.50×10^{-4} $-2.91°$
8	0	0	2.52×10^{-3} $-1.33°$	1.66×10^{-3} $-1.33°$
9	0	0	7.89×10^{-3} $-0.43°$	5.19×10^{-3} $-0.43°$
10	0	0	5.25×10^{-2} $-0.06°$	3.46×10^{-2} $-0.06°$
11	1	1	1.52 $0°$	1.0 $0°$

[†]Voltage differences as defined here represent the product of the incident electric field at the center of each segment and the corresponding segment length.

8.4 MOMENT METHOD SOLUTION

Equations (8-22), (8-25), and (8-28) each has the form of

$$F(g) = h \tag{8-34}$$

where F is a known linear operator, h is a known excitation function, and g is the response function. For (8-22) F is an integrodifferential operator while for (8-25) and (8-28) it is an integral operator. The objective here is to determine g once F and h are specified.

While the inverse problem is often intractable in closed form, the linearity of the operator F makes a numerical solution possible. One technique, known as the Moment Method [7]–[14] requires that the unknown response function be expanded as a linear combination of N terms and written as

$$g(z') \simeq a_1 g_1(z') + a_2 g_2(z') + \cdots + a_N g_N(z') = \sum_{n=1}^{N} a_n g_n(z') \tag{8-35}$$

Each a_n is an unknown constant and each $g_n(z')$ is a known function usually referred to as a *basis* or *expansion* function. The domain of the $g_n(z')$ functions is the same as that of $g(z')$. Substituting

(8-35) into (8-34) and using the linearity of the F operator reduces (8-34) to

$$\sum_{n=1}^{N} a_n F(g_n) = h \tag{8-36}$$

The basis functions g_n are chosen so that each $F(g_n)$ in (8-36) can be evaluated conveniently, preferably in closed form or at the very least numerically. The only task remaining then is to find the a_n unknown constants.

Expansion of (8-36) leads to one equation with N unknowns. It alone is not sufficient to determine the N unknown a_n ($n = 1, 2, \ldots, N$) constants. To resolve the N constants, it is necessary to have N linearly independent equations. This can be accomplished by evaluating (8-36) (e.g., applying boundary conditions) at N different points. This is referred to as *point-matching* (or *collocation*). Doing this, (8-36) takes the form of

$$\sum_{n=1}^{N} I_n F(g_n) = h_m, \quad m = 1, 2, \ldots, N \tag{8-37}$$

In matrix form, (8-37) can be expressed as

$$[Z_{mn}][I_n] = [V_m] \tag{8-38}$$

where

$$Z_{mn} = F(g_n) \tag{8-38a}$$

$$I_n = a_n \tag{8-38b}$$

$$V_m = h_m \tag{8-38c}$$

The unknown coefficients a_n can be found by solving (8-38) using matrix inversion techniques, or

$$[I_n] = [Z_{mn}]^{-1}[V_m] \tag{8-39}$$

8.4.1 Basis (Expansion) Functions

One very important step in any numerical solution is the choice of basis functions. In general, one chooses as basis functions the set that has the ability to accurately represent and resemble the anticipated unknown function, while minimizing the computational effort required to employ it [28]–[30]. *Do not choose basis functions with smoother properties than the unknown being represented.*

Theoretically, there are many possible basis sets. However, only a limited number are used in practice. These sets may be divided into two general classes. The first class consists of subdomain functions, which are nonzero only over a part of the domain of the function $g(x')$; its domain is the surface of the structure. The second class contains entire-domain functions that exist over the entire domain of the unknown function. The entire-domain basis function expansion is analogous to the well-known Fourier series expansion method.

A. Subdomain Functions

Of the two types of basis functions, subdomain functions are the most common. Unlike entire-domain bases, they may be used without prior knowledge of the nature of the function that they must represent.

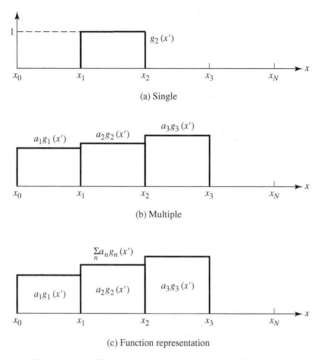

Figure 8.8 Piecewise constant subdomain functions.

The subdomain approach involves subdivision of the structure into N nonoverlapping segments, as illustrated on the axis in Figure 8.8(a). For clarity, the segments are shown here to be collinear and of equal length, although neither condition is necessary. The basis functions are defined in conjunction with the limits of one or more of the segments.

Perhaps the most common of these basis functions is the conceptually simple piecewise constant, or "pulse" function, shown in Figure 8.8(a). It is defined by

Piecewise Constant

$$g_n(x') = \begin{cases} 1 & x'_{n-1} \leq x' \leq x'_n \\ 0 & \text{elsewhere} \end{cases} \tag{8-40}$$

Once the associated coefficients are determined, this function will produce a staircase representation of the unknown function, similar to that in Figures 8.8(b) and (c).

Another common basis set is the piecewise linear, or "triangle," functions seen in Figure 8.9(a). These are defined by

Piecewise Linear

$$g_n(x') = \begin{cases} \dfrac{x' - x'_{n-1}}{x'_n - x'_{n-1}} & x'_{n-1} \leq x' \leq x'_n \\ \dfrac{x'_{n+1} - x'}{x'_{n+1} - x'_n} & x'_n \leq x' \leq x'_{n+1} \\ 0 & \text{elsewhere} \end{cases} \tag{8-41}$$

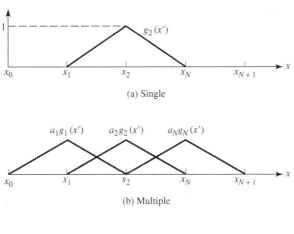

Figure 8.9 Piecewise linear subdomain functions.

and are seen to cover two segments, and overlap adjacent functions [Figure 8.9(b)]. The resulting representation [Figure 8.9(c)] is smoother than that for "pulses," but at the cost of increased computational complexity.

Increasing the sophistication of subdomain basis functions beyond the level of the "triangle" may not be warranted by the possible improvement in accuracy. However, there are cases where more specialized functions are useful for other reasons. For example, some integral operators may be evaluated without numerical integration when their integrands are multiplied by a $\sin(kx')$ or $\cos(kx')$ function, where x' is the variable of integration. In such examples, considerable advantages in computation time and resistance to errors can be gained by using basis functions like the piecewise sinusoid of Figure 8.10 or truncated cosine of Figure 8.11. These functions are defined, respectively, by

Piecewise Sinusoid

$$g_n(x') = \begin{cases} \dfrac{\sin[k(x' - x'_{n-1})]}{\sin[k(x'_n - x'_{n-1})]} & x'_{n-1} \leq x' \leq x'_n \\ \dfrac{\sin[k(x'_{n+1} - x')]}{\sin[k(x'_{n+1} - x'_n)]} & x'_n \leq x' \leq x'_{n+1} \\ 0 & \text{elsewhere} \end{cases} \quad (8\text{-}42)$$

Truncated Cosine

$$g_n(x') = \begin{cases} \cos\left[k\left(x' - \dfrac{x'_n - x'_{n-1}}{2}\right)\right] & x'_{n-1} \leq x' \leq x'_n \\ 0 & \text{elsewhere} \end{cases} \quad (8\text{-}43)$$

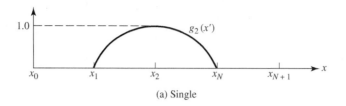
(a) Single

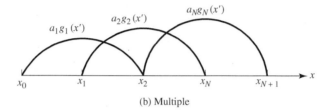
(b) Multiple

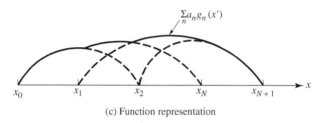
(c) Function representation

Figure 8.10 Piecewise sinusoids subdomain functions.

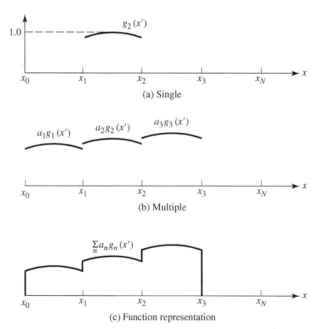

Figure 8.11 Truncated cosines subdomain functions.

B. Entire-Domain Functions

Entire-domain basis functions, as their name implies, are defined and are nonzero over the entire length of the structure being considered. Thus no segmentation is involved in their use.

A common entire-domain basis set is that of sinusoidal functions, where

Entire Domain

$$g_n(x') = \cos\left[\frac{(2n-1)\pi x'}{l}\right] \quad -\frac{l}{2} \leq x' \leq \frac{l}{2} \quad (8\text{-}44)$$

Note that this basis set would be particularly useful for modeling the current distribution on a wire dipole, which is known to have primarily sinusoidal distribution. The main advantage of entire-domain basis functions lies in problems where the unknown function is assumed *a priori* to follow a known pattern. Such entire-domain functions may render an acceptable representation of the unknown while using far fewer terms in the expansion of (8-35) than would be necessary for subdomain bases. Representation of a function by entire-domain cosine and/or sine functions is similar to the Fourier series expansion of arbitrary functions.

Because we are constrained to use a finite number of functions (or modes, as they are sometimes called), entire-domain basis functions usually have difficulty in modeling arbitrary or complicated unknown functions.

Entire-domain basis functions, sets like (8-44), can be generated using Tschebyscheff, Maclaurin, Legendre, and Hermite polynomials, or other convenient functions.

8.4.2 Weighting (Testing) Functions

To improve the point-matching solution of (8-37), (8-38), or (8-39) an inner product $\langle w, g \rangle$ can be defined which is a scalar operation satisfying the laws of

$$\langle \mathbf{w}, \mathbf{g} \rangle = \langle \mathbf{g}, \mathbf{w} \rangle \quad (8\text{-}45a)$$

$$\langle b\mathbf{f} + c\mathbf{g}, \mathbf{w} \rangle = b\langle \mathbf{f}, \mathbf{w} \rangle + c\langle \mathbf{g}, \mathbf{w} \rangle \quad (8\text{-}45b)$$

$$\langle \mathbf{g}^*, \mathbf{g} \rangle > 0 \quad \text{if } \mathbf{g} \neq 0 \quad (8\text{-}45c)$$

$$\langle \mathbf{g}^*, \mathbf{g} \rangle = 0 \quad \text{if } \mathbf{g} = 0 \quad (8\text{-}45d)$$

where b and c are scalars and the asterisk (*) indicates complex conjugation. A typical, but not unique, inner product is

$$\langle \mathbf{w}, \mathbf{g} \rangle = \iint_S \mathbf{w}^* \cdot \mathbf{g} \, ds \quad (8\text{-}46)$$

where the **w**'s are the *weighting (testing)* functions and S is the surface of the structure being analyzed. Note that the functions w and g can be vectors. This technique is better known as the *Moment Method* or *Method of Moments* (MM) [7], [8].

The collocation (point-matching) method is a numerical technique whose solutions satisfy the electromagnetic boundary conditions (e.g., vanishing tangential electric fields on the surface of an electric conductor) only at discrete points. Between these points the boundary conditions may not be satisfied, and we define the deviation as a *residual* (e.g., residual = $\Delta E|_{\text{tan}} = E(\text{scattered})|_{\text{tan}} + E(\text{incident})|_{\text{tan}} \neq 0$ on the surface of an electric conductor). For a half-wavelength dipole, a typical residual is shown in Figure 8.12(a) for pulse basis functions and point-matching and Figure 8.12(b) exhibits the residual for piecewise sinusoids-Galerkin method [31]. As expected, the pulse basis

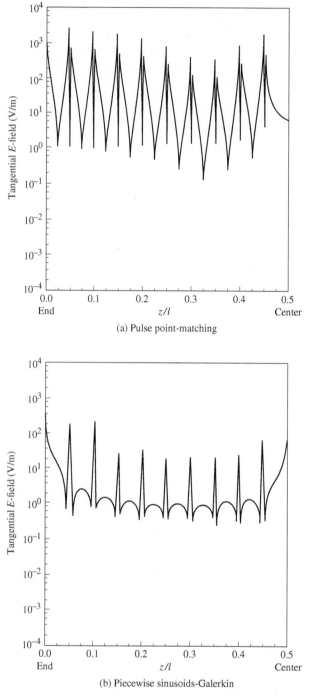

Figure 8.12 Tangential electric field on the conducting surface of a $\lambda/2$ dipole. (SOURCE: E. K. Miller and F. J. Deadrick, "Some computational aspects of thin-wire modeling" in *Numerical and Asymptotic Techniques in Electromagnetics*, 1975, Springer-Verlag).

point-matching exhibits the most ill-behaved residual and the piecewise sinusoids-Galerkin method indicates an improved residual. To minimize the residual in such a way that its overall average over the entire structure approaches zero, the method of *weighted residuals* is utilized in conjunction with the inner product of (8-46). This technique, referred to as the *Moment Method* (MM), does not lead to a vanishing residual at every point on the surface of a conductor, but it forces the boundary conditions to be satisfied in an average sense over the entire surface.

The choice of weighting functions is important in that the elements of $\{w_n\}$ must be linearly independent, [7]–[9], [29], [30]. Further, it will generally be advantageous to choose weighting functions that minimize the computations required to evaluate the inner product.

The condition of linear independence between elements and the advantage of computational simplicity are also important characteristics of basis functions. Because of this, similar types of functions are often used for both weighting and expansion. A particular choice of functions may be to let the weighting and basis function be the same, that is, $w_n = g_n$. This technique is known as *Galerkin's method* [32].

It should be noted that there are N^2 terms to be evaluated in (8-46), where N represents the number of weighting (or testing) functions. Each term usually requires two or more integrations; at least one to evaluate each $F(g_n)$ and one to perform the inner products of (8-46). When these integrations are to be done numerically, as is often the case, vast amounts of computation time may be necessary. There is, however, a unique set of weighting functions that reduce the number of required integrations. This is the set of Dirac delta weighting functions

$$[w_m] = [\delta(p - p_m)] = [\delta(p - p_1), \delta(p - p_2), \ldots] \quad (8\text{-}47)$$

where p specifies a position with respect to some reference (origin) and p_m represents a point at which the boundary condition is enforced.

Using (8-47) reduces (8-46) to one integration; the one represented by $F(g_n)$. This simplification may lead to solutions that would be impractical if other weighting functions were used. Physically, the use of Dirac delta weighting functions is seen as the relaxation of boundary conditions so that they are enforced only at discrete points on the surface of the structure, hence the name point-matching.

An important consideration when using point-matching is the positioning of the N matching points (p_m). While equally-spaced points often yield good results, much depends on the basis functions used. When using subsectional basis functions in conjunction with point-matching, one match point should be placed on each segment (to maintain linear independence). Placing the points at the center of the segments usually produces the best results. It is important that a match point does not coincide with the "peak" of a triangle or a similar discontinuous function, where the basis function is not differentiably continuous. This may cause errors in some situations.

8.5 SELF-IMPEDANCE

The input impedance of an antenna is a very important parameter, and it is used to determine the efficiency of the antenna. In Section 4.5 the real part of the impedance (referred either to the current at the feed terminals or to the current maximum) was found. At that time, because of mathematical complexities, no attempt was made to find the imaginary part (reactance) of the impedance. In this section the self-impedance of a linear element will be examined using both the Integral Equation-Moment Method and the induced emf method. The real and imaginary parts of the impedance will be found using both methods.

8.5.1 Integral Equation-Moment Method

To use this method to find the self-impedance of a dipole, the first thing to do is to solve the integral equation for the current distribution. This is accomplished using either Pocklington's Integral

456 INTEGRAL EQUATIONS, MOMENT METHOD, AND SELF AND MUTUAL IMPEDANCES

equation of (8-22) or (8-25) or Hallén's integral equation of (8-28). For Pocklington's integral equation you can use either the delta-gap voltage excitation of (8-29) or the magnetic-frill model of (8-32) or (8-33). Hallén's integral equation is based on the delta-gap model of (8-29).

Once the current distribution is found, using either or both of the integral equations, then the self (input) impedance is determined using the ratio of the voltage to current, or

$$Z_{in} = \frac{V_{in}}{I_{in}} \qquad (8\text{-}48)$$

A computer program **MOM** (Method of Moments) has been developed based on Pocklington's and Hallén's integral equations, and it is found in the publisher's website for the book. Pocklington's uses both the delta-gap and magnetic-frill models while Hallén's uses only the delta-gap feed model. Both, however, use piecewise constant subdomain functions and point-matching. The program computes the current distribution, normalized amplitude radiation pattern, and input impedance. The user must specify the length of the wire, its radius (both in wavelengths), and the type of feed modeling (delta-gap or magnetic-frill) and the number of segments.

To demonstrate the procedure and compare the results using the two-source modelings (delta-gap and magnetic-frill generator) for Pocklington's integral equation, an example is performed.

Example 8.4

Assume a center-fed linear dipole of $l = 0.47\lambda$ and $a = 0.005\lambda$. This is the same element of Example 8.3.

1. Determine the normalized current distribution over the length of the dipole using $N = 21$ segments to subdivide the length. Plot the current distribution.
2. Determine the input impedance using segments of $N = 7, 11, 21, 29, 41, 51, 61, 71,$ and 79.

Use Pocklington's integrodifferential equation (8-25) with piecewise constant subdomain basis functions and point-matching to solve the problem, model the gap with one segment, and use both the delta-gap and magnetic-frill generator to model the excitation. Use (8-33) for the magnetic-frill generator. Because the current at the ends of the wire vanishes, the piecewise constant subdomain basis functions are not the most judicious choice. However, because of their simplicity, they are chosen here to illustrate the principles even though the results are not the most accurate. Assume that the characteristic impedance of the annular aperture is 50 ohms and the excitation voltage V_i is 1 V.

Solution:

1. The voltage distribution was found in Example 8.3, and it is listed in Table 8.1. The corresponding normalized currents obtained using (8-25) with piecewise constant pulse functions and point-matching technique for both the delta-gap and magnetic-frill generator are shown plotted in Figure 8.13(a). It is apparent that the two distributions are almost identical in shape, and they resemble that of the ideal sinusoidal current distribution which is more valid for very thin wires and very small gaps. The distributions obtained using Pocklington's integral equation do not vanish at the ends because of the use of piecewise constant subdomain basis functions.

SELF-IMPEDANCE

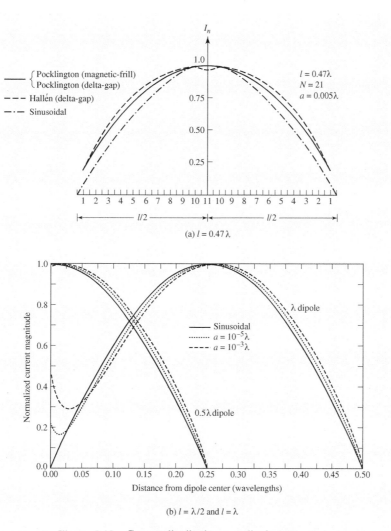

Figure 8.13 Current distribution on a dipole antenna.

2. The input impedances using both the delta-gap and the magnetic-frill generator are shown listed in Table 8.2. It is evident that the values begin to stabilize and compare favorably to each other once 61 or more segments are used.

TABLE 8.2 Dipole Input Impedance for Delta-Gap and Magnetic-Frill Generator Using Pocklington's Integral Equation ($l = 0.47\lambda, a = 0.005\lambda$)

N	Delta Gap	Magnetic Frill
7	$122.8 + j113.9$	$26.8 + j24.9$
11	$94.2 + j49.0$	$32.0 + j16.7$
21	$77.7 - j0.8$	$47.1 - j0.2$
29	$75.4 - j6.6$	$57.4 - j4.5$
41	$75.9 - j2.4$	$68.0 - j1.0$
51	$77.2 + j2.4$	$73.1 + j4.0$
61	$78.6 + j6.1$	$76.2 + j8.5$
71	$79.9 + j7.9$	$77.9 + j11.2$
79	$80.4 + j8.8$	$78.8 + j12.9$

458   INTEGRAL EQUATIONS, MOMENT METHOD, AND SELF AND MUTUAL IMPEDANCES

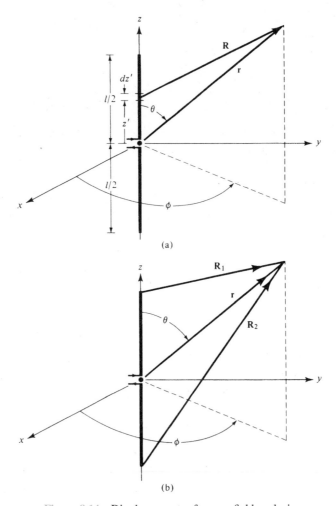

Figure 8.14   Dipole geometry for near-field analysis.

To further illustrate the point on the variation of the current distribution on a dipole, it has been computed by Moment Method and plotted in Figure 8.13(b) for $l = \lambda/2$ and $l = \lambda$ for wire radii of $a = 10^{-5}\lambda$ and $10^{-3}\lambda$ where it is compared with that based on the sinusoidal distribution. It is apparent that the radius of the wire does not influence to a large extent the distribution of the $l = \lambda/2$ dipole. However it has a profound effect on the current distribution of the $l = \lambda$ dipole at and near the feed point. Therefore the input impedance of the $l = \lambda$ dipole is quite different for the three cases of Figure 8.13(b), since the zero current at the center of the sinusoidal distribution predicts an infinite impedance. In practice, the impedance is not infinite but is very large.

8.5.2   Induced EMF Method

The induced emf method is a classical method to compute the self and mutual impedances [1]–[6], [33]. The method is basically limited to straight, parallel, and in echelon elements, and it is more difficult to account accurately for the radius of the wires as well as the gaps at the feeds. However it leads to closed-form solutions which provide very good design data. From the analysis of the infinitesimal dipole in Section 4.2, it was shown that the imaginary part of the power density, which contributes to the imaginary power, is dominant in the near-zone of the element and becomes negligible in the far-field. Thus, near-fields of an antenna are required to find its input reactance.

A. Near-Field of Dipole

In Chapter 4 the far-zone electric and magnetic fields radiated by a finite length dipole with a sinusoidal current distribution were found. The observations were restricted in the far-field in order to reduce the mathematical complexities. The expressions of these fields were used to derive the radiation resistance and the input resistance of the dipole. However, when the input reactance and/or the mutual impedance between elements are desired, the near-fields of the element must be known. It is the intent here to highlight the derivation.

The fields are derived based on the geometry of Figure 8.14. The procedure is identical to that used in Section 4.2.1 for the infinitesimal dipole. The major difference is that the integrations are much more difficult. To minimize long derivations involving complex integrations, only the procedure will be outlined and the final results will be given. The derivation is left as an end of the chapter problems. The details can also be found in the first edition of this book.

To derive the fields, the first thing is to specify the sinusoidal current distribution for a finite dipole which is that of (4-56). Once that is done, then the vector potential **A** of (4-2) is determined. Then the magnetic field is determined using (3-2a), or

$$\mathbf{H} = \frac{1}{\mu}\nabla \times \mathbf{A} = -\hat{\mathbf{a}}_\phi \frac{1}{\mu}\frac{\partial A_z}{\partial \rho} \tag{8-49}$$

It is recommended that cylindrical coordinates are used. By following this procedure and after some lengthy analytical details, it can be shown by referring to Figure 8.14(b) that the magnetic field radiated by the dipole is

$$\mathbf{H} = \hat{\mathbf{a}}_\phi H_\phi = -\hat{\mathbf{a}}_\phi \frac{I_0}{4\pi j}\frac{1}{y}\left[e^{-jkR_1} + e^{-jkR_2} - 2\cos\left(\frac{kl}{2}\right)e^{-jkr}\right] \tag{8-50}$$

where

$$r = \sqrt{x^2 + y^2 + z^2} = \sqrt{\rho^2 + z^2} \tag{8-50a}$$

$$R_1 = \sqrt{x^2 + y^2 + \left(z - \frac{l}{2}\right)^2} = \sqrt{\rho^2 + \left(z - \frac{l}{2}\right)^2} \tag{8-50b}$$

$$R_2 = \sqrt{x^2 + y^2 + \left(z + \frac{l}{2}\right)^2} = \sqrt{\rho^2 + \left(z + \frac{l}{2}\right)^2} \tag{8-50c}$$

The corresponding electric field is found using Maxwell's equation of

$$\mathbf{E} = \frac{1}{j\omega\varepsilon}\nabla \times \mathbf{H} \tag{8-51}$$

Once this is done, it can be shown that the electric field radiated by the dipole is

$$\mathbf{E} = \hat{\mathbf{a}}_\rho E_\rho + \hat{\mathbf{a}}_z E_z = -\hat{\mathbf{a}}_\rho \frac{1}{j\omega\varepsilon}\frac{\partial H_\phi}{\partial z} + \hat{\mathbf{a}}_z \frac{1}{j\omega\varepsilon}\frac{1}{\rho}\frac{\partial}{\partial \rho}(\rho H_\phi) \tag{8-52}$$

where

$$E_\rho = E_y = j\frac{\eta I_0}{4\pi y}\left[\left(z - \frac{l}{2}\right)\frac{e^{-jkR_1}}{R_1} + \left(z + \frac{l}{2}\right)\frac{e^{-jkR_2}}{R_2} - 2z\cos\left(\frac{kl}{2}\right)\frac{e^{-jkr}}{r}\right] \tag{8-52a}$$

$$E_z = -j\frac{\eta I_0}{4\pi}\left[\frac{e^{-jkR_1}}{R_1} + \frac{e^{-jkR_2}}{R_2} - 2\cos\left(\frac{kl}{2}\right)\frac{e^{-jkr}}{r}\right] \tag{8-52b}$$

460 INTEGRAL EQUATIONS, MOMENT METHOD, AND SELF AND MUTUAL IMPEDANCES

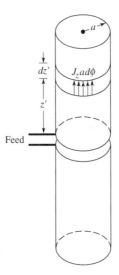

Figure 8.15 Uniform linear current density over cylindrical surface of wire.

It should be noted that the last term in (8-50), (8-52a), and (8-52b) vanishes when the overall length of the element is an integral number of odd half-wavelengths ($l = n\lambda/2, n = 1, 3, 5, ...$) because $\cos(kl/2) = \cos(n\pi/2) = 0$ for $n = 1, 3, 5, ...$.

The fields of (8-50), (8-52a), and (8-52b) were derived assuming a zero radius wire. In practice all wire antennas have a finite radius which in most cases is very small electrically (typically less than $\lambda/200$). Therefore the fields of (8-50), (8-52a), and (8-52b) are good approximations for finite, but small, radius dipoles.

B. Self-Impedance

The technique, which is used in this chapter to derive closed-form expressions for the self- and driving-point impedances of finite linear dipoles, is known as the *induced emf method*. The general approach of this method is to form the Poynting vector using (8-50), (8-52a), and (8-52b), and to integrate it over the surface that coincides with the surface of the antenna (linear dipole) itself. However, the same results can be obtained using a slightly different approach, as will be demonstrated here. The expressions derived using this method are more valid for small radii dipoles. Expressions, which are more accurate for larger radii dipoles, were derived in the previous section based on the Integral Equation-Moment Method.

To find the input impedance of a linear dipole of finite length and radius, shown in Figure 8.15, the tangential electric-field component on the surface of the wire is needed. This was derived previously and is represented by (8-52b). Based on the current distribution and tangential electric field along the surface of the wire, the induced potential developed at the terminals of the dipole based on the maximum current is given by

$$V_m = \int_{-l/2}^{+l/2} dV_m = -\frac{1}{I_m} \int_{-l/2}^{+l/2} I_z(\rho = a, z = z') E_z(\rho = a, z = z') \, dz' \qquad (8\text{-}53)$$

where I_m is the maximum current. The input impedance (*referred to at the current maximum* I_m) is defined as

$$Z_m = \frac{V_m}{I_m} \qquad (8\text{-}54)$$

and can be expressed using (8-53) as

$$Z_m = -\frac{1}{I_m^2} \int_{-l/2}^{l/2} I_z(\rho = a, z = z') E_z(\rho = a, z = z') \, dz' \quad (8\text{-}54a)$$

Equation (8-54a) can also be obtained by forming the complex power density, integrating it over the surface of the antenna, and then relating the complex power to the terminal and induced voltages [2]. The integration can be performed either over the gap at the terminals or over the surface of the conducting wire.

For a wire dipole, the total current I_z is uniformly distributed around the surface of the wire, and it forms a linear current sheet J_z. The current is concentrated primarily over a very small thickness of the conductor, as shown in Figure 8.15, and it is given, based on (4-56), by

$$I_z = 2\pi a J_z = I_m \sin\left[k\left(\frac{l}{2} - |z'|\right)\right] \quad (8\text{-}55)$$

Therefore (8-54a) can be written as

$$Z_m = -\frac{1}{I_m} \int_{-l/2}^{l/2} \sin\left[k\left(\frac{l}{2} - |z'|\right)\right] E_z(\rho = a, z = z') \, dz' \quad (8\text{-}56)$$

For simplicity, it is assumed that the E-field produced on the surface of the wire by a current sheet is the same as if the current were concentrated along a filament placed along the axis of the wire. Then the E-field used in (8-56) is the one obtained along a line parallel to the wire at a distance $\rho = a$ from the filament.

Letting $I_m = I_o$ and substituting (8-52b) into (8-56) it can be shown, after some lengthy but straightforward manipulations, that the real and imaginary parts of the input impedance (referred to at the *current maximum*) can be expressed as

$$R_r = R_m = \frac{\eta}{2\pi} \left\{ C + \ln(kl) - C_i(kl) + \frac{1}{2}\sin(kl)[S_i(2kl) - 2S_i(kl)] \right.$$
$$\left. + \frac{1}{2}\cos(kl)[C + \ln(kl/2) + C_i(2kl) - 2C_i(kl)] \right\} \quad (8\text{-}57a)$$

$$X_m = \frac{\eta}{4\pi} \left\{ 2S_i(kl) + \cos(kl)[2S_i(kl) - S_i(2kl)] \right.$$
$$\left. - \sin(kl)\left[2C_i(kl) - C_i(2kl) - C_i\left(\frac{2ka^2}{l}\right)\right] \right\} \quad (8\text{-}57b)$$

where $C = 0.5772$ (Euler's constant), and $S_i(x)$ and $C_i(x)$ are the sine and cosine integrals of Appendix III. Equation (8-57a) is identical to (4-70). In deriving (8-57a) it was assumed that the radius of the wire is negligible (in this case set to zero), and it has little effect on the overall answer. This is a valid assumption provided $l \gg a$, and it has been confirmed by other methods.

The input resistance and input reactance (*referred to at the current at the input terminals*) can be obtained by a transfer relation given by (4-79), or

$$R_{in} = \left(\frac{I_0}{I_{in}}\right)^2 R_r = \frac{R_r}{\sin^2(kl/2)} \quad (8\text{-}58a)$$

$$X_{in} = \left(\frac{I_0}{I_{in}}\right)^2 X_m = \frac{X_m}{\sin^2(kl/2)} \quad (8\text{-}58b)$$

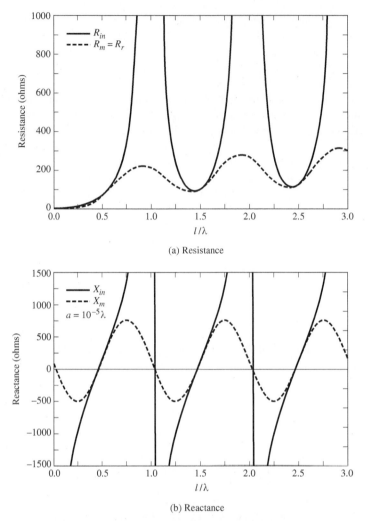

Figure 8.16 Self-resistance and self-reactance of dipole antenna with wire radius of $10^{-5}\lambda$.

For a small dipole the input reactance is given by [34]

$$X_{in} = X_m = -120\frac{[\ln(l/2a) - 1]}{\tan(kl/2)} \qquad (8\text{-}59)$$

while its input resistance and radiation resistance are given by (4-37). Plots of the self-impedance, both resistance and reactance, based on (8-57a), (8-57b) and (8-58a), (8-58b) for $0 \le l \le 3\lambda$ are shown in Figures 8.16(a,b). The radius of the wire is $10^{-5}\lambda$. It is evident that when the length of the wire is multiples of a wavelength the resistances and reactances become infinite; in practice they are large.

Ideally the radius of the wire does not affect the input resistance, as is indicated by (8-57a). However in practice it does have an effect, although it is not as significant as it is for the input reactance. To examine the effect the radius has on the values of the reactance, its values as given by (8-57b) have been plotted in Figure 8.17 for $a = 10^{-5}\lambda$, $10^{-4}\lambda$, $10^{-3}\lambda$, and $10^{-2}\lambda$. The overall length of the wire is taken to be $0 < l \le 3\lambda$. It is apparent that the reactance can be reduced to zero provided

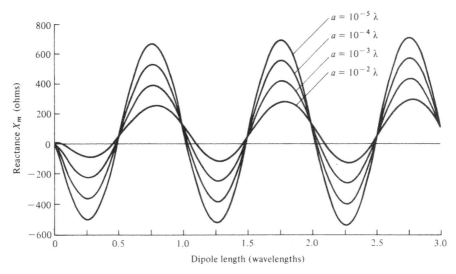

Figure 8.17 Reactance (referred to the current maximum) of linear dipole with sinusoidal current distribution for different wire radii.

the overall length is slightly less than $n\lambda/2, n = 1, 3, \ldots$, or slightly greater than $n\lambda/2, n = 2, 4, \ldots$. This is commonly done in practice for $l \simeq \lambda/2$ because the input resistance is close to 50 ohms, an almost ideal match for the widely used 50-ohm lines. For small radii, the reactance for $l = \lambda/2$ is equal to 42.5 ohms.

From (8-57b) it is also evident that when $l = n\lambda/2, n = 1, 2, 3, \ldots$, the terms within the last bracket do not contribute because $\sin(kl) = \sin(n\pi) = 0$. Thus for dipoles whose overall length is an integral number of half-wavelengths, the radius has no effect on the antenna reactance. This is illustrated in Figure 8.17 by the intersection points of the curves.

Example 8.5

Using the induced emf method, compute the input reactance for a linear dipole whose lengths are $n\lambda/2$, where $n = 1 - 6$.

Solution: The input reactance for a linear dipole based on the induced emf method is given by (8-57b) whose values are equal to 42.5 for $\lambda/2$, 125.4 for λ, 45.5 for $3\lambda/2$, 133.1 for 2λ, 46.2 for $5\lambda/2$, and 135.8 for 3λ.

8.6 MUTUAL IMPEDANCE BETWEEN LINEAR ELEMENTS

In the previous section, the input impedance of a linear dipole was derived when the element was radiating into an unbounded medium. The presence of an obstacle, which could be another element, would alter the current distribution, the field radiated, and in turn the input impedance of the antenna. Thus the antenna performance depends not only on its own current but also on the current of neighboring elements. For resonant elements with no current excitation of their own, there could be a substantial current induced by radiation from another source. These are known as parasitic elements, as in the case of a Yagi-Uda antenna (see Section 10.3.3), and play an important role in the

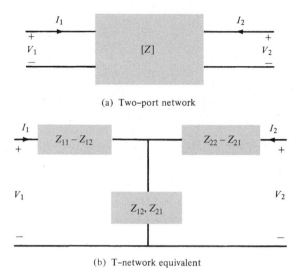

Figure 8.18 Two-port network and its T-equivalent.

overall performance of the entire antenna system. The antenna designer, therefore, must take into account the interaction and mutual effects between elements. The input impedance of the antenna in the presence of the other elements or obstacles, which will be referred to as *driving-point impedance*, depends upon the self-impedance (input impedance in the absence of any obstacle or other element) and the mutual impedance between the driven element and the other obstacles or elements.

To simplify the analysis, it is assumed that the antenna system consists of two elements. The system can be represented by a two-port (four-terminal) network, as shown in Figure 8.18, and by the voltage-current relations

$$\left.\begin{array}{l} V_1 = Z_{11}I_1 + Z_{12}I_2 \\ V_2 = Z_{21}I_1 + Z_{22}I_2 \end{array}\right\} \tag{8-60}$$

where

$$Z_{11} = \left.\frac{V_1}{I_1}\right|_{I_2=0} \tag{8-60a}$$

is the input impedance at port #1 with port #2 open-circuited,

$$Z_{12} = \left.\frac{V_1}{I_2}\right|_{I_1=0} \tag{8-60b}$$

is the mutual impedance at port #1 due to a current at port #2 (with port #1 open-circuited),

$$Z_{21} = \left.\frac{V_2}{I_1}\right|_{I_2=0} \tag{8-60c}$$

is the mutual impedance at port #2 due to a current in port #1 (with port #2 open-circuited),

$$Z_{22} = \frac{V_2}{I_2}\bigg|_{I_1=0} \tag{8-60d}$$

is the input impedance at port #2 with port #1 open-circuited. For a reciprocal network, $Z_{12} = Z_{21}$.

The impedances Z_{11} and Z_{22} are the input impedances of antennas 1 and 2, respectively, when each is radiating in an unbounded medium. The presence of another element modifies the input impedance and the extent and nature of the effects depends upon (1) the antenna type, (2) the relative placement of the elements, and (3) the type of feed used to excite the elements.

Equation (8-60) can also be expressed as

$$Z_{1d} = \frac{V_1}{I_1} = Z_{11} + Z_{12}\left(\frac{I_2}{I_1}\right) \tag{8-61a}$$

$$Z_{2d} = \frac{V_2}{I_2} = Z_{22} + Z_{21}\left(\frac{I_1}{I_2}\right) \tag{8-61b}$$

Z_{1d} and Z_{2d} represent the driving-point impedances of antennas 1 and 2, respectively. Each driving-point impedance depends upon the current ratio I_1/I_2, the mutual impedance, and the self-input impedance (when radiating into an unbounded medium). When attempting to match any antenna, it is the driving-point impedance that must be matched. It is, therefore, apparent that the mutual impedance plays an important role in the performance of an antenna and should be investigated. However, the analysis associated with it is usually quite complex and only simplified models can be examined with the induced emf method. Integral Equation-Moment Method techniques can be used for more complex geometries, including skewed arrangements of elements.

Referring to Figure 8.19, the induced open-circuit voltage in antenna 2, *referred to its current at the input terminals*, due to radiation from antenna 1 is given by

$$V_{21} = -\frac{1}{I_{2i}} \int_{-l_2/2}^{l_2/2} E_{z21}(z') I_2(z') \, dz' \tag{8-62}$$

where

$E_{z21}(z') = $ E-field component radiated by antenna 1, which is parallel to antenna 2

$I_2(z') = $ current distribution along antenna 2

Therefore the mutual impedance of (8-60c), (referred to at the *input current I_{1i} of antenna 1*), is expressed as

$$Z_{21i} = \frac{V_{21}}{I_{1i}} = -\frac{1}{I_{1i}I_{2i}} \int_{-l_2/2}^{l_2/2} E_{z21}(z') I_2(z') \, dz' \tag{8-63}$$

8.6.1 Integral Equation-Moment Method

To use this method to find the mutual impedance based on (8-63), an integral equation must be formed to find E_{z21}, which is the field radiated by antenna 1 at any point on antenna 2. This integral equation must be a function of the unknown current on antenna 1, and it can be derived using a procedure similar to that used to form Pocklington's Integral Equation of (8-22) or (8-25), or Hallén's Integral Equation of (8-28). The unknown current of antenna 1 can be represented by a series of finite number of terms with N unknown coefficients and a set of known (chosen) basis functions. The current $I_2(z)$ must also be expanded into a finite series of N terms with N unknown coefficients

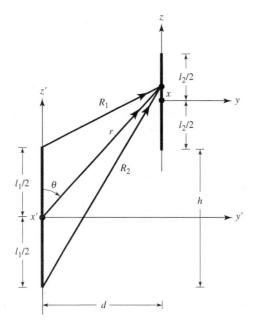

Figure 8.19 Dipole positioning for mutual coupling.

and a set of N chosen basis functions. Once each of them is formulated, then they can be used interactively to reduce (8-63) into an $N \times N$ set of linearly independent equations to find the mutual impedance.

To accomplish this requires a lengthy formulation and computer programming. The process usually requires numerical integrations or special functions for the evaluation of the impedance matrices of E_{z21} and the integral of (8-63). There are national computer codes, such as the Numerical Electromagnetics Code (NEC) and the simplified version Mini Numerical Electromagnetics Code (MININEC), for the evaluation of the radiation characteristics, including impedances, of wire antennas [35]–[37]. Both of these are based on an Integral Equation-Moment Method formulation. Information concerning these two codes follows. There are other codes; however, these two seem to be popular, especially for wire type antennas.

Another procedure that has been suggested to include mutual effects in arrays of linear elements is to use a convergent iterative algorithm [38], [39]. This method can be used in conjunction with a calculator [38], and it has been used to calculate impedances, patterns, and directivities of arrays of dipoles [39].

A. Numerical Electromagnetics Code (NEC)
The Numerical Electromagnetics Code (NEC) is a user-oriented program developed at Lawrence Livermore National Laboratory. It is a moment method code for analyzing the interaction of electromagnetic waves with arbitrary structures consisting of conducting wires and surfaces. It combines an integral equation for smooth surfaces with one for wires to provide convenient and accurate modeling for a wide range of applications. The code can model nonradiating networks and transmission lines, perfect and imperfect conductors, lumped element loading, and perfect and imperfect conducting ground planes. It uses the electric-field integral equation (EFIE) for thin wires and the magnetic field integral equation (MFIE) for surfaces. The excitation can be either an applied voltage source or an incident plane wave. The program computes induced currents and charges, near- and far-zone electric and magnetic fields, radar cross section, impedances or admittances, gain and directivity, power budget, and antenna-to-antenna coupling.

B. Mini-Numerical Electromagnetics Code (MININEC)

The Mini-Numerical Electromagnetics Code (MININEC) [36], [37] is a user-oriented compact version of the NEC developed at the Naval Ocean Systems Center (NOSC) [now Space and Naval Warfare Systems Command (SPAWAR)]. It is also a moment method code, but coded in BASIC, and has retained the most frequently used options of the NEC. It is intended to be used in mini, micro, and personal computers, as well as work stations, and it is most convenient to analyze wire antennas. It computes currents, and near- and far-field patterns. It also optimizes the feed excitation voltages that yield desired radiation patterns.

8.6.2 Induced EMF Method

The induced emf method is also based on (8-63) except that $I_2(z')$ is assumed to be the ideal current distribution of (4-56) or (8-55) while $E_{z21}(z')$ is the electric field of (8-52b). Using (8-55) and (8-52b), we can express (8-63) as

$$Z_{21i} = \frac{V_{21}}{I_{1i}} = j\frac{\eta I_{1m}I_{2m}}{4\pi I_{1i}I_{2i}} \int_{-l_2/2}^{l_2/2} \sin\left[k\left(\frac{l_2}{2} - |z'|\right)\right] \left[\frac{e^{-jkR_1}}{R_1} \right. $$
$$\left. + \frac{e^{-jkR_2}}{R_2} - 2\cos\left(k\frac{l_1}{2}\right) \frac{e^{-jkr}}{r}\right] dz' \qquad (8\text{-}64)$$

where $r, R_1,$ and R_2 are given, respectively, by (8-50a), (8-50b) and (8-50c) but with $y = d$ and $l = l_1$. I_{1m}, I_{2m} and I_{1i}, I_{2i} represent, respectively, the maximum and input currents for antennas 1 and 2. By referring each of the maximum currents to those at the input using (4-78) and assuming free-space, we can write (8-64) as

$$Z_{21i} = j\frac{30}{\sin\left(\frac{kl_1}{2}\right)\sin\left(\frac{kl_2}{2}\right)} \int_{-l_2/2}^{l_2/2} \sin\left[k\left(\frac{l_2}{2} - |z'|\right)\right] \left[\frac{e^{-jkR_1}}{R_1}\right.$$
$$\left. + \frac{e^{-jkR_2}}{R_2} - 2\cos\left(k\frac{l_1}{2}\right) \frac{e^{-jkr}}{r}\right] dz' \qquad (8\text{-}65)$$

The mutual impedance referred to the current maxima is related to that at the input of (8-65) by

$$Z_{21m} = Z_{21i} \sin\left(\frac{kl_1}{2}\right) \sin\left(\frac{kl_2}{2}\right) \qquad (8\text{-}66)$$

which for identical elements ($l_1 = l_2 = l$) reduces to

$$Z_{21m} = Z_{21i} \sin^2\left(\frac{kl}{2}\right) \qquad (8\text{-}67)$$

whose real and imaginary parts are related by

$$R_{21m} = R_{21i} \sin^2\left(\frac{kl}{2}\right) \qquad (8\text{-}67a)$$

$$X_{21m} = X_{21i} \sin^2\left(\frac{kl}{2}\right) \qquad (8\text{-}67b)$$

468   INTEGRAL EQUATIONS, MOMENT METHOD, AND SELF AND MUTUAL IMPEDANCES

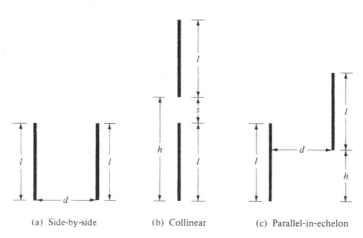

(a) Side-by-side   (b) Collinear   (c) Parallel-in-echelon

Figure 8.20 Dipole configuration of two identical elements for mutual impedance computations.

For a two-element array of linear dipoles, there are three classic configurations for which closed-form solutions for (8-65), in terms of sine and cosine integrals, are obtained [33]. These are shown in Figure 8.20, and they are referred to as the *side-by-side* [Figure 8.20(a)], *collinear* [Figure 8.20(b)], and *parallel-in-echelon* [Figure 8.20(c)]. For two identical elements (each with *odd* multiples of $\lambda/2, l = n\lambda/2, n = 1, 3, 5, \ldots$) (8-67) reduces for each arrangement to the expressions that follow. Expressions for linear elements of any length are much more complex and can be found in [33].

A MATLAB and FORTRAN computer program referred to as **Impedances**, based on (8-68a)–(8-70i), is included in the publisher's website for the book.

Side-by-Side Configuration [Figure 8.20(a)]

$$R_{21m} = \frac{\eta}{4\pi}[2C_i(u_0) - C_i(u_1) - C_i(u_2)] \tag{8-68a}$$

$$X_{21m} = -\frac{\eta}{4\pi}[2S_i(u_0) - S_i(u_1) - S_i(u_2)] \tag{8-68b}$$

$$u_0 = kd \tag{8-68c}$$

$$u_1 = k(\sqrt{d^2 + l^2} + l) \tag{8-68d}$$

$$u_2 = k(\sqrt{d^2 + l^2} - l) \tag{8-68e}$$

Collinear Configuration [Figure 8-20(b)]

$$R_{21m} = -\frac{\eta}{8\pi}\cos(v_0)[-2C_i(2v_0) + C_i(v_2) + C_i(v_1) - \ln(v_3)]$$
$$+ \frac{\eta}{8\pi}\sin(v_0)[2S_i(2v_0) - S_i(v_2) - S_i(v_1)] \tag{8-69a}$$

$$X_{21m} = -\frac{\eta}{8\pi}\cos(v_0)[2S_i(2v_0) - S_i(v_2) - S_i(v_1)]$$
$$+ \frac{\eta}{8\pi}\sin(v_0)[2C_i(2v_0) - C_i(v_2) - C_i(v_1) - \ln(v_3)] \tag{8-69b}$$

$$v_0 = kh \tag{8-69c}$$

$$v_1 = 2k(h+l) \tag{8-69d}$$

$$v_2 = 2k(h-l) \tag{8-69e}$$

$$v_3 = (h^2 - l^2)/h^2 \tag{8-69f}$$

Parallel-in-Echelon Configuration [Figure 8.20(c)]

$$\begin{aligned}R_{21m} = &-\frac{\eta}{8\pi}\cos(w_0)[-2C_i(w_1) - 2C_i(w'_1) + C_i(w_2)\\ &+ C_i(w'_2) + C_i(w_3) + C_i(w'_3)]\\ &+ \frac{\eta}{8\pi}\sin(w_0)[2S_i(w_1) - 2S_i(w'_1) - S_i(w_2)\\ &+ S_i(w'_2) - S_i(w_3) + S_i(w'_3)]\end{aligned} \tag{8-70a}$$

$$\begin{aligned}X_{21m} = &-\frac{\eta}{8\pi}\cos(w_0)[2S_i(w_1) + 2S_i(w'_1) - S_i(w_2)\\ &- S_i(w'_2) - S_i(w_3) - S_i(w'_3)]\\ &+ \frac{\eta}{8\pi}\sin(w_0)[2C_i(w_1) - 2C_i(w'_1) - C_i(w_2) + C_i(w'_2)\\ &- C_i(w_3) + C_i(w'_3)]\end{aligned} \tag{8-70b}$$

$$w_0 = kh \tag{8-70c}$$

$$w_1 = k(\sqrt{d^2 + h^2} + h) \tag{8-70d}$$

$$w'_1 = k(\sqrt{d^2 + h^2} - h) \tag{8-70e}$$

$$w_2 = k[\sqrt{d^2 + (h-l)^2} + (h-l)] \tag{8-70f}$$

$$w'_2 = k[\sqrt{d^2 + (h-l)^2} - (h-l)] \tag{8-70g}$$

$$w_3 = k[\sqrt{d^2 + (h+l)^2} + (h+l)] \tag{8-70h}$$

$$w'_3 = k[\sqrt{d^2 + (h+l)^2} - (h+l)] \tag{8-70i}$$

The mutual impedance, referred to the current maximum, based on the induced emf method of a side-by-side and a collinear arrangement of two half-wavelength dipoles is shown plotted in Figure 8.21. It is apparent that the side-by-side arrangement exhibits larger mutual effects since the antennas are placed in the direction of maximum radiation. The data is compared with those based on the Moment Method/NEC [35] using a wire with a radius of $10^{-5}\lambda$. A very good agreement is indicated between the two sets because a wire with a radius of $10^{-5}\lambda$ for the MM/NEC is considered very thin. Variations as a function of the radius of the wire for both the side-by-side and collinear arrangements using the MM/NEC are shown, respectively, in Figures 8.22(a,b). Similar sets of data were computed for the parallel-in-echelon arrangement of Figure 8.20(c), and they are shown, respectively, in Figures 8.23(a) and 8.23(b) for $d = \lambda/2$, $0 \le h \le \lambda$ and $h - \lambda/2$, $0 < d < \lambda$ for wire radii of $10^{-5}\lambda$. Again a very good agreement between the induced emf and Moment Method/NEC data is indicated.

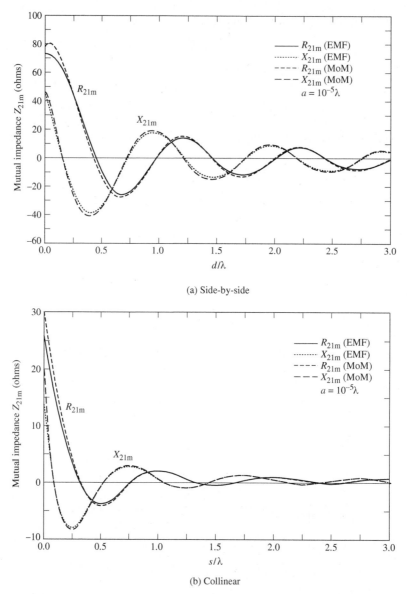

Figure 8.21 Mutual impedance of two side-by-side and collinear $\lambda/2$ dipoles using the moment method and induced emf method.

Example 8.6

Two identical linear half-wavelength dipoles are placed in a side-by-side arrangement, as shown in Figure 8.20(a). Assuming that the separation between the elements is $d = 0.35\lambda$, find the driving-point impedance of each.

Solution: Using (8-61a)

$$Z_{1d} = \frac{V_1}{I_1} = Z_{11} + Z_{12}\left(\frac{I_2}{I_1}\right)$$

Since the dipoles are identical, $I_1 = I_2$. Thus

$$Z_{1d} = Z_{11} + Z_{12}$$

From Figure 8.21(a)

$$Z_{12} \simeq 25 - j38$$

Since

$$Z_{11} = 73 + j42.5$$

Z_{1d} reduces to

$$Z_{1d} \simeq 98 + j4.5$$

which is also equal to Z_{2d} of (8-61b).

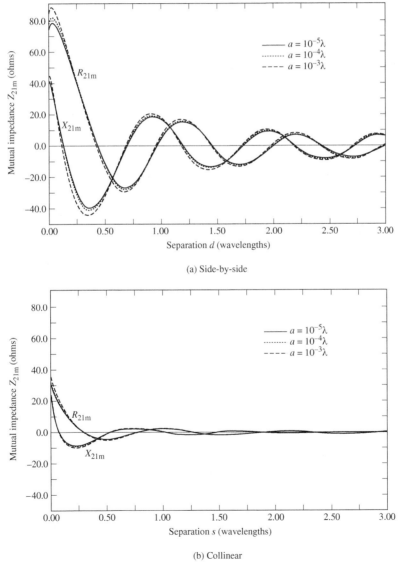

(a) Side-by-side

(b) Collinear

Figure 8.22 Variations of mutual impedance as a function of wire radius for side-by-side and collinear $\lambda/2$ dipole arrangements.

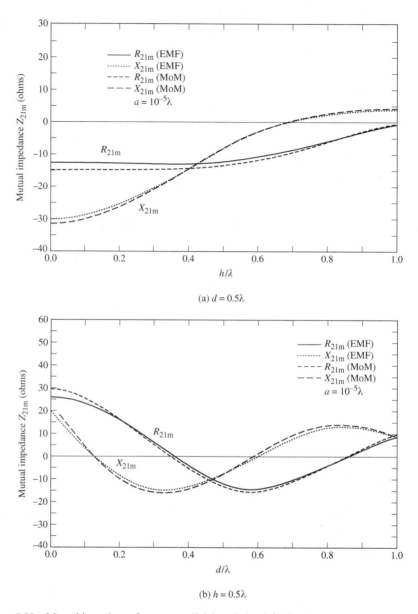

Figure 8.23 Mutual impedance for two parallel-in-echelon $\lambda/2$ dipoles with wire radii of $10^{-5}\,\lambda$.

As discussed in Chapter 2, Section 2.13, maximum power transfer between the generator and the transmitting antenna occurs when their impedances are conjugate-matched. The same is necessary for the receiver and receiving antenna. This ensures maximum power transfer between the transmitter and receiver, when there is no interaction between the antennas. In practice, the input impedance of one antenna depends on the load connected to the other antenna. Under those conditions, the matched loads and maximum coupling can be computed using the Linville method [40], which is used in rf amplifier design. This technique has been incorporated into the NEC [35]. Using this method, the maximum coupling C_{max} is computed using [35]

$$C_{max} = \frac{1}{L}[1 - (1 - L^2)^{1/2}] \tag{8-71}$$

where

$$L = \frac{|Y_{12}Y_{21}|}{2\text{Re}(Y_{11})\text{Re}(Y_{22}) - \text{Re}(Y_{12}Y_{21})} \tag{8-71a}$$

To ensure maximum coupling, the admittance of the matched load on the receiving antenna should be [35]

$$Y_L = \left[\frac{1-\rho}{1+\rho} + 1\right]\text{Re}(Y_{22}) - Y_{22} \tag{8-72}$$

where

$$\rho = \frac{C_{\max}(Y_{12}Y_{21})^*}{|Y_{12}Y_{21}|} \tag{8-72a}$$

and $C_{\max}$ computed using (8-71). The corresponding input admittance of the transmitting antenna is

$$Y_{in} = Y_{11} - \frac{Y_{21}Y_{12}}{Y_L + Y_{22}} \tag{8-73}$$

Based on (8-71)–(8-73), maximum coupling for two half-wavelength dipoles in side-by-side and collinear arrangements as a function of the element separation (d for side-by-side and s for collinear) was computed using the NEC, and it is shown in Figure 8.24. As expected, the side-by-side arrangement exhibits much stronger coupling, since the elements are placed along the direction of their respective maximum radiation. Similar curves can be computed for the parallel-in-echelon arrangement.

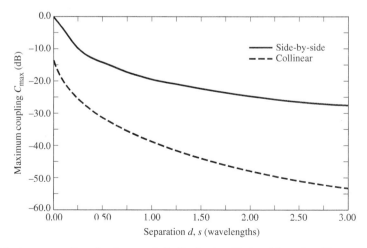

Figure 8.24 Maximum coupling for the two $\lambda/2$ dipoles in side-by-side and collinear arrangements as a function of separation.

8.7 MUTUAL COUPLING IN ARRAYS

When two antennas are near each other, whether one and/or both are transmitting or receiving, some of the energy that is primarily intended for one ends up at the other. The amount depends primarily on the

a. radiation characteristics of each
b. relative separation between them
c. relative orientation of each

There are many different mechanisms that can cause this interchange of energy. For example, even if both antennas are transmitting, some of the energy radiated from each will be received by the other because of the nonideal directional characteristics of practical antennas. Part of the incident energy on one or both antennas may be rescattered in different directions allowing them to behave as secondary transmitters. This interchange of energy is known as "mutual coupling," and in many cases it complicates the analysis and design of an antenna. Furthermore, for most practical configurations, mutual coupling is difficult to predict analytically but must be taken into account because of its significant contribution. Because the mutual effects of any antenna configuration cannot be generalized, in this section we want first to briefly introduce them in a qualitative manner and then examine their general influence on the behavior of the radiation characteristics of the antenna.

8.7.1 Coupling in the Transmitting Mode

To simplify the discussion, let us assume that two antennas m and n of an array are positioned relative to each other as shown in Figure 8.25(a). The procedure can be extended to a number of elements. If a source is attached to antenna n, the generated energy traveling toward the antenna labeled as (0) will be radiated into space (1) and toward the mth antenna (2). The energy incident on the mth antenna sets up currents which have a tendency to rescatter some of the energy (3) and allow the remaining to travel toward the generator of m (4). Some of the rescattered energy (3) may be redirected back toward antenna n (5). This process can continue indefinitely. The same process would take place if antenna m is excited and antenna n is the parasitic element. If both antennas, m and n, are excited simultaneously, the radiated and rescattered fields by and from each must be added vectorially to arrive at the total field at any observation point. Thus, "*the total contribution to the far-field pattern of a particular element in the array depends not only upon the excitation furnished by its own generator (the direct excitation) but upon the total parasitic excitation as well, which depends upon the couplings from and the excitation of the other generators* [41]."

The wave directed from the n to the m antenna and finally toward its generator (4) adds vectorially to the incident and reflected waves of the m antenna itself, thus enhancing the existing standing wave pattern within m. For coherent excitations, the coupled wave (4) due to source n differs from the reflected one in m only in phase and amplitude. The manner in which these two waves interact depends on the coupling between them and the excitation of each. It is evident then that the vector sum of these two waves will influence the input impedance looking in at the terminals of antenna m and will be a function of the position and excitation of antenna n. This coupling effect is commonly modeled as a change in the apparent driving impedance of the elements and it is usually referred to as *mutual impedance variation*.

To demonstrate the usefulness of the driving impedance variation, let us assume that a set of elements in an array are excited. For a given element in the array, the generator impedance that is optimum in the sense of maximizing the radiated power for that element is that which would be a conjugate impedance match at the element terminals. This is accomplished by setting a reflected wave which is equal in amplitude and phase to the backwards traveling waves induced due to the coupling. Even though this is not the generator impedance which is a match to a given element when all other elements are not excited, it does achieve maximum power transfer.

MUTUAL COUPLING IN ARRAYS **475**

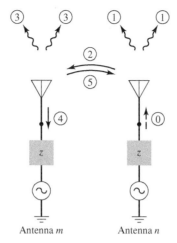

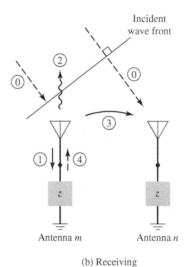

Figure 8.25 Transmitting mode coupling paths between antennas m and n. (Reprinted with permission of MIT Lincoln Laboratory, Lexington, MA.)

To minimize confusion, let us adopt the following terminology [41]:

1. *Antenna impedance*: The impedance looking into a single isolated element.
2. *Passive driving impedance*: The impedance looking into a single element of an array with all other elements of the array passively terminated in their normal generator impedance unless otherwise specified.
3. *Active driving impedance*: The impedance looking into a single element of an array with all other elements of the array excited.

Since the passive driving impedance is of minor practical importance and differs only slightly from the antenna impedance, the term *driving impedance* will be used to indicate *active* driving impedance, unless otherwise specified.

Since the driving impedance for a given element is a function of the placement and excitation of the other elements, then *optimum generator impedance that maximizes array radiation efficiency (gain) varies with array excitation.* These changes, with element excitation variations, constitute one of the principal aggravations in electronic scanning arrays.

8.7.2 Coupling in the Receiving Mode

To illustrate the coupling mechanism in the receiving mode, let us again assume that the antenna system consists of two passively loaded elements of a large number, as shown in Figure 8.25(b). Assume that a plane wave (0) is incident, and it strikes antenna m first where it causes current flow. Part of the incident wave will be rescattered into space as (2), the other will be directed toward antenna n as (3) where it will add vectorially with the incident wave (0), and part will travel into its feed as (1). It is then evident that the amount of energy received by each element of an antenna array is the vector sum of the direct waves and those that are coupled to it parasitically from the other elements.

The amount of energy that is absorbed and reradiated by each element depends on its match to its terminating impedance. Thus, the amount of energy received by any element depends upon its terminating impedance as well as that of the other elements. In order to maximize the amount of energy extracted from an incident wave, we like to minimize the total backscattered (2) energy into space by properly choosing the terminating impedance. This actually can be accomplished by mismatching the receiver itself relative to the antenna so that the reflected wave back to the antenna (4) is cancelled by the rescattered wave, had the receiver been matched to the actual impedance of each antenna.

As a result of the previous discussion, it is evident that mutual coupling plays an important role in the performance of an antenna. However, the analysis and understanding of it may not be that simple.

8.7.3 Mutual Coupling on Array Performance

The effects of the mutual coupling on the performance of an array depends upon the

a. antenna type and its design parameters
b. relative positioning of the elements in the array
c. feed of the array elements
d. scan volume of the array

These design parameters influence the performance of the antenna array by varying its element impedance, reflection coefficients, and overall antenna pattern. In a finite-element array, the multipath routes the energy follows because of mutual coupling will alter the pattern in the absence of these interactions. However, for a very large regular array (array with elements placed at regular intervals on a grid and of sufficient numbers so that edge effects can be ignored), the relative shape of the pattern will be the same with and without coupling interactions. It will only require a scaling up or down in amplitude while preserving the shape. This, however, is not true for irregular placed elements or for small regular arrays where edge effects become dominant.

8.7.4 Coupling in an Infinite Regular Array

The analysis and understanding of coupling can be considerably simplified by considering an infinite regular array. Although such an array is not physically realizable, it does provide an insight and, in many cases, answers of practical importance. For the infinite regular array we assume that

a. all the elements are placed at regular intervals
b. all the elements are identical

c. all the elements have uniform (equal) amplitude excitation
d. there can be a linear relative phasing between the elements in the two orthogonal directions

The geometry of such an array is shown in Figure 6.28 with its array factor given by (6-87) or (6-88). This simplified model will be used to analyze the coupling and describes fairly accurately the behavior of most elements in arrays of modest to large size placed on flat or slowly curved surfaces with smoothly varying amplitude and phase taper.

To assess the behavior of the element driving impedance as a function of scan angle, we can write the terminal voltage of any one element in terms of the currents flowing in the others, assuming a single-mode operation, as

$$V_{mn} = \sum_p \sum_q Z_{mn,pq} I_{pq} \tag{8-74}$$

where $Z_{mn,pq}$ defines the terminal voltage at antenna mn due to a unity current in element pq when the current in all the other elements is zero. Thus the $Z_{mn,pq}$ terms represent the *mutual impedances* when the indices mn and pq are not identical. The *driving impedance* of the mnth element is defined as

$$Z_{Dmn} = \frac{V_{mn}}{I_{mn}} = \sum_p \sum_q Z_{mn,pq} \frac{I_{pq}}{I_{mn}} \tag{8-75}$$

Since we assumed that the amplitude excitation of the elements of the array was uniform and the phase linear, we can write that

$$I_{pq} = I_{00} e^{j(p\beta_x + q\beta_y)} \tag{8-76a}$$

$$I_{mn} = I_{00} e^{j(m\beta_x + n\beta_y)} \tag{8-76b}$$

Thus, (8-75) reduces to

$$Z_{Dmn} = \sum_p \sum_q Z_{mn,pq} e^{j(p-m)\beta_x} e^{j(q-n)\beta_y} \tag{8-77}$$

It is evident that the driving-point impedance of mn element is equal to the vector sum of the element self impedance ($mn = pq$) and the phased mutual impedances between it and the other elements ($mn \neq pq$). The element self-impedance ($mn = pq$) is obtained when all other elements are open-circuited so that the current at their feed points is zero $[I_{pq}(pq \neq mn) = 0]$. For most practical antennas, physically this is almost equivalent to removing the pq elements and finding the impedance of a single isolated element.

A consequence of the mutual coupling problem is the change of the array input reflection coefficient, and thus the input impedance, with scan angle. To demonstrate the variations of the input reflection coefficient, and thus of the input impedance, of an infinite array as a function of scan angle, the input reflection coefficient of an infinite array of circular microstrip patches matched at broadside is shown in Figure 8.26 for the E-plane and H-plane [46]. The variations are due mainly to coupling between the elements. The variations are more pronounced for the E-plane than for the H-plane. For microstrip patches, coupling is attributed to space waves (with $1/r$ radial variations), higher order waves (with $1/\rho^2$ radial variations), surface waves (with $1/\rho^2$ radial variations), and leaky waves [with $\exp(-\lambda\rho)/\rho^{1/2}$ radial variations]. As is shown in Chapter 14 and Figures 14.41, 14.42, the variations of the reflection coefficient can be reduced by suppressing the surface waves supported by the substrate using cavities to back the patches [46]. The variations of the reflection coefficient as a function of scan angle can lead, due to large values of the reflection coefficient (ideally unity), to

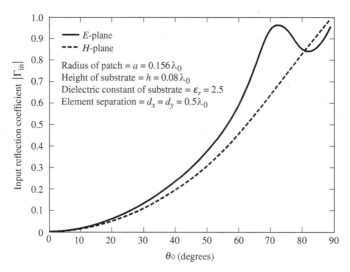

Figure 8.26 Typical magnitude of input reflection coefficient versus scan angle in E- and H-planes for infinite array of microstrip patches (courtesy J. T. Aberle and F. Zavosh).

what is usually referred as *array scan blindness* [47]–[50]. This is evident for the E-plane near 72°–73° and is due to excitation in that plane of a leaky-wave mode, which is not as strongly excited as the scan angle increases beyond those values. Scan blindness is reached at a scan angle of 90°. Also there can be degradation of side lobe level and main beam shape due to the large variations of the reflection coefficient.

Scan blindness is attributed to slow waves which are supported by the structure of the antenna array. These structures may take the form of dielectric layers (such as radomes, superstrates, and substrates) over the face of the array, or metallic grids or fence structures. The scan blindness has been referred to as a "forced surface wave" [47], [48], or a "leaky wave" [49], resonant response of the slow wave structure by the phased array. For the microstrip arrays, the substrate layer supports a slow surface wave which contributes to scan blindness [50].

8.7.5 Active Element Pattern in an Array

In Figure 8.26, it was demonstrated how the reflection coefficient, and thus the input impedance, in an array of microstrip patches is impacted by the mutual impedance, due primarily to the surface waves, as the array is scanned. In fact, if the array is not designed properly, the reflection coefficient can reach unity, leading to scan blindness.

Similarly, the pattern of a single element of an array of element is influenced by the presence of the other elements. In Chapter 6, we used the pattern multiplication rule to find the total field of an array of identical elements by simply multiplying the element pattern by the array factor, as represented by (6-5). The pattern multiplication rule does not take into account the mutual coupling between the elements, and it is only an approximation. It has been demonstrated in [51] that the pattern of an element in an array, in the presence of the other elements, referred to as the *active element pattern* is influenced by the presence of the other elements, and it is not necessarily the same as the pattern of the element in the absence of the others, referred to as the *isolated element pattern*.

In [51], the impact of the pattern of an element in an infinite array, by the presence of the other elements, is derived rigorously using scattering parameters. However, in the same article, the author derives the same relationship using a more simplistic and less rigorous but uses more intuitive arguments based on basic concepts, even those derived in Chapter 2. It will be the less rigorous but more intuitive derivation that we will outline here. The reader is referred to [51] to for the more rigorous derivation.

Let us consider an infinite uniform array of aperture elements. The maximum effective aperture, and in turn the maximum gain, from a single isolated rectangular aperture element with dimensions a and b, assuming no losses, can be written according to (2-110), (12-37), and (12-40) as

$$A_{em} = \varepsilon_{ap}A_p = \varepsilon_{ap}(ab) = \frac{\lambda^2}{4\pi}G_0 \tag{8-78}$$

$$G_0 = \left(\frac{4\pi}{\lambda^2}A_{em}\right) = \frac{4\pi}{\lambda^2}(\varepsilon_{ap}A_p) = \frac{4\pi}{\lambda^2}(\varepsilon_{ap}ab) \tag{8-78a}$$

If we consider a large uniform array of N elements, the maximum gain $G_0(\theta_0, \phi_0)$ available from a single isolated element scanned at an angle θ_0, ϕ_0 is, according to (6-100),

$$G_0(\theta_0, \phi_0) = \left(\frac{4\pi}{\lambda^2}A_{em}\right) = \frac{4\pi}{\lambda^2}(\varepsilon_{ap}A_p \cos\theta_0) = \frac{4\pi}{\lambda^2}\varepsilon_{ap}(ab)\cos\theta_0 \tag{8-79}$$

while the overall maximum gain $G_0^a(\theta_0, \phi_0)$ of the entire array of N elements can be expressed, assuming no coupling and using superposition, as

$$G_0^a(\theta_0, \phi_0) = N\left(\frac{4\pi}{\lambda^2}A_{em}\right) = N\frac{4\pi}{\lambda^2}(\varepsilon_{ap}A_p \cos\theta_0) = \frac{4\pi}{\lambda^2}N\varepsilon_{ap}(ab)\cos\theta_0 \tag{8-79a}$$

The realized maximum gain $G_{re0}(\theta_0, \phi_0)$ of the array, accounting for reflection/mismatch loss, is according to (2-49b) or (2-49c)

$$G_{re0}(\theta_0, \phi_0) = \frac{4\pi}{\lambda^2}N\left[1 - |\Gamma_i(\theta_0, \phi_0)|^2\right]ab\cos\theta_0 \tag{8-80}$$

Note that (8-80) assumes identical reflections at each element, and that $|\Gamma_i(\theta_0, \phi_0)|$ is the magnitude of the active reflection coefficient of the ith element of a fully excited phased array.

Now, if the maximum gain $G_0^e(\theta_0, \phi_0)$ of the active element is defined as the gain $G_0^i(\theta_0, \phi_0)$ of a singly excited isolated element of the array, with all the other elements match-terminated, we can rewrite it as

$$G_0^e(\theta_0, \phi_0) = G_0^i(\theta_0, \phi_0)\left[1 - |\Gamma_i(\theta_0, \phi_0)|^2\right] \tag{8-81}$$

After the superposition, the gain of the fully excited array in the direction of the main beam can now be written, taking into account coupling, as

$$G_{re0}^a(\theta_0, \phi_0) = NG_0^e(\theta_0, \phi_0) \tag{8-82}$$

Based on the above, the realized active element gain pattern $G_{re}^e(\theta, \phi)$ of an array element in the presence of the other array elements, taking into account coupling, can be related to the gain pattern $G^i(\theta, \phi)$ of an isolated array element pattern by

$$G_{re}^e(\theta, \phi) = G^i(\theta, \phi)\left[1 - |\Gamma_i(\theta, \phi)|^2\right] \tag{8-83}$$

Figure 8.27 displays the E-plane gain patterns of an isolated $G^i(\theta)$ and an active $G_{re}^e(\theta)$, based on relationship of (8-83), printed dipole element in an infinite microstrip array [51]. The active element

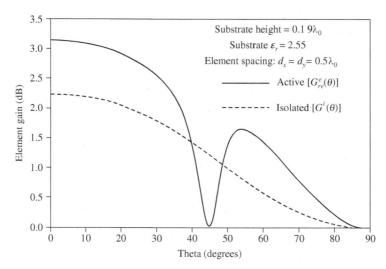

Figure 8.27 *E*-plane element gain patterns of an Infinite Planar Array of microstrip dipoles. (SOURCE [51] © 1944 IEEE).

pattern displays a null around 46°, due to scan blindness (reflection coefficient nearly unity) influenced primarily by the surface waves of the printed array, as was illustrated in Figure 8.26. The gain pattern of the isolated element does not display a null, and the isolated element pattern is quite different than the active element pattern. This comparison indicates that mutual coupling in the array, due to surface waves within the substrate of the microstrip array, impacts the performance of the array, and thus, it should be taken into account.

8.8 MULTIMEDIA

In the publisher's website for the book, the following multimedia resources are included for the review, understanding, and visualization of the material of this chapter:

a. **Java**-based **interactive questionnaire**, with answers.
b. **Matlab** and **Fortran** Method of Moment computer program, designated **MOM**, for computing and displaying the radiation characteristics of dipoles, using Hallén's and Pocklington's integral equations.
c. **Matlab** and **Fortran** computer program, designated **Impedance**, for computing the self and mutual impedance of linear elements.
d. **Power Point (PPT)** viewgraphs, in multicolor.

REFERENCES

1. R. E. Burgess, "Aerial Characteristics," *Wireless Engr.*, Vol. 21, pp. 154–160, April 1944.
2. J. D. Kraus, *Antennas*, McGraw-Hill, New York, 1988, Chapters 9, 10, pp. 359–434.
3. S. A. Schelkunoff and H. T. Friis, *Antennas: Theory and Practice*, Wiley, New York, 1952, pp. 213–242.
4. A. A. Pistolkors, "The Radiation Resistance of Beam Antennas," *Proc. IRE*, Vol. 17, pp. 562–579, March 1929.

5. R. Bechmann, "On the Calculation of Radiation Resistance of Antennas and Antenna Combinations," *Proc. IRE*, Vol. 19. pp. 461–466, March 1931.

6. P. S. Carter, "Circuit Relations in Radiation Systems and Applications to Antenna Problems," *Proc. IRE*, Vol. 20, pp. 1004–1041, June 1932.

7. R. F. Harrington, "Matrix Methods for Field Problems," *Proc. IEEE*, Vol. 55, No. 2, pp. 136–149, February 1967.

8. R. F. Harrington, *Field Computation by Moment Methods*, Macmillan, New York, 1968.

9. J. H. Richmond, "Digital Computer Solutions of the Rigorous Equations for Scattering Problems," *Proc. IEEE*, Vol. 53, pp. 796–804, August 1965.

10. L. L. Tsai, "Moment Methods in Electromagnetics for Undergraduates," *IEEE Trans. Educ.*, Vol. E–21, No. 1, pp. 14–22, February 1978.

11. R. Mittra (Ed.), *Computer Techniques for Electromagnetics*, Pergamon, New York, 1973.

12. J. Moore and R. Pizer, *Moment Methods in Electromagnetics*, John Wiley and Sons, New York, 1984.

13. J. J. H. Wang, *Generalized Moment Methods in Electromagnetics*, John Wiley and Sons, New York, 1991.

14. C. A. Balanis, *Advanced Engineering Electromagnetics*, John Wiley and Sons, Second Edition, New York, 2012.

15. J. D. Lilly, "Application of The Moment Method to Antenna Analysis," MSEE Thesis, Department of Electrical Engineering, West Virginia University, 1980.

16. J. D. Lilly and C. A. Balanis, "Current Distributions, Input Impedances, and Radiation Patterns of Wire Antennas," North American Radio Science Meeting of URSI, Université Laval, Quebec, Canada, June 2–6, 1980.

17. D. K. Cheng, *Field and Wave Electromagnetics*, Addison-Wesley, Reading, MA, 1989, p. 97.

18. H. C. Pocklington, "Electrical Oscillations in Wire," *Cambridge Philos. Soc. Proc.*, Vol. 9, pp. 324–332, 1897.

19. E. Hallén, "Theoretical Investigations into the Transmitting and Receiving Qualities of Antennae," *Nova Acta Regiae Soc. Sci. Upsaliensis*, Ser. IV, No. 4, pp. 1–44, 1938.

20. R. King and C. W. Harrison, Jr., "The Distribution of Current along a Symmetric Center-Driven Antenna," *Proc. IRE*, Vol. 31, pp. 548–567, October 1943.

21. J. H. Richmond, "A Wire-Grid Model for Scattering by Conducting Bodies," *IEEE Trans. Antennas Propagat.*, Vol. AP–14, No. 6, pp. 782–786, November 1966.

22. G. A. Thiele, "Wire Antennas," in *Computer Techniques for Electromagnetics*, R. Mittra (Ed.), Pergamon, New York, Chapter 2, pp. 7–70, 1973.

23. C. M. Butler and D. R. Wilton, "Evaluation of Potential Integral at Singularity of Exact Kernel in Thin-Wire Calculations," *IEEE Trans. Antennas Propagat.*, Vol. AP-23, No. 2, pp. 293–295, March 1975.

24. L. W. Pearson and C. M. Butler, "Inadequacies of Collocation Solutions to Pocklington-Type Models of Thin-Wire Structures," *IEEE Trans. Antennas Propagat.*, Vol. AP-23, No. 2, pp. 293–298, March 1975.

25. C. M. Butler and D. R. Wilton, "Analysis of Various Numerical Techniques Applied to Thin-Wire Scatterers," *IEEE Trans. Antennas Propagat.*, Vol. AP-23, No. 4, pp. 534–540, July 1975.

26. D. R. Wilton and C. M. Butler, "Efficient Numerical Techniques for Solving Pocklington's Equation and their Relationships to Other Methods," *IEEE Trans. Antennas Propagat.*, Vol. AP-24, No. 1, pp. 83–86, January 1976.

27. L. L. Tsai, "A Numerical Solution for the Near and Far Fields of an Annular Ring of Magnetic Current," *IEEE Trans. Antennas Propagat.*, Vol. AP-20, No. 5, pp. 569–576, September 1972.

28. R. Mittra and C. A. Klein, "Stability and Convergence of Moment Method Solutions," in *Numerical and Asymptotic Techniques in Electromagnetics*, R. Mittra (Ed.), Springer-Verlag, New York, 1975, Chapter 5, pp. 129–163.

29. T. K. Sarkar, "A Note on the Choice Weighting Functions in the Method of Moments," *IEEE Trans. Antennas Propagat.*, Vol. AP-33, No. 4, pp. 436–441, April 1985.

30. T. K. Sarkar, A. R. Djordjević and E. Arvas, "On the Choice of Expansion and Weighting Functions in the Numerical Solution of Operator Equations," *IEEE Trans. Antennas Propagat.*, Vol. AP-33, No. 9, pp. 988–996, September 1985.

31. E. K. Miller and F. J. Deadrick, "Some Computational Aspects of Thin-Wire Modeling," in *Numerical and Asymptotic Techniques in Electromagnetics*, R. Mittra (Ed.), Springer-Verlag, New York, 1975, Chapter 4, pp. 89–127.

32. L. Kantorovich and G. Akilov, *Functional Analysis in Normed Spaces*, Pergamon, Oxford, pp. 586–587, 1964.

33. H. E. King, "Mutual Impedance of Unequal Length Antennas in Echelon," *IRE Trans. Antennas Propagat.*, Vol. AP-5, pp. 306–313, July 1957.

34. R. C. Hansen, "Fundamental Limitations in Antennas," *Proc. IEEE*, Vol. 69, No. 2, pp. 170–182, February 1981.

35. G. J. Burke and A. J. Poggio, "Numerical Electromagnetics Code (NEC)-Method of Moments," Technical Document 11, Naval Ocean Systems Center, San Diego, Calif., January 1981.

36. A. J. Julian, J. M. Logan, and J. W. Rockway, "MININEC: A Mini-Numerical Electromagnetics Code," Technical Document 516, Naval Ocean Systems Center, San Diego, Calif, September 6, 1982.

37. J. Rockway, J. Logan, D. Tam, and S. Li, *The MININEC SYSTEM: Microcomputer Analysis of Wire Antennas*, Artech House, 1988.

38. J. A. G. Malherbe, "Calculator Program for Mutual Impedance," *Microwave Journal*, (Euro-Global Ed.), pp. 82-H–82-M, February 1984.

39. J. A. G. Malherbe, "Analysis of a Linear Antenna Array Including the Effects of Mutual Coupling," *IEEE Trans. Educ.*, Vol. 32, No. 1, pp. 29–34, February 1989.

40. D. Rubin, The Linville Method of High Frequency Transistor Amplifier Design, Naval Weapons Center, NWCCL TP 845, Corona Labs., Corona, CA, March 1969.

41. J. L. Allen and B. L. Diamond, "Mutual Coupling in Array Antennas," Technical Report EDS-66-443, Lincoln Lab., MIT, October 4, 1966.

42. H. A. Wheeler, "The Radiation Resistance of an Antenna in an Infinite Array or Waveguide," *Proc. IRE*, Vol. 48, pp. 478–487, April 1948.

43. S. Edelberg and A. A. Oliner, "Mutual Coupling Effects in Large Antenna Arrays, Part I," *IRE Trans. Antennas Propagat.*, Vol. AP-8, No. 3, pp. 286–297, May 1960.

44. S. Edelberg and A. A. Oliner, "Mutual Coupling Effects in Large Antenna Arrays," Part II," *IRE Trans. Antennas Propagat.*, Vol. AP-8, No. 4, pp. 360–367, July 1960.

45. L. Stark, "Radiation Impedance of a Dipole in an Infinite Planar Phased Array," *Radio Sci.*, Vol. 3, pp. 361–375, 1966.

46. F. Zavosh and J. T. Aberle, "Infinite Phased Arrays of Cavity-Backed Patches," *IEEE Trans. Antennas Propagat.*, Vol. 42, No. 3, pp. 390–394, March 1994.

47. N. Amitay, V. Galindo, and C. P. Wu, *Theory and Analysis of Phased Array Antennas*, John Wiley and Sons, New York, 1972.

48. L. Stark, "Microwave Theory of Phased-Array Antennas—A Review," *Proc. IEEE*, Vol. 62, pp. 1661–1701, December 1974.

49. G. H. Knittel, A. Hessel and A. A. Oliner, "Element Pattern Nulls in Phased Arrays and Their Relation to Guided Waves," *Proc. IEEE*, Vol. 56, pp. 1822–1836, November 1968.

50. D. M. Pozar and D. H. Schaubert, "Scan Blindness in Infinite Phased Arrays of Printed Dipoles," *IEEE Trans. Antennas Propagat.*, Vol. AP-32, No. 6, pp. 602–610, June 1984.

51. D. M. Pozar, "The Active Element Pattern," *IEEE Trans. Antennas Propagat.*, Vol. 42, No. 8, pp. 1176–1178, August 1994.

PROBLEMS

8.1. Derive Pocklington's integral equation (8-25) using (8-22)–(8-24c).

8.2. Show that the incident tangential electric field (E_z^i) generated on the surface of a wire of radius a by a magnetic field generator of (8-31) is given by (8-32).

8.3. Reduce (8-32) to (8-33) valid only along the z axis ($\rho = 0$).

8.4. For the center-fed dipole of Example 8.3 write the [Z] matrix for $N = 21$ using for the gap the delta-gap generator and the magnetic-frill generator. Use the computer program **MOM** (Pocklington) of this chapter.

8.5. For an infinitesimal center-fed dipole of $l = \lambda/50$ and radius $a = 0.005\lambda$, derive the input impedance using Pocklington's integral equation with piecewise constant subdomain basis functions and point-matching. Use $N = 21$ and model the gap as a delta-gap generator and as a magnetic-frill generator. Use the **MOM** (Pocklington) computer program of this chapter.

8.6. Using the **MOM** (Hallén) computer program at the end of the chapter, compute the input impedance of a $\lambda/4$ and $3\lambda/4$ dipole with an l/d ratio of $l/d = 50$ and 25. Use 20 subsections. Compare the results with the impedances of a dipole with $l/d = 10^9$. Plot the current distribution and the far-field pattern of each dipole.

8.7. Derive (8-50)-(8-52b) using (8-49), (3-2a), and (4-56).

8.8. For a linear dipole with sinusoidal current distribution, radiating in free-space, find, using tabulated data or subroutines for the sine and cosine integrals, the radiation Z_{im} and the input Z_{in} impedances when $a = \lambda/20$. Verify using the computer program **Impedance** of this chapter.
  (a) $l = \lambda/4$ (b) $l = \lambda/2$ (c) $l = 3\lambda/4$ (d) $l = \lambda$

8.9. A $\lambda/2$ dipole of finite radius is not self-resonant. However, if the dipole is somewhat less than $\lambda/2$, it becomes self-resonant. For a dipole with radius of $a = \lambda/200$ radiating in free-space, find the
  (a) nearest length by which the $\lambda/2$ dipole becomes self-resonant
  (b) radiation resistance (referred to the current maximum) of the new resonant dipole
  (c) input resistance   (d) VSWR when the dipole is connected to a 50-ohm line

8.10. Find the length, at the first resonance, of linear dipoles with wire radii of
  (a) $10^{-5}\lambda$ (b) $10^{-4}\lambda$ (c) $10^{-3}\lambda$ (d) $10^{-2}\lambda$

  Compute the radiation resistance of each.

8.11. A quarter-wavelength monopole of radius $a = 10^{-2}\lambda$ is placed upon an infinite ground plane. Determine the
  (a) impedance of the monopole
  (b) length by which it must be shortened to become self-resonant (first resonance)
  (c) impedance of the monopole when its length is that given in part b.
  (d) VSWR when the monopole of part b is connected to a 50-ohm line.

8.12. For two half-wavelength dipoles radiating in free-space, compute (using equations, *not* curves) the mutual impedance Z_{21m} referred to the current maximum for
  (a) side-by-side arrangement with $d = \lambda/4$ (b) collinear configuration with $s = \lambda/4$

  Verify using the computer program **Impedance** of this chapter.

8.13. Two identical linear $\lambda/2$ dipoles are placed in a collinear arrangement a distance $s = 0.35\lambda$ apart. Find the driving-point impedance of each. Verify using the computer program **Impedance** of this chapter.

8.14. Two identical linear $\lambda/2$ dipoles are placed in a collinear arrangement. Find the spacings between them so that the driving-point impedance of each has the smallest reactive part.

CHAPTER 9

Broadband Dipoles and Matching Techniques

9.1 INTRODUCTION

In Chapter 4 the radiation properties (pattern, directivity, input impedance, mutual impedance, etc.) of very thin-wire antennas were investigated by assuming that the current distribution, which in most cases is nearly sinusoidal, is known. In practice, infinitely thin (electrically) wires are not realizable but can be approximated. In addition, their radiation characteristics (such as pattern, impedance, gain, etc.) are very sensitive to frequency. The degree to which they change as a function of frequency depends on the antenna bandwidth. For applications that require coverage over a broad range of frequencies, such as television reception of all channels, wide-band antennas are needed. There are numerous antenna configurations, especially of arrays, that can be used to produce wide bandwidths. Some simple and inexpensive dipole configurations, including the conical and cylindrical dipoles, can be used to accomplish this to some degree.

For a finite diameter wire (usually $d > 0.05\lambda$) the current distribution may not be sinusoidal and its effect on the radiation pattern of the antenna is usually negligible. However, it has been shown that the current distribution has a pronounced effect on the input impedance of the wire antenna, especially when its length is such that a near null in current occurs at its input terminals. The effects are much less severe when a near current maximum occurs at the input terminals.

Historically there have been three methods that have been used to take into account the finite conductor thickness. The first method treats the problem as boundary-value problem [1], the second as a tapered transmission line or electromagnetic horn [2], and the third finds the current distribution on the wire from an integral equation [3]. The boundary-value approach is well suited for idealistic symmetrical geometries (e.g., ellipsoids, prolate spheroids) which cannot be used effectively to approximate more practical geometries such as the cylinder. The method expresses the fields in terms of an infinite series of free oscillations or natural modes whose coefficients are chosen to satisfy the conditions of the driving source. For the assumed idealized configurations, the method does lead to very reliable data, but it is very difficult to know how to approximate more practical geometries (such as a cylinder) by the more idealized configurations (such as the prolate spheroid). For these reasons the boundary-value method is not very practical and will not be pursued any further in this text.

In the second method Schelkunoff represents the antenna as a two-wire uniformly tapered transmission line, each wire of conical geometry, to form a biconical antenna. Its solution is obtained by applying transmission-line theory (incident and reflected waves), so well known to the average engineer. The analysis begins by first finding the radiated fields which in turn are used, in conjunction with transmission-line theory, to find the input impedance.

Antenna Theory: Analysis and Design, Fourth Edition. Constantine A. Balanis.
© 2016 John Wiley & Sons, Inc. Published 2016 by John Wiley & Sons, Inc.
Companion Website: www.wiley.com/go/antennatheory4e

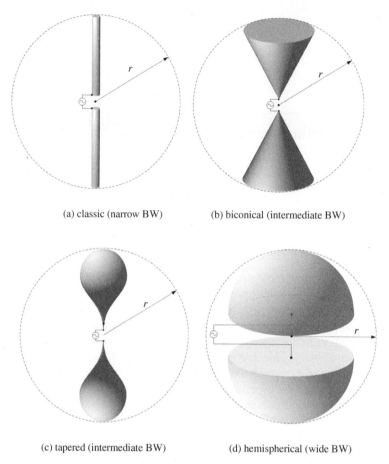

(a) classic (narrow BW) (b) biconical (intermediate BW)

(c) tapered (intermediate BW) (d) hemispherical (wide BW)

Figure 9.1 Dipole configurations and associated qualitative bandwidths (BW).

For the third technique, the main objectives are to find the current distribution on the antenna and in turn the input impedance. These were accomplished by Hallén by deriving an integral equation for the current distribution whose approximate solution, of different orders, was obtained by iteration and application of boundary conditions. Once a solution for the current is formed, the input impedance is determined by knowing the applied voltage at the feed terminals.

The details of the second method will follow in summary form. The integral equation technique of Hallén, along with that of Pocklington, form the basis of Moment Method techniques which were discussed in Chapter 8.

One of the main objectives in the design of an antenna is to broadband its characteristics—increase its bandwidth. Usually, this is a daunting task, especially if the specifications are quite ambitious. Typically, the response of each antenna, versus frequency, can be classified qualitatively into three categories: *narrowband, intermediate band*, and *wide band*. In Chapter 11, Section 11.5, it is pointed out that *the bandwidth of an antenna (which can be enclosed within a sphere of radius r) can be improved only if the antenna utilizes efficiently, with its geometrical configuration, the available volume within the sphere.* In Figure 9.1, we exhibit four different dipole configurations, starting with the classic dipole in Figure 9.1(a) and concluding with the hemispherical dipole of Figure 9.1(d). If we were to examine the frequency characteristics of a dipole on the basis of this fundamental principle, we can qualitatively categorize the frequency response of the different dipole configurations of Figure 9.1 into three groups; *narrowband, intermediate band*, and *wide band*. The same can be concluded for the geometries of the four monopole geometries of Figure 9.2. It should

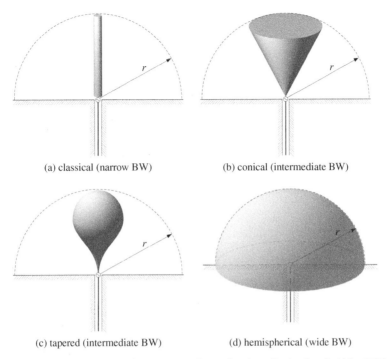

(a) classical (narrow BW)   (b) conical (intermediate BW)

(c) tapered (intermediate BW)   (d) hemispherical (wide BW)

Figure 9.2 Monopole configurations and associated qualitative bandwidths (BW).

be pointed out that the third configuration (*tapered*) in each figure will provide the best reflection (matching) efficiency when fed from traditional transmission lines. Although in each of the previous two figures, Figures 9.1 and 9.2, the last two configurations (*d* and *e* in each) exhibit the most broadband characteristics, usually these geometries are not as convenient and economical for practical implementation. However, any derivatives of these geometries, especially two-dimensional simulations, are configurations that may be used to broadband the frequency characteristics.

Since the biconical antenna of Figure 9.1(b) and derivatives of it are classic configurations with practical applications, they will be examined in some detail in the section that follows.

9.2 BICONICAL ANTENNA

One simple configuration that can be used to achieve broadband characteristics is the biconical antenna formed by placing two cones of infinite extent together, as shown in Figure 9.3(a). This can be thought to represent a uniformly tapered transmission line. The application of a voltage V_i at the input terminals will produce outgoing spherical waves, as shown in Figure 9.3(b), which in turn produce at any point $(r, \theta = \theta_c, \phi)$ a current I along the surface of the cone and voltage V between the cones (Figure 9.4). These can then be used to find the characteristic impedance of the transmission line, which is also equal to the input impedance of an infinite geometry. Modifications to this expression, to take into account the finite lengths of the cones, will be made using transmission-line analogy.

9.2.1 Radiated Fields

The analysis begins by first finding the radiated **E**- and **H**-fields between the cones, assuming dominant TEM mode excitation (**E** and **H** are transverse to the direction of propagation). Once these are

488 BROADBAND DIPOLES AND MATCHING TECHNIQUES

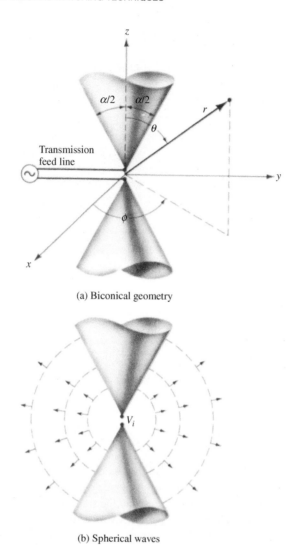

(a) Biconical geometry

(b) Spherical waves

Figure 9.3 Biconical antenna geometry and radiated spherical waves.

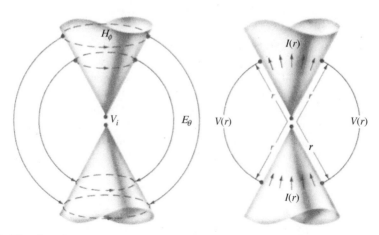

Figure 9.4 Electric and magnetic fields, and associated voltages and currents, for a biconical antenna.

determined for any point (r, θ, ϕ), the voltage V and current I at any point on the surface of the cone $(r, \theta = \theta_c, \phi)$ will be formed. From Faraday's law we can write that

$$\nabla \times \mathbf{E} = -j\omega\mu\mathbf{H} \tag{9-1}$$

which when expanded in spherical coordinates and assuming that the **E**-field has only an E_θ component independent of ϕ, reduces to

$$\nabla \times \mathbf{E} = \hat{\mathbf{a}}_\phi \frac{1}{r} \frac{\partial}{\partial r}(rE_\theta) = -j\omega\mu(\hat{\mathbf{a}}_r H_r + \hat{\mathbf{a}}_\theta H_\theta + \hat{\mathbf{a}}_\phi H_\phi) \tag{9-2}$$

Since **H** only has an H_ϕ component, necessary to form the TEM mode with E_θ, (9-2) can be written as

$$\frac{1}{r}\frac{\partial}{\partial r}(rE_\theta) = -j\omega\mu H_\phi \tag{9-2a}$$

From Ampere's law we have that

$$\nabla \times \mathbf{H} = +j\omega\varepsilon\mathbf{E} \tag{9-3}$$

which when expanded in spherical coordinates, and assuming only E_θ and H_ϕ components independent of ϕ, reduces to

$$\hat{\mathbf{a}}_r \frac{1}{r^2 \sin\theta}\left[\frac{\partial}{\partial\theta}(r\sin\theta H_\phi)\right] - \hat{\mathbf{a}}_\theta \frac{1}{r\sin\theta}\left[\frac{\partial}{\partial r}(r\sin\theta H_\phi)\right] = +j\omega\varepsilon(\hat{\mathbf{a}}_\theta E_\theta) \tag{9-4}$$

which can also be written as

$$\frac{\partial}{\partial\theta}(r\sin\theta H_\phi) = 0 \tag{9-4a}$$

$$\frac{1}{r\sin\theta}\frac{\partial}{\partial r}(r\sin\theta H_\phi) = -j\omega\varepsilon E_\theta \tag{9-4b}$$

Rewriting (9-4b) as

$$\frac{1}{r}\frac{\partial}{\partial r}(rH_\phi) = -j\omega\varepsilon E_\theta \tag{9-5}$$

and substituting it into (9-2a) we form a differential equation for H_ϕ as

$$-\frac{1}{j\omega\varepsilon r}\frac{\partial}{\partial r}\left[\frac{\partial}{\partial r}(rH_\phi)\right] = -j\omega\mu H_\phi \tag{9-6}$$

or

$$\frac{\partial^2}{\partial r^2}(rH_\phi) = -\omega^2\mu\varepsilon(rH_\phi) = -k^2(rH_\phi) \tag{9-6a}$$

A solution for (9-6a) must be obtained to satisfy (9-4a). To meet the condition of (9-4a), the θ variations of H_ϕ must be of the form

$$H_\phi = \frac{f(r)}{\sin\theta} \tag{9-7}$$

A solution of (9-6a), which also meets the requirements of (9-7) and represents an outward traveling wave, is

$$H_\phi = \frac{H_0}{\sin\theta} \frac{e^{-jkr}}{r} \quad (9\text{-}8)$$

where

$$f(r) = H_0 \frac{e^{-jkr}}{r} \quad (9\text{-}8a)$$

An inward traveling wave is also a solution but does not apply to the infinitely long structure.

Since the field is of TEM mode, the electric field is related to the magnetic field by the intrinsic impedance, and we can write it as

$$E_\theta = \eta H_\phi = \eta \frac{H_0}{\sin\theta} \frac{e^{-jkr}}{r} \quad (9\text{-}9)$$

In Figure 9.4(a) we have sketched the electric and magnetic field lines in the space between the two conical structures. The voltage produced between two corresponding points on the cones, a distance r from the origin, is found by

$$V(r) = \int_{\alpha/2}^{\pi-\alpha/2} \mathbf{E} \cdot d\mathbf{l} = \int_{\alpha/2}^{\pi-\alpha/2} (\hat{a}_\theta E_\theta) \cdot (\hat{a}_\theta r d\theta) = \int_{\alpha/2}^{\pi-\alpha/2} E_\theta r\, d\theta \quad (9\text{-}10)$$

or by using (9-9)

$$V(r) = \eta H_0 e^{-jkr} \int_{\alpha/2}^{\pi-\alpha/2} \frac{d\theta}{\sin\theta} = \eta H_0 e^{-jkr} \ln\left[\frac{\cot(\alpha/4)}{\tan(\alpha/4)}\right]$$

$$V(r) = 2\eta H_0 e^{-jkr} \ln\left[\cot\left(\frac{\alpha}{4}\right)\right] \quad (9\text{-}10a)$$

The current on the surface of the cones, a distance r from the origin, is found by using (9-8) as

$$I(r) = \int_0^{2\pi} H_\phi r \sin\theta\, d\phi = H_0 e^{-jkr} \int_0^{2\pi} d\phi = 2\pi H_0 e^{-jkr} \quad (9\text{-}11)$$

In Figure 9.4(b) we have sketched the voltage and current at a distance r from the origin.

9.2.2 Input Impedance

A. Infinite Cones

Using the voltage of (9-10a) and the current of (9-11), we can write the characteristic impedance as

$$Z_c = \frac{V(r)}{I(r)} = \frac{\eta}{\pi} \ln\left[\cot\left(\frac{\alpha}{4}\right)\right] \quad (9\text{-}12)$$

Since the characteristic impedance is not a function of the radial distance r, it also represents the input impedance at the antenna feed terminals of the infinite structure. For a free-space medium,

(9-12) reduces to

$$Z_c = Z_{in} = 120 \ln\left[\cot\left(\frac{\alpha}{4}\right)\right] \quad (9\text{-}12a)$$

which is a pure resistance. For small cone angles

$$Z_{in} = \frac{\eta}{\pi}\ln\left[\cot\left(\frac{\alpha}{4}\right)\right] = \frac{\eta}{\pi}\ln\left[\frac{1}{\tan(\alpha/4)}\right] \simeq \frac{\eta}{\pi}\ln\left(\frac{4}{\alpha}\right) \quad (9\text{-}12b)$$

Variations of Z_{in} as a function of the half-cone angle $\alpha/2$ are shown plotted in Figure 9.5(a) for $0° < \alpha/2 \leq 90°$ and in Figure 9.5(b) in an expanded scale for $0° < \alpha/2 \leq 2°$. Although the half-cone angle is not very critical in the design, it is usually chosen so that the characteristic impedance of the biconical configuration is nearly the same as that of the transmission line to which it will be attached. Small angle biconical antennas are not very practical but wide-angle configurations ($30° < \alpha/2 < 60°$) are frequently used as broadband antennas.

The radiation resistance of (9-12) can also be obtained by first finding the total radiated power

$$P_{rad} = \oiint_S \mathbf{W}_{av} \cdot d\mathbf{s} = \int_0^{2\pi}\int_{\alpha/2}^{\pi-\alpha/2} \frac{|E|^2}{2\eta} r^2 \sin\theta\, d\theta\, d\phi = \pi\eta|H_0|^2 \int_0^{\pi-\alpha/2} \frac{d\theta}{\sin\theta}$$

$$P_{rad} = 2\pi\eta|H_0|^2 \ln\left[\cot\left(\frac{\alpha}{4}\right)\right] \quad (9\text{-}13)$$

and by using (9-11) evaluated at $r = 0$ we form

$$R_r = \frac{2P_{rad}}{[I(0)]^2} = \frac{\eta}{\pi}\ln\left[\cot\left(\frac{\alpha}{4}\right)\right] \quad (9\text{-}14)$$

which is identical to (9-12).

B. Finite Cones

The input impedance of (9-12) or (9-14) is for an infinitely long structure. To take into account the finite dimensions in determining the input impedance, Schelkunoff [2] has devised an ingenious method where he assumes that for a finite length cone ($r = l/2$) some of the energy along the surface of the cone is reflected while the remaining is radiated. Near the equator most of the energy is radiated. This can be viewed as a load impedance connected across the ends of the cones. The electrical equivalent is a transmission line of characteristic impedance Z_c terminated in a load impedance Z_L. Values of the input impedance (resistance and reactance) have been computed, for small angle cones, in [4] where it is shown the antenna becomes more broadband (its resistance and reactance variations are less severe) as the cone angle increases.

The biconical antenna represents one of the canonical problems in antenna theory, and its model is well suited for examining general characteristics of dipole-type antennas.

C. Unipole

Whenever one of the cones is mounted on an infinite plane conductor (i.e., the lower cone is replaced by a ground plane), it forms a unipole and its input impedance is one-half of the two-cone structure, as is the case of a $\lambda/4$ monopole compared to that of a dipole, and it is represented by (4-106). Input impedances for unipoles of various cone angles as a function of the antenna length l have been

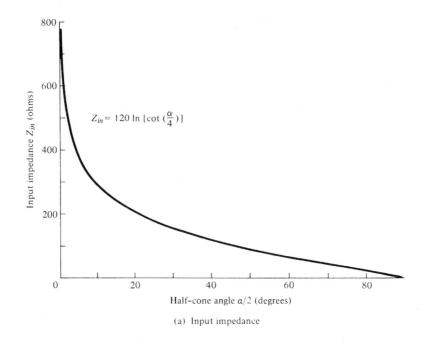

(a) Input impedance

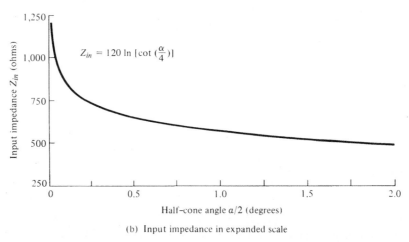

(b) Input impedance in expanded scale

Figure 9.5 Input impedance of an infinitely long biconical antenna radiating in free-space.

measured [5]. Radiation patterns of biconical dipoles fed by coaxial lines have been computed by Papas and King [6].

9.3 TRIANGULAR SHEET, FLEXIBLE AND CONFORMAL BOW-TIE, AND WIRE SIMULATION

Because of their broadband characteristics, biconical antennas have been employed for many years in the VHF and UHF frequency ranges. However, the solid or shell biconical structure is so massive for most frequencies of operation that it is impractical to use. Because of its attractive radiation characteristics, compared to those of other single antennas, realistic variations to its mechanical structure have been sought while retaining as many of the desired electrical features as possible.

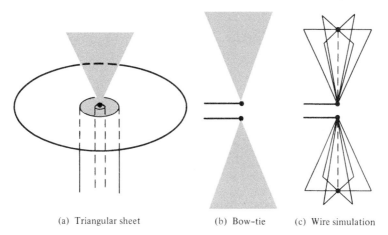

(a) Triangular sheet (b) Bow–tie (c) Wire simulation

Figure 9.6 Triangular sheet, bow-tie, and wire simulation of biconical antenna.

Geometrical approximations of the solid or shell conical unipole or biconical antenna are the triangular sheet and bow-tie antennas shown in Figures 9.6(a, b), respectively, each fabricated from sheet metal. The triangular sheet has been investigated experimentally by Brown and Woodward [5]. Each of these antennas can also be simulated by a wire along the periphery of its surface, which reduces significantly the weight and wind resistance of the structure. This was done in [7] using the Method of Moments. In order to simulate better the attractive surface of revolution of a biconical antenna of low-mass structures, multielement intersecting wire bow-ties were employed, as shown in Figure 9.6(c). It has been shown that eight or more intersecting wire-constructed bow-ties can approximate reasonably well the radiation characteristics of solid conical body-of-revolution antenna.

The bow-tie of Figure 9.6(b) was investigated more in depth in [8], [9] using a flexible thin substrate [10]. The substrate is a thin plastic (heat-stabilized PEN), allowing the antenna to be flexible. The substrate is covered with a very thin silicon nitride layer, which is the gate dielectric [11]. The conducting material used for the feed network, balun, and the antenna element is aluminum. Figure 9.7 illustrates a simplified model for the flexible substrate that best approximates the electrical properties of the actual substrate, also referred to as plastic in [10].

A bow-tie design, shown in Figure 9.8(a) along with its balun, was selected as the design because of its basic geometry, broadband characteristics, and variety of applications when compared to a linear wire and a printed dipole. Furthermore, bow-tie antennas are expected to be more directive than conventional dipole antennas because of the larger radiating area [12]. They are also used in size reduction applications of patch antennas to achieve lower operating frequencies without increasing the overall patch area.

The coplanar bow-tie [12]–[14] requires a balanced feed network, so that the antenna can be fed by a microstrip line or a coplanar waveguide with the use of a balun. The design procedure of a printed

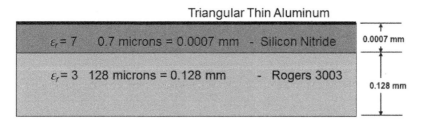

Figure 9.7 Simplified geometry PEN (polyethylene naphthalate) plastic. (SOURCE: [8] © 2011 IEEE).

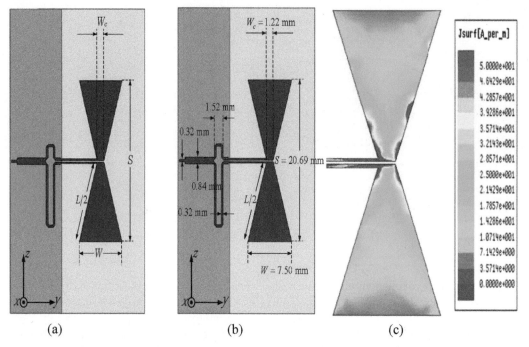

Figure 9.8 Flexible bow-tie antenna on a thin plastic (PEN) substrate at 7.66 GHz. (SOURCE: [8] © 2011 IEEE).

bow-tie antenna is similar to the design of rectangular microstrip patches. There is a set of design equations, which are obtained by modifying the semi-empirical design equations for rectangular patches of Chapter 14. The resonant frequency of a bow-tie patch, for the dominant TM_{10} mode, can be obtained using the equations that follow [8], [15], [16].

$$(f_r)_{10}^{TM} = \frac{1}{2\sqrt{\varepsilon_{reff}}L}\left(\frac{1.152}{R_t}\right) \qquad (9\text{-}15a)$$

$$R_t = \frac{L}{2}\frac{(W+2\Delta L)+(W_c+2\Delta L)}{(W+2\Delta L)+(S+2\Delta L)} \qquad (9\text{-}15b)$$

$$\frac{\Delta L}{h} = 0.412\frac{(\varepsilon_{reff}+0.3)\left(\frac{W_i}{h}+0.264\right)}{(\varepsilon_{reff}-0.258)\left(\frac{W_i}{h}+0.813\right)} \qquad (9\text{-}15c)$$

$$\varepsilon_{reff} = \frac{\varepsilon_r+1}{2}+\frac{\varepsilon_r-1}{2}\left[1+12\frac{h}{W_i}\right]^{-1/2} \qquad (9\text{-}15d)$$

$$W_i = \left(\frac{W+W_c}{2}\right) \qquad (9\text{-}15e)$$

In this set of design equations, the thickness, relative and effective permittivity of the substrate, are denoted by h, ε_r, and ε_{reff}, respectively. The other geometrical parameters are defined in Figure 9.8(a). Although these equations were primarily derived for microstrip patch type bow-ties, they were used to obtain an initial design of a coplanar bow-tie antenna. Afterward, the antenna design

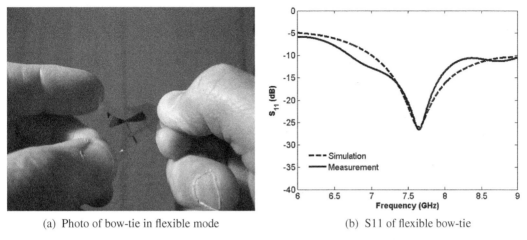

(a) Photo of bow-tie in flexible mode (b) S11 of flexible bow-tie

Figure 9.9 Photo and S_{11} of flexible bow-tie antenna. (SOURCE: [8] © 2011 IEEE).

was fine-tuned and finalized by numerical simulations [8]. There was a slight difference between the initial and the final values because of the approximate form of the design equations and the existence of the microstrip to coplanar feed network balun.

The balun introduces an 180° phase difference between the coupled microstrip lines near the center frequency. The length of the phase shifter is a very important parameter for the balun design. The lengths of the two branches of the microstrip line should be adjusted such that their difference is equal to one-quarter of the guided wavelength at the center frequency [16]. Another critical parameter is the gap between the coplanar strip lines. This gap can be adjusted to optimize the balun's performance [17].

Based on these equations and procedure, a bow-tie was designed, using the flexible substrate of Figure 9.7, to resonate at 7.66 GHz; the overall dimensions are indicated in Figure 9.8(b). A plot of the current density at 7.66 GHz is displayed in Figure 9.8(c), which is most intense along its edges. A photo of the bow tie, in its flexible mode, is shown in Figure 9.9(a). Its return loss is shown in Figure 9.9(b). It is worth to note that the bandwidth of the bow-tie, with respect to −15 dB (VSWR < 1.5:1) return loss, was decreased from 15% to 8.75% after the inclusion of the balun. This is due to the rapid change of the phase shift, introduced by the balun, with respect to frequency. This observation verifies that the balun is the critical device in the design of the bandwidth, and that the balun determines the bandwidth of the overall design.

In addition to return loss, amplitude radiation patterns of the antennas were simulated, both 3-D and 2-D, and they are displayed in Figure 9.10. The 2-D patterns, each normalized to its own maximum, were simulated in three planes: principal E-plane (y-z plane), secondary E-plane (x-z plane), and H-plane (x-y plane), and they were compared with measurements. The secondary E-plane (x-z plane) is defined as the one along which the E-field is parallel to it but does not pass through the overall field maximum. It can be seen that the measured radiation patterns are in excellent agreement with the simulated ones in all of the three planes. Although the pattern in the secondary E-plane is very close to the pattern of an ideal dipole, the patterns in the principal E- and H-planes are noticeably distorted. The back lobes of the patterns are approximately 10 dB lower compared to the forward lobes. This difference is due to the presence of the ground plane, which "pushes" the radiation pattern against itself and toward the bow-tie. This structure can also be considered as a 2-element Yagi-Uda antenna [18], which is composed of the bow-tie dipole and the ground plane. Yagi-Uda antennas are discussed in detail in Section 10.3.3. The ground plane acts as the reflector of the Yagi antenna, which has a decreased backward radiation. Therefore, the direction of the peak gain is away from the ground plane.

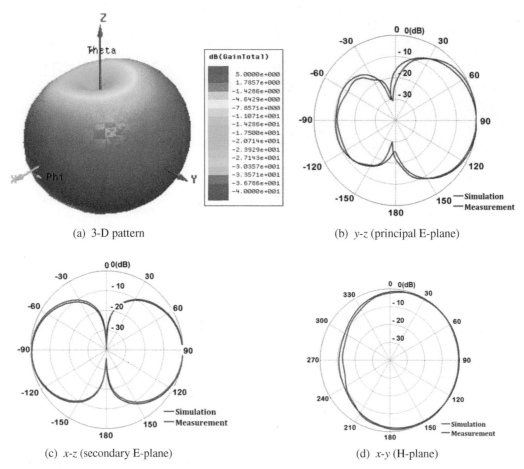

Figure 9.10 Normalized 3-D and 2-D amplitude patterns of bow-tie at 7.66 GHz. (SOURCE: [8] © 2011 IEEE).

In addition to the solid flexible bow-tie, an outline bow-tie was designed, simulated, and measured, where the center part of the metallic structure was removed [8], [9]. The outline bow-tie resonated at a lower frequency (7.4 GHz, instead of 7.66 GHz for the solid) because electrically was larger, due to the removal of the center part of the metallic structure where the current density is less intense and not critical to the performance of the radiating bow-tie. For more details, the reader is directed to [8], [9].

9.4 VIVALDI ANTENNA

The Vivaldi antenna is a broadband end-fire traveling wave type introduced by Gibson in 1979 [19]. Its basic geometry is shown in Figure 9.11, and it usually is implemented on a substrate with the Vivaldi design etched on the upper cladding of the substrate. The basic structure consists of a $\lambda_s/4$ uniform slot that is connected to an exponentially tapered slot; the subscript s is used to identify the slot. The slot is excited/fed by a microstrip transmission line from the undersurface of the substrate, as shown in Figure 9.11. An alternate design is to use a resonant area, typically either square or circular, instead of the $\lambda_s/4$ uniform slot, which is also usually excited by a microstrip line. The resonant area is connected to an exponentially tapered slot, with or without a transmission line.

The Vivaldi antenna is sometimes referred to as tapered slot antenna (TSA), flared or notch antenna, end-fire slot, and other related names. Because of its planar structure, it exhibits attractive

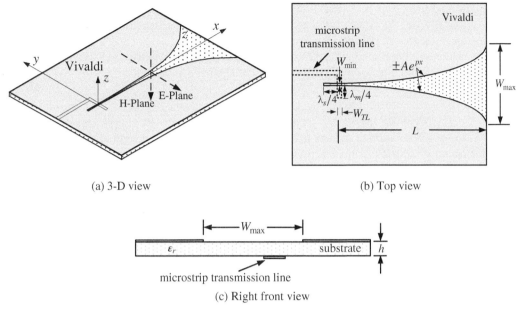

Figure 9.11 Vivaldi antenna geometry.

geometrical characteristics, especially to be integrated with MMICs (Monolithic Microwave Integrated Circuits). The Vivaldi design is usually low cost, and it possesses excellent radiation characteristics, such as high gain, broadband performance, constant beamwidth, and low side lobes. The directivity of Vivaldi antennas increases as the length L of the antenna increases, achieving gains up to 17 dB [20]. When first presented by Gibson [19], it had achieved, as a single element, an impedance bandwidth from below 2 GHz to above 40 GHz with a gain of 10 dB and side lobes of −20 dB. Subsequent wide bandwidth applications have used the Vivaldi antenna geometry in arrays [21], [22], including active electronically scanned arrays (AESAs).

One may question as to where the name came from, and why the name of Antonio Vivaldi, a composer from the Baroque period, is associated with an antenna. The question is best answered in [23], where according to an article on the Pharos JRA focal plane array, it "received its name from a resemblance to the shape of a cello or violin, instruments used by Antonio Vivaldi, the designer's favorite composer." Gibson himself was a musician, composer, and teacher of music.

Vivaldi antennas are categorized as broadband or frequency independent antennas, with bandwidth up to 6:1 [24]; even 10:1 or greater for VSWR < 2 ($S_{11} < -10$ dB). For arrays, the main lobe of the pattern is nearly proportional to $\cos(\theta)$, and it is basically maintained for scan angles up to about 50°–60° [22]. The bandwidth is limited by the opening width W_{min} and the aperture width W_{max} of the antenna.

This antenna is usually incorporated with MMICs; thus, the bandwidth is also limited by the transition between the microstrip line (which connects to the MMICs) and the slot line of the antenna, a subject of interest in [25]. The proper design of this transition results in a broadband performance that matches the antenna performance [24]. The simpler transition is a $\lambda_m/4$ open microstrip and uniform $\lambda_s/4$ slot line, where λ_m and λ_s are the wavelengths at the center frequency; the subscript m is used to identify the microstrip line. A coplanar waveguide feed can also be used, and it presents a wider bandwidth. Also, baluns can be used with Vivaldi antennas to make them more compatible when connected to strip lines and microstrips. A commonly used balun is that adapted from Knorr's microstrip-to-slot transition [22], [27]. Over the entire bandwidth, the beamwidth is nearly constant [24]. Furthermore, Vivaldi antennas exhibit a symmetric end-fire beam [26]; that is, the beamwidth is approximately the same in both the E-plane (parallel to the substrate) and the H-plane (perpendicular to the substrate). As the length of the antenna increases, the beamwidth narrows [20].

As can be observed in Figure 9.11, the antenna is an exponentially tapered slot cut in a thin film of metal that is supported by a substrate. The exponential taper can be defined by [19]

$$y(x) = \pm A e^{px} \tag{9-16}$$

where y is the half separation of the slot and x is the position across the length of the antenna, A is half of the opening width W_{min}, and p is the taper rate. Larger values of the rate of change p improve the low-frequency resistance but simultaneously create larger variations in resistance and reactance over the entire band. Therefore, for wide bandwidth applications, a compromise is usually required between the taper rate p and the square resonant/cavity area [22]. The taper rate has a significant impact on the bandwidth and beamwidth of the antenna. In general, as the taper rate increases, the beamwidth in the E-plane increases, the beamwidth in the H-plane decreases, and the bandwidth increases.

Parametric studies have shown that the optimal performance is achieved when the length L is greater than one wavelength at the lowest frequency. The opening width W_{min} is based on the highest frequency and the aperture width W_{max} influences the lower frequency [19]. Furthermore, the value of the aperture width W_{max} typically should, based on parametric examinations, be in the range W_{max1} and W_{max2}, where

$$W_{max1} \approx \lambda_0 \tag{9-17}$$

and

$$W_{max2} \approx \frac{\lambda_{min}}{2} \tag{9-18}$$

such that $W_{max1} < W_{max} < W_{max2}$, where λ_{min} is the wavelength at the minimum frequency and λ_0 is the wavelength at the center frequency [28].

The feed of Vivaldi antennas is usually a microstrip line that connects it to the MMICs, as illustrated in Figure 9.11. The microstrip line is printed on a different layer inside the substrate, and as mentioned before, the transition between the microstrip line and the slot line of the antenna should be designed properly so as not to introduce a significant limitation in the bandwidth.

Example 9.1

Design a Vivaldi antenna with a center frequency of 10 GHz using a substrate with permittivity of 2.33 and thickness of 0.508 mm. Assume that the minimum frequency is 4 GHz.

Solution:

With (9-17) and (9-18) used as a guideline, the range of values for the aperture width is

$$W_{max1} < W_{max} < W_{max2}$$

$$W_{max1} = \lambda_0 = \frac{3 \times 10^8}{(10 \times 10^9)\sqrt{2.33}} = 19.65 \approx 20 \text{ mm}$$

$$W_{max2} = \frac{\lambda_{min}}{2} = \frac{3 \times 10^8}{2(4 \times 10^9)\sqrt{2.33}} = 24.57 \approx 25 \text{ mm}$$

Through simulations, an aperture width of $W_{max} = 25$ mm led to the best performance and was used in the design. The length L was selected to be 27 mm, $W_{min} = 2A = 0.1$ mm, and $p = 0.204$. Based on these dimensions, the S_{11} of the design is illustrated in Figure 9.12. Using $S_{11} = -10$ dB as a criterion, the low and high frequencies are 8.6 GHz and 23.9 GHz, respectively, leading to a bandwidth of 2.77 : 1. However the antenna operates below 8.6 GHz but is not as well matched.

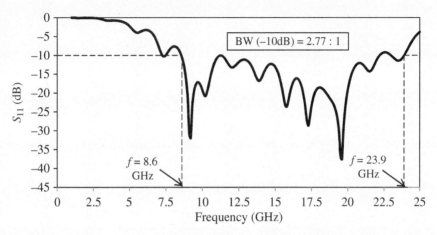

Figure 9.12 S_{11} for the Vivaldi antenna.

The 3-D radiation pattern and the 2-D patterns in the E- and H-planes, are shown in Figures 9.13 and 9.14, respectively.

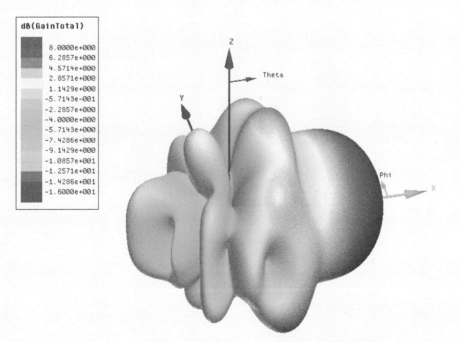

Figure 9.13 3-D radiation pattern for the Vivaldi antenna at 10 GHz.

500 BROADBAND DIPOLES AND MATCHING TECHNIQUES

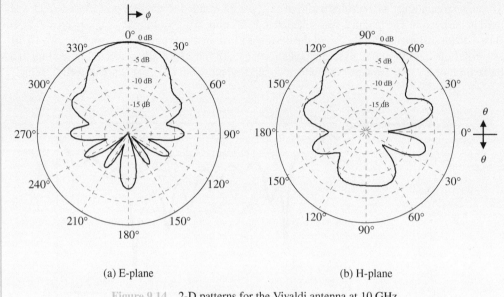

(a) E-plane

(b) H-plane

Figure 9.14   2-D patterns for the Vivaldi antenna at 10 GHz.

The current density distribution at 10 GHz, associated with the design of Example 9.1 is displayed in Figure 9.15.

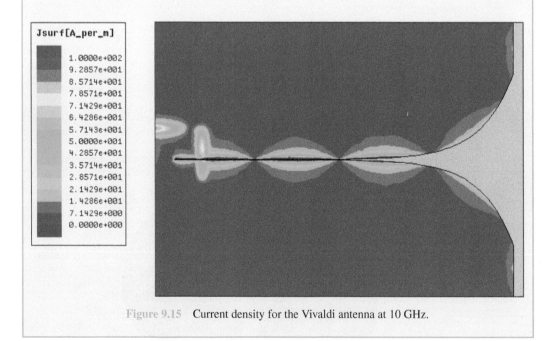

Figure 9.15   Current density for the Vivaldi antenna at 10 GHz.

9.5 CYLINDRICAL DIPOLE

Another simple and inexpensive antenna whose radiation characteristics are frequency dependent is a cylindrical dipole (i.e., a wire of finite diameter and length) of the form shown in Figure 9.16.

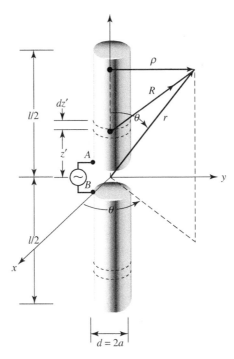

Figure 9.16 Center-fed cylindrical antenna configuration.

Thick dipoles are considered broadband while thin dipoles are more narrowband. This geometry can be considered to be a special form of the biconical antenna when $\alpha = 0°$. A thorough analysis of the current, impedance, pattern, and other radiation characteristics can be performed using the Moment Method. With that technique the antenna is analyzed in terms of integral formulations of the Hallén and Pocklington type which can be evaluated quite efficiently by the Moment Method. The analytical formulation of the Moment Method has been presented in Chapter 8. In this section we want to present, in summary form, some of its performance characteristics.

9.5.1 Bandwidth

As has been pointed out previously, a very thin linear dipole has very narrowband input impedance characteristics. Any small perturbations in the operating frequency will result in large changes in its operational behavior. One method by which its acceptable operational bandwidth can be enlarged will be to decrease the l/d ratio. For a given antenna, this can be accomplished by holding the length the same and increasing the diameter of the wire. For example, an antenna with a $l/d \simeq 5,000$ has an acceptable bandwidth of about 3%, which is a small fraction of the center frequency. An antenna of the same length but with a $l/d \simeq 260$ has a bandwidth of about 30%.

9.5.2 Input Impedance

The input impedance (resistance and reactance) of a very thin dipole of length l and diameter d can be computed using (8-60a)–(8-61b). As the radius of the wire increases, these equations become inaccurate. However, using integral equation analyses along with the Moment Method of Chapter 8, input impedances can be computed for wires with different l/d ratios. In general, it has been observed that for a given length wire its impedance variations become less sensitive as a function of frequency as the l/d ratio decreases. Thus more broadband characteristics can be obtained by increasing the

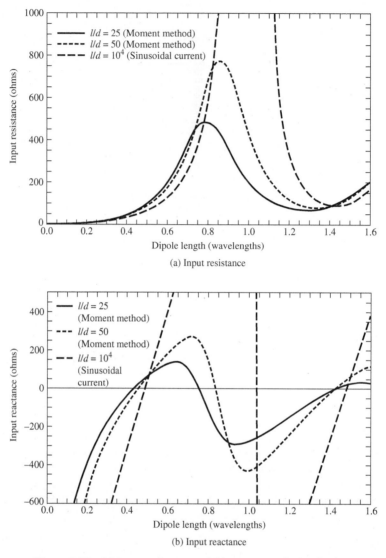

Figure 9.17 (a) Input resistance and (b) reactance of wire dipoles.

diameter of a given wire. To demonstrate this, in Figures 9.17(a) and (b) we have plotted, as a function of electrical length (as a function of frequency for a constant physical length), the input resistance and reactance of dipoles with $l/d = 10^4 (\Omega = 19.81)$, $50(\Omega = 9.21)$, and $25(\Omega = 7.824)$ where $\Omega = 2\ln(2l/d)$ and d is the diameter. For $l/d = 10^4$ the values were computed using (8-60a) and (8-61a) and then transferred to the input terminals by (8-60b) and (8-61b), respectively. The others were computed using the Moment Method techniques of Chapter 8. It is noted that the variations of each are less pronounced as the l/d ratio decreases, thus providing greater bandwidth.

Measured input resistances and reactances for a wide range of constant l/d ratios have been reported [29]. These curves are for a cylindrical antenna driven by a coaxial cable mounted on a large ground plane on the earth's surface. Thus they represent half of the input impedance of a center-fed cylindrical dipole radiating in free-space. The variations of the antenna's electrical length were obtained by varying the frequency while the length-to-diameter (l/d) ratio was held constant.

TABLE 9.1 Cylindrical Dipole Resonances

	First Resonance	Second Resonance	Third Resonance	Fourth Resonance
LENGTH	$0.48\lambda F$	$0.96\lambda F$	$1.44\lambda F$	$1.92\lambda F$
RESISTANCE (ohms)	67	$\dfrac{R_n^2}{67}$	95	$\dfrac{R_n^2}{95}$

$$F = \frac{l/2a}{1+l/2a}; \quad R_n = 150\log_{10}(l/2a)$$

9.5.3 Resonance and Ground Plane Simulation

The imaginary part of the input impedance of a linear dipole can be eliminated by making the total length, l, of the wire slightly less than an integral number of half-wavelengths (i.e., l slight less than $n\lambda/2, n = 1, 3, \ldots$) or slightly greater than an integral number of wavelengths (i.e., l slightly greater than $n\lambda$, $n = 1, 2, \ldots$). The amount of reduction or increase in length, is a function of the radius of the wire, and it can be determined for thin wires iteratively using (8-60b) and (8-61b). At the resonance length, the resistance can then be determined using (8-60a) and (8-61a). Empirical equations for approximating the length, impedance, and the order of resonance of the cylindrical dipoles are found in Table 9.1 [30]. R_n is called the natural resistance and represents the geometric mean resistance at an odd resonance and at the next higher even resonance. For a cylindrical stub above a ground plane, as shown in Figure 9.11, the corresponding values are listed in Table 9.2 [30].

To reduce the wind resistance, to simplify the design, and to minimize the costs, the ground plane of Figure 9.18(a) is often simulated, especially at low frequencies, by crossed wires as shown in Figure 9.18(b). Usually only two crossed wires (four radials) are employed. A larger number of radials results in a better simulation of the ground plane. Ground planes are also simulated by wire mesh. The spacing between the wires is usually selected to be equal or smaller than $\lambda/10$. The flat or shaped reflecting surfaces for UHF educational TV are usually realized approximately by using wire mesh.

9.5.4 Radiation Patterns

The theory for the patterns of infinitesimally thin wires was developed in Chapter 4. Although accurate patterns for finite diameter wires can be computed using current distributions obtained by the Moment Method of Chapter 8, the patterns calculated using ideal sinusoidal current distributions, valid for infinitely small diameters, provide a good first-order approximation even for relatively thick cylinders. To illustrate this, in Figure 9.19 we have plotted the relative patterns for $l = 3\lambda/2$ with $l/d = 10^4 (\Omega = 19.81), 50 (\Omega = 9.21), 25 (\Omega = 6.44)$, and $8.7 (\Omega = 5.71)$, where $\Omega = 2\ln(2l/d)$ and d is the diameter. For $l/d = 10^4$ the current distribution was assumed to be purely sinusoidal, as given by (4-56); for the others, the Moment Method techniques of Chapter 8 were used. The patterns were computed using the Moment Method formulations outlined in Section 8.4. It is noted that the

TABLE 9.2 Cylindrical Stub Resonances

	First Resonance	Second Resonance	Third Resonance	Fourth Resonance
LENGTH	$0.24\lambda F'$	$0.48\lambda F'$	$0.72\lambda F'$	$0.96\lambda F'$
RESISTANCE (ohms)	34	$\dfrac{(R_n')^2}{34}$	48	$\dfrac{(R_n')^2}{48}$

$$F' = \frac{l/a}{1+l/a}; \quad R_n' = 75\log_{10}(l/a)$$

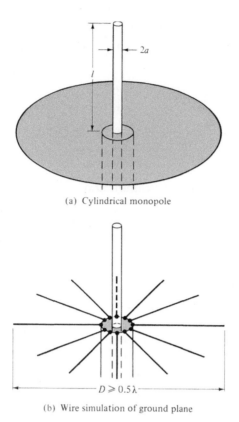

(a) Cylindrical monopole

(b) Wire simulation of ground plane

Figure 9.18 Cylindrical monopole above circular solid and wire-simulated ground planes.

pattern is essentially unaffected by the thickness of the wire in regions of intense radiation. However, as the radius of the wire increases, the minor lobes diminish in intensity and the nulls are filled by low-level radiation. The same characteristics have been observed for other length dipoles such as $l = \lambda/2, \lambda$ and 2λ. The input impedance for the $l = \lambda/2$ and $l = 3\lambda/2$ dipoles, with $l/d = 10^4$, 50, and 25, is equal to

$l = \lambda/2$	$l = 3\lambda/2$	
$Z_{in}(l/d = 10^4) = 73 + j42.5$	$Z_{in}(l/d = 10^4) = 105.49 + j45.54$	
$Z_{in}(l/d = 50) = 85.8 + j54.9$	$Z_{in}(l/d = 50) = 103.3 + j9.2$	(9-19)
$Z_{in}(l/d = 25) = 88.4 + j27.5$	$Z_{in}(l/d = 25) = 106.8 + j4.9$	

9.5.5 Equivalent Radii

Up to now, the formulations for the current distribution and the input impedance assume that the cross section of the wire is constant and of radius a. An electrical equivalent radius can be obtained for some uniform wires of noncircular cross section. This is demonstrated in Table 9.3 where the actual cross sections and their equivalent radii are illustrated.

The equivalent radius concept can be used to approximate the antenna or scattering characteristics of electrically small wires of arbitrary cross sections. It is accomplished by replacing the noncircular cross-section wire with a circular wire whose radius is the "equivalent" radius of the noncircular cross section. In electrostatics, the equivalent radius represents the radius of a circular wire whose

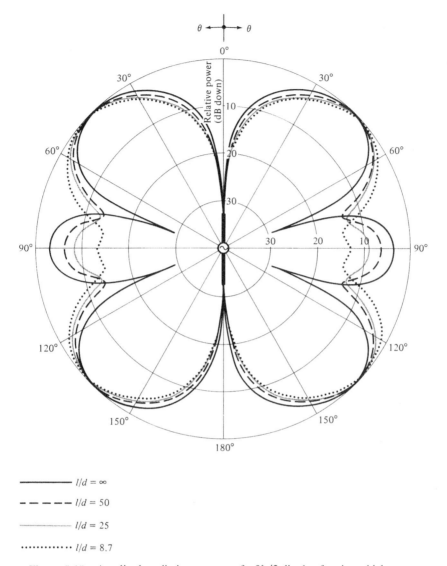

Figure 9.19 Amplitude radiation patterns of a $3\lambda/2$ dipole of various thicknesses.

capacitance is equal to that of the noncircular geometry. This definition can be used at all frequencies provided the wire remains electrically small. The circle with equivalent radius lies between the circles which circumscribe and inscribe the geometry and which together bound the noncircular cross section.

9.6 FOLDED DIPOLE

To achieve good directional pattern characteristics and at the same time provide good matching to practical coaxial lines with 50- or 75-ohm characteristic impedances, the length of a single wire element is usually chosen to be $\lambda/4 \leq l < \lambda$. The most widely used dipole is that whose overall length is $l \simeq \lambda/2$, and which has an input impedance of $Z_{in} \simeq 73 + j42.5$ and directivity of $D_0 \simeq 1.643$. In practice, there are other very common transmission lines whose characteristic impedance is much

TABLE 9.3 Conductor Geometrical Shapes and their Equivalent Circular Cylinder Radii

Geometrical Shape	Electrical Equivalent Radius
(flat strip, $b \simeq 0$)	$a_e = 0.25a$
(L-shape angle)	$a_e \simeq 0.2(a+b)$
(square cross-section)	$a_e = 0.59a$
(rectangular cross-section)	(plot of a_e/a vs b/a, ranging from about 0.2 at $b/a=0$ to about 0.55 at $b/a=1.0$)
(two cylinders, radii a and b)	$a_e = \tfrac{1}{2}(a+b)$
(two parallel conductors C_1, C_2 separated by s)	$\ln a_e \simeq \dfrac{1}{(S_1+S_2)^2} \times [S_1^2 \ln a_1 + S_2^2 \ln a_2 + 2S_1 S_2 \ln s]$ S_1, S_2 = peripheries of conductors C_1, C_2 a_1, a_2 = equivalent radii of conductors C_1, C_2

higher than 50 or 75 ohms. For example, a "twin-lead" transmission line (usually two parallel wires separated by about $\frac{5}{16}$ in. and embedded in a low-loss plastic material used for support and spacing) is widely used for TV applications and has a characteristic impedance of about 300 ohms.

In order to provide good matching characteristics, variations of the single dipole element must be used. One simple geometry that can achieve this is a folded wire which forms a very thin ($s \ll \lambda$) rectangular loop as shown in Figure 9.20(a). This antenna, when the spacing between the two larger sides is very small (usually $s < 0.05\lambda$), is known as a *folded dipole* and it serves as a step-up impedance transformer (approximately by a factor of 4 when $l = \lambda/2$) of the single-element impedance. Thus

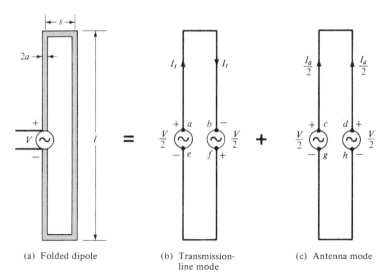

Figure 9.20 Folded dipole and its equivalent transmission-line and antenna mode models. (SOURCE: G. A. Thiele, E. P. Ekelman, Jr., and L. W. Henderson, "On the Accuracy of the Transmission Line Model for Folded Dipole," *IEEE Trans. Antennas Propagat.*, Vol. AP-28, No. 5, pp. 700–703, September 1980. © (1980) IEEE).

when $l = \lambda/2$ and the antenna is resonant, impedances on the order of about 300 ohms can be achieved, and it would be ideal for connections to "twin-lead" transmission lines.

A folded dipole operates basically as a balanced system, and it can be analyzed by assuming that its current is decomposed into two distinct modes: a transmission-line mode [Figure 9.20(b)] and an antenna mode [Figure 9.20(c)]. This type of an analytic model can be used to predict accurately the input impedance provided the longer parallel wires are close together electrically ($s \ll \lambda$).

To derive an equation for the input impedance, let us refer to the modeling of Figure 9.20. For the transmission-line mode of Figure 9.20(b), the input impedance at the terminals $a-b$ or $e-f$, looking toward the shorted ends, is obtained from the impedance transfer equation

$$Z_t = Z_0 \left[\frac{Z_L + jZ_0 \tan(kl')}{Z_0 + jZ_L \tan(kl')} \right]_{\substack{l'=l/2 \\ Z_L=0}} = jZ_0 \tan\left(k\frac{l}{2}\right) \tag{9-20}$$

where Z_0 is the characteristic impedance of a two-wire transmission line

$$Z_0 = \frac{\eta}{\pi} \cosh^{-1}\left(\frac{s/2}{a}\right) = \frac{\eta}{\pi} \ln\left[\frac{s/2 + \sqrt{(s/2)^2 - a^2}}{a}\right] \tag{9-21}$$

which can be approximated for $s/2 \gg a$ by

$$Z_0 = \frac{\eta}{\pi} \ln\left[\frac{s/2 + \sqrt{(s/2)^2 - a^2}}{a}\right] \simeq \frac{\eta}{\pi} \ln\left(\frac{s}{a}\right) = 0.733\eta \log_{10}\left(\frac{s}{a}\right) \tag{9-21a}$$

Since the voltage between the points a and b is $V/2$, and it is applied to a transmission line of length $l/2$, the transmission-line current is given by

$$I_t = \frac{V/2}{Z_t} \tag{9-22}$$

For the antenna mode of Figure 9.20(c), the generator points $c - d$ and $g - h$ are each at the same potential and can be connected, without loss of generality, to form a dipole. Each leg of the dipole is formed by a pair of closely spaced wires ($s \ll \lambda$) extending from the feed ($c - d$ or $g - h$) to the shorted end. Thus the current for the antenna mode is given by

$$I_a = \frac{V/2}{Z_d} \tag{9-23}$$

where Z_d is the input impedance of a linear dipole of length l and diameter d computed using (8-57a)–(8-58b). For the configuration of Figure 9.20(c), the radius that is used to compute Z_d for the dipole can be either the half-spacing between the wires ($s/2$) or an equivalent radius a_e. The equivalent radius a_e is related to the actual wire radius a by (from Table 9.3)

$$\ln(a_e) = \frac{1}{2}\ln(a) + \frac{1}{2}\ln(s) = \ln(a) + \frac{1}{2}\ln\left(\frac{s}{a}\right) = \ln\sqrt{as} \tag{9-24}$$

or

$$a_e = \sqrt{as} \tag{9-24a}$$

It should be expected that the equivalent radius yields the most accurate results.

The total current on the feed leg (left side) of the folded dipole of Figure 9.20(a) is given by

$$I_{in} = I_t + \frac{I_a}{2} = \frac{V}{2Z_t} + \frac{V}{4Z_d} = \frac{V(2Z_d + Z_t)}{4Z_t Z_d} \tag{9-25}$$

and the input impedance at the feed by

$$\boxed{Z_{in} = \frac{V}{I_{in}} = \frac{2Z_t(4Z_d)}{2Z_t + 4Z_d} = \frac{4Z_t Z_d}{2Z_d + Z_t}} \tag{9-26}$$

Based on (9-26), the folded dipole behaves as the equivalent of Figure 9.21(a) in which the antenna mode impedance is stepped up by a ratio of four. The transformed impedance is then placed in shunt with twice the impedance of the nonradiating (transmission-line) mode to result in the input impedance.

When $l = \lambda/2$, it can be shown that (9-26) reduces to

$$\boxed{Z_{in} = 4Z_d} \tag{9-27}$$

or that the impedance of the folded dipole is four times greater than that of an isolated dipole of the same length as one of its sides. This is left as an exercise for the reader (Prob. 9.9).

The impedance relation of (9-27) for the $l = \lambda/2$ can also be derived by referring to Figure 9.22. Since for a folded dipole the two vertical arms are closely spaced ($s \ll \lambda$), the current distribution in each is identical as shown in Figure 9.22(a). The equivalent of the folded dipole of Figure 9.22(a) is the ordinary dipole of Figure 9.22(b). Comparing the folded dipole to the ordinary dipole, it is apparent that the currents of the two closely spaced and identical arms of the folded dipole are equal to the one current of the ordinary dipole, or

$$2I_f = I_d \tag{9-28}$$

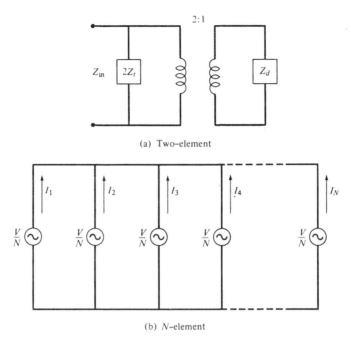

(a) Two-element

(b) N-element

Figure 9.21 Equivalent circuits for two-element and N-element (with equal radii elements) folded dipoles.

where I_f is the current of the folded dipole and I_d is the current of the ordinary dipole. Also the input power of the two dipoles are identical, or

$$P_f \equiv \frac{1}{2}I_f^2 Z_f = P_d \equiv \frac{1}{2}I_d^2 Z_d \tag{9-29}$$

Substituting (9-28) into (9-29) leads to

$$Z_f = 4Z_d \tag{9-30}$$

where Z_f is the impedance of the folded dipole while Z_d is the impedance of the ordinary dipole. Equation (9-30) is identical to (9-27).

To better understand the impedance transformation of closely spaced conductors (of equal diameter) and forming a multielement folded dipole, let us refer to its equivalent circuit in Figure 9.21(b).

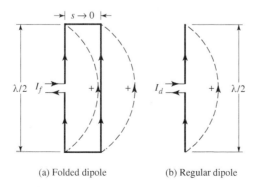

(a) Folded dipole   (b) Regular dipole

Figure 9.22 Folded dipole and equivalent regular dipole.

For N elements, the equivalent voltage at the center of each conductor is V/N and the current in each is $I_n, n = 1, 2, 3, \ldots, N$. Thus the voltage across the first conductor can be represented by

$$\frac{V}{N} = \sum_{n=1}^{N} I_n Z_{1n} \tag{9-31}$$

where Z_{1n} represents the self or mutual impedance between the first and nth element. Because the elements are closely spaced

$$I_n \simeq I_1 \quad \text{and} \quad Z_{1n} \simeq Z_{11} \tag{9-32}$$

for all values of $n = 1, 2, \ldots, N$. Using (9-32), we can write (9-31) as

$$\frac{V}{N} = \sum_{n=1}^{N} I_n Z_{1n} \simeq I_1 \sum_{n=1}^{N} Z_{1n} \simeq N I_1 Z_{11} \tag{9-33}$$

or

$$\boxed{Z_{in} = \frac{V}{I_1} \simeq N^2 Z_{11} = N^2 Z_r} \tag{9-33a}$$

since the self-impedance Z_{11} of the first element is the same as its impedance Z_r in the absence of the other elements. Additional impedance step-up of a single dipole can be obtained by introducing more elements. For a three-element folded dipole with elements of identical diameters and of $l \simeq \lambda/2$, the input impedance would be about nine times greater than that of an isolated element or about 650 ohms. Greater step-up transformations can be obtained by adding more elements; in practice, they are seldom needed. Many other geometrical configurations of a folded dipole can be obtained which would contribute different values of input impedances. Small variations in impedance can be obtained by using elements of slightly different diameters and/or lengths.

To test the validity of the transmission-line model for the folded dipole, a number of computations were made [31] and compared with data obtained by the Moment Method, which is considered to be more accurate. In Figures 9.23(a) and (b) the input resistance and reactance for a two-element folded dipole is plotted as a function of l/λ when the diameter of each wire is $d = 2a = 0.001\lambda$ and the spacing between the elements is $s = 0.00613\lambda$. The characteristic impedance of such a transmission line is 300 ohms. The equivalent radius was used in the calculations of Z_d. An excellent agreement is indicated between the results of the transmission-line model and the Moment Method. Computations and comparisons for other spacings ($s = 0.0213\lambda$, $Z_0 = 450$ ohms and $s = 0.0742\lambda$, $Z_0 = 600$ ohms) but with elements of the same diameter ($d = 0.001\lambda$) have been made [31]. It has been shown that as the spacing between the wires increased, the results of the transmission-line model began to disagree with those of the Moment Method. For a given spacing, the accuracy for the characteristic impedance, and in turn for the input impedance, can be improved by increasing the diameter of the wires. The characteristic impedance of a transmission line, as given by (9-21) or (9-21a), depends not on the spacing but on the spacing-to-diameter (s/d) ratio, which is more accurate for smaller s/d. Computations were also made whereby the equivalent radius was not used. The comparisons of these results indicated larger disagreements, thus concluding the necessity of the equivalent radius, especially for the larger wire-to-wire spacings.

A two-element folded dipole is widely used, along with "twin-lead" line, as feed element of TV antennas such as Yagi-Uda antennas, which are discussed in detail in Section 10.3.3. Although the impedance of an isolated folded dipole may be around 300 ohms, its value will be somewhat

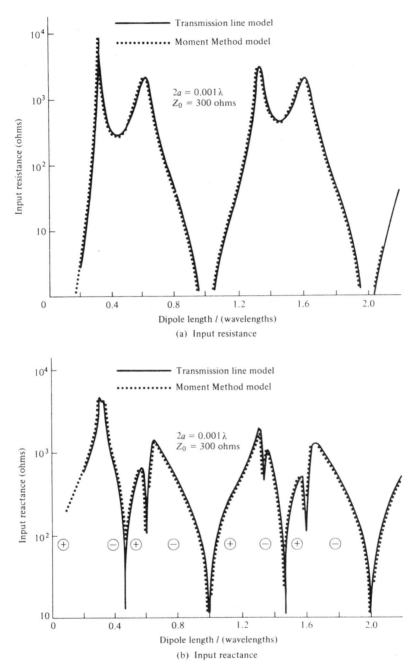

Figure 9.23 Input resistance and reactance of folded dipole. (SOURCE: G. A. Thiele, E. P. Ekelman, Jr., and L. W. Henderson, "On the Accuracy of the Transmission Line Model for Folded Dipole," *IEEE Trans. Antennas Propagat.*, Vol. AP-28, No. 5, pp. 700–703, September 1980. © (1980) IEEE).

different when it is used as an element in an array or with a reflector. The folded dipole has better bandwidth characteristics than a single dipole of the same size. Its geometrical arrangement tends to behave as a short parallel stub line which attempts to cancel the off resonance reactance of a single dipole. The folded dipole can be thought to have a bandwidth which is the same as that of a single dipole but with an equivalent radius ($a < a_e < s/2$).

Symmetrical and asymmetrical planar folded dipoles can also be designed and constructed using strips which can be fabricated using printed-circuit technology [32]. The input impedance can be varied over a wide range of values by adjusting the width of the strips. In addition, the impedance can be adjusted to match the characteristic impedance of printed-circuit transmission lines with four-to-one impedance ratios.

A MATLAB computer program, entitled **Folded**, has been developed to perform the design of a folded dipole. The description of the program is found in the corresponding READ ME file included in the publisher's website for the book.

9.7 DISCONE AND CONICAL SKIRT MONOPOLE

There are innumerable variations to the basic geometrical configurations of cones and dipoles, some of which have already been discussed, to obtain broadband characteristics. Two other common radiators that meet this characteristic are the conical skirt monopole and the discone antenna [33] shown in Figures 9.24(a) and (b), respectively.

For each antenna, the overall pattern is essentially the same as that of a linear dipole of length $l < \lambda$ (i.e., a solid of revolution formed by the rotation of a figure-eight) whereas in the horizontal (azimuthal) plane it is nearly omnidirectional. The polarization of each is vertical. Each antenna because of its simple mechanical design, ease of installation, and attractive broadband characteristics has wide applications in the VHF (30–300 MHz) and UHF (300 MHz–3 GHz) spectrum for broadcast, television, and communication applications.

The discone antenna is formed by a disk and a cone. The disk is attached to the center conductor of the coaxial feed line, and it is perpendicular to its axis. The cone is connected at its apex to the outer shield of the coaxial line. The geometrical dimensions and the frequency of operation of two designs [33] are shown in Table 9.4.

In general, the impedance and pattern variations of a discone as a function of frequency are much less severe than those of a dipole of fixed length l. The performance of this antenna as a function of frequency is similar to a high-pass filter. Below an effective cutoff frequency it becomes inefficient, and it produces severe standing waves in the feed line. At cutoff, the slant height of the cone is approximately $\lambda/4$.

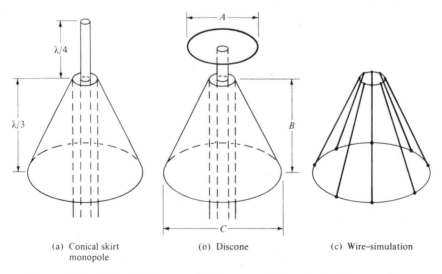

(a) Conical skirt monopole (b) Discone (c) Wire-simulation

Figure 9.24 Conical skirt monopole, discone, and wire-simulated cone surface.

TABLE 9.4 Frequency and Dimensions of Two Designs

Frequency (MHz)	A (cm)	B (cm)	C (cm)
90	45.72	60.96	50.80
200	22.86	31.75	35.56

Measured elevation (vertical) plane radiation patterns from 250 to 650 MHz, at 50-MHz intervals, have been published [33] for a discone with a cutoff frequency of 200 MHz. No major changes in the "figure-eight" shape of the patterns were evident other than at the high-frequency range where the pattern began to turn downward somewhat.

The conical skirt monopole is similar to the discone except that the disk is replaced by a monopole of length usually $\lambda/4$. Its general behavior also resembles that of the discone. Another way to view the conical skirt monopole is with a $\lambda/4$ monopole mounted above a finite ground plane. The plane has been tilted downward to allow more radiation toward and below the horizontal plane.

To reduce the weight and wind resistance of the cone, its solid surface can be simulated by radial wires, as shown in Figure 9.24(c). This is a usual practice in the simulation of finite size ground planes for monopole antennas. The lengths of the wires used to simulate the ground plane are on the order of about $\lambda/4$ or greater.

9.8 MATCHING TECHNIQUES

The operation of an antenna system over a frequency range is not completely dependent upon the frequency response of the antenna element itself but rather on the frequency characteristics of the transmission line–antenna element combination. In practice, the characteristic impedance of the transmission line is usually real whereas that of the antenna element is complex. Also the variation of each as a function of frequency is not the same. Thus efficient coupling-matching networks must be designed which attempt to couple-match the characteristics of the two devices over the desired frequency range.

There are many coupling-matching networks that can be used to connect the transmission line to the antenna element and which can be designed to provide acceptable frequency characteristics. Only a limited number will be introduced here.

9.8.1 Stub-Matching

Ideal matching at a given frequency can be accomplished by placing a short- or open-circuited shunt stub a distance s from the transmission-line–antenna element connection, as shown in Figure 9.25(a). Assuming a real characteristic impedance, the length s is controlled so as to make the real part of the antenna element impedance equal to the characteristic impedance. The length l of the shunt line is varied until the susceptance of the stub is equal in magnitude but opposite in phase to the line input susceptance at the point of the transmission line–shunt element connection. The matching procedure is illustrated best graphically with the use of a Smith chart. Analytical methods, on which the Smith chart graphical solution is based, can also be used. The short-circuited stub is more practical because an equivalent short can be created by a pin connection in a coaxial cable or a slider in a waveguide. This preserves the overall length of the stub line for matchings which may require longer length stubs.

A single stub with a variable length l cannot always match all antenna (load) impedances. A double-stub arrangement positioned a fixed distance s from the load, with the length of each stub variable and separated by a constant length d, will match a greater range of antenna impedances. However, a triple-stub configuration will always match all loads.

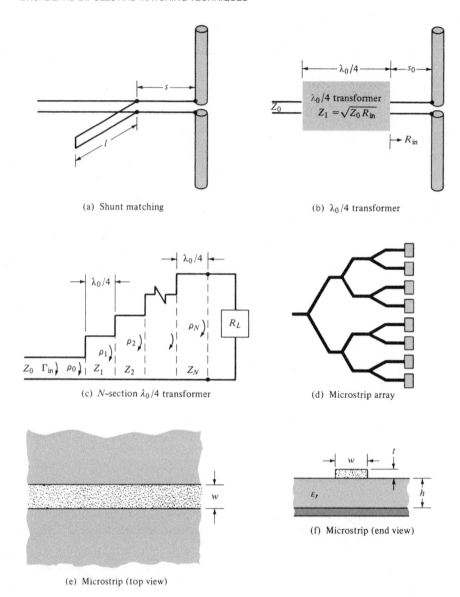

Figure 9.25 Matching and microstrip techniques.

Excellent treatments of the analytical and graphical methods for the single-, double-, triple-stub, and other matching techniques are presented in [34] and [35]. The higher-order stub arrangements provide more broad and less sensitive matchings (to frequency variations) but are more complex to implement. Usually a compromise is chosen, such as the double-stub.

9.8.2 Quarter-Wavelength Transformer

A. Single Section
Another technique that can be used to match the antenna to the transmission line is to use a $\lambda/4$ transformer. If the impedance of the antenna is real, the transformer is attached directly to the load. However if the antenna impedance is complex, the transformer is placed a distance s_0 away from the

antenna, as shown in Figure 9.25(b). The distance s_0 is chosen so that the input impedance toward the load at s_0 is real and designated as R_{in}. To provide a match, the transformer characteristic impedance Z_1 should be $Z_1 = \sqrt{R_{in}Z_0}$, where Z_0 is the characteristic impedance (real) of the input transmission line. The transformer is usually another transmission line with the desired characteristic impedance.

Because the characteristic impedances of most off-the-shelf transmission lines are limited in range and values, the quarter-wavelength transformer technique is most suitable when used with microstrip transmission lines. In microstrips, the characteristic impedance can be changed by simply varying the width of the center conductor.

B. Multiple Sections

Matchings that are less sensitive to frequency variations and that provide broader bandwidths, require multiple $\lambda/4$ sections. In fact the number and characteristic impedance of each section can be designed so that the reflection coefficient follows, within the desired frequency bandwidth, prescribed variations which are symmetrical about the center frequency. The antenna (load) impedance will again be assumed to be real; if not, the antenna element must be connected to the transformer at a point s_0 along the transmission line where the input impedance is real.

Referring to Figure 9.25(c), the total input reflection coefficient Γ_{in} for an N-section quarter-wavelength transformer with $R_L > Z_0$ can be written approximately as [34]

$$\Gamma_{in}(f) \simeq \rho_0 + \rho_1 e^{-j2\theta} + \rho_2 e^{-j4\theta} + \cdots + \rho_N e^{-j2N\theta}$$

$$= \sum_{n=0}^{N} \rho_n e^{-j2n\theta} \qquad (9\text{-}34)$$

where

$$\rho_n = \frac{Z_{n+1} - Z_n}{Z_{n+1} + Z_n} \qquad (9\text{-}34\text{a})$$

$$\theta = k\Delta l = \frac{2\pi}{\lambda}\left(\frac{\lambda_0}{4}\right) = \frac{\pi}{2}\left(\frac{f}{f_0}\right) \qquad (9\text{-}34\text{b})$$

In (9-34), ρ_n represents the reflection coefficient at the junction of two infinite lines with characteristic impedances Z_n and Z_{n+1}, f_0 represents the designed center frequency, and f the operating frequency. Equation (9-34) is valid provided the ρ_n's at each junction are small ($R_L \simeq Z_0$). If $R_L < Z_0$, the ρ_n's should be replaced by $-\rho_n$'s. For a real load impedance, the ρ_n's and Z_n's will also be real.

For a symmetrical transformer ($\rho_0 = \rho_N$, $\rho_1 = \rho_{N-1}$, etc.), (9-34) reduces to

$$\Gamma_{in}(f) \simeq 2e^{-jN\theta}[\rho_0 \cos N\theta + \rho_1 \cos(N-2)\theta + \rho_2 \cos(N-4)\theta + \cdots] \qquad (9\text{-}35)$$

The last term in (9-35) should be

$$\rho_{[(N-1)/2]} \cos\theta \quad \text{for } N = \text{odd integer} \qquad (9\text{-}35\text{a})$$

$$\tfrac{1}{2}\rho_{(N/2)} \quad \text{for } N = \text{even integer} \qquad (9\text{-}35\text{b})$$

C. Binomial Design

One technique, used to design an N-section $\lambda/4$ transformer, requires that the input reflection coefficient of (9-34) have maximally flat passband characteristics. For this method, the junction reflection coefficients (ρ_n's) are derived using the binomial expansion. Doing this, we can equate (9-34) to

$$\Gamma_{in}(f) = \sum_{n=0}^{N} \rho_n e^{-j2n\theta} = e^{-jN\theta} \frac{R_L - Z_0}{R_L + Z_0} \cos^N(\theta)$$

$$= 2^{-N} \frac{R_L - Z_0}{R_L + Z_0} \sum_{n=0}^{N} C_n^N e^{-j2n\theta} \qquad (9\text{-}36)$$

where

$$C_n^N = \frac{N!}{(N-n)!n!}, \quad n = 0, 1, 2, \ldots, N \qquad (9\text{-}36a)$$

From (9-34)

$$\rho_n = 2^{-N} \frac{R_L - Z_0}{R_L + Z_0} C_n^N \qquad (9\text{-}37)$$

For this type of design, the fractional bandwidth $\Delta f / f_0$ is given by

$$\frac{\Delta f}{f_0} = 2 \frac{(f_0 - f_m)}{f_0} = 2\left(1 - \frac{f_m}{f_0}\right) = 2\left(1 - \frac{2}{\pi}\theta_m\right) \qquad (9\text{-}38)$$

Since

$$\theta_m = \frac{2\pi}{\lambda_m}\left(\frac{\lambda_0}{4}\right) = \frac{\pi}{2}\left(\frac{f_m}{f_0}\right) \qquad (9\text{-}39)$$

(9-38) reduces using (9-36) to

$$\frac{\Delta f}{f_0} = 2 - \frac{4}{\pi} \cos^{-1}\left[\frac{\rho_m}{(R_L - Z_0)/(R_L + Z_0)}\right]^{1/N} \qquad (9\text{-}40)$$

where ρ_m is the maximum value of reflection coefficient which can be tolerated within the bandwidth. The usual design procedure is to specify the

1. load impedance (R_L)
2. input characteristic impedance (Z_0)
3. number of sections (N)
4. maximum tolerable reflection coefficient (ρ_m) [or fractional bandwidth ($\Delta f/f_0$)]

and to find the

1. characteristic impedance of each section
2. fractional bandwidth [or maximum tolerable reflection coefficient (ρ_m)]

To illustrate the principle, let us consider an example.

Example 9.2

A linear dipole with an input impedance of $70 + j37$ is connected to a 50-ohm line. Design a two-section $\lambda/4$ binomial transformer by specifying the characteristic impedance of each section to match the antenna to the line at $f = f_0$. If the input impedance (at the point the transformer is connected) is assumed to remain constant as a function of frequency, determine the maximum reflection coefficient and VSWR within a fractional bandwidth of 0.375.

Solution: Since the antenna impedance is not real, the antenna must be connected to the transformer through a transmission line of length s_0. Assuming a 50-ohm characteristic impedance for that section of the transmission line, the input impedance at $s_0 = 0.062\lambda$ is real and equal to 100 ohms. Using (9-36a) and (9-37)

$$\rho_n = 2^{-N} \frac{R_L - Z_0}{R_L + Z_0} C_n^N = 2^{-N} \frac{R_L - Z_0}{R_L + Z_0} \frac{N!}{(N-n)!n!}$$

which for $N = 2, R_L = 100, Z_0 = 50$

$$n = 0: \quad \rho_0 = \frac{Z_1 - Z_0}{Z_1 + Z_0} = \frac{1}{12} \Rightarrow Z_1 = 1.182 Z_0 = 59.09$$

$$n = 1: \quad \rho_1 = \frac{Z_2 - Z_1}{Z_2 + Z_1} = \frac{1}{6} \Rightarrow Z_2 = 1.399 Z_1 = 82.73$$

For a fractional bandwidth of 0.375 ($\theta_m = 1.276$ rad $= 73.12°$) we can write, using (9-40)

$$\frac{\Delta f}{f_0} = 0.375 = 2 - \frac{4}{\pi} \cos^{-1} \left[\frac{\rho_m}{(R_L - Z_0)/(R_L + Z_0)} \right]^{1/2}$$

which for $R_L = 100$ and $Z_0 = 50$ gives

$$\rho_m = 0.028$$

The maximum voltage standing wave ratio is

$$\text{VSWR}_m = \frac{1 + \rho_m}{1 - \rho_m} = 1.058$$

The magnitude of the reflection coefficient is given by (9-36) as

$$|\Gamma_{in}| = \rho_{in} = \left| \frac{R_L - Z_0}{R_L + Z_0} \right| \cos^2 \theta = \frac{1}{3} \cos^2 \left[\frac{2\pi}{\lambda} \left(\frac{\lambda_0}{4} \right) \right] = \frac{1}{3} \cos^2 \left[\frac{\pi}{2} \left(\frac{f}{f_0} \right) \right]$$

which is shown plotted in Figure 9.26, and it is compared with the response of a single-section $\lambda/4$ transformer and that of a two-section Tschebyscheff design whose maximum $\rho_m = 0.0147$.

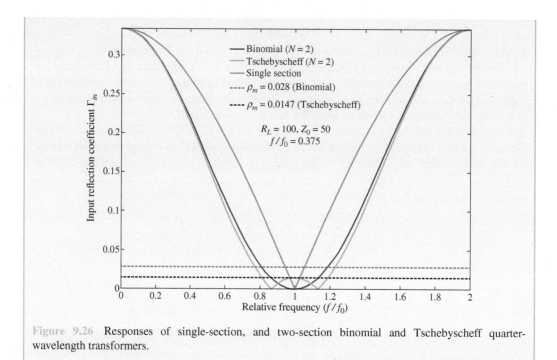

Figure 9.26 Responses of single-section, and two-section binomial and Tschebyscheff quarter-wavelength transformers.

Microstrip designs are ideally suited for antenna arrays, as shown in Figure 9.25(d). In general the characteristic impedance of a microstrip line, whose top and end views are shown in Figures 9.25(e) and (f), respectively, is given by [36]

$$Z_c = \frac{87}{\sqrt{\varepsilon_r + 1.41}} \ln\left(\frac{5.98h}{0.8w + t}\right) \qquad \text{for } h < 0.8w \qquad (9\text{-}41)$$

where

ε_r = dielectric constant of dielectric substrate (board material)

h = height of substrate

w = width of microstrip center conductor

t = thickness of microstrip center conductor

Thus for constant values of ε_r, h, and t, the characteristic impedance can be changed by simply varying the width (w) of the center conductor.

D. Tschebyscheff Design

The reflection coefficient can be made to vary within the bandwidth in an oscillatory manner and have equal-ripple characteristics. This can be accomplished by making Γ_{in} behave according to a Tschebyscheff polynomial. For the Tschebyscheff design, the equation that corresponds to (9-36) is

$$\Gamma_{in}(f) = e^{-jN\theta}\frac{R_L - Z_0}{R_L + Z_0}\frac{T_N(\sec\theta_m \cos\theta)}{T_N(\sec\theta_m)} \qquad (9\text{-}42)$$

where $T_N(x)$ is the Tschebyscheff polynomial of order N.

The maximum allowable reflection coefficient occurs at the edges of the passband where $\theta = \theta_m$ and $T_N(\sec\theta_m \cos\theta)|_{\theta=\theta_m} = 1$. Thus

$$\rho_m = \left|\frac{R_L - Z_0}{R_L + Z_0}\frac{1}{T_N(\sec\theta_m)}\right| \tag{9-43}$$

or

$$|T_N(\sec\theta_m)| = \left|\frac{1}{\rho_m}\frac{R_L - Z_0}{R_L + Z_0}\right| \tag{9-43a}$$

Using (9-43), we can write (9-42) as

$$\Gamma_{in}(f) = e^{-jN\theta}\rho_m T_N(\sec\theta_m \cos\theta) \tag{9-44}$$

and its magnitude as

$$|\Gamma_{in}(f)| = |\rho_m T_N(\sec\theta_m \cos\theta)| \tag{9-44a}$$

For this type of design, the fractional bandwidth $\Delta f/f_o$ is also given by (9-38).

To be physical, ρ_m must be smaller than the reflection coefficient when there is no matching. Therefore, from (9-43)

$$\rho_m = |\Gamma_m| = \left|\frac{R_L - Z_0}{R_L + Z_0}\frac{1}{T_N(\sec\theta_m)}\right| < \left|\frac{R_L - Z_0}{R_L + Z_0}\right| \Rightarrow |T_N(\sec\theta_m)| > 1 \tag{9-45}$$

The Tschebyscheff polynomial can be represented by either (6-71a) or (6-71b). Since $|T_N(\sec\theta_m)| > 1$, then using (6-71b) we can express it as

$$T_N(\sec\theta_m) = \cosh[N\cosh^{-1}(\sec\theta_m)] \tag{9-46}$$

or using (9-43a) as

$$|T_N(\sec\theta_m)| = |\cosh[N\cosh^{-1}(\sec\theta_m)]| = \left|\frac{1}{\rho_m}\frac{R_L - Z_0}{R_L + Z_0}\right| \tag{9-46a}$$

Thus

$$\sec\theta_m = \cosh\left[\frac{1}{N}\cosh^{-1}\left(\left|\frac{1}{\rho_m}\frac{R_L - Z_0}{R_L + Z_0}\right|\right)\right] \tag{9-47}$$

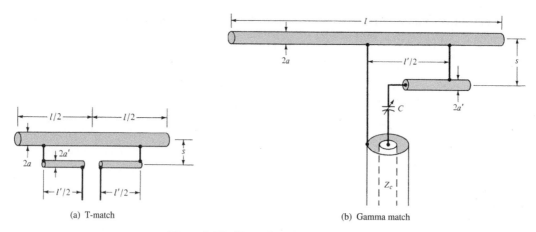

Figure 9.27 T-match and Gamma match.

or

$$\theta_m = \sec^{-1}\left\{\cosh\left[\frac{1}{N}\cosh^{-1}\left(\left|\frac{1}{\rho_m}\frac{R_L - Z_0}{R_L + Z_0}\right|\right)\right]\right\} \tag{9-47a}$$

Using (9-44), we can write the reflection coefficient of (9-35) as

$$\begin{aligned}\Gamma_{in}(\theta) &= 2e^{-jN\theta}[\rho_0 \cos(N\theta) + \rho_1 \cos(N-2)\theta + \cdots] \\ &= e^{-jN\theta}\rho_m T_N(\sec\theta_m \cos\theta)\end{aligned} \tag{9-48}$$

For a given N, replace $T_N(\sec\theta_m \cos\theta)$ by its polynomial series expansion of (6-69) and then match terms. The usual procedure for the Tschebyscheff design is the same as that of the binomial as outlined previously.

The first few Tschebyscheff polynomials are given by (6-69). For $z = \sec\theta_m \cos\theta$, the first three polynomials reduce to

$$\begin{aligned}T_1(\sec\theta_m \cos\theta) &= \sec\theta_m \cos\theta \\ T_2(\sec\theta_m \cos\theta) &= 2(\sec\theta_m \cos\theta)^2 - 1 = \sec^2\theta_m \cos 2\theta + (\sec^2\theta_m - 1) \\ T_3(\sec\theta_m \cos\theta) &= 4(\sec\theta_m \cos\theta)^3 - 3(\sec\theta_m \cos\theta) \\ &= \sec^3\theta_m \cos 3\theta + 3(\sec^3\theta_m - \sec\theta_m)\cos\theta\end{aligned} \tag{9-49}$$

The design of Example 9.2 using a Tschebyscheff transformer is assigned as an exercise to the reader (Prob. 9.24). However its response is shown plotted in Figure 9.26 for comparison whose maximum $\rho_m = 0.0147$ compared to $\rho_m = 0.028$ for the binomial design.

In general, the multiple sections (either binomial or Tschebyscheff) provide greater bandwidths than a single section. As the number of sections increases, the bandwidth also increases. The advantage of the binomial design is that the reflection coefficient values within the bandwidth monotonically decrease from both ends toward the center. Thus the values are always smaller than an acceptable and designed value that occurs at the "skirts" of the bandwidth. For the Tschebyscheff design, the reflection coefficient values within the designed bandwidth are equal or smaller than an acceptable and designed value. The number of times the reflection coefficient reaches the maximum ripple value within the bandwidth is determined by the number of sections. In fact, for an even number of sections the reflection coefficient at the designed center frequency is equal to its maximum allowable value, while for an odd number of sections it is zero. For a maximum tolerable reflection coefficient, the N-section Tschebyscheff transformer provides a larger bandwidth than a corresponding N-section binomial design, or for a given bandwidth the maximum tolerable reflection coefficient is smaller for a Tschebyscheff design.

A MATLAB computer program, entitled **Quarterwave**, has been developed to perform the design of binomial and Tschebyscheff quarter-wavelength impedance transformer designs. The description of the program is found in the corresponding READ ME file included in the publisher's website for the book.

Other matching techniques, especially for dipole type antennas, include the T-match and Gamma match. These are discussed in detail in the previous three editions of this book.

9.8.3 Baluns and Transformers

A twin-lead transmission line (two parallel-conductor line) is a symmetrical line whereas a coaxial cable is inherently unbalanced. Because the inner and outer (inside and outside parts of it) conductors of the coax are not coupled to the antenna in the same way, they provide the unbalance. The result is a net current flow to ground on the outside part of the outer conductor. This is shown in Figure 9.28(a) where an electrical equivalent is also indicated. The amount of current flow I_3 on the outside surface of the outer conductor is determined by the impedance Z_g from the outer shield to ground. If Z_g can be made very large, I_3 can be reduced significantly. Devices that can be used to balance inherently unbalanced systems, by canceling or choking the outside current, are known as *baluns* (*bal*ance to *un*balance).

One type of a balun is that shown in Figure 9.28(b), referred to usually as a *bazooka* balun. Mechanically it requires that a $\lambda/4$ in length metal sleeve, and shorted at its one end, encapsulates the coaxial line. Electrically the input impedance at the open end of this $\lambda/4$ shorted transmission line, which is equivalent to Z_g, will be very large (ideally infinity). Thus the current I_3 will be choked, if not completely eliminated, and the system will be nearly balanced.

Another type of a balun is that shown in Figure 9.28(c). It requires that one end of a $\lambda/4$ section of a transmission line be connected to the outside shield of the main coaxial line while the other is connected to the side of the dipole which is attached to the center conductor. This balun is used to cancel the flow of I_3. The operation of it can be explained as follows: In Figure 9.28(a) the voltages between each side of the dipole and the ground are equal in magnitude but 180° out of phase, thus producing a current flow on the outside of the coaxial line. If the two currents I_1 and I_2 are equal in magnitude, I_3 would be zero. Since arm #2 of the dipole is connected directly to the shield of the coax while arm #1 is weakly coupled to it, it produces a much larger current I_2. Thus there is relatively little cancellation in the two currents.

The two currents, I_1 and I_2, can be made equal in magnitude if the center conductor of the coax is connected directly to the outer shield. If this connection is made directly at the antenna terminals, the transmission line and the antenna would be short-circuited, thus eliminating any radiation. However, the indirect parallel-conductor connection of Figure 9.28(c) provides the desired current cancellation without eliminating the radiation. The current flow on the outer shield of the main line is canceled at the bottom end of the $\lambda/4$ section (where the two join together) by the equal in magnitude, but opposite in phase, current in the $\lambda/4$ section of the auxiliary line. Ideally then there is no current flow in the outer surface of the outer shield of the remaining part of the main coaxial line. It should be stated that the parallel auxiliary line need not be made $\lambda/4$ in length to achieve the balance. It is made $\lambda/4$ to prevent the upsetting of the normal operation of the antenna.

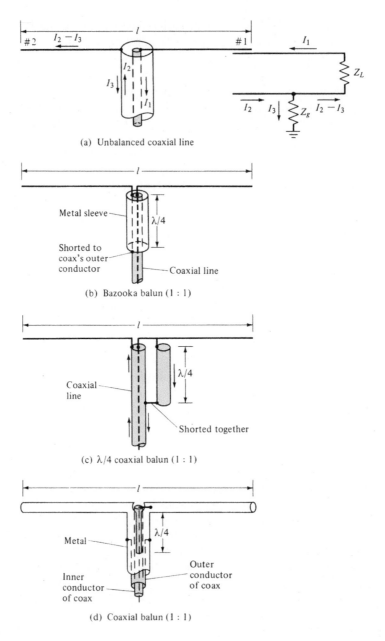

Figure 9.28 Balun configurations.

A compact construction of the balun in Figure 9.28(c) is that in Figure 9.28(d). The outside metal sleeve is split and a portion of it is removed on opposite sides. The remaining opposite parts of the outer sleeve represent electrically the two shorted $\lambda/4$ parallel transmission lines of Figure 9.28(c). Another balun is that of Figure 9.8(a,b) used to feed the flexible bow–tie antenna. All of the baluns shown in Figure 9.28 are narrowband devices.

Devices can be constructed which provide not only balancing but also step-up impedance transformations. One such device is the $\lambda/4$ coaxial balun, with a 4:1 impedance transformation, of Figure 9.29(a). The U-shaped section of the coaxial line must be $\lambda/2$ long [37].

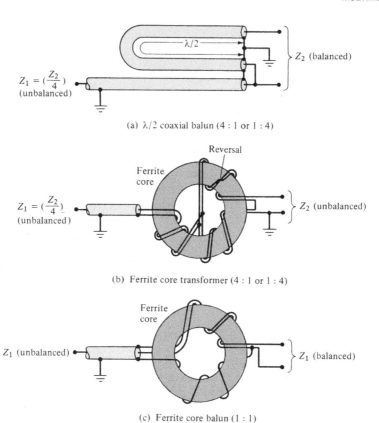

Figure 9.29 Balun and ferrite core transformers.

Because all the baluns-impedance transformers that were discussed so far are narrowband devices, the bandwidth can be increased by employing ferrite cores in their construction [38]. Two such designs, one a 4:1 or 1:4 transformer and the other a 1:1 balun, are shown in Figures 9.29(b,c). The ferrite core has a tendency to maintain high impedance levels over a wide frequency range [39]. A good design and construction can provide bandwidths of 8 or even 10 to 1. Coil coaxial baluns, constructed by coiling the coaxial line itself to form a balun [39], can provide bandwidths of 2 or 3 to 1.

9.9 MULTIMEDIA

In the publisher's website for this book, the following multimedia resources are included for the review, understanding, and visualization of the material of this chapter:

a. **Java**-based **interactive questionnaire**, with answers.
b. **Matlab** computer program, designated **Quarterwave**, for computing and displaying the characteristics of binomial and Tschebyscheff quarter-wavelength impedance matching designs.
c. **Matlab** computer programs, designated **Folded** for computing the characteristics of a folded dipole.
d. **Power Point (PPT)** viewgraphs, in multicolor.

REFERENCES

1. M. Abraham, "Die Electrischen Schwingungen um einen Stabformingen Leiter, Behandelt nach der Maxwelleschen Theorie," *Ann. Phys.*, Vol. 66, pp. 435–472, 1898.
2. S. A. Schelkunoff, *Electromagnetic Waves*, Van Nostrand, New York, 1943, Chapter 11.
3. E. Hallén, "Theoretical Investigations into the Transmitting and Receiving Qualities of Antennae," *Nova Acta Reg. Soc. Sci. Upsaliensis*, Ser. IV, Vol. 11, No. 4, pp. 1–44, 1938.
4. R. C. Johnson and H. Jasik (eds.), *Antenna Engineering Handbook*, McGraw-Hill, New York, 1984, Chapter 4.
5. G. H. Brown and O. M. Woodward Jr., "Experimentally Determined Radiation Characteristics of Conical and Triangular Antennas," *RCA Rev.*, Vol. 13, No. 4, p. 425, December 1952.
6. C. H. Papas and R. King, "Radiation from Wide-Angle Conical Antennas Fed by a Coaxial Line," *Proc. IRE*, Vol. 39, p. 1269, November 1949.
7. C. E. Smith, C. M. Butler, and K. R. Umashankar, "Characteristics of a Wire Biconical Antenna," *Microwave J.*, pp. 37–40, September 1979.
8. A. C. Durgun, C. A. Balanis, C. R. Birtcher and D. R. Allee, "Design, Simulation, Fabrication and Testing of Flexible Bow-Tie Antennas," *IEEE Trans. Antennas Propagat.*, Vol. 59, No. 12, pp. 4425–4435, December 2011.
9. A. C. Durgun, C. A. Balanis, C. R. Birtcher and D. A. Allee, "Radiation Characteristics of a Flexible Bow–tie Antenna," *IEEE AP-S International Symposium*, Spokane, WA, pp. 1–4, July 3-8, 2011.
10. G. B. Raupp, S. M. O'Rourke, D. R. Allee, S. Venugopal, E. J. Bawolek, D. E. Loy, S. K. Ageno, B. P. O'Brien, S. Rednour, and G. E. Jabbour, "Flexible Reflective and Emissive Display Integration and Manufacturing (invited paper)," in *Cockpit and Future Displays for Defense and Security*, Vol. 5801, No. 1, 2005, pp. 194–203. [Online]. Available: http://link.aip.org/link/?PSI/5801/194/l.
11. K. R. Wissmiller, J. E. Knudsen, T. J. Alward, Z. P. Li, D. R. Allee, and L. T. Clark, "Reducing Power in Flexible A-SI Digital Circuits While Preserving State," in *Proc. IEEE Custom Integrated Circuits Conf.*, Sept. 2005, pp. 219–222.
12. A. A. Eldek, A. Z. Elsherbeni, and C. E. Smith, "Wideband Microstrip Fed Printed Bow-Tie Antenna for Phased-Array Systems," *Microwave and Optical Technology Letters*, Vol. 43, No. 2, pp. 123–126, Oct. 2004.
13. M. Rahim, M. Abdul Aziz, and C. Goh, "Bow-Tie Microstrip Antenna Design," in *Networks, 2005. Jointly held with the 2005 IEEE 7th Malaysia International Conference on Communication., 2005 13th IEEE International Conference on*, Vol. 1, 2005, pp. 17–20.
14. Y.-D. Lin and S.-N. Tsai, "Coplanar Waveguide-Fed Uniplanar Bow-Tie Antenna," *IEEE Trans. Antennas Propag.*, Vol. 45, No. 2, pp. 305–306, Feb. 1997.
15. J. George, M. Deepukumar, C. Aanandan, P. Mohanan, and K. Nair, "New Compact Microstrip Antenna," *Electronics Letters*, Vol. 32, No. 6, pp. 508–509, March 1996.
16. B. Garibello and S. Barbin, "A Single Element Compact Printed Bowtie Antenna Enlarged Bandwidth," *2005 SBMO/IEEE MTT-S International Conference on Microwave and Optoelectronics*, July 2005, pp. 354–358.
17. Y. Qian and T. Itoh, "A Broadband Uniplanar Microstrip-to-CPW Transition," in *Microwave Conference Proceedings, APMC '97*, Vol. 2, Dec. 1997, pp. 609–612.
18. N. Kaneda, Y. Qian, and T. Itoh, "A Broad-Band Microstrip-to-Waveguide Transition Using Quasi-Yagi Antenna," *IEEE Trans. Microwave Theory Tech.*, vol. 47, no. 12, pp. 2562–2567, Dec. 1999.
19. P. J. Gibson, "The Vivaldi Aerial," *9th European Microwave Conference*, pp. 101–105, Sept. 17–20, 1979.
20. K. S. Yngvesson, T. L. Korzeniowski, Young-Sik Kim, E. L. Kollberg, J. F. Johansson, "The Tapered Slot Antenna - a New Integrated Element for Millimeter-Wave Applications," *IEEE Trans. Microwave Theory Tech*, Vol. 37, No. 2, pp. 365–374, Feb. 1989.
21. L. R. Lewis, M. Fasset and J. Hunt, "A Broadband Stripline Array Element," *IEEE Antennas Propagat. Symposium Digest*, pp. 335–337, 1974.
22. W. F. Croswell, T. Durham, M. Jones, D. Schaubert, P. Friedrich and J. G. Maloney, "Wideband Arrays," Chapter 12 in *Modern Antenna Handbook* (C. A. Balanis, editor), John Wiley & Sons, Publishers, 2008.

23. *Radio Astronomy Across Europe*. Alastair G. Gunn (editor), Springer, Dordrecht, The Netherlands, 2005.
24. E. Gazit, "Improved Design of the Vivaldi Antenna," *IEE Proceedings Microwaves, Antennas and Propagation*, Vol. 135, No. 2, pp. 89–92, April 1988.
25. B. Schuppert, "Micorstrip/Slotline Transitions: Modeling and Experimental Investigation," *IEEE Trans. Microwave Theory Tech.*, Vol. 36, No. 8, pp. 1272–1282, Aug. 1988.
26. J. B. Knorr, "Slot-Sine Transitions," *IEEE Trans. Microwave Theory Tech.*, Vol. 22, No. 5, pp. 548–554, May 1974.
27. K. S. Yngvesson, D. H. Schaubert, E. L. Kollberg, T. Korzeniowski, T. Thungren, J. J Johansson, "Endfire Tapered Slot Antennas on Dielectric Substrates," *IEEE Trans. Antennas Propagat.*, Vol. 33, No. 12, pp. 1392–1400, Dec. 1985.
28. N. Hamzah, and K. A. Othman, "Designing Vivaldi Antenna With Various Sizes Using CST Software," *Proceedings of the World Congress on Engineering 2011*, Vol. II, WCE 2011, London, UK, July 6–8, 2011.
29. G. H. Brown and O. M. Woodward Jr., "Experimentally Determined Impedance Characteristics of Cylindrical Antennas," *Proc. IRE*, Vol. 33, pp. 257–262, 1945.
30. J. D. Kraus, *Antennas*, McGraw-Hill, New York, 1950, pp. 276–278.
31. G. A. Thiele, E. P. Ekelman Jr., and L. W. Henderson, "On the Accuracy of the Transmission Line Model for Folded Dipole," *IEEE Trans. Antennas Propagat.*, Vol. AP-28, No. 5, pp. 700–703, September 1980.
32. R. W. Lampe, "Design Formulas for an Asymmetric Coplanar Strip Folded Dipole," *IEEE Trans. Antennas Propagat.*, Vol. AP-33, No. 9, pp. 1028–1031, September 1985.
33. A. G. Kandoian, "Three New Antenna Types and Their Applications," *Proc. IRE*, Vol. 34, pp. 70W–75W, February 1946.
34. R. E. Collin, *Foundations for Microwave Engineering*, McGraw-Hill, New York, 1992, Chapter 5, pp. 303–386.
35. D. M. Pozar, *Microwave Engineering*, John Wiley & Sons, Inc., NJ, 2004.
36. S. Y. Liao, *Microwave Devices and Circuits*, Prentice-Hall, Englewood Cliffs, N.J., 1980, pp. 418–422.
37. H. T. Tolles, "How To Design Gamma-Matching Networks," *Ham Radio*, pp. 46–55, May 1973.
38. O. M. Woodward Jr., "Balance Measurements on Balun Transformers," *Electronics*, Vol. 26, No. 9, pp. 188–191, September 1953.
39. C. L. Ruthroff, "Some Broad-Band Transformers," *Proc. IRE*, Vol. 47, pp. 1337–1342, August 1959.
40. W. L. Weeks, *Antenna Engineering*, McGraw-Hill, New York, 1968, pp. 161–180.

PROBLEMS

9.1. A 300-ohm "twin-lead" transmission line is attached to a biconical antenna.
   (a) Determine the cone angle that will match the line to an infinite length biconical antenna.
   (b) For the cone angle of part (a), determine the two smallest cone lengths that will resonate the antenna.
   (c) For the cone angle and cone lengths from part (b), what is the input VSWR?

9.2. Determine the first two resonant lengths, and the corresponding diameters and input resistances, for dipoles with $l/d = 25$, 50, and 10^4 using
   (a) the data in Figures 9.17(a) and 9.17(b)
   (b) Table 9.1

9.3. Design a resonant cylindrical stub monopole of length l, diameter d, and l/d of 50. Find the length (in λ), diameter (in λ), and the input resistance (in ohms) at the first four resonances.

9.4. A linear dipole of $l/d = 25$, 50, and 10^4 is attached to a 50-ohm line. Determine the VSWR of each l/d when
   (a) $l = \lambda/2$ (b) $l = \lambda$ (c) $l = 3\lambda/2$

9.5. Find the equivalent circular radius a_e for a
- (a) very thin flat plate of width $\lambda/10$
- (b) square wire with sides of $\lambda/10$
- (c) rectangular wire with sides of $\lambda/10$ and $\lambda/100$
- (d) elliptical wire with major and minor axes of $\lambda/10$ and $\lambda/20$
- (e) twin-lead transmission line with wire radii of 1.466×10^{-2} cm and separation of 0.8 cm

9.6. Compute the characteristic impedance of a two-wire transmission line with wire diameter of $d = 10^{-3}\lambda$ and center-to-center spacings of
  (a) $6.13\times10^{-3}\lambda$ (b) $2.13\times10^{-2}\lambda$ (c) $7.42\times10^{-2}\lambda$

9.7. From the expressions listed in Table 9.3, show that the "equivalent" radius a_e of the circular two-wire side-by-side parallel arrangemenet can be written as

$$\ln(a_e) \approx \frac{1}{(a' + a^2)}[(a')^2 \ln a' + (a)^2 \ln a + 2a'a \ln s]$$

where a and a' are, respectively, the radii of the two wires and s is the center-to-center sepration between the two parallel wires.

9.8. To increase its bandwidth, a $\lambda/4$ monopole antenna operating at 1 GHz is made of two side-by-side copper wires ($\sigma = 5.7\times10^7$ S/m) of circular cross section. The wires at each end of the arm are connected (shorted) electrically together. The radius of each wire is $\lambda/200$ and the separation between them is $\lambda/50$.
- (a) What is the effective radius (in meters) of the two wires? Compare with the physical radius of each wire (in meters).
- (b) What is the high-frequency loss resistance of each wire? What is the total loss resistance of the two together in a side-by-side shorted at the ends arrangement? What is the loss resistance based on the effective radius?
- (c) What is the radiation efficiency of one wire by itself? Compare with that of the two together in a side-by-side arrangement. What is the radiation efficiency based on the loss resistance of the effective radius?

9.9. Show that the input impedance of a two-element folded dipole of $l = \lambda/2$ is four times greater than that of an isolated element of the same length.

9.10. Design a two-element folded dipole with wire diameter of $10^{-3}\lambda$ and center-to-center spacing of $6.13\times10^{-3}\lambda$.
- (a) Determine its shortest length for resonance.
- (b) Compute the VSWR at the first resonance when it is attached to a 300-ohm line.

9.11. A two-element folded dipole of identical wires has an $l/d = 500$ and a center-to-center spacing of $6.13\times10^{-3}\lambda$ between the wires. Compute the
- (a) approximate length of a single wire at its first resonance
- (b) diameter of the wire at the first resonance
- (c) characteristic impedance of the folded dipole transmission line
- (d) input impedance of the transmission line mode model
- (e) input impedance of the folded dipole using as the radius of the antenna mode (1) the radius of the wire a, (2) the equivalent radius a_e of the wires, (3) half of the center-to-center spacing ($s/2$). Compare the results.

9.12. The input impedance of a 0.47λ folded dipole operating at 10 MHz is

$$Z_{in} = 306 + j75.3$$

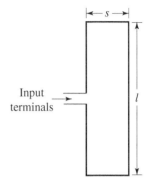

To resonate the element, it is proposed to place a lumped element in shunt (parallel) at the input terminals where the impedance is measured.
 (a) What kind of an element (capacitor or inductor) should be used to accomplish the requirement?
 (b) What is the value of the element (in Farads or Henries)?
 (c) What is the new value of the input impedance?
 (d) What is the VSWR when the resonant antenna is connected to a 300-ohm line?

Verify using the computer program **Folded**.

9.13. A half-wavelength, two-element symmetrical folded dipole is connected to a 300-ohm "twin-lead" transmission line. In order for the input impedance of the dipole to be real, an energy storage lumped element is placed across its input terminals. Determine, assuming $f = 100$ MHz, the
 (a) capacitance or inductance of the element that must be placed across the terminals.
 (b) VSWR at the terminals of the transmission line taking into account the dipole and the energy storage element.

Verify using the computer program **Folded**.

9.14. A half-wavelength, two-element symmetrical folded dipole whose each wire has a radius of $10^{-3}\lambda$ is connected to a 300-ohm "twin-lead" transmission line. The center-to-center spacing of the two wires is $4 \times 10^{-3}\lambda$. In order for the input impedance of the dipole to be real, determine, assuming $f = 100$ MHz, the
 (a) total capacitance C_T that must be placed in series at the input terminals.
 (b) capacitance C (two of them) that must be placed in series at each of the input terminals of the dipole in order to keep the antenna symmetrical.
 (c) VSWR at the terminals of the transmission line connected to the dipole through the two capacitances.

Verify using the computer program **Folded**.

9.15. An $l = 0.47\lambda$ folded dipole, whose wire radius is $5 \times 10^{-3}\lambda$, is fed by a "twin lead" transmission line with a 300-ohm characteristic impedance. The center-to-center spacing of the two side-by-side wires of the dipole is $s = 0.025\lambda$. The dipole is operating at $f = 10$ MHz. The input impedance of the "regular" dipole of $l = 0.47\lambda$ is $Z_d = 79 + j13$.
 (a) Determine the
  (i) Input impedance of the folded dipole.
  (ii) Amplification factor of the input impedance of the folded dipole, compared to the corresponding value of the regular dipole.
  (iii) Input reflection coefficient.
  (iv) Input VSWR.
 (b) To resonate the folded dipole and keep the system balanced, two capacitors (each C) are connected each symmetrically in series at the input terminals of the folded dipole.
  (i) What should C be to resonate the dipole?
  (ii) What is the new reflection coefficient at the input terminals of the "twin-lead" line?
  (iii) What is the new VSWR?

9.16. To increase the input impedance of a *regular* quarter-wavelength ($\lambda/4$) monopole, it was decided to design a *folded* quarter-wavelength ($\lambda/4$) monopole, as shown in the figure. Assuming that the radius of the monopole wire is very thin ($a \ll \lambda$) and the spacing between the wires is also very small ($s \ll \lambda$):

(a) Determine the input impedance of the regular monopole.

(b) Determine the input impedance of the folded monopole.

(c) What kind of a lumped element, *inductor or capacitor*, must be placed *in series* in order to resonate (*eliminate the imaginary part of the impedance*) the folded monopole?

(d) What is the value of the series inductor or capacitor at a frequency of 100 MHz?

(e) After the folded monopole is resonated, what is the input impedance of the folded monopole assuming it is connected to a coaxial line with a characteristic impedance of 75 ohms?

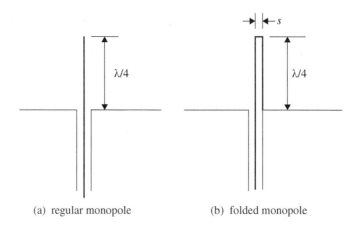

(a) regular monopole (b) folded monopole

9.17. A folded dipole of length $\lambda_0/2$ is composed of two wires, each of radii $\lambda_0/200$ and $\lambda_0/300$, respectively, and separated by a distance of $\lambda_0/100$. The folded dipole is connected to a 300-ohm transmission line. Determine the following:

(a) Input impedance of the dipole.

(b) What *capacitance or inductance* must be placed in *series* to the transmission line at the feed point so that the *dipole is resonant at* $f_o = 600$ *MHz*.

(c) Input reflection coefficient, assuming the resonant dipole is connected to a 300-ohm transmission line.

(d) Input VSWR of the resonant dipole when connected to the 300-ohm transmission line.

9.18. A folded dipole antenna operating at 100 MHz with identical wires in both arms, and with overall length of each arm being 0.48λ, is connected to a 300-ohm *twin-lead* line. The radius of each wire is $a = 5 \times 10^{-4}\lambda$ while the center-to-center separation is $s = 5 \times 10^{-3}\lambda$. Determine the

(a) Approximate length (*in* λ) of the regular dipole, *at the first resonance*, in the absence of the other wire.

(b) Input impedance of the single wire *resonant regular* dipole.

(c) Input impedance of the folded dipole at the length of the first resonance of the single element.

(d) Capacitance (*in F/m*) or inductance (*in H/m*), whichever is appropriate, that must be placed in series with the element at the feed to *resonate the folded dipole*. To keep the element balanced, place two elements, one in each arm.

(e) VSWR of the resonant folded dipole.

9.19. A folded dipole made of 3 *"legs" each with a total length* λ/2, as shown in the figure, is connected to a 300-ohm transmission line. Assuming *the wires are very thin and the spacing between the legs is small compared to the length*, determine the:
   (a) *Input impedance* of the folded dipole, *real and imaginary parts* (*in the absence of any resonant element at the input*).
   (b) Lumped element, *capacitor or inductor (whichever is appropriate)*, that must be placed *in parallel at the input (as shown in the figure dashed across the input)* to resonate *the folded dipole* at a frequency of 300 MHz.
   (c) *New input impedance* after the element is resonated; *take into account the capacitor/inductor*.
   (d) VSWR at the input of the *resonant folded dipole*.

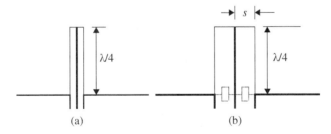

9.20. A folded monopole made up of 3 *"legs" each with a total length* λ/4, as shown in the figure, is connected to a 75-ohm transmission line. Assuming *the wires are very thin and the spacing between the legs is small compared to the length*, determine the:
   (a) Input impedance of the folded monopole, real and imaginary parts (*in the absence of any resonant element at the input*); figure (a) below.
   (b) *Total lumped capacitance/inductance, (whichever is appropriate)*, that must be placed *in parallel at the input* to resonate *the folded 3-wire* monopole *at 300 MHz*.
   (c) *Individual lumped capacitance (C) or inductance (L)* (*shown in the figure dashed across the input*) of *Each of two identical capacitance/inductance* elements *placed in parallel to resonate the folded monopole and keep it symmetrical at 300 MHz; figure (b) below. State the capacitance/inductance of Each of the 2 identical elements.*
   (d) *New input impedance* after the element is resonated; *take into account the presence of the 2 identical capacitors/inductors.*
   (e) VSWR at the input of the *resonant folded monopole, taking into account the presence of the two lumped elements. Input transmission line has 75-ohms characteristic impedance*.

9.21. A λ/2 folded dipole is used as the feed element for a Yagi-Uda array whose input impedance, *when the array is fed by a regular dipole*, is 30 + j3. The folded dipole used is *not symmetrical* (i.e., the two wires are not of the same radius). One of the wires has a radius of $10^{-4}\lambda$ and the other (*the one that is connected to the feed*) has a radius of $0.5 \times 10^{-4}\lambda$. The center-to-center spacing between the two dipoles is $2 \times 10^{-4}\lambda$. Determine the
   (a) *turns ratio* of the *equivalent-circuit impedance transformer*
   (b) *effective impedance* of the regular dipole when it is transferred from the *secondary* to the *primary* of the *equivalent-circuit impedance transformer*
   (c) *input impedance* of the folded dipole.

9.22. A three-element folded dipole, whose length of each element is λ/2, is used as the feed element of a Yagi-Uda array.
   (a) Determine the input impedance of the three-element folded dipole; real and imaginary parts.

(b) What is the *magnitude* of the input reflection coefficient and VSWR before the insert of the capacitors/inductors to resonate the antenna.

(c) In order *to resonate* the antenna (*cancel the imaginary part of the input impedance*), lumped element (*capacitor or inductor, whichever is appropriate*) is placed in *series (one in each arm to keep the system balanced), as shown in the box below*. Indicate which kind of an element, *capacitor or inductor*, will be appropriate. *Justify your selection.*

(d) For a frequency of *15 MHz*, determine the value of capacitor C or inductor L of Part c.

(e) What is the new input impedance after the antenna is resonated? *Account for the presence of the series capacitors or inductors when you are finding the new input impedance.*

(f) What is the VSWR if the resonated antenna is connected to a 300-ohm "twin-lead" transmission line?

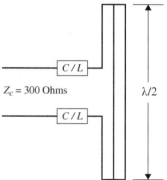

9.23. It is desired to design *two-section Tschebyscheff* impedance transformer to match an antenna, whose impedance is $Z_L = 185.56 + j50$, to a lossless input line whose characteristic impedance is $Z_0 = 50$ ohms. In order to accomplish this, the load impedance has to be real. Therefore, referring to page 516, Figure 9.25 (b) of the book, the *2-section Tschebyscheff impedance transformer* has to be connected at a distance s_0 from the load impedance [Figure 9.25(b) shows only one-section but in this problem we want two-sections]. Given that the required distance is $s_0 = 0.011475\lambda$, to convert the complex load impedance $Z_L = 186.56 + j50$ to a *real input impedance of 200 ohms* at $s_0 = 0.011475\lambda$, determine the:

(a) *Magnitude of the reflection coefficient and VSWR at a distance* $s_0 = 0.011475\lambda$, looking toward the load *before the insertion of the quarter-wavelength impedance transformer.* Assume a characteristic impedance of 50 ohms.

(b) *Fractional bandwidth* of the quarter-wavelength impedance transformer for a maximum tolerable reflection coefficient of $\rho_m = 0.3333$.

(c) *Intrinsic reflection coefficients* Γ_0, Γ_1 between the sections of the quarter-wavelength impedance transformer.

(d) *Characteristic impedances* Z_1, Z_2 of the two-section Tschcbyschcff quarter-wavelength impedance transformer (*in ohms*).

9.24. Repeat the design of Example 9.2 using a Tschebyscheff transformer.

9.25. Repeat the design of Example 9.2 for a three-section
(a) binomial transformer
(b) Tschebyscheff transformer
Verify using the computer program **Quarterwave**.

9.26. The *radiation resistance* of a *center-fed resonant* $l = 0.723\lambda$ linear dipole is $R_r = 175.35$ ohms. It is desired to connect the dipole to a transmission line so that the system meets certain specifications.

(a) Determine the magnitude of the *maximum reflection coefficient* so that the VSWR *does not exceed 3*.

(b) To achieve the specifications of *Part a*, determine the characteristic impedance of the input line. *Assume that* $R_L > Z_0$ where R_L is the load resistance.

(c) To reduce the reflection coefficient so and its *maximum tolerable value* is *one-half of that in Part (a)*, a *three-section quarter-wavelength binomial impedance transformer* is going to be placed between the load impedance and the input line. Determine the *fractional bandwidth (in %)* over which the system can now be operated so that it *does not exceed the new value of the maximum tolerable reflection coefficient* [*i.e.; not to exceed one-half of that of Part (a)*].

(d) Repeat *Part (c)* for a *3-section quarter-wavelength Tschebyscheff* design.

9.27. A self-resonant (*first resonance*) half-wavelength dipole of radius $a = 10^{-3}\lambda$ is connected to a 300-ohm "twin-lead" line through a three-section binomial impedance transformer. Determine the impedances of the three-section binomial transformer required to match the resonant dipole to the "twin-lead" line. Verify using the computer program **Quarterwave**. Assume maximum tolerable reflection coefficient of 1/3.

9.28. Design a *three-section binomial transformer* matching a 200-ohm load to a 100-ohm input transmission line. *The maximum tolerable reflection coefficient is 0.3*. Determine the:

(a) *Intrinsic reflection coefficients* at each intersection of one line to the other line.

(b) Characteristic impedances of each line between the input line and the load.

(c) VSWR at the input *in the absence of the impedance transformer*.

(d) Fractional bandwidth, *in the presence of the impedance transformer*, over which the system can operate without exceeding the maximum tolerable reflection coefficient.

9.29. Consider a center-fed thin-wire dipole with wire radius $a = 0.005\lambda$.

(a) Determine the shortest resonant length l (in wavelengths) and corresponding input resistance R_{in} of the antenna using assumed sinusoidal current distribution.

(b) Design a two-section Tschebyscheff quarter-wavelength transformer to match the antenna to a 75-ohm transmission line. Design the transformer to achieve an equal-ripple response to R_{in} over a fractional bandwidth of 0.25.

(c) Compare the performance of the matching network in Part (b) to an ideal transformer by plotting the input reflection coefficient magnitude versus normalized frequency for $0 \leq f/f_0 \leq 2$ for both cases.

Verify using the computer program **Quarterwave**.

CHAPTER 10

Traveling Wave and Broadband Antennas

10.1 INTRODUCTION

In the previous chapters we have presented the details of classical methods that are used to analyze the radiation characteristics of some of the simplest and most common forms of antennas (i.e., infinitely thin linear and circular wires, broadband dipoles, and arrays). In practice there is a myriad of antenna configurations, and it would be almost impossible to consider all of them in this book. In addition, many of these antennas have bizarre types of geometries and it would be almost impractical, if not even impossible, to investigate each in detail. However, the general performance behavior of some of them will be presented in this chapter with a minimum of analytical formulations. Today, comprehensive analytical formulations are available for most of them, but they would require so much space that it would be impractical to include them in this book.

10.2 TRAVELING WAVE ANTENNAS

In Chapter 4, center-fed linear wire antennas were discussed whose amplitude current distribution was

1. constant for infinitesimal dipoles ($l \leq \lambda/50$)
2. linear (triangular) for short dipoles ($\lambda/50 < l \leq \lambda/10$)
3. sinusoidal for long dipoles ($l > \lambda/10$)

In all cases the phase distribution was assumed to be constant. The sinusoidal current distribution of long open-ended linear antennas is a standing wave constructed by two waves of equal amplitude and 180° phase difference at the open end traveling in opposite directions along its length. The voltage distribution has also a standing wave pattern except that it has maxima (loops) at the end of the line instead of nulls (nodes) as the current. In each pattern, the maxima and minima repeat every integral number of half wavelengths. There is also a $\lambda/4$ spacing between a null and a maximum in each of the wave patterns. The current and voltage distributions on open-ended wire antennas are similar to the standing wave patterns on open-ended transmission lines. Linear antennas that exhibit current and voltage standing wave patterns formed by reflections from the open end of the wire are referred to as *standing wave* or *resonant* antennas.

Antenna Theory: Analysis and Design, Fourth Edition. Constantine A. Balanis.
© 2016 John Wiley & Sons, Inc. Published 2016 by John Wiley & Sons, Inc.
Companion Website: www.wiley.com/go/antennatheory4e

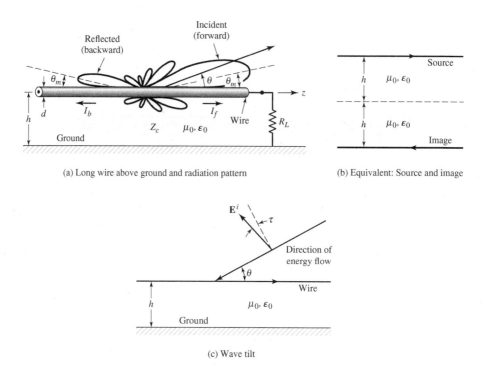

Figure 10.1 Beverage (long-wire) antenna above ground.

Antennas can be designed which have traveling wave (uniform) patterns in current and voltage. This can be achieved by properly terminating the antenna wire so that the reflections are minimized if not completely eliminated. An example of such an antenna is a long wire that runs horizontal to the earth, as shown in Figure 10.1. The input terminals consist of the ground and one end of the wire. This configuration is known as *Beverage or wave antenna*. There are many other configurations of traveling wave antennas. In general, all antennas whose current and voltage distributions can be represented by one or more traveling waves, usually in the same direction, are referred to as *traveling wave* or *nonresonant* antennas. A progressive phase pattern is usually associated with the current and voltage distributions.

Standing wave antennas, such as the dipole, can be analyzed as traveling wave antennas with waves propagating in opposite directions (forward and backward) and represented by traveling wave currents I_f and I_b in Figure 10.1(a). Besides the long-wire antenna there are many examples of traveling wave antennas such as dielectric rod (polyrod), helix, and various surface wave antennas. Aperture antennas, such as reflectors and horns, can also be treated as traveling wave antennas. In addition, arrays of closely spaced radiators (usually less than $\lambda/2$ apart) can also be analyzed as traveling wave antennas by approximating their current or field distribution by a continuous traveling wave. Yagi-Uda, log-periodic, and slots and holes in a waveguide are some examples of discrete-element traveling wave antennas. In general, a traveling wave antenna is usually one that is associated with radiation from a continuous source. An excellent book on traveling wave antennas is one by C. H. Walter [1].

A traveling wave may be classified as a *slow* wave if its phase velocity $v_p (v_p = \omega/k, \omega =$ wave angular frequency, $k =$ wave phase constant) is equal or smaller than the velocity of light c in free-space ($v_p/c \leq 1$). A fast wave is one whose phase velocity is greater than the speed of light ($v_p/c > 1$).

In general, there are two types of traveling wave antennas. One is the *surface wave* antenna defined as "an antenna which radiates power flow from discontinuities in the structure that interrupt a bound

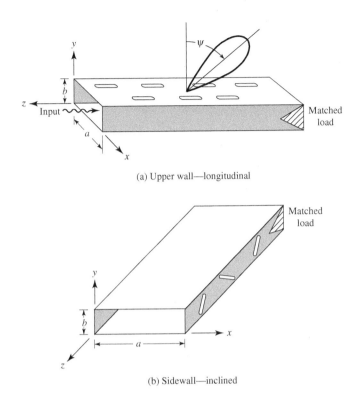

(a) Upper wall—longitudinal

(b) Sidewall—inclined

Figure 10.2 Leaky-wave waveguide slots; upper (broad) and side (narrow) walls.

wave on the antenna surface."[*] A surface wave antenna is, in general, a slow wave structure whose phase velocity of the traveling wave is equal to or less than the speed of light in free-space ($v_p/c \leq 1$).

For slow wave structures radiation takes place only at nonuniformities, curvatures, and discontinuities (see Figure 1.10). Discontinuities can be either discrete or distributed. One type of discrete discontinuity on a surface wave antenna is a transmission line terminated in an unmatched load, as shown in Figure 10.1(a). A distributed surface wave antenna can be analyzed in terms of the variation of the amplitude and phase of the current along its structure. In general, power flows parallel to the structure, except when losses are present, and for plane structures the fields decay exponentially away from the antenna. Most of the surface wave antennas are end-fire or near-end-fire radiators. Practical configurations include line, planar surface, curved, and modulated structures.

Another traveling wave antenna is a *leaky-wave* antenna defined as "an antenna that couples power in small increments per unit length, either continuously or discretely, from a traveling wave structure to free-space"[†] Leaky-wave antennas continuously lose energy due to radiation, as shown in Figure 10.2 by a slotted rectangular waveguide. The fields decay along the structure in the direction of wave travel and increase in others. Most of them are fast wave structures.

10.2.1 Long Wire

An example of a slow wave traveling antenna is a long wire, as shown in Figure 10.1. An antenna is usually classified as a *long* wire antenna if it is a straight conductor with a length

---

[*]"IEEE Standard Definitions of Terms for Antennas" (IEEE Std 145-1983), *IEEE Trans. Antennas and Propagat.*, Vol. AP-31, No. 6, Part II, Nov. 1983.
[†]Ibid.

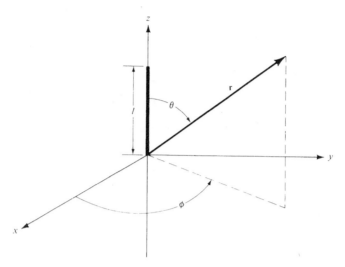

Figure 10.3 Long-wire antenna.

from one to many wavelengths. A long wire antenna has the distinction of being the first traveling wave antenna.

The long wire of Figure 10.1(a), in the presence of the ground, can be analyzed approximately using the equivalent of Figure 10.1(b) where an image is introduced to take into account the presence of the ground. The magnitude and phase of the image are determined using the reflection coefficient for horizontal polarization as given by (4-128). The height h of the antenna above the ground must be chosen so that the reflected wave (or wave from the image), which includes the phase due to reflection is in phase with the direct wave at the angles of desired maximum radiation. However, for typical electrical constitutive parameters of the earth, and especially for observation angles near grazing, the reflection coefficient for horizontal polarization is approximately -1. Therefore the total field radiated by the wire in the presence of the ground can be found by multiplying the field radiated by the wire in free space by the array factor of a two-element array, as was done in Section 4.8.2 and represented by (4-129).

The objective now is to find the field radiated by the long wire in free space. This is accomplished by referring to Figure 10.3. As the wave travels along the wire from the source toward the load, it continuously leaks energy. This can be represented by an attenuation coefficient. Therefore the current distribution of the forward traveling wave along the structure can be represented by

$$\mathbf{I}_f = \hat{\mathbf{a}}_z I_z(z')e^{-\gamma(z')z'} = \hat{\mathbf{a}}_z I_0 e^{-[\alpha(z')+jk_z(z')]z'} \tag{10-1}$$

where $\gamma(z')$ is the propagation coefficient [$\gamma(z') = \alpha(z') + jk_z(z')$ where $\alpha(z')$ is the attenuation constant (nepers/meter) while $k_z(z')$ is the phase constant (radians/meter) associated with the traveling wave]. In addition to the losses due to leakage, there are wire and ground losses. The attenuation factor $\alpha(z')$ can also be used to take into account the ohmic losses of the wire as well as ground losses. However, these, especially the ohmic losses, are usually very small and for simplicity are neglected. In addition, when the radiating medium is air, the loss of energy in a long wire ($l \gg \lambda$) due to leakage is also usually very small, and it can also be neglected. Therefore the current distribution of (10-1) can be approximated by

$$\mathbf{I} = \hat{\mathbf{a}}_z I(z')e^{-jk_z z'} = \hat{\mathbf{a}}_z I_0 e^{-jk_z z'} \tag{10-1a}$$

where $I(z') = I_0$ is assumed to be constant. Using techniques outlined and used in Chapter 4, it can be easily shown that in the far field

$$E_r \simeq E_\phi = H_r = H_\theta = 0 \tag{10-2a}$$

$$E_\theta \simeq j\eta \frac{kII_0 e^{-jkr}}{4\pi r} e^{-j(kl/2)(K-\cos\theta)} \sin\theta \frac{\sin[(kl/2)(\cos\theta - K)]}{(kl/2)(\cos\theta - K)} \tag{10-2b}$$

$$H_\phi \simeq \frac{E_\theta}{\eta} \tag{10-2c}$$

where K is used to represent the ratio of the phase constant of the wave along the transmission line (k_z) to that of free-space (k), or

$$K = \frac{k_z}{k} = \frac{\lambda}{\lambda_g} \tag{10-3}$$

λ_g = wavelength of the wave along the transmission line

Assuming a perfect electric conductor for the ground, the total field for Figure 10.1(a) is obtained by multiplying each of (10-2a)–(10-2c) by the array factor $\sin(kh\sin\theta)$.

For $k_z = k(K = 1)$ the time-average power density can be written as

$$\mathbf{W}_{av} = \mathbf{W}_{rad} = \hat{\mathbf{a}}_r \eta \frac{|I_0|^2}{8\pi^2 r^2} \frac{\sin^2\theta}{(\cos\theta - 1)^2} \sin^2\left[\frac{kl}{2}(\cos\theta - 1)\right] \tag{10-4}$$

which reduces to

$$\mathbf{W}_{av} = \mathbf{W}_{rad} = \hat{\mathbf{a}}_r \eta \frac{|I_0|^2}{8\pi^2 r^2} \cot^2\left(\frac{\theta}{2}\right) \sin^2\left[\frac{kl}{2}(\cos\theta - 1)\right] \tag{10-5}$$

From (10-5) it is evident that the power distribution of a wire antenna of length l is a multilobe pattern whose number of lobes depends upon its length. Assuming that l is very large such that the variations in the sine function of (10-5) are more rapid than those of the cotangent, the peaks of the lobes occur approximately when

$$\sin^2\left[\frac{kl}{2}(\cos\theta - 1)\right]_{\theta=\theta_m} = 1 \tag{10-6}$$

or

$$\frac{kl}{2}(\cos\theta_m - 1) = \pm\left(\frac{2m+1}{2}\right)\pi, \quad m = 0, 1, 2, 3, \ldots \tag{10-6a}$$

The angles where the peaks occur are given by

$$\theta_m = \cos^{-1}\left[1 \pm \frac{\lambda}{2l}(2m+1)\right], \quad m = 0, 1, 2, 3, \ldots \tag{10-7}$$

The angle where the maximum of the major lobe occurs is given by $m = 0$ (or $2m + 1 = 1$). As l becomes very large ($l \gg \lambda$), the angle of the maximum of the major lobe approaches zero degrees and the structure becomes a near-end-fire array.

In finding the values of the maxima, the variations of the cotangent term in (10-5) were assumed to be negligible (as compared to those of the sine term). If the effects of the cotangent term were to be included, then the values of the $2m + 1$ term in (10-7) should be

$$2m + 1 = 0.742, 2.93, 4.96, 6.97, 8.99, 11, 13, \ldots \tag{10-8}$$

(instead of $1, 3, 5, 7, 9, \ldots$) for the first, second, third, and so forth maxima. The approximate values approach those of the exact for the higher order lobes.

In a similar manner, the nulls of the pattern can be found and occur when

$$\sin^2\left[\frac{kl}{2}(\cos\theta - 1)\right]_{\theta=\theta_n} = 0 \tag{10-9}$$

or

$$\frac{kl}{2}(\cos\theta_n - 1) = \pm n\pi, \quad n = 1, 2, 3, 4, \ldots \tag{10-9a}$$

The angles where the nulls occur are given by

$$\theta_n = \cos^{-1}\left(1 \pm n\frac{\lambda}{l}\right), \quad n = 1, 2, 3, 4, \ldots \tag{10-10}$$

for the first, second, third, and so forth nulls.

The total radiated power can be found by integrating (10-5) over a closed sphere of radius r and reduces to

$$P_{\text{rad}} = \oiint_S \mathbf{W}_{\text{rad}} \cdot d\mathbf{s} = \frac{\eta}{4\pi}|I_0|^2\left[1.415 + \ln\left(\frac{kl}{\pi}\right) - C_i(2kl) + \frac{\sin(2kl)}{2kl}\right] \tag{10-11}$$

where $C_i(x)$ is the cosine integral of (4-68a). The radiation resistance is then found to be

$$R_r = \frac{2P_{\text{rad}}}{|I_0|^2} = \frac{\eta}{2\pi}\left[1.415 + \ln\left(\frac{kl}{\pi}\right) - C_i(2kl) + \frac{\sin(2kl)}{2kl}\right] \tag{10-12}$$

Using (10-5) and (10-11) the directivity can be written as

$$D_0 = \frac{4\pi U_{\max}}{P_{\text{rad}}} = \frac{2\cot^2\left[\frac{1}{2}\cos^{-1}\left(1 - \frac{0.371\lambda}{l}\right)\right]}{1.415 + \ln\left(\frac{2l}{\lambda}\right) - C_i(2kl) + \frac{\sin(2kl)}{2kl}} \tag{10-13}$$

A. Amplitude Patterns, Maxima, and Nulls

To verify some of the derivations and illustrate some of the principles, a number of computations were made. Shown in Figure 10.4(a) is the three-dimensional pattern of a traveling wire antenna with length $l = 5\lambda$ while in Figure 10.4(b) is the three-dimensional pattern for a standing wave wire antenna with length $l = 5\lambda$. The corresponding two-dimensional patterns are shown in Figure 10.5. The pattern of Figure 10.4(a) is formed by the forward traveling wave current $I_f = I_1 e^{-jkz}$ of Figure 10.1(a) while that of Figure 10.4(b) is formed by the forward I_f plus backward I_b traveling

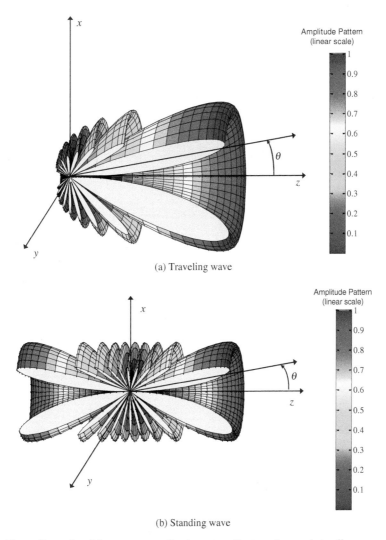

Figure 10.4 Three-dimensional free-space amplitude pattern for traveling and standing wave wire antennas of $l = 5\lambda$.

wave currents of Figure 10.1(a). The two currents I_f and I_b together form a standing wave; that is, $I_s = I_f + I_b = I_1 e^{-jkz} - I_2 e^{+jkz} = -2jI_0 \sin(kz)$ when $I_2 = I_1 = I_0$. As expected, for the traveling wave antenna of Figure 10.4(a) there is maximum radiation in the forward direction while for the standing wave antenna of Figure 10.4(b) there is maximum radiation in the forward and backward directions. The lobe near the axis of the wire in the directions of travel is the largest. The magnitudes of the other lobes from the main decrease progressively, with an envelope proportional to $\cot^2(\theta/2)$, toward the other direction. The traveling wave antenna is used when it is desired to radiate or receive predominantly from one direction. As the length of the wire increases, the maximum of the main lobe shifts closer toward the axis and the number of lobes increase. This is illustrated in Figure 10.6 for a traveling wave wire antenna with $l = 5\lambda$ and 10λ. The angles of the maxima of the first four lobes, computed using (10-8), are plotted in Figure 10.7(a) for $0.5\lambda \leq l \leq 10\lambda$. The corresponding angles of the first four nulls, computed using (10-10), are shown in Figure 10.7(b) for $0.5\lambda \leq l \leq 10\lambda$. These curves can be used effectively to design long wires when the direction of the maximum or null is desired.

540 TRAVELING WAVE AND BROADBAND ANTENNAS

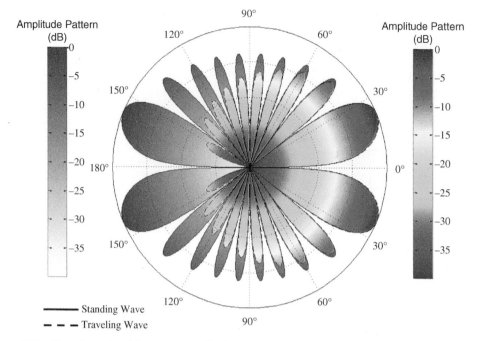

Figure 10.5 Two-dimensional free-space amplitude pattern for traveling and standing wave wire antennas of $l = 5\lambda$.

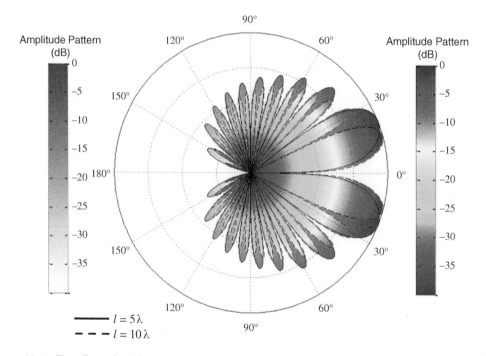

Figure 10.6 Two-dimensional free-space amplitude pattern for traveling wave wire antenna of $l = 5\lambda$ and 10λ.

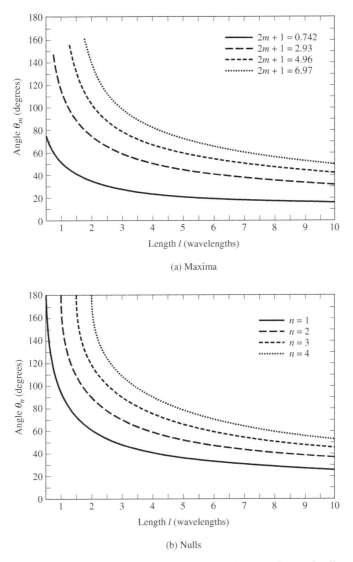

Figure 10.7 Angles versus length of wire antenna where maxima and nulls occur.

B. Input Impedance

For traveling wave wire antennas the radiation in the opposite direction from the maximum is suppressed by reducing, if not completely eliminating, the current reflected from the end of the wire. This is accomplished by increasing the diameter of the wire or more successfully by properly terminating it to the ground, as shown in Figure 10.1. Ideally a complete elimination of the reflections (perfect match) can only be accomplished if the antenna is elevated only at small heights (compared to the wavelength) above the ground, and it is terminated by a resistive load. The value of the load resistor, to achieve the impedance match, is equal to the characteristic impedance of the wire near the ground (which is found using image theory). For a wire with diameter d and height h above the ground, an approximate value of the termination resistance is obtained from

$$R_L = 138 \log_{10}\left(4\frac{h}{d}\right) \tag{10-14}$$

To achieve a reflection-free termination, the load resistor can be adjusted about this value (usually about 200–300 ohms) until there is no standing wave on the antenna wire. Therefore the input impedance is the same as the load impedance or the characteristic impedance of the line, as given by (10-14).

If the antenna is not properly terminated, waves reflected from the load traveling in the opposite direction from the incident waves create a standing wave pattern. Therefore the input impedance of the line is not equal to the load impedance. To find the input impedance, the transmission line impedance transfer equation of (9-18) can be used. Doing this we can write that the impedance at the input terminals of Figure 10.1(a) is

$$Z_{in}(l) = Z_c \left[\frac{R_L + jZ_c \tan(\beta l)}{Z_c + jR_L \tan(\beta l)} \right] \tag{10-15}$$

C. Polarization

A long-wire antenna is linearly polarized, and it is always parallel to the plane formed by the wire and radial vector from the center of the wire to the observation point. The direction of the linear polarization is not the same in all parts of the pattern, but it is perpendicular to the radial vector (and parallel to the plane formed by it and the wire). Thus the wire antenna of Figure 10.1, when its height above the ground is small compared to the wavelength and its main beam is near the ground, is not an effective element for horizontal polarization. Instead it is usually used to transmit or receive waves that have an appreciable vector component in the vertical plane. This is what is known as a *Beverage antenna* which is used more as a receiving rather than a transmitting element because of its poor radiation efficiency due to power absorbed in the load resistor.

When a TEM wave travels parallel to an air-conductor interface, it creates a forward wave tilt [2] which is determined by applying the boundary conditions on the tangential fields along the interface. The amount of tilt is a function of the constitutive parameters of the ground. If the conductor is a perfect electric conductor (PEC), then the wave tilt is zero because the tangential electric field vanishes along the PEC. The wave tilt increases with frequency and with ground resistivity. Therefore, for a Beverage wire antenna, shown in Figure 10.1(c) in the receiving mode, reception is influenced by the tilt angle of the incident vertically polarized wavefront, which is formed by the losses of the local ground. The electric-field vector of the incident wavefront produces an electric force that is parallel to the wire, which in turn induces a current in the wire. The current flows in the wire toward the receiver, and it is reinforced up to a certain point along the wire by the advancing wavefront. The wave along the wire is transverse magnetic.

D. Resonant Wires

Resonant wire antennas are formed when the load impedance of Figure 10.1(a) is not matched to the characteristic impedance of the line. This causes reflections which with the incident wave form a standing wave. Resonant antennas, including the dipole, were examined in detail in Chapter 4, and the electric and magnetic field components of a center-fed wire of total length l are given, respectively, by (4-62a) and (4-62b). Other radiation characteristics (including directivity, radiation resistance, maximum effective area, etc.) are found in Chapter 4.

Resonant antennas can also be formed using long wires. It can be shown that for *resonant* long wires with lengths odd multiple of half wavelength ($l = n\lambda/2, n = 1, 3, 5, \ldots$), the radiation resistance is given approximately (within 0.5 ohms) by [3], [4]

$$R_r = 73 + 69 \log_{10}(n) \tag{10-16}$$

This expression gives a very good answer even for $n = 1$. For the same elements, the angle of maximum radiation is given by

$$\theta_{max} = \cos^{-1}\left(\frac{n-1}{n}\right) \quad (10\text{-}17)$$

This formula is more accurate for small values of n, although it gives good results even for large values of n. It can also be shown that the maximum directivity is related to the radiation resistance by

$$D_0 = \frac{120}{R_r \sin^2 \theta_{max}} \quad (10\text{-}18)$$

The values based on (10-18) are within 0.5 dB from those based on (4-75). It is apparent that all three expressions, (10-16)–(10-18), lead to very good results for the half-wavelength dipole ($n = 1$).

Long-wire antennas (both resonant and nonresonant) are very simple, economical, and effective directional antennas with many uses for transmitting and receiving waves in the MF (300 KHz–3 MHz) and HF (3–30 MHz) ranges. Their properties can be enhanced when used in arrays.

A MATLAB computer program, entitled **Beverage**, has been developed to analyze the radiation characteristics of a long-wire antenna. The description of the program is found in the corresponding READ ME file included in the CD attached to the book.

10.2.2 V Antenna

For some applications a single long-wire antenna is not very practical because (1) its directivity may be low, (2) its side lobes may be high, and (3) its main beam is inclined at an angle, which is controlled by its length. These and other drawbacks of single long-wire antennas can be overcome by utilizing an array of wires.

One very practical array of long wires is the V antenna formed by using two wires each with one of its ends connected to a feed line as shown in Figure 10.8(a). In most applications, the plane formed by the legs of the V is parallel to the ground leading to a horizontal V array whose principal polarization is parallel to the ground and the plane of the V. Because of increased sidelobes, the directivity of ordinary linear dipoles begins to diminish for lengths greater than about 1.25λ, as shown in Figure 4.9. However by adjusting the included angle of a V-dipole, its directivity can be made greater and its side lobes smaller than those of a corresponding linear dipole. Designs for maximum directivity usually require smaller included angles for longer V's.

Most V antennas are symmetrical ($\theta_1 = \theta_2 = \theta_0$ and $l_1 = l_2 = l$). Also V antennas can be designed to have unidirectional or bidirectional radiation patterns, as shown in Figures 10.8(b) and (c), respectively. To achieve the unidirectional characteristics, the wires of the V antenna must be nonresonant which can be accomplished by minimizing if not completely eliminating reflections from the ends of the wire. The reflected waves can be reduced by making the inclined wires of the V relatively thick. In theory, the reflections can even be eliminated by properly terminating the open ends of the V leading to a purely traveling wave antenna. One way of terminating the V antenna will be to attach a load, usually a resistor equal in value to the open end characteristic impedance of the V-wire transmission line, as shown in Figure 10.9(a). The terminating resistance can also be divided in half and each half connected to the ground leading to the termination of Figure 10.9(b). If the length of each leg of the V is very long (typically $l > 5\lambda$), there will be sufficient leakage of the field along each leg that when the wave reaches the end of the V it will be sufficiently reduced that there will not necessarily be a need for a termination. Of course, termination with a load is not possible without a ground plane.

544 TRAVELING WAVE AND BROADBAND ANTENNAS

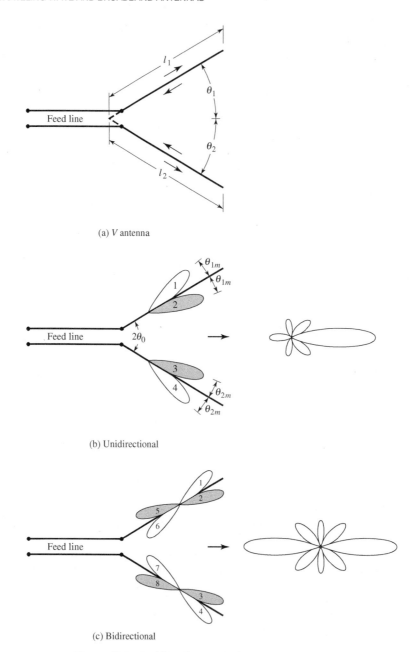

(a) V antenna

(b) Unidirectional

(c) Bidirectional

Figure 10.8 Unidirectional and bidirectional V antennas.

The patterns of the individual wires of the V antenna are conical in form and are inclined at an angle from their corresponding axes. The angle of inclination is determined by the length of each wire. For the patterns of each leg of a symmetrical V antenna to add in the direction of the line bisecting the angle of the V and to form one major lobe, the total included angle $2\theta_0$ of the V should be equal to $2\theta_m$, which is twice the angle that the cone of maximum radiation of each wire makes with its axis. When this is done, beams 2 and 3 of Figure 10.8(b) are aligned and add constructively. Similarly for Figure 10.8(c), beams 2 and 3 are aligned and add constructively in the forward direction, while beams 5 and 8 are aligned and add constructively in the backward direction. If the total included angle of the V is greater than $2\theta_m(2\theta_0 > 2\theta_m)$ the main lobe is split into two distinct beams. However, if

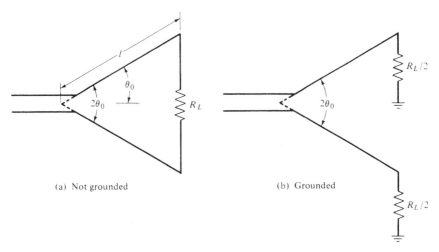

Figure 10.9 Terminated V antennas.

$2\theta_0 < 2\theta_m$, then the maximum of the single major lobe is still along the plane that bisects the V but it is tilted upward from the plane of the V. This may be a desired designed characteristic when the antenna is required to transmit waves upward toward the ionosphere for optimum reflection or to receive signals reflected downward by the ionosphere. For optimum operation, typically the included angle is chosen to be approximately $\theta_0 \simeq 0.8\theta_m$. When this is done, the reinforcement of the fields from the two legs of the V lead to a total directivity for the V of approximately twice the directivity of one leg of the V.

For a symmetrical V antenna with legs each of length l, there is an optimum included angle which leads to the largest directivity. Design data for optimum included angles of V dipoles were computed [5] using Moment Method techniques and are shown in Figure 10.10(a). The corresponding directivities are shown in Figure 10.10(b). In each figure the dots (·) represent values computed using the Moment Method while the solid curves represent second- or third-order polynomials fitted through the computed data. The polynomials for optimum included angles and maximum directivities are given by

$$2\theta_0 = \begin{cases} -149.3 \left(\frac{l}{\lambda}\right)^3 + 603.4 \left(\frac{l}{\lambda}\right)^2 - 809.5 \left(\frac{l}{\lambda}\right) + 443.6 & \text{(10-19a)} \\ \quad \text{for } 0.5 \leq l/\lambda \leq 1.5 \\ 13.39 \left(\frac{l}{\lambda}\right)^2 - 78.27 \left(\frac{l}{\lambda}\right) + 169.77 & \text{(10-19b)} \\ \quad \text{for } 1.5 \leq l/\lambda \leq 3 \end{cases}$$

$$D_0 = 2.94 \left(\frac{l}{\lambda}\right) + 1.15 \quad \text{for } 0.5 \leq l/\lambda \leq 3 \quad \text{(10-20)}$$

The dashed curves represent data obtained from empirical formulas [6]. The corresponding input impedances of the V's are slightly smaller than those of straight dipoles.

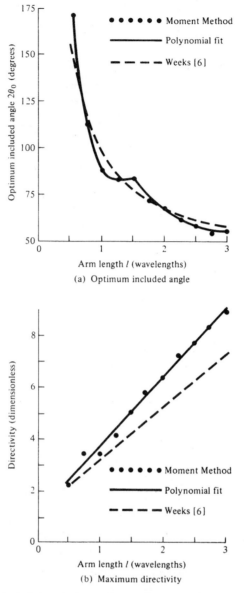

Figure 10.10 Optimum included angle for maximum directivity as a function of arm length for V dipoles. (SOURCE: G. A. Thiele and E. P. Ekelman, Jr., "Design Formulas for Vee Dipoles," *IEEE Trans. Antennas Propagat.*, Vol. AP-28, pp. 588–590, July 1980. © (1980) IEEE).

Another form of a V antenna is shown in the insert of Figure 10.11(a). The V is formed by a monopole wire, bent at an angle over a ground plane, and by its image shown dashed. The included angle of the V as well as the length can be used to tune the antenna. For included angles greater than 120° ($2\theta_0 > 120°$), the antenna exhibits primarily vertical polarization with radiation patterns almost identical to those of straight dipoles. As the included angle becomes smaller than about 120°, a horizontally polarized field component is excited which tends to fill the pattern toward the horizontal direction, making it a very attractive communication antenna for aircraft. The computed impedance of the ground plane and free-space V configurations obtained by the Moment Method [7] is shown plotted in Figure 10.11(a).

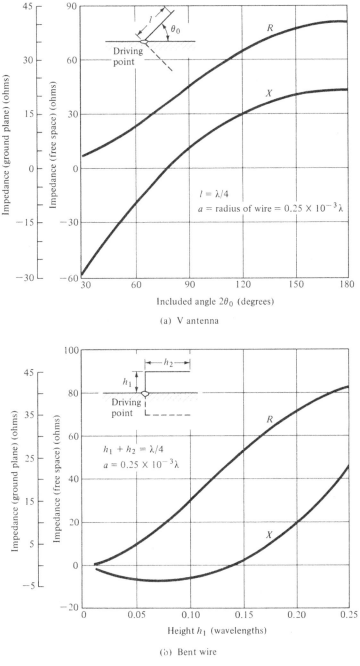

Figure 10.11 Computed impedance $(R + jX)$ of V and bent wire antennas above ground. (SOURCE: D. G. Fink (ed.), *Electronics Engineer's Handbook*, Chapter 18 (by W. F. Croswell), McGraw-Hill, New York, 1975).

Another practical form of a dipole antenna, particularly useful for airplane or ground-plane applications, is the 90° bent wire configuration of Figure 10.11(b). The computed impedance of the antenna, obtained also by the Moment Method [7], is shown plotted in Figure 10.11(b). This antenna can be tuned by adjusting its perpendicular and parallel lengths h_1 and h_2. The radiation pattern in the plane of the antenna is nearly omnidirectional for $h_1 \leq 0.1\lambda$. For $h_1 > 0.1\lambda$ the pattern approaches that of vertical $\lambda/2$ dipole.

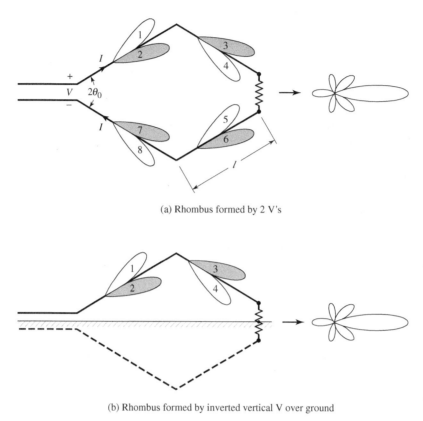

(a) Rhombus formed by 2 V's

(b) Rhombus formed by inverted vertical V over ground

Figure 10.12 Rhombic antenna configurations.

10.2.3 Rhombic Antenna

A. Geometry and Radiation Characteristics

Two V antennas can be connected at their open ends to form a diamond or rhombic antenna, as shown in Figure 10.12(a). The antenna is usually terminated at one end in a resistor, usually about 600–800 ohms, in order to reduce if not eliminate reflections. However, if each leg is long enough (typically greater than 5λ) sufficient leakage occurs along each leg that the wave that reaches the far end of the rhombus is sufficiently reduced that it may not be necessary to terminate the rhombus. To achieve the single main lobe, beams 2, 3, 6, and 7 are aligned and add constructively. The other end is used to feed the antenna. Another configuration of a rhombus is that of Figure 10.12(b) which is formed by an inverted V and its image (shown dashed). The inverted V is connected to the ground through a resistor. As with the V antennas, the pattern of rhombic antennas can be controlled by varying the element lengths, angles between elements, and the plane of the rhombus. Rhombic antennas are usually preferred over V's for nonresonant and unidirectional pattern applications because they are less difficult to terminate. Additional directivity and reduction in side lobes can be obtained by stacking, vertically or horizontally, a number of rhombic and/or V antennas to form arrays.

The field radiated by a rhombus can be found by adding the fields radiated by its four legs. For a symmetrical rhombus with equal legs, this can be accomplished using array theory and pattern multiplication. When this is done, a number of design equations can be derived [8]–[11]. For this design, the plane formed by the rhombus is placed parallel and a height h above a perfect electric conductor.

B. Design Equations

Let us assume that it is desired to design a rhombus such that the maximum of the main lobe of the pattern, in a plane which bisects the V of the rhombus, is directed at an angle ψ_0 above the ground plane. The design can be optimized if the height h is selected according to

$$\frac{h_m}{\lambda_0} = \frac{m}{4\cos(90° - \psi_0)}, \quad m = 1, 3, 5, \ldots \quad (10\text{-}21)$$

with $m = 1$ representing the minimum height.

The minimum optimum length of each leg of a symmetrical rhombus must be selected according to

$$\frac{l}{\lambda_0} = \frac{0.371}{1 - \sin(90° - \psi_0)\cos\theta_0} \quad (10\text{-}22)$$

The best choice for the included angle of the rhombus is selected to satisfy

$$\theta_0 = \cos^{-1}[\sin(90° - \psi_0)] \quad (10\text{-}23)$$

10.3 BROADBAND ANTENNAS

In Chapter 9 broadband dipole antennas were discussed. There are numerous other antenna designs that exhibit greater broadband characteristics than those of the dipoles. Some of these antenna can also provide circular polarization, a desired extra feature for many applications. In this section we want to discuss briefly some of the most popular broadband antennas.

10.3.1 Helical Antenna

Another basic, simple, and practical configuration of an electromagnetic radiator is that of a conducting wire wound in the form of a screw thread forming a helix, as shown in Figure 10.13. In most cases the helix is used with a ground plane. The ground plane can take different forms. One is for the ground to be flat, as shown in Figure 10.13. Typically the diameter of the ground plane should be at least $3\lambda/4$. However, the ground plane can also be cupped in the form of a cylindrical cavity (see Figure 10.17) or in the form of a frustrum cavity [8]. In addition, the helix is usually connected to the center conductor of a coaxial transmission line at the feed point with the outer conductor of the line attached to the ground plane.

The geometrical configuration of a helix consists usually of N turns, diameter D and spacing S between each turn. The total length of the antenna is $L = NS$ while the total length of the wire is $L_n = NL_0 = N\sqrt{S^2 + C^2}$ where $L_0 = \sqrt{S^2 + C^2}$ is the length of the wire between each turn and $C = \pi D$ is the circumference of the helix. Another important parameter is the pitch angle α which is the angle formed by a line tangent to the helix wire and a plane perpendicular to the helix axis. The pitch angle is defined by

$$\alpha = \tan^{-1}\left(\frac{S}{\pi D}\right) = \tan^{-1}\left(\frac{S}{C}\right) \quad (10\text{-}24)$$

When $\alpha = 0°$, then the winding is flattened and the helix reduces to a loop antenna of N turns. On the other hand, when $\alpha = 90°$ then the helix reduces to a linear wire. When $0° < \alpha < 90°$, then

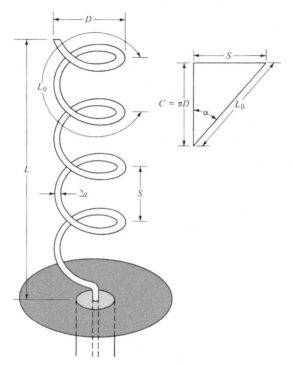

Figure 10.13 Helical antenna with ground plane.

a true helix is formed with a circumference greater than zero but less than the circumference when the helix is reduced to a loop ($\alpha = 0°$).

The radiation characteristics of the antenna can be varied by controlling the size of its geometrical properties compared to the wavelength. The input impedance is critically dependent upon the pitch angle and the size of the conducting wire, especially near the feed point, and it can be adjusted by controlling their values. The general polarization of the antenna is elliptical. However circular and linear polarizations can be achieved over different frequency ranges.

The helical antenna can operate in many modes; however the two principal ones are the *normal* (*broadside*) and the *axial* (*end-fire*) modes. The three-dimensional amplitude patterns representative of a helix operating, respectively, in the normal (broadside) and axial (end-fire) modes are shown in Figure 10.14. The one representing the normal mode, Figure 10.14(a), has its maximum in a plane normal to the axis and is nearly null along the axis. The pattern is similar in shape to that of a small dipole or circular loop. The pattern representative of the axial mode, Figure 10.14(b), has its maximum along the axis of the helix, and it is similar to that of an end-fire array. More details are in the sections that follow. The axial (end-fire) mode is usually the most practical because it can achieve circular polarization over a wider bandwidth (usually 2:1) and it is more efficient.

Because an elliptically polarized antenna can be represented as the sum of two orthogonal linear components in time-phase quadrature, a helix can always receive a signal transmitted from a rotating linearly polarized antenna. Therefore helices are usually positioned on the ground for space telemetry applications of satellites, space probes, and ballistic missiles to transmit or receive signals that have undergone Faraday rotation by traveling through the ionosphere.

A. Normal Mode

In the normal mode of operation the field radiated by the antenna is maximum in a plane normal to the helix axis and minimum along its axis, as shown sketched in Figure 10.14(a), which is a figure-eight rotated about its axis similar to that of a linear dipole of $l < \lambda_0$ or a small loop ($a \ll \lambda_0$).

BROADBAND ANTENNAS **551**

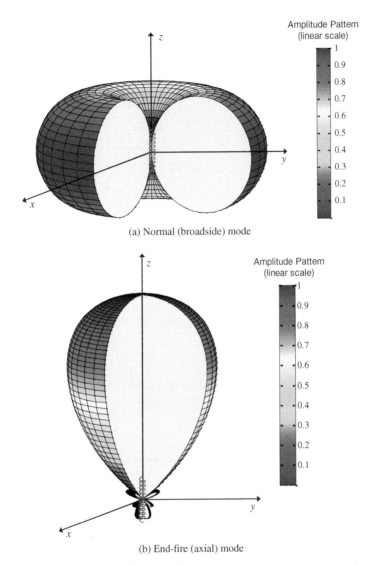

(a) Normal (broadside) mode

(b) End-fire (axial) mode

Figure 10.14 Three-dimensional normalized amplitude linear power patterns for *normal* and *end-fire* modes helical designs.

To achieve the normal mode of operation, the dimensions of the helix are usually small compared to the wavelength (i.e., $NL_0 \ll \lambda_0$).

The geometry of the helix reduces to a loop of diameter D when the pitch angle approaches zero and to a linear wire of length S when it approaches 90°. Since the limiting geometries of the helix are a loop and a dipole, the far field radiated by a small helix in the normal mode can be described in terms of E_θ and E_ϕ components of the dipole and loop, respectively. In the normal mode, the helix of Figure 10.15(a) can be simulated approximately by N small loops and N short dipoles connected together in series as shown in Figure 10.14(b). The fields are obtained by superposition of the fields from these elemental radiators. The planes of the loops are parallel to each other and perpendicular to the axes of the vertical dipoles. The axes of the loops and dipoles coincide with the axis of the helix.

Since in the normal mode the helix dimensions are small, the current throughout its length can be assumed to be constant and its relative far-field pattern to be independent of the number of loops and short dipoles. Thus its operation can be described accurately by the sum of the fields radiated

552 TRAVELING WAVE AND BROADBAND ANTENNAS

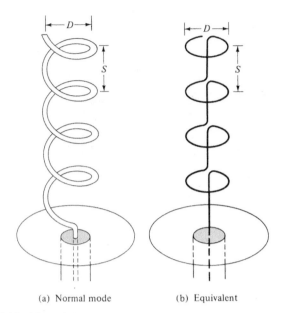

(a) Normal mode   (b) Equivalent

Figure 10.15 Normal (broadside) mode for helical antenna and its equivalent.

by a small loop of diameter D and a short dipole of length S, with its axis perpendicular to the plane of the loop, and each with the same constant current distribution.

The far-zone electric field radiated by a short dipole of length S and constant current I_0 is E_θ, and it is given by (4-26a) as

$$E_\theta = j\eta \frac{kI_0 S e^{-jkr}}{4\pi r} \sin\theta \qquad (10\text{-}25)$$

where l is being replaced by S. In addition the electric field radiated by a loop is E_ϕ, and it is given by (5-27b) as

$$E_\phi = \eta \frac{k^2 (D/2)^2 I_0 e^{-jkr}}{4r} \sin\theta \qquad (10\text{-}26)$$

where $D/2$ is substituted for a. A comparison of (10-25) and (10-26) indicates that the two components are in time-phase quadrature, a necessary but not sufficient condition for circular or elliptical polarization.

The ratio of the magnitudes of the E_θ and E_ϕ components is defined here as the axial ratio (AR), and it is given by

$$\boxed{\text{AR} = \frac{|E_\theta|}{|E_\phi|} = \frac{4S}{\pi k D^2} = \frac{2\lambda S}{(\pi D)^2}} \qquad (10\text{-}27)$$

By varying the D and/or S the axial ratio attains values of $0 \leq \text{AR} \leq \infty$. The value of AR = 0 is a special case and occurs when $E_\theta = 0$ leading to a linearly polarized wave of horizontal polarization (the helix is a loop). When AR = ∞, $E_\phi = 0$ and the radiated wave is linearly polarized with vertical polarization (the helix is a vertical dipole). Another special case is the one when AR is unity (AR = 1) and occurs when

$$\frac{2\lambda_0 S}{(\pi D)^2} = 1 \tag{10-28}$$

or

$$\boxed{C = \pi D = \sqrt{2S\lambda_0}} \tag{10-28a}$$

for which

$$\boxed{\tan\alpha = \frac{S}{\pi D} = \frac{\pi D}{2\lambda_0}} \tag{10-29}$$

When the dimensional parameters of the helix satisfy the above relation, the radiated field is circularly polarized in *all directions* other than $\theta = 0°$ where the fields vanish.

When the dimensions of the helix do not satisfy any of the above special cases, the field radiated by the antenna is not circularly polarized. The progression of polarization change can be described geometrically by beginning with the pitch angle of zero degrees ($\alpha = 0°$), which reduces the helix to a loop with linear horizontal polarization. As α increases, the polarization becomes elliptical with the major axis being horizontally polarized. When α, is such that $C/\lambda_0 = \sqrt{2S/\lambda_0}$, AR = 1 and we have circular polarization. For greater values of α, the polarization again becomes elliptical but with the major axis vertically polarized. Finally when $\alpha = 90°$ the helix reduces to a linearly polarized vertical dipole.

To achieve the normal mode of operation, it has been assumed that the current throughout the length of the helix is of constant magnitude and phase. This is satisfied to a large extent provided the total length of the helix wire NL_0 is very small compared to the wavelength ($L_n \ll \lambda_0$) and its end is terminated properly to reduce multiple reflections. Because of the critical dependence of its radiation characteristics on its geometrical dimensions, which must be very small compared to the wavelength, this mode of operation is very narrow in bandwidth and its radiation efficiency is very small. Practically this mode of operation is limited, and it is seldom utilized.

B. End-Fire Mode

A more practical mode of operation, which can be generated with great ease, is the end-fire or axial mode. In this mode of operation, there is only one major lobe and its maximum radiation intensity is along the axis of the helix, as shown in Figure 10.14(b). The minor lobes are at oblique angles to the axis.

To excite this mode, the diameter D and spacing S must be large fractions of the wavelength. To achieve circular polarization, primarily in the major lobe, the circumference of the helix must be in the $\frac{3}{4} < C/\lambda_0 < \frac{4}{3}$ range (with $C/\lambda_0 = 1$ near optimum), and the spacing about $S \simeq \lambda_0/4$. The pitch angle is usually $12° \leq \alpha \leq 14°$. Most often the antenna is used in conjunction with a ground plane, whose diameter is at least $\lambda_0/2$, and it is fed by a coaxial line. However, other types of feeds (such as waveguides and dielectric rods) are possible, especially at microwave frequencies. The dimensions of the helix for this mode of operation are not as critical, thus resulting in a greater bandwidth.

C. Design Procedure

The terminal impedance of a helix radiating in the end-fire mode is nearly resistive with values between 100 and 200 ohms. Smaller values, even near 50 ohms, can be obtained by properly designing the feed. Empirical expressions, based on a large number of measurements, have been derived [8], and they are used to determine a number of parameters. The input impedance (purely resistive) is obtained by

$$R \simeq 140 \left(\frac{C}{\lambda_0}\right) \qquad (10\text{-}30)$$

which is accurate to about ±20%, the half-power beamwidth by

$$\text{HPBW (degrees)} \simeq \frac{52\lambda_0^{3/2}}{C\sqrt{NS}} \qquad (10\text{-}31)$$

the beamwidth between nulls by

$$\text{FNBW (degrees)} \simeq \frac{115\lambda_0^{3/2}}{C\sqrt{NS}} \qquad (10\text{-}32)$$

the directivity by

$$D_0 \text{ (dimensionless)} \simeq 15N\frac{C^2 S}{\lambda_0^3} \qquad (10\text{-}33)$$

the axial ratio (for the condition of increased directivity) by

$$\text{AR} = \frac{2N+1}{2N} \qquad (10\text{-}34)$$

and the normalized far-field pattern by

$$E = \sin\left(\frac{\pi}{2N}\right) \cos\theta \, \frac{\sin[(N/2)\psi]}{\sin[\psi/2]} \qquad (10\text{-}35)$$

where

$$\psi = k_0 \left(S\cos\theta - \frac{L_0}{p}\right) \qquad (10\text{-}35\text{a})$$

$$p = \frac{L_0/\lambda_0}{S/\lambda_0 + 1} \qquad \text{For ordinary end-fire radiation} \qquad (10\text{-}35\text{b})$$

$$p = \frac{L_0/\lambda_0}{S/\lambda_0 + \left(\frac{2N+1}{2N}\right)} \qquad \text{For Hansen-Woodyard end-fire radiation} \qquad (10\text{-}35\text{c})$$

All these relations are approximately valid provided $12° < \alpha < 14°$, $\frac{3}{4} < C/\lambda_0 < \frac{4}{3}$, and $N > 3$.

The far-field pattern of the helix, as given by (10-35), has been developed by assuming that the helix consists of an array of N identical turns (each of nonuniform current and identical to that of the others), a uniform spacing S between them, and the elements are placed along the z-axis. The $\cos\theta$ term in (10-35) represents the field pattern of a single turn, and the last term in (10-35) is the array factor of a uniform array of N elements. The total field is obtained by multiplying the field from one turn with the array factor (pattern multiplication).

The value of p in (10-35a) is the ratio of the velocity with which the wave travels along the helix wire to that in free space, and it is selected according to (10-35b) for ordinary end-fire radiation or (10-35c) for Hansen-Woodyard end-fire radiation. These are derived as follows.

For ordinary end-fire the relative phase ψ among the various turns of the helix (elements of the array) is given by (6-7a), or

$$\psi = k_0 S \cos\theta + \beta \tag{10-36}$$

where $d = S$ is the spacing between the turns of the helix. For an end-fire design, the radiation from each one of the turns along $\theta = 0°$ must be in phase. Since the wave along the helix wire between turns travels a distance L_0 with a wave velocity $v = pv_0$ ($p < 1$ where v_0 is the wave velocity in free space) and the desired maximum radiation is along $\theta = 0°$, then (10-36) for *ordinary end-fire* radiation is equal to

$$\psi = (k_0 S \cos\theta - kL_0)_{\theta=0°} = k_0\left(S - \frac{L_0}{p}\right) = -2\pi m, \quad m = 0, 1, 2, \ldots \tag{10-37}$$

Solving (10-37) for p leads to

$$p = \frac{L_0/\lambda_0}{S/\lambda_0 + m} \tag{10-38}$$

For $m = 0$ and $p = 1$, $L_0 = S$. This corresponds to a straight wire ($\alpha = 90°$), and not a helix. Therefore the next value is $m = 1$, and it corresponds to the first transmission mode for a helix. Substituting $m = 1$ in (10-38) leads to

$$p = \frac{L_0/\lambda_0}{S/\lambda_0 + 1} \tag{10-38a}$$

which is that of (10-35b).

In a similar manner, it can be shown that for Hansen-Woodyard end-fire radiation (10-37) is equal to

$$\psi = (k_0 S \cos\theta - kL_0)_{\theta=0°} = k_0\left(S - \frac{L_0}{p}\right) = -\left(2\pi m + \frac{\pi}{N}\right), \quad m = 0, 1, 2, \ldots \tag{10-39}$$

which when solved for p leads to

$$p = \frac{L_0/\lambda_0}{S/\lambda_0 + \left(\dfrac{2mN+1}{2N}\right)} \tag{10-40}$$

For $m = 1$, (10-40) reduces to

$$p = \frac{L_0/\lambda_0}{S/\lambda_0 + \left(\frac{2N+1}{2N}\right)} \qquad (10\text{-}40a)$$

which is identical to (10-35c).

Example 10.1

Design a 10-turn helix to operate in the axial mode. For an optimum design,

1. Determine the:
   a. Circumference (in λ_o), pitch angle (*in degrees*), and separation between turns (in λ_o)
   b. Relative (to free space) wave velocity along the wire of the helix for:
      (i) Ordinary end-fire design
      (ii) Hansen-Woodyard end-fire design
   c. Half-power beamwidth of the main lobe (*in degrees*)
   d. Directivity (in dB) using:
      (i) A formula
      (ii) The computer program **Directivity** of Chapter 2
   e. Axial ratio (*dimensionless* and in *dB*)
2. Plot the normalized three-dimensional linear power pattern for the ordinary and Hansen-Woodyard designs.

Solution:

1. a. For an optimum design

$$C \simeq \lambda_o, \alpha \simeq 13° \Rightarrow S = C \tan \alpha = \lambda_o \tan(13°) = 0.231\lambda_o$$

   b. The length of a single turn is

$$L_o = \sqrt{S^2 + C^2} = \lambda_o \sqrt{(0.231)^2 + (1)^2} = 1.0263\lambda_o$$

   Therefore the relative wave velocity is:
   (i) Ordinary end-fire:

$$p = \frac{v_h}{v_o} = \frac{L_o/\lambda_o}{S_o/\lambda_o + 1} = \frac{1.0263}{0.231 + 1} = 0.8337$$

   (ii) Hansen-Woodyard end-fire:

$$p = \frac{v_h}{v_o} = \frac{L_o/\lambda_o}{S_o/\lambda_o + \left(\frac{2N+1}{2N}\right)} = \frac{1.0263}{0.231 + 21/20} = 0.8012$$

c. The half-power beamwidth according to (10-31) is

$$\text{HPBW} \simeq \frac{52\lambda_o^{3/2}}{C\sqrt{NS}} = \frac{52}{1\sqrt{10(0.231)}} = 34.2135°$$

d. The directivity is:
   (i) Using (10-33):

$$D_o \simeq 15N\frac{C^2 S}{\lambda_o^3} = 15(10)(1)^2(0.231) = 34.65 \ (dimensionless)$$
$$= 15.397 \text{ dB}$$

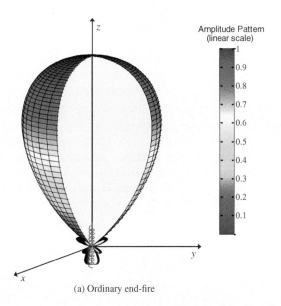

(a) Ordinary end-fire

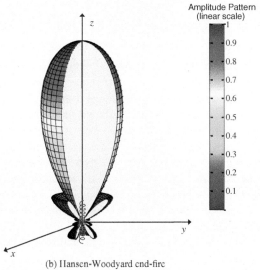

(b) Hansen-Woodyard end-fire

Figure 10.16 Three-dimensional normalized amplitude linear power patterns for helical *ordinary* ($p = 0.8337$) and *Hansen-Woodyard* ($p = 0.8012$) end-fire designs.

(ii) Using the computer program ***Directivity*** and (10-35):
  a. ordinary end-fire ($p = 0.8337$): $D_o = 12.678$ (*dimensionless*)
  $= 11.03$ dB
  b. H-W end-fire ($p = 0.8012$): $D_o = 26.36$ (*dimensionless*) $= 14.21$ dB
  e. The axial ratio according to (10-34) is:

$$AR = \frac{2N+1}{2N} = \frac{20+1}{20} = 1.05 \text{ (\textit{dimensionless})} = 0.424 \text{ dB}$$

2. The three-dimensional linear power patterns for the two end-fire designs, *ordinary* and *Hansen-Woodyard*, are shown in Figure 10.16.

D. Feed Design

The nominal impedance of a helical antenna operating in the axial mode, computed using (10-30), is 100–200 ohms. However, many practical transmission lines (such as a coax) have characteristic impedance of about 50 ohms. In order to provide a better match, the input impedance of the helix must be reduced to near that value. There may be a number of ways by which this can be accomplished. One way to effectively control the input impedance of the helix is to properly design the first 1/4 turn of the helix which is next to the feed [8], [12]. To bring the input impedance of the helix from nearly 150 ohms down to 50 ohms, the wire of the first 1/4 turn should be flat in the form of a strip and the transition into a helix should be very gradual. This is accomplished by making the wire from the feed, at the beginning of the formation of the helix, in the form of a strip of width w by flattening it and nearly touching the ground plane which is covered with a dielectric slab of height [2]

$$h = \frac{w}{\frac{377}{\sqrt{\varepsilon_r} Z_0} - 2} \quad (10\text{-}41)$$

where

w = width of strip conductor of the helix starting at the feed
ε_r = dielectric constant of the dielectric slab covering the ground plane
Z_0 = characteristic impedance of the input transmission line

Typically the strip configuration of the helix transitions from the strip to the regular circular wire and the designed pitch angle of the helix very gradually within the first 1/4–1/2 turn.

This modification decreases the characteristic impedance of the conductor-ground plane effective transmission line, and it provides a lower impedance over a substantial but reduced bandwidth. For example, a 50-ohm helix has a VSWR of less than 2:1 over a 40% bandwidth compared to a 70% bandwidth for a 140-ohm helix. In addition, the 50-ohm helix has a VSWR of less than 1.2:1 over a 12% bandwidth as contrasted to a 20% bandwidth for one of 140 ohms.

A simple and effective way of increasing the thickness of the conductor near the feed point will be to bond a thin metal strip to the helix conductor [12]. For example, a metal strip 70-mm wide was used to provide a 50-ohm impedance in a helix whose conducting wire was 13-mm in diameter and it was operating at 230.77 MHz.

A commercially available helix with a cupped ground plane is shown in Figure 10.17. It is right-hand circularly-polarized (RHCP) operating between 100–160 MHz with a gain of about 6 dB at 100 MHz and 12.8 dB at 160 MHz. The right-hand winding of the wire is clearly shown in the

Figure 10.17 Commercial helix with a cupped ground plane. (Courtesy: Seavey Engineering Associates, Inc, Pembroke, MA).

photo. The axial ratio is about 8 dB at 100 MHz and 2 dB at 160 MHz. The maximum VSWR in the stated operating frequency, relative to a 50-ohm line, does not exceed 3:1.

A MATLAB computer program, entitled **Helix**, has been developed to analyze and design a helical antenna. The description of the program is found in the corresponding READ ME file included in the CD attached to the book.

10.3.2 Electric-Magnetic Dipole

It has been shown in the previous section that the circular polarization of a helical antenna operating in the normal mode was achieved by assuming that the geometry of the helix is represented by a number of horizontal small loops and vertical infinitesimal dipoles. It would then seem reasonable that an antenna with only one loop and a single vertical dipole would, in theory, represent a radiator with an elliptical polarization. Ideally circular polarization, in all space, can be achieved if the current in each element can be controlled, by dividing the available power between the dipole and the loop, so that the magnitude of the field intensity radiated by each is equal.

Experimental models of such an antenna were designed and built [13] one operating around 350 MHz and the other near 1.2 GHz. A sketch of one of them is shown in Figure 10.18. The measured VSWR in the 1.15–1.32 GHz frequency range was less than 2:1.

This type of an antenna is very useful in UHF communication networks where considerable amount of fading may exist. In such cases the fading of the horizontal and vertical components are affected differently and will not vary in the same manner. Hopefully, even in severe cases, there will always be one component all the time which is being affected less than the other, thus providing continuous communication. The same results would apply in VHF and/or UHF broadcasting. In addition, a transmitting antenna of this type would also provide the versatility to receive with horizontally or vertically polarized elements, providing a convenience in the architectural design of the receiving station.

10.3.3 Yagi-Uda Array of Linear Elements

Another very practical radiator in the HF (3–30 MHz), VHF (30–300 MHz), and UHF (300–3,000 MHz) ranges is the Yagi-Uda antenna. This antenna consists of a number of linear dipole

560 TRAVELING WAVE AND BROADBAND ANTENNAS

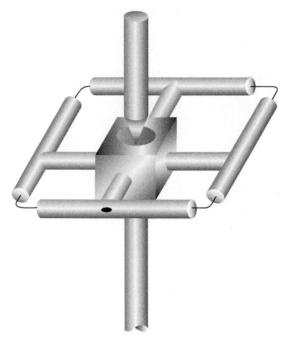

Figure 10.18 Electric-magnetic dipole configuration. (SOURCE: A. G. Kandoian, "Three New Antenna Types and Their Applications," *Proc. IRE*, Vol. 34, pp. 70W–75W, February 1946. © (1946) IEEE).

elements, as shown in Figure 10.19, one of which is energized directly by a feed transmission line while the others act as parasitic radiators whose currents are induced by mutual coupling. A common feed element for a Yagi-Uda antenna is a folded dipole. This radiator is exclusively designed to operate as an end-fire array, and it is accomplished by having the parasitic elements in the forward beam act as directors while those in the rear act as reflectors. Yagi designated the row of directors as a "wave canal." The Yagi-Uda array has been widely used as a home TV antenna; so it should be familiar to most of the readers, if not to the general public.

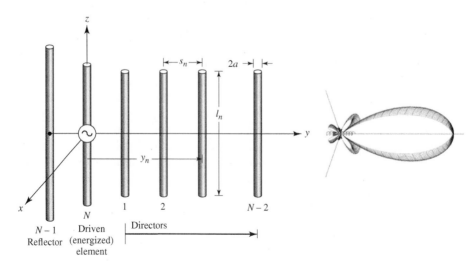

Figure 10.19 Yagi-Uda antenna configuration.

The original design and operating principles of this radiator were first described in Japanese in articles published in the Journal of I.E.E. of Japan by S. Uda of the Tohoku Imperial University in Japan [14]. In a later, but more widely circulated and read article [15], one of Professor Uda's colleagues, H. Yagi, described the operation of the same radiator in English. This paper has been considered a classic, and it was reprinted in 1984 in its original form in the *Proceedings of the IEEE* [15] as part of IEEE's centennial celebration. Despite the fact that Yagi in his English written paper acknowledged the work of Professor Uda on beam radiators at a wavelength of 4.4 m, it became customary throughout the world to refer to this radiator as a *Yagi* antenna, a generic term in the antenna dictionary. However, in order for the name to reflect more appropriately the contributions of both inventors, it should be called a *Yagi-Uda* antenna, a name that will be adopted in this book. Although the work of Uda and Yagi was done in the early 1920s and published in the middle 1920s, full acclaim in the United States was not received until 1928 when Yagi visited the United States and presented papers at meetings of the Institute of Radio Engineers (IRE) in New York, Washington, and Hartford. In addition, his work was published in the *Proceedings of IRE*, June 1928, where J. H. Dellinger, Chief of Radio Division, Bureau of Standards, Washington, D.C., and himself a pioneer of radio waves, characterized it as "exceptionally fundamental" and wrote "I have never listened to a paper that I felt so sure was destined to be a classic." So true!!

In 1984, IEEE celebrated its centennial year (1884–1984). Actually, IEEE was formed in 1963 when the IRE and AIEE united to form IEEE. During 1984, the *Proceedings of the IEEE* republished some classic papers, in their original form, in the different areas of electrical engineering that had appeared previously either in the *Proceeding of the IRE* or *IEEE*. In antennas, the only paper that was republished was that by Yagi [15]. Not only that, in 1997, the *Proceedings of the IEEE* republished for the second time the original paper by Yagi [15], [16]. That in itself tells us something of the impact this particular classic antenna design had on the electrical engineering profession.

The Yagi-Uda antenna has received exhaustive analytical and experimental investigations in the open literature and elsewhere. It would be impractical to list all the contributors, many of whom we may not be aware. However, we will attempt to summarize the salient point of the analysis, describe the general operation of the radiator, and present some design data.

To achieve the end-fire beam formation, the parasitic elements in the direction of the beam are somewhat smaller in length than the feed element. Typically the driven element is resonant with its length slightly less than $\lambda/2$ (usually $0.45-0.49\lambda$) whereas the lengths of the directors should be about 0.4 to 0.45λ. However, the directors are not necessarily of the same length and/or diameter. The separation between the directors is typically 0.3 to 0.4λ, and it is not necessarily uniform for optimum designs. It has been shown experimentally that for a Yagi-Uda array of 6λ total length the overall gain was independent of director spacing up to about 0.3λ. A significant drop (5–7 dB) in gain was noted for director spacings greater than 0.3λ. For that antenna, the gain was also independent of the radii of the directors up to about 0.024λ. The length of the reflector is somewhat greater than that of the feed. In addition, the separation between the driven element and the reflector is somewhat smaller than the spacing between the driven element and the nearest director, and it is found to be near optimum at 0.25λ.

Since the length of each director is smaller than its corresponding resonant length, the impedance of each is capacitive and its current leads the induced emf. Similarly the impedances of the reflectors is inductive and the phases of the currents lag those of the induced emfs. The total phase of the currents in the directors and reflectors is not determined solely by their lengths but also by their spacing to the adjacent elements. Thus, properly spaced elements with lengths slightly less than their corresponding resonant lengths (less than $\lambda/2$) act as directors because they form an array with currents approximately equal in magnitude and with equal progressive phase shifts which will reinforce the field of the energized element toward the directors. Similarly, a properly spaced element with a length of $\lambda/2$ or slightly greater will act as a reflector. Thus a Yagi-Uda array may be regarded as a structure supporting a traveling wave whose performance is determined by the current distribution in each element and the phase velocity of the traveling wave. It should be noted that the previous discussion on the lengths of the directors, reflectors, and driven elements is based on the first resonance. Higher resonances are available near lengths of λ, $3\lambda/2$, and so forth, but are seldom used.

In practice, the major role of the reflector is played by the first element next to the one energized, and very little in the performance of a Yagi-Uda antenna is gained if more than one (at the most two) elements are used as reflectors. However, considerable improvements can be achieved if more directors are added to the array. Practically there is a limit beyond which very little is gained by the addition of more directors because of the progressive reduction in magnitude of the induced currents on the more extreme elements. Usually most antennas have about 6 to 12 directors. However, many arrays have been designed and built with 30 to 40 elements. Array lengths on the order of 6λ have been mentioned [17] as typical. A gain (relative to isotropic) of about 5 to 9 per wavelength is typical for such arrays, which would make the overall gain on the order of about 30 to 54 (14.8–17.3 dB) typical.

The radiation characteristics that are usually of interest in a Yagi-Uda antenna are the *forward and backward gains, input impedance, bandwidth, front-to-back ratio*, and *magnitude of minor lobes*. The lengths and diameters of the directors and reflectors, as well as their respective spacings, determine the optimum characteristics. For a number of years optimum designs were accomplished experimentally. However, with the advent of high-speed computers many different numerical techniques, based on analytical formulations, have been utilized to derive the geometrical dimensions of the array for optimum operational performance. Usually Yagi-Uda arrays have low input impedance and relatively narrow bandwidth (on the order of about 2%). Improvements in both can be achieved at the expense of others (such as gain, magnitude of minor lobes, etc.). Usually a compromise is made, and it depends on the particular design. One way to increase the input impedance without affecting the performance of other parameters is to use an impedance step-up element as a feed (such as a two-element folded dipole with a step-up ratio of about 4). Front-to-back ratios of about 30 ($\simeq 15$ dB) can be achieved at wider than optimum element spacings, but they usually are compromised somewhat to improve other desirable characteristics.

The Yagi-Uda array can be summarized by saying that its performance can be considered in three parts:

1. the reflector-feeder arrangement
2. the feeder
3. the rows of directors

It has been concluded, numerically and experimentally, that the reflector spacing and size have (1) negligible effects on the forward gain and (2) large effects on the backward gain (front-to-back ratio) and input impedance, and they can be used to control or optimize antenna parameters without affecting the gain significantly. The feeder length and radius has a small effect on the forward gain but a large effect on the backward gain and input impedance. Its geometry is usually chosen to control the input impedance that most commonly is made real (resonant element). The size and spacing of the directors have a large effect on the forward gain, backward gain, and input impedance, and they are considered to be the most critical elements of the array.

Yagi-Uda arrays are quite common in practice because they are lightweight, simple to build, low-cost, and provide moderately desirable characteristics (including a unidirectional beam) for many applications. The design for a small number of elements (typically five or six) is simple but the design becomes quite critical if a large number of elements are used to achieve a high directivity. To increase the directivity of a Yagi-Uda array or to reduce the beamwidth in the E-plane, several rows of Yagi-Uda arrays can be used [18] to form a *curtain* antenna. To neutralize the effects of the feed transmission line, an odd number of rows is usually used.

A. Theory: Integral Equation-Moment Method

There have been many experimental [19], [20] investigations and analytical [21]–[30] formulations of the Yagi-Uda array. A method [25] based on rigorous integral equations for the electric field radiated by the elements in the array will be presented and it will be used to describe the complex current distributions on all the elements, the phase velocity, and the corresponding radiation patterns. The method is similar to that of [25], which is based on Pocklington's integral equation of (8-24) while the one presented here follows that of [25] but is based on Pocklington's integral equation of (8-22) and formulated by Tirkas [26]. Mutual interactions are also included and, in principle, there are no restrictions on the number of elements. However, for computational purposes, point-matching numerical methods, based on the techniques of Section 8.4, are used to evaluate and satisfy the integral equation at discrete points on the axis of each element rather than everywhere on the surface of every element. The number of discrete points where boundary conditions are matched must be sufficient in number to allow the computed data to compare well with experimental results.

The theory is based on Pocklington's integral equation of (8-22) for the total field generated by an electric current source radiating in an unbounded free-space, or

$$\int_{-l/2}^{+l/2} I(z') \left(\frac{\partial^2}{\partial z^2} + k^2 \right) \frac{e^{-jkR}}{R} dz' = j4\pi\omega\epsilon_0 E_z^t \tag{10-42}$$

where

$$R = \sqrt{(x-x')^2 + (y-y')^2 + (z-z')^2} \tag{10-42a}$$

Since

$$\frac{\partial^2}{\partial z^2}\left(\frac{e^{-jkR}}{R}\right) = \frac{\partial^2}{\partial z'^2}\left(\frac{e^{-jkR}}{R}\right) \tag{10-43}$$

(10-42) reduces to

$$\int_{-l/2}^{+l/2} I(z') \frac{\partial^2}{\partial z'^2}\left(\frac{e^{-jkR}}{R}\right) dz' + k^2 \int_{-l/2}^{+l/2} I(z') \frac{e^{-jkR}}{R} dz' = j4\pi\omega\epsilon_0 E_z^t \tag{10-44}$$

We will now concentrate in the integration of the first term of (10-44). Integrating the first term of (10-44) by parts where

$$u = I(z') \tag{10-45}$$

$$du = \frac{dI(z')}{dz'} dz' \tag{10-45a}$$

$$dv = \frac{\partial^2}{\partial z'^2}\left(\frac{e^{-jkR}}{R}\right) dz' = \frac{\partial}{\partial z'}\left[\frac{\partial}{\partial z'}\left(\frac{e^{-jkR}}{R}\right)\right] dz' \tag{10-46}$$

$$v = \frac{\partial}{\partial z'}\left(\frac{e^{-jkR}}{R}\right) \tag{10-46a}$$

reduces it to

$$\int_{-l/2}^{l/2} I(z') \frac{\partial^2}{\partial z'^2}\left(\frac{e^{-jkR}}{R}\right) dz' = I(z') \left[\frac{\partial}{\partial z'}\left(\frac{e^{-jkR}}{R}\right)\right]\bigg|_{-l/2}^{+l/2}$$

$$- \int_{-l/2}^{+l/2} \frac{\partial}{\partial z'}\left(\frac{e^{-jkR}}{R}\right) \frac{dI(z')}{dz'} dz' \tag{10-47}$$

564   TRAVELING WAVE AND BROADBAND ANTENNAS

Since we require that the current at the ends of each wire vanish [i.e., $I_z(z' = +l/2) = I_z(z' = -l/2) = 0$], (10-47) reduces to

$$\int_{-l/2}^{+l/2} I(z') \frac{\partial^2}{\partial z'^2}\left(\frac{e^{-jkR}}{R}\right) dz' = -\int_{-l/2}^{+l/2} \frac{\partial}{\partial z'}\left(\frac{e^{-jkR}}{R}\right) dz' \frac{dI(z')}{dz'} \qquad (10\text{-}48)$$

Integrating (10-48) by parts where

$$u = \frac{dI(z')}{dz'} \qquad (10\text{-}49)$$

$$du = \frac{d^2 I(z')}{dz'^2} dz' \qquad (10\text{-}49a)$$

$$dv = \frac{\partial}{\partial z'}\left(\frac{e^{-jkR}}{R}\right) dz' \qquad (10\text{-}50)$$

$$v = \frac{e^{-jkR}}{R} \qquad (10\text{-}50a)$$

reduces (10-48) to

$$\int_{-l/2}^{+l/2} I(z') \frac{\partial^2}{\partial z'^2}\left(\frac{e^{-jkR}}{R}\right) dz' = -\left.\frac{dI(z')}{dz'} \frac{e^{-jkR}}{R}\right|_{-l/2}^{+l/2}$$
$$+ \int_{-l/2}^{+l/2} \frac{d^2 I(z')}{dz'^2} \frac{e^{-jkR}}{R} dz' \qquad (10\text{-}51)$$

When (10-51) is substituted for the first term of (10-44) reduces it to

$$-\left.\frac{dI(z')}{dz'} \frac{e^{-jkR}}{R}\right|_{-l/2}^{+l/2} + \int_{-l/2}^{+l/2}\left[k^2 I(z') + \frac{d^2 I(z')}{dz'^2}\right] \frac{e^{-jkR}}{R} dz' = j4\pi\omega\varepsilon_0 E_z^t \qquad (10\text{-}52)$$

For small diameter wires the current on each element can be approximated by a finite series of odd-ordered even modes. Thus, the current on the nth element can be written as a Fourier series expansion of the form [26]

$$I_n(z') = \sum_{m=1}^{M} I_{nm} \cos\left[(2m-1)\frac{\pi z'}{l_n}\right] \qquad (10\text{-}53)$$

where I_{nm} represents the complex current coefficient of mode m on element n and l_n represents the corresponding length of the n element. Taking the first and second derivatives of (10-53) and substituting them, along with (10-53), into (10-52) reduces it to

$$\sum_{m=1}^{M} I_{nm} \left\{ \frac{(2m-1)\pi}{l_n} \sin\left[(2m-1)\frac{\pi z_n'}{l_n}\right] \left.\frac{e^{-jkR}}{R}\right|_{-l_n/2}^{+l_n/2} + \left[k^2 - \frac{(2m-1)^2 \pi^2}{l_n^2}\right] \right.$$
$$\left. \times \int_{-l_n/2}^{+l_n/2} \cos\left[(2m-1)\frac{\pi z_n'}{l_n}\right] \frac{e^{-jkR}}{R} dz_n' \right\} = j4\pi\omega\varepsilon_0 E_z^t \qquad (10\text{-}54)$$

Since the cosine is an even function, (10-54) can be reduced by integrating over only $0 \leq z' \leq l/2$ to

$$\sum_{m=1}^{M} I_{nm} \left\{ (-1)^{m+1} \frac{(2m-1)\pi}{l_n} G_2\left(x, x', y, y'/z, \frac{l_n}{2}\right) + \left[k^2 - \frac{(2m-1)^2 \pi^2}{l_n^2} \right] \right.$$

$$\left. \times \int_0^{l_n/2} G_2(x, x', y, y'/z, z'_n) \cos\left[\frac{(2m-1)\pi z'_n}{l_n} \right] dz'_n \right\} = j4\pi\omega\varepsilon_0 E_z^t$$

(10-55)

where

$$G_2(x, x', y, y'/z, z'_n) = \frac{e^{-jkR_-}}{R_-} + \frac{e^{-jkR_+}}{R_+}$$

(10-55a)

$$R_{\pm} = \sqrt{(x-x')^2 + (y-y')^2 + a^2 + (z \pm z')^2}$$

(10-55b)

$n = 1, 2, 3, \ldots, N$
$N = $ total number of elements

where

$R_{\pm}$ is the distance from the center of each wire radius to the center of any other wire, as shown in Figure 10.20(a).

The integral equation of (10-55) is valid for each element, and it assumes that the number M of current modes is the same for each element. To apply the Moment Method solution to the integral equation of (10-55), each wire element is subdivided in M segments. On each element, other than the driven element, the matching is done at the center of the wire, and it requires that E_z^t of (10-55) vanishes at each matching point of each segment [i.e., $E_z^t(z = z_i) = 0$], as shown in Figure 10.20(b). On the driven element the matching is done on the surface of the wire, and it requires that E_z^t of (10-55) vanishes at $M - 1$ points, even though there are m modes, and it excludes the segment at the feed as shown in Figure 10.20(c). This generates $M - 1$ equations. The Mth equation on the feed element is generated by the constraint that the normalized current for all M modes at the feed point ($z' = 0$) of the driven element is equal to unity [25], [27], or

$$\sum_{m=1}^{M} I_{nm}(z' = 0) \Big|_{n=N} = 1$$

(10-56)

Based on the above procedure, a system of linear equations is generated by taking into account the interaction of

a. each mode in each wire segment with each segment on the same wire.
b. each mode in each wire segment with each segment on the other wires.

This system of linear equations is then solved to find the complex amplitude coefficients of the current distribution in each wire as represented by (10-53). This is demonstrated in [25] for a three-element array (one director, one reflector, and the driven element) with two modes in each wire.

566 TRAVELING WAVE AND BROADBAND ANTENNAS

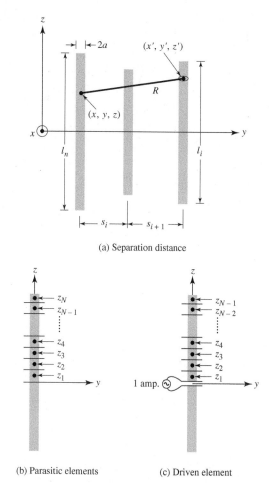

Figure 10.20 Geometry of Yagi-Uda array for Moment Method formulation (SOURCE: G. A. Thiele, "Yagi-Uda Type Antennas," *IEEE Trans. Antennas Propagat.*, Vol. AP-17, No. 1, pp. 24–31, January 1969. © (1969) IEEE).

B. Far-Field Pattern

Once the current distribution is found, the far-zone field generated by each element can be found using the techniques outlined in Chapter 3. The total field of the entire Yagi-Uda array is obtained by summing the contributions from each.

Using the procedure outlined in Chapter 3, Section 3.6, the far-zone electric field generated by the M modes of the nth element oriented parallel to the z axis is given by

$$E_{\theta n} \simeq -j\omega A_{\theta n} \tag{10-57}$$

$$A_{\theta n} \simeq -\frac{\mu e^{-jkr}}{4\pi r} \sin\theta \int_{-l_n/2}^{+l_n/2} I_n e^{jk(x_n \sin\theta\cos\phi + y_n \sin\theta\sin\phi + z'_n \cos\theta)} \, dz'_n$$

$$\simeq -\frac{\mu e^{-jkr}}{4\pi r} \sin\theta \left[e^{jk(x_n \sin\theta\cos\phi + y_n \sin\theta\sin\phi)} \int_{-l_n/2}^{+l_n/2} I_n e^{jkz'_n \cos\theta} \, dz'_n \right] \tag{10-57a}$$

where x_n, y_n represent the position of the nth element. The total field is then obtained by summing the contributions from each of the N elements, and it can be written as

$$E_\theta = \sum_{n=1}^{N} E_{\theta n} = -j\omega A_\theta \tag{10-58}$$

$$A_\theta = \sum_{n=1}^{N} A_{\theta n} = -\frac{\mu e^{-jkr}}{4\pi r} \sin\theta \sum_{n=1}^{N} \left\{ e^{jk(x_n \sin\theta \cos\phi + y_n \sin\theta \sin\phi)} \right.$$

$$\left. \times \left[\int_{-l_n/2}^{+l_n/2} I_n e^{jkz'_n \cos\theta} \, dz'_n \right] \right\} \tag{10-58a}$$

For each wire, the current is represented by (10-53). Therefore the last integral in (10-58a) can be written as

$$\int_{-l_n/2}^{+l_n/2} I_n e^{jkz'_n \cos\theta} \, dz'_n = \sum_{m=1}^{M} I_{nm} \int_{-l_n/2}^{+l_n/2} \cos\left[\frac{(2m-1)\pi z'_n}{l_n}\right] e^{jkz'_n \cos\theta} \, dz'_n \tag{10-59}$$

Since the cosine is an even function, (10-59) can also be expressed as

$$\int_{-l_n/2}^{+l_n/2} I_n e^{jkz'_n \cos\theta} \, dz'_n = \sum_{m=1}^{M} I_{nm} \int_{0}^{+l_n/2} 2\cos\left[\frac{(2m-1)\pi z'_n}{l_n}\right]$$

$$\times \left[\frac{e^{jkz'_n \cos\theta} + e^{-jkz'_n \cos\theta}}{2}\right] dz'_n$$

$$= \sum_{m=1}^{M} I_{nm} \int_{0}^{+l_n/2} 2\cos\left[\frac{(2m-1)\pi z'_n}{l_n}\right]$$

$$\times \cos(kz'_n \cos\theta) dz'_n \tag{10-60}$$

Using the trigonometric identity

$$2\cos(\alpha)\cos(\beta) = \cos(\alpha+\beta) + \cos(\alpha-\beta) \tag{10-61}$$

(10-60) can be rewritten as

$$\int_{-l_n/2}^{+l_n/2} I_n e^{jkz'_n \cos\theta} \, dz'_n = \sum_{m=1}^{M} I_{nm} \left\{ \int_{0}^{+l_n/2} \cos\left[\frac{(2m-1)\pi}{l_n} + k\cos\theta\right] z'_n \, dz'_n \right.$$

$$\left. + \int_{0}^{+l_n/2} \cos\left[\frac{(2m-1)\pi}{l_n} - k\cos\theta\right] z'_n \, dz'_n \tag{10-62}$$

Since

$$\int_0^{\alpha/2} \cos[(b \pm c)z]\, dz = \frac{\alpha}{2} \frac{\sin\left[(b \pm c)\frac{\alpha}{2}\right]}{(b \pm c)\frac{\alpha}{2}} \tag{10-63}$$

(10-62) can be reduced to

$$\int_{-l_n/2}^{+l_n/2} I_n e^{jkz'_n \cos\theta}\, dz'_n = \sum_{m=1}^{M} I_{nm} \left[\frac{\sin(Z^+)}{Z^+} + \frac{\sin(Z^-)}{Z^-}\right] \frac{l_n}{2} \tag{10-64}$$

$$Z^+ = \left[\frac{(2m-1)\pi}{l_n} + k\cos\theta\right] \frac{l_n}{2} \tag{10-64a}$$

$$Z^- = \left[\frac{(2m-1)\pi}{l_n} - k\cos\theta\right] \frac{l_n}{2} \tag{10-64b}$$

Thus, the total field represented by (10-58) and (10-58a) can be written as

$$E_\theta = \sum_{n=1}^{N} E_{\theta n} = -j\omega A \tag{10-65}$$

$$A_\theta = \sum_{n=1}^{N} A_{\theta n} = -\frac{\mu e^{-jkr}}{4\pi r} \sin\theta \sum_{n=1}^{N} \left\{ e^{jk(x_n \sin\theta \cos\phi + y_n \sin\theta \sin\phi)} \right.$$
$$\left. \times \sum_{m=1}^{M} I_{nm} \left[\frac{\sin(Z^+)}{Z^+} + \frac{\sin(Z^-)}{Z^-}\right]\right\} \frac{l_n}{2} \tag{10-65a}$$

There have been other analyses [28], [29] based on the integral equation formulation that allows the conversion to algebraic equations. In order not to belabor further the analytical formulations, which in call cases are complicated because of the antenna structure, it is appropriate at this time to present some results and indicate design procedures.

C. Computer Program and Results

Based on the preceding formulation, a MATLAB and FORTRAN computer program entitled **Yagi_Uda** has been developed [26] that computes the E- and H-plane patterns, their corresponding half-power beamwidths, and the directivity of the Yagi-Uda array. The program is described in the corresponding READ ME file, and both are included in the publisher's website for the book. The input parameters include the total number of elements (N), the number of current modes in each element (M), the length of each element, and the spacing between the elements. The program assumes one reflector, one driven element, and $N-2$ directors. For the development of the formulation and computer program, the numbering system ($n = 1, 2, \ldots, N$) for the elements begins with the first director ($n = 1$), second director ($n = 2$), etc. The reflector is represented by the next to the last element ($n = N-1$), while the driven element is designated as the last element ($n = N$), as shown in Figure 10.19.

One Yagi-Uda array design is considered here, which is the same as one of the two included in [25]; the other ones are assigned as end of the chapter problems. The patterns, beamwidths, and directivities were computed based on the computer program developed here.

Example 10.2

Design a Yagi-Uda array of 15 elements (13 directors, one reflector, and the exciter). Compute and plot the E- and H-plane patterns, normalized current at the center of each element, and directivity and front-to-back ratio as a function of reflector spacing and director spacing. Use the computer program **Yagi_Uda** of this chapter. The dimensions of the array are as follows:

$$N = \text{total number of elements} = 15$$
$$\text{number of directors} = 13$$
$$\text{number of reflectors} = 1$$
$$\text{number of exciters} = 1$$
$$\text{total length of reflector} = 0.5\lambda$$
$$\text{total length of feeder} = 0.47\lambda$$
$$\text{total length of each director} = 0.406\lambda$$
$$\text{spacing between reflector and feeder} = 0.25\lambda$$
$$\text{spacing between adjacent directors} = 0.34\lambda$$
$$a = \text{radius of wires} = 0.003\lambda$$

Solution: Using the computer program of this chapter, the computed E- and H-plane patterns of this design are shown in Figure 10.21. The corresponding beamwidths are: E-plane ($\Theta_e = 26.98°$), H-plane ($\Theta_h = 27.96°$) while the directivity is 14.64 dB. A plot of the current at the center of each element versus position of the element is shown in Figure 10.22; the current of the feed element at its center is unity, as required by (10-56). One important *figure-of-merit* in

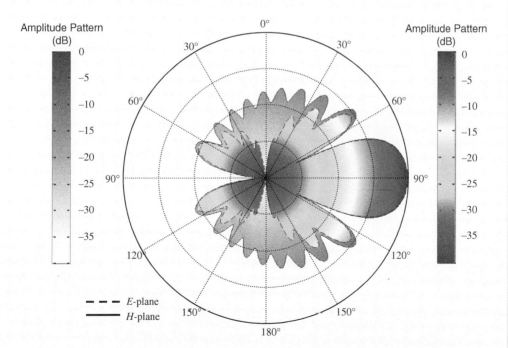

Figure 10.21 E- and H-plane amplitude patterns of 15-element Yagi-Uda array.

570 TRAVELING WAVE AND BROADBAND ANTENNAS

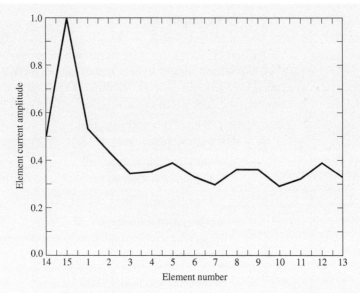

Figure 10.22 Normalized current at the center of each element of a 15-element Yagi-Uda array.

Yagi-Uda array is the *front-to-back* ratio of the pattern [$20 \log_{10} E(\theta = 90°, \phi = 90°)/E(\theta = 90°, \phi = 270°)$] as a function of the spacing of the reflector with respect to the feeder. This, along with the directivity, is shown in Figure 10.23 for spacing from 0.1λ–0.5λ. For this design, the maximum front-to-back ratio occurs for a reflector spacing of about 0.23λ while the directivity is monotonically decreasing very gradually, from about 15.2 dB at a spacing of 0.1λ down to about 10.4 dB at a spacing of 0.5λ.

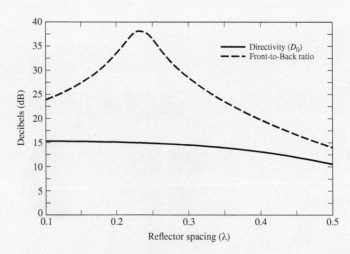

Figure 10.23 Directivity and front-to-back ratio, as a function of reflector spacing, of a 15-element Yagi-Uda array.

Another important parametric investigation is the variation of the front-to-back ratio and directivity as a function of director spacing. This is shown in Figure 10.24 for spacings from 0.1λ to 0.5λ. It is apparent that the directivity exhibits a slight increase from about 12 dB at a spacing of about 0.1λ to about 15.3 dB at a spacing of about 0.45λ. A steep drop in directivity occurs for spacings greater than about 0.45λ. This agrees with the conclusion arrived at in [19] and [28] that large reductions in directivity occur in Yagi-Uda array designs for spacings greater than about 0.4λ. For this design, the variations of the front-to-back ratio are much more sensitive as a function of director spacing, as shown in Figure 10.24; excursions on the order of 20–25 dB are evident for changes in spacing of about 0.05λ. Such variations in front-to-back ratio as shown in Figure 10.24, as a function of director spacing, are not evident in Yagi-Uda array designs with a smaller number of elements. However, they are even more pronounced for designs with a larger number of elements. Both of these are demonstrated in design problems assigned at the end of the chapter.

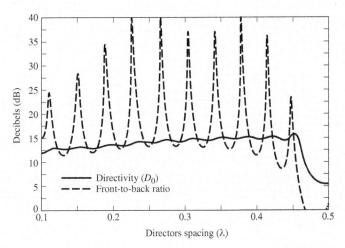

Figure 10.24 Directivity and front-to-back ratio, as a function of director spacing, for 15-element Yagi-Uda array.

D. Optimization

The radiation characteristics of the array can be adjusted by controlling the geometrical parameters of the array. This was demonstrated in Figures 10.23 and 10.24 for the 15-element array using uniform lengths and making uniform variations in spacings. However, these and other array characteristics can be optimized by using nonuniform director lengths and spacings between the directors. For example, the spacing between the directors can be varied while holding the reflector–exciter spacing and the lengths of all elements constant. Such a procedure was used by Cheng and Chen [28] to optimize the directivity of a six-element (four-director, reflector, exciter) array using a perturbational technique. The results of the initial and the optimized (perturbed) array are shown in Table 10.1. For the same array, they allowed all the spacings to vary while maintaining constant all other parameters. The results are shown in Table 10.2.

Another optimization procedure is to maintain the spacings between all the elements constant and vary the lengths so as to optimize the directivity. The results of a six-element array [29] are shown in Table 10.3. The ultimate optimization is to vary both the spacings and lengths. This was accomplished by Chen and Cheng [29] whereby they first optimized the array by varying the spacing, while maintaining the lengths constant. This was followed, on the same array, with perturbations in the lengths while maintaining the optimized spacings constant. The results of this procedure are shown in Table 10.4 with the corresponding H-plane ($\theta = \pi/2, \phi$) far-field patterns shown in Figure 10.25.

TABLE 10.1 Directivity Optimization for Six-Element Yagi-Uda Array (Perturbation of *Director Spacings*), $l_1 = 0.51\lambda$, $l_2 = 0.50\lambda$, $l_3 = l_4 = l_5 = l_6 = 0.43\lambda$, $a = 0.003369\lambda$

	s_{21}/λ	s_{32}/λ	s_{43}/λ	s_{54}/λ	s_{65}/λ	Directivity (dB)
Initial array	0.250	0.310	0.310	0.310	0.310	11.21
Optimized array	0.250	0.336	0.398	0.310	0.407	12.87

(SOURCE: D. K. Cheng and C. A. Chen, "Optimum Spacings for Yagi-Uda Arrays," *IEEE Trans. Antennas Propag.*, Vol. AP-21, pp. 615–623, September 1973. © (1973) IEEE).

TABLE 10.2 Directivity Optimization for Six-Element Yagi-Uda Array (Perturbation of *All Element Spacings*), $l_1 = 0.51\lambda$, $l_2 = 0.50\lambda$, $l_3 = l_4 = l_5 = l_6 = 0.43\lambda$, $a = 0.003369\lambda$

	s_{21}/λ	s_{32}/λ	s_{43}/λ	s_{54}/λ	s_{65}/λ	Directivity (dB)
Initial array	0.280	0.310	0.310	0.310	0.310	10.92
Optimized array	0.250	0.352	0.355	0.354	0.373	12.89

(SOURCE: D. K. Cheng and C. A. Chen, "Optimum Spacings for Yagi-Uda Arrays," *IEEE Trans. Antennas Propag.*, Vol. AP-21, pp. 615–623, September 1973. © (1973) IEEE).

In all, improvements in directivity and front-to-back ratio are noted. The ideal optimization will be to allow the lengths and spacings to vary simultaneously. Such an optimization was not performed in [28] or [29], although it could have been done iteratively by repeating the procedure.

Another parameter that was investigated for the directivity-optimized Yagi-Uda antenna was the frequency bandwidth [30]. The results of such a procedure are shown in Figure 10.26. The antenna was a six-element array optimized at a center frequency f_0. The array was designed, using space perturbations on all the elements, to yield an optimum directivity at f_0. The geometrical parameters are listed in Table 10.2. The 3-dB bandwidth seems to be almost the same for the initial and the optimized arrays. The rapid decrease in the directivity of the initial and optimized arrays at frequencies higher than f_0 and nearly constant values below f_0 may be attributed to the structure of the antenna which can support a "traveling wave" at $f < f_0$ but not at $f > f_0$. It has thus been suggested that an increase in the bandwidth can be achieved if the geometrical dimensions of the antenna are chosen slightly smaller than the optimum.

E. Input Impedance and Matching Techniques

The input impedance of a Yagi-Uda array, measured at the center of the driven element, is usually small and it is strongly influenced by the spacing between the reflector and feed element. For a 13-element array using a resonant driven element, the measured input impedances are listed in Table 10.5 [22]. Some of these values are low for matching to a 50-, 78-, or 300-ohm transmission lines.

TABLE 10.3 Directivity Optimization for Six-Element Yagi-Uda Array (Perturbation of *All Element Lengths*), $s_{21} = 0.250\lambda$, $s_{32} = s_{43} = s_{54} = s_{65} = 0.310\lambda$, $a = 0.003369\lambda$

	l_1/λ	l_2/λ	l_3/λ	l_4/λ	l_5/λ	l_6/λ	Directivity (dB)
Initial array	0.510	0.490	0.430	0.430	0.430	0.430	10.93
Length-perturbed array	0.472	0.456	0.438	0.444	0.432	0.404	12.16

(SOURCE: C. A. Chen and D. K. Cheng, "Optimum Element Lengths for Yagi-Uda Arrays," *IEEE Trans. Antennas Propag.*, Vol. AP-23, pp. 8–15, January 1975. © (1975) IEEE).

TABLE 10.4 Directivity Optimization for Six-Element Yagi-Uda Array (Perturbation of Director Spacings and All Element Lengths), $a = 0.003369\lambda$

	l_1/λ	l_2/λ	l_3/λ	l_4/λ	l_5/λ	l_6/λ	s_{21}/λ	s_{32}/λ	s_{43}/λ	s_{54}/λ	s_{65}/λ	Directivity (dB)
Initial array	0.510	0.490	0.430	0.430	0.430	0.430	0.250	0.310	0.310	0.310	0.310	10.93
Array after spacing perturbation	0.510	0.490	0.430	0.430	0.430	0.430	0.250	0.289	0.406	0.323	0.422	12.83
Optimum array after spacing and length perturbation	0.472	0.452	0.436	0.430	0.434	0.430	0.250	0.289	0.406	0.323	0.422	13.41

(SOURCE: C. A. Chen and D. K. Cheng, "Optimum Element Lengths for Yagi-Uda Arrays," *IEEE Trans. Antennas Propag.*, Vol. AP-23, pp. 8–15, January 1975. © (1975) IEEE).

574 TRAVELING WAVE AND BROADBAND ANTENNAS

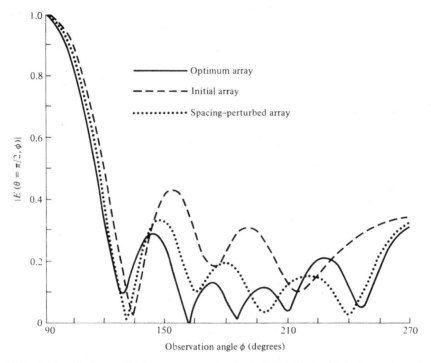

Figure 10.25 Normalized amplitude antenna patterns of initial, perturbed, and optimum six-element Yagi-Uda arrays (Table 10.4). (SOURCE: C. A. Chen and D. K. Cheng, "Optimum Element Lengths for Yagi-Uda Arrays," *IEEE Trans. Antennas Propagat.*, Vol. AP-23, pp. 8–15, January 1975. © (1975) IEEE).

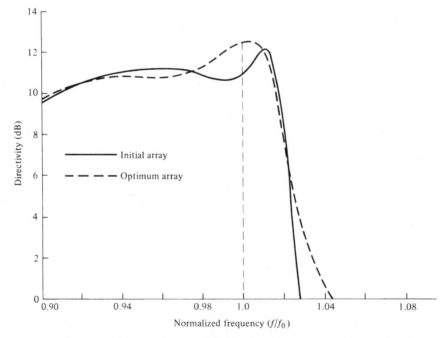

Figure 10.26 Bandwidth of initial and optimum six-element Yagi-Uda array with perturbation of all element spacings (Table 10.2). (SOURCE: N. K. Takla and L.-C. Shen, "Bandwidth of a Yagi Array with Optimum Directivity," *IEEE Trans. Antennas Propagat.*, Vol. AP-25, pp. 913–914, November 1977. © (1977) IEEE).

TABLE 10.5 Input Impedance of a 15-Element Yagi-Uda Array (Reflector Length = 0.5λ; Director Spacing = 0.34λ; Director Length = 0.406λ)

Reflector Spacing (s_{21}/λ)	Input Impedance (ohms)
0.25	62
0.18	50
0.15	32
0.13	22
0.10	12

There are many techniques that can be used to match a Yagi-Uda array to a transmission line and eventually to the receiver, which in many cases is a television set which has a large impedance (on the order of 300 ohms). Two common matching techniques are the use of the folded dipole, of Section 9.5, as a driven element and simultaneously as an impedance transformer, and the Gamma-match of Section 9.7.4. Which one of the two is used depends primarily on the transmission line from the antenna to the receiver.

The coaxial cable is now widely used as the primary transmission line for television, especially with the wide spread and use of cable TV; in fact, most television sets are already prewired with coaxial cable connections. Therefore, if the coax with a characteristic impedance of about 78 ohms is the transmission line used from the Yagi-Uda antenna to the receiver and since the input impedance of the antenna is typically 30–70 ohms (as illustrated in Table 10.5), the Gamma-match is the most prudent matching technique to use. This has been widely used in commercial designs where a clamp is usually employed to vary the position of the short to achieve a best match.

If, however, a "twin-lead" line with a characteristic impedance of about 300 ohms is used as the transmission line from the antenna to the receiver, as was used widely some years ago, then it would be most prudent to use a folded dipole as the driven element which acts as a step-up impedance transformer of about 4:1 (4:1) when the length of the element is exactly λ/2. This technique is also widely used in commercial designs.

Another way to explain the end-fire beam formation and whether the parameters of the Yagi-Uda array are properly adjusted for optimum directivity is by drawing a vector diagram of the progressive phase delay from element to element. If the current amplitudes throughout the array are equal, the total phase delay for maximum directivity should be about 180°, as is required by the Hansen-Woodyard criteria for improved end-fire radiation. Since the currents in a Yagi-Uda array are not equal in all the elements, the phase velocity of the traveling wave along the antenna structure is not the same from element-to-element but it is always slower than the velocity of light and faster than the corresponding velocity for a Hansen-Woodyard design. For a Yagi-Uda array, the decrease in the phase velocity is a function of the increase in total array length.

In general then, the phase velocity, and in turn the phase shift, of a traveling wave in a Yagi-Uda array structure is controlled by the geometrical dimensions of the array and its elements, and it is not uniform from element to element.

F. Design Procedure

A government document [31] has been published which provides extensive data of experimental investigations carried out by the National Bureau of Standards to determine how parasitic element diameter, element length, spacings between elements, supporting booms of different cross-sectional areas, various reflectors, and overall length affect the measured gain. Numerous graphical data is included to facilitate the design of different length antennas to yield maximum gain. In addition, design criteria are presented for stacking Yagi-Uda arrays either one above the other or side by side.

TABLE 10.6 Optimized Uncompensated Lengths of Parasitic Elements for Yagi-Uda Antennas of Six Different Lengths

$d/\lambda = 0.0085$ $s_{12} = 0.2\lambda$		Length of Yagi-Uda (in wavelengths)					
		0.4	0.8	1.20	2.2	3.2	4.2
LENGTH OF REFLECTOR (l_1/λ)		**0.482**	**0.482**	**0.482**	**0.482**	**0.482**	**0.475**
LENGTH OF DIRECTORS, λ	l_3	0.442	0.428	0.428	0.432	0.428	0.424
	l_4		0.424	0.420	0.415	0.420	0.424
	l_5			0.428	0.420	0.407	0.420
	l_6				0.428	0.398	0.407
	l_7				0.390	0.394	0.403
	l_8				0.390	0.390	0.398
	l_9				0.390	0.386	0.394
	l_{10}				0.390	0.386	0.390
	l_{11}				0.398	0.386	0.390
	l_{12}				0.407	0.386	0.390
	l_{13}					0.386	0.390
	l_{14}					0.386	0.390
	l_{15}					0.386	0.390
	l_{16}					0.386	
	l_{17}					0.386	
SPACING BETWEEN DIRECTORS (s_{ik}/λ)		0.20	0.20	0.25	0.20	0.20	0.308
DIRECTIVITY RELATIVE TO HALF-WAVE DIPOLE (dB)		7.1	9.2	10.2	12.25	13.4	14.2
DESIGN CURVE (SEE FIGURE 10.27)		(A)	(B)	(B)	(C)	(B)	(D)

(SOURCE: Peter P. Viezbicke, *Yagi Antenna Design*, NBS Technical Note 688, December 1976).

A step-by-step design procedure has been established in determining the geometrical parameters of a Yagi-Uda array for a desired directivity (over that of a $\lambda/2$ dipole mounted at the same height above ground). The included graphs can only be used to design arrays with overall lengths (*from reflector element to last director*) of 0.4, 0.8, 1.2, 2.2, 3.2, and 4.2λ with corresponding directivities of 7.1, 9.2, 10.2, 12.25, 13.4, and 14.2 dB, respectively, and with a diameter-to-wavelength ratio of $0.001 \leq d/\lambda \leq 0.04$. Although the graphs do not cover all possible designs, they do accommodate most practical requests. The driven element used to derive the data was a $\lambda/2$ folded dipole, and the measurements were carried out at $f = 400$ MHz. To make the reader aware of the procedure, it will be outlined by the use of an example. The procedure is identical for all other designs at frequencies where included data can accommodate the specifications.

The basis of the design is the data included in

1. Table 10.6 which represents optimized antenna parameters for six different lengths and for a $d/\lambda = 0.0085$
2. Figure 10.27 which represents *uncompensated* director and reflector lengths for $0.001 \leq d/\lambda \leq 0.04$
3. Figure 10.28 which provides compensation length increase for all the parasitic elements (directors and reflectors) as a function of boom-to-wavelength ratio $0.001 \leq D/\lambda \leq 0.04$

BROADBAND ANTENNAS 577

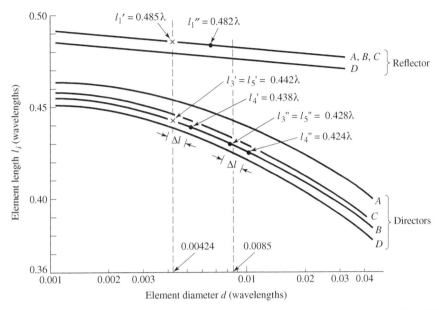

Figure 10.27 Design curves to determine element lengths of Yagi-Uda arrays. (SOURCE: P. P. Viezbicke, "Yagi Antenna Design," NBS Technical Note 688, U.S. Department of Commerce/National Bureau of Standards, December 1976).

The specified information is usually the center frequency, antenna directivity, d/λ and D/λ ratios, and it is required to find the optimum parasitic element lengths (directors and reflectors). The spacing between the directors is uniform but not the same for all designs. However, there is only one reflector and its spacing is $s = 0.2\lambda$ for all designs.

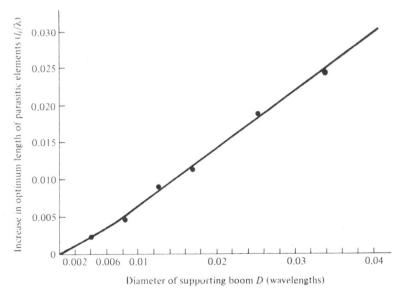

Figure 10.28 Increase in optimum length of parasitic elements as a function of metal boom diameter. (SOURCE: P. P. Viezbicke, "Yagi Antenna Design," NBS Technical Note 688, U.S. Department of Commerce/National Bureau of Standards, December 1976).

Example 10.3

Design a Yagi-Uda array with a directivity (relative to a $\lambda/2$ dipole at the same height above ground) of 9.2 dB at $f_0 = 50.1$ MHz. The desired diameter of the parasitic elements is 2.54 cm and of the metal supporting boom 5.1 cm. Find the element spacings, lengths, and total array length.

Solution:

a. At $f_0 = 50.1$ MHz the wavelength is $\lambda = 5.988$ m $= 598.8$ cm. Thus $d/\lambda = 2.54/598.8 = 4.24 \times 10^{-3}$ and $D/\lambda = 5.1/598.8 = 8.52 \times 10^{-3}$.

b. From Table 10.6, the desired array would have a total of five elements (three directors, one reflector, one feeder). For a $d/\lambda = 0.0085$ ratio the optimum uncompensated lengths would be those shown in the second column of Table 10.6 ($l_3 = l_5 = 0.428\lambda$, $l_4 = 0.424\lambda$, and $l_1 = 0.482\lambda$). The overall antenna length would be $L = (0.6 + 0.2)\lambda = 0.8\lambda$, the spacing between directors 0.2λ, and the reflector spacing 0.2λ. It is now desired to find the optimum lengths of the parasitic elements for a $d/\lambda = 0.00424$.

c. Plot the optimized lengths from Table 10.6 ($l_3'' = l_5'' = 0.428\lambda$, $l_4'' = 0.424\lambda$, and $l_1'' = 0.482\lambda$) on Figure 10.27 and mark them by a dot ($\cdot$).

d. In Figure 10.27 draw a vertical line through $d/\lambda = 0.00424$ intersecting curves (B) at director uncompensated lengths $l_3' = l_5' = 0.442\lambda$ and reflector length $l_1' = 0.485\lambda$. Mark these points by an x.

e. With a divider, measure the distance (Δl) along director curve (B) between points $l_3'' = l_5'' = 0.428\lambda$ and $l_4'' = 0.424\lambda$. Transpose this distance from the point $l_3' = l_5' = 0.442\lambda$ on curve (B), established in step (d) and marked by an x, downward along the curve and determine the uncompensated length $l_4' = 0.438\lambda$. Thus the boom uncompensated lengths of the array at $f_0 = 50.1$ MHz are

$$l_3' = l_5' = 0.442\lambda$$
$$l_4' = 0.438\lambda$$
$$l_1' = 0.485\lambda$$

f. Correct the element lengths to compensate for the boom diameter. From Figure 10.28, a boom diameter-to-wavelength ratio of 0.00852 requires a fractional length increase in each element of about 0.005λ. Thus the final lengths of the elements should be

$$l_3 = l_5 = (0.442 + 0.005)\lambda = 0.447\lambda$$
$$l_4 = (0.438 + 0.005)\lambda = 0.443\lambda$$
$$l_1 = (0.485 + 0.005)\lambda = 0.490\lambda$$

The design data were derived from measurements carried out on a nonconducting Plexiglas boom mounted 3λ above the ground. The driven element was a $\lambda/2$ folded dipole matched to a 50-ohm line by a double-stub tuner. All parasitic elements were constructed from aluminum tubing. Using Plexiglas booms, the data were repeatable and represented the same values as air-dielectric booms. However that was not the case for wooden booms because of differences in the moisture, which had a direct affect on the gain. Data on metal booms was also repeatable provided the element lengths were increased to compensate for the metal boom structure.

Figure 10.29 Commercial Yagi-Uda dipole TV array. (Courtesy: Winegard Company, Burlington, IA).

A commercial Yagi-Uda antenna is shown in Figure 10.29. It is a TV antenna designed primarily for channels 2–13. Its gain (over a dipole) ranges from 4.4 dB for channel 2 to 7.3 dB for channel 13, and it is designed for 300 ohms impedance.

10.3.4 Yagi-Uda Array of Loops

Aside from the dipole, the loop antenna is one of the most basic antenna elements. The pattern of a very small loop is similar to that of a very small dipole and in the far-field region it has a null along its axis. As the circumference of the loop increases, the radiation along its axis increases and reaches near maximum at about one wavelength [32]. Thus loops can be used as the basic elements, instead of the linear dipoles, to form a Yagi-Uda array as shown in Figure 10.30. By properly choosing the dimensions of the loops and their spacing, they can form a unidirectional beam along the axis of the loops and the array.

It has been shown that the radiation characteristics of a two-element loop array, one driven element and a parasitic reflector, resulted in the elimination of corona problems at high altitudes [33]. In addition, the radiation characteristics of loop arrays mounted above ground are less affected by the electrical properties of the soil, as compared with those of dipoles [34]. A two-element loop array also resulted in a 1.8 dB higher gain than a corresponding array of two dipoles [33]. A two-element

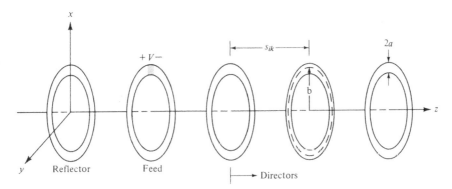

Figure 10.30 Yagi-Uda array of circular loops.

array of square loops (a feeder and a reflector) in a boxlike construction is called a "cubical quad" or simply a "quad" antenna, and it is very popular in amateur radio applications [35]. The sides of each square loop are $\lambda/4$ (perimeter of λ), and the loops are usually supported by a fiberglass or bamboo cross-arm assembly.

The general performance of a loop Yagi-Uda array is controlled by the same geometrical parameters (reflector, feeder, and director sizes, and spacing between elements), and it is influenced in the same manner as an array of dipoles [36]–[38].

In a numerical parametric study of coaxial Yagi-Uda arrays of circular loops [37] of 2 to 10 directors, it has been found that the optimum parameters for *maximum forward gain* were

1. circumference of feeder $2\pi b_2 \simeq 1.1\lambda$, where b_2 is its radius. This radius was chosen so that the input impedance for an isolated element is purely resistive.
2. circumference of the reflector $2\pi b_1 \simeq 1.05\lambda$, where b_1 is its radius. The size of the reflector does not strongly influence the forward gain but has a major effect on the backward gain and input impedance.
3. feeder–reflector spacing of about 0.1λ. Because it has negligible effect on the forward gain, it can be used to control the backward gain and/or the input impedance.
4. circumference of directors $2\pi b \simeq 0.7\lambda$, where b is the radius of any director and it was chosen to be the same for all. When the circumference approached a value of one wavelength, the array exhibited its cutoff properties.
5. spacing of directors of about 0.25λ, and it was uniform for all.

The radius a of all the elements was retained constant and was chosen to satisfy $\Omega = 2\ln(2\pi b_2/a) = 11$ where b_2 is the radius of the feeder.

While most of the Yagi-Uda designs have been implemented using dipoles, and some using loops, as the primary elements, there have been Yagi-Uda designs using slots [39] and microstrip patch elements [40]. The microstrip design was primarily configured for low-angle satellite reception for mobile communications [40].

10.4 MULTIMEDIA

In the publisher's website for this book, the following multimedia resources are included for the review, understanding, and visualization of the material of this chapter:

a. **Java**-based **interactive questionnaire**, with answers.
b. **Matlab** computer programs, designated
   - **Beverage**
   - **Helix**

   for computing and displaying the radiation characteristics of beverage and helical antennas.
c. **Matlab** and **Fortran** computer program, designated **Yagi_Uda**, for computing and displaying the radiation characteristics of a Yagi-Uda array design.
d. **Power Point (PPT)** viewgraphs, in multicolor.

REFERENCES

1. C. H. Walter, *Traveling Wave Antennas*, McGraw-Hill, New York, 1965.
2. J. D. Kraus, *Electromagnetics*, McGraw-Hill Book Co., New York, 1992.
3. J. G. Brainerd et al., *Ultra-High-Frequency Techniques*, Van Nostrand, New York, 1942.

4. L. V. Blake, *Antennas*, John Wiley and Sons, New York, 1966.
5. G. A. Thiele and E. P. Ekelman, Jr., "Design Formulas for Vee Dipoles," *IEEE Trans. Antennas Propagat.*, Vol. AP-28, No. 4, pp. 588–590, July 1980.
6. W. L. Weeks, *Antenna Engineering*, McGraw-Hill, New York, 1968, pp. 140–142.
7. D. G. Fink (ed.), *Electronics Engineers' Handbook*, Chapter 18 (by W. F. Croswell), McGraw-Hill, New York, 1975.
8. J. D. Kraus, *Antennas*, McGraw-Hill, New York, 1988.
9. A. A. de Carvallo, "On the Design of Some Rhombic Antenna Arrays," *IRE Trans. Antennas Propagat.*, Vol. AP-7, No. 1, pp. 39–46, January 1959.
10. E. Bruce, A. C. Beck, and L. R. Lowry, "Horizontal Rhombic Antennas," *Proc. IRE*, Vol. 23, pp. 24–26, January 1935.
11. R. S. Elliott, *Antenna Theory and Design*, Prentice-Hall, Englewood Cliffs, New Jersey, 1981.
12. J. D. Kraus, "A 50-Ohm Input Impedance for Helical Beam Antennas," *IEEE Trans. Antennas Propagat.*, Vol. AP-25, No. 6, p. 913, November 1977.
13. A. G. Kandoian, "Three New Antenna Types and Their Applications," *Proc. IRE*, Vol. 34, pp. 70W–75W, February 1946.
14. S. Uda, "Wireless Beam of Short Electric Waves," *J. IEE (Japan)*, pp. 273–282, March 1926, and pp. 1209–1219, November 1927.
15. H. Yagi, "Beam Transmission of Ultra Short Waves," *Proc. IRE*, Vol. 26, pp. 715–741, June 1928. Also *Proc. IEEE*, Vol. 72, No. 5, pp. 634–645, May 1984; *Proc. IEEE*, Vol. 85, No. 11, pp. 1864–1874, November 1997.
16. D. M. Pozar, "Beam Transmission of Ultra Short Waves: An Introduction to the Classic Paper by H. Yagi," *Proc. IEEE*, Vol. 85, No. 11, pp. 1857–1863, November 1997.
17. R. M. Fishender and E. R. Wiblin, "Design of Yagi Aerials," *Proc. IEE (London)*, pt. 3, Vol. 96, pp. 5–12, January 1949.
18. C. C. Lee and L.-C. Shen, "Coupled Yagi Arrays," *IEEE Trans. Antennas Propagat.*, Vol. AP-25, No. 6, pp. 889–891, November 1977.
19. H. W. Ehrenspeck and H. Poehler, "A New Method for Obtaining Maximum Gain from Yagi Antennas," *IRE Trans. Antennas Propagat.*, Vol. AP-7, pp. 379–386, October 1959.
20. H. E. Green, "Design Data for Short and Medium Length Yagi-Uda Arrays," *Elec. Engrg. Trans. Inst. Engrgs.* (Australia), pp. 1–8, March 1966.
21. W. Wilkinshaw, "Theoretical Treatment of Short Yagi Aerials," *Proc. IEE (London)*, pt. 3, Vol. 93, p. 598, 1946.
22. R. J. Mailloux, "The Long Yagi-Uda Array," *IEEE Trans. Antennas Propagat.*, Vol. AP-14, pp. 128–137, March 1966.
23. D. Kajfez, "Nonlinear Optimization Reduces the Sidelobes of Yagi Antennas," *IEEE Trans. Antennas Propagat.*, Vol. AP-21, No. 5, pp. 714–715, September 1973.
24. D. Kajfez, "Nonlinear Optimization Extends the Bandwidth of Yagi Antennas," *IEEE Trans. Antennas Propagat.*, Vol. AP-23, pp. 287–289, March 1975.
25. G. A. Thiele, "Analysis of Yagi-Uda Type Antennas," *IEEE Trans. Antennas Propagat.*, Vol. AP-17, No. 1, pp. 24–31, January 1969.
26. P. A. Tirkas, Private communication.
27. G. A. Thiele, "Calculation of the Current Distribution on a Thin Linear Antenna," *IEEE Trans. Antennas Propagat.*, Vol. AP-14, No. 5, pp. 648–649, September 1966.
28. D. K. Cheng and C. A. Chen, "Optimum Spacings for Yagi-Uda Arrays," *IEEE Trans. Antennas Propagat.*, Vol. AP-21, No. 5, pp. 615–623, September 1973.
29. C. A. Chen and D. K. Cheng, "Optimum Element Lengths for Yagi-Uda Arrays," *IEEE Trans. Antennas Propagat.*, Vol. AP-23, No. 1, pp. 8–15, January 1975.
30. N. K. Takla and L.-C. Shen, "Bandwidth of a Yagi Array with Optimum Directivity," *IEEE Trans. Antennas Propagat.*, Vol. AP-25, No. 6, pp. 913–914, November 1977.

31. P. P. Viezbicke, "Yagi Antenna Design," NBS Technical Note 688, U.S. Department of Commerce/National Bureau of Standards, December 1968.
32. S. Adachi and Y. Mushiake, "Studies of Large Circular Loop Antennas," *Sci. Rep. Research Institute of Tohoku University (RITU), B*, Vol. 9, No. 2, pp. 79–103, 1957.
33. J. E. Lindsay, Jr., "A Parasitic End-Fire Array of Circular Loop Elements," *IEEE Trans. Antennas Propagat.*, Vol. AP-15, No. 5, pp. 697–698, September 1967,
34. E. Ledinegg, W. Paponsek, and H. L. Brueckmann, "Low-Frequency Loop Antenna Arrays: Ground Reaction and Mutual Interaction," *IEEE Trans. Antennas Propagat.*, Vol. AP-21, No. 1, pp. 1–8, January 1973.
35. D. DeMaw (ed.), *The Radio Amateur's Handbook*, American Radio Relay League, Newington, CT, 56th ed., 1979.
36. A. Shoamanesh and L. Shafai, "Properties of Coaxial Yagi Loop Arrays," *IEEE Trans. Antennas Propagat.*, Vol. AP-26, No. 4, pp. 547–550, July 1978.
37. L. C. Shen and G. W. Raffoul, "Optimum Design of Yagi Array of Loops," *IEEE Trans. Antennas Propagat.*, Vol. AP-22, No. 11, pp. 829–830, November 1974.
38. A. Shoamanesh and L. Shafai, "Design Data for Coaxial Yagi Array of Circular Loops," *IEEE Trans. Antennas Propagat.*, Vol. AP-27, No. 5, pp. 711–713, September 1979.
39. R. J. Coe and G. Held, "A Parasitic Slot Array," *IEEE Trans. Antennas Propagat.*, Vol. AP-12, No. 1, pp. 10–16, January 1964.
40. J. Huang and A. C. Densmore, "Microstrip Yagi Antenna for Mobile Satellite Vehicle Application," *IEEE Trans. Antennas Propagat.*, Vol. AP-39, No. 6, pp. 1024–1030, July 1991.

PROBLEMS

10.1. Given the current distribution of (10-1a), show that the
   (a) far-zone electric field intensity is given by (10-2a) and (10-2b)
   (b) average power density is given by (10-4) and (10-5)
   (c) radiated power is given by (10-11)

10.2. Determine the phase velocity (compared to free-space) of the wave on a Beverage antenna (terminated long wire) of length $l = 50\lambda$ so that the maximum occurs at angles of
   (a) 10°   (b) 20°

   from the axis of the wire.

10.3. The current distribution on a terminated and matched long linear (traveling wave) antenna of length l, positioned along the x-axis and fed at its one end, is given by

$$\mathbf{I} = \hat{\mathbf{a}}_x I_0 e^{-jkx'}, \quad 0 \leq x' \leq l$$

   Find the far field electric and magnetic field components in *standard spherical coordinates*.

10.4. A long linear (traveling wave) antenna of length l, positioned along the z-axis and fed at the $z = 0$ end, is terminated in a load at the $z = l$ end. There is a nonzero reflection at the load such that the current distribution on the wire is given by

$$I(z) = I_0 \frac{e^{-jkz} + Re^{jkz}}{1 + R}, \quad 0 \leq z \leq l$$

   Determine as a function of R and l the
   (a) far-zone spherical electric-field components
   (b) radiation intensity in the $\theta = \pi/2$ direction

10.5. Design a Beverage antenna so that the first maximum occurs at 10° from its axis. Assuming the phase velocity of the wave on the line is the same as that of free-space, find the
  (a) lengths (exact and approximate) to accomplish that
  (b) angles (exact and approximate) where the next six maxima occur
  (c) angles (exact and approximate) where the nulls, between the maxima found in parts (a) and (b), occur
  (d) radiation resistance using the exact and approximate lengths
  (e) directivity using the exact and approximate lengths

  Verify using the computer program **Beverage**.

10.6. It is desired to place the first maximum of a long wire traveling wave antenna at an angle of 25° from the axis of the wire. For the wire antenna, find the
  (a) exact required length   (b) radiation resistance   (c) directivity (*in dB*)

  The wire is radiating into free space.

10.7. Compute the directivity of a long wire with lengths of $l = 2\lambda$ and 3λ.

  Verify using the computer program **Beverage**.

10.8. A long wire of diameter d is placed (in the air) at a height h above the ground.
  (a) Find its characteristic impedance assuming $h \gg d$.
  (b) Compare this value with (10-14).

10.9. A *long resonant* wire is placed vertically so that the axis of the wire is pointed along zenith ($\theta = 0°$). Design the wire so that its radiation resistance is *138.843 ohms*. Determine the *approximate*:
  (a) Length of the wire (*in odd number of* $\lambda/2$)
  (b) Angle (*in degrees*), measured from zenith, where the first maximum of the amplitude pattern will occur.
  (c) Maximum directivity (*in dB*).

10.10. Beverage (long-wire) antennas are used for over-the-horizon communication where the maximum of the main beam is pointed few degrees above the horizon. Assuming the wire antenna of length l and radius $\lambda/200$ is placed horizontally parallel to the z-axis a height $h = \lambda/20$ above a flat, perfect electric conducting plane of infinite extent (x-axis is perpendicular to the ground plane).
  (a) Derive the array factor for the equivalent two-element array.
  (b) What is the normalized total electric field of the wire in the presence of the conducting plane?
  (c) What value of load resistance should be placed at the terminating end to eliminate any reflections and not create a standing wave?

10.11. It is desired to design a very long resonant (standing wave) wire for over-the-horizon communication system, with a length equal to *odd multiple of a half of a wavelength*, so that its first maximum is 10° from the axis of the wire. To meet the design requirements, determine the resonant (standing wave) antenna's
  (a) approximate length (*in* λ).   (b) radiation resistance.   (c) directivity (*in dB*).

  To make the resonant (standing wave) antenna a *traveling wave antenna*, it will be terminated with a load resistance. If the wire antenna's diameter is $\lambda/400$ and it is placed horizontally a height $\lambda/20$ above an infinite perfectly conducting flat ground plane,
  (d) what should the load resistance be to accomplish this?

10.12. Compute the optimum directivities of a V antenna with leg lengths of $l = 2\lambda$ and $l = 3\lambda$. Compare these values with those of Problem 10.7.

10.13. Design a symmetrical V antenna so that its optimum directivity is 8 dB. Find the lengths of each leg (in λ) and the total included angle of the V (in degrees).

10.14. Repeat the design of Problem 10.13 for an optimum directivity of 5 dB.

10.15. It is desired to design a V-dipole with a maximized directivity. The length of each arm is 0.5λ (overall length of the entire V-dipole is λ). To meet the requirements of the design, what is the
(a) total included angle of the V-dipole (in degrees)?
(b) directivity (in dB)?
(c) gain (in dB) if the overall antenna efficiency is 35%?

10.16. It is desired to design a V-dipole with optimum dimensions so at to maximize its directivity. The desired maximum directivity is 5.257 dB. Determine the:
(a) Length (*in* λ) of each arm of the dipole.
(b) Included angle (*in degrees*) of the V-dipole.
(c) Approximate directivity (*In dB*) of a regular dipole (*with an included angle of* 180°) of the same overall length as the V-dipole.
(d) Which of the dipole designs (V or 180°) exhibits the larger directivity and by how many dB?

10.17. Ten identical elements of V antennas are placed along the z-axis to form a *uniform broadside* array. Each element is designed to have a maximum directivity of 9 dB. Assuming each element is placed so that its maximum is also broadside ($\theta = 90°$) and the elements are spaced $\lambda/4$ apart, find the
(a) arm length of each V (in λ)
(b) included angle (in degrees) of each V
(c) approximate total directivity of the array (in dB).

10.18. Design a resonant 90° bent, $\lambda/4$ long, $0.25 \times 10^{-3}\lambda$ radius wire antenna placed above a ground plane. Find the
(a) height where the bent must be made (b) input resistance of the antenna
(c) VSWR when the antenna is connected to a 50-ohm line.

10.19. Design a five-turn helical antenna which at 400 MHz operates in the normal mode. The spacing between turns is $\lambda_0/50$. It is desired that the antenna possesses circular polarization. Determine the
(a) circumference of the helix (in λ_0 and in meters)
(b) length of a single turn (in λ_0 and in meters)
(c) overall length of the entire helix (in λ_0 and in meters)
(d) pitch angle (in degrees).

10.20. A helical antenna of *4 turns* is operated in the *normal mode* at a frequency of 880 MHz and is used as an antenna for a wireless cellular telephone. The length L of the helical antenna is 5.7 cm and the diameter of each turn is 0.5 cm. Determine the:
(a) Spacing S (*in* λ_0) between the turns.
(b) Length L_0 (*in* λ_0) of each turn.
(c) Overall length L_n (*in* λ_0) of entire helix.

(d) Axial ratio of the helix (dimensionless).
(e) On the basis of the answer in Part *d*, the primary dominant component (E_θ or E_ϕ) of the far-zone field radiated by the helix.
(f) Primary polarization of the helix (*vertical* or *horizontal*) and why? Does the antenna primarily radiate as a linear vertical wire antenna or as a horizontal loop? Why? *Explain*.
(g) Radiation resistance of the helical antenna assuming that it can be determined using

$$R_r \approx 640 \left(\frac{L}{\lambda_0}\right)^2$$

(h) Radiation resistance of a single straight wire monopole of length *L* *(the same L as that of the helix)* mounted above an infinite ground plane.
(i) On the basis of the values of Parts *g* and *h*, what can you say about which antenna is preferable to be used as an antenna for a cellular telephone and why? Explain.

10.21. Helical antennas are often used for space applications where the polarization of the other antenna is not necessarily constant at all times. To assure uninterrupted communication, the helical antenna is designed to produce nearly circular polarization. For a 10-turn helix and axial mode operation, determine the following:
(a) radius (in λ_0), pitch angle (in degrees), and separation between turns (in λ) of the helix *for optimum operation*.
(b) half-power beamwidth (in degrees).
(c) directivity (in dB).
(d) axial ratio (dimensionless).
(e) minimum loss (in dB) of the received signal if the other antenna of the communication system is linearly polarized:
  1. in the vertical direction;
  2. at a 45° tilt relative to the vertical.

10.22. Design a nine-turn helical antenna operating in the axial mode so that the input impedance is about 110 ohms. The required directivity is 10 dB (above isotropic). For the helix, determine the approximate:
(a) circumference (*in λ_0*).   (b) spacing between the turns (*in λ_0*).
(c) half-power beamwidth (*in degrees*).

Assuming a symmetrical azimuthal pattern:
(d) determine the directivity (*in dB*) using *Kraus' formula*. Compare with the desired.

10.23. Design a helical antenna with *optimum* dimensions, maximum 17.16 dB (above isotropic) directivity for *optimized* end-fire radiation, and circular polarization. To achieve the desired specifications, determine the:
(a) Optimized pitch angle (*in degrees*), circumference (*in λ_o*) and spacing between turns (*in λ_o*)
(b) Nearest integer number of turns and corresponding axial ratio (*dimemionless*).
(c) Normalized phase velocity (*only one of them; the appropriate one*) of the wave as it travels on the wire of the helix.
(d) Assuming the helix can be approximated by the equivalent model of Figure 10.15(b), does the design meet the necessary Hansen-Woodyard (H-W) spacing between turns

condition for maximum end-fire radiation? Justify it by indicating whether the spacing between the turns meets or does not meet the desired conditions for optimized end-fire radiation. State the *necessary* spacing (in λ_o) for H-W conditions.

10.24. Design a helical antenna which operates in the axial (end-fire) mode with an axial ratio (AR) of 1.1. For *optimum design*, determine the:
(a) Number of turns (N), circumference (*in* λ) and α (*in degrees*).
(b) Half-power beamwidth (*in degrees*).
(c) Directivity (*dimensionless* and *in dB*).
(d) Directivity (*dimensionless* and *in dB*) using *another simple, closed form expression which is a function of the HPBW. Write the expression for this other formula and give its name.* Assume the amplitude pattern of the helical antenna operating in the axial (end-fire) mode is rotationally symmetric around the z-axis; not a function of ϕ when referring to Figure 10.14(b).
(e) Directivity (*dimensionless* and *in dB*) using *another simple, closed form expression which is a function of the HPBW; aside from those used in Parts d and e. Write the expression for this other formula and give its name.* Assume the amplitude pattern of the helical antenna operating in the axial (end-fire) mode is rotationally symmetric around the z-axis; not a function of ϕ when referring to Figure 10.14(b).
(f) Of the directivities found in parts *d* and *e*, which one agrees closer with the one in part *c*, and why?

10.25. It is desired to design an *optimum end-fire helical antenna* radiating in the axial mode at 100 MHz whose polarization axial ratio is 1.1. Determine the:
(a) directivity (*dimensionless* and *in dB*).
(b) half-power beamwidth (*in degrees*).
(c) input impedance.
(d) VSWR when connected to a 50-ohm line.
(e) wave velocity of the wave traveling along the helix for an ordinary end-fire radiation (*in m/sec*).

10.26. Design an *optimum configuration* of a helical antenna to operate in the *axial/end-fire* mode at a frequency of 500 MHz. The helix should have *10 turns* and operates in the *end-fire mode* with the *largest potential directivity*.
(a) *State the name of the end-fire design* that will accomplish the required specifications.
(b) Give the dimensions of the circumference (*in* λ_o), pitch angle (*in degrees*) and spacing between turns (*in* λ_o).
(c) Approximate side lobe level (*in dB*) of the *array factor part only* of the overall pattern.
(d) Phase velocity of wave along the helical wire (*in m/sec*).
(e) Half-power beamwidth (*in degrees*).
(f) Directivity (*in dB*) using *two different equations/methods* (i.e., two different values of directivity). *State the equations/methods you are using to do the calculation.*

10.27. Design a five-turn helical antenna which at 300 MHz operates in the axial mode and possesses circular polarization in the major lobe. Determine the
(a) near optimum circumference (in λ_0 and in meters)
(b) spacing (in λ_0 and in meters) for near optimum pitch angle design
(c) input impedance

(d) half-power beamwidth (in degrees), first-null beamwidth (in degrees), directivity (dimensionless and in dB), and axial ratio

(e) VSWR when the antenna is connected to 50- and 75-ohm coaxial lines

10.28. For civilian *L1* GPS applications (*f* = 1,575.42 MHz) it is desired to design and use an optimum configuration helical antenna with *right-hand circular polarization* and with an *Axial Ratio of 1.1*. The antenna should operate in the *end-fire/axial mode*. To accomplish this, *using optimum geometrical parameters,* determine the:

(a) Pitch angle (*in degrees*) and the number of turns.

(b) Circumference (*in* λ) and spacing between turns (*in* λ).

(c) Half-power beamwidth and first-null beamwidth (*both in degrees*).

(d) Directivity (*dimensionless and in dB*).

(e) Phase velocity (*in meters/sec*) the wave travels around the helix, *for both ordinary end-fire and Hansen-Woodyard designs.*

10.29. For Problem 10.27, plot the normalized polar amplitude pattern (in dB) assuming phasing for

(a) ordinary end-fire   (b) Hansen-Woodyard end-fire   (c) $p = 1$

10.30. For Problem 10.27, compute the directivity (in dB) using the computer program **Directivity** of Chapter 2 assuming phasing for

(a) ordinary end-fire   (b) Hansen-Woodyard end-fire   (c) $p = 1$

Compare with the value obtained using (10-33).

10.31. Repeat the design of Problem 10.27 at a frequency of 500 MHz.

10.32. Design an end-fire right-hand circularly polarized helix having a half-power beamwidth of 45°, pitch angle of 13°, and a circumference of 60 cm at a frequency of 500 MHz. Determine the

(a) turns needed   (b) directivity (in dB)   (c) axial ratio

(d) lower and upper frequencies of the bandwidth over which the required parameters remain relatively constant

(e) input impedance at the center frequency and the edges of the band from part (d).

10.33. An end-fire, *20-turn helical antenna, designed for increased directivity,* is used as a receiving antenna for a satellite communication system operating at *1 GHz*. The desired intercepted power to be delivered to the receiver is *1 watt* based on a *maximum incident power density of 1 mwatt/cm² of a linearly-polarized incident wave.* Assuming no other losses, determine the helix's:

(a) Axial ratio (*dimensionless*)

(b) Approximate directivity (*dimensionless* and *in dB*)

(c) Approximate gain (*dimensionless* and *in dB*)

(d) circumference (*in* λ_o), pitch angle (*in degrees*), spacing between turns (*in* λ_o), and length of each turn (*in* λ_o).

(e) normalized velocity, compared to that of free space.

10.34. Design a helical antenna with a directivity of 15 dB that is operating in the axial mode and whose polarization is nearly circular. The spacing between the turns is $\lambda_0/10$. Determine the

(a) Number of turns.

(b) Axial ratio, *both as an dimensionless quantity* and *in dB*.

(c) Directivity (in dB) based on *Kraus' formula* and *Tai & Pereira's formula*. How do they compare with the desired value?

(d) Progressive phase shifts (in degrees) between the turns to achieve the axial mode radiation.

10.35. Design a 10-turn helical antenna so that at the center frequency of 10 GHz, the circumference of each turn is $0.95\lambda_0$. Assuming a pitch angle of 14°, determine the

(a) mode in which the antenna operates (b) half-power beamwidth (in degrees)

(c) directivity (in dB). Compare this answer with what you get using *Kraus' formula*.

10.36. A *lossless* 10-turn helical antenna with a circumference of *one wavelength* is connected to a 78-ohm coaxial line, and it is used as a transmitting antenna in a 500-MHz spacecraft communication system. The spacing between turns is $\lambda_0/10$. The power in the coaxial line from the transmitter is 5 watts. Assuming the antenna is lossless:

(a) What is radiated power?

(b) If the antenna were isotropic, what would the power density (watts/m^2) be at a distance of 10 kilometers?

(c) What is the power density (in watts/m^2) at the same distance when the transmitting antenna is the 10-turn helix and the observations are made along the maximum of the major lobe?

(d) If at 10 kilometers along the maximum of the major lobe an identical 10-turn helix was placed as a receiving antenna, which was polarization-matched to the incoming wave, what is the maximum power (in watts) that can be received?

10.37. A 20-turn helical antenna operating in the axial mode is used as a transmitting antenna in a 500-MHz long distance communication system. The receiving antenna is a linearly polarized element. Because the transmitting and receiving elements are not matched in polarization, approximately how many dB of losses are introduced because of polarization mismatch?

10.38. A Yagi-Uda array of linear elements is used as a TV antenna receiving primarily channel 8 whose center frequency is 183 MHz. With a regular resonant $\lambda/2$ dipole as the feed element in the array, the input impedance is approximately 68 ohms. The antenna is to be connected to the TV using a "twin-lead" line with a characteristic impedance of nearly 300 ohms. At the center frequency of 183 MHz

(a) what should the smallest input impedance of the array be if it is desired to maintain a VSWR equal or less than 1.1?

(b) what is the best way to modify the present feed to meet the desired VSWR specifications? Be very specific in explaining why your recommendation will accomplish this.

10.39. The input impedance of a Yagi-Uda array design for TV reception using a folded dipole as the feed element is $300 + j25$. The antenna will be connected to the TV receiver with a lossless "twin-lead" line with a characteristic impedance of 300 ohms. To *eliminate the imaginary part* of the input impedance, a *shorted transmission line* with a characteristic impedance of 300 ohms will be connected *in parallel* to the antenna *at the feed points of the folded feed dipole element*.

(a) Determine the *reactance of the shorted transmission line*, which will be connected in parallel to the folded dipole at its feed, that will be required to accomplish this.

(b) State whether the reactance of the shorted transmission line *in part a* is *inductive or capacitive*.

(c) What is the *shortest length* (in λ) of the shorted transmission line?

(d) What is the VSWR in the "twin-lead" line from the antenna to the TV receiver?

10.40. Evaluate approximately the effect of the spacing between the director and driven element in the three-element Yagi-Uda array shown in the accompanying figure. Assume that the far-zone (radiated) field of the antenna is given by

$$E_\theta = \sin\theta[1 - e^{-j\pi/8}e^{-jks_{12}\sin\theta\sin\phi} - e^{j\pi/8}e^{+jks_{23}\sin\theta\sin\phi}]$$

where s_{12} is the spacing between the reflector and the driven element, and s_{23} is the spacing between the director and the driven element. For this problem, set $s_{12} = 0.2\lambda$ and let $s_{23} = 0.15\lambda, 0.20\lambda, 0.25\lambda$.

(a) Generate polar plots of the radiation power patterns in both *E*- and *H*-planes. Normalize the power pattern to its value for $\theta = \pi/2, \phi = \pi/2$. Generate two plots, one for *E*-plane and one for *H*-plane.

(b) Compute the front-to-back ratio (FBR) in the *E*-plane given by

$$\text{FBR}|_{E-plane} = \frac{P_n\left(\theta = \frac{\pi}{2}, \phi = \frac{\pi}{2}\right)}{P_n\left(\theta = \frac{\pi}{2}, \phi = \frac{-\pi}{2}\right)}$$

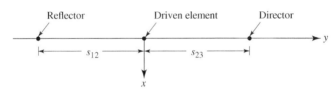

Leave your answers for both parts in terms of numbers, not dB.

10.41. Analyze a 27-element Yagi-Uda array, using the computer program **Yagi_Uda** of this chapter, having the following specifications:

N = total number of elements	= 27
Number of directors	= 25
Number of reflectors	= 1
Total length of reflector	= 0.5λ
Total length of feeder	= 0.47λ
Total length of each director	= 0.406λ
Spacing between reflector and feeder	= 0.125λ
Spacing between adjacent directors	= 0.34λ
a = radius of wires	= 0.003λ

Use 8 modes for each element. Compute the

(a) far-field *E*- and *H*-plane amplitude patterns (in dB).

(b) directivity of the array (in dB).

(c) *E*-plane half-power beamwidth (in degrees).

(d) *H*-plane half-power beamwidth (in degrees).

(e) *E*-plane front-to-back ratio (in dB).

(f) *H*-plane front to back ratio (in dB).

10.42. Repeat the analysis of Problem 10.41 for a three-element array having the following specifications:

N = total number of elements	= 3
Number of directors	= 1
Number of reflectors	= 1
Total length of reflector	= 0.5λ
Total length of feeder	= 0.475λ
Total length of director	= 0.45λ
Spacing between reflector and feeder	= 0.2λ
Spacing between feeder and director	= 0.16λ
a = radius of wires	= 0.005λ

Use 8 modes for each element.

10.43. Design a Yagi-Uda array of linear dipoles to cover all the VHF TV channels (starting with 54 MHz for channel 2 and ending with 216 MHz for channel 13. See Appendix IX). Perform the design at f_0 = 216 MHz. Since the gain is not affected appreciably at $f < f_0$, as Figure 10.26 indicates, this design should accommodate all frequencies below 216 MHz. The gain of the antenna should be 14.4 dB (above isotropic). The elements and the supporting boom should be made of aluminum tubing with outside diameters of $\frac{3}{8}$in.($\simeq$0.95 cm) and $\frac{3}{4}$in.($\simeq$1.90 cm), respectively. Find the number of elements, their lengths and spacings, and the total length of the array (in λ, meters, and feet).

10.44. Repeat the design of Problem 10.43 for each of the following:
(a) VHF-TV channels 2–6 (54–88 MHz. See Appendix IX)
(b) VHF-TV channels 7–13 (174–216 MHz. See Appendix IX)

10.45. Design a Yagi-Uda antenna to cover the entire FM band of 88–108 MHz (100 channels spaced at 200 KHz apart. See Appendix IX). The desired gain is 12.35 dB (above isotropic). Perform the design at f_0 = 108 MHz. The elements and the supporting boom should be made of aluminum tubing with outside diameters of $\frac{3}{8}$in.($\simeq$0.95 cm) and $\frac{3}{4}$in.($\simeq$1.90 cm), respectively. Find the number of elements, their lengths and spacings, and the total length of the array (in λ, meters, and feet).

10.46. Design a Yagi-Uda antenna to cover the UHF TV channels (512–806 MHz. See Appendix IX). The desired gain is 12.35 dB (above isotropic). Perform the design at f_0 = 806 MHz. The elements and the supporting boom should be made of wire with outside diameters of $\frac{3}{32}$ in.($\simeq$0.2375 cm) and $\frac{3}{16}$ in.($\simeq$0.475 cm), respectively. Find the number of elements, their lengths and spacings, and the total length of the array (in λ, meters, and feet).

CHAPTER 11

Frequency Independent Antennas, Antenna Miniaturization, and Fractal Antennas

11.1 INTRODUCTION

The numerous applications of electromagnetics to the advances of technology have necessitated the exploration and utilization of most of the electromagnetic spectrum. In addition, the advent of broadband systems have demanded the design of broadband radiators. The use of simple, small, lightweight, and economical antennas, designed to operate over the entire frequency band of a given system, would be most desirable. Although in practice all the desired features and benefits cannot usually be derived from a single radiator, most can effectively be accommodated. Previous to the 1950s, antennas with broadband pattern and impedance characteristics had bandwidths not greater than about 2:1. In the 1950s, a breakthrough in antenna evolution was made which extended the bandwidth to as great as 40:1 or more. The antennas introduced by the breakthrough were referred to as *frequency independent*, and they had geometries that were specified by angles. These antennas are primarily used in the 10–10,000 MHz region in a variety of practical applications such as TV, point-to-point communication, feeds for reflectors and lenses, and so forth.

In antenna scale modeling, characteristics such as impedance, pattern, polarization, and so forth, are invariant to a change of the physical size if a similar change is also made in the operating frequency or wavelength. For example, if *all* the physical dimensions are *reduced* by a factor of two, the performance of the antenna will remain unchanged if the operating frequency is *increased* by a factor of two. In other words, the performance is invariant if the electrical dimensions remain unchanged. This is the principle on which antenna scale model measurements are made. For a complete and thorough discussion of scaling, the reader is referred to Section 17.10 entitled "Scale Model Measurements."

The scaling characteristics of antenna model measurements also indicate that if the shape of the antenna were completely specified by angles, its performance would have to be independent of frequency [1]. The infinite biconical dipole of Figure 9.1 is one such structure. To make infinite structures more practical, the designs usually require that the current on the structure decrease with distance away from the input terminals. After a certain point the current is negligible, and the structure beyond that point to infinity can be truncated and removed. Practically then the truncated antenna has a lower cutoff frequency above which it radiation characteristics are the same as those of the infinite structure. The lower cutoff frequency is that for which the current at the point of truncation becomes negligible. The upper cutoff is limited to frequencies for which the dimensions of the feed

Antenna Theory: Analysis and Design, Fourth Edition. Constantine A. Balanis.
© 2016 John Wiley & Sons, Inc. Published 2016 by John Wiley & Sons, Inc.
Companion Website: www.wiley.com/go/antennatheory4e

transmission line cease to look like a "point" (usually about $\lambda_2/8$ where λ_2 is the wavelength at the highest desirable frequency). Practical bandwidths are on the order of about 40:1. Even higher ratios (i.e., 1,000:1) can be achieved in antenna design but they are not necessary, since they would far exceed the bandwidths of receivers and transmitters.

Even though the shape of the biconical antenna can be completely specified by angles, the current on its structure does not diminish with distance away from the input terminals, and its pattern does not have a limiting form with frequency. This can be seen by examining the current distribution as given by (9-11). It is evident that there are phase but no amplitude variations with the radial distance r. Thus the biconical structure cannot be truncated to form a frequency independent antenna. In practice, however, antenna shapes exist which satisfy the general shape equation, as proposed by Rumsey [1], to have frequency independent characteristics in pattern, impedance, polarization, and so forth, and with current distribution which diminishes rapidly.

Rumsey's general equation will first be developed, and it will be used as the unifying concept to link the major forms of frequency independent antennas. Classical shapes of such antennas include the equiangular geometries of planar and conical spiral structures [2]–[4], and the logarithmically periodic structures [5], [6].

Fundamental limitations in electrically small antennas will be discussed in Section 11.5. These will be derived using spherical mode theory, with the antenna enclosed in a virtual sphere. Minimum Q curves, which place limits on the achievable bandwidth, will be included. Fractal antennas, discussed in Section 11.5, is one class whose design is based on this fundamental principle.

11.2 THEORY

The analytical treatment of frequency independent antennas presented here parallels that introduced by Rumsey [1] and simplified by Elliott [7] for three-dimensional configurations.

We begin by assuming that an antenna, whose geometry is best described by the spherical coordinates (r, θ, ϕ), has both terminals infinitely close to the origin and each is symmetrically disposed along the $\theta = 0, \pi$-axes. It is assumed that the antenna is perfectly conducting, it is surrounded by an infinite homogeneous and isotropic medium, and its surface or an edge on its surface is described by a curve

$$r = F(\theta, \phi) \qquad (11\text{-}1)$$

where r represents the distance along the surface or edge. If the antenna is to be scaled to a frequency that is K times lower than the original frequency, the antenna's physical surface must be made K times greater to maintain the same electrical dimensions. Thus the new surface is described by

$$r' = KF(\theta, \phi) \qquad (11\text{-}2)$$

The new and old surfaces are identical; that is, not only are they similar but they are also congruent (if both surfaces are infinite). Congruence can be established only by rotation in ϕ. Translation is not allowed because the terminals of both surfaces are at the origin. Rotation in θ is prohibited because both terminals are symmetrically disposed along the $\theta = 0, \pi$-axes.

For the second antenna to achieve congruence with the first, it must be rotated by an angle C so that

$$KF(\theta, \phi) = F(\theta, \phi + C) \qquad (11\text{-}3)$$

The angle of rotation C depends on K but neither depends on θ or ϕ. Physical congruence implies that the original antenna electrically would behave the same at both frequencies. However the radiation pattern will be rotated azimuthally through an angle C. For unrestricted values of $K(0 \leq K \leq \infty)$, the pattern will rotate by C in ϕ with frequency, because C depends on K but its shape will be unaltered. Thus the impedance and pattern will be frequency independent.

To obtain the functional representation of $F(\theta, \phi)$, both sides of (11-3) are differentiated with respect to C to yield

$$\frac{d}{dC}[KF(\theta, \phi)] = \frac{dK}{dC} F(\theta, \phi) = \frac{\partial}{\partial C}[F(\theta, \phi + C)] = \frac{\partial}{\partial(\phi + C)}[F(\theta, \phi + C)] \quad (11\text{-}4)$$

and with respect to ϕ to give

$$\frac{\partial}{\partial \phi}[KF(\theta, \phi)] = K \frac{\partial F(\theta, \phi)}{\partial \phi} = \frac{\partial}{\partial \phi}[F(\theta, \phi + C)] = \frac{\partial}{\partial(\phi + C)}[F(\theta, \phi + C)] \quad (11\text{-}5)$$

Equating (11-5) to (11-4) yields

$$\frac{dK}{dC} F(\theta, \phi) = K \frac{\partial F(\theta, \phi)}{\partial \phi} \quad (11\text{-}6)$$

Using (11-1) we can write (11-6) as

$$\frac{1}{K} \frac{dK}{dC} = \frac{1}{r} \frac{\partial r}{\partial \phi} \quad (11\text{-}7)$$

Since the left side of (11-7) is independent of θ and ϕ, a general solution for the surface $r = F(\theta, \phi)$ of the antenna is

$$r = F(\theta, \phi) = e^{a\phi} f(\theta) \quad (11\text{-}8)$$

$$\text{where} \quad a = \frac{1}{K} \frac{dK}{dC} \quad (11\text{-}8a)$$

and $f(\theta)$ is a completely arbitrary function.

Thus for any antenna to have frequency independent characteristics, its surface must be described by (11-8). This can be accomplished by specifying the function $f(\theta)$ or its derivatives. Subsequently, interesting, practical, and extremely useful antenna configurations will be introduced whose surfaces are described by (11-8).

11.3 EQUIANGULAR SPIRAL ANTENNAS

The equiangular spiral is one geometrical configuration whose surface can be described by angles. It thus fulfills all the requirements for shapes that can be used to design frequency independent antennas. Since a curve along its surface extends to infinity, it is necessary to designate the length of the arm to specify a finite size antenna. The lowest frequency of operation occurs when the total arm length is comparable to the wavelength [2]. For all frequencies above this, the pattern and impedance characteristics are frequency independent.

11.3.1 Planar Spiral

The shape of an equiangular plane spiral curve can be derived by letting $f(\theta)$ in (11-8) be

$$f(\theta) = A\delta\left(\frac{\pi}{2} - \theta\right) \tag{11-9}$$

where A is a constant and δ is the Dirac delta function. Using (11-9) reduces (11-8) to

$$r|_{\theta=\pi/2} = \rho = \begin{cases} Ae^{a\phi} = \rho_0 e^{a(\phi-\phi_0)} & \theta = \pi/2 \\ 0 & \text{elsewhere} \end{cases} \tag{11-10}$$

where

$$A = \rho_0 e^{-a\phi_0} \tag{11-10a}$$

In wavelengths, (11-10) can be written as

$$\rho_\lambda = \frac{\rho}{\lambda} = \frac{A}{\lambda}e^{a\phi} = Ae^{a[\phi - \ln(\lambda)/a]} = Ae^{a(\phi-\phi_1)} \tag{11-11}$$

where

$$\phi_1 = \frac{1}{a}\ln(\lambda) \tag{11-11a}$$

Another form of (11-10) is

$$\phi = \frac{1}{a}\ln\left(\frac{\rho}{A}\right) = \tan\psi \ln\left(\frac{\rho}{A}\right) = \tan\psi(\ln\rho - \ln A) \tag{11-12}$$

where $1/a$ is the rate of expansion of the spiral and ψ is the angle between the radial distance ρ and the tangent to the spiral, as shown in Figure 11.1(a).

It is evident from (11-11) that changing the wavelength is equivalent to varying ϕ_0 which results in nothing more than a pure rotation of the infinite structure pattern. Within limitations imposed by the arm length, similar characteristics have been observed for finite structures. The same result can be concluded by examining (11-12). Increasing the logarithm of the frequency ($\ln f$) by C_0 is equivalent to rotating the structure by $C_0 \tan \psi$. As a result, the pattern is merely rotated but otherwise unaltered. Thus we have frequency independent antennas.

The total length L of the spiral can be calculated by

$$L = \int_{\rho_0}^{\rho_1}\left[\rho^2\left(\frac{d\phi}{d\rho}\right)^2 + 1\right]^{1/2} d\rho \tag{11-13}$$

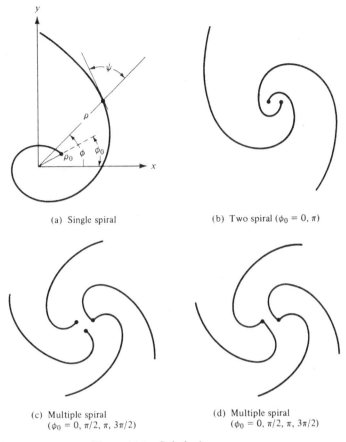

(a) Single spiral

(b) Two spiral ($\phi_0 = 0, \pi$)

(c) Multiple spiral
($\phi_0 = 0, \pi/2, \pi, 3\pi/2$)

(d) Multiple spiral
($\phi_0 = 0, \pi/2, \pi, 3\pi/2$)

Figure 11.1 Spiral wire antennas.

which reduces, using (11-10), to

$$L = (\rho_1 - \rho_0)\sqrt{1 + \frac{1}{a^2}} \tag{11-14}$$

where ρ_0 and ρ_1 represent the inner and outer radii of the spiral.

Various geometrical arrangements of the spiral have been used to form different antenna systems. If ϕ_0 in (11-10) is 0 and π, the spiral wire antenna takes the form of Figure 11.1(b). The arrangements of Figures 11.1(c) and 11.1(d) are each obtained when $\phi_0 = 0$, $\pi/2$, π, and $3\pi/2$. Numerous other combinations are possible.

An equiangular metallic solid surface, designated as P, can be created by defining the curves of its edges, using (11-10), as

$$\rho_2 = \rho_2' e^{a\phi} \tag{11-15a}$$

$$\rho_3 = \rho_3' e^{a\phi} = \rho_2' e^{a(\phi-\delta)} \tag{11-15b}$$

where

$$\rho_3' = \rho_2' e^{-a\delta} \tag{11-15c}$$

such that

$$K = \frac{\rho_3}{\rho_2} = e^{-a\delta} < 1 \qquad (11\text{-}16)$$

The two curves, which specify the edges of the conducting surface, are of identical relative shape with one magnified relative to the other or rotated by an angle δ with respect to the other. The magnification or rotation allows the arm of conductor P to have a finite width, as shown in Figure 11.2(a).

The metallic arm of a second conductor, designated as Q, can be defined by

$$\rho_4 = \rho_4' e^{a\phi} = \rho_2' e^{a(\phi-\pi)} \qquad (11\text{-}17)$$

where

$$\rho_4' = \rho_2' e^{-a\pi} \qquad (11\text{-}17\text{a})$$

$$\rho_5 = \rho_5' e^{a\phi} = \rho_4' e^{a(\phi-\delta)} = \rho_2' e^{a(\phi-\pi-\delta)} \qquad (11\text{-}18)$$

where

$$\rho_5' = \rho_4' e^{-a\delta} = \rho_2' e^{-a(\pi+\delta)} \qquad (11\text{-}18\text{a})$$

The system composed of the two conducting arms, P and Q, constitutes a balanced system, and it is shown in Figure 11.2(a). The finite size of the structure is specified by the fixed spiraling length L_0 along the centerline of the arm. The entire structure can be completely specified by the rotation angle δ, the arm length L_0, the rate of spiral $1/a$, and the terminal size ρ_2'. However, it has been found that most characteristics can be described adequately by only three; that is, L_0, ρ_2', and $K = e^{-a\delta}$ as given by (11-16). In addition each arm is usually tapered at its end, shown by dashed lines in Figure 11.2(a), to provide a better matching termination.

The previous analytical formulations can be used to describe two different antennas. One antenna would consist of two metallic arms suspended in free-space, as shown in Figure 11.2(a), and the other of a spiraling slot on a large conducting plane, as shown in Figure 11.2(b). The second is also usually tapered to provide better matching termination. The slot antenna is the most practical, because it can be conveniently fed by a balanced coaxial arrangement [2] to maintain its overall balancing. The antenna in Figure 11.2(a) with $\delta = \pi/2$ is self-complementary, as defined by Babinet's principle [8], and its input impedance for an infinite structure should be $Z_s = Z_c = 188.5 \simeq 60\pi$ ohms (for discussion of Babinet's Principle see Section 12.8). Experimentally, measured mean input

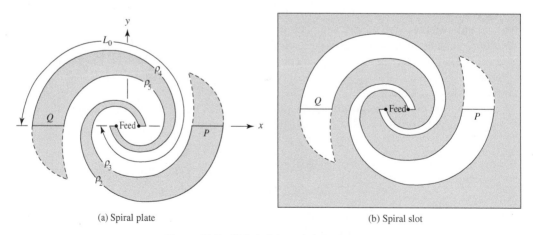

(a) Spiral plate (b) Spiral slot

Figure 11.2 Spiral plate and slot antennas.

impedances were found to be only about 164 ohms. The difference between theory and experiment is attributed to the finite arm length, finite thickness of the plate, and nonideal feeding conditions.

Spiral slot antennas, with good radiation characteristics, can be built with one-half to three turns. The most optimum design seems to be that with 1.25 to 1.5 turns with an overall length equal to or greater than one wavelength. The rate of expansion should not exceed about 10 per turn. The patterns are bidirectional, single lobed, broadside (maximum normal to the plane), and must vanish along the directions occupied by the infinite structure. The wave is circularly polarized near the axis of the main lobe over the usable part of the bandwidth. For a fixed cut, the beamwidth will vary with frequency since the pattern rotates. Typical variations are on the order of 10°. In general, however, slot antennas with more broad arms and/or more tightly wound spirals exhibit smoother and more uniform patterns with smaller variations in beamwidth with frequency. For symmetrical structures, the pattern is also symmetrical with no tilt to the lobe structure.

To maintain the symmetrical characteristics, the antenna must be fed by an electrically and geometrically balanced line. One method that achieves geometrical balancing requires that the coax is embedded into one of the arms of the spiral. To maintain symmetry, a dummy cable is usually placed into the other arm. No appreciable currents flow on the feed cables because of the rapid attenuation of the fields along the spiral. If the feed line is electrically unbalanced, a balun must be used. This limits the bandwidth of the system.

The polarization of the radiated wave is controlled by the length of the arms. For very low frequencies, such that the total arm length is small compared to the wavelength, the radiated field is linearly polarized. As the frequency increases, the wave becomes elliptically polarized and eventually achieves circular polarization. Since the pattern is essentially unaltered through this frequency range, the polarization change with frequency can be used as a convenient criterion to select the lower cutoff frequency of the usable bandwidth. In many practical cases, this is chosen to be the point where the axial ratio is equal or less than 2 to 1, and it occurs typically when the overall arm length is about one wavelength. A typical variation in axial ratio of the on-axis field as a function of frequency for a one-turn slot antenna is shown in Figure 11.3. The off-axis radiated field has nearly circular polarization over a smaller part of the bandwidth. In addition to the limitation imposed on the bandwidth by the overall length of the arms, another critical factor that can extend or reduce the bandwidth is the construction precision of the feed.

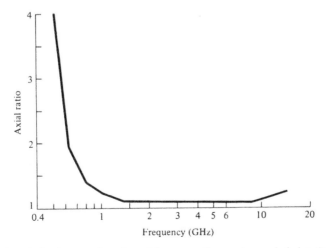

Figure 11.3   On-axis polarization as a function of frequency for one-turn spiral slot. (SOURCE: J. D. Dyson, "The Equiangular Spiral Antenna," *IRE Trans. Antennas Propagat.*, Vol. AP-7, pp. 181–187, April 1959. © (1959) IEEE).

The input impedance of a balanced equiangular slot antenna converges rapidly as the frequency is increased, and it remains reasonably constant for frequencies for which the arm length is greater than about one wavelength. Measured values for a 700–2,500 MHz antenna [2] were about 75–100 ohms with VSWR's of less than 2 to 1 for 50-ohm lines.

For slot antennas radiating in free-space, without dielectric material or cavity backing, typical measured efficiencies are about 98% for arm lengths equal to or greater than one wavelength. Rapid decreases are observed for shorter arms.

11.3.2 Conical Spiral

The shape of a nonplanar spiral can be described by defining the derivative of $f(\theta)$ to be

$$\frac{df}{d\theta} = f'(\theta) = A\delta(\beta - \theta) \tag{11-19}$$

in which β is allowed to take any value in the range $0 \leq \beta \leq \pi$. For a given value of β, (11-19) in conjunction with (11-8) describes a spiral wrapped on a conical surface. The edges of one conical spiral surface are defined by

$$r_2 = r_2' e^{(a \sin \theta_0)\phi} = r_2' e^{b\phi} \tag{11-20a}$$

$$r_3 = r_3' e^{a \sin \theta_0 \phi} = r_2' e^{a \sin \theta_0 (\phi - \delta)} \tag{11-20b}$$

where

$$r_3' = r_2' e^{-(a \sin \theta_0)\delta} \tag{11-20c}$$

and θ_0 is half of the total included cone angle. Larger values of θ_0 in $0 \leq \theta \leq \pi/2$ represent less tightly wound spirals. These equations correspond to (11-15a)–(11-15c) for the planar surface. The second arm of a balanced system can be defined by shifting each of (11-20a)–(11-20c) by 180°, as was done for the planar surface by (11-17)–(11-18a). A conical spiral metal strip antenna of elliptical polarization is shown in Figure 11.4.

The conducting conical spiral surface can be constructed conveniently by forming, using printed-circuit techniques, the conical arms on the dielectric cone which is also used as a support. The feed cable can be bonded to the metal arms which are wrapped around the cone. Symmetry can be preserved by observing the same precautions, like the use of a dummy cable, as was done for the planar surface.

A distinct difference between the planar and conical spirals is that the latter provides unidirectional radiation (single lobe) toward the apex of the cone with the maximum along the axis. Circular polarization and relatively constant impedances are preserved over large bandwidths. Smoother patterns have been observed for unidirectional designs. Conical spirals can be used in conjunction with a ground plane, with a reduction in bandwidth when they are flush mounted on the plane.

11.4 LOG-PERIODIC ANTENNAS

Another type of an antenna configuration, which closely parallels the frequency independent concept, is the log-periodic structure introduced by DuHamel and Isbell [4]. Because the entire shape of it cannot be solely specified by angles, it is not truly frequency independent.

Figure 11.4 Conical spiral metal strip antenna. (SOURCE: *Antennas, Antenna Masts and Mounting Adaptors*, American Electronic Laboratories, Inc., Lansdale, Pa., Catalog 7.5M-7-79. Courtesy of American Electronic Laboratories, Inc., Montgomeryville, PA 18936 USA).

11.4.1 Planar and Wire Surfaces

A planar log-periodic structure is shown in Figure 11.5(a). It consists of a metal strip whose edges are specified by the angle $\alpha/2$. However, in order to specify the length from the origin to any point on the structure, a distance characteristic must be included.

In spherical coordinates (r, θ, ϕ) the shape of the structure can be written as

$$\theta = \text{periodic function of } [b \ln(r)] \qquad (11\text{-}21)$$

An example of it would be

$$\theta = \theta_0 \sin\left[b \ln\left(\frac{r}{r_0}\right)\right] \qquad (11\text{-}22)$$

It is evident from (11-22) that the values of θ are repeated whenever the logarithm of the radial frequency $\ln(\omega) = \ln(2\pi f)$ differs by $2\pi/b$. The performance of the system is then periodic as a function of the logarithm of the frequency; thus the name *logarithmic-periodic* or *log-periodic*.

A typical log-periodic antenna configuration is shown in Figure 11.5(b). It consists of two coplanar arms of the Figure 11.5(a) geometry. The pattern is unidirectional toward the apex of the cone

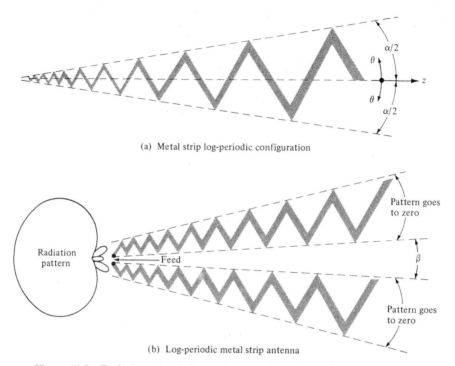

(a) Metal strip log-periodic configuration

(b) Log-periodic metal strip antenna

Figure 11.5 Typical metal strip log-periodic configuration and antenna structure.

formed by the two arms, and it is linearly polarized. Although the patterns of this and other log-periodic structures are not completely frequency independent, the amplitude variations of certain designs are very slight. Thus, practically, they are frequency independent.

Log-periodic wire antennas were introduced by DuHamel [5]. While investigating the current distribution on log-periodic surface structures of the form shown in Figure 11.6(a), he discovered that the fields on the conductors attenuated very sharply with distance. This suggested that perhaps

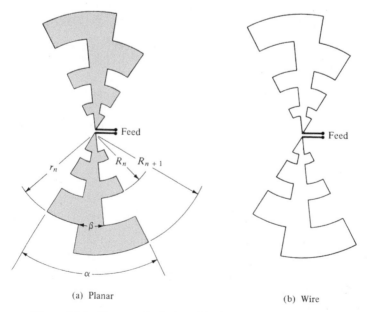

(a) Planar   (b) Wire

Figure 11.6 Planar and wire logarithmically periodic antennas.

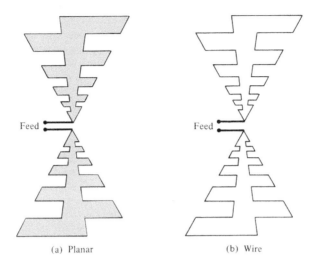

Figure 11.7 Planar and wire trapezoidal toothed log-periodic antennas.

there was a strong current concentration at or near the edges of the conductors. Thus, removing part of the inner surface to form a wire antenna as shown in Figure 11.6(b), should not seriously degrade the performance of the antenna. To verify this, a wire antenna, with geometrical shape identical to the pattern formed by the edges of the conducting surface, was built and it was investigated experimentally. As predicted, it was found that the performance of this antenna was almost identical to that of Figure 11.6(a); thus the discovery of a much simpler, lighter in weight, cheaper, and less wind resistant antenna. The fact that, for a planar bow-tie antenna, the removal of the inner surface of Figure 11.6(a), and also of Figure 11.7(a), does not impact significantly the radiation characteristics was also verified through simulations and measurements in [9]. The current distribution on the surface of a solid planar bow-tie is shown in Figure 9.8. As is seen in Figure 9.8, the most intense current density is along the perimeter edges; thus, the removal of the inner surface, to form an "outline" bow tie, did not affect significantly the performance of the bow-tie [9]. Nonplanar geometries in the form of a V, formed by bending one arm relative to the other, are also widely used.

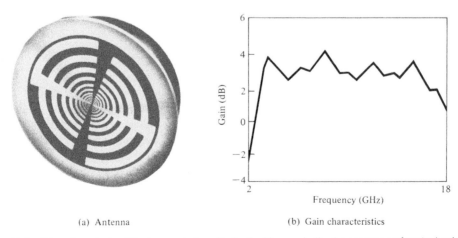

Figure 11.8 Linearly polarized flush-mounted cavity-backed log-periodic slot antenna and typical gain characteristics. (SOURCE: *Antennas, Antenna Masts and Mounting Adaptors*, American Electronic Laboratories, Inc., Lansdale, Pa., Catalog 7.5M-7-79. Courtesy of American Electronic Laboratories, Inc., Montgomeryville, PA 18936 USA).

If the wires or the edges of the plates are linear (instead of curved), the geometries of Figure 11.6 reduce, respectively, to the trapezoidal tooth log-periodic structures of Figure 11.7. These simplifications result in more convenient fabrication geometries with basically no loss in operational performance. There are numerous other bizarre but practical configurations of log-periodic structures, including log-periodic arrays.

If the geometries of Figure 11.6 use uniform periodic teeth, we define the geometric ratio of the log-periodic structure by

$$\tau = \frac{R_n}{R_{n+1}} \qquad (11\text{-}23)$$

and the width of the antenna slot by

$$\chi = \frac{r_n}{R_{n+1}} \qquad (11\text{-}24)$$

The geometric ratio τ of (11-23) defines the period of operation. For example, if two frequencies f_1 and f_2 are one period apart, they are related to the geometric ratio τ by

$$\tau = \frac{f_1}{f_2}, \quad f_2 > f_1 \qquad (11\text{-}25)$$

Extensive studies on the performance of the antenna of Figure 11.6(b) as a function of α, β, τ, and χ, have been performed [10]. In general, these structures performed almost as well as the planar and conical structures. The only major difference is that the log-periodic configurations are linearly polarized instead of circular.

A commercial lightweight, cavity-backed, linearly polarized, flush-mounted log-periodic slot antenna and its associated gain characteristics are shown in Figures 11.8(a) and (b). Typical electrical characteristics are: VSWR—2:1; E-plane beamwidth—70°; H-plane beamwidth—70°. The maximum diameter of the cavity is about 2.4 in. (6.1 cm), the depth is 1.75 in. (4.445 cm), and the weight is near 5 oz (0.14 kg).

11.4.2 Dipole Array

To the layman, the most recognized log-periodic antenna structure is the configuration introduced by Isbell [5] which is shown in Figure 11.9(a). It consists of a sequence of side-by-side parallel linear dipoles forming a coplanar array. Although this antenna has slightly smaller directivities than the Yagi–Uda array (7–12 dB), they are achievable and maintained over much wider bandwidths. There are, however, major differences between them.

While the geometrical dimensions of the Yagi–Uda array elements do not follow any set pattern, the lengths (l_n's), spacings (R_n's), diameters (d_n's), and even gap spacings at dipole centers (s_n's) of the log-periodic array increase logarithmically as defined by the inverse of the geometric ratio τ. That is,

$$\frac{1}{\tau} = \frac{l_2}{l_1} = \frac{l_{n+1}}{l_n} = \frac{R_2}{R_1} = \frac{R_{n+1}}{R_n} = \frac{d_2}{d_1} = \frac{d_{n+1}}{d_n} = \frac{s_2}{s_1} = \frac{s_{n+1}}{s_n} \qquad (11\text{-}26)$$

LOG-PERIODIC ANTENNAS

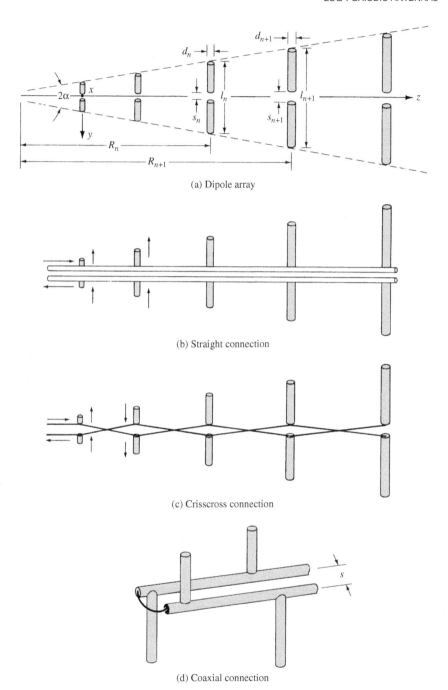

Figure 11.9 Log-periodic dipole array and associated connections.

Another parameter that is usually associated with a log-periodic dipole array is the spacing factor σ defined by

$$\sigma = \frac{R_{n+1} - R_n}{2l_{n+1}} \qquad (11\text{-}26a)$$

Straight lines through the dipole ends meet to form an angle 2α which is a characteristic of frequency independent structures.

Because it is usually very difficult to obtain wires or tubing of many different diameters and to maintain tolerances of very small gap spacings, constant dimensions in these can be used. These relatively minor factors will not sufficiently degrade the overall performance.

While only one element of the Yagi–Uda array is directly energized by the feed line, while the others operate in a parasitic mode, all the elements of the log-periodic array are connected. There are two basic methods, as shown in Figures 11.9(b) and 11.9(c), which could be used to connect and feed the elements of a log-periodic dipole array. In both cases the antenna is fed at the small end of the structure.

The currents in the elements of Figure 11.9(b) have the same phase relationship as the terminal phases. If in addition the elements are closely spaced, the phase progression of the currents is to the right. This produces an end-fire beam in the direction of the longer elements and interference effects to the pattern result.

It was recognized that by mechanically crisscrossing or transposing the feed between adjacent elements, as shown in Figure 11.9(c), a 180° phase is added to the terminal of each element. Since the phase between the adjacent closely spaced short elements is almost in opposition, very little energy is radiated by them and their interference effects are negligible. However, at the same time, the longer and larger spaced elements radiate. The mechanical phase reversal between these elements produces a phase progression so that the energy is beamed end fire in the direction of the shorter elements. The most active elements for this feed arrangement are those that are near resonant with a combined radiation pattern toward the vertex of the array.

The feed arrangement of Figure 11.9(c) is convenient provided the input feed line is a balanced line like the two-conductor transmission line. Using a coaxial cable as a feed line, a practical method to achieve the 180° phase reversal between adjacent elements is shown in Figure 11.9(d). This feed arrangement provides a built-in broadband balun resulting in a balanced overall system. The elements and the feeder line of this array are usually made of piping. The coaxial cable is brought to the feed through the hollow part of one of the feeder-line pipes. While the outside conductor of the coax is connected to that conductor at the feed, its inner conductor is extended and it is connected to the other pipe of the feeder line.

If the geometrical pattern of the log-periodic array, as defined by (11-26), is to be maintained to achieve a truly log-periodic configuration, an infinite structure would result. However, to be useful as a practical broadband radiator, the structure is truncated at both ends. This limits the frequency of operation to a given bandwidth.

The cutoff frequencies of the truncated structure can be determined by the electrical lengths of the longest and shortest elements of the structure. The lower cutoff frequency occurs approximately when the longest element is $\lambda/2$; however, the high cutoff frequency occurs when the shortest element is nearly $\lambda/2$ only when the active region is very narrow. Usually it extends beyond that element. The active region of the log-periodic dipole array is near the elements whose lengths are nearly or slightly smaller than $\lambda/2$. The role of active elements is passed from the longer to the shorter elements as the frequency increases. Also the energy from the shorter active elements traveling toward the longer inactive elements decreases very rapidly so that a negligible amount is reflected from the truncated end. The movement of the active region of the antenna, and its associated phase center, is an undesirable characteristic in the design of feeds for reflector antennas (see Chapter 15). For this reason, log-periodic arrays are not widely used as feeds for reflectors.

The decrease of energy toward the longer inactive elements is demonstrated in Figure 11.10(a). The curves represent typical computed and measured transmission-line voltages (amplitude and phase) on a log-periodic dipole array [11] as a function of distance from its apex. These are feeder-line voltages at the base of the elements of an array with $\tau = 0.95$, $\sigma = 0.0564$, $N = 13$, and $l_n/d_n = 177$. The frequency of operation is such that element No. 10 is $\lambda/2$. The amplitude voltage is nearly constant from the first (the feed) to the eighth element while the corresponding phase is uniformly progressive. Very rapid decreases in amplitude and nonlinear phase variations are noted beyond the eighth element.

The region of constant voltage along the structure is referred to as the *transmission region*, because it resembles that of a matched transmission line. Along the structure, there is about 150° phase change for every $\lambda/4$ free-space length of transmission line. This indicates that the phase velocity of the wave traveling along the structure is $v_p = 0.6v_0$, where v_0 is the free-space velocity. The smaller velocity results from the shunt capacitive loading of the line by the smaller elements. The loading is almost constant per unit length because there are larger spacings between the longer elements.

The corresponding current distribution is shown in Figure 11.10(b). It is noted that the rapid decrease in voltage is associated with strong current excitation of elements 7–10 followed by a rapid decline. The region of high current excitation is designated as the *active region*, and it encompasses 4 to 5 elements for this design. The voltage and current excitations of the longer elements (beyond the ninth) are relatively small, reassuring that the truncated larger end of the structure is not affecting the performance. The smaller elements, because of their length, are not excited effectively. As the frequency changes, the relative voltage and current patterns remain essentially the same, but they move toward the direction of the active region.

There is a linear increase in current phase, especially in the active region, from the shorter to the longer elements. This phase shift progression is opposite in direction to that of an unloaded line. It suggests that on the log-periodic antenna structure there is a wave that travels toward the feed forming a unidirectional end-fire pattern toward the vertex.

The radiated wave of a single log-periodic dipole array is linearly polarized, and it has horizontal polarization when the plane of the antenna is parallel to the ground. Bidirectional patterns and circular polarization can be obtained by phasing multiple log-periodic dipole arrays. For these, the overall effective phase center can be maintained at the feed.

If the input impedance of a log-periodic antenna is plotted as a function of frequency, it will be repetitive. However, if it is plotted as a function of the *logarithm* of the frequency, it will be *periodic* (not necessarily sinusoidal) with each cycle being exactly identical to the preceding one. Hence the name *log-periodic*, because the variations are *periodic* with respect to the *logarithm* of the frequency. A typical variation of the impedance as a function of frequency is shown in Figure 11.11. Other parameters that undergo similar variations are the pattern, directivity, beamwidth, and side lobe level.

The periodicity of the structure does not ensure broadband operation. However, if the variations of the impedance, pattern, directivity, and so forth within one cycle are made sufficiently small and acceptable for the corresponding bandwidth of the cycle, broadband characteristics are ensured within acceptable limits of variation. The total bandwidth is determined by the number of repetitive cycles for the given truncated structure.

The relative frequency span Δ of each cycle is determined by the geometric ratio as defined by (11-25) and (11-26).* Taking the logarithm of both sides in (11-25) reduces to

$$\Delta = \ln(f_2) - \ln(f_1) = \ln\left(\frac{1}{\tau}\right) \qquad (11\text{-}27)$$

The variations that occur within a given cycle ($f_1 \leq f \leq f_2 = f_1/\tau$) will repeat identically at other cycles of the bandwidth defined by $f_1/\tau^{n-1} \leq f \leq f_1/\tau^n$, $n = 1, 2, 3, \ldots$.

Typical designs of log-periodic dipole arrays have apex half angles of $10° \leq \alpha \leq 45°$ and $0.95 \geq \tau \geq 0.7$. There is a relation between the values of α and τ. As α increases, the corresponding τ values decrease, and vice versa. Larger values of α or smaller values of τ result in more compact designs which require smaller number of elements separated by larger distances. In contrast, smaller values

---

*In some cases, the impedance (but not the pattern) may vary with a period which is one-half of (11-27). That is, $\Delta = \frac{1}{2}\ln(1/\tau)$.

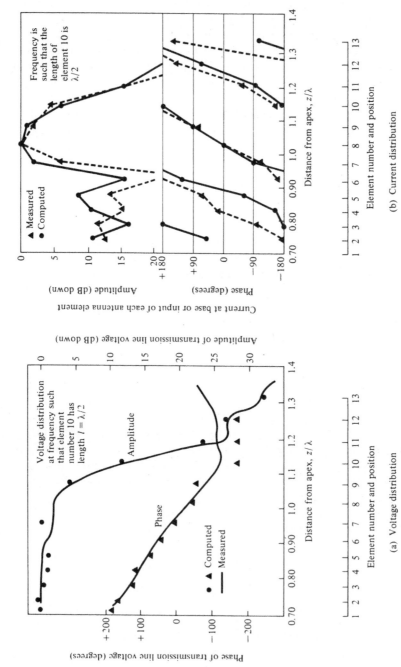

Figure 11.10 Measured and computed voltage and current distributions on a log-periodic dipole array of 13 elements with frequency such that $l_{10} = \lambda/2$. (SOURCE: R. L. Carrel, "Analysis and Design of the Log-Periodic Dipole Antenna," Ph.D. Dissertation, Elec. Eng. Dept., University of Illinois, 1961, University Microfilms, Inc., Ann Arbor, Michigan).

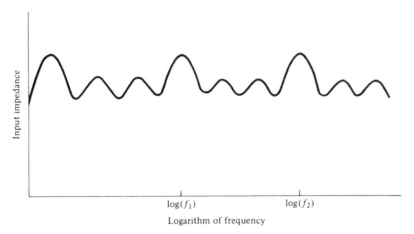

Figure 11.11 Typical input impedance variation of a log-periodic antenna as a function of the logarithm of the frequency.

of α or larger values of τ require a larger number of elements that are closer together. For this type of a design, there are more elements in the active region which are nearly $\lambda/2$. Therefore the variations of the impedance and other characteristics as a function of frequency are smaller, because of the smoother transition between the elements, and the gains are larger.

Experimental models of log-periodic dipole arrays have been built and measurements were made [6]. The input impedances (purely resistive) and corresponding directivities (*above isotropic*) for three different designs are listed in Table 11.1. Larger directivities can be achieved by arraying multiple log-periodic dipole arrays. There are other configurations of log-periodic dipole array designs, including those with V instead of linear elements [12]. This array provides moderate bandwidths with good directivities at the higher frequencies, and it is widely used as a single TV antenna covering the entire frequency spectrum from the lowest VHF channel (54 MHz) to the highest UHF (806 MHz). Typical gain, VSWR, and E- and H-plane half-power beamwidths of commercial log-periodic dipole arrays are shown in Figures 11.12(a), (b), (c), respectively. The overall length of each of these antennas is about 105 in. (266.70 cm) while the largest element in each has an overall length of about 122 in. (309.88 cm). The weight of each antenna is about 31 lb ($\simeq$14 kg).

TABLE 11.1 Input Resistances (R_{in} in ohms) and Directivities (dB above isotropic) for Log-Periodic Dipole Arrays

	$\tau = 0.81$		$\tau = 0.89$		$\tau = 0.95$	
α	R_{in}(ohms)	D_0(dB)	R_{in}(ohms)	D_0(dB)	R_{in}(ohms)	D_0(dB)
10	98	—	82	9.8	77.5	10.7
12.5	—	—	77	—	—	—
15	—	7.2	—	—	—	—
17.5	—	—	76	7.7	62	8.8
20	—	—	74	—	—	—
25	—	—	63	7.2	—	8.0
30	80	—	64	—	54	—
35	—	—	56	6.5	—	—
45	65	5.2	59	6.2	—	—

(SOURCE: D. E. Isbell, "Log Periodic Dipole Arrays," *IRE Trans. Antennas Propagat.*, Vol. AP-8, pp. 260–267, May 1960. © (1960) IEEE.)

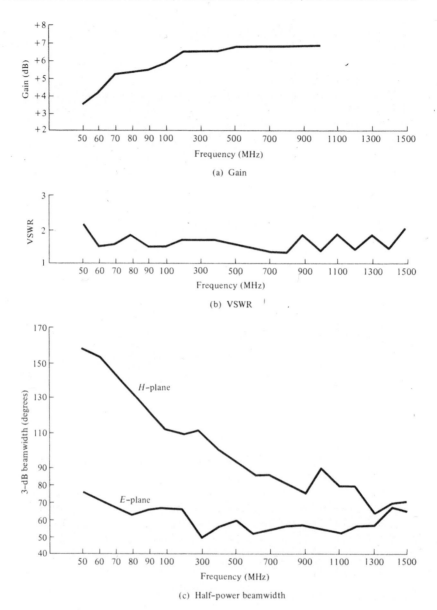

Figure 11.12 Typical gain, VSWR, and half-power beamwidth of commercial log-periodic dipole arrays. (SOURCE: *Antennas, Antenna Masts and Mounting Adaptors*, American Electronic Laboratories, Inc., Lansdale, Pa., Catalog 7.5M-7-79. Courtesy of American Electronic Laboratories, Inc., Montgomeryville, PA 18936 USA).

11.4.3 Design of Dipole Array

The ultimate goal of any antenna configuration is the design that meets certain specifications. Probably the most introductory, complete, and practical design procedure for a log-periodic dipole array is that by Carrel [11]. To aid in the design, he has a set of curves and nomographs. The general configuration of a log-periodic array is described in terms of the design parameters τ, α, and σ related by

$$\alpha = \tan^{-1}\left[\frac{1-\tau}{4\sigma}\right] \tag{11-28}$$

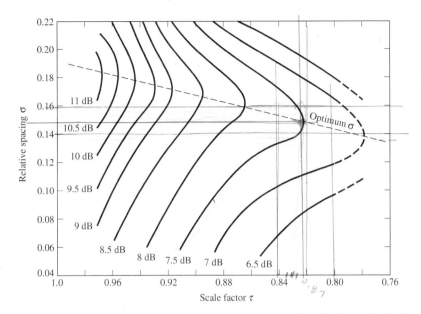

Figure 11.13 Computed contours of constant directivity versus σ and τ for log-periodic dipole arrays. (SOURCE: R. L. Carrel, "Analysis and Design of the Log-Periodic Dipole Antenna," Ph.D. Dissertation, Elec. Eng. Dept., University of Illinois, 1961, University Microfilms, Inc., Ann Arbor Michigan). *Note:* The initial curves led to designs whose directivities are 1–2 dB too high. They have been reduced by an average of 1 dB (see P. C. Butson and G. T. Thompson, "A Note on the Calculation of the Gain of Log-Periodic Dipole Antennas," *IEEE Trans. Antennas Propagat.*, AP-24, pp. 105–106, January 1976).

Once two of them are specified, the other can be found. Directivity (in dB) contour curves as a function of τ for various values of σ are shown in Figure 11.13.

The original directivity contour curves in [11] are in error because the expression for the E-plane field pattern in [11] is in error. To correct the error, the leading $\sin(\theta)$ function in front of the summation sign of equation 47 in [10] should be in the denominator and not in the numerator [i.e., replace $\sin\theta$ by $1/\sin(\theta)$] [11]. The influence of this error in the contours of Figure 11.13 is variable and leads to 1–2 dB higher directivities. However it has been suggested that, as an average, the directivity of each original contour curve be reduced by about 1 dB. This has been implemented already, and the curves in Figure 11.13 are more accurate as they now appear.

A. Design Equations

In this section a number of equations will be introduced that can be used to design a log-periodic dipole array.

While the bandwidth of the system determines the lengths of the shortest and longest elements of the structure, the width of the active region depends on the specific design. Carrel [11] has introduced a semiempirical equation to calculate the bandwidth of the active region B_{ar} related to α and τ by

$$B_{ar} = 1.1 + 7.7(1 - \tau)^2 \cot \alpha \tag{11-29}$$

In practice a slightly larger bandwidth (B_s) is usually designed than that which is required (B). The two are related by

$$B_s = BB_{ar} = B[1.1 + 7.7(1 - \tau)^2 \cot \alpha] \tag{11-30}$$

where

B_s = designed bandwidth
B = desired bandwidth
B_{ar} = active region bandwidth

The total length of the structure L, from the shortest (l_{min}) to the longest (l_{max}) element, is given by

$$L = \frac{\lambda_{max}}{4}\left(1 - \frac{1}{B_s}\right)\cot\alpha \tag{11-31}$$

where

$$\lambda_{max} = 2l_{max} = \frac{v}{f_{min}} \tag{11-31a}$$

From the geometry of the system, the number of elements are determined by

$$N = 1 + \frac{\ln(B_s)}{\ln(1/\tau)} \tag{11-32}$$

The center-to-center spacing s of the feeder-line conductors can be determined by specifying the required input impedance (assumed to be real), and the diameter of the dipole elements and the feeder-line conductors. To accomplish this, we first define an average characteristic impedance of the elements given by

$$Z_a = 120\left[\ln\left(\frac{l_n}{d_n}\right) - 2.25\right] \tag{11-33}$$

where l_n/d_n is the length-to-diameter ratio of the nth element of the array. For an ideal log-periodic design, this ratio should be the same for all the elements of the array. Practically, however, the elements are usually divided into one, two, three, or more groups with all the elements in each group having the same diameter but not the same length. The number of groups is determined by the total number of elements of the array. Usually three groups (for the small, middle, and large elements) should be sufficient.

The effective loading of the dipole elements on the input line is characterized by the graphs shown in Figure 11.14 where

$\sigma' = \sigma/\sqrt{\tau}$ = relative mean spacing
Z_a = average characteristic impedance of the elements
R_{in} = input impedance (real)
Z_0 = characteristic impedance of the feeder line

The center-to-center spacing s between the two rods of the feeder line, each of identical diameter d, is determined by

$$s = d\cosh\left(\frac{Z_0}{120}\right) \tag{11-34}$$

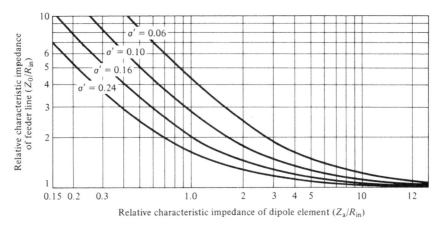

Figure 11.14 Relative characteristic impedance of a feeder line as a function of relative characteristic impedance of dipole element. (SOURCE: R. L. Carrel, "Analysis and Design of the Log-Periodic Dipole Antenna," Ph.D. Dissertation, Elec. Eng. Dept., University of Illinois, 1961, University Microfilms, Inc., Ann Arbor, Michigan).

B. Design Procedure

A design procedure is outlined here, based on the equations introduced above and in the previous page, and assumes that the directivity (in dB), input impedance R_{in} (real), diameter of elements of feeder line (d), and the lower and upper frequencies ($B = f_{max}/f_{min}$) of the bandwidth are specified. It then proceeds as follows:

1. Given D_0 (dB), determine σ and τ from Figure 11.13.
2. Determine α using (11-28).
3. Determine B_{ar} using (11-29) and B_s using (11-30).
4. Find L using (11-31) and N using (11-32).
5. Determine Z_a using (11-33) and $\sigma' = \sigma/\sqrt{\tau}$.
6. Determine Z_0/R_{in} using Figure 11.14.
7. Find s using (11-34).

Example 11.1

Design a log-periodic dipole antenna, of the form shown in Figure 11.9(d), to cover all the VHF TV channels (starting with 54 MHz for channel 2 and ending with 216 MHz for channel 13. See Appendix IX.) The desired directivity is 8 dB and the input impedance is 50 ohms (ideal for a match to 50-ohm coaxial cable). The elements should be made of aluminum tubing with $\frac{3}{4}$ in. (1.9 cm) outside diameter for the largest element and the feeder line and $\frac{3}{16}$ in. (0.48 cm) for the smallest element. These diameters yield identical l/d ratios for the smallest and largest elements.

Solution:

1. From Figure 11.13, for $D_0 = 8$ dB the optimum σ is $\sigma = 0.157$ and the corresponding τ is $\tau = 0.865$.
2. Using (11-28)

$$\alpha = \tan^{-1}\left[\frac{1 - 0.865}{4(0.157)}\right] = 12.13° \simeq 12°$$

3. Using (11-29)

$$B_{ar} = 1.1 + 7.7(1 - 0.865)^2 \cot(12.13°) = 1.753$$

and from (11-30)

$$B_s = BB_{ar} = \frac{216}{54}(1.753) = 4(1.753) = 7.01$$

4. Using (11-31a)

$$\lambda_{max} = \frac{v}{f_{min}} = \frac{3 \times 10^8}{54 \times 10^6} = 5.556 \text{ m (18.227 ft)}$$

From (11-31)

$$L = \frac{5.556}{4}\left(1 - \frac{1}{7.01}\right)\cot(12.13°) = 5.541 \text{ m (18.178 ft)}$$

and from (11-32)

$$N = 1 + \frac{\ln(7.01)}{\ln(1/0.865)} = 14.43 \text{ (14 or 15 elements)}$$

5. $\sigma' = \frac{\sigma}{\sqrt{\tau}} = \frac{0.157}{\sqrt{0.865}} = 0.169$

At the lowest frequency

$$l_{max} = \frac{\lambda_{max}}{2} = \frac{18.227}{2} = 9.1135 \text{ ft}$$

$$\frac{l_{max}}{d_{max}} = \frac{9.1135(12)}{0.75} = 145.816$$

Using (11-33)

$$Z_a = 120[\ln(145.816) - 2.25] = 327.88 \text{ ohms}$$

Thus

$$\frac{Z_a}{R_{in}} = \frac{327.88}{50} = 6.558$$

6. From Figure 11.14

$$Z_0 \simeq 1.2 R_{in} = 1.2(50) = 60 \text{ ohms}$$

7. Using (11-34), assuming the feeder line conductor is made of the same size tubing as the largest element of the array, the center-to-center spacing of the feeder conductors is

$$s = \frac{3}{4}\cosh\left(\frac{60}{120}\right) = 0.846 \simeq 0.85 \text{ in.}$$

which allows for a 0.1 in. separation between their conducting surfaces.

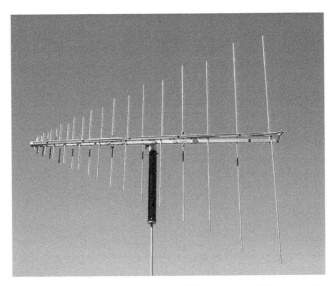

Figure 11.15 Commercial log-periodic dipole antenna of 21 elements. (Courtesy: Antenna Research Associates, Inc., Beltsville, MD).

For such a high-gain antenna, this is obviously a good practical design. If a lower gain is specified and designed for, a smaller length will result.

A commercial log-periodic dipole antenna of 21 elements is shown in Figure 11.15. The antenna is designed to operate in the 100–1,100 MHz with a gain of about 6 dBi, a VSWR (to a 50-ohm line) of 2:1 maximum, a front-to-back ratio of about 20 dB, and with typical half-power beamwidths of about: *E*-plane (75°) and *H*-plane (120°).

C. Design and Analysis Computer Program

A computer program Log-Periodic Dipole Array (**log_perd**) has been developed in FORTRAN and translated in MATLAB, based primarily on the design equations of (11-28)–(11-34), and Figures 11.13 and 11.14, to design a log-periodic dipole array whose geometry is shown in Figure 11.9(a). Although most of the program is based on the same design equations as outlined in the design subsection, this program takes into account more design specifications than those included in the previous design procedure, and it is more elaborate. Once the design is completed, the computer program can be used to analyze the design of the antenna. The description is found in the computer disc available with this book. The program has been developed based on input specifications, which are listed in the program. It can be used as a design tool to determine the geometry of the array (including the number of elements and their corresponding lengths, diameters, and positions) along with the radiation characteristics of the array (including input impedance, VSWR, directivity, front-to-back ratio, *E*- and *H*-plane patterns, etc.) based on desired specifications. The input data includes the desired directivity, lower and upper frequency of the operating band, length-to-diameter ratio of the elements, characteristic impedance of the input transmission line, desired input impedance, termination (load) impedance, etc. These and others are listed in the program included in the CD.

The program assumes that the current distribution on each antenna element is sinusoidal. This approximation would be very accurate if the elements were very far from each other. However, in the active region the elements are usually separated by a distance of about 0.1λ when $\alpha = 15°$ and $\tau = 0.9$. Referring to Figure 8.21, one can see that two $\lambda/2$ dipoles separated by 0.1λ have a mutual impedance (almost real) of about 70 ohms. If this mutual impedance is high compared to the resistance of the transmission line (not the characteristic impedance), then the primary method of coupling energy to each antenna will be through the transmission line. If the mutual impedance is high compared to the self-impedance of each element, then the effect on the radiation pattern should be small. In practice, this is usually the case, and the approximation is relatively good. However, an integral equation formulation with a Moment Method numerical solution would be more accurate. The program uses (8-60a) for the self-resistance and (8-60b) for the self-reactance. It uses (8-68) for the mutual impedance, which for the side-by-side configuration reduces to the sine and cosine integrals in [14], similar in form to (8-71a)–(8-71e) for the $l = \lambda/2$ dipole.

The geometry of the designed log-periodic dipole array is that of Figure 11.9, except that the program also allows for an input transmission line (connected to the first/shortest element), a termination transmission line (extending beyond the last/longest element), and a termination (load) impedance. The length of the input transmission line changes the phase of computed data (such as voltage, current, reflection coefficient, etc.) while its characteristic impedance is used to calculate the VSWR, which in turn affects the input impedance measured at the source. The voltages and currents are found based on the admittance method of Kyle [15]. The termination transmission line and the termination (load) impedance allow for the insertion of a matching section whose primary purpose is to absorb any energy which manages to continue past the active region. Without the termination (load) impedance, this energy would be reflected along the transmission line back into the active region where it would affect the radiation characteristics of the array design and performance.

In designing the array, the user has the choice to select σ and τ (but not the directivity) or to select the directivity (but not σ and τ). In the latter case, the program finds σ and τ by assuming an *optimum design* as defined by the dashed line of Figure 11.13. For the geometry of the array, the program assumes that the elements are placed along the z-axis (with the shortest at $z = 0$ and the longest along the positive z-axis). Each linear element of the array is directed along the y-axis (i.e., the array lies on the yz-plane). The angle θ is measured from the z axis toward the xy-plane while angle ϕ is measured from the x-axis (which is normal to the plane of the array) toward the y-axis along the xy-plane. The E-plane of the array is the yz-plane ($\phi = 90°, 270°; 0° \le \theta \le 180°$) while the H-plane is the xz-plane ($\phi = 0°, 180°; 0° \le \theta \le 180°$).

11.5 FUNDAMENTAL LIMITS OF ELECTRICALLY SMALL ANTENNAS

In all areas of electrical engineering, especially in electronic devices and computers, the attention has been shifted toward miniaturization. Electromagnetics, and antennas in particular, are of no exception. A large emphasis in the last few years has been placed toward electrically small antennas, including printed board designs. However, there are fundamental limits as to how small the antenna elements can be made. The basic limitations are imposed by the free-space wavelength to which the antenna element must couple to, which has not been or is expected to be miniaturized [16].

An excellent paper on the fundamental limits in antennas has been published [16], and revisited in [17], and most of the material in this section is drawn from [16]. It reviews the limits of electrically small, superdirective, super-resolution, and high-gain antennas. The limits on electrically small antennas are derived by assuming that the entire antenna structure (with a largest linear dimension of $2a$), and its transmission line and oscillator are all enclosed within a sphere of radius a as shown in Figure 11.16(a). Because of the arbitrary current or source distribution of the antenna inside the sphere, its radiated field outside the sphere is expressed as a complete set of orthogonal spherical vector waves or modes. For vertically polarized omnidirectional antennas, only TM_{m0} circularly symmetric (no azimuthal variations) modes are required. Each mode is used to represent a spherical wave which propagates in the outward radial direction. This approach was introduced first by Chu [18], and it was followed by Harrington [19]. Earlier papers on the fundamental limitations and performance of small antennas were published by Wheeler [20], [21]. He derived the limits of a small dipole and a small loop (used as a magnetic dipole) from the limitations of a capacitor and an inductor, respectively. The capacitor and inductor were chosen to occupy, respectively, volumes equal to those of the dipole and the loop.

Using the mathematical formulation introduced by Chu [18], the source or current distribution of the antenna system inside the sphere is not uniquely determined by the field distribution outside the sphere. Since it is possible to determine an infinite number of different source or current distributions inside the sphere, for a given field configuration outside the sphere, Chu [18] confined his interest to the most favorable source distribution and its corresponding antenna structure that could exist within the sphere. This approach was taken to minimize the details and to simplify the task of identifying the antenna structure. It was also assumed that the desired current or source distribution minimizes the amount of energy stored inside the sphere so that the input impedance at a given frequency is resistive.

Because the spherical wave modes outside the sphere are orthogonal, the total energy (electric or magnetic) outside the sphere and the complex power transmitted across the closed spherical surface are equal, respectively, to the sum of the energies and complex powers associated with each corresponding spherical mode. Therefore there is no coupling, in energy or power, between any two modes outside the sphere. As a result, the space outside the sphere can be replaced by a number of independent equivalent circuits as shown in Figure 11.16(b). The number of equivalent circuits is equal to the number of spherical wave modes outside the sphere, plus one. The terminals of each equivalent circuit are connected to a box which represents the inside of the sphere, and from inside the box a pair of terminals are drawn to represent the input terminals. Using this procedure, the antenna space problem has been reduced to one of equivalent circuits.

The radiated power of the antenna is calculated from the propagating modes while all modes contribute to the reactive power. When the sphere (which encloses the antenna element) becomes

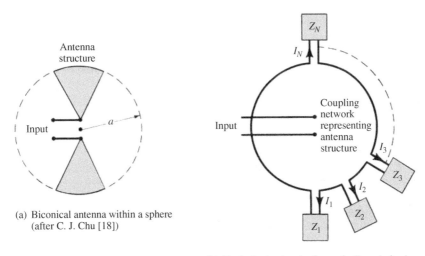

(a) Biconical antenna within a sphere (after C. J. Chu [18])

(b) Equivalent network of a vertically-polarized omnidirectional antenna (after C. J. Chu [18])

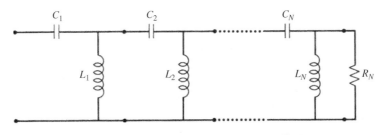

(c) Equivalent circuit for N spherical modes (after C. J. Chu [18])

Figure 11.16 Antenna within a sphere of radius a, and its equivalent circuit modeling. (SOURCE: C. J. Chu, "Physical Limitations of Omnidirectional Antennas," *J. Appl. Phys.*, Vol. 19, pp. 1163–1175, December 1948).

very small, there exist no propagating modes. Therefore the Q of the system becomes very large since all modes are evanescent (below cutoff) and contribute very little power. However, unlike closed waveguides, each evanescent mode here has a real part (even though it is very small).

For a lossless antenna (radiation efficiency $e_{cd} = 100\%$), the equivalent circuit of each spherical mode is a single network section with a series C and a shunt L. The total circuit is a ladder network of $L-C$ sections (one for each mode) with a final shunt resistive load, as shown in Figure 11.16(c). The resistive load is used to represent the normalized antenna radiation resistance.

From this circuit structure, the input impedance is found. The Q of each mode is formed by the ratio of its stored to its radiated energy. When several modes are supported, the Q is formed from the contributions of all the modes.

It has been shown that the higher order modes within a sphere of radius a become evanescent when $ka < 1$. Therefore the Q of the system, for the lowest order TM_{01} spherical mode, reduces to [16]

$$Q = \frac{1 + 2(ka)^2}{(ka)^3[1 + (ka)^2]} \stackrel{ka \ll 1}{\simeq} \frac{1}{(ka)^3} \quad (11\text{-}35a)$$

or for the lowest TM_{01} spherical mode, according to [17], to

$$Q = \frac{1}{(ka)^3} + \frac{1}{ka} = \frac{[1 + (ka)^2]}{(ka)^3} \stackrel{ka \ll 1}{\simeq} \frac{1}{(ka)^3} \quad (11\text{-}35b)$$

Both (11-35a) and (11-35b) lead to the same results when $ka \ll 1$.

When two modes are excited, one TE_{01} and the other TM_{01}, the values of Q are halved. Equations (11-35a) and (11-35b), which relate the lowest achievable Q to the largest linear dimension of an electrically small antenna, are independent of the geometrical configuration of the antenna within the sphere of radius a. The shape of the radiating element within the bounds of the sphere only determines whether TE, TM, or TE and TM modes are excited. *Therefore (11-35a) and (11-35b) represent the fundamental limit on the electrical size of an antenna.* In practice, this limit is only approached but is never exceeded or even equaled.

The losses of an antenna can be taken into account by including a loss resistance in series with the radiation resistance, as shown by the equivalent circuits of Figures 2.27(b) and 2.28(b). This influences the Q of the system and the antenna radiation efficiency as given by (2-90).

Computed values of Q versus kr for idealized antennas enclosed within a sphere of radius a, and with radiation efficiencies of $e_{cd} = 100, 50, 10,$ and 5, are shown plotted in Figure 11.17. These curves represent the minimum values of Q that can be obtained from an antenna whose structure can be enclosed within a sphere of radius a and whose radiated field, outside the sphere, can be represented by a single spherical wave mode.

For antennas with equivalent circuits of fixed values, the fractional bandwidth is related to the Q of the system by

$$\text{fractional bandwidth} = \text{FBW} = \frac{\Delta f}{f_0} = \frac{1}{Q} \quad (11\text{-}36)$$

where

f_0 = center frequency
Δf = bandwidth

The relationship of (11-36) is valid for $Q \gg 1$ since the equivalent resonant circuit with fixed values is a good approximation for an antenna. For values of $Q < 2$, (11-36) is not accurate.

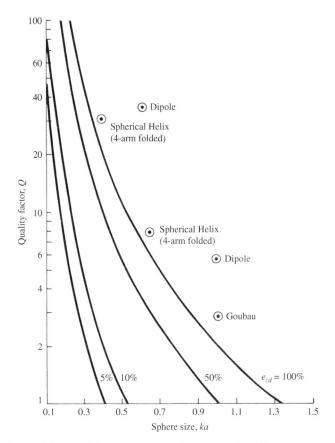

Figure 11.17 Fundamental limits of Q versus antenna size (enclosed within a sphere of radius a) for single-mode antennas of various radiation efficiencies. (SOURCE: R. C. Hansen, "Fundamental Limitations in Antennas," *Proc. IEEE*, Vol. 69, No. 2, February 1981. © (1981) IEEE).

To compare the results of the minimum Q curves of Figure 11.17 with values of practical antenna structures, data points for a small linear dipole and a Goubau [22] antenna are included in the same figure. For a small linear dipole of length l and wire radius b, its impedance is given by (4-37) and (8-62) [16]

$$Z_{in} \simeq 20\pi^2 \left(\frac{l}{\lambda}\right)^2 - j120\frac{\left[\ln\left(\frac{l}{2b}\right) - 1\right]}{\tan\left(\pi\frac{l}{\lambda}\right)} \qquad (11\text{-}37)$$

and its corresponding Q by

$$Q \simeq \frac{\left[\ln\left(\frac{l}{2b}\right) - 1\right]}{\left(\pi\frac{l}{\lambda}\right)^2 \tan\left(\pi\frac{l}{\lambda}\right)} \qquad (11\text{-}38)$$

The computed Q values of the small dipole were for $kl/2 = ka \simeq 0.62$ and 1.04 with $l/2b = l/d = 50$, and of the Goubau antenna were for $ka \simeq 1.04$.

It is apparent that the Q's of the dipole are much higher than the corresponding values of the minimum Q curves even for the 100% efficient antennas. However the Goubau antenna, of the same

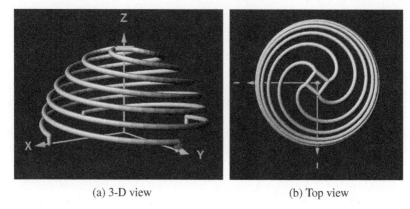

(a) 3-D view   (b) Top view

Figure 11.18   Three-dimensional and top views of a four-arm folded helix: (SOURCE: [23] © 2004 IEEE).

radius sphere, demonstrates a lower value of Q and approaches the values of the 100% minimum Q curve. This indicates that the fractional bandwidth of the Goubau antenna, which is inversely proportional to its Q as defined by (11-36), is higher than that of a dipole enclosed within the same radius sphere. In turn, the bandwidth of an idealized antenna, enclosed within the same sphere, is even larger.

From the above, it is concluded that *the bandwidth of an antenna (which can be closed within a sphere of radius a) can be improved only if the antenna utilizes efficiently, with its geometrical configuration, the available volume within the sphere.* The dipole, being a one-dimensional structure, is a poor utilizer of the available volume within the sphere. However a Goubau antenna, being a clover leaf dipole with coupling loops over a ground plane (or a double cover leaf dipole without a ground plane), is a more effective design for utilizing the available three-dimensional space within the sphere. A design, such as that of a spiral, that utilizes the space even more efficiently than the Goubau antenna would possess a lower Q and a higher fractional bandwidth. Ultimately, the values would approach the minimum Q curves. In practice, these curves are only approached but are never exceeded or even equaled.

A geometry of a radiating element that utilizes efficiently the available space within a sphere of radius a, and improves the bandwidth, is that of a folded spherical helix. Three-dimensional and top views of a four-arm folded helix is shown in Figure 11.18 [23].

The resonant properties of a four-arm folded spherical helical antenna, of the form shown in Figure 11.18, are listed on Table 11.2. Two points of its corresponding Q, from [23], are indicated in Figure 11.17 where they have approached the fundamental limit curve. It has been concluded in [23] that the radiation properties of a folded spherical helical antenna are functions of:

- Number of turns
- Number of helical arms

TABLE 11.2   **Resonant performance properties of a four-arm folded spherical helix antenna**

No. of Turns	Arm Length (cm)	f_R (MHz)	R_A (Ohms)	Efficiency (%)	Q
$1/2$	17	515.8	87.6	99.6	5.6
1	30.9	300.3	43.1	98.6	32
$1^1/_2$	45.07	210	23.62	97.6	88

(SOURCE: [23] © 2004 IEEE).

Increasing the number of arms, while maintaining the same sphere radius, decreases the corresponding Q and increases the bandwidth. By properly selecting these parameters, the spherical helices can be designed to [23]:

- Be electrically small ($ka < 0.5$)
- Be self-resonant
- Have high radiation efficiency
- Have Qs within 1.5 times the fundamental limit
- Have radiation resistance, at resonance, near 50 ohm

While the Qs are within 1.5 the fundamental limit, the absolute value of the Qs can be high because of the small values of ka at which these antennas are self-resonant. Any increase in the bandwidth or decrease in the Q beyond their corresponding limits, while maintaining the same volume, would be at the expense of reducing the antenna efficiency, as indicated in Figure 11.17.

Other examples of dipole and monopole antennas that illustrate the principle of efficient utilization of the space within a sphere in order to increase the bandwidth (lower the Q) are shown in Figures 9.1 and 9.2. The qualitative bandwidths are also indicated in these figures.

11.6 ANTENNA MINIATURIZATION

Antenna miniaturization is a process by, the antenna designer can exercise to reduce the dimensions of the antenna while maintaining, within reasonable limits, the original antenna radiation characteristics. The effort to shrink the physical dimensions of an antenna, without significant performance degradation, has continued for over half a century. However, this is not an easy task, and most often some compromises have had to be accepted between reduction in dimensions and antenna performance. Over the years, many creative techniques have been developed to miniaturize such basic antenna elements as dipoles, monopoles, loops, and slots microstrip/patches, as well as many other radiating elements. An excellent review paper of these attempts is [24]. This section follows and reports on some of the miniaturization experimentations. In the examples reported here and in [24], miniaturization involves modifying the geometry of the structure, by adding components, or by altering the material characteristics. It should be mentioned that at this time, although the performance characteristics of the miniaturized design, most likely is not as good as that of the original antenna, the result is still better than that of an unmodified antenna reduced by the same size; otherwise, miniaturization is really not achieved, even though the dimensions are reduced but the radiation characteristics are considerably degraded. This is why *small antennas*, such as the *infinitesimal linear dipole* of Figure 4.1 ($l \leq \lambda/50$), have radiation characteristics (e.g., impedance, efficiency, gain, bandwidth) that are not very attractive. Therefore, *small antennas* are differentiated from *miniaturized antennas*, and as is defined in [24], for an antenna to be miniaturized, it must be: "*An antenna from a well-established category that has been reduced in size while preserving the fidelity of at least one performance characteristic.*"

Miniaturization, as already mentioned, is accomplished by altering the geometry of the structure, and possibly adding components or integrating materials into the structure. The performance of the revised design does not need to be as good as that of the original version, but it must be better than a version of the original antenna that had been reduced in size by the same amount as the miniaturized antenna. Therefore, in the design techniques reported in [24] and some here, no active elements are considered that will enhance the performance of the designs.

A recent technique that has received considerable attention for miniaturizing antenna design is the integration of metamaterials into the antenna structures. However, the many papers that have reported on such designs base their results only on simulations. Numerous references pertaining to such work are provided in [24], and the designs reported there are those that have been fabricated and their radiation characteristics have been verified with experimental data. Nevertheless, because metamaterial techniques are beyond the scope of this book, they will not be covered here. The reader is encouraged to refer to [24] for more details and references.

Some of the antenna characteristics that can be used to judge the merits of miniaturization, not all inclusive, are as follows:

- Antenna input impedance
- Return loss
- Impedance bandwidth
- Fractional bandwidth
- Impedance matching/reflection
- Radiation efficiency
- Directivity
- Gain
- Realized gain
- Resonant Q

Numerous techniques have been devised to miniaturize existing types of antennas have varying levels of complexity and ingenuity. The examples of miniaturization included here, particularly those related to dipole and monopole antennas, are also reported in [24]; a discussion of patch antennas as reported in [24] is provided later in this book, in Section 14.9.

11.6.1 Monopole Antenna

Monopole antennas, as well as dipoles, are basic elements whose analysis and radiation characteristics are discussed in Chapter 4.

A monopole antenna is resonant (real input impedance) for a length (height) slightly smaller than $l = \lambda_o/4$. For this reason, miniaturization is can succeed for designs in which the length is significantly smaller than $\lambda_o/4$. A few such basic designs will be reported here.

A. Impedance Loading
One of the simplest designs used to miniaturize monopoles and dipoles is to add lumped impedance elements along the antenna's length. This is referred to as *impedance loading*. One such approach, reported in 1963 by Harrison [25], was to add lumped inductor coils at various points along a monopole, as shown in Figure 11.19. It is reported in [24], [25] that by inserting a coil of a given Q, the antenna was made considerable shorter. However, the incorporation and integration of coil(s) in

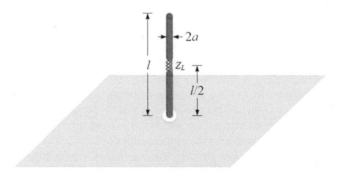

Figure 11.19 Cylindrical monopole loaded with lumped inductor [24].

the design impacted the antenna's performance, decreasing the radiation resistance, reflection coefficient, radiation efficiency, and realized gain. It was in fact discovered that, as the radiating element is made smaller, due to the insertion of a coil, the radiation resistance decreases, the bandwidth narrows and the radiation efficiency decreases. To illustrate the trade-offs of miniaturization due to impedance loading, we give an example below.

Example 11.2

Consider an electrically small resonant monopole of length (height) of $l = \lambda_o/(10\pi) = 3.183 \times 10^{-2}\lambda_o$, radius $a = 2.229 \times 10^{-4}\lambda_o$ $[2\ln(2l/a) = 12.5]$ that is connected to a 50-ohm line. A coil, with a $Q = 300$, is placed in its middle, as in Figure 11.19. In accord with [25], the coil's:

- radiation resistance is $R_r = 20.15$ ohms
- radiation efficiency e_{cd} (accounting for the coil's losses) is $e_{cd} = 19.85\%$. (No other losses on the element itself are included.)

Assuming that the maximum directivity D_o of the shortened monopole, loaded with the coil, is the same as that of a regular short monopole ($D_o = 3$ dimensionless = 4.77 dB, $l \ll \lambda_o$), the objective is to determine the following components:

a. Reflection coefficient Γ
b. Return loss S_{11} (in dB)
c. Maximum realized gain G_{re} (in dB)
d. Maximum realized gain G_{re} (in dB) when the coil is lossless ($Q = \infty$), $e_{cd} = 100\%$, and whose radiation resistance is $R_r = 2.37$ ohms [25]
e. Compare the realized gain of parts c and d with those of a regular $\lambda_o/4$ resonant, lossless monopole

Solution:

a. $\Gamma = \frac{20.15 - 50}{20.15 + 50} = -0.4255$

b. $S_{11} = 20 \log_{10} |\Gamma| = -20 \log_{10} |-0.4255| = -7.422 \, \text{dB}$

c. The maximum directivity of a regular infinitesimal ($l \ll \lambda_o$) monopole is $D_o = 2(1.5) = 3$. Therefore the realized gain of the short monopole, with the coil in its middle, is

$$G_{re0} = 10 \log_{10}\left[e_{cd}\left(1 - |\Gamma|^2\right) D_o\right] = 10 \log_{10}\left[0.1985\left(1 - |-0.4255|^2\right) 3\right] = -3.119 \, \text{dB}$$

d. For a very short ($l \ll \lambda_o$) monopole with a lossless coil ($Q = \infty, e_{cd} = 100\%$), $R_r = 2.37$

$$\Gamma = \frac{2.37 - 50}{2.37 + 50} = -0.9095$$

$$G_{re0} = 10 \log_{10}\left[e_{cd}\left(1 - |\Gamma|^2\right) D_o\right] = 10 \log_{10}\left[1\left(1 - |-0.9095|^2\right) 3\right] = -2.853 \, \text{dB}$$

e. The realized gain of a regular lossless ($e_{cd} = 1$) resonant $\lambda_o/4$ monopole, based on the radiation resistance of $R_r = 73/2 = 36.5$, is

$$\Gamma = \frac{R_r - 50}{R_r + 50} = \frac{36.5 - 50}{36.5 + 50} = -0.1561$$

$$G_{re0} = 10 \log_{10}\left[e_{cd}\left(1 - |\Gamma|^2\right) D_o\right] = 10 \log_{10}\left[1\left(1 - |-0.1561|^2\right) 3\right] = 4.664 \, \text{dB}$$

Therefore, while a coil-loaded monopole of length (height) $l = \lambda_o/(10\pi)$ is physically shortened, compared to a regular $\lambda_o/4$ monopole, by a factor of

$$\frac{\lambda_o/4}{\lambda_o/(10\pi)} = \frac{10\pi}{4} = 7.85$$

its realized gain has been reduced by:

a. $l = \lambda_o/(10\pi)$ monopole with coil of $Q = 300$:

$$|4.664 - (-3.119)| = 7.783 \text{ dB}$$

b. $l = \lambda_o/(10\pi)$ monopole with coil of $Q = \infty$:

$$|4.664 - (-2.853)| = 7.517 \text{ dB}$$

So, there is a trade-off of small size, by a factor of 7.85, at the expense of reduction in realized gain by 7.783 dB and 7.517 dB, respectively. Whether such a trade-off is acceptable will depend on the application and antenna engineer. However, the example illustrates the trade-offs of miniaturization; length reduction versus reduction in realized gain.

B. Materials Loading

Several authors have attempted at miniaturizating dipole and monopole antennas by covering them with dielectric sleeves [26] and spheres [27]. James and Henderson [28] provide a detailed account of such material loading and the impact on the antenna's pattern, bandwidth, input impedance, and radiation efficiency. To illustrate the trade-offs of this type of miniaturization, we select an example from [28], [24] of a monopole loaded with a cylindrical dielectric cover, as shown in Figure 11.20.

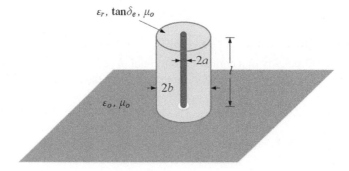

Figure 11.20 Cylindrical monopole loaded covered with a cylindrical dielectric cover [24].

Example 11.3

The resonant monopole, shown in Figure 11.20, has length (height) $l = 23$ mm and radius $a = 0.75$ mm, and it is surrounded by a dielectric cylinder of the same length (height) and radius $b = 23.5$ mm. The dielectric cylinder is filled with pure water ($\varepsilon_r = 73$, $\mu_r = 1$, $\tan \delta_e = 0.02$), and it was measured to be resonant at 381 MHz ($\lambda_o = 30 \times 10^{10}/381 \times 10^6 = 787.4$ mm; $\lambda_o/4 = 196.85$ mm, $\lambda/4 = 23$ mm; a reduction in length/height by a factor of $\sqrt{73} = 8.5$). The radiation resistance and radiation efficiency were measured to be, respectively, $R_r = 0.4$ ohm and $e_r = 12.6\%$ [28]. Assuming that the monopole is connected to a 50-ohm line and the directivity of the dielectric loaded monopole is the same as a regular monopole, determine for the dielectric loaded monopole the following components:

a. Reflection coefficient Γ
b. Return loss S_{11} (in dB)
c. Maximum realized gain G_{re} (in dB)
d. Reduction of maximum realized gain G_{re} (in dB) of resonant dielectric loaded monopole compared to a resonant lossless regular infinitesimal monopole
e. Reduction of length (height) factor of dielectric loaded monopole
f. Increased of volume of dielectric loaded monopole

Solution:

a. Since the radiation resistance of the dielectric covered resonant monopole is 0.4 ohm, its reflection coefficient is $\Gamma = (0.4 - 50)/(0.4 + 50) = -0.9841$.
b. $S_{11} = 20 \log_{10} |\Gamma| = -20 \log_{10} |-0.9841| = -0.139 \text{dB}$.
c. Since the length (height) of the monopole is $l = 23 = \lambda/4$ mm, the maximum directivity is $D_o = 2(1.643) = 3.286$. Therefore, the realized gain of the dielectric monopole is

$$G_{re0} = 10 \log_{10} \left[\varepsilon_r \left(1 - |\Gamma|^2\right) D_o \right] = 10 \log_{10} \left[0.126 \left(1 - |-0.9841|^2\right) 3.286 \right] = -18.85 \text{ dB}$$

d. The realized gain of a regular lossless resonant $\lambda_o/4$ monopole is

$$\Gamma = \frac{36.5 - 50}{36.5 + 50} = -0.1561$$

$$G_{re0} = 10 \log_{10} \left[e_{cd} \left(1 - |\Gamma|^2\right) D_o \right] = 10 \log_{10} \left[1 \left(1 - |-0.1561|^2\right) 3.286 \right] = 5.06 \text{ dB}$$

Therefore, the reduction of the realized gain of the dielectric loaded monopole, compared to an infinitesimal monopole is

$$5.06 - (-18.85) = 24.01 \text{ dB}$$

e. Since the dielectric constant of the dielectric cover is 73, the wavelength λ in the dielectric, compared to the free-space wavelength λ_o, has be reduced by a factor of

$$\frac{\lambda_o}{\lambda} = \sqrt{73} = 8.544 \text{ factor}$$

Thus, the dielectric-loaded monopole has been shortened by a factor of 8.544.

f. The volume of the dielectric monopole, compared to the volume of the free-space monopole, is

$$\frac{V}{V_o} = \frac{\pi b^2(l)}{\pi a^2(l_o)} = \left(\frac{23.5}{0.74}\right)^2 \left(\frac{1}{\sqrt{73}}\right) = 118$$

Therefore, while the height of the dielectric loaded monopole has been reduced by a factor of 8.544, its volume has been increased by a factor of 118, and its realized gain has been reduced by 23.9 dB. This does not seem to be a good trade-off of miniaturization, considering the large reduction in realized gain.

A second example in [28] concerns a monopole of length (height) of $l = 160$ mm and radius $a = 0.6$ mm, resonant at 253 MHz ($l_o = \lambda_o/4 = 1{,}185.77/4 = 296.44$ mm; $l = 160$ mm; reduction in length (height) $= 296.44/160 = 1.85$) surrounded by a cylinder of water of radius $b = 11.5$ mm. For this example, the reduction of length (height) is only 1.85, but there is a considerable increase in volume by a factor of 199:

$$\frac{V}{V_o} = \left(\frac{11.5}{0.6}\right)^2 \times \left(\frac{l}{l_o}\right) = \left(\frac{11.5}{0.6}\right)^2 \times \left(\frac{160}{296.44}\right) = 198.3 \approx 199$$

which is much greater than the 118 factor of Example 11.3. However, because the measured [28] radiation efficiency was $e_{cd} = 96.6\%$ and the radiation resistance was measured to be 8.5 ohms ($|\Gamma| = 0.7094$, $S_{11} = -2.9822$ dB), the realized gain is

$$G_{re0} = 10 \log_{10}\left[e_{cd}\left(1 - |\Gamma|^2\right)D_o\right] = 10 \log_{10}\left[0.966\left(1 - |-0.7094|^2\right)3.286\right] = 1.978 \text{ dB}$$

In this case, the radiation efficiency is much higher, which leads a positive realized gain. Therefore, for this example, miniaturization leads to more positive trade-offs.

C. Folding

Another technique that can be used to miniaturize a dipole is to fold its arm and create a compact structure. There are two ways to do this:

- The arm's linear physical length could be shortened and its current path length maintained; this way, its resonance is maintained.
- The arm's linear length could be increased and its current path length also increased; this way, its resonance is decreased.

However, in both approaches, the radiation characteristics (radiation resistance, bandwidth, efficiency) might be compromised. As a result, many of the attempts to miniaturize dipoles and monopoles were by bending the wire into different compact shapes, including zig-zag, which results in a reduction in length but an increase in area [29]. In [30], there are discussed a number of folding shapes that were tried and the trade-offs between length reduction and radiation efficiency that ensued.

In [31], there is report of a meandering shape of a dipole that included measurements of radiation resistance, bandwidth, and efficiency. This was accomplished by folding the monopole N times into U shapes with 90° corners to form planar structures of width $2W$ and length l, as shown in Figure 11.21.

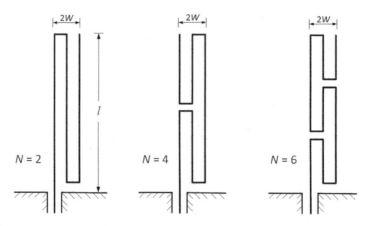

Figure 11.21 Monopole with meandering pattern. (SOURCE: [31] © 1999 IEEE).

One such example reported in [31] involved a resonant linear monopole of length $l \approx \lambda_o/4 \approx$ 135 mm, radius $a = 0.4$ mm, resonant frequency of 545 MHz ($\lambda_o = 550.46$ mm), radiation resistance of 36.5 ohms ($|\Gamma| = 0.156$, based on a line characteristic impedance of 50 ohms), radiation efficiency of $e_{cd} = 99.1\%$, and a fractional bandwidth 9.5% [based on $S_{11} = -10$ dB (VSWR = 2)]. This monopole was physically miniaturized by folding it into meandering patterns, of the form shown in Figure 11.21, but with each of the various values of N occupying a surface of width $W = 2.7$ mm and length $\ell = 45$ mm.

To judge the merits of the folding, a *reduction factor* β was introduced and defined as $\beta = L/l$ (where L is the length of a conventional monopole and l is the length of the length of the folded monopole). It was determined, through experimental data, that with $N = 2$ the antenna resonated at 922 MHz ($\lambda = 325.38$ mm; $\lambda/4 = 81.34$ mm). This folding corresponded to a physical length reduction of about $\beta = L/l = 81.34/45 = 1.8$ but an increase in width by a factor of $2W/2a = 5.4/0.8 = 6.75$. For the reduced folding geometry, the measured radiation resistance was $R_r = 13$ ohms ($|\Gamma| = 0.5873$, $S_{11} = -4.62$ dB, based a 50-ohm line), the radiation efficiency was 96.7%, and the fractional bandwidth was 3%. Based on these figures of merit, it can be stated that the radiation efficiency had likely not been drastically reduced, but both the radiation resistance and the bandwidth, were significantly compromised. Ultimately, and the realized gain was reduced to

$$G_{re0} = 10 \log_{10} \left[e_r \left(1 - |\Gamma|^2 \right) D_o \right] = 10 \log_{10} \left[0.967 \left(1 - |-0.5873|^2 \right) 2(1.643) \right] = 3.18 \text{ dB}$$

Compared to the realized gain of $G_{re0} = 5.02$ dB of a resonant $\lambda/4$ monopole ($R_r = 36.5$, $e_{cd} = 99.1\%$, $|\Gamma| = 0.1561$, 50-omh line), the realized gain of the $N = 2$ folded monopole was reduced by about 2 dB. Since the radiation efficiency was reduced slightly (from 99.1% to 96.7%) and the realized gain was reduced by about 2 dB, this could be considered a miniaturized design, based on the definition of miniaturization stated in this section.

It is shown in [31] that as N increases, as indicated in Figure 11.21, the meandering pattern becomes more complex and the reduction factor is minimized; however, the radiation efficiency and bandwidth increase. Another example of folding geometry from [31], for the case where $N = 14$, is as follows:

- Length reduction factor is only $\beta = 1/0.75 = 1.33$.
- Radiation efficiency is 98%.
- Fractional bandwidth is 8%.
- Radiation resistance is 23.5 ohms ($|\Gamma| = 0.3605$, $S_{11} = -8.86$ dB; based on 50-ohm line).

Its corresponding realized gain is

$$G_{re0} = 10 \log_{10} \left[e_r \left(1 - |\Gamma|^2 \right) D_o \right] = 10 \log_{10} \left[0.98 \left(1 - |-0.3605|^2 \right) 2(1.643) \right] = 4.47 \text{ dB}$$

which is reduced by only 0.6 dB compared to an ideal resonant lossless $\lambda/4$ monopole.

Two additional folding geometries are shown in Figure 11.22 [32]. As is evident in these two meandering patternsshown in the figure, the currents in the horizontal lengths oppose each other, basically canceling each other, whereas the vertical currents reinforce each other. Thus the radiation due to the horizontal sections basically cancels and that of the vertical section reinforces. This design usually leads to a reduction of the effective or net-inductance and in the overall electrical length; thus, resonance occurs at a higher frequency. Also, it should be stated that the vertical sections are primarily responsible for the radiation resistance and bandwidth, which are usually small, as in small vertical straight-wire monopoles discussed in Section 4.7. However, in Figure 11.22(b), the currents in the helical meandering pattern reinforce each other in all parts, leading to increases of the effective or net-inductance and the overall electrical length; thus, resonance occurs at a lower frequency, as discussed in the design of helical antennas in Section 10.3.1. For example, it is stated in [32], that for designs of Figure 11.22 with same overall height (10 cm), same total wire length

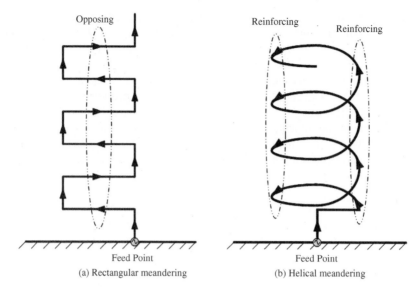

Figure 11.22 Monopole with meandering pattern. ([32] Reprinted with permission from John Wiley & Sons, Inc.).

(30 cm), same conductor diameter (1 mm), and same overall cylindrical diameter (3.3 cm), the geometry of Figure 11.22(a) resonates at 361 MHz while that of the helix Figure 11.22(b) (operating in the normal/broadside mode) resonates at 312.6 MHz; a clear indication that the geometry of Figure 11.22(b) is electrically longer; resonates at a lower frequency. It should be further noted that the spacing between the turns plays a key role in the reinforcing or opposing current vectors. Because of its compactness, despite some of its shortcomings, the design of Figure 11.22(a) is an attractive design that has had wide use in wireless mobile units and in external cell phone antennas. The frequency ranges of both designs in Figure 11.22 are considered to be electrically small and their radiation patterns donut shaped (null along the axis), compared to the patterns of Figure 4.15 (of a small monopole) and of Figure 10.14(a) (of the broadside/normal mode helical antenna). For more details on the electrical performances of these and other designs, including some other coupling and decoupling configurations that result in resonances at higher frequencies, the reader is referred to [32].

11.6.2 Patch Antennas

There are numerous techniques that are used to miniaturize planar type antennas, such as microstrips. One such technique is to use a shorting pin/strip, as shown in Figure 14.45, leading to a microstrip of length nearly $\lambda/4$ (instead of $\lambda/2$); is a physical reduction by a factor of 2. Similar methods are utilized in the design of PIFA and IFA antennas (as discussed later, respectively, in Sections 14.9.1 and 14.9.3). The reader is also referred to [24] for more on patch/microstrip antenna miniaturization techniques.

11.6.3 Antenna Miniaturization Using Metamaterials

Artificial impedance surfaces, also known as *engineered electromagnetic surfaces*, have been developed over the last few decades to alter the impedance boundary conditions of the surface and thus control the radiation characteristics, such as radiation efficiency and pattern, of antenna elements placed at or near them, or the scattering of impinging electromagnetic waves. When electromagnetic waves interact with surfaces that exhibit geometrical periodicity, they result in some interesting and exciting characteristics, which typically have numerous applications that have captured the attention and imagination of engineers and scientists. Using a "broad brush" designation, these *metamaterials* [33] have included applications of dielectric and metallic structures of waveguides, resonators, filters, antennas, and other devices.

Another antenna miniaturization technique that appears to have much potential has involved attempts to incorporate metamaterials into the structures. Even though many papers have reported on such attempts, the reported designs are based only on simulations. The reader is referred to [24], for the numerous references on such designs, but only reported are designs that have been fabricated and their radiation characteristics have verified with experimental data. Otherwise, because metamaterial techniques are beyond the scope of this book, they will not be reported here.

11.7 FRACTAL ANTENNAS

One of the main objectives in wireless communication systems is the design of wideband, or even multiband, low profile, small antennas. Applications of such antennas include, but are not limited to, personal communication systems, small satellite communication terminals, unmanned aerial vehicles, and many more. In order to meet the specification that the antenna be small, some severe limitations are placed on the design, which must meet the fundamental limits of electrically small antennas discussed in the previous section. It was pointed out in the previous section as well as in Section 9.1 that *the bandwidth of an antenna enclosed in a sphere of radius a can be improved only if the antenna utilizes efficiently, with its geometrical configuration, the available volume within the sphere*. For planar geometries, the bandwidth of the antenna can be improved as the geometry of the antenna best utilizes the available planar area of a circle of radius a that encloses the antenna. It was illustrated in Figure 11.17 that a Goubau antenna exhibits a lower Q, and thus a larger bandwidth, than a small linear cylindrical dipole that can be enclosed in a sphere of the same radius. An even lower Q is achieved using a 4-arm folded spherical helix [23] of Figure 11.18, also indicated in Figure 11.17. Both fo these were achieved because a Goubau antenna utilizes the available volume of the sphere more efficiently. Another antenna that accomplishes the same objective, compared to a cylindrical dipole, is a biconical antenna, as illustrated in Figure 11.16. Ideally a biconical antenna of Figure 11.16 that would exhibit the smallest Q (largest bandwidth) would be one whose included angle of each cone is 180°, or a hemispherical dipole. Other dipole and monopole configurations, as well as their qualitative bandwidths, are shown in Figures 9.1 and 9.2.

Another antenna that can meet the requirements of utilizing the available space within a sphere of radius a more efficiently is a fractal antenna. Fractal antennas are based on the concept of a fractal, which is a recursively generated geometry that has fractional dimensions, as pioneered and advanced by Benoit B. Mandelbrot [34]. Mandelbrot coined the term *fractal* and investigated the relationship between fractals and nature using discoveries made by Gaston Julia, Pierre Fatou, and Felix Hausdorff [35]–[40]. He was able to show that many fractals exist in nature and can be used to accurately model certain phenomena. In addition, he was able to introduce new fractals to model more complex structures, including trees and mountains, that possess an inherent self-similarity and self-affinity in their geometrical shape.

Fractal concepts have been applied to many branches of science and engineering, including *fractal electrodynamics* for radiation, propagation, and scattering [41]–[45]. These fractal concepts have been extended to antenna theory and design, and there have been many studies and implementations of different fractal antenna elements and arrays [46]–[57], and many others. The theory of the basics and advances of fractal antennas is well documented in some of the stated references and others, and it will not be repeated here because of space limitations. Instead, a qualitative discussion will be given here. The interested reader is directed to the references for more details.

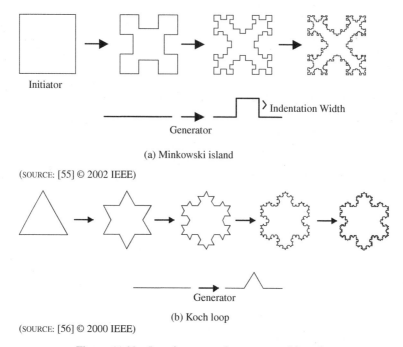

Figure 11.23 Iterative generation process of fractals.

Fractals can be classified in two categories: *deterministic* and *random*. *Deterministic*, such as the von Koch snowflake and the Sierpinski gaskets, are those that are generated of several scaled-down and rotated copies of themselves [40]. Such fractals can be generated using computer graphics requiring particular mapping that is repeated over and over using a recursive algorithm. *Random* fractals also contain elements of randomness that allow simulation of natural phenomena. Procedures and algorithms for generating fractals, both deterministic and random, can be found in [40].

Fractal geometries can best be described and generated using an iterative process that leads to self-similar and self-affinity structures as outlined in [55], [56]. The process can best be illustrated graphically as shown for the two different geometries in Figure 11.23(a,b). Figure 11.23(a) exhibits what is referred to as the Minkowski island fractal [54], while Figure 11.23(b) illustrates the Koch fractal loop [55]. The geometry generating process of a fractal begins with a basic geometry referred to as the *initiator*, which in Figure 11.23(a) is a Euclidean square while that of Figure 11.23(b) is a Euclidean triangle. In Figure 11.23(a), each of the four straight sides of the square is replaced with a generator that is shown at the bottom of the figure. The first three generated iterations are displayed. In Figure 11.23(b), the middle third of each side of the triangle is replaced with its own generator. The first four generated iterations are displayed. The process and generator of Figure 11.23(b) can also be used to generate the Koch dipole that will be discussed later in this section. The trend of the fractal antenna geometry can be deduced by observing several iterations of the process. The final fractal geometry is a curve with an infinitely intricate underlying structure such that, no matter how closely the structure is viewed, the fundamental building blocks cannot be differentiated because they are scaled versions of the initiator. Fractal geometries are used to represent structures in nature, such as trees, plants, mountain ranges, clouds, waves, and so on. Mathematically generated fractal geometries of plants are shown in Figure 11.24. The theory and design of fractal antenna arrays is described in [52].

Another classic fractal is the Sierpinski gasket [49], [50]. This fractal can be generated with the Pascal triangle of (6-63) using the following procedure [42], [46]. Consider an equilateral triangular grid of nodes, as shown in Figure 11.25(a). Starting from the top, each row is labeled $n = 1, 2, 3, \ldots,$

FRACTAL ANTENNAS **629**

Figure 11.24 Fractals used to represent plants in nature.

and each row contains n nodes. A number is assigned to each node for identification purposes. If all the nodes whose number is divisible by a prime number $p(p = 2, 3, 5, \ldots)$ are deleted, the result is a self-similar fractal referred to as the Sierpinski gasket of *Mod-p*. For example, if the nodes in Figure 11.25(a) whose numbers are divisible by 2 are deleted, the result is a Sierpinski gasket of *Mod-2*. If the nodes whose number is divisible by 3 are deleted, a Sierpinski gasket of *Mod-3* is

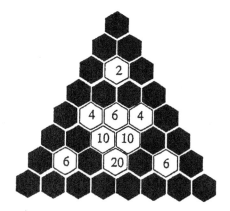

(a) Pascal's triangle and Sierpinski gasket Mod-2

(b) Mod-3   (c) Mod-5

Figure 11.25 Pascal's triangle and Sierpinski gaskets (Mod-2, 3, 5). (SOURCE: [54] © 2001 IEEE).

630 FREQUENCY INDEPENDENT ANTENNAS, ANTENNA MINIATURIZATION, AND FRACTAL ANTENNAS

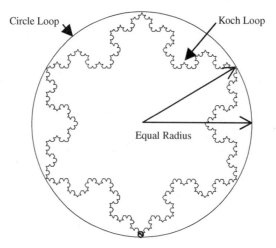

Figure 11.26 Circular and Koch loops of equal radii. (SOURCE: [55] © 2002 IEEE).

obtained, as shown in Figure 11.25(b). Sierpinski gaskets can be used as elements in monopoles and dipoles having geometries whose peripheries are similar to the cross section of conical monopoles and biconical dipoles. The Sierpinski gaskets exhibit favorable radiation characteristics in terms of resonance, impedance, directivity, pattern, and so on, just like the other fractals.

Fractals antennas exhibit space-filling properties that can be used to miniaturize classic antenna elements, such as dipoles and loops, and overcome some of the limitations of small antennas. The line that is used to represent the fractal can meander in such a way as to effectively fill the available space, leading to curves that are electrically long but compacted into a small physical space. This is part of the fundamental limit of small antennas, discussed in Section 11.5 and represented by (11-35a) and (11-35b) (see Figure 11.17), which leads to smaller Qs/larger bandwidths. It also results in antenna elements that, although are compacted in small space, can be resonate and exhibit input resistances that are much greater than the classic geometries of dipoles, loops, etc.

To demonstrate this, let us consider two classic geometries: a circular loop and a linear dipole. Both of these are discussed in Chapters 5 and 4, respectively, and both, when they are small electrically, exhibit small input resistances and must be large in order to resonate. For an ideal circular loop, as discussed in detail in Chapter 5, the input impedance is very small (usually around 1 ohm) if the loop is electrically small. This can be overcome by using a Koch loop of Figure 11.23(b). In fact, in Figure 11.26 we display a Koch loop of four iterations, shown also in Figure 11.23(b), circumscribed by a circular loop of equal radius. In Figure 11.55, we compare the input resistance below resonance of these two geometries for a circumference ranging from 0.15λ to 0.27λ (for the circular loop) and 0.39λ to 0.7λ for the Koch loop. While the input impedance of a small circular loop is very small (around 1.33 ohms at about 0.265λ in circumference for the circular loop), the input resistance of the Koch fractal loop of equal radius (radius $\approx 0.04218\lambda$) is 35 ohms. Both antennas are below resonance and would require matching for the reactive part.

For an ideal linear dipole, the first resonance occurs when the overall length is $\lambda/2$ (as shown in Figure 8.17), which for some frequencies can be physically large. The length can be miniaturized using fractal dipoles, such as the Koch dipole and other similar geometries. To generate the Koch dipole, we apply the iterative generating procedure using the generator of Figure 11.23(b), as shown on the top part of Figure 11.28. This procedure can be extended to generate quasi-fractal tree dipoles and 3D quasi-fractal tree dipoles, as also illustrated in the second and third parts of Figure 11.28, respectively.

In Figure 11.29, we exhibit the resonant frequency for the first five iterations of each fractal dipole of Figure 11.28. It is apparent that the higher iterative geometries exhibit lower resonant frequencies, as if the overall length of the dipole was large electrically. The resonant frequencies plotted in

FRACTAL ANTENNAS 631

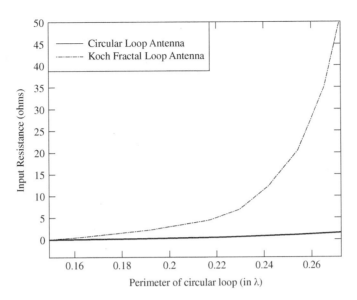

Figure 11.27 Input resistance of circular loop and Koch fractal loop vs. perimeter. (SOURCE: [56] © 2000 IEEE).

Figure 11.29, which correspond to decrease in the length of the dipole, were computed when the reactance was zero and the input impedance was approximately 50 ohms at equal frequencies. It should also be pointed out that the major decrease of the resonant frequency occurs for the 3D fractal dipole, corresponding to about 40% reduction, after five iterations over the equivalent tree dipole [56]. It should also be stated that most of the miniaturization benefits of the fractal dipole occur within the first five iterations with very little changes in the characteristics of the dipole occurring with minimal increases in the complexity afterwards.

As was indicated in Section 11.6, the theoretical limit on the lowest Q of an antenna of any shape is represented by (11-35a) and (11-35b) and plotted in Figure 11.17. Because of the space-filling properties of fractal antennas, their Q can be lower than that of a classic cylindrical dipole, as was

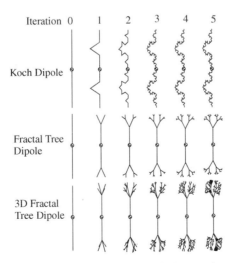

Figure 11.28 Fractal dipole geometries for Koch, tree, and 3D-tree fractal dipoles for five iterations. (SOURCE: [55] © 2002 IEEE).

632 FREQUENCY INDEPENDENT ANTENNAS, ANTENNA MINIATURIZATION, AND FRACTAL ANTENNAS

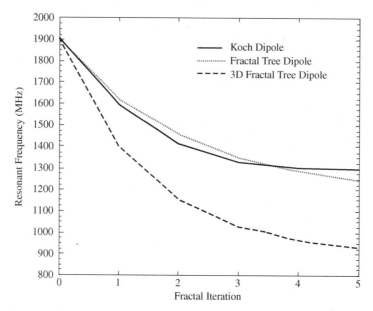

Figure 11.29 Resonant frequency for first five iterations of Koch, tree, and 3D-tree fractal dipoles. (SOURCE: [55] © 2002 IEEE).

the case for Goubau and spherical helix antennas illustrated in Figure 11.17. To demonstrate that, we have plotted in Figure 11.30 the Q of a Koch dipole of the first five iterations, including the zero-order linear dipole. The curve representing the fundamental limit is also included. It is apparent that each of the Koch dipoles exhibits lower Qs/higher bandwidths, compared to the classic linear dipole, as the order of iteration increases. The Qs can be lowered even more using the fractal tree and especially the 3D fractal dipole that exhibits the most space-filling properties [55]. The Qs shown in Figure 11.30 were obtained using Method of Moments [56].

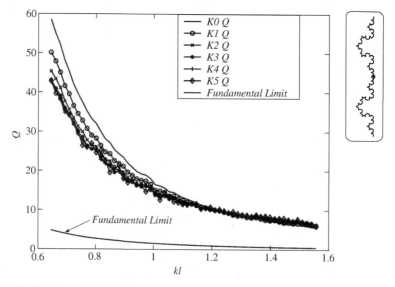

Figure 11.30 Quality factor Q for Koch dipole of up to five iterations as a function of kh (h is maximum length/height of dipole). (SOURCE: [55] © 2002 IEEE).

The concepts discussed above for loops and dipoles can be extended to other antenna elements, including microstrip/patch antennas [55]. The reader is directed to the references.

11.8 MULTIMEDIA

In the publisher's website for this book the following multimedia resources are included for the review, understanding, and visualization of the material of this chapter:

a. **Java**-based **interactive questionnaire**, with answers.
b. **Matlab** and **Fortran** computer program, designated **log_perd**, for computing and displaying the radiation characteristics of a log-periodic linear dipole array design.
c. **Power Point (PPT)** viewgraphs, in multicolor.

REFERENCES

1. V. H. Rumsey, "Frequency Independent Antennas," *1957 IRE National Convention Record*, pt. 1, pp. 114–118.
2. J. D. Dyson, "The Equiangular Spiral Antenna," *IRE Trans. Antennas Propagat.*, Vol. AP-7, pp. 181–187, April 1959.
3. J. D. Dyson, "The Unidirectional Equiangular Spiral Antenna," *IRE Trans. Antennas Propagat.*, Vol. AP-7, pp. 329–334, October 1959.
4. D. S. Filipovic and J. L. Volakis, "Novel Slot Spiral Antenna Designs for Dual-Band/Multiband Operation," *IEEE Trans. Antennas Propagat.*, Vol. 51, No. 3, pp. 430–440, March 2003.
5. R. H. DuHamel and D. E. Isbell, "Broadband Logarithmically Periodic Antenna Structures," *1957 IRE National Convention Record*, pt. 1, pp. 119–128.
6. D. E. Isbell, "Log Periodic Dipole Arrays," *IRE Trans. Antennas Propagat.*, Vol. AP-8, pp. 260–267, May 1960.
7. R. S. Elliott, "A View of Frequency Independent Antennas," *Microwave J.*, pp. 61–68, December 1962.
8. H. G. Booker, "Slot Aerials and Their Relation to Complementary Wire Aerials," *J. IEE (London)*, Vol. 93, Pt. IIIA, April 1946.
9. A. C. Durgan, C. A. Balanis, C. R. Birtcher and D. R. Allee, "Design, Simulation, Fabrication and Testing of Flexible Bow-Tie Antennas," *IEEE Trans. Antennas propagat.*, Vol 59, No. 12, pp. 4425–4435. December 2011.
10. R. H. DuHamel and F. R. Ore, "Logarithmically Periodic Antenna Designs," *IRE National Convention Record*, pt. 1, pp. 139–152, 1958.
11. R. L. Carrel, "Analysis and Design of the Log-Periodic Dipole Antenna," Ph.D. Dissertation, Elec. Eng. Dept., University of Illinois, University Microfilms, Inc., Ann Arbor, MI, 1961.
12. P. E. Mayes and R. L. Carrel, *"Log-Periodic Resonant-V Arrays,"* Presented at WESCON, San Francisco, CA, August 22–25, 1961.
13. P. C. Budson and G. T. Thompson, "A Note on the Calculation of the Gain of Log-Periodic Dipole Antennas," *IEEE Trans. Antennas Propagat.*, AP-24, pp. 105–106, January 1976.
14. H. E. King, "Mutual Impedance of Unequal Length Antennas in Echelon," *IRE Trans. Antennas Propagat.*, Vol. AP-5, pp. 306–313, July 1957.
15. R. H. Kyle, "Mutual Coupling Between Log-Periodic Dipole Antennas," General Electric Tech. Info. Series, Report No. R69ELS 3, December 1968.
16. R. C. Hansen, "Fundamental Limitations in Antennas," *Proc. IEEE*, Vol. 69, No. 2, February 1981.
17. J. S. McLean, "A Re-examination of the Fundamental Limits on the Radiation Q of Electrically Small Antennas," *IEEE Trans. Antennas Propagat.*, Vol. 44, No. 5, pp. 672–676, May 1996.

18. L. J. Chu, "Physical Limitations of Omnidirectional Antennas," *J. Appl. Phys.*, Vol. 19, pp. 1163–1175, December 1948.
19. R. F. Harrington, "Effect of Antenna Size on Gain, Bandwidth, and Efficiency," *J. Res. Nat. Bur. Stand.-D, Radio Propagat.*, Vol. 64D, pp. 1–12, January–February 1960.
20. H. A. Wheeler, "Fundamental Limitations of Small Antennas," *Proc. IRE*, pp. 1479–1488, December 1947.
21. H. A. Wheeler, "Small Antennas," *IEEE Trans. Antennas Propagat.*, Vol. AP-23, No. 4, pp. 462–469, July 1975.
22. G. Goubau, "Multi-element Monopole Antennas," Proc. Workshop on Electrically Small Antennas ECOM, Ft. Monmouth, NJ, pp. 63–67, May 1976.
23. S. R. Best, "The Radiation Properties of Electrically Small Folded Spherical Helix Antennas," *IEEE Trans. Antennas Propagat.*, Vol. 52, No. 4, pp. 953–960, April 2004.
24. E. J. Rothwell and R. O. Ouedraogo, "Antenna Miniaturization: Definitions, Concepts and a Review with Emphasis on Metamaterials," *J. Electromagnetic Waves Appl.*, Vol. 28, No. 17, pp. 2089–2123, 2014.
25. C. W. Harrison, "Monopole with Inductive Loading," *IEEE Trans. Antennas Propagat.*, Vol. 11, No. 4, pp. 394–400, July 1963.
26. J. L. Birchfield and W. R. Free, "Dielectrically Loaded Short Antennas," *IEEE Trans. Antennas Propagat.*, Vol. 22, No. 3, pp. 471–472, May 1974.
27. M. S. Smith, "Properties of Dielectrically Loaded Antennas," *Proc. IEE*, Vol. 124, No. 10, pp. 837–839, 1977.
28. J. R. James and A. Henderson, "Electrically Short Monopole Antennas with Dielectric or Ferrite Coatings," *Proc. IEE*, Vol. 125, No. 9, pp. 793–803, 1978.
29. H. Nakano, H. Togami, A. Yoshizawa and J. Yamauchi, "Shortening Ratios of Modified Dipole Antennas," *IEEE Trans. Antennas Propagat.*, Vol. 32, No. 4, pp. 385–386, April 1984.
30. J. M. Gonzalez-Arbesu, S. Blanch, and J. Romeu, "Shortening Ratios of Modified Dipole Antenna," *IEEE Antennas Wireless Propagat. Lett.*, Vol. 2, pp. 147–150, 2003.
31. R. Rashed and C.-T. Tai, "A New Class of Resonant Antennas," *IEEE Trans. Antennas Propagat.*, Vol. 39, No. 9, pp. 1428–1430, Sept. 1991.
32. S. S. Best, "Small and Fractal Antennas," Chapter 10, pp. 475–528 in *Modern Antenna Handbook* (C. A. Balanis, editor), John Wiley & Sons, Inc., 2008.
33. C. A. Balanis, *Advanced Engineering Electromagnetics*, 2nd edition, John Wiley & Sons, 2012.
34. B. B. Mandelbrot, *The Fractal Geometry of Nature*, Freeman, 1983.
35. X. Yang, J. Chiochetti, D. Papadopoulos, and L. Susman, "Fractal Antenna Elements and Arrays," *Appl. Microwave Wireless*, Vol. 11, No. 5, pp. 34–46, May 1999.
36. K. Falconer, *Fractal Geometry: Mathematical Foundations and Applications*, John Wiley & Sons, New York, 1990.
37. H. Lauwerier, *Fractals: Endlessly Repeated Geometrical Figures*, Princeton University Press, Princeton, NJ, 1991.
38. J.-F. Gouyet, *Physics and Fractal Structures*, Springer, New York, 1996.
39. H.-O. Peitgen and P. H. Richter, *The Beauty of Fractals; Images of Complex Dynamical Systems*, Springer-Verlag, Berlin/New York, 1986.
40. E.-O. Peitgen and D. Saupe (eds.), *The Science of Fractal Images*, Springer-Verlag, Berlin/New York, 1988.
41. D. L. Jaggard, "On Fractal Electrodynamics," in H. N. Kritikos and D. L. Jaggard (eds.), *Recent Advances in Electromagnetics Theory*, Springer-Verlag, New York, pp. 183–224, 1990.
42. D. L. Jaggard, "Fractal Electrodynamics and Modeling," in H. L. Bertoni and L. B. Felsen (eds.), *Directions in Electromagnetics Wave Modeling*, Plenum Publishing, New York, pp. 435–446, 1991.
43. D. L. Jaggard, "Fractal Electrodynamics: Wave Interactions with Discretely Self-Similar Structures," in C. Baum and H. Kritikos (eds.), *Electromagnetic Symmetry*, Taylor & Francis Publishers, Washington, DC, pp. 231–281, 1995.

44. D. H. Werner, "An Overview of Fractal Electrodynamics Research," *Proceedings of the 11th Annual Review of Progress in Applied Computational Electromagnetics (ACES)*, Vol. II, Naval Postgraduate School, Monterey, CA, pp. 964–969, March 1955.
45. D. L. Jaggard, "Fractal Electrodynamics: From Super Antennas to Superlattices," in J. L. Vehel, E. Lutton, and C. Tricot (eds.), *Fractals in Engineering*, Springer-Verlag, New York, pp. 204–221, 1997.
46. Y. Kim and D. L. Jaggard, "The Fractal Random Array," *Proc. IEEE*, Vol. 74, No. 9, pp. 1278–1280, 1986.
47. N. Cohen, "Fractal Antennas Part 1: Introduction and the Fractal Quad," *Communications Quarterly*, Summer, pp. 7–22, Summer 1995.
48. N. Cohen, "Fractal Antennas Part 2: A Discussion of Relevant, but Disparate Qualities," *Communications Quarterly*, Summer, pp. 53–66, Summer 1996.
49. C. Puente, J. Romeu, and A. Cardama, "Fractal Antennas," in D. H. Werner and R. Mittra (eds.), *Frontiers in Electromagnetics*, IEEE Press, Piscataway, NJ, pp. 48–93, 2000.
50. C. Puente, J. Romeu, R. Pous, and A. Cardama, "On the Behavior of the Sierpinski Multiband Antenna," *IEEE Trans. Antennas Propagat.*, Vol. 46, No. 4, pp. 517–524, April 1998.
51. C. Puente, J. Romeu, R. Pous, J. Ramis, and A. Hijazo, "Small but Long Koch Fractal Monopole," *Electronic Letters*, Vol. 34, No. 1, pp. 9–10, 1998.
52. D. H. Werner, R. L. Haupt, and P. L. Werner, "Fractal Antenna Engineering: The Theory and Design of Fractal Antenna Arrays," *IEEE Antennas Propagation Magazine*, Vol. 41, No. 5, pp. 37–59, October 1999.
53. C. Puente Baliarda, J. Romeu, and A. Cardama, "The Koch Monopole: A Small Fractal Antenna," *IEEE Trans. Antennas Propagat.*, Vol. 48, No. 11, pp. 1773–1781, November 2000.
54. J. Romeu and J. Soler, "Generalized Sierpinski Fractal Multiband Antenna," *IEEE Trans. Antennas Propagat.*, Vol. 49, No. 8, pp. 1237–1239, August 2001.
55. J. P. Gianvittorio and Y. Rahmat-Samii, "Fractal Antennas: A Novel Antenna Miniaturization Technique, and Applications," *IEEE Antennas Propagation Magazine*, Vol. 44, No. 1, pp. 20–36, February 2002.
56. J. P. Gianvittorio and Y. Rahmat-Samii, "Fractal Element Antennas: A Compilation of Configurations with Novel Characteristics," 2000 IEEE Antennas and Propagation Society International Symposium, Vol. 3, Salt Lake City, Utah, pp. 1688–1691, July 16–21, 2000.
57. C. Borja and J. Romeu, "Multiband Sierpinski Fractal Patch Antenna," 2000 IEEE Antennas and Propagation Society International Symposium, Vol. 3, Salt Lake City, Utah, pp. 1708–1711, July 16–21, 2000.
58. N. S. Holter, A. Lakhatakia, V. K. Varadan, V. V. Varadan, and R. Messier, "A New Class of Fractals: The Pascal-Sierpinski Gaskets," *Journal of Physics A: Math. Gen.*, Vol. 19, pp. 1753–1759, 1986.

PROBLEMS

11.1. Design a symmetrical two-wire plane spiral ($\phi_0 = 0, \pi$) at $f = 10$ MHz with total feed terminal separation of $10^{-3}\lambda$. The total length of each spiral should be one wavelength and each wire should be of one turn.

(a) Determine the rate of spiral of each wire.

(b) Find the radius (in λ and in meters) of each spiral at its terminal point.

(c) Plot the geometric shape of one wire. Use meters for its length.

11.2. Verify (11-28).

11.3. Design log-periodic dipole arrays, of the form shown in Figure 11.9(d), each with directivities of 9 dB, input impedance of 75 ohms, and each with the following additional specifications: Cover the (see Appendix IX)

(a) VHF TV channels 2–13 (54–216 MHz). Use aluminum tubing with outside diameters of $\frac{3}{4}$ in. (1.905 cm) and $\frac{3}{16}$ in. (0.476 cm) for the largest and smallest elements, respectively.

(b) VHF TV channels 2–6 (54–88 MHz). Use diameters of 1.905 and 1.1169 cm for the largest and smallest elements, respectively.

(c) VHF TV channels 7–13 (174–216 MHz). Use diameters of 0.6 and 0.476 cm for the largest and smallest elements, respectively.

(d) UHF TV channels (512–806 MHz). The largest and smallest elements should have diameters of 0.2 and 0.128 cm, respectively.

(e) FM band of 88–108 MHz (100 channels at 200 KHz apart). The largest and smallest elements should have diameters of 1.169 and 0.9525 cm, respectively.

In each design, the feeder line should have the same diameter as the largest element.

11.4. For each design in Problem 11.3, determine the

(a) span of each period over which the radiation characteristics will vary slightly

(b) number of periods (cycles) within the desired bandwidth

11.5. Using the **log_perd** computer program and Appendix IX, design an array which covers the VHF television band. Design the antenna for 7 dBi gain optimized in terms of $\sigma - \tau$. The antenna should be matched to a 75-ohm coaxial input cable. For this problem, set the input line length to 0 meters, the source resistance to 0 ohms, the termination line length to 0 meters, the termination impedance to 100 Kohms, the length-to-diameter ratio to 40, and the boom diameter to 10 cm. To make the actual input impedance 75 ohms, one must iteratively find the optimal desired input impedance.

(a) Plot the gain, magnitude of the input impedance, and VSWR versus frequency from 30 MHz to 400 MHz.

(b) Based on the ripples in the plot of gain versus frequency, what is τ? Compare this value to the value calculated by the computer program.

(c) Why does the gain decrease rapidly for frequencies less than the lower design frequency yet decrease very slowly for frequencies higher than the upper design frequency?

11.6. For the antenna of Problem 11.5, replace the 100-Kohm load with a 75-ohm resistor.

(a) Plot the gain, magnitude of the input impedance, and VSWR versus frequency from 30 MHz to 400 MHz.

(b) What does the termination resistor do which makes this antenna an improvement over the antenna of Problem 11.5?

11.7. For the antenna of Problem 11.5, replace the 100-Kohm termination (load) with a 75-ohm resistor and make the source resistance 10 ohms. This resistance represents the internal resistance of the power supply as well as losses in the input line.

(a) Plot the gain versus frequency from 30 MHz to 400 MHz.

(b) What is the antenna efficiency of this antenna?

(c) Based on your result from parts (a) and (b), what should the gain versus frequency plot look like for Problem 11.6?

11.8. Design a log-periodic dipole array which operates from 470 MHz to 806 MHz (UHF band) with 8 dBi gain. This antenna should be matched to a 50-ohm cable of length 2 meters with no source resistance. The termination should be left open. Select the length-to-diameter ratio to be 25. At 600 MHz, do the following. Use the computer program **log_perd** of this chapter.

(a) Plot the E- and H-plane patterns.

(b) Calculate the E- and H-plane half-power beamwidths.

(c) Find the front-to-back ratio.

(d) Why does the E-plane pattern have deep nulls while the H-plane pattern does not?

11.9. The overall length of a small linear dipole antenna (like a biconical antenna, or cylindrical dipole, or any other) is λ/π. Assuming the antenna is 100% efficient, what is:
   (a) The smallest possible value of Q for an antenna of such a length? Practically it will be larger than that value.
   (b) The largest fractional bandwidth ($\Delta f/f_0$ where f_0 is the center frequency)?

11.10. For a small ($l = \lambda/10$) dipole with radius $a = \lambda/500$ operating at a center frequency of 500 MHz:
   (a) Approximate *maximum possible* fractional bandwidth.
   (b) Lower and upper frequencies (in MHz) of the operating bandwidth.
   (c) The radius of the sphere (*in* λ) which encloses the dipole and which separates the region where the energy within the sphere is basically imaginary and outside the sphere is basically read.

11.11. It is desired to design a 100% efficient biconical dipole antenna whose overall length is $\lambda/20$. The design guidelines specify a need to optimize the frequency response (bandwidth). To accomplish this, the quality factor Q of the antenna should be minimized. In order to get some indications as to the fundamental limits of the design:
   (a) What is the lowest possible limit of the Q for this size antenna?
   (b) In order to approach this lower fundamental limit, should the included angle of the biconical antenna be made larger or smaller, and why?

11.12. It is desired to design a very small, *100% efficient* spiral antenna whose *largest* fractional bandwidth is 10%. Determine the following:
   (a) The *smallest possible* quality factor (Q) that the spiral can ever have.
   (b) The largest possible diameter (in λ) the spiral can ever have.

11.13. Small antennas are very highly reactive radiating elements and exhibit very high quality factors and very small bandwidths. Assuming the antenna is 50% efficient and its fractional bandwidth is 1%
   (a) determine the largest *overall* linear dimension (in λ) of the antenna;
   (b) would, in practice, the measured.
      1. Q be smaller or larger than the one designed based on the 1% fractional bandwidth, *and why?*
      2. fractional bandwidth be smaller or larger than the designed one of 1%, *and why?*

11.14. Consider a *small* biconical antenna with 100% efficiency and a fractional bandwidth of 2%.
   (a) Determine the largest possible *overall* dimension (in λ) of the antenna.
   (b) Does the Q become smaller or larger with increasing cone angle? Why?
   (c) Does the fractional bandwidth become smaller or larger with increasing cone angle? Why?

CHAPTER 12

Aperture Antennas

12.1 INTRODUCTION

Aperture antennas are most common at microwave frequencies. There are many different geometrical configurations of an aperture antenna with some of the most popular shown in Figure 1.4. They may take the form of a waveguide or a horn whose aperture may be square, rectangular, circular, elliptical, or any other configuration. Aperture antennas are very practical for space applications, because they can be flush mounted on the surface of the spacecraft or aircraft. Their opening can be covered with a dielectric material to protect them from environmental conditions. This type of mounting does not disturb the aerodynamic profile of the craft, which in high-speed applications is critical.

In this chapter, the mathematical tools will be developed to analyze the radiation characteristics of aperture antennas. The concepts will be demonstrated by examples and illustrations. Because they are the most practical, emphasis will be given to the rectangular and circular configurations. Due to mathematical complexities, the observations will be restricted to the far-field region. The edge effects, due to the finite size of the ground plane to which the aperture is mounted, can be taken into account by using diffraction methods such as the Geometrical Theory of Diffraction, better known as GTD. This is discussed briefly and only qualitatively in Section 12.10.

The radiation characteristics of wire antennas can be determined once the current distribution on the wire is known. For many configurations, however, the current distribution is not known exactly and only physical intuition or experimental measurements can provide a reasonable approximation to it. This is even more evident in aperture antennas (slits, slots, waveguides, horns, reflectors, lenses). It is therefore expedient to have alternate methods to compute the radiation characteristics of antennas. Emphasis will be placed on techniques that for their solution rely primarily not on the current distribution but on reasonable approximations of the fields on or in the vicinity of the antenna structure. One such technique is the *Field Equivalence Principle*.

12.2 FIELD EQUIVALENCE PRINCIPLE: HUYGENS' PRINCIPLE

The *field equivalence* is a principle by which actual sources, such as an antenna and transmitter, are replaced by equivalent sources. The fictitious sources are said to be *equivalent within a region*

Antenna Theory: Analysis and Design, Fourth Edition. Constantine A. Balanis.
© 2016 John Wiley & Sons, Inc. Published 2016 by John Wiley & Sons, Inc.
Companion Website: www.wiley.com/go/antennatheory4e

because *they produce the same fields within that region*. The formulations of scattering and diffraction problems by the equivalence principle are more suggestive to approximations.

The field equivalence was introduced in 1936 by S. A. Schelkunoff [1], [2], and it is a more rigorous formulation of Huygens' principle [3] which states that "*each point on a primary wavefront can be considered to be a new source of a secondary spherical wave and that a secondary wavefront can be constructed as the envelope of these secondary spherical waves* [4]." The equivalence principle is based on the *uniqueness theorem* which states that "*a field in a lossy region is uniquely specified by the sources within the region plus the tangential components of the electric field over the boundary, or the tangential components of the magnetic field over the boundary, or the former over part of the boundary and the latter over the rest of the boundary* [2], [5]." The field in a lossless medium is considered to be the limit, as the losses go to zero, of the corresponding field in a lossy medium. Thus if the tangential electric and magnetic fields are completely known over a closed surface, the fields in the source-free region can be determined.

By the equivalence principle, the fields outside an imaginary closed surface are obtained by placing over the closed surface suitable electric- and magnetic-current densities which satisfy the boundary conditions. The current densities are selected so that the fields inside the closed surface are zero and outside they are equal to the radiation produced by the actual sources. Thus the technique can be used to obtain the fields radiated outside a closed surface by sources enclosed within it. The formulation is exact but requires integration over the closed surface. The degree of accuracy depends on the knowledge of the tangential components of the fields over the closed surface.

In most applications, the closed surface is selected so that most of it coincides with the conducting parts of the physical structure. This is preferred because the vanishing of the tangential electric field components over the conducting parts of the surface reduces the physical limits of integration.

The equivalence principle is developed by considering an actual radiating source, which electrically is represented by current densities $\mathbf{J}_1$ and $\mathbf{M}_1$, as shown in Figure 12.1(a). The source radiates fields $\mathbf{E}_1$ and $\mathbf{H}_1$ everywhere. However, it is desired to develop a method that will yield the fields outside a closed surface. To accomplish this, a closed surface S is chosen, shown dashed in Figure 12.1(a), which encloses the current densities $\mathbf{J}_1$ and $\mathbf{M}_1$. The volume within S is denoted by V_1 and outside S by V_2. *The primary task will be to replace the original problem, shown in Figure 12.1(a), by an equivalent one which yields the same fields* $\mathbf{E}_1$ *and* $\mathbf{H}_1$ *outside S (within* V_2*)*. The formulation of the problem can be aided eminently if the closed surface is judiciously chosen so that fields over most, if not the entire surface, are known *a priori*.

An equivalent problem of Figure 12.1(a) is shown in Figure 12.1(b). The original sources $\mathbf{J}_1$ and $\mathbf{M}_1$ are removed, and we assume that there exist fields $\mathbf{E}$ and $\mathbf{H}$ inside S and fields $\mathbf{E}_1$ and $\mathbf{H}_1$ outside of S. For these fields to exist within and outside S, they must satisfy the boundary conditions on the tangential electric and magnetic field components. Thus on the imaginary surface S there must exist

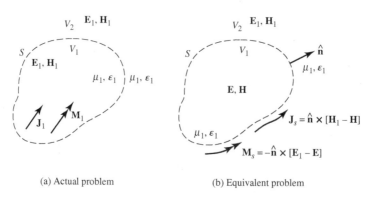

(a) Actual problem (b) Equivalent problem

Figure 12.1 Actual and equivalent models.

the *equivalent sources*

$$\mathbf{J}_s = \hat{\mathbf{n}} \times [\mathbf{H}_1 - \mathbf{H}] \tag{12-1}$$

$$\mathbf{M}_s = -\hat{\mathbf{n}} \times [\mathbf{E}_1 - \mathbf{E}] \tag{12-2}$$

and they radiate into an *unbounded space* (same medium everywhere). *The current densities of (12-1) and (12-2) are said to be equivalent only within V_2, because they produce the original fields $(\mathbf{E}_1, \mathbf{H}_1)$ only outside S. Fields $\mathbf{E}, \mathbf{H}$, different from the originals $(\mathbf{E}_1, \mathbf{H}_1)$, result within V_1.* Since the currents of (12-1) and (12-2) radiate in an unbounded space, the fields can be determined using (3-27)–(3-30a) and the geometry of Figure 12.2(a). In Figure 12.2(a), R is the distance from any point on the surface S, where $\mathbf{J}_s$ and $\mathbf{M}_s$ exist, to the observation point.

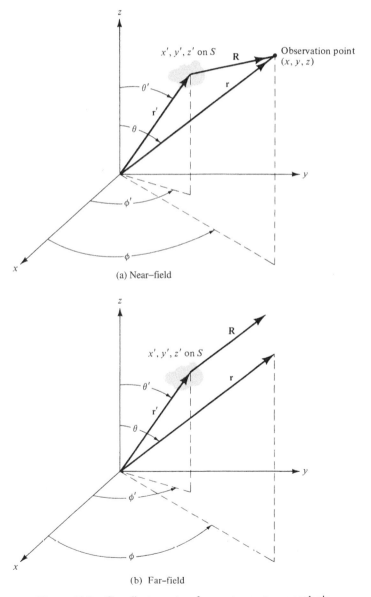

Figure 12.2 Coordinate system for aperture antenna analysis.

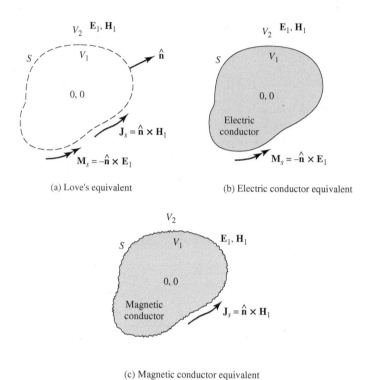

Figure 12.3 Equivalence principle models.

So far, the tangential components of *both* **E** and **H** have been used in setting up the equivalent problem. From electromagnetic uniqueness concepts, it is known that the tangential components of only **E** or **H** are needed to determine the fields. It will be demonstrated that equivalent problems can be found which require only the magnetic-current densities (tangential **E**) or only electric current densities (tangential **H**). This requires modifications to the equivalent problem of Figure 12.1(b).

Since the fields **E, H** within S can be anything (this is not the region of interest), it can be assumed that they are zero. In that case the equivalent problem of Figure 12.1(b) reduces to that of Figure 12.3(a) with the equivalent current densities being equal to

$$\mathbf{J}_s = \hat{\mathbf{n}} \times (\mathbf{H}_1 - \mathbf{H})|_{\mathbf{H}=0} = \hat{\mathbf{n}} \times \mathbf{H}_1 \tag{12-3}$$

$$\mathbf{M}_s = -\hat{\mathbf{n}} \times (\mathbf{E}_1 - \mathbf{E})|_{\mathbf{E}=0} = -\hat{\mathbf{n}} \times \mathbf{E}_1 \tag{12-4}$$

This form of the field equivalence principle is known as *Love's Equivalence Principle* [2], [6]. Since the current densities of (12-3) and (12-4) radiate in an unbounded medium (same μ, ε everywhere), they can be used in conjunction with (3-27)–(3-30a) to find the fields everywhere.

Love's Equivalence Principle of Figure 12.3(a) produces a null field within the imaginary surface S. Since the value of the $\mathbf{E} = \mathbf{H} = 0$ within S cannot be disturbed if the properties of the medium within it are changed, let us assume that it is replaced by a perfect electric conductor ($\sigma = \infty$). The introduction of the perfect conductor will have an effect on the equivalent source $\mathbf{J}_s$, and it will prohibit the use of (3-27)–(3-30a) because the current densities no longer radiate into an unbounded medium. Imagine that the geometrical configuration of the electric conductor is identical to the profile of the imaginary surface S, over which $\mathbf{J}_s$ and $\mathbf{M}_s$ exist. As the electric conductor takes its place,

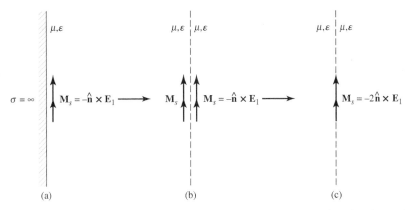

Figure 12.4 Equivalent models for magnetic source radiation near a perfect electric conductor.

as shown in Figure 12.3(b), according to the *uniqueness theorem* [2], the equivalent of Figure 12.3(a) reduces to that of Figure 12.3(b). Only a magnetic current density $\mathbf{M}_s$ (tangential components of the electric field) are necessary over the entire S, and it radiates in the presence of the electric conductor producing outside S the original fields $\mathbf{E}_1, \mathbf{H}_1$. Within S the fields are zero but, as before, this is not a region of interest. The difficulty in trying to use the equivalent problem of Figure 12.3(b) is that (3-27)–(3-30a) cannot be used, because the current densities do not radiate into an unbounded medium. The problem of a magnetic current density radiating in the presence of an electric conducting surface must be solved. So it seems that the equivalent problem is just as difficult as the original problem itself.

Before some special simple geometries are considered and some suggestions are made for approximating complex geometries, let us introduce another equivalent problem. Referring to Figure 12.3(a), let us assume that instead of placing a perfect electric conductor within S we introduce a perfect magnetic conductor (PMC). Again, *according to the uniqueness theorem* [2], the equivalent problem of Figure 12.3(a) reduces to that shown in Figure 12.3(c) (requires only a $\mathbf{J}_s$ over the entire surface S, i.e., tangential components of the magnetic field). As was with the equivalent problem of Figure 12.3(b), (3-27)–(3-30a) cannot be used with Figure 12.3(c) and the problem is just as difficult as that of Figure 12.3(b) or the original of Figure 12.1(a).

To begin to see the utility of the field equivalence principle, especially that of Figure 12.3(b), let us assume that the surface of the electric conductor is flat and extends to infinity as shown in Figure 12.4(a). For this geometry, the problem is to determine how a magnetic source radiates in the presence of a flat electric conductor. From image theory, this problem reduces to that of Figure 12.4(b) where an imaginary magnetic source is introduced on the side of the conductor and takes its place (remove conductor). Since the imaginary source is in the same direction as the equivalent source, the equivalent problem of Figure 12.4(b) reduces to that of Figure 12.4(c). The magnetic current density is doubled, it radiates in an unbounded medium, and (3-27)–(3-30a) can be used. The equivalent problem of Figure 12.4(c) yields the correct $\mathbf{E}, \mathbf{H}$ fields to the right side of the interface. If the surface of the obstacle is not flat and infinite, but its curvature is large compared to the wavelength, a good approximation is the equivalent problem of Figure 12.3(c).

SUMMARY

In the analysis of electromagnetic problems, many times it is easier to form equivalent problems that yield the same solution within a region of interest. This is the case for scattering, diffraction, and

644 APERTURE ANTENNAS

aperture antenna problems. In this chapter, the main interest is in aperture antennas. The concepts will be demonstrated with examples.

The steps that must be used to form an equivalent and solve an aperture problem are as follows:

1. Select an imaginary surface that encloses the actual sources (the aperture). The surface must be judiciously chosen so that the tangential components of the electric and/or the magnetic field are known, exactly or approximately, over its entire span. In many cases this surface is a flat plane extending to infinity.
2. Over the imaginary surface form equivalent current densities $\mathbf{J}_s, \mathbf{M}_s$ which take one of the following forms:
    a. $\mathbf{J}_s$ and $\mathbf{M}_s$ over S assuming that the $\mathbf{E}$- and $\mathbf{H}$-fields within S are not zero.
    b. or $\mathbf{J}_s$ and $\mathbf{M}_s$ over S assuming that the $\mathbf{E}$- and $\mathbf{H}$-fields within S are zero (Love's theorem)
    c. or $\mathbf{M}_s$ over S ($\mathbf{J}_s = 0$) assuming that within S the medium is a perfect electric conductor
    d. or $\mathbf{J}_s$ over S ($\mathbf{M}_s = 0$) assuming that within S the medium is a perfect magnetic conductor.
3. Solve the equivalent problem. For forms (a) and (b), (3-27)–(3-30a) can be used. For form (c), the problem of a magnetic current source next to a perfect electric conductor must be solved [(3-27)–(3-30a) cannot be used directly, because the current density does not radiate into an unbounded medium]. If the electric conductor is an infinite flat plane the problem can be solved exactly by image theory. For form (d), the problem of an electric current source next to a perfect magnetic conductor must be solved. Again (3-27)–(3-30a) cannot be used directly. If the magnetic conductor is an infinite flat plane, the problem can be solved exactly by image theory.

To demonstrate the usefulness and application of the field equivalence theorem to aperture antenna theory, an example is considered.

Example 12.1

A waveguide aperture is mounted on an infinite ground plane, as shown in Figure 12.5(a). Assuming that the tangential components of the electric field over the aperture are known, and are given by $\mathbf{E}_a$, find an equivalent problem that will yield the same fields $\mathbf{E}, \mathbf{H}$ radiated by the aperture to the right side of the interface.

Solution: First an imaginary closed surface is chosen. For this problem it is appropriate to select a flat plane extending from minus infinity to plus infinity, as shown in Figure 12.5(b). Over the infinite plane, the equivalent current densities $\mathbf{J}_s$ and $\mathbf{M}_s$ are formed. Since the tangential components of $\mathbf{E}$ do not exist outside the aperture, because of vanishing boundary conditions, the magnetic current density $\mathbf{M}_s$ is only nonzero over the aperture. The electric current density $\mathbf{J}_s$ is nonzero everywhere and is yet unknown. Now let us assume that an imaginary flat electric conductor approaches the surface S, and it shorts out the current density $\mathbf{J}_s$ everywhere. $\mathbf{M}_s$ exists only over the space occupied originally by the aperture, and it radiates in the presence of the conductor [see Figure 12.5(c)]. By image theory, the conductor can be removed and replaced by an imaginary (equivalent) source $\mathbf{M}_s$ as shown in Figure 12.5(d), which is analogous to Figure 12.4(b). Finally, the equivalent problem of Figure 12.5(d) reduces to that of Figure 12.5(e), which is analogous to that of Figure 12.4(c). The original problem has been reduced to a very simple equivalent and (3-27)–(3-30a) can be utilized for its solution.

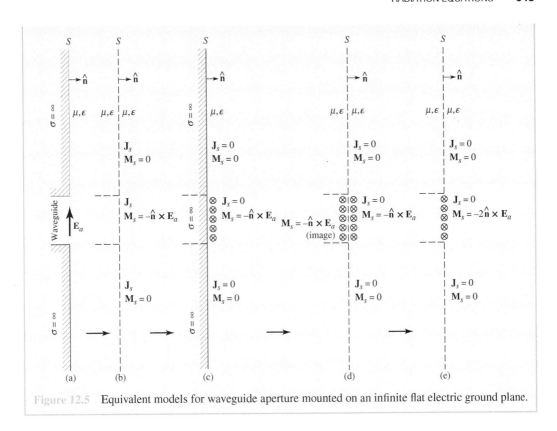

Figure 12.5 Equivalent models for waveguide aperture mounted on an infinite flat electric ground plane.

In this chapter the theory will be developed to compute the fields radiated by an aperture, like that shown in Figure 12.5(a), making use of its equivalent of Figure 12.5(e). For other problems, their equivalent forms will not necessarily be the same as that shown in Figure 12.5(e).

12.3 RADIATION EQUATIONS

In Chapter 3 and in the previous section it was stated that the fields radiated by sources $\mathbf{J}_s$ and $\mathbf{M}_s$ in an unbounded medium can be computed by using (3-27)–(3-30a) where the integration must be performed over the entire surface occupied by $\mathbf{J}_s$ and $\mathbf{M}_s$. These equations yield valid solutions for all observation points [2], [7]. For most problems, the main difficulty is the inability to perform the integrations in (3-27) and (3-28). However for far-field observations, the complexity of the formulation can be reduced.

As was shown in Section 4.4.1, for far-field observations R can most commonly be approximated by

$$R \simeq r - r' \cos\psi \quad \text{for phase variations} \tag{12-5a}$$

$$R \simeq r \quad \text{for amplitude variations} \tag{12-5b}$$

where ψ is the angle between the vectors $\mathbf{r}$ and $\mathbf{r}'$, as shown in Figure 12.2(b). The primed coordinates $(x', y', z', \text{or } r', \theta', \phi')$ indicate the space occupied by the sources $\mathbf{J}_s$ and $\mathbf{M}_s$, over which integration must be performed. The unprimed coordinates $(x, y, z \text{ or } r, \theta, \phi)$ represent the observation point. Geometrically the approximation of (12-5a) assumes that the vectors $\mathbf{R}$ and $\mathbf{r}$ are parallel, as shown in Figure 12.2(b).

Using (12-5a) and (12-5b), (3-27) and (3-28) can be written as

$$\mathbf{A} = \frac{\mu}{4\pi} \iint_S \mathbf{J}_s \frac{e^{-jkR}}{R} \, ds' \simeq \frac{\mu e^{-jkr}}{4\pi r} \mathbf{N} \tag{12-6}$$

$$\mathbf{N} = \iint_S \mathbf{J}_s e^{jkr' \cos\psi} \, ds' \tag{12-6a}$$

$$\mathbf{F} = \frac{\varepsilon}{4\pi} \iint_S \mathbf{M}_s \frac{e^{-jkR}}{R} \, ds' \simeq \frac{\varepsilon e^{-jkr}}{4\pi r} \mathbf{L} \tag{12-7}$$

$$\mathbf{L} = \iint_S \mathbf{M}_s e^{jkr' \cos\psi} \, ds' \tag{12-7a}$$

In Section 3.6 it was shown that in the far-field only the θ and ϕ components of the **E**- and **H**-fields are dominant. Although the radial components are not necessarily zero, they are negligible compared to the θ and ϕ components. Using (3-58a)–(3-59b), the $\mathbf{E}_A$ of (3-29) and $\mathbf{H}_F$ of (3-30) can be written as

$$(E_A)_\theta \simeq -j\omega A_\theta \tag{12-8a}$$

$$(E_A)_\phi \simeq -j\omega A_\phi \tag{12-8b}$$

$$(H_F)_\theta \simeq -j\omega F_\theta \tag{12-8c}$$

$$(H_F)_\phi \simeq -j\omega F_\phi \tag{12-8d}$$

and the $\mathbf{E}_F$ of (3-29) and $\mathbf{H}_A$ of (3-30), with the aid of (12-8a)–(12-8d), as

$$(E_F)_\theta \simeq +\eta(H_F)_\phi = -j\omega\eta F_\phi \tag{12-9a}$$

$$(E_F)_\phi \simeq -\eta(H_F)_\theta = +j\omega\eta F_\theta \tag{12-9b}$$

$$(H_A)_\theta \simeq -\frac{(E_A)_\phi}{\eta} = +j\omega \frac{A_\phi}{\eta} \tag{12-9c}$$

$$(H_A)_\phi \simeq +\frac{(E_A)_\theta}{\eta} = -j\omega \frac{A_\theta}{\eta} \tag{12-9d}$$

Combining (12-8a)–(12-8d) with (12-9a)–(12-9d), and making use of (12-6)–(12-7a) the total **E**- and **H**-fields can be written as

$$E_r \simeq 0 \tag{12-10a}$$

$$E_\theta \simeq -\frac{jke^{-jkr}}{4\pi r}(L_\phi + \eta N_\theta) \tag{12-10b}$$

$$E_\phi \simeq +\frac{jke^{-jkr}}{4\pi r}(L_\theta - \eta N_\phi) \tag{12-10c}$$

$$H_r \simeq 0 \tag{12-10d}$$

$$H_\theta \simeq \frac{jke^{-jkr}}{4\pi r}\left(N_\phi - \frac{L_\theta}{\eta}\right) \tag{12-10e}$$

$$H_\phi \simeq -\frac{jke^{-jkr}}{4\pi r}\left(N_\theta + \frac{L_\phi}{\eta}\right) \tag{12-10f}$$

The $N_\theta, N_\phi, L_\theta$, and L_ϕ can be obtained from (12-6a) and (12-7a). That is,

$$\mathbf{N} = \iint_S \mathbf{J}_s e^{+jkr' \cos\psi} \, ds' = \iint_S (\hat{\mathbf{a}}_x J_x + \hat{\mathbf{a}}_y J_y + \hat{\mathbf{a}}_z J_z) e^{+jkr' \cos\psi} \, ds' \qquad (12\text{-}11a)$$

$$\mathbf{L} = \iint_S \mathbf{M}_s e^{+jkr' \cos\psi} \, ds' = \iint_S (\hat{\mathbf{a}}_x M_x + \hat{\mathbf{a}}_y M_y + \hat{\mathbf{a}}_z M_z) e^{+jkr' \cos\psi} \, ds' \qquad (12\text{-}11b)$$

Using the rectangular-to-spherical component transformation, obtained by taking the inverse (in this case also the transpose) of (4-5), (12-11a) and (12-11b) reduce for the θ and ϕ components to

$$N_\theta = \iint_S [J_x \cos\theta \cos\phi + J_y \cos\theta \sin\phi - J_z \sin\theta] e^{+jkr' \cos\psi} \, ds' \qquad (12\text{-}12a)$$

$$N_\phi = \iint_S [-J_x \sin\phi + J_y \cos\phi] e^{+jkr' \cos\psi} \, ds' \qquad (12\text{-}12b)$$

$$L_\theta = \iint_S [M_x \cos\theta \cos\phi + M_y \cos\theta \sin\phi - M_z \sin\theta] e^{+jkr' \cos\psi} \, ds' \qquad (12\text{-}12c)$$

$$L_\phi = \iint_S [-M_x \sin\phi + M_y \cos\phi] e^{+jkr' \cos\psi} \, ds' \qquad (12\text{-}12d)$$

SUMMARY

To summarize the results, the procedure that must be followed to solve a problem using the radiation integrals will be outlined. Figures 12.2(a) and 12.2(b) are used to indicate the geometry.

1. Select a closed surface over which the total electric and magnetic fields $\mathbf{E}_a$ and $\mathbf{H}_a$ are known.
2. Form the equivalent current densities $\mathbf{J}_s$ and $\mathbf{M}_s$ over S using (12-3) and (12-4) with $\mathbf{H}_1 = \mathbf{H}_a$ and $\mathbf{E}_1 = \mathbf{E}_a$.
3. Determine the $\mathbf{A}$ and $\mathbf{F}$ potentials using (12-6)–(12-7a) where the integration is over the closed surface S.
4. Determine the radiated $\mathbf{E}$- and $\mathbf{H}$-fields using (3-29) and (3-30).

The above steps are valid for all regions (near-field and far-field) outside the surface S. If, however, the observation point is in the far-field, steps 3 and 4 can be replaced by 3' and 4'. That is,

3'. Determine $N_\theta, N_\phi, L_\theta$ and L_ϕ using (12-12a)–(12-12d).
4'. Determine the radiated $\mathbf{E}$- and $\mathbf{H}$-fields using (12-10a)–(12-10f).

Some of the steps outlined above can be reduced by a judicious choice of the equivalent model. In the remaining sections of this chapter, the techniques will be applied and demonstrated with examples of rectangular and circular apertures.

12.4 DIRECTIVITY

The directivity of an aperture can be found in a manner similar to that of other antennas. The primary task is to formulate the radiation intensity $U(\theta, \phi)$, using the far-zone electric and magnetic field components, as given by (2-12a) or

$$U(\theta, \phi) = \frac{1}{2}\text{Re}[(\hat{\mathbf{a}}_\theta E_\theta + \hat{\mathbf{a}}_\phi E_\phi) \times (\hat{\mathbf{a}}_\theta H_\theta + \hat{\mathbf{a}}_\phi H_\phi)^*] = \frac{1}{2\eta}(|E_\theta^0|^2 + |E_\phi^0|^2) \qquad (12\text{-}13)$$

which in normalized form reduces to

$$U_n(\theta, \phi) = (|E_\theta^0(\theta, \phi)|^2 + |E_\phi^0(\theta, \phi)|^2) = B_0 F(\theta, \phi) \qquad (12\text{-}13a)$$

The directive properties can then be found using (2-19)–(2-22).

Because the radiation intensity $U(\theta, \phi)$ for each aperture antenna will be of a different form, a general equation for the directivity cannot be formed. However, a general MATLAB and FORTRAN computer program, designated as **Directivity**, has been written to compute the directivity of any antenna, including an aperture, once the radiation intensity is specified. The program is based on the formulations of (12-13a), (2-19)–(2-20), and (2-22), and it is included in Chapter 2. In the main program, it requires the lower and upper limits on θ and ϕ. The radiation intensity for the antenna in question must be specified in the subroutine $U(\theta, \phi, F)$ of the program.

Expressions for the directivity of some simple aperture antennas, rectangular and circular, will be derived in later sections of this chapter.

12.5 RECTANGULAR APERTURES

In practice, the rectangular aperture is probably the most common microwave antenna. Because of its configuration, the rectangular coordinate system is the most convenient system to express the fields at the aperture and to perform the integration. Shown in Figure 12.6 are the three most common and convenient coordinate positions used for the solution of an aperture antenna. In Figure 12.6(a) the aperture lies on the y-z plane, in Figure 12.6(b) on the x-z plane, and in Figure 12.6(c) on the x-y plane. For a given field distribution, the analytical forms for the fields for each of the arrangements are not the same. However the computed values will be the same, since the physical problem is identical in all cases.

For each of the geometries shown in Figure 12.6, the only difference in the analysis is in the formulation of

1. the components of the equivalent current densities $(J_x, J_y, J_z, M_x, M_y, M_z)$
2. the difference in paths from the source to the observation point $(r' \cos \psi)$
3. the differential area ds'

In general, the nonzero components of $\mathbf{J}_s$ and $\mathbf{M}_s$ are

$$J_y, J_z, M_y, M_z \quad [\text{Figure 12.6(a)}] \qquad (12\text{-}14a)$$

$$J_x, J_z, M_x, M_z \quad [\text{Figure 12.6(b)}] \qquad (12\text{-}14b)$$

$$J_x, J_y, M_x, M_y \quad [\text{Figure 12.6(c)}] \qquad (12\text{-}14c)$$

RECTANGULAR APERTURES **649**

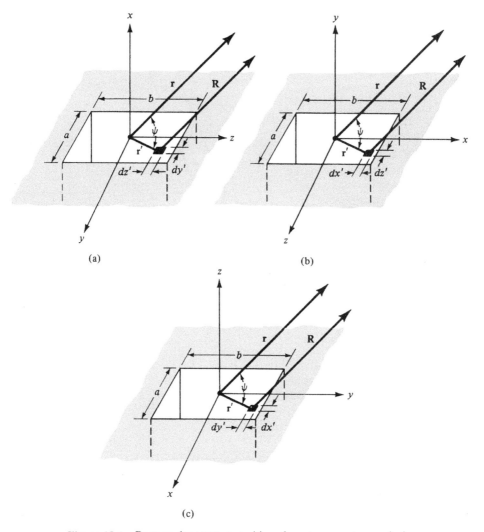

Figure 12.6 Rectangular aperture positions for antenna system analysis.

The differential paths take the form of

$$r' \cos \psi = \mathbf{r}' \cdot \hat{\mathbf{a}}_r = (\hat{\mathbf{a}}_y y' + \hat{\mathbf{a}}_z z') \cdot (\hat{\mathbf{a}}_x \sin\theta \cos\phi + \hat{\mathbf{a}}_y \sin\theta \sin\phi + \hat{\mathbf{a}}_z \cos\theta)$$
$$= y' \sin\theta \sin\phi + z' \cos\theta \quad \text{[Figure 12.6(a)]} \quad (12\text{-}15\text{a})$$

$$r' \cos \psi = \mathbf{r}' \cdot \hat{\mathbf{a}}_r = (\hat{\mathbf{a}}_x x' + \hat{\mathbf{a}}_z z') \cdot (\hat{\mathbf{a}}_x \sin\theta \cos\phi + \hat{\mathbf{a}}_y \sin\theta \sin\phi + \hat{\mathbf{a}}_z \cos\theta)$$
$$= x' \sin\theta \cos\phi + z' \cos\theta \quad \text{[Figure 12.6(b)]} \quad (12\text{-}15\text{b})$$

$$r' \cos \psi = \mathbf{r}' \cdot \hat{\mathbf{a}}_r = (\hat{\mathbf{a}}_x x' + \hat{\mathbf{a}}_y y') \cdot (\hat{\mathbf{a}}_x \sin\theta \cos\phi + \hat{\mathbf{a}}_y \sin\theta \sin\phi + \hat{\mathbf{a}}_z \cos\theta)$$
$$= x' \sin\theta \cos\phi + y' \sin\theta \sin\phi \quad \text{[Figure 12.6(c)]} \quad (12\text{-}15\text{c})$$

650 APERTURE ANTENNAS

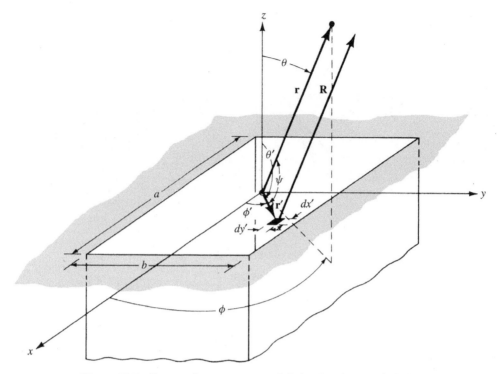

Figure 12.7 Rectangular aperture on an infinite electric ground plane.

and the differential areas are represented by

$$ds' = dy' \, dz' \quad \text{[Figure 12.6(a)]} \tag{12-16a}$$

$$ds' = dx' \, dz' \quad \text{[Figure 12.6(b)]} \tag{12-16b}$$

$$ds' = dx' \, dy' \quad \text{[Figure 12.6(c)]} \tag{12-16c}$$

12.5.1 Uniform Distribution on an Infinite Ground Plane

The first aperture examined is a rectangular aperture mounted on an infinite ground plane, as shown in Figure 12.7. To reduce the mathematical complexities, initially the field over the opening is assumed to be constant and given by

$$\mathbf{E}_a = \hat{\mathbf{a}}_y E_0 \quad -a/2 \le x' \le a/2, \quad -b/2 \le y' \le b/2 \tag{12-17}$$

where E_0 is a constant. The task is to find the fields radiated by it, the pattern beamwidths, the side lobe levels of the pattern, and the directivity. To accomplish these, the equivalent will be formed first.

A. Equivalent
To form the equivalent, a closed surface is chosen which extends from $-\infty$ to $+\infty$ on the x-y plane. Since the physical problem of Figure 12.7 is identical to that of Figure 12.5(a), its equivalents are

those of Figures 12.5(a)–(e). Using the equivalent of Figure 12.5(e)

$$\mathbf{M}_s = \begin{cases} -2\hat{\mathbf{n}} \times \mathbf{E}_a = -2\hat{\mathbf{a}}_z \times \hat{\mathbf{a}}_y E_0 = +\hat{\mathbf{a}}_x 2E_0 & \begin{array}{l} -a/2 \leq x' \leq a/2 \\ -b/2 \leq y' \leq b/2 \end{array} \\ 0 & \text{elsewhere} \end{cases} \quad (12\text{-}18)$$

$$\mathbf{J}_s = 0 \qquad \text{everywhere}$$

B. Radiation Fields: Element and Space Factors

The far-zone fields radiated by the aperture of Figure 12.7 can be found by using (12-10a)–(12-10f), (12-12a)–(12-12d), (12-14c), (12-15c), (12-16c), and (12-18). Thus,

$$N_\theta = N_\phi = 0 \qquad (12\text{-}19)$$

$$L_\theta = \int_{-b/2}^{+b/2} \int_{-a/2}^{+a/2} [M_x \cos\theta \cos\phi] e^{jk(x'\sin\theta\cos\phi + y'\sin\theta\sin\phi)} \, dx' \, dy'$$

$$L_\theta = \cos\theta \cos\phi \left[\int_{-b/2}^{+b/2} \int_{-a/2}^{+a/2} M_x e^{jk(x'\sin\theta\cos\phi + y'\sin\theta\sin\phi)} \, dx' \, dy' \right] \qquad (12\text{-}19\text{a})$$

In (12-19a), the integral within the brackets represents the *space factor* for a two-dimensional distribution. It is analogous to the space factor of (4-58a) for a line source (one-dimensional distribution). For the L_θ component of the vector potential $\mathbf{F}$, the *element factor* is equal to the product of the factor outside the brackets in (12-19a) and the factor outside the brackets in (12-10c). The total field is equal to the product of the element and space factors, as defined by (4-59), and expressed in (12-10b) and (12-10c).

Using the integral

$$\int_{-c/2}^{+c/2} e^{j\alpha z} \, dz = c \left[\frac{\sin\left(\frac{\alpha}{2}c\right)}{\frac{\alpha}{2}c} \right] \qquad (12\text{-}20)$$

(12-19a) reduces to

$$L_\theta = 2abE_0 \left[\cos\theta \cos\phi \left(\frac{\sin X}{X}\right) \left(\frac{\sin Y}{Y}\right) \right] \qquad (12\text{-}21)$$

where

$$X = \frac{ka}{2} \sin\theta \cos\phi \qquad (12\text{-}21\text{a})$$

$$Y = \frac{kb}{2} \sin\theta \sin\phi \qquad (12\text{-}21\text{b})$$

Similarly it can be shown that

$$L_\phi = -2abE_0 \left[\sin\phi \left(\frac{\sin X}{X}\right) \left(\frac{\sin Y}{Y}\right) \right] \qquad (12\text{-}22)$$

Substituting (12-19), (12-21), and (12-22) into (12-10a)–(12-10f), the fields radiated by the aperture can be written as

$$E_r = 0 \tag{12-23a}$$

$$E_\theta = j\frac{abkE_0 e^{-jkr}}{2\pi r}\left[\sin\phi\left(\frac{\sin X}{X}\right)\left(\frac{\sin Y}{Y}\right)\right] \tag{12-23b}$$

$$E_\phi = j\frac{abkE_0 e^{-jkr}}{2\pi r}\left[\cos\theta\cos\phi\left(\frac{\sin X}{X}\right)\left(\frac{\sin Y}{Y}\right)\right] \tag{12-23c}$$

$$H_r = 0 \tag{12-23d}$$

$$H_\theta = -\frac{E_\phi}{\eta} \tag{12-23e}$$

$$H_\phi = +\frac{E_\theta}{\eta} \tag{12-23f}$$

Equations (12-23a)–(12-23f) represent the three-dimensional distributions of the far-zone fields radiated by the aperture. Experimentally only two-dimensional plots can be measured. To reconstruct experimentally a three-dimensional plot, a series of two-dimensional plots must be made. In many applications, however, only a pair of two-dimensional plots are usually sufficient. These are the principal E- and H-plane patterns whose definition was stated in Section 2.2.3 and illustrated in Figure 2.3.

For the problem in Figure 12.7, the E-plane pattern is on the y-z plane ($\phi = \pi/2$) and the H-plane is on the x-z plane ($\phi = 0$). Thus

E–Plane ($\phi = \pi/2$)

$$E_r = E_\phi = 0 \tag{12-24a}$$

$$E_\theta = j\frac{abkE_0 e^{-jkr}}{2\pi r}\left[\frac{\sin\left(\frac{kb}{2}\sin\theta\right)}{\frac{kb}{2}\sin\theta}\right] \tag{12-24b}$$

H–Plane ($\phi = 0$)

$$E_r = E_\theta = 0 \tag{12-25a}$$

$$E_\phi = j\frac{abkE_0 e^{-jkr}}{2\pi r}\left\{\cos\theta\left[\frac{\sin\left(\frac{ka}{2}\sin\theta\right)}{\frac{ka}{2}\sin\theta}\right]\right\} \tag{12-25b}$$

To demonstrate the techniques, three-dimensional patterns have been plotted in Figures 12.8 and 12.9. The dimensions of the aperture are indicated in each figure. Multiple lobes appear, because the dimensions of the aperture are greater than one wavelength. The number of lobes increases as the dimensions increase. For the aperture whose dimensions are $a = 3\lambda$ and $b = 2\lambda$ (Figure 12.8), there are a total of five lobes in the principal H-plane and three lobes in the principal E-plane. The pattern in the H-plane is only a function of the dimension a whereas that in the E-plane is only influenced

RECTANGULAR APERTURES 653

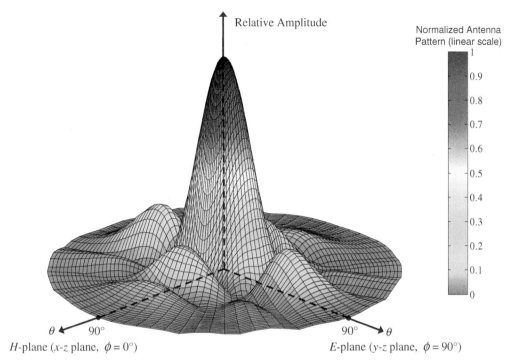

Figure 12.8 Three-dimensional field pattern of a constant field rectangular aperture mounted on an infinite ground plane ($a = 3\lambda, b = 2\lambda$).

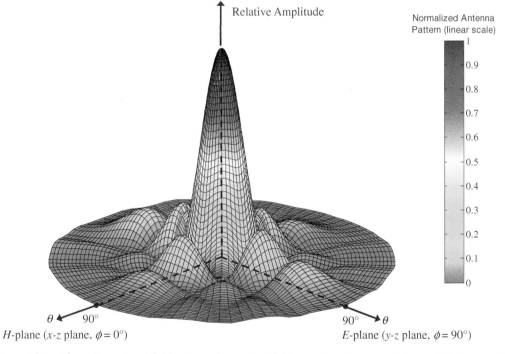

Figure 12.9 Three-dimensional field pattern of a constant field square aperture mounted on an infinite ground plane ($a = b = 3\lambda$).

654 APERTURE ANTENNAS

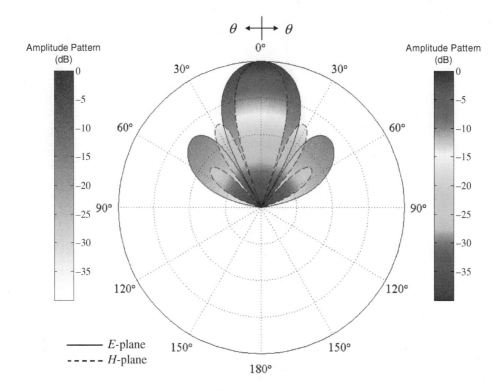

Figure 12.10 E- and H-plane amplitude patterns for uniform distribution aperture mounted on an infinite ground plane ($a = 3\lambda, b = 2\lambda$).

by b. In the E-plane, the side lobe formed on each side of the major lobe is a result of $\lambda < b \leq 2\lambda$. In the H-plane, the first minor lobe on each side of the major lobe is formed when $\lambda < a \leq 2\lambda$ and the second side lobe when $2\lambda < a \leq 3\lambda$. Additional lobes are formed when one or both of the aperture dimensions increase. This is illustrated in Figure 12.9 for an aperture with $a = b = 3\lambda$.

The two-dimensional principal plane patterns for the aperture with $a = 3\lambda, b = 2\lambda$ are shown in Figure 12.10. For this, and for all other size apertures mounted on an infinite ground plane, the H-plane patterns along the ground plane vanish. This is dictated by the boundary conditions. The E-plane patterns, in general, do not have to vanish along the ground plane, unless the dimension of the aperture in that plane (in this case b) is a multiple of a wavelength.

The patterns computed above assumed that the aperture was mounted on an infinite ground plane. In practice, infinite ground planes are not realizable, but they can be approximated by large structures. Edge effects, on the patterns of apertures mounted on finite size ground planes, can be accounted for by diffraction techniques. They will be introduced and illustrated in Section 12.9. Computed results, which include diffractions, agree extremely well with measurements [8]–[10].

C. Beamwidths

For the E-plane pattern given by (12-24b), the maximum radiation is directed along the z-axis ($\theta = 0$). The nulls (zeros) occur when

$$\frac{kb}{2} \sin \theta |_{\theta=\theta_n} = n\pi, \quad n = 1, 2, 3, \ldots \quad (12\text{-}26)$$

or at the angles of

$$\theta_n = \sin^{-1}\left(\frac{2n\pi}{kb}\right) = \sin^{-1}\left(\frac{n\lambda}{b}\right) \text{ rad} \qquad (12\text{-}26a)$$
$$= 57.3 \sin^{-1}\left(\frac{n\lambda}{b}\right) \text{ degrees}, \quad n = 1, 2, 3, \ldots$$

If $b \gg n\lambda$, (12-26a) reduces approximately to

$$\theta_n \simeq \frac{n\lambda}{b} \text{ rad} = 57.3 \left(\frac{n\lambda}{b}\right) \text{ degrees}, \quad n = 1, 2, 3, \ldots \qquad (12\text{-}26b)$$

The total *beamwidth between nulls* is given by

$$\Theta_n = 2\theta_n = 2\sin^{-1}\left(\frac{n\lambda}{b}\right) \text{ rad} \qquad (12\text{-}27)$$
$$= 114.6 \sin^{-1}\left(\frac{n\lambda}{b}\right) \text{ degrees}, \quad n = 1, 2, 3, \ldots$$

or approximately (for large apertures, $b \gg n\lambda$) by

$$\Theta_n \simeq \frac{2n\lambda}{b} \text{ rad} = 114.6 \left(\frac{n\lambda}{b}\right) \text{ degrees}, \quad n = 1, 2, 3, \ldots \qquad (12\text{-}27a)$$

The *first-null beamwidth* (FNBW) is obtained by letting $n = 1$.

The half-power point occurs when (see Appendix I)

$$\frac{kb}{2} \sin\theta\big|_{\theta=\theta_h} = 1.391 \qquad (12\text{-}28)$$

or at an angle of

$$\theta_h = \sin^{-1}\left(\frac{2.782}{kb}\right) = \sin^{-1}\left(\frac{0.443\lambda}{b}\right) \text{ rad} \qquad (12\text{-}28a)$$
$$= 57.3 \sin^{-1}\left(\frac{0.443\lambda}{b}\right) \text{ degrees}$$

If $b \gg 0.443\lambda$, (12-28a) reduces approximately to

$$\theta_h \simeq \left(0.443\frac{\lambda}{b}\right) \text{ rad} = 25.38 \left(\frac{\lambda}{b}\right) \text{ degrees} \qquad (12\text{-}28b)$$

Thus the total *half-power beamwidth* (HPBW) is given by

$$\Theta_h = 2\theta_h = 2\sin^{-1}\left(\frac{0.443\lambda}{b}\right) \text{ rad} = 114.6 \sin^{-1}\left(\frac{0.443\lambda}{b}\right) \text{ degrees} \qquad (12\text{-}29)$$

or approximately (when $b \gg 0.443\lambda$) by

$$\Theta_h \simeq \left(0.886\frac{\lambda}{b}\right) \text{ rad} = 50.8\left(\frac{\lambda}{b}\right) \text{ degrees} \qquad (12\text{-}29\text{a})$$

The maximum of the first side lobe occurs when (see Appendix I)

$$\frac{kb}{2}\sin\theta\big|_{\theta=\theta_s} = 4.494 \qquad (12\text{-}30)$$

or at an angle of

$$\theta_s = \sin^{-1}\left(\frac{8.988}{kb}\right) = \sin^{-1}\left(\frac{1.43\lambda}{b}\right) \text{ rad} = 57.3\sin^{-1}\left(\frac{1.43\lambda}{b}\right) \text{ degrees} \qquad (12\text{-}30\text{a})$$

If $b \gg 1.43\lambda$, (12-30a) reduces to

$$\theta_s \simeq 1.43\left(\frac{\lambda}{b}\right) \text{ rad} = 81.9\left(\frac{\lambda}{b}\right) \text{ degrees} \qquad (12\text{-}30\text{b})$$

The total beamwidth between first side lobes (FSLBW) is given by

$$\Theta_s = 2\theta_s = 2\sin^{-1}\left(\frac{1.43\lambda}{b}\right) \text{ rad} = 114.6\sin^{-1}\left(\frac{1.43\lambda}{b}\right) \text{ degrees} \qquad (12\text{-}30\text{c})$$

or approximately (when $b \gg 1.43\lambda$) by

$$\Theta_s \simeq 2.86\left(\frac{\lambda}{b}\right) \text{ rad} = 163.8\left(\frac{\lambda}{b}\right) \text{ degrees} \qquad (12\text{-}30\text{d})$$

D. Side Lobe Level

The maximum of (12-24b) at the first side lobe is given by (see Appendix I)

$$|E_\theta(\theta = \theta_s)| = \left|\frac{\sin(4.494)}{4.494}\right| = 0.217 = -13.26 \text{ dB} \qquad (12\text{-}31)$$

which is 13.26 dB down from the maximum of the main lobe.

An approximate value of the maximum of the first side lobe can be obtained by assuming that the maximum of (12-24b) occurs when its numerator is maximum. That is, when

$$\frac{kb}{2}\sin\theta\big|_{\theta=\theta_s} \simeq \frac{3\pi}{2} \qquad (12\text{-}32)$$

Thus,

$$|E_\theta(\theta = \theta_s)| \simeq \frac{1}{3\pi/2} = 0.212 = -13.47 \text{ dB} \qquad (12\text{-}33)$$

These values are very close to the exact ones given by (12-31).

A similar procedure can be followed to find the nulls, 3-dB points, beamwidth between nulls and 3-dB points, angle where the maximum of first side lobe occurs, and its magnitude at that point for the *H*-plane pattern of (12-25b). A comparison between the *E*- and *H*-plane patterns of (12-24b) and (12-25b) shows that they are similar in form except for the additional $\cos\theta$ term that appears in (12-25b). An examination of the terms in (12-25b) reveals that the $\cos\theta$ term is a much slower varying function than the $\sin(ka\sin\theta/2)/(ka\sin\theta/2)$ term, especially when *a* is large.

As a first approximation, (12-26)–(12-33), with *b* replaced by *a*, can also be used for the *H*-plane. More accurate expressions can be obtained by also including the $\cos\theta$ term. In regions well removed from the major lobe, the inclusion of the $\cos\theta$ term becomes more essential for accurate results.

E. Directivity

The directivity for the aperture can be found using (12-23a)–(12-23c), (12-13)–(12-13a), and (2-19)–(2-22). The analytical details using this procedure, especially the integration to compute the radiated power (P_{rad}), are more cumbersome.

Because the aperture is mounted on an infinite ground plane, an alternate and much simpler method can be used to compute the radiated power. The average power density is first formed using the fields at the aperture, and it is then integrated over the physical bounds of the opening. The integration is confined to the physical bounds of the opening. Using Figure 12.7 and assuming that the magnetic field at the aperture is given by

$$\mathbf{H}_a = -\hat{\mathbf{a}}_x \frac{E_0}{\eta} \qquad (12\text{-}34)$$

where η is the intrinsic impedance, the radiated power reduces to

$$P_{\text{rad}} = \oiint_S \mathbf{W}_{\text{av}} \cdot d\mathbf{s} = \frac{|E_0|^2}{2\eta} \iint_{S_a} ds = ab\frac{|E_0|^2}{2\eta} \qquad (12\text{-}35)$$

The maximum radiation intensity (U_{max}), using the fields of (12-23a)–(12-23b), occurs toward $\theta = 0°$ and it is equal to

$$U_{\text{max}} = \left(\frac{ab}{\lambda}\right)^2 \frac{|E_0|^2}{2\eta} \qquad (12\text{-}36)$$

Thus the directivity is equal to

$$\boxed{D_0 = \frac{4\pi U_{\text{max}}}{P_{\text{rad}}} = \frac{4\pi}{\lambda^2}ab = \frac{4\pi}{\lambda^2}A_p = \frac{4\pi}{\lambda^2}A_{em}} \qquad (12\text{-}37)$$

where

A_p = physical area of the aperture
A_{em} = maximum effective area of the aperture

Using the definition of (2-110), it is shown that *the physical and maximum effective areas of a constant distribution aperture are equal.*

The beamwidths, side lobe levels, and directivity of this and other apertures are summarized in Table 12.1.

TABLE 12.1 Equivalents, Fields, Beamwidths, Side Lobe Levels, and Directivities of Rectangular Apertures

	Uniform Distribution Aperture on Ground Plane	Uniform Distribution Aperture in Free-Space	TE$_{10}$-Mode Distribution Aperture on Ground Plane
Aperture distribution of tangential components (analytical)	$\mathbf{E}_a = \hat{\mathbf{a}}_y E_0 \left.\begin{array}{l}-a/2 \leq x' \leq a/2 \\ -b/2 \leq y' \leq b/2\end{array}\right\}$	$\mathbf{E}_a = \hat{\mathbf{a}}_y E_0$ $\left.\begin{array}{l}-a/2 \leq x' \leq a/2 \\ -b/2 \leq y' \leq b/2\end{array}\right\}$ $\mathbf{H}_a = -\hat{\mathbf{a}}_x \dfrac{E_0}{\eta}$	$\mathbf{E}_a = \hat{\mathbf{a}}_y E_0 \cos\left(\dfrac{\pi}{a}x'\right) \left.\begin{array}{l}-a/2 \leq x' \leq a/2 \\ -b/2 \leq y' \leq b/2\end{array}\right\}$
Aperture distribution of tangential components (graphical)			
Equivalent	$\mathbf{M}_s = \begin{cases}-2\hat{\mathbf{n}} \times \mathbf{E}_a & \begin{array}{l}-a/2 \leq x' \leq a/2 \\ -b/2 \leq y' \leq b/2\end{array} \\ 0 & \text{elsewhere} \\ & \text{everywhere}\end{cases}$ $\mathbf{J}_s = 0$	$\mathbf{M}_s = -\hat{\mathbf{n}} \times \mathbf{E}_a \left.\begin{array}{l}-a/2 \leq x' \leq a/2 \\ -b/2 \leq y' \leq b/2\end{array}\right\}$ $\mathbf{J}_s = \hat{\mathbf{n}} \times \mathbf{H}_a$ $\mathbf{M}_s \simeq \mathbf{J}_s \simeq 0$ elsewhere	$\mathbf{M}_s = \begin{cases}-2\hat{\mathbf{n}} \times \mathbf{E}_a & \begin{array}{l}-a/2 \leq x' \leq a/2 \\ -b/2 \leq y' \leq b/2\end{array} \\ 0 & \text{elsewhere} \\ & \text{everywhere}\end{cases}$ $\mathbf{J}_s = 0$
Far-zone fields	$E_r = H_r = 0$ $E_\theta = C \sin\phi \dfrac{\sin X}{X} \dfrac{\sin Y}{Y}$ $E_\phi = C \cos\theta \cos\phi \dfrac{\sin X}{X} \dfrac{\sin Y}{Y}$ $H_\theta = -E_\phi/\eta$ $H_\phi = E_\theta/\eta$	$E_r = H_r = 0$ $E_\theta = \dfrac{C}{2} \sin\phi(1+\cos\theta) \dfrac{\sin X}{X} \dfrac{\sin Y}{Y}$ $E_\phi = \dfrac{C}{2} \cos\phi(1+\cos\theta) \dfrac{\sin X}{X} \dfrac{\sin Y}{Y}$ $H_\theta = -E_\phi/\eta$ $H_\phi = E_\theta/\eta$	$E_r = H_r = 0$ $E_\theta = -\dfrac{\pi}{2} C \sin\phi \dfrac{\cos X}{(X)^2 - \left(\dfrac{\pi}{2}\right)^2} \dfrac{\sin Y}{Y}$ $E_\phi = -\dfrac{\pi}{2} C \cos\theta \cos\phi \dfrac{\cos X}{(X)^2 - \left(\dfrac{\pi}{2}\right)^2} \dfrac{\sin Y}{Y}$ $H_\theta = -E_\phi/\eta$ $H_\phi = E_\theta/\eta$
$X = \dfrac{ka}{2} \sin\theta \cos\phi$			
$Y = \dfrac{kb}{2} \sin\theta \sin\phi$			
$C = j\dfrac{abkE_0 e^{-jkr}}{2\pi r}$			

Half-power beamwidth (degrees)	E-plane $b \gg \lambda$	$\dfrac{50.8}{b/\lambda}$	$\dfrac{50.8}{b/\lambda}$	$\dfrac{50.8}{b/\lambda}$
	H-plane $a \gg \lambda$	$\dfrac{50.8}{a/\lambda}$	$\dfrac{50.8}{a/\lambda}$	$\dfrac{68.8}{a/\lambda}$
First null beamwidth (degrees)	E-plane $b \gg \lambda$	$\dfrac{114.6}{b/\lambda}$	$\dfrac{114.6}{b/\lambda}$	$\dfrac{114.6}{b/\lambda}$
	H-plane $a \gg \lambda$	$\dfrac{114.6}{a/\lambda}$	$\dfrac{114.6}{a/\lambda}$	$\dfrac{171.9}{a/\lambda}$
First side lobe max. (to main max.) (dB)	E-plane	-13.26	-13.26	-13.26
	H-plane	-13.26 $a \gg \lambda$	-13.26 $a \gg \lambda$	-23 $a \gg \lambda$
Directivity D_0 (dimensionless)		$\dfrac{4\pi}{\lambda^2}(\text{area}) = 4\pi\left(\dfrac{ab}{\lambda^2}\right)$	$\dfrac{4\pi}{\lambda^2}(\text{area}) = 4\pi\left(\dfrac{ab}{\lambda^2}\right)$	$\dfrac{8}{\pi^2}\left[4\pi\left(\dfrac{ab}{\lambda^2}\right)\right] = 0.81\left[4\pi\left(\dfrac{ab}{\lambda^2}\right)\right]$

Example 12.2

A rectangular aperture with a constant field distribution, with $a = 3\lambda$ and $b = 2\lambda$, is mounted on an infinite ground plane. Compute the

a. FNBW in the E-plane
b. HPBW in the E-plane
c. FSLBW in the E-plane
d. FSLMM in the E-plane
e. directivity using (12-37)
f. directivity using the computer program **Directivity** at the end of Chapter 2, the fields of (12-23a)–(12-23f), and the formulation of Section 12.4

Solution:

a. Using (12-27)

$$\Theta_1 = 114.6 \sin^{-1}(\tfrac{1}{2}) = 114.6(0.524) = 60°$$

b. Using (12-29)

$$\Theta_h = 114.6 \sin^{-1}\left(\frac{0.443}{2}\right) = 114.6(0.223) = 25.6°$$

c. Using (12-30c)

$$\Theta_s = 2\theta_s = 114.6 \sin^{-1}\left(\frac{1.43}{2}\right) = 114.6(0.796) = 91.3°$$

d. Using (12-31)

$$|E_\theta|_{\theta=\theta_s} = 0.217 \simeq -13.26 \text{ dB}$$

e. Using (12-37)

$$D_0 = 4\pi(3)(2) = 75.4 = 18.77 \text{ dB}$$

f. Using the computer program at the end of Chapter 2

$$D_0 \simeq 80.4 = 19.05 \text{ dB}$$

The difference in directivity values using (12-37) and the computer program is not attributed to the accuracy of the numerical method. The main contributor is the aperture tangential magnetic field of (12-34), which was assumed to be related to the aperture tangential electric field by the intrinsic impedance. Although this is a good assumption for large size apertures, it is not exact. Therefore the directivity value computed using the computer program should be considered to be the more accurate.

12.5.2 Uniform Distribution in Space

The second aperture examined is that of Figure 12.7 when it is *not* mounted on an infinite ground plane. The field distribution is given by

$$\left.\begin{array}{l} \mathbf{E}_a = \hat{\mathbf{a}}_y E_0 \\ \mathbf{H}_a = -\hat{\mathbf{a}}_x \dfrac{E_0}{\eta} \end{array}\right\} \quad \begin{array}{l} -a/2 \le x' \le a/2 \\ -b/2 \le y' \le b/2 \end{array} \tag{12-38}$$

where E_0 is a constant. The geometry of the opening for this problem is identical to the previous one. However the equivalents and radiated fields are different, because this time the aperture is not mounted on an infinite ground plane.

A. Equivalent

To form the equivalent, a closed surface is chosen which again extends from $-\infty$ to $+\infty$ on the *x-y* plane. Over the entire surface $\mathbf{J}_s$ and $\mathbf{M}_s$ are formed. The difficulty encountered in this problem is that both $\mathbf{J}_s$ and $\mathbf{M}_s$ are not zero outside the opening, and expressions for them are not known there. The replacement of the semi-infinite medium to the left of the boundary (negative z) by an imaginary electric or magnetic conductor only eliminates one or the other current densities ($\mathbf{J}_s$ or $\mathbf{M}_s$) but not both. Thus, even though an exact equivalent for this problem exists in principle, it cannot be used practically because the fields outside the opening are not known *a priori*. We are therefore forced to adopt an approximate equivalent.

The usual and most accurate relaxation is to assume that both $\mathbf{E}_a$ and $\mathbf{H}_a$ (and in turn $\mathbf{M}_s$ and $\mathbf{J}_s$) exist over the opening but are zero outside it. It has been shown, by comparison with measurements and other available data, that this approximate equivalent yields the best results.

B. Radiated Fields

Using a procedure similar to that of the previous section, the radiation characteristics of this aperture can be derived. A summary of them is shown in Table 12.1.

The field components of this aperture are identical in form to those of the aperture when it is mounted on an infinite ground plane if the $(1 + \cos\theta)$ factor is replaced in the E_θ component by 2 and in the E_ϕ component by $2\cos\theta$. Thus for small values of θ (in the main lobe and especially near its maximum), the patterns of the two apertures are almost identical. This procedure can be used, in general, to relate the fields of an aperture when it is and it is not mounted on an infinite ground plane. However, the coordinate system chosen must have the z-axis perpendicular to the aperture.

A three-dimensional pattern for an aperture with $a = 3\lambda, b = 2\lambda$ was computed, and it is shown in Figure 12.11. The dimensions of this aperture are the same as those of Figure 12.8. However the angular limits over which the radiated fields now exist have been extended to $0° \le \theta \le 180°$. Although the general structures of the two patterns are similar, they are not identical. Because of the enlarged space over which fields now exist, additional minor lobes are formed.

C. Beamwidths and Side Lobe Levels

To find the beamwidths and the angle at which the maximum of the side lobe occurs, it is usually assumed that the $(1 + \cos\theta)$ term is a much slower varying function than the $\sin(ka\sin\theta/2)/(ka\sin\theta/2)$ or the $\sin(kb\sin\theta/2)/(kb\sin\theta/2)$ terms. This is an approximation, and it is more valid for large apertures (large a and/or b) and for angles near the main maximum. More accurate results can be obtained by considering the $(1 + \cos\theta)$ term. Thus (12-26)–(12-33) can be used, to a good approximation, to compute the beamwidths and side lobe level. A summary is included in Table 12.1.

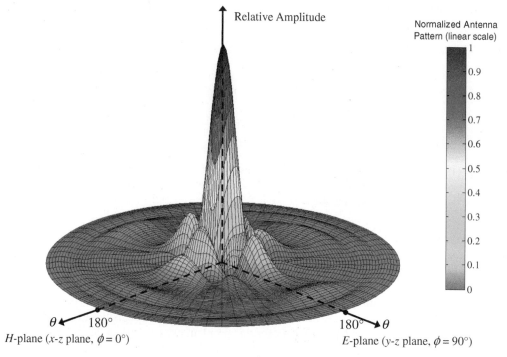

Figure 12.11 Three-dimensional field pattern of a constant field rectangular aperture ($a = 3\lambda, b = 2\lambda$).

D. Directivity

Although the physical geometry of the opening of this problem is identical to that of Section 12.5.1, their directivities are not identical. This is evident by examining their far-zone field expressions or by realizing that the fields outside the aperture along the x-y plane are not exactly the same.

To derive an exact expression for the directivity of this aperture would be a very difficult task. Since the patterns of the apertures are nearly the same, especially at the main lobe, their directivities are almost the same. To verify this, an example is taken.

Example 12.3

Repeat the problem of Example 12.2 for an aperture that is not mounted on an infinite ground plane.

Solution: Since the E-plane patterns of the two apertures are identical, the FNBW, HPBW, FSLBW, and FSLMM are the same. The directivities as computed by (12-37), are also the same. Since the fields radiated by the two apertures are not identical, their directivities computed using the far-zone fields will not be exactly the same. Therefore for this problem

$$D_0 \simeq 81.16 \text{(dimensionless)} = 19.09 \text{ dB}$$

As with Example 12.2, the directivities computed using (12-37) and the computer program do not agree exactly. For this problem, however, neither one is exact. For (12-37), it has been assumed that the aperture tangential magnetic field is related to the aperture tangential electric field by the intrinsic impedance η. This relationship is good but not exact. For the computer program, the formulation is based on the equivalent of this section where the fields outside the aperture were assumed to be negligible. Again this is a good assumption for some problems, but it is not exact.

A summary of the radiation characteristics of this aperture is included in Table 12.1 where it is compared with that of other apertures.

12.5.3 TE$_{10}$-Mode Distribution on an Infinite Ground Plane

In practice, a commonly used aperture antenna is that of a rectangular waveguide mounted on an infinite ground plane. At the opening, the field is usually approximated by the dominant TE$_{10}$-mode. Thus

$$\mathbf{E}_a = \hat{\mathbf{a}}_y E_0 \cos\left(\frac{\pi}{a}x'\right) \quad \begin{cases} -a/2 \leq x' \leq +a/2 \\ -b/2 \leq y' \leq +b/2 \end{cases} \quad (12\text{-}39)$$

A. Equivalent, Radiated Fields, Beamwidths, and Side Lobe Levels
Because the physical geometry of this antenna is identical to that of Figure 12.7, their equivalents and the procedure to analyze each one are identical. They differ only in the field distribution over the aperture.

The details of the analytical formulation are not included. However, a summary of its radiation characteristics is included in Table 12.1. The *E*-plane pattern of this aperture is identical in form (with the exception of a normalization factor) to the *E*-plane of the aperture of Section 12.5.1. This is expected, since the TE$_{10}$-mode field distribution along the *E*-plane (*y-z* plane) is also a constant. That is not the case for the *H*-plane or at all other points removed from the principal planes. To demonstrate that, a three-dimensional pattern for the TE$_{10}$-mode aperture with $a = 3\lambda, b = 2\lambda$ was computed and it is shown in Figure 12.12. This pattern should be compared with that of Figure 12.8.

The expressions for the beamwidths and side lobe level in the *E*-plane are identical to those given by (12-26)–(12-33). However those for the *H*-plane are more complex, and a simple procedure is

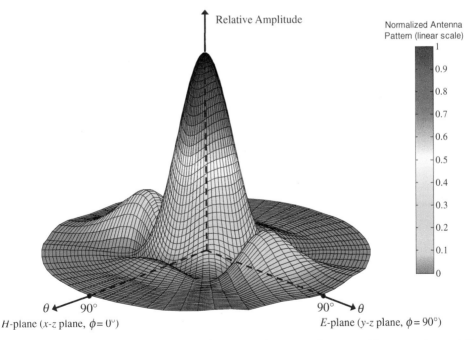

Figure 12.12 Three-dimensional field pattern of a TE$_{10}$-mode rectangular waveguide mounted on an infinite ground plane ($a = 3\lambda, b = 2\lambda$).

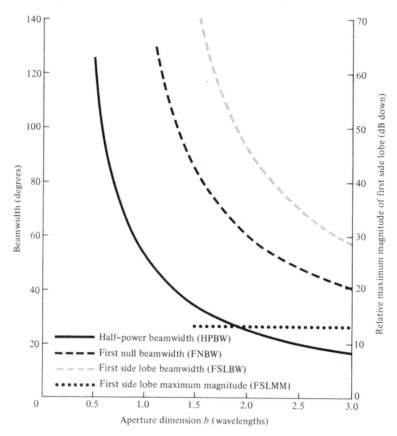

Figure 12.13 E-plane beamwidths and first side lobe relative maximum magnitude for TE_{10}-mode rectangular waveguide mounted on an infinite ground plane.

not available. Computations for the HPBW, FNBW, FSLBW, FSLMM in the E- and H-planes were made, and they are shown graphically in Figures 12.13 and 12.14.

When the same aperture is not mounted on a ground plane, the far-zone fields do not have to be re-derived but rather can be written by inspection. This is accomplished by introducing appropriately, in each of the field components (E_θ and E_ϕ) of the fourth column of Table 12.1, a $(1 + \cos\theta)/2$ factor, as is done for the fields of the two apertures in the second and third columns. This factor is appropriate when the z-axis is perpendicular to the plane of the aperture. Other similar factors will have to be used when either the x-axis or y-axis is perpendicular to the plane of the aperture.

B. Directivity and Aperture Efficiency

The directivity of this aperture is found in the same manner as that of the uniform distribution aperture of Section 12.5.1. Using the aperture electric field of (12-39), and assuming that the aperture magnetic field is related to the electric field by the intrinsic impedance η, the radiated power can be written as

$$P_{rad} = \oiint_S \mathbf{W}_{av} \cdot ds = ab\frac{|E_0|^2}{4\eta} \qquad (12\text{-}39a)$$

The maximum radiation intensity occurs at $\theta = 0°$, and it is given by

$$U_{max} = \frac{8}{\pi^2}\left(\frac{ab}{\lambda}\right)^2 \frac{|E_0|^2}{4\eta} \qquad (12\text{-}39b)$$

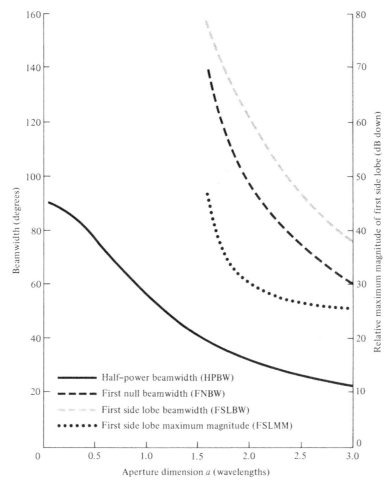

Figure 12.14 *H*-plane beamwidths and first side lobe relative maximum magnitude for TE_{10}-mode rectangular waveguide mounted on an infinite ground plane.

Thus the directivity is equal to

$$D_0 = \frac{8}{\pi^2}\left[ab\left(\frac{4\pi}{\lambda^2}\right)\right] = 0.81\left[ab\left(\frac{4\pi}{\lambda^2}\right)\right] = 0.81 A_p\left(\frac{4\pi}{\lambda^2}\right) = A_{em}\left(\frac{4\pi}{\lambda^2}\right) \qquad (12\text{-}39c)$$

In general, the maximum effective area A_{em} is related to the physical area A_p by

$$A_{em} = \varepsilon_{ap} A_p, \quad 0 \le \varepsilon_{ap} \le 1 \qquad (12\text{-}40)$$

where ε_{ap} is the aperture efficiency. For this problem $\varepsilon_{ap} = 8/\pi^2 \simeq 0.81$. The aperture efficiency is a figure of merit which indicates how efficiently the physical area of the antenna is utilized. Typically, aperture antennas have aperture efficiencies from about 30% to 90%, horns from 35% to 80% (optimum gain horns have $\varepsilon_{ap} \simeq 50\%$), and circular reflectors from 50% to 80%.

For reflectors, the aperture efficiency is a function of many factors. The most prominent are the spillover, amplitude taper, phase distribution, polarization uniformity, blockage, and surface random errors. These are discussed in detail in Section 15.4.1 of Chapter 15.

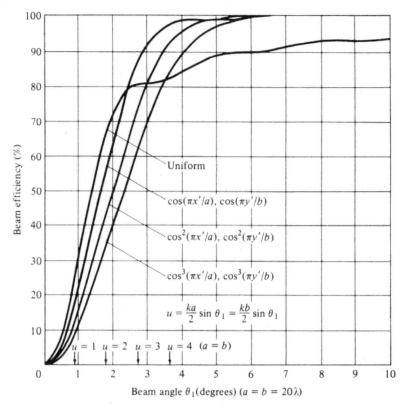

Figure 12.15 Beam efficiency versus half-cone angle θ_1, for a square aperture with different field distributions. The aperture is not mounted on an infinite ground plane. (SOURCE: D. G. Fink (ed.), *Electronics Engineers' Handbook*, Section 18 (by W. F. Croswell), McGraw-Hill, New York, 1975).

12.5.4 Beam Efficiency

The beam efficiency for an antenna was introduced in Section 2.10 and was defined by (2-53). When the aperture is mounted on the *x-y* plane, the beam efficiency can be calculated using (2-54). The beam efficiency can be used to judge the ability of the antenna to discriminate between signals received through its main lobe and those through the minor lobes. Beam efficiencies for rectangular apertures with different aperture field distributions are plotted, versus the half-cone angle θ_1, in Figure 12.15 [11]. The uniform field distribution aperture has the least ability to discriminate between main lobe and minor lobe signals. The aperture radiates in an unbounded medium, and it is not mounted on an infinite ground plane. The lower abscissa scale is in terms of θ_1 (*in degrees*), and it should be used only when $a = b = 20\lambda$. The upper abscissa scale is in terms of $u[u = (ka/2)\sin\theta_1 = (kb/2)\sin\theta_1]$, and it should be used for any square aperture.

Example 12.4

Determine the beam efficiency, within a cone of half-angle $\theta_1 = 10°$, for a square aperture with uniform field distribution and with

a. $a = b = 20\lambda$
b. $a = b = 3\lambda$

Solution: The solution is carried out using the curves of Figure 12.15.

a. When $a = b = 20\lambda$, the lower abscissa scale can be used. For $\theta_1 = 10°$, the efficiency for the uniform aperture is about 94%.
b. For $a = b = 3\lambda$ and $\theta_1 = 10°$

$$u = \frac{ka}{2} \sin\theta_1 = 3\pi \sin(10°) = 1.64$$

Using the upper abscissa scale, the efficiency for the uniform aperture at $u = 1.64$ is about 58%.

An antenna array of slotted rectangular waveguides used for the AWACS airborne system is shown in Figure 6.27. It utilizes waveguide sticks, with slits on their narrow wall.

A MATLAB computer program, designated as **Aperture**, has been developed to compute and display different radiation characteristics of rectangular and circular apertures. The description of the program is found in the corresponding READ ME file included in the publisher's website for this book.

12.6 CIRCULAR APERTURES

A widely used microwave antenna is the circular aperture. One of the attractive features of this configuration is its simplicity in construction. In addition, closed form expressions for the fields of all the modes that can exist over the aperture can be obtained.

The procedure followed to determine the fields radiated by a circular aperture is identical to that of the rectangular, as summarized in Section 12.3. The primary differences lie in the formulation of the equivalent current densities $(J_x, J_y, J_z, M_x, M_y, M_z)$, the differential paths from the source to the observation point ($r' \cos\psi$), and the differential area (ds'). Before an example is considered, these differences will be reformulated for the circular aperture.

Because of the circular profile of the aperture, it is often convenient and desirable to adopt cylindrical coordinates for the solution of the fields. In most cases, therefore, the electric and magnetic field components over the circular opening will be known in cylindrical form; that is, $E_\rho, E_\phi, E_z, H_\rho, H_\phi$, and H_z. Thus the components of the equivalent current densities $\mathbf{M}_s$ and $\mathbf{J}_s$ would also be conveniently expressed in cylindrical form $(M_\rho, M_\phi, M_z, J_\rho, J_\phi, J_z)$. In addition, the required integration over the aperture to find $N_\theta, N_\phi, L_\theta$, and L_ϕ of (12-12a)–(12-12d) should also be done in cylindrical coordinates. It is then desirable to reformulate $r' \cos\psi$ and ds', as given by (12-15a)–(12-16c).

The most convenient position for placing the aperture is that shown in Figure 12.16 (aperture on *x-y* plane). The transformation between the rectangular and cylindrical components of $\mathbf{J}_s$ is given by (see Appendix VII)

$$\begin{bmatrix} J_x \\ J_y \\ J_z \end{bmatrix} = \begin{bmatrix} \cos\phi' & -\sin\phi' & 0 \\ \sin\phi' & \cos\phi' & 0 \\ 0 & 0 & 1 \end{bmatrix} \begin{bmatrix} J_\rho \\ J_\phi \\ J_z \end{bmatrix} \quad (12\text{-}41\text{a})$$

A similar transformation exists for the components of $\mathbf{M}_s$. The rectangular and cylindrical coordinates are related by (see Appendix VII)

$$x' = \rho' \cos\phi'$$
$$y' = \rho' \sin\phi' \quad (12\text{-}41\text{b})$$
$$z' = z'$$

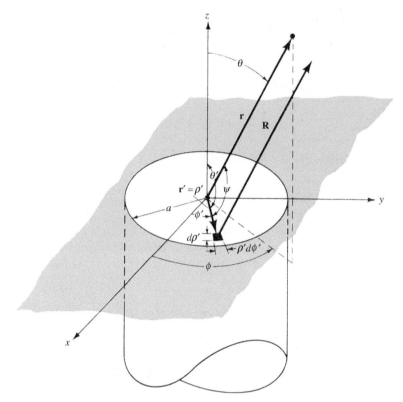

Figure 12.16 Circular aperture mounted on an infinite ground plane.

Using (12-41a), (12-12a)–(12-12d) can be written as

$$N_\theta = \iint_S [J_\rho \cos\theta \cos(\phi - \phi') + J_\phi \cos\theta \sin(\phi - \phi') - J_z \sin\theta] \\ \times e^{+jkr' \cos\psi} \, ds' \quad (12\text{-}42a)$$

$$N_\phi = \iint_S [-J_\rho \sin(\phi - \phi') + J_\phi \cos(\phi - \phi')] e^{+jkr' \cos\psi} \, ds' \quad (12\text{-}42b)$$

$$L_\theta = \iint_S [M_\rho \cos\theta \cos(\phi - \phi') + M_\phi \cos\theta \sin(\phi - \phi') - M_z \sin\theta] \\ \times e^{+jkr' \cos\psi} \, ds' \quad (12\text{-}42c)$$

$$L_\phi = \iint_S [-M_\rho \sin(\phi - \phi') + M_\phi \cos(\phi - \phi')] e^{+jkr' \cos\psi} \, ds' \quad (12\text{-}42d)$$

where $r' \cos\psi$ and ds' can be written, using (12-15c) and (12-41b), as

$$r' \cos\psi = x' \sin\theta \cos\phi + y' \sin\theta \sin\phi = \rho' \sin\theta \cos(\phi - \phi') \quad (12\text{-}43a)$$

$$ds' = dx' \, dy' = \rho' \, d\rho' \, d\phi' \quad (12\text{-}43b)$$

In summary, for a circular aperture antenna the fields radiated can be obtained by *either* of the following:

1. If the fields over the aperture are known in *rectangular components*, use the same procedure as for the rectangular aperture with (12-43a) and (12-43b) substituted in (12-12a)–(12-12d).
2. If the fields over the aperture are known in *cylindrical components*, use the same procedure as for the rectangular aperture with (12-42a)–(12-42d), along with (12-43a) and (12-43b), taking the place of (12-12a)–(12-12d).

12.6.1 Uniform Distribution on an Infinite Ground Plane

To demonstrate the methods, the field radiated by a circular aperture mounted on an infinite ground plane will be formulated. To simplify the mathematical details, the field over the aperture is assumed to be constant and given by

$$\mathbf{E}_a = \hat{\mathbf{a}}_y E_0 \quad \rho' \leq a \tag{12-44}$$

where E_0 is a constant.

A. Equivalent and Radiation Fields

The equivalent problem of this is identical to that of Figure 12.7. That is,

$$\mathbf{M}_s = \begin{cases} -2\hat{\mathbf{n}} \times \mathbf{E}_a = \hat{\mathbf{a}}_x 2E_0 & \rho' \leq a \\ 0 & \text{elsewhere} \end{cases}$$
$$\mathbf{J}_s = 0 \quad \text{everywhere} \tag{12-45}$$

Thus,

$$N_\theta = N_\phi = 0 \tag{12-46}$$

$$L_\theta = 2E_0 \cos\theta \cos\phi \int_0^a \rho' \left[\int_0^{2\pi} e^{+jk\rho' \sin\theta \cos(\phi-\phi')} \, d\phi' \right] d\rho' \tag{12-47}$$

Because

$$\int_0^{2\pi} e^{+jk\rho' \sin\theta \cos(\phi-\phi')} \, d\phi' = 2\pi J_0(k\rho' \sin\theta) \tag{12-48}$$

(12-47) can be written as

$$L_\theta = 4\pi E_0 \cos\theta \cos\phi \int_0^a J_0(k\rho' \sin\theta) \rho' \, d\rho' \tag{12-49}$$

where $J_0(t)$ is the Bessel function of the first kind of order zero. Making the substitution

$$t = k\rho' \sin\theta$$
$$dt = k \sin\theta \, d\rho' \tag{12-49a}$$

670 APERTURE ANTENNAS

reduces (12-49) to

$$L_\theta = \frac{4\pi E_0 \cos\theta \cos\phi}{(k\sin\theta)^2} \int_0^{ka\sin\theta} tJ_0(t)\,dt \tag{12-49b}$$

Since

$$\int_0^\beta zJ_0(z)\,dz = zJ_1(z)\Big|_0^\beta = \beta J_1(\beta) \tag{12-50}$$

where $J_1(\beta)$ is the Bessel function of order one, (12-49b) takes the form of

$$L_\theta = 4\pi a^2 E_0 \left\{ \cos\theta \cos\phi \left[\frac{J_1(ka\sin\theta)}{ka\sin\theta}\right] \right\} \tag{12-51}$$

Similarly

$$L_\phi = -4\pi a^2 E_0 \sin\phi \left[\frac{J_1(ka\sin\theta)}{ka\sin\theta}\right] \tag{12-52}$$

Using (12-46), (12-51), and (12-52), the electric field components of (12-10a)–(12-10c) can be written as

$$E_r = 0 \tag{12-53a}$$

$$E_\theta = j\frac{ka^2 E_0 e^{-jkr}}{r} \left\{ \sin\phi \left[\frac{J_1(ka\sin\theta)}{ka\sin\theta}\right] \right\} \tag{12-53b}$$

$$E_\phi = j\frac{ka^2 E_0 e^{-jkr}}{r} \left\{ \cos\theta \cos\phi \left[\frac{J_1(ka\sin\theta)}{ka\sin\theta}\right] \right\} \tag{12-53c}$$

In the principal *E*- and *H*-planes, the electric field components simplify to

E-Plane ($\phi = \pi/2$)

$$E_r = E_\phi = 0 \tag{12-54a}$$

$$E_\theta = j\frac{ka^2 E_0 e^{-jkr}}{r} \left[\frac{J_1(ka\sin\theta)}{ka\sin\theta}\right] \tag{12-54b}$$

H-Plane ($\phi = 0$)

$$E_r = E_\theta = 0 \tag{12-55a}$$

$$E_\phi = j\frac{ka^2 E_0 e^{-jkr}}{r} \left\{ \cos\theta \left[\frac{J_1(ka\sin\theta)}{ka\sin\theta}\right] \right\} \tag{12-55b}$$

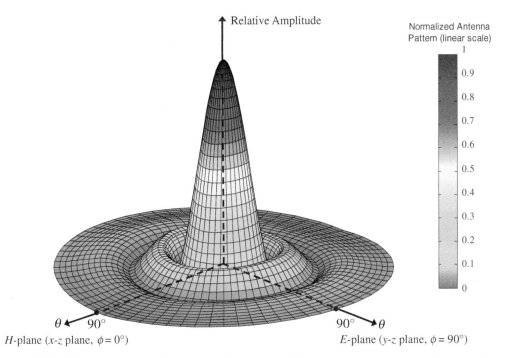

| θ ← 90° | | 90° → θ |
| H-plane (x-z plane, φ = 0°) | | E-plane (y-z plane, φ = 90°) |

Figure 12.17 Three-dimensional field pattern of a constant field circular aperture mounted on an infinite ground plane ($a = 1.5\lambda$).

A three-dimensional pattern has been computed for the constant field circular aperture of $a = 1.5\lambda$, and it is shown in Figure 12.17. The pattern of Figure 12.17 seems to be symmetrical. However closer observation, especially through the two-dimensional E- and H-plane patterns, will reveal that not to be the case. It does, however, possess characteristics that are almost symmetrical.

B. Beamwidth, Side Lobe Level, and Directivity

Exact expressions for the beamwidths and side lobe levels cannot be obtained easily. However approximate expressions are available, and they are shown tabulated in Table 12.2. More exact data can be obtained by numerical methods.

Since the field distribution over the aperture is constant, the directivity is given by

$$D_0 = \frac{4\pi}{\lambda^2} A_{em} = \frac{4\pi}{\lambda^2} A_p = \frac{4\pi}{\lambda^2}(\pi a^2) = \left(\frac{2\pi a}{\lambda}\right)^2 = \left(\frac{C}{\lambda}\right)^2 \quad (12\text{-}56)$$

since the maximum effective area A_{em} is equal to the physical area A_p of the aperture [as shown for the rectangular aperture in (12-37)].

A summary of the radiation parameters of this aperture is included in Table 12.2.

12.6.2 TE$_{11}$-Mode Distribution on an Infinite Ground Plane

A very practical antenna is a circular waveguide of radius a mounted on an infinite ground plane, as shown in Figure 12.16. However, the field distribution over the aperture is usually that of the

TABLE 12.2 Equivalents, Fields, Beamwidths, Side Lobe Levels, and Directivities of Circular Apertures

	Uniform Distribution Aperture on Ground Plane	TE_{11}-Mode Distribution Aperture on Ground Plane
Aperture distribution of tangential components (analytical)	$\mathbf{E}_a = \hat{\mathbf{a}}_y E_0 \quad \rho' \leq a$	$\mathbf{E}_a = \hat{\mathbf{a}}_\rho E_\rho + \hat{\mathbf{a}}_\phi E_\phi$ $E_\rho = E_0 J_1(\chi'_{11}\rho'/a)\sin\phi'/\rho' \quad \left. \begin{array}{l} \rho' \leq a \\ \chi'_{11} = 1.841 \\ ' = \dfrac{\partial}{\partial \rho'} \end{array} \right.$ $E_\phi = E_0 J'_1(\chi'_{11}\rho'/a)\cos\phi'$
Aperture distribution of tangential components (graphical)		
Equivalent	$\mathbf{M}_s = \begin{cases} -2\hat{\mathbf{n}} \times \mathbf{E}_a & \rho' \leq a \\ 0 & \text{elsewhere} \end{cases}$ $\mathbf{J}_s = 0 \quad \text{everywhere}$	$\mathbf{M}_s = \begin{cases} -2\hat{\mathbf{n}} \times \mathbf{E}_a & \rho' \leq a \\ 0 & \text{elsewhere} \end{cases}$ $\mathbf{J}_s = 0 \quad \text{everywhere}$
Far-zone fields $Z = ka\sin\theta$ $C_1 = j\dfrac{ka^2 E_0 e^{-jkr}}{r}$ $C_2 = j\dfrac{kaE_0 J_1(\chi'_{11})e^{-jkr}}{r}$ $\chi'_{11} = 1.841$	$E_r = H_r = 0$ $E_\theta = jC_1 \sin\phi \dfrac{J_1(Z)}{Z}$ $E_\phi = jC_1 \cos\theta \cos\phi \dfrac{J_1(Z)}{Z}$ $H_\theta = -E_\phi/\eta$ $H_\phi = E_\theta/\eta$	$E_r = H_r = 0$ $E_\theta = C_2 \sin\phi \dfrac{J_1(Z)}{Z}$ $E_\phi = C_2 \cos\theta \cos\phi \dfrac{J'_1(Z)}{1 - (Z/\chi'_{11})^2}$ $H_\theta = -E_\phi/\eta$ $H_\phi = E_\theta/\eta$ $J'_1(Z) = J_0(Z) - J_1(Z)/Z$

Half-power beamwidth (degrees)	E-plane $a \gg \lambda$	$\dfrac{29.2}{a/\lambda}$	$\dfrac{29.2}{a/\lambda}$
	H-plane $a \gg \lambda$	$\dfrac{29.2}{a/\lambda}$	$\dfrac{37.0}{a/\lambda}$
First null beamwidth (degrees)	E-plane $a \gg \lambda$	$\dfrac{69.9}{a/\lambda}$	$\dfrac{69.9}{a/\lambda}$
	H-plane $a \gg \lambda$	$\dfrac{69.9}{a/\lambda}$	$\dfrac{98.0}{a/\lambda}$
First side lobe max. (to main max.) (dB)	E-plane	-17.6	-17.6
	H-plane	-17.6	-26.2
Directivity D_0 (dimensionless)		$\dfrac{4\pi}{\lambda^2}(\text{area}) = \dfrac{4\pi}{\lambda^2}(\pi a^2) = \left(\dfrac{2\pi a}{\lambda}\right)^2$	$0.836\left(\dfrac{2\pi a}{\lambda}\right)^2 = 10.5\pi\left(\dfrac{a}{\lambda}\right)^2$

674 APERTURE ANTENNAS

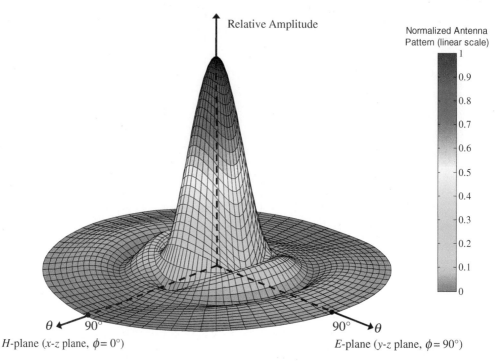

H-plane (*x*-*z* plane, $\phi = 0°$)   *E*-plane (*y*-*z* plane, $\phi = 90°$)

Figure 12.18 Three-dimensional field pattern of a TE_{11}-mode circular waveguide mounted on an infinite ground plane ($a = 1.5\lambda$).

dominant TE_{11}-mode for a circular waveguide given by

$$\left.\begin{array}{l} E_\rho = \dfrac{E_0}{\rho'} J_1\left(\dfrac{\chi'_{11}}{a}\rho'\right) \sin\phi' \\[6pt] E_\phi = E_0 \dfrac{\partial}{\partial \rho'}\left[J_1\left(\dfrac{\chi'_{11}}{a}\rho'\right)\right] \cos\phi' \\[6pt] E_z = 0 \\[4pt] \chi'_{11} = 1.841 \end{array}\right\} \qquad (12\text{-}57)$$

The analysis of this problem is assigned at the end of this chapter as an exercise to the reader (Problem 12.35). However, a three-dimensional pattern for $a = 1.5\lambda$ was calculated, and it is shown in Figure 12.18. This pattern should be compared with that of Figure 12.17 for the constant aperture field distribution.

The beamwidths and the side lobe levels in the *E*- and *H*-planes are different, and exact closed-form expressions cannot be obtained. However, they can be calculated using iterative methods, and the data are shown in Figures 12.19 and 12.20 for the *E*- and *H*-planes, respectively.

A summary of all the radiation characteristics is included in Table 12.2. When the same apertures of Table 12.2 are not mounted on a ground plane, the far-zone fields do not have to be re-derived but rather can be written by inspection. This is accomplished by introducing appropriately, in each of the field components (E_θ and E_ϕ) of the second and third columns of Table 12.2, a $(1 + \cos\theta)/2$ factor, as was done for the fields of the two apertures in the second and third columns of Table 12.1.

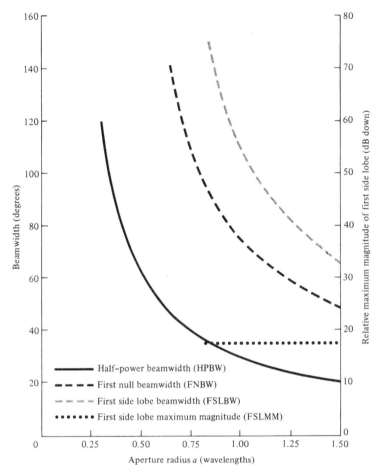

Figure 12.19 E-plane beamwidths and first side lobe relative maximum magnitude for TE_{11}-mode circular aperture mounted on an infinite ground plane.

12.6.3 Beam Efficiency

Beam efficiency, as defined by (2-53) and calculated by (2-54), for circular apertures not mounted on infinite ground planes is shown in Figure 12.21 [11]. The lower abscissa scale (in degrees) is in terms of the half-cone angle θ_1 (in degrees), and it should be used only when the radius of the aperture is $20\lambda (a = 20\lambda)$. The upper abscissa scale is in terms of $u (u = ka \sin \theta_1)$, and it should be used for any radius circular aperture.

The procedure for finding the beam efficiency of a circular aperture is similar to that of a rectangular aperture as discussed in Section 12.5.4, illustrated in Figure 12.15, and demonstrated by Example 12.4.

A MATLAB computer program, designated as **Aperture**, has been developed to compute and display different radiation characteristics of rectangular and circular apertures. The description of the program is found in the corresponding READ ME file included in the publisher's website for this book.

12.7 DESIGN CONSIDERATIONS

As is the case for arrays, aperture antennas can be designed to control their radiation characteristics. Typically the level of the minor lobes can be controlled by tapering the distribution across the

676 APERTURE ANTENNAS

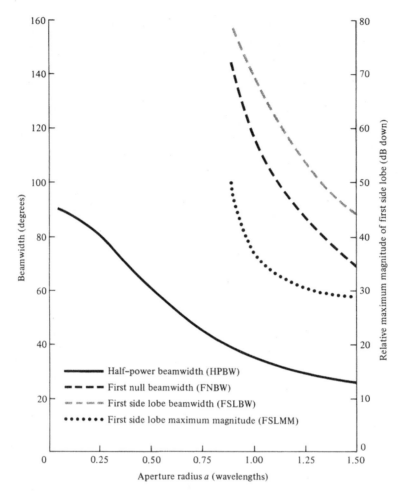

Figure 12.20 *H*-plane beamwidths and first side lobe relative maximum magnitude for TE_{11}-mode circular waveguide mounted on an infinite ground plane.

aperture; the smoother the taper from the center of the aperture toward the edge, the lower the side lobe level and the larger the half-power beamwidth, and conversely. Therefore a very smooth taper, such as that represented by a binomial distribution or others, would result in very low side lobes but larger half-power beamwidths. In contrast, an abrupt distribution, such as that of uniform illumination, exhibits the smaller half-power beamwidth but the highest side lobe level (about - 13.5 dB). Therefore if it is desired to achieve simultaneously both a very low sidelobe level, as well as a small half-power beamwidth, a compromise has to be made. Typically an intermediate taper, such as that of a Tschebyscheff distribution or any other similar one, will have to be selected. This has been discussed in detail both in Chapter 6 for arrays and in Chapter 7 for continuous sources. These can be used to design continuous distributions for apertures.

Aperture antennas, both rectangular and circular, can also be designed for satellite applications where the beamwidth can be used to determine the "footprint" area of the coverage. In such designs, it is important to relate the beamwidth to the size of the aperture. In addition, it is also important to maximize the directivity of the antennas within a desired angular sector defined by the beamwidth, especially at the edge of coverage (EOC) [12]. This can be accomplished, using approximate closed-form expressions, as outlined in [12]. This procedure was used in Section 6.11 of Chapter 6 for arrays, and it is applicable for apertures, both rectangular and circular.

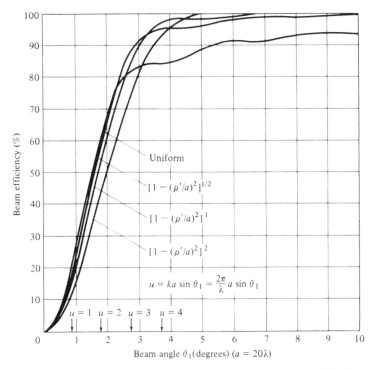

Figure 12.21 Beam efficiency versus half-cone angle θ_1, for a circular aperture with different field distributions. The aperture is not mounted on an infinite ground plane. (SOURCE: D. G. Fink (ed.), *Electronics Engineers' Handbook*, Section 18 (by W. F. Croswell), McGraw-Hill, New York, 1975).

12.7.1 Rectangular Aperture

For a rectangular aperture, of dimensions a and b, and with a uniform distribution, the procedure to determine the optimum aperture dimensions a,b to maximize the directivity at an edge angle θ_c of a given angular sector ($0 \leq \theta \leq \theta_c$) is identical to that outlined in Section 6.11. Thus to determine the optimum dimension b of the aperture so that the directivity is maximum at an edge-of-coverage angle θ_{ce} of an angular sector $0 \leq \theta \leq \theta_{ce}$ in the E-plane is given by (6-105a), or

$$E\text{-Plane:} \quad b = \frac{\lambda}{2 \sin \theta_{ce}} \tag{12-58a}$$

Similarly for the H-plane, the optimum dimension a is determined by

$$H\text{-Plane:} \quad a = \frac{\lambda}{2 \sin \theta_{ch}} \tag{12-58b}$$

where θ_{ch} is the angle, in the H-plane, at the edge-of-coverage (EOC) angular sector where the directivity needs to be maximized.

Since the aperture antenna is uniformly illuminated, the directivity of (6-103) based on the optimum dimensions of (12-58a) and (12-58b) is

$$D_0 = \frac{4\pi}{\lambda^2} A_{em} = \frac{4\pi}{\lambda^2} A_p = \frac{4\pi}{\lambda^2} \left(\frac{\lambda}{2 \sin \theta_{ce}} \right) \left(\frac{\lambda}{2 \sin \theta_{ch}} \right) \tag{12-59}$$

12.7.2 Circular Aperture

A procedure similar to that for the rectangular aperture can be used for the circular aperture. In fact, it can be used for circular apertures with uniform distributions as well as tapered (parabolic or parabolic with a pedestal) [12].

For a circular aperture with uniform distribution, the normalized power pattern multiplied by the maximum directivity can be written as

$$P(\theta) = (2\pi a)^2 \left\{ \frac{2J_1(ka \sin \theta)}{ka \sin \theta} \right\}^2 \qquad (12\text{-}60)$$

The maximum value of (12-60) occurs when $\theta = 0$. However, for any other angle $\theta = \theta_c$, the maximum of the pattern occurs when

$$ka \sin \theta_c = 1.841 \qquad (12\text{-}61)$$

or

$$a = \frac{1.841\lambda}{2\pi \sin \theta_c} = \frac{\lambda}{3.413 \sin \theta_c} \qquad (12\text{-}61a)$$

Therefore to maximize the directivity at the edge $\theta = \theta_c$ of a given angular sector $0 \leq \theta \leq \theta_c$, the optimum radius of the uniformly illuminated circular aperture must be chosen according to (12-61a).

The maximum value of (12-60), which occurs at $\theta = 0$, is equal to

$$P(\theta = 0)|_{max} = (2\pi a)^2 \qquad (12\text{-}62)$$

while at the edge of the angular sector $(\theta = \theta_c)$ is equal to

$$P(\theta = \theta_c) = (2\pi a)^2 \left\{ \frac{2(0.5818)}{1.841} \right\}^2 = (2\pi a)^2 (0.3995) \qquad (12\text{-}63)$$

Therefore the value of the directivity at the edge of the desired coverage $(\theta = \theta_c)$, relative to its maximum value at $\theta = 0$, is

$$\frac{P(\theta = \theta_c)}{P(\theta = 0)} = 0.3995 = -3.985 \text{ dB} \qquad (12\text{-}64)$$

Since the aperture is uniformly illuminated, the directivity based on the optimum radius of (12-61a) is

$$D_0 = \frac{4\pi}{\lambda^2} A_p = \frac{4\pi}{\lambda^2} \pi \left(\frac{1.841\lambda}{2\pi \sin \theta_c} \right)^2 = \frac{3.3893}{\sin^2 \theta_c} = \frac{1.079\pi}{\sin^2 \theta_c} \qquad (12\text{-}65)$$

A similar procedure can be followed for circular apertures with radial taper (parabolic) and radial taper squared of Table 7.2, as well as radial taper (parabolic) with pedestal. The characteristics of these, along with those of the uniform, are listed in Table 12.3.

TABLE 12.3 Edge-of-Coverage (EOC) Designs for Square and Circular Apertures

Aperture	Distribution	Size Square: Side Circular: Radius	Directivity	EOC Directivity (relative to peak)
Square	Uniform	$\dfrac{\lambda}{2\sin(\theta_c)}$	$\dfrac{\pi}{\sin^2(\theta_c)}$	−3.920 dB
Circular	Uniform	$\dfrac{\lambda}{3.413\sin(\theta_c)}$	$\dfrac{1.086\pi}{\sin^2(\theta_c)}$	−3.985 dB
Circular	Parabolic taper	$\dfrac{\lambda}{2.732\sin(\theta_c)}$	$\dfrac{1.263\pi}{\sin^2(\theta_c)}$	−4.069 dB
Circular	Parabolic taper with −10 dB pedestal	$\dfrac{\lambda}{3.064\sin(\theta_c)}$	$\dfrac{1.227\pi}{\sin^2(\theta_c)}$	−4.034 dB

(SOURCE: K. Praba, "Optimal Aperture for Maximum Edge-of-Coverage (EOC) Directivity," *IEEE Antennas & Propagation Magazine*, Vol. 36, No. 3, pp. 72–74, June 1994. © (1994) IEEE)

Example 12.5

It is desired to design an aperture antenna, with uniform illumination, so that the directivity is maximized at an angle 30° from the normal to the aperture. Determine the optimum dimension and its associated directivity when the aperture is

a. Square
b. Circular

Solution: For a square aperture $\theta_{ce} = \theta_{ch}$. Therefore the optimum dimension, according to (12-58a) or (12-58b), is

$$a = b = \frac{\lambda}{2\sin(30°)} = \lambda$$

while the directivity, according to (12-59), is

$$D_0 = \frac{\pi}{\sin^2\theta_c} = \frac{\pi}{\sin^2(30°)} = 12.5664 = 10.992 \text{ dB}$$

The directivity at $\theta = 30°$ is −3.920 dB from the maximum at $\theta = 0°$, or 7.072 dB.

For a circular aperture the optimum radius, according to (12-61a), is

$$a = \frac{\lambda}{3.413\sin(30°)} = \frac{\lambda}{3.413(0.5)} = 0.586\lambda$$

while the directivity, according to (12-65), is

$$D_0 = \frac{1.079\pi}{\sin^2\theta_c} = \frac{1.079\pi}{\sin^2(30°)} = 13.559 = 11.32 \text{ dB}$$

The directivity at $\theta = 30°$ is −3.985 dB from the maximum at $\theta = 0°$, or 7.365 dB.

680 APERTURE ANTENNAS

12.8 BABINET'S PRINCIPLE

Now that wire and aperture antennas have been analyzed, one may inquire as to whether there is any relationship between them. This can be answered better by first introducing *Babinet's principle* which in optics states that *when the field behind a screen with an opening is added to the field of a complementary structure, the sum is equal to the field when there is no screen.* Babinet's principle in optics does not consider polarization, which is so vital in antenna theory; it deals primarily with absorbing screens. An extension of Babinet's principle, which includes polarization and the more practical conducting screens, was introduced by Booker [13], [14]. Referring to Figure 12.22(a), let us assume that an electric source **J** radiates into an unbounded medium of intrinsic impedance $\eta = (\mu/\varepsilon)^{1/2}$ and produces at point P the fields $\mathbf{E}_0, \mathbf{H}_0$. The same fields can be obtained by combining the fields when the electric source radiates in a medium with intrinsic impedance $\eta = (\mu/\varepsilon)^{1/2}$ in the presence of

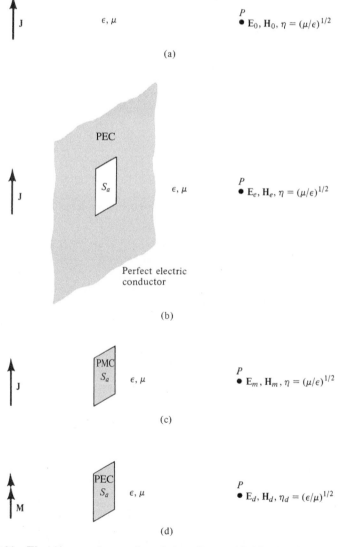

Figure 12.22 Electric source in an unbounded medium and Babinet's principle equivalents.

1. an infinite, planar, very thin, perfect electric conductor with an opening S_a, which produces at P the fields $\mathbf{E}_e, \mathbf{H}_e$ [Figure 12.22(b)]
2. a flat, very thin, perfect magnetic conductor S_a, which produces at P the fields $\mathbf{E}_m, \mathbf{H}_m$ [Figure 12.22(c)].

That is,

$$\mathbf{E}_0 = \mathbf{E}_e + \mathbf{E}_m$$
$$\mathbf{H}_0 = \mathbf{H}_e + \mathbf{H}_m \quad (12\text{-}66a)$$

The field produced by the source in Figure 12.22(a) can also be obtained by combining the fields of

1. an electric source $\mathbf{J}$ radiating in a medium with intrinsic impedance $\eta = (\mu/\varepsilon)^{1/2}$ in the presence of an infinite, planar, very thin, perfect electric conductor S_a, which produces at P the fields $\mathbf{E}_e, \mathbf{H}_e$ [Figure 12.22(b)]
2. a magnetic source $\mathbf{M}$ radiating in a medium with intrinsic impedance $\eta_d = (\varepsilon/\mu)^{1/2}$ in the presence of a flat, very thin, perfect electric conductor S_a, which produces at P the fields $\mathbf{E}_d, \mathbf{H}_d$ [Figure 12.22(d)]

That is,

$$\mathbf{E}_0 = \mathbf{E}_e + \mathbf{H}_d$$
$$\mathbf{H}_0 = \mathbf{H}_e - \mathbf{E}_d \quad (12\text{-}66b)$$

The dual of Figure 12.22(d) is more easily realized in practice than that of Figure 12.22(c).

To obtain Figure 12.22(d) from Figure 12.22(c), $\mathbf{J}$ is replaced by $\mathbf{M}$, $\mathbf{E}_m$ by $\mathbf{H}_d$, $\mathbf{H}_m$ by $-\mathbf{E}_d$, ε by μ, and μ by ε. This is a form of duality often used in electromagnetics (see Section 3.7, Table 3.2). The electric screen with the opening in Figure 12.22(b) and the electric conductor of Figure 12.22(d) are also dual. They are usually referred to as *complementary structures*, because when combined they form a single solid screen with no overlaps. A proof of Babinet's principle and its extension can be found in the literature [5].

Using Booker's extension it can be shown [13], [14] by referring to Figure 12.23, that if a screen and its complement are immersed in a medium with an intrinsic impedance η and have terminal

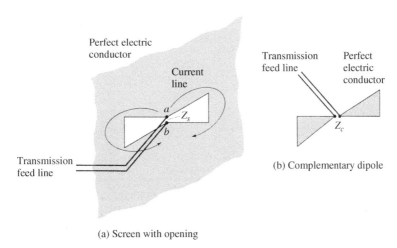

(a) Screen with opening

Figure 12.23 Opening on a screen and its complementary dipole.

impedances of Z_s and Z_c, respectively, the impedances are related by

$$Z_s Z_c = \frac{\eta^2}{4} \qquad (12\text{-}67)$$

To obtain the impedance Z_c of the complement (dipole) in a practical arrangement, a gap must be introduced to represent the feed points. In addition, the far-zone fields radiated by the opening on the screen ($E_{\theta s}, E_{\phi s}, H_{\theta s}, H_{\phi s}$) are related to the far-zone fields of the complement ($E_{\theta c}, E_{\phi c}, H_{\theta c}, H_{\phi c}$) by

$$E_{\theta s} = H_{\theta c}, \quad E_{\phi s} = H_{\phi c}, \quad H_{\theta s} = -\frac{E_{\theta c}}{\eta_0^2}, \quad H_{\phi s} = -\frac{E_{\phi c}}{\eta_0^2} \qquad (12\text{-}68)$$

Infinite, flat, very thin conductors are not realizable in practice but can be closely approximated. If a slot is cut into a plane conductor that is large compared to the wavelength and the dimensions of the slot, the behavior predicted by Babinet's principle can be realized to a high degree. The impedance properties of the slot may not be affected as much by the finite dimensions of the plane as would be its pattern. The slot of Figure 12.23(a) will also radiate on both sides of the screen. Unidirectional radiation can be obtained by placing a backing (box or cavity) behind the slot, forming a so-called *cavity-backed slot* whose radiation properties (impedance and pattern) are determined by the dimensions of the cavity.

To demonstrate the application of Babinet's principle, an example is considered.

Example 12.6
A very thin half-wavelength slot is cut on an infinite, planar, very thin, perfectly conducting electric screen as shown in Figure 12.24(a). Find its input impedance. Assume it is radiating into free-space.

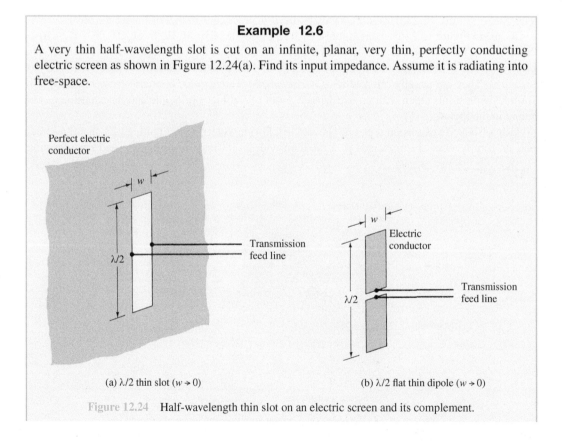

(a) $\lambda/2$ thin slot ($w \to 0$) (b) $\lambda/2$ flat thin dipole ($w \to 0$)

Figure 12.24 Half-wavelength thin slot on an electric screen and its complement.

Solution: From Babinet's principle and its extension we know that a very thin half-wavelength dipole, shown in Figure 12.24(b), is the complementary structure to the slot. From Chapter 4, the terminal (input) impedance of the dipole is $Z_c = 73 + j42.5$. Thus the terminal (input) impedance of the slot, using (12-67), is given by

$$Z_s = \frac{\eta_0^2}{4Z_c} \simeq \frac{(376.7)^2}{4(73 + j42.5)} \simeq \frac{35{,}475.72}{73 + j42.5}$$

$$Z_s \simeq 362.95 - j211.31$$

The slot of Figure 12.24(a) can be made to resonate by choosing the dimensions of its complement (dipole) so that it is also resonant. The pattern of the slot is identical in shape to that of the dipole except that the **E**- and **H**-fields are interchanged. When a vertical slot is mounted on a vertical screen, as shown in Figure 12.25(a), its electric field is horizontally polarized while that of the dipole is

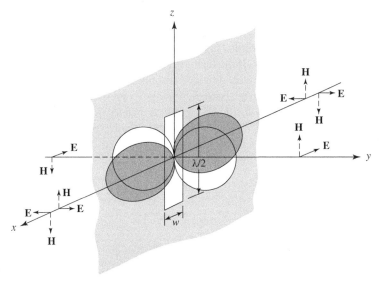

(a) $\lambda/2$ slot on a screen

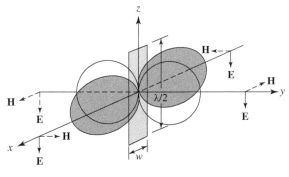

(b) $\lambda/2$ flat dipole

Figure 12.25 Radiation fields of a $\lambda/2$ slot on a screen and of a $\lambda/2$ flat dipole. (SOURCE: J. D. Kraus, *Antennas*, McGraw-Hill, New York, 1988, Chapter 13).

684 APERTURE ANTENNAS

vertically polarized [Fig. 12.25(b)]. Changing the angular orientation of the slot or screen will change the polarization.

The slot antenna, as a cavity-backed design, has been utilized in a variety of law enforcement applications. Its main advantage is that it can be fabricated and concealed within metallic objects, and with a small transmitter it can provide covert communications. There are various methods of feeding a slot antenna [15]. For proper operation, the cavity depth must be equal to odd multiples of $\lambda_g/4$, where λ_g is the guide wavelength.

12.9 FOURIER TRANSFORMS IN APERTURE ANTENNA THEORY

Previously the spatial domain analysis of aperture antennas was introduced, and it was applied to rectangular and circular apertures radiating in an infinite, homogeneous, lossless medium. The analysis of aperture antennas mounted on infinite ground planes, covered with lossless and/or lossy dielectric media, becomes too complex when it is attempted in the spatial domain. Considerable simplification can result with the utility of the frequency (*spectral*) domain.

12.9.1 Fourier Transforms-Spectral Domain

From Fourier series analysis, any periodic function $f(x)$ with a period T can be represented by a Fourier series of cosine and sine terms. If the function $f(x)$ is aperiodic and exists only in the interval of $0 < x < T$, a Fourier series can be formed by constructing, in a number of ways, a periodic function. The Fourier series for the constructed periodic function represents the actual aperiodic function $f(x)$ only in the interval $0 < x < T$. Outside this space, the aperiodic function $f(x)$ is zero and the series representation is not needed. A Fourier series for $f(x)$ converges to the values of $f(x)$ at each point of continuity and to the midpoint of its values at each discontinuity.

In addition, $f(x)$ can also be represented as a superposition of discrete complex exponentials of the form

$$f(x) = \sum_{n=-\infty}^{+\infty} c_n e^{-j(2n\pi/T)x} \qquad (12\text{-}69)$$

$$c_n = \frac{1}{T} \int_0^T f(x) e^{+j(2n\pi/T)x} \, dx \qquad (12\text{-}69a)$$

or of continuous complex exponentials of the form

$$f(x) = \frac{1}{2\pi} \int_{-\infty}^{+\infty} \mathcal{F}(\omega) e^{-jx\omega} \, d\omega \quad -\infty < \omega < +\infty \qquad (12\text{-}70a)$$

whose inverse is given by

$$\mathcal{F}(\omega) = \int_{-\infty}^{+\infty} f(x) e^{+j\omega x} \, dx \quad -\infty < x < +\infty \qquad (12\text{-}70b)$$

The integral operation in (12-70a) is referred to as the *direct transformation* and that of (12-70b) as the *inverse transformation* and both form a *transform pair*.

Another useful identity is *Parseval's theorem*, which for the transform pair, can be written as

$$\int_{-\infty}^{+\infty} f(x) g^*(x) \, dx = \frac{1}{2\pi} \int_{-\infty}^{+\infty} \mathcal{F}(\omega) \mathcal{G}^*(\omega) \, d\omega \qquad (12\text{-}71)$$

where * indicates complex conjugate.

From the definitions of (12-70a), (12-70b) and (12-71), the Fourier transforms can be expanded to two dimensions and can be written as

$$f(x,y) = \frac{1}{4\pi^2} \int_{-\infty}^{+\infty} \int_{-\infty}^{+\infty} \mathcal{F}(\omega_1, \omega_2) e^{-j(\omega_1 x + \omega_2 y)} \, d\omega_1 \, d\omega_2 \qquad (12\text{-}72\text{a})$$

$$\mathcal{F}(\omega_1, \omega_2) = \int_{-\infty}^{+\infty} \int_{-\infty}^{+\infty} f(x,y) e^{+j(\omega_1 x + \omega_2 y)} \, dx \, dy \qquad (12\text{-}72\text{b})$$

$$\int_{-\infty}^{+\infty} \int_{-\infty}^{+\infty} f(x,y) g^*(x,y) \, dx \, dy$$
$$= \frac{1}{4\pi^2} \int_{-\infty}^{+\infty} \int_{-\infty}^{+\infty} \mathcal{F}(\omega_1, \omega_2) \mathcal{G}^*(\omega_1, \omega_2) \, d\omega_1 \, d\omega_2 \qquad (12\text{-}72\text{c})$$

The process can be continued to n dimensions.

The definitions, theorems, and principles introduced will be utilized in the sections that follow to analyze the radiation characteristics of aperture antennas mounted on infinite ground planes.

12.9.2 Radiated Fields

To apply Fourier transforms (*spectral techniques*) to the analysis of aperture antennas, let us consider a rectangular aperture of dimensions a and b mounted on an infinite ground plane, as shown in Figure 12.26. In the source-free region ($z > 0$), the field $\mathbf{E}(x, y, z)$ of a monochromatic wave radiated by the aperture can be written as a superposition of plane waves (all of the same frequency, different amplitudes, and traveling in different directions) of the form $\mathbf{f}(k_x, k_y) e^{-j\mathbf{k}\cdot\mathbf{r}}$ [16], [17]. The function $\mathbf{f}(k_x, k_y)$ is the vector amplitude of the wave, and k_x and k_y are the spectral frequencies which extend over the entire frequency spectrum ($-\infty \leq k_x, k_y \leq \infty$). Thus the field $\mathbf{E}(x, y, z)$ can be written as

$$\mathbf{E}(x, y, z) = \frac{1}{4\pi^2} \int_{-\infty}^{+\infty} \int_{-\infty}^{+\infty} \mathbf{f}(k_x, k_y) e^{-j\mathbf{k}\cdot\mathbf{r}} \, dk_x \, dk_y \qquad (12\text{-}73)$$

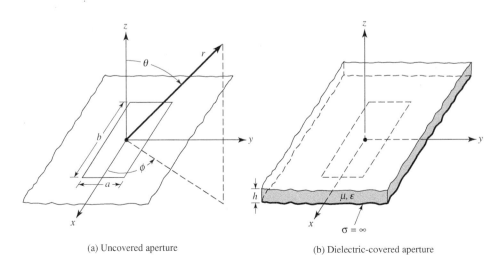

(a) Uncovered aperture

(b) Dielectric-covered aperture

Figure 12.26 Rectangular apertures mounted on infinite ground planes.

according to the definition of (12-72a). The object of a plane wave expansion is to determine the unknown amplitudes $\mathbf{f}(k_x, k_y)$ and the direction of propagation of the plane waves. Since

$$\mathbf{r} = \hat{\mathbf{a}}_x x + \hat{\mathbf{a}}_y y + \hat{\mathbf{a}}_z z \quad (12\text{-}74)$$

and the propagation factor $\mathbf{k}$ (often referred to as the *vector wavenumber*) can be defined as

$$\mathbf{k} = \hat{\mathbf{a}}_x k_x + \hat{\mathbf{a}}_y k_y + \hat{\mathbf{a}}_z k_z \quad (12\text{-}75)$$

(12-73) can be written as

$$\mathbf{E}(x, y, z) = \frac{1}{4\pi^2} \int_{-\infty}^{+\infty} \int_{-\infty}^{+\infty} [\mathbf{f}(k_x, k_y) e^{-jk_z z}] e^{-j(k_x x + k_y y)} \, dk_x \, dk_y. \quad (12\text{-}76)$$

The part of the integrand within the brackets can be regarded as the transform of $\mathbf{E}(x, y, z)$. This allows us to write the transform pair as

$$\boxed{\mathbf{E}(x, y, z) = \frac{1}{4\pi^2} \int_{-\infty}^{+\infty} \int_{-\infty}^{+\infty} \mathscr{E}(k_x, k_y, z) e^{-j(k_x x + k_y y)} \, dk_x \, dk_y} \quad (12\text{-}77\text{a})$$

$$\boxed{\mathscr{E}(k_x, k_y, z) = \int_{-\infty}^{+\infty} \int_{-\infty}^{+\infty} \mathbf{E}(x, y, z) e^{+j(k_x x + k_y y)} \, dx \, dy} \quad (12\text{-}77\text{b})$$

where

$$\boxed{\mathscr{E}(k_x, k_y, z) = \mathbf{f}(k_x, k_y) e^{-jk_z z}} \quad (12\text{-}77\text{c})$$

In principle then, according to (12-77a) and (12-77b) the fields radiated by an aperture $\mathbf{E}(x, y, z)$ can be found provided its transform $\mathscr{E}(k_x, k_y, z)$ is known. To this point the transform field $\mathscr{E}(k_x, k_y, z)$ can only be found provided the actual field $\mathbf{E}(x, y, z)$ is known *a priori*. In other words, the answer must be known beforehand! However, as it will be seen from what follows, if the transform field at $z = 0$

$$\mathscr{E}(k_x, k_y, z = 0) = \mathbf{f}(k_x, k_y) \quad (12\text{-}78)$$

is formed, it will be sufficient to determine $\mathbf{E}(x, y, z)$. To form the transform $\mathscr{E}(k_x, k_y, z = 0) = \mathbf{f}(k_x, k_y)$, it will be necessary and sufficient to know only the tangential components of the E-field at $z = 0$. For the problem of Figure 12.26(a), the tangential components of the E-field at $z = 0$ exist only over the bounds of the aperture (they vanish outside it because of the presence of the infinite ground plane).

In general

$$\mathbf{f}(k_x, k_y) = \hat{\mathbf{a}}_x f_x(k_x, k_y) + \hat{\mathbf{a}}_y f_y(k_x, k_y) + \hat{\mathbf{a}}_z f_z(k_x, k_y) \quad (12\text{-}79)$$

which can also be written as

$$\mathbf{f}(k_x, k_y) = \mathbf{f}_t(k_x, k_y) + \hat{\mathbf{a}}_z f_z(k_x, k_y) \quad (12\text{-}79\text{a})$$

$$\mathbf{f}_t(k_x, k_y) = \hat{\mathbf{a}}_x f_x(k_x, k_y) + \hat{\mathbf{a}}_y f_y(k_x, k_y) \quad (12\text{-}79\text{b})$$

For aperture antennas positioned along the x-y plane, the only components of $\mathbf{f}(k_x, k_y)$ that need to be found are f_x and f_y. As will be shown in what follows, f_z can be found once f_x and f_y are known. This is a further simplification of the problem. The functions f_x and f_y are found, using (12-77a) and (12-77b), provided the tangential components of the **E**-field over the aperture (E_{xa} and E_{ya}) are specified. The solution of (12-77c) is valid provided the z variations of $\mathbf{E}(k_x, k_y, z)$ are separable. In addition, in the source-free region the field $\mathbf{E}(x, y, z)$ of (12-77a) must satisfy the homogeneous vector wave equation. These allow us to relate the propagation constant k_z to k_x, k_y and $k = (\omega\sqrt{\mu\epsilon})$, by

$$k_z^2 = k^2 - (k_x^2 + k_y^2) \tag{12-80}$$

or

$$k_z = \begin{cases} +[k^2 - (k_x^2 + k_y^2)]^{1/2} & \text{when } k^2 \geq k_x^2 + k_y^2 \tag{12-80a} \\ -j[(k_x^2 + k_y^2) - k^2]^{1/2} & \text{when } k^2 < k_x^2 + k_y^2 \tag{12-80b} \end{cases}$$

This is left as an exercise to the reader. The form of k_z as given by (12-80a) contributes to the propagating waves (radiation field) of (12-76) and (12-77a) whereas that of (12-80b) contributes to the evanescent waves. Since the field in the far zone of the antenna is of the radiation type, its contribution comes from the part of the k_x, k_y spectrum which satisfies (12-80a). The values of k_x and k_y in (12-80)–(12-80b) are analogous to the eigenvalues for the fields inside a rectangular waveguide [12]. In addition, k_z is analogous to the propagation constant for waveguides which is used to define cutoff.

To find the relation between f_z and f_x, f_y, we proceed as follows. In the source-free region ($z > 0$) the field $\mathbf{E}(x, y, z)$, in addition to satisfying the vector wave equation, must also be solenoidal so that

$$\nabla \cdot \mathbf{E}(x, y, z) = \nabla \cdot \left\{ \frac{1}{4\pi^2} \int_{-\infty}^{+\infty} \int_{-\infty}^{+\infty} \mathbf{f}(k_x, k_y) e^{-j\mathbf{k}\cdot\mathbf{r}} \, dk_x \, dk_y \right\} = 0 \tag{12-81}$$

Interchanging differentiation with integration and using the vector identity

$$\nabla \cdot (\alpha \mathbf{A}) = \alpha \nabla \cdot \mathbf{A} + \mathbf{A} \cdot \nabla \alpha \tag{12-82}$$

reduces (12-81) to

$$\frac{1}{4\pi^2} \int_{-\infty}^{+\infty} \int_{-\infty}^{+\infty} [\mathbf{f} \cdot \nabla(e^{-j\mathbf{k}\cdot\mathbf{r}})] \, dk_x \, dk_y = 0 \tag{12-83}$$

since $\nabla \cdot \mathbf{f}(k_x, k_y) = 0$. Equation (12-83) is satisfied provided that

$$\mathbf{f} \cdot \nabla e^{-j\mathbf{k}\cdot\mathbf{r}} = -j\mathbf{f} \cdot \mathbf{k} e^{-j\mathbf{k}\cdot\mathbf{r}} = 0 \tag{12-84}$$

or

$$\mathbf{f} \cdot \mathbf{k} = (\mathbf{f}_t + \hat{\mathbf{a}}_z f_z) \cdot \mathbf{k} = 0 \tag{12-84a}$$

or

$$f_z = \frac{\mathbf{f}_t \cdot \mathbf{k}}{k_z} = -\frac{(f_x k_x + f_y k_y)}{k_z} \tag{12-84b}$$

From (12-84b) it is evident that f_z can be formed once f_x and f_y are known.

688 APERTURE ANTENNAS

All three components of $\mathbf{f}$ (f_x, f_y and f_z) can be found, using (12-77b) and (12-78), provided the two components of $\mathbf{E}$ (E_x, E_y) at $z = 0$, which is the plane of the aperture and ground plane of Figure 12.26(a), are known. Because E_x and E_y along the $z = 0$ plane are zero outside the bounds of the aperture ($|x| > a/2$, $|y| > b/2$), (12-77b) and (12-78) reduce for f_x and f_y to

$$f_x(k_x, k_y) = \int_{-b/2}^{+b/2} \int_{-a/2}^{+a/2} E_{xa}(x', y', z' = 0) e^{+j(k_x x' + k_y y')} \, dx' \, dy' \tag{12-85a}$$

$$f_y(k_x, k_y) = \int_{-b/2}^{+b/2} \int_{-a/2}^{+a/2} E_{ya}(x', y', z' = 0) e^{+j(k_x x' + k_y y')} \, dx' \, dy' \tag{12-85b}$$

where primes indicate source points. $E_{xa}(x', y', z' = 0)$ and $E_{ya}(x', y', z' = 0)$, which represent the tangential components of the electric field over the aperture, are the only fields that need to be known. Once f_x and f_y are found by using (12-85a) and (12-85b), f_z and $\mathscr{E}(k_x, k_y, z)$ can be formed using (12-84b) and (12-77c), respectively. Thus, the solution for $\mathbf{E}(x, y, z)$ for the aperture in Figure 12.26(a) is given by

$$\mathbf{E}(x, y, z) = \frac{1}{4\pi^2} \Bigg\{ \iint_{\substack{k_x^2 + k_y^2 \le k^2 \\ k_z = [k^2 - (k_x^2 + k_y^2)]^{1/2}}} \mathscr{E}(k_x, k_y, z) e^{-j(k_x x + k_y y)} \, dk_x \, dk_y$$

$$+ \iint_{\substack{k_x^2 + k_y^2 > k^2 \\ k_z = -j[(k_x^2 + k_y^2) - k^2]^{1/2}}} \mathscr{E}(k_x, k_y, z) e^{-j(k_x x + k_y y)} \, dk_x \, dk_y \Bigg\} \tag{12-86}$$

$$\mathscr{E}(k_x, k_y, z) = \left[\hat{\mathbf{a}}_x f_x + \hat{\mathbf{a}}_y f_y - \hat{\mathbf{a}}_z \left(\frac{f_x k_x + f_y k_y}{k_z} \right) \right] e^{-jk_z z} \tag{12-86a}$$

where f_x and f_y are given by (12-85a) and (12-85b).

In summary, the field radiated by the aperture of Figure 12.26(a) can be found by the following procedure:

1. Specify the tangential components of the **E**-field (E_{xa} and E_{ya}) over the bounds of the aperture.
2. Find f_x and f_y using (12-85a) and (12-85b), respectively.
3. Find f_z using (12-84b).
4. Find $\mathscr{E}(k_x, k_y, z)$ using (12-86a).
5. Formulate $\mathbf{E}(x, y, z)$ using (12-86).

This completes the solution for $\mathbf{E}(x, y, z)$. However, as is evident from (12-86), the integration is quite difficult even for the simplest of problems. However, if the observations are restricted in the far-field region, many simplifications in performing the integrations can result. This was apparent in Chapters 4, 5 and in others. In many practical problems, the far zone is usually the region of greatest importance. Since it is also known that for all antennas the fields in the far zone are primarily of the radiated type (*propagating waves*), then only the first integral in (12-86) contributes in that region.

In the next section, our attention is directed toward the evaluation of (12-86a) or (12-73) in the far-zone region (large values of kr). This is accomplished by evaluating (12-73) asymptotically for large values of kr by the method of *Stationary Phase* [18], [19].

To complete the formulation of the radiated fields in all regions, let us outline the procedure to find $\mathbf{H}(x, y, z)$. From Maxwell's equations

$$\mathbf{H}(x,y,z) = -\frac{1}{j\omega\mu} \nabla \times \mathbf{E}(x,y,z)$$

$$= -\frac{1}{j\omega\mu} \nabla \times \left[\frac{1}{4\pi^2} \int_{-\infty}^{+\infty} \int_{-\infty}^{+\infty} \mathbf{f}(k_x, k_y) e^{-j\mathbf{k}\cdot\mathbf{r}} \, dk_x \, dk_y \right] \quad (12\text{-}87)$$

Interchanging integration with differentiation and using the vector identity

$$\nabla \times (\alpha \mathbf{A}) = \alpha \nabla \times \mathbf{A} + (\nabla \alpha) \times \mathbf{A} \quad (12\text{-}88)$$

reduces (12-87) to

$$\mathbf{H}(x,y,z) = -\frac{1}{4\pi^2 k\eta} \int_{-\infty}^{+\infty} \int_{-\infty}^{+\infty} (\mathbf{f} \times \mathbf{k}) e^{-j\mathbf{k}\cdot\mathbf{r}} \, dk_x \, dk_y \quad (12\text{-}89)$$

since $\nabla \times \mathbf{f}(k_x, k_y) = 0$ and $\nabla(e^{-j\mathbf{k}\cdot\mathbf{r}}) = -j\mathbf{k}e^{-j\mathbf{k}\cdot\mathbf{r}}$ from (12-84).

12.9.3 Asymptotic Evaluation of Radiated Field

The main objective in this section is the evaluation of (12-73) or (12-86a) for observations made in the far-field. For most practical antennas, the field distribution on the aperture is such that an exact evaluation of (12-73) in closed form is not possible. However, if the observations are restricted to the far-field region (large kr), the integral evaluation becomes less complex. This was apparent in Chapters 4, 5, and others. The integral of (12-73) will be evaluated asymptotically for large values of kr using the method of *Stationary Phase* (Appendix VIII) [18], [19].

The stationary phase method assumes that the main contribution to the integral of (12-73) comes from values of k_x and k_y where $\mathbf{k} \cdot \mathbf{r}$ does not change for first order changes in k_x and k_y. That is to say $\mathbf{k} \cdot \mathbf{r}$ remains stationary at those points. For the other values of k_x and k_y, $\mathbf{k} \cdot \mathbf{r}$ changes very rapidly and the function $e^{-j\mathbf{k}\cdot\mathbf{r}}$ oscillates very rapidly between the values of $+1$ and -1. Assuming that $\mathbf{f}(k_x, k_y)$ is a slowly varying function of k_x and k_y, the integrand of (12-73) oscillates very rapidly outside the stationary points so that the contribution to the integral from that region is negligible. As the observation point approaches infinity, the contributions to the integral from the region outside the stationary points is zero. For practical applications, the observation point cannot be at infinity. However, it will be assumed to be far enough such that the major contributions come from the stationary points.

The first task in the asymptotic evaluation of (12-73) is to find the stationary points of $\mathbf{k} \cdot \mathbf{r}$. For that, $\mathbf{k} \cdot \mathbf{r}$ is written as

$$\mathbf{k} \cdot \mathbf{r} = (\hat{\mathbf{a}}_x k_x + \hat{\mathbf{a}}_y k_y + \hat{\mathbf{a}}_z k_z) \cdot \mathbf{a}_r r \quad (12\text{-}90)$$

Using the inverse transformation of (4-5), (12-90) can be written as

$$\mathbf{k} \cdot \mathbf{r} = r(k_x \sin\theta \cos\phi + k_y \sin\theta \sin\phi + k_z \cos\theta) \quad (12\text{-}91)$$

which reduces, using (12-80a) to

$$\mathbf{k} \cdot \mathbf{r} = r[k_x \sin\theta \cos\phi + k_y \sin\theta \sin\phi + \sqrt{k^2 - k_x^2 - k_y^2} \cos\theta] \tag{12-92}$$

The stationary points can be found by

$$\frac{\partial(\mathbf{k} \cdot \mathbf{r})}{\partial k_x} = 0 \tag{12-93a}$$

$$\frac{\partial(\mathbf{k} \cdot \mathbf{r})}{\partial k_y} = 0 \tag{12-93b}$$

Using (12-92) and (12-80), (12-93a) and (12-93b) reduce to

$$\frac{\partial(\mathbf{k} \cdot \mathbf{r})}{\partial k_x} = r\left(\sin\theta \cos\phi - \frac{k_x}{k_z}\cos\theta\right) = 0 \tag{12-94a}$$

$$\frac{\partial(\mathbf{k} \cdot \mathbf{r})}{\partial k_y} = r\left(\sin\theta \sin\phi - \frac{k_y}{k_z}\cos\theta\right) = 0 \tag{12-94b}$$

whose solutions are given, respectively, by

$$k_x = k_z \frac{\sin\theta \cos\phi}{\cos\theta} \tag{12-95a}$$

$$k_y = k_z \frac{\sin\theta \sin\phi}{\cos\theta} \tag{12-95b}$$

Using (12-95a) and (12-95b), (12-80) can be written as

$$k^2 = k_z^2 + k_x^2 + k_y^2 = k_z^2\left(1 + \frac{\sin^2\theta}{\cos^2\theta}\right) \tag{12-96}$$

which reduces for k_z to

$$k_z = k\cos\theta \tag{12-97}$$

With the aid of (12-97), the stationary point of (12-95a) and (12-95b) simplify to

$$k_x = k\sin\theta \cos\phi = k_1 \tag{12-98a}$$

$$k_y = k\sin\theta \sin\phi = k_2 \tag{12-98b}$$

The function $\mathbf{k} \cdot \mathbf{r}$ can be expanded into a Taylor series, about the stationary point k_1, k_2, and it can be approximated by the zero, first, and second order terms. That is,

$$\mathbf{k} \cdot \mathbf{r} \simeq \mathbf{k} \cdot \mathbf{r}\bigg|_{k_1,k_2} + \frac{\partial(\mathbf{k} \cdot \mathbf{r})}{\partial k_x}\bigg|_{k_1,k_2}(k_x - k_1) + \frac{\partial(\mathbf{k} \cdot \mathbf{r})}{\partial k_y}\bigg|_{k_1,k_2}(k_y - k_2)$$

$$+ \frac{1}{2}\frac{\partial^2(\mathbf{k} \cdot \mathbf{r})}{\partial k_x^2}\bigg|_{k_1,k_2}(k_x - k_1)^2 + \frac{1}{2}\frac{\partial^2(\mathbf{k} \cdot \mathbf{r})}{\partial k_y^2}\bigg|_{k_1,k_2}(k_y - k_2)^2$$

$$+ \frac{\partial^2(\mathbf{k} \cdot \mathbf{r})}{\partial k_x \partial k_y}\bigg|_{k_1,k_2}(k_x - k_1)(k_y - k_2) \tag{12-99}$$

Since the second and third terms vanish at the stationary point $k_x = k_1$ and $k_y = k_2$, (12-99) can be expressed as

$$\mathbf{k} \cdot \mathbf{r} = \mathbf{k} \cdot \mathbf{r}\Big|_{k_1,k_2} - A\xi^2 - B\eta^2 - C\xi\eta \tag{12-100}$$

where

$$A = -\frac{1}{2}\frac{\partial^2 (\mathbf{k} \cdot \mathbf{r})}{\partial k_x^2}\Big|_{k_1,k_2} \tag{12-100a}$$

$$B = -\frac{1}{2}\frac{\partial^2 (\mathbf{k} \cdot \mathbf{r})}{\partial k_y^2}\Big|_{k_1,k_2} \tag{12-100b}$$

$$C = -\frac{\partial^2 (\mathbf{k} \cdot \mathbf{r})}{\partial k_x \partial k_y}\Big|_{k_1,k_2} \tag{12-100c}$$

$$\xi = (k_x - k_1) \tag{12-100d}$$

$$\eta = (k_y - k_2) \tag{12-100e}$$

Using (12-97)-(12-98b), (12-90) reduces to

$$\mathbf{k} \cdot \mathbf{r}\Big|_{k_1,k_2} = kr \tag{12-101}$$

Similarly, with the aid of (12-92), A, B, and C can be written, after a few manipulations, as

$$A = -\frac{1}{2}\frac{\partial^2 (\mathbf{k} \cdot \mathbf{r})}{\partial k_x^2}\Big|_{k_1,k_2} = \frac{r}{2k}\left(1 + \frac{\sin^2\theta \cos^2\phi}{\cos^2\theta}\right) \tag{12-102a}$$

$$B = -\frac{1}{2}\frac{\partial^2 (\mathbf{k} \cdot \mathbf{r})}{\partial k_y^2}\Big|_{k_1,k_2} = \frac{r}{2k}\left(1 + \frac{\sin^2\theta \sin^2\phi}{\cos^2\theta}\right) \tag{12-102b}$$

$$C = -\frac{\partial^2 (\mathbf{k} \cdot \mathbf{r})}{\partial k_x \partial k_y}\Big|_{k_1,k_2} = \frac{r}{k}\frac{\sin^2\theta}{\cos^2\theta}\cos\phi \sin\phi \tag{12-102c}$$

Thus (12-73) can be approximated around the stationary point $k_x = k_1$ and $k_y = k_2$, which contributes mostly to the integral, by

$$\mathbf{E}(x,y,z) \simeq \frac{1}{4\pi^2} \iint_{S_{1,2}} \mathbf{f}(k_x = k_1, k_y = k_2) e^{-j(kr - A\xi^2 - B\eta^2 - C\xi\eta)} \, d\xi \, d\eta \tag{12-103}$$

or

$$\mathbf{E}(x,y,z) \simeq \frac{1}{4\pi^2} \mathbf{f}(k_1, k_2) e^{-jkr} \iint_{S_{1,2}} e^{+j(A\xi^2 + B\eta^2 + C\xi\eta)} \, d\xi \, d\eta \tag{12-103a}$$

where $S_{1,2}$ is the surface near the stationary point.

The integral of (12-103a) can be evaluated with the method of *Stationary Phase*. That is, (see Appendix VIII)

$$\iint_{S_{1,2}} e^{j(A\xi^2 + B\eta^2 + C\xi\eta)} \, d\xi \, d\eta = j \frac{2\pi\delta}{\sqrt{|4AB - C^2|}} \tag{12-104}$$

$$\delta = \begin{cases} +1 & \text{if } 4AB > C^2 \text{ and } A > 0 \\ -1 & \text{if } 4AB > C^2 \text{ and } A < 0 \\ -j & \text{if } 4AB < C^2 \end{cases} \tag{12-104a}$$

With the aid of (12-102a)-(12-102c), the factor $4AB - C^2$ is

$$4AB - C^2 = \left(\frac{r}{k \cos\theta}\right)^2 \tag{12-105}$$

Since $4AB > C^2$ and $A > 0$, (12-103) reduces to

$$\iint_{S_{1,2}} e^{j(A\xi^2 + B\eta^2 + C\xi\eta)} \, d\xi \, d\eta = j \frac{2\pi k}{r} \cos\theta \tag{12-106}$$

and (12-103a) to

$$\mathbf{E}(r, \theta, \phi) \simeq j \frac{k e^{-jkr}}{2\pi r} [\cos\theta \mathbf{f}(k_1 = k\sin\theta\cos\phi, k_2 = k\sin\theta\sin\phi)] \tag{12-107}$$

In the far-field region, only the θ and ϕ components of the electric and magnetic fields are dominant. Therefore, the E_θ and E_ϕ components of (12-107) can be written in terms of f_x and f_y. With the aid of (12-84b), $\mathbf{f}$ can be expressed as

$$\mathbf{f} = \hat{\mathbf{a}}_x f_x + \hat{\mathbf{a}}_y f_y + \hat{\mathbf{a}}_z f_z = \left[\hat{\mathbf{a}}_x f_x + \hat{\mathbf{a}}_y f_y - \hat{\mathbf{a}}_z \frac{(f_x k_x + f_y k_y)}{k_z} \right] \tag{12-108}$$

At the stationary point ($k_x = k_1 = k\sin\theta\cos\phi$, $k_y = k_2 = k\sin\theta\sin\phi$, $k_z = k\cos\theta$), (12-108) reduces to

$$\mathbf{f}(k_1, k_2) = \left[\hat{\mathbf{a}}_x f_x + \hat{\mathbf{a}}_y f_y - \hat{\mathbf{a}}_z \frac{\sin\theta}{\cos\theta}(f_x \cos\phi + f_y \sin\phi) \right] \tag{12-109}$$

Using the inverse transformation of (4-5), the θ and ϕ components of $\mathbf{f}$ can be written as

$$f_\theta = \frac{f_x \cos\phi + f_y \sin\phi}{\cos\theta} \tag{12-110a}$$

$$f_\phi = -f_x \sin\phi + f_y \cos\phi \tag{12-110b}$$

The **E**-field of (12-107) reduces, for the θ and ϕ components, to

$$\mathbf{E}(r,\theta,\phi) \simeq j\frac{ke^{-jkr}}{2\pi r}[\hat{\mathbf{a}}_\theta(f_x \cos\phi + f_y \sin\phi) + \hat{\mathbf{a}}_\phi \cos\theta(-f_x \sin\phi + f_y \cos\phi)]$$

(12-111)

and the **H**-field to

$$\mathbf{H}(r,\theta,\phi) = \sqrt{\frac{\varepsilon}{\mu}}[\hat{\mathbf{a}}_r \times \mathbf{E}(r,\theta,\phi)]$$

(12-112)

where from (12-85a) and (12-85b)

$$f_x(k_x = k_1, k_y = k_2)$$
$$= \int_{-b/2}^{+b/2}\int_{-a/2}^{+a/2} E_{xa}(x',y',z'=0)e^{jk(x'\sin\theta\cos\phi + y'\sin\theta\sin\phi)}\,dx'\,dy'$$

(12-113a)

$$f_y(k_x = k_1, k_y = k_2)$$
$$= \int_{-b/2}^{+b/2}\int_{-a/2}^{+a/2} E_{ya}(x',y',z'=0)e^{jk(x'\sin\theta\cos\phi + y'\sin\theta\sin\phi)}\,dx'\,dy'$$

(12-113b)

To illustrate the frequency domain (*spectral*) techniques, the problem of a uniform illuminated aperture, which was previously analyzed in Section 12.5.1 using spatial methods, will be solved again using transform methods.

Example 12.7

A rectangular aperture of dimensions a and b is mounted on an infinite ground plane, as shown in Figure 12.26(a). Find the field radiated by it assuming that over the opening the electric field is given by

$$\mathbf{E}_a = \hat{\mathbf{a}}_y E_0, \quad \begin{array}{l} -a/2 \leq x' \leq a/2 \\ -b/2 \leq y' \leq b/2 \end{array}$$

where E_0 is a constant.
Solution: From (12-113a) and (12-113b)

$$f_x = 0$$

$$f_y = E_0 \int_{-b/2}^{+b/2} e^{jky'\sin\theta\sin\phi}\,dy' \int_{-a/2}^{+a/2} e^{jkx'\sin\theta\cos\phi}\,dx'$$

which, when integrated, reduces to

$$f_y = abE_0 \left(\frac{\sin X}{X}\right)\left(\frac{\sin Y}{Y}\right)$$

$$X = \frac{ka}{2} \sin\theta \cos\phi$$

$$Y = \frac{kb}{2} \sin\theta \sin\phi$$

The θ and ϕ components of (12-111) can be written as

$$E_\theta = j\frac{abkE_0 e^{-jkr}}{2\pi r}\left\{\sin\phi \left[\frac{\sin X}{X}\right]\left[\frac{\sin Y}{Y}\right]\right\}$$

$$E_\phi = j\frac{abkE_0 e^{-jkr}}{2\pi r}\left\{\cos\theta \cos\phi \left[\frac{\sin X}{X}\right]\left[\frac{\sin Y}{Y}\right]\right\}$$

which are identical to those of (12-23b) and (12-23c), respectively.

12.9.4 Dielectric-Covered Apertures

The transform (*spectral*) technique can easily be extended to determine the field radiated by dielectric-covered apertures [20], [21]. For the sake of brevity, the details will not be included here. However, it can be shown that for a single lossless dielectric sheet cover of thickness h, dielectric constant ε_r, unity relative permeability, and free-space phase constant k_0, the far-zone radiated field E_θ, E_ϕ of the covered aperture of Figure 12.26(b) are related to E_θ^0, E_ϕ^0 of the uncovered aperture of Figure 12.26(a) by

$$E_\theta(r,\theta,\phi) = f(\theta)E_\theta^0(r,\theta,\phi) \tag{12-114a}$$

$$E_\phi(r,\theta,\phi) = g(\theta)E_\phi^0(r,\theta,\phi) \tag{12-114b}$$

where

E_θ, E_ϕ = field components of dielectric-covered aperture [Fig. 12.26(b)]
E_θ^0, E_ϕ^0 = field components of uncovered aperture [Fig. 12.26(a)]

$$f(\theta) = \frac{e^{jk_0 h \cos\theta}}{\cos\psi + jZ_h \sin\psi} \tag{12-114c}$$

$$g(\theta) = \frac{e^{jk_0 h \cos\theta}}{\cos\psi + jZ_e \sin\psi} \tag{12-114d}$$

$$\psi = k_0 h \sqrt{\varepsilon_r - \sin^2\theta} \tag{12-114e}$$

$$Z_e = \frac{\cos\theta}{\sqrt{\varepsilon_r - \sin^2\theta}} \qquad (12\text{-}114\text{f})$$

$$Z_h = \frac{\sqrt{\varepsilon_r - \sin^2\theta}}{\varepsilon_r \cos\theta} \qquad (12\text{-}114\text{g})$$

The above relations do not include surface wave contributions which can be taken into account but are beyond the scope of this section [20].

To investigate the effect of the dielectric sheet, far-zone principal E- and H-plane patterns were computed for a rectangular waveguide shown in Figure 12.26(b). The waveguide was covered with a single dielectric sheet, was operating in the dominant TE_{10} mode, and was mounted on an infinite ground plane. The E- and H-plane patterns are shown in Figure 12.27(a) and 12.27(b), respectively. In the E-plane patterns, it is evident that the surface impedance of the modified ground plane forces the normal electric field component to vanish along the surface ($\theta = \pi/2$). This is similar to the effects experienced by the patterns of the vertical dipole above ground shown in Figure 4.31. Since the H-plane patterns have vanishing characteristics when the aperture is radiating in free-space, the presence of the dielectric sheet has a very small overall effect. This is similar to the effects experienced by the patterns of a horizontal dipole above ground shown in Figure 4.32. However, both the E- and H-plane patterns become more broad near the surface, and more narrow elsewhere, as the thickness increases.

12.9.5 Aperture Admittance

Another parameter of interest, especially when the antenna is used as a diagnostic tool, is its terminating impedance or admittance. In this section, using Fourier transform (*spectral*) techniques, the admittance of an aperture antenna mounted on an infinite ground plane and radiating into free-space will be formulated. Computations will be presented for a parallel-plate waveguide. The techniques can best be presented by considering a specific antenna configuration and field distribution. Similar steps can be used for any other geometry and field distribution.

The geometrical arrangement of the aperture antenna under consideration is shown in Figure 12.26(a). It consists of a rectangular waveguide mounted on an infinite ground plane. It is assumed that the field distribution, *above cutoff*, is that given by the TE_{10} mode, or

$$\mathbf{E}_a = \hat{\mathbf{a}}_y E_0 \cos\left(\frac{\pi}{a}x'\right) \quad \begin{array}{l} -a/2 \leq x' \leq a/2 \\ -b/2 \leq y' \leq b/2 \end{array} \qquad (12\text{-}115)$$

where E_0 is a constant. The aperture admittance is defined as

$$Y_a = \frac{2P^*}{|V|^2} \qquad (12\text{-}116)$$

where

P^* = conjugate of complex power transmitted by the aperture

V = aperture reference voltage.

696 APERTURE ANTENNAS

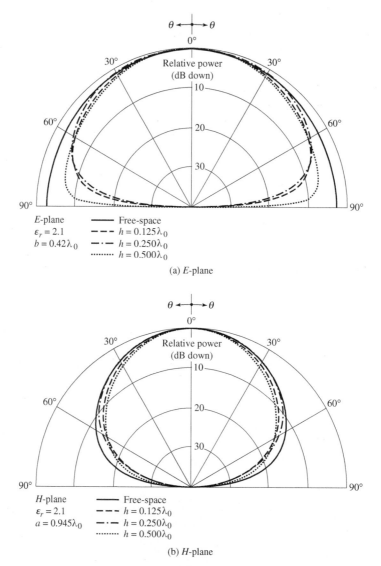

Figure 12.27 Amplitude radiation patterns of a dielectric-covered waveguide mounted on an infinite ground plane and with a TE_{10}-mode aperture field distribution.

The complex power transmitted by the aperture can be written as

$$P = \frac{1}{2} \iint_{S_a} [\mathbf{E}(x',y',z'=0) \times \mathbf{H}^*(x',y',z'=0)] \cdot \hat{\mathbf{a}}_z \, dx' \, dy' \qquad (12\text{-}117)$$

where S_a is the aperture of the antenna. $\mathbf{E}(x',y',z'=0)$ and $\mathbf{H}(x',y',z'=0)$ represent the total electric and magnetic fields at the aperture including those of the modes which operate below cutoff and contribute to the imaginary power. For the field distribution given by (12-115), (12-117) reduces to

$$P = -\frac{1}{2} \iint_{S_a} [E_y(x',y',z'=0) H_x^*(x',y',z'=0)] \, dx' \, dy' \qquad (12\text{-}117a)$$

The amplitude coefficients of all modes that can exist within the waveguide, propagating and nonpropagating, can be evaluated provided the total tangential **E**- and/or **H**-field at any point within the waveguide is known. Assuming that (12-115) represents the total tangential **E**-field, it allows the determination of all mode coefficients. Even though this can be accomplished, the formulation of (12-117a) in the spatial domain becomes rather complex [22].

An alternate and simpler method in the formulation of the aperture admittance is to use Fourier transforms. By Parseval's theorem of (12-72c), (12-117a) can be written as

$$P = -\frac{1}{2} \int_{-\infty}^{+\infty} \int_{-\infty}^{+\infty} E_y(x', y', z' = 0) H_x^*(x', y', z' = 0)\, dx'\, dy'$$

$$= -\frac{1}{8\pi^2} \int_{-\infty}^{+\infty} \int_{-\infty}^{+\infty} \mathcal{E}_y(k_x, k_y) \mathcal{H}_x^*(k_x, k_y)\, dk_x\, dk_y \qquad (12\text{-}118)$$

where the limits of the first integral have been extended to infinity since $E_y(x', y', z' = 0)$ vanishes outside the physical bounds of the aperture. $\mathcal{E}_y(k_x, k_y)$ and $\mathcal{H}_x(k_x, k_y)$ are the Fourier transforms of the aperture *E*- and *H*-fields, respectively.

The transform $\mathcal{E}(k_x, k_y, z = 0)$ is obtained from (12-78) while $\mathcal{H}(k_x, k_y, z = 0)$ can be written, by referring to (12-89), as

$$\mathcal{H}(k_x, k_y, z = 0) = -\frac{1}{k\eta}(\mathbf{f} \times \mathbf{k}) \qquad (12\text{-}119)$$

For the problem at hand, the transforms $\mathcal{E}_y$ and $\mathcal{H}_x$ are given by

$$\mathcal{E}_y(k_x, k_y) = f_y(k_x, k_y) \qquad (12\text{-}120)$$

$$\mathcal{H}_x(k_x, k_y) = -\frac{1}{k\eta}\left(k_z + \frac{k_y^2}{k_z}\right) f_y = -\frac{1}{k\eta}\left(\frac{k^2 - k_x^2}{k_z}\right) f_y \qquad (12\text{-}121)$$

Using (12-77b) and (12-115), (12-120) reduces to

$$f_y(k_x, k_y) = E_0 \int_{-b/2}^{+b/2} \int_{-a/2}^{+a/2} \cos\left(\frac{\pi}{a}x'\right) e^{j(k_x x' + k_y y')}\, dx'\, dy'$$

$$f_y(k_x, k_y) = \left(\frac{\pi a b}{2}\right) E_0 \left[\frac{\cos X}{(\pi/2)^2 - (X)^2}\right]\left[\frac{\sin Y}{Y}\right] \qquad (12\text{-}122)$$

where

$$X = \frac{k_x a}{2} \qquad (12\text{-}122\text{a})$$

$$Y = \frac{k_y b}{2} \qquad (12\text{-}122\text{b})$$

Substituting (12-120)–(12-122b) into (12-118) leads to

$$P = \frac{(\pi ab E_0)^2}{32\pi^2 k\eta} \int_{-\infty}^{+\infty} \int_{-\infty}^{+\infty} \left\{ \frac{(k^2 - k_x^2)}{k_z^*} \left[\frac{\cos X}{(\pi/2)^2 - (X)^2}\right]^2 \left[\frac{\sin Y}{Y}\right]^2 \right\} dk_x\, dk_y \qquad (12\text{-}123)$$

If the reference aperture voltage is given by

$$V = \frac{ab}{\sqrt{2}} E_0 \qquad (12\text{-}124)$$

the aperture admittance can be written as

$$Y_a = \frac{2P^*}{|V|^2} = \frac{1}{8k\eta} \int_{-\infty}^{+\infty} \left\{ \left[\frac{\sin\left(\frac{k_y b}{2}\right)}{\frac{k_y b}{2}}\right]^2 \int_{-\infty}^{+\infty} \frac{(k^2 - k_x^2)}{k_z} \right. $$
$$\left. \times \left[\frac{\cos\left(\frac{k_x a}{2}\right)}{\left(\frac{\pi}{2}\right)^2 - \left(\frac{k_x a}{2}\right)^2}\right]^2 dk_x \right\} dk_y \qquad (12\text{-}125)$$

where k_z is given by (12-80a) and (12-80b). As stated before, the values of k_z as given by (12-80a) contribute to the radiated (real) power and those of (12-80b) contribute to the reactive (imaginary) power. Referring to Figure 12.28, values of k_x and k_y within the circle contribute to the aperture conductance, and the space is referred to as the *visible region*. Values of k_x and k_y outside the circle contribute to the aperture susceptance and constitute the *invisible region*. Thus (12-125) can be separated into its real and imaginary parts, and it can be written as

$$Y_a = G_a + jB_a \qquad (12\text{-}126)$$

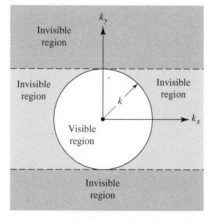

Figure 12.28 Visible and invisible regions in k-space.

$$G_a = \frac{1}{4k\eta} \int_0^k \left[\frac{\sin\left(\frac{k_y b}{2}\right)}{\frac{k_y b}{2}} \right]^2 \left\{ \int_0^{\sqrt{k^2-k_y^2}} \frac{(k^2 - k_x^2)}{[k^2 - (k_x^2 + k_y^2)]^{1/2}} \right.$$

$$\left. \times \left[\frac{\cos\left(\frac{k_x a}{2}\right)}{\left(\frac{\pi}{2}\right)^2 - \left(\frac{k_x a}{2}\right)^2} \right]^2 dk_x \right\} dk_y \quad (12\text{-}126a)$$

$$B_a = -\frac{1}{4k\eta} \left\{ \int_0^k \left[\frac{\sin\left(\frac{k_y b}{2}\right)}{\frac{k_y b}{2}} \right]^2 \left\{ \int_{\sqrt{k^2-k_y^2}}^{\infty} \frac{(k_x^2 - k^2)}{[(k_x^2 + k_y^2) - k^2]^{1/2}} \right. \right.$$

$$\left. \times \left[\frac{\cos\left(\frac{k_x a}{2}\right)}{\left(\frac{\pi}{2}\right)^2 - \left(\frac{k_x a}{2}\right)^2} \right]^2 dk_x \right\} dk_y$$

$$+ \int_k^{\infty} \left[\frac{\sin\left(\frac{k_y b}{2}\right)}{\frac{k_y b}{2}} \right]^2 \left\{ \int_0^{\infty} \frac{(k_x^2 - k^2)}{[(k_x^2 + k_y^2) - k^2]^{1/2}} \right.$$

$$\left. \left. \times \left[\frac{\cos\left(\frac{k_x a}{2}\right)}{\left(\frac{\pi}{2}\right)^2 - \left(\frac{k_x a}{2}\right)^2} \right]^2 dk_x \right\} dk_y \right\} \quad (12\text{-}126b)$$

The first term in (12-126b) takes into account the contributions from the strip outside the circle for which $k_y < k$, and the second term includes the remaining space outside the circle.

The numerical evaluation of (12-126a) and (12-126b) is complex and will not be attempted here. Computations for the admittance of rectangular apertures radiating into lossless and lossy half spaces have been carried out and appear in the literature [23]–[27]. Various ingenious techniques have been used to evaluate these integrals.

Because of the complicated nature of (12-126a) and (12-126b) to obtain numerical data, a simpler configuration will be considered as an example.

Example 12.8

A parallel plate waveguide (slot) is mounted on an infinite ground plane, as shown in Figure 12.29. Assuming the total electric field at the aperture is given by

$$\mathbf{E}_a = \hat{\mathbf{a}}_y E_0 \quad -b/2 \leq y' \leq b/2$$

where E_0 is a constant, find the aperture admittance assuming the aperture voltage is given by $V = bE_0$.

Solution: This problem bears a very close similarity to that of Figure 12.26(a), and most of the results of this example can be obtained almost directly from the previous formulation. Since the problem is two-dimensional, (12-120)–(12-122) reduce to

$$\mathscr{E}_y(k_y) = f_y(k_y) = bE_0 \frac{\sin\left(\frac{k_y b}{2}\right)}{\frac{k_y b}{2}}$$

$$\mathscr{H}_x(k_y) = \frac{k}{\eta}\frac{f_y}{k_z} = \frac{kbE_0}{\eta k_z} \frac{\sin\left(\frac{k_y b}{2}\right)}{\frac{k_y b}{2}}$$

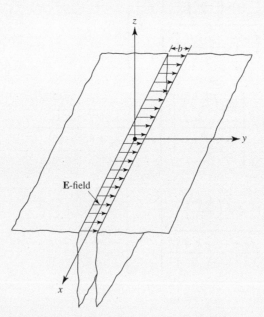

Figure 12.29 Parallel-plate waveguide geometry and aperture field distribution.

and (12-123) to

$$P = \frac{(bE_0)^2 k}{4\pi\eta} \int_{-\infty}^{+\infty} \frac{1}{k_z^*} \left[\frac{\sin\left(\frac{k_y b}{2}\right)}{\frac{k_y b}{2}}\right]^2 dk_y$$

Since the aperture voltage is given by $V = bE_0$, the aperture slot admittance (*per unit length along the x-direction*) of (12-125) can be written as

$$Y_a = \frac{k}{2\pi\eta} \int_{-\infty}^{+\infty} \frac{1}{k_z} \left[\frac{\sin\left(\frac{k_y b}{2}\right)}{\frac{k_y b}{2}} \right]^2 dk_y$$

and the aperture slot conductance and susceptance of (12-126a) and (12-126b) as

$$G_a = \frac{k}{2\pi\eta} \int_{-k}^{k} \frac{1}{\sqrt{k^2 - k_y^2}} \left[\frac{\sin\left(\frac{k_y b}{2}\right)}{\frac{k_y b}{2}} \right]^2 dk_y$$

$$= \frac{k}{\pi\eta} \int_{0}^{k} \frac{1}{\sqrt{k^2 - k_y^2}} \left[\frac{\sin\left(\frac{k_y b}{2}\right)}{\frac{k_y b}{2}} \right]^2 dk_y$$

$$B_a = \frac{k}{2\pi\eta} \left\{ \int_{-\infty}^{-k} \frac{1}{\sqrt{k_y^2 - k^2}} \left[\frac{\sin\left(\frac{k_y b}{2}\right)}{\frac{k_y b}{2}} \right]^2 dk_y \right.$$

$$\left. + \int_{k}^{\infty} \frac{1}{\sqrt{k_y^2 - k^2}} \left[\frac{\sin\left(\frac{k_y b}{2}\right)}{\frac{k_y b}{2}} \right]^2 dk_y \right\}$$

$$B_a = \frac{k}{\pi\eta} \int_{k}^{\infty} \frac{1}{\sqrt{k_y^2 - k^2}} \left[\frac{\sin\left(\frac{k_y b}{2}\right)}{\frac{k_y b}{2}} \right]^2 dk_y$$

If

$$w = \frac{b}{2} k_y$$

the expressions for the slot conductance and susceptance reduce to

$$G_a = \frac{2}{\eta\lambda} \int_{0}^{kb/2} \frac{1}{\sqrt{(kb/2)^2 - w^2}} \left(\frac{\sin w}{w}\right)^2 dw$$

$$B_a = \frac{2}{\eta\lambda} \int_{kb/2}^{\infty} \frac{1}{\sqrt{w^2 - (kb/2)^2}} \left(\frac{\sin w}{w}\right)^2 dw$$

The admittance will always be capacitive since B_a is positive.

The expressions for the slot conductance and susceptance (*per unit length along the x-direction*) reduce for small values of kb to [5]

$$\left. \begin{array}{l} G_a \simeq \dfrac{\pi}{\eta\lambda}\left[1 - \dfrac{(kb)^2}{24}\right] \\[1em] B_a \simeq \dfrac{\pi}{\eta\lambda}[1 - 0.636 \ln(kb)] \end{array} \right\} \quad \dfrac{b}{\lambda} < \dfrac{1}{10}$$

and for large values of kb to

$$\left. \begin{array}{l} G_a \simeq \dfrac{1}{\eta b} \\[1em] B_a \simeq \dfrac{\lambda}{\eta}\left(\dfrac{1}{\pi b}\right)^2\left[1 - \dfrac{1}{2}\sqrt{\dfrac{\lambda}{b}}\cos\left(\dfrac{2b}{\lambda} + \dfrac{1}{4}\right)\pi\right] \end{array} \right\} \quad \dfrac{b}{\lambda} > 1$$

Normalized values of λG_a and λB_a as a function of b/λ for an aperture radiating into free space are shown plotted in Figure 12.30.

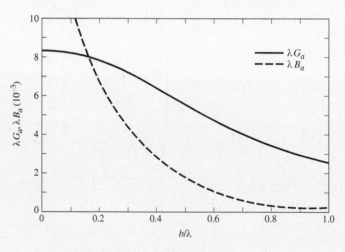

Figure 12.30 Normalized values of conductance and susceptance of narrow slot.

12.10 GROUND PLANE EDGE EFFECTS: THE GEOMETRICAL THEORY OF DIFFRACTION

Infinite size (physically and/or electrically) ground planes are not realizable in practice, but they can be approximated closely by very large structures. The radiation characteristics of antennas (current distribution, pattern, impedance, etc.) mounted on finite size ground planes can be modified considerably, especially in regions of very low intensity, by the effects of the edges. The ground plane edge diffractions for an aperture antenna are illustrated graphically in Figure 12.31. For these

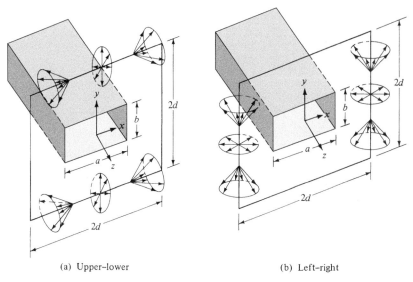

Figure 12.31 Diffraction mechanisms for an aperture mounted on a finite size ground plane (diffractions at upper-lower and left-right edges of the ground plane).

problems, rigorous solutions do not exist unless the object's surface can be described by curvilinear coordinates. Presently there are two methods that can be used conveniently to account for the edge effects. One technique is the *Moment Method* (MM) [28] discussed in Chapter 8 and the other is the *Geometrical Theory of Diffraction* (GTD) [29].

The Moment Method describes the solution in the form of an integral, and it can be used to handle arbitrary shapes. It mostly requires the use of a digital computer for numerical computations and, because of capacity limitations of computers, it is most computationally efficient for objects that are small electrically. Therefore, it is usually referred to as a *low-frequency asymptotic method*.

When the dimensions of the radiating object are large compared to the wavelength, *high-frequency asymptotic techniques* can be used to analyze many otherwise not mathematically tractable problems. One such technique, which has received considerable attention in the past few years, is the Geometrical *T*heory of *D*iffraction (GTD) which was originally developed by Keller [29]. The GTD is an extension of the classical *G*eometrical *O*ptics (GO; direct, reflected, and refracted rays), and it overcomes some of the limitations of GO by introducing a diffraction mechanism [2].

The diffracted field, which is determined by a generalization of Fermat's principle [2], [30], is initiated at points on the surface of the object where there is a discontinuity in the incident GO field (incident and reflected shadow boundaries). The phase of the field on a diffracted ray is assumed to be equal to the product of the optical length of the ray (from some reference point) and the phase constant of the medium. Appropriate phase jumps must be added as a ray passes through caustics.*
The amplitude is assumed to vary in accordance with the principle of conservation of energy in a narrow tube of rays. The initial value of the field on a diffracted ray is determined from the incident field with the aid of an appropriate diffraction coefficient (which, in general, is a dyadic for electromagnetic fields). The diffraction coefficient is usually determined from the asymptotic solutions of the simplest boundary-value problems which have the same local geometry at the points of diffraction as the object(s) of investigation. Geometries of this type are referred to as *canonical* problems.

---

*A caustic is a point or a line through which all the rays of a wave pass. Examples of it are the focal point of a paraboloid (parabola of revolution) and the focal line of a parabolic cylinder. The field at the caustic is infinite because, in principle, an infinite number of rays pass through it.

704 APERTURE ANTENNAS

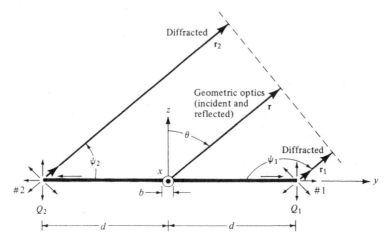

Figure 12.32 Aperture geometry in principal E-plane ($\phi = \pi/2$).

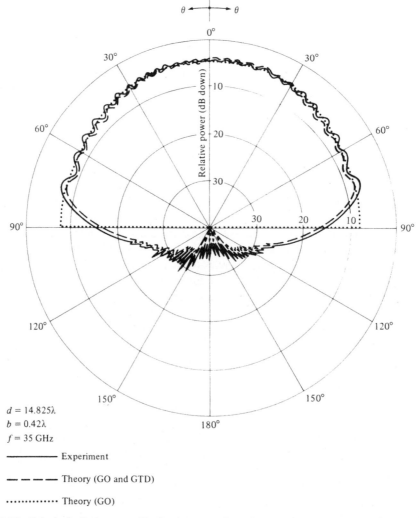

$d = 14.825\lambda$
$b = 0.42\lambda$
$f = 35$ GHz

——— Experiment

– – – – Theory (GO and GTD)

············ Theory (GO)

Figure 12.33 Principal E-plane amplitude patterns of an aperture antenna mounted on a finite size ground plane.

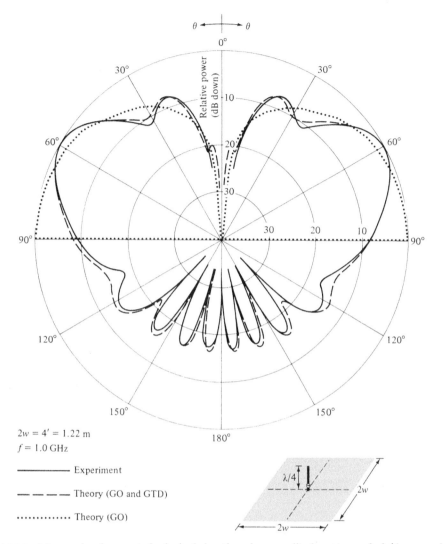

Figure 12.34 Measured and computed principal elevation plane amplitude patterns of a $\lambda/4$ monopole above infinite and finite square ground planes.

One of the simplest geometries is a conducting wedge [31], [32]. Another is that of a conducting, smooth, and convex surface [33]–[35].

The primary objective in using the GTD to solve complicated geometries is to resolve each such problem into smaller components [8]–[10], [35]. The partitioning is made so that each smaller component represents a canonical geometry of a known solution. These techniques have also been applied for the modeling and analysis of antennas on airplanes [36], and they have combined both wedge and smooth conducting surface diffractions [33], [35]. The ultimate solution is a superposition of the contributions from each canonical problem.

Some of the advantages of GTD are

1. It is simple to use.
2. It can be used to solve complicated problems that do not have exact solutions.
3. It provides physical insight into the radiation and scattering mechanisms from the various parts of the structure.

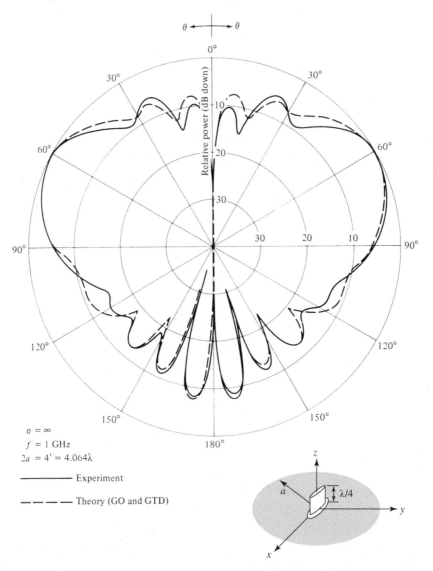

Figure 12.35 Measured and computed principal elevation plane amplitude patterns of a $\lambda/4$ monopole (blade) above a circular ground plane.

4. It yields accurate results which compare extremely well with experiments and other methods.
5. It can be combined with other techniques such as the Moment Method [37].

The derivation of the diffraction coefficients for a conducting wedge and their application are lengthy, and will not be repeated here. An extensive and detailed treatment of over 100 pages, for both antennas and scattering, can be found in [2]. However, to demonstrate the versatility and potential of the GTD, three examples are considered. The first is the E-plane pattern of a rectangular aperture of dimensions a,b mounted on a finite size ground plane, as shown in Figure 12.31. The GTD formulation along the E-plane includes the direct radiation and the fields diffracted by the two edges of the ground plane, as shown in Figure 12.32. The computed E-plane pattern along with the measured one are shown in Figure 12.33; an excellent agreement is indicated.

The two other examples considered here are the elevation pattern of a $\lambda/4$ monopole mounted on square and circular ground planes. The diffraction mechanism on the principal planes for these is the same as that of the aperture, which is shown in Figure 12.32. The corresponding principal elevation plane pattern of the monopole on the square ground plane is displayed in Figure 12.34 while that on the circular one is exhibited in Figure 12.35. For each case an excellent agreement is indicated with the measurements. It should be noted that the minor lobes near the symmetry axis ($\theta = 0°$ and $\theta = 180°$) for the circular ground plane of Figure 12.35 are more intense than the corresponding ones for the square ground plane of Figure 12.34. These effects are due to the ring-source radiation by the rim of the circular ground plane toward the symmetry axis [2], [10].

12.11 MULTIMEDIA

In the publisher's website for this book, the following multimedia resources are included for the review, understanding, and visualization of the material of this chapter:

a. **Java**-based **interactive questionnaire**, with answers.
b. **Matlab** computer program, designated as **Aperture**, for computing and displaying the radiation characteristics of rectangular and circular apertures.
c. **Power Point (PPT)** viewgraphs, in multicolor.

REFERENCES

1. S. A. Schelkunoff, "Some Equivalence Theorems of Electromagnetics and Their Application to Radiation Problems," *Bell Syst. Tech. J.* Vol. 15, pp. 92–112, 1936.
2. C. A. Balanis, *Advanced Engineering Electromagnetics*, 2nd edition, John Wiley and Sons, New York, 2012.
3. C. Huygens, *Traite De La Lumiere*, Leyden, 1690. Translated into English by S. P. Thompson, London, 1912, Reprinted by The University of Chicago Press.
4. J. D. Kraus and K. R. Carver, *Electromagnetics* (Second edition), McGraw-Hill, New York, 1973, pp. 464–467.
5. R. F. Harrington, *Time-Harmonic Electromagnetic Fields*, McGraw-Hill, New York, 1961, pp. 100–103, 143–263, 365–367.
6. A. E. H. Love, "The Integration of the Equations of Propagation of Electric Waves," *Philos. Trans. R. Soc. London, Ser. A*, Vol. 197, pp. 1–45, 1901.
7. R. Mittra (ed.), *Computer Techniques for Electromagnetics*, Pergamon Press, New York, 1973, pp. 9–13.
8. C. A. Balanis and L. Peters, Jr., "Equatorial Plane Pattern of an Axial-TEM Slot on a Finite Size Ground Plane," *IEEE Trans. Antennas Propagat.*, Vol. AP-17, No. 3, pp. 351–353, May 1969.
9. C. A. Balanis, "Pattern Distortion Due to Edge Diffractions," *IEEE Trans. Antennas Propagat.*, Vol. AP-18, No. 4, pp. 561–563, July 1970.
10. C. R. Cockrell and P. H. Pathak, "Diffraction Theory Techniques Applied to Aperture Antennas on Finite Circular and Square Ground Planes," *IEEE Trans. Antennas Propagat.*, Vol. AP-22, No. 3, pp. 443–448, May 1974.
11. D. G. Fink (ed.), *Electronics Engineers' Handbook*, Section 18 (Antennas by W. F. Croswell), McGraw-Hill, New York, 1975.
12. K. Praba, "Optimal Aperture for Maximum Edge-of-Coverage (EOC) Directivity," *IEEE Antennas Propagation Magazine*, Vol. 36, No. 3, pp. 72–74, June 1994.
13. H. G. Booker, "Slot Aerials and Their Relation to Complementary Wire Aerials," *J. Inst. Elect. Eng.*, part III A, pp. 620–626, 1946.

14. E. C. Jordan and K. G. Balmain, *Electromagnetic Waves and Radiating Systems*, Prentice-Hall, Inc., Englewood Cliffs, NJ, 1968.
15. J. D. Kraus, *Antennas*, Chapter 13, McGraw-Hill, New York, 1988.
16. H. G. Booker and P. C. Clemmow, "The Concept of an Angular Spectrum of Plane Waves, and its Relation to that of Polar Diagram and Aperture Distribution," *Proc. IEE (London)*, Vol. 97, part III, pp. 11–17, January 1950.
17. G. Borgiotti, "Fourier Transforms Method of Aperture Antennas," *Alta Freq.*, Vol. 32, pp. 196–204, November 1963.
18. L. B. Felsen and N. Marcuvitz, *Radiation and Scattering of Waves*, Prentice-Hall, Englewood Cliffs, NJ, 1973.
19. R. E. Collin and F. J. Zucker, *Antenna Theory: Part 1*, Chapter 3, McGraw-Hill Book Co., New York, 1969.
20. C. M. Knop and G. I. Cohn, "Radiation from an Aperture in a Coated Plane," *Radio Sci. J. Res.*, Vol. 68D, No. 4, pp. 363–378, April 1964.
21. F. L. Whetten, "Dielectric Coated Meandering Leaky-Wave Long Slot Antennas," Ph.D. Dissertation, Arizona State University, May 1993.
22. M. H. Cohen, T. H. Crowley, and C. A. Lewis, "The Aperture Admittance of a Rectangular Waveguide Radiating Into Half-Space," Antenna Lab., Ohio State University, Rept. 339-22, Contract W 33-038 ac21114, November 14, 1951.
23. R. T. Compton, "The Admittance of Aperture Antennas Radiating Into Lossy Media," Antenna Lab., Ohio State University Research Foundation, Rept. 1691-5, March 15, 1964.
24. A. T. Villeneuve, "Admittance of a Waveguide Radiating into a Plasma Environment," *IEEE Trans. Antennas Propagat.*, Vol. AP-13, No. 1, pp. 115–121, January 1965.
25. J. Galejs, "Admittance of a Waveguide Radiating Into a Stratified Plasma," *IEEE Trans. Antennas Propagat.*, Vol. AP-13, No. 1, pp. 64–70, January 1965.
26. M. C. Bailey and C. T. Swift, "Input Admittance of a Circular Waveguide Aperture Covered by a Dielectric Slab," *IEEE Trans. Antennas Propagat.*, Vol. AP-16, No. 4, pp. 386–391, July 1968.
27. W. F. Croswell, W. C. Taylor, C. T. Swift, and C. R. Cockrell, "The Input Admittance of a Rectangular Waveguide-Fed Aperture Under an Inhomogeneous Plasma: Theory and Experiment," *IEEE Trans. Antennas Propagat.*, Vol. AP-16, No. 4, pp. 475–487, July 1968.
28. R. F. Harrington, *Field Computation by Moment Methods*, Macmillan Co., New York, 1968.
29. J. B. Keller, "Geometrical Theory of Diffraction," *Journal Optical Society of America*, Vol. 52, No. 2, pp. 116–130, February 1962.
30. R. G. Kouyoumjian, "The Geometrical Theory of Diffraction and Its Applications," in *Numerical and Asymptotic Techniques in Electromagnetics* (R. Mittra, ed.), Springer-Verlag, New York, 1975.
31. R. G. Kouyoumjian and P. H. Pathak, "A Uniform Geometrical Theory of Diffraction for an Edge in a Perfectly Conducting Surface," *Proc. IEEE*, Vol. 62, No. 11, pp. 1448–1461, November 1974.
32. D. L. Hutchins, "Asymptotic Series Describing the Diffraction of a Plane Wave by a Two-Dimensional Wedge of Arbitrary Angle," Ph.D. Dissertation, The Ohio State University, Dept. of Electrical Engineering, 1967.
33. P. H. Pathak and R. G. Kouyoumjian, "An Analysis of the Radiation from Apertures on Curved Surfaces by the Geometrical Theory of Diffraction, *Proc. IEEE*, Vol. 62, No. 11, pp. 1438–1447, November 1974.
34. G. L. James, *Geometrical Theory of Diffraction for Electromagnetic Waves*, Peter Peregrinus, Ltd., Stevenage, Herts., England, 1976.
35. C. A. Balanis and L. Peters, Jr., "Analysis of Aperture Radiation from an Axially Slotted Circular Conducting Cylinder Using GTD," *IEEE Trans. Antennas Propagat.*, Vol. AP-17, No. 1, pp. 93–97, January 1969.
36. C. A. Balanis and Y.-B. Cheng, "Antenna Radiation and Modeling for Microwave Landing System," *IEEE Trans. Antennas Propagat.*, Vol. AP-24, No. 4, pp. 490–497, July 1976.

PROBLEMS

12.1. A uniform plane wave traveling in the $+z$ direction, whose magnetic field is expressed as

$$\mathbf{H}^i = \hat{\mathbf{a}}_y H_0 e^{-jkz} \quad z \leq 0$$

impinges upon an aperture on an infinite, flat, perfect electric conductor whose cross section is indicated in the figure.

(a) State the equivalent that must be used to determine the field radiated by the aperture to the right of the conductor ($z > 0$).

(b) Assuming the aperture dimension in the y direction is b, determine the far-zone fields for $z > 0$.

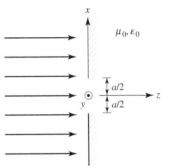

12.2. Repeat Problem 12.1 when the incident magnetic field is polarized in the x direction.

12.3. Repeat Problem 12.1 when the incident electric field is polarized in the y direction.

12.4. Repeat Problem 12.1 when the incident electric field is polarized in the x direction.

12.5. A perpendicularly polarized plane wave is obliquely incident upon an aperture, with dimension a and b, on a perfectly electric conducting ground plane of infinite extent, as shown in the figure. Assuming the field over the aperture is given by the incident field (ignore diffractions from the edges of the aperture), find the far-zone spherical components of the fields for $x > 0$.

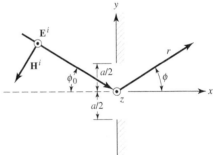

12.6. Repeat Problem 12.5 for a parallelly polarized plane wave (when the incident magnetic field is polarized in the z direction, i.e., the incident magnetic field is perpendicular to the x-y plane while the incident electric field is parallel to the x-y plane).

12.7. A narrow rectangular slot of size L by W is mounted on an infinite ground plane that covers the x-y plane. The tangential field over the aperture is given by

$$\mathbf{E}_a = \hat{\mathbf{a}}_y E_0 e^{-jk_0 x' \sqrt{2}/2}$$

Using the equivalence principle and image theory, we can replace the aperture and infinite ground plane with an equivalent magnetic current radiating in free-space. Determine the

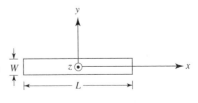

(a) appropriate equivalent
(b) far-zone spherical electric field components for $z > 0$
(c) direction (θ, ϕ) in which the radiation intensity is maximum

12.8. A rectangular aperture, of dimensions a and b, is mounted on an infinite ground plane, as shown in Figure 12.6(a). Assuming the tangential field over the aperture is given by

$$\mathbf{E}_a = \hat{\mathbf{a}}_z E_0 \quad -a/2 \leq y' \leq a/2, \quad -b/2 \leq z' \leq b/2$$

find the far-zone spherical electric and magnetic field components radiated by the aperture.

12.9. Repeat Problem 12.8 when the same aperture is analyzed using the coordinate system of Figure 12.6(b). The tangential aperture field distribution is given by

$$\mathbf{E}_a = \hat{\mathbf{a}}_x E_0 \quad -b/2 \leq x' \leq b/2, \quad -a/2 \leq z' \leq a/2$$

12.10. Repeat Problem 12.8 when the aperture field is given by

$$\mathbf{E}_a = \hat{\mathbf{a}}_z E_0 \cos\left(\frac{\pi}{a} y'\right) \quad -a/2 \leq y' \leq a/2, \quad -b/2 \leq z' \leq b/2$$

12.11. Repeat Problem 12.9 when the aperture field distribution is given by

$$\mathbf{E}_a = \hat{\mathbf{a}}_x E_0 \cos\left(\frac{\pi}{a} z'\right) \quad -b/2 \leq x' \leq b/2, \quad -a/2 \leq z' \leq a/2$$

12.12. Find the fields radiated by the apertures of Problems
(a) 12.8 (b) 12.9 (c) 12.10 (d) 12.11
when each of the apertures with their associated field distributions are *not* mounted on a ground plane. Assume the tangential **H**-field at the aperture is related to the **E**-field by the intrinsic impedance.

12.13. Find the fields radiated by the rectangular aperture of Section 12.5.3 when it is not mounted on an infinite ground plane.

12.14. For the rectangular aperture of Section 12.5.3 (with $a = 4\lambda$, $b = 3\lambda$), compute the
(a) E-plane beamwidth (*in degrees*) between the maxima of the *second* minor lobe
(b) E-plane amplitude (*in dB*) of the maximum of the second minor lobe (relative to the maximum of the major lobe)
(c) approximate directivity of the antenna using Kraus' formula. Compare it with the value obtained using the expression in Table 12.1.

12.15. A rectangular X-band (8.2–12.4 GHz) waveguide (with inside dimensions of 0.9 in by 0.4 in) operating in the dominant TE_{10} mode at 10 GHz is mounted on an infinite ground plane and used as a receiving antenna. This antenna is connected to a matched lossless transmission line and a matched load is attached to the transmission line. Determine the:
(a) Directivity (*dimensionless* and *in dB*) using:
  1. the most accurate formula that is available to you in class.
  2. Kraus' formula.
(b) Maximum power (*in watts*) that can be delivered to the load when a uniform plane wave with a power density of 10 mW/cm^2 is incident upon the antenna at normal incidence. Neglect any losses.

12.16. A lossless aperture antenna has a gain of 11 dB and overall physical area of $2\lambda^2$.
(a) What is the aperture efficiency of this antenna (*in %*)?
(b) Assuming the antenna is matched to a lossless transmission line that in turn is connected to a load that is also matched to the transmission line, what is the maximum power that

can be delivered to the load if the incident power density at the antenna aperture is 10×10^{-3} watts/cm^2? The frequency of operation is 10 GHz.

12.17. For the rectangular aperture of Section 12.5.1 with $a = b = 3\lambda$, compute the directivity using (12-37) and the computer program **Aperture**.

12.18. For the rectangular aperture of Section 12.5.2 with $a = b = 3\lambda$, compute the directivity using (12-37) and the computer program **Aperture**.

12.19. Compute the directivity of the aperture of Section 12.5.3, using the computer program **Aperture**, when
(a) $a = 3\lambda, b = 2\lambda$ (b) $a = b = 3\lambda$

12.20. Repeat Problem 12.19 when the aperture is not mounted on an infinite ground plane.

12.21. For the rectangular aperture of Section 12.5.3 with $a = 3\lambda, b = 2\lambda$, compute the
(a) E-plane half-power beamwidth
(b) H-plane half-power beamwidth
(c) E-plane first-null beamwidth
(d) H-plane first-null beamwidth
(e) E-plane first side lobe maximum (relative to main maximum)
(f) H-plane first side lobe maximum (relative to main maximum)

using the formulas of Table 12.1. Compare the results with the data from Figures 12.13 and 12.14. Verify using the computer program **Aperture**.

12.22. A square waveguide aperture, of dimensions $a = b$ and lying on the x-y plane, is radiating into free-space. Assuming a $\cos(\pi x'/a)$ by $\cos(\pi y'/b)$ distribution over the aperture, find the dimensions of the aperture (in wavelengths) so that the beam efficiency within a 37° total included angle cone is 90%.

12.23. Verify (12-39a), (12-39b), (12-39c), and (12-40).

12.24. A rectangular aperture mounted on an infinite ground plane has aperture electric field distributions and corresponding efficiencies of

Field Distribution	Aperture Efficiency
(a) Triangular	75%
(b) Cosine square	66.67%

What are the corresponding directives (in dB) if the dimensions of the aperture are $a = \lambda/2$ and $b = \lambda/4$?

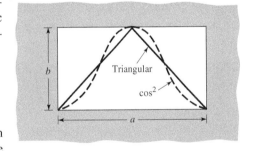

12.25. The physical area of an aperture antenna operating at 10 GHz is 200 cm^2 while its directivity is 23 dB. Assuming the antenna has an overall radiation efficiency of 90% and it is perfectly matched to the input transmission line, find the aperture efficiency of the antenna.

12.26. Two X-band (8.2–12.4 GHz) rectangular waveguides, each operating in the dominant TE$_{10}$-mode, are used, respectively, as transmitting and receiving antennas in a long distance communication system. The dimensions of each waveguide are $a = 2.286$ cm (0.9 in.) and $b = 1.016$ cm (0.4 in.) and the center frequency of operation is 10 GHz. Assuming the waveguides are separated by 10 kilometers and they are positioned for maximum radiation and reception toward each other, and the radiated power is 1 watt, find the:
(a) Incident power density at the receiving antenna
(b) Maximum power that can be delivered to a matched load

712 APERTURE ANTENNAS

Assume the antennas are lossless, are polarization matched, and each is mounted on an infinite ground plane.

12.27. The normalized far-zone electric field radiated in the E-plane (x-z plane; $\phi = 0°$) by a waveguide aperture antenna of dimensions a and b, mounted on an infinite ground plane as shown in the figure, is given by

$$\mathbf{E} = -\hat{\mathbf{a}}_\theta j \frac{\omega\mu b I_0 e^{-jkr}}{4\pi r} \frac{\sin\left(\frac{kb}{2}\cos\theta\right)}{\frac{kb}{2}\cos\theta}$$

Determine in the E-plane the:
(a) Vector effective length of the antenna.
(b) Maximum value of the effective length.
State the value of θ (*in degrees*) which maximizes the effective length.

12.28. A uniform plane wave is incident upon an X-band rectangular waveguide, with dimensions of 2.286 cm and 1.016 cm, mounted on an infinite ground plane. Assuming the waveguide is operating in the dominant TE_{10} mode, determine the maximum power that can be delivered to a matched load. The frequency is 10 GHz and the power density of the incident plane wave is 10^{-4} watts/m².

12.29. Compute the aperture efficiency of a rectangular aperture, mounted on an infinite ground plane as shown in Figure 12.7, with an **E**-field aperture distribution directed toward y but with variations

(a) triangular in the x and uniform in the y
(b) cosine-squared in the x and uniform in the y
(c) cosine in the x and cosine in the y
(d) cosine-squared in both the x and y directions.

How do they compare with those of a cosine distribution?

12.30. An X-band (8.2–12.4 GHz) WR 90 rectangular waveguide, with inner dimensions of 0.9 in. (2.286 cm) and 0.4 in. (1.016 cm), is mounted on an infinite ground plane. Assuming the waveguide is operating in the dominant TE_{10}-mode, find its directivity at $f = 10$ GHz using the

(a) computer program **Aperture**   (b) formula in Table 12.1

Compare the answers.

12.31. An X-band (8.2–12.4 GHz) rectangular waveguide, with internal dimensions of $a = 2.286$ cm and $b = 1.016$ cm, and mounted on an infinite PEC, is used as a ground-based receiving antenna for a satellite borne communication system. The power radiated by the satellite antenna is *10 watts*, the satellite is at a distance of *100 kilometers* from the ground-based antenna, and its antenna is assumed to have isotropic radiation characteristics. Assuming an operating frequency of 10 GHz and dominant TE_{10} mode field configuration, determine the maximum power (*in watts*) that can be intercepted by the receiving antenna when the incident wave is perpendicularly (normal to the aperture) incident on the receiving antenna and its polarization is:

(a) *Linearly polarized. Assume no other losses.*
(b) *Circularly polarized. Assume no other losses.*

The maximum radiation of the ground-based receiving antenna is directed toward the satellite borne communication system.

12.32. Repeat Problem 12.30 at $f = 20$ GHz for a K-band (18–26.5 GHz) WR 42 rectangular waveguide with inner dimensions of 0.42 in. (1.067 cm) and 0.17 in. (0.432 cm).

12.33. A rectangular waveguide, of dimensions a and b [as shown in Figure 12.6(c)], is mounted on an *infinite in extent planar Perfect Magnetic Conductor (PMC) ground plane*. The aperture lies on the x-y plane, with the z-axis perpendicular to the aperture. To simplify the formulation, the *tangential* electric and magnetic field components over the aperture are given by:

$$\left.\begin{array}{l}\mathbf{E}_a = \hat{\mathbf{a}}_y E_o \\ \mathbf{H}_a = -\hat{\mathbf{a}}_x \dfrac{E_o}{\eta}\end{array}\right\} \begin{array}{l} -a/2 \leq x' \leq +a/2 \\ -b/2 \leq y' \leq +b/2 \end{array}$$

where η is the intrinsic impedance of free space.

(a) Based on the *Field Equivalence Principle (FEP)*, also known as *Huygen's Principle*, determine the *exact vector* surface equivalent current densities $\mathbf{J}_s$ and $\mathbf{M}_s$ that will provide *a unique and exact solution to the actual physical radiation problem*.
You do NOT have to derive the equivalent surface current densities $\mathbf{J}_s$ and $\mathbf{M}_s$.
Also you must *give a proper explanation how you arrived at them using the FEP and any other theorem(s)*.

(b) Determine the far-field *electric fields* radiated by the aperture.
The electric fields *must* be expressed in *simplified form* (using $\sin X/X$, $\sin Y/Y$ and sine/cosine functions, as is done in the text book).

12.34. Four rectangular X-band waveguides of dimensions $a = 0.9$ in. (2.286 cm) and $b = 0.4$ in. (1.016 cm) and each operating on the dominant TE_{10}-mode, are mounted on an infinite ground plane so that their apertures and the ground plane coincide with the x-y plane. The apertures form a linear array, are placed along the x-axis with a center-to-center separation of $d = 0.85\lambda$ apart, and they are fed so that they form a broadside Dolph–Tschebyscheff array of -30 dB minor lobes. Assuming a center frequency of 10 GHz, determine the overall directivity of the array in decibels.

12.35. Sixty-four (64) X-band rectangular waveguides are mounted so that the aperture of each is mounted on an infinite ground plane that coincides with the x-y plane, and all together form an $8 \times 8 = 64$ planar array. Each waveguide has dimensions of $a = 0.9$ in. (2.286 cm), $b = 0.4$ in. (1.086 cm) and the center-to-center spacing between the waveguides is $d_x = d_y = 0.85\lambda$. Assuming a TE_{10}-mode operation for each waveguide, a center frequency of 10 GHz, and the waveguides are fed to form a *uniform* broadside planar array, find the directivity of the total array.

12.36. Find the far-zone fields radiated when the circular aperture of Section 12.6.1 is not mounted on an infinite ground plane.

12.37. Derive the far-zone fields when the circular aperture of Section 12.6.2
(a) is; (b) is not
mounted on an infinite ground plane.

12.38. A circular waveguide (not mounted on a ground plane), operating in the dominant TE_{11} mode, is used as an antenna radiating in free-space. Write in simplified form the normalized far-zone electric field components radiated by the waveguide antenna. You do not have to derive them.

12.39. A circular waveguide of radius $a = 1.125$ cm, NOT mounted on an infinite PEC ground plane and operating in the dominant TE_{11} mode at a frequency of 10 GHz, is used as a receiving antenna. On the basis of the *approximate equivalent*, determine the following:

(a) Far-zone electric and magnetic radiated fields (*you do not have to derive them*). Specify the angular limits (lower and higher *in degrees*) on the observation angles θ and ϕ.

(b) Maximum power (*in watts*) that can be delivered to a receiver (load) assuming the receiver (load) is matched to the transmission line that connects the antenna and the receiver (load). Assume that the transmission line has a characteristic impedance of 300 ohms while the antenna has an input impedance of $350 + j\,400$ ohms. Assume no other losses. The maximum power density of the wave impinging upon the antenna is 100 *watts*/m^2.

12.40. A ground-based and airborne communication system operates at X-Band ($f = 10\,GHz$). The *receiving ground-based system* utilizes a *circular waveguide, with a radius of $a = 2$ cm*, and it mounted on an infinite PEC ground plane. The *transmitting airborne system* utilizes a *rectangular waveguide with dimensions of $a = 2.286$ cm and $b = 1.016$ cm*, and it is also mounted on an infinite PEC ground plane. Assuming that the two systems are aligned so that the maximum radiation from airborne system is directed toward the maximum reception of the ground-based system, the power radiated by the airborne system is 100 watts, the distance between the two systems is 100 kilometers, and assuming *no conduction/dielectric or reflection/mismatch losses* in either of the two systems, determine the:

(a) *Maximum* directivity (*dimensionless* and *in dB*) of the transmitting antenna.

(b) *Maximum* power density (in watts/cm^2) radiated by the airborne system and impinging at the ground-based system, which is at 100 kilometers.

(c) *Maximum* effective area (*in cm^2*) of the ground-based system; do not include any polarization mismatch at this point.

(d) *Maximum* power (*in watts*) delivered to the ground-based system if the:
  - Incident wave and the ground based antenna are polarization matched (no other losses).
  - Incident wave is circularly polarized while the ground based antenna is linearly polarized (no other losses).

12.41. A *lossless* circular aperture antenna has a gain of 15 dB and overall physical area of 25 cm^2. The frequency of operation is 10 GHz.

(a) What is the aperture efficiency of the antenna (*in %*)?

(b) What is the power P_L delivered to a matched load given that the power density of the incident wave at the antenna aperture is 30 mW/cm^2? Assume ideal conditions (no losses).

12.42. A *lossless* circular aperture antenna operating on the *dominant TE_{11}-mode* and mounted on an infinite ground plane has an overall *gain of 9 dB*. Determine the following:

(a) Physical area (*in λ^2*) of the antenna.

(b) Maximum effective/equivalent area (*in λ^2*) of the antenna.

(c) Aperture efficiency (*in percent*).

(d) How much *more or less* efficient (*in percent*) is this antenna with a TE_{11}-mode distribution compared with the same antenna but with a *uniform field distribution* over its aperture. State which one is more or less efficient.

12.43. For the circular aperture of Section 12.6.1, compute its directivity, using the computer program **Aperture** of this chapter, when its radius is

(a) $a = 0.5\lambda$ (b) $a = 1.5\lambda$ (c) $a = 3.0\lambda$

Compare the results with data from Table 12.2.

12.44. Repeat Problem 12.43 when the circular aperture of Section 12.6.1 is not mounted on an infinite ground plane. Compare the results with those of Problem 12.43.

12.45. For the circular aperture of Problem 12.37, compute the directivity, using the computer program **Aperture** of this chapter, when its radius is
(a) $a = 0.5\lambda$ (b) $a = 1.5\lambda$ (c) $a = 3.0\lambda$

Compare the results with data from Table 12.2.

12.46. For the circular aperture of Section 12.6.2 with $a = 1.5\lambda$, compute the
(a) *E*-plane half-power beamwidth
(b) *H*-plane half-power beamwidth
(c) *E*-plane first-null beamwidth
(d) *H*-plane first-null beamwidth
(e) *E*-plane first side lobe maximum (relative to main maximum)
(f) *H*-plane first side lobe maximum (relative to main maximum) using the formulas of Table 12.2. Compare the results with the data from Figures 12.19 and 12.20. Verify using the computer program **Aperture**.

12.47. A circular waveguide, of radius a [as shown in Figure 12.16], is *not* mounted on a PEC ground plane; *it is radiating in free-space with no ground plane*. The aperture lies on the *x-y* plane, with the *z*-axis perpendicular to the aperture. To simplify the formulation, the *tangential* electric and magnetic field components over the aperture are given by:

$$\left. \begin{array}{l} \mathbf{E}_a = \hat{\mathbf{a}}_y E_o \\ \mathbf{H}_a = -\hat{\mathbf{a}}_x \dfrac{E_o}{\eta} \end{array} \right\} 0 \leq \rho' \leq a$$

where η is the intrinsic impedance of free space.

(a) Detrmine/write the *approximate* far-zone electric and magnetic field components radiated by the aperture, based on the *approximate Field Equivalence Principle (FEP)*, also known as *Huygen's Principle*. You do not have to derive the expressions for the electric and magnetic fields, but you must indicate how you got them.

The expressions for the fields must be written in simplified form, just as the examples in the book.

(b) For an aperture radius of $a = 3\lambda$, determine the:
- E-plane half-power beamwidth (*in degrees*).
- H-plane half-power beamwidth (*in degrees*).
- Directivity (*dimensions* and *in dB*).
- Directivity (*dimensionless* and *in dB*) using *an appropriate* alternate/another expression *that uses the half-power beamwidth information.*

State which alternate expression you are using and why?

12.48. A circular aperture of radius a is mounted on an infinite electric ground plane. Assuming the opening is on the *x-y* plane and its field distribution is given by

(a) $\mathbf{E}_a = \hat{\mathbf{a}}_y E_0 \left[1 - \left(\dfrac{\rho'}{a} \right)^2 \right]$, $\rho' \leq a$

(b) $\mathbf{E}_a = \hat{\mathbf{a}}_y E_0 \left[1 - \left(\dfrac{\rho'}{a}\right)^2\right]^2$, $\rho' \leq a$

find the far-zone electric and magnetic field components radiated by the antenna.

12.49. Repeat Problem 12.48 when the electric field is given by

$$\mathbf{E}_a = \hat{\mathbf{a}}_y E_0 [1 - (\rho'/a)], \quad \rho' \leq a$$

Find only the radiation vectors **L** and **N**. Work as far as you can. If you find you cannot complete the solution in closed form, state clearly why you cannot. Simplify as much as possible.

12.50. A coaxial line of inner and outer radii a and b, respectively, is mounted on an infinite electric ground plane. Assuming that the electric field over the aperture of the coax is

$$\mathbf{E}_a = -\hat{\mathbf{a}}_\rho \dfrac{V}{\varepsilon \ln(b/a)} \dfrac{1}{\rho'}, \quad a \leq \rho' \leq b$$

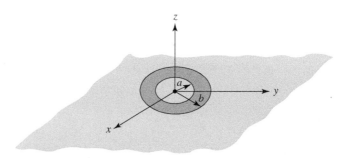

where V is the applied voltage and ε is the permittivity of the coax medium, find the far-zone spherical electric and magnetic field components radiated by the antenna.

12.51. It is desired to design a circular aperture antenna with a field distribution over its opening of

$$E = C[1 - (\rho'/a)^2]$$

where C is a constant, a its radius, and ρ' any point on the aperture, such that its beam efficiency within a 60° total included angle cone is 90%. Find its radius in wavelengths.

12.52. For the antenna of Problem 12.51, find its efficiency within a 40° total included angle cone when its radius is 2λ.

12.53. Design square apertures with uniform illumination so that the directivity at 60° from the normal is maximized relative to that at $\theta = 0°$. Determine the:
(a) Dimensions of the aperture (*in* λ)
(b) Maximum directivity (*in dB*)
(c) Directivity (*in dB*) at 60° from the maximum

12.54. Design a circular aperture with uniform illumination so that the directivity at 60° from the normal is maximized relative to that at $\theta = 0°$. Determine the
(a) Radius of the aperture (*in* λ)   (b) Maximum directivity (*in dB*)
(c) Directivity (*in dB*) at 60° from the maximum

12.55. Repeat Problem 12.50 for a circular aperture with a parabolic distribution.

12.56. Repeat Problem 12.50 for a circular aperture with a parabolic taper on 10 dB pedestal.

12.57. Derive the edge-of-coverage (EOC) design characteristics for a circular aperture with a parabolic taper.

12.58. Design a rectangular aperture of uniform illumination so that its directivity in the E- and H-planes is maximized (relative to its maximum value), respectively, at angles of 30° and 45° from the normal to the aperture.
   (a) Determine the optimum dimensions (*in* λ) of the aperture.
   (b) What is the *maximum* directivity (*in dB*) of the aperture and at what angle(s) (*in degrees*) will this occur?
   (c) What is the directivity (*in dB*) of the aperture along the E-plane at 30° from the normal to the aperture?
   (d) What is the directivity (*in dB*) of the aperture along the H-plane at 45° from the normal to the aperture?

12.59. Design a circular aperture with *uniform* distribution so that its directivity at an angle $\theta = 35°$ from the normal to the aperture is maximized relative to its maximum value at $\theta = 0°$. Specifically,
   (a) find the optimum radius (*in* λ) of the aperture.
   (b) what is the maximum directivity (*in dB*) at $\theta = 0°$?
   (c) what is the directivity (*in dB*) at $\theta = 35°$?

12.60. A vertical dipole is radiating into a free-space medium and produces fields $\mathbf{E}_0$ and $\mathbf{H}_0$. Illustrate alternate methods for obtaining the same fields using Babinet's principle and extensions of it.

12.61. (a) (1) Sketch the six principal-plane patterns, and (2) define the direction of $\mathbf{E}$ and $\mathbf{H}$ along the three principal axes and at 45° to the axes, for a thin slot one-half wavelength long, cut in a conducting sheet which has infinite conductivity and extending to infinity, and open on both sides. Inside dimensions of the slot are approximately 0.5λ by 0.1λ. Assume that the width (0.1λ) of the slot is small compared to a wavelength. Assume a coordinate system such that the conducting plane lies on the x-y plane with the larger dimension of the slot parallel to the y-axis.
   (b) Sketch the six approximate principal-plane patterns $E_\theta(\phi = 0°), E_\phi(\phi = 0°), E_\theta(\phi = 90°), E_\phi(\phi = 90°), E_\theta(\theta = 90°), E_\phi(\theta = 90°)$.

12.62. A very thin circular annular slot with circumference of one wavelength is cut on a very thin, infinite, flat, perfectly electric conducting plate. The slot is radiating into free-space. What is the impedance (real and imaginary parts) of the slot?

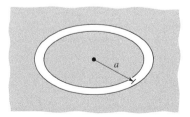

12.63. Repeat Example 12.7 for a rectangular aperture with an electric field distribution of

$$\mathbf{E}_a = \hat{\mathbf{a}}_y E_0 \cos\left(\frac{\pi}{a}x'\right), \quad \begin{array}{l} -a/2 \leq x' \leq a/2 \\ -b/2 \leq y' \leq b/2 \end{array}$$

12.64. Two identical very thin ($b = \lambda/20$) parallel-plate waveguides (slots), *each mounted on an infinite ground plane*, as shown in Figure 12.29 of the book for one of them, are separated by a distance of $\lambda_g/2$ where λ_g is the parallel-plate waveguide (transmission line) that is connecting the two slots. Assuming each slot is of width $W = 10$ cm, the parallel-plate waveguide (transmission line) is filled with air, the slots are radiating in free-space and are operating at *10 GHz*:

(a) What is the admittance Y_a of *one slot in the absence of the other (both real and imaginary parts)*?

(b) What is the total input impedance Z_{in} of *both slots together when looking in at the input of one of them in the presence of the other (both real and imaginary parts)*?

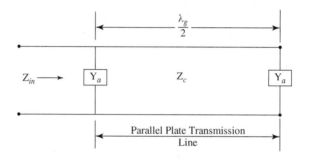

CHAPTER 13

Horn Antennas

13.1 INTRODUCTION

One of the simplest and probably the most widely used microwave antenna is the horn. Its existence and early use dates back to the late 1800s. Although neglected somewhat in the early 1900s, its revival began in the late 1930s from the interest in microwaves and waveguide transmission lines during the period of World War II. Since that time a number of articles have been written describing its radiation mechanism, optimization design methods, and applications. Many of the articles published since 1939 which deal with the fundamental theory, operating principles, and designs of a horn as a radiator can be found in a book of reprinted papers [1] and chapters in handbooks [2], [3].

The horn is widely used as a feed element for large radio astronomy, satellite tracking, and communication dishes found installed throughout the world. In addition to its utility as a feed for reflectors and lenses, it is a common element of phased arrays and serves as a universal standard for calibration and gain measurements of other high-gain antennas. Its widespread applicability stems from its simplicity in construction, ease of excitation, versatility, large gain, and preferred overall performance.

An electromagnetic horn can take many different forms, four of which are shown in Figure 13.1. The horn is nothing more than a hollow pipe of different cross sections, which has been tapered (flared) to a larger opening. The type, direction, and amount of taper (flare) can have a profound effect on the overall performance of the element as a radiator. In this chapter, the fundamental theory of horn antennas will be examined. In addition, data will be presented that can be used to understand better the operation of a horn and its design as an efficient radiator.

13.2 *E*-PLANE SECTORAL HORN

The *E*-plane sectoral horn is one whose opening is flared in the direction of the *E*-field, and it is shown in Figure 13.2(a). A more detailed geometry is shown in Figure 13.2(b).

13.2.1 Aperture Fields

The horn can be treated as an aperture antenna. To find its radiation characteristics, the equivalent principle techniques developed in Chapter 12 can be utilized. To develop an exact equivalent of it,

---

Portions of this chapter on aperture-matched horns, multimode horns, and dielectric-loaded horns were first published by the author in [2], Copyright 1988, reprinted by permission of Van Nostrand Reinhold Co.

Antenna Theory: Analysis and Design, Fourth Edition. Constantine A. Balanis.
© 2016 John Wiley & Sons, Inc. Published 2016 by John Wiley & Sons, Inc.
Companion Website: www.wiley.com/go/antennatheory4e

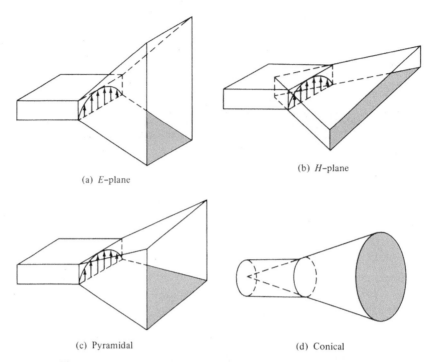

Figure 13.1 Typical electromagnetic horn antenna configurations.

it is necessary that the tangential electric and magnetic field components over a closed surface are known. The closed surface that is usually selected is an infinite plane that coincides with the aperture of the horn. When the horn is not mounted on an infinite ground plane, the fields outside the aperture are not known and an exact equivalent cannot be formed. However, the usual approximation is to assume that the fields outside the aperture are zero, as was done for the aperture of Section 12.5.2.

The fields at the aperture of the horn can be found by treating the horn as a radial waveguide [4]–[6]. The fields within the horn can be expressed in terms of cylindrical TE and TM wave functions which include Hankel functions. This method finds the fields not only at the aperture of the horn but also within the horn. The process is straightforward but laborious, and it will not be included here. However, it is assigned as an exercise at the end of the chapter (Problem 13.1).

It can be shown that if the (1) fields of the feed waveguide are those of its dominant TE_{10} mode and (2) horn length is large compared to the aperture dimensions, the lowest order mode fields at the aperture of the horn are given by

$$E'_z = E'_x = H'_y = 0 \tag{13-1a}$$

$$E'_y(x', y') \simeq E_1 \cos\left(\frac{\pi}{a}x'\right) e^{-j[ky'^2/(2\rho_1)]} \tag{13-1b}$$

$$H'_z(x', y') \simeq jE_1 \left(\frac{\pi}{ka\eta}\right) \sin\left(\frac{\pi}{a}x'\right) e^{-j[ky'^2/(2\rho_1)]} \tag{13-1c}$$

$$H'_x(x', y') \simeq -\frac{E_1}{\eta} \cos\left(\frac{\pi}{a}x'\right) e^{-j[ky'^2/(2\rho_1)]} \tag{13-1d}$$

$$\rho_1 = \rho_e \cos\psi_e \tag{13-1e}$$

where E_1 is a constant. The primes are used to indicate the fields at the aperture of the horn. The expressions are similar to the fields of a TE_{10}-mode for a rectangular waveguide with aperture

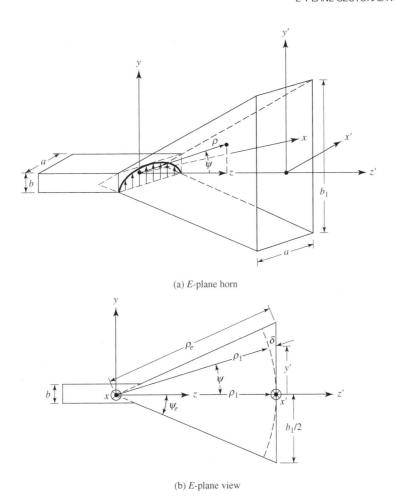

(a) *E*-plane horn

(b) *E*-plane view

Figure 13.2 *E*-plane horn and coordinate system.

dimensions of a and $b_1 (b_1 > a)$. The only difference is the complex exponential term which is used here to represent the quadratic phase variations of the fields over the aperture of the horn.

The necessity of the quadratic phase term in (13-1b)–(13-1d) can be illustrated geometrically. Referring to Figure 13.2(b), let us assume that at the imaginary apex of the horn (shown dashed) there exists a line source radiating cylindrical waves. As the waves travel in the outward radial direction, the constant phase fronts are cylindrical. At any point y' at the aperture of the horn, the phase of the field will not be the same as that at the origin ($y' = 0$). The phase is different because the wave has traveled different distances from the apex to the aperture. The difference in path of travel, designated as $\delta(y')$, can be obtained by referring to Figure 13.2(b). For any point y'

$$[\rho_1 + \delta(y')]^2 = \rho_1^2 + (y')^2 \tag{13-2}$$

or

$$\delta(y') = -\rho_1 + [\rho_1^2 + (y')^2]^{1/2} = -\rho_1 + \rho_1 \left[1 + \left(\frac{y'}{\rho_1}\right)^2\right]^{1/2} \tag{13-2a}$$

which is referred to as the *spherical* phase term.

Using the binomial expansion and retaining only the first two terms of it, (13-2a) reduces to

$$\delta(y') \simeq -\rho_1 + \rho_1 \left[1 + \frac{1}{2}\left(\frac{y'}{\rho_1}\right)^2\right] = \frac{1}{2}\left(\frac{y'^2}{\rho_1}\right) \qquad (13\text{-}2\text{b})$$

when (13-2b) is multiplied by the phase factor k, the result is identical to the *quadratic* phase term in (13-1b)–(13-1d).

The quadratic phase variation for the fields of the dominant mode at the aperture of a horn antenna has been a standard for many years, and it has been chosen because it yields in most practical cases very good results. Because of its simplicity, it leads to closed form expressions, in terms of sine and cosine Fresnel integrals, for the radiation characteristics (far-zone fields, directivity, etc.) of the horn. It has been shown recently [7] that using the more accurate expression of (13-2a) for the phase, error variations and numerical integration yield basically the same directivities as using the approximate expression of (13-2b) for large aperture horns (b_1 of Figures 13.2 or a_1 of Figure 13.10 greater than 50λ) or small peak aperture phase error ($S = \rho_e - \rho_1$ of Figure 13.2 or $T = \rho_h - \rho_2$ of Figure 13.10 less than 0.2λ). However, for intermediate aperture sizes (5$\lambda \leq b_1$ or $a_1 \leq 8\lambda$) or intermediate peak aperture phase errors (0.2$\lambda \leq S$ or $T \leq 0.6\lambda$) the more accurate expression of (13-2a) for the phase variation yields directivities which are somewhat higher (by as much as a few tenths of a decibel) than those obtained using (13-2b). Also it has been shown using a full-wave Moment Method analysis of the horn [8] that as the horn dimensions become large the amplitude distribution at the aperture of the horn contains higher-order modes than the TE_{10} mode and the phase distribution at the aperture approaches the parabolic phase front.

Example 13.1

Design an E-plane sectoral horn so that the maximum phase deviation at the aperture of the horn is 56.72°. The dimensions of the horn are $a = 0.5\lambda$, $b = 0.25\lambda$, $b_1 = 2.75\lambda$.

Solution: Using (13-2b)

$$\Delta\phi|_{max} = k\delta(y')|_{y'=b_1/2} = \frac{k(b_1/2)^2}{2\rho_1} = 56.72\left(\frac{\pi}{180}\right)$$

or

$$\rho_1 = \left(\frac{2.75}{2}\right)^2 \frac{180}{56.72}\lambda = 6\lambda$$

The total flare angle of the horn should be equal to

$$2\psi_e = 2\tan^{-1}\left(\frac{b_1/2}{\rho_1}\right) = 2\tan^{-1}\left(\frac{2.75/2}{6}\right) = 25.81°$$

13.2.2 Radiated Fields

To find the fields radiated by the horn, only the tangential components of the **E**- and/or **H**-fields over a closed surface must be known. The closed surface is chosen to coincide with an infinite plane passing through the mouth of the horn. To solve for the fields, the approximate equivalent of Section 12.5.2 is used. That is,

$$J_y = -\frac{E_1}{\eta} \cos\left(\frac{\pi}{a}x'\right) e^{-jk\delta(y')} \quad \left.\begin{array}{l} -a/2 \le x' \le a/2 \\ \\ M_x = E_1 \cos\left(\frac{\pi}{a}x'\right) e^{-jk\delta(y')} \end{array}\right\} \quad -b_1/2 \le y' \le b_1/2 \tag{13-3}$$

and

$$\mathbf{J}_s = \mathbf{M}_s = 0 \quad \text{elsewhere} \tag{13-3a}$$

Using (12-12a)

$$N_\theta = -\frac{E_1}{\eta} \cos\theta \sin\phi I_1 I_2 \tag{13-4}$$

where

$$I_1 = \int_{-a/2}^{+a/2} \cos\left(\frac{\pi}{a}x'\right) e^{jkx'\sin\theta\cos\phi} \, dx' = -\left(\frac{\pi a}{2}\right) \left[\frac{\cos\left(\frac{ka}{2}\sin\theta\cos\phi\right)}{\left(\frac{ka}{2}\sin\theta\cos\phi\right)^2 - \left(\frac{\pi}{2}\right)^2}\right] \tag{13-4a}$$

$$I_2 = \int_{-b_1/2}^{+b_1/2} e^{-jk[\delta(y') - y'\sin\theta\sin\phi]} \, dy' \tag{13-4b}$$

The integral of (13-4b) can also be evaluated in terms of cosine and sine Fresnel integrals. To do this, I_2 can be written, by completing the square, as

$$I_2 = \int_{-b_1/2}^{+b_1/2} e^{-j[ky'^2/(2\rho_1) - k_y y']} \, dy' = e^{j(k_y^2 \rho_1/2k)} \int_{-b_1/2}^{+b_1/2} e^{-j[(ky' - k_y \rho_1)^2/2k\rho_1]} \, dy' \tag{13-5}$$

where

$$k_y = k \sin\theta \sin\phi \tag{13-5a}$$

Making a change of variable

$$\sqrt{\frac{\pi}{2}} t = \sqrt{\frac{1}{2k\rho_1}} (ky' - k_y \rho_1) \tag{13-6a}$$

$$t = \sqrt{\frac{1}{\pi k \rho_1}} (ky' - k_y \rho_1) \tag{13-6b}$$

$$dt = \sqrt{\frac{k}{\pi \rho_1}} \, dy' \tag{13-6c}$$

reduces (13-5) to

$$I_2 = \sqrt{\frac{\pi \rho_1}{k}} e^{j(k_y^2 \rho_1/2k)} \int_{t_1}^{t_2} e^{-j(\pi/2)t^2} \, dt$$

$$= \sqrt{\frac{\pi \rho_1}{k}} e^{j(k_y^2 \rho_1/2k)} \int_{t_1}^{t_2} \left[\cos\left(\frac{\pi}{2}t^2\right) - j \sin\left(\frac{\pi}{2}t^2\right) \right] dt \quad (13\text{-}7)$$

and takes the form of

$$I_2 = \sqrt{\frac{\pi \rho_1}{k}} e^{j(k_y^2 \rho_1/2k)} \{[C(t_2) - C(t_1)] - j[S(t_2) - S(t_1)]\} \quad (13\text{-}8)$$

where

$$t_1 = \sqrt{\frac{1}{\pi k \rho_1}} \left(-\frac{kb_1}{2} - k_y \rho_1 \right) \quad (13\text{-}8a)$$

$$t_2 = \sqrt{\frac{1}{\pi k \rho_1}} \left(\frac{kb_1}{2} - k_y \rho_1 \right) \quad (13\text{-}8b)$$

$$C(x) = \int_0^x \cos\left(\frac{\pi}{2}t^2\right) dt \quad (13\text{-}8c)$$

$$S(x) = \int_0^x \sin\left(\frac{\pi}{2}t^2\right) dt \quad (13\text{-}8d)$$

$C(x)$ and $S(x)$ are known as the cosine and sine Fresnel integrals and are well tabulated [9] (see Appendix IV). Computer subroutines are also available for efficient numerical evaluation of each [10], [11].

Using (13-4a) and (13-8), (13-4) can be written as

$$N_\theta = E_1 \frac{\pi a}{2} \sqrt{\frac{\pi \rho_1}{k}} e^{j(k_y^2 \rho_1/2k)} \times \left\{ \frac{\cos\theta \sin\phi}{\eta} \left[\frac{\cos\left(\frac{k_x a}{2}\right)}{\left(\frac{k_x a}{2}\right)^2 - \left(\frac{\pi}{2}\right)^2} \right] F(t_1, t_2) \right\} \quad (13\text{-}9)$$

where

$$k_x = k \sin\theta \cos\phi \quad (13\text{-}9a)$$

$$k_y = k \sin\theta \sin\phi \quad (13\text{-}9b)$$

$$F(t_1, t_2) = [C(t_2) - C(t_1)] - j[S(t_2) - S(t_1)] \quad (13\text{-}9c)$$

E-PLANE SECTORAL HORN   725

In a similar manner, N_ϕ, L_θ, L_ϕ of (12-12b)–(12-12d) reduce to

$$N_\phi = E_1 \frac{\pi a}{2} \sqrt{\frac{\pi \rho_1}{k}} e^{j(k_y^2 \rho_1/2k)} \left\{ \frac{\cos \phi}{\eta} \left[\frac{\cos\left(\frac{k_x a}{2}\right)}{\left(\frac{k_x a}{2}\right)^2 - \left(\frac{\pi}{2}\right)^2} \right] F(t_1, t_2) \right\} \quad (13\text{-}10a)$$

$$L_\theta = E_1 \frac{\pi a}{2} \sqrt{\frac{\pi \rho_1}{k}} e^{j(k_y^2 \rho_1/2k)} \times \left\{ -\cos\theta \cos\phi \left[\frac{\cos\left(\frac{k_x a}{2}\right)}{\left(\frac{k_x a}{2}\right)^2 - \left(\frac{\pi}{2}\right)^2} \right] F(t_1, t_2) \right\} \quad (13\text{-}10b)$$

$$L_\phi = E_1 \frac{\pi a}{2} \sqrt{\frac{\pi \rho_1}{k}} e^{j(k_y^2 \rho_1/2k)} \left\{ \sin\phi \left[\frac{\cos\left(\frac{k_x a}{2}\right)}{\left(\frac{k_x a}{2}\right)^2 - \left(\frac{\pi}{2}\right)^2} \right] F(t_1, t_2) \right\} \quad (13\text{-}10c)$$

The electric field components radiated by the horn can be obtained by using (12-10a)–(12-10c), and (13-9)–(13-10c). Thus,

$$E_r = 0 \quad (13\text{-}11a)$$

$$E_\theta = -j\frac{a\sqrt{\pi k \rho_1} E_1 e^{-jkr}}{8r}$$
$$\times \left\{ e^{j(k_y^2 \rho_1/2k)} \sin\phi (1 + \cos\theta) \left[\frac{\cos\left(\frac{k_x a}{2}\right)}{\left(\frac{k_x a}{2}\right)^2 - \left(\frac{\pi}{2}\right)^2} \right] F(t_1, t_2) \right\} \quad (13\text{-}11b)$$

$$E_\phi = -j\frac{a\sqrt{\pi k \rho_1} E_1 e^{-jkr}}{8r}$$
$$\times \left\{ e^{j(k_y^2 \rho_1/2k)} \cos\phi (\cos\theta + 1) \left[\frac{\cos\left(\frac{k_x a}{2}\right)}{\left(\frac{k_x a}{2}\right)^2 - \left(\frac{\pi}{2}\right)^2} \right] F(t_1, t_2) \right\} \quad (13\text{-}11c)$$

where t_1, t_2, k_x, k_y, and $F(t_1, t_2)$ are given, respectively, by (13-8a), (13-8b), (13-9a), (13-9b), and (13-9c). The corresponding **H**-field components are obtained using (12-10d)–(12-10f).

In the principal E- and H-planes, the electric field reduces to

E-Plane ($\phi = \pi/2$)

$$E_r = E_\phi = 0 \tag{13-12a}$$

$$E_\theta = -j\frac{a\sqrt{\pi k\rho_1}E_1 e^{-jkr}}{8r}\left\{-e^{j(k\rho_1\sin^2\theta/2)}\left(\frac{2}{\pi}\right)^2(1+\cos\theta)F(t_1',t_2')\right\} \tag{13-12b}$$

$$t_1' = \sqrt{\frac{k}{\pi\rho_1}}\left(-\frac{b_1}{2} - \rho_1\sin\theta\right) \tag{13-12c}$$

$$t_2' = \sqrt{\frac{k}{\pi\rho_1}}\left(+\frac{b_1}{2} - \rho_1\sin\theta\right) \tag{13-12d}$$

H-Plane ($\phi = 0$)

$$E_r = E_\theta = 0 \tag{13-13a}$$

$$E_\phi = -j\frac{a\sqrt{\pi k\rho_1}E_1 e^{-jkr}}{8r}\left\{(1+\cos\theta)\left[\frac{\cos\left(\frac{ka}{2}\sin\theta\right)}{\left(\frac{ka}{2}\sin\theta\right)^2 - \left(\frac{\pi}{2}\right)^2}\right]F(t_1'',t_2'')\right\} \tag{13-13b}$$

$$t_1'' = -\frac{b_1}{2}\sqrt{\frac{k}{\pi\rho_1}} \tag{13-13c}$$

$$t_2'' = +\frac{b_1}{2}\sqrt{\frac{k}{\pi\rho_1}} \tag{13-13d}$$

To better understand the performance of an E-plane sectoral horn and gain some insight into its performance as an efficient radiator, a three-dimensional normalized field pattern has been plotted in Figure 13.3 utilizing (13-11a)–(13-11c). As expected, the E-plane pattern is much narrower than the H-plane because of the flaring and larger dimensions of the horn in that direction. Figure 13.3 provides an excellent visual view of the overall radiation performance of the horn. To display additional details, the corresponding normalized E- and H-plane patterns (*in dB*) are illustrated in Figure 13.4. These patterns also illustrate the narrowness of the E-plane and provide information on the relative levels of the pattern in those two planes.

To examine the behavior of the pattern as a function of flaring, the E-plane patterns for a horn antenna with $\rho_1 = 15\lambda$ and with flare angles of $20° \leq 2\psi_e \leq 35°$ are plotted in Figure 13.5. A total of four patterns is illustrated. Since each pattern is symmetrical, only half of each pattern is displayed. For small included angles, the pattern becomes narrower as the flare increases. Eventually the pattern begins to widen, becomes flatter around the main lobe, and the phase tapering at the aperture is such that even the main maximum does not occur on axis. This is illustrated in Figure 13.5 by the pattern with $2\psi_e = 35°$. As the flaring is extended beyond that point, the flatness (with certain allowable ripple) increases and eventually the main maximum returns again on axis. It is also observed that as the flaring increases, the pattern exhibits much sharper cutoff characteristics. In practice, to compensate for the phase taper at the opening, a lens is usually placed at the aperture making the pattern of the horn always narrower as its flare increases.

E-PLANE SECTORAL HORN 727

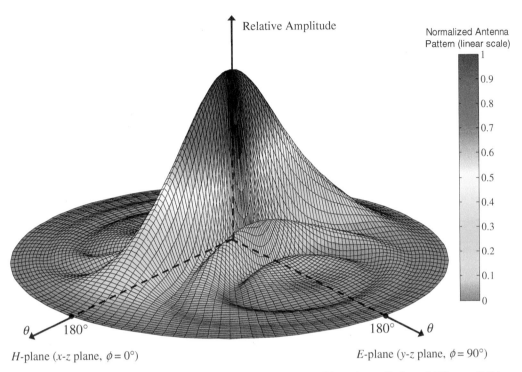

H-plane (x-z plane, $\phi = 0°$)   E-plane (y-z plane, $\phi = 90°$)

Figure 13.3 Three-dimensional field pattern of E-plane sectoral horn ($\rho_1 = 6\lambda$, $b_1 = 2.75\lambda$, $a = 0.5\lambda$).

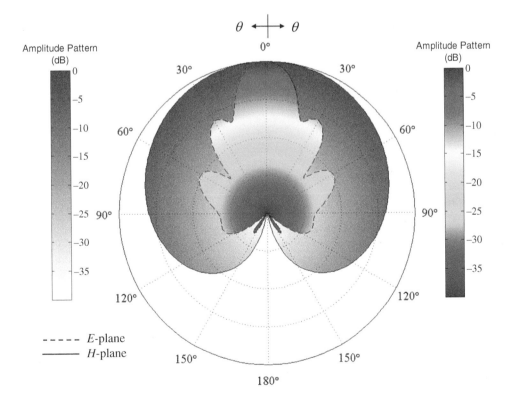

Figure 13.4 E- and H-plane patterns of an E-plane sectoral horn.

728 HORN ANTENNAS

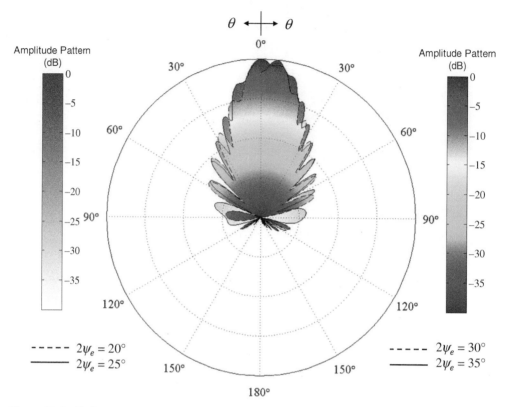

Figure 13.5 E-plane patterns of E-plane sectoral horn for constant length and different included angles.

Similar pattern variations occur as the length of the horn is varied while the flare angle is held constant. As the length increases, the pattern begins to broaden and eventually becomes flatter (with a ripple). Beyond a certain length, the main maximum does not even occur on axis, and the pattern continues to broaden and to become flatter (within an allowable ripple) until the maximum returns on axis. The process continues indefinitely.

13.2.3 Directivity

The directivity is one of the parameters that is often used as a figure of merit to describe the performance of an antenna. To find the directivity, the maximum radiation is formed. That is,

$$U_{max} = U(\theta, \phi)|_{max} = \frac{r^2}{2\eta}|\mathbf{E}|^2_{max} \tag{13-14}$$

For most horn antennas $|\mathbf{E}|_{max}$ is directed nearly along the z-axis ($\theta = 0°$). Thus,

$$|\mathbf{E}|_{max} = \sqrt{|E_\theta|^2_{max} + |E_\phi|^2_{max}} = \frac{2a\sqrt{\pi k \rho_1}}{\pi^2 r}|E_1||F(t)| \tag{13-15}$$

Using (13-11b), (13-11c), and (13-9c)

$$|E_\theta|_{\max} = \frac{2a\sqrt{\pi k \rho_1}}{\pi^2 r}|E_1 \sin\phi F(t)| \tag{13-15a}$$

$$|E_\phi|_{\max} = \frac{2a\sqrt{\pi k \rho_1}}{\pi^2 r}|E_1 \cos\phi F(t)| \tag{13-15b}$$

$$F(t) = [C(t) - jS(t)] \tag{13-15c}$$

$$t = \frac{b_1}{2}\sqrt{\frac{k}{\pi \rho_1}} = \frac{b_1}{\sqrt{2\lambda \rho_1}} \tag{13-15d}$$

since

$$k_x = k_y = 0 \tag{13-15e}$$

$$t_1 = -t = -\frac{b_1}{2}\sqrt{\frac{k}{\pi \rho_1}} = -\frac{b_1}{\sqrt{2\lambda \rho_1}} \tag{13-15f}$$

$$t_2 = +t = +\frac{b_1}{2}\sqrt{\frac{k}{\pi \rho_1}} = \frac{b_1}{\sqrt{2\lambda \rho_1}} \tag{13-15g}$$

$$C(-t) = -C(t) \tag{13-15h}$$

$$S(-t) = -S(t) \tag{13-15i}$$

Thus

$$U_{\max} = \frac{r^2}{2\eta}|\mathbf{E}|_{\max}^2 = \frac{2a^2 k \rho_1}{\eta \pi^3}|E_1|^2|F(t)|^2 = \frac{4a^2 \rho_1 |E_1|^2}{\eta \lambda \pi^2}|F(t)|^2 \tag{13-16}$$

where

$$|F(t)|^2 = \left[C^2\left(\frac{b_1}{\sqrt{2\lambda \rho_1}}\right) + S^2\left(\frac{b_1}{\sqrt{2\lambda \rho_1}}\right)\right] \tag{13-16a}$$

The total power radiated can be found by simply integrating the average power density over the aperture of the horn. Using (13-1a)–(13-1d)

$$P_{\text{rad}} = \frac{1}{2}\iint_{S_0} \text{Re}(\mathbf{E}' \times \mathbf{H}'^*) \cdot d\mathbf{s} = \frac{1}{2\eta}\int_{-b_1/2}^{+b_1/2}\int_{-a/2}^{+a/2}|E_1|^2 \cos^2\left(\frac{\pi}{a}x'\right)dx' \, dy' \tag{13-17}$$

which reduces to

$$P_{\text{rad}} = |E_1|^2 \frac{b_1 a}{4\eta} \tag{13-17a}$$

Using (13-16) and (13-17a), the directivity for the E-plane horn can be written as

$$D_E = \frac{4\pi U_{\max}}{P_{\text{rad}}} = \frac{64a\rho_1}{\pi \lambda b_1}|F(t)|^2$$
$$= \frac{64a\rho_1}{\pi \lambda b_1}\left[C^2\left(\frac{b_1}{\sqrt{2\lambda \rho_1}}\right) + S^2\left(\frac{b_1}{\sqrt{2\lambda \rho_1}}\right)\right] \quad (13\text{-}18)$$

The overall performance of an antenna system can often be judged by its beamwidth and/or its directivity. The half-power beamwidth (HPBW), as a function of flare angle, for different horn lengths is shown in Figure 13.6. In addition, the directivity (normalized with respect to the constant aperture dimension a) is displayed in Figure 13.7. For a given length, the horn exhibits a monotonic decrease in half-power beamwidth and an increase in directivity up to a certain flare. Beyond that point a monotonic increase in beamwidth and decrease in directivity is indicated followed by rises and falls. The increase in beamwidth and decrease in directivity beyond a certain flare indicate the broadening of the main beam.

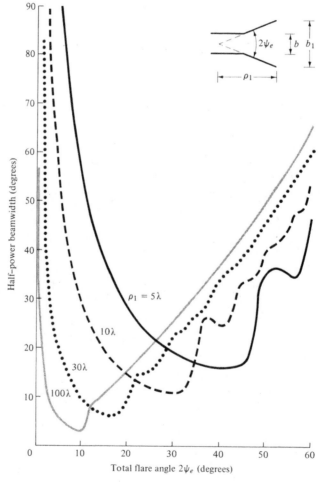

Figure 13.6 Half-power beamwidth of E-plane sectoral horn as a function of included angle and for different lengths.

E-PLANE SECTORAL HORN

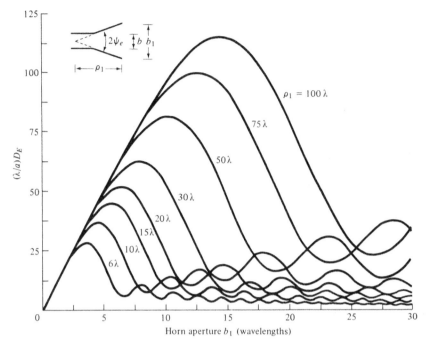

Figure 13.7 Normalized directivity of E-plane sectoral horn as a function of aperture size and for different lengths.

If the values of b_1 (in λ), which correspond to the maximum directivities in Figure 13.7, are plotted versus their corresponding values of ρ_1 (in λ), it can be shown that each optimum directivity occurs when

$$b_1 \simeq \sqrt{2\lambda\rho_1} \tag{13-18a}$$

with a corresponding value of s equal to

$$s\big|_{b_1 = \sqrt{2\lambda\rho_1}} = s_{op} = \frac{b_1^2}{8\lambda\rho_1}\bigg|_{b_1 = \sqrt{2\lambda\rho_1}} = \frac{1}{4} \tag{13-18b}$$

The classic expression of (13-18) for the directivity of an E-plane horn has been the standard for many years. However, it has been shown that this expression may not always yield very accurate values for the on-axis directivity. A more accurate expression for the maximum on-axis directivity based on an exact open-ended parallel-plate waveguide analysis has been derived, and it yields a modification to the on-axis value of (13-18), which provides sufficient accuracy for most designs [12], [13]. Using (13-18a), the modified formula for the on-axis value of (13-18) can be written as [12], [13]

$$D_E(\max) = \frac{16ab_1}{\lambda^2(1+\lambda_g/\lambda)} \left[C^2\left(\frac{b_1}{\sqrt{2\lambda\rho_1}}\right) + S^2\left(\frac{b_1}{\sqrt{2\lambda\rho_1}}\right) \right] e^{\frac{\pi a}{\lambda}\left(1-\frac{\lambda}{\lambda_g}\right)} \tag{13-18c}$$

where λ_g is the guide wavelength in the feed waveguide for the dominant TE_{10} mode. Predicted values based on (13-18) and (13-18c) have been compared with measurements and it was found that (13-18c) yielded results which were closer to the measured values [12].

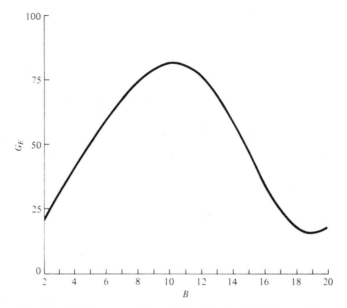

Figure 13.8 G_E as a function of B. (SOURCE: Adopted from data by E. H. Braun, "Some Data for the Design of Electromagnetic Horns," *IRE Trans. Antennas Propagat.*, Vol. AP-4, No. 1, January 1956. © 1956 IEEE).

The directivity of an *E*-plane sectoral horn can also be computed by using the following procedure [14].

1. Calculate B by

$$B = \frac{b_1}{\lambda}\sqrt{\frac{50}{\rho_e/\lambda}} \qquad (13\text{-}19\text{a})$$

2. Using this value of B, find the corresponding value of G_E from Figure 13.8. If, however, the value of B is smaller than 2, compute G_E using

$$G_E = \frac{32}{\pi}B \qquad (13\text{-}19\text{b})$$

3. Calculate D_E by using the value of G_E from Figure 13.8 or from (13-19b). Thus

$$\boxed{D_E = \frac{a}{\lambda}\frac{G_E}{\sqrt{\dfrac{50}{\rho_e/\lambda}}}} \qquad (13\text{-}19\text{c})$$

Example 13.2

An *E*-plane sectoral horn has dimensions of $a = 0.5\lambda$, $b = 0.25\lambda$, $b_1 = 2.75\lambda$, and $\rho_1 = 6\lambda$. Compute the directivity using (13-18) and (13-19c). Compare the answers.

Solution: For this horn

$$\frac{b_1}{\sqrt{2\lambda\rho_1}} = \frac{2.75}{\sqrt{2(6)}} = 0.794$$

Therefore (from Appendix IV)
$$[C(0.794)]^2 = (0.72)^2 = 0.518$$
$$[S(0.794)]^2 = (0.24)^2 = 0.0576$$

Using (13-18)
$$D_E = \frac{64(0.5)6}{2.75\pi}(0.518 + 0.0576) = 12.79 = 11.07 \text{ dB}$$

To compute the directivity using (13-19c), the following parameters are evaluated:
$$\rho_e = \lambda\sqrt{(6)^2 + \left(\frac{2.75}{2}\right)^2} = 6.1555\lambda$$

$$\sqrt{\frac{50}{\rho_e/\lambda}} = \sqrt{\frac{50}{6.1555}} = 2.85$$

$$B = 2.75(2.85) = 7.84$$

For $B = 7.84$, $G_E = 73.5$ from Figure 13.8. Thus, using (13-19c)
$$D_E = \frac{0.5(73.5)}{2.85} = 12.89 = 11.10 \text{ dB}$$

Obviously an excellent agreement between the results of (13-18) and (13-19c).

13.3 H-PLANE SECTORAL HORN

Flaring the dimensions of a rectangular waveguide in the direction of the **H**-field, while keeping the other constant, forms an *H*-plane sectoral horn shown in Figure 13.1(b). A more detailed geometry is shown in Figure 13.9.

The analysis procedure for this horn is similar to that for the *E*-plane horn, which was outlined in the previous section. Instead of including all the details of the formulation, a summary of each radiation characteristic will be given.

13.3.1 Aperture Fields

The fields at the aperture of the horn can be found by treating the horn as a radial waveguide forming an imaginary apex shown dashed in Figure 13.9. Using this method, it can be shown that at the aperture of the horn

$$E'_x = H'_y = 0 \tag{13-20a}$$

$$E'_y(x') = E_2 \cos\left(\frac{\pi}{a_1}x'\right) e^{-jk\delta(x')} \tag{13-20b}$$

$$H'_x(x') = -\frac{E_2}{\eta} \cos\left(\frac{\pi}{a_1}x'\right) e^{-jk\delta(x')} \tag{13-20c}$$

$$\delta(x') = \frac{1}{2}\left(\frac{x'^2}{\rho_2}\right) \tag{13-20d}$$

$$\rho_2 = \rho_h \cos\psi_h \tag{13-20e}$$

734 HORN ANTENNAS

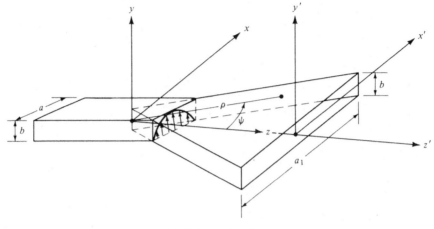

(a) *H*-plane sectoral horn

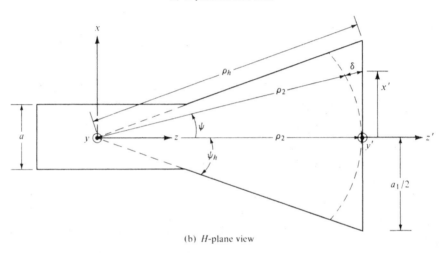

(b) *H*-plane view

Figure 13.9 *H*-plane sectoral horn and coordinate system.

13.3.2 Radiated Fields

The fields radiated by the horn can be found by first formulating the equivalent current densities $\mathbf{J}_s$ and $\mathbf{M}_s$. Using (13-20a)–(13-20c), it can be shown that over the aperture of the horn

$$J_x = J_z = M_y = M_z = 0 \tag{13-21a}$$

$$J_y = -\frac{E_2}{\eta} \cos\left(\frac{\pi}{a_1}x'\right) e^{-jk\delta(x')} \tag{13-21b}$$

$$M_x = E_2 \cos\left(\frac{\pi}{a_1}x'\right) e^{-jk\delta(x')} \tag{13-21c}$$

and they are assumed to be zero elsewhere. Thus (12-11a) can be expressed as

$$N_\theta = \iint_S J_y \cos\theta \sin\phi \, e^{+jkr' \cos\psi} \, ds' = -\frac{E_2}{\eta} \cos\theta \sin\phi I_1 I_2 \tag{13-22}$$

where

$$I_1 = \int_{-b/2}^{+b/2} e^{+jky' \sin\theta \sin\phi} \, dy' = b \left[\frac{\sin\left(\frac{kb}{2}\sin\theta\sin\phi\right)}{\frac{kb}{2}\sin\theta\sin\phi} \right] \quad (13\text{-}22a)$$

$$I_2 = \int_{-a_1/2}^{+a_1/2} \cos\left(\frac{\pi}{a_1}x'\right) e^{-jk[\delta(x')-x'\sin\theta\cos\phi]} \, dx' \quad (13\text{-}22b)$$

By rewriting $\cos[(\pi/a_1)x']$ as

$$\cos\left(\frac{\pi}{a_1}x'\right) = \left[\frac{e^{j(\pi/a_1)x'} + e^{-j(\pi/a_1)x'}}{2} \right] \quad (13\text{-}23)$$

(13-22b) can be expressed as

$$I_2 = I_2' + I_2'' \quad (13\text{-}24)$$

where

$$I_2' = \frac{1}{2}\sqrt{\frac{\pi\rho_2}{k}} e^{j(k_x'^2 \rho_2/2k)} \{[C(t_2') - C(t_1')] - j[S(t_2') - S(t_1')]\} \quad (13\text{-}25)$$

$$t_1' = \sqrt{\frac{1}{\pi k \rho_2}} \left(-\frac{ka_1}{2} - k_x' \rho_2 \right) \quad (13\text{-}25a)$$

$$t_2' = \sqrt{\frac{1}{\pi k \rho_2}} \left(+\frac{ka_1}{2} - k_x' \rho_2 \right) \quad (13\text{-}25b)$$

$$k_x' = k \sin\theta \cos\phi + \frac{\pi}{a_1} \quad (13\text{-}25c)$$

$$I_2'' = \frac{1}{2}\sqrt{\frac{\pi\rho_2}{k}} e^{j(k_x''^2 \rho_2/2k)} \{[C(t_2'') - C(t_1'')] - j[S(t_2'') - S(t_1'')]\} \quad (13\text{-}26)$$

$$t_1'' = \sqrt{\frac{1}{\pi k \rho_2}} \left(-\frac{ka_1}{2} - k_x'' \rho_2 \right) \quad (13\text{-}26a)$$

$$t_2'' = \sqrt{\frac{1}{\pi k \rho_2}} \left(+\frac{ka_1}{2} - k_x'' \rho_2 \right) \quad (13\text{-}26b)$$

$$k_x'' = k \sin\theta \cos\phi - \frac{\pi}{a_1} \quad (13\text{-}26c)$$

736 HORN ANTENNAS

$C(x)$ and $S(x)$ are the cosine and sine Fresnel integrals of (13-8c) and (13-8d), and they are well tabulated (see Appendix IV).

With the aid of (13-22a), (13-24), (13-25), and (13-26), (13-22) reduces to

$$N_\theta = -E_2 \frac{b}{2}\sqrt{\frac{\pi \rho_2}{k}} \left\{ \frac{\cos\theta \sin\phi}{\eta} \frac{\sin Y}{Y}[e^{jf_1}F(t'_1,t'_2) + e^{jf_2}F(t''_1,t''_2)] \right\} \tag{13-27}$$

$$F(t_1,t_2) = [C(t_2) - C(t_1)] - j[S(t_2) - S(t_1)] \tag{13-27a}$$

$$f_1 = \frac{k'^2_x \rho_2}{2k} \tag{13-27b}$$

$$f_2 = \frac{k''^2_x \rho_2}{2k} \tag{13-27c}$$

$$Y = \frac{kb}{2}\sin\theta\sin\phi \tag{13-27d}$$

In a similar manner, N_ϕ, L_θ, and L_ϕ of (12-12b)–(12-12d) can be written as

$$N_\phi = -E_2 \frac{b}{2}\sqrt{\frac{\pi \rho_2}{k}} \left\{ \frac{\cos\phi}{\eta} \frac{\sin Y}{Y}[e^{jf_1}F(t'_1,t'_2) + e^{jf_2}F(t''_1,t''_2)] \right\} \tag{13-28a}$$

$$L_\theta = E_2 \frac{b}{2}\sqrt{\frac{\pi \rho_2}{k}} \left\{ \cos\theta\cos\phi \frac{\sin Y}{Y}[e^{jf_1}F(t'_1,t'_2) + e^{jf_2}F(t''_1,t''_2)] \right\} \tag{13-28b}$$

$$L_\phi = -E_2 \frac{b}{2}\sqrt{\frac{\pi \rho_2}{k}} \left\{ \sin\phi \frac{\sin Y}{Y}[e^{jf_1}F(t'_1,t'_2) + e^{jf_2}F(t''_1,t''_2)] \right\} \tag{13-28c}$$

The far-zone electric field components of (12-10a)–(12-10c) can then be expressed as

$$E_r = 0 \tag{13-29a}$$

$$E_\theta = jE_2 \frac{b}{8}\sqrt{\frac{k\rho_2}{\pi}} \frac{e^{-jkr}}{r}$$
$$\times \left\{ \sin\phi(1+\cos\theta)\frac{\sin Y}{Y}[e^{jf_1}F(t'_1,t'_2) + e^{jf_2}F(t''_1,t''_2)] \right\} \tag{13-29b}$$

$$E_\phi = jE_2 \frac{b}{8}\sqrt{\frac{k\rho_2}{\pi}} \frac{e^{-jkr}}{r}$$
$$\times \left\{ \cos\phi(\cos\theta+1)\frac{\sin Y}{Y}[e^{jf_1}F(t'_1,t'_2) + e^{jf_2}F(t''_1,t''_2)] \right\} \tag{13-29c}$$

The electric field in the principal E- and H-planes reduces to

E-Plane ($\phi = \pi/2$)

$$E_r = E_\phi = 0 \qquad (13\text{-}30\text{a})$$

$$E_\theta = jE_2\frac{b}{8}\sqrt{\frac{k\rho_2}{\pi}}\frac{e^{-jkr}}{r} \times \left\{(1+\cos\theta)\frac{\sin Y}{Y}[e^{jf_1}F(t_1', t_2') + e^{jf_2}F(t_1'', t_2'')]\right\} \qquad (13\text{-}30\text{b})$$

$$Y = \frac{kb}{2}\sin\theta \qquad (13\text{-}30\text{c})$$

$$k_x' = \frac{\pi}{a_1} \qquad (13\text{-}30\text{d})$$

$$k_x'' = -\frac{\pi}{a_1} \qquad (13\text{-}30\text{e})$$

H-Plane ($\phi = 0$)

$$E_r = E_\theta = 0 \qquad (13\text{-}31\text{a})$$

$$E_\phi = jE_2\frac{b}{8}\sqrt{\frac{k\rho_2}{\pi}}\frac{e^{-jkr}}{r} \times \{(\cos\theta + 1)[e^{jf_1}F(t_1', t_2') + e^{jf_2}F(t_1'', t_2'')]\} \qquad (13\text{-}31\text{b})$$

$$k_x' = k\sin\theta + \frac{\pi}{a_1} \qquad (13\text{-}31\text{c})$$

$$k_x'' = k\sin\theta - \frac{\pi}{a_1} \qquad (13\text{-}31\text{d})$$

with $f_1, f_2, F(t_1', t_2'), F(t_1'', t_2''), t_1', t_2', t_1''$, and t_2'' as defined previously.

Computations similar to those for the E-plane sectoral horn were also performed for the H-plane sectoral horn. A three-dimensional field pattern of an H-plane sectoral horn is shown in Figure 13.10.

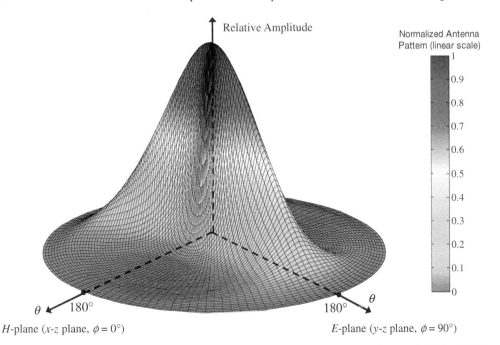

Figure 13.10 Three-dimensional field pattern of an H-plane sectoral horn ($\rho_2 = 6\lambda$, $a_1 = 5.5\lambda$, $b = 0.25\lambda$).

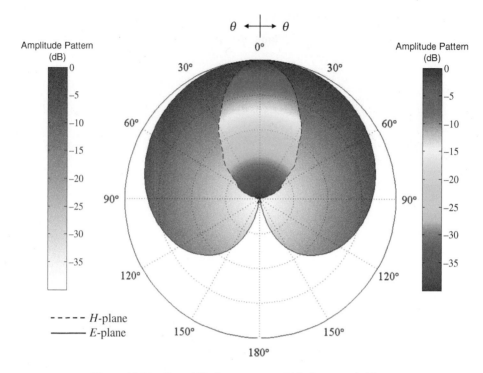

Figure 13.11 E- and H-plane patterns of H-plane sectoral horn.

Its corresponding E- and H-plane patterns are displayed in Figure 13.11. This horn exhibits narrow pattern characteristics in the flared H-plane.

Normalized H-plane patterns for a given length horn ($\rho_2 = 12\lambda$) and different flare angles are shown in Figure 13.12. A total of four patterns is illustrated. Since each pattern is symmetrical, only half of each pattern is displayed. As the included angle is increased, the pattern begins to become narrower up to a given flare. Beyond that point the pattern begins to broaden, attributed primarily to the phase taper (phase error) across the aperture of the horn. To correct this, a lens is usually placed at the horn aperture, which would yield narrower patterns as the flare angle is increased. Similar pattern variations are evident when the flare angle of the horn is maintained fixed while its length is varied.

13.3.3 Directivity

To find the directivity of the H-plane sectoral horn, a procedure similar to that for the E-plane is used. As for the E-plane sectoral horn, the maximum radiation is directed nearly along the z-axis ($\theta = 0°$). Thus

$$|E_\theta|_{\max} = |E_2|\frac{b}{4r}\sqrt{\frac{2\rho_2}{\lambda}} |\sin \phi\{[C(t'_2) + C(t''_2) - C(t'_1) - C(t''_1)] - j[S(t'_2) + S(t''_2) - S(t'_1) - S(t''_1)]\}| \quad (13\text{-}32)$$

$$t'_1 = \sqrt{\frac{1}{\pi k \rho_2}}\left(-\frac{ka_1}{2} - \frac{\pi}{a_1}\rho_2\right) \quad (13\text{-}32a)$$

$$t'_2 = \sqrt{\frac{1}{\pi k \rho_2}}\left(+\frac{ka_1}{2} - \frac{\pi}{a_1}\rho_2\right) \quad (13\text{-}32b)$$

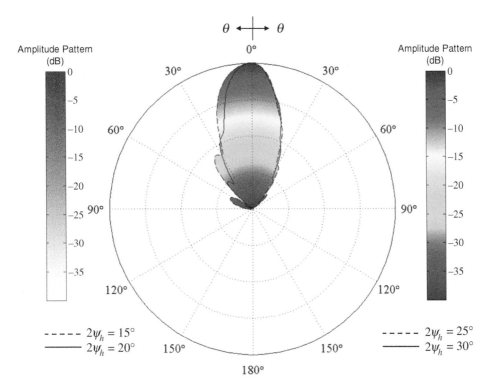

Figure 13.12 H-plane patterns of H-plane sectoral horn for constant length and different included angles.

$$t_1'' = \sqrt{\frac{1}{\pi k \rho_2}} \left(-\frac{ka_1}{2} + \frac{\pi}{a_1} \rho_2 \right) = -t_2' = v \tag{13-32c}$$

$$t_2'' = \sqrt{\frac{1}{\pi k \rho_2}} \left(+\frac{ka_1}{2} + \frac{\pi}{a_1} \rho_2 \right) = -t_1' = u \tag{13-32d}$$

Since

$$C(-x) = -C(x) \tag{13-33a}$$

$$S(-x) = -S(x) \tag{13-33b}$$

$$|E_\theta|_{\max} = |E_2| \frac{b}{r} \sqrt{\frac{\rho_2}{2\lambda}} |\sin\phi \{[C(u) - C(v)] - j[S(u) - S(v)]\}| \tag{13-34}$$

$$u = t_2'' = -t_1' = \sqrt{\frac{1}{\pi k \rho_2}} \left(+\frac{ka_1}{2} + \frac{\pi}{a_1} \rho_2 \right) = \frac{1}{\sqrt{2}} \left(\frac{\sqrt{\lambda \rho_2}}{a_1} + \frac{a_1}{\sqrt{\lambda \rho_2}} \right) \tag{13-34a}$$

$$v = t_1'' = -t_2' = \sqrt{\frac{1}{\pi k \rho_2}} \left(-\frac{ka_1}{2} + \frac{\pi}{a_1} \rho_2 \right) = \frac{1}{\sqrt{2}} \left(\frac{\sqrt{\lambda \rho_2}}{a_1} - \frac{a_1}{\sqrt{\lambda \rho_2}} \right) \tag{13-34b}$$

Similarly

$$|E_\phi|_{\max} = |E_2| \frac{b}{r} \sqrt{\frac{\rho_2}{2\lambda}} |\cos\phi \{[C(u) - C(v)] - j[S(u) - S(v)]\}| \tag{13-35}$$

Thus

$$|\mathbf{E}|_{max} = \sqrt{|E_\theta|^2_{max} + |E_\phi|^2_{max}} = |E_2|\frac{b}{r}\sqrt{\frac{\rho_2}{2\lambda}}\{[C(u)-C(v)]^2 + [S(u)-S(v)]^2\}^{1/2} \quad (13\text{-}36)$$

$$U_{max} = |E_2|^2 \frac{b^2 \rho_2}{4\eta\lambda}\{[C(u)-C(v)]^2 + [S(u)-S(v)]^2\} \quad (13\text{-}37)$$

The total power radiated can be obtained by simply integrating the average power density over the mouth of the horn, and it is given by

$$P_{rad} = |E_2|^2 \frac{ba_1}{4\eta} \quad (13\text{-}38)$$

Using (13-37) and (13-38), the directivity for the H-plane sectoral horn can be written as

$$\boxed{D_H = \frac{4\pi U_{max}}{P_{rad}} = \frac{4\pi b \rho_2}{a_1 \lambda} \times \{[C(u)-C(v)]^2 + [S(u)-S(v)]^2\}} \quad (13\text{-}39)$$

where

$$\boxed{u = \frac{1}{\sqrt{2}}\left(\frac{\sqrt{\lambda\rho_2}}{a_1} + \frac{a_1}{\sqrt{\lambda\rho_2}}\right)} \quad (13\text{-}39a)$$

$$\boxed{v = \frac{1}{\sqrt{2}}\left(\frac{\sqrt{\lambda\rho_2}}{a_1} - \frac{a_1}{\sqrt{\lambda\rho_2}}\right)} \quad (13\text{-}39b)$$

The half-power beamwidth (HPBW) as a function of flare angle is plotted in Figure 13.13. The normalized directivity (relative to the constant aperture dimension b) for different horn lengths, as a function of aperture dimension a_1, is displayed in Figure 13.14. As for the E-plane sectoral horn, the HPBW exhibits a monotonic decrease and the directivity a monotonic increase up to a given flare; beyond that, the trends are reversed.

If the values of a_1 (in λ), which correspond to the maximum directivities in Figure 13.14, are plotted versus their corresponding values of ρ_2 (in λ), it can be shown that each optimum directivity occurs when

$$\boxed{a_1 \simeq \sqrt{3\lambda\rho_2}} \quad (13\text{-}39c)$$

with a corresponding value of t equal to

$$\boxed{t|_{a_1=\sqrt{3\lambda\rho_2}} = t_{op} = \frac{a_1^2}{8\lambda\rho_2}\bigg|_{a_1=\sqrt{3\lambda\rho_2}} = \frac{3}{8}} \quad (13\text{-}39d)$$

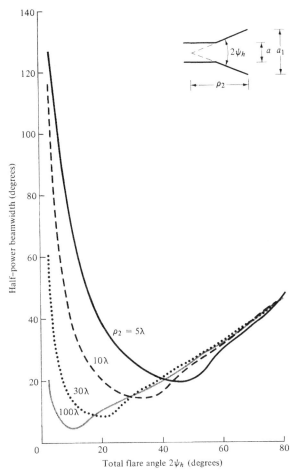

Figure 13.13 Half-power beamwidth of H-plane sectoral horn as a function of included angle and for different lengths.

The directivity of an H-plane sectoral horn can also be computed by using the following procedure [14].

1. Calculate A by

$$A = \frac{a_1}{\lambda}\sqrt{\frac{50}{\rho_h/\lambda}} \qquad (13\text{-}40\text{a})$$

2. Using this value of A, find the corresponding value of G_H from Figure 13.15. *If the value of A is smaller than 2, then compute G_H using*

$$G_H = \frac{32}{\pi}A \qquad (13\text{-}40\text{b})$$

3. Calculate D_H by using the value of G_H from Figure 13.15 or from (13-40b). Thus

$$\boxed{D_H = \frac{b}{\lambda}\frac{G_H}{\sqrt{\dfrac{50}{\rho_h/\lambda}}}} \qquad (13\text{-}40\text{c})$$

This is the actual directivity of the horn.

742 HORN ANTENNAS

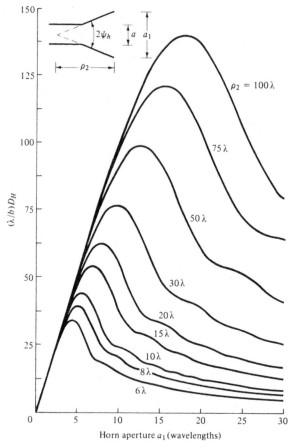

Figure 13.14 Normalized directivity of H-plane sectoral horn as a function of aperture size and for different lengths.

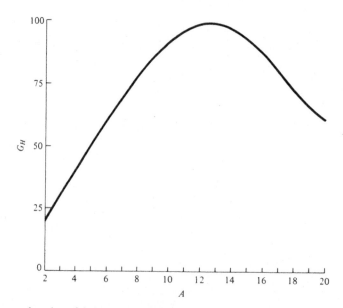

Figure 13.15 G_H as a function of A. (SOURCE: Adopted from data by E. H. Braun, "Some Data for the Design of Electromagnetic Horns," *IRE Trans. Antennas Propagat.*, Vol. AP-4, No. 1, January 1956. © 1956 IEEE).

Example 13.3

An H-plane sectoral horn has dimensions of $a = 0.5\lambda$, $b = 0.25\lambda$, $a_1 = 5.5\lambda$, and $\rho_2 = 6\lambda$. Compute the directivity using (13-39) and (13-40c). Compare the answers.

Solution: For this horn

$$u = \frac{1}{\sqrt{2}}\left(\frac{\sqrt{6}}{5.5} + \frac{5.5}{\sqrt{6}}\right) = 1.9$$

$$v = \frac{1}{\sqrt{2}}\left(\frac{\sqrt{6}}{5.5} - \frac{5.5}{\sqrt{6}}\right) = -1.273$$

Therefore (from Appendix IV)

$$C(1.9) = 0.394$$
$$C(-1.273) = -C(1.273) = -0.659$$
$$S(1.9) = 0.373$$
$$S(-1.273) = -S(1.273) = -0.669$$

Using (13-39)

$$D_H = \frac{4\pi(0.25)6}{5.5}[(0.394 + 0.659)^2 + (0.373 + 0.669)^2]$$

$$D_H = 7.52 = 8.763 \text{ dB}$$

To compute the directivity using (13-40c), the following parameters are computed:

$$\rho_h = \lambda\sqrt{(6)^2 + (5.5/2)^2} = 6.6\lambda$$

$$\sqrt{\frac{50}{\rho_h/\lambda}} = \sqrt{\frac{50}{6.6}} = 2.7524$$

$$A = 5.5(2.7524) = 15.14$$

For $A = 15.14$, $G_H = 91.8$ from Figure 13.15. Thus, using (13-40c)

$$D_H = \frac{0.25(91.8)}{2.7524} = 8.338 = 9.21 \text{ dB}$$

Although there is a good agreement between the results of (13-39) and (13-40c), they do not compare as well as those of Example 13.2.

13.4 PYRAMIDAL HORN

The most widely used horn is the one which is flared in both directions, as shown in Figure 13.16. It is widely referred to as a pyramidal horn, and its radiation characteristics are essentially a combination of the E- and H-plane sectoral horns.

744 HORN ANTENNAS

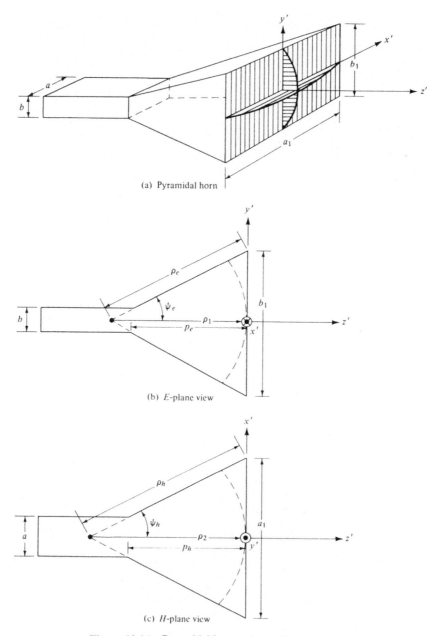

Figure 13.16 Pyramidal horn and coordinate system.

13.4.1 Aperture Fields, Equivalent, and Radiated Fields

To simplify the analysis and to maintain a modeling that leads to computations that have been shown to correlate well with experimental data, the tangential components of the E- and H-fields over the aperture of the horn are approximated by

$$E'_y(x', y') = E_0 \cos\left(\frac{\pi}{a_1}x'\right) e^{-j[k(x'^2/\rho_2 + y'^2/\rho_1)/2]} \qquad (13\text{-}41\text{a})$$

$$H'_x(x', y') = -\frac{E_0}{\eta} \cos\left(\frac{\pi}{a_1}x'\right) e^{-j[k(x'^2/\rho_2 + y'^2/\rho_1)/2]} \qquad (13\text{-}41\text{b})$$

and the equivalent current densities by

$$J_y(x', y') = -\frac{E_0}{\eta} \cos\left(\frac{\pi}{a_1}x'\right) e^{-j[k(x'^2/\rho_2 + y'^2/\rho_1)/2]} \quad (13\text{-}42a)$$

$$M_x(x', y') = E_0 \cos\left(\frac{\pi}{a_1}x'\right) e^{-j[k(x'^2/\rho_2 + y'^2/\rho_1)/2]} \quad (13\text{-}42b)$$

The above expressions contain a cosinusoidal amplitude distribution in the x' direction and quadratic phase variations in both the x' and y' directions, similar to those of the sectoral E- and H-plane horns.

The N_θ, N_ϕ, L_θ and L_ϕ can now be formulated as before, and it can be shown that they are given by

$$N_\theta = -\frac{E_0}{\eta} \cos\theta \sin\phi I_1 I_2 \quad (13\text{-}43a)$$

$$N_\phi = -\frac{E_0}{\eta} \cos\phi I_1 I_2 \quad (13\text{-}43b)$$

$$L_\theta = E_0 \cos\theta \cos\phi I_1 I_2 \quad (13\text{-}43c)$$

$$L_\phi = -E_0 \sin\phi I_1 I_2 \quad (13\text{-}43d)$$

where

$$I_1 = \int_{-a_1/2}^{+a_1/2} \cos\left(\frac{\pi}{a_1}x'\right) e^{-jk[x'^2/(2\rho_2) - x'\sin\theta\cos\phi]} \, dx' \quad (13\text{-}43e)$$

$$I_2 = \int_{-b_1/2}^{+b_1/2} e^{-jk[y'^2/(2\rho_1) - y'\sin\theta\sin\phi]} \, dy' \quad (13\text{-}43f)$$

Using (13-22b), (13-24), (13-25), and (13-26), (13-43e) can be expressed as

$$I_1 = \frac{1}{2}\sqrt{\frac{\pi\rho_2}{k}} (e^{j(k_x'^2 \rho_2/2k)}\{[C(t_2') - C(t_1')] - j[S(t_2') - S(t_1')]\} \\ + e^{j(k_x''^2 \rho_2/2k)}\{[C(t_2'') - C(t_1'')] - j[S(t_2'') - S(t_1'')]\}) \quad (13\text{-}44)$$

where $t_1', t_2', k_x', t_1'', t_2''$, and k_x'' are given by (13-25a)–(13-25c) and (13-26a)–(13-26c). Similarly, using (13-5)–(13-8d), I_2 of (13-43f) can be written as

$$I_2 = \sqrt{\frac{\pi\rho_1}{k}} e^{j(k_y^2 \rho_1/2k)}\{[C(t_2) - C(t_1)] - j[S(t_2) - S(t_1)]\} \quad (13\text{-}45)$$

where k_y, t_1, and t_2 are given by (13-5a), (13-8a), and (13-8b).

Combining (13-43a)–(13-43d), the far-zone **E**- and **H**-field components of (12-10a)–(12-10c) reduce to

$$E_r = 0 \tag{13-46a}$$

$$\begin{aligned}E_\theta &= -j\frac{ke^{jkr}}{4\pi r}[L_\phi + \eta N_\theta] \\ &= j\frac{kE_0 e^{-jkr}}{4\pi r}[\sin\phi(1+\cos\theta)I_1 I_2]\end{aligned} \tag{13-46b}$$

$$\begin{aligned}E_\phi &= +j\frac{ke^{-jkr}}{4\pi r}[L_\theta - \eta N_\phi] \\ &= j\frac{kE_0 e^{-jkr}}{4\pi r}[\cos\phi(\cos\theta+1)I_1 I_2]\end{aligned} \tag{13-46c}$$

where I_1 and I_2 are given by (13-44) and (13-45), respectively.

The fields radiated by a pyramidal horn, as given by (13-46a)–(13-46c), are valid for all angles of observation. An examination of these equations reveals that the principal E-plane pattern ($\phi = \pi/2$) of a pyramidal horn, aside from a normalization factor, is identical to the E-plane pattern of an E-plane sectoral horn. Similarly the H-plane ($\phi = 0$) is identical to that of an H-plane sectoral horn. Therefore the pattern of a pyramidal horn is very narrow in both principal planes and, in fact, in all planes. This is illustrated in Figure 13.17. The corresponding E-plane pattern is shown in Figure 13.4 and the H-plane pattern in Figure 13.11.

To demonstrate that the maximum radiation for a pyramidal horn is not necessarily directed along its axis, the three-dimensional field pattern for a horn with $\rho_1 = \rho_2 = 6\lambda$, $a_1 = 12\lambda$, $b_1 = 6\lambda$, $a = 0.50\lambda$ and $b = 0.25\lambda$ is displayed in Figure 13.18. The corresponding two-dimensional E- and H-plane patterns are shown in Figure 13.19. The maximum does not occur on

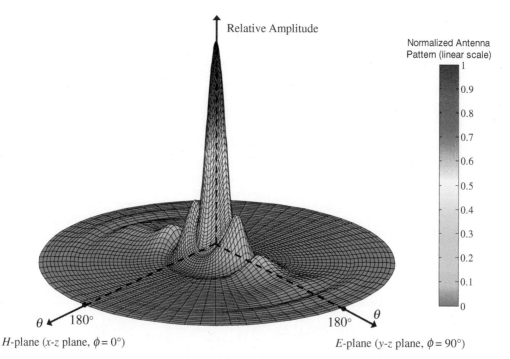

Figure 13.17 Three-dimensional field pattern of a pyramidal horn ($\rho_1 = \rho_2 = 6\lambda$, $a_1 = 5.5\lambda$, $b_1 = 2.75\lambda$, $a = 0.5\lambda$, $b = 0.25\lambda$).

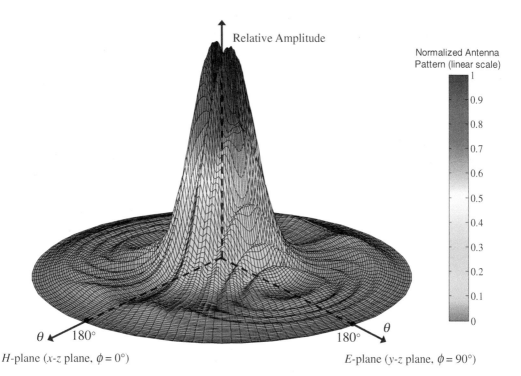

H-plane (x-z plane, $\phi = 0°$) E-plane (y-z plane, $\phi = 90°$)

Figure 13.18 Three-dimensional field pattern of a pyramidal horn with maximum not on axis ($\rho_1 = \rho_2 = 6\lambda$, $a_1 = 12\lambda$, $b_1 = 6\lambda$, $a = 0.5\lambda$, $b = 0.25\lambda$).

axis because the phase error taper at the aperture is such that the rays emanating from the different parts of the aperture toward the axis are not in phase and do not add constructively.

To physically construct a pyramidal horn, the dimension p_e of Figure 13.16(b) given by

$$p_e = (b_1 - b)\left[\left(\frac{\rho_e}{b_1}\right)^2 - \frac{1}{4}\right]^{1/2} \tag{13-47a}$$

should be equal to the dimension p_h of Figure 13.16(c) given by

$$p_h = (a_1 - a)\left[\left(\frac{\rho_h}{a_1}\right)^2 - \frac{1}{4}\right]^{1/2} \tag{13-47b}$$

The dimensions chosen for Figures 13.17 and 13.18 do satisfy these requirements. For the horn of Figure 13.17, $\rho_e = 6.1555\lambda$, $\rho_h = 6.6\lambda$, and $p_e = p_h = 5.4544\lambda$, whereas for that of Figure 13.18, $\rho_e = 6.7082\lambda$, $\rho_h = 8.4853\lambda$, and $p_e = p_h = 5.75\lambda$. The fields of (13-46a)–(13-46c) provide accurate patterns for angular regions near the main lobe and its closest minor lobes. To accurately predict the field intensity of the pyramidal and other horns, especially in the minor lobes, diffraction techniques can be utilized [15]–[18]. These methods take into account diffractions that occur near the aperture edges of the horn. The diffraction contributions become more dominant in regions where the radiation of (13-46a)–(13-46c) is of very low intensity.

In addition to the previous methods, the horn antenna has been examined using full-wave analyses, such as the Method of Moments (MoM) [8] and the Finite-Difference Time-Domain (FDTD) [19]. These methods yield more accurate results in all regions, and they are able to include many of the

748 HORN ANTENNAS

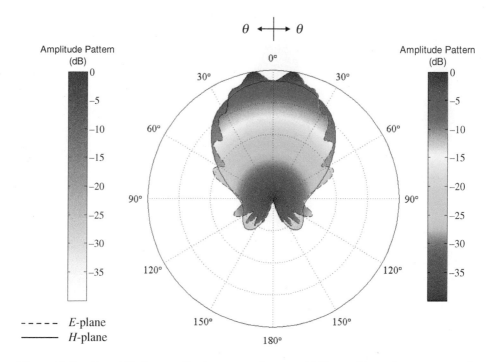

Figure 13.19 E- and H-plane amplitude patterns of a pyramidal horn with maximum not on axis.

other features of the horn, such as its wall thickness, etc. Predicted patterns based on these methods compare extremely well with measurements, even in regions of very low intensity (such as the back lobes). An example of such a comparison is made in Figure 13.20(a,b) for the E- and H-plane patterns of a 20-dB standard-gain horn whose Method of Moment predicted values are compared with measured patterns and with predicted values based on (13-46a) and (13-46c), which in Figure 13.20 are labeled as *approximate*. It is apparent that the MoM predicted patterns compare extremely well with the measured data.

All of the patterns presented previously represent the main polarization of the field radiated by the antenna (referred to as *copolarized* or *copol*). If the horn is symmetrical and it is excited in the dominant mode, ideally there should be no field component radiated by the antenna which is orthogonal to the main polarization (referred to as *cross polarization* or *cross-pol*), especially in the principal planes. However, in practice, either because of nonsymmetries, defects in construction and/or excitation of higher-order modes, all antennas exhibit cross-polarized components. These cross-pol components are usually of very low intensity compared to those of the primary polarization. For good designs, these should be 30 dB or more below the copolarized fields and are difficult to measure accurately or be symmetrical, as they should be in some cases.

13.4.2 Directivity

As for the E- and H-plane sectoral horns, the directivity of the pyramidal configuration is vital to the antenna designer. The maximum radiation of the pyramidal horn is directed nearly along the z-axis ($\theta = 0°$). It is a very simple exercise to show that $|E_\theta|_{\max}$, $|E_\phi|_{\max}$, and in turn $U_{\max}$ can be written, using (13-46b) and (13-46c), as

$$|E_\theta|_{\max} = |E_0 \sin\phi| \frac{\sqrt{\rho_1 \rho_2}}{r} \{[C(u) - C(v)]^2 + [S(u) - S(v)]^2\}^{1/2}$$

$$\times \left\{ C^2\left(\frac{b_1}{\sqrt{2\lambda\rho_1}}\right) + S^2\left(\frac{b_1}{\sqrt{2\lambda\rho_1}}\right) \right\}^{1/2} \quad \text{(13-48a)}$$

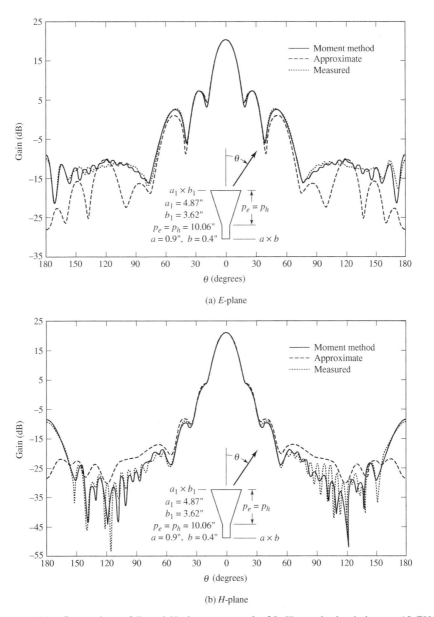

Figure 13.20 Comparison of E- and H-plane patterns for 20-dB standard-gain horn at 10 GHz.

$$|E_\phi|_{\max} = |E_0 \cos\phi| \frac{\sqrt{\rho_1 \rho_2}}{r} \{[C(u) - C(v)]^2 + [S(u) - S(v)]^2\}^{1/2}$$

$$\times \left\{ C^2\left(\frac{b_1}{\sqrt{2\lambda\rho_1}}\right) + S^2\left(\frac{b_1}{\sqrt{2\lambda\rho_1}}\right) \right\}^{1/2} \quad (13\text{-}48\text{b})$$

$$U_{\max} = \frac{r^2}{2\eta}|\mathbf{E}|^2_{\max} = |E_0|^2 \frac{\rho_1 \rho_2}{2\eta} \{[C(u) - C(v)]^2 + [S(u) - S(v)]^2\}$$

$$\times \left\{ C^2\left(\frac{b_1}{\sqrt{2\lambda\rho_1}}\right) + S^2\left(\frac{b_1}{\sqrt{2\lambda\rho_1}}\right) \right\} \quad (13\text{-}48\text{c})$$

where u and v are defined by (13-39a) and (13-39b).

Since

$$P_{rad} = |E_0|^2 \frac{a_1 b_1}{4\eta} \tag{13-49}$$

the directivity of the pyramidal horn can be written as

$$D_p = \frac{4\pi U_{max}}{P_{rad}} = \frac{8\pi \rho_1 \rho_2}{a_1 b_1} \{[C(u) - C(v)]^2 + [S(u) - S(v)]^2\}$$
$$\times \left\{ C^2\left(\frac{b_1}{\sqrt{2\lambda\rho_1}}\right) + S^2\left(\frac{b_1}{\sqrt{2\lambda\rho_1}}\right) \right\} \tag{13-50}$$

which reduces to

$$D_p = \frac{\pi \lambda^2}{32ab} D_E D_H \tag{13-50a}$$

where D_E and D_H are the directivities of the E- and H-plane sectoral horns as given by (13-18) and (13-39), respectively. This is a well-known relationship and has been used extensively in the design of pyramidal horns.

The directivity (*in dB*) of a pyramidal horn, over isotropic, can also be approximated by

$$D_p(\text{dB}) = 10\left[1.008 + \log_{10}\left(\frac{a_1 b_1}{\lambda^2}\right)\right] - (L_e + L_h) \tag{13-51}$$

where L_e and L_h represent, respectively, the losses (*in dB*) due to phase errors in the E- and H-planes of the horn which are found plotted in Figure 13.21.

The directivity of a pyramidal horn can also be calculated by doing the following [14].

1. Calculate

$$A = \frac{a_1}{\lambda} \sqrt{\frac{50}{\rho_h/\lambda}} \tag{13-52a}$$

$$B = \frac{b_1}{\lambda} \sqrt{\frac{50}{\rho_e/\lambda}} \tag{13-52b}$$

2. Using A and B, find G_H and G_E, respectively, from Figures 13.15 and 13.8. If the values of either A or B or both are smaller than 2, then calculate G_E and/or G_H by

$$G_E = \frac{32}{\pi} B \tag{13-52c}$$

$$G_H = \frac{32}{\pi} A \tag{13-52d}$$

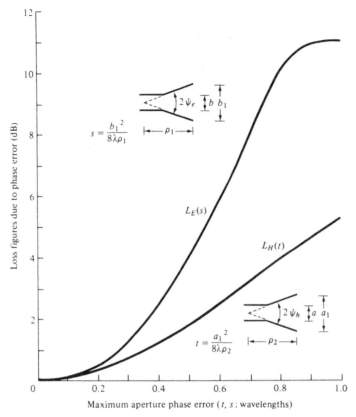

Figure 13.21 Loss figures for E- and H-planes due to phase errors. (SOURCE: W. C. Jakes, in H. Jasik (ed.), *Antenna Engineering Handbook*, McGraw-Hill, New York, 1961).

3. Calculate D_p by using the values of G_E and G_H from Figures 13.8 and 13.15 or from (13-52c) and (13-52d). Thus

$$D_p = \frac{G_E G_H}{\dfrac{32}{\pi}\sqrt{\dfrac{50}{\rho_e/\lambda}}\sqrt{\dfrac{50}{\rho_h/\lambda}}} = \frac{G_E G_H}{10.1859\sqrt{\dfrac{50}{\rho_e/\lambda}}\sqrt{\dfrac{50}{\rho_h/\lambda}}} \qquad (13\text{-}52\text{e})$$

$$= \frac{\lambda^2 \pi}{32ab} D_E D_H$$

where D_E and D_H are, respectively, the directivities of (13-19c) and (13-40c). This is the actual directivity of the horn. The above procedure has led to results accurate to within 0.01 dB for a horn with $\rho_e = \rho_h = 50\lambda$.

A commercial X-band (8.2–12.4 GHz) horn is that shown in Figure 13.22. It is a lightweight precision horn antenna, which is usually cast of aluminum, and it can be used as a

1. standard for calibrating other antennas
2. feed for reflectors and lenses
3. pickup (probe) horn for sampling power
4. receiving and/or transmitting antenna.

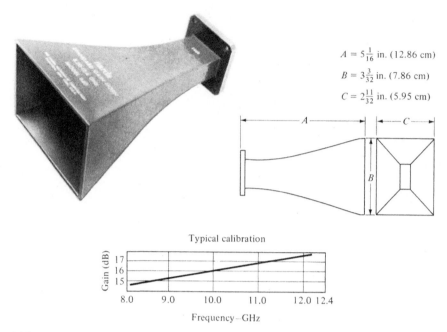

Figure 13.22 Typical standard gain X-band (8.2–12.4 GHz) pyramidal horn and its gain characteristics. (Courtesy of The NARDA Microwave Corporation).

It possesses an exponential taper, and its dimensions and typical gain characteristics are indicated in the figure. The half-power beamwidth in both the E- and H-planes is about 28° while the side lobes in the E- and H-planes are, respectively, about 13 and 20 dB down.

Gains of the horn antenna which were measured, predicted, and provided by the manufacturer, whose amplitude patterns are shown in Figure 13.20, are displayed in Figure 13.23. A very good agreement amongst all three sets is indicated.

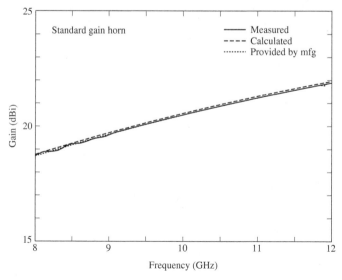

Figure 13.23 Gains of the pyramidal horn which were measured, computed, and provided by the manufacturer. The amplitude patterns of the horn are shown in Figure 13.22.

Example 13.4

A pyramidal horn has dimensions of $\rho_1 = \rho_2 = 6\lambda$, $a_1 = 5.5\lambda$, $b_1 = 2.75\lambda$, $a = 0.5\lambda$, and $b = 0.25\lambda$.

a. Check to see if such a horn can be constructed physically.
b. Compute the directivity using (13-50a), (13-51), and (13-52e).

Solution: From Examples 13.2 and 13.3.

$$\rho_e = 6.1555\lambda$$
$$\rho_h = 6.6\lambda$$

Thus

$$p_e = (2.75 - 0.25)\lambda\sqrt{\left(\frac{6.1555}{2.75}\right)^2 - \frac{1}{4}} = 5.454\lambda$$

$$p_h = (5.5 - 0.5)\lambda\sqrt{\left(\frac{6.6}{5.5}\right)^2 - \frac{1}{4}} = 5.454\lambda$$

Therefore the horn can be constructed physically.

The directivity can be computed by utilizing the results of Examples 13.2 and 13.3. Using (13-50a) with the values of D_E and D_H computed using, respectively, (13-18) and (13-39) gives

$$D_p = \frac{\pi\lambda^2}{32ab}D_E D_H = \frac{\pi}{32(0.5)(0.25)}(12.79)(7.52) = 75.54 = 18.78 \text{ dB}$$

Utilizing the values of D_E and D_H computed using, respectively, (13-19c) and (13-40c), the directivity of (13-52e) is equal to

$$D_p = \frac{\pi\lambda^2}{32ab}D_E D_H = \frac{\pi}{32(0.5)0.25}(12.89)(8.338) = 84.41 = 19.26 \text{ dB}$$

For this horn

$$s = \frac{b_1^2}{8\lambda\rho_1} = \frac{(2.75)^2}{8(6)} = 0.1575$$

$$t = \frac{a_1^2}{8\lambda\rho_2} = \frac{(5.5)^2}{8(6)} = 0.63$$

For these values of s and t

$$L_E = 0.20 \text{ dB}$$
$$L_H = 2.75 \text{ dB}$$

from Figure 13.21. Using (13-51)

$$D_p = 10\{1.008 + \log_{10}[5.5(2.75)]\} - (0.20 + 2.75) = 18.93 \text{ dB}$$

The agreement is best between the directivities of (13-50a) and (13-51).

A MATLAB and FORTRAN computer program entitled **Analysis** has been developed to analyze the radiation characteristics of a pyramidal horn and the directivities of the corresponding E- and H-plane sectoral horns. The program and the READ ME file are found in the publisher's website for this book.

13.4.3 Design Procedure

The pyramidal horn is widely used as a standard to make gain measurements of other antennas (see Section 17.4, Chapter 17), and as such it is often referred to as a *standard gain horn*. To design a pyramidal horn, one usually knows the desired gain G_0 and the dimensions a, b of the rectangular feed waveguide. The objective of the design is to determine the remaining dimensions ($a_1, b_1, \rho_e, \rho_h, P_e$, and p_h) that will lead to an optimum gain. The procedure that follows can be used to accomplish this [2], [3].

The design equations are derived by first selecting values of b_1 and a_1 that lead, respectively, to optimum directivities for the E- and H-plane sectoral horns using (13-18a) and (13-39c). Since the overall efficiency (including both the antenna and aperture efficiencies) of a horn antenna is about 50% [2], [3], the gain of the antenna can be related to its physical area. Thus it can be written using (12-37c), (12-38), and (13-18a), (13-39c) as

$$G_0 = \frac{1}{2}\frac{4\pi}{\lambda^2}(a_1 b_1) = \frac{2\pi}{\lambda^2}\sqrt{3\lambda\rho_2}\sqrt{2\lambda\rho_1} \simeq \frac{2\pi}{\lambda^2}\sqrt{3\lambda\rho_h}\sqrt{2\lambda\rho_e} \quad (13\text{-}53)$$

since for long horns $\rho_2 \simeq \rho_h$ and $\rho_1 \simeq \rho_e$. For a pyramidal horn to be physically realizable, P_e and P_h of (13-47a) and (13-47b) must be equal. Using this equality, it can be shown that (13-53) reduces to

$$\left(\sqrt{2\chi} - \frac{b}{\lambda}\right)^2 (2\chi - 1) = \left(\frac{G_0}{2\pi}\sqrt{\frac{3}{2\pi}}\frac{1}{\sqrt{\chi}} - \frac{a}{\lambda}\right)^2 \left(\frac{G_0^2}{6\pi^3}\frac{1}{\chi} - 1\right) \quad (13\text{-}54)$$

where

$$\frac{\rho_e}{\lambda} = \chi \quad (13\text{-}54a)$$

$$\frac{\rho_h}{\lambda} = \frac{G_0^2}{8\pi^3}\left(\frac{1}{\chi}\right) \quad (13\text{-}54b)$$

Equation (13-54) is the horn-design equation.

1. As a first step of the design, find the value of χ which satisfies (13-54) for a desired gain G_0 (dimensionless). Use an iterative technique and begin with a trial value of

$$\chi(\text{trial}) = \chi_1 = \frac{G_0}{2\pi\sqrt{2\pi}} \quad (13\text{-}55)$$

2. Once the correct χ has been found, determine ρ_e and ρ_h using (13-54a) and (13-54b), respectively.

3. Find the corresponding values of a_1 and b_1 using (13-18a) and (13-39c) or

$$a_1 = \sqrt{3\lambda\rho_2} \simeq \sqrt{3\lambda\rho_h} = \frac{G_0}{2\pi}\sqrt{\frac{3}{2\pi\chi}}\lambda \qquad (13\text{-}56\text{a})$$

$$b_1 = \sqrt{2\lambda\rho_1} \simeq \sqrt{2\lambda\rho_e} = \sqrt{2\chi}\lambda \qquad (13\text{-}56\text{b})$$

4. The values of p_e and p_h can be found using (13-47a) and (13-47b).

A MATLAB and FORTRAN computer program entitled **Design** has been developed to accomplish this, and it is included in the publisher's website for this book.

Example 13.5

Design an optimum gain X-band (8.2–12.4 GHz) pyramidal horn so that its gain (above isotropic) at $f = 11$ GHz is 22.6 dB. The horn is fed by a WR 90 rectangular waveguide with inner dimensions of $a = 0.9$ in. (2.286 cm) and $b = 0.4$ in. (1.016 cm).

Solution: Convert the gain G_0 from dB to a dimensionless quantity. Thus

$$G_0(\text{dB}) = 22.6 = 10\log_{10} G_0 \Rightarrow G_0 = 10^{2.26} = 181.97$$

Since $f = 11$ GHz, $\lambda = 2.7273$ cm and

$$a = 0.8382\lambda$$
$$b = 0.3725\lambda$$

1. The initial value of χ is taken, using (13-55), as

$$\chi_1 = \frac{181.97}{2\pi\sqrt{2\pi}} = 11.5539$$

which does not satisfy (13-54) for the desired design specifications. After a few iterations, a more accurate value is $\chi = 11.1157$.

2. Using (13-54a) and (13-54b)

$$\rho_e = 11.1157\lambda = 30.316 \text{ cm} = 11.935 \text{ in.}$$
$$\rho_h = 12.0094\lambda = 32.753 \text{ cm} = 12.895 \text{ in.}$$

3. The corresponding values of a_1 and b_1 are

$$a_1 = 6.002\lambda = 16.370 \text{ cm} = 6.445 \text{ in.}$$
$$b_1 = 4.715\lambda = 12.859 \text{ cm} = 5.063 \text{ in.}$$

4. The values of p_e and p_h are equal to

$$p_e = p_h = 10.005\lambda = 27.286 \text{ cm} = 10.743 \text{ in.}$$

The same values are obtained using the computer program at the end of this chapter.

The derived design parameters agree closely with those of a commercial gain horn available in the market. As a check, the gain of the designed horn was computed using (13-50a) and (13-51), assuming an antenna efficiency e_t of 100%, and (13-53). The values were

$$G_0 \simeq D_0 = 22.4 \text{ dB} \quad \text{for (13-52a)}$$

$$G_0 \simeq D_0 = 22.1 \text{ dB} \quad \text{for (13-53)}$$

$$G_0 = 22.5 \text{ dB} \quad \text{for (13-55)}$$

All three computed values agree closely with the designed value of 22.6 dB.

The previous formulations for all three horn configurations (E-plane, H-plane, and pyramidal) are based on the use of the *quadratic* phase term of (13-2b) instead of the *spherical* phase term of (13-2a). This was necessary so that the integrations can be performed and expressed in terms of cosine and sine Fresnel integrals. In order to examine the differences using the *spherical* phase term instead of the *quadratic*, especially as it relates to the directivity, a numerical integration was used in [20]. It was found that the directivity of a *pyramidal* horn of Figure 13.18 using the spherical phase term, instead of the quadratic

- *is basically the same for*
  1. large aperture horns ($a_1, b_1 > 50\lambda$);
  2. small peak aperture phase errors ($S = \rho_e - \rho_1 < 0.2\lambda$ or $T = \rho_h - \rho_2 < 0.2\lambda$ of Figure 13.16).
- *is slightly higher (more improved) by as much as 0.6 dB for*
  1. intermediate apertures
     ($5\lambda \leq a_1 \leq 8\lambda$ or $5\lambda \leq b_1 \leq 8\lambda$);
  2. intermediate peak aperture phase errors
     ($0.2\lambda \leq S = \rho_e - \rho_1 \leq 0.6\lambda$ or $0.2\lambda \leq T = \rho_h - \rho_2 \leq 0.6\lambda$ of Figure 13.16).
- *can be lower, especially for E-plane horns, for*
  1. large peak aperture phase errors
     ($S = \rho_e - \rho_1 > 0.6\lambda$ or $T = \rho_h - \rho_2 > 0.6\lambda$ of Figure 13.16).

13.5 CONICAL HORN

Another very practical microwave antenna is the conical horn shown in Figure 13.24 with a photo of one in Figure 13.25. While the pyramidal, E-, and H-plane sectoral horns are usually fed by a rectangular waveguide, the feed of a conical horn is often a circular waveguide.

The first rigorous treatment of the fields radiated by a conical horn is that of Schorr and Beck [21]. The modes within the horn are found by introducing a spherical coordinate system and are in terms of spherical Bessel functions and Legendre polynomials. The analysis is too involved and will not be attempted here. However data, in the form of curves have been presented [22] which give a qualitative description of the performance of a conical horn.

Despite its popularity and wide range of applications, the directivity and amplitude patterns of a conical horn have not received the same attention as those of the pyramidal horn. For the classic curves by Gray and Schelkunoff, which were reported in [22] and included in as a reference in many books and papers, there has not been a clear documentation whether the reported directivity curves were computed based on quadratic or spherical phase distributions. This and other issues related

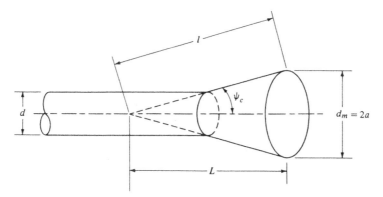

Figure 13.24 Geometry of conical horn.

to the conical horn, like loss figures and amplitude patterns, are addressed in [23]. Figure 13.26 displays a set of directivity curves, versus diameter of horn aperture, for different axial lengths L. The curves were computed based on quadratic (QPD) and spherical (SPD) phase distributions, and they are compared with those based on modal solution (MS) as well as those reported in [22]. It is clear that the directivities obtained based on the SPD are in closer agreement with those of the MS as well as those of Gray and Schelkunoff, which were reported in [22].

Based on the numerical data of Figure 13.26 and using linear curve fitting in the least square sense, a set of approximate equations were obtained for the optimum maximum directivity $(D_c)_{opt}$ and axial length L based on the optimum horn line of Figure 13.26 [23]. These equations can be used for designing optimum directivity conical horn, and they are represented by

$$(D_c)_{opt} \approx 15.9749 \left(\frac{L}{\lambda}\right) + 1.7209 \tag{13-57a}$$

$$(D_c)_{opt} \approx 5.1572 \left(\frac{d_m}{\lambda}\right)^2 - 0.6451 \left(\frac{d_m}{\lambda}\right) + 1.3645 \tag{13-57b}$$

$$L \approx 0.3232 \left(\frac{d_m}{\lambda}\right)^2 - 0.0475 \left(\frac{d_m}{\lambda}\right) + 0.0052 \tag{13-57c}$$

Figure 13.25 Photo of an X-band commercial conical horn ($L = 7.147\lambda$, $2\psi_c = 35°$).

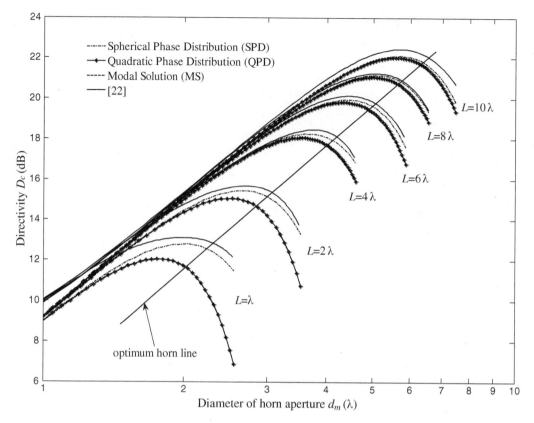

Figure 13.26 Directivity of a conical horn as a function of aperture diameter for different axial horn lengths. (SOURCE: [23] © 2013 IEEE).

These expressions are used to determine the dimensions of the conical horn once the optimum directivity of the conical horn is specified.

Referring to Figure 13.26, it is apparent that the behavior of a conical horn is similar to that of a pyramidal or a sectoral horn. As the flare angle increases, the directivity for a given length horn increases until it reaches a maximum beyond which it begins to decrease. The decrease is a result of the dominance of the quadratic phase error at the aperture. In the same figure, an optimum directivity line is indicated.

The results of Figure 13.26 behave as those of Figures 13.7 and 13.14. When the horn aperture (d_m) is held constant and its length (L) is allowed to vary, the maximum directivity is obtained when the flare angle is zero ($\psi_c = 0$ or $L = \infty$). This is equivalent to a circular waveguide of diameter d_m. As for the pyramidal and sectoral horns, a lens is usually placed at the aperture of the conical horn to compensate for its quadratic phase error. The result is a narrower pattern as its flare increases.

The directivity (*in dB*) of a conical horn, with an aperture efficiency of ε_{ap} and aperture circumference C, can be computed using

$$D_c(\text{dB}) = 10 \log_{10}\left[\varepsilon_{ap}\frac{4\pi}{\lambda^2}(\pi a^2)\right] = 10 \log_{10}\left(\frac{C}{\lambda}\right)^2 - L(s) \quad (13\text{-}58)$$

where a is the radius of the horn at the aperture and

$$L(s) = -10\log_{10}(\varepsilon_{ap}) \tag{13-58a}$$

The first term in (13-58) represents the directivity of a uniform circular aperture whereas the second term, represented by (13-58a), is a correction figure to account for the loss in directivity due to the aperture efficiency. Usually the term in (13-58a) is referred to as *loss figure* which can be computed (*in dB*) using the advanced expressions of [23]

$$L(s) \approx \begin{cases} 0.5030 + 5.1123s - 7.1138s^2 + 23.1401s^3, & L \leq 3\lambda \\ 0.7853 - 0.3976s + 13.112s^2 + 3.901s^3, & L > 3\lambda \end{cases} \tag{13-58b} \tag{13-58c}$$

where s is the maximum phase deviation (*in wavelengths*), and it is given by

$$s = \frac{d_m^2}{8\lambda l} \tag{13-58d}$$

The directivity of a conical horn is optimum when its diameter is equal to

$$d_m \simeq \sqrt{3l\lambda} \tag{13-59}$$

which corresponds to a maximum aperture phase deviation of $s = 3/8$ (*wavelengths*) and a loss figure of about 2.9 dB (or an aperture efficiency of about 51%).

A set of figures based on (13-58b) and (13-58c), and their impact on the maximum directivity of a conical horn, are displayed in Figure 13.27(a,b), respectively, where they are compared with exact data. It is apparent that the advanced expressions of (13-58b) and (13-58c) provide a good approximation to both the loss figure and directivity of the conical horn over an extended range of aperture phase deviation represented by s.

Far-field amplitude patterns, E-plane and H-plane, over an extended dynamic range of 0–60 dB for the X-band conical horn of Figure 13.25, are displayed in Figure 13.28 where they are compared with predictions based on GTD/UTD, simulations and measurements [23]. Excellent agreement is indicated everywhere, including in the back lobe region.

A summary of all the pertinent formulas and equation numbers that can be used to compute the directivity of E-plane, H-plane, pyramidal, and conical horns is listed in Table 13.1.

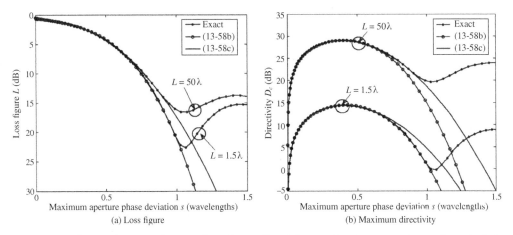

Figure 13.27 Conical horn loss factor and maximum directivity as a function of maximum phase deviations ($L = 50\lambda, L = 1.5\lambda$). (SOURCE: [23] © 2013 IEEE).

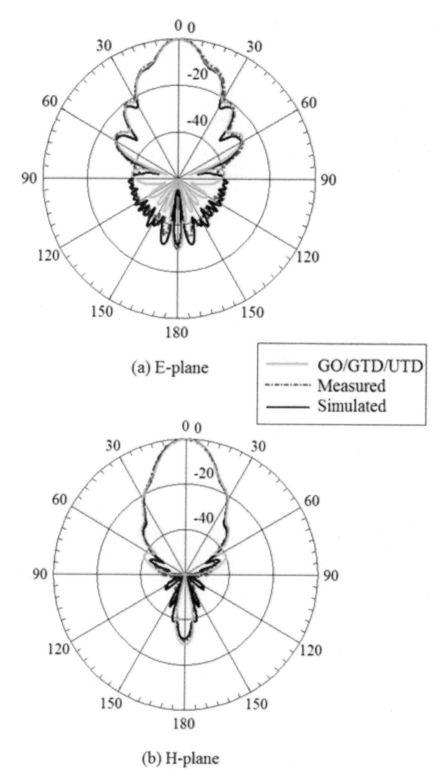

Figure 13.28 Normalized far-field amplitude patterns of X-band conical horn, of Figure 13.25, at $f = 10.5$ GHz ($L = 7.147\lambda, 2\psi_c = 35°$). (SOURCE: [23] © 2013 IEEE).

TABLE 13.1 Directivity Formulas and Equation Numbers for Horns

Horn Type	Directivity	Equation Number
E-plane	$D_E = \dfrac{64a\rho_1}{\pi\lambda b_1}\left[C^2\left(\dfrac{b_1}{\sqrt{2\lambda\rho_1}}\right) + S^2\left(\dfrac{b_1}{\sqrt{2\lambda\rho_1}}\right)\right]$	(13-18)
E-plane (alternate)	$D_E = \dfrac{a}{\lambda}\dfrac{G_E}{\sqrt{\dfrac{50}{\rho_e/\lambda}}}, \quad B = \dfrac{b_1}{\lambda}\sqrt{\dfrac{50}{\rho_e/\lambda}}$	(13-19c), (13-19a)
	$G_E = \dfrac{32}{\pi}B \quad \text{if } B < 2$ $G_E \text{ from Figure 13.8} \quad \text{if } B \geq 2$	(13-19b) Figure 13.8
H-plane	$D_H = \dfrac{4\pi b\rho_2}{a_1\lambda}\{[C(u) - C(v)]^2 + [S(u) - S(v)]^2\}$	(13-39)
	$u = \dfrac{1}{\sqrt{2}}\left(\dfrac{\sqrt{\lambda\rho_2}}{a_1} + \dfrac{a_1}{\sqrt{\lambda\rho_2}}\right), \quad v = \dfrac{1}{\sqrt{2}}\left(\dfrac{\sqrt{\lambda\rho_2}}{a_1} - \dfrac{a_1}{\sqrt{\lambda\rho_2}}\right)$	(13-39a), (13-39b)
H-plane (alternate)	$D_H = \dfrac{b}{\lambda}\dfrac{G_H}{\sqrt{\dfrac{50}{\rho_h/\lambda}}}, \quad A = \dfrac{a_1}{\lambda}\sqrt{\dfrac{50}{\rho_h/\lambda}}$	(13-40c), (13-40a)
	$G_H = \dfrac{32}{\pi}A \quad \text{if } A < 2$ $G_H \text{ from Figure 13.15} \quad \text{if } A \geq 2$	(13-40b) Figure 13.15
Pyramidal	$D_P = \dfrac{\pi\lambda^2}{32ab}D_E D_H$	(13-50a), (13-18), (13-39)
Pyramidal (alternate)	$D_P(dB) = 10\left[1.008 + \log_{10}\left(\dfrac{a_1 b_1}{\lambda^2}\right)\right] - (L_e + L_h)$ L_e, L_h	(13-51) Figure 13.21
Pyramidal (alternate)	$D_P = \dfrac{G_E G_H}{\dfrac{32}{\pi}\sqrt{\dfrac{50}{\rho_e/\lambda}}\sqrt{\dfrac{50}{\rho_h/\lambda}}} = \dfrac{\lambda^2\pi}{32ab}D_E D_H$	(13-52e), (13-50a) (13-19c), (13-40c)
Conical	$D_c(dB) = 10\log_{10}\left[\varepsilon_{ap}\dfrac{4\pi}{\lambda^2}(\pi a^2)\right] = 10\log_{10}\left(\dfrac{C}{\lambda}\right)^2 - L(s)$	(13-58)
	$L(s) = -10\log_{10}(\varepsilon_{ap})$ $\simeq LF(s) \approx \begin{cases} 0.5030 + 5.1123s - 7.1138s^2 + 23.1401s^3, & L \leq 3\lambda \\ 0.7853 - 0.3976s + 13.112s^2 + 3.901s^3, & L > 3\lambda \end{cases}$	(13-58b), (13-58c)
	$s = \dfrac{d_m^2}{8\lambda l}, \quad d_m \simeq \sqrt{3l\lambda}$	(13-58d), (13-59)

13.6 CORRUGATED HORN

The large emphasis placed on horn antenna research in the 1960s was inspired by the need to reduce spillover efficiency and cross-polarization losses and increase aperture efficiencies of large reflectors used in radio astronomy and satellite communications. In the 1970s, high-efficiency and rotationally symmetric antennas were needed in microwave radiometry. Using conventional feeds, aperture

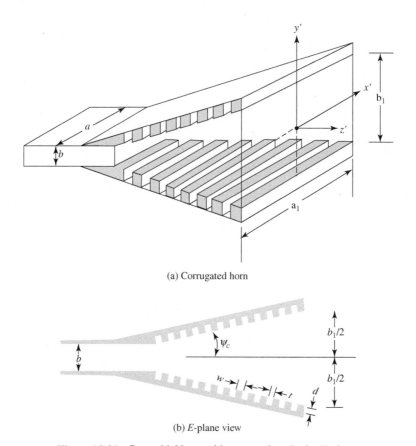

Figure 13.29 Pyramidal horn with corrugations in the E-plane.

efficiencies of 50–60% were obtained. However, efficiencies of the order of 75–80% can be obtained with improved feed systems utilizing corrugated horns.

The aperture techniques introduced in Chapter 12 can be used to compute the pattern of a horn antenna and would yield accurate results only around the main lobe and the first few minor lobes. The antenna pattern structure in the back lobe region is strongly influenced by diffractions from the edges, especially from those that are perpendicular to the E-field at the horn aperture. The diffractions lead to undesirable radiation not only in the back lobes but also in the main lobe and in the minor lobes. However, they dominate only in low-intensity regions.

In 1964, Kay [24] realized that grooves on the walls of a horn antenna would present the same boundary conditions to all polarizations and would taper the field distribution at the aperture in all the planes. The creation of the same boundary conditions on all four walls would eliminate the spurious diffractions at the edges of the aperture. For a square aperture, this would lead to an almost rotationally symmetric pattern with equal E- and H-plane beamwidths. A *corrugated (grooved)* pyramidal horn, with corrugations in the E-plane walls, is shown in Figure 13.29(a) with a side view in Figure 13.29(b). Since diffractions at the edges of the aperture in the H-plane are minimal, corrugations are usually not placed on the walls of that plane. Corrugations can also be placed in a conical horn forming a *conical corrugated* horn, also referred to in [24] as a *scalar horn*. However, instead of the corrugations being formed as shown in Figure 13.30(a), practically it is much easier to machine them to have the profile shown in Figure 13.30(b).

A photograph of a corrugated conical horn, often referred to as a *scalar* horn, is shown in Figure 13.31. This type of a horn is widely used as a feed of reflector antennas, especially of the Cassegrain (dual reflector) configuration of Figures 15.9 and 15.30.

CORRUGATED HORN 763

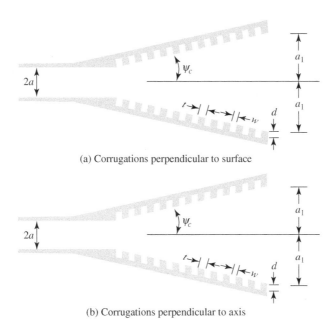

(a) Corrugations perpendicular to surface

(b) Corrugations perpendicular to axis

Figure 13.30 Side view profiles of conical corrugated horns.

To form an effective corrugated surface, it usually requires 10 or more slots (corrugations) per wavelength [25]. To simplify the analysis of an infinite corrugated surface, the following assumptions are usually required:

1. The teeth of the corrugations are vanishingly thin.
2. Reflections from the base of the slot are only those of a TEM mode.

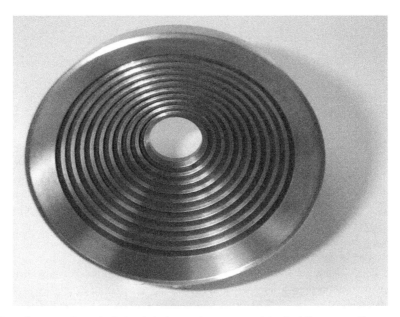

Figure 13.31 Corrugated conical (scalar) horn. (COURTESY: March Microwave Systems, B.V., The Netherlands).

The second assumption is satisfied provided the width of the corrugation (w) is small compared to the free-space wavelength (λ_0) and the slot depth (d) (usually $w < \lambda_0/10$). For a corrugated surface satisfying the above assumptions, its approximate surface reactance is given by [26], [27]

$$X = \frac{w}{w+t}\sqrt{\frac{\mu_0}{\varepsilon_0}}\tan(k_0 d) \qquad (13\text{-}60)$$

when

$$\frac{w}{w+t} \simeq 1 \qquad (13\text{-}60\text{a})$$

which can be satisfied provided $t \leq w/10$.

The surface reactance of a corrugated surface, used on the walls of a horn, must be capacitive in order for the surface to force to zero the tangential magnetic field parallel to the edge at the wall. Such a surface will not support surface waves, will prevent illumination of the E-plane edges, and will diminish diffractions. This can be accomplished, according to (13-60), if $\lambda_0/4 < d < \lambda_0/2$ or more generally when $(2n+1)\lambda_0/4 < d < (n+1)\lambda_0/2$. Even though the cutoff depth is also a function of the slot width w, its influence is negligible if $w < \lambda_0/10$ and $\lambda_0/4 < d < \lambda_0/2$.

The effect of the corrugations on the walls of a horn is to modify the electric field distribution in the E-plane from uniform (at the waveguide-horn junction) to cosine (at the aperture). Through measurements, it has been shown that the transition from uniform to cosine distribution takes place almost at the onset of the corrugations. For a horn of about 45 corrugations, the cosine distribution has been established by the fifth corrugation (from the onset) and the spherical phase front by the fifteenth [28].

Referring to Figure 13.29(a), the field distribution at the aperture can be written as

$$E'_y(x',y') = E_0 \cos\left(\frac{\pi}{a_1}x'\right)\cos\left(\frac{\pi}{b_1}y'\right) e^{-j[k(x'^2/\rho_2 + y'^2/\rho_1)/2]} \qquad (13\text{-}61\text{a})$$

$$H'_x(x',y') = -\frac{E_0}{\eta} \cos\left(\frac{\pi}{a_1}x'\right)\cos\left(\frac{\pi}{b_1}y'\right) e^{-j[k(x'^2/\rho_2 + y'^2/\rho_1)/2]} \qquad (13\text{-}61\text{b})$$

corresponding to (13-41a) and (13-41b) of the uncorrugated pyramidal horn. Using the above distributions, the fields radiated by the horn can be computed in a manner analogous to that of the pyramidal horn of Section 13.4. Patterns have been computed and compare very well with measurements [28].

In Figure 13.32(a) the measured E-plane patterns of an uncorrugated square pyramidal horn (*referred to as the control horn*) and a corrugated square pyramidal horn are shown. The aperture size on each side was 3.5 in. (2.96λ at 10 GHz) and the total flare angle in each plane was 50°. It is evident that the levels of the minor lobes and back lobes are much lower for the corrugated horn than those of the control horn. However the corrugated horn also exhibits a wider main beam for small angles; thus a larger 3-dB beamwidth (HPBW) but a lower 10-dB beamwidth. This is attributed to the absence of the diffracted fields from the edges of the corrugated horn which, for nearly on-axis observations, add to the direct wave contribution because of their in-phase relationship. The fact that the on-axis far-fields of the direct and diffracted fields are nearly in phase is also evident from the pronounced on-axis maximum of the control horn. The E- and H-plane patterns of the corrugated horn are almost identical to those of Figure 13.32(a) over the frequency range from 8 to 14 GHz. These suggest that the main beam in the E-plane can be obtained from known H-plane patterns of horn antennas.

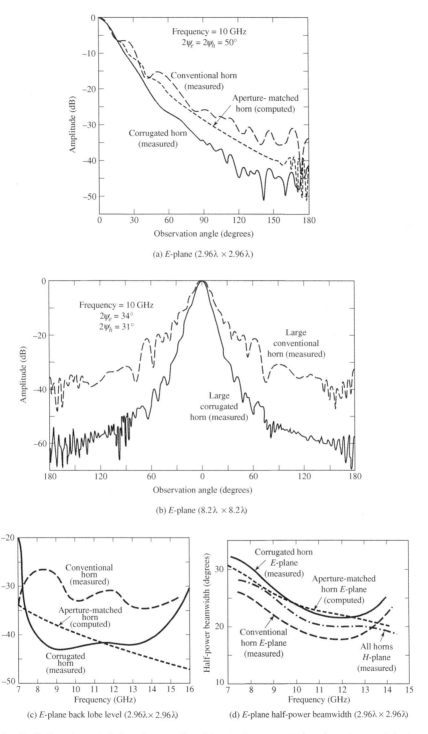

Figure 13.32 Radiation characteristics of conventional (control), corrugated, and aperture-matched pyramidal horns. (SOURCE: (a), (c), (d). W. D. Burnside and C. W. Chuang, "An Aperture-Matched Horn Design," *IEEE Trans. Antennas Propagat.*, Vol. AP-30, No. 4, pp. 790–796, July 1982. © 1982 IEEE (b) R. E. Lawrie and L. Peters, Jr., "Modifications of Horn Antennas for Low Side Lobe Levels," *IEEE Trans. Antennas Propagat.*, Vol. AP-14, No. 5, pp. 605–610, September 1966. © 1966 IEEE).

In Figure 13.32(b) the measured E-plane patterns of larger control and corrugated square pyramidal horns, having an aperture of 9.7 in. on each side (8.2λ at 10 GHz) and included angles of 34° and 31° in the E- and H-planes, respectively, are shown. For this geometry, the pattern of the corrugated horn is narrower and its minor and back lobes are much lower than those of the corresponding control horn. The saddle formed on the main lobe of the control horn is attributed to the out-of-phase relations between the direct and diffracted rays. The diffracted rays are nearly absent from the corrugated horn and the minimum on-axis field is eliminated. The control horn is a thick-edged horn which has the same interior dimensions as the corrugated horn. The H-plane pattern of the corrugated horn is almost identical to the H-plane pattern of the corresponding control horn.

In Figures 13.32(c) and 13.32(d) the back lobe level and the 3-dB beamwidth for the smaller size control and corrugated horns, whose E-plane patterns are shown in Figure 13.32(a), are plotted as a function of frequency. All the observations made previously for that horn are well evident in these figures.

The presence of the corrugations, especially near the waveguide-horn junction, can affect the impedance and VSWR of the antenna. The usual practice is to begin the corrugations at a small distance away from the junction. This leads to low VSWR's over a broad band. Previously it was indicated that the width w of the corrugations must be small (usually $w < \lambda_0/10$) to approximate a corrugated surface. This would cause corona and other breakdown phenomena. However the large corrugated horn, whose E-plane pattern is shown in Figure 13.32(b), has been used in a system whose peak power was 20 kW at 10 GHz with no evidence of any breakdown phenomena.

The design concepts of the pyramidal corrugated horn can be extended to include circumferentially corrugated conical horns, as shown in Figures 13.30 and 13.31. Several designs of conical corrugated horns were investigated in terms of pattern symmetry, low cross polarization, low side lobe levels, circular polarization, axial ratio, and phase center [29]–[38]. For small flare angles (ψ_c less than about 20° to 25°) the slots can be machined perpendicular to the axis of the horn, as shown in Figure 13.30(b), and the grooves can be considered sections of parallel-plate TEM-mode waveguides of depth d. For large flare angles, however, the slots should be constructed perpendicular to the surface of the horn, as shown in Figure 13.30(a) and implemented in the design of Figure 13.31. The groove arrangement of Figure 13.30(b) is usually preferred because it is easier to fabricate.

13.7 APERTURE-MATCHED HORNS

A horn which provides significantly better performance than an ordinary horn (in terms of pattern, impedance, and frequency characteristics) is that shown in Figure 13.33(a), which is referred to as an *aperture-matched* horn [39]. The main modification to the ordinary (conventional) horn, which we refer to here as the *control* horn, consists of the attachment of curved-surface sections to the outside of the aperture edges, which reduces the diffractions that occur at the sharp edges of the aperture and provides smooth matching sections between the horn modes and the free-space radiation.

In contrast to the corrugated horn, which is complex and costly and reduces the diffractions at the edges of the aperture by minimizing the incident field, the aperture-matched horn reduces the diffractions by modifying the structure (without sacrificing size, weight, bandwidth, and cost) so that the diffraction coefficient is minimized. The basic concepts were originally investigated using elliptic cylinder sections, as shown in Figure 13.33(b); however, other convex curved surfaces, which smoothly blend to the ordinary horn geometry at the attachment point, will lead to similar improvements. This modification in geometry can be used in a wide variety of horns, and includes E-plane, H-plane, pyramidal, and conical horns. Bandwidths of 2:1 can be attained easily with aperture-matched horns having elliptical, circular, or other curved surfaces. The radii of curvature of the curved surfaces used in experimental models [39] ranged over $1.69\lambda \leq a \leq 8.47\lambda$ with $a = b$ and $b = 2a$. Good results can be obtained by using circular cylindrical surfaces with $2.5\lambda \leq a \leq 5\lambda$.

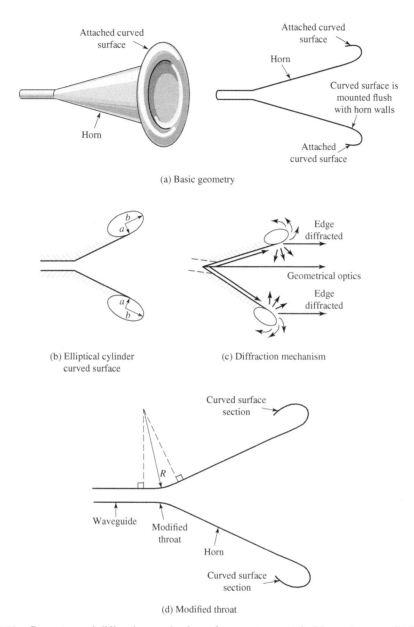

Figure 13.33 Geometry and diffraction mechanism of an aperture-matched horn. (SOURCE: W. D. Burnside and C. W. Chuang, "An Aperture-Matched Horn Design," *IEEE Trans. Antennas Propagat.*, Vol. AP-30, No. 4, pp. 790–796, July 1982. © 1982 IEEE).

The basic radiation mechanism of such a horn is shown in Figure 13.33(c). The introduction of the curved sections at the edges does not eliminate diffractions; instead it substitutes edge diffractions by curved-surface diffractions which have a tendency to provide an essentially undisturbed energy flow across the junction, around the curved surface, and into free-space. Compared with conventional horns, this radiation mechanism leads to smoother patterns with greatly reduced black lobes and negligible reflections back into the horn. The size, weight, and construction costs of the aperture-matched horn are usually somewhat larger and can be held to a minimum if half (one-half sections of an ellipse) or quadrant (one-fourth sections of an ellipse) sections are used instead of the complete closed surfaces.

To illustrate the improvements provided by the aperture-matched horns, the E-plane pattern, back lobe level, and half-power beamwidth of a pyramidal $2.96\lambda \times 2.96\lambda$ horn were computed and compared with the measured data of corresponding control and corrugated horns. The data are shown in Figures 13.32(a,c,d). It is evident by examining the patterns of Figure 13.32(a) that the aperture-matched horn provides a smoother pattern and lower back lobe level than conventional horns (referred to here as control horn); however, it does not provide, for the wide minor lobes, the same reduction as the corrugated horn. To achieve nearly the same E-plane pattern for all three horns, the overall horn size must be increased. If the modifications for the aperture-matched and corrugated horns were only made in the E-plane edges, the H-plane patterns for all three horns would be virtually the same except that the back lobe level of the aperture-matched and corrugated horns would be greatly reduced.

The back lobe level of the same three horns (control, corrugated, and aperture matched) are shown in Figure 13.32(c). The corrugated horn has lower back lobe intensity at the lower end of the frequency band, while the aperture-matched horn exhibits superior performance at the high end. However, both the corrugated and aperture-matched horns exhibit superior back lobe level characteristics to the control (conventional) horn throughout almost the entire frequency band. The half-power beamwidth characteristics of the same three horns are displayed in Figure 13.32(d). Because the control (conventional) horn has uniform distribution across the complete aperture plane, compared with the tapered distributions for the corrugated and aperture-matched horns, it possesses the smallest beamwidth almost over the entire frequency band.

In a conventional horn the VSWR and antenna impedance are primarily influenced by the throat and aperture reflections. Using the aperture-matched horn geometry of Figure 13.33(a), the aperture reflections toward the inside of the horn are greatly reduced. Therefore the only remaining dominant factors are the throat reflections. To reduce the throat reflections it has been suggested that a smooth curved surface be used to connect the waveguide and horn walls, as shown in Figure 13.33(d). Such a transition has been applied in the design and construction of a commercial X-band (8.2–12.4 GHz) pyramidal horn (see Figure 13.22), whose tapering is of an exponential nature. The VSWRs measured in the 8–12 GHz frequency band using the conventional exponential X-band horn (shown in Figure 13.22), with and without curved sections at its aperture, are shown in Figure 13.34.

The matched sections used to create the aperture-matched horn were small cylinder sections. The VSWR's for the conventional horn are very small (less than 1.1) throughout the frequency band because the throat reflections are negligible compared with the aperture reflections. It is evident, however, that the VSWR's of the corresponding aperture-matched horn are much superior to those of the conventional horn because both the throat and aperture reflections are very minimal.

The basic design of the aperture-matched horn can be extended to include corrugations on its inside surface [31]. This type of design enjoys the advantages presented by both the aperture-matched

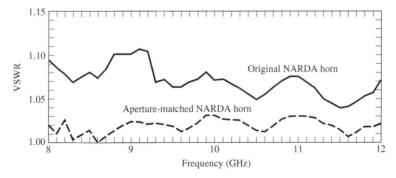

Figure 13.34 Measured VSWR for exponentially tapered pyramidal horns (conventional and aperture matched). (SOURCE: W. D. Burnside and C. W. Chuang, "An Aperture-Matched Horn Design," *IEEE Trans. Antennas Propagat.*, Vol. AP-30, No. 4, pp. 790–796, July 1982. © 1982 IEEE).

and corrugated horns with cross-polarized components of less than −45 dB over a significant part of the bandwidth. Because of its excellent cross-polarization characteristics, this horn is recommended for use as a reference and for frequency reuse applications in both satellite and terrestrial applications.

13.8 MULTIMODE HORNS

Over the years there has been a need in many applications for horn antennas which provide symmetric patterns in all planes, phase center coincidence for the electric and magnetic planes, and side lobe suppression. All of these are attractive features for designs of optimum reflector systems and monopulse radar systems. Side lobe reduction is a desired attribute for horn radiators utilized in antenna range, anechoic chamber, and standard gain applications, while pattern plane symmetry is a valuable feature for polarization diversity.

Pyramidal horns have traditionally been used over the years, with good success, in many of these applications. Such radiators, however, possess nonsymmetric beamwidths and undesirable side lobe levels, especially in the E-plane. Conical horns, operating in the dominant TE_{11} mode, have a tapered aperture distribution in the E-plane. Thus, they exhibit more symmetric electric- and magnetic-plane beamwidths and reduced side lobes than do the pyramidal horns. One of the main drawbacks of a conical horn is its relative incompatibility with rectangular waveguides.

To remove some of the deficiencies of pyramidal and conical horns and further improve some of their attractive characteristics, corrugations were introduced on the interior walls of the waveguides, which lead to the corrugated horns that were discussed in a previous section of this chapter. In some other cases designs were suggested to improve the beamwidth equalization in all planes and reduce side lobe levels by utilizing horn structures with multiple-mode excitations. These have been designated as *multimode* horns, and some of the designs will be discussed briefly here. For more information the reader should refer to the cited references.

One design of a multimode horn is the "diagonal" horn [40], shown in Figure 13.35, all of whose cross sections are square and whose internal fields consist of a superposition of TE_{10} and TE_{01} modes in a square waveguide. For small flare angles, the field structure within the horn is such that the **E**-field

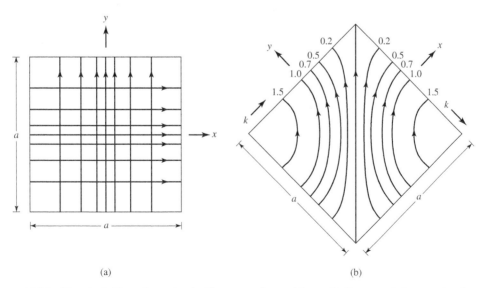

Figure 13.35 Electric field configuration inside square diagonal horn. (a) Two coexisting equal orthogonal modes. (b) Result of combining the two modes shown in (a). (*After Love* [39] reprinted with permission of *Microwave Journal*, Vol. V, No. 3, March 1962).

vector is parallel to one of the diagonals. Although it is not a multimode horn in the true sense of the word, because it does not make use of higher-order TE and TM modes, it does possess the desirable attributes of the usual multimode horns, such as equal beamwidths and suppressed beamwidths and side lobes in the E- and H-planes which are nearly equal to those in the principal planes. These attractive features are accomplished, however, at the expense of pairs of cross-polarized lobes in the intercardinal planes which make such a horn unattractive for applications where a high degree of polarization purity is required.

Diagonal horns have been designed, built, and tested [40] such that the 3-, 10-, and 30-dB beamwidths are nearly equal not only in the principal E- and H-planes, but also in the 45° and 135° planes. Although the theoretical limit of the side lobe level in the principal planes is 31.5 dB down, side lobes of at least 30 dB down have been observed in those planes. Despite a theoretically predicted level of −19.2 dB in the ±45° planes, side lobes with level of −23 to −27 dB have been observed. The principal deficiency in the side lobe structure appears in the ±45°-plane cross-polarized lobes whose intensity is only 16 dB down; despite this, the overall horn efficiency remains high. Compared with diagonal horns, conventional pyramidal square horns have H-plane beamwidths which are about 35% wider than those in the E-plane, and side lobe levels in the E-plane which are only 12 to 13 dB down (although those in the H-plane are usually acceptable).

For applications which require optimum performance with narrow beamwidths, lenses are usually recommended for use in conjunction with diagonal horns. Diagonal horns can also be converted to radiate circular polarization by inserting a differential phase shifter inside the feed guide whose cross section is circular and adjusted so that it produces phase quadrature between the two orthogonal modes.

Another multimode horn which exhibits suppressed side lobes, equal beamwidths, and reduces cross polarization is the *dual-mode* conical horn [41]. Basically this horn is designed so that diffractions at the aperture edges of the horn, especially those in the E-plane, are minimized by reducing the fields incident on the aperture edges and consequently the associated diffractions. This is accomplished by utilizing a conical horn which at its throat region is excited in both the dominant TE_{11} and higher-order TM_{11} mode. A discontinuity is introduced at a position within the horn where two modes exist. The horn length is adjusted so that the superposition of the relative amplitudes of the two modes at the edges of the aperture is very small compared with the maximum aperture field magnitude. In addition, the dimensions of the horn are controlled so that the total phase at the aperture is such that, in conjunction with the desired amplitude distribution, it leads to side lobe suppression, beamwidth equalization, and phase center coincidence.

Qualitatively the pattern formation of a dual-mode conical horn operating in the TE_{11} and TM_{11} modes is accomplished by utilizing a pair of modes which have radiation functions with the same argument. However, one of the modes, in this case the TM_{11} mode, contains an additional envelope factor which varies very rapidly in the main beam region and remains relatively constant at large angles. Thus, it is possible to control the two modes in such a way that their fields cancel in all directions except within the main beam. The TM_{11} mode exhibits a null in its far-field pattern. Therefore a dual-mode conical horn possesses less axial gain than a conventional dominant-mode conical horn of the same aperture size. Because of that, dual-mode horns render better characteristics and are more attractive for applications where pattern plane symmetry and side lobe reduction are more important than maximum aperture efficiency. A most important application of a dual-mode horn is as a feed of Cassegrain reflector systems.

Dual-mode conical horns have been designed, built, and tested [41] with relatively good success in their performance. Generally, however, diagonal horns would be good competitors for the dual-mode horns if it were not for the undesirable characteristics (especially the cross-polarized components) that they exhibit in the very important 45° and 135° planes. Improved performance can be obtained from dual-mode horns if additional higher-order modes (such as the TE_{12}, TE_{13} and TM_{12}) are excited [42] and if their relative amplitudes and phases can be properly controlled. Computed maximum aperture efficiencies of paraboloidal reflectors, using such horns as feeds, have reached 90% contrasted with efficiencies of about 76% for reflector systems using conventional dominant-mode horn feeds. In practice the actual maximum efficiency achieved is determined by the number of modes that can be excited and the degree to which their relative amplitudes and phases can be controlled.

The techniques of the dual-mode and multimode conical horns can be extended to the design of horns with rectangular cross sections. In fact a multimode pyramidal horn has been designed, built, and tested to be used as a feed in a low-noise Cassegrain monopulse system [43]. This rectangular pyramidal horn utilizes additional higher-order modes to provide monopulse capability, side lobe suppression in both the E- and H-planes, and beamwidth equalization. Specifically the various pattern modes for the monopulse system are formed in a single horn as follows:

a. *Sum:* Utilizes $TE_{10} + TE_{30}$ instead of only TE_{10}. When the relative amplitude and phase excitations of the higher-order TE_{30} mode are properly adjusted, they provide side lobe cancellation at the second minor lobe of the TE_{10}-mode pattern
b. *E-Plane Difference:* Utilizes $TE_{11} + TM_{11}$ modes
c. *H-Plane Difference:* Utilizes TE_{20} mode

In its input, the horn of [43] contained a four-guide monopulse bridge circuitry, a multimode matching section, a difference mode phasing section, and a sum mode excitation and control section. To illustrate the general concept, in Figure 13.36(a–c) are plots of three-dimensional patterns of the sum, E-plane difference, and H-plane difference modes which utilize, respectively, the $TE_{10} + TE_{30}$, $TE_{11} + TM_{11}$, and TE_{20} modes. The relative excitation between the modes has been controlled so that each pattern utilizing multiple modes in its formation displays its most attractive features for its function.

13.9 DIELECTRIC-LOADED HORNS

Over the years much effort has been devoted to enhance the antenna and aperture efficiencies of aperture antennas, especially for those that serve as feeds for reflectors (such as the horn). One technique that was proposed and then investigated was to use dielectric guiding structures, referred to as *Dielguides* [44], between the primary feed and the reflector (or subreflector). The technique is simple and inexpensive to implement and provides broadband, highly efficient, and low-noise antenna feeds. The method negates the compromise between taper and spillover efficiencies, and it is based on the principle of internal reflections, which has been utilized frequently in optics. Its role bears a very close resemblance to that of a lens, and it is an extension of the classical parabolic-shaped lens to other geometrical shapes.

Another method that has been used to control the radiation pattern of electromagnetic horns is to insert totally within them various shapes of dielectric material (wedges, slabs, etc.) [45]–[53] to control in a predictable manner not only the phase distribution over the aperture, as is usually done by using the classical parabolic lenses, but also to change the power (amplitude) distribution over the aperture. The control of the amplitude and phase distributions over the aperture are essential in the design of very low side lobe antenna patterns.

Symmetrical loading of the H-plane walls has also been utilized, by proper parameter selection, to create a dominant longitudinal section electric (LSE) mode and to enhance the aperture efficiency and pattern-shaping capabilities of symmetrically loaded horns [45]. The method is simple and inexpensive, and it can also be utilized to realize high efficiency from small horns which can be used in limited scan arrays. Aperture efficiencies on the order of 92 to 96% have been attained, in contrast to values of 81% for unloaded horns.

A similar technique has been suggested to symmetrically load the E-plane walls of rectangular horns [47]–[53] and eventually to line all four of its walls with dielectric slabs. Other similar

772 HORN ANTENNAS

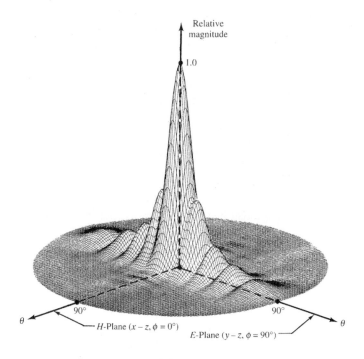

(a) Sum ($TE_{10} + TE_{30}$ modes)

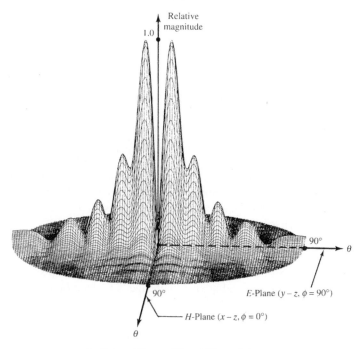

(b) E-plane difference ($TE_{11} + TM_{11}$ modes)

Figure 13.36 Three-dimensional sum and difference (E- and H-planes) field patterns of a monopulse pyramidal horn. (SOURCE: C. A. Balanis, "Horn Antennas," in *Antenna Handbook* (Y. T. Lo and S. W. Lee, eds.), © 1988, Van Nostrand Reinhold Co., Inc.).

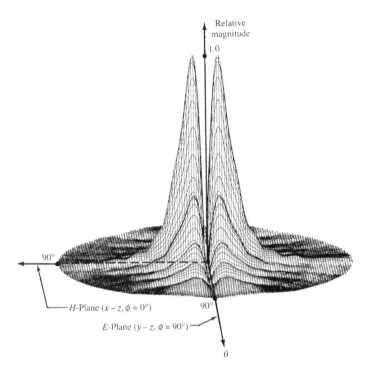

(c) *H*-plane difference (TE$_{20}$ mode)

Figure 13.36 (*Continued*)

techniques have been suggested, and a summary of these and other classical papers dealing with dielectric-loaded horns can be found in [1].

13.10 PHASE CENTER

Each far-zone field component radiated by an antenna can be written, in general, as

$$\mathbf{E}_u = \hat{\mathbf{u}} E(\theta, \phi) e^{j\psi(\theta,\phi)} \frac{e^{-jkr}}{r} \quad (13\text{-}62)$$

where $\hat{\mathbf{u}}$ is a unit vector. The terms $E(\theta, \phi)$ and $\psi(\theta, \phi)$ represent, respectively, the (θ, ϕ) variations of the amplitude and phase.

In navigation, tracking, homing, landing, and other aircraft and aerospace systems it is usually desirable to assign to the antenna a reference point such that for a given frequency, $\psi(\theta, \phi)$ of (13-63) is independent of θ and ϕ (i.e., $\psi(\theta, \phi)$ = constant). The reference point which makes $\psi(\theta, \phi)$ independent of θ and ϕ is known as the *phase center* of the antenna [54]–[58]. When referenced to the phase center, the fields radiated by the antenna are spherical waves with ideal spherical wave fronts or equiphase surfaces. Therefore a phase center is a reference point from which radiation is said to emanate, and radiated fields measured on the surface of a sphere whose center coincides with the phase center have the same phase.

For practical antennas such as arrays, reflectors, and others, a single unique phase center valid for all values of θ and ϕ does not exist; for most, however, their phase center moves along a surface, and its position depends on the observation point. However, in many antenna systems a reference point can be found such that $\psi(\theta, \phi)$ = constant, or nearly so, over most of the angular space, especially over the main lobe. When the phase center position variation is sufficiently small, that point is usually referred to as the *apparent phase center*.

The need for the phase center can best be explained by examining the radiation characteristics of a paraboloidal reflector (parabola of revolution). Plane waves incident on a paraboloidal reflector focus at a single point which is known as the *focal point*. Conversely, spherical waves emanating from the focal point are reflected by the paraboloidal surface and form plane waves. Thus in the receiving mode all the energy is collected at a single point. In the transmitting mode, ideal plane waves are formed if the radiated waves have spherical wavefronts and emanate from a single point.

In practice, no antenna is a point source with ideal spherical equiphases. Many of them, however, contain a point from which their radiation, over most of the angular space, seems to have spherical wavefronts. When such an antenna is used as a feed for a reflector, its phase center must be placed at the focal point. Deviations of the feed from the focal point of the reflector lead to phase errors which result in significant gain reductions of the antenna, as illustrated in Section 15.4.1(G) of Chapter 15.

The analytical formulations for locating the phase center of an antenna are usually very laborious and exist only for a limited number of configurations [54]–[56]. Experimental techniques [57], [58] are available to locate the phase center of an antenna. The one reported in [57] is also discussed in some detail in [2], and it will not be repeated here. The interested reader is referred to [2] and [57].

The horn is a microwave antenna that is widely used as a feed for reflectors [59]. To perform as an efficient feed for reflectors, it is imperative that its phase center is known and it is located at the focal point of the reflector. Instead of presenting analytical formulations for the phase center of a horn, graphical data will be included to illustrate typical phase centers.

Usually the phase center of a horn is not located at its mouth (throat) or at its aperture but mostly between its imaginary apex point and its aperture. The exact location depends on the dimensions of the horn, especially on its flare angle. For large flare angles the phase center is closer to the apex. As the flare angle of the horn becomes smaller, the phase center moves toward the aperture of the horn.

Computed phase centers for an *E*-plane and an *H*-plane sectoral horn are displayed in Figure 13.37(a,b). It is apparent that for small flare angles the *E*- and *H*-plane phase centers are identical. Although each specific design has its own phase center, the data of Figure 13.37(a,b) are typical. If the *E*- and *H*-plane phase centers of a pyramidal horn are not identical, its phase center can be taken to be the average of the two.

Phase center nomographs for conical corrugated and uncorrugated (TE_{11}-mode) horns are available [55], and they can be found in [2] and [55]. The procedure to use these in order to locate a phase center is documented in [2] and [55], and it is not repeated here. The interested reader is referred to [2] where examples are also illustrated.

13.11 MULTIMEDIA

In the publisher's website for this book, the following multimedia resources are included for the review, understanding, and visualization of the material of this chapter:

a. **Java**-based **interactive questionnaire**, with answers.
b. **Matlab** and **Fortran** computer programs, designated
   - **Analysis**
   - **Design**

   for computing and displaying the analysis and design characteristics of a pyramidal horn.
c. **Matlab**-based **animation-visualization** program, designated **te_horn**, that can be used to animate and visualize the radiation of a two-dimensional horn. A detailed description of this program is provided in Section 1.3.4 and in the corresponding READ ME file of Chapter 1.
d. **Power Point (PPT)** viewgraphs, in multicolor.

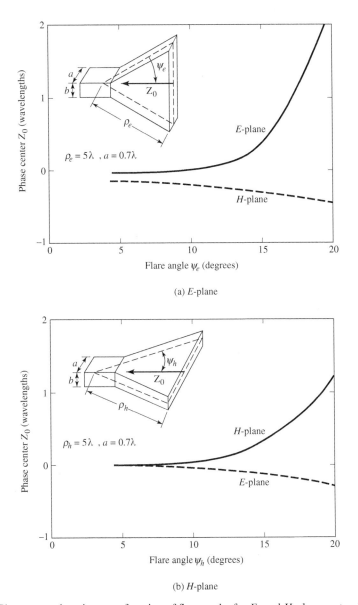

Figure 13.37 Phase center location, as a function of flare angle, for *E*- and *H*-plane sectoral horns. (*Adapted from Hu* [53]).

REFERENCES

1. A. W. Love, *Electromagnetic Horn Antennas*, IEEE Press, New York, 1976.
2. C. A. Balanis, "Horn Antennas," Chapter 8 in *Antenna Handbook: Theory, Applications and Design* (Y. T. Lo and S. W. Lee, eds.), Van Nostrand Reinhold Co., New York, 1988.
3. A. W. Love, "Horn Antennas," Chapter 15 in *Antenna Engineering Handbook* (R. C. Johnson and H. Jasik, eds.), New York, 1984.
4. R. F. Harrington, *Time-Harmonic Electromagnetic Fields*, McGraw-Hill, New York, 1961, pp. 208–213.
5. S. Silver (ed.), *Microwave Antenna Theory and Design*, MIT Radiation Laboratory Series, Vol. 12, McGraw-Hill, New York, 1949, pp. 349–376.

6. C. A. Balanis, *Advanced Engineering Electromagnetics*, Second edition, John Wiley and Sons, New York, 2012.
7. M. J. Maybell and P. S. Simon, "Pyramidal Horn Gain Calculation with Improved Accuracy," *IEEE Trans. Antennas Propagat.*, Vol. 41, No. 7, pp. 884–889, July 1993.
8. K. Liu, C. A. Balanis, C. R. Birtcher and G. C. Barber, "Analysis of Pyramidal Horn Antennas Using Moment Method," *IEEE Trans. Antennas Propagat.*, Vol. 41, No. 10, pp. 1379–1389, October 1993.
9. M. Abramowitz and I. A. Stegun (eds.), Handbook of Mathematical Functions, National Bureau of Standards, United States Dept. of Commerce, June 1964.
10. J. Boersma, "Computation of Fresnel Integrals," *Math. Comp.*, Vol. 14, p. 380, 1960.
11. Y.-B. Cheng, "Analysis of Aircraft Antenna Radiation for Microwave Landing System Using Geometrical Theory of Diffraction," MSEE Thesis, Dept. of Electrical Engineering, West Virginia University, 1976, pp. 208–211.
12. E. V. Jull, "Gain of an E-Plane Sectoral Horn—A Failure of the Kirchoff Theory and a New Proposal," *IEEE Trans. Antennas Propagat.*, Vol. AP-22, No. 2, pp. 221–226, March 1974.
13. E. V. Jull, *Aperture Antennas and Diffraction Theory*, Peter Peregrinus Ltd., London, United Kingdom, 1981, pp. 55–65.
14. E. H. Braun, "Some Data for the Design of Electromagnetic Horns," *IRE Trans. Antennas Propagat.*, Vol. AP-4, No. 1, pp. 29–31, January 1956.
15. P. M. Russo, R. C. Rudduck, and L. Peters, Jr., "A Method for Computing E-Plane Patterns of Horn Antennas," *IEEE Trans. Antennas Propagat.*, Vol. AP-13, No. 2, pp. 219–224, March 1965.
16. J. S. Yu, R. C. Rudduck, and L. Peters, Jr., "Comprehensive Analysis for E-Plane of Horn Antennas by Edge Diffraction Theory," *IEEE Trans. Antennas Propagat.*, Vol. AP-14, No. 2, pp. 138–149, March 1966.
17. M. A. K. Hamid, "Diffraction by a Conical Horn," *IEEE Trans. Antennas Propagat.*, Vol. AP-16, No. 5, pp. 520–528, September 1966.
18. M. S. Narasimhan and M. S. Shehadri, "GTD Analysis of the Radiation Patterns of Conical Horns," *IEEE Trans. Antennas Propagat.*, Vol. AP-26, No. 6, pp. 774–778, November 1978.
19. P. A. Tirkas and C. A. Balanis, "Contour Path FDTD Method for Analysis of Pyramidal Horns with Composite Inner E-Plane Walls," *IEEE Trans. Antennas Propagat.*, Vol. AP-42, No. 11, pp. 1476–1483, November 1994.
20. M. J. Maybell and P. S. Simon, "Pyramidal Horn Gain Calculation with Improved Accuracy," *IEEE Trans. Antennas Propagat.*, Vol. 41, No. 7, pp. 884–889, July 1993.
21. M. G. Schorr and F. J. Beck, Jr., "Electromagnetic Field of a Conical Horn," *J. Appl. Phys.*, Vol. 21, pp. 795–801, August 1950.
22. A. P. King, "The Radiation Characteristics of Conical Horn Antennas," *Proc. IRE*, Vol. 38, pp. 249–251, March 1950.
23. N. A. Aboserwal, C. A. Balanis and C. R. Birtcher, "Conical Horn: Gain and Amplitude Patterns," *IEEE Trans. Antenna Propagat.*, Vol. 61, No. 7, pp. 3427–3433, July 2013.
24. A. F. Kay, "The Scalar Feed," AFCRL Rep. 64–347, AD601609, March 1964.
25. R. E. Lawrie and L. Peters, Jr., "Modifications of Horn Antennas for Low Side Lobe Levels," *IEEE Trans. Antennas Propagat.*, Vol. AP-14, No. 5, pp. 605–610, September 1966.
26. R. S. Elliott, "On the Theory of Corrugated Plane Surfaces," *IRE Trans. Antennas Propagat.*, Vol. AP-2, No. 2, pp. 71–81, April 1954.
27. C. A. Mentzer and L. Peters, Jr., "Properties of Cutoff Corrugated Surfaces for Corrugated Horn Design," *IEEE Trans. Antennas Propagat.*, Vol. AP-22, No. 2, pp. 191–196, March 1974.
28. C. A. Mentzer and L. Peters, Jr., "Pattern Analysis of Corrugated Horn Antennas," *IEEE Trans. Antennas Propagat.*, Vol. AP-24, No. 3, pp. 304–309, May 1976.
29. M. J. Al-Hakkak and Y. T. Lo, "Circular Waveguides and Horns with Anisotropic and Corrugated Boundaries," *Antenna Laboratory Report No. 73-3*, Department of Electrical Engineering, University of Illinois, Urbana, January 1973.
30. B. MacA. Thomas, G. L. James and K. J. Greene, "Design of Wide-Band Corrugated Conical Horns for Cassegrain Antennas," *IEEE Trans. Antennas Propagat.*, Vol. AP-34, No. 6, pp. 750–757, June 1986.

31. B. MacA. Thomas, "Design of Corrugated Conical Horns," *IEEE Trans. Antennas Propagat.*, Vol. AP-26, No. 2, pp. 367–372, March 1978.
32. B. MacA. Thomas and K. J. Greene, "A Curved-Aperture Corrugated Horn Having Very Low Cross-Polar Performance," *IEEE Trans. Antennas Propagat.*, Vol. AP-30, No. 6, pp. 1068–1072, November 1982.
33. G. L. James, "TE_{11}-to-HE_{11} Mode Converters for Small-Angle Corrugated Horns," *IEEE Trans. Antennas Propagat.*, Vol. AP-30, No. 6, pp. 1057–1062, November 1982.
34. K. Tomiyasu, "Conversion of TE_{11} mode by a Large-Diameter Conical Junction," *IEEE Trans. Microwave Theory Tech.*, Vol. MTT-17, No. 5, pp. 277–279, May 1969.
35. B. MacA. Thomas, "Mode Conversion Using Circumferentially Corrugated Cylindrical Waveguide," *Electron. Lett.*, Vol. 8, pp. 394–396, 1972.
36. J. K. M. Jansen and M. E. J. Jeuken, "Surface Waves in Corrugated Conical Horn," *Electronic Letters*, Vol. 8, pp. 342–344, 1972.
37. Y. Tacheichi, T. Hashimoto, and F. Takeda, "The Ring-Loaded Corrugated Waveguide," *IEEE Trans. Microwave Theory Tech.*, Vol. MTT-19, No. 12, pp. 947–950, December 1971.
38. F. Takeda and T. Hashimoto, "Broadbanding of Corrugated Conical Horns by Means of the Ring-Loaded Corrugated Waveguide Structure," *IEEE Trans. Antennas Propagat.*, Vol. AP-24, No. 6, pp. 786–792, November 1976.
39. W. D. Burnside and C. W. Chuang, "An Aperture-Matched Horn Design," *IEEE Trans. Antennas Propagat.*, Vol. AP-30, No. 4, pp. 790–796, July 1982.
40. A. W. Love, "The Diagonal Horn Antenna," *Microwave Journal*, Vol. V, pp. 117–122, March 1962.
41. P. D. Potter, "A New Horn Antenna with Suppressed Side Lobes and Equal Beamwidths," *Microwave Journal*, pp. 71–78, June 1963.
42. P. D. Potter and A. C. Ludwig, "Beamshaping by Use of Higher-Order Modes in Conical Horns," *Northeast Electron. Res. Eng. Mtg.*, pp. 92–93, November 1963.
43. P. A. Jensen, "A Low-Noise Multimode Cassegrain Monopulse with Polarization Diversity," *Northeast Electron. Res. Eng. Mtg.*, pp. 94–95, November 1963.
44. H. E. Bartlett and R. E. Moseley, "Dielguides—Highly Efficient Low-Noise Antenna Feeds," *Microwave Journal*, Vol. 9, pp. 53–58, December 1966.
45. L. L. Oh, S. Y. Peng, and C. D. Lunden, "Effects of Dielectrics on the Radiation Patterns of an Electromagnetic Horn," *IEEE Trans. Antennas Propagat.*, Vol. AP-18, No. 4, pp. 553–556, July 1970.
46. G. N. Tsandoulas and W. D. Fitzgerald, "Aperture Efficiency Enhancement in Dielectrically Loaded Horns," *IEEE Trans. Antennas Propagat.*, Vol. AP-20, No. 1, pp. 69–74, January 1972.
47. R. Baldwin and P. A. McInnes, "Radiation Patterns of Dielectric Loaded Rectangular Horns," *IEEE Trans. Antennas Propagat.*, Vol. AP-21, No. 3, pp. 375–376, May 1973.
48. C. M. Knop, Y. B. Cheng, and E. L. Osterlag, "On the Fields in a Conical Horn Having an Arbitrary Wall Impedance," *IEEE Trans. Antennas Propagat.*, Vol. AP-34, No. 9, pp. 1092–1098, September 1986.
49. J. J. H. Wang, V. K. Tripp, and R. P. Zimmer, "Magnetically Coated Horn for Low Sidelobes and Low Cross-Polarization," *IEE Proc.*, Vol. 136, pp. 132–138, April 1989.
50. J. J. H. Wang, V. K. Tripp, and J. E. Tehan, "The Magnetically Coated Conducting Surface as a Dual Conductor and Its Application to Antennas and Microwaves," *IEEE Trans. Antennas Propagat.*, Vol. AP-38, No. 7, pp. 1069–1077, July 1990.
51. K. Liu and C. A. Balanis, "Analysis of Horn Antennas with Impedance Walls," 1990 IEEE Antennas and Propagation Symposium Digest, Vol. I, pp. 1184–1187, May 7–11, 1990, Dallas, TX.
52. K. Liu and C. A. Balanis, "Low-Loss Material Coating for Horn Antenna Beam Shaping," 1991 IEEE Antennas and Propagation Symposium Digest, Vol. 3, pp. 1664–1667, June 24–28, 1991, London, Ontario, Canada.
53. P. A. Tirkas, "Finite-Difference Time-Domain for Aperture Antenna Radiation," PhD Dissertation, Dept. of Electrical Engineering, Arizona State University, December 1993.
54. Y. Y. Hu, "A Method of Determining Phase Centers and Its Applications to Electromagnetic Horns," *Journal of the Franklin Institute*, Vol. 271, pp. 31–39, January 1961.

55. E. R. Nagelberg, "Fresnel Region Phase Centers of Circular Aperture Antennas," *IEEE Trans. Antennas Propagat.*, Vol. AP-13, No. 3, pp. 479–480, May 1965.
56. I. Ohtera and H. Ujiie, "Nomographs for Phase Centers of Conical Corrugated and TE_{11} Mode Horns," *IEEE Trans. Antennas Propagat.*, Vol. AP-23, No. 6, pp. 858–859, November 1975.
57. J. D. Dyson, "Determination of the Phase Center and Phase Patterns of Antennas," in *Radio Antennas for Aircraft and Aerospace Vehicles*, W. T. Blackband (ed.), AGARD Conference Proceedings, No. 15, Slough, England Technivision Services, 1967.
58. M. Teichman, "Precision Phase Center Measurements of Horn Antennas," *IEEE Trans. Antennas Propagat.*, Vol. AP-18, No. 5, pp. 689–690, September 1970.
59. W. M. Truman and C. A. Balanis, "Optimum Design of Horn Feeds for Reflector Antennas," *IEEE Trans. Antennas Propagat.*, Vol. AP-22, No. 4, pp. 585–586, July 1974.

PROBLEMS

13.1. Derive (13-1a)–(13-1e) by treating the *E*-plane horn as a radial waveguide.

13.2. Design an *E*-plane horn such that the maximum phase difference between two points at the aperture, one at the center and the other at the edge, is 120°. Assuming that the maximum length along its wall (ρ_e), measured from the aperture to its apex, is 10λ, find the
(a) maximum total flare angle of the horn
(b) largest dimension of the horn at the aperture
(c) directivity of the horn (*dimensionless* and *in dB*)
(d) gain of the antenna (*in dB*) when the reflection coefficient within the waveguide feeding the horn is 0.2. Assume only mismatch losses. The waveguide feeding the horn has dimensions of 0.5λ and 0.25λ

13.3. For an *E*-plane horn with $\rho_1 = 6\lambda$, $b_1 = 3.47\lambda$, and $a = 0.5\lambda$,
(a) compute (*in dB*) its pattern at $\theta = 0°$, 10°, and 20°.
(b) compute its directivity using (13-18) and (13-19c). Compare the answers.

13.4. Repeat Problem 13.3 for $\rho_1 = 6\lambda$, $b_1 = 6\lambda$, and $a = 0.5\lambda$.

13.5. For an *E*-plane sectoral horn, plot b_1 (*in* λ) versus ρ_1 (*in* λ) using (13-18a). Verify, using the data of Figure 13.7, that the maximum directivities occur when (13-18a) is satisfied.

13.6. For an *E*-plane sectoral horn with $\rho_1 = 20\lambda$, $a = 0.5\lambda$
(a) find its optimum aperture dimensions for maximum normalized directivity
(b) compute the total flare angle of the horn
(c) compute its directivity, using (13-18), and compare it with the graphical answer
(d) find its half-power beamwidth (*in degrees*)
(e) compute the directivity using (13-19c)

13.7. An *E*-plane horn is fed by an *X*-band WR 90 rectangular waveguide with inner dimensions of 0.9 in. (2.286 cm) and $b = 0.4$ in. (1.016 cm). Design the horn so that its maximum directivity at $f = 11$ GHz is 30 (14.77 dB).

13.8. Design an optimum directivity *E*-plane sectoral horn whose axial length is $\rho_1 = 10\lambda$. The horn is operating at *X*-band with a desired center frequency equal to $f = 10$ GHz. The dimensions of the feed waveguide are $a = 0.9$ in. (2.286 cm) and $b = 0.4$ in. (1.016 cm). Assuming a 100% efficient horn ($e_0 = 1$), find the
(a) horn aperture dimensions b_1 and ρ_e (*in* λ), and flare half-angle ψ_e (*in degrees*)

(b) directivity D_E (*in dB*) using (13-19c)

(c) aperture efficiency

(d) largest phase difference (*in degrees*) between center of horn at the aperture and any point on the horn aperture along the principal *E*-plane.

13.9. Derive (13-20a)–(13-20e) by treating the *H*-plane horn as a radial waveguide.

13.10. For an *H*-plane sectoral horn with $\rho_2 = 6\lambda, a_1 = 6\lambda$, and $b = 0.25\lambda$ compute the

(a) directivity (*in dB*) using (13-39), (13-40c) and compare the answers

(b) normalized field strength (*in dB*) at $\theta = 30°, 45°$, and $90°$.

13.11. For an *H*-plane sectoral horn, plot a_1 (*in* λ) versus ρ_2 (*in* λ) using (13-39c). Verify, using the data of Figure 13.14, that the maximum directivities occur when (13-39c) is satisfied.

13.12. Design an *H*-plane horn so that its maximum directivity at $f = 10$ GHz is 13.25 dB. The horn is fed with a standard X-band waveguide with dimensions $a = 2.286$ cm and $b = 1.016$ cm. Determine:

(a) the horn aperture dimension a_1 (*in cm*).

(b) the axial length ρ_2 of the horn (*in cm*).

(c) the flare angle of the horn (*in degrees*).

13.13. An *H*-plane sectoral horn is fed by an *X*-band WR 90 rectangular waveguide with dimensions of $a = 0.9$ in. (2.286 cm) and $b = 0.4$ in. (1.016 cm). Design the horn so that its maximum directivity at $f = 11$ GHz is 16.3 (12.12 dB).

13.14. Repeat the design of Problem 13.8 for an *H*-plane sectoral horn where axial length is also $\rho_2 = 10\lambda$. The feed waveguide dimensions and center frequency of operation are the same as in Problem 13.8. Assuming an 100% efficient horn ($e_0 = 1$), find the

(a) horn aperture dimensions a_1 and ρ_h (*in* λ), and the flare half-angle ψ_h (*in degrees*)

(b) directivity D_H (*in dB*) using (13-40c)

(c) aperture efficiency

(d) largest phase difference (*in degrees*) between the center of the horn at the aperture and any point on the horn aperture along the principal *H*-plane.

13.15. Show that (13-47a) and (13-47b) must be satisfied in order for a pyramidal horn to be physically realizable.

13.16. A standard-gain X-band (8.2–12.4 GHz) pyramidal horn has dimensions of $\rho_1 \simeq 13.5$ in. (34.29 cm), $\rho_2 \simeq 14.2$ in. (36.07 cm), $a_1 = 7.65$ in. (19.43 cm), $b_1 = 5.65$ in. (14.35 cm), $a = 0.9$ in. (2.286 cm), and $b = 0.4$ in. (1.016 cm).

(a) Check to see if such a horn can be constructed physically.

(b) Compute the directivity (*in dB*) at $f = 8.2, 10.3, 12.4$ GHz using for each (13-50a), (13-51), and (13-52e). Compare the answers. Verify with the computer program **Analysis** of this chapter.

13.17. A standard-gain X-band (8.2–12.4 GHz) pyramidal horn has dimensions of $\rho_1 \simeq 5.3$ in. (13.46 cm), $\rho_2 \simeq 6.2$ in. (15.75 cm), $a_1 = 3.09$ in. (7.85 cm), $b_1 = 2.34$ in. (5.94 cm), $a = 0.9$ in. (2.286 cm), and $b = 0.4$ in. (1.016 cm).

(a) Check to see if such a horn can be constructed physically.

(b) Compute the directivity (in dB) at $f = 8.2, 10.3, 12.4$ GHz using for each (13-50a), (13-51), and (13-52e). Compare the computed answers with the gains of Figure 13.22. Verify with the computer program **Analysis** of this chapter.

13.18. A *lossless linearly polarized* pyramidal horn antenna is used as a receiver in a microwave communications system operating at *10 GHz*. Over the aperture of the horn, the incident wave of the communications system is uniform *and circularly polarized* with a total power density of 10 mW/λ^2. *The pyramidal horn has been designed for optimum gain*. The dimensions of the horn at the aperture are $4\lambda_o$ by $2.5\lambda_o$. Determine the
(a) Approximate aperture efficiency of the *optimum gain horn (in %)*.
(b) Maximum directivity of the horn *(in dB)*.
(c) Maximum power *(in mW)* that can be delivered to the receiver that is assumed to be matched to the transmission line that connects the antenna to the receiver. *Assume no other losses.*

13.19. Repeat the design of the optimum X-band pyramidal horn of Example 13.4 so that the gain at $f = 11$ GHz is 17.05 dB.

13.20. It is desired to design an *optimum* directivity pyramidal horn antenna. The length of the horn from its interior apex is $\rho_1 = \rho_2 = 9\lambda$. The horn is fed by an X-band waveguide whose interior dimensions are 0.5λ by 0.22λ.
(a) To accomplish this, what should the aperture dimensions *(in λ)* of the horn be?
(b) What is the directivity *(in dB)* of the horn?

13.21. Design a pyramidal horn antenna with optimum gain at a frequency of 10 GHz. The overall length of the antenna from the imaginary vertex of the horn to the center of the aperture is 10λ and is nearly the same in both planes. Determine the
(a) Aperture dimensions of the horn *(in cm)*.   (b) Gain of the antenna *(in dB)*
(c) Aperture efficiency of the antenna *(in %)*. Assume the reflection, conduction, and dielectric losses of the antenna are negligible.
(d) Power delivered to a matched load when the incident power density is 10 μwatts/m².

13.22. Design an optimum gain C-band (3.95–5.85 GHz) pyramidal horn so that its gain at $f = 4.90$ GHz is 20.0 dBi. The horn is fed by a WR 187 rectangular waveguide with inner dimensions of $a = 1.872$ in. (4.755 cm) and $b = 0.872$ in. (2.215 cm). Refer to Figure 13.16 for the horn geometry. Determine in cm, the remaining dimensions of the horn: $\rho_e, \rho_h, a_1, b_1, p_e$, and p_h. Verify using the computer program **Design** of this chapter.

13.23. It is desired to design a *pyramidal horn for optimum directivity* with *total* flare angles of 43° in the E-plane and 50° in the H-plane. Assuming the dimensions of the feed rectangular waveguide are $a = 0.5\lambda$ and $b = 0.25\lambda$, determinet the
(a) Horn dimensions at its aperture (a_1 and b_1; both *in λ*).
(b) Other horn dimensions ρ_1, ρ_2, ρ_e and ρ_h *(all in λ)*.
(c) Verify that this horn *is physically realizable*.

13.24. For a conical horn, plot d_m *(in λ)* versus l *(in λ)* using (13-59). Verify, using the data of Figure 13.26, that the maximum directivities occur when (13-59) is satisfied.

13.25. A conical horn has dimensions of $L = 19.5$ in. (49.53 cm), $d_m = 15$ in. (38.10 cm), and $d = 2.875$ in. (7.3025 cm).
(a) Find the frequency *(in GHz)* which will result in maximum directivity for this horn. What is that directivity *(in dB)*?
(b) Find the directivity *(in dB)* at 2.5 and 5 GHz.
(c) Compute the cutoff frequency *(in GHz)* of the TE_{11}-mode which can exist inside the circular waveguide that is used to feed the horn.

13.26. It is desired to design an *optimum directivity* conical horn antenna of circular cross section whose overall *slanted* length l is 10λ. Determine the
  (a) geometrical dimensions of the conical horn [*radius (in λ), diameter (in λ), total flare angle (in degrees)*].
  (b) aperture efficiency of the horn *(in %)*.
  (c) directivity of the horn *(dimensionless and in dB)*.

13.27. Design an *optimum directivity* conical horn, using (13-58)–(13-59), so that its directivity *(above isotropic)* at $f = 11$ GHz is 22.6 dB. Check your design with the data in Figure 13.26. Compare the design dimensions with those of the pyramidal horn of Example 13.5.

13.28. Design an *optimum directivity* conical horn so that its maximum directivity is 20 dB *(above isotropic)*. Determine the
  (a) Diameter *(in λ)* of the horn at its aperture.
  (b) Length *(in λ)* of the horn from its virtual apex to the edge of the aperture.
  (c) Length l *(in λ)* of the horn from its virtual apex to the edge of the aperture.
  (d) *Total* flare angle *(in degrees)* of the horn.
  (e) Aperture efficiency *(in %)* of the horn.
  Refer to Figure 13.24 for the geometry of the problem.

13.29. Design an optimum directivity conical horn so that its directivity at 10 GHz *(above a standard-gain horn of 15 dB directivity)* is 5 dB. Determine the horn diameter *(in cm)* and its flare angle *(in degrees)*.

13.30. Design a conical horn, *for an optimum directivity*, when the *total* included angle is 50°. Calculate
  (a) The diameter of the horn at its aperture *(in λ)*.
  (b) The aperture efficiency *(in %)*.
  (c) What should the expected aperture efficiency *(in %)* be for maximum directivity? How does the answer in Part b compares with the expected value?
  (d) The directivity of the horn *(dimensionless and in dB)*.

13.31. It is desired to design *optimum gain, lossless* ($e_{cd} = 1$) pyramidal (as shown in Fig. 13.16a) and conical (as shown in Fig. 13.24) horns. Assuming *the area* of each horn at its respective aperture is $20\,\pi\lambda^2$, determine the gain of the *optimum gain* horn *(dimensionless and in dB)* for the
  (a) Pyramidal horn of Figure 13.16a.
  (b) (b) Conical horn of Figure 13.24.

13.32. As part of a 10-GHz microwave communication system, you purchase a horn antenna that is said to have a directivity of 75 (dimensionless). The conduction and dielectric losses of the antenna are negligible, and the horn is polarization matched to the incoming signal. A standing wave meter indicates a voltage reflection coefficient of 0.1 at the antenna-waveguide junction.
  (a) Calculate the maximum effective aperture of the horn.
  (b) If an impinging wave with a uniform power density of 1 µwatts/m^2 is incident upon the horn, what is the maximum power delivered to a load which is connected and matched to the lossless waveguide?

13.33. For an X-band pyramidal corrugated horn operating at 10.3 GHz, find the
(a) smallest lower and upper limits of the corrugation depths (*in cm*)
(b) width w of each corrugation (*in cm*) (c) width t of each corrugation tooth (*in cm*)

13.34. Find the E- and H-plane phase centers (*in* λ) of
(a) an E-plane ($\rho_e = 5\lambda, a = 0.7\lambda$) (b) an H-plane ($\rho_h = 5\lambda, a = 0.7\lambda$)
sectoral horn with a total included angle of 30°.

CHAPTER 14

Microstrip and Mobile Communications Antennas

14.1 INTRODUCTION

In high-performance aircraft, spacecraft, satellite, and missile applications, where size, weight, cost, performance, ease of installation, and aerodynamic profile are constraints, low-profile antennas may be required. Presently there are many other government and commercial applications, such as mobile radio and wireless communications, that have similar specifications. To meet these requirements, microstrip antennas [1]–[38] can be used. These antennas are low profile, conformable to planar and nonplanar surfaces, simple and inexpensive to manufacture using modern printed-circuit technology, mechanically robust when mounted on rigid surfaces, compatible with MMIC designs, and when the particular patch shape and mode are selected, they are very versatile in terms of resonant frequency, polarization, pattern, and impedance. In addition, by adding loads between the patch and the ground plane, such as pins and varactor diodes, adaptive elements with variable resonant frequency, impedance, polarization, and pattern can be designed [18], [39]–[44].

Major operational disadvantages of microstrip antennas are their low efficiency, low power, high Q (sometimes in excess of 100), poor polarization purity, poor scan performance, spurious feed radiation and very narrow frequency bandwidth, which is typically only a fraction of a percent or at most a few percent. In some applications, such as in government security systems, narrow bandwidths are desirable. However, there are methods, such as increasing the height of the substrate, that can be used to extend the efficiency (to as large as 90 percent if surface waves are not included) and bandwidth (up to about 35 percent) [38]. However, as the height increases, surface waves are introduced which usually are not desirable because they extract power from the total available for direct radiation (space waves). The surface waves travel within the substrate and they are scattered at bends and surface discontinuities, such as the truncation of the dielectric and ground plane [45]–[49], and degrade the antenna pattern and polarization characteristics. Surface waves can be eliminated, while maintaining large bandwidths, by using cavities [50], [51]. Stacking, as well as other methods, of microstrip elements can also be used to increase the bandwidth [13], [52]–[62]. In addition, microstrip antennas also exhibit large electromagnetic signatures at certain frequencies outside the operating band, are rather large physically at VHF and possibly UHF frequencies, and in large arrays there is a trade-off between bandwidth and scan volume [63]–[65].

Many commercial substrates are available for use in the design and fabrication of microstrip type antennas. Some common substrates are listed in Table 14.1, along with the most pertinent parameters (name of company, substrate name, thickness, frequency range, dielectric constant, and loss tangent).

Antenna Theory: Analysis and Design, Fourth Edition. Constantine A. Balanis.
© 2016 John Wiley & Sons, Inc. Published 2016 by John Wiley & Sons, Inc.
Companion Website: www.wiley.com/go/antennatheory4e

TABLE 14.1 Typical Substrates and Their Parameters

Company	Substrate	Thickness (mm)	Frequency (GHz)	ε_r	tanδ
Rogers Corporation	Duroid® 5880	0.127	0 – 40	2.20	0.0009
	RO 3003	1.575	0 – 40	3.00	0.0010
	RO 3010	3.175	0 – 10	10.2	0.0022
	RO 4350	0.168	0 – 10	3.48	0.0037
		0.508			
		1.524			
—	FR4	0.05 – 100	0.001	4.70	—
DuPont	HK 04J	0.025	0.001	3.50	0.005
Isola	IS 410	0.05 – 3.2	0.1	5.40	0.035
Arlon	DiClad 870	0.091	0 – 10	2.33	0.0013
Polyflon	Polyguide	0.102	0 – 10	2.32	0.0005
Neltec	NH 9320	3.175	0 – 10	3.20	0.0024
Taconic	RF-60A	0.102	0 – 10	6.15	0.0038

This is only a small sample; some companies, especially the Rogers Corporation, have a variety of substrate types to choose from. The Rogers substrates are usually referred to as PTFE (polytetrafluorethylene), which are woven glass laminates, and are very popular for microstrip designs. FR4 is another very popular substrate.

14.1.1 Basic Characteristics

Microstrip antennas received considerable attention starting in the 1970s, although the idea of a microstrip antenna can be traced to 1953 [1] and a patent in 1955 [2]. Microstrip antennas, as shown in Figure 14.1(a), consist of a very thin ($t \ll \lambda_0$, where λ_0 is the free-space wavelength) metallic strip (patch) placed a small fraction of a wavelength ($h \ll \lambda_0$, usually $0.003\lambda_0 \leq h \leq 0.05\lambda_0$) above

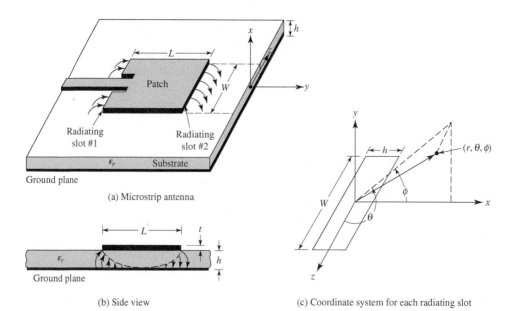

(a) Microstrip antenna

(b) Side view

(c) Coordinate system for each radiating slot

Figure 14.1 Microstrip antenna and coordinate system.

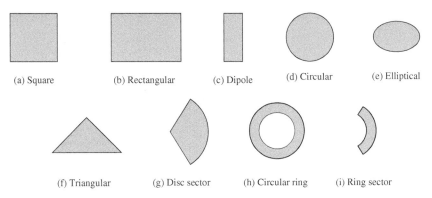

Figure 14.2 Representative shapes of microstrip patch elements.

a ground plane. The microstrip patch is designed so its pattern maximum is normal to the patch (broadside radiator). This is accomplished by properly choosing the mode (field configuration) of excitation beneath the patch. End-fire radiation can also be accomplished by judicious mode selection. For a rectangular patch, the length L of the element is usually $\lambda_0/3 < L < \lambda_0/2$. The strip (patch) and the ground plane are separated by a dielectric sheet (referred to as the substrate), as shown in Figure 14.1(a).

There are numerous substrates that can be used for the design of microstrip antennas, and their dielectric constants are usually in the range of $2.2 \leq \varepsilon_r \leq 12$. The ones that are most desirable for good antenna performance are thick substrates whose dielectric constant is in the lower end of the range because they provide better efficiency, larger bandwidth, loosely bound fields for radiation into space, but at the expense of larger element size [38]. Thin substrates with higher dielectric constants are desirable for microwave circuitry because they require tightly bound fields to minimize undesired radiation and coupling, and lead to smaller element sizes; however, because of their greater losses, they are less efficient and have relatively smaller bandwidths [38]. Since microstrip antennas are often integrated with other microwave circuitry, a compromise has to be reached between good antenna performance and circuit design.

Often microstrip antennas are also referred to as *patch* antennas. The radiating elements and the feed lines are usually photoetched on the dielectric substrate. The radiating patch may be square, rectangular, thin strip (dipole), circular, elliptical, triangular, or any other configuration. These and others are illustrated in Figure 14.2. Square, rectangular, dipole (strip), and circular are the most common because of ease of analysis and fabrication, and their attractive radiation characteristics, especially low cross-polarization radiation. Microstrip dipoles are attractive because they inherently possess a large bandwidth and occupy less space, which makes them attractive for arrays [14], [22], [30], [31]. Linear and circular polarizations can be achieved with either single elements or arrays of microstrip antennas. Arrays of microstrip elements, with single or multiple feeds, may also be used to introduce scanning capabilities and achieve greater directivities. These will be discussed in later sections.

14.1.2 Feeding Methods

There are many configurations that can be used to feed microstrip antennas. The four most popular are the microstrip line, coaxial probe, aperture coupling, and proximity coupling [15], [16], [30], [35], [38], [66]–[68]. These are displayed in Figure 14.3. One set of equivalent circuits for each one of these is shown in Figure 14.4. The microstrip feed line is also a conducting strip, usually of much smaller width compared to the patch. The microstrip-line feed is easy to fabricate, simple to match by controlling the inset position and rather simple to model. However as the substrate thickness

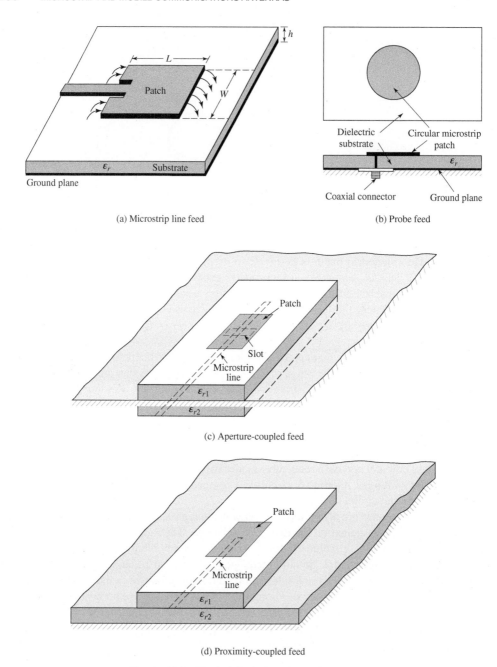

Figure 14.3 Typical feeds for microstrip antennas.

increases, surface waves and spurious feed radiation increase, which for practical designs limit the bandwidth (typically 2–5%).

Coaxial-line feeds, where the inner conductor of the coax is attached to the radiation patch while the outer conductor is connected to the ground plane, are also widely used. The coaxial probe feed is also easy to fabricate and match, and it has low spurious radiation. However, it also has narrow bandwidth and it is more difficult to model, especially for thick substrates ($h > 0.02\lambda_0$).

Both the microstrip feed line and the probe possess inherent asymmetries which generate higher order modes which produce cross-polarized radiation. To overcome some of these problems,

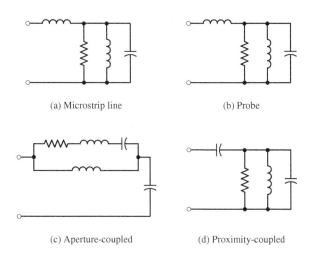

(a) Microstrip line (b) Probe
(c) Aperture-coupled (d) Proximity-coupled

Figure 14.4 Equivalent circuits for typical feeds of Figure 14.3.

noncontacting aperture-coupling feeds, as shown in Figures 14.3(c,d), have been introduced. The aperture coupling of Figure 14.3(c) is the most difficult of all four to fabricate and it also has narrow bandwidth. However, it is somewhat easier to model and has moderate spurious radiation. The aperture coupling consists of two substrates separated by a ground plane. On the bottom side of the lower substrate there is a microstrip feed line whose energy is coupled to the patch through a slot on the ground plane separating the two substrates. This arrangement allows independent optimization of the feed mechanism and the radiating element. Typically a high dielectric material is used for the bottom substrate, and thick low dielectric constant material for the top substrate. The ground plane between the substrates also isolates the feed from the radiating element and minimizes interference of spurious radiation for pattern formation and polarization purity. For this design, the substrate electrical parameters, feed line width, and slot size and position can be used to optimize the design [38]. Typically matching is performed by controlling the width of the feed line and the length of the slot. The coupling through the slot can be modeled using the theory of Bethe [69], which is also used to account for coupling through a small aperture in a conducting plane. This theory has been successfully used to analyze waveguide couplers using coupling through holes [70]. In this theory the slot is represented by an equivalent normal electric dipole to account for the normal component (to the slot) of the electric field, and an equivalent horizontal magnetic dipole to account for the tangential component (to the slot) magnetic field. If the slot is centered below the patch, where ideally for the dominant mode the electric field is zero while the magnetic field is maximum, the magnetic coupling will dominate. Doing this also leads to good polarization purity and no cross-polarized radiation in the principal planes [38]. Of the four feeds described here, the proximity coupling has the largest bandwidth (as high as 13 percent), is somewhat easy to model and has low spurious radiation. However its fabrication is somewhat more difficult. The length of the feeding stub and the width-to-line ratio of the patch can be used to control the match [61].

14.1.3 Methods of Analysis

There are many methods of analysis for microstrip antennas. The most popular models are the *transmission-line* [16], [35], *cavity* [12], [16], [18], [35], and *full wave* (which include primarily integral equations/Moment Method) [22], [26], [71]–[74]. The transmission-line model is the easiest of all, it gives good physical insight, but is less accurate and it is more difficult to model coupling [75]. Compared to the transmission-line model, the cavity model is more accurate but at the same time more complex. However, it also gives good physical insight and is rather difficult to model coupling, although it has been used successfully [8], [76], [77]. In general when applied properly, the full-wave models are very accurate, very versatile, and can treat single elements, finite and infinite arrays, stacked elements, arbitrary shaped elements, and coupling. However they are the most complex models and usually give less physical insight. In this chapter we will cover the transmission-line and cavity models only. However results and design curves from full-wave models will also be included. Since they are the most popular and practical, in this chapter the only two patch configurations that will be considered are the rectangular and circular. Representative radiation characteristics of some other configurations will be included.

14.2 RECTANGULAR PATCH

The rectangular patch is by far the most widely used configuration. It is very easy to analyze using both the transmission-line and cavity models, which are most accurate for thin substrates [78]. We begin with the transmission-line model because it is easier to illustrate.

14.2.1 Transmission-Line Model

It was indicated earlier that the transmission-line model is the easiest of all but it yields the least accurate results and it lacks the versatility. However, it does shed some physical insight. As it will be demonstrated in Section 14.2.2 using the cavity model, a rectangular microstrip antenna can be represented as an array of two *radiating* narrow apertures (slots), each of width W and height h, separated by a distance L. Basically the transmission-line model represents the microstrip antenna by two slots, separated by a low-impedance Z_c transmission line of length L.

A. Fringing Effects
Because the dimensions of the patch are finite along the length and width, the fields at the edges of the patch undergo fringing. This is illustrated along the length in Figures 14.1(a,b) for the two radiating slots of the microstrip antenna. The same applies along the width. The amount of fringing is a function of the dimensions of the patch and the height of the substrate. For the principal E-plane (xy-plane) fringing is a function of the ratio of the length of the patch L to the height h of the substrate (L/h) and the dielectric constant ε_r of the substrate. Since for microstrip antennas $L/h \gg 1$, fringing is reduced; however, it must be taken into account because it influences the resonant frequency of the antenna. The same applies for the width.

For a microstrip line shown in Figure 14.5(a), typical electric field lines are shown in Figure 14.5(b). This is a nonhomogeneous line of two dielectrics; typically the substrate and air. As can be seen, most of the electric field lines reside in the substrate and parts of some lines exist in air. As $W/h \gg 1$ and $\varepsilon_r \gg 1$, the electric field lines concentrate mostly in the substrate. Fringing in this case makes the microstrip line look wider electrically compared to its physical dimensions. Since some of the waves travel in the substrate and some in air, an *effective dielectric constant* $\varepsilon_{\text{reff}}$ is introduced to account for fringing and the wave propagation in the line.

To introduce the effective dielectric constant, let us assume that the center conductor of the microstrip line with its original dimensions and height above the ground plane is embedded into one dielectric, as shown in Figure 14.5(c). The effective dielectric constant is defined *as the dielectric constant of the uniform dielectric material so that the line of Figure 14.5(c) has identical electrical characteristics, particularly propagation constant, as the actual line of Figure 14.5(a).* For a line with air above the substrate, the effective dielectric constant has values in the range of $1 < \varepsilon_{\text{reff}} < \varepsilon_r$. For most applications where the dielectric constant of the substrate is much greater than unity ($\varepsilon_r \gg 1$), the value of $\varepsilon_{\text{reff}}$ will be closer to the value of the actual dielectric constant ε_r of the substrate. The effective dielectric constant is also a function of frequency. As the frequency of operation increases, most of the electric field lines concentrate in the substrate. Therefore the microstrip line behaves

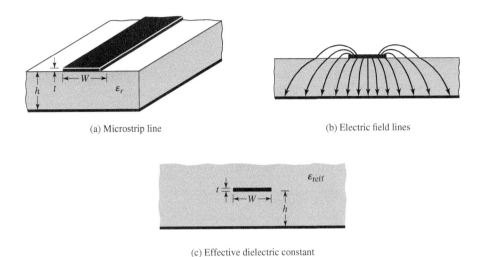

(a) Microstrip line   (b) Electric field lines

(c) Effective dielectric constant

Figure 14.5 Microstrip line and its electric field lines, and effective dielectric constant geometry.

more like a homogeneous line of one dielectric (only the substrate), and the effective dielectric constant approaches the value of the dielectric constant of the substrate. Typical variations, as a function of frequency, of the effective dielectric constant for a microstrip line with three different substrates are shown in Figure 14.6.

For low frequencies the effective dielectric constant is essentially constant. At intermediate frequencies its values begin to monotonically increase and eventually approach the values of the dielectric constant of the substrate. The initial values (at low frequencies) of the effective dielectric constant are referred to as the *static values*, and they are given by [79]

$$\underline{W/h > 1}$$

$$\varepsilon_{\text{reff}} = \frac{\varepsilon_r + 1}{2} + \frac{\varepsilon_r - 1}{2} \left[1 + 12\frac{h}{W}\right]^{-1/2} \tag{14-1}$$

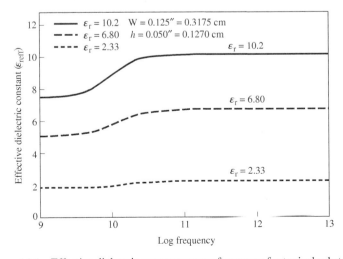

Figure 14.6 Effective dielectric constant versus frequency for typical substrates.

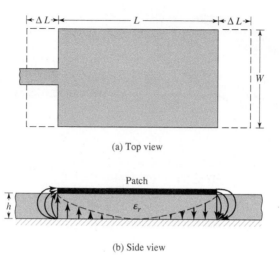

Figure 14.7 Physical and effective lengths of rectangular microstrip patch.

B. Effective Length, Resonant Frequency, and Effective Width

Because of the fringing effects, electrically the patch of the microstrip antenna looks greater than its physical dimensions. For the principal *E*-plane (*xy*-plane), this is demonstrated in Figure 14.7 where the dimensions of the patch along its length have been extended on each end by a distance ΔL, which is a function of the effective dielectric constant $\varepsilon_{\text{reff}}$ and the width-to-height ratio (W/h). A very popular and practical approximate relation for the normalized extension of the length is [80]

$$\frac{\Delta L}{h} = 0.412 \frac{(\varepsilon_{\text{reff}} + 0.3)\left(\frac{W}{h} + 0.264\right)}{(\varepsilon_{\text{reff}} - 0.258)\left(\frac{W}{h} + 0.8\right)} \tag{14-2}$$

Since the length of the patch has been extended by ΔL on each side, the effective length of the patch is now ($L = \lambda/2$ for dominant TM_{010} mode with no fringing)

$$L_{\text{eff}} = L + 2\Delta L \tag{14-3}$$

For the dominant TM_{010} mode, the resonant frequency of the microstrip antenna is a function of its length. Usually it is given by

$$(f_r)_{010} = \frac{1}{2L\sqrt{\varepsilon_r}\sqrt{\mu_0\varepsilon_0}} = \frac{v_0}{2L\sqrt{\varepsilon_r}} \tag{14-4}$$

where v_0 is the speed of light in free space. Since (14-4) does not account for fringing, it must be modified to include edge effects and should be computed using

$$(f_{rc})_{010} = \frac{1}{2L_{\text{eff}}\sqrt{\varepsilon_{\text{reff}}}\sqrt{\mu_0\varepsilon_0}} = \frac{1}{2(L + 2\Delta L)\sqrt{\varepsilon_{\text{reff}}}\sqrt{\mu_0\varepsilon_0}}$$

$$= q\frac{1}{2L\sqrt{\varepsilon_r}\sqrt{\mu_0\varepsilon_0}} = q\frac{v_0}{2L\sqrt{\varepsilon_r}} \tag{14-5}$$

where

$$q = \frac{(f_{rc})_{010}}{(f_r)_{010}} \tag{14-5a}$$

The q factor is referred to as the *fringe factor* (length reduction factor). As the substrate height increases, fringing also increases and leads to larger separations between the radiating edges and lower resonant frequencies. The designed resonant frequency, based on fringing, is lower as the patch looks longer, as indicated in Figure 14.7. The resonant frequency decrease due to fringing is usually 2–6%.

C. Design

Based on the simplified formulation that has been described, a design procedure is outlined which leads to practical designs of rectangular microstrip antennas. The procedure assumes that the specified information includes the dielectric constant of the substrate (ε_r), the resonant frequency (f_r), and the height of the substrate h. The procedure is as follows:

Specify:

$$\varepsilon_r, f_r \text{ (in Hz), and } h$$

Determine:

$$W, L$$

Design procedure:

1. For an efficient radiator, a practical width that leads to good radiation efficiencies is [15]

$$W = \frac{1}{2f_r\sqrt{\mu_0\varepsilon_0}}\sqrt{\frac{2}{\varepsilon_r+1}} = \frac{v_0}{2f_r}\sqrt{\frac{2}{\varepsilon_r+1}} \tag{14-6}$$

   where v_0 is the free-space velocity of light.
2. Determine the effective dielectric constant of the microstrip antenna using (14-1).
3. Once W is found using (14-6), determine the extension of the length ΔL using (14-2).
4. The actual length of the patch can now be determined by solving (14-5) for L, or

$$L = \frac{1}{2f_r\sqrt{\varepsilon_{\text{reff}}}\sqrt{\mu_0\varepsilon_0}} - 2\Delta L \tag{14-7}$$

Typical lengths of microstrip patches vary between

$$L \approx (0.47 - 0.49)\frac{\lambda_o}{\sqrt{\varepsilon_r}} = (0.47 - 0.49)\lambda_d \tag{14-7a}$$

where λ_d is the wavelength in the dielectric. The smaller the dielectric constant of the substrate, the larger is the fringing; thus, the length of the microstrip patch is smaller. In contrast, the larger the dielectric constant, the more tightly the fields are held within the substrate; thus, the fringing is smaller and the length is longer and closer to half-wavelength in the dielectric.

Example 14.1

Design a rectangular microstrip antenna using a substrate (RT/duroid 5880) with dielectric constant of 2.2, $h = 0.1588$ cm (0.0625 inches) so as to resonate at 10 GHz.

Solution: Using (14-6), the width W of the patch is

$$W = \frac{30}{2(10)} \sqrt{\frac{2}{2.2+1}} = 1.186 \text{ cm } (0.467 \text{ in})$$

The effective dielectric constant of the patch is found using (14-1), or

$$\varepsilon_{\text{reff}} = \frac{2.2+1}{2} + \frac{2.2-1}{2} \left(1 + 12\frac{0.1588}{1.186}\right)^{-1/2} = 1.972$$

The extended incremental length of the patch ΔL is, using (14-2)

$$\Delta L = 0.1588(0.412) \frac{(1.972 + 0.3)\left(\frac{1.186}{0.1588} + 0.264\right)}{(1.972 - 0.258)\left(\frac{1.186}{0.1588} + 0.8\right)}$$

$$= 0.081 \text{ cm } (0.032 \text{ in})$$

The actual length L of the patch is found using (14-3), or

$$L = \frac{\lambda}{2} - 2\Delta L = \frac{30}{2(10)\sqrt{1.972}} - 2(0.081) = 0.906 \text{ cm } (0.357 \text{ in})$$

Finally the effective length is

$$L_e = L + 2\Delta L = \frac{\lambda}{2} = 1.068 \text{ cm } (0.421 \text{ in})$$

An experimental rectangular patch based on this design was built and tested. It is probe fed from underneath by a coaxial line and is shown in Figure 14.8(a). Its principal E- and H-plane patterns are displayed in Figure 14.21(c,d) for $f_o = 9.8$ GHz.

(a) Rectangular

(b) Circular

Figure 14.8 Experimental models of rectangular and circular patches based, respectively, on the designs of Examples 14.1 and 14.4.

D. Conductance

Each radiating slot is represented by a parallel equivalent admittance Y (with conductance G and susceptance B). This is shown in Figure 14.9. The slots are labeled as #1 and #2. The equivalent admittance of slot #1, based on an infinitely wide, uniform slot, is derived in Example 12.8 of Chapter 12, and it is given by [81]

$$Y_1 = G_1 + jB_1 \qquad (14\text{-}8)$$

where for a slot of finite width W

$$G_1 = \frac{W}{120\lambda_0}\left[1 - \frac{1}{24}(k_0 h)^2\right] \qquad \frac{h}{\lambda_0} < \frac{1}{10} \qquad (14\text{-}8a)$$

$$B_1 = \frac{W}{120\lambda_0}[1 - 0.636\ln(k_0 h)] \qquad \frac{h}{\lambda_0} < \frac{1}{10} \qquad (14\text{-}8b)$$

Since slot #2 is identical to slot #1, its equivalent admittance is

$$Y_2 = Y_1, \quad G_2 = G_1, \quad B_2 = B_1 \qquad (14\text{-}9)$$

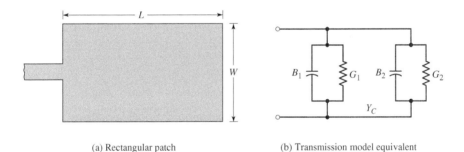

(a) Rectangular patch   (b) Transmission model equivalent

Figure 14.9 Rectangular microstrip patch and its equivalent circuit transmission-line model.

The conductance of a single slot can also be obtained by using the field expression derived by the cavity model. In general, the conductance is defined as

$$G_1 = \frac{2P_{\text{rad}}}{|V_0|^2} \qquad (14\text{-}10)$$

Using the electric field of (14-41), the radiated power is written as

$$P_{\text{rad}} = \frac{|V_0|^2}{2\pi\eta_0}\int_0^\pi \left[\frac{\sin\left(\frac{k_0 W}{2}\cos\theta\right)}{\cos\theta}\right]^2 \sin^3\theta\, d\theta \qquad (14\text{-}11)$$

Therefore the conductance of (14-10) can be expressed as

$$G_1 = \frac{I_1}{120\pi^2} \qquad (14\text{-}12)$$

where

$$I_1 = \int_0^\pi \left[\frac{\sin\left(\frac{k_0 W}{2} \cos\theta\right)}{\cos\theta} \right]^2 \sin^3\theta \, d\theta$$

$$= -2 + \cos(X) + X S_i(X) + \frac{\sin(X)}{X} \qquad (14\text{-}12a)$$

$$X = k_0 W \qquad (14\text{-}12b)$$

Asymptotic values of (14-12) and (14-12a) are

$$G_1 = \begin{cases} \dfrac{1}{90}\left(\dfrac{W}{\lambda_0}\right)^2 & W \ll \lambda_0 \\[2mm] \dfrac{1}{120}\left(\dfrac{W}{\lambda_0}\right) & W \gg \lambda_0 \end{cases} \qquad (14\text{-}13)$$

The values of (14-13) for $W \gg \lambda_0$ are identical to those given by (14-8a) for $h \ll \lambda_0$. A plot of G as a function of W/λ_0 is shown in Figure 14.10.

E. Resonant Input Resistance

The total admittance at slot #1 (input admittance) is obtained by transferring the admittance of slot #2 from the output terminals to input terminals using the admittance transformation equation of transmission lines [16], [70], [79]. Ideally the two slots should be separated by $\lambda/2$ where λ is the wavelength in the dielectric (substrate). However, because of fringing the length of the patch is electrically longer than the actual length. Therefore the actual separation of the two slots is slightly less than $\lambda/2$. If the reduction of the length is properly chosen using (14-2) (typically $0.47\lambda < L < 0.49\lambda$), the transformed admittance of slot #2 becomes

$$\tilde{Y}_2 = \tilde{G}_2 + j\tilde{B}_2 = G_1 - jB_1 \qquad (14\text{-}14)$$

or

$$\tilde{G}_2 = G_1 \qquad (14\text{-}14a)$$

$$\tilde{B}_2 = -B_1 \qquad (14\text{-}14b)$$

Therefore the total resonant input admittance is real and is given by

$$Y_{in} = Y_1 + \tilde{Y}_2 = 2G_1 \qquad (14\text{-}15)$$

Since the total input admittance is real, the resonant input impedance is also real, or

$$Z_{in} = \frac{1}{Y_{in}} = R_{in} = \frac{1}{2G_1} \qquad (14\text{-}16)$$

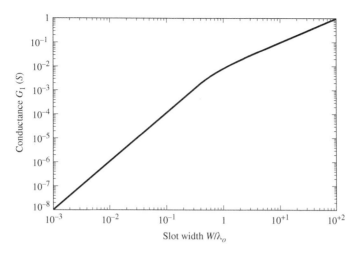

Figure 14.10 Slot conductance as a function of slot width.

The resonant input resistance, as given by (14-16), does not take into account mutual effects between the slots. This can be accomplished by modifying (14-16) to [8]

$$R_{in} = \frac{1}{2(G_1 \pm G_{12})} \quad (14\text{-}17)$$

where the plus (+) sign is used for modes with odd (antisymmetric) resonant voltage distribution beneath the patch and between the slots while the minus (−) sign is used for modes with even (symmetric) resonant voltage distribution. The mutual conductance is defined, in terms of the far-zone fields, as

$$G_{12} = \frac{1}{|V_0|^2} \text{Re} \iint_S \mathbf{E}_1 \times \mathbf{H}_2^* \cdot d\mathbf{s} \quad (14\text{-}18)$$

where $\mathbf{E}_1$ is the electric field radiated by slot #1, $\mathbf{H}_2$ is the magnetic field radiated by slot #2, V_0 is the voltage across the slot, and the integration is performed over a sphere of large radius. It can be shown that G_{12} can be calculated using [8], [34]

$$G_{12} = \frac{1}{120\pi^2} \int_0^\pi \left[\frac{\sin\left(\frac{k_0 W}{2} \cos\theta\right)}{\cos\theta} \right]^2 J_0(k_0 L \sin\theta) \sin^3\theta \, d\theta \quad (14\text{-}18a)$$

where J_0 is the Bessel function of the first kind of order zero. For typical microstrip antennas, the mutual conductance obtained using (14-18a) is small compared to the self conductance G_1 of (14-8a) or (14-12).

An alternate approximate expression for the input impedance, R_{in}, for a resonant patch [82] is

$$R_{in} = 90 \frac{(\varepsilon_r)^2}{\varepsilon_r - 1} \left(\frac{L}{W}\right) \qquad (14\text{-}18b)$$

This expression is valid for thin substrates ($h \ll \lambda_0$). It was derived on approximations of rigorously developed formulations, including Sommerfeld integrals, for thin grounded substrates. Equation (14-17), and (14-18b) give reasonable results for the input resistance R_{in}, although they are not identical.

As shown by (14-8a) and (14-17), the input resistance is not strongly dependent upon the substrate height h. In fact for very small values of h, such that $k_0 h \ll 1$, the input resistance is not dependent on h. Modal-expansion analysis also reveals that the input resistance is not strongly influenced by the substrate height h. It is apparent from (14-8a) and (14-17) that the resonant input resistance can be decreased by increasing the width W of the patch. This is acceptable as long as the ratio of W/L does not exceed 2 because the aperture efficiency of a single patch begins to drop, as W/L increases beyond 2.

F. Matching Techniques

The resonant input resistance, as calculated by (14-17), is referenced at slot #1. However, it has been shown that the resonant input resistance can be changed by using an inset feed, recessed a distance y_0 from slot #1, as shown in Figure 14.11(a). This technique can be used effectively to match the

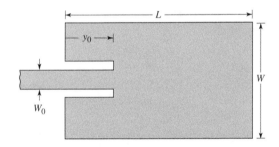

(a) Recessed microstrip-line feed

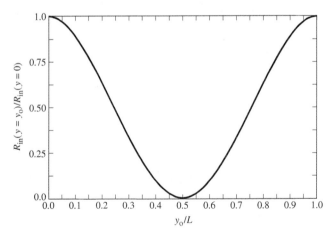

(b) Normalized input resistance

Figure 14.11 Recessed microstrip-line feed and variation of normalized input resistance.

patch antenna using a microstrip-line feed whose characteristic impedance is given by [79]

$$Z_c = \begin{cases} \dfrac{60}{\sqrt{\varepsilon_{\text{reff}}}} \ln\left[\dfrac{8h}{W_0} + \dfrac{W_0}{4h}\right], & \dfrac{W_0}{h} \leq 1 \quad (14\text{-}19a) \\[2ex] \dfrac{120\pi}{\sqrt{\varepsilon_{\text{reff}}}\left[\dfrac{W_0}{h} + 1.393 + 0.667 \ln\left(\dfrac{W_0}{h} + 1.444\right)\right]}, & \dfrac{W_0}{h} > 1 \quad (14\text{-}19b) \end{cases}$$

where W_0 is the width of the microstrip line, as shown in Figure 14.11. Using modal-expansion analysis, the input resistance for the inset feed is given approximately by [8], [16]

$$R_{in}(y = y_0) = \dfrac{1}{2(G_1 \pm G_{12})} \left[\cos^2\left(\dfrac{\pi}{L} y_0\right) \right.$$
$$\left. + \dfrac{G_1^2 + B_1^2}{Y_c^2} \sin^2\left(\dfrac{\pi}{L} y_0\right) - \dfrac{B_1}{Y_c} \sin\left(\dfrac{2\pi}{L} y_0\right)\right] \quad (14\text{-}20)$$

where $Y_c = 1/Z_c$. Since for most typical microstrips $G_1/Y_c \ll 1$ and $B_1/Y_c \ll 1$, (14-20) reduces to

$$R_{in}(y = y_0) = \dfrac{1}{2(G_1 \pm G_{12})} \cos^2\left(\dfrac{\pi}{L} y_0\right)$$
$$= R_{in}(y = 0) \cos^2\left(\dfrac{\pi}{L} y_0\right) \quad (14\text{-}20a)$$

A plot of the normalized value of (14-20a) is shown in Figure 14.11(b).

The values obtained using (14-20) agree fairly well with experimental data. However, the inset feed introduces a physical notch, which in turn introduces a junction capacitance. The physical notch and its corresponding junction capacitance influence slightly the resonance frequency, which typically may vary by about 1%. It is apparent from (14-20a) and Figure 14.11(b) that the maximum value occurs at the edge of the slot ($y_0 = 0$) where the voltage is maximum and the current is minimum; typical values are in the 150–300 ohms. The minimum value (zero) occurs at the center of the patch ($y_0 = L/2$) where the voltage is zero and the current is maximum. As the inset feed point moves from the edge toward the center of the patch the resonant input impedance decreases monotonically and reaches zero at the center. When the value of the inset feed point approaches the center of the patch ($y_0 = L/2$), the $\cos^2(\pi y_0/L)$ function varies very rapidly; therefore the input resistance also changes rapidly with the position of the feed point. To maintain very accurate values, a close tolerance must be preserved.

Other matching techniques, aside from the recessed microstrip of Figure 14.11, are the coupled recessed microstrip and the $\lambda/4$ impedance transformer of Figure 14.12(a,b). The R_{in} in Figure 14.12(b) is the input resistance at the leading edge of the resonant patch; it must be real.

Example 14.2

A microstrip antenna with overall dimensions of $L = 0.906$ cm (0.357 inches) and $W = 1.186$ cm (0.467 inches), substrate with height $h = 0.1588$ cm (0.0625 inches) and dielectric constant of $\varepsilon_r = 2.2$, is operating at 10 GHz. Find:

a. The input impedance.
b. The position of the inset feed point where the input impedance is 50 ohms.

Solution:

$$\lambda_0 = \frac{30}{10} = 3 \text{ cm}$$

Using (14-12) and (14-12a)

$$G_1 = 0.00157 \text{ siemens}$$

which compares with $G_1 = 0.00328$ using (14-8a). Using (14-18a)

$$G_{12} = 6.1683 \times 10^{-4}$$

Using (14-17) with the (+) sign because of the odd field distribution between the radiating slots for the dominant TM_{010} mode

$$R_{in} = 228.3508 \text{ ohms}.$$

Since the input impedance at the leading radiating edge of the patch is 228.3508 ohms while the desired impedance is 50 ohms, the inset feed point distance y_0 is obtained using (14-20a). Thus

$$50 = 228.3508 \cos^2\left(\frac{\pi}{L} y_0\right)$$

or

$$y_0 = 0.3126 \text{ cm } (0.123 \text{ inches})$$

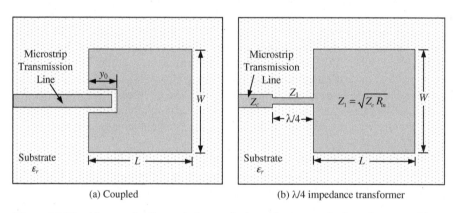

Figure 14.12 Alternate feeding techniques of microstrip antenna for impedance matching.

14.2.2 Cavity Model

Microstrip antennas resemble dielectric-loaded cavities, and they exhibit higher order resonances. The normalized fields within the dielectric substrate (between the patch and the ground plane) can be found more accurately by treating that region as a cavity bounded by electric conductors (above and below it) and by magnetic walls (to simulate an open circuit) along the perimeter of the patch. This

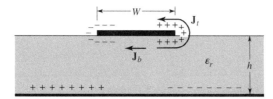

Figure 14.13 Charge distribution and current density creation on microstrip patch.

is an approximate model, which in principle leads to a reactive input impedance (of zero or infinite value of resonance), and it does not radiate any power. However, assuming that the actual fields are approximate to those generated by such a model, the computed pattern, input admittance, and resonant frequencies compare well with measurements [12], [16], [18]. This is an accepted approach, and it is similar to the perturbation methods which have been very successful in the analysis of waveguides, cavities, and radiators [81].

To shed some insight into the cavity model, let us attempt to present a physical interpretation into the formation of the fields within the cavity and radiation through its side walls. When the microstrip patch is energized, a charge distribution is established on the upper and lower surfaces of the patch, as well as on the surface of the ground plane, as shown in Figure 14.13. The charge distribution is controlled by two mechanisms; an *attractive* and a *repulsive* mechanism [34]. The *attractive* mechanism is between the corresponding opposite charges on the bottom side of the patch and the ground plane, which tends to maintain the charge concentration on the bottom of the patch. The *repulsive* mechanism is between like charges on the bottom surface of the patch, which tends to push some charges from the bottom of the patch, around its edges, to its top surface. The movement of these charges creates corresponding current densities $\mathbf{J}_b$ and $\mathbf{J}_t$, at the bottom and top surfaces of the patch, respectively, as shown in Figure 14.13. Since for most practical microstrips the height-to-width ratio is very small, the attractive mechanism dominates and most of the charge concentration and current flow remain underneath the patch. A small amount of current flows around the edges of the patch to its top surface. However, this current flow decreases as the height-to-width ratio decreases. In the limit, the current flow to the top would be zero, which ideally would not create any tangential magnetic field components to the edges of the patch. This would allow the four side walls to be modeled as perfect magnetic conducting surfaces which ideally would not disturb the magnetic field and, in turn, the electric field distributions beneath the patch. Since in practice there is a finite height-to-width ratio, although small, the tangential magnetic fields at the edges would not be exactly zero. However, since they will be small, a good approximation to the cavity model is to treat the side walls as perfectly magnetic conducting. This model produces good normalized electric and magnetic field distributions (modes) beneath the patch.

The fringing field is another mechanism that can be used to explain why the microstrip antenna radiates. Consider the side view of a patch antenna shown in Figure 14.13. Because the current at the end of the patch is zero (open circuit), the current is maximum at the center of the half-wave patch and ideally zero at the beginning of the patch. This low current value at the feed explains in part why the impedance is high when fed at the leading edge, as shown in Figure 14.11(a). Because the patch antenna can be ideally viewed as an open-circuited transmission line, the voltage reflection coefficient of the voltage is +1; therefore the voltage and current are out of phase. Hence, at the end of the patch, the voltage is maximum. At the center of the patch (a quarter-wavelength away from the edges), the voltage must be minimum. Basically, taking into account the finite dimensions of the cavity, the fields underneath the patch resemble those of Figure 14.12(b), which roughly displays the fringing of the fields around the edges.

Based on the field structure in Figure 14.14, it is the fringing fields that are responsible for the radiation. Note that the fringing fields at the leading and trailing edges (left and right of Figure 14.14) of the patch antenna are both in the same direction (in this case y direction). Hence, the fringing

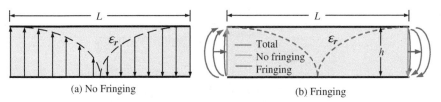

Figure 14.14 Side view of microstrip antenna, without and with fringing.

E-fields on the leading and trailing edges of the microstrip antenna add up in phase and produce the radiation of the microstrip antenna. The current also adds up in phase on the patch; however, an equal current but with opposite direction is on the ground plane (as illustrated in Figure 14.13), which cancels the radiation. This also explains why the microstrip antenna radiates but the microstrip transmission line does not. The microstrip antenna's radiation arises from the fringing fields, which are due to the favorable voltage distribution; hence, the radiation is attributed due to the voltage and not the current. The patch antenna is, therefore, a "voltage radiator," as opposed to wire antennas, which are "current radiators."

Furthermore, note that the smaller the dielectric constant ε_r of the substrate is, the fringing field is greater; that is, the field "bows" away further from the patch. Therefore, using a smaller permittivity for the microstrip yields more efficient radiation. In contrast, when no radiation is desired, as in transmission lines, a high value of ε_r is desired. The high ε_r causes the fields to become more tightly bound (less fringing), resulting in a desired smaller radiation efficiency (ideally zero).

If the microstrip antenna were treated only as a cavity, it would not be sufficient to find the absolute amplitudes of the electric and magnetic fields. In fact by treating the walls of the cavity, as well as the material within it as lossless, the cavity would not radiate and its input impedance would be purely reactive. Also the function representing the impedance would only have real poles. To account for radiation, a loss mechanism has to be introduced. In Figures 2.27 and 2.28 of Chapter 2, this was taken into account by the radiation resistance R_r and loss resistance R_L. These two resistances allow the input impedance to be complex and for its function to have complex poles; the imaginary poles representing, through R_r and R_L, the radiation and conduction-dielectric losses. To make the microstrip lossy using the cavity model, which would then represent an antenna, the loss is taken into account by introducing an effective loss tangent δ_{eff}. The effective loss tangent is chosen appropriately to represent the loss mechanism of the cavity, which now behaves as an antenna and is taken as the reciprocal of the antenna quality factor Q ($\delta_{\text{eff}} = 1/Q$).

Because the thickness of the microstrip is usually very small, the waves generated within the dielectric substrate (between the patch and the ground plane) undergo considerable reflections when they arrive at the edge of the patch. Therefore only a small fraction of the incident energy is radiated; thus the antenna is considered to be very inefficient. The fields beneath the patch form standing waves that can be represented by cosinusoidal wave functions. Since the height of the substrate is very small ($h \ll \lambda$ where λ is the wavelength within the dielectric), the field variations along the height will be considered constant. In addition, because of the very small substrate height, the fringing of the fields along the edges of the patch are also very small whereby the electric field is nearly normal to the surface of the patch. Therefore only TM^x field configurations will be considered within the cavity. While the top and bottom walls of the cavity are perfectly electric conducting, the four side walls will be modeled as perfectly conducting magnetic walls (tangential magnetic fields vanish along those four walls).

A. Field Configurations (modes) — TM^x

The field configurations within the cavity can be found using the vector potential approach described in detail in Chapter 8 of [79]. Referring to Figure 14.15, the volume beneath the patch can be treated as a rectangular cavity loaded with a dielectric material with dielectric constant ε_r. The dielectric

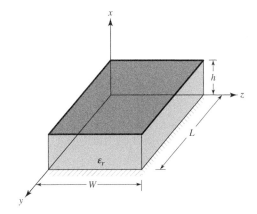

Figure 14.15 Rectangular microstrip patch geometry.

material of the substrate is assumed to be truncated and not extended beyond the edges of the patch. The vector potential A_x must satisfy the homogeneous wave equation of

$$\nabla^2 A_x + k^2 A_x = 0 \tag{14-21}$$

whose solution is written in general, using the separation of variables, as [79]

$$A_x = [A_1 \cos(k_x x) + B_1 \sin(k_x x)][A_2 \cos(k_y y) + B_2 \sin(k_y y)]$$
$$\cdot [A_3 \cos(k_z z) + B_3 \sin(k_z z)] \tag{14-22}$$

where k_x, k_y and k_z are the wavenumbers along the x, y, and z directions, respectively. These will be determined subject to the boundary conditions. The electric and magnetic fields within the cavity are related to the vector potential A_x by [79]

$$E_x = -j\frac{1}{\omega\mu\varepsilon}\left(\frac{\partial^2}{\partial x^2} + k^2\right) A_x \quad H_x = 0$$

$$E_y = -j\frac{1}{\omega\mu\varepsilon}\frac{\partial^2 A_x}{\partial x \partial y} \qquad H_y = \frac{1}{\mu}\frac{\partial A_x}{\partial z} \tag{14-23}$$

$$E_z = -j\frac{1}{\omega\mu\varepsilon}\frac{\partial^2 A_x}{\partial x \partial z} \qquad H_z = -\frac{1}{\mu}\frac{\partial A_x}{\partial y}$$

subject to the boundary conditions of

$$E_y(x' = 0, 0 \le y' \le L, 0 \le z' \le W)$$
$$= E_y(x' = h, 0 \le y' \le L, 0 \le z' \le W) = 0$$
$$H_y(0 \le x' \le h, 0 \le y' \le L, z' = 0)$$
$$= H_y(0 \le x' \le h, 0 \le y' \le L, z' = W) = 0 \tag{14-24}$$
$$H_z(0 \le x' \le h, y' = 0, 0 \le z' \le W)$$
$$= H_z(0 \le x' \le h, y' = L, 0 \le z' \le W) = 0$$

The primed coordinates x', y', z' are used to represent the fields within the cavity.

Applying the boundary conditions $E_y(x' = 0, 0 \leq y' \leq L, 0 \leq z' \leq W) = 0$ and $E_y(x' = h, 0 \leq y' \leq L, 0 \leq z' \leq W) = 0$, it can be shown that $B_1 = 0$ and

$$k_x = \frac{m\pi}{h}, \quad m = 0, 1, 2, \ldots \tag{14-25}$$

Similarly, applying the boundary conditions $H_y(0 \leq x' \leq h, 0 \leq y' \leq L, z' = 0) = 0$ and $H_y(0 \leq x' \leq h, 0 \leq y' \leq L, z' = W) = 0$, it can be shown that $B_3 = 0$ and

$$k_z = \frac{p\pi}{W}, \quad p = 0, 1, 2, \ldots \tag{14-26}$$

Finally, applying the boundary conditions $H_z(0 \leq x' \leq h, y' = 0, 0 \leq z' \leq W) = 0$ and $H_z(0 \leq x' \leq h, y' = L, 0 \leq z' \leq W) = 0$, it can be shown that $B_2 = 0$ and

$$k_y = \frac{n\pi}{L}, \quad n = 0, 1, 2, \ldots \tag{14-27}$$

Thus the final form for the vector potential A_x within the cavity is

$$A_x = A_{mnp} \cos(k_x x') \cos(k_y y') \cos(k_z z') \tag{14-28}$$

where A_{mnp} represents the amplitude coefficients of each mnp mode. The wavenumbers k_x, k_y, k_z are equal to

$$\left. \begin{array}{l} k_x = \left(\dfrac{m\pi}{h}\right), \quad m = 0, 1, 2, \ldots \\[4pt] k_y = \left(\dfrac{n\pi}{L}\right), \quad n = 0, 1, 2, \ldots \\[4pt] k_z = \left(\dfrac{p\pi}{W}\right), \quad p = 0, 1, 2, \ldots \end{array} \right\} m = n = p \neq 0 \tag{14-29}$$

where m, n, p represent, respectively, the number of half-cycle field variations along the x, y, z directions.

Since the wavenumbers k_x, k_y, and k_z are subject to the constraint equation

$$k_x^2 + k_y^2 + k_z^2 = \left(\frac{m\pi}{h}\right)^2 + \left(\frac{n\pi}{L}\right)^2 + \left(\frac{p\pi}{W}\right)^2 = k_r^2 = \omega_r^2 \mu \varepsilon \tag{14-30}$$

the resonant frequencies for the cavity are given by

$$(f_r)_{mnp} = \frac{1}{2\pi\sqrt{\mu\varepsilon}} \sqrt{\left(\frac{m\pi}{h}\right)^2 + \left(\frac{n\pi}{L}\right)^2 + \left(\frac{p\pi}{W}\right)^2} \tag{14-31}$$

Substituting (14-28) into (14-23), the electric and magnetic fields within the cavity are written as

$$E_x = -j\frac{(k^2 - k_x^2)}{\omega\mu\varepsilon}A_{mnp}\cos(k_x x')\cos(k_y y')\cos(k_z z')$$

$$E_y = -j\frac{k_x k_y}{\omega\mu\varepsilon}A_{mnp}\sin(k_x x')\sin(k_y y')\cos(k_z z')$$

$$E_z = -j\frac{k_x k_z}{\omega\mu\varepsilon}A_{mnp}\sin(k_x x')\cos(k_y y')\sin(k_z z') \quad (14\text{-}32)$$

$$H_x = 0$$

$$H_y = -\frac{k_z}{\mu}A_{mnp}\cos(k_x x')\cos(k_y y')\sin(k_z z')$$

$$H_z = \frac{k_y}{\mu}A_{mnp}\cos(k_x x')\sin(k_y y')\cos(k_z z')$$

To determine the dominant mode with the lowest resonance, we need to examine the resonant frequencies. The mode with the lowest order resonant frequency is referred to as the *dominant* mode. Placing the resonant frequencies in ascending order determines the order of the modes of operation. For all microstrip antennas $h \ll L$ and $h \ll W$. If $L > W > h$, the mode with the lowest frequency (dominant mode) is the TM_{010}^x whose resonant frequency is given by

$$(f_r)_{010} = \frac{1}{2L\sqrt{\mu\varepsilon}} = \frac{v_0}{2L\sqrt{\varepsilon_r}} \quad (14\text{-}33)$$

where v_0 is the speed of light in free-space. If in addition $L > W > L/2 > h$, the next higher order (second) mode is the TM_{001}^x whose resonant frequency is given by

$$(f_r)_{001} = \frac{1}{2W\sqrt{\mu\varepsilon}} = \frac{v_0}{2W\sqrt{\varepsilon_r}} \quad (14\text{-}34)$$

If, however, $L > L/2 > W > h$, the second order mode is the TM_{020}^x, instead of the TM_{001}^x, whose resonant frequency is given by

$$(f_r)_{020} = \frac{1}{L\sqrt{\mu\varepsilon}} = \frac{v_0}{L\sqrt{\varepsilon_r}} \quad (14\text{-}35)$$

If $W > L > h$, the dominant mode is the TM_{001}^x whose resonant frequency is given by (14-34) while if $W > W/2 > L > h$ the second order mode is the TM_{002}^x. Based upon (14-32), the distribution of the tangential electric field along the side walls of the cavity for the TM_{010}^x, TM_{001}^x, TM_{020}^x and TM_{002}^x is as shown, respectively, in Figure 14.16.

In all of the preceding discussion, it was assumed that there is no fringing of the fields along the edges of the cavity. This is not totally valid, but it is a good assumption. However, fringing effects and their influence were discussed previously, and they should be taken into account in determining the resonant frequency. This was done in (14-5) for the dominant TM_{010}^x mode.

B. Equivalent Current Densities

It has been shown using the cavity model that the microstrip antenna can be modeled reasonably well by a dielectric-loaded cavity with two perfectly conducting electric walls (top and bottom), and four perfectly conducting magnetic walls (sidewalls). It is assumed that the material of the substrate

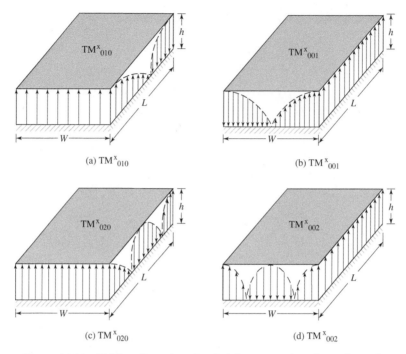

Figure 14.16 Field configurations (modes) for rectangular microstrip patch.

is truncated and does not extend beyond the edges of the patch. The four sidewalls represent four narrow apertures (slots) through which radiation takes place. Using the Field Equivalence Principle (Huygens' Principle) of Section 12.2 of Chapter 12, the microstrip patch is represented by an equivalent electric current density $\mathbf{J}_t$ at the top surface of the patch to account for the presence of the patch (there is also a current density $\mathbf{J}_b$ at the bottom of the patch which is not needed for this model). The four side slots are represented by the equivalent electric current density $\mathbf{J}_s$ and equivalent magnetic current density $\mathbf{M}_s$, as shown in Figure 14.17(a), each represented by

$$\mathbf{J}_s = \hat{\mathbf{n}} \times H_a \qquad (14\text{-}36)$$

and

$$\mathbf{M}_s = -\hat{\mathbf{n}} \times E_a \qquad (14\text{-}37)$$

where $\mathbf{E}_a$ and $\mathbf{H}_a$ represent, respectively, the electric and magnetic fields at the slots.

Because it was shown for microstrip antennas with very small height-to-width ratio that the current density $\mathbf{J}_t$ at the top of the patch is much smaller than the current density $\mathbf{J}_b$ at the bottom of the patch, it will be assumed it is negligible here and it will be set to zero. Also it was argued that the tangential magnetic fields along the edges of the patch are very small, ideally zero. Therefore the corresponding equivalent electric current density $\mathbf{J}_s$ will be very small (ideally zero), and it will be set to zero here. Thus the only nonzero current density is the equivalent magnetic current density $\mathbf{M}_s$ of (14-37) along the side periphery of the cavity radiating in the presence of the ground plane, as shown in Figure 14.17(b). The presence of the ground plane can be taken into account by image theory which will double the equivalent magnetic current density of (14-37). Therefore the final equivalent is a magnetic current density of twice (14-37) or

$$\mathbf{M}_s = -2\hat{\mathbf{n}} \times E_a \qquad (14\text{-}38)$$

around the side periphery of the patch radiating into free-space, as shown in Figure 14.17(c).

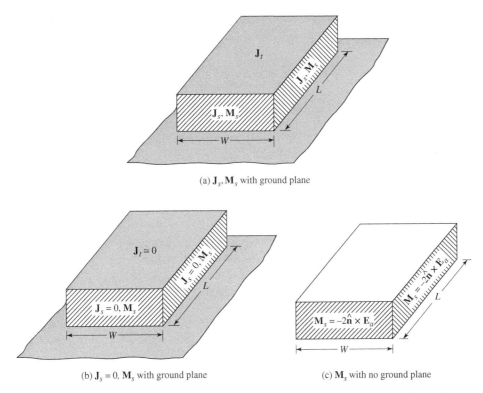

Figure 14.17 Equivalent current densities on four sides of rectangular microstrip patch.

It was shown, using the transmission-line model, that the microstrip antenna can be represented by two radiating slots along the length of the patch (each of width W and height h). Similarly it will be shown here also that while there are a total of four slots representing the microstrip antenna, only two (the radiating slots) account for most of the radiation; the fields radiated by the other two, which are separated by the width W of the patch, cancel along the principal planes. Therefore the same two slots, separated by the length of the patch, are referred to here also as *radiating* slots. The slots are separated by a very low-impedance parallel-plate transmission line of length L, which acts as a transformer. The length of the transmission line is approximately $\lambda/2$, where λ is the guide wavelength in the substrate, in order for the fields at the aperture of the two slots to have opposite polarization. This is illustrated in Figures 14.1(a) and 14.16(a). The two slots form a two-element array with a spacing of $\lambda/2$ between the elements. It will be shown here that in a direction perpendicular to the ground plane the components of the field add in phase and give a maximum radiation normal to the patch; thus it is a broadside antenna.

Assuming that the dominant mode within the cavity is the TM^x_{010} mode, the electric and magnetic field components reduce from (14-32) to

$$E_x = E_0 \cos\left(\frac{\pi}{L}y'\right)$$
$$H_z = H_0 \sin\left(\frac{\pi}{L}y'\right) \quad (14\text{-}39)$$
$$E_y = E_z = H_x = H_y = 0$$

where $E_0 = -j\omega A_{010}$ and $H_0 = (\pi/\mu L)A_{010}$. The electric field structure within the substrate and between the radiating element and the ground plane is sketched in Figures 14.1(a,b) and 14.16(a).

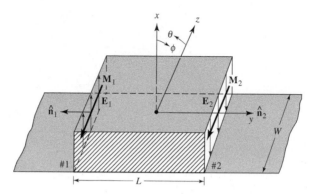

Figure 14.18 Rectangular microstrip patch radiating slots and equivalent magnetic current densities.

It undergoes a phase reversal along the length but it is uniform along its width. The phase reversal along the length is necessary for the antenna to have broadside radiation characteristics.

Using the equivalence principle of Section 12.2, each slot radiates the same fields as a magnetic dipole with current density $\mathbf{M}_s$ equal to (14-38). By referring to Figure 14.18 the equivalent magnetic current densities along the two slots, each of width W and height h, are both of the same magnitude and of the same phase. Therefore these two slots form a two-element array with the sources (current densities) of the same magnitude and phase, and separated by L. Thus these two sources will add in a direction normal to the patch and ground plane forming a broadside pattern. This is illustrated in Figures 14.19(a) where the normalized radiation pattern of each slot in the principal E-plane is sketched individually along with the total pattern of the two. In the H-plane, the normalized pattern of each slot and of the two together is the same, as shown in Figure 14.19(b).

The equivalent current densities for the other two slots, each of length L and height h, are shown in Figure 14.20. Since the current densities on each wall are of the same magnitude but of opposite direction, the fields radiated by these two slots cancel each other in the principal H-plane. Also since corresponding slots on opposite walls are 180° out of phase, the corresponding radiations cancel each other in the principal E-plane. This will be shown analytically. The radiation from these two side walls in nonprincipal planes is small compared to the other two side walls. Therefore these two slots are usually referred to as *nonradiating* slots.

C. Fields Radiated—TM^x_{010} Mode

To find the fields radiated by each slot, we follow a procedure similar to that used to analyze the aperture in Section 12.5.1. The total field is the sum of the two-element array with each element

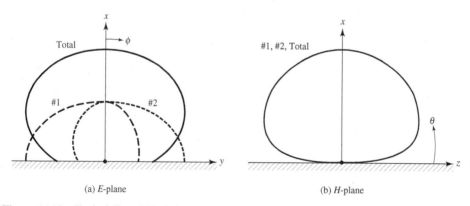

(a) E-plane (b) H-plane

Figure 14.19 Typical E- and H-plane patterns of each microstrip patch slot, and of the two together.

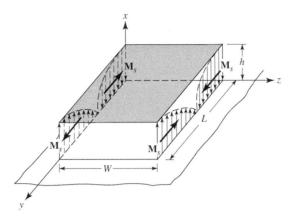

Figure 14.20 Current densities on nonradiating slots of rectangular microstrip patch.

representing one of the slots. Since the slots are identical, this is accomplished by using an array factor for the two slots.

Radiating Slots Following a procedure similar to that used to analyze the aperture in Section 12.5.1, the far-zone electric fields radiated by each slot, using the equivalent current densities of (14-38), are written as

$$E_r \simeq E_\theta \simeq 0 \tag{14-40a}$$

$$E_\phi = +j\frac{k_0 h W E_0 e^{-jk_0 r}}{2\pi r}\left\{\sin\theta\frac{\sin(X)}{X}\frac{\sin(Z)}{Z}\right\} \tag{14-40b}$$

where

$$X = \frac{k_0 h}{2}\sin\theta\cos\phi \tag{14-40c}$$

$$Z = \frac{k_0 W}{2}\cos\theta \tag{14-40d}$$

For very small heights ($k_0 h \ll 1$), (14-40b) reduces to

$$E_\phi \simeq +j\frac{V_0 e^{-jk_0 r}}{\pi r}\left\{\sin\theta\frac{\sin\left(\frac{k_0 W}{2}\cos\theta\right)}{\cos\theta}\right\} \tag{14-41}$$

where $V_0 = hE_0$.

According to the theory of Chapter 6, the array factor for the two elements, of the same magnitude and phase, separated by a distance L_e along the y direction is

$$(AF)_y = 2\cos\left(\frac{k_0 L_e}{2}\sin\theta\sin\phi\right) \tag{14-42}$$

where L_e is the effective length of (14-3). Thus, the total electric field for the two slots (also for the microstrip antenna) is

$$E_\phi^t = +j\frac{k_0 h W E_0 e^{-jk_0 r}}{\pi r}\left\{\sin\theta\frac{\sin(X)}{X}\frac{\sin(Z)}{Z}\right\}$$
$$\times \cos\left(\frac{k_0 L_e}{2}\sin\theta\sin\phi\right) \quad (14\text{-}43)$$

where

$$X = \frac{k_0 h}{2}\sin\theta\cos\phi \quad (14\text{-}43a)$$

$$Z = \frac{k_0 W}{2}\cos\theta \quad (14\text{-}43b)$$

For small values of h ($k_0 h \ll 1$), (14-43) reduces to

$$E_\phi^t \simeq +j\frac{2V_0 e^{-jk_0 r}}{\pi r}\left\{\sin\theta\frac{\sin\left(\frac{k_0 W}{2}\cos\theta\right)}{\cos\theta}\right\}\cos\left(\frac{k_0 L_e}{2}\sin\theta\sin\phi\right) \quad (14\text{-}44)$$

where $V_0 = hE_0$ is the voltage across the slot.

E-Plane ($\theta = 90°, 0° \leq \phi \leq 90°$ and $270° \leq \phi \leq 360°$)

For the microstrip antenna, the x-y plane ($\theta = 90°, 0° \leq \phi \leq 90°$ and $270° \leq \phi \leq 360°$) is the principal E-plane. For this plane, the expressions for the radiated fields of (14-43)–(14-43b) reduce to

$$E_\phi^t = +j\frac{k_0 W V_0 e^{-jk_0 r}}{\pi r}\left\{\frac{\sin\left(\frac{k_0 h}{2}\cos\phi\right)}{\frac{k_0 h}{2}\cos\phi}\right\}\cos\left(\frac{k_0 L_e}{2}\sin\phi\right) \quad (14\text{-}45)$$

H-Plane ($\phi = 0°, 0° \leq \theta \leq 180°$)

The principal H-plane of the microstrip antenna is the x-z plane ($\phi = 0°, 0° \leq \theta \leq 180°$), and the expressions for the radiated fields of (14-43)–(14-43b) reduce to

$$E_\phi^t \simeq +j\frac{k_0 W V_0 e^{-jk_0 r}}{\pi r}\left\{\sin\theta\frac{\sin\left(\frac{k_0 h}{2}\sin\theta\right)}{\frac{k_0 h}{2}\sin\theta}\frac{\sin\left(\frac{k_0 W}{2}\cos\theta\right)}{\frac{k_0 W}{2}\cos\theta}\right\} \quad (14\text{-}46)$$

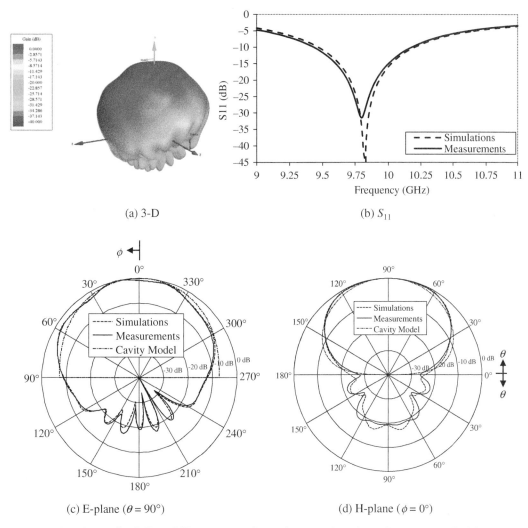

Figure 14.21 Normalized 3D and 2D patterns and S_{11} of rectangular microstrip patch ($L = 0.906$ cm, $W = 1.186$ cm, $h = 0.1588$ cm, $y_0 = 0.203$ cm, $\varepsilon_r = 2.2, f_0 = 9.8$ GHz).

To illustrate the modeling of the microstrip using the cavity model, the 3-D, S_{11} and 2-D principal E- and H-plane patterns have been computed at $f_0 = 9.8$ GHz for the rectangular microstrip of Example 14.1 and Figure 14.8(a). These are displayed in Figure 14.21 where they are compared with measurements. A good agreement is indicated. The ground plane was 10 cm × 10 cm. Edge effects can be taken into account using diffraction theory [48], [79]. The noted asymmetry in the measured and simulated patterns is due to the feed which is not symmetrically positioned along the E-plane. The simulation accounts for the position of the feed, while the cavity model does not account for it. The pattern for $0° \le \phi \le 180°$ [left half in Figure 14.21(c)] corresponds to observation angles which lie on the same side of the patch as does the feed probe.

The presence of the dielectric-covered ground plane modifies the reflection coefficient, which influences the magnitude and phase of the image. This is similar to the ground effects discussed in Section 4.8 of Chapter 4. To account for the dielectric, the reflection coefficient for vertical polarization of +1 must be replaced by the reflection coefficient of (4-125) while the reflection coefficient for horizontal polarization of −1 must be replaced by the reflection coefficient of (4-128).

Basically the introduction of the reflection coefficients of (4-125) and (4-128) to account for the dielectric cover of the ground plane is to modify the boundary conditions of the perfect conductor to one with an impedance surface. The result is for (4-125) to modify the shape of the pattern in the E-plane of the microstrip antenna, primarily for observation angles near grazing (near the ground plane), as was done in Figure 4.35 for the lossy earth. Similar changes are expected for the microstrip antenna. The changes in the pattern near grazing come from the fact that, for the perfect conductor, the reflection coefficient for vertical polarization is $+1$ for all observation angles. However for the dielectric-covered ground plane (impedance surface), the reflection coefficient of (4-125) is nearly $+1$ for observation angles far away from grazing but begins to change very rapidly near grazing and becomes -1 at grazing [79]; thus the formation of an ideal null at grazing.

Similarly the reflection coefficient of (4-128) should basically control the pattern primarily in the H-plane. However, because the reflection coefficient for horizontal polarization for a perfect conductor is -1 for all observation angles while that of (4-128) for the dielectric-covered ground plane is nearly -1 for all observation angles, the shape of the pattern in the H-plane is basically unaltered by the presence of the dielectric cover [79]. This is illustrated in Figure 4.37 for the earth. The pattern also exhibits a null along the ground plane. Similar changes are expected for the microstrip antenna.

Nonradiating Slots The fields radiated by the so-called nonradiating slots, each of effective length L_e and height h, are found using the same procedure as for the two radiating slots. Using the fields of (14-39), the equivalent magnetic current density of one of the nonradiating slots facing the $+z$ axis is

$$\mathbf{M}_s = -2\hat{\mathbf{n}} \times \mathbf{E}_a = \hat{\mathbf{a}}_y 2E_0 \cos\left(\frac{\pi}{L_e}y'\right) \quad (14\text{-}47)$$

and it is sketched in Figure 14.20. A similar one is facing the $-z$ axis. Using the same procedure as for the radiating slots, the normalized far-zone electric field components radiated by each slot are given by

$$E_\theta = -\frac{k_0 h L_e E_0 e^{-jk_0 r}}{2\pi r} \left\{ Y \cos\phi \frac{\sin X}{X} \frac{\cos Y}{(Y)^2 - (\pi/2)^2} \right\} e^{j(X+Y)} \quad (14\text{-}48a)$$

$$E_\phi = \frac{k_0 h L_e E_0 e^{-jk_0 r}}{2\pi r} \left\{ Y \cos\theta \sin\phi \frac{\sin X}{X} \frac{\cos Y}{(Y)^2 - (\pi/2)^2} \right\} e^{j(X+Y)} \quad (14\text{-}48b)$$

where

$$X = \frac{k_0 h}{2} \sin\theta \cos\phi \quad (14\text{-}48c)$$

$$Y = \frac{k_0 L_e}{2} \sin\theta \sin\phi \quad (14\text{-}48d)$$

Since the two nonradiating slots form an array of two elements, of the same magnitude but of opposite phase, separated along the z axis by a distance W, the array factor is

$$(AF)_z = 2j \sin\left(\frac{k_0 W}{2} \cos\theta\right) \quad (14\text{-}49)$$

Therefore the total far-zone electric field is given by the product of each of (14-48a) and (14-48b) with the array factor of (14-49).

RECTANGULAR PATCH

In the H-plane ($\phi = 0°, 0° \leq \theta \leq 180°$), (14-48a) and (14-48b) are zero because the fields radiated by each quarter cycle of each slot are cancelled by the fields radiated by the other quarter. Similarly in the E-plane ($\theta = 90°, 0° \leq \phi \leq 90°$ and $270° \leq \phi \leq 360°$) the total fields are also zero because (14-49) vanishes. This implies that the fields radiated by each slot are cancelled by the fields radiated by the other. The nonradiation in the principal planes by these two slots was discussed earlier and demonstrated by the current densities in Figure 14.20. However, these two slots do radiate away from the principal planes, but their field intensity in these other planes is small compared to that radiated by the two radiating slots such that it is usually neglected. Therefore they are referred to as *nonradiating* slots.

14.2.3 Directivity

As for every other antenna, the directivity is one of the most important figures-of-merit whose definition is given by (2-16a) or

$$D_0 = \frac{U_{max}}{U_0} = \frac{4\pi U_{max}}{P_{rad}} \qquad (14\text{-}50)$$

Single Slot ($k_0 h \ll 1$) Using the electric field of (14-41), the maximum radiation intensity and radiated power can be written, respectively, as

$$U_{max} = \frac{|V_0|^2}{2\eta_0 \pi^2} \left(\frac{\pi W}{\lambda_0} \right)^2 \qquad (14\text{-}51)$$

$$P_{rad} = \frac{|V_0|^2}{2\eta_0 \pi} \int_0^\pi \left[\frac{\sin\left(\frac{k_0 W}{2} \cos\theta\right)}{\cos\theta} \right]^2 \sin^3\theta \, d\theta \qquad (14\text{-}52)$$

Therefore, the directivity of a single slot can be expressed as

$$D_0 = \left(\frac{2\pi W}{\lambda_0} \right)^2 \frac{1}{I_1} \qquad (14\text{-}53)$$

where

$$I_1 = \int_0^\pi \left[\frac{\sin\left(\frac{k_0 W}{2} \cos\theta\right)}{\cos\theta} \right]^2 \sin^3\theta \, d\theta$$

$$= \left[-2 + \cos(X) + X S_i(X) + \frac{\sin(X)}{X} \right] \qquad (14\text{-}53a)$$

$$X = k_0 W \qquad (14\text{-}53b)$$

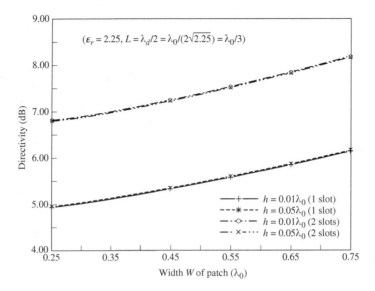

Figure 14.22 Computed directivity of one and two slots as a function of the slot width.

Asymptotically the values of (14-53) vary as

$$D_0 = \begin{cases} 3.3 \text{(dimensionless)} = 5.2 \text{ dB} & W \ll \lambda_0 \\ 4\left(\dfrac{W}{\lambda_0}\right) & W \gg \lambda_0 \end{cases} \quad (14\text{-}54)$$

The directivity of a single slot can be computed using (14-53) and (14-53a). In addition, it can also be computed using (14-41) and the computer program **Directivity** of Chapter 2. Since both are based on the same formulas, they should give the same results. Plots of the directivity of a single slot for $h = 0.01\lambda_0$ and $0.05\lambda_0$ as a function of the width of the slot are shown in Figure 14.22. It is evident that the directivity of a single slot is not influenced strongly by the height of the substrate, as long as it is maintained electrically small.

Two Slots ($k_0 h \ll 1$) For two slots, using (14-44), the directivity can be written as

$$D_2 = \left(\frac{2\pi W}{\lambda_0}\right)^2 \frac{\pi}{I_2} = \frac{2}{15 G_{\text{rad}}}\left(\frac{W}{\lambda_0}\right)^2 \quad (14\text{-}55)$$

where G_{rad} is the radiation conductance and

$$I_2 = \int_0^\pi \int_0^\pi \left[\frac{\sin\left(\dfrac{k_0 W}{2}\cos\theta\right)}{\cos\theta}\right]^2 \sin^3\theta \cos^2\left(\frac{k_0 L_e}{2}\sin\theta\sin\phi\right) d\theta\, d\phi \quad (14\text{-}55\text{a})$$

The total broadside directivity D_2 for the two radiating slots, separated by the dominant TM_{010}^x mode field (antisymmetric voltage distribution), can also be written as [8], [83]

$$D_2 = D_0 D_{AF} = D_0 \frac{2}{1 + g_{12}} \quad (14\text{-}56)$$

$$D_{AF} = \frac{2}{1 + g_{12}} \overset{g_{12} \ll 1}{\simeq} 2 \quad (14\text{-}56a)$$

where

D_0 = directivity of single slot [as given by (14-53) and (14-53a)]
D_{AF} = directivity of array factor AF

$$\left[AF = \cos\left(\frac{k_0 L_e}{2} \sin\theta \sin\phi \right) \right]$$

g_{12} = normalized mutual conductance = G_{12}/G_1

This can also be justified using the array theory of Chapter 6. The normalized mutual conductance g_{12} can be obtained using (14-12), (14-12a), and (14-18a). Computed values based on (14-18a) show that usually $g_{12} \ll 1$; thus (14-56a) is usually a good approximation to (14-56).

Asymptotically the directivity of two slots (microstrip antenna) can be expressed as

$$D_2 = \begin{cases} 6.6 \text{(dimensionless)} = 8.2 \text{ dB} & W \ll \lambda_0 \\ 8\left(\dfrac{W}{\lambda_0}\right) & W \gg \lambda_0 \end{cases} \quad (14\text{-}57)$$

The directivity of the microstrip antenna can now be computed using (14-55) and (14-55a). In addition, it can also be computed using (14-44) and the computer program **Directivity** of Chapter 2. Since they are based on the same formulas, they should give the same results. Plots of directivity of a microstrip antenna, modeled by two slots, for $h = 0.01\lambda_0$ and $0.05\lambda_0$ are shown plotted as a function of the width of the patch (W/λ_0) in Figure 14.22. It is evident that the directivity is not a strong function of the height, as long as the height is maintained electrically small. About 2 dB difference is indicated between the directivity of one and two slots. A typical plot of the directivity of a patch for a fixed resonant frequency as a function of the substrate height (h/λ_0), for two different dielectrics, is shown in Figure 14.23.

The directivity of the slots also can be approximated by Kraus's, (2-26), and Tai & Pereira's, (2-30a), formulas in terms of the E- and H-plane beamwidths, which can be approximated by [36]

$$\Theta_E \simeq 2 \sin^{-1} \sqrt{\frac{7.03\lambda_0^2}{4(3L_e^2 + h^2)\pi^2}} \quad (14\text{-}58)$$

$$\Theta_H \simeq 2 \sin^{-1} \sqrt{\frac{1}{2 + k_0 W}} \quad (14\text{-}59)$$

The values of the directivities obtained using (14-58) and (14-59) along with either (2-26) or (2-30a) will not be very accurate since the beamwidths, especially in the E-plane, are very large. However, they can serve as guidelines.

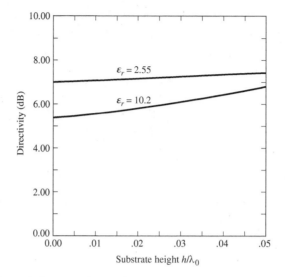

Figure 14.23 Directivity variations as a function of substrate height for a square microstrip patch antenna. (Courtesy of D. M. Pozar).

Example 14.3

For the rectangular microstrip antenna of Examples 14.1 and 14.2, with overall dimensions of $L = 0.906$ cm and $W = 1.186$ cm, substrate height $h = 0.1588$ cm, and dielectric constant of $\varepsilon_r = 2.2$, center frequency of 10 GHz, find the directivity based on (14-56) and (14-56a). Compare with the values obtained using (14-55) and (14-55a).

Solution: From the solution of Example 14.2

$$G_1 = 0.00157 \text{ Siemens}$$
$$G_{12} = 6.1683 \times 10^{-4} \text{ Siemens}$$
$$g_{12} = G_{12}/G_1 = 0.3921$$

Using (14-56a)

$$D_{AF} = \frac{2}{1+g_{12}} = \frac{2}{1+0.3921} = 1.4367 = 1.5736 \text{ dB}$$

Using (14-53) and (14-53a)

$$I_1 = 1.863$$
$$D_0 = \left(\frac{2\pi W}{\lambda_0}\right)^2 \frac{1}{I_1} = 3.312 = 5.201 \text{ dB}$$

According to (14-56)

$$D_2 = D_0 D_{AF} = 3.312(1.4367) = 4.7584 = 6.7746 \text{ dB}$$

Using (14-55a)

$$I_2 = 3.59801$$

Finally, using (14-55)

$$D_2 = \left(\frac{2\pi W}{\lambda_0}\right)^2 \frac{\pi}{I_2} = 5.3873 \text{ (dimensionless)} = 7.314 \text{ dB}$$

There is about 0.5 dB difference between the directivities computed using (14-55) and (14-56); the one based on (14-55) is more accurate.

A MATLAB and FORTRAN computer program, designated as **Microstrip**, has been developed to design and compute the radiation characteristics of rectangular and circular microstrip patch antennas. The description of the program is found in the corresponding READ ME file included in the publisher's website for this book.

14.3 CIRCULAR PATCH

Other than the rectangular patch, the next most popular configuration is the circular patch or disk, as shown in Figure 14.24. It also has received a lot of attention not only as a single element [6], [10], [13], [46], [47], [51], but also in arrays [65] and [74]. The modes supported by the circular patch antenna can be found by treating the patch, ground plane, and the material between the two as a circular cavity. As with the rectangular patch, the modes that are supported primarily by a circular microstrip antenna whose substrate height is small ($h \ll \lambda$) are TM^z where z is taken perpendicular to the patch. As far as the dimensions of the patch, there are two degrees of freedom to control (length and width) for the rectangular microstrip antenna. Therefore the order of the modes can be changed by changing the relative dimensions of the width and length of the patch (width-to-length ratio). However, for the circular patch there is only one degree of freedom to control (radius of the patch). Doing this does not change the order of the modes; however, it does change the absolute value of the resonant frequency of each [79].

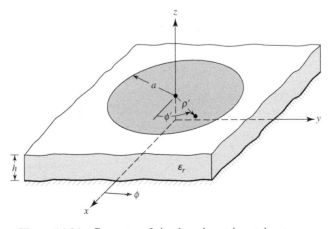

Figure 14.24 Geometry of circular microstrip patch antenna.

816 MICROSTRIP AND MOBILE COMMUNICATIONS ANTENNAS

Other than using full-wave analysis [51], [65], [74], the circular patch antenna can only be analyzed conveniently using the cavity model [10], [46], [47]. This can be accomplished using a procedure similar to that for the rectangular patch but now using cylindrical coordinates [79]. The cavity is composed of two perfect electric conductors at the top and bottom to represent the patch and the ground plane, and by a cylindrical perfect magnetic conductor around the circular periphery of the cavity. The dielectric material of the substrate is assumed to be truncated beyond the extent of the patch.

14.3.1 Electric and Magnetic Fields — TM^z_{mnp}

To find the fields within the cavity, we use the vector potential approach. For TM^z we need to first find the magnetic vector potential A_z, which must satisfy, in cylindrical coordinates, the homogeneous wave equation of

$$\nabla^2 A_z(\rho,\phi,z) + k^2 A_z(\rho,\phi,z) = 0. \tag{14-60}$$

It can be shown that for TM^z modes, whose electric and magnetic fields are related to the vector potential A_z by [79]

$$E_\rho = -j\frac{1}{\omega\mu\varepsilon}\frac{\partial^2 A_z}{\partial\rho\partial z} \qquad H_\rho = \frac{1}{\mu}\frac{1}{\rho}\frac{\partial A_z}{\partial \phi}$$

$$E_\phi = -j\frac{1}{\omega\mu\varepsilon}\frac{1}{\rho}\frac{\partial^2 A_z}{\partial\phi\partial z} \qquad H_\phi = -\frac{1}{\mu}\frac{\partial A_z}{\partial \rho} \tag{14-61}$$

$$E_z = -j\frac{1}{\omega\mu\varepsilon}\left(\frac{\partial^2}{\partial z^2} + k^2\right) A_z \qquad H_z = 0$$

subject to the boundary conditions of

$$E_\rho(0 \le \rho' \le a, 0 \le \phi' \le 2\pi, z' = 0) = 0$$
$$E_\rho(0 \le \rho' \le a, 0 \le \phi' \le 2\pi, z' = h) = 0$$
$$H_\phi(\rho' = a, 0 \le \phi' \le 2\pi, 0 \le z' \le h) = 0 \tag{14-62}$$

the magnetic vector potential A_z reduces to [79]

$$A_z = B_{mnp} J_m(k_\rho \rho')[A_2 \cos(m\phi') + B_2 \sin(m\phi')]\cos(k_z z') \tag{14-63}$$

with the constraint equation of

$$(k_\rho)^2 + (k_z)^2 = k_r^2 = \omega_r^2 \mu\varepsilon \tag{14-63a}$$

The primed cylindrical coordinates ρ', ϕ', z' are used to represent the fields within the cavity while $J_m(x)$ is the Bessel function of the first kind of order m, and

$$k_\rho = \chi'_{mn}/a \tag{14-63b}$$

$$k_z = \frac{p\pi}{h} \tag{14-63c}$$

$$m = 0, 1, 2, \ldots \tag{14-63d}$$

$$n = 1, 2, 3, \ldots \tag{14-63e}$$

$$p = 0, 1, 2, \ldots \tag{14-63f}$$

In (14-63b) χ'_{mn} represents the zeroes of the derivative of the Bessel function $J_m(x)$, and they determine the order of the resonant frequencies. The first four values of χ'_{mn}, in ascending order, are

$$\begin{aligned} \chi'_{11} &= 1.8412 \\ \chi'_{21} &= 3.0542 \\ \chi'_{01} &= 3.8318 \\ \chi'_{31} &= 4.2012 \end{aligned} \tag{14-64}$$

14.3.2 Resonant Frequencies

The resonant frequencies of the cavity, and thus of the microstrip antenna, are found using (14-63a)–(14-63f). Since for most typical microstrip antennas the substrate height h is very small (typically $h < 0.05\lambda_0$), the fields along z are essentially constant and are presented in (14-63f) by $p = 0$ and in (14-63c) by $k_z = 0$. Therefore the resonant frequencies for the TM^z_{mn0} modes can be written using (14-63a) as

$$(f_r)_{mn0} = \frac{1}{2\pi\sqrt{\mu\varepsilon}} \left(\frac{\chi'_{mn}}{a} \right) \tag{14-65}$$

Based on the values of (14-64), the first four modes, in ascending order, are TM^z_{110}, TM^z_{210}, TM^z_{010}, and TM^z_{310}. The dominant mode is the TM^z_{110} whose resonant frequency is

$$(f_r)_{110} = \frac{1.8412}{2\pi a\sqrt{\mu\varepsilon}} = \frac{1.8412 v_0}{2\pi a\sqrt{\varepsilon_r}} \tag{14-66}$$

where v_0 is the speed of light in free-space.

The resonant frequency of (14-66) does not take into account fringing. As was shown for the rectangular patch, and illustrated in Figure 14.7, fringing makes the patch look electrically larger and it was taken into account by introducing a length correction factor given by (14-2). Similarly for the circular patch a correction is introduced by using an effective radius a_e, to replace the actual radius a, given by [6]

$$a_e = a \left\{ 1 + \frac{2h}{\pi a \varepsilon_r} \left[\ln\left(\frac{\pi a}{2h}\right) + 1.7726 \right] \right\}^{1/2} \tag{14-67}$$

Therefore the resonant frequency of (14-66) for the dominant TM^z_{110} should be modified by using (14-67) and expressed as

$$(f_{rc})_{110} = \frac{1.8412 v_0}{2\pi a_e \sqrt{\varepsilon_r}} = \frac{8.791 \times 10^9}{a_e \sqrt{\varepsilon_r}} \Rightarrow a_e = \frac{8.791 \times 10^9}{f_{rc} \sqrt{\varepsilon_r}} \quad (14\text{-}68)$$

14.3.3 Design

Based on the cavity model formulation, a design procedure is outlined which leads to practical designs of circular microstrip antennas for the dominant TM^z_{110} mode. The procedure assumes that the specified information includes the dielectric constant of the substrate (ε_r), the resonant frequency (f_r) and the height of the substrate h. The procedure is as follows:

Specify

$$\varepsilon_r, f_r (\text{in Hz}), \text{ and } h \text{ (in cm)}$$

Determine The actual radius a of the patch.

Design Procedure A first-order approximation to the solution of (14-67) for a is to find a_e. This is accomplished by substituting (14-68) for a_e on the left side of (14-67) and for a only within the brackets {..} on the right side of (14-67). Doing this and solving for a leads to

$$a = \frac{F}{\left\{ 1 + \frac{2h}{\pi \varepsilon_r F} \left[\ln\left(\frac{\pi F}{2h}\right) + 1.7726 \right] \right\}^{1/2}} \quad (14\text{-}69)$$

where

$$F = \frac{8.791 \times 10^9}{f_r \sqrt{\varepsilon_r}} \quad (14\text{-}69\text{a})$$

Remember that h in (14-69) must be in cm.

Example 14.4

Design a circular microstrip antenna using a substrate (RT/duroid 5880) with a dielectric constant of 2.2, $h = 0.1588$ cm (0.0625 in.) so as to resonate at 10 GHz.
  Solution: Using (14-69a)

$$F = \frac{8.791 \times 10^9}{10 \times 10^9 \sqrt{2.2}} = 0.593$$

Therefore using (14-69)

$$a = \frac{F}{\left\{ 1 + \frac{2h}{\pi \varepsilon_r F} \left[\ln\left(\frac{\pi F}{2h}\right) + 1.7726 \right] \right\}^{1/2}} = 0.525 \text{ cm } (0.207 \text{ in.})$$

An experimental circular patch based on this design was built and tested. It is probe fed from underneath by a coaxial line and is shown in Figure 14.8(b). Its 3-D, S_{11} and principal E- and H-plane patterns are displayed in Figure 14.25, where they are compared with simulations, measurements and cavity model. A very good agreement is indicated.

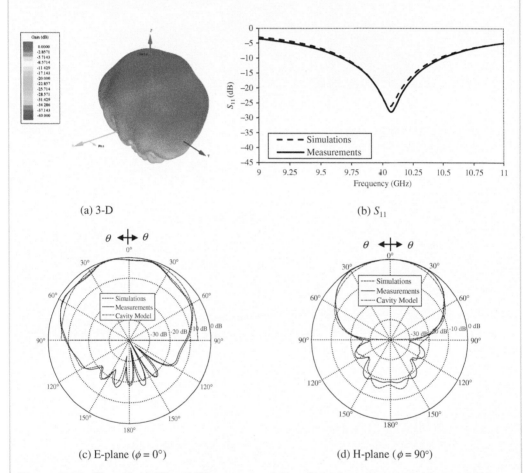

Figure 14.25 Normalized 3D and 2D patterns and S_{11} of circular microstrip patch ($a = 0.525$ cm, $a_e = 0.598$ cm, $h = 0.1588$ cm, $\rho_f = 0.2$ cm, $\varepsilon_r = 2.2, f_o = 10$ GHz).

14.3.4 Equivalent Current Densities and Fields Radiated

As was done for the rectangular patch using the cavity model, the fields radiated by the circular patch can be found by using the Equivalence Principle whereby the circumferential wall of the cavity is replaced by an equivalent magnetic current density of (14-38) as shown in Figure 14.26. Based on (14-61)–(14-63) and assuming a TM_{110}^z mode field distribution beneath the patch, the normalized electric and magnetic fields within the cavity for the cosine azimuthal variations can be written as

$$E_\rho = E_\phi = H_z = 0 \tag{14-70a}$$

$$E_z = E_0 J_1(k\rho') \cos \phi' \tag{14-70b}$$

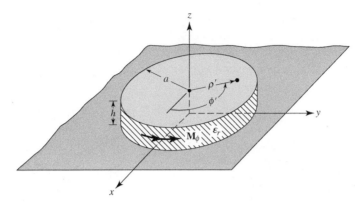

Figure 14.26 Cavity model and equivalent magnetic current density for circular microstrip patch antenna.

$$H_\rho = j\frac{E_0}{\omega\mu_0}\frac{1}{\rho}J_1(k\rho')\sin\phi' \tag{14-70c}$$

$$H_\phi = j\frac{E_0}{\omega\mu_0}J_1'(k\rho')\cos\phi' \tag{14-70d}$$

where $' = \partial/\partial\rho$ and ϕ' is the azimuthal angle along the perimeter of the patch.

Based on (14-70b) evaluated at the electrical equivalent edge of the disk ($\rho' = a_e$), the magnetic current density of (14-38) can be written as

$$\mathbf{M}_s = -2\hat{\mathbf{n}} \times \mathbf{E}_a|_{\rho'=a_e} = \hat{\mathbf{a}}_\phi 2E_0 J_1(ka_e)\cos\phi' \tag{14-71}$$

Since the height of the substrate is very small and the current density of (14-71) is uniform along the z direction, we can approximate (14-71) by a filamentary magnetic current of

$$\mathbf{I}_m = h\mathbf{M}_s = \hat{\mathbf{a}}_\phi 2hE_0 J_1(ka_e)\cos\phi' = \hat{\mathbf{a}}_\phi 2V_0\cos\phi' \tag{14-71a}$$

where $V_0 = hE_0 J_1(ka_e)$ at $\phi' = 0$.

Using (14-71a) the microstrip antenna can be treated as a circular loop. Referring to Chapter 5 for the loop and using the radiation equations of Sections 12.3 and 12.6, we can write that [10], [83]

$$E_r = 0 \tag{14-72a}$$

$$E_\theta = -j\frac{k_0 a_e V_0 e^{-jk_0 r}}{2r}\{\cos\phi J_{02}'\} \tag{14-72b}$$

$$E_\phi = j\frac{k_0 a_e V_0 e^{-jk_0 r}}{2r}\{\cos\theta \sin\phi J_{02}\} \tag{14-72c}$$

$$J_{02}' = J_0(k_0 a_e \sin\theta) - J_2(k_0 a_e \sin\theta) \tag{14-72d}$$

$$J_{02} = J_0(k_0 a_e \sin\theta) + J_2(k_0 a_e \sin\theta) \tag{14-72e}$$

where a_e is the effective radius as given by (14-67). The fields in the principal planes reduce to:

E-plane ($\phi = 0°, 180°, 0° \leq \theta \leq 90°$)

$$E_\theta = j\frac{k_0 a_e V_0 e^{-jk_0 r}}{2r}[J'_{02}] \tag{14-73a}$$

$$E_\phi = 0 \tag{14-73b}$$

H-plane ($\phi = 90°, 270°, 0° \leq \theta \leq 90°$)

$$E_\theta = 0 \tag{14-74a}$$

$$E_\phi = j\frac{k_0 a_e V_0 e^{-jk_0 r}}{2r}[\cos\theta J_{02}] \tag{14-74b}$$

Patterns have been computed for the circular patch of Example 14.4, Figure 14.8(b) based on (14-73a)–(14-74b), and they are shown in Figure 14.25 where they are compared with measurements and simulated patterns. The noted asymmetry in the measured and simulated patterns is due to the feed which is not symmetrically positioned along the E-plane. The simulations accounts for the position of the feed, while the cavity model does not account for it. The pattern for the left half of Figure 14.25(c) corresponds to observation angles which lie on the same side of the patch as does the feed probe. The ground plane was 10 cm × 10 cm.

14.3.5 Conductance and Directivity

The conductance due to the radiated power and directivity of the circular microstrip patch antenna can be computed using their respective definitions of (14-10) and (14-50). For each we need the radiated power, which based on the fields of (14-72b) and (14-72c) of the cavity model can be expressed as

$$P_{\text{rad}} = |V_0|^2 \frac{(k_0 a_e)^2}{960} \int_0^{\pi/2} [J'^2_{02} + \cos^2\theta J^2_{02}] \sin\theta \, d\theta \tag{14-75}$$

Therefore the conductance across the gap between the patch and the ground plane at $\phi' = 0°$ based on (14-10) and (14-75) can be written as

$$G_{\text{rad}} = \frac{(k_0 a_e)^2}{480} \int_0^{\pi/2} [J'^2_{02} + \cos^2\theta J^2_{02}] \sin\theta \, d\theta \tag{14-76}$$

A plot of the conductance of (14-76) for the TM^z_{110} mode is shown in Figure 14.27. While the conductance of (14-76) accounts for the losses due to radiation, it does not take into account losses due to conduction (ohmic) and dielectric losses, which each can be expressed as [10]

$$G_c = \frac{\varepsilon_{mo}\pi(\pi\mu_0 f_r)^{-3/2}}{4h^2\sqrt{\sigma}}[(ka_e)^2 - m^2] \tag{14-77}$$

$$G_d = \frac{\varepsilon_{mo}\tan\delta}{4\mu_0 h f_r}[(ka_e)^2 - m^2] \tag{14-78}$$

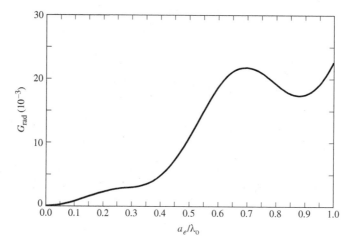

Figure 14.27 Radiation conductance versus effective radius for circular microstrip patch operating in dominant TM^z_{110} mode.

where $\varepsilon_{mo} = 2$ for $m = 0$, $\varepsilon_{mo} = 1$ for $m \neq 0$, and f_r represents the resonant frequency of the $mn0$ mode. Thus, the total conductance can be written as

$$G_t = G_{rad} + G_c + G_d \qquad (14\text{-}79)$$

Based on (14-50), (14-72b), (14-72c), (14-75) and (14-76), the directivity for the slot at $\theta = 0°$ can be expressed as

$$D_0 = \frac{(k_0 a_e)^2}{120 G_{rad}} \qquad (14\text{-}80)$$

A plot of the directivity of the dominant TM^z_{110} mode as a function of the radius of the disk is shown plotted in Figure 14.28. For very small values of the radius the directivity approaches 3 (4.8 dB), which is equivalent of that of a slot above a ground plane and it agrees with the value of (14-54) for $W \ll \lambda_0$.

14.3.6 Resonant Input Resistance

As was the case for the rectangular patch antenna, the input impedance of a circular patch at resonance is real. The input power is independent of the feed-point position along the circumference. Taken the reference of the feed at $\phi' = 0°$, the input resistance at any radial distance $\rho' = \rho_0$ from the center of the patch, for the dominant TM_{11} mode (the one that does not have a zero in the amplitude pattern normal to the patch), can be written as

$$R_{in}(\rho' = \rho_0) = \frac{1}{G_t} \frac{J_1^2(k\rho_0)}{J_1^2(ka_e)} \qquad (14\text{-}81)$$

where G_t is the total conductance due to radiation, conduction (ohmic) and dielectric losses, as given by (14-79). As was the case with the rectangular patch, the resonant input resistance of a circular

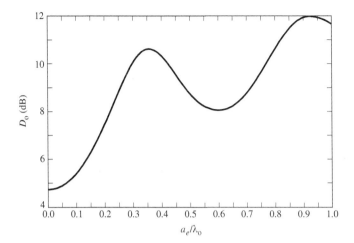

Figure 14.28 Directivity versus effective radius for circular microstrip patch antenna operating in dominant TM_{110}^z mode.

patch with an inset feed, which is usually a probe, can be written as

$$R_{in}(\rho' = \rho_0) = R_{in}(\rho' = a_e)\frac{J_1^2(k\rho_0)}{J_1^2(ka_e)} \tag{14-82}$$

$$R_{in}(\rho' = a_e) = \frac{1}{G_t} \tag{14-82a}$$

where G_t is given by (14-79). This is analogous to (14-20a) for the rectangular patch.

A MATLAB and FORTRAN computer program, designated as **Microstrip**, has been developed to design and compute the radiation characteristics of rectangular and circular microstrip patch antennas. The description of the program is found in the corresponding READ ME file included in the publisher's website for this book.

14.4 QUALITY FACTOR, BANDWIDTH, AND EFFICIENCY

The quality factor, bandwidth, and efficiency are antenna figures-of-merit, which are interrelated, and there is no complete freedom to independently optimize each one. Therefore there is always a trade-off between them in arriving at an optimum antenna performance. Often, however, there is a desire to optimize one of them while reducing the performance of the other.

The quality factor is a figure-of-merit that is representative of the antenna losses. Typically there are radiation, conduction (ohmic), dielectric and surface wave losses. Therefore the total quality factor Q_t is influenced by all of these losses and is, in general, written as [16]

$$\frac{1}{Q_t} = \frac{1}{Q_{rad}} + \frac{1}{Q_c} + \frac{1}{Q_d} + \frac{1}{Q_{sw}} \tag{14-83}$$

where

Q_t = total quality factor
Q_{rad} = quality factor due to radiation (space wave) losses

Q_c = quality factor due to conduction (ohmic) losses
Q_d = quality factor due to dielectric losses
Q_{sw} = quality factor due to surface waves

For very thin substrates, the losses due to surface waves are very small and can be neglected. However, for thicker substrates they need to be taken into account [84]. These losses can also be eliminated by using cavities [50] and [51].

For very thin substrates ($h \ll \lambda_0$) of arbitrary shapes (including rectangular and circular), there are approximate formulas to represent the quality factors of the various losses [16], [85]. These can be expressed as

$$Q_c = h\sqrt{\pi f \mu \sigma} \quad (14\text{-}84)$$

$$Q_d = \frac{1}{\tan \delta} \quad (14\text{-}85)$$

$$Q_{rad} = \frac{2\omega \varepsilon_r}{hG_t/l} K \quad (14\text{-}86)$$

where $\tan \delta$ is the loss tangent of the substrate material, σ is the conductivity of the conductors associated with the patch and ground plane, G_t/l is the total conductance per unit length of the radiating aperture and

$$K = \frac{\iint_{\text{area}} |E|^2 \, dA}{\oint_{\text{perimeter}} |E|^2 \, dl} \quad (14\text{-}86a)$$

For a rectangular aperture operating in the dominant TM_{010}^x mode

$$K = \frac{L}{4} \quad (14\text{-}87a)$$

$$G_t/l = \frac{G_{rad}}{W} \quad (14\text{-}87b)$$

The Q_{rad} as represented by (14-86) is inversely proportional to the height of the substrate, and for very thin substrates is usually the dominant factor.

The fractional bandwidth of the antenna is inversely proportional to the Q_t of the antenna, and it is defined by (11-36) or

$$\frac{\Delta f}{f_0} = \frac{1}{Q_t} \quad (14\text{-}88)$$

However, (14-88) may not be as useful because it does not take into account impedance matching at the input terminals of the antenna. A more meaningful definition of the fractional bandwidth is over a band of frequencies where the VSWR at the input terminals is equal to or less than a desired maximum value, assuming that the VSWR is unity at the design frequency. A modified form of (14-88) that takes into account the impedance matching is [16]

$$\frac{\Delta f}{f_0} = \frac{\text{VSWR} - 1}{Q_t \sqrt{\text{VSWR}}} \qquad (14\text{-}88a)$$

In general the bandwidth is proportional to the volume, which for a rectangular microstrip antenna at a constant resonant frequency can be expressed as

$$\text{BW} \sim \text{volume} = \text{area} \cdot \text{height} = \text{length} \cdot \text{width} \cdot \text{height}$$

$$\sim \frac{1}{\sqrt{\varepsilon_r}} \frac{1}{\sqrt{\varepsilon_r}} \sqrt{\varepsilon_r} = \frac{1}{\sqrt{\varepsilon_r}} \qquad (14\text{-}89)$$

An approximate expression for the bandwidth (for VSWR ≤ 2, $|\Gamma| \leq 1/3$), provided that the surface wave power is much smaller than the radiated (space-wave) power [82], is

$$\text{BW} = \frac{16}{3\sqrt{2}} \left[\frac{\varepsilon_r - 1}{(\varepsilon_r)^2} \right] \frac{h}{\lambda_o} \left(\frac{W}{L} \right) = 3.771 \left[\frac{\varepsilon_r - 1}{(\varepsilon_r)^2} \right] \frac{h}{\lambda_o} \left(\frac{W}{L} \right) \qquad (14\text{-}89a)$$

The expression is valid for $h \ll \lambda_o$ as it is derived [82] based on approximations of rigorously developed formulas, including Sommerfeld integrals for thin grounded substrates.

Therefore, according to (14-89), the bandwidth is inversely proportional to the square root of the dielectric constant of the substrate. A typical variation of the bandwidth for a microstrip antenna as a function of the normalized height of the substrate, for two different substrates, is shown in Figure 14.29. It is evident that the bandwidth increases as the substrate height increases.

The radiation efficiency of an antenna is expressed by (2-90), and it is defined as the power radiated over the input power. It can also be expressed in terms of the quality factors, which for

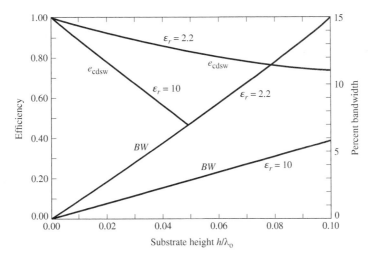

Figure 14.29 Efficiency and bandwidth versus substrate height at constant resonant frequency for rectangular microstrip patch for two different substrates. (SOURCE: D. M. Pozar, "Microstrip Antennas," *Proc. IEEE*, Vol. 80, No. 1, January 1992. © 1992 IEEE).

a microstrip antenna can be written as

$$e_{cdsw} = \frac{1/Q_{rad}}{1/Q_t} = \frac{Q_t}{Q_{rad}} \qquad (14\text{-}90)$$

where Q_t is given by (14-83). Typical variations of the efficiency as a function of the substrate height for a microstrip antenna, with two different substrates, are shown in Figure 14.29.

14.5 INPUT IMPEDANCE

In the previous sections of this chapter, we derived approximate expressions for the resonant input resistance for both rectangular and circular microstrip antennas. Also, approximate expressions were stated which describe the variation of the resonant input resistance as a function of the inset-feed position, which can be used effectively to match the antenna element to the input transmission line. In general, the input impedance is complex and it includes both a resonant and a nonresonant part which is usually reactive. Both the real and imaginary parts of the impedance vary as a function of frequency, and a typical variation is shown in Figure 14.30. Ideally both the resistance and reactance exhibit symmetry about the resonant frequency, and the reactance at resonance is equal to the average of sum of its maximum value (which is positive) and its minimum value (which is negative).

Typically the feed reactance is very small, compared to the resonant resistance, for very thin substrates. However, for thick elements the reactance may be significant and needs to be taken into account in impedance matching and in determining the resonant frequency of a loaded element [34]. The variations of the feed reactance as a function of position can be intuitively explained by considering the cavity model for a rectangular patch with its four side perfect magnetic conducting walls [34], [85]. As far as the impedance is concerned, the magnetic walls can be taken into account by introducing multiple images with current flow in the same direction as the actual feed. When the

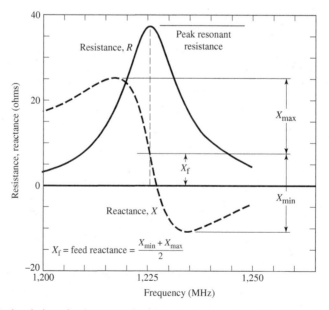

Figure 14.30 Typical variation of resistance and reactance of rectangular microstrip antenna versus frequency (*Electromagnetics*, Vol. 3, Nos. 3 and 4, p. 33, W. F. Richards, J. R. Zinecker, and R. D. Clark, Taylor & Francis, Washington, D.C. Reproduced by permission. All rights reserved).

feed point is far away from one of the edges, the magnetic field associated with the images and that of the actual feed do not overlap strongly. Therefore the inductance associated with the magnetic energy density stored within a small testing volume near the feed will be primarily due to the current of the actual feed. However, when the feed is at one of the edges, the feed and one of the images, which accounts for the magnetic wall at that edge, coincide. Thus, the associated magnetic field stored energy of the equivalent circuit doubles while the respective stored magnetic energy density quadruples. However, because the volume in the testing region of the patch is only half from that when the feed was far removed from the edge, the net stored magnetic density is only double of that of the feed alone. Thus, the associated inductance and reactance, when the feed is at the edge, is twice that when the feed is far removed from the edge. When the feed is at a corner, there will be three images in the testing volume of the patch, in addition to the actual feed, to take into account the edges that form the corner. Using the same argument as above, the associated inductance and reactance for a feed at a corner is four times that when the feed is removed from an edge or a corner. Thus, the largest reactance (about a factor of four larger) is when the feed is at or near a corner while the smallest is when the feed is far removed from an edge or a corner.

Although such an argument predicts the relative variations (trends) of the reactance as a function of position, they do predict very accurately the absolute values especially when the feed is at or very near an edge. In fact it overestimates the values for feeds right on the edge; the actual values predicted by the cavity model with perfect magnetic conducting walls are smaller [34]. A formula that has been suggested to approximate the feed reactance, which does not take into account any images, is

$$x_f \simeq -\frac{\eta k h}{2\pi}\left[\ln\left(\frac{kd}{4}\right) + 0.577\right] \qquad (14\text{-}91)$$

where d is the diameter of the feed probe. More accurate predictions of the input impedance, based on full-wave models, have been made for circular patches where an attachment current mode is introduced to match the current distribution of the probe to that of the patch [74].

14.6 COUPLING

The coupling between two or more microstrip antenna elements can be taken into account easily using full-wave analyses. However, it is more difficult to do using the transmission-line and cavity models, although successful attempts have been made using the transmission-line model [75] and the cavity model [76], [77]. It can be shown that coupling between two patches, as is coupling between two aperture or two wire antennas, is a function of the position of one element relative to the other. This has been demonstrated in Figure 4.23 for a vertical half-wavelength dipole above a ground plane and in Figure 4.33 for a horizontal half-wavelength dipole above a ground plane. From these two, the ground effects are more pronounced for the horizontal dipole. Also, mutual effects have been discussed in Chapter 8 for the three different arrangements of dipoles, as shown in Figure 8.20 whose side-by-side arrangement exhibits the largest variations of mutual impedance.

For two rectangular microstrip patches the coupling for two side-by-side elements is a function of the relative alignment. When the elements are positioned collinearly along the E-plane, this arrangement is referred to as the *E-plane*, as shown in Figure 14.31(a); when the elements are positioned collinearly along the H-plane, this arrangement is referred to as the *H-plane*, as shown in Figure 14.31(b). For an edge-to-edge separation of s, the E-plane exhibits the smallest coupling isolation for very small spacing (typically $s < 0.10\lambda_0$) while the H plane exhibits the smallest coupling for large spacing (typically $s > 0.10\lambda_0$). The spacing at which one plane coupling overtakes the other one depends on the electrical properties and geometrical dimensions of the microstrip antenna. Typical variations are shown in Figure 14.32.

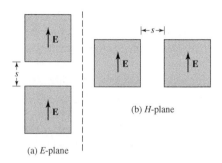

Figure 14.31 *E*- and *H*-plane arrangements of microstrip patch antennas.

In general, mutual coupling is primarily attributed to the fields that exist along the air-dielectric interface. The fields can be decomposed to space waves (with $1/\rho$ radial variations), higher order waves (with $1/\rho^2$ radial variations), surface waves (with $1/\rho^{1/2}$ radial variations), and leaky waves [with $\exp(-\lambda\rho)/\rho^{1/2}$ radial variations] [23], [87]. Because of the spherical radial variation, space ($1/\rho$) and higher order waves ($1/\rho^2$) are most dominant for very small spacing while surface waves, because of their $1/\rho^{1/2}$ radial variations are dominant for large separations. Surface waves exist and propagate within the dielectric, and their excitation is a function of the thickness of the substrate [79]. In a given direction, the lowest order (dominant) surface wave mode is TM(odd) with zero cutoff frequency followed by a TE(even), and alternatively by TM(odd) and TE(even) modes. For a rectangular microstrip patch, the fields are TM in a direction of propagation along the *E*-plane and TE in a direction of propagation along the *H*-plane. Since for the *E*-plane arrangement of Figure 14.31(a) the elements are placed collinearly along the *E*-plane where the fields in the space between the elements are primarily TM, there is a stronger surface wave excitation (based on a single dominant surface wave mode) between the elements, and the coupling is larger. However for the *H*-plane arrangement of Figure 14.31(b), the fields in the space between the elements are primarily TE and there is not a strong dominant mode surface wave excitation; therefore there is less coupling between the elements. This does change as the thickness of the substrate increases which allows higher order TE surface wave excitation.

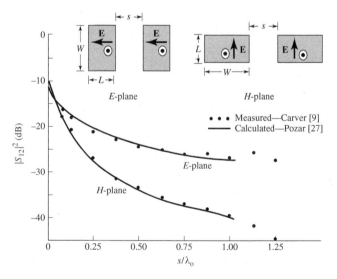

Figure 14.32 Measured and calculated mutual coupling between two coax-fed microstrip antennas, for both *E*-plane and *H*-plane coupling, ($W = 10.57$ cm, $L = 6.55$ cm, $h = 0.1588$ cm, $\varepsilon_r = 2.55$, $f_r = 1{,}410$ MHz). (SOURCE: D. M. Pozar, "Input Impedance and Mutual Coupling of Rectangular Microstrip Antennas," *IEEE Trans. Antennas Propagat.*, Vol. AP-30, No. 6, November 1982. © 1982 IEEE).

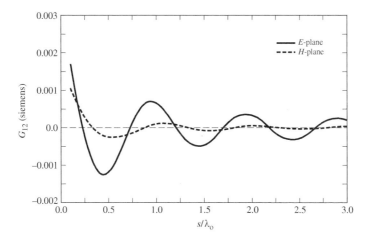

Figure 14.33 E- and H-plane mutual conductance versus patch separation for rectangular microstrip patch antennas ($W = 1.186$ cm, $L = 0.906$ cm, $\varepsilon_r = 2.2, \lambda_0 = 3$ cm).

The mutual conductance between two rectangular microstrip patches has also been found using the basic definition of conductance given by (14-18), the far fields based on the cavity model, and the array theory of Chapter 6. For the E-plane arrangement of Figure 14.31(a) and for the odd mode field distribution beneath the patch, which is representative of the dominant mode, the mutual conductance is [8]

$$G_{12} = \frac{1}{\pi}\sqrt{\frac{\varepsilon}{\mu}} \int_0^\pi \left[\frac{\sin\left(\frac{k_0 W}{2}\cos\theta\right)}{\cos\theta}\right]^2 \sin^3\theta \left\{ 2J_0\left(\frac{Y}{\lambda_0} 2\pi \sin\theta\right) \right.$$

$$\left. + J_0\left(\frac{Y+L}{\lambda_0} 2\pi \sin\theta\right) + J_0\left(\frac{Y-L}{\lambda_0} 2\pi \sin\theta\right) \right\} d\theta \qquad (14\text{-}92)$$

where Y is the center-to-center separation between the slots and J_0 is the Bessel function of the first kind of order zero. The first term in (14-92) represents the mutual conductance of two slots separated by a distance X along the E-plane while the second and third terms represent, respectively, the conductances of two slots separated along the E-plane by distances $Y+L$ and $Y-L$. Typical normalized results are shown by the solid curve in Figure 14.33.

For the H-plane arrangement of Figure 14.31(b) and for the odd mode field distribution beneath the patch, which is representative of the dominant mode, the mutual conductance is [8]

$$G_{12} = \frac{2}{\pi}\sqrt{\frac{\varepsilon}{\mu}} \int_0^\pi \left[\frac{\sin\left(\frac{k_0 W}{2}\cos\theta\right)}{\cos\theta}\right]^2 \sin^3\theta \cos\left(\frac{Z}{\lambda_0} 2\pi \cos\theta\right)$$

$$\cdot \left\{ 1 + J_0\left(\frac{L}{\lambda_0} 2\pi \sin\theta\right) \right\} d\theta \qquad (14\text{-}93)$$

where Z is the center-to-center separation between the slots and J_0 is the Bessel function of the first kind of order zero. The first term in (14-93) represents twice the mutual conductance of two slots separated along the H-plane by a distance Z while the second term represents twice the conductance between two slots separated along the E-plane by a distance L and along the H-plane by a distance Z. Typical normalized results are shown by the dashed curve in Figure 14.31. By comparing the results of Figure 14.31 it is clear that the mutual conductance for the H-plane arrangement, as expected, decreases with distance faster than that of the E-plane. Also it is observed that the mutual conductance for the E-plane arrangement is higher for wider elements while it is lower for wider elements for the H-plane arrangement.

14.7 CIRCULAR POLARIZATION

The patch elements that we discussed so far, both the rectangular and the circular, radiate primarily linearly polarized waves if conventional feeds are used with no modifications. However, circular and elliptical polarizations can be obtained using various feed arrangements or slight modifications made to the elements. We will discuss here some of these arrangements.

Circular polarization can be obtained if two orthogonal modes are excited with a 90° time-phase difference between them. This can be accomplished by adjusting the physical dimensions of the patch and using either single, or two, or more feeds. There have been some suggestions made and reported in the literature using single patches. For a square patch element, the easiest way to excite ideally circular polarization is to feed the element at two adjacent edges, as shown in Figures 14.34(a,b), to excite the two orthogonal modes; the TM_{010}^{x} with the feed at one edge and the TM_{001}^{x} with the feed at the other edge. The quadrature phase difference is obtained by feeding the element with a 90° power divider or 90° hybrid. Examples of arrays of linear elements that generate circular polarization are discussed in [88].

For a circular patch, circular polarization for the TM_{110}^{z} mode is achieved by using two feeds with proper angular separation. An example is shown in Figure 14.34(c) using two coax feeds separated by 90° which generate fields that are orthogonal to each other under the patch, as well as outside the patch. Also with this two-probe arrangement, each probe is always positioned at a point where the field generated by the other probe exhibits a null; therefore there is very little mutual coupling between the two probes. To achieve circular polarization, it is also required that the two feeds are fed in such a manner that there is 90° time-phase difference between the fields of the two; this is achieved through the use of a 90° hybrid, as shown in Figure 14.34(c). The shorting pin is placed at the center of the patch to ground the patch to the ground plane which is not necessary for circular polarization but is used to suppress modes with no ϕ variations and also may improve the quality of circular polarization.

For higher order modes, the spacing between the two feeds to achieve circular polarization is different. This is illustrated in Figure 14.34(d) and tabulated in Table 14.2, for the TM_{110}^{z} [same as in Figure 14.32(c)], TM_{210}^{z}, TM_{310}^{z}, and TM_{410}^{z} modes [89]. However to preserve symmetry and minimize cross polarization, especially for relatively thick substrates, two additional feed probes located diametrically opposite of the original poles are usually recommended. The additional probes are used to suppress the neighboring (adjacent) modes which usually have the next highest magnitudes [88]. For the even modes (TM_{210}^{z} and TM_{410}^{z}), the four feed probes should have phases of 0°, 90°, 0° and 90° while the odd modes (TM_{110}^{z} and TM_{310}^{z}) should have phases of 0°, 90°, 180° and 270°, as shown in Figure 14.34(d) [89].

To overcome the complexities inherent in dual-feed arrangements, circular polarization can also be achieved with a single feed. One way to accomplish this is to feed the patch at a single point to excite two orthogonal degenerate modes (of some resonant frequency) of ideally equal amplitudes. By introducing then a proper asymmetry in the cavity, the degeneracy can be removed with one mode increasing with frequency while the orthogonal mode will be decreasing with frequency by the same amount. Since the two modes will have slightly different frequencies, by proper design the field of one mode can lead by 45° while that of the other can lag by 45° resulting in a 90° phase difference necessary for circular polarization [16]. To achieve this, several arrangements have been suggested. This can be illustrated with a square patch fed as shown in Figure 14.35 [34].

The resonant frequencies f_1 and f_2 of the bandwidth of (14-88a) associated with the two lengths L and W of a rectangular microstrip are [90]

$$f_1 = \frac{f_0}{\sqrt{1 + 1/Q_t}} \tag{14-94a}$$

$$f_2 = f_0 \sqrt{1 + 1/Q_t} \tag{14-94b}$$

where f_0 is the center frequency. Feeding the element along the diagonal starting at the lower left corner toward the upper right corner, shown dashed in Figure 14.35(b), yields ideally left-hand circular polarization at broadside. Right-hand circular polarization can be achieved by feeding along the opposite diagonal, which starts at the lower right corner and proceeds toward the upper left corner, shown dashed in Figure 14.35(c). Instead of moving the feed point each time to change the modes in order to change the type of circular polarization, varactor diodes can be used to adjust the capacitance and bias, which effectively shifts by electrical means the apparent physical location of the feed point.

Example 14.5

The fractional bandwidth at a center frequency of 10 GHz of a rectangular patch antenna whose substrate is RT/duroid 5880 ($\varepsilon_r = 2.2$) with height $h = 0.1588$ cm is about 5% for a VSWR of 2:1. Within that bandwidth, find resonant frequencies associated with the two lengths of the rectangular patch antenna, and the relative ratio of the two lengths.

Solution: The total quality factor Q_t of the patch antenna is found using (14-88a) or

$$Q_t = \frac{1}{0.05\sqrt{2}} = 14.14$$

Using (14-94a) and (14-94b)

$$f_1 = \frac{10 \times 10^9}{\sqrt{1 + 1/14.14}} = 9.664 \text{ GHz}$$

$$f_2 = 10 \times 10^9 \sqrt{1 + 1/14.14} = 10.348 \text{ GHz}$$

The relative ratio of the two lengths from Figure 14.35(a)

$$\frac{L}{W} = 1 + \frac{1}{Q_t} = 1 + \frac{1}{14.14} = 1.07$$

which makes the patch nearly square.

There are some other practical ways of achieving nearly circular polarization. For a square patch, this can be accomplished by cutting very thin slots as shown in Figures 14.36(a,b) with dimensions

$$c = \frac{L}{2.72} = \frac{W}{2.72} \tag{14-95a}$$

$$d = \frac{c}{10} = \frac{L}{27.2} = \frac{W}{27.2} \tag{14-95b}$$

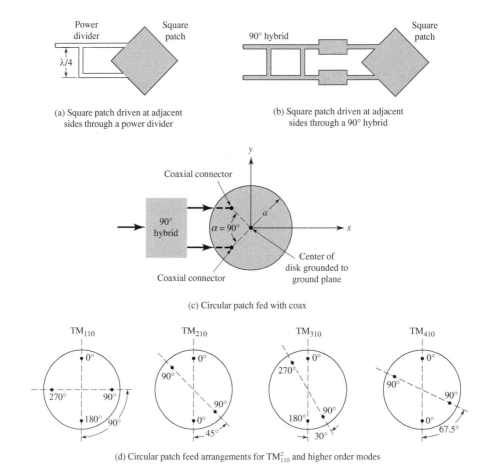

Figure 14.34 Rectangular and circular patch arrangements for circular polarization. (SOURCE: J. Huang, "Circularly Polarized Conical Patterns from Circular Microstrip Antennas," *IEEE Trans. Antennas Propagat.*, Vol. AP-32, No. 9, Sept. 1984. © 1984 IEEE).

An alternative way is to trim the ends of two opposite corners of a square patch and feed at points 1 or 3, as shown in Figure 14.37(a). Circular polarization can also be achieved with a circular patch by making it slightly elliptical or by adding tabs, as shown in Figure 14.37(b).

TABLE 14.2 Feed Probe Angular Spacing of Different Modes for Circular Polarization (after [89])

	TM_{110}	TM_{210}	TM_{310}	TM_{410}	TM_{510}	TM_{610}
α	90°	45° or 135°	30° or 90°	22.5° or 67.5°	18°, 54° or 90°	15°, 45° or 75°

14.8 ARRAYS AND FEED NETWORKS

Microstrip antennas are used not only as single elements but are very popular in arrays [17], [23], [30], [31], [50], [51], [54], [63]–[65], and [74]–[77]. As discussed in Chapter 6, arrays are very versatile and are used, among other things, to synthesize a required pattern that cannot be achieved with a single element. In addition, they are used to scan the beam of an antenna system, increase

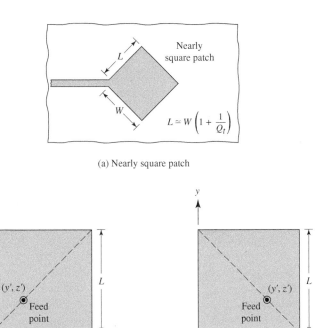

Figure 14.35 Single-feed arrangements for circular polarization of rectangular microstrip patches.

the directivity, and perform various other functions which would be difficult with any one single element. The elements can be fed by a single line, as shown in Figure 14.38(a), or by multiple lines in a feed network arrangement, as shown in Figure 14.38(b). The first is referred to as a *series-feed network* while the second is referred to as a *corporate-feed network*.

The corporate-feed network is used to provide power splits of 2^n (i.e., $n = 2, 4, 8, 16, 32$, etc.). This is accomplished by using either tapered lines, as shown in Figure 14.39(a), to match 100-ohm patch elements to a 50-ohm input or using quarter-wavelength impedance transformers, as shown in Figure 14.39(b) [3]. The design of single- and multiple-section quarter-wavelength impedance transformers is discussed in Section 9.8.

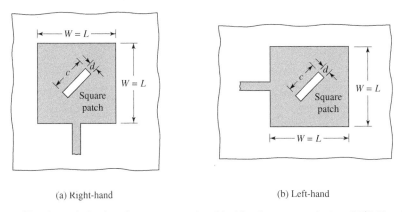

Figure 14.36 Circular polarization for square patch with thin slots on patch ($c = W/2.72 = L/2.72$, $d = c/10 = W/27.2 = L/27.2$).

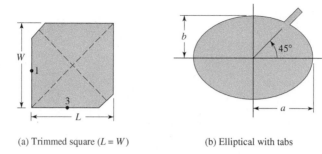

(a) Trimmed square ($L = W$)   (b) Elliptical with tabs

Figure 14.37 Circular polarization by trimming opposite corners of a square patch and by making circular patch slightly elliptical and adding tabs.

Series-fed arrays can be conveniently fabricated using photolithography for both the radiating elements and the feed network. However, this technique is limited to arrays with a fixed beam or those which are scanned by varying the frequency, but it can be applied to linear and planar arrays with single or dual polarization. Also any changes in one of the elements or feed lines affects the performance of the others. Therefore in a design it is important to be able to take into account these and other effects, such as mutual coupling, and internal reflections.

Corporate-fed arrays are general and versatile. With this method the designer has more control of the feed of each element (amplitude and phase) and it is ideal for scanning phased arrays, multi-beam arrays, or shaped-beam arrays. As discussed in Chapter 6, the phase of each element can be controlled using phase shifters while the amplitude can be adjusted using either amplifiers or attenuators. An electronically-steered phased array (ATDRSS) of 10×10 rectangular microstrip elements, operating in the 2–2.3 GHz frequency range and used for space-to-space communications, is shown in Figure 14.40.

Those who have been designing and testing microstrip arrays indicate that radiation from the feed line, using either a series or corporate-feed network, is a serious problem that limits the cross-polarization and side lobe level of the arrays [38]. Both cross-polarization and side lobe levels can be improved by isolating the feed network from the radiating face of the array. This can be accomplished using either probe feeds or aperture coupling.

Arrays can be analyzed using the theory of Chapter 6. However, such an approach does not take into account mutual coupling effects, which for microstrip patches can be significant. Therefore for more accurate results, full-wave solutions must be performed. In microstrip arrays [63], as in any other array [91], mutual coupling between elements can introduce *scan blindness* which limits, for a certain maximum reflection coefficient, the angular volume over which the arrays can be scanned. For microstrip antennas, this scan limitation is strongly influenced by surface waves within the substrate. This scan angular volume can be extended by eliminating surface waves. One way to do this is to use

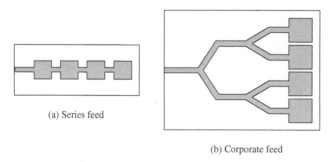

(a) Series feed

(b) Corporate feed

Figure 14.38 Feed arrangements for microstrip patch arrays.

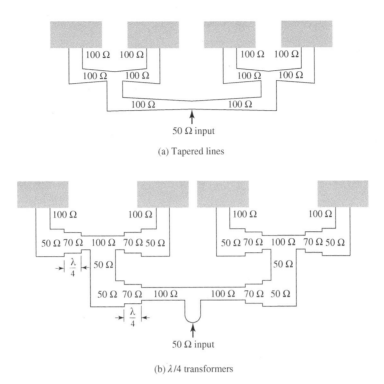

Figure 14.39 Tapered and λ/4 impedance transformer lines to match 100-ohm patches to a 50-ohm line. (SOURCE: R. E. Munson, "Conformal Microstrip Antennas and Microstrip Phased Arrays," *IEEE Trans. Antennas Propagat.*, Vol. AP-22, No. 1, January 1974. © 1974 IEEE).

cavities in conjunction with microstrip elements [50], [51]. Figure 14.41 shows an array of circular patches backed by either circular or rectangular cavities. It has been shown that the presence of cavities, either circular or rectangular, can have a pronounced enhancement in the E-plane scan volume, especially for thicker substrates [51]. The H-plane scan volume is not strongly enhanced. However the shape of the cavity, circular or rectangular, does not strongly influence the results. Typical results for broadside-matched reflection coefficient infinite array of circular patches, with a

Figure 14.40 Antenna array of 10 × 10 rectangular microstrip patches, 2–2.3 GHz, for space-to-space communications. (Courtesy: Ball Aerospace & Technologies Corp.).

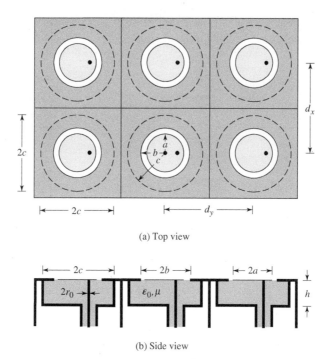

Figure 14.41 Array of circular patches backed by circular cavities. (Courtesy J. T. Aberle and F. Zavosh).

substrate $0.08\lambda_0$ thick and backed by circular and rectangular cavities, are shown in Figure 14.42 for the E-plane and H-plane. The broadside-matched reflection coefficient $\Gamma(\theta, \phi)$ is defined as

$$\Gamma(\theta, \phi) = \frac{Z_{in}(\theta, \phi) - Z_{in}(0, 0)}{Z_{in}(\theta, \phi) + Z_{in}^*(0, 0)} \qquad (14\text{-}96)$$

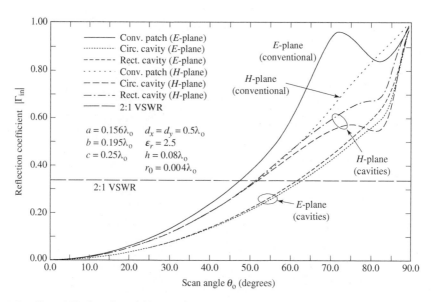

Figure 14.42 E- and H-plane broadside-matched input reflection coefficient versus scan angle for infinite array of circular microstrip patches with and without cavities. (Courtesy J. T. Aberle and F. Zavosh).

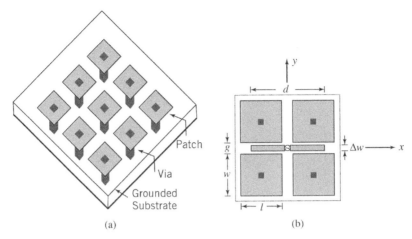

Figure 14.43 Dipole phased array on a PBG surface (SOURCE: [92] © 2004 IEEE). (a) 3 × 3 array. (b) Dipole between 2 × 2 unit cell.

where $Z_{in}(\theta, \phi)$ is the input impedance when the main beam is scanned toward an angle (θ, ϕ). The results are compared with those of a conventional cavity (noncavity backed). It is apparent that there is a significant scan enhancement for the E-plane, especially for a VSWR of about 2:1. H-plane enhancement occurs for reflection coefficients greater than about 0.60. For the conventional array, the E-plane response exhibits a large reflection coefficient, which approaches unity, near a scan angle of $\theta_0 = 72.5°$. This is evidence of scan blindness which ideally occurs when the reflection coefficient is unity, and it is attributed to the coupling between the array elements due to leaky waves [63]. Scan blindness occurs for both the E- and H-planes at grazing incidence ($\theta_0 = 90°$).

Another way to minimize the surface waves, without the use of cavities, is to mount the patches on EBG textured surfaces with vias, as shown in Figure 4.13(a) and Figure 14.43(a) for a 3 × 3 array. A 2 × 2 unit cell of EBG surface, with a dipole in its middle, is displayed in Figure 14.43(b) [92]. Such surfaces have the ability to control and minimize surface waves in substrates. A microstrip dipole element, placed within a textured high-impedance surface, was designed, simulated, fabricated, and measured [92]. It consists of a dipole patch of length 9.766 mm placed in the middle of 4 × 4 EBG unit cells; each cell had dimensions of $w = l = 1.22$ mm and a separation gap between them of $g = 1.66$ mm. The substrate had a dielectric constant of $\varepsilon_r = 2.2$ and a height of $h = 4.771$ mm. The scanned magnitude of the simulated reflection coefficient at 13 GHz of such a design is shown in Figure 14-44. The band-gap frequency range was 9.7–15.1 GHz. The E-plane curves are presented by the E curves, the H-plane by the H curves and the diagonal (45°) plane by the D curves. It is clear that using conventional substrates, the reflection coefficient varies as a function of the scan angle, and in fact creates scan bindness around 50°. However, when the same elements were placed on a textured high-impendance surface, the reflection coefficient was reduced, especially in the E-plane, and the scan blindness was eliminated.

14.9 ANTENNAS FOR MOBILE COMMUNICATIONS

For conventional rectangular microstrips, of the form shown in Figure 14.1(a), their length L has to be slightly less than $\lambda/2$ for them to operate properly. For many applications, especially at lower frequencies and mobile units, this length may be too long. However, the length of the microstrip, whether it is fed by a microstrip line or a probe, can be reduced to basically one-half of its size of the

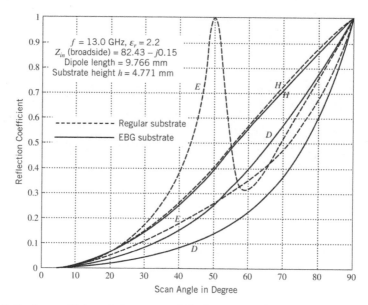

Figure 14.44 Reflection coefficient of phase array of dipoles on a regular and 4 × 4 until cell PBG substrate. (SOURCE: [92] © 2004 IEEE).

conventional design, by using a shorting sheet or pin at its end, as shown in Figure 14.45(a,b), respectively, and they may be referred to as $\lambda/4$ microstrips. Their radiation characteristics are similar to those of conventional $\lambda/2$ microstrip designs.

14.9.1 Planar Inverted-F Antenna (PIFA)

Planar Inverted-F Antenna (PIFA) is a very popular design in mobile communications [93]–[95]. Its name is due to the resemblance of a letter F with its face down in its side view. The intrinsic PIFA is basically a derivative of the $\lambda/4$ rectangular microstrip of Figure 14.45.

Among its advantages are the following:

- Low backward radiation reduces the specific absorption rate (SAR) compared to other antenna types when used in mobile applications.
- Both vertical and horizontal polarizations can be received.

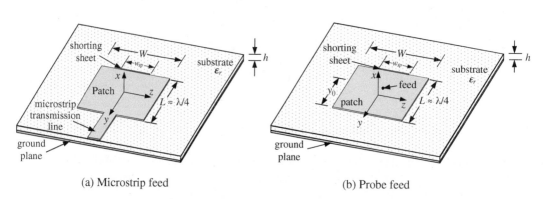

Figure 14.45 Rectangular $\lambda/4$ microstrips.

- It can be easily integrated in mobile devices.
- It is light weight.
- It is conformal.
- It is easy to design.
- It is low cost.
- It is reliable.

To understand its performance, it is important to understand the $\lambda/2$ patch antenna. The current distribution of a $\lambda/2$ antenna has a peak on the center of the patch, and it goes to zero at the edges. The voltage is out of phase from the current, starting from a positive at one edge to the negative at the other edge or vice versa. Therefore, at the center of the $\lambda/2$ patch, the voltage is zero and the current reaches its maximum. The zero voltage at the center of the $\lambda/2$ patch means a short circuit from the patch to the ground plane. If a shorting pin is physically placed at the center of a $\lambda/2$ microstrip patch and half of the antenna is removed, the current and voltage will be basically the same as the ones for a $\lambda/2$ antenna; thus, the antenna will radiate [96]. By reducing the size to $\lambda/4$, the disadvantages, compared to the $\lambda/2$ patch, include a reduction in its bandwidth and gain. The PIFA design is shown in Figure 14.46.

The resonant frequency of the PIFA antenna is determined by the length of the patch L, the width of the patch W, the width of the shorting sheet w_s, and the height of the substrate h. Basically, the height h is small, and it can be neglected. If the height is neglected, the main restriction in the design is a quarter-wavelength distance between the pin and the opposite edge of the patch. Thus, if the shorting sheet extends along the entire width of the patch, the distance L should be equal to quarter wavelength. Alternatively, if the shorting sheet is only a pin, then $L + W \approx \lambda/4$. In summary, the design equation is [96]

$$L + W - w_s = \frac{\lambda}{4} + h \quad \Rightarrow \quad L = -W + w_s + \frac{\lambda}{4} + h \qquad (14\text{-}97)$$

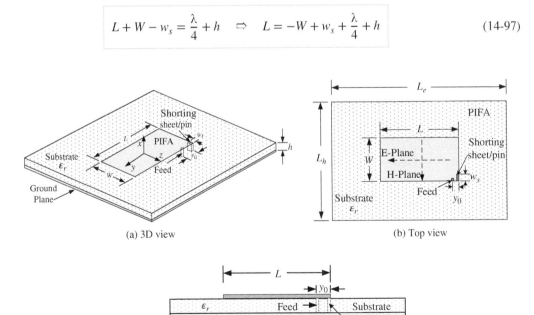

Figure 14.46 Planar Inverted-F Antenna (PIFA) design.

where λ is the wavelength in the dielectric. For the special cases, (14-97) reduces to

$$L = \frac{\lambda}{4} + h \qquad \text{for } w_s = W \qquad (14\text{-}97a)$$

$$L \approx -W + \frac{\lambda}{4} + h \qquad \text{for } w_s \approx 0 \qquad (14\text{-}97b)$$

Equation (14-97) can be solved for the resonant frequency and expressed as

$$\boxed{f = \frac{v_0}{4(L + W - w_s - h)\sqrt{\varepsilon_r}}} \qquad (10\text{-}98)$$

where v_0 is the free-space speed of light and ε_r is the relative permittivity (dielectric constant) of the substrate.

The input impedance of the PIFA is controlled by the distance from the shorting sheet (pin) to the feed point. The magnitude of the input impedance decreases as this distance y_0 decreases, and vice versa. To match the antenna to a 50-ohm transmission line, the feed should be closer to the shorting sheet (pin) than to the open end of the PIFA (i.e., $y_0 < L/2$).

The radiation pattern is a hybrid between a patch antenna and an Inverted-F Antenna (IFA). Several modifications, such as U-slots, can be incorporated to the PIFA to obtain multiple resonances required in mobile communications [97], [98].

Example 14.6

Design a PIFA antenna using an air substrate ($\varepsilon_r = 1$) and height $h = 0.448$ cm to resonate at 1.8 GHz. Assume $W = 0.773L$, $w_s = 0.428W = 0.331L$.

- Determine L, W, w_s (in cm).
- Simulate the design for S_{11}, 3-D, and principal plane 2-D amplitude patterns.

Solution: Using (14-97), write

$$L + 0.773L - 0.331L = \frac{\lambda}{4} + h \Rightarrow L = \frac{1}{1.442}\left(\frac{\lambda}{4} + h\right)$$

$$L = \frac{1}{1.442}\left(\frac{30 \times 10^9}{4\sqrt{1}(1.8 \times 10^9)} + 0.448\right) = \frac{4.6147}{1.442} = 3.2 \text{ cm} = 0.192\lambda$$

$$W = 0.773L = 2.474 \text{ cm} = 0.1484\lambda$$

$$w_s = 0.331L = 1.059 \text{ cm} = 0.0636\lambda$$

A PIFA design, based on this example, was simulated. The S_{11}, 3-D and 2-D E-plane and H-plane patterns are shown in Figures 14.47 and 14.48, respectively.

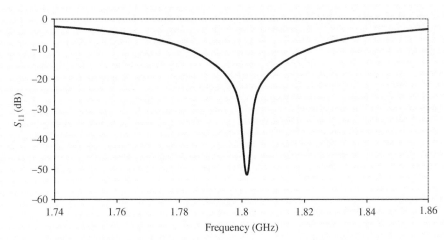

Figure 14.47 S_{11} for a PIFA design: $f = 1.8$ GHz, $h = 0.048$ cm, $L = 0.192\lambda = 3.2$ cm, $W = 0.148\lambda = 2.474$ cm, $w_s = 0.0636\lambda = 1.06$ cm, $y_0 = 0.57$ cm.

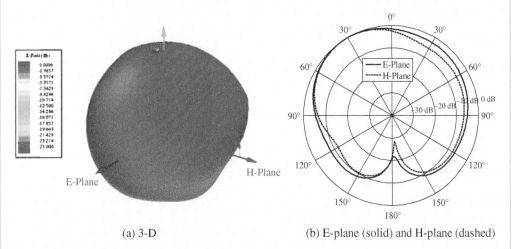

(a) 3-D   (b) E-plane (solid) and H-plane (dashed)

Figure 14.48 Normalized 3-D and 2-D (E-plane; xy and H-plane; yz) radiation patterns for a PIFA design of Example 14.6: $f = 1.8$ GHz, $h = 0.448$ cm, $L = 0.192\lambda = 3.2$ cm, $W = 0.1484\lambda = 2.474$ cm, $w_s = 0.0636\lambda = 1.059$ cm, $y_0 = 0.57$ cm.

14.9.2 Slot Antenna

A slot antenna is a radiating element which typically is formed by cutting an opening on a ground plane, as is indicated in Figures 12.6 and 12.7, and repeated here in Figure 14.49.

Usually, this opening is referred to as a slot instead of an aperture, and the length L in Figure 14.49 should be much longer than the width W. For proper radiation, the length L of the slot should be around half a wavelength ($L \approx \lambda/2$) while the width W typically should be $W \leq (0.05 - 0.1)\lambda$. The slot is usually excited by a voltage source placed symmetrically in the middle of the slot, at $\lambda/4$ from each of its edges. For such symmetrical excitation, the voltage reaches its maximum at the center ($\lambda/4$ from the edges), and it is minimum at the edges, as shown in Figure 14.50(a). In contrast, the current is negative at one edge, reaches zero at the center ($\lambda/4$ from the edges), and it

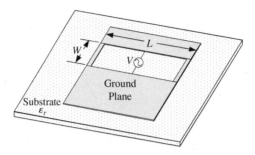

Figure 14.49 Slot antenna with voltage excitation.

is positive at the other edge, as shown in Figure 14.50(b) or vice-versa in polarity. The center of the slot antenna, where the current is zero, can be seen as an open circuit.

A slot antenna is the complement of a dipole type of a radiator, where the arms of the dipole are PEC strips, as shown in Figure 12.24 for a $L = \lambda/2$ length slot and dipole, and where the slot excitation at the center is replaced by that of the dipole. The impedance of the dipole in free space, Z_d, is related to that of the slot in free space, Z_s, by Babinet's principle of Section 12.8, and it is equal to (12-67) or

$$Z_d Z_s = \frac{\eta_o^2}{4} \tag{14-99}$$

where η_o is the free-space intrinsic impedance. Knowing either Z_d or Z_s, you can determine the other by (14-99).

The slot antenna of Figure 14.49 is a bidirectional radiator; that is, it radiates both in the forward and backward directions of the ground plane. Typically, to convert it to a unidirectional radiator in the

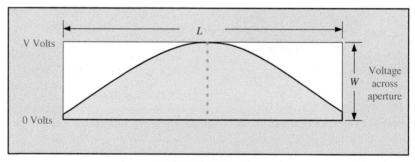

(a) Voltage distribution

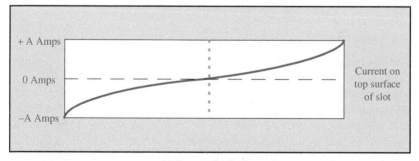

(b) Current distribution

Figure 14.50 Voltage and current distribution of a $L = \lambda/2$ slot antenna.

forward direction, a cavity is inserted in its back side, reducing it to what it is usually referred to as a *Cavity-Backed Slot* (CBS) antenna [99]–[102]. If the slot antenna is cut to half, to physically place an open circuit, it reduces to an IFA, which will be discussed next, whose performance is ideally similar to the original slot antenna.

Example 14.7

A very thin slot ($W = \lambda/10$) with length $L = \lambda/2$ is cut on a PEC ground plane. Determine analytically its approximate input impedance Z_s. Verify your result by simulations using a CEM software. Assume that the slot is radiating in free space.

Solution:

The slot ($L = \lambda/2$, $W = \lambda/10$) is, according to Figure 12.24, the complementary structure of a thin strip half-wavelength ($l = \lambda/2$) dipole, whose ideal thin-wire input impedance is, according to (4-93a),

$$Z_d = 73 + j42.5$$

Using Babinet's principle of Section 12.8 and (12-67) or (14-99), the input impedance of the slot Z_s can be represented approximately by

$$Z_s \approx \frac{\eta_o}{4Z_d} = \frac{376.7}{4(73 + j42.5)} = 362.95 - j211.31$$

Using a commercial CEM software, the simulated input impedance of a slot Z_s with $L = \lambda/2$, $W = \lambda/10$ on a $5\lambda \times 5\lambda$ PEC ground plane is

$$Z_s = 339.38 - j78.85$$

While the input resistance of 339.38 ohms is close to the analytical value of 362.95 ohms, the simulated reactance (capacitive) of −78.85 ohms is considerably different from the analytical of −211.31 ohms. However, as the width W of the slot decreases, the simulated capacitive reactance becomes more negative and approaches the analytical one.

14.9.3 Inverted-F Antenna (IFA)

A planar Inverted-F Antenna (IFA) consists of a thin arm or wire shorted at one of the ends to the ground plane [96], [103], [104], as shown in Figure 14.51. The length of the arm should be nearly $\lambda/4$. The position of the feed with respect to the shorting pin controls the input impedance. As the feed becomes closer to the shorting pin, the input impedance is reduced. Thus, to obtain a 50-ohm input impedance (used in most practical designs), the feed should be closer to the shorting pin than to the open end.

Although the IFA looks like the PIFA (with a wire or thin arm instead of a planar patch), its performance is different. The performance of the IFA is similar to that of the slot antenna [96]. The length of the IFA should be half of a slot antenna, as mentioned in the previous section. If the slot antenna is halved, and physically create an open circuit, it reduces to an IFA whose performance is ideally similar to that of the original slot antenna. The shorting pin introduces to the input impedance an inductance L_s while the open end introduces a capacitance C_s [105]. To obtain the impedance at resonance, the two reactive components should cancel out, leaving only the radiation resistance R_s. The width W of the shorting pin is very small compared to the wavelength ($W \ll \lambda$), usually $W \leq (0.05 - 0.1)\lambda$.

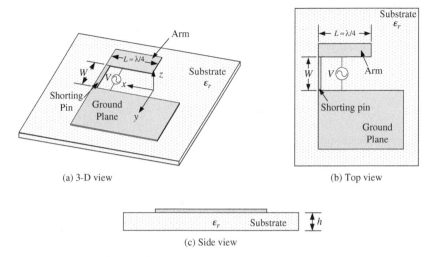

(a) 3-D view (b) Top view

(c) Side view

Figure 14.51 Inverted-F Antenna (IFA) design.

Example 14.8

Design a planar IFA antenna using Rogers RT/duroid 5870 substrate (dielectric constant of 2.33) and thickness of 0.1524 cm to resonate at 1.8 GHz.

- Determine the length L (in cm)
- Simulate the design for S_{11}, and 3-D and 2-D amplitude patterns.

Solution:
The length of the IFA antenna should be equal to

$$L = \frac{\lambda}{4} = \frac{30 \times 10^9}{4\sqrt{2.33}(1.8 \times 10^9)} = 2.73 \text{ cm}$$

An IFA based on this design was simulated using $W = 0.06\lambda = 0.655$ cm. The S_{11}, 3-D and 2-D E-plane and H-plane patterns are shown in Figures 14.52 and 14.53, respectively. It is apparent

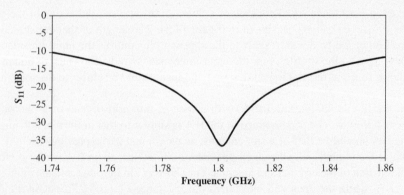

Figure 14.52 S_{11} for the IFA design of Example 14.8: $f = 1.8$ GHz, $W = 0.006\lambda = 0.655$ cm, $L = 0.25\lambda = 2.73$ cm. Rogers RT/duroid 5870 substrate, $h = 0.1524$ cm.

that the pattern of the IFA is nearly ominidirectional, as it should be, since the slot is the complement of the dipole and their patterns (slot and dipole) are, according to Babinet's principle, ideally identical as indicated in Figure 12.25 with the **E** and **H** fields reversed.

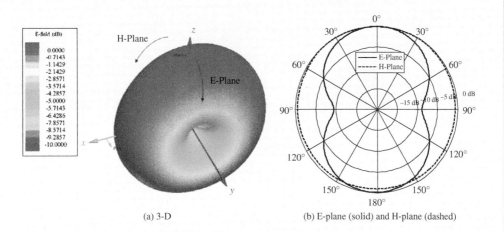

(a) 3-D   (b) E-plane (solid) and H-plane (dashed)

Figure 14.53 Normalized 3-D and 2-D (E-plane;yz and H-plane;xz) radiation patterns for the IFA design of Example 14.8: $f = 1.8$ GHz, $W = 0.06\lambda = 0.655$ cm, $L = 0.25\lambda = 2.73$ cm. Rogers RT/duroid 5870 substrate, $h = 0.1524$ cm.

While the IFA configuration of Figure 14.51 is considered a planar design, which is most widely used for mobile devices, there is another IFA design, which is wire-type and nonplanar [106], as shown in Figure 14.54.

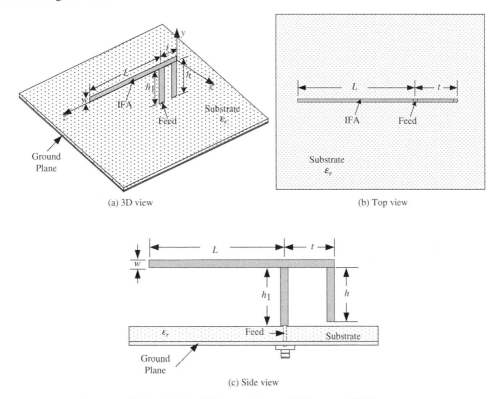

Figure 14.54 Wire-type Inverted-F Antenna (IFA).

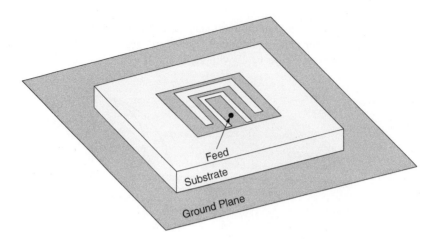

Figure 14.55 Three-band U-shot microstrip/patch antenna.

14.9.4 Multiband Antennas for Mobile Units

Today's mobile units, such as the smartphone, provide a plethora of services, from phone calls, emails, messaging, news, sports, music, to maps, just to name a few. Also these units have become smaller and slimmer, and less room is available to accommodate efficiently all these services. Since it is impossible to dedicate an individual antenna design for each of the services, antenna design engineers face major challenges to come up with designs whereby each individual antenna structure can accommodate multiple services. This requires that each physical antenna structure resonate at multiple frequencies, leading to multiband designs.

One planar antenna structure that can resonate at multiple frequencies is a rectangular microstrip with multiple slots, in the form of a U, placed on its patch. Each U slot is designed to resonate at a given frequency, and this is accomplished by making the overall length of each slot nearly $\lambda/2$ [107]. A three-band U-shape microstrip slot design is shown in Figure 14.55. The first resonance is attributed to the rectangular patch and the other two to the two U slots. Observe that the slots are separated from each other and the edges of the patch so that there is no major coupling between them.

A U-shape two-band planar antenna is shown in Figure 14.56. It is designed to operate at two bands and to accommodate two cellular services; the outer/longer slot to service the GSM band

GSM: 874–958 MHz
BW = 84 MHz

DCS: 1,711–1,943 MHz
BW = 232 MHz

Figure 14.56 Two-band (GSM and DCS) U-shot antenna (courtesy of Prof. Seong-Ook Park).

(874–958 MHz), with a bandwidth of 84 MHz (about 9%), and the inner/smaller one to accommodate the DCS band (1,711–1,943 MHz), with a bandwidth of 232 MHz (nearly 13%).

14.10 DIELECTRIC RESONATOR ANTENNAS

The dielectric resonator antenna followed the introduction and development of dielectric resonators, which are nonmetalized dielectric objects (spheres, hemispheres, cylinders, disks, parallelepipeds, etc.) that can function as energy storage devices due to a high dielectric constant (usually ceramic) and a high quality factor Q. Electromagnetic energy is injected into the block of dielectric material creating electromagnetic waves that bounce back and forth between the walls of the cavity creating standing waves. For the dielectric resonator to function as a resonant cavity, the dielectric constant of the material must be large (usually 50 or greater). Under these conditions, the dielectric–air interface acts almost as an open circuit and causes internal reflections that confine the energy in the dielectric material, thus creating a resonant structure. This effect can be expressed as an approximation, using the plane wave reflection coefficient Γ at a dielectric–air interface of [79]

$$\Gamma = \frac{\eta_0 - \eta_1}{\eta_0 + \eta_1} \overset{\mu_1 = \mu_0}{=} \frac{\sqrt{\mu_0/\varepsilon_0} - \sqrt{\mu_0/\varepsilon_1}}{\sqrt{\mu_0/\varepsilon_0} + \sqrt{\mu_0/\varepsilon_1}} = \frac{\sqrt{\varepsilon_1/\varepsilon_0} - 1}{\sqrt{\varepsilon_1/\varepsilon_0} + 1} = \frac{\sqrt{\varepsilon_r} - 1}{\sqrt{\varepsilon_r} + 1} \overset{\varepsilon_r \to \text{large}}{=} +1 \qquad (14\text{-}100)$$

Thus, the coefficient Γ approaches +1 as the dielectric constant becomes very large. Given these conditions, the dielectric–air interface can further be approximated by an ideal nonphysical perfect magnetic conductor (PMC) surface, which requires that the tangential components of the magnetic field to vanish (in contrast to the perfect electric conductor, PEC, which requires that the tangential electric field components to vanish). This is a well-known and widely used technique in solving boundary-value electromagnetic problems. It is, however, a first-order approximation, although it usually leads to reasonable results. The magnetic wall model has been used to analyze both dielectric waveguides and dielectric resonant cavities, and was used in Sections 14.2.2 and 14.3 to analyze rectangular and circular microstrip antennas.

Dielectric resonators were introduced in 1939 by Richtmyer [108], but for almost 25 years his theoretical work failed to generate any continuous and prolonged interest. The introduction in the 1960s of material, such as rutile, of high dielectric constant (around 100) renewed the interest in dielectric resonators [109]–[114], and the work of Van Bladel introduced a detailed theory on the modes of the dielectric resonator [115]. However, the poor temperature stability of rutile resulted in large resonant frequency changes and prevented the development of practical microwave components. In the 1970s, low-loss and temperature-stable ceramics, such as barium tetratitanate and $(Zr–Sn)TiO_4$, were introduced and were used for the design of high performance microwave components such as filters and oscillators. Because dielectric resonators are small, lightweight, temperature stable, of high Q, and low cost, they are ideal for the design and fabrication of monolithic microwave integrated circuits (MMICs) and general semiconductor devices. Thus, dielectric resonators have replaced traditional waveguide resonators, especially in MIC applications, and have implementations in the millimeter wave region.

Dielectric Resonator Antennas (DRAs) also consist of blocks of ceramic materials. DRAs are mounted on a PEC ground plane, and are commonly used at microwave frequencies and above. They were reported initially and introduced as practical radiating elements in the early 1980s by Long et al. in [116]–[118] for the cylindrical, rectangular, and hemispherical geometries, respectively. Since then, there have been numerous advances and publications on DRAs both as single elements and as elements in arrays, which have been documented in [119]–[124], and many others. For the blocks of dielectric material to act as radiators, their dielectric constants must be moderate, in the range of 5–30, to allow some of the energy from inside the cavity to *escape* through the walls of the cavity and *create* radiation.

In general, DRAs have a number of desirable features:

- Are efficient radiators, especially when compared to microstrips antennas; they do not generate surface waves and they are open on all sides, other than the side mounted on the PEC ground plane, whereas microstrip antennas radiate basically through their leading and trailing sides, as modeled in Figures 14.9 and 14.18.
- Are proportional in size to the wavelength in the dielectric, which, as is the case with microstrip antennas, decreases as the dielectric constant increases [$\lambda = \lambda_o/\sqrt{\varepsilon_r}$, λ_o = free space wavelength, ε_r = dielectric constant].
- Are wideband (typically BW $\approx$ 10%), especially as compared to microstrip antennas, with some geometries achieving bandwidths of 50% or even higher. In cases where the Q is very high, their bandwidth can be reduced to single digits.
- Have high power capability, as the dielectric constant is high.
- Are versatile, as different shapes can be used to excite various modes with different radiation characteristics.
- Can be excited by different feeds, such as probes (e.g., coaxial line), microstrip lines, slots, and coplanar waveguides, as is also the case for microstrip antennas; some of them are illustrated in Figure 14.3.
- Can be of low profile, as for some geometries their height h can be as small as 0.05λ.
- Can be used in a wide range of applications, including mobile wireless communication, satellite systems, GPS, and indoor communications.
- Can range in frequency of operation from about 1 to 60 GHz, and higher. The low frequency is limited by the size and weight of the DRAs.
- Exhibit lower losses at the millimeter wavelength region because at these frequencies metallic radiators become more inefficient due to skin effect while DRAs are made purely of dielectrics with, ideally, no losses.
- Are low cost and lightweight.

14.10.1 Basic DRA Geometries

Four basic DRA geometries are the rectangular, cylindrical, hemicylindrical, and hemispherical ones shown, respectively, in Figures 14.57 to 14.60. In each illustration, a probe feed line is also indicated whose position y_o, ρ_o, r_o is relative to the center of the cavity, and can be adjusted to improve the matching of the DRA to the feed line; for a coaxial line, this is typically 50 ohms. Although there are other configurations, these four are considered basic and are most commonly used. They also can be analyzed for their resonant frequencies and radiated fields by using the cavity model.

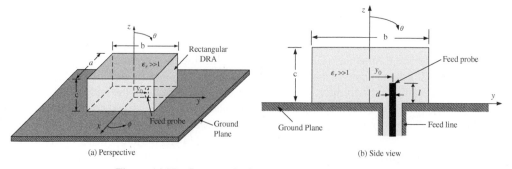

Figure 14.57 Rectangular DRA with a probe feeding line.

DIELECTRIC RESONATOR ANTENNAS   849

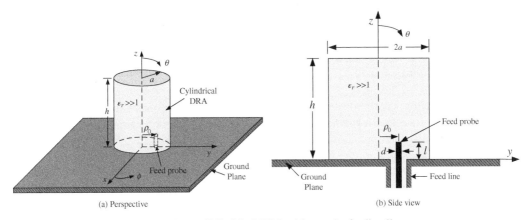

Figure 14.58   Cylindrical DRA with a probe feeding line.

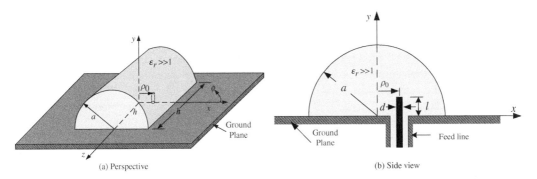

Figure 14.59   Hemicylindrical DRA with a probe feeding line.

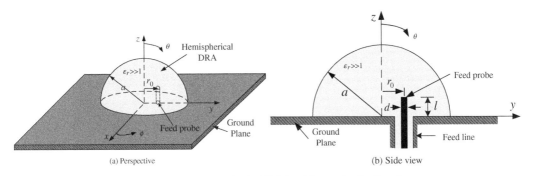

Figure 14.60   Hemispherical DRA with a probe feeding line.

14.10.2 Methods of Analysis and Design

These, and similar geometries, can be analyzed and designed using various methods such as:

- Cavity modal model where the open sides are treated as PMC while the ground plane is treated as a PEC.
- Full-wave solutions, such as:

  1. Integral equations (IE)/MoM
  2. Finite-difference time-domain (FDTD)

850 MICROSTRIP AND MOBILE COMMUNICATIONS ANTENNAS

3. Finite element method (FEM)
4. Transmission line method (TLM)
5. Hybrid methods
6. Perturbation methods

Although the cavity modal model (magnetic wall modeling of the open sides) does not always lead to the most accurate data, it will be utilized initially here, as a first-order approximation, because it is simple, instructive, and provides physical insight. It was used, as a first-order approximation, for the rectangular and circular microstrip antennas in Sections 14.2.2 and 14.3. Refinements upon the cavity modal model can be implemented to improve the designs. It has been demonstrated that this model is accurate for high dielectric constants, especially for resonant frequencies and radiation patterns [116], though it may not be necessarily accurate for the input impedance. Design equations based on other hybrid mode models will also be included.

14.10.3 Cavity Model Resonant Frequencies (TE and TM Modes)

The procedure for predicting the radiation characteristics using the cavity model was outlined in Section 14.2.2 for the TM^x modes for the rectangular microstrip antenna, and in Section 14.3 for the TM^z modes for the circular microstrip antenna. A similar procedure can be used for TE modes, but it is beyond the scope of this book. A detailed procedure for all modes, TEM, TM, TE, and hybrid modes, is provided in Chapter 6 of [79]. The TE and TM mode resonant frequencies, for the rectangular, cylindrical, hemicylindrical, and hemispherical are summarized here.

A. Rectangular DRA of Figure 14.57 (mnp modes: m for x variations; n for y variations; p for z variations).

$$(f_r)_{mnp}^{TE^z} = \frac{1}{2\pi\sqrt{\mu_d\varepsilon_d}}\sqrt{\left(\frac{m\pi}{a}\right)^2 + \left(\frac{n\pi}{b}\right)^2 + \left[(2p-1)\left(\frac{\pi}{2c}\right)\right]^2} \quad \begin{matrix} m=1,2,3\ldots \\ n=1,2,3,\ldots \\ p=1,2,3,\ldots \end{matrix} \quad (14\text{-}101a)$$

$$(f_r)_{mnp}^{TM^z} = \frac{1}{2\pi\sqrt{\mu_d\varepsilon_d}}\sqrt{\left(\frac{m\pi}{a}\right)^2 + \left(\frac{n\pi}{b}\right)^2 + \left[(2p-1)\left(\frac{\pi}{2c}\right)\right]^2} \quad \begin{matrix} m=0,1,2\ldots \\ n=0,1,2,\ldots \\ p=1,2,3,\ldots \end{matrix} \Bigg\} m=n\neq 0$$

$$(14\text{-}101b)$$

- For $a > b > 2c$ and $c > a > b$, the TE^z_{mnp} or TM^z_{mnp} mode with the smallest resonant frequency $(f_r)_{mnp}$ is the TM^z_{101} ($m=1, n=0, p=1$) with $(f_r)^{TM^z}_{101}$ of

$$(f_r)^{TM^z}_{101} = \frac{1}{2\sqrt{\mu_d\varepsilon_d}}\sqrt{\left(\frac{1}{a}\right)^2 + \left(\frac{1}{2c}\right)^2} = \frac{v_o}{2\sqrt{\mu_r\varepsilon_r}}\sqrt{\left(\frac{1}{a}\right)^2 + \left(\frac{1}{2c}\right)^2} \quad (14\text{-}102a)$$

where v_o is the speed of light.
- However for $c > b > a$, the lowest order TE^z_{mnp} or TM^z_{mnp} mode with the smallest resonant $(f_r)_{mnp}$ is the TM^z_{011} ($m=0, n=1, p=1$) with $(f_r)^{TM^z}_{011}$ of

$$(f_r)^{TM^z}_{011} = \frac{1}{2\sqrt{\mu_d\varepsilon_d}}\sqrt{\left(\frac{1}{b}\right)^2 + \left(\frac{1}{2c}\right)^2} = \frac{v_o}{2\sqrt{\mu_r\varepsilon_r}}\sqrt{\left(\frac{1}{b}\right)^2 + \left(\frac{1}{2c}\right)^2} \quad (14\text{-}102b)$$

B. Cylindrical DRA of Figure 14.58 *(mnp modes: m for azimuthal φ variations, n for radial ρ variations, p for height z variations)*

$$(f_r)_{mnp}^{TE^z} = \frac{1}{2\pi\sqrt{\mu_d\varepsilon_d}}\sqrt{\left(\frac{\chi_{mn}}{a}\right)^2 + \left[(2p+1)\left(\frac{\pi}{2h}\right)\right]^2} \quad \begin{array}{l} m = 0,1,2\ldots \\ n = 1,2,3,\ldots \\ p = 0,1,2,3,\ldots \end{array} \quad (14\text{-}103\text{a})$$

$$(f_r)_{mnp}^{TM^z} = \frac{1}{2\pi\sqrt{\mu_d\varepsilon_d}}\sqrt{\left(\frac{\chi'_{mn}}{a}\right)^2 + \left[(2p+1)\left(\frac{\pi}{2h}\right)\right]^2} \quad \begin{array}{l} m = 0,1,2\ldots \\ n = 1,2,3,\ldots \\ p = 0,1,2,3,\ldots \end{array} \quad (14\text{-}103\text{b})$$

where χ_{mn} and χ'_{mn} are the $n = 1, 2, \ldots$, zeros of the Bessel function J_m and its derivative J'_m, respectively. These are listed, respectively in Tables 9.2 and 9.1 of Chapter 9 of [79]. The five smallest values of each are

$$\chi_{01} = 2.4049, \quad \chi_{11} = 3.8318, \quad \chi_{21} = 5.1357, \quad \chi_{02} = 5.5201, \quad \chi_{31} = 6.3802$$
$$\chi'_{11} = 1.8412, \quad \chi'_{21} = 3.0542, \quad \chi'_{01} = 3.8318, \quad \chi'_{31} = 4.2012, \quad \chi'_{31} = 5.3175$$

- Based on these values of χ_{mn} and χ'_{mn}, the TE^z_{mnp} and TM^z_{mnp} modes with the smallest resonant frequency $(f_r)_{mnp}$ are the

$$(f_r)_{010}^{TE^z} = \frac{1}{2\pi\sqrt{\mu_d\varepsilon_d}}\sqrt{\left(\frac{2.4049}{a}\right)^2 + \left(\frac{\pi}{2h}\right)^2} \quad (14\text{-}104\text{a})$$

$$(f_r)_{110}^{TM^z} = \frac{1}{2\pi\sqrt{\mu_d\varepsilon_d}}\sqrt{\left(\frac{1.8412}{a}\right)^2 + \left(\frac{\pi}{2h}\right)^2} \quad (14\text{-}104\text{b})$$

Of these two resonant frequencies, the smaller of the two is the TM^z_{110} ($m = 1, n = 1, p = 0$) mode whose resonant frequency is that of (14-104b).

C. Hemicylindrical DRA of Figure 14.59 *(mnp modes: m for azimuthal φ variations, n for radial ρ variations, p for height z variations)*

$$(f_r)_{mnp}^{TE^z} = \frac{1}{2\pi\sqrt{\mu_d\varepsilon_d}}\sqrt{\sqrt{\left(\frac{\chi_{mn}}{a}\right)^2 + \left(\frac{p\pi}{h}\right)^2}} \quad \begin{array}{l} m = 0,1,2\ldots \\ n = 1,2,3,\ldots \\ p = 0,1,2,3,\ldots \end{array} \quad (14\text{-}105\text{a})$$

$$(f_r)_{mnp}^{TM^z} = \frac{1}{2\pi\sqrt{\mu_d\varepsilon_d}}\sqrt{\sqrt{\left(\frac{\chi'_{mn}}{a}\right)^2 + \left(\frac{p\pi}{h}\right)^2}} \quad \begin{array}{l} m = 1,2,3\ldots \\ n = 1,2,3,\ldots \\ p = 1,2,3,\ldots \end{array} \quad (14\text{-}105\text{b})$$

- Based on these values of χ_{mn} and χ'_{mn}, the TE^z_{mnp} and TM^z_{mnp} modes with the smallest resonant frequency $(f_r)_{mnp}$ are the

$$(f_r)_{010}^{TE^z} = \frac{1}{2\pi\sqrt{\mu_d\varepsilon_d}}\left(\frac{\chi_{01}}{a}\right) = \frac{2.4049}{2\pi a\sqrt{\mu_d\varepsilon_d}} \quad (14\text{-}106\text{a})$$

$$(f_r)_{111}^{TM^z} = \frac{1}{2\pi\sqrt{\mu_d\varepsilon_d}}\sqrt{\left(\frac{\chi'_{11}}{a}\right)^2 + \left(\frac{\pi}{h}\right)^2} = \frac{1}{2\pi\sqrt{\mu_d\varepsilon_d}}\sqrt{\left(\frac{1.841}{a}\right)^2 + \left(\frac{\pi}{h}\right)^2} \quad (14\text{-}106\text{b})$$

The TE^z_{mnp} or TM^z_{mnp} mode with the smallest resonant frequency is the:

- For $h/a < 2.031$: $\text{TE}^z_{010}(m = 0, n = 1, p = 0)$ with a resonant frequency of (14-106a).
- For $h/a > 2.031$: $\text{TM}^z_{111}(m = 1, n = 1, p = 1)$ with a resonant frequency of (14-106b).

D. **Hemispherical DRA of Figure 14.60** *(mnp modes: m for azimuthal ϕ variations, n for elevation θ variations; p for radial r variations)*

$$(f_r)^{\text{TE}^r}_{mnp}(\text{even, odd}) = \frac{\zeta_{np}}{2\pi a \sqrt{\mu_d \varepsilon_d}} \tag{14-107a}$$

$$(f_r)^{\text{TM}^r}_{mnp}(\text{even, odd}) = \frac{\zeta'_{np}}{2\pi a \sqrt{\mu_d \varepsilon_d}} \tag{14-107b}$$

$$(f_r)^{\text{TE}^r}_{011}(\text{even}) = (f_r)^{\text{TE}^r}_{111}(\text{even, odd}) = \frac{\zeta_{11}}{2\pi a \sqrt{\mu_d \varepsilon_d}} = \frac{4.493}{2\pi a \sqrt{\mu_d \varepsilon_d}} \tag{14-107c}$$

$$(f_r)^{\text{TM}^r}_{011}(\text{even}) = (f_r)^{\text{TM}^r}_{111}(\text{even, odd}) = \frac{\zeta'_{11}}{2\pi a \sqrt{\mu_d \varepsilon_d}} = \frac{2.744}{2\pi a \sqrt{\mu_d \varepsilon_d}} \tag{14-107d}$$

where ζ_{np} and ζ'_{np} are, for a given n, the $p = 1, 2, \ldots$ zeroes of the spherical Bessel function $\hat{J}_n$ and its derivative $\hat{J}'_n$, respectively. These are listed, respectively, in Tables 10.1 and 10.2 of Chapter 10 of [79].

The five smallest values of each are:

$$\zeta_{11} = 4.493, \; \zeta_{21} = 5.763, \; \zeta_{31} = 6.988, \; \zeta_{12} = 7.725, \; \zeta_{41} = 8.183$$
$$\zeta'_{11} = 2.744, \; \zeta'_{21} = 3.870, \; \zeta'_{31} = 4.973, \; \zeta'_{41} = 6.062, \; \zeta'_{12} = 6.117$$

- Based upon these values of ζ_{mn} and ζ'_{mn}, the TE^r_{mnp} or TM^r_{mnp} mode with the smallest resonant frequency $(f_r)_{mnp}$, having a three-fold degeneracy, is the TM^r_{011} (even: $m = 0, n = 1, p = 1$) = TM^r_{111} (even: $m = 1, n = 1, p = 1$) = TM^r_{111} (odd: $m = 1, n = 1, p = 1$) whose resonant frequency $(f_r)^{\text{TM}^r}_{011}$ (even) = $(f_r)^{\text{TM}^r}_{111}$ (even) = $(f_r)^{\text{TM}^r}_{111}$ (odd) is that of (14-107d).

14.10.4 Hybrid Modes: Resonant Frequencies and Quality Factors

As indicated previously, the formulations based on the cavity PMC/PEC models of Section 14.10.3 are first-order approximations, especially for the resonant frequency and far-zone radiated fields. More accurate expressions for the resonant frequencies (f_r) and radiation quality factors (Q) of the *cylindrical* DRA of Figure 14.58 have been derived for $\text{TE}_{mn\delta}$, $\text{TM}_{mn\delta}$ and hybrid $\text{HE}_{mn\delta}$ modes, which are the most common modes for this DRA geometry. The notation here is equivalent to that of Section 9.5.2 of [79], adopted based on the notation of [123], where the first integer subscript m ($m = 0, 1, 2, \ldots$) represents azimuthal ϕ field variations, the second integer subscript n ($n = 1, 2, 3, \ldots$) represents radial ρ field variations, while the third subscript δ (δ = noninteger) represents z field variations. The noninteger value of δ indicates that the field variations along the z direction are not full periods or the length of the DRA is not integer multiple of $\lambda_d/2$, where λ_d is the wavelength in the DRA dielectric. In [123], the dielectric resonator has been modeled using two PEC flat plates, each placed a distance h from each end of the *cylindrical* dielectric resonator, as shown in Figure 9.15 of [79]. The value of δ depends on the dielectric constant ε_r of the DRA and the proximity h_1, h_2 of each of the two PEC plates from the ends of the DRA [122], as illustrated in Figure 9.15 of [79].

It has been shown in [122], [123] that the field variations within the cylindrical DRA for the $\text{TE}_{mn\delta}$ and $\text{TM}_{mn\delta}$ are azimuthally symmetric (not ϕ dependent) while those of the hybrid $\text{HE}_{mn\delta}$ are ϕ dependent, as shown, respectively, in Figures 14.61 (a,b,c).

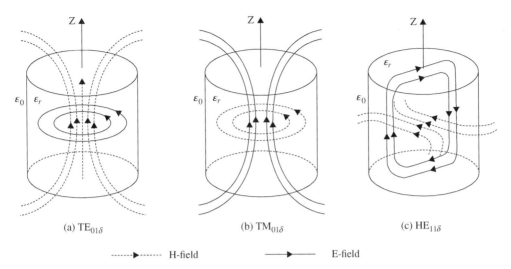

(a) TE$_{01\delta}$ (b) TM$_{01\delta}$ (c) HE$_{11\delta}$

------▶------ H-field   ———▶——— E-field

Figure 14.61 Field configurations within a cylindrical DRA for the TE$_{01\delta}$, TM$_{01\delta}$, and hybrid HE$_{11\delta}$ modes [122] [SOURCE: A. A. Kishk and Y. M. M. Antar, "Dielectric Resonator Antennas," Chapter 17 in *Antenna Engineering Handbook*, J. L. Volakis (ed.), © McGraw-Hill Book Co., 2007].

Based on these more accurate TE$_{mn\delta}$, TM$_{mn\delta}$, and hybrid HE$_{mn\delta}$ modes, closed-form expressions for the resonant frequencies and Q have been derived, for the *cylindrical* DRA of Figure 14.58, through extensive numerical and experimental investigations and curve fitting routines. One set of these expressions, based on [121], [124], is listed here; another set can be found in [122].

TE$_{01\delta}$ Mode :

$$f_r = \frac{v_o}{2\pi a}\left(\frac{2.327}{\sqrt{\varepsilon_r + 1}}\right)\left[1 + 0.2123\left(\frac{a}{h}\right) - 0.00898\left(\frac{a}{h}\right)^2\right]; \quad 0.125 \leq \frac{a}{h} \leq 5 \quad (14\text{-}108a)$$

$$Q = 0.078192\left(\varepsilon_r\right)^{1.27}\left[\begin{array}{l}1 + 17.31\left(\frac{h}{a}\right) - 21.57\left(\frac{h}{a}\right)^2 \\ + 10.86\left(\frac{h}{a}\right)^3 - 1.98\left(\frac{h}{a}\right)^4\end{array}\right]; \quad 0.5 \leq \frac{a}{h} \leq 5 \quad (14\text{-}108b)$$

TM$_{01\delta}$ Mode :

$$f_r = \frac{v_o}{2\pi a}\left(\frac{1}{\sqrt{\varepsilon_r + 2}}\right)\sqrt{\left[(03.83)^2 + \left(\frac{\pi}{2}\right)^2\left(\frac{a}{h}\right)^2\right]}; \quad 0.125 \leq \frac{a}{h} \leq 5 \quad (14\text{-}109a)$$

$$Q = 0.00872\left(\varepsilon_r\right)^{0.888} e^{0.03975\varepsilon_r}\left[1 - (0.3 - 0.2z)\left(\frac{38 - \varepsilon_r}{28}\right)\right]$$

$$\times \left(9.498z + 2{,}058.33 z^{4.3226} e^{-3.501z}\right) \quad 0.125 \leq \frac{a}{h} \leq 5 \quad (14\text{-}109b)$$

HE$_{11\delta}$ Mode :

$$f_r = \frac{v_o}{2\pi a}\left(\frac{6.324}{\sqrt{\varepsilon_r + 2}}\right)\left(0.27 + 0.18z + 0.005z^2\right); \quad 0.125 \leq \frac{a}{h} \leq 5 \quad (14\text{-}110a)$$

$$Q = 0.01z\left(\varepsilon_r\right)^{1.3}\left[1 + 100 e^{-2.05\left(0.5z - 0.0125z^2\right)}\right]; \quad 0.5 \leq \frac{a}{h} \leq 5 \quad (14\text{-}110b)$$

TABLE 14.3 Updated Predicted and Measured Resonant Frequencies and Qs [121]

		$\varepsilon_r = 38, a = 0.6415$ cm, $h = 0.2810$ cm				
Mode	Measured f_r (GHz)	Theoretical f_r (GHz)	% Difference	Measured Q-Factor	Theoretical Q-Factor	% Difference
$TE_{01\delta}$	3.97	3.988	0.45%	46.2	41.92	9.3%
$TM_{01\delta}$	6.13	6.175	0.73%	72.1	46.34	35.7%
$HE_{11\delta}$	5.18	5.262	1.58%	30.2	31.24	3.4%
		$\varepsilon_r = 79, a = 0.5145$ cm, $h = 0.2255$ cm				
Mode	Measured f_r (GHz)	Theoretical f_r (GHz)	% Difference	Measured Q-Factor	Theoretical Q-Factor	% Difference
$TE_{01\delta}$	3.48	3.47	2.9%	114.7	106.21	7.4%
$TM_{01\delta}$	5.41	5.41	0.0%	336.7	349.61	3.8%
$HE_{11\delta}$	4.56	4.61	1.1%	76.4	80.94	5.9%

It should be pointed out that the Qs based on the equations above are small, near the lower limits of a/h; larger values are attained at the middle and upper range of a/h ratios. Table 14.2 compares predicted, based on (14-108a) and (14-110b), and measured, based on [121], resonant frequencies and Qs of two isolated cylindrical DRAs, of dielectric constants $\varepsilon_r = 38$ and $\varepsilon_r = 79$. It should be noted that the predicted values in Table 14.3 are slightly different than those from Table 4.2 of [121]. The original data for the measured Qs are based on radar cross sections measurements from [125]. The resonant frequencies based on (14-108a), (14-109a), and (14-110a) are less than 3% from measured ones while the differences for the Qs, based on (14-108b), (14-109b), and (14-110b), are somewhat larger than the corresponding ones based on measurements but, except for one case, always smaller than 10%.

A Matlab computer program, **DRA_Analysis_Design**, has been developed and it can be used to:

- *Analyze* the four DRAs of Figures 14.57–14.60
- *Design* the cylindrical DRA of Figure 14.58

....

- In the *analysis* part of the program, once the radius a, height h and dielectric constant ε_r are specified, the following are computed:

  1. Dominant mode resonant frequencies, of any of the four DRA geometries of Figures 14.57 to 14.60, based on the modal solution expressions of (14-101a)–(14-107d). The computed resonant frequencies are:
     - Lowest-order 5 TE^z, 5 TM^z, and 5 TE^z/TM^z modes of each of the cubic, cylindrical, and hemicylindrical DRAs of Figures 14.57 to 14.59.
     - Lowest-order 3 degenerate TE^r modes of the hemispherical DRA of Figure 14.60.
  2. The resonant frequencies and Qs of the *cylindrical* resonator of Figure 14.58 for all three modes ($TE_{01\delta}$, $TM_{01\delta}$, or $HE_{11\delta}$), based on (14-108a)–(14-110b). Example 14.9 is an analysis exercise.

- The *design* part of the program is for a *cylindrical* DRA of Figure 14.58 which, once the following are specified:

  1. Mode ($TE_{01\delta}$, $TM_{01\delta}$, or $HE_{11\delta}$)
  2. Fractional bandwidth (BW, in %)
  3. VSWR
  4. Resonant frequency (f_r, in GHz)

the program performs the following:

1. Computes the Q of the cavity [based on (14-88a)].
2. Prompts the user to select a desired dielectric constant from the following:
   - From a range of values for the $TE_{01\delta}$ mode.
   - Greater than some minimum value for the $TM_{01\delta}$ mode.
   - From a range of values for the $HE_{11\delta}$ mode.

   These ranges and values for the dielectric constant are determined, once the Q is computed using (14-88a), by solving using a nonlinear procedure as follows:
   - $TE_{01\delta}$ mode: (14-108b) for $0.5 \leq a/h \leq 5$.
   - $TM_{01\delta}$ mode: (14-109b) for $0.125 \leq a/h \leq 5$.
   - $HE_{11\delta}$ mode: (14-110b) for $0.5 \leq a/h \leq 5$.

   Otherwise, the stated design specifications cannot be met.
3. At this point, the Q, resonant frequency f_r and dielectric constant ε_r have been decided step by step. To complete the design, based on the stated specifications, the computer program (using a nonlinear procedure) determines the dimensions of the cylindrical DRA (radius a and height h, both in cm) by solving simultaneously:
   - $TE_{01\delta}$ mode: (14-108a) and (14-108b).
   - $TM_{01\delta}$ mode: (14-109a) and (14-109b).
   - $HE_{11\delta}$ mode: (14-110a) and (14-110b).

   Example 14.10 provides a design exercise based on this procedure.

14.10.5 Radiated Fields

The far-zone radiated fields, for each of the configurations in Figures 14.57 to 14.60, can be found by formulating the equivalent surface magnetic current density $\mathbf{M}_s$ on the open surface of each of the four geometries and then by using the radiation equations. This process was implemented in Chapter 12. In Chapter 12 and this chapter the procedure is used to determine the far-zone fields radiated of the following geometries:

- Rectangular apertures in Section 12.3
- Circular apertures in Section 12.6
- Rectangular microstrip antennas in Section 14.2.2(C)
- Circular microstrips in 14.3.4

The same method was applied successfully in [116] for cylindrical DRAs where the predicted patterns compared well with measurements.

As has been shown in [122], two-dimensional normalized amplitude patterns, in the respective principal planes, of the electric fields radiated by the cylindrical DRA of Figure 14.58 for the $HE_{11\delta}$ and $M_{01\delta}$ modes resemble, respectively, those in Figure 14.62(a,b). The patterns of Figure 14.62(a) for the $HE_{11\delta}$ were found to be typical of those of a horizontal magnetic dipole, and those of Figure 14.62(b) for the $TM_{01\delta}$ mode to be typical of those of an infinitesimal vertical electric monopole [122], as represented in Figure 4.15 of [122].

In addition, it has been shown in [122] that typical ideal two-dimensional normalized amplitude patterns, along the principle planes, of the fields radiated by the hemicylindrical DRA of Figure 14.59, for the $TE_{01\delta}$ mode, look like those in Figure 14.63. The patterns of Figure 14.63 for the $TE_{01\delta}$ mode were found to be typical of those radiated by a short horizontal magnetic dipole.

The radiation characteristics (resonant frequency, radiation pattern, and input impedance) of a cylindrical DRA were examined analytically, numerically, and experimentally in [116]. In [116], the dominant mode, as computed using (14-103a)–(14-104b), was determined to be the TM_{110}^z followed

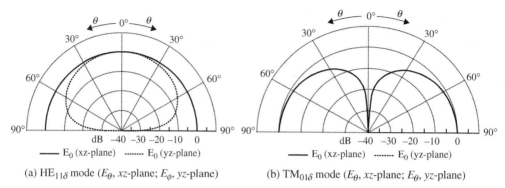

(a) $HE_{11\delta}$ mode (E_θ, xz-plane; E_ϕ, yz-plane)

(b) $TM_{01\delta}$ mode (E_θ, xz-plane; E_θ, yz-plane)

Figure 14.62 Normalized amplitude patterns for the cylindrical DRA of Figure 14.58 for the $HE_{11\delta}$ and $TM_{01\delta}$ modes [122] [SOURCE: A. A. Kishk and Y. M. M. Antar, "Dielectric Resonator Antennas," Chapter 17 in *Antenna Engineering Handbook*, J. L. Volakis (ed.), © McGraw-Hill Book Co., 2007].

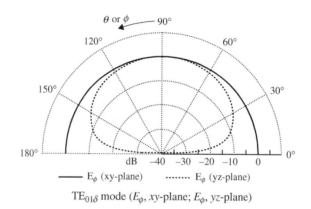

$TE_{01\delta}$ mode (E_ϕ, xy-plane; E_ϕ, yz-plane)

Figure 14.63 Normalized amplitude patterns for the hemicylindrical DRA of Figure 14.58 for the $TE_{01\delta}$ mode [122] [SOURCE: A. A. Kishk and Y. M. M. Antar, "Dielectric Resonator Antennas," Chapter 17 in *Antenna Engineering Handbook*, J. L. Volakis (ed.), © McGraw-Hill Book Co., 2007].

by either the TM^z_{111} for small a/h ratios or the TE^z_{010} for large h/a ratios. For the cylindrical DRA of Figure 14.58, the far-field patterns for large h/a ratios match those of the $HE_{11\delta}$ hybrid mode of Figure 14.62(a), with the maximum radiation along the z axis ($\theta = 0°$). However, for small h/a ratios, a null begins to form along $\theta = 0°$.

Example 14.9: Analysis

Given a cylindrical DRA of Figure 14.58 with $a = 0.5$ cm, $h = 0.3$ cm ($a/h = 1.67$), $l = 0.38$ cm, $d = 0.05$ cm, and probe position $\phi' = 90°$, $\rho_o = 0.36$ cm.

a. Using commercial software, simulate and plot versus frequency the:
   - S_{11}
   - Input impedance (real and imaginary parts)
b. Compute the resonant frequencies of the dominant TE^z and TM^z modes based on (14-104a) and (14-104b) for:
   - $\varepsilon_r = 8.9$
   - $\varepsilon_r = 89$

c. Compute the resonant frequencies of the $TE_{01\delta}$ and $TM_{01\delta}$ modes, and hybrid $HE_{11\delta}$ mode for:
   - $\varepsilon_r = 8.9$
   - $\varepsilon_r = 89$
d. Compute the Qs of the $TE_{01\delta}$ and $TM_{01\delta}$ modes, and hybrid $HE_{11\delta}$ mode for:
   - $\varepsilon_r = 8.9$
   - $\varepsilon_r = 89$
e. Compute the 3-D and 2-D principal E-plane and H-plane patterns for $\varepsilon_r = 8.9$ based on the resonant frequency of the lowest S_{11} of part a.

Solution:

a. The DRA, with the subject dimensions, has been simulated and the S_{11} and input impedance vs. frequency are plotted in Figure 14.64(a,b). The resonant frequency, based on the lowest S_{11}, is 10.69 GHz.
b. Based on (14-104a) and (14-104b), the resonant frequencies are:
   - $\varepsilon_r = 8.9 : f_r(TE_{010}) = 11.38$ GHz, $f_r(TM_{110}) = 10.24$ GHz
   - $\varepsilon_r = 89 : f_r(TE_{010}) = 3.60$ GHz, $f_r(TM_{110}) = 3.24$ GHz

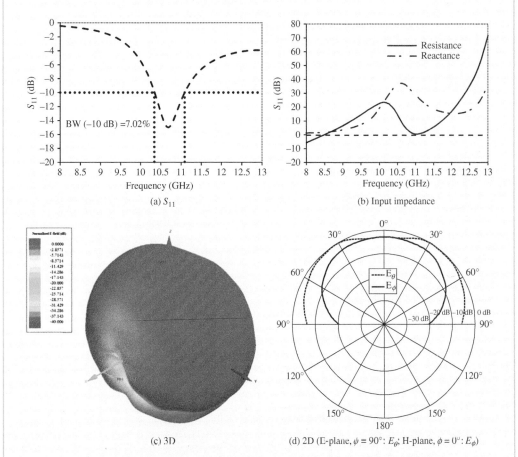

Figure 14.64 S_{11}, input impedance, and 3-D and 2-D normalized amplitude patterns ($f = 10.69$ GHz) of Example 4.9.

c. Based on (14-108a), (14-109a), and (14-110a), the resonant frequencies are:
   - $\varepsilon_r = 8.9$: $f_r(TE_{01\delta}) = 9.385$ GHz, $f_r(TM_{01\delta}) = 13.419$ GHz, $f_r(HE_{11\delta}) = 8.564$ GHz
   - $\varepsilon_r = 89$: $f_r(TE_{01\delta}) = 3.113$ GHz, $f_r(TM_{01\delta}) = 4.644$ GHz, $f_r(HE_{11\delta}) = 2.964$ GHz
d. Based on (14-108b), (14-109b), and (14-110b), the Qs are:
   - $\varepsilon_r = 8.9$: $Q = (TE_{01\delta}) = 7.17$, $Q(TM_{01\delta}) = 6.325$, $Q(HE_{11\delta}) = 5.8864$
   - $\varepsilon_r = 89$: $Q(TE_{01\delta}) = 133.512$, $Q(TM_{01\delta}) = 1,071.95$, $Q(HE_{11\delta}) = 117.45$
e. The corresponding 3-D and 2-D amplitude patterns are displayed, respectively, in Figure 14.64(c,d), and they match those of [116]. Note that the feed in the present model of Figure 14.57 is at $\phi' = 90°$ while that of Figure 1 of [116] is at $\phi' = 0°$; therefore the E-plane and H-plane are interchanged, as they should be.

Example 14.10: Design

Design a cylindrical DRA, with geometry of Figure 14.58, to operate on the $TE_{01\delta}$ mode and to exhibit a fractional bandwidth of 2.887%, VSWR = 3 and resonant frequency $f_r = 10$ GHz. Use the *Design* part of the Matlab computer program **DRA_Analysis_Design** to:

- Determine the Q.
- Select a dielectric constant ε_r (greater than some minimum value).
- Determine the dimensions a and h (both in cm).

Solution:
The Matlab computer program **DRA_Analysis_Design** computes the following:

- $Q = 40$ based on (14-88a)
- Dielectric constant range: $34 < \varepsilon_r < 48.65$. Select $\varepsilon_r = 38$.
- Based on the answers for the Q and dielectric constant ε_r, the radius and height are $a = 0.2653$ cm, $h = 0.1020$ cm

When the dimensions $a = 0.2653$ cm, $h = 0.1020$ cm, and $\varepsilon_r = 38$ are used in the *Analysis* part of the Matlab program, they lead, for the $TE_{01\delta}$ mode, to a resonant frequency $f_r = 10$ GHz and a $Q = 40$, which verifies the design procedure.

A summary of the pertinent parameters, and associated formulas and equation numbers for this chapter are listed in Table 14.4.

14.11 MULTIMEDIA

In the publisher's website for this book, the following multimedia resources are included for the review, understanding, and visualization of the material of this chapter:

a. **Java**-based **interactive questionnaire**, with answers.
b. **Matlab** and **Fortran** computer program, designated **Microstrip**, for computing and displaying the radiation characteristics of rectangular and circular microstrip antennas.
c. **Matlab** computer program, designated **DRA_Analysis_Design**, which:
   - Analyzes a DRA cavity, as in Figures 14.57 to 14.60, once the dimensions and dielectric constant are specified.

- Designs a cylindrical DRA, as in Figure 14.58, once the mode ($TE_{01\delta}$, $TM_{01\delta}$, or $HE_{11\delta}$), fractional bandwidth (BW), VSWR, and resonant frequency are specified.
  d. **Power Point (PPT)** viewgraphs, in multicolor.

TABLE 14.4 Summary of Important Parameters and Associated Formulas and Equation Numbers

Parameter	Formula	Equation Number
	Transmission-Line Model-Rectangular Patch	
Effective dielectric constant ε_{reff} ($W/h \gg 1$)	$\varepsilon_{reff} = \dfrac{\varepsilon_r + 1}{2} + \dfrac{\varepsilon_r - 1}{2}\left[1 + 12\dfrac{h}{W}\right]^{-1/2}$	(14-1)
Effective length L_{eff}	$L_{eff} = L + 2\Delta L$	(14-3)
Normalized extension length $\Delta L/h$	$\dfrac{\Delta L}{h} = 0.412 \dfrac{(\varepsilon_{reff} + 0.3)\left[\dfrac{W}{h} + 0.264\right]}{(\varepsilon_{reff} - 0.258)\left[\dfrac{W}{h} + 0.8\right]}$	(14-2)
Resonant frequency; dominant mode ($L > W$) (*no fringing*)	$(f_r)_{010} = \dfrac{1}{2L\sqrt{\varepsilon_r}\sqrt{\mu_o \varepsilon_o}}$	(14-4)
Resonant frequency; dominant mode ($L > W$) (*with fringing*)	$(f_{rc})_{010} = \dfrac{1}{2L_{eff}\sqrt{\varepsilon_{reff}}\sqrt{\mu_o \varepsilon_o}}$	(14-5)
Slot conductance G_1	$G_1 = \dfrac{W}{120\lambda_o}\left[1 - \dfrac{1}{24}(k_o h)^2\right], \quad \dfrac{h}{\lambda_o} < \dfrac{1}{10}$	(14-8a)
Slot susceptance B_1	$B_1 = \dfrac{W}{120\lambda_o}[1 - 0.636\ln(k_o h)], \quad \dfrac{h}{\lambda_o} < \dfrac{1}{10}$	(14-8b)
Input slot resistance R_{in} (*at resonance; no coupling*)	$R_{in} = \dfrac{1}{2G_1}$	(14-16)
Input slot resistance R_{in} (*at resonance; with coupling*)	$R_{in} = \dfrac{1}{2(G_1 \pm G_{12})}$ + for modes with *odd* symmetry − for modes with *even* symmetry	(14-17)
Input slot resistance R_{in} (*at resonance; with coupling*)	$R_{in} = 90\dfrac{(\varepsilon_r)^2}{\varepsilon_r - 1}\left(\dfrac{L}{W}\right)$	(14-18b)
Input resistance $R_{in}(y = y_o)$ (*no coupling*)	$R_{in}(y = y_o) = R_{in}(y = 0)\cos^2\left(\dfrac{\pi}{L}y_o\right)$ $= \dfrac{1}{2G_1}\cos^2\left(\dfrac{\pi}{L}y_o\right)$	(14-20a)
Input resistance $R_{in}(y = y_o)$ (*with coupling*)	$R_{in}(y = y_o) = R_{in}(y = 0)\cos^2\left(\dfrac{\pi}{L}y_o\right)$ $= \dfrac{1}{2(G_1 \pm G_{12})}\cos^2\left(\dfrac{\pi}{L}y_o\right)$	(14-20a)

(*continued overleaf*)

TABLE 14.4 (continued)

Parameter	Formula	Equation Number
Cavity Model-Rectangular Patch		
Resonant frequency $(f_{rc})_{010}$; dominant mode $(L > W)$ (*no fringing*)	$(f_{rc})_{010} = \dfrac{1}{2L\sqrt{\varepsilon_r}\sqrt{\mu_o \varepsilon_o}}$	(14-33)
Resonant frequency $(f_r)_{010}$; dominant mode $(L > W)$ (*with fringing*)	$(f_r)_{010} = \dfrac{1}{2L_{eff}\sqrt{\varepsilon_{reff}}\sqrt{\mu_o \varepsilon_o}}$	(14-5)
Resonant frequency $(f_r)_{001}$; dominant mode $(L > W > L/2 > h)$ (*no fringing*)	$(f_r)_{001} = \dfrac{1}{2W\sqrt{\varepsilon_r}\sqrt{\mu_o \varepsilon_o}}$	(14-34)
Resonant frequency $(f_r)_{020}$; dominant mode $(L > L/2 > h)$; (*no fringing*)	$(f_r)_{020} = \dfrac{1}{L\sqrt{\varepsilon_r}\sqrt{\mu_o \varepsilon_o}}$	(14-35)
Total electric field E_ϕ^t	$E_\phi^t = E_\phi(\text{single slot}) \times \text{AF}$	(14-40a)–(14-41), (14-43)
Array factor $(\text{AF})_y$	$(\text{AF})_y = 2\cos\left(\dfrac{k_o L_e}{2}\sin\theta\sin\phi\right)$	(14-42)
Directivity D_o (*single slot*)	$D_o = \begin{cases} 3.3\ (dimensionless) = 5.2\ \text{dB}; & W \ll \lambda_o \\ 4\left(\dfrac{W}{\lambda_o}\right); & W \gg \lambda_o \end{cases}$	(14-54)
Directivity D_o (*two slots*)	$D_o = \begin{cases} 6.6\ (dimensionless) = 8.2\ \text{dB}; & W \ll \lambda_o \\ 8\left(\dfrac{W}{\lambda_o}\right); & W \gg \lambda_o \end{cases}$	(14-57)
Cavity Model-Circular Patch		
Resonant frequency $(f_r)_{110}$; dominant mode TM_{110} mode; (*no fringing*)	$(f_r)_{110} = \dfrac{1.8412}{2\pi a \sqrt{\varepsilon_r}\sqrt{\mu_o \varepsilon_o}}$	(14-66)
Resonant frequency $(f_{rc})_{110}$; dominant mode TM_{110} mode; (*with fringing*)	$(f_{rc})_{110} = \dfrac{1.8412}{2\pi a_e \sqrt{\varepsilon_r}\sqrt{\mu_o \varepsilon_o}}$	(14-68)
Effective radius a_e	$a_e = a\left\{1 + \dfrac{2h}{\pi a \varepsilon_r}\left[\ln\left(\dfrac{\pi a}{2h}\right) + 1.7726\right]\right\}^{1/2}$	(14-68)

(*continued*)

TABLE 14.4 (*continued*)

Parameter	Formula	Equation Number
Physical radius a	$a = \dfrac{F}{\left\{1 + \dfrac{2h}{\pi \varepsilon_r F}\left[\ln\left(\dfrac{\pi F}{2h}\right) + 1.7726\right]\right\}^{1/2}}$	(14-69)
	$F = \dfrac{8.791 \times 10^9}{f_r \sqrt{\varepsilon_r}}$; ($h$ in cm)	(14-69a)
Directivity D_o	$D_o = \dfrac{(k_o a_e)^2}{120 G_{\text{rad}}}$	(14-80)
Radiation conductance G_{rad}	$G_{\text{rad}} = \dfrac{(k_o a_e)^2}{480} \displaystyle\int_0^{\pi/2} [(J'_{02})^2 + \cos^2\theta (J_{02})^2] \sin\theta \, d\theta$	(14-76)
	$J'_{02} = J_o(k_o a_e \sin\theta) - J_2(k_o a_e \sin\theta)$	(14-72d)
	$J_{02} = J_o(k_o a_e \sin\theta) + J_2(k_o a_e \sin\theta)$	(14-72e)
Input resistance $R_{in}(\rho' = \rho_o)$	$R_{in}(\rho' = \rho_o) = R_{in}(\rho' = a_e) \dfrac{J_1^2(k\rho_o)}{J_1^2(ka_e)}$	(14-82)
	$R_{in}(\rho' = a_e) = \dfrac{1}{G_t}$	(14-82a)
	$G_t = G_{\text{rad}} + G_c + G_d$	(14-79)
	$G_c = \dfrac{\varepsilon_{mo}\pi(\pi\mu_o f_r)^{-3/2}}{4h^2 \sqrt{\sigma}}[(ka_e)^2 - m^2]$	(14-77)
	$G_d = \dfrac{\varepsilon_{mo} \tan\delta}{4\mu_o h f_r}[(ka_e)^2 - m^2]$	(14-78)
	where for $mn0$ mode ($m = n = 1$ for dominant mode) $\varepsilon_{mo} = 2$ for $m = 0$ $\varepsilon_{mo} = 1$ for $m \neq 0$	
Total quality factor Q_t	$\dfrac{1}{Q_t} = \dfrac{1}{Q_{\text{rad}}} + \dfrac{1}{Q_c} + \dfrac{1}{Q_d} + \dfrac{1}{Q_{sw}}$	(14-83)
	For $h \ll \lambda_o$ $Q_c = h\sqrt{\pi f \mu \sigma}$; $Q_d = \dfrac{1}{\tan\delta}$	(14-84), (14-85)
	$Q_{\text{rad}} = \dfrac{2\omega\varepsilon_r}{hG_t/l}K$; $K = \dfrac{\displaystyle\iint_{\text{area}} |E|^2 \, dA}{\displaystyle\oint_{\text{perimeter}} |E|^2 \, dl}$	(14-86), (14-86a)
Fractional bandwidth $\dfrac{\Delta f}{f_o}$	$\dfrac{\Delta f}{f_o} = \dfrac{VSWR - 1}{Q_t \sqrt{VSWR}}$	(14-88a)

REFERENCES

1. G. A. Deschamps, "Microstrip Microwave Antennas," Presented at the Third USAF Symposium on Antennas, 1953.
2. H. Gutton and G. Baissinot, "Flat Aerial for Ultra High Frequencies," French Patent No. 703 113, 1955.
3. R. E. Munson, "Conformal Microstrip Antennas and Microstrip Phased Arrays," *IEEE Trans. Antennas Propagat.*, Vol. AP-22, No. 1, pp. 74–78, January 1974.
4. J. Q. Howell, "Microstrip Antennas," *IEEE Trans. Antennas Propagat.*, Vol. AP-23, No. 1, pp. 90–93, January 1975.
5. A. G. Derneryd, "Linearly Polarized Microstrip Antennas," *IEEE Trans. Antennas Propagat.*, Vol. AP-24, No. 6, pp. 846–851, November 1976.
6. L. C. Shen, S. A. Long, M. R. Allerding, and M. D. Walton, "Resonant Frequency of a Circular Disc, Printed-Circuit Antenna," *IEEE Trans. Antennas Propagat.*, Vol. AP-25, No. 4, pp. 595–596, July 1977.
7. P. K. Agrawal and M. C. Bailey, "An Analysis Technique for Microstrip Antennas," *IEEE Trans. Antennas Propagat.*, Vol. AP-25, No. 6, pp. 756–759, November 1977.
8. A. G. Derneryd, "A Theoretical Investigation of the Rectangular Microstrip Antenna Element," *IEEE Trans. Antennas Propagat.*, Vol. AP-26, No. 4, pp. 532–535, July 1978.
9. *Proc. of the Workshop on Printed-Circuit Antenna Technology*, October 17–19, 1979, New Mexico State Univ., Las Cruces, NM.
10. A. G. Derneryd, "Analysis of the Microstrip Disk Antenna Element," *IEEE Trans. Antennas Propagat.*, Vol. AP-27, No. 5, pp. 660–664, September 1979.
11. A. G. Derneryd, "Extended Analysis of Rectangular Microstrip Resonator Antennas," *IEEE Trans. Antennas Propagat.*, Vol. AP-27, No. 6, pp. 846–849, November 1979.
12. Y. T. Lo, D. Solomon, and W. F. Richards, "Theory and Experiment on Microstrip Antennas," *IEEE Trans. Antennas Propagat.*, Vol. AP-27, No. 2, pp. 137–145, March 1979.
13. S. A. Long and M. D. Walton, "A Dual-Frequency Stacked Circular-Disc Antenna," *IEEE Trans. Antennas Propagat.*, Vol. AP-27, No. 2, pp. 270–273, March 1979.
14. N. K. Uzunoglu, N. G. Alexopoulos, and J. G. Fikioris, "Radiation Properties of Microstrip Dipoles," *IEEE Trans. Antennas Propagat.*, Vol. AP-27, No. 6, pp. 853–858, November 1979.
15. I. J. Bahl and P. Bhartia, *Microstrip Antennas*, Artech House, Dedham, MA, 1980.
16. K. R. Carver and J. W. Mink, "Microstrip Antenna Technology," *IEEE Trans. Antennas Propagat.*, Vol. AP-29, No. 1, pp. 2–24, January 1981.
17. R. J. Mailloux, J. F. McIlvenna, and N. P. Kernweis, "Microstrip Array Technology," *IEEE Trans. Antennas Propagat.*, Vol. AP-29, No. 1, pp. 25–27, January 1981.
18. W. F. Richards, Y. T. Lo, and D. D. Harrison, "An Improved Theory of Microstrip Antennas with Applications," *IEEE Trans. Antennas Propagat.*, Vol. AP-29, No. 1, pp. 38–46, January 1981.
19. E. H. Newman and P. Tylyathan, "Analysis of Microstrip Antennas Using Moment Methods," *IEEE Trans. Antennas Propagat.*, Vol. AP-29, No. 1, pp. 47–53, January 1981.
20. D. C. Chang, "Analytical Theory of an Unloaded Rectangular Microstrip Patch," *IEEE Trans. Antennas Propagat.*, Vol. AP-29, No. 1, pp. 54–62, January 1981.
21. T. Itoh and W. Menzel, "A Full-Wave Analysis Method for Open Microstrip Structures," *IEEE Trans. Antennas Propagat.*, Vol. AP-29, No. 1, pp. 63–68, January 1981.
22. I. E. Rana and N. G. Alexopoulos, "Current Distribution and Input Impedance of Printed Dipoles," *IEEE Trans. Antennas Propagat.*, Vol. AP-29, No. 1, pp. 99–105, January 1981.
23. N. G. Alexopoulos and I. E. Rana, "Mutual Impedance Computation Between Printed Dipoles," *IEEE Trans. Antennas Propagat.*, Vol. AP-29, No. 1, pp. 106–111, January 1981.
24. J. R. James, P. S. Hall, C. Wood, and A. Henderson, "Some Recent Developments in Microstrip Antenna Design," *IEEE Trans. Antennas Propagat.*, Vol. AP-29, No. 1, pp. 124–128, January 1981.

25. M. D. Deshpande and M. C. Bailey, "Input Impedance of Microstrip Antennas," *IEEE Trans. Antennas Propagat.*, Vol. AP-30, No. 4, pp. 645–650, July 1982.
26. M. C. Bailey and M. D. Deshpande, "Integral Equation Formulation of Microstrip Antennas," *IEEE Trans. Antennas Propagat.*, Vol. AP-30, No. 4, pp. 651–656, July 1982.
27. D. M. Pozar, "Input Impedance and Mutual Coupling of Rectangular Microstrip Antenna," *IEEE Trans. Antennas Propagat.*, Vol. AP-30, No. 6, pp. 1191–1196, November 1982.
28. D. M. Pozar, "Considerations for Millimeter-Wave Printed Antennas," *IEEE Trans. Antennas Propagat.*, Vol. AP-31, No. 5, pp. 740–747, September 1983.
29. E. F. Kuester and D. C. Chang, "A Geometrical Theory for the Resonant Frequencies and Q- Factors of Some Triangular Microstrip Patch Antennas," *IEEE Trans. Antennas Propagat.*, Vol. AP-31, No. 1, pp. 27–34, January 1983.
30. P. B. Katehi and N. G. Alexopoulos, "On the Modeling of Electromagnetically Coupled Microstrip Antennas-The Printed Strip Dipole," *IEEE Trans. Antennas Propagat.*, Vol. AP-32, No. 11, pp. 1179–1186, November 1984.
31. D. M. Pozar, "Analysis of Finite Phased Arrays of Printed Dipoles," *IEEE Trans. Antennas Propagat.*, Vol. AP-33, No. 10, pp. 1045–1053, October 1985.
32. J. R. James, P. S. Hall, and C. Wood, *Microstrip Antenna Theory and Design*, Peter Peregrinus, London, UK, 1981.
33. R. E. Munson, "Microstrip Antennas," Chapter 7 in *Antenna Engineering Handbook* (R. C. Johnson and H. Jasik, eds.), McGraw-Hill Book Co., New York, 1984.
34. W. F. Richards, "Microstrip Antennas," Chapter 10 in *Antenna Handbook: Theory, Applications and Design* (Y. T. Lo and S. W. Lee, eds.), Van Nostrand Reinhold Co., New York, 1988.
35. J. R. James and P. S. Hall, *Handbook of Microstrip Antennas*, Vols. 1 and 2, Peter Peregrinus, London, UK, 1989.
36. P. Bhartia, K. V. S. Rao, and R. S. Tomar, *Millimeter-Wave Microstrip and Printed Circuit Antennas*, Artech House, Boston, MA, 1991.
37. J. R. James, "What's New In Antennas," *IEEE Antennas Propagat. Mag.*, Vol. 32, No. 1, pp. 6–18, February 1990.
38. D. M. Pozar, "Microstrip Antennas," *Proc. IEEE*, Vol. 80, No. 1, pp. 79–81, January 1992.
39. D. H. Schaubert, F. G. Farrar, A. Sindoris, and S. T. Hayes, "Microstrip Antennas with Frequency Agility and Polarization Diversity," *IEEE Trans. Antennas Propagat.*, Vol. AP-29, No. 1, pp. 118–123, January 1981.
40. P. Bhartia and I. J. Bahl, "Frequency Agile Microstrip Antennas," *Microwave Journal*, pp. 67–70, October 1982.
41. W. F. Richards and Y. T. Lo, "Theoretical and Experimental Investigation of a Microstrip Radiator with Multiple Lumped Linear Loads," *Electromagnetics*, Vol. 3, No. 3–4, pp. 371–385, July–December 1983.
42. W. F. Richards and S. A. Long, "Impedance Control of Microstrip Antennas Utilizing Reactive Loading," *Proc. Intl. Telemetering Conf.*, pp. 285–290, Las Vegas, 1986.
43. W. F. Richards and S. A. Long, "Adaptive Pattern Control of a Reactively Loaded, Dual-Mode Microstrip Antenna," *Proc. Intl. Telemetering Conf.*, pp. 291–296, Las Vegas, 1986.
44. M. P. Purchine and J. T. Aberle, "A Tunable L-Band Circular Microstrip Patch Antenna," *Microwave Journal*, pp. 80, 84, 87, and 88, October 1994.
45. C. M. Krowne, "Cylindrical-Rectangular Microstrip Antenna," *IEEE Trans. Antennas Propagat.*, Vol. AP-31, No. 1, pp. 194–199, January 1983.
46. S. B. De Assis Fonseca and A. J. Giarola, "Microstrip Disk Antennas, Part I: Efficiency of Space Wave Launching," *IEEE Trans. Antennas Propagat.*, Vol. AP-32, No. 6, pp. 561–567, June 1984.
47. S. B. De Assis Fonseca and A. J. Giarola, "Microstrip Disk Antennas, Part II: The Problem of Surface Wave Radiation by Dielectric Truncation," *IEEE Trans. Antennas Propagat.*, Vol. AP-32, No. 6, pp. 568–573, June 1984.

48. J. Huang, "The Finite Ground Plane Effect on the Microstrip Antenna Radiation Patterns," *IEEE Trans. Antennas Propagat.*, Vol. AP-31, No. 7, pp. 649–653, July 1983.

49. I. Lier and K. R. Jakobsen, "Rectangular Microstrip Patch Antennas with Infinite and Finite Ground-Plane Dimensions," *IEEE Trans. Antennas Propagat.*, Vol. AP-31, No. 6, pp. 978–984, November 1983.

50. R. J. Mailloux, "On the Use of Metallized Cavities in Printed Slot Arrays with Dielectric Substrates," *IEEE Trans. Antennas Propagat.*, Vol. AP-35, No. 5, pp. 477–487, May 1987.

51. J. T. Aberle and F. Zavosh, "Analysis of Probe-Fed Circular Microstrip Patches Backed by Circular Cavities," *Electromagnetics*, Vol. 14, pp. 239–258, 1994.

52. A. Henderson, J. R. James, and C. M. Hall, "Bandwidth Extension Techniques in Printed Conformal Antennas," *Military Microwaves*, Vol. MM 86, pp. 329–334, 1986.

53. H. F. Pues and A. R. Van de Capelle, "An Impedance Matching Technique for Increasing the Bandwidth of Microstrip Antennas," *IEEE Trans. Antennas Propagat.*, Vol. AP-37, No. 11, pp. 1345–1354, November 1989.

54. J. J. Schuss, J. D. Hanfling, and R. L. Bauer, "Design of Wideband Patch Radiator Phased Arrays," *IEEE Antennas Propagat. Symp. Dig.*, pp. 1220–1223, 1989.

55. C. H. Tsao, Y. M. Hwang, F. Kilburg, and F. Dietrich, "Aperture-Coupled Patch Antennas with Wide-Bandwidth and Dual Polarization Capabilities," *IEEE Antennas Propagat. Symp. Dig.*, pp. 936–939, 1988.

56. A. Ittipiboon, B. Clarke, and M. Cuhaci, "Slot-Coupled Stacked Microstrip Antennas," *IEEE Antennas Propagat. Symp. Dig.*, pp. 1108–1111, 1990.

57. S. Sabban, "A New Broadband Stacked Two-Layer Microstrip Antenna," *IEEE Antennas Propagat. Symp. Dig.*, pp. 63–66, 1983.

58. C. H. Chen, A. Tulintseff, and M. Sorbello, "Broadband Two-Layer Microstrip Antenna," *IEEE Antennas Propagat. Symp. Dig.*, pp. 251–254, 1984.

59. R. W. Lee, K. F. Lee, and J. Bobinchak, "Characteristics of a Two-Layer Electromagnetically Coupled Rectangular Patch Antenna," *Electron. Lett.*, Vol. 23, pp. 1070–1072, September 1987.

60. W. F. Richards, S. Davidson, and S. A. Long, "Dual-Band, Reactively Loaded Microstrip Antennas," *IEEE Trans. Antennas Propagat.*, Vol. AP-33, No. 5, pp. 556–561, May 1985.

61. D. M. Pozar and B. Kaufman, "Increasing the Bandwidth of a Microstrip Antenna by Proximity Coupling," *Electronic Letters*, Vol. 23, pp. 368–369, April 1987.

62. N. W. Montgomery, "Triple-Frequency Stacked Microstrip Element," *IEEE Antennas Propagat. Symp. Dig.*, pp. 255–258, Boston, MA, 1984.

63. D. M. Pozar and D. H. Schaubert, "Scan Blindness in Infinite Phased Arrays of Printed Dipoles," *IEEE Trans. Antennas Propagat.*, Vol. AP-32, No. 6, pp. 602–610, June 1984.

64. D. M. Pozar, "Finite Phased Arrays of Rectangular Microstrip Antennas," *IEEE Trans. Antennas Propagat.*, Vol. AP-34, No. 5, pp. 658–665, May 1986.

65. F. Zavosh and J. T. Aberle, "Infinite Phased Arrays of Cavity-Backed Patches," Vol. AP-42, No. 3, pp. 390–398, March 1994.

66. H. G. Oltman and D. A. Huebner, "Electromagnetically Coupled Microstrip Dipoles," *IEEE Trans. Antennas Propagat.*, Vol. AP-29, No. 1, pp. 151–157, January 1981.

67. D. M. Pozar, "A Microstrip Antenna Aperture Coupled to a Microstrip Line," *Electronic Letters*, Vol. 21, pp. 49–50, January 1985.

68. G. Gronau and I. Wolff, "Aperture-Coupling of a Rectangular Microstrip Resonator," *Electronic Letters*, Vol. 22, pp. 554–556, May 1986.

69. H. A. Bethe, "Theory of Diffractions by Small Holes," *Physical Review*, Vol. 66, pp. 163–182, 1944.

70. R. E. Collin, *Foundations for Microwave Engineering*, Chapter 6, McGraw-Hill Book Co., New York, 1992.

71. J. R. Mosig and F. E. Gardiol, "General Integral Equation Formulation for Microstrip Antennas and Scatterers," *Proc. Inst. Elect. Eng.*, Pt. H, Vol. 132, pp. 424–432, 1985.

72. N. G. Alexopoulos and D. R. Jackson, "Fundamental Superstrate (Cover) Effects on Printed Circuit Antennas," *IEEE Trans. Antennas Propagat.*, Vol. AP-32, No. 8, pp. 807–816, August 1984.
73. C. C. Liu, A. Hessel, and J. Shmoys, "Performance of Probe-Fed Rectangular Microstrip Patch Element Phased Arrays," *IEEE Trans. Antennas Propagat.*, Vol. AP-36, No. 11, pp. 1501–1509, November 1988.
74. J. T. Aberle and D. M. Pozar, "Analysis of Infinite Arrays of One- and Two-Probe-Fed Circular Patches," *IEEE Trans. Antennas Propagat.*, Vol. AP-38, No. 4, pp. 421–432, April 1990.
75. E. H. Van Lil and A. R. Van de Capelle, "Transmission-Line Model for Mutual Coupling Between Microstrip Antennas," *IEEE Trans. Antennas Propagat.*, Vol. AP-32, No. 8, pp. 816–821, August 1984.
76. K. Malkomes, "Mutual Coupling Between Microstrip Patch Antennas," *Electronic Letters*, Vol. 18, No. 122, pp. 520–522, June 1982.
77. E. Penard and J.-P. Daniel, "Mutual Coupling Between Microstrip Antennas," *Electronic Letters*, Vol. 18, No. 4, pp. 605–607, July 1982.
78. D. H. Schaubert, D. M. Pozar, and A. Adrian, "Effect of Microstrip Antenna Substrate Thickness and Permittivity: Comparison of Theories and Experiment," *IEEE Trans. Antennas Propagat.*, Vol. AP-37, No. 6, pp. 677–682, June 1989.
79. C. A. Balanis, *Advanced Engineering Electromagnetics*, Second Edition, John Wiley & Sons, New York, 2012.
80. E. O. Hammerstad, "Equations for Microstrip Circuit Design," *Proc. Fifth European Microwave Conf.*, pp. 268–272, September 1975.
81. R. F. Harrington, *Time-Harmonic Electromagnetic Fields*, McGraw-Hill Book Co., p. 183, 1961.
82. D. R. Jackson and N. Alexopoulos, "Simple Approximate Formulas for Input Resistance, Bandwidth and Efficiency of a Resonant Rectangular Patch," *IEEE Trans. Antennas Propagat.*, Vol. AP-39, No. 3, pp. 407–410, March 1991.
83. R. E. Collin and F. J. Zucker, *Antenna Theory*, Part I, Chapter 5, McGraw-Hill Book Co., New York, 1969.
84. E. J. Martin, "Radiation Fields of Circular Loop Antennas by a Direct Integration Process," *IRE Trans. Antennas Propagat.*, Vol. AP-8, pp. 105–107, January 1960.
85. R. J. Collier and P. D. White, "Surface Waves in Microstrip Circuits," *Proc. 6th European Microwave Conference*, 1976, pp. 632–636.
86. W. F. Richards, J. R. Zinecker, R. D. Clark, and S. A. Long, "Experimental and Theoretical Investigation of the Inductance Associated with a Microstrip Antenna Feed," *Electromagnetics*, Vol. 3, No. 3–4, pp. 327–346, July–December 1983.
87. L. B. Felsen and N. Marcuvitz, *Radiation and Scattering of Waves*, Prentice-Hall, Englewood Cliffs, NJ, 1973.
88. J. Huang, "A Technique for an Array to Generate Circular Polarization with Linearly Polarized Elements," *IEEE Trans. Antennas Propagat.*, Vol. AP-34, No. 9, pp. 1113–1124, September 1986.
89. J. Huang, "Circularly Polarized Conical Patterns from Circular Microstrip Antennas," *IEEE Trans. Antennas Propagat.*, Vol. AP-32, No. 9, pp. 991–994, September 1984.
90. T. A. Milligan, *Modern Antenna Design*, McGraw-Hill Book Co., New York, 1985.
91. R. J. Mailloux, "Phase Array Theory and Technology," *Proc. IEEE*, Vol. 70, No. 3, pp. 246–291, March 1982.
92. L. Zhang, J. A. Casteneda, and N. G. Alexopoulos, "Scan Blindness Free Phased Array Design Using PBG Materials," *IEEE Trans. Antennas Propagat.*, Vol. 52, No. 8, pp. 2000–2007, Aug. 2004.
93. T. Taga, K. Tsunekawa, and A. Sasaki, "Antennas for Detachable Mobile Radio Units," Review of the ECL, NTT, Japan, Vol. 35, No. 1, pp. 59–65, Jan. 1987.
94. Taga, T., and K. Tsunekawa, "Performance Analysis of a Built-In Planar Inverted-F Antenna for 800 MHz Band Portable Radio Units," *IEEE Journal on Selected Areas in Communications*, Vol. 5, No. 5, pp. 921–929, 1987.

95. K. L. Melde, H.-J. Park, H.-H. Yeh, B. Fankem, Z. Zhou, and W. R. Eisenstadt., "Software Defined Match Control Circuit Integrated with a Planar Inverted F Antenna," *IEEE Trans. Antennas Propagat.,* Vol. 58, No. 12, pp. 3884–3890, Oct. 2010.

96. P. J. Bevelacqua (2014, Aug. 14). *Antenna Theory.* Retrieved from http://www.antenna-theory.com

97. D. M. Nashaat, H. A. Elsadek, and H. Ghali, "Single Feed Compact Quad-Band PIFA Antenna for Wireless Communication Applications," *IEEE Trans. Antennas Propagat.*, Vol. 53, No. 8, pp. 2631–2635, Aug. 2005.

98. A. Cabedo, J. Anguera, C. Picher, M. Ribo, and C. Puente, "Multiband Handset Antenna Combining a PIFA, Slots, and Ground Plane Modes," *IEEE Trans. Antennas Propagat.*, Vol. 57, No. 9, pp. 2526–2533, Sept. 2009.

99. A. C. Polycarpou, C. A. Balanis, J. T. Aberle, and C. R. Birtcher, "Radiation and Scattering from Ferrite-Tuned Cavity-Backed Slot Antennas: Theory and Experiment," *IEEE Trans. Antennas Propagat.*, Vol. 46, No. 9, pp. 1297–1306, Sept. 1998.

100. V. G. Kononov, C. A. Balanis, and C. R. Birtcher, "Analysis, Simulation and Measurements of CBS Antennas Loaded with Non-Uniformly Biased Ferrite Material," *IEEE Trans. Antennas Propagat.*, Vol. 60, No. 4, pp. 1717–1726, Apr. 2012.

101. V. G. Kononov, C. A. Balanis and C. R. Birtcher, "The Impact of the Non-Uniform Bias Field on the Radiation Patterns of the Ferrite-Loaded CBS Antennas," *IEEE Trans. Antennas Propagat.*, Vol. 61, No. 8, pp. 4367–4371, Aug. 2013.

102. M. Askarian Amiri, C. A. Balanis, and C. R. Birtcher, "Gain and Bandwidth Enhancement of Ferrite-Loaded CBS Antenna Using Material Shaping and Positioning," *IEEE Antennas and Wireless Propagation Letters*, Vol. 61, pp. 611–614, 2013.

103. C. Soras, M. Karaboikis, G. Tsachtsiris, and V. Makios, "Analysis and Design of an Inverted-F Antenna Printed on a PCMCIA Card for the 2.4 GHz ISM Band," *IEEE Antennas Propagat. Magazine*, Vol. 44, No. 1, pp. 37–44, Feb. 2002.

104. Z. Zhang, *Antenna Design for Mobile Devices*, John Wiley & Sons, Asia, 2011.

105. H.-Y. D. Yang, "Miniaturized Printed Wire Antenna for Wireless Communications," *IEEE Antennas Wireless Propag. Lett.,* Vol. 4, pp. 358–361, 2005.

106. D. K. Karmokar, and K. M. Morshed, "Analysis of Inverted-F and Loaded Inverted-F Antennas for 2.4 GHz ISM Band Applications," *Journal of Electrical Engineering*, Vol. 36, No. 2, pp. 4–9, 2009.

107. K. F. Lee, S. L. S. Yang, A. A. Kishk, and K. M. Luk, "The Versatile U-Slot Patch Antenna," *IEEE Antennas Propagat. Magazine*, Vol. 52, No. 1, pp. 71–88, Feb. 2010.

108. R. D. Richtmyer, "Dielectric Resonator," *J. Appl. Phys.*, Vol. 10, pp. 391–398, June 1939.

109. A. Okaya, "The Rutile Microwave Resonator," *Proc. IRE*, Vol. 48, p. 1921, Nov. 1960.

110. A. Okaya and L. F. Barash, "The Dielectric Microwave Resonator," *Proc. IRE*, Vol. 50, pp. 2081–2092, October 1962.

111. H. Y. Yee, "Natural Resonant Frequencies of Microwave Dielectric Resonators," *IEEE Trans. Microwave Theory Tech.*, Vol. MTT-13, p. 256, Mar. 1965.

112. S. J. Fiedziuszko, "Microwave Dielectric Resonators," *Microwave J.*, pp. 189–200, September 1980.

113. M. W. Pospieszalski, "Cylindrical Dielectric Resonators and Their Applications in TEM Line Microwave Circuits," *IEEE Trans. Microwave Theory Tech.*, Vol. MTT-27, No. 3, pp. 233–238, Mar. 1979.

114. K. A. Zaki and C. Chen, "Loss Mechanisms in Dielectric-Loaded Resonators," *IEEE Trans. Microwave Theory Tech.*, Vol. MTT-33, No. 12, pp. 1448–1452, Dec. 1985.

115. J. Van Bladel, "The Excitation of Dielectric Resonators of Very High Permittivity," *IEEE Trans. Microwave Theory Tech.*, Vol. MTT-23, pp. 208–217, 1975.

116. S. A. Long, M. W. McAllister, and L. C. Shen, "The Resonant Cylindrical Dielectric Cavity Antenna," *IEEE Trans. Antennas Propagat.*, Vol. 19, pp. 406–412, May 1983.

117. M. W. McAllister, S. A. Long, and G. L. Conway, "Rectangular Dielectric Resonator Antenna," *Electronic Letters*, Vol. 19, pp. 218–219, March 1983.

118. M. W. McAllister and S. A. Long, "Resonant Hemispherical Dielectric Antenna," *Electronic Letters*, Vol. 20, pp. 657–659, Aug. 1984.

119. K. M. Luk and K. W. Leung (editors), *Dielectric Resonator Antennas*, Research Studies Press Ltd., Baldock, Hertfordshire, England, 2003.
120. R. K. Mongia and P. Bhartia, "Dielectric Resonator Antennas: A Review and General Design Relations for Resonant Frequency and Bandwidth," *International J. Microwave and Millimeter-Wave Computer-Aided Engineering*, John Wiley & Sons, Inc., Vol. 4, No. 3, pp. 230–247, 1994.
121. A. Petosa, *Dielectric Resonator Antenna Handbook*, Artech House, Norwood, MA, 2007.
122. A. A. Kishk and Y. M. M. Antar, "Dielectric Resonator Antennas," Chapter 17 in *Antenna Engineering Handbook* (J. L. Volakis, editor), McGraw-Hill, 2007.
123. Y. Kobayashi and S. Tanaka, "Resonant Modes of a Dielectric Rod Resonator Short-Circuited at Both Ends by Parallel Conducting Plates," *IEEE Trans. Microwave Theory Tech.*, Vol. MTT-28, pp. 1077–1085, Oct. 1980.
124. A. A. Kishk, A. W. Glisson, and J. P. Junker, "Bandwidth Enhancement for Split Cylindrical Dielectric Resonator Antennas," *PIERS 33*, pp. 97–118, 2001.
125. R. K. Mongia, C. L. Larose, S. R. Mishra, and P. Bhartia, "Accurate Measurement of *Q*-Factors of Isolated Dielectric Resonators," *IEEE Trans. MTT*, Vol. 42, No. 8, pp. 1463–1467, Aug. 1994.

PROBLEMS

14.1. A microstrip line is used as a feed line to a microstrip patch. The substrate of the line is alumina ($\varepsilon_r \simeq 10$) while the dimensions of the line are $w/h = 1.2$ and $t/h = 0$. Determine the effective dielectric constant and characteristic impedance of the line. Compare the computed characteristic impedance to that of a 50-ohm line.

14.2. A microstrip transmission line of beryllium oxide ($\varepsilon_r \simeq 6.8$) has a width-to-height ratio of $w/h = 1.5$. Assuming that the thickness-to-height ratio is $t/h = 0$, determine:
(a) effective dielectric constant
(b) characteristic impedance of the line

14.3. A microstrip line, *which is open at one end and extends to infinity toward the other end*, has a center conductor *width* $= 0.4\lambda_o$, substrate height of $0.05\lambda_o$, and it is operating at *10 GHz*. The dielectric constant of the substrate is *2.25*. This type of microstrip line is used to construct rectangular patch antennas. Determine the following:
(a) The *input admittance (real and imaginary parts)* of the microstrip line at the leading open edge. *Is it capacitive or inductive?*
(b) What kind of a lumped element *(capacitor or inductor)* can be placed at the leading open edge between the center conductor of the line and its ground plane *to resonate the admittance? What is the value of the lumped element?*
(c) The new *input impedance, taking into account the presence of the lumped element.*

14.4. Design a *rectangular* microstrip antenna so that it will resonate at *2 GHz*. The idealistic lossless substrate (RT/Duroid 6010.2) has a dielectric constant of *10.2* and a height of *0.05 in. (0.127 cm)*.
(a) Determine the *physical dimensions (width* and *length)* of the patch *(in cm)*.
(b) *Approximate range of lengths (in cm)* between the two radiating slots of the rectangular patch, *if we want the input impedance (taking into account both radiating slots) to be real*.
(c) What is the *real input impedance of Part b? Neglect coupling*.
(d) Location *(in cm from the leading radiating slot)* of a coaxial feed so that the *total input impedance is 150 ohms*.

14.5. Design a rectangular microstrip antenna to resonate at 9 GHz using a substrate with a dielectric constant of 2.56. Determine the following:
   (a) Directivity of a single radiating slot (*dimensionless* and in *dB*). Use the cavity model.
   (b) Approximate directivity of the entire patch (*dimensions* and in *dB*). Use the cavity model and neglect coupling between the two slots.

14.6. It is desired to design rectangular microstrip patch antenna to resonate at a frequency of $f = 10$ GHz. Using a substrate with a dielectric constant of 4 ($\varepsilon_r = 4$) and a height of $h = 0.25$ cm, determine the:
   (a) Width of the microstrip patch (*in cm*).
   (b) Effective dielectric constant (ε_{reff}) of the substrate.
   (c) Effective length L_e of the patch (*in cm*).
   (d) Physical length L of the patch (*in cm*).

14.7. A rectangular microstrip antenna was designed, without taking into account fringing effects from any of the four edges of the patch, to operate at a center frequency of 4.6 GHz. The width of the patch was chosen to be $W = 1.6046$ cm and the substrate had a height of 0.45 cm and a dielectric constant of 6.8. However, when the patch was tested, it was found to resonate at a frequency of 4.046 GHz!
   (a) Find the physical length L of the patch (*in cm*).
   (b) Why did the patch resonate at 4.046 GHz, instead of the designed frequency of 4.6 GHz? Verify the new resonant frequency. Must justify your answer mathematically. Show that the measured resonant frequency is correct.

14.8. Cellular and mobile telephony, using earth-based repeaters, has received wide acceptance and has become an essential means of communication for business, even for the household. Cellular telephony by satellites is the wave of the future and communication systems are being designed for that purpose. The present allocated frequency band for satellites is at *L*-band ($\simeq 1.6$ GHz). Various antennas are being examined for that purpose; one candidate is the microstrip patch antenna. Design a rectangular microstrip patch antenna, based on the dominant mode, that can be mounted on the roof of a car to be used for satellite cellular telephone. The designed center frequency is 1.6 GHz, the dielectric constant of the substrate is 10.2 (i.e., RT/duroid), and the thickness of the substrate is 0.127 cm. Determine the
   (a) dimensions of the rectangular patch (in cm)
   (b) resonant input impedance, assuming no coupling between the two radiating slots
   (c) mutual conductance between the two radiating slots of the patch
   (d) resonant input impedance, taking into account coupling
   (e) position of the feed to match the patch antenna to a 75-ohm line

14.9. Repeat the design of Problem 14.8 using a substrate with a dielectric constant of 2.2 (i.e., RT/duroid 5880) and with a height of 0.1575 cm. Are the new dimensions of the patch realistic for the roof of a personal car?

14.10. Design a rectangular microstrip patch with dimensions W and L, over a single substrate, whose center frequency is 10 GHz. The dielectric constant of the substrate is 10.2 and the height of the substrate is 0.127 cm (0.050 in.). Determine the physical dimensions W and L (in cm) of the patch, taking into account field fringing.

14.11. Using the transmission-line model of Figure 14.9(b), derive (14-14)–(14-15).

14.12. To take into account coupling between the two radiating slots of a rectangular microstrip patch, the resonant input resistance is represented by (14-17). Justify, explain, and/or show why the plus (+) sign is used for modes with odd (antisymmetric) resonant voltage distributions beneath the patch while the minus (−) sign is used for modes with even (symmetric) resonant voltage distributions.

14.13. Show that for typical rectangular microstrip patches $G_1/Y_c \ll 1$ and $B_1/Y_c \ll 1$ so that (14-20) reduces to (14-20a).

14.14. A rectangular microstrip patch antenna is operating at 10 GHz with $\varepsilon_r = 10.2$ and dimensions of length $L = 0.4097$ cm, width $W = 0.634$ cm, and substrate height $h = 0.127$ cm. It is desired to feed the patch using a probe feed. Neglecting mutual coupling, calculate:
(a) What is the input impedance of the patch at one of the radiating edges based on the transmission-line model?
(b) At what distance y_0 (in cm) from one of the radiating edges should the coax feed be placed so that the input impedance is 50 ohms?

14.15. A rectangular microstrip patch antenna, whose input impedance is 152.44 ohms at its leading radiating edge, is fed by a microstrip line as shown in Figure 14.11. Assuming the width of the feeding line is $W_0 = 0.2984$ cm, the height of the substrate is 0.1575 cm and the dielectric constant of the substrate is 2.2, at what distance y_0 should the microstrip patch antenna be fed so as to have a perfect match between the line and the radiating element? The overall microstrip patch element length is 0.9068 cm.

14.16. The rectangular microstrip patch of Example 14.2 is fed by a microstrip transmission line of Figure 14.5. In order to reduce reflections at the inset feed point between the line and the patch element, design the microstrip line so that its characteristic impedance matches that of the radiating element.

14.17. Repeat the design of Example 14.2 so that the input impedance of the radiating patch at the feed point is:
(a) 75 ohms
(b) 100 ohms

Then, assuming the feed line is a microstrip line, determine the dimensions of the line so that its characteristic impedance matches that of the radiating patch.

14.18. A rectangular microstrip patch antenna has dimensions of $L = 0.906$ cm, $W = 1.186$ cm, and $h = 0.1575$ cm. The dielectric constant of the substrate is $\varepsilon_r = 2.2$. Using the geometry of Figure 14.15 and assuming no fringing, determine the resonant frequency of the first 4 TM^z_{0np} modes, in order of ascending resonant frequency.

14.19. Derive the TM^z_{mnp} field configurations (modes) for the rectangular microstrip patch based on the geometry of Figure P14.19. Determine the:
(a) eigenvalues
(b) resonant frequency $(f_r)_{mnp}$ for the mnp mode.
(c) dominant mode if $L > W > h$
(d) resonant frequency of the dominant mode.

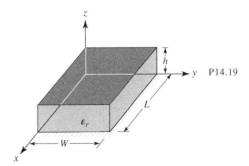

P14.19

14.20. A microstrip antenna with a rectangular patch, with dimensions $L = 5$ cm and $W = 2$ cm, and a substrate with $h = 0.1568$ cm and a dielectric constant of $\varepsilon_r = 2.2$, is used in a wireless communications system. *Accounting for fringing:*

(a) Identify the second-order TM$^x_{mnp}$ mode. Selection of the wrong mode will make all the other answers wrong, and no credit will be given for answers of the other parts of this problem for the wrong mode.

(b) Compute the resonant frequency of the 2nd-order TM$^x_{mnp}$ mode.

(c) Calculate the resonant input resistance (*assume no coupling*) of the 2nd-order TM$^x_{mnp}$ mode.

(d) Compute the resonant input resistance, of the 2nd-order TM$^x_{mnp}$ mode, when the mutual conductance of this mode is 0.24921×10^{-3} Siemens.

14.21. Repeat Problem 14.19 for the TM$^y_{mnp}$ modes based on the geometry of Figure P14.21.

14.22. Derive the array factor of (14-42).

14.23. Assuming the coordinate system for the rectangular microstrip patch is that of Problem 14.19 (Figure P14.19), derive based on the cavity model the

(a) far-zone electric field radiated by one of the radiating slots of the patch

(b) array factor for the two radiating slots of the patch

(c) far-zone total electric field radiated by both of the radiating slots

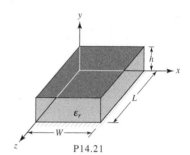

P14.21

14.24. Repeat Problem 14.23 for the rectangular patch geometry of Problem 14.21 (Figure P14.21).

14.25. Determine the directivity (in dB) of the rectangular microstrip patch of Example 14.3 using

(a) Kraus' approximate formula

(b) Tai & Pereira's approximate formula

14.26. Compute the directivity (in dB) of the rectangular microstrip patch of Problem 14.8.

14.27. Derive the directivity (in dB) of the rectangular microstrip patch of Problem 14.9.

14.28. For a circular microstrip patch antenna operating in the dominant TM$^z_{110}$ mode, derive the far-zone electric fields radiated by the patch based on the cavity model.

14.29. Using the cavity model, derive the TM$^z_{mnp}$ resonant frequencies for a microstrip patch whose shape is that of a half of a circular patch (semicircle).

14.30. Repeat Problem 14.29 for a 90° circular disc (angular sector of 90°) microstrip patch.

14.31. Repeat Problem 14.29 for the circular sector microstrip patch antenna whose geometry is shown in Figure P14.31.

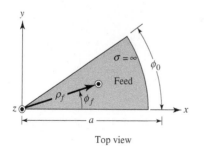

Top view

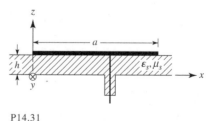

P14.31

Side view

14.32. Repeat Problem 14.29 for the annular microstrip patch antenna whose geometry is shown in Figure P14.32.

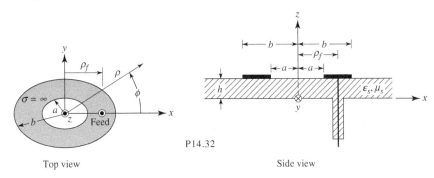

P14.32

Top view   Side view

14.33. Repeat Problem 14.29 for the annular sector microstrip patch antenna whose geometry is shown in Figure P14.33.

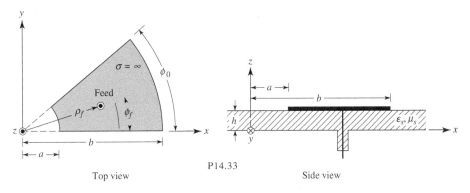

P14.33

Top view   Side view

14.34. Repeat the design of Problem 14.8 for a circular microstrip patch antenna operating in the *dominant* TM_{110}^z mode. Use $\sigma = 10^7$ S/m and $\tan\delta = 0.0018$.

14.35. Repeat the design of Problem 14.9 for a circular microstrip patch antenna operating in the *dominant* TM_{110}^z mode. Use $\sigma = 10^7$ S/m and $\tan\delta = 0.0018$.

14.36. For ground-based cellular telephony, the desired pattern coverage is omnidirectional and similar to that of a monopole (*with a null toward zenith*, $\theta = 0°$). This can be accomplished using a *circular microstrip patch* antenna operating in a higher order mode, such as the TM_{210}^z. Assuming the desired resonant frequency is 900 MHz, design a circular microstrip patch antenna operating in the TM_{210}^z mode. Assuming a substrate with a dielectric constant of 10.2 and a height of 0.127 cm:

(a) Derive an expression for the resonant frequency of the TM_{210}^z mode;
(b) Determine the radius of the circular patch (*in cm*). Neglect fringing.

14.37. Microstrip (patch) antennas are usually designed so that the maximum of the amplitude radiation pattern is perpendicular to the patch. However for ground-based cellular telephony, the pattern of the antenna should usually match that of a vertical monopole with a null towards zenith ($\theta = 0°$). This can be accomplished if a circular patch is selected and it is excited at a higher-order mode, such as the TM_{210}^z mode.

Assuming a TM_{210}^z mode (NOT dominant TM_{210}^z mode), the desired operating frequency, *without taking into account fringing*, is 1.9 GHz, the substrate has a dielectric constant of 10.2 and its height is 0.127 cm. Determine the:

(a) *Physical* radius of the circular patch (*in cm*). *Neglect fringing*.

(b) *Effective* radius of the circular patch (*in cm*). *Account for fringing.*

(c) New resonant frequency (*in GHz*) *taking into account fringing.*

P.S. *The design procedure of Section 14.3.3 is for the* TM_{110}^z *and should not be used. Other equations should be used. However Equation (14-67) is still valid for the* TM_{210}^z.

14.38. For ground-based cellular telephony, the desired pattern coverage is omnidirectional and similar to that of a monopole (with a null toward zenith). This can be accomplished using circular microstrip patch antennas operating in higher order modes, such as the TM_{210}^z, TM_{310}^z, TM_{410}^z, etc. Assuming that the desired resonant frequency is 900 MHz, design a circular microstrip patch antenna operating in the TM_{210}^z mode. Assuming a substrate with a dielectric constant of 10.2 and a height of 0.127 cm:

(a) Derive an expression for the resonant frequency.

(b) Determine the radius of the circular patch. Neglect fringing.

(c) Derive expressions for the far-zone radiated fields.

(d) Plot the normalized *E*- and *H*-plane amplitude patterns (in dB).

(e) Plot the normalized azimuthal (*x-y* plane) amplitude pattern (in dB).

(f) Determine the directivity (in dB) using the **Directivity** computer program of Chapter 2.

14.39. Repeat Problem 14.38 for the TM_{310}^z mode.

14.40. Repeat Problem 14.38 for the TM_{410}^z mode.

14.41. The diameter of a typical probe feed for a microstrip patch antenna is $d = 0.1$ cm. At $f = 10$ GHz, determine the feed reactance assuming a substrate with a dielectric constant of 2.2 and height of 0.1575 cm.

14.42. Determine the impedance of a single-section quarter-wavelength impedance transformer to match a 100-ohm patch element to a 50-ohm microstrip line. Determine the dimensions of the line assuming a substrate with a dielectric constant of 2.2 and a height of 0.1575 cm.

14.43. Repeat the design of Problem 14.42 using a two-section binomial transformer. Determine the dimensions of each section of the transformer.

14.44. Repeat the design of Problem 14.42 using a two-section Tschebyscheff transformer. Determine the dimensions of each section of the transformer.

14.45. A very thin ($w \ll \lambda$) half-guide wavelength slot, as shown in the Figure P14.45, is fed by a dipole of length slightly less than $\lambda_g/2$ (the ends of the dipole do not touch the ground plane forming the slot). Determine the impedance of the slot.

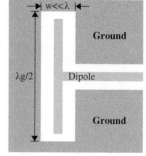

P14.45

14.46. For a cubic dielectric resonator of Figure 14.57, with dimensions of $a = 1$ cm, $b = 1$ cm, and $c = 0.3$ cm, determine the resonant frequencies for the first five TE^z and/or TM^z modes when: (a) $\varepsilon_r = 8.9$ (b) $\varepsilon_r = 89$. Use Matlab computer program **DRA_Analysis_Design**.

14.47. Repeat Problem 14.46 for a hemicylindrical dielectric resonator of Figure 14.59 with dimensions of $a = 0.3$ cm and $h = 1$ cm.

14.48. Repeat Problem 14.46 for the three dominant degenerate modes of a hemispherical dielectric resonator of Figure 14.60 with radius of 0.3 cm.

14.49. Design a cylindrical resonator of Figure 14.58, to operate in the $TM_{01\delta}$ mode and to exhibit a fractional bandwidth of 2.887%, VSWR = 3 and resonant frequency $f_r = 10$ GHz. Determine the:

(a) Q of the cavity
(b) Minimum dielectric constant to achieve the design
(c) Choose a dielectric constant of 38
(d) Determine the dimensions a and h (both in cm) based on a dielectric constant of 38. Use Matlab computer program **DRA_Analysis_Design.**

14.50. Design a cylindrical resonator of Figure 14.58, to operate in the $HE_{11\delta}$ mode and to exhibit a fractional bandwidth of 2.887%, VSWR = 3 and resonant frequency $f_r = 10$ GHz. Determine the:

(a) Q of the cavity
(b) Range of dielectric constants to achieve the design
(c) Choose a dielectric constant of 38
(d) Determine the dimensions a and h (both in cm) based on a dielectric constant of 38. Use Matlab computer program **DRA_Analysis_Design.**

CHAPTER 15

Reflector Antennas

15.1 INTRODUCTION

Reflector antennas, in one form or another, have been in use since the discovery of electromagnetic wave propagation in 1888 by Hertz. However the fine art of analyzing and designing reflectors of many various geometrical shapes did not forge ahead until the days of World War II when numerous radar applications evolved. Subsequent demands of reflectors for use in radio astronomy, microwave communication, and satellite tracking resulted in spectacular progress in the development of sophisticated analytical and experimental techniques in shaping the reflector surfaces and optimizing illumination over their apertures so as to maximize the gain. The use of reflector antennas for deep-space communication, such as in the space program and especially their deployment on the surface of the moon, resulted in establishing the reflector antenna almost as a household word during the 1960s. Although reflector antennas take many geometrical configurations, some of the most popular shapes are the plane, corner, and curved reflectors (especially the paraboloid), as shown in Figure 15.1, each of which will be discussed in this chapter. Many articles on various phases of the analysis and design of curved reflectors have been published and some of the most referenced can be found in a book of reprinted papers [1].

15.2 PLANE REFLECTOR

The simplest type of reflector is a plane reflector introduced to direct energy in a desired direction. The arrangement is that shown in Figure 15.1(a) which has been extensively analyzed in Section 4.7 when the radiating source is a vertical or horizontal linear element. It has been clearly demonstrated that the polarization of the radiating source and its position relative to the reflecting surface can be used to control the radiating properties (pattern, impedance, directivity) of the overall system. Image theory has been used to analyze the radiating characteristics of such a system. Although the infinite dimensions of the plane reflector are idealized, the results can be used as approximations for electrically large surfaces. The perturbations introduced by keeping the dimensions finite can be accounted for by using special methods such as the Geometrical Theory of Diffraction [2]–[5] which was introduced in Section 12.10.

Antenna Theory: Analysis and Design, Fourth Edition. Constantine A. Balanis.
© 2016 John Wiley & Sons, Inc. Published 2016 by John Wiley & Sons, Inc.
Companion Website: www.wiley.com/go/antennatheory4e

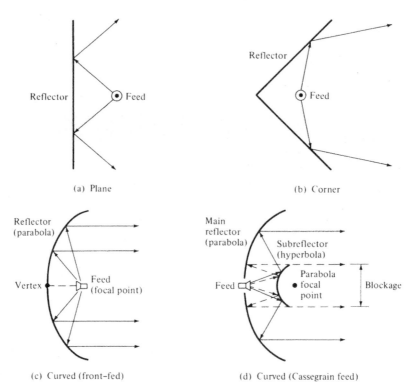

Figure 15.1 Geometrical configuration for some reflector systems.

15.3 CORNER REFLECTOR

To better collimate the energy in the forward direction, the geometrical shape of the plane reflector itself must be changed so as to prohibit radiation in the back and side directions. One arrangement which accomplishes that consists of two plane reflectors joined so as to form a corner, as shown in Figures 15.1(b) and in 15.2(a). This is known as the corner reflector. Because of its simplicity in construction, it has many unique applications. For example, if the reflector is used as a passive target for radar or communication applications, it will return the signal exactly in the same direction as it received it when its included angle is 90°. This is illustrated geometrically in Figure 15.2(b). Because of this unique feature, military ships and vehicles are designed with minimum sharp corners to reduce their detection by enemy radar. Corner reflectors are also widely used as receiving elements for home television.

In most practical applications, the included angle formed by the plates is usually 90°; however other angles are sometimes used. To maintain a given system efficiency, the spacing between the vertex and the feed element must increase as the included angle of the reflector decreases, and vice versa. For reflectors with infinite sides, the gain increases as the included angle between the planes decreases. This, however, may not be true for finite size plates. For simplicity, in this chapter it will be assumed that the plates themselves are infinite in extent ($l = \infty$). However, since in practice the dimensions must be finite, guidelines on the size of the aperture (D_a), length (l), and height (h) will be given.

The feed element for a corner reflector is almost always a dipole or an array of collinear dipoles placed parallel to the vertex a distance s away, as shown in a perspective view in Figure 15.2(c). Greater bandwidth is obtained when the feed elements are cylindrical or biconical dipoles instead of thin wires. In many applications, especially when the wavelength is large compared to tolerable

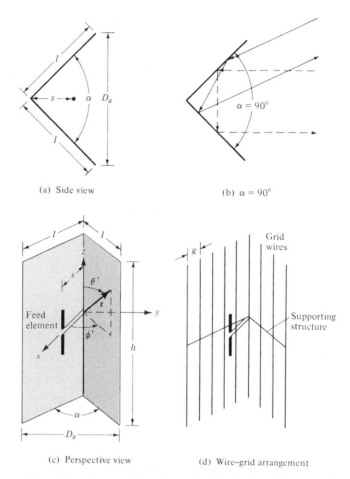

Figure 15.2 Side and perspective views of solid and wire-grid corner reflectors.

physical dimensions, the surfaces of the corner reflector are frequently made of grid wires rather than solid sheet metal, as shown in Figure 15.2(d). One of the reasons for doing that is to reduce wind resistance and overall system weight. The spacing (g) between wires is made a small fraction of a wavelength (usually $g \leq \lambda/10$). For wires that are parallel to the length of the dipole, as is the case for the arrangement of Figure 15.2(d), the reflectivity of the grid-wire surface is as good as that of a solid surface.

In practice, the aperture of the corner reflector (D_a) is usually made between one and two wavelengths ($\lambda < D_a < 2\lambda$). The length of the sides of a 90° corner reflector is most commonly taken to be about twice the distance from the vertex to the feed ($l \simeq 2s$). For reflectors with smaller included angles, the sides are made larger. The feed-to-vertex distance (s) is usually taken to be between $\lambda/3$ and $2\lambda/3$ ($\lambda/3 < s < 2\lambda/3$). For each reflector, there is an optimum feed-to-vertex spacing. If the spacing becomes too small, the radiation resistance decreases and becomes comparable to the loss resistance of the system which leads to an inefficient antenna. For very large spacing, the system produces undesirable multiple lobes, and it loses its directional characteristics. It has been experimentally observed that increasing the size of the sides does not greatly affect the beamwidth and directivity, but it increases the bandwidth and radiation resistance. The main lobe is somewhat broader for reflectors with finite sides compared to that of infinite dimensions. The height (h) of the reflector is usually taken to be about 1.2 to 1.5 times greater than the total length of the feed element, in order to reduce radiation toward the back region from the ends.

878 REFLECTOR ANTENNAS

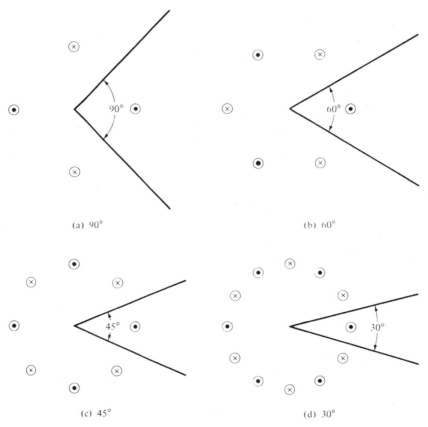

Figure 15.3 Corner reflectors and their images (with perpendicularly polarized feeds) for angles of 90°, 60°, 45°, and 30°.

The analysis for the field radiated by a source in the presence of a corner reflector is facilitated when the included angle (α) of the reflector is $\alpha = \pi/n$, where n is an integer ($\alpha = \pi, \pi/2, \pi/3, \pi/4$, etc.). For those cases ($\alpha = 180°, 90°, 60°, 45°$, etc.) it is possible to find a system of images, which when properly placed in the absence of the reflector plates, form an array that yields the same field within the space formed by the reflector plates as the actual system. The number of images, polarity, and position of each is controlled by the included angle of the corner reflector and the polarization of the feed element. In Figure 15.3 we display the geometrical and electrical arrangement of the images for corner reflectors with included angles of 90°, 60°, 45°, and 30° and a feed with perpendicular polarization. The procedure for finding the number, location, and polarity of the images is demonstrated graphically in Figure 15.4 for a corner reflector with a 90° included angle. It is assumed that the feed element is a linear dipole placed parallel to the vertex. A similar procedure can be followed for all other reflectors with an included angle of $\alpha = 180°/n$, where n is an integer.

15.3.1 90° Corner Reflector

The first corner reflector to be analyzed is the one with an included angle of 90°. Because its radiation characteristics are the most attractive, it has become the most popular.

Referring to the reflector of Figure 15.2(c) with its images in Figure 15.4(b), the total field of the system can be derived by summing the contributions from the feed and its images. Thus

$$\mathbf{E}(r, \theta, \phi) = \mathbf{E}_1(r_1, \theta, \phi) + \mathbf{E}_2(r_2, \theta, \phi) + \mathbf{E}_3(r_3, \theta, \phi) + \mathbf{E}_4(r_4, \theta, \phi) \qquad (15\text{-}1)$$

CORNER REFLECTOR

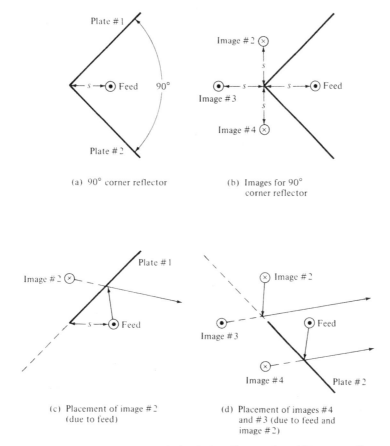

Figure 15.4 Geometrical placement and electrical polarity of images for a 90° corner reflector with a parallel polarized feed.

In the far-zone, the normalized scalar field can be written as

$$E(r, \theta, \phi) = f(\theta, \phi)\frac{e^{-jkr_1}}{r_1} - f(\theta, \phi)\frac{e^{-jkr_2}}{r_2} + f(\theta, \phi)\frac{e^{-jkr_3}}{r_3} - f(\theta, \phi)\frac{e^{-jkr_4}}{r_4}$$

$$E(r, \theta, \phi) = [e^{+jks\cos\psi_1} - e^{+jks\cos\psi_2} + e^{+jks\cos\psi_3} - e^{+jks\cos\psi_4}]f(\theta, \phi)\frac{e^{-jkr}}{r} \quad (15\text{-}2)$$

where

$$\cos\psi_1 = \hat{\mathbf{a}}_x \cdot \hat{\mathbf{a}}_r = \sin\theta\cos\phi \quad (15\text{-}2a)$$

$$\cos\psi_2 = \hat{\mathbf{a}}_y \cdot \hat{\mathbf{a}}_r = \sin\theta\sin\phi \quad (15\text{-}2b)$$

$$\cos\psi_3 = -\hat{\mathbf{a}}_x \cdot \hat{\mathbf{a}}_r = -\sin\theta\cos\phi \quad (15\text{-}2c)$$

$$\cos\psi_4 = -\hat{\mathbf{a}}_y \cdot \hat{\mathbf{a}}_r = -\sin\theta\sin\phi \quad (15\text{-}2d)$$

since $\hat{\mathbf{a}}_r = \hat{\mathbf{a}}_x \sin\theta\cos\phi + \hat{\mathbf{a}}_y \sin\theta\sin\phi + \hat{\mathbf{a}}_z \cos\theta$. Equation (15-2) can also be written, using (15-2a)–(15-2d), as

$$E(r, \theta, \phi) = 2[\cos(ks\sin\theta\cos\phi) - \cos(ks\sin\theta\sin\phi)]f(\theta, \phi)\frac{e^{-jkr}}{r} \quad (15\text{-}3)$$

where for $\alpha = \pi/2 = 90°$

$$0 \leq \theta \leq \pi, \quad \begin{array}{c} 0 \leq \phi \leq \alpha/2 \\ 2\pi - \alpha/2 \leq \phi \leq 2\pi \end{array} \tag{15-3a}$$

Letting the field of a single isolated (radiating in free-space) element to be

$$E_0 = f(\theta, \phi) \frac{e^{-jkr}}{r} \tag{15-4}$$

(15-3) can be rewritten as

$$\frac{E}{E_0} = \text{AF}(\theta, \phi) = 2[\cos(ks \sin\theta \cos\phi) - \cos(ks \sin\theta \sin\phi)] \tag{15-5}$$

Equation (15-5) represents not only the ratio of the total field to that of an isolated element at the origin but also the array factor of the entire reflector system. In the azimuthal plane ($\theta = \pi/2$), (15-5) reduces to

$$\frac{E}{E_0} = \text{AF}(\theta = \pi/2, \phi) = 2[\cos(ks \cos\phi) - \cos(ks \sin\phi)] \tag{15-6}$$

To gain some insight into the performance of a corner reflector, in Figure 15.5 we display the normalized patterns for an $\alpha = 90°$ corner reflector for spacings of $s = 0.1\lambda, 0.7\lambda, 0.8\lambda, 0.9\lambda$, and 1.0λ. It is evident that for the small spacings the pattern consists of a single major lobe whereas multiple lobes appear for the larger spacings ($s > 0.7\lambda$). For $s = \lambda$ the pattern exhibits two lobes separated by a null along the $\phi = 0°$ axis.

Another parameter of performance for the corner reflector is the field strength along the symmetry axis ($\theta = 90°, \phi = 0°$) as a function of feed-to-vertex distance s [6]. The normalized (relative to the field of a single isolated element) absolute field strength $|E/E_0|$ as a function of $s/\lambda (0 \leq s \leq 10\lambda)$ for $\alpha = 90°$ is shown plotted in Figure 15.6. It is apparent that the first field strength peak is achieved when $s = 0.5\lambda$, and it is equal to 4. The field is also periodic with a period of $\Delta s/\lambda = 1$.

15.3.2 Other Corner Reflectors

A similar procedure can be used to derive the array factors and total fields for all other corner reflectors with included angles of $\alpha = 180°/n$. Referring to Figure 15.3, it can be shown that the array factors for $\alpha = 60°, 45°$, and $30°$ can be written as

$$\boxed{\begin{array}{c} \alpha = 60° \\ \text{AF}(\theta, \phi) = 4\sin\left(\frac{X}{2}\right)\left[\cos\left(\frac{X}{2}\right) - \cos\left(\sqrt{3}\frac{Y}{2}\right)\right] \end{array}} \tag{15-7}$$

$$\boxed{\begin{array}{c} \alpha = 45° \\ \text{AF}(\theta, \phi) = 2\left[\cos(X) + \cos(Y) - 2\cos\left(\frac{X}{\sqrt{2}}\right)\cos\left(\frac{Y}{\sqrt{2}}\right)\right] \end{array}} \tag{15-8}$$

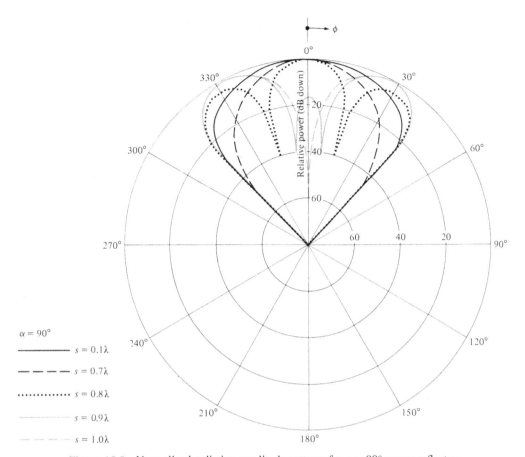

Figure 15.5 Normalized radiation amplitude patterns for $\alpha = 90°$ corner reflector.

$$\underline{\alpha = 30°}$$
$$\text{AF}(\theta, \phi) = 2\left[\cos(X) - 2\cos\left(\frac{\sqrt{3}}{2}X\right)\cos\left(\frac{Y}{2}\right) - \cos(Y) + 2\cos\left(\frac{X}{2}\right)\cos\left(\frac{\sqrt{3}}{2}Y\right)\right]$$

(15-9)

where

$$X = ks\sin\theta\cos\phi \qquad (15\text{-}9a)$$
$$Y = ks\sin\theta\sin\phi \qquad (15\text{-}9b)$$

These are assigned, at the end of the chapter, as exercises to the reader (Problem 15.2).

For a corner reflector with an included angle of $\alpha = 180°/n, n = 1, 2, 3, \ldots$, the number of images is equal to $N = (360/\alpha) - 1 = 2n - 1$.

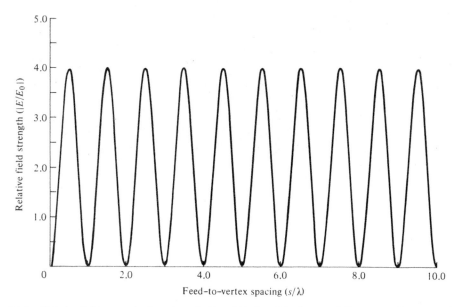

Figure 15.6 Relative field strength along the axis ($\theta = 90°$, $\phi = 0°$) of an $\alpha = 90°$ corner reflector as a function of feed-to-vertex spacing.

It has also been shown [7] by using long filament wires as feeds, that the azimuthal plane ($\theta = \pi/2$) array factor for corner reflectors with $\alpha = 180°/n$, where n is an integer, can also be written as

$n =$ even ($n = 2, 4, 6, \ldots$)

$$\text{AF}(\phi) = 4n(-1)^{n/2}[J_n(ks)\cos(n\phi) + J_{3n}(ks)\cos(3n\phi) + J_{5n}(ks)\cos(5n\phi) + \cdots] \quad (15\text{-}10a)$$

$n =$ odd ($n = 1, 3, 5, \ldots$)

$$\text{AF}(\phi) = 4nj(-1)^{(n-1)/2}[J_n(ks)\cos(n\phi) - J_{3n}(ks)\cos(3n\phi) + J_{5n}(ks)\cos(5n\phi) + \cdots] \quad (15\text{-}10b)$$

where $J_m(x)$ is the Bessel function of the first kind of order m (see Appendix V).

When n is not an integer, the field must be found by retaining a sufficient number of terms of the infinite series. It has also been shown [7] that for all values of $n = m$ (integral or fractional) that the field can be written as

$$\text{AF}(\phi) = 4m[e^{jm\pi/2}J_m(ks)\cos(m\phi) + e^{j3m\pi/2}J_{3m}(ks)\cos(3m\phi) + \cdots] \quad (15\text{-}11)$$

The array factor for a corner reflector, as given by (15-10a)–(15-11), has a form that is similar to the array factor for a uniform circular array, as given by (6-121). This should be expected since the feed sources and their images in Figure 15.3 form a circular array. The number of images increase as the included angle of the corner reflector decreases.

Patterns have been computed for corner reflectors with included angles of 60°, 45°, and 30°. It has been found that these corner reflectors have also single-lobed patterns for the smaller values of s, and they become narrower as the included angle decreases. Multiple lobes begin to appear when

$$s \simeq 0.95\lambda \quad \text{for} \quad \alpha = 60°$$

$$s \simeq 1.2\lambda \quad \text{for} \quad \alpha = 45°$$

$$s \simeq 2.5\lambda \quad \text{for} \quad \alpha = 30°$$

The field strength along the axis of symmetry ($\theta = 90°, \phi = 0°$) as a function of the feed-to-vertex distance s, has been computed for reflectors with included angles of $\alpha = 60°, 45°$, and $30°$. The results for $\alpha = 45°$ are shown in Figure 15.7 for $0 \leq s \leq 10\lambda$.

For reflectors with $\alpha = 90°$ and $60°$, the normalized field strength is periodic with periods of λ and 2λ, respectively. However, for the 45° and 30° reflectors the normalized field is not periodic but rather "*almost periodic*" or "*pseudoperiodic*" [8]. For the 45° and 30° reflectors the arguments of the trigonometric functions representing the arrays factors, and given by (15-8)–(15-9b), are related by irrational numbers and therefore the arrays factors do not repeat. However, when plotted they look very similar. Therefore when examined only graphically, the observer erroneously may conclude that the patterns are periodic (because they look so much the same). However, when the array factors are examined analytically it is concluded that the functions are not periodic but rather nearly periodic. The field variations are "nearly similar" in form in the range $\Delta s \simeq 16.69\lambda$ for the $\alpha = 45°$ and $\Delta s \simeq 30\lambda$ for the $\alpha = 30°$. Therefore the array factors of (15-8) and (15-9) belong to the class of *nearly periodic functions* [8].

It has also been found that the maximum field strength increases as the included angle of the reflector decreases. This is expected since a smaller angle reflector exhibits better directional characteristics because of the narrowness of its angle. The maximum values of $|E/E_0|$ for $\alpha = 60°, 45°$,

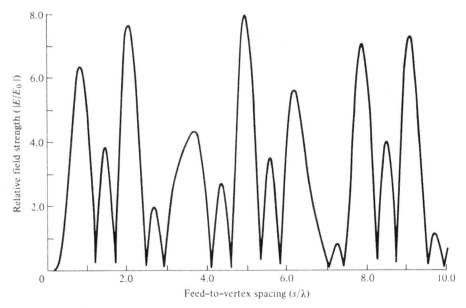

Figure 15.7 Relative field strength along the axis ($\theta = 90°, \phi = 0°$) for an $\alpha = 45°$ corner reflector as a function of feed-to-vertex spacing.

and 30° are approximately 5.2, 8, and 9, respectively. The first field strength peak, but not necessarily the ultimate maximum, is achieved when

$$s \simeq 0.65\lambda \quad \text{for} \quad \alpha = 60°$$
$$s \simeq 0.85\lambda \quad \text{for} \quad \alpha = 45°$$
$$s \simeq 1.20\lambda \quad \text{for} \quad \alpha = 30°$$

15.4 PARABOLIC REFLECTOR

The overall radiation characteristics (antenna pattern, antenna efficiency, polarization discrimination, etc.) of a reflector can be improved if the structural configuration of its surface is upgraded. It has been shown by geometrical optics that if a beam of parallel rays is incident upon a reflector whose geometrical shape is a parabola, the radiation will converge (focus) at a spot which is known as the *focal point*. In the same manner, if a point source is placed at the focal point, the rays reflected by a parabolic reflector will emerge as a parallel beam. This is one form of the principle of reciprocity, and it is demonstrated geometrically in Figure 15.1(c). The symmetrical point on the parabolic surface is known as the *vertex*. Rays that emerge in a parallel formation are usually said to be *collimated*. In practice, collimation is often used to describe the highly directional characteristics of an antenna even though the emanating rays are not exactly parallel. Since the transmitter (receiver) is placed at the focal point of the parabola, the configuration is usually known as *front fed*.

The disadvantage of the front-fed arrangement is that the transmission line from the feed must usually be long enough to reach the transmitting or the receiving equipment, which is usually placed behind or below the reflector. This may necessitate the use of long transmission lines whose losses may not be tolerable in many applications, especially in low-noise receiving systems. In some applications, the transmitting or receiving equipment is placed at the focal point to avoid the need for long transmission lines. However, in some of these applications, especially for transmission that may require large amplifiers and for low-noise receiving systems where cooling and weatherproofing may be necessary, the equipment may be too heavy and bulky and will provide undesirable blockage.

Another arrangement that avoids placing the feed (transmitter and/or receiver) at the focal point is that shown in Figure 15.1(d), and it is known as the *Cassegrain feed*. Through geometrical optics, Cassegrain, a famous astronomer (hence its name), showed that incident parallel rays can be focused to a point by utilizing two reflectors. To accomplish this, the main (primary) reflector must be a parabola, the secondary reflector (subreflector) a hyperbola, and the feed placed along the axis of the parabola usually at or near the vertex. Cassegrain used this scheme to construct optical telescopes, and then its design was copied for use in radio frequency systems. For this arrangement, the rays that emanate from the feed illuminate the subreflector and are reflected by it in the direction of the primary reflector, as if they originated at the focal point of the parabola (primary reflector). The rays are then reflected by the primary reflector and are converted to parallel rays, provided the primary reflector is a parabola and the subreflector is a hyperbola. Diffractions occur at the edges of the subreflector and primary reflector, and they must be taken into account to accurately predict the overall system pattern, especially in regions of low intensity [9]–[11]. Even in regions of high intensity, diffractions must be included if an accurate formation of the fine ripple structure of the pattern is desired. With the Cassegrain-feed arrangement, the transmitting and/or receiving equipment can be placed behind the primary reflector. This scheme makes the system relatively more accessible for servicing and adjustments.

A parabolic reflector can take two different forms. One configuration is that of the parabolic right cylinder, shown in Figure 15.8(a), whose energy is collimated at a line that is parallel to the axis of the cylinder through the focal point of the reflector. The most widely used feed for this type of a

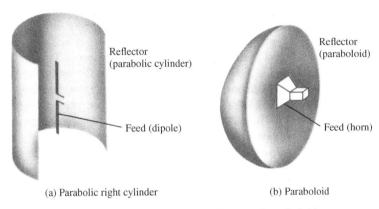

(a) Parabolic right cylinder (b) Paraboloid

Figure 15.8 Parabolic right cylinder and paraboloid.

reflector is a linear dipole, a linear array, or a slotted waveguide. The other reflector configuration is that of Figure 15.8(b) which is formed by rotating the parabola around its axis, and it is referred to as a *paraboloid* (parabola of revolution). A pyramidal or a conical horn has been widely utilized as a feed for this arrangement.

There are many other types of reflectors whose analysis is widely documented in the literature [12]–[14]. The spherical reflector, for example, has been utilized for radioastronomy and small earth-station applications, because its beam can be efficiently scanned by moving its feed. An example of that is the 1,000-ft (305-m) diameter spherical reflector at Arecibo, Puerto Rico [12] whose primary surface is built into the ground and scanning of the beam is accomplished by movement of the feed. For spherical reflectors a substantial blockage may be provided by the feed leading to unacceptable minor lobe levels, in addition to the inherent reduction in gain and less favorable cross-polarization discrimination.

To eliminate some of the deficiencies of the symmetric configurations, offset-parabolic reflector designs have been developed for single- and dual-reflector systems [13]. Because of the asymmetry of the system, the analysis is more complex. However the advent and advances of computer technology have made the modeling and optimization of the offset-reflector designs available and convenient. Offset-reflector designs reduce aperture blocking and VSWR. In addition, they lead to the use of larger f/d ratios while maintaining acceptable structural rigidity, which provide an opportunity for improved feed pattern shaping and better suppression of cross-polarized radiation emanating from the feed. However, offset-reflector configurations generate cross-polarized antenna radiation when illuminated by a linearly polarized primary feed. Circularly polarized feeds eliminate depolarization, but they lead to squinting of the main beam from boresight. In addition, the structural asymmetry of the system is usually considered a major drawback.

Paraboloidal reflectors are the most widely used large aperture ground-based antennas [14]. At the time of its construction, the world's largest fully steerable reflector was the 100-m diameter radio telescope [15] of the Max Planck Institute for Radioastronomy at Effelsberg, West Germany, while the largest in the United States was the 64-m diameter [16] reflector at Goldstone, California built primarily for deep-space applications. When fed efficiently from the focal point, paraboloidal reflectors produce a high-gain pencil beam with low side lobes and good cross-polarization discrimination characteristics. This type of an antenna is widely used for low-noise applications, such as in radioastronomy, and it is considered a good compromise between performance and cost. To build a large reflector requires not only a large financial budget but also a difficult structural undertaking, because it must withstand severe weather conditions.

Cassegrain designs, employing dual-reflector surfaces, are used in applications where pattern control is essential, such as in satellite ground-based systems, and have efficiencies of 65–80%. They

Figure 15.9 Shaped 10-m earth-station dual-reflector antenna. (Courtesy Andrew Corp).

supersede the performance of the single-reflector front-fed arrangement by about 10%. Using geometrical optics, the classical Cassegrain configuration, consisting of a paraboloid and a hyperboloid, is designed to achieve a uniform phase front in the aperture of the paraboloid. By employing good feed designs, this arrangement can achieve lower spillover and more uniform illumination of the main reflector. In addition, slight shaping of one or both of the dual-reflector's surfaces can lead to an aperture with almost uniform amplitude and phase with a substantial enhancement in gain [14]. These are referred to as *shaped* reflectors. Shaping techniques have been employed in dual-reflectors used in earth-station applications. An example is the 10-m earth-station dual-reflector antenna, shown in Figure 15.9, whose main reflector and subreflector are shaped.

For many years horns or waveguides, operating in a single mode, were used as feeds for reflector antennas. However because of radioastronomy and earth-station applications, considerable efforts have been placed in designing more efficient feeds to illuminate either the main reflector or the subreflector. It has been found that corrugated horns that support hybrid mode fields (combination of TE and TM modes) can be used as desirable feeds. Such feed elements match efficiently the fields of the feeds with the desired focal distribution produced by the reflector, and they can reduce cross-polarization. Dielectric cylinders and cones are other antenna structures that support hybrid modes [14]. Their structural configuration can also be used to support the subreflector and to provide attractive performance figures.

There are primarily two techniques that can be used to analyze the performance of a reflector system [17]. One technique is the *aperture distribution method* and the other the *current distribution method*. Both techniques will be introduced to show the similarities and differences.

15.4.1 Front-Fed Parabolic Reflector

Parabolic cylinders have widely been used as high-gain apertures fed by line sources. The analysis of a parabolic cylinder (single curved) reflector is similar, but considerably simpler than that of a paraboloidal (double curved) reflector. The principal characteristics of aperture amplitude, phase, and polarization for a parabolic cylinder, as contrasted to those of a paraboloid, are as follows:

1. The amplitude taper, due to variations in distance from the feed to the surface of the reflector, is proportional to $1/\rho$ in a cylinder compared to $1/r^2$ in a paraboloid.
2. The focal region, where incident plane waves converge, is a line source for a cylinder and a point source for a paraboloid.
3. When the fields of the feed are linearly polarized parallel to the axis of the cylinder, no cross-polarized components are produced by the parabolic cylinder. That is not the case for a paraboloid.

Generally, parabolic cylinders, as compared to paraboloids, (1) are mechanically simpler to build, (2) provide larger aperture blockage, and (3) do not possess the attractive characteristics of a paraboloid. In this chapter, only paraboloidal reflectors will be examined.

A. Surface Geometry

The surface of a paraboloidal reflector is formed by rotating a parabola about its axis. Its surface must be a paraboloid of revolution so that rays emanating from the focus of the reflector are transformed into plane waves. The design is based on optical techniques, and it does not take into account any deformations (diffractions) from the rim of the reflector. Referring to Figure 15.10 and choosing a plane perpendicular to the axis of the reflector through the focus, it follows that

$$OP + PQ = \text{constant} = 2f \tag{15-12}$$

Since

$$OP = r' \tag{15-13}$$
$$PQ = r' \cos \theta'$$

(15-12) can be written as

$$r'(1 + \cos \theta') = 2f \tag{15-14}$$

or

$$\boxed{r' = \frac{2f}{1 + \cos \theta'} = f \sec^2\left(\frac{\theta'}{2}\right) \quad \theta \leq \theta_0} \tag{15-14a}$$

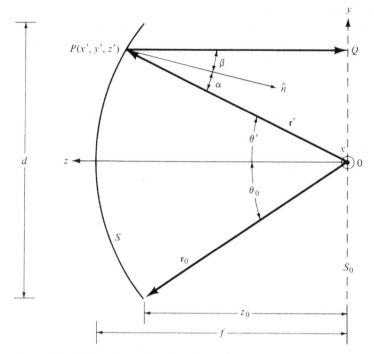

Figure 15.10 Two-dimensional configuration of a paraboloidal reflector.

Since a paraboloid is a parabola of revolution (about its axis), (15-14a) is also the equation of a paraboloid in terms of the spherical coordinates r', θ', ϕ'. Because of its rotational symmetry, there are no variations with respect to ϕ'.

Equation (15-14a) can also be written in terms of the rectangular coordinates x', y', z'. That is,

$$r' + r' \cos \theta' = \sqrt{(x')^2 + (y')^2 + (z')^2} + z' = 2f \tag{15-15}$$

or

$$(x')^2 + (y')^2 = 4f(f - z') \quad \text{with } (x')^2 + (y')^2 \leq (d/2)^2 \tag{15-15a}$$

In the analysis of parabolic reflectors, it is desirable to find a unit vector that is normal to the local tangent at the surface reflection point. To do this, (15-14a) is first expressed as

$$f - r' \cos^2\left(\frac{\theta'}{2}\right) = S = 0 \tag{15-16}$$

and then a gradient is taken to form a normal to the surface. That is,

$$\mathbf{N} = \nabla\left[f - r' \cos^2\left(\frac{\theta'}{2}\right)\right] = \hat{\mathbf{a}}_r' \frac{\partial S}{\partial r'} + \hat{\mathbf{a}}_\theta' \frac{1}{r'} \frac{\partial S}{\partial \theta'}$$

$$= -\hat{\mathbf{a}}_r' \cos^2\left(\frac{\theta'}{2}\right) + \hat{\mathbf{a}}_\theta' \cos\left(\frac{\theta'}{2}\right) \sin\left(\frac{\theta'}{2}\right) \tag{15-17}$$

A unit vector, normal to S, is formed from (15-17) as

$$\hat{\mathbf{n}} = \frac{\mathbf{N}}{|\mathbf{N}|} = -\hat{\mathbf{a}}'_r \cos\left(\frac{\theta'}{2}\right) + \hat{\mathbf{a}}'_\theta \sin\left(\frac{\theta'}{2}\right) \qquad (15\text{-}18)$$

To find the angle between the unit vector $\hat{\mathbf{n}}$ which is normal to the surface at the reflection point, and a vector directed from the focus to the reflection point, we form

$$\cos\alpha = -\hat{\mathbf{a}}'_r \cdot \hat{\mathbf{n}} = -\hat{\mathbf{a}}'_r \cdot \left[-\hat{\mathbf{a}}'_r \cos\left(\frac{\theta'}{2}\right) + \hat{\mathbf{a}}'_\theta \sin\left(\frac{\theta'}{2}\right)\right]$$
$$= \cos\left(\frac{\theta'}{2}\right) \qquad (15\text{-}19)$$

In a similar manner we can find the angle between the unit vector $\hat{\mathbf{n}}$ and the z-axis. That is,

$$\cos\beta = -\hat{\mathbf{a}}_z \cdot \hat{\mathbf{n}} = -\hat{\mathbf{a}}_z \cdot \left[-\hat{\mathbf{a}}'_r \cos\left(\frac{\theta'}{2}\right) + \hat{\mathbf{a}}'_\theta \sin\left(\frac{\theta'}{2}\right)\right] \qquad (15\text{-}20)$$

Using the transformation of (4-5), (15-20) can be written as

$$\cos\beta = -(\hat{\mathbf{a}}'_r \cos\theta' - \hat{\mathbf{a}}'_\theta \sin\theta') \cdot \left[-\hat{\mathbf{a}}'_r \cos\left(\frac{\theta'}{2}\right) + \hat{\mathbf{a}}'_\theta \sin\left(\frac{\theta'}{2}\right)\right]$$
$$= \cos\left(\frac{\theta'}{2}\right) \qquad (15\text{-}21)$$

which is identical to α of (15-19). This is nothing more than a verification of Snell's law of reflection at each differential area of the surface, which has been assumed to be flat locally.

Another expression that is usually very prominent in the analysis of reflectors is that relating the subtended angle θ_0 to the f/d ratio. From the geometry of Figure 15.10

$$\theta_0 = \tan^{-1}\left(\frac{d/2}{z_0}\right) \qquad (15\text{-}22)$$

where z_0 is the distance along the axis of the reflector from the focal point to the edge of the rim. From (15-15a)

$$z_0 = f - \frac{x_0^2 + y_0^2}{4f} = f - \frac{(d/2)^2}{4f} = f - \frac{d^2}{16f} \qquad (15\text{-}23)$$

Substituting (15-23) into (15-22) reduces it to

$$\theta_0 = \tan^{-1}\left|\frac{\frac{d}{2}}{f - \frac{d^2}{16f}}\right| = \tan^{-1}\left|\frac{\frac{1}{2}\left(\frac{f}{d}\right)}{\left(\frac{f}{d}\right)^2 - \frac{1}{16}}\right| \qquad (15\text{-}24)$$

It can also be shown that another form of (15-24) is

$$f = \left(\frac{d}{4}\right)\cot\left(\frac{\theta_0}{2}\right) \tag{15-25}$$

B. Induced Current Density

To determine the radiation characteristics (pattern, gain, efficiency, polarization, etc.) of a parabolic reflector, the current density induced on its surface must be known.

The current density $\mathbf{J}_s$ can be determined by using

$$\mathbf{J}_s = \hat{\mathbf{n}} \times \mathbf{H} = \hat{\mathbf{n}} \times (\mathbf{H}^i + \mathbf{H}^r) \tag{15-26}$$

where $\mathbf{H}^i$ and $\mathbf{H}^r$ represent, respectively, the incident and reflected magnetic field components evaluated at the surface of the conductor, and $\hat{\mathbf{n}}$ is a unit vector normal to the surface. If the reflecting surface can be approximated by an *infinite plane surface* (this condition is met locally for a parabola), then by the method of images

$$\hat{\mathbf{n}} \times \mathbf{H}^i = \hat{\mathbf{n}} \times \mathbf{H}^r \tag{15-27}$$

and (15-26) reduces to

$$\mathbf{J}_s = \hat{\mathbf{n}} \times (\mathbf{H}^i + \mathbf{H}^r) = 2\hat{\mathbf{n}} \times \mathbf{H}^i = 2\hat{\mathbf{n}} \times \mathbf{H}^r \tag{15-28}$$

The current density approximation of (15-28) is known as the *physical-optics* approximation, and it is valid when the transverse dimensions of the reflector, radius of curvature of the reflecting object, and the radius of curvature of the incident wave are large compared to a wavelength.

If the reflecting surface is in the far-field of the source generating the incident waves, then (15-28) can also be written as

$$\mathbf{J}_s = 2\hat{\mathbf{n}} \times \mathbf{H}^i \simeq \frac{2}{\eta}[\hat{\mathbf{n}} \times (\hat{\mathbf{s}}_i \times \mathbf{E}^i)] \tag{15-29}$$

or

$$\mathbf{J}_s = 2\hat{\mathbf{n}} \times \mathbf{H}^r \simeq \frac{2}{\eta}[\hat{\mathbf{n}} \times (\hat{\mathbf{s}}_r \times \mathbf{E}^r)] \tag{15-29a}$$

where η is the intrinsic impedance of the medium, $\hat{\mathbf{s}}_i$ and $\hat{\mathbf{s}}_r$ are radial unit vectors along the ray paths of the incident and reflected waves (as shown in Figure 15.11), and $\mathbf{E}^i$ and $\mathbf{E}^r$ are the incident and reflected electric fields.

C. Aperture Distribution Method

It was pointed out earlier that the two most commonly used techniques in analyzing the radiation characteristics of reflectors are the *aperture distribution* and the *current distribution* methods.

For the aperture distribution method, the field reflected by the surface of the paraboloid is first found over a plane which is normal to the axis of the reflector. Geometrical optics techniques (ray tracing) are usually employed to accomplish this. In most cases, the plane is taken through the focal point, and it is designated as the *aperture plane*, as shown in Figure 15.12. Equivalent sources are

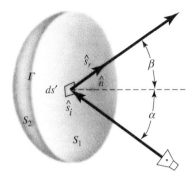

Figure 15.11 Reflecting surface with boundary Γ.

then formed over that plane. Usually it is assumed that the equivalent sources are zero outside the projected area of the reflector on the aperture plane. These equivalent sources are then used to compute the radiated fields utilizing the aperture techniques of Chapter 12.

For the *current distribution method*, the physical optics approximation of the induced current density $\mathbf{J}_s$ given by (15-28) ($\mathbf{J}_s \simeq 2\hat{\mathbf{n}} \times \mathbf{H}^i$ where $\mathbf{H}^i$ is the incident magnetic field and $\hat{\mathbf{n}}$ is a unit vector normal to the reflector surface) is formulated over the illuminated side of the reflector (S_1) of Figure 15.11. This current density is then integrated over the surface of the reflector to yield the far-zone radiation fields.

For the reflector of Figure 15.11, approximations that are common to both methods are:

1. The current density is zero on the shadow side (S_2) of the reflector.
2. The discontinuity of the current density over the rim (Γ) of the reflector is neglected.
3. Direct radiation from the feed and aperture blockage by the feed are neglected.

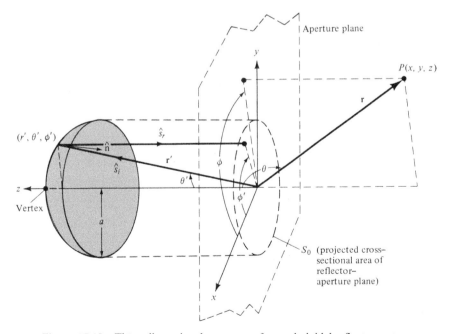

Figure 15.12 Three-dimensional geometry of a paraboloidal reflector system.

These approximations lead to accurate results, using either technique, for the radiated fields on the main beam and nearby minor lobes. To predict the pattern more accurately in all regions, especially the far minor lobes, geometrical diffraction techniques [9]–[11] can be applied. Because of the level of the material, it will not be included here. The interested reader can refer to the literature.

The advantage of the aperture distribution method is that the integration over the aperture plane can be performed with equal ease for any feed pattern or feed position [18]. The integration over the surface of the reflector as required for the current distribution method, becomes quite complex and time consuming when the feed pattern is asymmetrical and/or the feed is placed off-axis.

Let us assume that a y-polarized source with a gain function of $G_f(\theta', \phi')$ is placed at the focal point of a paraboloidal reflector. The radiation intensity of this source is given by

$$U(\theta', \phi') = \frac{P_t}{4\pi} G_f(\theta', \phi') \quad (15\text{-}30)$$

where P_t is the total radiated power. Referring to Figure 15.12, at a point r' in the far-zone of the source

$$U(\theta', \phi') = \frac{1}{2}\operatorname{Re}[\mathbf{E}^\circ(\theta', \phi') \times \mathbf{H}^{\circ*}(\theta', \phi')] = \frac{1}{2\eta} |\mathbf{E}^\circ(\theta', \phi')|^2 \quad (15\text{-}31)$$

or

$$|\mathbf{E}^\circ(\theta', \phi')| = [2\eta U(\theta', \phi')]^{1/2} = \left[\eta \frac{P_t}{2\pi} G_f(\theta', \phi')\right]^{1/2} \quad (15\text{-}31\text{a})$$

The incident field, with a direction perpendicular to the radial distance, can then be written as

$$\mathbf{E}^i(r', \theta', \phi') = \hat{\mathbf{e}}_i \left[\sqrt{\frac{\mu}{\varepsilon}} \frac{P_t}{2\pi} G_f(\theta', \phi')\right]^{1/2} \frac{e^{-jkr'}}{r'} = \hat{\mathbf{e}}_i C_1 \sqrt{G_f(\theta', \phi')} \frac{e^{-jkr'}}{r'} \quad (15\text{-}32)$$

$$C_1 = \left(\frac{\mu}{\varepsilon}\right)^{1/4} \left(\frac{P_t}{2\pi}\right)^{1/2} \quad (15\text{-}32\text{a})$$

where $\hat{\mathbf{e}}_i$ is a unit vector perpendicular to $\hat{\mathbf{a}}'_r$ and parallel to the plane formed by $\hat{\mathbf{a}}'_r$ and $\hat{\mathbf{a}}_y$, as shown in Figure 15.13.

It can be shown [19] that on the surface of the reflector

$$\mathbf{J}_s = 2\sqrt{\frac{\varepsilon}{\mu}} [\hat{\mathbf{n}} \times (\hat{\mathbf{s}}_i \times \mathbf{E}^i)] = 2\sqrt{\frac{\varepsilon}{\mu}} C_1 \sqrt{G_f(\theta', \phi')} \frac{e^{-jkr'}}{r'} \mathbf{u} \quad (15\text{-}33)$$

where

$$\mathbf{u} = \hat{\mathbf{n}} \times (\hat{\mathbf{a}}'_r \times \hat{\mathbf{e}}_i) = (\hat{\mathbf{n}} \cdot \hat{\mathbf{e}}_i)\hat{\mathbf{a}}'_r - (\hat{\mathbf{n}} \cdot \hat{\mathbf{a}}'_r)\hat{\mathbf{e}}_i \quad (15\text{-}33\text{a})$$

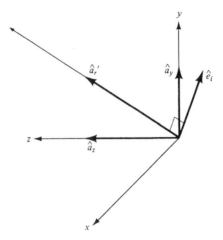

Figure 15.13 Unit vector alignment for a paraboloidal reflector system.

which reduces to

$$\mathbf{u} = \left[-\hat{\mathbf{a}}_x \sin\theta' \sin\left(\frac{\theta'}{2}\right) \sin\phi' \cos\phi' \right.$$
$$+ \hat{\mathbf{a}}_y \cos\left(\frac{\theta'}{2}\right) (\sin^2\phi' \cos\theta' + \cos^2\phi')$$
$$\left. - \hat{\mathbf{a}}_z \cos\theta' \sin\phi' \sin\left(\frac{\theta'}{2}\right) \right] \bigg/ \sqrt{1 - \sin^2\theta' \sin^2\phi'} \quad (15\text{-}34)$$

To find the aperture field $\mathbf{E}_{ap}$ at the plane through the focal point, due to the reflector currents of (15-33), the reflected field $\mathbf{E}^r$ at r' (the reflection point) is first found. This is of the form

$$\mathbf{E}^r = \hat{\mathbf{e}}_r C_1 \sqrt{G_f(\theta',\phi')} \frac{e^{-jkr'}}{r'} \quad (15\text{-}35)$$

where $\hat{\mathbf{e}}_r$ is a unit vector depicting the polarization of the reflected field. From (15-29a)

$$\mathbf{J}_s = 2\sqrt{\frac{\varepsilon}{\mu}} [\hat{\mathbf{n}} \times (\hat{\mathbf{s}}_r \times \mathbf{E}^r)] \quad (15\text{-}36)$$

Because $\hat{\mathbf{s}}_r = -\hat{\mathbf{a}}_z$, (15-36) can be written, using (15-35), as

$$\mathbf{J}_s = 2\sqrt{\frac{\varepsilon}{\mu}} C_1 \sqrt{G_f(\theta',\phi')} \frac{e^{-jkr'}}{r'} \mathbf{u} \quad (15\text{-}37)$$

where

$$\mathbf{u} = \hat{\mathbf{n}} \times (-\hat{\mathbf{a}}_z \times \hat{\mathbf{e}}_r) = -\hat{\mathbf{a}}_z(\hat{\mathbf{n}} \cdot \hat{\mathbf{e}}_r) - \hat{\mathbf{e}}_r \cos\left(\frac{\theta'}{2}\right) \quad (15\text{-}37a)$$

Since **u** in (15-37) and (15-37a) is the same as that of (15-33)–(15-34), it can be shown [19] through some extensive mathematical manipulations that

$$\hat{\mathbf{e}}_r = \frac{\hat{\mathbf{a}}_x \sin\phi' \cos\phi' (1 - \cos\theta') - \hat{\mathbf{a}}_y (\sin^2\phi' \cos\theta' + \cos^2\phi')}{\sqrt{1 - \sin^2\theta' \sin^2\phi'}} \quad (15\text{-}38)$$

Thus the field $\mathbf{E}^r$ at the point of reflection r' is given by (15-35) where $\hat{\mathbf{e}}_r$ is given by (15-38). At the plane passing through the focal point, the field is given by

$$\mathbf{E}_{ap} = \hat{\mathbf{e}}_r C_1 \sqrt{G_f(\theta', \phi')} \frac{e^{-jkr'(1+\cos\theta')}}{r'} = \hat{\mathbf{a}}_x E_{xa} + \hat{\mathbf{a}}_y E_{ya} \quad (15\text{-}39)$$

where E_{xa} and E_{ya} represent the x- and y-components of the reflected field over the aperture. Since the field from the reflector to the aperture plane is a plane wave, no correction in amplitude is needed to account for amplitude spreading.

Using the reflected electric field components (E_{xa} and E_{ya}) as given by (15-39), an equivalent is formed at the aperture plane. That is,

$$\mathbf{J}'_s = \hat{\mathbf{n}} \times \mathbf{H}_a = -\hat{\mathbf{a}}_z \times \left(\hat{\mathbf{a}}_x \frac{E_{ay}}{\eta} - \hat{\mathbf{a}}_y \frac{E_{ax}}{\eta} \right) = -\hat{\mathbf{a}}_x \frac{E_{ax}}{\eta} - \hat{\mathbf{a}}_y \frac{E_{ay}}{\eta} \quad (15\text{-}40a)$$

$$\mathbf{M}'_s = -\hat{\mathbf{n}} \times \mathbf{E}_a = +\hat{\mathbf{a}}_z \times (\hat{\mathbf{a}}_x E_{ax} + \hat{\mathbf{a}}_y E_{ay}) = -\hat{\mathbf{a}}_x E_{ay} + \hat{\mathbf{a}}_y E_{ax} \quad (15\text{-}40b)$$

The radiated field can be computed using the (15-40a), (15-40b), and the formulations of Section 12.3. The integration is restricted only over the projected cross-sectional area S_0 of the reflector at the aperture plane shown dashed in Figure 15.12. That is,

$$E_{\theta s} = \frac{jke^{-jkr}}{4\pi r}(1 - \cos\theta) \iint_{S_0} (-E_{ax}\cos\phi - E_{ay}\sin\phi) \\ \times e^{jk(x'\sin\theta\cos\phi + y'\sin\theta\sin\phi)} \, dx' \, dy' \quad (15\text{-}41a)$$

$$E_{\phi s} = \frac{jke^{-jkr}}{4\pi r}(1 - \cos\theta) \iint_{S_0} (-E_{ax}\sin\phi + E_{ay}\cos\phi) \\ \times e^{jk(x'\sin\theta\cos\phi + y'\sin\theta\sin\phi)} \, dx' \, dy' \quad (15\text{-}41b)$$

The aperture distribution method has been used to compute, using efficient numerical integration techniques, the radiation patterns of paraboloidal [18] and spherical [20] reflectors. The fields given by (15-41a) and (15-41b) represent only the secondary pattern due to scattering from the reflector. The total pattern of the system is represented by the sum of secondary pattern and the primary pattern of the feed element. For most feeds (such as horns), the primary pattern in the boresight (forward) direction of the reflector is of very low intensity and usually can be neglected.

To demonstrate the utility of the techniques, the principal E- and H-plane secondary patterns of a 35 GHz reflector, with an $f/d \simeq 0.82$ [$f = 8.062$ in. (20.48 cm), $d = 9.84$ in. (24.99 cm)] and fed by a conical dual-mode horn, were computed and they are displayed in Figure 15.14. Since the feed horn has identical E- and H-plane patterns and the reflector is fed symmetrically, the reflector E- and H-plane patterns are also identical and do not possess any cross-polarized components.

To simultaneously display the field intensity associated with each point in the aperture plane of the reflector, a computer generated plot was developed [20]. The field point locations, showing quantized contours of constant amplitude in the aperture plane, are illustrated in Figure 15.15. The

PARABOLIC REFLECTOR **895**

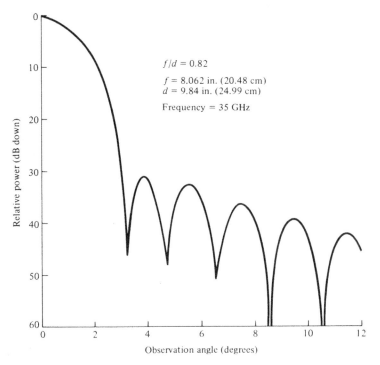

Figure 15.14 Principal E- or H-plane pattern of a symmetrical front-fed paraboloidal reflector. (Courtesy M. C. Bailey, NASA Langley Research Center).

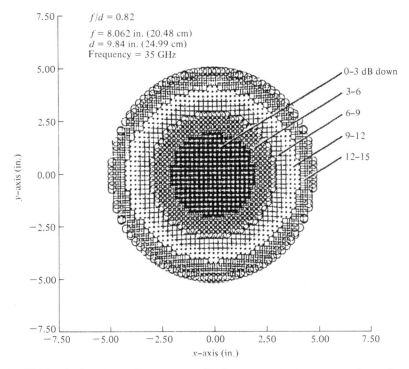

Figure 15.15 Field point locations of constant amplitude contours in the aperture plane of a symmetrical front-fed paraboloidal reflector. (Courtesy M. C. Bailey, NASA Langley Research Center).

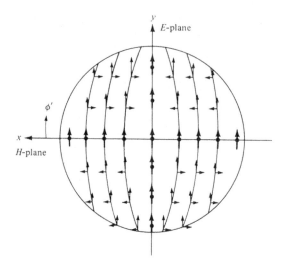

Figure 15.16 Principal (y-direction) and cross-polarization (x-direction) components of a paraboloidal reflector. (SOURCE: S. Silver (ed.), *Microwave Antenna Theory and Design* (MIT Radiation Lab. Series, Vol. 12), McGraw-Hill, New York, 1949).

reflector system has an $f/d \simeq 0.82$ with the same physical dimensions [f = 8.062 in. (20.48 cm), d = 9.84 in. (24.99 cm)] and the same feed as the principal pattern of Figure 15.14. One symbol is used to represent the amplitude level of each 3-dB region. The field intensity within the bounds of the reflector aperture plane is within the 0–15 dB range.

D. Cross-Polarization

The field reflected by the paraboloid, as represented by (15-35) and (15-38) of the aperture distribution method, contains x- and y-polarized components when the incident field is y-polarized. The y-component is designated as the *principal polarization* and the x-component as the *cross-polarization*. This is illustrated in Figure 15.16. It is also evident that symmetrical (with respect to the principal planes) cross-polarized components are 180° out of phase with one another. However for very narrow beam reflectors or for angles near the boresight axis ($\theta' \simeq 0°$), the cross-polarized x-component diminishes and it vanishes on axis ($\theta' = 0°$). A similar procedure can be used to show that for an incident x-polarized field, the reflecting surface decomposes the wave to a y-polarized field, in addition to its x-polarized component.

An interesting observation about the polarization phenomenon of a parabolic reflector can be made if we first assume that the feed element is an infinitesimal electric dipole ($l \ll \lambda$) with its length along the y-axis. For that feed, the field reflected by the reflector is given by (15-35) where from (4-114)

$$C_1\sqrt{G_f(\theta', \phi')} = j\eta\frac{kI_0 l}{4\pi}\sin\psi = j\eta\frac{kI_0 l}{4\pi}\sqrt{1 - \cos^2\psi}$$
$$= j\eta\frac{kI_0 l}{4\pi}\sqrt{1 - \sin^2\theta'\sin^2\phi'} \qquad (15\text{-}42)$$

The angle ψ is measured from the y-axis toward the observation point.

When (15-42) is inserted in (15-35), we can write with the aid of (15-38) that

$$\mathbf{E}^r = [\hat{\mathbf{a}}_x \sin\phi'\cos\phi'(1 - \cos\theta') - \hat{\mathbf{a}}_y(\sin^2\phi'\cos\theta' + \cos^2\phi')] \times j\eta\frac{kI_0 l}{4\pi}\frac{e^{-jkr'}}{r'} \qquad (15\text{-}43)$$

Now let us assume that an infinitesimal magnetic dipole, with its length along the x-axis (or a small loop with its area parallel to the y-z plane) and with a magnetic moment of $-\hat{\mathbf{a}}_x M$, is placed at

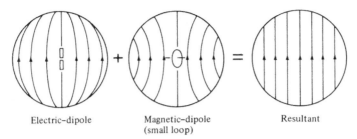

Figure 15.17 Electric and magnetic dipole fields combined to form a Huygens' source with ideal feed polarization for reflector. (SOURCE: A. W. Love, "Some Highlights in Reflector Antenna Development," *Radio Science*, Vol. 11, Nos. 8, 9, August–September 1976).

the focal point and used as a feed. It can be shown [21]–[23] that the field reflected by the reflector has *x*- and *y*-components. However the *x*-component has a reverse sign to the *x*-component of the electric dipole feed. By adjusting the ratio of the electric to the magnetic dipole moments to be equal to $\sqrt{\mu/\varepsilon}$, the two cross-polarized reflected components (*x*-components) can be made equal in magnitude and for their sum to vanish (because of reverse signs). Thus a cross electric and magnetic dipole combination located at the focal point of a paraboloid can be used to induce currents on the surface of the reflector which are parallel everywhere. This is illustrated graphically in Figure 15.17.

The direction of the induced current flow determines the far-field polarization of the antenna. Thus for the crossed electric and magnetic dipole combination feed, the far-field radiation is free of cross-polarization. This type of feed is "ideal" in that it does not require that the surface of the reflector be solid but can be formed by closely spaced parallel conductors. Because of its ideal characteristics, it is usually referred to as a *Huygens' source*.

E. Current Distribution Method

The current distribution method was introduced as a technique that can be used to better approximate, as compared to the geometrical optics (ray-tracing) method, the field scattered from a surface. Usually the main difficulty in applying this method is the approximation of the current density over the surface of the scatterer.

To analyze the reflector using this technique, we refer to the radiation integrals and auxiliary potential functions formulations of Chapter 3. While the two-step procedure of Figure 3.1 often simplifies the solution of most problems, the one-step formulation of Figure 3.1 is most convenient for the reflectors.

Using the potential function methods outlined in Chapter 3, and referring to the coordinate system of Figure 12.2(a), it can be shown [17] that the **E**- and **H**-fields radiated by the sources **J** and **M** can be written as

$$\mathbf{E} = \mathbf{E}_A + \mathbf{E}_F = -j\frac{1}{4\pi\omega\varepsilon}\int_V [(\mathbf{J}\cdot\nabla)\nabla + k^2\mathbf{J} + j\omega\varepsilon\mathbf{M}\times\nabla]\frac{e^{-jkR}}{R}\,dv' \quad (15\text{-}44\text{a})$$

$$\mathbf{H} = \mathbf{H}_A + \mathbf{H}_F = -j\frac{1}{4\pi\omega\mu}\int_V [(\mathbf{M}\cdot\nabla)\nabla + k^2\mathbf{M} - j\omega\mu\mathbf{J}\times\nabla]\frac{e^{-jkR}}{R}\,dv' \quad (15\text{-}44\text{b})$$

which for far-field observations reduce, according to the coordinate system of Figure 12.2(b), to

$$\mathbf{E} \simeq -j\frac{\omega\mu}{4\pi r}e^{-jkr}\int_V \left[\mathbf{J} - (\mathbf{J}\cdot\hat{\mathbf{a}}_r)\hat{\mathbf{a}}_r + \sqrt{\frac{\varepsilon}{\mu}}\mathbf{M}\times\hat{\mathbf{a}}_r\right]e^{+jk\mathbf{r'}\cdot\hat{\mathbf{a}}_r}\,dv' \quad (15\text{-}45\text{a})$$

$$\mathbf{H} \simeq -j\frac{\omega\varepsilon}{4\pi r}e^{-jkr}\int_V \left[\mathbf{M} - (\mathbf{M}\cdot\hat{\mathbf{a}}_r)\hat{\mathbf{a}}_r - \sqrt{\frac{\mu}{\varepsilon}}\mathbf{J}\times\hat{\mathbf{a}}_r\right]e^{+jk\mathbf{r'}\cdot\hat{\mathbf{a}}_r}\,dv' \quad (15\text{-}45\text{b})$$

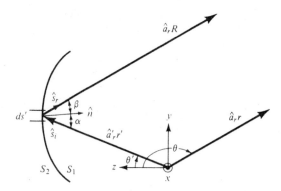

Figure 15.18 Geometrical arrangement of reflecting surface.

If the current distributions are induced by electric and magnetic fields incident on a perfect electric conducting ($\sigma = \infty$) surface shown in Figure 15.18, the fields created by these currents are referred to as *scattered fields*. If the conducting surface is closed, the far-zone fields are obtained from (15-45a) and (15-45b) by letting $\mathbf{M} = 0$ and reducing the volume integral to a surface integral with the surface current density $\mathbf{J}$ replaced by the linear current density $\mathbf{J}_s$. Thus

$$\mathbf{E}_s = -j\frac{\omega\mu}{4\pi r}e^{-jkr}\oiint_S [\mathbf{J}_s - (\mathbf{J}_s \cdot \hat{\mathbf{a}}_r)\hat{\mathbf{a}}_r]e^{+jk\mathbf{r}'\cdot\hat{\mathbf{a}}_r}\,ds' \quad (15\text{-}46\text{a})$$

$$\mathbf{H}_s = +j\frac{\omega\sqrt{\mu\varepsilon}}{4\pi r}e^{-jkr}\oiint_S [\mathbf{J}_s \times \hat{\mathbf{a}}_r]e^{+jk\mathbf{r}'\cdot\hat{\mathbf{a}}_r}\,ds' \quad (15\text{-}46\text{b})$$

The electric and magnetic fields scattered by the closed surface of the reflector of Figure 15.11, and given by (15-46a) and (15-46b), are valid provided the source-density functions (current and charge) satisfy the equation of continuity. This would be satisfied if the scattering object is a smooth closed surface. For the geometry of Figure 15.11, the current distribution is discontinuous across the boundary Γ (being zero over the shadow area S_2) which divides the illuminated (S_1) and shadow (S_2) areas. It can be shown [17] that the equation of continuity can be satisfied if an appropriate line-source distribution of charge is introduced along the boundary Γ. Therefore the total scattered field would be the sum of the (1) surface currents over the illuminated area, (2) surface charges over the illuminated area, and (3) line-charge distribution over the boundary Γ.

The contributions from the surface charge density are taken into account by the current distribution through the equation of continuity. However it can be shown [17] that in the far-zone the contribution due to line-charge distribution cancels out the longitudinal component introduced by the surface current and charge distributions. Since in the far-zone the field components are predominantly transverse, the contribution due to the line-charge distribution need not be included and (15-46a)–(15-46b) can be applied to an open surface.

In this section, (15-46a) and (15-46b) will be used to calculate the field scattered from the surface of a parabolic reflector. Generally the field radiated by the currents on the shadow region of the reflector is very small compared to the total field, and the currents and field can be set equal to zero. The field scattered by the illuminated (concave) side of the parabolic reflector can be formulated, using the current distribution method, by (15-46a) and (15-46b) when the integration is restricted over the illuminated area.

The total field of the system can be obtained by a superposition of the radiation from the primary source in directions greater than $\theta_0 (\theta > \theta_0)$ and that scattered by the surface as obtained by using either the aperture distribution or the current distribution method.

Generally edge effects are neglected. However the inclusion of diffracted fields [9]–[11] from the rim of the reflector not only introduce fields in the shadow region of the reflector, but also modify those present in the transition and lit regions. Any discontinuities introduced by geometrical optics methods along the transition region (between lit and shadow regions) are removed by the diffracted components.

The far-zone electric field of a parabolic reflector, neglecting the direct radiation, is given by (15-46a). When expanded, (15-46a) reduces, by referring to the geometry of Figure 15.18, to the two components of

$$E_\theta = -j\frac{\omega\mu}{4\pi r} e^{-jkr} \iint_{S_1} \hat{\mathbf{a}}_\theta \cdot \mathbf{J}_s e^{+jk\mathbf{r}'\cdot\hat{\mathbf{a}}_r} \, ds' \tag{15-47a}$$

$$E_\phi = -j\frac{\omega\mu}{4\pi r} e^{-jkr} \iint_{S_1} \hat{\mathbf{a}}_\phi \cdot \mathbf{J}_s e^{+jk\mathbf{r}'\cdot\hat{\mathbf{a}}_r} \, ds' \tag{15-47b}$$

According to the geometry of Figure 15.19

$$ds' = dW \, dN = (r' \sin\theta' \, d\phi') \left[r' \sec\left(\frac{\theta'}{2}\right) d\theta' \right]$$

$$= (r')^2 \sin\theta' \sec\left(\frac{\theta'}{2}\right) d\theta' \, d\phi' \tag{15-48}$$

since

$$dW = r' \sin\theta' \, d\phi' \tag{15-48a}$$

$$dH = -\hat{\mathbf{a}}_r' \cdot d\mathbf{N} = -\hat{\mathbf{a}}_r' \cdot \hat{\mathbf{n}} \, dN$$

$$= -\hat{\mathbf{a}}_r' \cdot \left[-\hat{\mathbf{a}}_r' \cos\left(\frac{\theta'}{2}\right) + \hat{\mathbf{a}}_\theta' \sin\left(\frac{\theta'}{2}\right) \right] dN = \cos\left(\frac{\theta'}{2}\right) dN \tag{15-48b}$$

$$dN = \sec\left(\frac{\theta'}{2}\right) dH = \sec\left(\frac{\theta'}{2}\right) r' \, d\theta' = r' \sec\left(\frac{\theta'}{2}\right) d\theta' \tag{15-48c}$$

Therefore, it can be shown that (15-47a) and (15-47b) can be expressed, with the aid of (15-37), (15-37a), and (15-48), as

$$\begin{bmatrix} E_\theta \\ E_\phi \end{bmatrix} = -j\frac{\omega\mu}{2\pi r}\sqrt{\frac{\varepsilon}{\mu}} C_1 e^{-jkr} \begin{bmatrix} \hat{\mathbf{a}}_\theta \cdot \mathbf{I} \\ \hat{\mathbf{a}}_\phi \cdot \mathbf{I} \end{bmatrix} = -j\frac{\omega\mu e^{-jkr}}{2\pi r} \left[\sqrt{\frac{\varepsilon}{\mu}} \frac{P_t}{2\pi} \right]^{1/2} \begin{bmatrix} \hat{\mathbf{a}}_\theta \cdot \mathbf{I} \\ \hat{\mathbf{a}}_\phi \cdot \mathbf{I} \end{bmatrix} \tag{15-49}$$

where

$$\mathbf{I} = \mathbf{I}_t + \mathbf{I}_z \tag{15-49a}$$

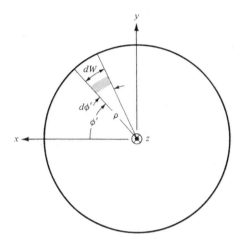

(a) Projected cross section

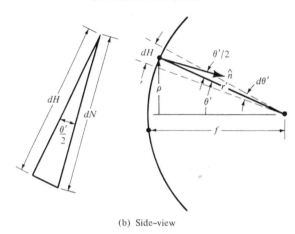

(b) Side-view

Figure 15.19 Projected cross section and side view of reflector.

$$\mathbf{I}_t = -\int_0^{2\pi}\int_0^{\theta_0} \hat{\mathbf{e}}_r \cos\left(\frac{\theta'}{2}\right) \frac{\sqrt{G_f(\theta',\phi')}}{r'} e^{-jkr'[1-\sin\theta'\sin\theta\cos(\phi'-\phi)-\cos\theta'\cos\theta]}$$
$$\times (r')^2 \sin\theta' \sec\left(\frac{\theta'}{2}\right) d\theta'\, d\phi' \qquad (15\text{-}49\text{b})$$

$$\mathbf{I}_z = -\hat{\mathbf{a}}_z \int_0^{2\pi}\int_0^{\theta_0} (\hat{\mathbf{n}}\cdot\hat{\mathbf{e}}_r) \frac{\sqrt{G_f(\theta',\phi')}}{r'} e^{-jkr'[1-\sin\theta'\sin\theta\cos(\phi'-\phi)-\cos\theta'\cos\theta]}$$
$$\times (r')^2 \sin\theta' \sec\left(\frac{\theta'}{2}\right) d\theta'\, d\phi' \qquad (15\text{-}49\text{c})$$

By comparing (15-49) with (15-35), the radiated field components formulated by the aperture distribution and current distribution methods lead to similar results provided the $\mathbf{I}_z$ contribution of (15-49c) is neglected. As the ratio of the aperture diameter to wavelength (d/λ) increases, the current distribution method results reduce to those of the aperture distribution and the angular pattern becomes more narrow.

For variations near the $\theta = \pi$ region, the $\mathbf{I}_z$ contribution becomes negligible because

$$\hat{\mathbf{a}}_\theta \cdot [-\hat{\mathbf{a}}_z(\hat{\mathbf{n}} \cdot \hat{\mathbf{e}}_r)] = [\hat{\mathbf{a}}_x \sin\theta\cos\phi + \hat{\mathbf{a}}_y \cos\theta\sin\phi - \hat{\mathbf{a}}_z \sin\theta]$$
$$\cdot [-\hat{\mathbf{a}}_z(\hat{\mathbf{n}} \cdot \hat{\mathbf{e}}_r)] = (\hat{\mathbf{n}} \cdot \hat{\mathbf{e}}_r)\sin\theta \quad (15\text{-}50\text{a})$$

$$\hat{\mathbf{a}}_\phi \cdot [-\hat{\mathbf{a}}_z(\hat{\mathbf{n}} \cdot \hat{\mathbf{e}}_r)] = [-\hat{\mathbf{a}}_x \sin\phi + \hat{\mathbf{a}}_y \cos\phi] \cdot [-\hat{\mathbf{a}}_z(\hat{\mathbf{n}} \cdot \hat{\mathbf{e}}_r)] = 0 \quad (15\text{-}50\text{b})$$

F. Directivity and Aperture Efficiency

In the design of antennas, the directivity is a very important figure of merit. The purpose of this section will be to examine the dependence of the directivity and aperture efficiency on the primary-feed pattern $G_f(\theta', \phi')$ and f/d ratio (or the included angle $2\theta_0$) of the reflector. To simplify the analysis, it will be assumed that the feed pattern $G_f(\theta', \phi')$ is circularly symmetric (not a function of ϕ') and that $G_f(\theta') = 0$ for $\theta' > 90°$.

The secondary pattern (formed by the surface of the reflector) is given by (15-49). Approximating the **I** of (15-49a) by $\mathbf{I}_t$, the total **E**-field in the $\theta = \pi$ direction is given by either E_θ or E_ϕ of (15-49). Assuming the feed is circularly symmetric, linearly polarized in the y-direction, and by neglecting cross-polarized contributions, it can be shown with the aid of (15-14a) that (15-49) reduces to

$$E(r, \theta = \pi) = -j\frac{2\omega\mu f}{r}\left[\sqrt{\frac{\varepsilon}{\mu}}\frac{P_t}{2\pi}\right]^{1/2} e^{-jk(r+2f)} \int_0^{\theta_0} \sqrt{G_f(\theta')} \tan\left(\frac{\theta'}{2}\right) d\theta' \quad (15\text{-}51)$$

The power intensity (power/unit solid angle) in the forward direction $U(\theta = \pi)$ is given by

$$U(\theta = \pi) = \frac{1}{2}r^2\sqrt{\frac{\varepsilon}{\mu}}|E(r, \theta = \pi)|^2 \quad (15\text{-}52)$$

which by using (15-51) reduces to

$$U(\theta = \pi) = \frac{16\pi^2}{\lambda^2}f^2\frac{P_t}{4\pi}\left|\int_0^{\theta_0} \sqrt{G_f(\theta')} \tan\left(\frac{\theta'}{2}\right) d\theta'\right|^2 \quad (15\text{-}52\text{a})$$

The antenna directivity in the forward direction can be written, using (15-52a), as

$$D_0 = \frac{4\pi U(\theta = \pi)}{P_t} = \frac{U(\theta = \pi)}{P_t/4\pi} = \frac{16\pi^2}{\lambda^2}f^2\left|\int_0^{\theta_0} \sqrt{G_f(\theta')} \tan\left(\frac{\theta'}{2}\right) d\theta'\right|^2 \quad (15\text{-}53)$$

The focal length is related to the angular spectrum and aperture diameter d by (15-25). Thus (15-53) reduces to

$$\boxed{D_0 = \left(\frac{\pi d}{\lambda}\right)^2\left\{\cot^2\left(\frac{\theta_0}{2}\right)\left|\int_0^{\theta_0} \sqrt{G_f(\theta')} \tan\left(\frac{\theta'}{2}\right) d\theta'\right|^2\right\}} \quad (15\text{-}54)$$

The factor $(\pi d/\lambda)^2$ is the directivity of a uniformly illuminated constant phase aperture; the remaining part is the aperture efficiency defined as

$$\boxed{\varepsilon_{ap} = \cot^2\left(\frac{\theta_0}{2}\right)\left|\int_0^{\theta_0} \sqrt{G_f(\theta')} \tan\left(\frac{\theta'}{2}\right) d\theta'\right|^2} \quad (15\text{-}55)$$

It is apparent by examining (15-55) that the aperture efficiency is a function of the subtended angle (θ_0) and the feed pattern $G_f(\theta')$ of the reflector. Thus for a given feed pattern, all paraboloids with the same f/d ratio have identical aperture efficiency.

To illustrate the variation of the aperture efficiency as a function of the feed pattern and the angular extent of the reflector, Silver [17] considered a class of feeds whose patterns are defined by

$$G_f(\theta') = \begin{cases} G_0^{(n)} \cos^n(\theta') & 0 \le \theta' \le \pi/2 \\ 0 & \pi/2 < \theta' \le \pi \end{cases} \quad (15\text{-}56)$$

where $G_0^{(n)}$ is a constant for a given value of n. Although idealistic, these patterns were chosen because (1) closed form solutions can be obtained, and (2) they often are used to represent a major part of the main lobe of many practical antennas. The intensity in the back region ($\pi/2 < \theta' \le \pi$) was assumed to be zero in order to avoid interference between the direct radiation from the feed and scattered radiation from the reflector.

The constant $G_0^{(n)}$ can be determined from the relation

$$\oiint_S G_f(\theta') \, d\Omega = \oiint_S G_f(\theta') \sin\theta' \, d\theta' \, d\phi' = 4\pi \quad (15\text{-}57)$$

which for (15-56) becomes

$$G_0^{(n)} \int_0^{\pi/2} \cos^n \theta' \sin \theta' \, d\theta' = 2 \Rightarrow G_0^{(n)} = 2(n+1) \quad (15\text{-}58)$$

Substituting (15-56) and (15-58) into (15-55) leads, for the even values of $n = 2$ through $n = 8$, to

$$\varepsilon_{ap}(n=2) = 24 \left\{ \sin^2\left(\frac{\theta_0}{2}\right) + \ln\left[\cos\left(\frac{\theta_0}{2}\right)\right] \right\}^2 \cot^2\left(\frac{\theta_0}{2}\right) \quad (15\text{-}59a)$$

$$\varepsilon_{ap}(n=4) = 40 \left\{ \sin^4\left(\frac{\theta_0}{2}\right) + \ln\left[\cos\left(\frac{\theta_0}{2}\right)\right] \right\}^2 \cot^2\left(\frac{\theta_0}{2}\right) \quad (15\text{-}59b)$$

$$\varepsilon_{ap}(n=6) = 14 \left\{ 2\ln\left[\cos\left(\frac{\theta_0}{2}\right)\right] + \frac{[1-\cos(\theta_0)]^3}{3} + \frac{1}{2}\sin^2(\theta_0) \right\}^2 \cot^2\left(\frac{\theta_0}{2}\right) \quad (15\text{-}59c)$$

$$\varepsilon_{ap}(n=8) = 18 \left\{ \frac{1-\cos^4(\theta_0)}{4} - 2\ln\left[\cos\left(\frac{\theta_0}{2}\right)\right] - \frac{[1-\cos(\theta_0)]^3}{3} - \frac{1}{2}\sin^2(\theta_0) \right\}^2 \cot^2\left(\frac{\theta_0}{2}\right)$$

$$(15\text{-}59d)$$

The variations of (15-59a)–(15-59d), as a function of the angular aperture of the reflector θ_0 or the f/d ratio, are shown plotted in Figure 15.20. It is apparent, from the graphical illustration, that for a given feed pattern (n = constant)

1. There is only one reflector with a given angular aperture or f/d ratio which leads to a maximum aperture efficiency.
2. Each maximum aperture efficiency is in the neighborhood of 82–83%.

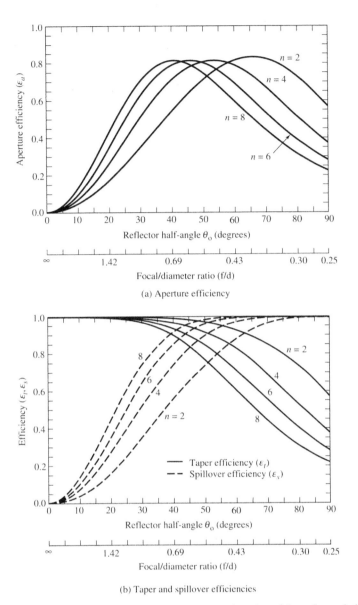

(a) Aperture efficiency

(b) Taper and spillover efficiencies

Figure 15.20 Aperture, and taper and spillover efficiencies as a function of the reflector half-angle θ_0 (or f/d ratio) for different feed patterns.

3. Each maximum aperture efficiency, for any one of the given patterns, is almost the same as that of any of the others.
4. As the feed pattern becomes more directive (n increases), the angular aperture of the reflector that leads to the maximum efficiency is smaller.

The aperture efficiency is generally the product of the

1. fraction of the total power that is radiated by the feed, intercepted, and collimated by the reflecting surface (generally known as *spillover efficiency* ε_s)
2. uniformity of the amplitude distribution of the feed pattern over the surface of the reflector (generally known as *taper efficiency* ε_t)

3. phase uniformity of the field over the aperture plane (generally known as *phase efficiency* ε_p)
4. polarization uniformity of the field over the aperture plane (generally known as *polarization efficiency* ε_x)
5. *blockage efficiency* ε_b
6. *random error efficiency* ε_r over the reflector surface

Thus in general

$$\varepsilon_{ap} = \varepsilon_s \varepsilon_t \varepsilon_p \varepsilon_x \varepsilon_b \varepsilon_r \tag{15-60}$$

For feeds with symmetrical patterns

$$\varepsilon_s = \frac{\int_0^{\theta_0} G_f(\theta') \sin\theta' \, d\theta'}{\int_0^{\pi} G_f(\theta') \sin\theta' \, d\theta'} \tag{15-61}$$

$$\varepsilon_t = 2\cot^2\left(\frac{\theta_0}{2}\right) \frac{\left|\int_0^{\theta_0} \sqrt{G_f(\theta')} \tan\left(\frac{\theta'}{2}\right) d\theta'\right|^2}{\int_0^{\theta_0} G_f(\theta') \sin\theta' \, d\theta'} \tag{15-62}$$

which by using (15-25) can also be written as

$$\varepsilon_t = 32\left(\frac{f}{d}\right)^2 \frac{\left|\int_0^{\theta_0} \sqrt{G_f(\theta')} \tan\left(\frac{\theta'}{2}\right) d\theta'\right|^2}{\int_0^{\theta_0} G_f(\theta') \sin\theta' \, d\theta'} \tag{15-62a}$$

Thus

1. $100(1 - \varepsilon_s)$ = percent power loss due to energy from feed spilling past the main reflector.
2. $100(1 - \varepsilon_t)$ = percent power loss due to nonuniform amplitude distribution over the reflector surface.
3. $100(1 - \varepsilon_p)$ = percent power loss if the field over the aperture plane is not in phase everywhere.
4. $100(1 - \varepsilon_x)$ = percent power loss if there are cross-polarized fields over the antenna aperture plane.
5. $100(1 - \varepsilon_b)$ = percent power loss due to blockage provided by the feed or supporting struts (also by subreflector for a dual reflector).
6. $100(1 - \varepsilon_r)$ = percent power loss due to random errors over the reflector surface.

An additional factor that reduces the antenna gain is the attenuation in the antenna feed and associated transmission line.

For feeds with

1. symmetrical patterns
2. aligned phase centers
3. no cross-polarized field components
4. no blockage
5. no random surface error

the two main factors that contribute to the aperture efficiency are the spillover and nonuniform amplitude distribution losses. Because these losses depend primarily on the feed pattern, a compromise between spillover and taper efficiency must emerge. Very high spillover efficiency can be achieved by a narrow beam pattern with low minor lobes at the expense of a very low taper efficiency.

Example 15.1

Show that a parabolic reflector using a point source as a feed, as shown in Figure 15.21, creates an amplitude taper proportional to

$$\left(\frac{f}{r'}\right) = \cos^4\left(\frac{\theta'}{2}\right)$$

and that the feed pattern has to be equal to

$$G_f(\theta') = G_0 \sec^4\left(\frac{\theta'}{2}\right) = \sec^4\left(\frac{\theta'}{2}\right)$$

to produce uniform illumination.

Solution: The field radiated by a point source varies as

$$\frac{e^{-jkr}}{r}$$

Therefore, a field radiated by a point source at the focal point of a parabolic reflector, which acts as a feed, creates an amplitude taper across the reflector due to the space factor (r'/f), by referring to Figure 15.21, which compares the field at the vertex to that at any other point θ' on the reflector surface.

According to (15-14) or (15-14a)

$$\frac{r'}{f} = \frac{2}{1+\cos(\theta')} = \sec^2\left(\frac{\theta'}{2}\right) \Rightarrow \left(\frac{f}{r'}\right) = \frac{1}{\sec^2(\theta'/2)} = \cos\left(\frac{\theta'}{2}\right)$$

The power amplitude taper created across the reflector is therefore

$$\left(\frac{f}{r'}\right)^2 = \cos^4\left(\frac{\theta'}{2}\right)$$

To compensate for this and obtain uniform illumination, the feed power must be

$$G_f(\theta') = \frac{1}{\cos^4(\theta'/2)} = \sec^4\left(\frac{\theta'}{2}\right), \qquad \theta' \leq \theta_0$$

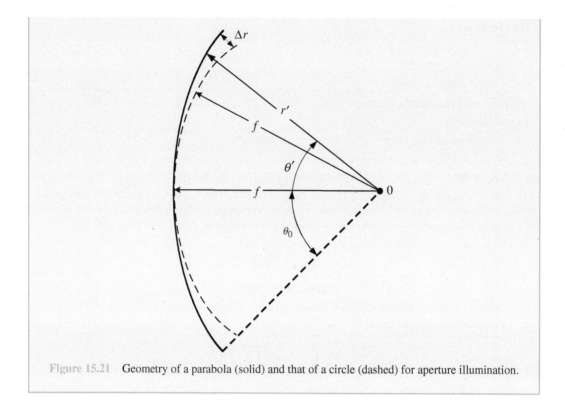

Figure 15.21 Geometry of a parabola (solid) and that of a circle (dashed) for aperture illumination.

Through Example 15.1, uniform illumination and ideal taper, spillover and aperture efficiencies can be obtained when the feed power pattern is

$$G_f(\theta') = \begin{cases} G_0 \sec^4\left(\dfrac{\theta'}{2}\right) & 0 \le \theta' \le \theta_0 \\ 0 & \theta' > \theta_0 \end{cases} \quad (15\text{-}63)$$

whose normalized distribution is shown plotted in Figure 15.22. To accomplish this, the necessary value of G_0 is derived in Example 15.2. Although such a pattern is "ideal" and impractical to achieve, much effort has been devoted to develop feed designs which attempt to approximate it [14].

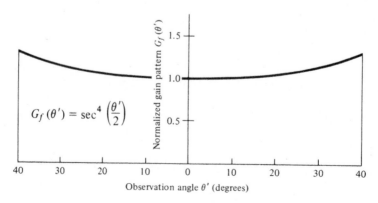

Figure 15.22 Normalized gain pattern of feed for uniform amplitude illumination of paraboloidal reflector with a total subtended angle of 80°.

Example 15.2

Show that a feed pattern of

$$G_f(\theta') = \begin{cases} G_0 \sec^4\left(\dfrac{\theta'}{2}\right) & 0 \le \theta' \le \theta_0 \\ 0 & \theta' > \theta_0 \end{cases}$$

in conjunction with a parabolic reflector, leads to an ideal aperture efficiency of $\varepsilon_{ap} = 1$. Determine the value of G_0 that will accomplish this.

Solution: The aperture efficiency of (15-55) reduces, using the given feed pattern, to

$$\varepsilon_{ap} = \cot^2\left(\frac{\theta_0}{2}\right) \left| \int_0^{\theta_0} \sqrt{G_f(\theta')} \tan\left(\frac{\theta'}{2}\right) d\theta' \right|^2$$

$$= \cot^2\left(\frac{\theta_0}{2}\right) G_0 \left| \int_0^{\theta_0} \sec^2\left(\frac{\theta'}{2}\right) \tan\left(\frac{\theta'}{2}\right) d\theta' \right|^2$$

$$\varepsilon_{ap} = \cot^2\left(\frac{\theta_0}{2}\right) G_0 |I|^2$$

where

$$I = \int_0^{\theta_0} \sec^2\left(\frac{\theta'}{2}\right) \tan\left(\frac{\theta'}{2}\right) d\theta' = \int_0^{\theta_0} \frac{1}{\cos^2(\theta'/2)} \frac{\sin(\theta'/2)}{\cos(\theta'/2)} d\theta' = \int_0^{\theta_0} \frac{\sin(\theta'/2)}{\cos^3(\theta'/2)} d\theta'$$

$$I = \left[\frac{1}{\cos^2(\theta_0/2)} - 1\right] = \left[\sec^2\left(\frac{\theta_0}{2}\right) - 1\right] = \tan^2\left(\frac{\theta_0}{2}\right)$$

Thus

$$\varepsilon_{ap} = G_0 \cot^2\left(\frac{\theta_0}{2}\right) \tan^4\left(\frac{\theta_0}{2}\right) = G_0 \tan^2\left(\frac{\theta_0}{2}\right)$$

To insure also an ideal spillover efficiency ($\varepsilon_s = 1$), all the power that is radiated by the feed must be contained within the angular region subtended by the edges of the reflector. Using (15-57), and assuming symmetry in the pattern with respect to ϕ, we can write that

$$\int_0^{2\pi} \int_0^{\pi} G_f(\theta') \underbrace{\sin\theta' \, d\theta' \, d\phi'}_{d\Omega} = 2\pi \int_0^{\pi} G_f(\theta') \sin\theta' \, d\theta' = 4\pi$$

$$\int_0^{\pi} G_f(\theta') \sin\theta' \, d\theta' = 2$$

Using the given feed pattern

$$G_f(\theta') = G_0 \sec^4\left(\frac{\theta'}{2}\right) \quad 0 \le \theta' \le \theta_0$$

we can write that

$$\int_0^{\theta_0} G_0 \sec^4\left(\frac{\theta'}{2}\right) \sin\theta' \, d\theta' = G_0 \int_0^{\theta_0} \sec^4\left(\frac{\theta'}{2}\right) \sin\theta' \, d\theta' = 2$$

$$G_0 = \frac{2}{\int_0^{\theta_0} \sec^4(\theta'/2) \sin\theta' \, d\theta'} = \frac{2}{2\tan^2(\theta_0/2)} = \cot^2\left(\frac{\theta_0}{2}\right)$$

Thus the total aperture efficiency, taking into account uniform illumination and total spillover control, is

$$\varepsilon_{ap} = G_0 \cot^2\left(\frac{\theta_0}{2}\right) \tan^4\left(\frac{\theta_0}{2}\right) = G_0 \tan^2\left(\frac{\theta_0}{2}\right) = \cot^2\left(\frac{\theta_0}{2}\right) \tan^2\left(\frac{\theta_0}{2}\right) = 1$$

To develop guidelines for designing practical feeds which result in high aperture efficiencies, it is instructive to examine the relative field strength at the edges of the reflector's bounds ($\theta' = \theta_0$) for patterns that lead to optimum efficiencies. For the patterns of (15-56), when used with reflectors that result in optimum efficiencies as demonstrated graphically in Figure 15.20, the relative field strength at the edges of their angular bounds ($\theta' = \theta_0$) is shown plotted in Figure 15.23. Thus for $n = 2$ the field strength of the pattern at $\theta' = \theta_0$ is 8 dB down from the maximum. As the pattern becomes more narrow (n increases), the relative field strength at the edges for maximum efficiency is further reduced as illustrated in Figure 15.20. Since for $n = 2$ through $n = 10$ the field strength is between 8 to 10.5 dB down, for most practical feeds the figure used is 9–10 dB.

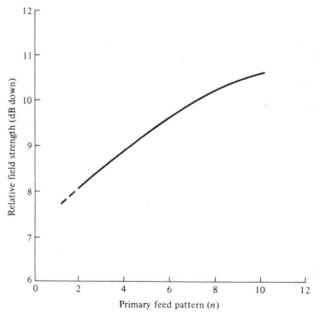

Figure 15.23 Relative field strength of feed pattern along reflector edge bounds as a function of primary-feed pattern number ($\cos^n \theta$). (SOURCE: S. Silver (ed.), *Microwave Antenna Theory and Design* (MIT Radiation Lab. Series, Vol. 12), McGraw-Hill, New York, 1949).

TABLE 15.1 Aperture Efficiency and Field Strength at the Edge of Reflector, Relative to That at the Vertex, due to the Feed Pattern (Feed) and Path Length (Path) between the Edge and the Vertex

n	ε_{ap}	θ_0 (deg)	f/d	Feed (dB) $10\log[G(\theta_0)]$	Path (dB) $10\log(f/r')^2$	Total (dB) Feed + Path
2	0.829	66	0.385	−7.8137	−3.056	−10.87
4	0.8196	53.6	0.496	−8.8215	−1.959	−10.942
6	0.8171	46.2	0.5861	−9.4937	−1.439	−10.93
8	0.8161	41.2	0.6651	−9.7776	−1.137	−10.914

Another parameter to examine for the patterns of (15-56), when used with reflectors that lead to optimum efficiency, is the amplitude taper or illumination of the main aperture of the reflector which is defined as *the ratio of the field strength at the edge of the reflector surface to that at the vertex*. The aperture illumination is a function of the feed pattern and the *f/d* ratio of the reflector. To obtain that, the ratio of the angular variation of the pattern toward the two points $[G_f(\theta' = \theta_0)/G_f(\theta' = 0)]$ is multiplied by the space attenuation factor $(f/r_0)^2$, where f is the focal distance of the reflector and r_0 is the distance from the focal point to the edge of the reflector. For each of the patterns, the reflector edge illumination for maximum efficiency is nearly 11 dB down from that at the vertex. The details for $2 \leq n \leq 8$ are shown in Table 15.1.

The results obtained with the idealized patterns of (15-56) should only be taken as typical, because it was assumed that

1. the field intensity for $\theta' > 90°$ was zero
2. the feed was placed at the phase center of the system
3. the patterns were symmetrical
4. there were no cross-polarized field components
5. there was no blockage
6. there were no random errors at the surface of the reflector

Each factor can have a significant effect on the efficiency, and each has received much attention which is well documented in the open literature [1].

In practice, maximum reflector efficiencies are in the 65–80% range. To demonstrate that, paraboloidal reflector efficiencies for square corrugated horns feeds were computed, and they are shown plotted in Figure 15.24. The corresponding amplitude taper and spillover efficiencies for the aperture efficiencies of Figures 15.20(a) and 15.24 are displayed, respectively, in Figures 15.20(b) and 15.25. For the data of Figures 15.24 and 15.25, each horn had aperture dimensions of $8\lambda \times 8\lambda$, their patterns were assumed to be symmetrical (by averaging the *E*- and *H*-planes), and they were computed using the techniques of Section 13.6. From the plotted data, it is apparent that the maximum aperture efficiency for each feed pattern is in the range of 74–79%, and that the product of the taper and spillover efficiencies is approximately equal to the total aperture efficiency.

We would be remiss if we left the discussion of this section without reporting the gain of some of the largest reflectors that exist around the world [23]. The gains are shown in Figure 15.26 and include the 1,000-ft (305-m) diameter spherical reflector [12] at Arecibo, Puerto Rico, the 100-m radio telescope [15] at Effelsberg, West Germany, the 64-m reflector [16] at Goldstone, California, the 22-m reflector at Krim, USSR, and the 12-m telescope at Kitt Peak, Arizona. The dashed portions of the curves indicate extrapolated values. For the Arecibo reflector, two curves are shown. The 215-m diameter curve is for a reduced aperture of the large reflector (305-m) for which a line feed at 1,415 MHz was designed [12].

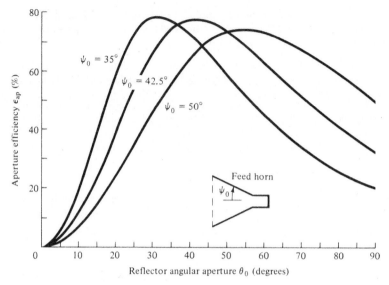

Figure 15.24 Parabolic reflector aperture efficiency as a function of angular aperture for $8\lambda \times 8\lambda$ square corrugated horn feed with total flare angles of $2\psi_0 = 70°, 85°,$ and $100°$.

G. Phase Errors

Any departure of the phase, over the aperture of the antenna, from uniform can lead to a significant diminution of its directivity [24]. For a paraboloidal reflector system, phase errors result from [17]

1. displacement (defocusing) of the feed phase center from the focal point
2. deviation of the reflector surface from a parabolic shape or random errors at the surface of the reflector
3. departure of the feed wavefronts from spherical shape

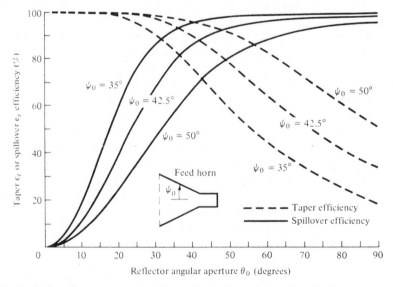

Figure 15.25 Parabolic reflector taper and spillover efficiencies as a function of reflector aperture for different corrugated horn feeds.

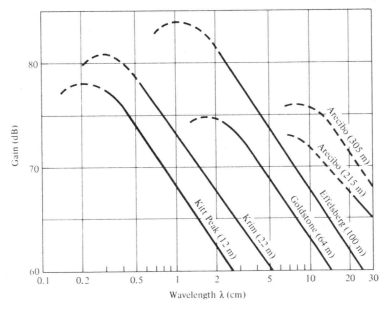

Figure 15.26 Gains of some worldwide large reflector antennas. (SOURCE: A. W. Love, "Some Highlights in Reflector Antenna Development," *Radio Science*, Vol. 11, Nos. 8, 9, August–September 1976).

The defocusing effect can be reduced by first locating the phase center of the feed antenna and then placing it at the focal point of the reflector. In Chapter 13 (Section 13.10) it was shown that the phase center for horn antennas, which are widely utilized as feeds for reflectors, is located between the aperture of the horn and the apex formed by the intersection of the inclined walls of the horn [25].

Very simple expressions have been derived [24] to predict the loss in directivity for rectangular and circular apertures when the peak values of the aperture phase deviation is known. When the phase errors are assumed to be relatively small, it is not necessary to know the exact amplitude or phase distribution function over the aperture.

Assuming the maximum radiation occurs along the axis of the reflector, and that the maximum phase deviation over the aperture of the reflector can be represented by

$$|\Delta\phi(\rho)| = |\phi(\rho) - \overline{\phi(\rho)}| \leq m, \quad \rho \leq a \tag{15-64}$$

where $\phi(\rho)$ is the aperture phase function and $\overline{\phi(\rho)}$ is its average value, then the ratio of the directivity with (D) and without (D_0) phase errors can be written as [24]

$$\frac{D}{D_0} = \frac{\text{directivity with phase error}}{\text{directivity without phase error}} \geq \left(1 - \frac{m^2}{2}\right)^2 \tag{15-65}$$

and the maximum fractional reduction in directivity as

$$\frac{\Delta D}{D_0} = \frac{D_0 - D}{D_0} \leq m^2\left(1 - \frac{m^2}{4}\right) \tag{15-66}$$

Relatively simple expressions have also been derived [24] to compute the maximum possible change in half-power beamwidth.

Example 15.3

A 10-m diameter reflector, with an f/d ratio of 0.5, is operating at $f = 3$ GHz. The reflector is fed with an antenna whose primary pattern is symmetrical and which can be approximated by $G_f(\theta') = 6\cos^2\theta'$, $0° \leq \theta \leq 90°$ and zero elsewhere. Find its

(a) aperture efficiency
(b) overall directivity
(c) spillover and taper efficiencies
(d) directivity when the maximum aperture phase deviation is $\pi/8$ rad

Solution: Using (15-24), half of the subtended angle of the reflector is equal to

$$\theta_0 = \tan^{-1}\left[\frac{0.5(0.5)}{(0.5)^2 - \frac{1}{16}}\right] = 53.13°$$

(a) The aperture efficiency is obtained using (15-59a). Thus

$$\varepsilon_{ap} = 24\{\sin^2(26.57°) + \ln[\cos(26.57°)]\}^2 \cot^2(26.57°)$$
$$= 0.75 = 75\%$$

which agrees with the data of Figure 15.20.

(b) The overall directivity is obtained by (15-54), or

$$D = 0.75[\pi(100)]^2 = 74{,}022.03 = 48.69 \text{ dB}$$

(c) The spillover efficiency is computed using (15-61) where the upper limit of the integral in the denominator has been replaced by $\pi/2$. Thus

$$\varepsilon_s = \frac{\int_0^{53.13°} \cos^2\theta' \sin\theta'\, d\theta'}{\int_0^{90°} \cos^2\theta' \sin\theta'\, d\theta'} = \frac{2\cos^3\theta'\big|_0^{53.13°}}{2\cos^3\theta'\big|_0^{90°}} = 0.784 = 78.4\%$$

In a similar manner, the taper efficiency is computed using (15-62). Since the numerator in (15-62) is identical in form to the aperture efficiency of (15-55), the taper efficiency can be found by multiplying (15-59a) by 2 and dividing by the denominator of (15-62). Thus

$$\varepsilon_t = \frac{2(0.75)}{1.568} = 0.9566 = 95.66\%$$

The product of ε_s and ε_t is equal to

$$\varepsilon_s\varepsilon_t = 0.784(0.9566) = 0.75$$

and it is identical to the total aperture efficiency computed above.

(d) The directivity for a maximum phase error of $m = \pi/8 = 0.3927$ rad can be computed using (15-65). Thus

$$\frac{D}{D_0} \geq \left(1 - \frac{m^2}{2}\right)^2 = \left[1 - \frac{(0.3927)^2}{2}\right]^2 = 0.8517 = -0.69 \text{ dB}$$

or $D \geq 0.8517 D_0 = 0.8517(74{,}022.03) = 63{,}046.94 = 48.0$ dB.

Surface roughness effects on the directivity of the antenna were first examined by Ruze [26] where he indicated that for any reflector antenna there is a wavelength ($\lambda_{\max}$) at which the directivity reaches a maximum. This wavelength depends on the RMS deviation (σ) of the reflector surface from an ideal paraboloid. For a random roughness of Gaussian distribution, with correlation interval large compared to the wavelength, they are related by

$$\lambda_{\max} = 4\pi\sigma \tag{15-67}$$

Thus the directivity of the antenna, given by (15-54), is modified to include surface roughness and can be written as

$$D = \left(\frac{\pi d}{\lambda}\right)^2 \varepsilon_{ap} e^{-(4\pi\sigma/\lambda)^2} \tag{15-68}$$

Using (15-67), the maximum directivity of (15-68) can be written as

$$D_{\max} = 10^{2q} \varepsilon_{ap} \left(\frac{e^{-1}}{16}\right) \tag{15-69}$$

where q is the index of smoothness defined by

$$\frac{d}{\sigma} = 10^{+q} \tag{15-70}$$

In decibels, (15-69) reduces to

$$D_{\max}(\text{dB}) = 20q - 16.38 + 10\log_{10}(\varepsilon_{ap}) \tag{15-71}$$

For an aperture efficiency of unity ($\varepsilon_{ap} = 1$), the directivity of (15-68) is plotted in Figure 15.27, as a function of (d/λ), for values of $q = 3.5$, 4.0, and 4.5. It is apparent that for each value of q and a given reflector diameter d, there is a maximum wavelength where the directivity reaches a maximum value. This maximum wavelength is given by (15-67).

H. Feed Design

The widespread use of paraboloidal reflectors has stimulated interest in the development of feeds to improve the aperture efficiency and to provide greater discrimination against noise radiation from the ground. This can be accomplished by developing design techniques that permit the synthesis of feed patterns with any desired distribution over the bounds of the reflector, rapid cutoff at its edges, and very low minor lobes in all the other space. In recent years, the two main problems that concerned feed designers were aperture efficiency and low cross-polarization.

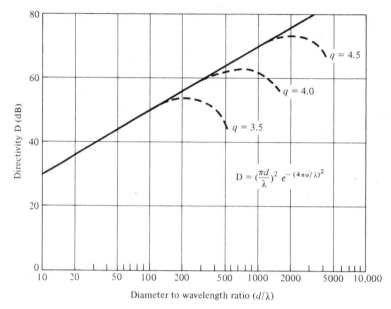

Figure 15.27 Reflector surface roughness effects on antenna directivity. (SOURCE: A. W. Love, "Some Highlights in Reflector Antenna Development," *Radio Science*, Vol. 11, Nos. 8, 9, August–September 1976).

In the receiving mode, an ideal feed and a matched load would be one that would absorb all the energy intercepted by the aperture when uniform and linearly polarized plane waves are normally incident upon it. The feed field structure must be made to match the focal region field structure formed by the reflecting, scattering, and diffracting characteristics of the reflector. By reciprocity, an ideal feed in the transmitting mode would be one that radiates only within the solid angle of the aperture and establishes within it an identical outward traveling wave. For this ideal feed system, the transmitting and receiving mode field structures within the focal region are identical with only the direction of propagation reversed.

An optical analysis indicates that the focal region fields, formed by the reflection of linearly polarized plane waves incident normally on an axially symmetric reflector, are represented by the well-known Airy rings described mathematically by the amplitude distribution intensity of $[2J_1(u)/u]^2$. This description is incomplete, because it is a scalar solution, and it does not take into account polarization effects. In addition, it is valid only for reflectors with large f/d ratios, which are commonly used in optical systems, and it would be significantly in error for reflectors with f/d ratios of 0.25 to 0.5, which are commonly used in microwave applications.

A vector solution has been developed [27] which indicates that the fields near the focal region can be expressed by hybrid TE and TM modes propagating along the axis of the reflector. This representation provides a clear physical picture for the understanding of the focal region field formation. The boundary conditions of the hybrid modes indicate that these field structures can be represented by a spectrum of hybrid waves that are linear combinations of TE_{1n} and TM_{1n} modes of circular waveguides.

A single hollow pipe cannot simultaneously satisfy both the TE and TM modes because of the different radial periodicities. However, it has been shown that $\lambda/4$ deep annular slots on the inner surface of a circular pipe force the boundary conditions on **E** and **H** to be the same and provide a single anisotropic reactance surface which satisfies the boundary conditions on both TE and TM modes. This provided the genesis of hybrid mode waveguide radiators [28] and corrugated horns [29]. Corrugated horns, whose aperture size and flare angle are such that at least 180° phase error over their aperture is assured, are known as "scalar" horns [30]. Design data for uncorrugated horns that can be used to maximize the aperture efficiency or to produce maximum power transmission to the feed have been calculated [31] and are shown in graphical form in Figure 15.28.

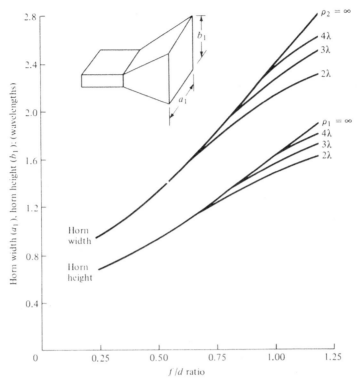

Figure 15.28 Optimum pyramidal horn dimensions versus f/d ratio for various horn lengths.

A FORTRAN software designated **Reflector** for computer-aided analysis and design of reflector antennas has been developed [32]. A converted MATLAB version of **Reflector** is included in the publisher's website for this book. The program computes the radiation of a parabolic reflector. In addition it provides spatial and spectral methods to compute radiation due to an aperture distribution. The software package can also be used to investigate the directivity, sidelobe level, polarization, and near-to-far-zone fields.

15.4.2 Cassegrain Reflectors

To improve the performance of large ground-based microwave reflector antennas for satellite tracking and communication, it has been proposed that a two-reflector system be utilized. The arrangement suggested was the Cassegrain dual-reflector system [33] of Figure 15.1(d), which was often utilized in the design of optical telescopes and it was named after its inventor. To achieve the desired collimation characteristics, the larger (main) reflector must be a paraboloid and the smaller (secondary) a hyperboloid. The use of a second reflector, which is usually referred to as the subreflector or subdish, gives an additional degree of freedom for achieving good performance in a number of different applications. For an accurate description of its performance, diffraction techniques must be used to take into account diffractions from the edges of the subreflector, especially when its diameter is small [34].

In general, the Cassegrain arrangement provides a variety of benefits, such as the

1. ability to place the feed in a convenient location
2. reduction of spillover and minor lobe radiation
3. ability to obtain an equivalent focal length much greater than the physical length
4. capability for scanning and/or broadening of the beam by moving one of the reflecting surfaces

To achieve good radiation characteristics, the subreflector or subdish must be several, at least a few, wavelengths in diameter. However, its presence introduces shadowing which is the principal limitation of its use as a microwave antenna. The shadowing can significantly degrade the gain of the system, unless the main reflector is several wavelengths in diameter. Therefore the Cassegrain is usually attractive for applications that require gains of 40 dB or greater. There are, however, a variety of techniques that can be used to minimize aperture blocking by the subreflector. Some of them are [33] (1) minimum blocking with simple Cassegrain, and (2) twisting Cassegrains for least blocking.

The first comprehensive published analysis of the Cassegrain arrangement as a microwave antenna is that by Hannan [33]. He uses geometrical optics to derive the geometrical shape of the reflecting surfaces, and he introduces the equivalence concepts of the virtual feed and the equivalent parabola. Although his analysis does not predict fine details, it does give reasonably good results. Refinements to his analysis have been introduced [34]–[36].

To improve the aperture efficiency, suitable modifications to the geometrical shape of the reflecting surfaces have been suggested [37]–[39]. The reshaping of the reflecting surfaces is used to generate desirable amplitude and phase distributions over one or both of the reflectors. The resultant system is usually referred to as *shaped* dual reflector. The reflector antenna of Figure 15.9 is such a system. Shaped reflector surfaces, generated using analytical models, are illustrated in [39]. It also has been suggested [35] that a flange is placed around the subreflector to improve the aperture efficiency.

Since many reflectors have dimensions and radii of curvature large compared to the operating wavelength, they were traditionally designed based on geometrical optics (GO) [39]. Both the single- and double-reflector (Cassegrain) systems were designed to convert a spherical wave at the source (focal point) into a plane wave. Therefore the reflecting surfaces of both reflector systems were primarily selected to convert the phase of the wavefront from spherical to planar. However, because of the variable radius of curvature at each point of reflection, the magnitude of the reflected field is also changed due to spatial attenuation or amplitude spreading factor [40] or divergence factor (4-131) of Section 4.8.3, which are functions of the radius of curvature of the surface at the point of reflection. This ultimately leads to amplitude taper of the wavefront at the aperture plane. This is usually undesirable, and it can sometimes be compensated to some extent by the design of the pattern of the feed element or of the reflecting surface.

For a shaped-dual reflector system, there are two surfaces or degrees of freedom that can be utilized to compensate for any variations in the phase and amplitude of the field at the aperture plane. To determine how each surface may be reshaped to control the phase and/or amplitude of the field at the aperture plane, let us use geometrical optics and assume that the field radiated by the feed (pattern) is represented, both in amplitude and phase, by a bundle of rays which has a well-defined periphery. This bundle of rays is initially intercepted by the subreflector and then by the main reflector. Ultimately the output, after two reflections, is also a bundle of rays with prescribed phase and amplitude distributions, and a prescribed periphery, as shown in Figure 15.29(a) [41]. It has been shown in [42] that for a two-reflector system with high magnification (i.e., large ratio of main reflector diameter to subreflector diameter) that over the aperture plane the

a. amplitude distribution is controlled largely by the subreflector curvature.
b. phase distribution is controlled largely by the curvature of the main reflector.

Therefore in a Cassegrain two-reflector system the reshaping of the paraboloid main reflector can be used to optimize the phase distribution while the reshaping of the hyperboloid subreflector can be used to control the amplitude distribution. This was used effectively in [42] to design a shaped two-reflector system whose field reflected by the subreflector had a nonspherical phase but a $\csc^4(\theta/2)$ amplitude pattern. However, the output from the main reflector had a perfect plane wave phase front and a very good approximate uniform amplitude distribution, as shown in Figure 15.29(b).

Because a comprehensive treatment of this arrangement can be very lengthy, only a brief introduction of the system will be presented here. The interested reader is referred to the referenced literature.

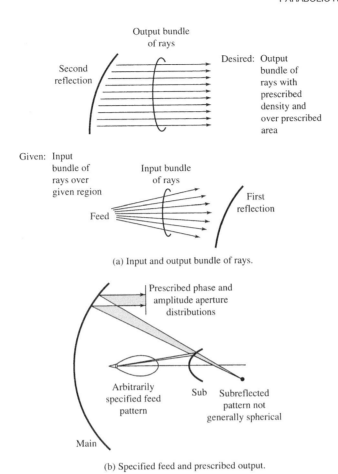

Figure 15.29 Geometrical optics for the reshaping and synthesis of the reflectors of a Cassegrain system. (SOURCE: R. Mittra and V. Galindo-Israel "Shaped Dual Reflector Synthesis," *IEEE Antennas and Propagation Society Newsletter*, Vol. 22, No. 4, pp. 5–9, August 1980. © (1980) IEEE).

A. Classical Cassegrain Form

The operation of the Cassegrain arrangement can be introduced by referring to Figure 15.1(d) and assuming the system is operating in the receiving or transmitting mode. To illustrate the principle, a receiving mode is adopted.

Let us assume that energy, in the form of parallel rays, is incident upon the reflector system. Energy intercepted by the main reflector, which has a large concave surface, is reflected toward the subreflector. Energy collected by the convex surface of the subdish is reflected by it, and it is directed toward the vertex of the main dish. If the incident rays are parallel, the main reflector is a paraboloid, and the subreflector is a hyperboloid, then the collected bundle of rays is focused at a single point. The receiver is then placed at this focusing point.

A similar procedure can be used to describe the system in the transmitting mode. The feed is placed at the focusing point, and it is usually sufficiently small so that the subdish is located in its far-field region. In addition, the subreflector is large enough that it intercepts most of the radiation from the feed. Using the geometrical arrangement of the paraboloid and the hyperboloid, the rays reflected by the main dish will be parallel. The amplitude taper of the emerging rays is determined by the feed pattern and the tapering effect of the geometry.

The geometry of the classical Cassegrain system, employing a concave paraboloid as the main dish and a convex hyperboloid as the subreflector, is simple and it can be described completely

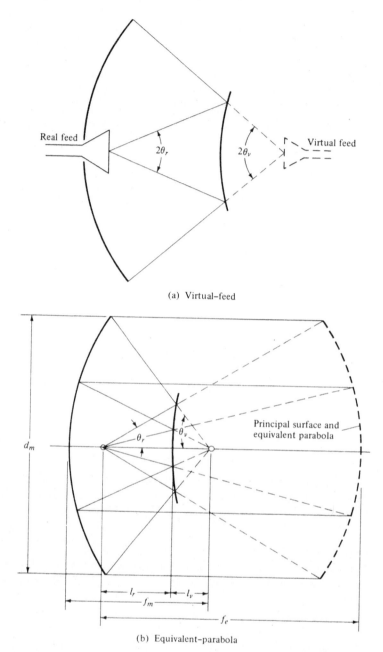

Figure 15.30 Virtual-feed and equivalent parabola concepts. (SOURCE: P. W. Hannan, "Microwave Antennas Derived from the Cassegrain Telescope," *IRE Trans. Antennas Propagat.* Vol. AP-9, No. 2, March 1961. © (1961) IEEE).

by only four independent parameters (two for each reflector). The analytical details can be found in [33].

To aid in the understanding and in predicting the essential performance of a Cassegrain, the concept of *virtual feed* [33] is useful. By this principle, the real feed and the subreflector are replaced by an equivalent system which consists of a virtual feed located at the focal point of the main reflector, as shown by the dashed lines of Figure 15.30(a). For analysis purposes then, the new system is a single-reflector arrangement with the original main dish, a different feed, and no subreflector.

The configuration of the virtual feed can be determined by finding the optical image of the real feed. This technique is only accurate when examining the effective aperture of the feed and when the dimensions of the real and virtual feeds are larger than a wavelength. In fact, for the classical Cassegrain arrangement of Figure 15.30(a), the virtual feed has a smaller effective aperture, and a corresponding broader beamwidth, than the real feed. The increase in beamwidth is a result of the convex curvature of the subreflector, and it can be determined by equating the ratio of the virtual to real-feed beamwidths to the ratio of the angles θ_v/θ_r.

The ability to obtain a different effective aperture for the virtual feed as compared to that of the real feed is useful in many applications such as in a monopulse antenna [33]. To maintain efficient and wideband performance and effective utilization of the main aperture, this system requires a large feed aperture, a corresponding long focal length, and a large antenna structure. The antenna dimensions can be maintained relatively small by employing the Cassegrain configuration which utilizes a large feed and a short focal length for the main reflector.

Although the concept of virtual feed can furnish useful qualitative information for a Cassegrain system, it is not convenient for an accurate quantitative analysis. Some of the limitations of the virtual-feed concept can be overcome by the concept of the *equivalent parabola* [33].

By the technique of the equivalent parabola, the main dish and the subreflector are replaced by an equivalent focusing surface at a certain distance from the real focal point. This surface is shown dashed in Figure 15.30(b), and it is defined as [33] "the locus of intersection of incoming rays parallel to the antenna axis with the extension of the corresponding rays converging toward the real focal point." Based on simple geometrical optics ray tracing, the equivalent focusing surface for a Cassegrain configuration is a paraboloid whose focal length equals the distance from its vertex to the real focal point. This equivalent system also reduces to a single-reflector arrangement, which has the same feed but a different main reflector, and it is accurate when the subreflector is only a few wavelengths in diameter. More accurate results can be obtained by including diffraction patterns. It also has the capability to focus toward the real focal point an incoming plane wave, incident from the opposite direction, in exactly the same manner as the actual main dish and the subreflector.

B. Cassegrain and Gregorian Forms

In addition to the classical Cassegrain forms, there are other configurations that employ a variety of main reflector and subreflector surfaces and include concave, convex, and flat shapes [33]. In one form, the main dish is held invariant while its feed beamwidth progressively increases and the axial dimensions of the antenna progressively decrease. In another form, the feed beamwidth is held invariant while the main reflector becomes progressively flatter and the axial dimensions progressively increase.

One of the other reflector arrangements is the classical Gregorian design of Figure 15.31 where the main reflector is a parabola while the subreflector is a concave ellipse. The focal point is

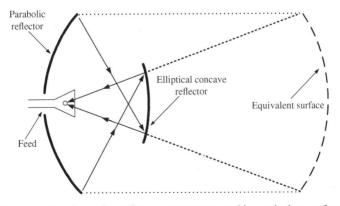

Figure 15.31 Gregorian reflector arrangement and its equivalent surface.

between the main reflector and subreflector. Its equivalent parabola is shown dashed in Figure 15.31. When the overall size and the feed beamwidth of the classical Gregorian, of Figure 15.31, are identical to those of the classical Cassegrain, of Figure 15.30, the Gregorian requires a shorter focal length for the main dish [31].

15.5 SPHERICAL REFLECTOR

The discussion and results presented in the previous sections illustrate that a paraboloidal reflector is an ideal collimating device. However, despite all of its advantages and many applications it is severely handicapped in angular scanning. Although scanning can be accomplished by (1) mechanical rotation of the entire structure, and (2) displacement of the feed alone, it is thwarted by the large mechanical moment of inertia in the first case and by the large coma and astigmatism in the second. By contrast, the spherical reflector can make an ideal wide-angle scanner because of its perfectly symmetrical geometrical configuration. However, it is plagued by poor inherent collimating properties. If, for example, a point source is placed at the focus of the sphere, it does not produce plane waves. The departure of the reflected wavefront from a plane wave is known as *spherical aberration*, and it depends on the diameter and focal length of the sphere. By reciprocity, plane waves incident on a spherical reflector surface parallel to its axis do not converge at the focal point. However a spherical reflector has the capability of focusing plane waves incident at various angles by translating and orientating the feed and by illuminating different parts of the structural geometry. The 1,000-ft diameter reflector [12] at Arecibo, Puerto Rico is a spherical reflector whose surface is built into the earth and the scanning is accomplished by movement of the feed.

The focusing characteristics of a typical spherical reflector is illustrated in Figure 15.32 for three rays. The point F in the figure is the paraxial focus, and it is equal to one-half the radius of the

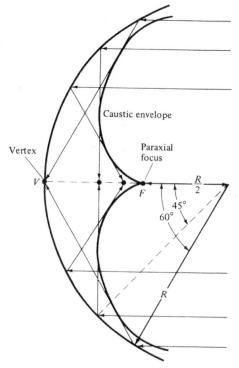

Figure 15.32 Spherical reflector geometry and rays that form a caustic.

sphere. The caustic* surface is an epicycloid and is generated by the reflection of parallel rays. A degenerate line FV of this caustic is parallel to the incident rays and extends from the paraxial focus to the vertex of the reflector. If one draws a ray diagram of plane waves incident within a 120° cone, it will be shown that all energy must pass through the line FV. Thus, the line FV can be used for the placement of the feed for the interception of plane waves incident parallel to the axial line. It can thus be said that a spherical reflector possesses a line focus instead of a point. However, amplitude and phase corrections must be made in order to realize maximum antenna efficiency.

Ashmead and Pippard [43] proposed to reduce spherical aberration and to minimize path error by placing a point-source feed not at the paraxial focus F (half radius of the sphere), as taught in optics, but displaced slightly toward the reflector. For an aperture of diameter d, the correct location for placing a point source is a distance f_0 from the vertex such that the maximum path error value is [43]

$$\Delta_{max} \simeq \frac{d^4}{2000 f_0^3} \qquad (15\text{-}72)$$

and the maximum phase error does not differ from a paraboloid by more than one-eighth of a wavelength. This, however, leads to large f/d and to poor area utilization. A similar result was obtained by Li [44]. He stated that the total phase error (sum of maximum absolute values of positive and negative phase errors) over an aperture of radius a is least when the phase error at the edge of the aperture is zero. Thus the optimum focal length is

$$f_{op} = \frac{1}{4}(R + \sqrt{R^2 - a^2}) \qquad (15\text{-}73)$$

where

R = radius of the spherical reflector
a = radius of the utilized aperture

Thus when $R = 2a$, the optimum focal length is $0.4665R$ and the corresponding total phase error, using the formula found in [44], is $0.02643(R/\lambda)$ rad. Even though the optimum focal length leads to minimum total phase error over a prescribed aperture, it does not yield the best radiation pattern when the illumination is not uniform. For a tapered amplitude distribution, the focal length that yields the best radiation pattern will be somewhat longer, and in practice, it is usually determined by experiment. Thus for a given maximum aperture size there exists a maximum value of total allowable phase error, and it is given by [44]

$$\left(\frac{a}{R}\right)^4_{max} = 14.7 \frac{(\Delta/\lambda)_{total}}{(R/\lambda)} \qquad (15\text{-}74)$$

where (Δ/λ) is the total phase error in wavelengths.

---

*A caustic is a point, a line, or a surface through which all the rays in a bundle pass and where the intensity is infinite. The caustic also represents the geometrical loci of all the centers of curvature of the wave surfaces. Examples of it include the focal line for cylindrical parabolic reflector and the focal point of a paraboloidal reflector.

> **Example 15.4**
>
> A spherical reflector has a 10-ft diameter. If at 11.2 GHz the maximum allowable phase error is $\lambda/16$, find the maximum permissible aperture.
>
> *Solution*: At $f = 11.2$ GHz
>
> $$\lambda = 0.08788 \text{ ft}$$
>
> $$\left(\frac{a}{R}\right)^4_{\max} = 14.7 \left(\frac{1/16}{56.8957}\right) = 0.01615$$
>
> $$a^4 \simeq 10.09$$
>
> $$a = 1.78 \text{ ft}$$

To overcome the shortcoming of a point feed and minimize spherical aberration, Spencer, Sletten, and Walsh [45] were the first to propose the use of a line-source feed. Instead of a continuous line source, a set of discrete feed elements can be used to reduce spherical aberration when they are properly placed along the axis in the vicinity of the paraxial focus. The number of elements, their position, and the portion of the reflector surface which they illuminate is dictated by the allowable wavefront distortion, size, and curvature of the reflector. This is shown in Figure 15.33 [46]. A single feed located near the paraxial focus will illuminate the central portion of the reflector. If the reflector is large, additional feed elements along the axis toward the vertex will be needed to minimize the phase errors in the aperture. The ultimate feed design will be a compromise between a single element and a line-source distribution.

An extensive effort has been placed on the analysis and experiment of spherical reflectors, and most of it can be found well documented in a book of reprinted papers [1]. In addition, a number of two-dimensional patterns and aperture plane constant amplitude contours, for symmetrical and offset feeds, have been computed [20].

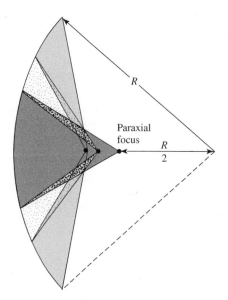

Figure 15.33 Reflector illumination by feed sections placed between paraxial focus and vertex. (SOURCE: A. C. Schell, "The Diffraction Theory of Large-Aperture Spherical Reflector Antennas," *IRE Trans. Antennas Propagat.*, Vol. AP-11, No. 4, July 1963. © (1963) IEEE).

15.6 MULTIMEDIA

In the publisher's website for this book, the following multimedia resources are included for the review, understanding, and visualization of the material of this chapter:

a. **Java**-based **interactive questionnaire**, with answers.
b. **Matlab** computer program, designated **Reflector**, for computing and displaying the radiation characteristics of a paraboloidal reflector.
c. **Power Point (PPT)** viewgraphs, in multicolor.

REFERENCES

1. A. W. Love (ed.), *Reflector Antennas*, IEEE Press, New York, 1978.
2. Y. Obha, "On the Radiation of a Corner Reflector Finite in Width," *IEEE Trans. Antennas Propagat.*, Vol. AP-11, No. 2, pp. 127–132, March 1963.
3. C. A. Balanis and L. Peters, Jr., "Equatorial Plane Pattern of an Axial-TEM Slot on a Finite Size Ground Plane," *IEEE Trans. Antennas Propagat.*, Vol. AP-17, No. 3, pp. 351–353, May 1969.
4. C. A. Balanis, "Pattern Distortion Due to Edge Diffractions," *IEEE Trans. Antennas Propagat.*, Vol. AP-18, No. 4, pp. 551–563, July 1970.
5. C. A. Balanis, "Analysis of an Array of Line Sources Above a Finite Ground Plane," *IEEE Trans. Antennas Propagat.*, Vol. AP-19, No. 2, pp. 181–185, March 1971.
6. D. Proctor, "Graphs Simplify Corner Reflector Antenna Design," *Microwaves*, Vol. 14, No. 7, pp. 48–52, July 1975.
7. E. B. Moullin, *Radio Aerials*, Oxford University Press, 1949, Chapters 1 and 3.
8. R. E. Paley and N. Wiener, *Fourier Transforms in the Complex Domain*, American Mathematical Society, Providence, R.I., p. 116, 1934.
9. P. A. J. Ratnasiri, R. G. Kouyoumjian, and P. H. Pathak, "The Wide Angle Side Lobes of Reflector Antennas," ElectroScience Laboratory, The Ohio State University, Technical Report 2183-1, March 23, 1970.
10. G. L. James and V. Kerdemelidis, "Reflector Antenna Radiation Pattern Analysis by Equivalent Edge Currents," *IEEE Trans. Antennas Propagat.*, Vol. AP-21, No. 1, pp. 19–24, January 1973.
11. C. A. Mentzer and L. Peters, Jr., "A GTD Analysis of the Far-out Side Lobes of Cassegrain Antennas," *IEEE Trans. Antennas Propagat.*, Vol. AP-23, No. 5, pp. 702–709, September 1975.
12. L. M. LaLonde and D. E. Harris, "A High Performance Line Source Feed for the AIO Spherical Reflector," *IEEE Trans. Antennas Propagat.*, Vol. AP-18, No. 1, pp. 41–48, January 1970.
13. A. W. Rudge, "Offset-Parabolic-Reflector Antennas: A Review," *Proc. IEEE*, Vol. 66, No. 12, pp. 1592–1618, December 1978.
14. P. J. B. Clarricoats and G. T. Poulton, "High-Efficiency Microwave Reflector Antennas—A Review," *Proc. IEEE*, Vol. 65, No. 10, pp. 1470–1502, October 1977.
15. O. Hachenberg, B. H. Grahl, and R. Wielebinski, "The 100-Meter Radio Telescope at Effelsberg," *Proc. IEEE*, Vol. 69, No. 9, pp. 1288–1295, 1973.
16. P. D. Potter, W. D. Merrick, and A. C. Ludwig, "Big Antenna Systems for Deep-Space Communications," *Astronaut. Aeronaut.*, pp. 84–95, October 1966.
17. S. Silver (ed.), *Microwave Antenna Theory and Design*, McGraw-Hill, New York, 1949 (MIT Radiation Lab. Series, Vol. 12).
18. J. F. Kauffman, W. F. Croswell, and L. J. Jowers, "Analysis of the Radiation Patterns of Reflector Antennas," *IEEE Trans. Antennas Propagat.*, Vol. AP-24, No. 1, pp. 53–65, January 1976.
19. R. E. Collin and F. J. Zucker (eds.), *Antenna Theory Part II*, McGraw-Hill, New York, pp. 36–48, 1969.

20. P. K. Agrawal, J. F. Kauffman, and W. F. Croswell, "Calculated Scan Characteristics of a Large Spherical Reflector Antenna," *IEEE Trans. Antennas Propagat.*, Vol. AP-27, No. 3, pp. 430–431, May 1979.
21. E. M. T. Jones, "Paraboloid Reflector and Hyperboloid Lens Antennas," *IRE Trans. Antennas Propagat.*, Vol. AP-2, No. 3, pp. 119–127, July 1954.
22. I. Koffman, "Feed Polarization for Parallel Currents in Reflectors Generated by Conic Sections," *IEEE Trans. Antennas Propagat.*, Vol. AP-14, No. 1, pp. 37–40, January 1966.
23. A. W. Love, "Some Highlights in Reflector Antenna Development," *Radio Sci.*, Vol. 11, Nos. 8, 9, pp. 671–684, August–September 1976.
24. D. K. Cheng, "Effect of Arbitrary Phase Errors on the Gain and Beamwidth Characteristics of Radiation Pattern," *IRE Trans. Antennas Propagat.*, Vol. AP-3, No. 3, pp. 145–147, July 1955.
25. Y. Y. Hu, "A Method of Determining Phase Centers and Its Application to Electromagnetic Horns," *J. Franklin Inst.*, pp. 31–39, January 1961.
26. J. Ruze, "The Effect of Aperture Errors on the Antenna Radiation Pattern," *Nuevo Cimento Suppl.*, Vol. 9, No. 3, pp. 364–380, 1952.
27. H. C. Minnett and B. MacA. Thomas, "Fields in the Image Space of Symmetrical Focusing Reflectors," *Proc. IEE*, Vol. 115, pp. 1419–1430, October 1968.
28. G. F. Koch, "Coaxial Feeds for High Aperture Efficiency and Low Spillover of Paraboloidal Reflector Antennas," *IEEE Trans. Antennas Propagat.*, Vol. AP-21, No. 2, pp. 164–169, March 1973.
29. R. E. Lawrie and L. Peters, Jr., "Modifications of Horn Antennas for Low Side Lobe Levels," *IEEE Trans. Antennas Propagat.*, Vol. AP-14, No. 5, pp. 605–610, September 1966.
30. A. J. Simmons and A. F. Kay, "The Scalar Feed-A High Performance Feed for Large Paraboloid Reflectors," *Design and Construction of Large Steerable Aerials*, IEE Conf. Publ. 21, pp. 213–217, 1966.
31. W. M. Truman and C. A. Balanis, "Optimum Design of Horn Feeds for Reflector Antennas," *IEEE Trans. Antennas Propagat.*, Vol. AP-22, No. 4, pp. 585–586, July 1974.
32. B. Houshmand, B. Fu, and Y. Rahmat-Samii, "Reflector Antenna Analysis Software," Vol. II, Chapter 11, CAEME Center for Multimedia Education, University of Utah, pp. 447–465, 1995.
33. P. W. Hannan, "Microwave Antennas Derived from the Cassegrain Telescope," *IRE Trans. Antennas Propagat.*, Vol. AP-9, No. 2, pp. 140–153, March 1961.
34. W. V. T. Rusch, "Scattering from a Hyperboloidal Reflector in a Cassegrain Feed System," *IEEE Trans. Antennas Propagat.*, Vol. AP-11, No. 4, pp. 414–421, July 1963.
35. P. D. Potter, "Application of Spherical Wave Theory to Cassegrainian-Fed Paraboloids," *IEEE Trans. Antennas Propagat.*, Vol. AP-15, No. 6, pp. 727–736, November 1967.
36. W. C. Wong, "On the Equivalent Parabola Technique to Predict the Performance Characteristics of a Cassegrain System with an Offset Feed," *IEEE Trans. Antennas Propagat.*, Vol. AP-21, No. 3, pp. 335–339, May 1973.
37. V. Galindo, "Design of Dual-Reflector Antennas with Arbitrary Phase and Amplitude Distributions," *IEEE Trans. Antennas Propagat.*, Vol. AP-12, No. 4, pp. 403–408, July 1964.
38. W. F. Williams, "High Efficiency Antenna Reflector," *Microwave Journal*, Vol. 8, pp. 79–82, July 1965.
39. G. W. Collins, "Shaping of Subreflectors in Cassegrainian Antennas for Maximum Aperture Efficiency," *IEEE Trans. Antennas Propagat.*, Vol. AP-21, No. 3, pp. 309–313, May 1973.
40. C. A. Balanis, *Advanced Engineering Electromagnetics*, John Wiley & Sons, Inc., New York, pp. 744–764, 1989.
41. R. Mittra and V. Galindo-Israel, "Shaped Dual Reflector Synthesis," *IEEE Antennas Propagation Society Newsletter*, Vol. 22, No. 4, pp. 5–9, August 1980.
42. K. A. Green, "Modified Cassegrain Antenna for Arbitrary Aperture Illumination," *IEEE Trans. Antennas Propagat.*, Vol. AP-11, No. 5, pp. 589–590, September 1963.
43. J. Ashmead and A. B. Pippard, "The Use of Spherical Reflectors as Microwave Scanning Aerials," *J. Inst. Elect. Eng.*, Vol. 93, Part III-A, pp. 627–632, 1946.
44. T. Li, "A Study of Spherical Reflectors as Wide-Angle Scanning Antennas," *IRE Trans. Antennas Propagat.*, Vol. AP-7, No. 3, pp. 223–226, July 1959.

PROBLEMS

45. R. C. Spencer, C. J. Sletten, and J. E. Walsh, "Correction of Spherical Aberration by a Phased Line Source," *Proceedings National Electronics Conference*, Vol. 5, pp. 320–333, 1949.
46. A. C. Schell, "The Diffraction Theory of Large-Aperture Spherical Reflector Antennas," *IEEE Trans. Antennas Propagat.*, Vol. AP-11, No. 4, pp. 428–432, July 1963.

PROBLEMS

15.1. An infinite line source, of constant electric current I_0, is placed a distance s above a flat and infinite electric ground plane. Derive the array factor.

15.2. For corner reflectors with included angles of $\alpha = 60°, 45°$, and $30°$:
   (a) Derive the array factors of (15-7)–(15-9b).
   (b) Plot the field strength along the axis ($\theta = 90°, \phi = 0°$) as a function of the feed-to-vertex spacing, $0 \leq s/\lambda \leq 10$.

15.3. Consider a corner reflector with an included angle of $\alpha = 36°$.
   (a) Derive the array factor.
   (b) Plot the relative field strength along the axis ($\theta = 90°, \phi = 0°$) as a function of the feed-to-vertex spacing s, for $0 \leq s/\lambda \leq 10$.
   (c) Determine the spacing that yields the first maximum possible field strength along the axis. For this spacing, what is the ratio of the field strength of the corner reflector along the axis to the field strength of the feed element alone?
   (d) For the spacing in part c, plot the normalized power pattern in the azimuthal plane ($\theta = 90°$).

15.4. A 60° corner reflector, in conjunction with a $\lambda/2$ dipole feed, is used in a radar tracking system. One of the requirements for such a system is that the antenna, in one of its modes of operation, has a null along the forward symmetry axis. In order to accomplish this, what should be the feed spacing from the vertex (in wavelengths)? Give all the possible values of the feed-to-vertex spacing.

15.5. For a parabolic reflector, derive (15-25) which relates the f/d ratio to its subtended angle θ_0.

15.6. Show that for a parabolic reflector
   (a) $0 \leq f/d \leq 0.25$ relates to $180° \geq \theta_0 \geq 90°$
   (b) $0.25 \leq f/d \leq \infty$ relates to $90° \geq \theta_0 \geq 0°$

15.7. The diameter of a paraboloidal reflector antenna (dish), used for public television stations, is 10 meters. Find the far-zone distance if the antenna is used at 2 and 4 GHz.

15.8. Show that the directivity of a uniformly illuminated circular aperture of diameter d is equal to $(\pi d/\lambda)^2$.

15.9. Verify (15-33) and (15-33a).

15.10. The field radiated by a paraboloidal reflector with an f/d ratio of 0.5 is given by
$$\mathbf{E} = (\hat{\mathbf{a}}_x + \hat{\mathbf{a}}_y \sin\phi \cos\phi) f(r,\theta,\phi)$$
where the x-component is the co-pol and the y-component is the cross-pol.
   (a) At what observation angle(s) (in degrees) (0°–180°) is the cross-pol minimum? What is the minimum value?
   (b) At what observation angle(s) (in degrees) (0°–180°) is the cross-pol maximum? What is the maximum value?

(c) What is the polarization loss factor when the receiving antenna is linearly polarized in the x-direction.

(d) What is the polarization loss factor when the receiving antenna is linearly polarized in the y-direction.

(e) What should the polarization of the receiving antenna be in order to eliminate the losses due to polarization? Write an expression for the polarization of the receiving antenna to achieve this.

15.11. Verify (15-49) and (15-54).

15.12. A small parabolic reflector (dish) of revolution, referred to as a paraboloid, is now being advertised as a TV antenna for direct broadcast. Assuming the diameter of the reflector is 1 meter, determine at 3 GHz the directivity (in dB) of the antenna if the feed is such that

(a) the illumination over the aperture is uniform (ideal)

(b) the taper efficiency is 80% while the spillover efficiency is 85%. Assume no other losses. What is the total aperture efficiency of the antenna (in dB)?

15.13. The 140-ft (42.672-m) paraboloidal reflector at the National Radio Astronomy Observatory, Green Bank, W. Va, has an f/d ratio of 0.4284. Determine the

(a) subtended angle of the reflector

(b) aperture efficiency assuming the feed pattern is symmetrical and its gain pattern is given by $2\cos^2(\theta'/2)$, where θ' is measured from the axis of the reflector

(c) directivity of the entire system when the antenna is operating at 10 GHz, and it is illuminated by the feed pattern of part (b)

(d) directivity of the entire system at 10 GHz when the reflector is illuminated by the feed pattern of part (b) and the maximum aperture phase deviation is $\pi/16$ rad

15.14. A paraboloidal reflector has an f/d ratio of 0.38. Determine

(a) which $\cos^n\theta'$ symmetrical feed pattern will maximize its aperture efficiency

(b) the directivity of the reflector when the focal length is 10λ

(c) the value of the feed pattern in dB (relative to the main maximum) along the edges of the reflector.

15.15. The diameter of an educational TV station reflector is 10 meters. It is desired to design the reflector at 1 GHz with a f/d ratio of 0.5. The pattern of the feed is given by

$$G_f = \begin{cases} 3.4286\cos^4(\theta'/2) & 0 \le \theta' \le \pi/2 \\ 0 & \text{Elsewhere} \end{cases}$$

Assume a symmetrical pattern in the ϕ' direction. Determine the following:

(a) Total subtended angle of the reflector (*in degrees*)

(b) Aperture efficiency (*in %*)

(c) Directivity of the reflector (*dimensionless* and *in dB*).

15.16. The feed pattern of a paraboloidal reflector, with a *diameter of 10 meters* and a *f/d ratio of 0.433*, is rotationally symmetric and it is given by:

$$G_f(\theta') = \begin{cases} 2.667\cos^2(\theta'/2) & 0° \le \theta' \le 90° \\ 0 & \text{Elsewhere} \end{cases}$$

Calculate the:
(a) Total subtended angle of the reflector (*in degrees*).
(b) Aperture efficiency (*in %*).
(c) Maximum power (*in Watts*) that can be delivered, in the receiving mode, to a load connected to the reflector when the power density of an incident wave is 10^{-6} Watts/cm^2.

15.17. A 10-m diameter earth-based paraboloidal reflector antenna, which is used for satellite television, has a f/d ratio of 0.433. Assuming the amplitude gain pattern of the feed is rotationally symmetric in ϕ' and is given by

$$G_f(\theta', \phi') = \begin{cases} G_0 \cos^2(\theta'/2) & 0° \leq \theta' \leq 90° \\ 0 & \text{Elsewhere} \end{cases}$$

where G_0 is a constant, determine the
(a) *total* subtended angle of the reflector (*in degrees*)   (b) value of G_0.
(c) aperture efficiency (*in %*).

15.18. A 10-m diameter earth-based paraboloidal reflector antenna, which is used for satellite television, has a total subtended angle of 120°. Assuming the amplitude gain pattern of the feed is rotationally symmetric in ϕ and is given by

$$G_f(\theta') = \begin{cases} G_0 \cos^2(\theta'/2) & 0° \leq \theta' \leq 90° \\ 0 & \text{Elsewhere} \end{cases}$$

where G_0 is a constant, determine the
(a) value of G_0 (b) taper efficiency (*in %*).
Compare the aperture efficiency of the above pattern with that of

$$G_f(\theta') = \begin{cases} G_0 \cos^2(\theta') & 0° \leq \theta' \leq 90° \\ 0 & \text{Elsewhere} \end{cases}$$

Which of the two patterns' aperture efficiency is lower, *and why*. Explain the *why* by comparing *qualitatively* the corresponding *taper* and *spillover* efficiencies, and thus the aperture efficiency, of the two feed patterns; do this without calculating the taper and spillover efficiencies.

15.19. The symmetrical feed pattern for a paraboloidal reflector is given by

$$G_f = \begin{cases} G_0 \cos^4\left(\dfrac{\theta'}{2}\right) & 0 \leq \theta' \leq \pi/2 \\ 0 & \text{Elsewhere} \end{cases}$$

where G_0 is a constant.
(a) Evaluate the constant G_0.   (b) Derive an expression for the aperture efficiency.
(c) Find the subtended angle of the reflector that will maximize the aperture efficiency. What is the maximum aperture efficiency?

15.20. A paraboloidal reflector is operating at a frequency of 5 GHz. It is 8 meters in diameter, with an f/d ratio of 0.25. It is fed with an antenna whose primary pattern is symmetrical and which can be approximated by

$$G_f = \begin{cases} 10\cos^4\theta' & 0 \le \theta' \le \pi/2 \\ 0 & \text{Elsewhere} \end{cases}$$

Find its
(a) aperture efficiency (b) overall directivity
(c) spillover efficiency (d) taper efficiency

15.21. A paraboloidal reflector operating at 10 GHz, with a diameter of 10 meters and a *total* subtended angle of 80°, is fed with an antenna whose gain pattern is given by:

$$G_f(\theta') = \begin{cases} G_0\cos^8(\theta') & 0° \le \theta' \le 90° \\ 0 & \text{Elsewhere} \end{cases}$$

(a) Calculate the:
 1. Aperture efficiency (*in %*). No graphical solution.
 2. Value of G_0.
(b) Determine the (you can use graphical solution):
 1. Taper efficiency (*in %*).
 2. Spillover efficiency (*in %*).
(c) Calculate the directivity (*dimensionless* and *in dB*).

15.22. A parabolic reflector has a diameter of 10 meters and has an included angle of $\theta_0 = 30°$. The directivity at the operating frequency of 25 GHz is 5,420,000. The phase efficiency, polarization efficiency, blockage efficiency, and random error efficiency are all 100%. The feed has a phi-symmetric pattern given by

$$G_f = \begin{cases} G_0\cos^{10}\theta' & 0° \le \theta' \le 90° \\ 0 & \text{Elsewhere} \end{cases}$$

Find the taper, spillover, and overall efficiencies.

15.23. A 10-meter diameter paraboloidal reflector is used as a TV satellite antenna. The focus-to-diameter ratio of the reflector is 0.536 and the pattern of the feed in the forward region can be approximated by $\cos^2(\theta')$. Over the area of the reflector, the incident power density from the satellite can be approximated by a uniform plane wave with a power density of 10 μwatts/m². At a center frequency of 9 GHz:
(a) What is the maximum directivity of the reflector (*in dB*)?
(b) Assuming no losses of any kind, what is the maximum power that can be delivered to a TV receiver connected to the reflector through a lossless transmission line?

15.24. A 3-meter diameter parabolic reflector is used as a receiving antenna for satellite television reception at 5 GHz. The reflector is connected to the television receiver through a 78-ohm coaxial cable. The aperture efficiency is approximately 75%. Assuming the maximum incident power density from the satellite is 10 microwatts/square meter and the incident wave is polarization-matched to the reflector antenna, what is the:
(a) Directivity of the antenna (*in dB*)

(b) Maximum power (*in watts*) that can be delivered to the receiving TV? Assume no losses.

(c) Power (*in watts*) delivered to the receiving TV if the reflection coefficient at the transmission line/receiving TV terminal junction is 0.2. Assume no other losses.

15.25. A reflector antenna with a total subtended angle of 120° is illuminated at 10 GHz with a specially designed feed so that its aperture efficiency is nearly unity. The focal distance of the reflector is 5 meters. Assuming the radiation pattern is nearly symmetric, determine the:

(a) Half-power beamwidth (*in degrees*).   (b) Sidelobe level (*in dB*).

(c) Directivity (*in dB*).

(d) Directivity (*in dB*) based on Kraus' formula and Tai & Pereira's formula.

(e) Loss in directivity (*in dB*) if the surface has rms random roughness of 0.64 mm.

15.26. A front-fed paraboloidal reflector with an f/d radio of 0.357, whose diameter is 10 meters and which operates at 10 GHz, is fed by an antenna whose power pattern is rotationally symmetric and it is represented by $\cos^2(\theta/2)$. All the power is radiated in the forward region ($0° \leq \theta' \leq 90°$) of the feed. Determine the

(a) spillover efficiency   (b) taper efficiency   (c) overall aperture efficiency

(d) directivity of the reflector (*in dB*)

(e) directivity of the reflector (*in dB*) if the RMS, reflector surface deviation from an ideal paraboloid is $\lambda/100$.

15.27. A three-dimensional paraboloidal reflector, as shown in Figure 15.12 with radius a at its aperture, has a aperture field distribution $A(\rho')$ which is represented by

$$A(\rho') = A_0 \left[1 - \left(\frac{\rho'}{a} \right)^2 \right]$$

where

A_0 = constant

ρ' = radial distance from the center of aperture ($0 \leq \rho' \leq a$)

a = radius of the aperture

Determine, for an aperture radius of $a = 50\lambda$, the:

(a) Aperture efficiency (*in %*).   (b) Directivity (*dimensionless* and *in dB*) of the reflector.

(c) Half-power beamwidth (*in degrees*).

(d) Approximate directivity (*dimensionless* and *in dB*) using another *appropriate* formula at your availability.

15.28. A uniform plane wave at a frequency of *10 GHz* and with a power density of *10 μwatts/cm²* is incident upon a three-dimensional paraboloidal reflector, with a $f/d = 0.43$, which is used as a receiving antenna. The antenna used as a feed, which is placed at the focal point of the reflector, has a rotationally symmetric pattern (*not a function of ϕ*) and it is given by

$$G_f(\theta') = \begin{cases} \dfrac{\cot^2\left(\dfrac{\theta_0}{2}\right)}{\cos^4\left(\dfrac{\theta'}{2}\right)} & 0 \leq \theta' \leq \theta_0 \\ 0 & \theta' \geq \theta_0 \end{cases}$$

Assuming the reflector has a *diameter of 10 meters*, determine the:
(a) *Total* subtended angle (*in degrees*) of the reflector.
(b) Spillover efficiency (*in %*); *do not have to derive*.
(c) Taper efficiency (*in %*) *do not have to derive*.
(d) Aperture efficiency (*in %*); *do not have to derive*.
(e) *Maximum directivity* (*in dB*) of the reflector.
(f) *Maximum* power (*in watts*) delivered to a load connected to the feed. *Assume no other losses.*

Refer to Figure 15.10 for the geometry of the problem.

15.29. Design pyramidal horn antennas that will maximize the aperture efficiency or produce maximum power transmission to the feed, for paraboloidal reflectors with *f/d* ratios of
(a) 0.50   (b) 0.75   (c) 1.00

15.30. A three-dimensional paraboloidal reflector, like the one shown in Figure 15.12, is fed in such a way that it creates, after the field radiated by the feed if reflected by the reflector surface, over its aperture (over the area of the reflector which includes the rim with radius *a*, as shown in Figure 15.12), a normalized tapered electric field distribution given by:

$$\text{Normalized Aperture Electric Field Distribution} = \left[1 - \left(\frac{\rho'}{a}\right)^2\right], \quad 0 \leq \rho' \leq a$$

where ρ' = distance from the center of the aperture towards the reflector circumference/rim/edge, as shown in Figure 15.12. What is the:

(a) Aperture efficiency ε_{ap} (*in %*) of the aperture of radius *a* with the above aperture field distribution.
(b) Directivity of the reflector (*dimensionless* and *in dB*) when its radius is 20λ.

CHAPTER 16

Smart Antennas

16.1 INTRODUCTION

Over the last decade, wireless technology has grown at a formidable rate, thereby creating new and improved services at lower costs. This has resulted in an increase in airtime usage and in the number of subscribers. The most practical solution to this problem is to use spatial processing. As Andrew Viterbi, founder of Qualcomm Inc., clearly stated: "*Spatial processing remains as the most promising, if not the last frontier, in the evolution of multiple access systems*" [1].

Spatial processing is the central idea of adaptive antennas or smart-antenna systems. Although it might seem that adaptive antennas have been recently discovered, they date back to World War II with the conventional Bartlett beamformer [2]. It is only of today's advancement in powerful low-cost digital signal processors, general-purpose processors (and ASICs—Application-Specific Integrated Circuits), as well as innovative software-based signal-processing techniques (algorithms), that smart-antenna systems have received enormous interest worldwide. In fact, many overviews and tutorials [1]–[14] have emerged, and a great deal of research is being done on the adaptive and direction-of-arrival (DOA) algorithms for smart-antenna systems [15], [16]. As the number of users and the demand for wireless services increases at an exponential rate, the need for wider coverage area and higher transmission quality rises. Smart-antenna systems provide a solution to this problem.

This chapter presents an introduction and general overview of smart-antenna systems. First, it gives the reader an insight on smart-antenna systems using the human auditory system as an analogy. Then, it presents the purpose for smart antennas by introducing the cellular radio system and its evolution. A brief overview of signal propagation is given to emphasize the need for smart-antenna systems. These topics are followed by antenna array theory, time of arrival and adaptive digital processing algorithms, mutual coupling, mobile *ad hoc* networks, antenna design, and simulation and impact of antenna designs on network capacity/*throughput* and communication channel bit-error-rate (BER). Material of this chapter is extracted from [17]–[22]. More extensive discussion and details on each topic can be found in [17] and other sources.

16.2 SMART-ANTENNA ANALOGY

The functionality of many engineering systems is readily understood when it is related to our human body system [3]. Therefore, to give an insight into how a smart-antenna system works, let us imagine

---

Antenna Theory: Analysis and Design, Fourth Edition. Constantine A. Balanis.
© 2016 John Wiley & Sons, Inc. Published 2016 by John Wiley & Sons, Inc.
Companion Website: www.wiley.com/go/antennatheory4e

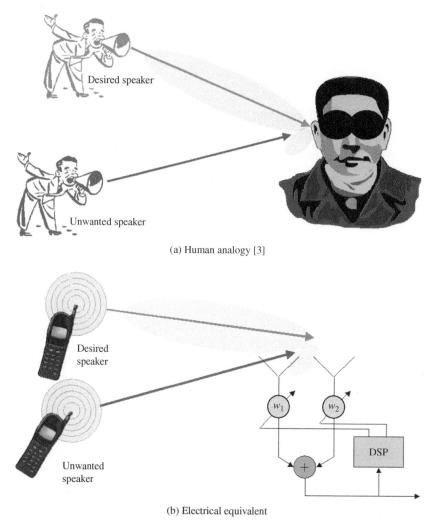

Figure 16.1 Smart-antenna analogy. (a) Human analogy [3]; (b) Electrical equivalent.

two persons carrying on a conversation inside a dark room [refer to Figure 16.1(a)]. The listener among the two persons is capable of determining the location of the speaker as he moves about the room because the voice of the speaker arrives at each acoustic sensor, the ear, at a different time. The human signal processor, the brain, computes the direction of the speaker from the time differences or delays of the voice received by the two ears. Afterward, the brain adds the strength of the signals from each ear so as to focus on the sound of the computed direction. Furthermore, if additional speakers join in the conversation, the brain can tune out unwanted interferers and concentrate on one conversation at a time. Conversely, the listener can respond back to the same direction of the desired speaker by orienting the transmitter (mouth) toward the speaker.

Electrical smart-antenna systems work the same way using two antennas instead of the two ears and a digital signal processor instead of a brain [refer to Figure 16.1(b)]. Therefore, after the digital signal processor measures the time delays from each antenna element, it computes the direction of arrival (DOA) of the signal-of-interest (SOI), and then it adjusts the excitations (gains and phases of the signals) to produce a radiation pattern that focuses on the SOI while, ideally, tuning out any signal-not-of-interest (SNOI).

16.3 CELLULAR RADIO SYSTEMS EVOLUTION

Maintaining capacity has always been a challenge as the number of services and subscribers increased. To achieve the capacity demand required by the growing number of subscribers, cellular radio systems had to evolve throughout the years. To justify the need for smart-antenna systems in the current cellular system structure, a brief history on the evolution of the cellular radio systems is presented. For more in-depth details refer to [23]–[25].

16.3.1 Omnidirectional Systems

Since the early days, system designers knew that capacity was going to be a problem, especially when the number of channels or frequencies allotted by the Federal Communications Commission (FCC) was limited. Therefore, to achieve the capacity required for thousands of subscribers, a suitable cellular structure had to be designed; an example of the resulting structure is depicted in Figure 16.2.

Each shaded hexagonal area in Figure 16.2 represents a small geographical area named *cell* with maximum radius R [24]. At the center of each cell resides a base station equipped with an omnidirectional antenna with a given band of frequencies. Base stations in adjacent cells are assigned frequency bands that contain different frequencies compared to the neighboring cells. By limiting the coverage area to within the boundaries of a cell, the same band of frequencies may be used to cover different cells that are separated from one another by distances large enough to keep interference levels below the threshold of the others. The design process of selecting and allocating the same bands of frequencies to different cells of cellular base stations within a system is referred to as *frequency reuse* [23]. This is shown in Figure 16.2 by the repeating shaded pattern or *clusters* [23]; cells having the same shaded pattern use the same frequency spectrum. In the first cellular radio systems deployed, each base station was equipped with an omnidirectional antenna with a typical amplitude pattern as that shown in Figure 16.3. Because only a small percentage of the total energy reached the desired user, the remaining energy was radiated in undesired directions. As the number of users increased, so did the interference, thereby reducing capacity. An immediate solution to this problem was to subdivide a cell into smaller cells; this technique is referred to as *cell splitting* [24].

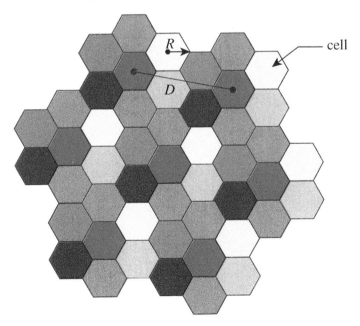

Figure 16.2 Typical cellular structure with 7 cells reuse pattern.

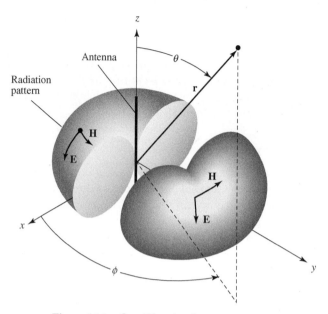

Figure 16.3 Omnidirectional antenna pattern.

A. Cell Splitting

Cell splitting [24], as shown in Figure 16.4, subdivides a congested cell into smaller cells called *microcells*, each with its own base station and a corresponding reduction in antenna height and transmitter power. Cell splitting improves capacity by decreasing the cell radius R and keeping the D/R ratio unchanged; D is the distance between the centers of the clusters. The disadvantages of cell splitting are costs incurred from the installation of new base stations, the increase in the number of *handoffs* (the process of transferring communication from one base station to another when the mobile unit travels from one cell to another), and a higher processing load per subscriber.

B. Sectorized Systems

As the demand for wireless service grew even higher, the number of frequencies assigned to a cell eventually became insufficient to support the required number of subscribers. Thus, a cellular design technique was needed to provide more frequencies per coverage area. This technique is referred to as *cell sectoring* [23], where a single omnidirectional antenna is replaced at the base station with several directional antennas. Typically, a cell is sectorized into three sectors of 120° each [3], as shown in Figure 16.5.

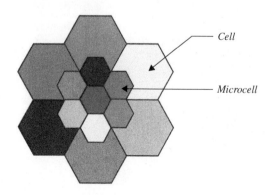

Figure 16.4 Cell splitting.

Figure 16.5 Sectorized base-station antenna.

In sectoring, capacity is improved while keeping the cell radius unchanged and reducing the D/R ratio. In other words, by reducing the number of cells in a cluster and thus increasing the frequency reuse, capacity improvement is achieved. However, in order to accomplish this, it is necessary to reduce the relative interference without decreasing the transmitting power. The cochannel interference in such cellular systems is reduced since only two neighboring cells interfere instead of six for the omnidirectional case [24], [26] (see Figure 16.6). The penalty for improved signal-to-interference (S/I) ratio and capacity is an increase in the number of antennas at the base station, and a decrease in *trunking efficiency* [23], [26] due to channel sectoring at the base station [23]–[26]. Trunking

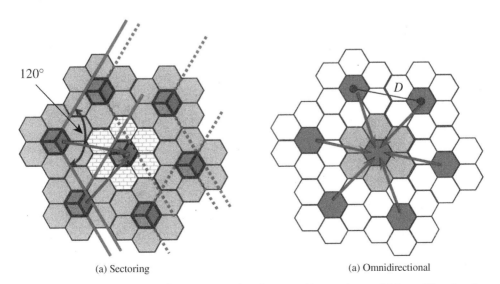

(a) Sectoring   (a) Omnidirectional

Figure 16.6 Cochannel interference comparison between (a) sectoring, and (b) omnidirectional.

efficiency is a measure of the number of users that can be offered service with a particular configuration of fixed number of frequencies.

16.3.2 Smart-Antenna Systems

Despite its benefits, cell sectoring did not provide the solution needed for the capacity problem. Therefore, the system designers began to look into a system that could dynamically sectorize a cell. Hence, they began to examine *smart antennas*.

Many refer to smart-antenna systems as smart antennas, but in reality antennas are not smart; it is the digital signal processing, along with the antennas, which makes the system smart. Although it might seem that smart-antenna systems is a new technology, the fundamental theory of smart (adaptive) antennas is not new [5], [6]. In fact, it has been applied in defense-related systems since World War II. Until recent years, with the emergence of powerful low-cost digital signal processors (DSPs), general-purpose processors (and Application-Specific Integrated Circuits—ASICs), as well as innovative signal-processing algorithms, smart-antenna systems have become practical for commercial use [3].

Smart-antenna systems are basically an extension of *cell sectoring* in which the sector coverage is composed of multiple beams [26]. This is achieved by the use of antenna arrays, and the number of beams in the sector (e.g., 120°) is a function of the array geometry. Because smart antennas can focus their radiation pattern toward the desired users while rejecting unwanted interferences, they can provide greater coverage area for each base station. Moreover, because smart antennas have a higher rejection interference, and therefore lower bit error rate (BER), they can provide a substantial capacity improvement. These systems can generally be classified as either *Switched-Beam* or *Adaptive Array* [5], [6], [27].

A. Switched-Beam Systems
A *switched-beam system* is a system that can choose from one of many predefined patterns in order to enhance the received signal (see Figure 16.7), and it is obviously an extension of cell sectoring as each sector is subdivided into smaller sectors. As the mobile unit moves throughout the cell, the switched-beam system detects the signal strength, chooses the appropriate predefined beam pattern,

Figure 16.7 Switched-beam system.

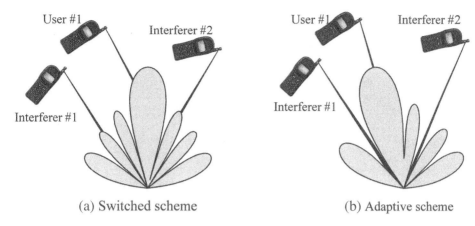

Figure 16.8 Comparison of (a) switched-beam scheme, and (b) adaptive array scheme.

and continually switches the beams as necessary. The overall goal of the switched-beam system is to increase the gain according to the location of the user. However, since the beams are fixed, the intended user may not be in the center of any given main beam. If there is an interferer near the center of the active beam, it may be enhanced more than the desired user [3].

B. Adaptive Array Systems
Adaptive array systems [5], [6] provide more degrees of freedom since they have the ability to adapt in real time the radiation pattern to the RF signal environment. In other words, they can direct the main beam toward the pilot signal or SOI while suppressing the antenna pattern in the direction of the interferers or SNOIs. To put it simply, adaptive array systems can customize an appropriate radiation pattern for each individual user. This is far superior to the performance of a switched-beam system, as shown in Figure 16.8. This figure shows that not only the switched-beam system may not able to place the desired signal at the maximum of the main lobe but also it exhibits the inability to fully reject the interferers. Because of the ability to control the overall radiation pattern in a greater coverage area for each cell site, as illustrated in Figure 16.9, adaptive array systems greatly increase capacity. Figure 16.9 shows a comparison, in terms of relative coverage area, of conventional sectorized switched-beam and adaptive arrays. In the presence of a low-level interference, both types

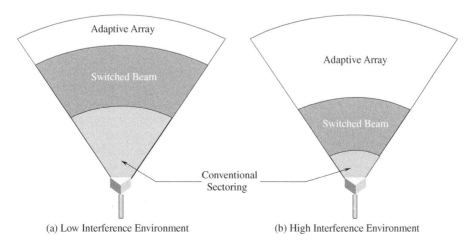

Figure 16.9 Relative coverage area comparison among sectorized systems, switched-beam systems, and adaptive array systems in (a) low interference environment, and (b) high interference environment [3].

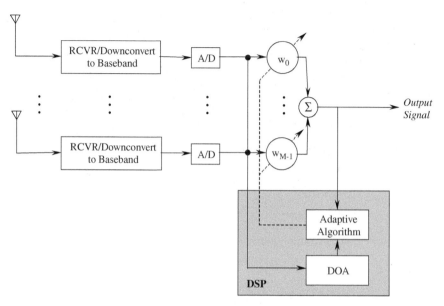

Figure 16.10 Functional block diagram of an adaptive array system.

of smart antennas provide significant gains over the conventional sectored systems. However, when a high-level interference is present, the interference rejection capability of the adaptive systems provides significantly more coverage than either the conventional or switched-beam system [3].

Adaptive array systems can locate and track signals (users and interferers) and dynamically adjust the antenna pattern to enhance reception while minimizing interference using signal-processing algorithms. A functional block diagram of such a system is shown in Figure 16.10. This figure shows that after the system downconverts the received signals to baseband and digitizes them, it locates the SOI using the direction-of-arrival (DOA) algorithm, and it continuously tracks the SOI and SNOIs by dynamically changing the weights (amplitudes and phases of the signals). Basically, the DOA computes the direction of arrival of all signals by computing the time delays between the antenna elements, and afterward the adaptive algorithm, using a cost function, computes the appropriate weights that result in an optimum radiation pattern. The details of how the time delays and the weights are computed are discussed later in this chapter. Because adaptive arrays are generally more digital-processing intensive and require a complete RF portion of the transceiver behind each antenna element, they tend to be more costly than switched-beam systems.

C. Spatial Division Multiple Access (SDMA)

The ultimate goal in the development of cellular radio systems is SDMA [3], [27]. SDMA is among the most-sophisticated utilization of smart-antenna technology; its advanced spatial-processing capability enables it to locate many users, creating different beam for each user, as shown in Figure 16.11. This means that more than one user can be allocated to the same physical communication channel in the same cell simultaneously, with only an angle separation. This is accomplished by having N parallel beamformers at the base station operating independently, where each beamformer has its own adaptive beamforming algorithm to control its own set of weights and its own direction-of-arrival algorithm (DOA) to determine the time delay of each user's signal (see Figure 16.12) [28], [29]. Each beamformer creates a maximum toward its desired user while nulling or attenuating the other users. This technology dramatically improves the interference suppression capability while greatly increasing frequency reuse, resulting in increased capacity and reduced infrastructure cost. Basically, capacity is increased not only through intercell frequency reuse but also through intracell frequency reuse [27].

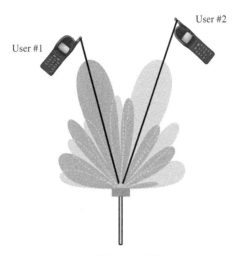

Figure 16.11   SDMA multibeam system.

16.4 SIGNAL PROPAGATION

Up until now, the problem of capacity has been associated solely with cochannel interference and with the depletion of channels due to the high number of users. However, multipath fading and delay spread also play a role in reducing system capacity [3], [30]. Fortunately, because of the ability of smart-antenna systems to adapt to the signal environment, they are able to

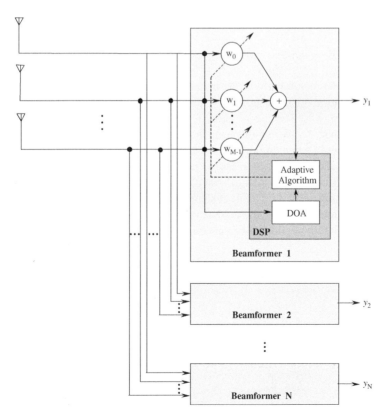

Figure 16.12   SDMA system block diagram [28], [29].

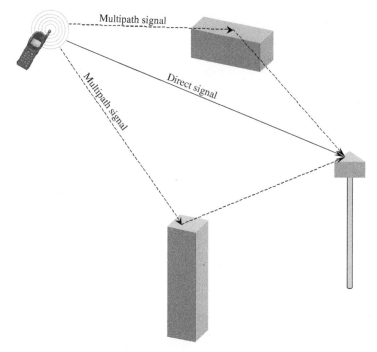

Figure 16.13 Multipath environment.

considerably reduce delay spread and multipath fading, thereby increasing capacity. This section gives a brief overview on signal propagation; for an in-depth study of the subject, the reader is referred to [3], [31], [32].

The signal generated by the user mobile device is omnidirectional in nature; therefore, it causes the signal to be reflected by structures, such as buildings. Ultimately, this results in the arrival of multiple delayed versions (multipath) of the main (direct) signals at the base station, as depicted in Figure 16.13. This condition is referred to as *multipath* [3], [32]. In general, these multiple delayed signals do not match in phase because of the difference in path length at the base station, as shown by the example in Figure 16.14 [3]. Because smart-antenna systems can tailor themselves to the signal environment, they can exploit or reject the reflected signals depending whether the signals are delayed copies of the SOI or the SNOIs. This is an advantage because smart antennas are not only capable of extracting information from the direct path of the SOI but they can also extract information from the reflected version of the SOI while rejecting all interferers or SNOIs. Therefore, because of this ability to manage multipath signals, smart-antenna systems improve link quality. As the signals are delayed, the phases of the multipath signal components can combine destructively over a narrow

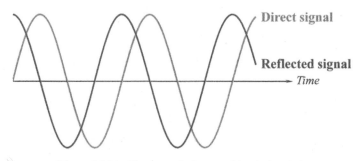

Figure 16.14 Two out-of-phase multipath signals [3].

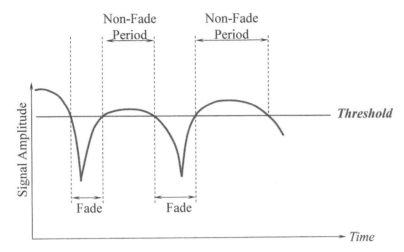

Figure 16.15 Fade effect on a user signal [3].

bandwidth, leading to *fading* of the received signal level. This results in a reduction of the signal strength, and a representative example is illustrated in Figure 16.15 [3].

One type of fading is *Rayleigh fading* or *fast fading* [32], [33]. Fading is constantly changing, and it is a three-dimensional (3-D) phenomenon that creates *fade zones* [30]. These fade zones are usually small, and they tend to periodically attenuate the received signal (i.e., degrade it in quality) as the users pass through them. Although fading, in general, is a difficult problem, smart-antenna systems perform better than earlier systems, unless fading is severe. Figure 16.16 shows lighter shaded area as a representation of fade zones in a multipath environment.

Occasionally, the multipath signals are 180° out of phase, as shown in Figure 16.17. This multipath problem is called *phase cancellation* [3]. When this happens, a call cannot be maintained for a long period of time, and it is dropped. In digital signals, the effect of multipath causes a condition called *delay spread*. In other words, the symbols representing the bits collide with one another causing intersymbol interference (ISI), as shown in Figure 16.18. When this occurs, the bit error rate (BER) rises, and a noticeable degradation in quality is observed [31].

Finally, another signal propagation problem is *cochannel interference* [30]. This occurs when a user's signal interferes with a cell having the same set of frequencies. Omnidirectional cells suffer more from cochannel interference than do sectorized cells, and smart-antenna systems, because of their ability to tune out cochannel interference, perform at their best under such environment [24]; this comparison is shown in Figure 16.19.

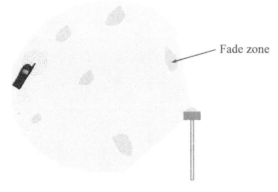

Figure 16.16 Fade zones [3].

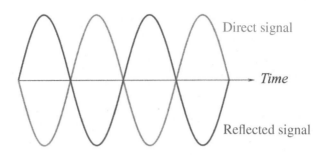

Figure 16.17 Phase cancellation.

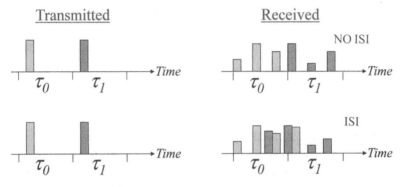

Figure 16.18 Delay spread and ISI.

16.5 SMART ANTENNAS' BENEFITS

The primary reason for the growing interest in smart-antenna systems is the *capacity* increase. In densely populated areas, mobile systems are usually interference-limited, meaning that the interference from other users is the main source of noise in the system. This means that the signal-to-interference ratio (SIR) is much smaller than the signal-to-noise ratio (SNR). In general, smart antennas will, by simultaneously increasing the useful received signal level and lowering the interference level, increase the SIR.

Another benefit that smart-antenna systems provide is *range* increase. Because smart antennas are more directional than omnidirectional and sectorized antennas, a range increase potential is available

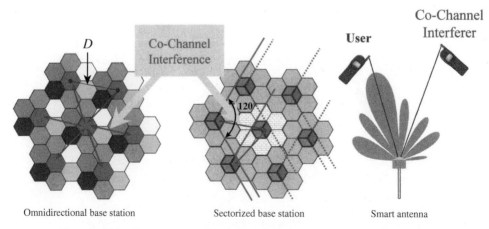

Figure 16.19 Cochannel interference comparison of the different systems [24].

[27]. In other words, smart antennas are able to focus their energy toward the intended users, instead of directing it in other unnecessary directions (wasting) like omnidirectional antennas do. This means that base stations can be placed further apart, leading to a more cost-efficient development. Therefore, in rural and sparsely populated areas, where radio coverage rather than capacity is more important, smart-antenna systems are also well suited [27].

Another added advantage of smart-antenna systems is *security*. In a society that becomes more dependent on conducting business and transmitting personal information, security is an important issue. Smart antennas make it more difficult to tap a connection because the intruder must be positioned in the same direction as the user as seen from the base station to successfully tap a connection [27].

Finally, because of the spatial detection nature of smart-antenna systems, they can be used to locate humans in emergencies or for any other location-specific service [27].

16.6 SMART ANTENNAS' DRAWBACKS

While smart antennas provide many benefits, they do suffer from certain drawbacks. For example, their transceivers are much more complex than traditional base station transceivers. The antenna needs separate transceiver chains for each array antenna element and accurate real-time calibration for each of them [27]. Moreover, the antenna beamforming is computationally intensive, which means that smart-antenna base stations must be equipped with very powerful digital signal processors. This tends to increase the system costs in the short term, but since the benefits outweigh the costs, it will be less expensive in the long run. For a smart antenna to have pattern-adaptive capabilities and reasonable gain, an array of antenna elements is necessary.

16.7 ANTENNA

One essential component of a smart-antenna system is its sensors or antenna elements. Just as in humans the ears are the transducers that convert acoustic waves into electrochemical impulses, antenna elements convert electromagnetic waves into electrical impulses. These antenna elements play an important role in shaping and scanning the radiation pattern and constraining the adaptive algorithm used by the digital signal processor. There are a plethora of antenna elements [34] that can be selected to form an adaptive array. This includes classic radiators such as dipoles, monopoles, loops, apertures, horns, reflectors, microstrips, and so on. Thus, a good antenna designer should consider the type of antenna element that is best suited for the application.

One element that meets the requirements and capabilities of a mobile device is that of an array of printed elements. There are a number of printed element geometries, *patches* as they are usually referred to, some shown in Figure 14.2. The two most popular are the rectangular, discussed in detail in Section 14.2, and the circular, discussed in detail in Section 14.3. Such an array possesses the attributes to provide the necessary bandwidth, scanning capabilities, beamwidth, and sidelobe level. Furthermore, it is a low-cost technology suitable for commercial products, in addition to being lightweight and conformal to surfaces. The analysis and design of microstrip/patch antennas is discussed in detail in Chapter 14. There are a number of analysis methods as well as software packages; one is that of [35].

16.7.1 Array Design

The main beam of a larger array can resolve, because of its narrower beamwidth, the signals-of-interest (SOIs) more accurately and allows the smart-antenna system to reject more signals-not-of-interest (SNOIs). Although this may seem attractive for a smart-antenna system, it has two main

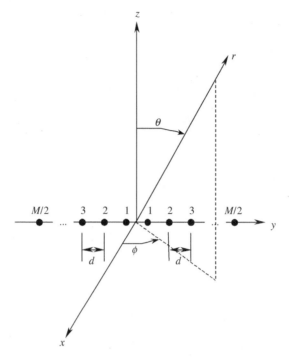

Figure 16.20 Linear array with elements along the y-axis.

disadvantages. One disadvantage is that it tends to increase the cost and the complexity of the hardware implementation, and the other is that it increases the convergence time for the adaptive algorithms, thereby reducing valuable bandwidth. Therefore, this issue is resolved on the basis of the analysis of the network *throughput* that will be discussed later in this chapter.

The array configuration that is well suited for mobile communication is usually a planar array. The linear array configuration is not as attractive because of its inability to scan in 3-D space. On the other hand, a planar array can scan the main beam in any direction of θ (elevation) and ϕ (azimuth), as discussed in Chapter 6. Initially, a linear array will be analyzed to demonstrate some of the basic principles of array theory; eventually, most of the effort will be on planar arrays.

16.7.2 Linear Array

The array factor of a linear array of M (even) identical elements with uniform spacing positioned symmetrically along the *y-axis*, as shown in Figure 16.20, can be written on the basis of the theory of Chapter 6, as

$$(AF)_M = w_1 e^{+j(1/2)\psi_1} + w_2 e^{+j(3/2)\psi_2} + \cdots + w_{M/2} e^{+j[(M-1)/2]\psi_{M/2}}$$
$$+ w_1 e^{-j(1/2)\psi_1} + w_2 e^{-j(3/2)\psi_2} + \cdots + w_{M/2} e^{-j[(M-1)/2]\psi_{M/2}} \quad (16\text{-}1)$$

Simplifying and normalizing (16-1), the array factor for an even number of elements with uniform spacing along the *y-axis* reduces to

$$(AF)_M = \sum_{n=1}^{M/2} w_n \cos[(2n-1)\psi_n] \quad (16\text{-}2)$$

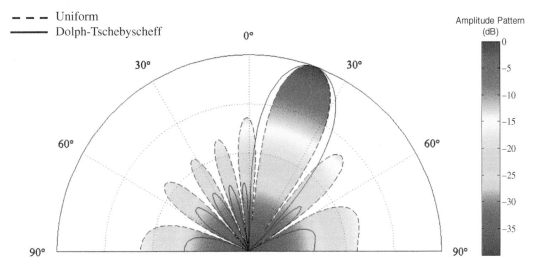

Figure 16.21 Array factors of an eight-element linear array (interelement spacing of 0.5λ).

where

$$\psi_n = \frac{\pi d}{\lambda} \sin\theta \sin\phi + \beta_n \tag{16-2a}$$

In (16-2) and (16-2a), w_n and β_n represent, respectively, the amplitude and phase excitations of the individual elements. While in Chapter 6 the amplitude coefficients were represented by a_n, in signal-processing beamforming it is most common to represent them by w_n, and it will be adopted in this chapter. The amplitude coefficients w_n control primarily the shape of the pattern and the major-to-minor lobe level while the phase excitations control primarily the scanning capabilities of the array. Tapered amplitude distributions exhibit wider beamwidths but lower sidelobes. Therefore, an antenna designer can choose different amplitude distributions to conform to the application specifications. This is shown in Figure 16.21, where the array factor of a uniform linear array is compared with the array factor of a Dolph–Tchebyscheff [34] linear array. As discussed in Chapter 6, Dolph–Tchebyscheff arrays maintain all their minor lobes at the same level while compromising slightly on a wider half-power beamwidth.

16.7.3 Planar Array

As mentioned earlier, linear arrays lack the ability to scan in 3-D space, and since it is necessary for portable devices to scan the main beam in any direction of θ (elevation) and ϕ (azimuth), planar arrays are more attractive for these mobile devices. Let us assume that we have $M \times N$ identical elements, M and N being even, with uniform spacing positioned symmetrical in the *xy-plane* as shown in Figure 16.22.

The array factor for this planar array with its maximum along θ_0, ϕ_0, for an even number of elements in each direction can be written as

$$[AF(\theta, \phi)]_{M \times N} = 4 \sum_{m=1}^{M/2} \sum_{n=1}^{N/2} w_{mn} \cos[(2m-1)u] \cos[(2n-1)v] \tag{16-3}$$

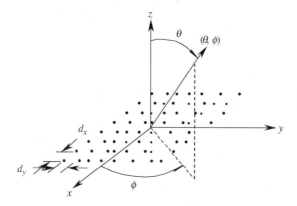

Figure 16.22 Planar array with uniformly spaced elements.

where

$$u = \frac{\pi d_x}{\lambda}(\sin\theta \cos\phi - \sin\theta_0 \cos\phi_0) \quad (16\text{-}3a)$$

$$v = \frac{\pi d_y}{\lambda}(\sin\theta \sin\phi - \sin\theta_0 \sin\phi_0) \quad (16\text{-}3b)$$

In (16-3), w_{mn} is the amplitude excitation of the individual element. For separable distributions $w_{mn} = w_m w_n$. However, for nonseparable distributions, $w_{mn} \neq w_m w_n$. This means that for an $M \times N$ planar array, only $M + N$ excitation values need to be computed from a separable distribution, while $M \times N$ values are needed from a nonseparable distribution [36].

16.8 ANTENNA BEAMFORMING

Intelligence, based on the definition of the Webster's dictionary, is the ability to apply knowledge and to manipulate one's environment. Consequently, the amount of intelligence a system has depends on the information collected, how it gains knowledge from the processed information, and its ability to apply this knowledge. In smart-antenna systems, this knowledge is gained and applied via algorithms processed by a digital signal processor (DSP) as shown in Figure 16.10. The objectives of a DSP are to estimate

1. the direction of arrival (DOA) of all impinging signals, and
2. the appropriate weights to ideally steer the maximum radiation of the antenna pattern toward the SOI and to place nulls toward the SNOI.

Hence, the work on smart antennas promotes research in adaptive signal-processing algorithms such as DOA and adaptive beamforming. The DOA estimation involves a correlation analysis of the array signals followed by eigenanalysis and signal/noise subspace formation while in adaptive beamforming the goal is to adapt the beam by adjusting the magnitude and phase of each antenna element such that a desirable pattern is formed [37].

This section presents a brief overview of direction-of-arrival algorithms followed by a review of adaptive beamforming algorithms. The adaptive beamforming begins with a simple example to give the reader an insight on the basics of adaptive beamforming.

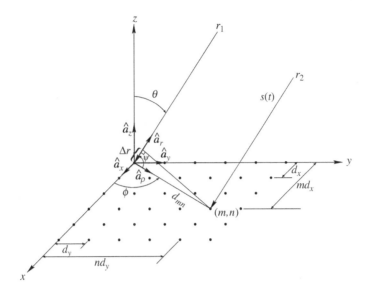

Figure 16.23 $M \times N$ planar array with graphical representation of the time delay.

16.8.1 Overview of Direction-Of-Arrival (DOA) Algorithms

After the antenna array receives the incoming signals from all directions, the DOA algorithm determines the directions of all incoming signals based on the *time delays*. To compute the time delays, consider an $M \times N$ planar array with interelement spacing d_x along the x-axis and d_y along the y-axis as shown in Figure 16.23. When an incoming wave, carrying a baseband signal $s(t)$ impinges at an angle (θ, ϕ) on the antenna array, it produces time delays relative to the other antenna elements. These time delays depend on the antenna geometry, number of elements, and interelement spacing. For the rectangular grid array of Figure 16.23, the time delay of the signal $s(t)$ at the (m,n)th element, relative to the reference element $(0,0)$ at the origin, is

$$\tau_{mn} = \frac{\Delta r}{v_0} \qquad (16\text{-}4)$$

where Δr and v_0 represent respectively the differential distance and the speed of the light in free-space. The differential distance, Δr, is computed using

$$\Delta r = d_{mn} \cos(\psi) \qquad (16\text{-}5)$$

where d_{mn} is the distance from the origin and the (m,n)th element, and ψ is the angle between the radial unit vector from the origin to the (m,n)th element and the radial unit vector in the direction of the incoming signal $s(t)$. Subsequently, d_{mn} and $\cos(\psi)$ are determined using

$$d_{mn} = \sqrt{m^2 d_x^2 + n^2 d_y^2} \qquad (16\text{-}6)$$

$$\cos(\psi) = \frac{\hat{\mathbf{a}}_r \cdot \hat{\mathbf{a}}_\rho}{|\hat{\mathbf{a}}_r| |\hat{\mathbf{a}}_\rho|} \qquad (16\text{-}7)$$

where $\hat{\mathbf{a}}_r$ and $\hat{\mathbf{a}}_\rho$ are, respectively, the unit vectors along the direction of the incoming signal $s(t)$ and along the distance d_{mn} to the (m, n)th element. Thus, the unit vectors (i.e., $\hat{\mathbf{a}}_r$ and $\hat{\mathbf{a}}_\rho$) are expressed as

$$\hat{\mathbf{a}}_r = \hat{\mathbf{a}}_x \sin\theta \cos\phi + \hat{\mathbf{a}}_y \sin\theta \sin\phi + \hat{\mathbf{a}}_z \cos\theta \qquad (16\text{-}8a)$$

$$\hat{\mathbf{a}}_\rho = \frac{\hat{\mathbf{a}}_x m d_x + \hat{\mathbf{a}}_y n d_y}{\sqrt{m^2 d_x^2 + n^2 d_y^2}} \qquad (16\text{-}8b)$$

where $\hat{\mathbf{a}}_x$, $\hat{\mathbf{a}}_y$, and $\hat{\mathbf{a}}_z$ are, respectively, the unit vectors along the x-, y-, and z-axis. Finally, substituting (16-5)–(16-8b) into (16-4), the *time delay* of the (m, n)th element, with respect to the element at the origin [i.e., $(0, 0)$], is written as

$$\tau_{mn} = \frac{m d_x \sin\theta \cos\phi + n d_y \sin\theta \sin\phi}{v_0} \qquad (16\text{-}9)$$

DOA estimation techniques can be categorized on the basis of the data analysis and implementation into four different areas: *conventional methods, subspace-based methods, maximum likelihood methods*, and *integrated methods*, which combine property-restoral techniques with subspace-based approach [27].

Conventional methods for DOA estimation are based on the concepts of beamforming and null steering and do not exploit the statistics of the received signal. In this technique, the DOA of all the signals is determined from the peaks of the output power spectrum obtained from steering the beam in all possible directions. Examples of conventional methods are the delay-and-sum method (classical beamformer method or Fourier method) and Capon's minimum variance method. One major disadvantage of the delay-and-sum method is its poor resolution; that is, the width of the main beam and the height of the sidelobes limit its ability to separate closely spaced signals [27]. On the other hand, Capon's minimum variance technique tries to overcome the poor resolution problem associated with the delay-and-sum method, and in fact, it gives a significant improvement. Although it provides better resolution, Capon's method fails when the SNOIs are correlated with the SOI.

Unlike conventional methods, *subspace methods* exploit the structure of the received data, resulting in a dramatic improvement in resolution. Two main algorithms that fall into this category are the MUltiple SIgnal Classification (MUSIC) algorithm and the Estimation of Signal Parameters via Rotational Invariance Technique (ESPRIT). In 1979, Schmidt proposed the conventional MUSIC algorithm that exploited the eigenstructure of the input covariance matrix [38]. This algorithm provides information about the number of incident signals, DOA of each signal, strengths and cross correlations between incident signals, and noise powers. Like many algorithms, the conventional MUSIC possesses drawbacks. One of the drawbacks is that it requires very precise and accurate array calibration. Another drawback is that, if the impinging signals are highly correlated, it fails because the covariance matrix of the received signals becomes singular. And lastly, it is computationally intensive.

To improve the conventional MUSIC algorithm further, several attempts were made to increase its resolution performance and decrease its computational complexity. In 1983, Barabell developed the Root-MUSIC algorithm based on polynomial rooting and provided higher resolution; its drawback was that it was applicable only to uniformly spaced linear arrays [39]. In 1989, Schmidt proposed the Cyclic MUSIC, a selective direction finding algorithm, which exploited the spectral coherence properties of the received signal and made it possible to resolve signals spaced more closely than the resolution threshold of the array. Moreover, the Cyclic MUSIC also avoids the requirements that the total number of signals impinging on the array must be less than the number of sensor elements [40]. Then, in 1994, Xu presented the Fast Subspace Decomposition (FSD) technique to decrease the computational complexity of the MUSIC algorithm [41]. In a signal environment with multipath, where the signals received are highly correlated, the performance of MUSIC degrades severely. To overcome such a detriment, a technique called spatial smoothing was applied to the covariance matrix [42], [43].

The ESPRIT algorithm is another subspace-based DOA estimation technique originally proposed by Roy [44]. Because ESPRIT has several advantages over MUSIC, such as that it

1. is less computationally intensive,
2. requires much less storage,
3. does not involve an exhaustive search through all possible steering vectors to estimate the DOA, and
4. does not require the calibration of the array,

it has become the algorithm of choice. It is also used in the computer program, designated as **DOA**, which is found in the publisher's website for this book, to determine the directions of arrival for linear and planar array designs of isotropic sources. Since its conception, the ESPRIT has evolved into the 2-D Unitary ESPRIT [45] and the Equirotational Stack ESPRIT (ES-ESPRIT) [46], a more accurate version of the ESPRIT. The corresponding READ ME file explains the details of the program.

Maximum Likelihood (ML) techniques were some of the first techniques to be investigated for DOA estimation, but they are less popular than suboptimal subspace techniques because ML methods are computationally intensive. However, ML techniques outperform the subspace-based techniques in low SNR and in correlated signal environment [47].

The final category of DOA algorithms is the *integrated technique* that combines the property-restoral method with the subspace-based approach. A property-restoral technique is the Iterative Least Squares Projection Based Constant Modulus Algorithm (ILSP-CMA), a data-efficient and cost-efficient approach that is used to detect the envelope of the received signals and overcome many of the problems associated with the Multistage CMA algorithms [48].

In 1995, Xu and Lin [49] proposed a new scheme that integrated ILSP-CMA and the subspace DOA approach. With an M-element antenna array, this scheme can estimate up to $2M^2/3$ DOAs of direct path and multipath signals while a conventional DOA can resolve no more than M DOAs. In 1996, Muhamed and Rappaport [50] showed improvement in performance using an integrated DOA over the conventional ESPRIT when they combined the subspace-based techniques, such as ESPRIT and MUSIC, with the ILSP-CMA.

To show that the DOA can be determined on the basis of the time delays, the DOA of a two-element array will be derived.

Example 16.1

Derive the DOA of a two-element array. Show that the angle of arrival/incidence can be determined on the basis of time delays and geometry of the system.

Solution: On the basis of the geometry of Figure 16.24, the time difference of the signal arriving at the two elements can be written as

$$\Delta t = (t_1 - t_2) = \frac{\Delta d}{v_0} = \frac{d \cos(\theta)}{v_0}$$

where v_0 is the speed of light in free-space. This equation can be rewritten as

$$\cos(\theta) = \frac{v_0}{d} \Delta t = \frac{v_0}{d}(t_1 - t_2)$$

or

$$\theta = \cos^{-1}\left(\frac{d}{v_0}\Delta t\right) = \cos^{-1}\left[\frac{d}{v_0}(t_1 - t_2)\right]$$

This clearly demonstrates that the angle of incidence θ (direction of arrival) can be determined knowing the time delay between the two elements ($\Delta t = t_1 - t_2$), and the geometry of the antenna array (in this case a linear array of two elements with a spacing d between the elements).

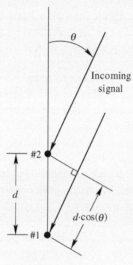

Figure 16.24 Incoming signal on a two-element array.

16.8.2 Adaptive Beamforming

As depicted in Figure 16.10, the information supplied by the DOA algorithm is processed by means of an adaptive algorithm to ideally steer the maximum radiation of the antenna pattern toward the SOI and place nulls in the pattern toward the SNOIs. *This is only necessary for DOA-based adaptive beamforming algorithms. However, for reference (or training) based adaptive beamforming algorithms, like the Least Mean Square (LMS) [51], [52] that is used in this chapter, the adaptive beamforming algorithm does not need the DOA information but instead uses the reference signal, or training sequence, to adjust the magnitudes and phases of each weight to match the time delays created by the impinging signals into the array.* In essence, this requires solving a linear system of normal equations. The main reason why it is generally undesirable to solve the normal equations directly is because the signal environment is constantly changing. Before reviewing the most common adaptive algorithm used in smart antennas, an example is given, based on [53], to illustrate the basic concept of how the weights are computed to satisfy certain requirements of the pattern, especially the formation of nulls.

Example 16.2

Determine the complex weights of a two-element linear array, half-wavelength apart, to receive a desired signal of certain magnitude (unity) at $\theta_0 = 0°$ while tuning out an interferer (SNOI) at $\theta_1 = 30°$, as shown in Figure 16.25. The elements of the array in Figure 16.25 are assumed to be, for simplicity, isotropic and the impinging signals are sinusoids.

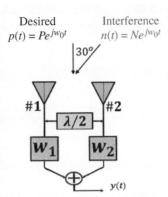

Figure 16.25 Two-element array receiving a desired signal at $\theta_0 = 0°$ and an SNOI at $\theta_0 = 30°$.

Solution: The output $y(t)$ of the array due to the desired signal $p(t)$ is analyzed first, followed by the output due to the interferer $n(t)$. Thus, the output $y(t)$ of the array due to the desired signal $p(t)$ is

$$y(t) = Pe^{j\omega_0 t}(\dot{w}_1 + \dot{w}_2)$$

For the output $y(t)$ to be equal (unity) only to the desired signal, $p(t)$, it is necessary that

$$\dot{w}_1 + \dot{w}_2 = 1$$

On the other hand, the output $y(t)$ due to the interfering signal $n(t)$ is given as

$$y(t) = Ne^{j(\omega_0 t - \pi/4)}\dot{w}_1 + Ne^{j(\omega_0 t + \pi/4)}\dot{w}_2$$

where $-\pi/4$ and $+\pi/4$ terms in the above equation are due to the phase lag and lead, respectively, in reference to the array midpoint of the impinging interfering signal (see Figure 16.26 for details). Because

$$e^{j(\omega_0 t - \pi/4)} = \frac{e^{j\omega_0 t}}{\sqrt{2}}(1-j)$$

$$e^{j(\omega_0 t + \pi/4)} = \frac{e^{j\omega_0 t}}{\sqrt{2}}(1+j),$$

the output $y(t)$ can be rewritten as

$$y(t) = Ne^{j\omega_0 t}\left[\frac{\sqrt{2}}{2}(1-j)\dot{w}_1 + \frac{\sqrt{2}}{2}(1+j)\dot{w}_2\right]$$

Therefore, for the output response $y(t)$ to be zero (i.e., reject totally the interference), it is necessary that

$$\frac{\sqrt{2}}{2}(1-j)\dot{w}_1 + \frac{\sqrt{2}}{2}(1+j)\dot{w}_2 = 0$$

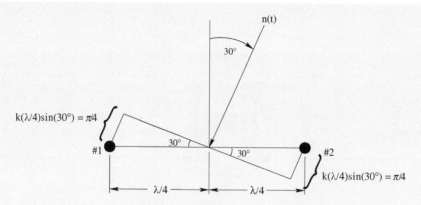

Figure 16.26 Phase lag and lead computations due to signal $n(t)$.

Solving simultaneously the linear system of two complex equations, the second one of this example and the previous one, for $\dot{w}_1$ and $\dot{w}_2$, yields

$$\dot{w}_1 = w'_1 + jw''_1 = \frac{1}{2} - j\frac{1}{2}$$

$$\dot{w}_2 = w'_2 + jw''_2 = \frac{1}{2} + j\frac{1}{2}$$

Thus, the above values of $\dot{w}_1$ and $\dot{w}_2$ are the optimum weights that guarantee the maximum signal-to-interference ratio (SIR) for a desired signal at $\theta_0 = 0°$ and an interferer at $\theta_1 = 30°$. Figure 16.27 shows, *by the solid line*, the array factor obtained on the basis of the weights derived above.

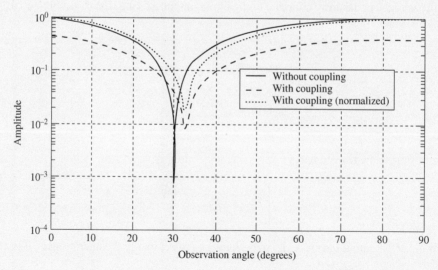

Figure 16.27 Comparison of array factors in the absence and in the presence of mutual coupling.

If the signal environment is stationary, the weights are easily computed by solving the normal equations as shown in this example. However, in practice, the signal environment is dynamic or time varying, and therefore, the weights need to be computed with adaptive methods. In Section 16.8.4, one of the optimal beamforming techniques and adaptive algorithms (Least Mean Square—LMS) used in smart-antenna systems [51], [52], is reviewed; it is the one used in this chapter. There are others, and the interested reader is referred to the literature [17], [53]–[65].

16.8.3 Mutual Coupling

When the radiating elements in the array are in the vicinity of each other, the radiation characteristics, such as the impedance and radiation pattern, of an excited antenna element is influenced by the presence of the others. This effect is known as *mutual coupling*, and it can have a deleterious impact on the performance of a smart-antenna array. Mutual coupling usually causes the maximum and nulls of the radiation pattern to shift and to fill the nulls; consequently, the DOA algorithm and the beamforming algorithm produce inaccurate results unless this effect is taken into account. Furthermore, this detrimental effect intensifies as the interelement spacing is reduced [34], [66]. This will not be discussed in this chapter. The interested reader is referred to the literature [66], [67]. However, a simple example follows to illustrate the effects of coupling on adaptive beamforming.

Example 16.3

To illustrate the effects of mutual coupling in beamforming, Example 16.2 is repeated here. However, this time, mutual coupling is considered. Let us reconsider the two-element linear array, with half-wavelength spacing, receiving a desired signal at $\theta_0 = 0°$ while tuning out an interferer (SNOI) at $\theta_1 = 30°$ in the presence of mutual coupling. The elements of the array in Figure 16.28 are assumed to be, for simplicity, isotropic and the impinging signals are sinusoids.

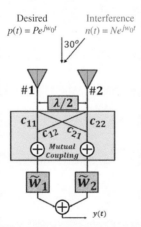

Figure 16.28 Two-element array receiving an SOI at $\theta_0 = 0°$ and an SNOI at $\theta_1 = 30°$ in the presence of mutual coupling.

Solution: The output $y(t)$ of the array due to the desired signal $p(t)$ is analyzed first, followed by the output due to the interferer $n(t)$. Thus, the output $y(t)$ of the array due to $p(t)$ is

$$y(t) = Pe^{jw_0 t}[(c_{11} + c_{12})\tilde{w}_1 + (c_{22} + c_{21})\tilde{w}_2]$$

where $c_{11}, c_{12}, c_{21},$ and c_{22} represent, respectively, the mutual coefficients that describe how an element is affected because of the presence of another. Therefore, for the output $y(t)$ to be equal only to the desired signal, $p(t)$, it is necessary that

$$(c_{11} + c_{12})\dot{w}_1 + (c_{22} + c_{21})\dot{w}_2 = 1$$

On the other hand, the output $y(t)$ due to the interfering signal $n(t)$ is given as

$$y(t) = Ne^{j(\omega_0 t - \pi/4)}(c_{11}\dot{w}_1 + c_{21}\dot{w}_2) + Ne^{j(\omega_0 t + \pi/4)}(c_{12}\dot{w}_1 + c_{22}\dot{w}_2)$$

where $-\pi/4$ and $+\pi/4$ terms are due to the phase lag and lead, respectively, in reference to the array midpoint of the impinging interfering signal (see Figure 16.26 for details). Because

$$e^{j(\omega_0 t - \pi/4)} = \frac{e^{j\omega_0 t}}{\sqrt{2}}(1 - j)$$

$$e^{j(\omega_0 t + \pi/4)} = \frac{e^{j\omega_0 t}}{\sqrt{2}}(1 + j),$$

the output $y(t)$ can be rewritten as

$$y(t) = Ne^{j\omega_0 t}\left\{\left[\frac{\sqrt{2}}{2}(1-j)c_{11} + \frac{\sqrt{2}}{2}(1+j)c_{12}\right]\dot{w}_1 + \left[\frac{\sqrt{2}}{2}(1-j)c_{21} + \frac{\sqrt{2}}{2}(1+j)c_{22}\right]\dot{w}_2\right\}$$

Therefore, for the output response to be zero (i.e., reject totally the interference), it is necessary that

$$\left[\frac{\sqrt{2}}{2}(1-j)c_{11} + \frac{\sqrt{2}}{2}(1+j)c_{12}\right]\dot{w}_1 + \left[\frac{\sqrt{2}}{2}(1-j)c_{21} + \frac{\sqrt{2}}{2}(1+j)c_{22}\right]\dot{w}_2 = 0$$

Solving simultaneously the linear system of complex equations, the second one of this example and the previous one, for $\dot{w}_1$ and $\dot{w}_2$, yields

$$\dot{w}_1 = \frac{1}{2}\frac{c_{22} - c_{21}}{c_{22}c_{11} - c_{12}c_{21}} - j\frac{1}{2}\frac{c_{22} + c_{21}}{c_{22}c_{11} - c_{12}c_{21}}$$

$$\dot{w}_2 = \frac{1}{2}\frac{c_{11} - c_{12}}{c_{22}c_{11} - c_{12}c_{21}} + j\frac{1}{2}\frac{c_{11} + c_{12}}{c_{22}c_{11} - c_{12}c_{21}}$$

Note that the above computed weights in the presence of mutual coupling are related to those weights in the absence of mutual coupling by

$$\dot{w}_1 = \dot{w}_1\left(\frac{c_{22}}{c_{22}c_{11} - c_{12}c_{21}} - j\frac{c_{21}}{c_{22}c_{11} - c_{12}c_{21}}\right)$$

$$\dot{w}_2 = \dot{w}_2\left(\frac{c_{11}}{c_{22}c_{11} - c_{12}c_{21}} + j\frac{c_{12}}{c_{22}c_{11} - c_{12}c_{21}}\right)$$

where $\dot{w}_1$ and $\dot{w}_2$ are the computed weights in the absence of mutual coupling as derived in Example 16.2. Let us assume that the values for c_{11}, c_{12}, c_{21}, and c_{22} are given, on the basis of the formulation in [66], by

$$c_{11} = c_{22} = 2.37 + j0.340$$
$$c_{12} = c_{21} = -0.130 - j0.0517$$

Then, the computed altered weights in the presence of mutual coupling, using the mutual coupling coefficients above and the weights of Example 16.2, are

$$\dot{\tilde{w}}_1 = \tilde{w}'_1 + j\tilde{w}''_1 = 0.189 - j0.223$$
$$\dot{\tilde{w}}_2 = \tilde{w}'_2 + j\tilde{w}''_2 = 0.250 + j0.167$$

On the basis of these new weights, $\dot{\tilde{w}}_1$ and $\dot{\tilde{w}}_2$, the computed patterns with coupling are displayed and compared with that without coupling in Figure 16.27. The one with coupling is also normalized so that its value at $\theta = 0°$ is unity. It is apparent that mutual coupling plays a significant role in the formation of the pattern. For example, in the presence of coupling, the pattern minimum is at $\theta_1 = 32.5°$ at a level of nearly -35 dB (from the maximum), while in the absence of coupling the null is at $\theta_1 = 30°$ at a level nearly $-\infty$ dB.

16.8.4 Optimal Beamforming Techniques

In optimal beamforming techniques, a weight vector that minimizes a cost function is determined. Typically, this cost function, related with a performance measure, is inversely associated with the quality of the signal at the array output, so that when the cost function is minimized, the quality of the signal is maximized at the array output [27]. The most commonly used optimally beamforming techniques or performance measures are the Minimum Mean Square Error (MMSE), Maximum Signal-to-Noise Ratio (MSNR), and Minimum (noise) Variance (MV).

A. Minimum Mean Square Error (MMSE) Criterion
One of the most widely used performance measures in computing the optimum weights is to minimize the MSE cost function. The solution of this function leads to a special class of optimum filters called *Wiener filters* [51], [52]. In order to derive the weights based on the MMSE criterion, the error, ε_k, between the desired signal, d_k, and the output signal of the array, y_k, is written as

$$\varepsilon_k = d_k - \mathbf{w}^H \mathbf{x}_k \qquad (16\text{-}10)$$

Therefore, the MSE based cost function can be written as

$$J_{\text{MSE}}(E[\varepsilon_k^2]) = E[(d_k - \mathbf{w}^H \mathbf{x}_k)^2]$$
$$= E[d_k^2 - 2d_k \mathbf{w}^H \mathbf{x}_k + \mathbf{w}^H \mathbf{x}_k \mathbf{x}_k^H \mathbf{w}] \qquad (16\text{-}11)$$
$$= d_k^2 - 2\mathbf{w}^H E[d_k \mathbf{x}_k] + \mathbf{w}^H E[\mathbf{x}_k \mathbf{x}_k^H] \mathbf{w}$$

where $E[\cdot]$ represents the expected value of $[\cdot]$. Because $E[d_k \mathbf{x}_k]$ and $E[\mathbf{x}_k \mathbf{x}_k^T]$ in (16-11) are the cross correlation, $\mathbf{r}_{xd}$, and the covariance, $\mathbf{R}_{xx}$, respectively, (16-11) can be rewritten as

$$J_{\text{MSE}}(E[\varepsilon_k^2]) = d_k^2 - 2\mathbf{w}^H \mathbf{r}_{xd} + \mathbf{w}^H \mathbf{R}_{xx} \mathbf{w} \qquad (16\text{-}12)$$

In order to minimize the cost function (i.e., to minimize the MSE) in (16-12) with respect to the weights, one must compute the gradient, which is achieved by taking the derivative with respect to the weights and setting it equal to zero; that is,

$$\min_{\mathbf{w}}\{J_{\text{MSE}}(E[\varepsilon_k^2])\} \Rightarrow \frac{\partial}{\partial \mathbf{w}}\{J_{\text{MSE}}(E[\varepsilon_k^2])\} = 0 \qquad (16\text{-}13)$$

Taking the derivative in (16-13) and solving in terms of the weights, $\mathbf{w}$, yields

$$\mathbf{w}_{\text{opt}} = \mathbf{R}_{xx}^{-1} \mathbf{r}_{xd} \qquad (16\text{-}14)$$

Equation (16-14) is the so-called *Wiener solution*, which is the optimal antenna array weight vector, $\mathbf{w}_{\text{opt}}$, in the MMSE sense.

B. Least Mean Square (LMS) Algorithm

One of the simplest algorithms that is commonly used to adapt the weights is the Least Mean Square (LMS) algorithm [51]. The LMS algorithm is a low complexity algorithm that requires no direct matrix inversion and no memory. Moreover, it is an approximation of the steepest descent method using an estimator of the gradient instead of the actual value of the gradient, since computation of the actual value of the gradient is impossible because it would require knowledge of the incoming signals *a priori*. Therefore, at each iteration in the adaptive process, the estimate of the gradient is of the form

$$\widehat{\nabla}\,[J(\mathbf{w})]_k = \begin{bmatrix} \dfrac{\partial J(\mathbf{w})}{\partial w_0} \\ \vdots \\ \dfrac{\partial J(\mathbf{w})}{\partial w_L} \end{bmatrix} \qquad (16\text{-}15)$$

where the $J(\mathbf{w})$ is the cost function of (16-12) to be minimized. Hence, according to the method of steepest descent, the iterative equation that updates the weights at each iteration is

$$\mathbf{w}_{k+1} = \mathbf{w}_k - \mu \widehat{\nabla}\,[J(\mathbf{w})]_k \qquad (16\text{-}16)$$

where μ is the step size related to the rate of convergence. This simplifies the calculations to be performed considerably and allows algorithms, like the LMS algorithm, to be used in real-time applications.

To summarize, the LMS algorithm minimizes the MSE cost function, and it solves the Wiener–Hopf equation, represented by (16-14), iteratively without the need for matrix inversion. Thus, the LMS algorithm computes the weights iteratively as

$$\mathbf{w}_{k+1} = \mathbf{w}_k + 2\mu \mathbf{x}_k (d_k - \mathbf{x}_k^T \mathbf{w}_k) \qquad (16\text{-}17)$$

In order to assure convergence of the weights, $\mathbf{w}_k$, the step size μ is bounded by the condition

$$0 < \mu < \frac{1}{\lambda_{\text{max}}} \qquad (16\text{-}18)$$

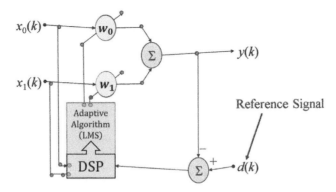

Figure 16.29 Implementation of the LMS algorithm based on the Wiener-Hopf equation [51].

where λ_{max} is the maximum eigenvalue of the covariance matrix, $\mathbf{R}_{xx}$. The main disadvantage of the LMS algorithm is that it tends to converge slowly, especially in noisy environments. A block diagram of the implementation of this (LMS) algorithm is shown in Figure 16.29 where $d(k)$ is the reference/desired signal.

An interactive MATLAB computer program entitled **Smart** is included in the publisher's website for this book, and it can be used to perform beamforming based on the LMS algorithm. One part of the program, designated **Linear**, is used for linear arrays of N isotropic elements while the other part, designated **Planar**, is used for planar arrays of $M \times N$ isotropic elements. The description of each is found in the READ ME file of the corresponding program.

To demonstrate the beamforming capabilities of the LMS algorithm, let us consider two examples.

Example 16.4

For an eight-element ($M = 8$) linear array of isotropic elements with a spacing of $d = 0.5\lambda$ between them, as shown in Figure 16.20, it is desired to form a pattern where the pattern maximum (SOI) is at $\theta_0 = 20°$. There are no requirements on any desired nulls (SNOIs) at any specific angles. Determine the relative amplitude (w's) and phase (β's) excitation coefficients of the elements using the following:

1. LMS beamforming algorithm
2. Classical method described in Chapter 6. Compare the results (amplitude, phase, and pattern) of the two methods.

Solution: Using the LMS algorithm, the obtained normalized amplitude and phase excitation coefficients are listed in Table 16.1 while the corresponding pattern, after 55 iterations, is exhibited in Figure 16.30. The beamformed pattern does meet the desired requirement of having a maximum at $\theta_0 = 20°$. Using the classical method of a scanning array detailed in Section 6.3.3, the obtained amplitude and phase excitation coefficients are also listed in Table 16.1 while the corresponding pattern is also exhibited in Figure 16.30.

By comparing the results of the two methods, it is apparent that the two methods lead to basically identical results in amplitude and phase excitation and corresponding patterns (they are basically indistinguishable). This indicates that the LMS algorithm, using only an SOI, converges to the element excitations, and corresponding pattern, of a *uniform* linear array.

TABLE 16.1 Amplitude (w's) and Phase (β's) Excitation Coefficients of an Eight-Element ($M = 8$) Array Using the Classical Method and LMS Algorithm ($d = 0.5\lambda$, SOI = 20°, $\mu = 0.01$, 55 iterations)

Element	Classical		LMS ($i = 55$)	
	w	β (degrees)	w	β (degrees)
1	1.0000	0	1.0000	0
2	1.0000	−61.56	1.0000	−61.56
3	1.0000	−123.12	1.0000	−123.13
4	1.0000	−184.69	1.0000	−184.69
5	1.0000	−246.25	1.0000	−246.25
6	1.0000	−307.82	1.0000	−307.82
7	1.0000	−369.38	1.0000	−369.38
8	1.0000	−430.95	1.0000	−430.95

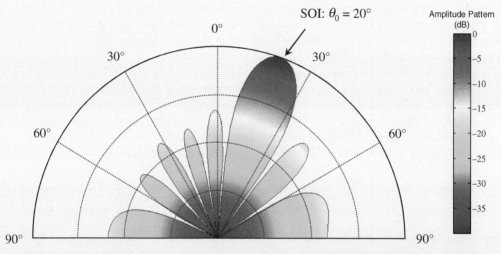

Figure 16.30 Normalized pattern of an eight-element ($M = 8$) linear array of isotropic elements with a spacing of $d = 0.5\lambda$ using both the classical method and *LMS* algorithm ($\mu = 0.01$, 55 iterations).

The requirements of the beamformed pattern of the linear array of Example 16.4 were that only a maximum (SOI) was placed at a specific angle, with no requirements on forming nulls (SNOIs) at specific angles. This was accomplished using both the LMS algorithm and the classical method of Section 6.3.3. However, if both a maximum and a null are desired *simultaneously* at specific angles, the classical method of Section 6.3.3 cannot be used. In contrast, the beamforming method using the LMS algorithm, or any other algorithm, has that versatility.

To demonstrate the capability of the LMS algorithm for pattern beamforming, with both maxima and nulls, let us consider another example.

Example 16.5

For the eight-element linear array of isotropic elements of Example 16.4, with $d = 0.5\lambda$ spacing between them, it is now desired to place a maximum (SOI) at $\theta_0 = 20°$ (as was the case for Example 16.4) and also to simultaneously place a null (SNOI) at $\theta_1 = 45°$. This cannot be accomplished with the classical method of Section 6.3.3. Therefore, perform this using the LMS algorithm.

1. Determine the normalized amplitude (w's) and phase (β's) excitation coefficients of the elements.
2. Plot the beamformed pattern.

Solution: Using the LMS algorithm, the obtained normalized excitation amplitude and phase coefficients, after 81 iterations, are listed in Table 16.2. The corresponding beamformed pattern is shown in Figure 16.31.

It is evident that the obtained pattern does meet the desired specifications. It is also observed that the amplitude excitation coefficients are symmetrical about the center of the array (4–5th element). A similar symmetry has also been observed in beamforming using circular arrays.

TABLE 16.2 **Amplitude (w's) and Phase (β's) Excitation Coefficients of an Eight-Element ($M = 8$) Array Using the LMS Algorithm ($d = 0.5\lambda$, SOI = 20°, SNOI = 45°, $\mu = 0.01$, 81 Iterations)**

Element	LMS ($i = 81$)	
	w	β (degrees)
1	1.0000	−11.62
2	0.8982	−57.05
3	1.1384	−109.98
4	1.3760	−178.77
5	1.3760	−252.21
6	1.1384	−321.01
7	0.8982	−373.94
8	1.0000	−419.37

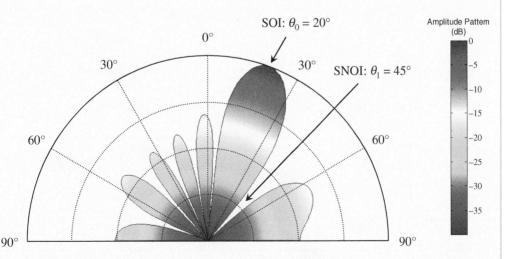

Figure 16.31 Normalized pattern of an eight-element ($M = 8$) linear array of isotropic elements with a spacing of $d = 0.5\lambda$ using both the classical method and *LMS* algorithm ($\mu = 0.01$, 81 iterations).

16.9 MOBILE AD HOC NETWORKS (MANETs)

As the necessity of exchanging and sharing data increases in our daily life, users demand ubiquitous, easy connectivity, and fast networks whether they are at work, at home, or on the move. Moreover, these users are interested in interconnecting all their personal electronic devices (PEDs) in an *ad hoc* fashion. This type of network is referred to as *Mobile Ad hoc NETwork (MANET)*. However, MANETs are not only limited to civilian use but also to disaster recovery (such as fire, flood, and earthquake), law enforcement (such as crowd control, search, and rescue), and tactical communications (such as soldiers coordinating moves in a battlefield) [68]. Thus, research in the area of mobile *ad hoc* networking has received wide interest. The ultimate objective is to design high capacity MANETs employing smart antennas.

This section gives a brief overview of MANETs and discusses MANETs employing smart-antenna systems. In particular, a description of the network layout used for the results and simulations of Sections 16.10.5 and 16.10.6 is presented. Moreover, the protocol used by the MANETs employing smart antennas is described.

16.9.1 Overview of Mobile Ad hoc NETworks (MANETs)

A MANET consists of a collection of wireless mobile stations (nodes) forming a dynamic network whose topology changes continuously and randomly, and its internodal connectivities are managed without the aid of any centralized administration. In contrast, cellular networks are managed by a centralized administration or base station controller (BSC) where each node is connected to a fixed base station. Moreover, cellular networks provide *single hop* connectivity between a node and a fixed base station while MANETs provide *multihop* connectivity between Node A (source) and Node B (destination), as illustrated in Figure 16.32. This figure shows an example of two nodes, Node A and Node B, that desire to exchange data and are at some distance apart. Because they are out of radio range with each other, it is necessary for Node A to use the neighboring or intermediate nodes in forwarding its data packets to Node B. In other words, the data packets from Node A are passed onto the neighboring nodes in a series of single hops until the data packets reach Node B. This type of interaction among nodes is referred to as *peer to peer* and follows a set of rules for communication, referred to as *protocol*. Because the responsibilities of organizing and controlling this network are distributed among the nodes, efficient routing protocols must operate in a distributed

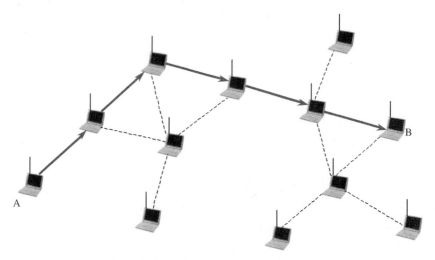

Figure 16.32 Multihop example of a MANET.

manner and must be topology independent [69]. Communication among the nodes takes place using one of the multiple access techniques: *Time Division Multiple Access* (*TDMA*), *Frequency Division Multiple Access* (*FDMA*), *Code Division Multiple Access* (*CDMA*), and *Space Division Multiple Access* (*SDMA*) [23].

The main advantage of MANETs is that they do not rely on extensive and expensive installations of fixed base stations throughout the usage area. Moreover, with the availability of multiple routes from the source node to the destination node, MANETs can perform route selection based on various metrics such as robustness and energy cost, thus optimizing network capacity and energy consumption of the individual node [70]. In other words, if the route with the smallest distances among the intermediate nodes from the source node to the destination node is considered, it would result in a low-capacity network because the many hops produce a long network delay and a higher possibility of a link failure (less robust network). However, because of the small distances among the intermediate nodes, each node or wireless device requires a small amount of energy to propagate its signal to its neighbor, thereby saving battery power. On the other hand, if the route with the farthest distances among the intermediate nodes was selected, it would result in a higher capacity network but with the most energy consumption. Therefore, this trade-off promotes research on algorithms that optimize network capacity, represented by *throughput*, and energy consumption based on the route selection. The main disadvantage of MANETs is that the network capacity drops considerably as the network size (number of nodes) increases. For a more in-depth study on the area of MANETs, the reader is referred to [71], [72].

The next section discusses mobile *ad hoc* networks employing smart-antenna systems to improve network *throughput*. Furthermore, it discusses the protocol, based on the Medium Access Control (MAC) protocol of the IEEE 802.11 Wireless Local Area Network (WLAN) standard, used to allow the nodes to access the channel in a TDMA environment.

16.9.2 MANETs Employing Smart-Antenna Systems

The ability of smart antennas to direct their radiation energy toward the direction of the intended node while suppressing interference can significantly increase the network capacity compared to a network equipped with omnidirectional antennas because they allow the communication channel to be reused. In other words, nodes with omnidirectional antennas keep the neighboring nodes on standby during their transmission while nodes with smart antennas focus only on the desired nodes and allow the neighboring nodes to communicate (refer to Figure 16.33). Therefore, smart antennas together with efficient access protocols can provide high capacity as well as robustness and reliability to mobile *ad hoc* networks. This section discusses the MANET layout used for the design of a high network *throughput* smart-antenna system. Also, included in this section is a discussion of the MAC protocol used by the MANET.

A. The Wireless Network
A MANET of 55 nodes uniformly distributed in an area of $1{,}000 \times 600$ m^2, as shown in Figure 16.34, is implemented using OPNET^{TM} Modeler/Radio$^®$ tool [73], a simulation software package developed by OPNET Technologies, Inc. to analyze, design, and implement communication networks, devices, and protocols. The nodes of the wireless network are equipped with four planar subarrays to cover all possible directions as shown in Figure 16.35. Specifically, each planar subarray in a node covers a sector of $90°$ (from $-45°$ to $+45°$) relative to broadside ($\theta = 0°$). For example, in the multihop network of Figure 16.35, *Node A* communicates with *Node B* with a beam scanned at θ_0, and *Node B* communicates with *Node C* with a beam scanned at θ_1. The data traffic through each of these nodes is modeled to follow a Poisson distribution, and each node changes position at random every time it transfers two consecutive packets (payload), to model the nodes' mobility. The length of each payload packet is 1024 bits. For more details on the MANET layout refer to [69], while

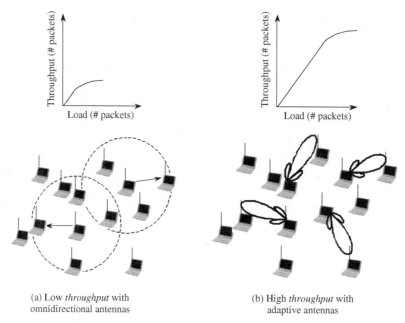

(a) Low *throughput* with omnidirectional antennas

(b) High *throughput* with adaptive antennas

Figure 16.33   Capacity comparison of a network employing omnidirectional antennas and a network with smart antennas.

the design of the smart-antenna system (i.e., antenna and adaptive signal processing) is presented in Section 16.10.

B. The Protocol

The protocol used for the MANET of Figure 16.35 is based on the MAC protocol of IEEE 802.11 Wireless Local Area Network standard and incorporates all necessary features to allow nodes to

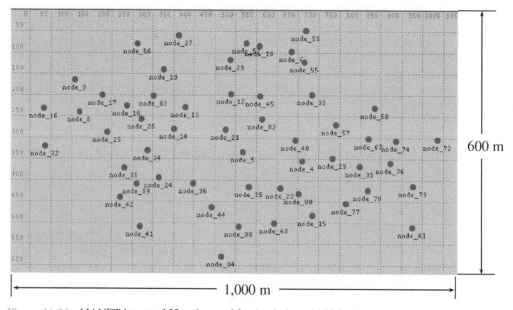

Figure 16.34   MANET layout of 55 nodes used for the design of a high *throughput* smart-antenna system.

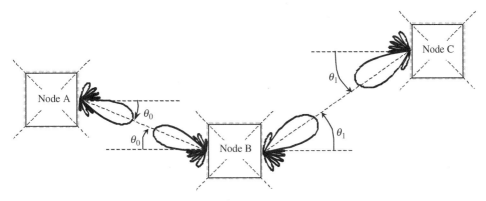

Figure 16.35 Example of three nodes communicating with each other.

access the channel in a TDMA environment. Moreover, it facilitates the use of smart antennas and increases the spatial reuse of TDMA slots, thereby increasing the capacity of the network. The protocol timing diagram is shown in Figure 16.36. A brief explanation of the protocol follows.

a. *Channel Access*—When a *source node* (SRC) has a packet to transmit to a *destination node* (DEST), which is within radio range in a single hop, it first senses the state of the communication channel. If the communication channel is busy, SRC waits until it becomes idle. Once the communication channel is idle, SRC sends a *Request-To-Send* (RTS) signal using the antenna array in nearly omnidirectional mode. This is possible by having only one active element of each subarray of Figure 16.35. When DEST receives successfully the RTS signal, it transmits in nearly omnidirectional mode a *Clear-To-Send* (CTS) signal to SRC.

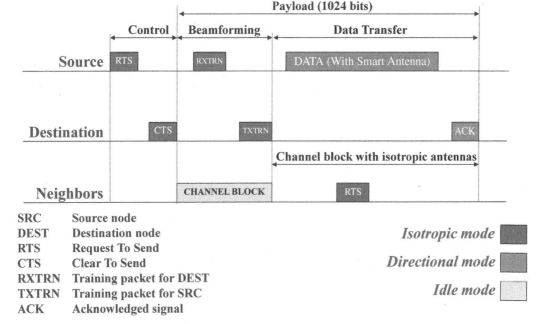

SRC	Source node
DEST	Destination node
RTS	Request To Send
CTS	Clear To Send
RXTRN	Training packet for DEST
TXTRN	Training packet for SRC
ACK	Acknowledged signal

Figure 16.36 MAC Protocol timing diagram (based on IEEE 802.11).

b. *Beamforming*—When all neighboring nodes receive the CTS signal sent by DEST, they wait or standby until beamforming is complete before accessing the channel. After SRC receives the CTS signal, it transmits the *training packet for* DEST (RXTRN) in nearly omnidirectional mode. DEST receives the RXTRN, determines the angle of arrival of the RXTRN signal using the DOA algorithm and computes the complex weights to steer the beam toward SRC. Then, DEST sends the *training packet for* SRC (TXTRN) in smart-antenna mode (i.e., using the computed weights for the array directed towards DEST). Finally, after SRC receives the TXTRN signal, it also computes the angle of arrival of the TXTRN signal using the DOA algorithm and the complex weights to steer the beam toward DEST.

c. *Data Transfer*—Once SRC and DEST have directed their radiation pattern towards one another, the beamforming period is complete, and *data transferring* (DATA) begins. At this time, the channel is freed for the other neighboring nodes to start transmitting their RTS signals. If the neighboring nodes' transmission causes interference above a predefined threshold to SRC and DEST, a new set of weights are computed by SRC and DEST to place nulls toward the direction of the interfering nodes (i.e., SNOIs). At the completion of the DATA packet, the *acknowledged* (ACK) signal is transmitted by DEST, acknowledging the successful reception of the DATA packet.

Note that the training packets, RXTRN and TXTRN, are variable and are part of the payload (refer to Figure 16.36); thus, their lengths affect the length of the DATA packet, which is important because it is the packet that carries information. In other words, if the DATA packet is short, less information is transmitted and thus low network throughput results. Conversely, if the DATA packet is long, more information is transmitted and thus high network throughput is achieved. The influence of the beamforming period or training sequence length on the overall network *throughput* is investigated in Section 16.10.6.

Another point to clarify is that if omnidirectional antennas were used instead of smart antennas, the communication channel would be blocked throughout DATA transmission (refer to Figure 16.36). In other words, no node within radio range of SRC and DEST would be allowed to transmit in order to avoid packet collision (i.e., interference). Thus, this infers that smart antennas increase network capacity by allowing the communication channel to be reused.

Simulation results of MANETs using smart antennas, described in this section, are presented in Sections 16.10.5 and 16.10.6.

16.10 SMART-ANTENNA SYSTEM DESIGN, SIMULATION, AND RESULTS

This section illustrates the design process of a smart-antenna system. The first step of this process is to choose and design an antenna element (in this chapter, a microstrip patch). Then, these designed antenna elements are combined in a particular configuration to form an array. The array configuration, the interelement spacing, and reference signal or training sequence are used in the LMS algorithm to compute the appropriate complex weights. Finally, these weights are tested using the *Ensemble*® [35] software to verify the overall design. The design process just described is documented below with the aid of some preliminary results. Network capacity for various antenna patterns is evaluated in order to guide the antenna design for high network capacity.

16.10.1 Design Process

The objective in this chapter is to introduce a process to design a smart-antenna system for a portable device. The proposed design process is composed of several steps or objectives. The first step of this process is to design an antenna suitable for the network/communication requirements such as the required beamwidth to maintain a tolerable bit error rate (BER) or *throughput*. The antenna design constitutes of a single-element design (e.g., dimensions, material, and geometry), and array design (e.g., configuration, interelement spacing, and number of elements). To optimize the antenna design, the *Ensemble*® [35], a MoM (Method of Moments) simulation software package, analyzes the design and computes the S-parameters, Z-parameters, and far-field amplitude patterns.

After the antenna design step is complete, the second step is to select an adaptive algorithm that minimizes the MSE. In addition, a sidelobe control technique is implemented to prevent environmental noise or *clutter noise* from being received by the high sidelobes and reduce the overall system performance. Because of its complexity, the details of this step will be omitted. The interested reader is referred to [17], [18].

After the adaptive algorithm has determined the complex weights that scan the beam toward the direction of the SOI, and place the nulls toward the direction of the SNOIs, these complex weights are entered in *Ensemble*® [35] as the final step of the design process to verify the overall design. Specifically, the final step consists of comparing the far-field amplitude patterns produced by *Ensemble*® [35] using the computed complex weights with the cavity model far-field patterns (i.e., using the array factor produced by the LMS algorithm's weights and the single-element pattern).

16.10.2 Single Element — Microstrip Patch Design

The first step of the design process is to design a single element. As mentioned earlier, because microstrip patch antennas are inexpensive, lightweight, conformal, easy to manufacture, and versatile, they are the most suitable type of elements for portable devices. A rectangular microstrip patch antenna is considered in this chapter. It is designed to operate at a frequency of 20 GHz (i.e., $f_r = 20$ GHz) using silicon as a substrate material with a dielectric constant of 11.7 (i.e., $\varepsilon_r = 11.7$), a loss tangent of 0.04 (i.e., $\tan \delta = 0.04$), a thickness of 0.300 mm (i.e., $h = 0.300$ mm), and an input impedance of 50 ohms (i.e., $R_{in} = 50$ ohms).

16.10.3 Rectangular Patch

The rectangular patch is by far the most widely used configuration. It has very attractive radiation characteristics and low cross-polarization radiation [34]. In this chapter, the rectangular patch is analyzed and designed using the cavity model, and then its design is optimized using the IE/MoM with the aid of *Ensemble*® [35] that computes S-parameters for microstrip and planar microwave structures.

The first step of the design procedure of a rectangular patch antenna is to compute its physical dimensions. The physical width, W, is calculated using (14-6) while its physical length, L, is computed using (14-7) and (14-1) to (14-2). Afterward, the probe location or excitation feed point is determined using (14-20a) in order to match the impedance (i.e., $R_{in} = 50$ ohms). The computed geometry values obtained from the cavity model are summarized in Figure 16.37.

After the rectangular patch antenna is designed using the cavity model, its design is verified with *Ensemble*® [35]. As expected, because the cavity model is not as accurate as the IE/MoM, the location and the physical dimensions of the probe of the rectangular patch need to be taken into account and optimized in order for the element to resonate at 20 GHz with an input resistance of 50 ohms. The optimized design parameters of the rectangular patch antenna using the *Ensemble*® are also shown in Figure 16.37.

The magnitude of S_{11} versus frequency (also referred to as *return loss*) is shown, as a verification of the design, in Figure 16.38. It shows that the rectangular patch antenna is indeed resonating at 20 GHz with a return loss of −21.5 dB, and its −3 dB and −10 dB bandwidths are 0.74 GHz and 0.25 GHz, respectively. Because the radio's power amplifier tends to reduce its output power, or

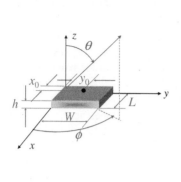

	Cavity Model	Optimized Ensemble®
W	2.976 mm	2.247 mm
L	2.129 mm	2.062 mm
y_o	1.488 mm	1.164 mm
x_o	1.256 mm	0.794 mm
h	0.300 mm	0.300 mm
ε_r	11.7, Si	11.7, Si
f_r	20 GHz	20 GHz

Figure 16.37 Rectangular patch antenna design using the cavity model.

worse become unstable if the VSWR is too large, a stricter definition of the antenna bandwidth is recommended. Thus, less than the −10 dB bandwidth is used.

The far-field amplitude patterns along the *E-plane* (i.e., E_θ when $\phi = 0°$) and the *H-plane* (i.e., E_ϕ when $\phi = 90°$) from *Ensemble*® [35] are compared with the cavity model patterns, and they are shown in Figure 16.39. The *H-plane* far-field patterns from the two models match almost perfectly; however, the *E-plane* far-field pattern, based on *Ensemble*® [35], shows a maximum at $\theta = 5°$ and a null at $\theta = 90°$. The reasons for these discrepancies between the two patterns are that the cavity model (the least accurate of the two methods)

1. does not take into account the probe location, and that it
2. assumes the dielectric material of the substrate is truncated and does not cover the ground plane beyond the edges of the patch (see Chapter 14, Section 14.2.2).

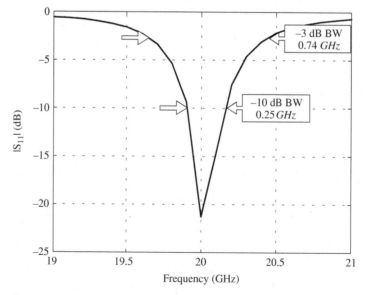

Figure 16.38 S_{11} (*return loss*) for the 20 GHz rectangular patch antenna. (SOURCE: [18] © 2002 IEEE).

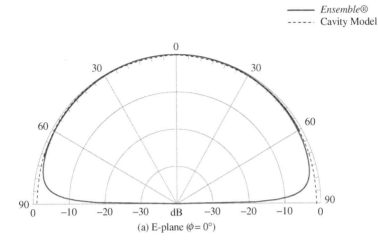

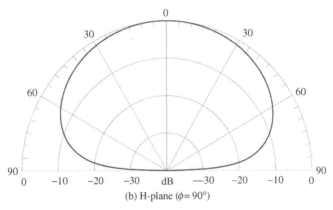

Figure 16.39 Rectangular patch antenna amplitude patterns using the cavity model and *Ensemble®* [35]. (SOURCE: [18] © 2002 IEEE).

16.10.4 Array Design

To electronically scan a radiation pattern in a given direction, an array of elements arranged in a specific configuration is essential. Two configurations are chosen here: the *linear array* and the *planar array*. Although linear arrays lack the ability to scan in 3-D space, planar arrays can scan the main beam in any direction of θ (elevation) and ϕ (azimuth). Consequently, a planar array is best suited for portable devices that require to communicate in any direction. However, in this chapter, a linear array is designed and analyzed initially for simplicity to illustrate some important features. Then, a planar array is designed to be eventually incorporated in the smart-antenna system for a wireless and mobile device.

A. Linear Array Design
Following the design of the individual rectangular patch antenna, a linear array of eight microstrip patches with interelement spacing of $\lambda/2$ (half wavelength), where λ is 1.5 cm (based on the resonance frequency of 20 GHz), is designed (see Figure 16.40). The reasons for choosing interelement spacing of $\lambda/2$ are as follows. To combat fading, the interelement spacing of at least $\lambda/2$ is necessary

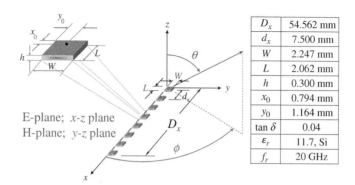

Figure 16.40 Design of an eight-element linear array with rectangular patches.

so that the signals received from different antenna elements are (almost) independent in a rich scattering environment (more precisely, in a uniform scattering environment [74], [75]). In such cases, the antenna arrays provide performance improvement through spatial diversity. However, to avoid *grating lobes* (multiple maxima), the interelement spacing should not exceed one wavelength. But most important, to avoid *aliasing* and causing of nulls to be misplaced, the interelement spacing should be less or equal to $\lambda/2$ (the *Nyquist rate*) [76]. Thus, to satisfy all three conditions, the interelement spacing of $\lambda/2$ (half wavelength) is chosen.

The total amplitude radiation patterns of the eight-element linear array based on the cavity model are represented, neglecting coupling, by the product of the element pattern (static pattern) and the array factor (dynamic pattern). In Chapter 6 this is referred to as pattern multiplication, and it is represented by (6-5). The array factor is dynamic in the sense that it can be controlled by the complex weights, w_n. The amplitudes of these complex weights control primarily the shape of the pattern and the major-to-minor lobe ratio while the phases control primarily the scanning capability of the array.

B. Planar Array Design

The planar arrays designed in this section are composed of only rectangular patches. Planar arrays of 4×4 and 8×8 elements of rectangular microstrip patches with interelement spacing of $\lambda/2$ (half

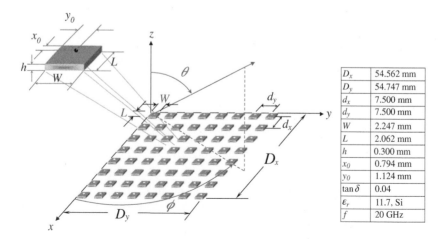

Figure 16.41 Design of an 8×8 planar array with rectangular patches. (SOURCE: [18] © 2002 IEEE).

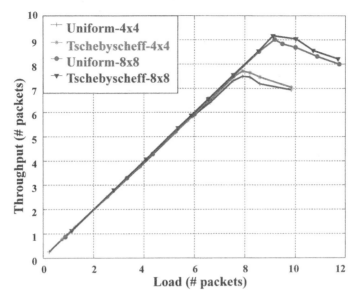

Figure 16.42 Average network *throughput* versus network *load* for different array sizes and excitations. (SOURCE: [18] © 2002 IEEE).

wavelength), where λ is 1.5 cm, are designed. The design and dimensions of the 8 × 8 planar array are given in Figure 16.41. Those of the 4 × 4 design are similar.

16.10.5 4 × 4 Planar Array versus 8 × 8 Planar Array

After the design and analysis of the 4 × 4 and 8 × 8 planar arrays with rectangular patch elements, it is necessary to determine which array configuration is most attractive for a wireless device in a Mobile Ad hoc NETwork (MANET) described in Section 16.9. As observed in Chapter 6, the number of array elements affects the beamwidth of a radiation pattern. That is, when more elements are used in an array (larger size array), the narrower is the main beam. Furthermore, the narrower beamwidth will resolve more accurately the SOIs and the SNOIs. However, these results and inferences do not quantify the overall performance of the network. Therefore, an *ad hoc* network of 55 nodes (see Figure 16.34) equipped with the 4 × 4 and 8 × 8 planar arrays is simulated using OPNET^{TM} Modeler/Radio$^{®}$ tool [73], and the average network *throughput* (G_{avg}) is measured.

Figure 16.42 shows the average network *throughput* versus network *load* (i.e., G_{avg} vs. L_{avg}) for the different nonadaptive antenna patterns; both uniform and Tschebyscheff (−26 dB). This figure indicates that the *throughput* for the case of the 8 × 8 array size is greater compared to the 4 × 4 array size, and also the Tschebyscheff arrays provide slightly greater *throughput* than their respective uniform arrays. These are attributed, respectively, to smaller beamwidths of the 8 × 8 arrays (compared to the 4 × 4 arrays) and lower sidelobes of the Tschebyscheff arrays (compared to the uniform arrays). In both cases, the smaller beamwidths and lower sidelobes lead to lower cochannel interference. Thus, on the basis of the network *throughput* results, the 8 × 8 array configuration is chosen for the remaining part of this chapter.

16.10.6 Adaptive Beamforming

The previous section described nonadaptive beamforming where the array factors of the array were obtained from complex weights that did not depend on the signal environment. However, the array factors of this section are produced from complex weights or excitations that depend on the signal

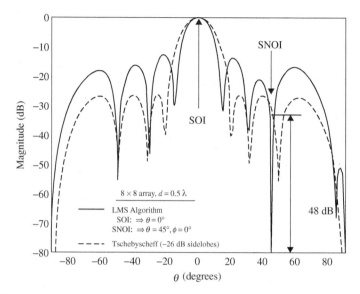

Figure 16.43 *E-plane* ($\phi = 0°$) radiation pattern comparison for the nonadaptive beamformer and adaptive beamformer.

environment. This technique is referred to as *adaptive beamforming* where a digital signal processor (DSP) computes the complex weights using an adaptive algorithm that generate an array factor for an optimal signal-to-interference ratio (SIR). Specifically, this results in an array pattern where ideally the maximum of the pattern is placed toward the source or SOI while nulling the interferers or SNOIs. Because of this, adaptive beamformers tend to be more costly owing to the extra hardware compared to nonadaptive beamformers. Thus, the question may be, how much improvement do adaptive beamformers provide over nonadaptive ones?

In order to answer this question, the overall network *throughput* of an adaptive beamformer is compared to the *throughput* of a nonadaptive beamformer. Each beamformer is equipped with an 8 × 8 planar array and receives an SOI at broadside (i.e., $\theta_0 = 0°$) and an SNOI at $\theta_1 = 45°, \phi_1 = 0°$. The radiation pattern used for the nonadaptive beamformer is a Tschebyscheff design of -26 dB and for the adaptive beamformer it is the LMS algorithm [discussed in Section 16.8.4(B)] generated pattern. The *E-plane* ($\phi = 0°$) of each beamformer is compared in Figure 16.43. The adaptive beamformer or the LMS algorithm generated pattern shown in Figure 16.43 has the maximum toward the SOI and the null toward the SNOI while the nonadaptive Tschebyscheff design beamformer has only the maximum toward the SOI. Also, the intensity toward the SNOI for the Tschebyscheff design is 48 dB higher than that of the LMS algorithm generated pattern. Thus, by attenuating the strength of the interferer more, it would be expected that the *throughput* of the adaptive beamformer will be higher than the *throughput* of the nonadaptive beamformer since the cochannel interference is reduced and the SIR is higher. In fact, as shown in Figure 16.44, where the *throughput* of the adaptive beamformer is compared to the *throughput* of the Tschebyscheff (-26 dB sidelobe level) nonadapted antenna pattern, the LMS beamforming algorithm pattern leads to a higher *throughput* even though the Tschebyscheff pattern exhibits much lower minor lobes.

To investigate the influence of the beamforming period or training sequence length (refer to Figure 16.36) on the network efficiency, the network *throughput* and the network *delay* are plotted, respectively, in Figure 16.45 and Figure 16.46 for various beamforming periods or training sequence lengths. It is apparent that the network *throughput* drops and the network *delay* increases significantly with increasing beamforming period or training sequence length. These figures show that, for this

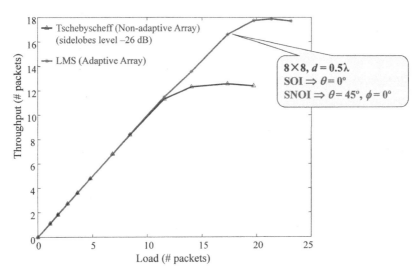

Figure 16.44 Network *throughput* comparison of the adaptive and nonadaptive beamformer. (SOURCE: [18] © 2002 IEEE).

particular network, for beamforming periods or training sequence lengths larger than 20% of the payload, results in a network *throughput* and network *delay* similar to a network of omnidirectional antennas. Therefore, this suggests that smart antennas may be not as effective for networks with long beamforming periods or large training sequence lengths.

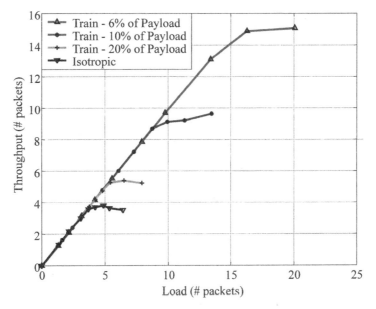

Figure 16.45 Beamforming period or training sequence length effects on the network *throughput*. (SOURCE: [18] © 2002 IEEE).

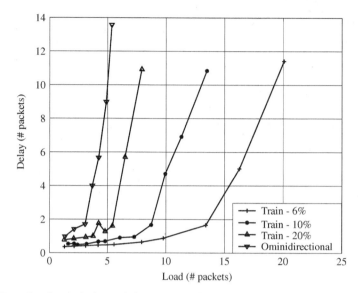

Figure 16.46 Beamforming period or training sequence length effects on the network delay. (SOURCE: [18] © 2002 IEEE).

16.11 BEAMFORMING, DIVERSITY COMBINING, RAYLEIGH-FADING, AND TRELLIS-CODED MODULATION

In this section, adaptive antenna arrays using the LMS algorithm are used to evaluate the performance of a communication channel using the Bit Error Rate (BER) as criterion. The adaptive array used is an 8×8 planar array whose patterns are beamformed so that the SOIs and SNOIs are in directions listed in Table 16.3. The BER of the channel is evaluated initially assuming only an SOI and then an SOI plus an SNOI; then the performance of the two is compared. *For all the cases considered in this section, the desired (SOI) and interfering signals (SNOI) are assumed to be synchronized, which is considered as a worst-case assumption. It is also assumed that the desired and interfering signals have equal power.* For the LMS algorithm, the length of the training sequence was selected to be 60 symbols, and it is transmitted every data sequence of length 940 symbols (total 1,000 symbols; i.e., 6% overhead). The symbol rate chosen was 100 Hz (symbol duration of $T = 1/100 = 0.01 = 10$ msec) [18], [75].

The BER of a communication channel with Binary Phase Shift Keying (BPSK) modulation is first evaluated where the received signal is corrupted with an Additive White Gaussian Noise (AWGN). The results are shown in Figure 16.47 with the set of curves marked uncoded. It is apparent that the adaptive antenna array using the LMS algorithm can suppress one interferer without any performance loss over an AWGN channel.

To improve the performance of the system, trellis-coded modulation (TCM) [74] schemes are used together with adaptive arrays [75]–[77]. In this scheme, the source bits are mapped to channel symbols using a TCM scheme, and the symbols are interleaved using a pseudorandom interleaver

TABLE 16.3   Signals and Their Directions

	SOI		SNOI	
DOA	θ_0	ϕ_0	θ_1	ϕ_1
	0°	0°	45°	0°

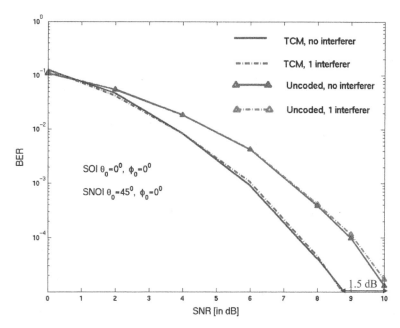

Figure 16.47 BER for uncoded BPSK over AWGN channel and trellis-coded QPSK modulation based on eight-state trellis encoder. (SOURCE: [18] © 2002 IEEE).

to uncorrelate the consecutive symbols to prevent bursty errors. The signal received by the adaptive antenna array consists of a faded version of the desired signal and a number of interfering signals plus AWGN. The receiver combines the signals from each antenna element using the LMS algorithm. A trellis-coded Quadrature Phase Shift Keying (QPSK) modulation scheme based on an eight-state trellis encoder was considered [75]. The performance of the TCM QPSK system in terms of BER is also displayed in Figure 16.47 with the set of curves indicated as TCM. It is again observed that the adaptive antenna array using the LMS algorithm can suppress one interferer without loss in performance. By comparing the two sets of data, uncoded versus TCM coded, it is apparent that the QPSK TCM system over AWGN is better than that of the uncoded BPSK system over AWGN channel by about 1.5 dB at a BER of 10^{-5}.

Wireless communication systems are characterized by time-varying multipath channels, which are typically modeled as "fading channels." Fading, if not mitigated by powerful signal-processing and communication techniques, degrades the performance of a wireless system dramatically. In order to combat fading, the receiver is typically provided with multiple replicas of the transmitted signal [31]. If the replicas fade (almost) independently of each other, then the transmitted information will be recovered with high probability since all the replicas will not typically fade simultaneously. An effective diversity technique is space diversity where multiple transmit and/or receive antennas are used. At the receiver, if the separation between antenna elements is at least $\lambda/2$, the signal received from different antenna elements are (almost) independent in a rich scattering environment (more precisely, in a uniform scattering environment) [78] and hence, in such cases, the antenna arrays provide performance improvement through spatial diversity. For the channels considered in this section, a Rayleigh-fading channel is assumed.

In order to simulate the fading channel, a filtered Gaussian model is used with a first-order lowpass filter. The BER results of the LMS algorithm of the uncoded system over a Rayleigh flat fading AWGN channel are displayed in Figure 16.48. The BER results indicate that when the Doppler spread of the channel is $f_m = 0.1$ Hz ($f_m T = 0.001$), the performance of the system degraded about 4 dB if one equal power interferer is present compared to the case of no interferers. If the channel

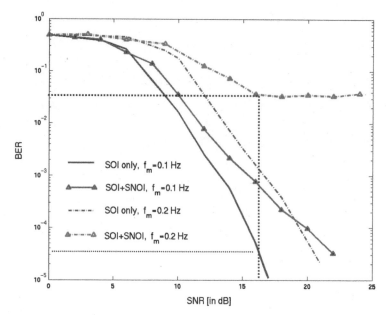

Figure 16.48 BER over *uncoded* Rayleigh-fading channel with Doppler spreads of $f_m = 0.1$ Hz ($f_m T = 0.001$) and $f_m = 0.2$ Hz ($f_m T = 0.002$). (SOURCE: [18] © 2002 IEEE).

faded more rapidly, it is observed that the LMS algorithm performs poorly. For example, the performance of the system over the channel with $f_m = 0.2$ Hz ($f_m T = 0.002$) Doppler spread degraded about 4 dB at a BER of 10^{-4} compared to the case when the Doppler spread was 0.1 Hz. An error floor for the BER was observed for SNRs larger than about 18 dB. For a relatively faster fading in the presence of an equal power interferer, the performance of the system degrades dramatically, implying that the performance of the adaptive algorithm depends highly on the fading rate. Furthermore,

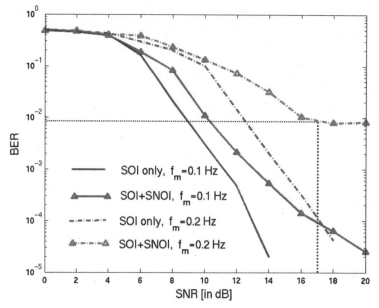

Figure 16.49 BER for *trellis-coded* QPSK modulation over Rayleigh-fading channel with Doppler spreads of $f_m = 0.1$ Hz ($f_m T = 0.001$) and $f_m = 0.2$ Hz ($f_m T = 0.002$). (SOURCE: [18] © 2002 IEEE).

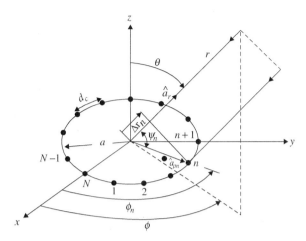

Figure 16.50 Geometry of uniform circular array of N elements.

if the convergence rate of the LMS algorithm is not sufficiently high to track the variations over a rapidly fading channel, adaptive algorithms with faster convergence should be employed.

The same system but with TCM coding was then analyzed over a Rayleigh-fading channel and the BER results for Doppler spreads of $f_m = 0.1$ Hz ($f_m T = 0.001$) and $f_m = 0.2$ Hz ($f_m T = 0.002$), are shown in Figure 16.49. When the TCM coded data of Figure 16.49 is compared with that of uncoded system of Figure 16.48, it is observed that when the Doppler spread is 0.2 Hz and there is one interferer, there is still a reducible error floor on the BER. However, the error floor is reduced compared to the uncoded BPSK case. It can be concluded that the TCM scheme provides some coding advantage in addition to spatial diversity advantage.

16.12 OTHER GEOMETRIES

Until now, the investigation of smart antennas suitable for wireless communication systems has involved primarily rectilinear arrays: *uniform linear arrays* (RLAs) and *uniform rectangular arrays* (URAs). Different algorithms have been proposed for the estimation of the *direction of arrivals*

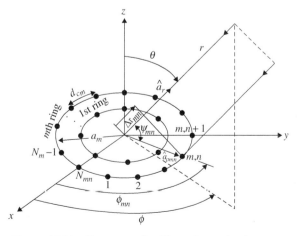

Figure 16.51 Geometry of uniform planar circular array.

(DOAs) of signals arriving to the array and several adaptive techniques have been examined for the shaping of the radiation pattern under different constraints imposed by the wireless environment. Another general antenna configuration that can be used for pattern beamforming is that of circular topologies. Two such geometries are shown in Figures 16.50 and 16.51: a *uniform circular array* (UCA) of Figure 16.50 and a *uniform planar circular array* (UPCA) of Figure 16.51. In the literature for adaptive antennas, not as much attention has been devoted so far to circular configurations despite their ability to offer some advantages. An obvious advantage results from the azimuthal symmetry of the UCA geometry. Because of the fact that a UCA does not have edge elements, directional patterns synthesized by this geometry can be electronically scanned in the azimuthal plane without a significant change in beam shape.

Because of space limitations, the analysis and beamforming capabilities of these configurations will not be pursued further here. The reader is referred to the literature [79]–[82] and others.

16.13 MULTIMEDIA

In the publisher's website for this book, the following multimedia resources are included for the review, understanding, and visualization of the material of this chapter:

a. **Java**-based **interactive questionnaire**, with answers.
b. **Matlab** computer program, designated **DOA**, for computing the direction of arrival of linear and planar arrays.
c. **Matlab** computer program, designated **Smart**, for computing and displaying the radiation characteristics of linear and planar smart-antenna beamforming designs.
d. **Power Point (PPT)** viewgraphs, in multicolor.

REFERENCES

1. R. H. Roy, "An Overview of Smart Antenna Technology: The Next Wave in Wireless Communications," *1998 IEEE Aerospace Conference*, Vol. 3, pp. 339–345, May 1998.
2. H. Krim and M. Viberg, "Two Decades of Array Signal Processing Research: The Parametric Approach," *IEEE Signal Process. Mag.*, pp. 67–94, July 1996.
3. International Engineering Consortium, *Smart Antenna Systems*, A On-line Tutorial Found on http://www.iec.org/online/tutorials/smart_ant/index.html.
4. M. Chryssomallis, "Smart Antennas," *IEEE Antennas Propagat. Mag.*, Vol. 42, No. 3, pp. 129–136, June 2000.
5. Special issue, *IEEE Trans. Antennas Propagat.*, Vol. 24, No. 5, Sept. 1976.
6. Special issue, *IEEE Trans. Antennas Propagat.*, Vol. 34, No. 3, Mar. 1986.
7. G. V. Tsoulos and G. E. Athanasiadou, "Adaptive Antenna Arrays for Mobile Communications: Performance/System Considerations and Challenges," *COMCON '99*, Athens, Greece, 1999.
8. A. O. Boukalov and S. G. Häggman, "System Aspects of Smart-Antenna Technology in Cellular Wireless Communications—An Overview," *IEEE Trans. Microwave Theory Tech.*, Vol. 48, No. 6, pp. 919–928, June 2000.
9. G. V. Tsoulos, "Smart Antennas for Mobile Communication Systems: Benefits and Challenges," *Electron. Commun. Eng. J.*, pp. 84–94, April 1999.
10. T. Sarkar, M. C. Wicks, M. Salazar-Palma and R. J. Bonneau, *Smart Antennas*, John Wiley—IEEE Press, 2003.

11. G. V. Tsoulos, G. E. Athanasiadou, M. A. Beach and S. C. Swales, "Adaptive Antennas for Microcellular and Mixed Cell Environments with DS-CDMA," *Wireless Personal Communications 7*, pp. 147–169, Kluwer Academic Publishers, Netherlands, 1998.
12. R. Kohno, "Spatial and Temporal Communication Theory Using Adaptive Antenna Array," *IEEE Personal Commun.*, Vol. 51, pp. 28–35, Feb. 1998.
13. T. S. Rappaport, *Smart Antennas: Adaptive Arrays, Algorithms, & Wireless Position Locations—Selected Readings*, IEEE, NJ, 1998.
14. G. V. Tsoulos, *Adaptive Antennas for Wireless Communications*, IEEE, NJ, 2001.
15. P. Strobach, "Bi-Iteration Multiple Invariance Subspace Tracking and Adaptive ESPRIT," *IEEE Trans. Signal Process.*, Vol. 48, No. 2, Feb. 2000.
16. B. D. V. Veen and K. M. Buckley, "Beamforming: A Versatile Approach to Spatial Filtering," *IEEE ASSP Mag.*, pp. 4–24, Apr. 1988.
17. S. Bellofiore, "Smart Antennas for Mobile Communications and Network Systems," Ph.D. Dissertation, Arizona State University, Dec. 2002.
18. S. Bellofiore, J. Foutz, R. Govindarajula, I. Bahceci, C. A. Balanis, A. S. Spanias, J. M. Capone, and T. M. Duman, "Smart Antenna System Analysis, Integration and Performance for Mobile Ad-Hoc Networks (MANETS)," *IEEE Trans. Antennas Propagat.*, Vol. 50, No. 5, pp. 571–581, May 2002.
19. S. Bellofiore, C. A. Balanis, J. Foutz, and A. S. Spanias, "Smart Antenna System for Mobile Communication Networks, Part I: Overview and Antenna Design," *IEEE Antennas Propagat. Mag.*, Vol. 44, No. 4, June 2002.
20. S. Bellofiore, J. Foutz, C. A. Balanis, and A. S. Spanias, "Smart Antenna System for Mobile Communication Networks, Part II: Beamforming and Network Throughput," *IEEE Antennas Propagat. Mag.*, Vol. 44, No. 5, August 2002.
21. S. Bellofiore, J. Foutz, C. A. Balanis, and A. S. Spanias, "Smart Antennas for Wireless Communications," *2001 IEEE Antennas and Propagation Society International Symposium*, Boston, MA, Vol. IV, pp. 26–29, July 9–13, 2001.
22. S. Bellofiore, C. A. Balanis, J. Foutz and A. S. Spanias, "Impact of Smart Antenna Designs on Network Capacity," *2002 IEEE Antennas and Propagation Society International Symposium*, San Antonio, TX, Vol. III, pp. 210–213, June 16–21, 2002.
23. T. S. Rappaport, *Wireless Communications: Principles & Practice*, Prentice Hall PTR, Upper Saddle River, NJ, 1999.
24. V. K. Garg and J. E. Wilkes, *Wireless and Personal Communications Systems*, Prentice Hall PTR, Upper Saddle River, NJ, 1996.
25. P. M. Shankar, *Introduction to Wireless Systems*, John Wiley & Sons, 2002.
26. B. Pattan, *Robust Modulation Methods & Smart Antennas in Wireless Communications*, Prentice Hall PTR, Upper Saddle River, NJ, 2000.
27. J. C. Liberti, Jr., and T. S. Rappaport, *Smart Antennas for Wireless Communications: IS-95 and Third Generation CDMA Applications,* Prentice Hall PTR, Upper Saddle River, NJ, 1999.
28. F. Shad, T. D. Todd, V. Kezys, and J. Litva, "Dynamic Slot Allocation (DSA) in Indoor SDMA/TDMA Using a Smart Antenna Basestation," *IEEE/ACM Trans. Networking*, Vol. 9, No. 1, pp. 69–81, Feb. 2001.
29. C. Ung and R. H. Johnston, "A Space Division Multiple Access Receiver," *2001 IEEE International Symposium on Antennas and Propagation*, Vol. 1, pp. 422–425, 2001.
30. R. Janaswamy, *Radiowave Propagation and Smart Antennas for Wireless Communications*, Kluwer Academic Publishers, Norwell, MA, 2001.
31. J. G. Proakis, *Digital Communications*, McGraw-Hill, Inc., New York, 1995.
32. H. L. Bertoni, *Radio Propagation for Modern Wireless Systems*, Prentice Hall, Upper Saddle River, NJ, 2000.
33. P. M. Shankar, *Introduction to Wireless Systems*, John Wiley & Sons, New York, 2002.
34. C. A. Balanis, "Antenna Theory: A Review," *Proc. IEEE*, Vol. 80, No. 1, pp. 7–23, Jan. 1992.
35. Ansoft Corp., *Ensemble*®, http://www.ansoft.com/products/hf/ensemble/.

36. Y. T. Lo and S. W. Lee (Eds), *Antenna Handbook, Theory, Applications, and Design*, Chapter 11, pp. 49–52, Van Nostrand Reinhold Company, New York, 1988.
37. B. Widrow, P. E. Mantey, L. J. Griffiths, and B. B. Goode, "Adaptive Antenna Systems," *Proc. IEEE*, Vol. 55, No. 12, pp. 2143–2159, Aug. 1967.
38. R. O. Schmidt, "Multiple Emitter Location and Signal Parameter Estimation," *Proceedings of RADC Spectrum Estimation Workshop*, Griffiss AFB, New York, pp. 243–258, 1979.
39. A. J. Barabell, "Improving the Resolution Performance of Eigenstructure-based Direction Finding Algorithms," *Proceedings of the IEEE International Conference on Acoustics, Speech, and Signal Processing*, pp. 336–339, 1983.
40. W. A. Gardner, "Simplification of MUSIC and ESPRIT by Exploitation of Cyclostationarity", *Proc. IEEE*, Vol. 76, pp. 845–847, July 1988.
41. G. Xu and T. Kailath, "Fast Subspace Decomposition," *IEEE Trans. Signal Process.*, Vol. 42, No. 3, pp. 539–550, Mar. 1994.
42. J. E. Evans, J. R. Johnson, and D. F. Sun, "High Resolution Angular Spectrum Estimation Techniques for Terrain Scattering Analysis and Angle of Arrival Estimation in ATC Navigation and Surveillance System," MIT Lincoln Lab., Lexington, MA, Rep. 582, 1982.
43. S. U. Pillai and B. H. Kwon, "Forward/Backward Spatial Smoothing Techniques for Coherent Signal Identification," *IEEE Trans. Acoustics, Speech, and Signal Process.*, Vol. 37, No. 1, pp. 8–15, Jan. 1989.
44. R. H. Roy and T. Kailath, "ESPRIT—Estimation of Signal Parameters Via Rotational Invariance Techniques," *IEEE Trans. Acoust., Speech, Signal Process.*, Vol. 37, pp. 984–995, July 1989.
45. M. D. Zoltowski, M. Haardt, and C. P. Mathews, "Closed-Form 2-D Angle Estimation with Rectangular Arrays in Element Space or Beam Space via Unitary ESPRIT," *IEEE Trans. Signal Process.*, Vol. 44, pp. 316–328, Feb. 1996.
46. P. Strobach, "Equirotational Stack Parameterization in Subspace Estimation and Fitting," *IEEE Trans. Signal Process.*, Vol. 48, pp. 713–722, Mar. 2000.
47. I. Ziskind and M. Wax, "Maximum Likelihood Localization of Multiple Sources by Alternating Projection," *IEEE Trans. Acoustics, Speech, and Signal Process.*, Vol. 36, No. 10, pp. 1553–1560, Oct. 1988.
48. I. Parra, G. Xu, and H. Liu, "A Least Squares Projective Constant Modulus Approach," *Proceedings of Sixth IEEE International Symposium on Personal, Indoor and Mobile Radio Communications*, Vol. 2, pp. 673–676, 1995.
49. G. Xu and H. Liu, "An Effective Transmission Beamforming Scheme for Frequency-Division-Duplex Digital Wireless Communication Systems," *1995 International Conference on Acoustics, Speech, and Signal Processing*, Vol. 3, pp. 1729–1732, 1995.
50. R. Muhamed and T. S. Rappaport, "Comparison of Conventional Subspace Based DOA Estimation Algorithms with those Employing Property-Restoral Techniques: Simulation and Measurements," *1996 5th IEEE International Conference on Universal Personal Communications*, Vol. 2, pp. 1004–1008, 1996.
51. B. Widrow and S. D. Stearns, *Adaptive Signal Processing*, Prentice Hall, Englewood Cliffs, NJ, 1985.
52. S. Haykin, *Adaptive Filter Theory*, Prentice Hall PTR, Upper Saddle River, NJ, 1996.
53. R. A. Monzingo and T. W. Miller, *Introduction to Adaptive Arrays*, John Wiley & Sons, New York, 1980.
54. J. Litva and T. K.-Y. Lo, *Digital Beamforming in Wireless Communications*, Artech House, Norwood, MA, 1996.
55. L. C. Godara, "Application of Antenna Arrays to Mobile Communications, Part II: Beam-Forming and Direction-of-Arrival Considerations," *Proc. IEEE*, Vol. 85, No. 8, pp. 1195–1245, Aug. 1997.
56. S. Choi and T. K. Sarkar, "Adaptive Antenna Array Utilizing the Conjugate Gradient Method for Multipath Mobile Communication," *Signal Process.*, Vol. 29, pp. 319–333, 1992.
57. G. D. Mandyam, N. Ahmed, and M. D. Srinath, "Adaptive Beamforming Based on the Conjugate Gradient Algorithm," *IEEE Trans. Aerospace Electron. Syst.*, Vol. 33, No. 1, pp. 343–347, Jan. 1997.
58. S. Haykin, *Neural Networks: A Comprehensive Foundation*, Prentice Hall PTR, Upper Saddle River, NJ, 1999.

59. B. Widrow and M. A. Lehr, "30 Years of Adaptive Neural Networks: Perceptron, Madaline, and Backpropagation," *Proc. IEEE*, Vol. 78, No. 9, pp. 1415–1442, Sept. 1990.
60. A. H. El-Zooghby, C. G. Christodoulou, and M. Georgiopoulos, "Neural Network-Based Adaptive Beamforming for One- and Two-Dimensional Antenna Arrays," *IEEE Trans. Antennas Propagat.*, Vol. 46, No. 12, pp. 1891–1893, Dec. 1998.
61. C. Christodoulou, *Applications of Neural Networks in Electromagnetics*, Artech House, Norwood, MA, 2001.
62. R. L. Haupt and S. E. Haupt, *Practical Genetic Algorithms*, John Wiley & Sons, New York, 1998.
63. Y. Rahmat-Samii and E. Michielssen (Eds.), *Electromagnetic Optimization by Generic Algorithms*, John Wiley & Sons, 1999.
64. D. S. Weile and E. Michielssen, "The Control of Adaptive Antenna Arrays with Genetic Algorithms using Dominance and Diploidy," *IEEE Trans. Antennas Propagat.*, Vol. 49, No. 10, pp. 1424–1433, Oct. 2001.
65. K.-K. Yan and Y. Lu, "Sidelobe Reduction in Array-Pattern Synthesis using Genetic Algorithm," *IEEE Trans. Antennas Propagat.*, Vol. 45, Vo. 7, July 1997.
66. I. J. Gupta and A. A. Ksienski, "Effect of Mutual Coupling on the Performance of Adaptive Arrays," *IEEE Trans. Antennas Propagat.*, Vol. 31, No. 5, Sept. 1983.
67. H. Steyskal and J. S. Herd, "Mutual Coupling Compensation in Small Array Antennas," *IEEE Trans. Antennas Propagat.*, Vol. 38, No. 12, Dec. 1990.
68. H. Xiaoyan, M. Gerla, Y. Yunjung, X. Kaixin, and T. J. Kwon, "Scalable Ad Hoc Routing in Large, Dense Wireless Networks using Clustering and Landmarks," *2002 IEEE International Conference on Communications*, Vol. 5, pp. 3179–3185, Apr. 2002.
69. R. Govindarajula, "Multiple Access Techniques for Mobile Ad Hoc Networks," MS Thesis, Arizona State University, May 2001.
70. S. Agarwal, R. H. Katz, S. V. Krishnamurthy, and S. K. Dao, "Distributed Power Control in Ad-Hoc Wireless Networks," *2001 12th IEEE International Symposium on Personal, Indoor and Mobile Radio Communications*, Vol. 2, pp. F-59–F-66, 2001.
71. C. K. Toh, *Wireless Networks: Protocols and Systems*, Prentice Hall PTR, Upper Saddle River, NJ, 2002.
72. Special issue, *IEEE J. Select. Areas Commun.*, Vol. 17, Vo. 8, Aug. 1999.
73. OPNET Technologies, Inc.^{TM}, *OPTNET*$^{®}$, http://www.opnet.com.
74. Y. Bahçeci and T. M. Duman, "Combined Turbo Coding and Unitary Space-Time Modulation," *IEEE Trans. Commun.*, Vol. 50, No. 8, pp. 1244–1249, Aug. 2002.
75. Y. Bahçeci, "Trellis- and Turbo-Coded Modulation for Multiple Antennas Over Fading Channels," MS Thesis, Arizona State University, Aug. 2001.
76. D. H. Johnson and D. E. Dudgeon, *Array Signal Processing: Concepts and Techniques*, Prentice Hall PTR, Englewood Cliffs, NJ, 1993.
77. G. Ungerboeck, "Channel Coding with Multilevel/Phase Signals," *IEEE Trans. Inform. Theory*, Vol. 28, No. 1, pp. 55–67, 1982.
78. G. L. Stuber, *Principles of Mobile Communications*, Kluwer Publishers, 1996.
79. D. E. N. Davies, "Circular Arrays," *The Handbook of Antenna Design*, Vol. 2, Chapter 12, pp. 299–329, Steven Peregrinus, Stevenage, 1983.
80. P. I. Ioannides, "Uniform Circular Arrays for Smart Antenna Systems," MS Thesis, Department of Electrical Engineering, Arizona State University, Aug. 2004.
81. P. Ioannides and C. A. Balanis, "Uniform Circular Arrays for Smart Antennas," *IEEE Antennas and Propagation Society International Symposium*, Monterey, CA, June 20–25, 2004, Vol. 3, pp. 2796–2799.
82. P. Ioannides and C. A. Balanis, "Mutual Coupling in Adaptive Circular Arrays," *IEEE Antennas and Propagation Society International Symposium*, Monterey, CA, June 20–25, 2004, Vol. 1, pp. 403–406.

PROBLEMS

16.1. For a linear array of 8 elements with a spacing of $d = 0.5\lambda$, determine the DOA, using the **DOA** computer program, assuming an SOI at $\theta_0 = 0°$ and no SNOIs.

16.2. Repeat Problem 16.1 assuming an SOI at $\theta_0 = 0°$ and an SNOI at $\theta_1 = 60°$.

16.3. For a planar array of 8×8 elements with a spacing of $d_x = d_y = 0.5\lambda$, determine the DOA, using the **DOA** computer program, assuming an SOI at $\theta_0 = 20°$, $\phi_0 = 90°$ and no SNOIs.

16.4. Repeat Problem 16.3 assuming an SOI at $\theta_0 = 20°$, $\phi_0 = 90°$ and two SNOIs, one at $\theta_1 = 60°$, $\phi_1 = 180°$ and the other at $\theta_2 = 45°$, $\phi_2 = 270°$.

16.5. Repeat Example 16.2 for a desired signal at $\theta_0 = 0°$ and an interference signal at $45°$. Assume no coupling. Plot the pattern.

16.6. Repeat Example 16.2 for a desired signal at $\theta_0 = 0°$ and an interference signal at $60°$. Assume no coupling. Plot the pattern.

16.7. Repeat Example 16.3 for a desired signal at $\theta_0 = 0°$ and an interference signal at $45°$. Assume coupling and use the same coupling coefficients. Plot the pattern.

16.8. Repeat Example 16.3 for a desired signal at $\theta_0 = 0°$ and an interference signal at $60°$. Assume coupling and use the same coupling coefficients. Plot the pattern.

16.9. Using the LMS algorithm, beamform the pattern of the array factor of a linear array of 10 elements, with a uniform $\lambda/2$ spacing between them, so the maximum (SOI) of the array is broadside ($\theta_0 = 0°$). Assume no requirements on the nulls (SNOIs).

16.10. Repeat Problem 16.9 but with the maximum (SOI) of the array toward $\theta_0 = 30°$. Assume no requirements on the nulls (SNOIs).

16.11. Repeat Problem 16.9 with the maximum (SOI) of the array toward $\theta_0 = 0°$ and a null (SNOI) toward $\theta_1 = 30°$.

16.12. Repeat Problem 16.10 with the maximum (SOI) of the array toward $\theta_0 = 30°$ and a null (SNOI) toward $\theta_1 = 60°$.

16.13. Using the LMS algorithm, beamform the pattern of the array factor of a planar array of 10×10 elements, with a uniform $\lambda/2$ spacing between them so that the maximum (SOI) of the array is broadside ($\theta_0 = 0°$). Assume no requirements on the nulls (SNOIs).

16.14. Repeat Problem 16.13 but with the maximum (SOI) of the array toward $\phi_0 = 45°$, $\theta_0 = 30°$. Assume no requirements on the nulls (SNOIs).

16.15. Repeat Problem 16.13 with the maximum (SOI) of the array toward $\theta_0 = 0°$ and a null (SNOI) toward $\phi_1 = 45°$, $\theta_1 = 30°$.

16.16. Repeat Problem 16.14 with the maximum (SOI) of the array toward $\phi_0 = 45°$, $\theta_0 = 30°$ and a null (SNOI) toward $\phi_1 = 45°$, $\theta_1 = 60°$.

CHAPTER 17

Antenna Measurements

17.1 INTRODUCTION

In the previous sixteen chapters, analytical methods have been outlined which can be used to analyze, synthesize, and numerically compute the radiation characteristics of antennas. Often many antennas, because of their complex structural configurations and excitation methods, cannot be investigated analytically. Although the number of radiators that fall into this category has diminished, because special analytical methods (such as the GTD, Moment Method, Finite-Difference Time-Domain and Finite Element) have been developed during the past few years, there are still a fair number that have not been examined analytically. In addition, experimental results are often needed to validate theoretical data.

As was discussed in Chapter 3, Section 3.8.2, it is usually most convenient to perform antenna measurements with the test antenna in its receiving mode. If the test antenna is reciprocal, the receiving mode characteristics (gain, radiation pattern, etc.) are identical to those transmitted by the antenna. The ideal condition for measuring far-field radiation characteristics then, is the illumination of the test antenna by plane waves: uniform amplitude and phase. Although this ideal condition is not achievable, it can be approximated by separating the test antenna from the illumination source by a large distance on an outdoor range. At large radii, the curvature of the spherical phasefront produced by the source antenna is small over the test antenna aperture. If the separation distance is equal to the inner boundary of the far-field region, $2D^2/\lambda$, then the maximum phase error of the incident field from an ideal plane wave is about 22.5°, as shown in Figure 17.1. In addition to phasefront curvature due to finite separation distances, reflections from the ground and nearby objects are possible sources of degradation of the test antenna illumination.

Experimental investigations suffer from a number of drawbacks such as:

1. For pattern measurements, the distance to the far-field region ($r > 2D^2/\lambda$) is too long even for outside ranges. It also becomes difficult to keep unwanted reflections from the ground and the surrounding objects below acceptable levels.
2. In many cases, it may be impractical to move the antenna from the operating environment to the measuring site.
3. For some antennas, such as phased arrays, the time required to measure the necessary characteristics may be enormous.

Antenna Theory: Analysis and Design, Fourth Edition. Constantine A. Balanis.
© 2016 John Wiley & Sons, Inc. Published 2016 by John Wiley & Sons, Inc.
Companion Website: www.wiley.com/go/antennatheory4e

982 ANTENNA MEASUREMENTS

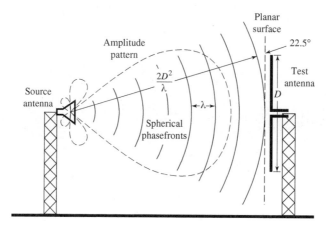

Figure 17.1 Phase error at the edges of a test antenna in the far-field when illuminated by a spherical wave.

4. Outside measuring systems provide an uncontrolled environment, and they do not possess an all-weather capability.
5. Enclosed measuring systems usually cannot accommodate large antenna systems (such as ships, aircraft, large spacecraft, etc.).
6. Measurement techniques, in general, are expensive.

Some of the above shortcomings can be overcome by using special techniques, such as indoor measurements, far-field pattern prediction from near-field measurements [1]–[4], scale model measurements, and automated commercial equipment specifically designed for antenna measurements and utilizing computer assisted techniques.

Because of the accelerated progress made in aerospace/defense related systems (with increasingly small design margins), more accurate measurement methods were necessary. To accommodate these requirements, improved instrumentation and measuring techniques were developed which include tapered anechoic chambers [5], compact and extrapolation ranges [2], near-field probing techniques [2]–[4], improved polarization techniques and swept-frequency measurements [6], indirect measurements of antenna characteristics, and automated test systems.

The parameters that often best describe an antenna system's performance are the pattern (amplitude and phase), gain, directivity, efficiency, impedance, current distribution, and polarization. Each of these topics will be addressed briefly in this chapter. A more extensive and exhaustive treatment of these and other topics can be found in the *IEEE Standard Test Procedures for Antennas* [7], in a summarized journal paper [8], and in a book on microwave antenna measurements [6]. Most of the material in this chapter is drawn from these three sources. The author recommends that the IEEE publication on test procedures for antennas becomes part of the library of every practicing antenna and microwave engineer.

17.2 ANTENNA RANGES

The testing and evaluation of antennas are performed in antenna ranges. Antenna facilities are categorized as *outdoor* and *indoor* ranges, and limitations are associated with both of them. Outdoor ranges are not protected from environmental conditions whereas indoor facilities are limited by space restrictions. Because some of the antenna characteristics are measured in the receiving mode and require far-field criteria, the ideal field incident upon the test antenna should be a uniform plane wave. To meet this specification, a large space is usually required and it limits the value of indoor facilities.

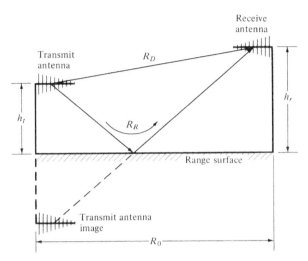

Figure 17.2 Geometrical arrangement for reflection range. (SOURCE: L. H. Hemming and R. A. Heaton, "Antenna Gain Calibration on a Ground Reflection Range," *IEEE Trans. Antennas Propagat.*, Vol. AP-21, No. 4, pp. 532–537, July 1973. © (1973) IEEE).

17.2.1 Reflection Ranges

In general, there are two basic types of antenna ranges: the *reflection* and the *free-space* ranges. The reflection ranges, if judiciously designed [9], can create a constructive interference in the region of the test antenna which is referred to as the "quiet zone." This is accomplished by designing the ranges so that specular reflections from the ground, as shown in Figure 17.2, combine constructively with direct rays.

Usually it is desirable for the illuminating field to have a small and symmetric amplitude taper. This can be achieved by adjusting the transmitting antenna height while maintaining constant that of the receiving antenna. These ranges are of the outdoor type, where the ground is the reflecting surface, and they are usually employed in the UHF region for measurements of patterns of moderately broad antennas. They are also used for systems operating in the UHF to the 16-GHz frequency region.

17.2.2 Free-Space Ranges

Free-space ranges are designed to suppress the contributions from the surrounding environment and include *elevated ranges, slant ranges* [10], *anechoic chambers, compact ranges* [2], and *near-field ranges* [4].

A. Elevated Ranges
Elevated ranges are usually designed to operate mostly over smooth terrains. The antennas are mounted on towers or roofs of adjacent buildings. These ranges are used to test physically large antennas. A geometrical configuration is shown in Figure 17.3(a). The contributions from the surrounding environment are usually reduced or eliminated by [7]

1. carefully selecting the directivity and side lobe level of the source antenna
2. clearing the line-of-sight between the antennas
3. redirecting or absorbing any energy that is reflected from the range surface and/or from any obstacles that cannot be removed
4. utilizing special signal-processing techniques such as modulation tagging of the desired signal or by using short pulses

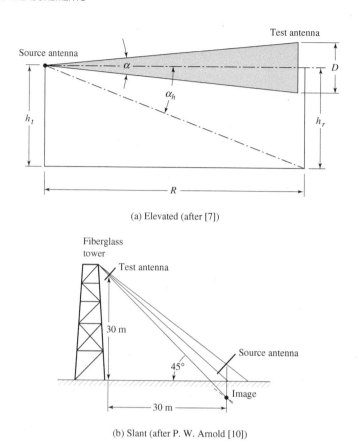

(a) Elevated (after [7])

(b) Slant (after P. W. Arnold [10])

Figure 17.3 Geometries of elevated and slant ranges. (SOURCES: *IEEE Standard Test Procedures for Antennas*, IEEE Std 149–1979, published by IEEE, Inc., 1979, distributed by Wiley; and P. W. Arnold, "The 'Slant' Antenna Range," *IEEE Trans. Antennas Propagat.*, Vol. AP-14, No. 5, pp. 658–659, September 1966. © (1966) IEEE).

In some applications, such as between adjacent mountains or hilltops, the ground terrain may be irregular. For these cases, it is more difficult to locate the specular reflection points (points that reflect energy toward the test antenna). To take into account the irregular surface, scaled drawings of the vertical profile of the range are usually constructed from data obtained from the U.S. Geological Survey. The maps show ground contours [11], and they give sufficient details which can be used to locate the specular reflection points, determine the level of energy reflected toward the test antenna, and make corrections if it is excessive.

B. Slant Ranges

Slant ranges [10] are designed so that the test antenna, along with its positioner, are mounted at a fixed height on a nonconducting tower while the source (transmitting) antenna is placed near the ground, as shown in Figure 17.3(b). The source antenna is positioned so that the pattern maximum, of its free-space radiation, is oriented toward the center of the test antenna. The first null is usually directed toward the ground specular reflection point to suppress reflected signals. Slant ranges, in general, are more compact than elevated ranges in that they require less land.

C. Anechoic Chambers

To provide a controlled environment, an all-weather capability, and security, and to minimize electromagnetic interference, indoor anechoic chambers have been developed as an alternative to outdoor testing. By this method, the testing is performed inside a chamber having walls that are covered with RF absorbers. The availability of commercial high-quality RF absorbing material, with improved electrical characteristics, has provided the impetus for the development and proliferation of anechoic chambers. Anechoic chambers are mostly utilized in the microwave region, but materials have been developed [12] which provide a reflection coefficient of −40 dB at normal incidence at frequencies as low as 100 MHz. In general, as the operating frequency is lowered, the thickness of RF absorbing material must be increased to maintain a given level of reflectivity performance. An RF absorber that meets the minimum electrical requirements at the lower frequencies usually possesses improved performance at higher frequencies.

Presently there are two basic types of anechoic chamber designs: the *rectangular* and the *tapered chamber*. The design of each is based on geometrical optics techniques, and each attempts to reduce or to minimize specular reflections. The geometrical configuration of each, with specular reflection points depicted, is shown in Figures 17.4(a) and 17.4(b).

The rectangular chamber [13] is usually designed to simulate free-space conditions and maximize the volume of the quiet zone. The design takes into account the pattern and location of the source, the frequency of operation, and it assumes that the receiving antenna at the test point is isotropic. Reflected energy is minimized by the use of high-quality RF absorbers. Despite the use of RF absorbing material, significant specular reflections can occur, especially at large angles of incidence.

Tapered anechoic chambers [14] take the form of a pyramidal horn. They begin with a tapered chamber which leads to a rectangular configuration at the test region, as shown in Figure 17.4(b). At the lower end of the frequency band at which the chamber is designed, the source is usually placed near the apex so that the reflections from the side walls, which contribute to the illuminating fields

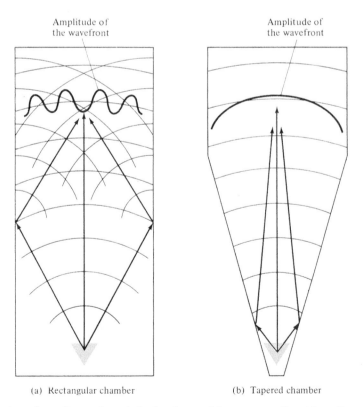

(a) Rectangular chamber   (b) Tapered chamber

Figure 17.4 Rectangular and tapered anechoic chambers and the corresponding side-wall specular reflections. (SOURCE: W. H. Kummer and E. S. Gillespie, "Antenna Measurements—1978," *Proc. IEEE*, Vol. 66, No. 4, pp. 483–507, April 1978. © (1978) IEEE).

in the region of the test antenna, occur near the source antenna. For such paths, the phase difference between the direct radiation and that reflected from the walls near the source can be made very small by properly locating the source antenna near the apex. Thus the direct and reflected rays near the test antenna region add vectorially and provide a relatively smooth amplitude illumination taper. This can be illustrated by ray-tracing techniques.

As the frequency of operation increases, it becomes increasingly difficult to place the source sufficiently close to the apex that the phase difference between the direct and specularly reflected rays can be maintained below an acceptable level. For such applications, reflections from the walls of the chamber are suppressed by using high-gain source antennas whose radiation toward the walls is minimal. In addition, the source is moved away from the apex, and it is placed closer to the end of the tapering section so as to simulate a rectangular chamber.

17.2.3 Compact Ranges

Microwave antenna measurements require that the radiator under test be illuminated by a uniform plane wave. This is usually achieved only in the far-field region, which in many cases dictates very large distances. The requirement of an ideal plane wave illumination can be nearly achieved by utilizing a compact range.

A *Compact Antenna Test Range* (CATR) is a collimating device which generates nearly planar wavefronts in a very short distance (typically 10–20 meters) compared to the $2D^2/\lambda$ (minimum) distance required to produce the same size test region using the standard system configuration of testing shown in Figure 17.1. Some attempts have been made to use dielectric lenses as collimators [15], but generally the name compact antenna test range refers to one or more curved metal reflectors which perform the collimating function. Compact antenna test ranges are essentially very large reflector antennas designed to optimize the planar characteristics of the fields in the near field of the aperture. Compact range configurations are often designated according to their analogous reflector antenna configurations: parabolic, Cassegrain, Gregorian, and so forth.

One compact range configuration is that shown in Figure 17.5 where a source antenna is used as an offset feed that illuminates a paraboloidal reflector, which converts the impinging spherical waves to plane waves [2]. Geometrical Optics (GO) is used in Figure 17.5 to illustrate general CATR operation. The rays from a feed antenna can, over the main beam, be viewed as emanating from a point at its phase center. When the phase center of the feed is located at the prime focus of a parabolic reflector, all rays that are reflected by the reflector and arrive at a plane transverse to the axis of the parabola have traveled an equal distance. See Chapter 15, Section 15.4 for details. Therefore, the field at the aperture of the reflector has a uniform phase; i.e., that of a plane wave. In addition to Geometrical

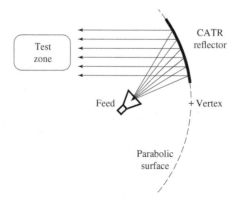

Figure 17.5 A Compact Antenna Test Range (CATR) synthesizes planar phase fronts by collimating spherical waves with a section of paraboloidal reflector.

Optics, analysis and design of CATRs have been performed with a number of other analytical methods. Compact range test zone fields have been predicted by the Method of Moments (MoM), but at high frequencies, the large electrical size of the CATR system makes the use of MoM, Finite-Difference Time-Domain (FD-TD), and Finite Element Method (FEM) impractical. High-frequency techniques, however, are well suited for compact range analysis because the fields of interest are near the specular reflection direction, and the reflector is electrically large. The Geometrical Theory of Diffraction (GTD) is, in principle, an appropriate technique, but it is difficult to implement for serrated-edge reflectors due to the large number of diffracting edges. To date, Physical Optics (PO) is probably the most practical and efficient method of predicting the performance of CATRs [16], [17].

The major drawbacks of compact ranges are aperture blockage, direct radiation from the source to the test antenna, diffractions from the edges of the reflector and feed support, depolarization coupling between the two antennas, and wall reflections. The use of an offset feed eliminates aperture blockage and reduces diffractions. Direct radiation and diffractions can be reduced further if a reflector with a long focal length is chosen. With such a reflector, the feed can then be mounted below the test antenna and the depolarization effects associated with curved surfaces are reduced. Undesirable radiation toward the test antenna can also be minimized by the use of high-quality absorbing material. These and other concerns will be discussed briefly.

A. CATR Performance

A perfect plane wave would be produced by a CATR if the reflector has an ideal parabolic curvature, is infinite in size and is fed by a point source located at its focus. Of course CATR reflectors are of finite size, and their surfaces have imperfections; thus the test zone fields they produce can only approximate plane waves. Although there are different configurations of CATR, their test zone fields have some common characteristics. The usable portion of the test zone consists of nearly planar wavefronts and is referred to as the "quiet zone." Outside the quiet zone, the amplitude of the fields decreases rapidly as a function of distance transverse to the range axis. The size of the quiet zone is typically about 50%–60% of the dimensions of the main reflector. Although the electromagnetic field in the quiet zone is often a very good approximation, it is not a "perfect" plane wave. The imperfections of the fields in the quiet zone from an ideal plane wave are usually represented by phase errors, and ripple and taper amplitude components. These discrepancies from an ideal plane wave, that occur over a specified test zone dimension, are the primary figures of merit of CATRs. For most applications phase deviations of less than 10°, peak-to-peak amplitude ripples of less than 1 dB, and amplitude tapers of less than 1 dB are considered adequate. More stringent quiet-zone specifications may be required to measure, within acceptable error levels, low-side lobe antennas and low-observable scatterers. The sources of quiet-zone taper and ripple are well known, but their minimization is a source of much debate.

Amplitude taper across the quiet zone can be attributed to two sources: the feed pattern and space-attenuation. That portion of the radiation pattern of the feed antenna which illuminates the CATR reflector is directly mirrored into the quiet zone. For example, if the 3-dB beamwidth of the feed is equal to about 60% of the angle formed by lines from the reflector edges to the focal point, then the feed will contribute 3 dB of quiet-zone amplitude taper. In general, as the directivity of the feed antenna increases, quiet-zone amplitude taper increases. Usually, low-gain feed antennas are designed to add less than a few tenths of a dB of amplitude taper. The $1/r^2$ space-attenuation occurs with the spherical spreading of the uncollimated radiation from the feed. Although the total path from the feed to the quiet zone is a constant, the distance from the feed to the reflector varies. These differences in the propagation distances from the feed to various points across the reflector surface cause amplitude taper in the quiet zone due to space-attenuation. This taper is asymmetric in the plane of the feed offset.

Amplitude and phase ripple are primarily caused by diffractions from the edges of the reflector. The diffracted fields are spread in all directions which, along with the specular reflected signal, form constructive and destructive interference patterns in the quiet zone, as shown in Figure 17.6(a).

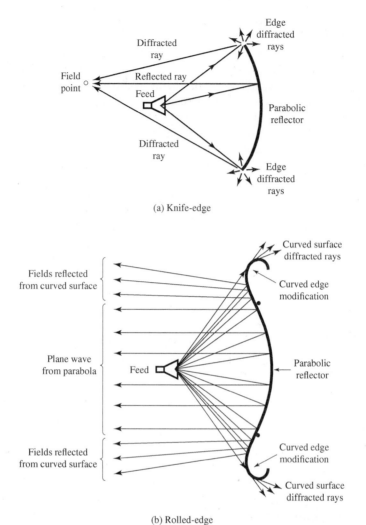

Figure 17.6 Amplitude and phase ripple in the quiet-zone fields produced by a compact antenna test range caused by the phasor sum of the reflected and diffracted rays from the reflector. (SOURCE: W. D. Burnside, M. C. Gilreath, B. M. Kent and G. L. Clerici, "Curved Edge Modification of Compact Range Reflectors," *IEEE Trans. Antennas Propagat.*, Vol. AP-35, No. 2, pp. 176–182, February 1987. © (1987) IEEE).

Considerable research has been done on reflector edge terminations in an effort to minimize quiet-zone ripple. Reflector edge treatments are the physical analogs of windowing functions used in Fourier transforms. Edge treatments reduce the discontinuity of the reflector/free-space boundary, caused by the finite size of the reflector, by providing a gradually tapered transition. Common reflector edge treatments include serrations and rolled edges, as shown in Figure 17.7(a,b). The serrated edge of a reflector tapers the amplitude of the reflected fields near the edge. An alternate interpretation of the effects of serrations is based on edge diffraction. Serrations produce many low-amplitude diffractions as opposed to, for example, the large-amplitude diffractions that would be generated by the four straight edges and corners of a rectangular knife-edged reflector. These small diffractions are quasi-randomized in location and direction; hence, they are likely to have cancellations in the quiet zone. Although most serrated-edge CATRs have triangular serrations, curving the edges of each serration can result in improved performance at high frequencies [18]. A number of blended, rolled-edge treatments have been suggested as alternatives to serrations, and have been implemented

ANTENNA RANGES

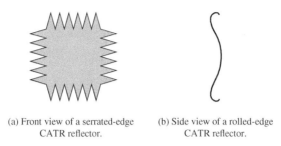

(a) Front view of a serrated-edge CATR reflector.

(b) Side view of a rolled-edge CATR reflector.

Figure 17.7 Two common CATR reflector edge treatments that are used to reduce the diffracted fields in the quiet zone.

to gradually redirect energy away from the quiet zone, as shown in Figure 17.6(b) [19]–[21]. In these designs, the concave parabolic surface of the reflector is blended into a convex surface which wraps around the edges of the reflector and terminates behind it. The predicted quiet-zone fields produced by a knife-edged reflector compared to those produced by a rolled-edged reflector are shown, respectively, in Figures 17.8(a,b) and demonstrate the effectiveness of this edge treatment. Another method of reducing quiet-zone ripple is to taper the illumination amplitude near the reflector edges. This can be accomplished with a high-gain feed or the feed can consist of an array of small elements designed so that a null in the feed pattern occurs at the reflector edges [22]–[25]. Finally, the surface currents on the reflector can be terminated gradually at the edges by tapering the conductivity and/or the impedance of the reflector via the application of lossy material.

The frequency of operation of a CATR is determined by the size of the reflector and its surface accuracy. The low-frequency limit is usually encountered when the reflector is about 25 to 30 wavelengths in diameter [26]. Quiet-zone ripple becomes large at the low-frequency limit. At high frequencies, reflector surface imperfections contribute to the quiet-zone ripple. A rule of thumb used in the design of CATRs is that the surface must deviate less than about 0.007λ from that of a true paraboloid [27]. Since the effects of reflector surface imperfections are additive, dual reflector systems must maintain twice the surface precision of a single reflector system to operate at the same frequency. Many CATR systems operate typically from 1 GHz to 100 GHz.

B. CATR Designs

Four reflector configurations that have been commercially developed will be briefly discussed: *the single paraboloid, the dual parabolic-cylinder, the dual shaped-reflector*, and *the single parabolic-cylinder systems*. The first three configurations are relatively common fully collimating compact ranges; the fourth is a hybrid approach which combines aspects of compact range technology with near-field/far-field (NF/FF) techniques.

The single paraboloidal reflector CATR design was illustrated in Figure 17.5. As with all compact range designs, the feed antenna is offset by some angle from the propagation direction of the collimated energy. This is done to eliminate blockage and to reduce scattering of the collimated fields by the feed. To achieve this offset, the reflector is a sector of a paraboloid that does not include the vertex. This design is referred to as a "*virtual vertex*" compact range. With only one reflector, the paraboloidal CATR has a minimum number of surfaces and edges that can be sources of quiet-zone ripple. Feed spillover into the quiet zone is also low with this design since the feed antenna is pointed almost directly away from the test zone. On the other hand, it is more difficult and costly to produce a high-precision surface that is curved in two planes (three-dimensional) compared to producing a reflector that is curved in only one plane (two-dimensional). In addition, it has been reported that the single paraboloidal reflector design depolarizes the incident fields to a greater degree than other CATR designs. This is due to the relatively low *f/d* ratio needed to simultaneously maintain the feed antenna between the test zone and the reflector while keeping the test zone as close as possible to the reflector aperture [28].

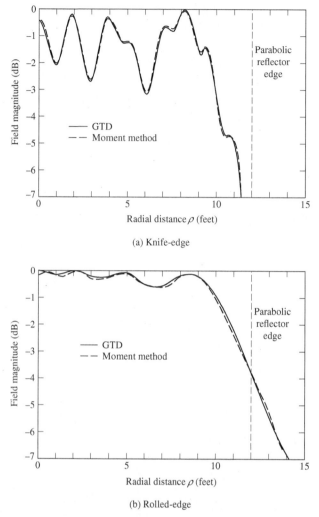

Figure 17.8 Predicted quiet-zone field amplitude versus transverse distance for knife-edge and rolled-edge reflectors. (SOURCE: W. D. Burnside, M. C. Gilreath, B. M. Kent, and G. L. Clerici, "Curved Edge Modification of Compact Range Reflectors," *IEEE Trans. Antennas Propagat.*, Vol. AP-35, No. 2, pp. 176–182, February 1987. © (1987) IEEE).

The dual parabolic-cylinder reflector concept is illustrated in Figure 17.9, and it consists of two parabolic cylinders arranged so that one is curved in one plane (vertical or horizontal) while the other is curved in the orthogonal plane. The spherical phase fronts radiated by the feed antenna are collimated first in the horizontal or vertical plane by the first reflector, then are collimated in the orthogonal plane by the second reflector [29]. Because the boresight of the feed antenna is directed at almost 90° to the plane wave propagation direction, direct illumination of the test zone by the feed can be relatively high. In practice, quiet-zone contamination from feed spillover is virtually eliminated through the use of range gating. Relatively low cross polarization is produced with this design because the doubly folded optics results in a long focal length main reflector.

The dual shaped-reflector CATR, shown schematically in Figure 17.10, is similar in design to a Cassegrain antenna, but the reflector surfaces are altered from the classical parabolic/hyperbolic shapes. An iterative design process is used to determine the shapes of the subreflector and main reflector needed to yield the desired quiet-zone performance. The shape of the subreflector maps the

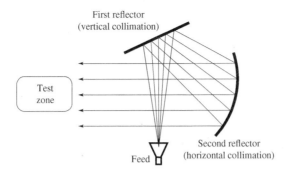

Figure 17.9 Dual parabolic-cylinder compact range collimates the fields in one plane with first reflector and then collimates the fields in the orthogonal plane with second reflector.

high-gain feed pattern into a nearly optimum illumination of the main reflector. An almost uniform energy density illuminates the central part of the main reflector while the amplitude tapers toward the reflector edges. This design results in a very high illumination efficiency (the power of the collimated quiet-zone fields relative to the system input power) [30]. Two of the consequences of this high illumination efficiency are (1) the reduction of spillover into the chamber reduces range clutter, and (2) the increased RF power delivered to the target increases system sensitivity.

The single parabolic-cylinder reflector system is essentially half of the dual parabolic-cylinder CATR. The reflector has a parabolic curvature in the vertical plane and is flat in the horizontal plane. This semicompact antenna test range collimates the fields only in the vertical plane, producing a quiet zone which consists of cylindrical waves, as shown in Figure 17.11 [31]–[33]. Such a compact range configuration is utilized in the ElectroMagnetic Anechoic Chamber (EMAC) at Arizona State University [32], [33].

This *Single-Plane Collimating Range* (SPCR) approach results in a number of advantages and compromises compared to conventional CATR systems and near-field/far-field (NF/FF) systems. For antennas that are small compared to the curvature of the cylindrical phasefront, far-field radiation patterns can be measured directly. Because of the folded optics, the radius of the cylindrical phasefront produced by the SPCR is larger than the radius of the spherical phasefront obtainable by separating the source antenna from the test antenna in a direct illumination configuration within the same anechoic chamber. Thus, with the SPCR it is possible to measure, directly, the far-field patterns of larger antennas compared to those directly measurable on an indoor far-field range. When the size of the antenna is significant relative to the curvature of the cylindrical phasefront, a NF/FF transformation is used to obtain the far-field pattern. However, because the fields are collimated in the vertical plane, only a one-dimensional transformation is required. This greatly simplifies the transformation algorithm. Most importantly, there is a one-to-one correlation between a single azimuthal pattern cut measured in the near-field, and the predicted far-field pattern. The data acquisition time is identical to that of conventional CATRs, and the NF/FF calculation time is nearly negligible. Another advantage

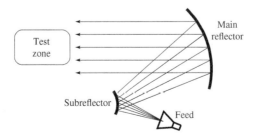

Figure 17.10 Dual shaped-reflector compact range analogous to a Cassegrain system.

992 ANTENNA MEASUREMENTS

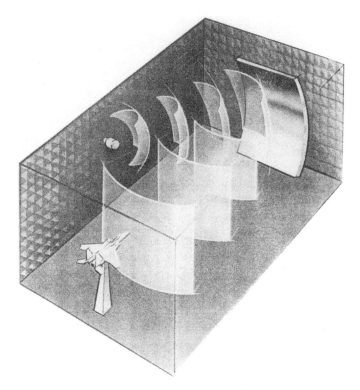

Figure 17.11   ASU Single-Plane Collimating Range (SPCR) produces a cylindrical wave in the quiet zone (artist rendering by Michael Hagelberg).

of the SPCR is the size of the quiet zone. In the vertical plane, the quiet-zone dimension compared to the SPCR reflector is similar to that of conventional CATRs (about 50% to 60%). However, in the horizontal plane, the quiet zone is nearly 100% of the horizontal dimension of the reflector. For a given size anechoic chamber and reflector, targets having much larger horizontal dimensions (yaw patterns of aircraft, for example) can be measured using the SPCR than is possible using a conventional CATR. The SPCR system is relatively inexpensive; the manufacturer estimates that its cost is about 60% of conventional CATR systems.

In addition to the added complexity of NF/FF transformation considerations, this cylindrical wave approach has other disadvantages compared to conventional CATR designs. Because the quiet-zone fields are expanding cylindrically as they propagate along the axis of the range, a large portion of the anechoic chamber is directly illuminated. This should be carefully considered in the design of the side walls of the anechoic chamber to control range clutter. Also, some measurement sensitivity is sacrificed for the same reason.

Compact antenna test ranges enable the measurement of full-sized antennas in very short distances, usually within the controlled environment of an anechoic chamber. A compact antenna test range can be used to accomplish any type of antenna testing (including radiation patterns, gain, efficiency, etc.) that can be performed on an outdoor facility.

17.2.4   Near-Field/Far-Field Methods

The dimensions of a conventional test range can be reduced by making measurements in the near field, and then using analytical methods to transform the measured near-field data to compute the far-field radiation characteristics [2]–[4], [34]. These are referred to as *near-field to far-field (NF/FF)* methods. Such techniques are usually used to measure patterns, and they are often performed indoors. Therefore, they provide a controlled environment and an all-weather capability, the measuring system is time and cost effective, and the computed patterns are as accurate as those measured in a far-field range. However, such methods require more complex and expensive systems, more extensive calibration procedures, more sophisticated computer software, and the patterns are not obtained in real time.

The near-field measured data (usually amplitude and phase distributions) are measured by a scanning field probe over a preselected surface which may be a *plane*, a *cylinder*, or a *sphere*. The measured data are then transformed to the far-field using analytical Fourier transform methods. The complexity of the analytical transformation increases from the planar to the cylindrical, and from the cylindrical to the spherical surfaces. The choice is primarily determined by the antenna to be measured.

In general, the planar system is better suited for high-gain antennas, especially planar phased arrays, and it requires the least amount of computations and no movement of the antenna. Although the cylindrical system requires more computations than the planar, for many antennas its measuring, positioning, and probe equipment are the least expensive. The spherical system requires the most expensive computation, and antenna and probe positioning equipment, which can become quite significant for large antenna systems. This system is best suited for measurements of low-gain and omnidirectional antennas.

Generally, implementation of NF/FF transformation techniques begins with measuring the magnitude and phase of the tangential electric field components radiated by the test antenna at regular intervals over a well-defined surface in the near field. By the principle of *modal expansion*, the sampled E-field data is used to determine the amplitude and phase of an angular spectrum of plane, cylindrical, or spherical waves. Expressing the total field of the test antenna in terms of a modal expansion allows the calculation of the field at any distance from the antenna. Solving for the fields at an infinite distance results in the far-field pattern.

A consideration of the general case of scanning with ideal probes over an arbitrary surface [34] reveals that the choice of scanning surfaces is limited. Morse and Feshbach [35] show that derivation of the far-zone vector field from the near-field depends on vector wave functions that are orthogonal to that surface. Planar, circular cylindrical, spherical, elliptic cylindrical, parabolic cylindrical, and conical are the six coordinate systems that support orthogonal vector wave solutions. The first three coordinate systems are conducive to convenient data acquisition, but the last three require scanning on an elliptic cylinder, a parabolic cylinder, or a sphere in conical coordinates [34]. Thus, the three NF/FF techniques that have been developed and are widely used are based on *planar, cylindrical*, and *spherical* near-field scanning surfaces.

Acquisition of planar near-field data is usually conducted over a rectangular x-y grid, as shown in Figure 17.12(a), with a maximum near-field sample spacing of $\Delta x = \Delta y = \lambda/2$ [36]. It is also

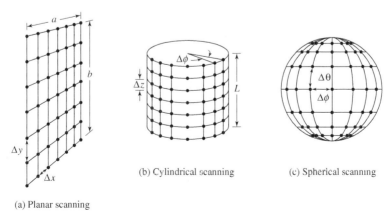

Figure 17.12 Three near-field scanning surfaces that permit convenient data acquisition (planar, cylindrical, and spherical).

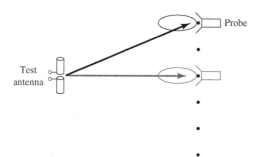

Figure 17.13 Probe compensation of near-field measurements due to nonisotropic radiation pattern of the probe.

possible to acquire the near-field measurements on a plane-polar grid [37] or a bipolar grid [38]. The test antenna is held stationary while the probe (typically an open-ended waveguide or small horn) is moved to each grid location on the plane. As the probe location varies, its orientation relative to the test antenna changes, as illustrated in Figure 17.13. This directive property of the probe, as well as its polarization, must be taken into account using the technique of *probe compensation* [3], [4]. Probe compensation methods use the well-known Lorentz reciprocity theorem to couple the far-zone fields of the test antenna to those of the measuring probe.

The principal advantage of the planar near-field to far-field transformation, over the cylindrical and spherical techniques, is its mathematical simplicity. Furthermore, the planar transformation is suitable for applying the computationally efficient Fast Fourier Transform (FFT) algorithm [39]. Assuming that the number of near-field data points is 2^n (or artificially padded to that number with points of zero value) where n is a positive integer, the full planar far-field transformation can be computed in a time proportional to $(ka)^2 \log_2(ka)$ where a is the radius of the smallest circle that inscribes the test antenna [34]. Planar NF/FF techniques are well suited for measuring antennas which have low backlobes. These include directional antennas such as horns, reflector antennas, planar arrays, and so forth. The primary disadvantage of probing the near-field on a planar surface to calculate the far-field is that the resulting far-field pattern is over a limited angular span. If the planar scanning surface is of infinite extent, one complete hemisphere of the far-field can be computed.

A complete set of near-field measurements over a *cylindrical* surface includes the information needed to compute complete azimuthal patterns for all elevation angles, excluding the conical regions at the top and bottom of the cylinder axis. Since the numerical integrations can be performed with the FFT, the cylindrical transformation exhibits numerical efficiencies and proportional computation times similar to those of the planar transformation. The angular modal expansion, however, is expressed in terms of Hankel functions, which can be more difficult to calculate, especially for large orders.

The cylindrical scanning grid is shown in Figure 17.12(b). The maximum angular and vertical sample spacing is

$$\Delta \phi = \frac{\lambda}{2(a + \lambda)} \tag{17-1}$$

and

$$\Delta z = \lambda/2 \tag{17-2}$$

where λ is the wavelength and a is the radius of the smallest cylinder that encloses the test antenna.

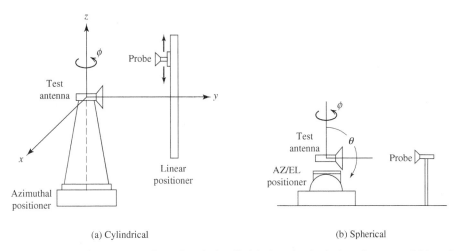

Figure 17.14 Schematic representation of typical cylindrical and spherical surface near-field positioning systems.

A typical cylindrical scanning system is illustrated in Figure 17.14(a). The azimuthal location of the antenna is held constant while the fields are probed at discrete locations in the vertical direction at some fixed distance from the antenna. At the completion of each vertical scan, the test antenna is rotated to the next angular position. The orientation of the probe with respect to the test antenna changes as the vertical location of the probe changes, thus a probe correction is generally required as in the planar case. In addition to directional antennas, the radiation patterns of antennas with narrow patterns along the vertical axis (horizontal fan beam antennas and vertical dipoles for example) can be predicted efficiently with the cylindrical NF/FF technique.

The information obtained by scanning the near-field radiation over a *spherical* surface enclosing a test antenna makes possible the most complete prediction of the far-field radiation pattern. The spherical scanning grid is illustrated in Figure 17.12(c). When sampled at the rate of

$$\Delta\theta = \frac{\lambda}{2(a+\lambda)} \qquad (17\text{-}3)$$

and

$$\Delta\phi = \frac{\lambda}{2(a+\lambda)} \qquad (17\text{-}4)$$

all of the spatial radiation characteristics of the test antenna are included in the transformation. Any far-field pattern cut can be computed from a complete near-field measurement with the spherical scanning scheme. Typically, a spherical scan is accomplished by fixing the location and orientation of the probe and varying the angular orientation of the test antenna with a dual-axis positioner, as shown in Figure 17.14(b). Since the probe is always pointed directly toward the test antenna, probe correction can be neglected for sufficiently large scan radii [34]. However, in general, probe correction is necessary.

The primary drawback of the spherical scanning technique lies in the mathematical transformation. A significant portion of the transformation cannot be accomplished via FFTs. Numerical integrations, matrix operations, and simultaneous solution of equations are required. This increases the computational time and difficulty of the transformation considerably over those of the planar and cylindrical transformations.

A. Modal-Expansion Method for Planar Systems

The mathematical formulations of the planar NF/FF system are based on the plane wave (modal) expansion using Fourier transform (spectral) techniques. Simply stated, any monochromatic, but otherwise arbitrary, wave can be represented as a superposition of plane waves traveling in different directions, with different amplitudes, but all of the same frequency. The objective of the plane wave expansion is to determine the unknown amplitudes and directions of propagation of the plane waves. The results comprise what is referred to as a *modal expansion* of the arbitrary wave. Similarly, cylindrical wave and spherical wave expansions are used to determine far-field patterns from fields measured in the near field over cylindrical and spherical surfaces, respectively.

The relationships between the near-zone E-field measurements and the far-zone fields for planar systems follow from the transform (spectral) techniques of Chapter 12, Section 12.9, represented by (12-73)–(12-75), or

$$\mathbf{E}(x, y, z) = \frac{1}{4\pi^2} \int_{-\infty}^{\infty} \int_{-\infty}^{\infty} \mathbf{f}(k_x, k_y) e^{-j\mathbf{k} \cdot \mathbf{r}} \, dk_x \, dk_y \tag{17-5}$$

where

$$\mathbf{f}(k_x, k_y) = \hat{\mathbf{a}}_x f_x(k_x, k_y) + \hat{\mathbf{a}}_y f_y(k_x, k_y) + \hat{\mathbf{a}}_z f_z(k_x, k_y) \tag{17-5a}$$

$$\mathbf{k} = \hat{\mathbf{a}}_x k_x + \hat{\mathbf{a}}_y k_y + \hat{\mathbf{a}}_z k_z \tag{17-5b}$$

$$\mathbf{r} = \hat{\mathbf{a}}_x x + \hat{\mathbf{a}}_y y + \hat{\mathbf{a}}_z z \tag{17-5c}$$

where $\mathbf{f}(k_x, k_y)$ represents the plane wave spectrum of the field. The x and y components of the electric field measured over a plane surface ($z = 0$) from (17-5) are

$$E_{xa}(x, y, z = 0) = \frac{1}{4\pi^2} \int_{-\infty}^{\infty} \int_{-\infty}^{\infty} f_x(k_x, k_y) e^{-j(k_x x + k_y y)} \, dk_x \, dk_y \tag{17-6a}$$

$$E_{ya}(x, y, z = 0) = \frac{1}{4\pi^2} \int_{-\infty}^{\infty} \int_{-\infty}^{\infty} f_y(k_x, k_y) e^{-j(k_x x + k_y y)} \, dk_x \, dk_y \tag{17-6b}$$

The x and y components of the *plane wave spectrum*, $f_x(k_x, k_y)$ and $f_y(k_x, k_y)$, are determined in terms of the near-zone electric field from the Fourier transforms of (17-6a) and (17-6b) as given by (12-85a), (12-85b), or

$$\boxed{f_x(k_x, k_y) = \int_{-b/2}^{+b/2} \int_{-a/2}^{+a/2} E_{xa}(x', y', z' = 0) e^{+j(k_x x' + k_y y')} \, dx' \, dy'} \tag{17-7a}$$

$$\boxed{f_y(k_x, k_y) = \int_{-b/2}^{+b/2} \int_{-a/2}^{+a/2} E_{ya}(x', y', z' = 0) e^{+j(k_x x' + k_y y')} \, dx' \, dy'} \tag{17-7b}$$

The far-field pattern of the antenna, in terms of the plane wave spectrum function $\mathbf{f}$, is then that of (12-107)

$$\boxed{\mathbf{E}(r, \theta, \phi) \simeq j \frac{k e^{-jkr}}{2\pi r} [\cos \theta \mathbf{f}(k_x, k_y)]} \tag{17-8}$$

or (12-111)

$$E_\theta(r,\theta,\phi) \simeq j\frac{ke^{-jkr}}{2\pi r}(f_x \cos\phi + f_y \sin\phi) \quad (17\text{-}9a)$$

$$E_\phi(r,\theta,\phi) \simeq j\frac{ke^{-jkr}}{2\pi r}\cos\theta(-f_x \sin\phi + f_y \cos\phi) \quad (17\text{-}9b)$$

The procedure then, to determine the far-zone field from near-field measurements, is as follows:

1. Measure the electric field components $E_{xa}(x',y',z'=0)$ and $E_{ya}(x',y',z'=0)$ in the near field.
2. Find the plane wave spectrum functions f_x and f_y using (17-7a) and (17-7b).
3. Determine the far-zone electric field using (17-8) or (17-9a) and (17-9b).

Similar procedures are used for cylindrical and spherical measuring systems except that the constant surfaces are, respectively, cylinders and spheres. However, their corresponding analytical expressions have different forms.

It is apparent once again, from another application problem, that if the tangential field components are known along a plane, the plane wave spectrum can be found, which in turn permits the evaluation of the field at any point. The computations become more convenient if the evaluation is restricted to the far-field region.

B. Measurements and Computations

The experimental procedure requires that a plane surface, a distance z_0 from the test antenna, be selected where measurements are made as shown in Figure 17.12(a). The distance z_0 should be at least two or three wavelengths away from the test antenna to be out of its reactive near-field region. The plane over which measurements are made is usually divided into a rectangular grid of M × N points spaced Δx and Δy apart and defined by the coordinates $(m\Delta x, n\Delta y, 0)$ where $-M/2 \leq m \leq M/2 - 1$ and $-N/2 \leq n \leq N/2 - 1$. The values of M and N are determined by the linear dimensions of the sampling plane divided by the sampling space. To compute the far-field pattern, it requires that both polarization components of the near field are measured. This can be accomplished by a simple rotation of a linear probe about the longitudinal axis or by the use of a dual-polarized probe. The probe used to make the measurements must not be circularly polarized, and it must not have nulls in the angular region of space over which the test antenna pattern is determined because the probe correction coefficients become infinite.

The measurements are carried out until the signal at the edges of the plane is of very low intensity, usually about 45 dB below the largest signal level within the measuring plane. Defining a and b the width and height, respectively, of the measuring plane, M and N are determined using

$$M = \frac{a}{\Delta x} + 1 \quad (17\text{-}10a)$$

$$N = \frac{b}{\Delta y} + 1 \quad (17\text{-}10b)$$

The sampling points on the measuring grid are chosen to be less than $\lambda/2$ in order to satisfy the Nyquist sampling criterion. If the plane $z = 0$ is located in the far field of the source, the sample spacings can increase to their maximum value of $\lambda/2$. Usually the rectangular lattice points are separated by the grid spacings of

$$\Delta x = \frac{\pi}{k_{xo}} \qquad (17\text{-}11a)$$

$$\Delta y = \frac{\pi}{k_{yo}} \qquad (17\text{-}11b)$$

where k_{xo} and k_{yo} are real numbers and represent the largest magnitudes of k_x and k_y, respectively, such that $\mathbf{f}(k_x, k_y) \simeq 0$ for $|k_x| > k_{xo}$ or $|k_y| > k_{yo}$.

At the grid sample points, the tangential electric field components, E_x and E_y, are recorded. The subscripts x and y represent, respectively, the two polarizations of the probe. The procedure for probe compensation is neglected here. A previously performed characterization of the probe is used to compensate for its directional effects in what is essentially an application of its "transfer function." The electric field components over the entire plane can be reconstructed from the samples taken at the grid points, and each is given by

$$E_{xa}(x, y, z = 0) \simeq \sum_{n=-N/2}^{N/2-1} \sum_{m=-M/2}^{M/2-1} E_x(m\Delta x, n\Delta y, 0) \frac{\sin(k_{xo}x - m\pi)}{k_{xo}x - m\pi} \frac{\sin(k_{yo}y - n\pi)}{k_{yo}y - n\pi} \qquad (17\text{-}12a)$$

$$E_{ya}(x, y, z = 0) \simeq \sum_{n=-N/2}^{N/2-1} \sum_{m=-M/2}^{M/2-1} E_y(m\Delta x, n\Delta y, 0) \frac{\sin(k_{xo}x - m\pi)}{k_{xo}x - m\pi} \frac{\sin(k_{yo}y - n\pi)}{k_{yo}y - n\pi} \qquad (17\text{-}12b)$$

Using (17-12a) and (17-12b), f_x and f_y of (17-7a) and (17-7b) can be evaluated, using a FFT algorithm, at the set of wavenumbers explicitly defined by the discrete Fourier transform and given by

$$k_x = \frac{2\pi m}{M\Delta x}, \qquad -\frac{M}{2} \leq m \leq \frac{M}{2} - 1 \qquad (17\text{-}13a)$$

$$k_y = \frac{2\pi n}{N\Delta y}, \qquad -\frac{N}{2} \leq n \leq \frac{N}{2} - 1 \qquad (17\text{-}13b)$$

The wavenumber spectrum points are equal to the number of points in the near-field distribution, and the maximum wavenumber coordinate of the wavenumber spectrum is inversely proportional to the near-field sampling spacing. While the maximum sampling spacing is $\lambda/2$, there is no minimum spacing restrictions. However, there are no advantages to increasing the near-field sample points by decreasing the sample spacing. The decreased sample spacing will increase the limits of the wavenumber spectrum points, which are in the large evanescent mode region, and do not contribute to increased resolution of the far-field pattern.

Increased resolution in the far-field power pattern can be obtained by adding artificial data sampling points (with zero value) at the outer extremities of the near-field distribution. This artificially increases the number of sample points without decreasing the sample spacing. Since the sample spacing remains fixed, the wavenumber limits also stay fixed. The additional wavenumber spectrum points are all within the original wavenumber limits and lead to increased resolution in the computed far-field patterns.

To validate the techniques, numerous comparisons between computed far-field patterns, from near-field measurements, and measured far-field patterns have been made. In Figure 17.15, the computed and measured sum and difference far-field azimuth plane patterns for a four-foot diameter

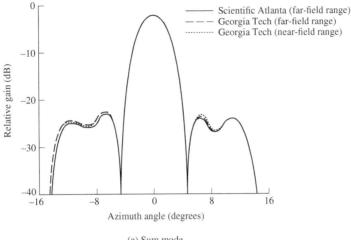

(a) Sum mode

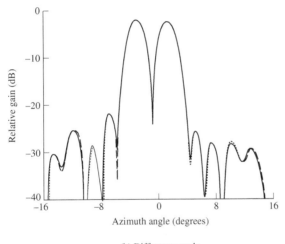

(b) Difference mode

Figure 17.15 Measured and computed sum and difference mode principal-plane far-field patterns for a four-foot parabolic reflector. (SOURCE: E. D. Joy, W. M. Leach, Jr., G. P. Rodrique, and D. T. Paris, "Applications of Probe Compensated Near-Field Measurements," *IEEE Trans. Antennas Propagat.*, Vol. AP-26, No. 3, pp. 379–389, May 1978. © (1978) IEEE).

parabolic reflector with a nominal gain of 30 dB are displayed [4]. Two measured far-field patterns were obtained on two different high-quality far-field ranges, one at the Georgia Institute of Technology and the other at Scientific-Atlanta. The third trace represents the computed far-field pattern from near-field measurements made at Georgia Tech. There are some minor discrepancies between the two measured far-field patterns which were probably caused by extraneous range reflections. The best agreement is between the computed far-field pattern and the one measured at Scientific-Atlanta. Many other comparisons have been made with similar success. The limited results shown here, and the many others published in the literature [4], [40]–[42] clearly demonstrate the capability of the near-field technique.

The near-field technique provides the antenna designers information not previously available to them. For example, if a given far-field pattern does not meet required specifications, it is possible to use near-field data to pinpoint the cause [43]. Near-field measurements can be applied also to antenna analysis and diagnostic tasks [44], and it is most attractive when efficient near-field data collection and transformation methods are employed.

17.3 RADIATION PATTERNS

The radiation patterns (amplitude and phase), polarization, and gain of an antenna, which are used to characterize its radiation capabilities, are measured on the surface of a constant radius sphere. Any position on the sphere is identified using the standard spherical coordinate system of Figure 17.16. Since the radial distance is maintained fixed, only the two angular coordinates (θ, ϕ) are needed for positional identification. A representation of the radiation characteristics of the radiator as a function of θ and ϕ for a constant radial distance and frequency is defined as the *pattern* of the antenna.

In general, the pattern of an antenna is three-dimensional. Because it is impractical to measure a three-dimensional pattern, a number of two-dimensional patterns, as defined in Section 2.2, are measured. They are used to construct a three-dimensional pattern. The number of two-dimensional patterns needed to construct faithfully a three-dimensional graph is determined by the functional requirements of the description, and the available time and funds. The minimum number of two-dimensional patterns is two, and they are usually chosen to represent the orthogonal principal E- and H-plane patterns, as defined in Section 2.2.

A two-dimensional pattern is also referred to as a *pattern cut*, and it is obtained by fixing one of the angles (θ or ϕ) while varying the other. For example, by referring to Figure 17.16, pattern cuts can be obtained by fixing $\phi_j (0 \leq \phi_j \leq 2\pi)$ and varying $\theta (0 \leq \theta \leq \pi)$. These are referred to as elevation patterns, and they are also displayed in Figure 2.19. Similarly, θ can be maintained fixed $(0 \leq \theta_i \leq \pi)$ while ϕ is varied $(0 \leq \phi \leq 2\pi)$. These are designated as azimuthal patterns. Part $(0 \leq \phi \leq \pi/2)$ of the $\theta_i = \pi/2$ azimuthal pattern is displayed in Figure 2.19.

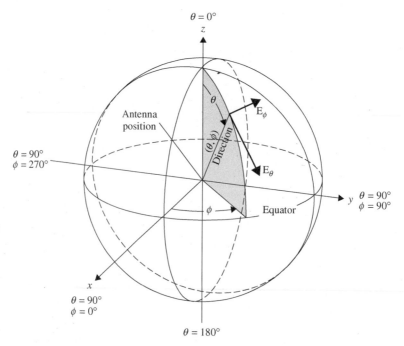

Figure 17.16 Spherical coordinate system geometry. (SOURCE: *IEEE Standard Test Procedures for Antennas*, IEEE Std 149–1979, published by IEEE, Inc., 1979, distributed by Wiley).

The patterns of an antenna can be measured in the transmitting or receiving mode. The mode is dictated by the application. However, if the radiator is reciprocal, as is the case for most practical antennas, then either the transmitting or receiving mode can be utilized. For such cases, the receiving mode is selected. The analytical formulations upon which an amplitude pattern is based, along with the advantages and disadvantages for making measurements in the transmitting or receiving mode, are found in Section 3.8.2. The analytical basis of a phase pattern is discussed in Section 13.10. Unless otherwise specified, it will be assumed here that the measurements are performed in the receiving mode.

17.3.1 Instrumentation

The instrumentation required to accomplish a measuring task depends largely on the functional requirements of the design. An antenna-range instrumentation must be designed to operate over a wide range of frequencies, and it usually can be classified into five categories [7]:

1. source antenna and transmitting system
2. receiving system
3. positioning system
4. recording system
5. data-processing system

A block diagram of a system that possesses these capabilities is shown in Figure 17.17.

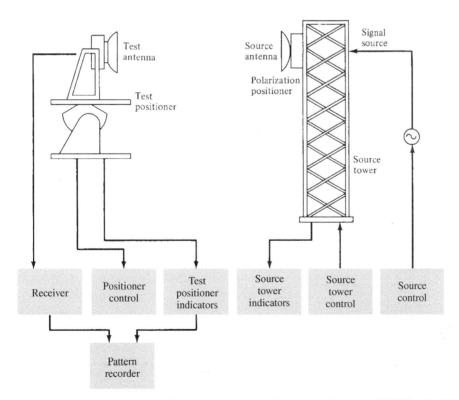

Figure 17.17 Instrumentation for typical antenna-range measuring system. (SOURCE: *IEEE Standard Test Procedures for Antennas*, IEEE Std 149–1979, published by IEEE, Inc., 1979, distributed by Wiley).

The source antennas are usually log-periodic antennas for frequencies below 1 GHz, families of parabolas with broadband feeds for frequencies above 400 MHz, and even large horn antennas. The system must be capable of controlling the polarization. Continuous rotation of the polarization can be accomplished by mounting a linearly polarized source antenna on a polarization positioner. Antennas with circular polarization can also be designed, such as crossed log-periodic arrays, which are often used in measurements.

The transmitting RF source must be selected so that it has [7] frequency control, frequency stability, spectral purity, power level, and modulation. The receiving system could be as simple as a bolometer detector, followed possibly by an amplifier, and a recorder. More elaborate and expensive receiving systems that provide greater sensitivity, precision, and dynamic range can be designed. One such system is a heterodyne receiving system [7], which uses double conversion and phase locking, which can be used for amplitude measurements. A dual-channel heterodyne system design is also available [7], and it can be used for phase measurements.

To achieve the desired plane cuts, the mounting structures of the system must have the capability to rotate in various planes. This can be accomplished by utilizing rotational mounts (pedestals), two of which are shown in Figure 17.18. Tower-model elevation-over-azimuth pedestals are also available [7].

There are primarily two types of recorders; one that provides a linear (rectangular) plot and the other a polar plot. The polar plots are most popular because they provide a better visualization of the radiation distribution in space. Usually the recording equipment is designed to graph the relative pattern. Absolute patterns are obtained by making, in addition, gain measurements which will be discussed in the next section. The recording instrumentation is usually calibrated to record relative field or power patterns. Power pattern calibrations are in decibels with dynamic ranges of 0–60 dB. For most applications, a 40-dB dynamic range is usually adequate and it provides sufficient resolution to examine the pattern structure of the main lobe and the minor lobes.

In an indoor antenna range, the recording equipment is usually placed in a room that adjoins the anechoic chamber. To provide an interference free environment, the chamber is closed during

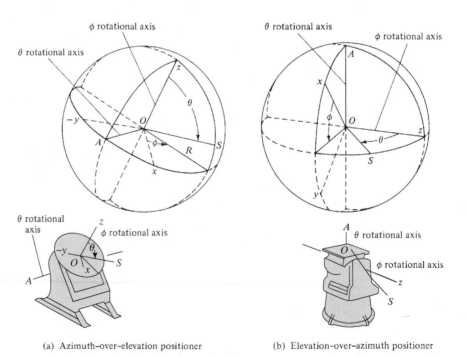

(a) Azimuth–over–elevation positioner   (b) Elevation–over–azimuth positioner

Figure 17.18 Azimuth-over-elevation and elevation-over-azimuth rotational mounts. (SOURCE: *IEEE Standard Test Procedures for Antennas*, IEEE Std 149–1979, published by IEEE, Inc., 1979, distributed by Wiley).

measurements. To monitor the procedures, windows or closed-circuit TVs are utilized. In addition, the recording equipment is connected, through synchronous servo-amplifier systems, to the rotational mounts (pedestals) using the traditional system shown in Figure 17.19(a). The system can record rectangular or polar plots. Position references are recorded simultaneously with measurements, and they are used for angular positional identification. As the rotational mount moves, the pattern is graphed simultaneously by the recorder on a moving chart. One of the axes of the chart is used to record the amplitude of the pattern while the other identifies the relative position of the radiator. A modern configuration to measure antenna and RCS patterns, using a network analyzer and being computer automated, is shown in Figure 17.19(b).

17.3.2 Amplitude Pattern

The total amplitude pattern of an antenna is described by the vector sum of the two orthogonally polarized radiated field components. The pattern on a conventional antenna range can be measured using the system of Figure 17.17 or Figure 17.19 with an appropriate detector. The receiver may be a simple bolometer (followed possibly by an amplifier), a double-conversion phase-locking heterodyne system [7, Fig. 14], or any other design.

In many applications, the movement of the antenna to the antenna range can significantly alter the operational environment. Therefore, in some cases, antenna pattern measurements must be made *in situ* to preserve the environmental performance characteristics. A typical system arrangement that can be used to accomplish this is shown in Figure 17.20. The source is mounted on an airborne vehicle, which is maneuvered through space around the test antenna and in its far field, to produce a plane wave and to provide the desired pattern cuts. The tracking device provides to the recording equipment the angular position data of the source relative to a reference direction. The measurements can be conducted either by a point-by-point or by a continuous method. Usually the continuous technique is preferred.

17.3.3 Phase Measurements

Phase measurements are based on the analytical formulations of Section 13.10. The phase pattern of the field, in the direction of the unit vector $\hat{\mathbf{u}}$, is given by the $\psi(\theta, \phi)$ phase function of (13-63). For linear polarization, $\hat{\mathbf{u}}$ is real, and it may represent $\hat{\mathbf{a}}_\theta$ or $\hat{\mathbf{a}}_\phi$ in the direction of θ or ϕ.

The phase of an antenna is periodic, and it is defined in multiples of 360°. In addition, the phase is a relative quantity, and a reference must be provided during measurements for comparison.

Two basic system techniques that can be used to measure phase patterns at short and long distances from the antenna are shown respectively, in Figures 17.21(a) and 17.21(b). For the design of Figure 17.21(a), a reference signal is coupled from the transmission line, and it is used to compare, in an appropriate network, the phase of the received signal. For large distances, this method does not permit a direct comparison between the reference and the received signal. In these cases, the arrangement of Figure 17.21(b) can be used in which the signal from the source antenna is received simultaneously by a fixed antenna and the antenna under test. The phase pattern is recorded as the antenna under test is rotated while the fixed antenna serves as a reference. The phase measuring circuit may be the dual-channel heterodyne system [7, Figure 15].

17.4 GAIN MEASUREMENTS

The most important figure of merit that describes the performance of a radiator is the gain. There are various techniques and antenna ranges that are used to measure the gain. The choice of either depends largely on the frequency of operation.

Usually free-space ranges are used to measure the gain above 1 GHz. In addition, microwave techniques, which utilize waveguide components, can be used. At lower frequencies, it is more

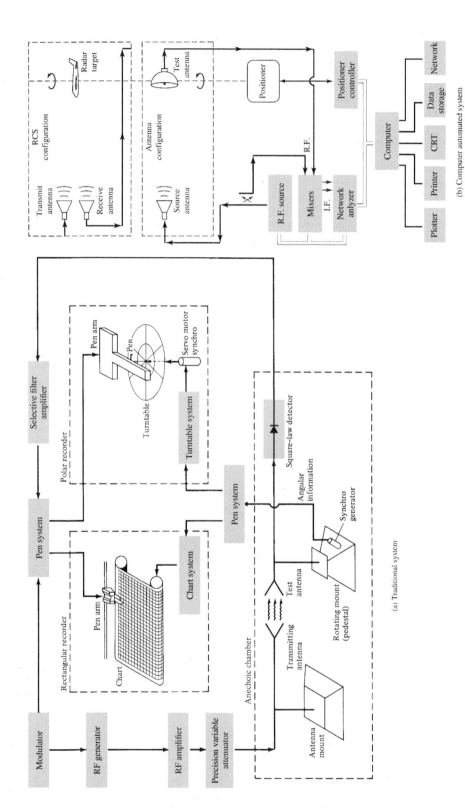

Figure 17.19 Block diagrams of typical instrumentations for measuring rectangular and polar antenna and RCS patterns.

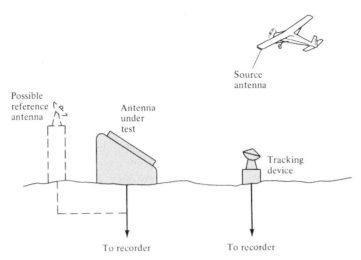

Figure 17.20 System arrangement for *in situ* antenna pattern measurements. (SOURCE: *IEEE Standard Test Procedures for Antennas*, IEEE Std 149–1979, published by IEEE, Inc., 1979, distributed by Wiley).

difficult to simulate free-space conditions because of the longer wavelengths. Therefore between 0.1–1 GHz, ground-reflection ranges are utilized. Scale models can also be used in this frequency range. However, since the conductivity and loss factors of the structures cannot be scaled conveniently, the efficiency of the full scale model must be found by other methods to determine the gain of the antenna. This is accomplished by multiplying the directivity by the efficiency to result in the gain. Below 0.1 GHz, directive antennas are physically large and the ground effects become increasingly pronounced. Usually the gain at these frequencies is measured *in situ*. Antenna gains are not usually measured at frequencies below 1 MHz. Instead, measurements are conducted on the field strength of the ground wave radiated by the antenna.

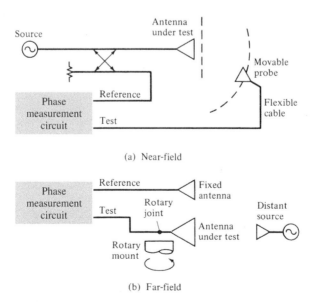

Figure 17.21 Near-field and far-field phase pattern measuring systems. (SOURCE: *IEEE Standard Test Procedures for Antennas*, IEEE Std 149–1979, published by IEEE, Inc., 1979, distributed by Wiley).

Usually there are two basic methods that can be used to measure the gain of an electromagnetic radiator: *realized-gain* and *gain-transfer* (or *gain-comparison*) measurements. The realized-gain method is used to calibrate antennas that can then be used as standards for gain measurements, and it requires no *a priori* knowledge of the gains of the antennas. Gain-transfer methods must be used in conjunction with standard gain antennas to determine the realized gain of the antenna under test.

The two antennas that are most widely used and universally accepted as gain standards are the resonant $\lambda/2$ dipole (with a gain of about 2.1 dB) and the pyramidal horn antenna (with a gain ranging from 12–25 dB). Both antennas possess linear polarizations. The dipole, in free-space, exhibits a high degree of polarization purity. However, because of its broad pattern, its polarization may be suspect in other than reflection-free environments. Pyramidal horns usually possess, in free-space, slightly elliptical polarization (axial ratio of about 40 to infinite dB). However, because of their very directive patterns, they are less affected by the surrounding environment.

17.4.1 Realized-Gain Measurements

There are a number of techniques that can be employed to make realized-gain measurements. A very brief review of each will be included here. More details can be found in [6]–[8]. All of these methods are based on the Friis transmission formula [as given by (2-118)] which assumes that the measuring system employs, each time, two antennas (as shown in Figure 2.31). The antennas are separated by a distance R, and it must satisfy the far-field criterion of each antenna. For polarization matched antennas, aligned for maximum directional radiation, (2-118) reduces to (2-119).

A. Two-Antenna Method

Equation (2-119) can be written in a logarithmic decibel form as

$$(G_{0t})_{\text{dB}} + (G_{0r})_{\text{dB}} = 20\log_{10}\left(\frac{4\pi R}{\lambda}\right) + 10\log_{10}\left(\frac{P_r}{P_t}\right) \qquad (17\text{-}14)$$

where

$(G_{0t})_{\text{dB}}$ = gain of the transmitting antenna (dB)
$(G_{0r})_{\text{dB}}$ = gain of the receiving antenna (dB)
P_r = received power (W)
P_t = transmitted power (W)
R = antenna separation (m)
λ = operating wavelength (m)

If the transmitting and receiving antennas are identical ($G_{0t} = G_{0r}$), (17-14) reduces to

$$(G_{0t})_{\text{dB}} = (G_{0r})_{\text{dB}} = \frac{1}{2}\left[20\log_{10}\left(\frac{4\pi R}{\lambda}\right) + 10\log_{10}\left(\frac{P_r}{P_t}\right)\right] \qquad (17\text{-}15)$$

By measuring R, λ, and the ratio of P_r/P_t, the gain of the antenna can be found. At a given frequency, this can be accomplished using the system of Figure 17.22(a). The system is simple and the procedure straightforward. For continuous multifrequency measurements, such as for broadband antennas, the swept-frequency instrumentation of Figure 17.22(b) can be utilized.

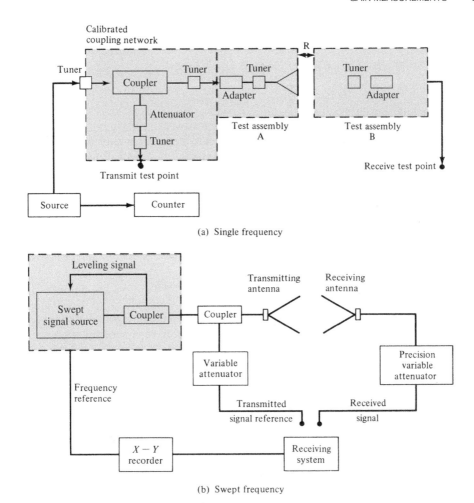

Figure 17.22 Typical two- and three-antenna measuring systems for single- and swept-frequency measurements. (SOURCE: J. S. Hollis, T. J. Lyon, and L. Clayton, Jr., *Microwave Antenna Measurements*, Scientific-Atlanta, Inc., Atlanta, Georgia, July 1970).

B. Three-Antenna Method

If the two antennas in the measuring system are not identical, three antennas (a, b, c) must be employed and three measurements must be made (using all combinations of the three) to determine the gain of each of the three. Three equations (one for each combination) can be written, and each takes the form of (17-14). Thus

(a-b Combination)

$$(G_a)_{\text{dB}} + (G_b)_{\text{dB}} = 20 \log_{10}\left(\frac{4\pi R}{\lambda}\right) + 10 \log_{10}\left(\frac{P_{rb}}{P_{ta}}\right) \qquad (17\text{-}16a)$$

(a-c Combination)

$$(G_a)_{\text{dB}} + (G_c)_{\text{dB}} = 20 \log_{10}\left(\frac{4\pi R}{\lambda}\right) + 10 \log_{10}\left(\frac{P_{rc}}{P_{ta}}\right) \qquad (17\text{-}16b)$$

(b-c Combination)

$$(G_b)_{\text{dB}} + (G_c)_{\text{dB}} = 20\log_{10}\left(\frac{4\pi R}{\lambda}\right) + 10\log_{10}\left(\frac{P_{rc}}{P_{tb}}\right) \qquad (17\text{-}16\text{c})$$

From these three equations, the gains $(G_a)_{\text{dB}}, (G_b)_{\text{dB}}$, and $(G_c)_{\text{dB}}$ can be determined provided R, λ, and the ratios of $P_{rb}/P_{ta}, P_{rc}/P_{ta}$, and P_{rc}/P_{tb} are measured.

The two- and three-antenna methods are both subject to errors. Care must be utilized so

1. the system is frequency stable
2. the antennas meet the far-field criteria
3. the antennas are aligned for boresight radiation
4. all the components are impedance and polarization matched
5. there is a minimum of proximity effects and multipath interference

Impedance and polarization errors can be accounted for by measuring the appropriate complex reflection coefficients and polarizations and then correcting accordingly the measured power ratios. The details for these corrections can be found in [7], [8]. There are no rigorous methods to account for proximity effects and multipath interference. These, however, can be minimized by maintaining the antenna separation by at least a distance of $2D^2/\lambda$, as is required by the far-field criteria, and by utilizing RF absorbers to reduce unwanted reflections. The interference pattern that is created by the multiple reflections from the antennas themselves, especially at small separations, is more difficult to remove. It usually manifests itself as a cyclic variation in the measured antenna gain as a function of separation.

C. Extrapolation Method

The extrapolation method is an realized-gain method, which can be used with the three-antenna method, and it was developed [15] to rigorously account for possible errors due to proximity, multipath, and nonidentical antennas. If none of the antennas used in the measurements are circularly polarized, the method yields the gains and polarizations of all three antennas. If only one antenna is circularly polarized, this method yields only the gain and polarization of the circularly polarized antenna. The method fails if two or more antennas are circularly polarized.

The method requires both amplitude and phase measurements when the gain and the polarization of the antennas are to be determined. For the determination of gains, amplitude measurements are sufficient. The details of this method can be found in [8], [45].

D. Ground-Reflection Range Method

A method that can be used to measure the gain of moderately broad-beam antennas, usually for frequencies below 1 GHz, has been reported [46]. The method takes into account the specular reflections from the ground (using the system geometry of Figure 17.2), and it can be used with some restrictions and modifications with the two- or three-antenna methods. As described here, the method is applicable to linear antennas that couple only the electric field. Modifications must be made for loop radiators. Using this method, it is recommended that the linear vertical radiators be placed in a horizontal position when measurements are made. This is desired because the reflection coefficient of the earth, as a function of incidence angle, varies very rapidly for vertically polarized waves. Smoother variations are exhibited for horizontally polarized fields. Circularly and elliptically polarized antennas are excluded, because the earth exhibits different reflective properties for vertical and horizontal fields.

To make measurements using this technique, the system geometry of Figure 17.2 is utilized. Usually it is desirable that the height of the receiving antenna h_r be much smaller than the range

R_0 ($h_r \ll R_0$). Also the height of the transmitting antenna is adjusted so that the field of the receiving antenna occurs at the first maximum nearest to the ground. Doing this, each of the gain equations of the two- or three-antenna methods take the form of

$$(G_a)_{\text{dB}} + (G_b)_{\text{dB}} = 20 \log_{10}\left(\frac{4\pi R_D}{\lambda}\right) + 10 \log_{10}\left(\frac{P_r}{P_t}\right)$$

$$- 20 \log_{10}\left(\sqrt{D_A D_B} + \frac{rR_D}{R_R}\right) \quad (17\text{-}17)$$

where D_A and D_B are the directivities (relative to their respective maximum values) along R_D, and they can be determined from amplitude patterns measured prior to the gain measurements. R_D, R_R, λ, and P_r/P_t are also measured. The only quantity that needs to be determined is the factor r which is a function of the radiation patterns of the antennas, the frequency of operation, and the electrical and geometrical properties of the antenna range.

The factor r can be found by first repeating the above measurements but with the transmitting antenna height adjusted so that the field at the receiving antenna is minimized. The quantities measured with this geometry are designated by the same letters as before but with a prime (′) to distinguish them from those of the previous measurement.

By measuring or determining the parameters

1. R_R, R_D, P_r, D_A, and D_B at a height of the transmitting antenna such that the receiving antenna is at the first maximum of the pattern
2. R'_R, R'_D, P'_r, D'_A, and D'_B at a height of the transmitting antenna such that the receiving antenna is at a field minimum

it can be shown [46] that r can be determined from

$$r = \left(\frac{R_R R'_R}{R_D R'_D}\right)\left[\frac{\sqrt{(P_r/P'_r)(D'_A D'_B)}R_D - \sqrt{D_A D_B} R'_D}{\sqrt{(P_r/P'_r)}R_R + R'_R}\right] \quad (17\text{-}18)$$

Now all parameters included in (17-17) can either be measured or computed from measurements. The free-space range system of Figure 17.22(a) can be used to perform these measurements.

17.4.2 Gain-Transfer (Gain-Comparison) Measurements

The method most commonly used to measure the gain of an antenna is the gain-transfer method. This technique utilizes a gain standard (with a known gain) to determine realized gains. Initially relative gain measurements are performed, which when compared with the known gain of the standard antenna, yield realized values. The method can be used with free-space and reflection ranges, and for *in situ* measurements.

The procedure requires two sets of measurements. In one set, using the test antenna as the receiving antenna, the received power (P_T) into a matched load is recorded. In the other set, the test antenna is replaced by the standard gain antenna and the received power (P_S) into a matched load is recorded. In both sets, the geometrical arrangement is maintained intact (other than replacing the receiving antennas), and the input power is maintained the same.

Writing two equations of the form of (17-14) or (17-17), for free-space or reflection ranges, it can be shown that they reduce to [7]

$$(G_T)_{\text{dB}} = (G_S)_{\text{dB}} + 10 \log_{10}\left(\frac{P_T}{P_S}\right) \qquad (17\text{-}19)$$

where $(G_T)_{\text{dB}}$ and $(G_S)_{\text{dB}}$ are the gains (in dB) of the test and standard gain antennas.

System disturbance during replacement of the receiving antennas can be minimized by mounting the two receiving antennas back-to-back on either side of the axis of an azimuth positioner and connecting both of them to the load through a common switch. One antenna can replace the other by a simple, but very precise, 180° rotation of the positioner. Connection to the load can be interchanged by proper movement of the switch.

If the test antenna is not too dissimilar from the standard gain antenna, this method is less affected by proximity effects and multipath interference. Impedance and polarization mismatches can be corrected by making proper complex reflection coefficient and polarization measurements [8].

If the test antenna is circularly or elliptically polarized, gain measurements using the gain-transfer method can be accomplished by at least two different methods. One way would be to design a standard gain antenna that possesses circular or elliptical polarization. This approach would be attractive in mass productions of power-gain measurements of circularly or elliptically polarized antennas.

The other approach would be to measure the gain with two orthogonal linearly polarized standard gain antennas. Since circularly and elliptically polarized waves can be decomposed to linear (vertical and horizontal) components, the total power of the wave can be separated into two orthogonal linearly polarized components. Thus the total gain of the circularly or elliptically polarized test antenna can be written as

$$(G_T)_{\text{dB}} = 10 \log_{10}(G_{TV} + G_{TH}) \qquad (17\text{-}20)$$

G_{TV} and G_{TH} are, respectively, the partial power gains with respect to vertical-linear and horizontal-linear polarizations.

G_{TV} is obtained, using (17-19), by performing a gain-transfer measurement with the standard gain antenna possessing vertical polarization. The measurements are repeated with the standard gain antenna oriented for horizontal polarization. This allows the determination of G_{TH}. Usually a single linearly polarized standard gain antenna (a linear $\lambda/2$ resonant dipole or a pyramidal horn) can be used, by rotating it by 90°, to provide both vertical and horizontal polarizations. This approach is very convenient, especially if the antenna possesses good polarization purity in the two orthogonal planes.

The techniques outlined above yield good results provided the transmitting and standard gain antennas exhibit good linear polarization purity. Errors will be introduced if either one of them possesses a polarization with a finite axial ratio. In addition, these techniques are accurate if the tests can be performed in a free-space, a ground-reflection, or an extrapolation range. These requirements place a low-frequency limit of 50 MHz.

Below 50 MHz, the ground has a large effect on the radiation characteristics of the antenna, and it must be taken into account. It usually requires that the measurements are performed on full scale models and *in situ*. Techniques that can be used to measure the gain of large HF antennas have been devised [47]–[49].

17.5 DIRECTIVITY MEASUREMENTS

If the directivity of the antenna cannot be found using solely analytical techniques, it can be computed using measurements of its radiation pattern. One of the methods is based on the approximate expressions of (2-27) by Kraus or (2-30b) by Tai and Pereira, whereas the other relies on the numerical techniques that were developed in Section 2.7. The computations can be performed very efficiently and economically with modern computational facilities and numerical techniques.

The simplest, but least accurate, method requires that the following procedure is adopted:

1. Measure the two principal E- and H-plane patterns of the test antenna.
2. Determine the half-power beamwidths (in degrees) of the E- and H-plane patterns.
3. Compute the directivity using either (2-27) or (2-30b).

The method is usually employed to obtain rough estimates of directivity. It is more accurate when the pattern exhibits only one major lobe, and its minor lobes are negligible.

The other method requires that the directivity be computed using (2-35) where P_{rad} is evaluated numerically using (2-43). The $F(\theta_i, \phi_j)$ function represents the radiation intensity or radiation pattern, as defined by (2-42), and it will be obtained by measurements. U_{max} in (2-35) represents the maximum radiation intensity of $F(\theta, \phi)$ in all space, as obtained by the measurements.

The radiation pattern is measured by sampling the field over a sphere of radius r. The pattern is measured in two-dimensional plane cuts with ϕ_j constant ($0 \leq \phi_j \leq 2\pi$) and θ variable ($0 \leq \theta \leq \pi$), as shown in Figure 2.19, or with θ_i fixed ($0 \leq \theta_i \leq \pi$) and ϕ variable ($0 \leq \phi \leq 2\pi$). The first are referred to as elevation or great-circle cuts, whereas the second represent azimuthal or conical cuts. Either measuring method can be used. Equation (2-43) is written in a form that is most convenient for elevation or great-circle cuts. However, it can be rewritten to accommodate azimuthal or conical cuts.

The spacing between measuring points is largely controlled by the directive properties of the antenna and the desired accuracy. The method is most accurate for broad-beam antennas. However, with the computer facilities and the numerical methods now available, this method is very attractive even for highly directional antennas. To maintain a given accuracy, the number of sampling points must increase as the pattern becomes more directional. The pattern data is recorded digitally on tape, and it can be entered into a computer at a later time. If on-line computer facilities are available, the measurements can be automated to provide essentially real-time computations.

The above discussion assumes that all the radiated power is contained in a single polarization, and the measuring probe possesses that polarization. If the antenna is polarized such that the field is represented by both θ and ϕ components, the *partial directivities* $D_\theta(\theta, \phi)$ and $D_\phi(\theta, \phi)$ of (2-17)–(2-17b) must each be found. This is accomplished from pattern measurements with the probe positioned, respectively, to sample the θ and ϕ components. The *total directivity* is then given by (2-17)–(2-17b), or

$$D_0 = D_\theta + D_\phi \tag{17-21}$$

where

$$D_\theta = \frac{4\pi U_\theta}{(P_{rad})_\theta + (P_{rad})_\phi} \tag{17-21a}$$

$$D_\phi = \frac{4\pi U_\phi}{(P_{rad})_\theta + (P_{rad})_\phi} \tag{17-21b}$$

$U_\theta, (P_{rad})_\theta$ and $U_\phi, (P_{rad})_\phi$ represent the radiation intensity and radiated power as contained in the two orthogonal θ and ϕ field components, respectively.

The same technique can be used to measure the field intensity and to compute the directivity of any antenna that possesses two orthogonal polarizations. Many antennas have only one polarization (θ or ϕ). This is usually accomplished by design and/or proper selection of the coordinate system. In this case, the desired polarization is defined as the *primary polarization*. Ideally, the other polarization should be zero. However, in practice, it is nonvanishing, but it is very small. Usually it is referred to as the *cross polarization*, and for good designs it is usually below −40 dB.

The directivity of circularly or elliptically polarized antennas can also be measured. Precautions must be taken [7] as to which component represents the primary polarization and which the cross-polarization contribution.

17.6 RADIATION EFFICIENCY

The radiation efficiency is defined as the ratio of the total power radiated by the antenna to the total power accepted by the antenna at its input terminals during radiation. System factors, such as impedance and/or polarization mismatches, do not contribute to the radiation efficiency because it is an inherent property of the antenna.

The radiation efficiency can also be defined, using the direction of maximum radiation as reference, as

$$\text{radiation efficiency} = \frac{\text{gain}}{\text{directivity}} \quad (17\text{-}22)$$

where directivity and gain, as defined in Sections 2.6 and 2.9, imply that they are measured or computed in the direction of maximum radiation. Using techniques that were outlined in Sections 17.4 and 17.5 for the measurements of the gain and directivity, the radiation efficiency can then be computed using (17-22).

If the antenna is very small and simple, it can be represented as a series network as shown in Figures 2.27(b) or 2.28(b). For antennas that can be represented by such a series network, the radiation efficiency can also be defined by (2-90) and it can be computed by another method [50]. For these antennas, the real part of the input impedance is equal to the total antenna resistance which consists of the radiation resistance and the loss resistance.

The radiation resistance accounts for the radiated power. For many simple antennas (dipoles, loops, etc.), it can be found by analytically or numerically integrating the pattern, relating it to the radiated power and to the radiation resistance by a relation similar to (4-18). The loss resistance accounts for the dissipated power, and it is found by measuring the input impedance (input resistance − radiation resistance = loss resistance).

Because the loss resistance of antennas coated with lossy dielectrics or antennas over lossy ground cannot be represented in series with the radiation resistance, this method cannot be used to determine their radiation efficiency. The details of this method can be found in [50].

17.7 IMPEDANCE MEASUREMENTS

Associated with an antenna, there are two types of impedances: a *self* and a *mutual* impedance. When the antenna is radiating into an unbounded medium and there is no coupling between it and other antennas or surrounding obstacles, the self-impedance is also the driving-point impedance of the antenna. If there is coupling between the antenna under test and other sources or obstacles, the driving-point impedance is a function of its self-impedance and the mutual impedances between it and the other sources or obstacles. In practice, the driving-point impedance is usually referred to as the input impedance. The definitions and the analytical formulations that underlie the self, mutual, and input impedances are presented in Chapter 8.

To attain maximum power transfer between a source or a source-transmission line and an antenna (or between an antenna and a receiver or transmission line-receiver), a conjugate match is usually desired. In some applications, this may not be the most ideal match. For example, in some receiving systems, minimum noise is attained if the antenna impedance is lower than the load impedance.

However, in some transmitting systems, maximum power transfer is attained if the antenna impedance is greater than the load impedance. If conjugate matching does not exist, the power lost can be computed [7] using

$$\frac{P_{\text{lost}}}{P_{\text{available}}} = \left|\frac{Z_{\text{ant}} - Z_{\text{cct}}^*}{Z_{\text{ant}} + Z_{\text{cct}}}\right|^2 \qquad (17\text{-}23)$$

where

Z_{ant} = input impedance of the antenna

Z_{cct} = input impedance of the circuits which are connected to the antenna at its input terminals

When a transmission line is associated with the system, as is usually the case, the matching can be performed at either end of the line. In practice, however, the matching is performed near the antenna terminals, because it usually minimizes line losses and voltage peaks in the line and maximizes the useful bandwidth of the system.

In a mismatched system, the degree of mismatch determines the amount of incident or available power which is reflected at the input antenna terminals into the line. The degree of mismatch is a function of the antenna input impedance and the characteristic impedance of the line. These are related to the input reflection coefficient and the input VSWR at the antenna input terminals by the standard transmission-line relationships of

$$\frac{P_{\text{refl}}}{P_{\text{inc}}} = |\Gamma|^2 = \frac{|Z_{\text{ant}} - Z_c|^2}{|Z_{\text{ant}} + Z_c|^2} = \left|\frac{\text{VSWR} - 1}{\text{VSWR} + 1}\right|^2 \qquad (17\text{-}24)$$

where

$\Gamma = |\Gamma|e^{j\gamma}$ = voltage reflection coefficient at the antenna input terminals

VSWR = voltage standing wave ratio at the antenna input terminals

Z_c = characteristic impedance of the transmission line

Equation (17-24) shows a direct relationship between the antenna input impedance (Z_{ant}) and the VSWR. In fact, if Z_{ant} is known, the VSWR can be computed using (17-24). In practice, however, that is not the case. What is usually measured is the VSWR, and it alone does not provide sufficient information to uniquely determine the complex input impedance. To overcome this, the usual procedure is to measure the VSWR, and to compute the magnitude of the reflection coefficient using (17-24). The phase of the reflection coefficient can be determined by locating a voltage maximum or a voltage minimum (from the antenna input terminals) in the transmission line. Since in practice the minima can be measured more accurately than the maxima, they are usually preferred. In addition, the first minimum is usually chosen unless the distance from it to the input terminals is too small to measure accurately. The phase γ of the reflection coefficient is then computed using [51]

$$\gamma = 2\beta x_n \pm (2n-1)\pi = \frac{4\pi}{\lambda_g} x_n \pm (2n-1)\pi, \quad n = 1, 2, 3, \ldots \qquad (17\text{-}25)$$

where

n = the voltage minimum from the input terminals (i.e., $n = 1$ is used to locate the first voltage minimum)

x_n = distance from the input terminals to the nth voltage minimum

λ_g = wavelength measured inside the input transmission line (it is twice the distance between two voltage minima or two voltage maxima)

Once the reflection coefficient is completely described by its magnitude and phase, it can be used to determine the antenna impedance by

$$Z_{ant} = Z_c \left[\frac{1+\Gamma}{1-\Gamma}\right] = Z_c \left[\frac{1+|\Gamma|e^{j\gamma}}{1-|\Gamma|e^{j\gamma}}\right] \tag{17-26}$$

Other methods, utilizing impedance bridges, slotted lines, and broadband swept-frequency network analyzers, can be utilized to determine the antenna impedance [51]–[53].

The input impedance is generally a function of frequency, geometry, method of excitation, and proximity to its surrounding objects. Because of its strong dependence on the environment, it should usually be measured *in situ* unless the antenna possesses very narrow beam characteristics.

Mutual impedances, which take into account interaction effects, are usually best described and measured by the cross-coupling coefficients S_{mn} of the device's (antenna's) scattering matrix. The coefficients of the scattering matrix can then be related to the coefficients of the impedance matrix [54].

17.8 CURRENT MEASUREMENTS

The current distribution along an antenna is another very important antenna parameter. A complete description of its amplitude and phase permit the calculation of the radiation pattern.

There are a number of techniques that can be used to measure the current distribution [55]–[58]. One of the simplest methods requires that a small sampling probe, usually a small loop, be placed near the radiator. On the sampling probe, a current is induced which is proportional to the current of the test antenna.

The indicating meter can be connected to the loop in many different ways [55]. If the wavelength is very long, the meter can be consolidated into one unit with the measuring loop. At smaller wavelengths, the meter can be connected to a crystal rectifier. In order not to disturb the field distribution near the radiator, the rectifier is attached to the meter using long leads. To reduce the interaction between the measuring instrumentation and the test antenna and to minimize induced currents on the leads, the wires are wound on a dielectric support rod to form a helical choke. Usually the diameter of each turn, and spacing between them, is about $\lambda/50$. The dielectric rod can also be used as a support for the loop. To prevent a dc short circuit on the crystal rectifier, a bypass capacitor is placed along the circumference of the loop.

There are many other methods, some of them more elaborate and accurate, and the interested reader can refer to the literature [55]–[58].

17.9 POLARIZATION MEASUREMENTS

The polarization of a wave was defined in Section 2.12 as *the curve traced by the instantaneous electric field, at a given frequency, in a plane perpendicular to the direction of wave travel*. The far-field polarization of an antenna is usually measured at distances where the field radiated by the antenna forms, in a small region, a plane wave that propagates in the outward radial direction.

In a similar manner, the polarization of the antenna is defined as *the curve traced by the instantaneous electric field radiated by the antenna in a plane perpendicular to the radial direction*, as shown in Figure 17.23(a). The locus is usually an ellipse. In a spherical coordinate system, which is usually adopted in antennas, the polarization ellipse is formed by the orthogonal electric field components of E_θ and E_ϕ. The sense of rotation, also referred to as the sense of polarization, is defined by the sense of rotation of the wave as it is observed along the direction of propagation [see Figure 17.23(b)].

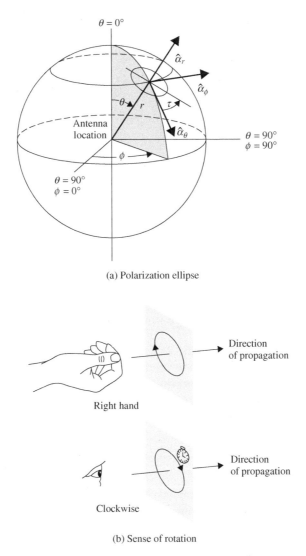

(a) Polarization ellipse

(b) Sense of rotation

Figure 17.23 Polarization ellipse and sense of rotation for antenna coordinate system. (SOURCE: *IEEE Standard Test Procedures for Antennas*, IEEE Std 149–1979, published by IEEE, Inc., 1979, distributed by Wiley).

The general polarization of an antenna is characterized by the axial ratio (AR), the sense of rotation (CW or CCW, RH or LH), and the tilt angle τ. The tilt angle is used to identify the spatial orientation of the ellipse, and it is usually measured clockwise from the reference direction. This is demonstrated in Figure 17.23(a) where τ is measured clockwise with respect to $\hat{\mathbf{a}}_\theta$, for a wave traveling in the outward radial direction.

Care must be exercised in the characterization of the polarization of a receiving antenna. If the tilt angle of an incident wave that is polarization matched to the receiving antenna is τ_m, it is related to the tilt angle τ_t of a wave transmitted by the same antenna by

$$\tau_t = 180° - \tau_m \tag{17-27}$$

if a single coordinate system and one direction of view are used to characterize the polarization. If the receiving antenna has a polarization that is different from that of the incident wave, the polarization loss factor (PLF) of Section 2.12.2 can be used to account for the polarization mismatch losses.

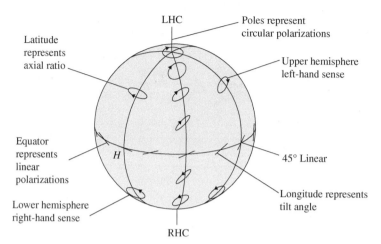

Figure 17.24 Polarization representation on the Poincaré sphere. (SOURCE: W. H. Kummer and E. S. Gillespie, "Antenna Measurements—1978," *Proc. IEEE*, Vol. 66, No. 4, pp. 483–507, April 1978. © (1978) IEEE).

The polarization of a wave and/or an antenna can best be displayed and visualized on the surface of a Poincaré sphere [59]. Each polarization occupies a unique point on the sphere, as shown in Figure 17.24. If one of the two points on the Poincaré sphere is used to define the polarization of the incident wave and the other the polarization of the receiving antenna, the angular separation can be used to determine the polarization losses. The procedure requires that the complex polarization ratios of each are determined, and they are used to compute the polarization efficiency in a number of different ways. The details of this procedure are well documented, and they can be found in [7], [8].

Practically it is very difficult to design radiators that maintain the same polarization state in all parts of their pattern. A complete description requires a number of measurements in all parts of the pattern. The number of measurements is determined by the required degree of polarization description.

There are a number of techniques that can be used to measure the polarization state of a radiator [7], [8], and they can be classified into three main categories:

1. Those that yield partial polarization information. They do not yield a unique point on the Poincaré sphere.
2. Those that yield complete polarization information but require a polarization standard for comparison. They are referred to as *comparison methods*.
3. Those that yield complete polarization information and require no *a priori* polarization knowledge or no polarization standard. They are designated as *absolute methods*.

The method selected depends on such factors as the type of antenna, the required accuracy, and the time and funds available. A complete description requires not only the polarization ellipse (axial ratio and tilt angle), but also its sense of rotation (CW or CCW, RH or LH).

In this text, a method will be discussed which can be used to determine the polarization ellipse (axial ratio and tilt angle) of an antenna but not its sense of rotation. This technique is referred to as the *polarization-pattern method*. The sense of polarization or rotation can be found by performing auxiliary measurements or by using other methods [7].

To perform the measurements, the antenna under test can be used either in the transmitting or in the receiving mode. Usually the transmitting mode is adopted. The method requires that a linearly polarized antenna, usually a dipole, be used to probe the polarization in the plane that contains the direction of the desired polarization. The arrangement is shown in Figure 17.25(a). The dipole is

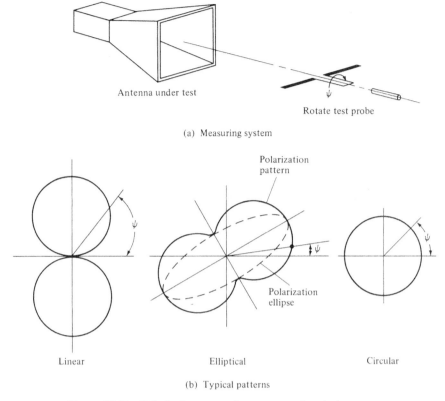

Figure 17.25 Polarization measuring system and typical patterns.

rotated in the plane of the polarization, which is taken to be normal to the direction of the incident field, and the output voltage of the probe is recorded.

If the test antenna is linearly polarized, the output voltage response will be proportional to $\sin\psi$ (which is the far-zone field pattern of an infinitesimal dipole). The pattern forms a figure-eight, as shown in Figure 17.25(b), where ψ is the rotation angle of the probe relative to a reference direction. For an elliptically polarized test antenna, the nulls of the figure-eight are filled and a dumbbell polarization curve (usually referred to as a *polarization pattern*) is generated, as shown in Figure 17.25(b). The dashed curve represents the polarization ellipse.

The polarization ellipse is tangent to the polarization pattern, and it can be used to determine the axial ratio and the tilt angle of the test antenna. The polarization pattern will be a circle, as shown in Figure 17.25(b), if the test antenna is circularly polarized. Ideally, this process must be repeated at every point of the antenna pattern. Usually it is performed at a number of points that describe sufficiently well the polarization of the antenna at the major and the minor lobes.

In some cases the polarization needs to be known over an entire plane. The axial ratio part of the polarization state can be measured using the arrangement of Figure 17.25(a) where the test probe antenna is used usually as a source while the polarization pattern of the test antenna is being recorded while the test antenna is rotated over the desired plane. This arrangement does not yield the tilt angle or sense of rotation of the polarization state. In order to obtain the desired polarization pattern, the rate of rotation of the linear probe antenna (usually a dipole) is much greater than the rotation rate of the positioner over which the test antenna is mounted and rotated to allow, ideally, the probe antenna to measure the polarization response of the test antenna at that direction before moving to another angle. When this is performed over an entire plane, a typical pattern recorded in decibels is shown in Figure 17.26 [60], and it is referred as the *axial ratio pattern*. It is apparent that the axial ratio pattern can be inscribed by inner and outer envelopes. At any given angle, the ratio of the outer and inner

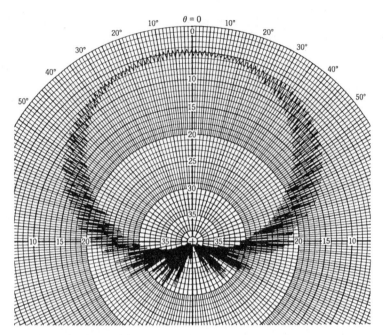

Figure 17.26 Pattern of a circularly polarized test antenna taken with a rotating, linearly polarized, source antenna. (SOURCE: E. S. Gillespie, "Measurement of Antenna Radiation Characteristics on Far-Field Ranges," in *Antenna Handbook* (Y. T. Lo & S. W. Lee, eds.), 1988, © Van Nostrand Reinhold Co., Inc).

envelope responses represent the axial ratio. If the pattern is recorded in decibels, the axial ratio is the difference between the outer and inner envelopes (in dB); zero dB difference represents circular polarization (axial ratio of unity). Therefore the polarization pattern of Figure 17.26 indicates that the test antenna it represents is nearly circularly polarized (within 1 dB; axial ratio less than 1.122) at and near $\theta = 0°$ and deviates from that almost monotonically at greater angles (typically by about 7 dB maximum; maximum axial ratio of about 2.24).

The sense of rotation can be determined by performing auxiliary measurements. One method requires that the responses of two circularly polarized antennas, one responsive to CW and the other to CCW rotation, be compared [55]. The rotation sense of the test antenna corresponds to the sense of polarization of the antenna which produced the more intense response.

Another method would be to use a dual-polarized probe antenna, such as a dual-polarized horn, and to record simultaneously the amplitude polarization pattern and the relative phase between the two orthogonal polarizations. This is referred to as the *phase-amplitude* method, and it can be accomplished using the instrumentation of Figure 17.27. Double-conversion phase-locked receivers can be used to perform the amplitude and phase comparison measurements.

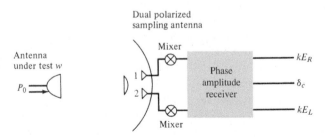

Figure 17.27 System configuration for measurements of polarization amplitude and phase. (SOURCE: W. H. Kummer and E. S. Gillespie, "Antenna Measurements—1978," *Proc. IEEE*, Vol. 66, No. 4, pp. 483–507, April 1978. © (1978) IEEE).

Another absolute polarization method, which can be used to completely describe the polarization of a test antenna, is referred to as the *three-antenna* method [7], [8]. As its name implies, it requires three antennas, two of which must not be circularly polarized. There are a number of transfer methods [7], [8], but they require calibration standards for complete description of the polarization state.

17.10 SCALE MODEL MEASUREMENTS

In many applications (such as with antennas on ships, aircraft, large spacecraft, etc.), the antenna and its supporting structure are so immense in weight and/or size that they cannot be moved or accommodated by the facilities of a measuring antenna range. In addition, a movement of the structure to an antenna range can eliminate or introduce environmental effects. To satisfy these system requirements, *in situ* measurements are usually executed.

A technique that can be used to perform antenna measurements associated with large structures is *geometrical scale modeling*. Geometrical modeling is employed to

1. physically accommodate, within small ranges or enclosures, measurements that can be related to large structures
2. provide experimental control over the measurements
3. minimize costs associated with large structures and corresponding experimental parametric studies

Geometrical scale modeling by a factor of n (n smaller or greater than unity) requires the scaling indicated in Table 17.1. The primed parameters represent the scaled model while the unprimed represent the full scale model. For a geometrical scale model, all the linear dimensions of the antenna and its associated structure are divided by n whereas the operating frequency and the conductivity of the antenna material and its structure are multiplied by n. In practice, the scaling factor n is usually chosen greater than unity.

Ideal scale modeling for antenna measurements requires exact replicas, both physically and electrically, of the full scale structures. In practice, however, this is closely approximated. The most difficult scaling is that of the conductivity. If the full scale model possesses excellent conductors, even better conductors will be required in the scaled models. At microwave and millimeter wave frequencies this can be accomplished by utilizing clean polished surfaces, free of films and other residues.

Geometrical scaling is often used for pattern measurements. However, it can also be employed to measure gain, directivity, radiation efficiency, input and mutual impedances, and so forth. For gain measurements, the inability to properly scale the conductivity can be overcome by measuring the directivity and the antenna efficiency and multiplying the two to determine the gain. Scalings that permit additional parameter changes are available [61]. The changes must satisfy the *theorem of similitude*.

TABLE 17.1 Geometrical Scale Model

Scaled Parameters		Unchanged Parameters	
Length:	$l' = l/n$	Permittivity:	$\varepsilon' = \varepsilon$
Time:	$t' = t/n$	Permeability:	$\mu' = \mu$
Wavelength:	$\lambda' = \lambda/n$	Velocity:	$v' = v$
Capacitance:	$C' = C/n$	Impedance:	$Z' = Z$
Inductance:	$L' = L/n$	Antenna gain:	$G_0' = G_0$
Echo area (RCS):	$A_e' = A_e/n^2$		
Frequency:	$f' = nf$		
Conductivity:	$\sigma' = n\sigma$		

1020 ANTENNA MEASUREMENTS

To illustrate the scaling for *Gain (Amplitude)* and *Echo Area (RCS)*, two examples are used.

17.10.1 Gain (Amplitude) Measurements, Simulations and Comparisons

For the gain, the absolute amplitude radiation patterns of a $\lambda/4$ monopole placed at the belly (bottom side) of a generic scale model helicopter [see Figure 17.28(a)], whose dimensions are about 1/10 the size of a full-scale helicopter. The absolute amplitude patterns of the $\lambda/4$ monopole on the scale model generic helicopter were measured at 7 GHz along the three principal planes; pitch, roll and yaw. In addition, the same patterns were simulated on a full-scale helicopter (10 times larger) but at a frequency of 700 MHz (1/10 the measured frequency) of the same helicopter geometry. The geometry

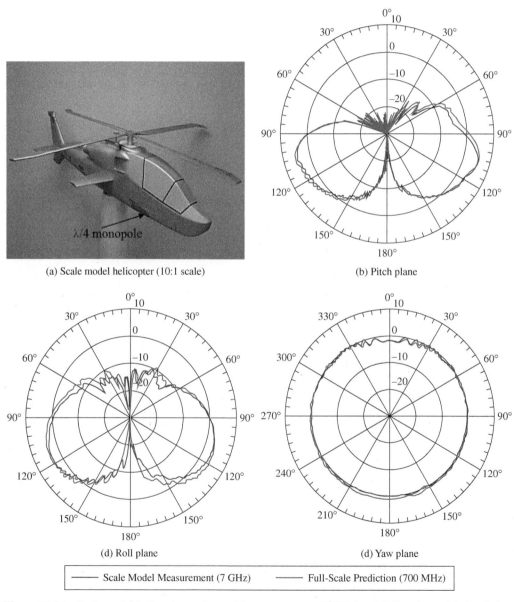

Figure 17.28 Scale model helicopter, scale model measurements (7 GHz) and full-scale model simulations (700 MHz) of a $\lambda/4$ monopole on the belly of airframe.

of the generic scale helicopter, along with the measured and simulated patterns, is indicated in Figure 17.28. It is evident that there is, as expected, a correct scaling between the measured amplitude (gain) patterns on the 1/10 scale model but at a frequency of 7 GHz (increased by a factor of 10 since the size of the scale model was 1/10 of the full-scale) and the simulated patterns at 700 MHz (a factor of 1/10 of the measured) but on a full-scale model (larger by a factor of 10). The maximum gain is about 6 dB, which is basically what is expected from a $\lambda/4$ monopole. In addition, there is an excellent comparison between the respective two sets of patterns (measured and simulated) considering the complexity of the airframe.

17.10.2 Echo Area (RCS) Measurements, Simulations and Comparisons

In addition to the scaled and full-scale models' gain (amplitude) radiation patterns, measured and simulated, the procedure was repeated for scaled and full-scale Echo Area (RCS) monostatic patterns, both measured and simulated. The geometry of the radar target to perform the monostatic RCS measurements and simulations is a square ($a = b$) PEC flat plate, whose configuration and polarization designations are shown in Figure 17.29. The scaling factor used for the RCS was 3:1 for both the measurements and simulations; the frequencies for both the measurements and simulations were 15 GHz (scaled) and 5 GHz (full-scale). The dimensions of the full-scale (large) square plate were 30 cm × 30 cm while for the scaled (small) square plate were 10 cm × 10 cm.

The vertical polarization [Figure 17.29(b)], measured and simulated, monostatic RCS patterns (in dBsm) of the scaled and full-scale RCS patterns along the principal y-z plane are displayed in Figure 17.30(a), while those for the horizontal polarization [Figure 17.29(c)] along the principal y-z plane are displayed in Figure 17.30(b). An excellent agreement is indicated between the measured and simulated patterns, for both the full-scale (large) and scaled (small) plates. In fact, all the curves for the full-scale and for the scaled plates, for both the vertical and horizontal polarizations, are so close to each other that they are indistinguishable from one another. The horizontal (soft) polarization patterns of Figure 14.30(b) follow, as they should, a nearly $\sin(q)/q$ distribution based on physical optics [62] and due to the very weak first-order diffractions from the edges of the plate for this polarization [62]. For the vertical polarization patterns of Figure 17.30(a) the agreement between measurements and simulations is also excellent. However for the vertical polarization the RCS patterns do not follow the $\sin(q)/q$ distribution, especially at the far minor lobes, because the first-order diffractions for this polarization (vertical; hard) are more intense and impact the overall distribution to be different from $\sin(q)/q$.

The maximum monostatic RCS of a flat plate of any geometry, for either vertical or horizontal polarization (both are identical based on physical optics), occurs at normal incidence and it is represented by

$$\text{RCS}_{\max} = 4\pi \left(\frac{\text{Area of plate}}{\lambda} \right)^2 \quad (17\text{-}28)$$

Based on (17-28), the maximum monostatic RCS of the full-scale (large) square plate of 30 cm × 30 cm at 5 GHz is:

$$\text{RCS (full-scale)}_{\max} = 4\pi \left[\frac{30(30)}{6(100)} \right]^2 = 28.274 \text{ (m}^2\text{)} = 14.51 \text{ dBsm} \quad (17\text{-}28a)$$

while the simulated maximum of Figures 17.30(a) and 17.30(b) is 14.6 dBsm [basically the same as that of (17-28a)]. For the scaled (small) square plate, of 10 cm × 10 cm at 15 GHz, the maximum

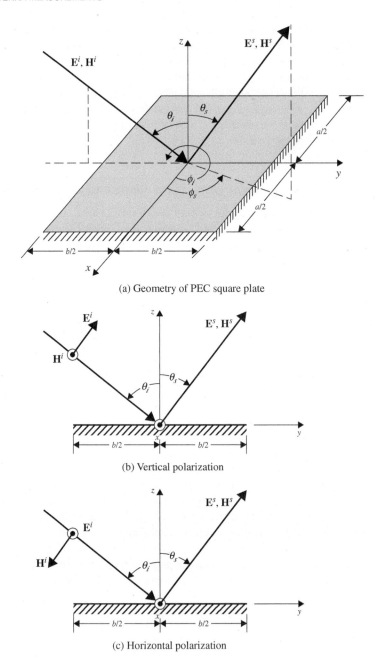

Figure 17.29 Flat PEC plate geometry and polarization designations for RCS (echo area) measurements and simulations [62].

monostatic RCS based on (17-28) is:

$$\text{RCS (scaled)}_{max} = 4\pi \left[\frac{10(10)}{2(100)}\right]^2 = 3.142 \text{ (m}^2\text{)} = 4.97 \text{ dBsm} \qquad (17\text{-}28b)$$

while the simulated maximum of Figures 17.30(a) and 17.30(b) is 5.10 dBsm [again basically the same as that of (17.28b)].

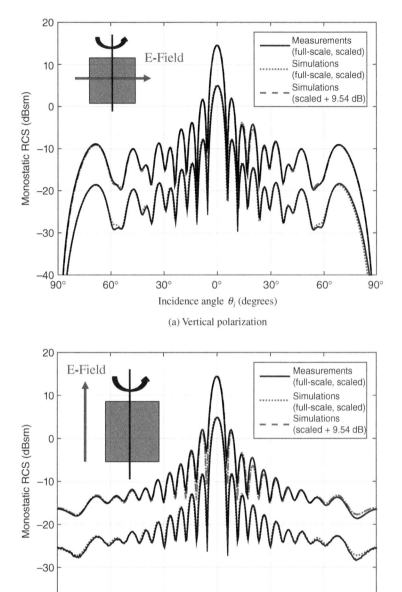

Figure 17.30 Vertical and horizontal polarizations measured and simulated monostatic RCS (echo area) patterns for full-scale (large; 30 cm × 30 cm) and scaled (small; 10 cm × 10 cm) square flat PEC plates of Figure 17.29.

Thus, according to (17-28a) and (17-28b), the RCS difference between the two PEC flat square plates, full-scale (large) and scale (small), is $28.274/3.142 = 9$ (dimensionless) or $14.5 - 4.97 = 9.54$ dB. This confirms the scaling of the echo area (RCS) of Table 7.1 which indicates that the scaling factor between the full-scale and scaled models is n^2, which for this example is $n^2 = 3^2 = 9$ (dimensionless), or $10 \log_{10}(9) = 9.54$ dB.

It is also apparent from the vertical polarization monostatic RCS patterns in Figure 17.30(a) and the horizontal polarization of 17.30(b) that there is a difference of $n^2 = 3^2 = 9$ (dimensionless) or $10 \log_{10}(9) = 9.54$ dB, between the scaled and full-scale (both measured and simulated) RCS patterns; i.e. the full-scale measured and simulated monostatic RCS patterns are 9.54 dB greater than those of the scaled, as they should be according the scaling listed on Table 7.1. In fact, if 9.54 dB is added to the measured and simulated monostatic RCS patterns of the scaled (small) plate monostatic RCS patterns, the adjusted (by +9.54 dB) patterns match those of the full-scale (large) plate, as shown in Figures 17.30(a) and 17.30(b). Again, the agreement is so good that it is difficult to distinguish any differences between any of the patterns for the full-scale and scaled plates. Such comparisons and agreements illustrate and validate the scaling principle for Echo Area (RCS).

It should be pointed out that in both the gain (amplitude) and RCS (echo area), the conductivity was not scaled, but should be by a factor of n [the conductivity of the scale (small) models should be greater by a factor of n compared to that of the full-scale (large) models] to properly account for the scaling according to the listing in Table 17.1. For the measurements, of both the helicopter and radar target, the conductivity was finite but very large (about 5×10^7 S/m) while for the simulations the conductivity of the airframe and plate was assumed infinity. Because the conductivity of the physical targets was very large (on the order of 10^7 S/m), it did not impact, within measuring accuracy, the amplitude (gain) and RCS (echo area) scaling illustrations and validations.

REFERENCES

1. J. Brown and E. V. Jull, "The Prediction of Aerial Patterns from Near-Field Measurements," *IEE (London)*, Paper No. 3469E, pp. 635–644, November 1961.
2. R. C. Johnson, H. A. Ecker, and J. S. Hollis, "Determination of Far-Field Antenna Patterns from Near-Field Measurements," *Proc. IEEE*, Vol. 61, No. 12, pp. 1668–1694, December 1973.
3. D. T. Paris, W. M. Leach, Jr., and E. B. Joy, "Basic Theory of Probe-Compensated Near-Field Measurements," *IEEE Trans. Antennas Propagat.*, Vol. AP-26, No. 3, pp. 373–379, May 1978.
4. E. B. Joy, W. M. Leach, Jr., G. P. Rodrigue, and D. T. Paris, "Applications of Probe-Compensated Near-Field Measurements," *IEEE Trans. Antennas Propagat.*, Vol. AP-26, No. 3, pp. 379–389, May 1978.
5. E. F. Buckley, *"Modern Microwave Absorbers and Applications,"* Emerson & Cuming, Inc., Canton, MA.
6. J. S. Hollis, T. J. Lyon, and L. Clayton, Jr., *Microwave Antenna Measurements*, Scientific-Atlanta, Inc., Atlanta, GA, July 1970.
7. *IEEE Standard Test Procedures for Antennas*, IEEE Std 149-1979, Published by IEEE, Inc., 1979, Distributed by Wiley-Interscience.
8. W. H. Kummer and E. S. Gillespie, "Antenna Measurements—1978," *Proc. IEEE*, Vol. 66, No. 4, pp. 483–507, April 1978.
9. L. H. Hemming and R. A. Heaton, "Antenna Gain Calibration on a Ground Reflection Range," *IEEE Trans. Antennas Propagat.*, Vol. AP-21, No. 4, pp. 532–537, July 1973.
10. P. W. Arnold, "The 'Slant' Antenna Range," *IEEE Trans. Antennas Propagat.*, Vol. AP-14, No. 5, pp. 658–659, September 1966.
11. A. W. Moeller, "The Effect of Ground Reflections on Antenna Test Range Measurements," *Microwave Journal*, Vol. 9, pp. 47–54, March 1966.
12. W. H. Emerson, "Electromagnetic Wave Absorbers and Anechoic Chambers Through the Years," *IEEE Trans. Antennas Propagat.*, Vol. AP-21, No. 4, pp. 484–490, July 1973.
13. M. R. Gillette and P. R. Wu, "RF Anechoic Chamber Design Using Ray Tracing," *1977 Int. IEEE/AP-S Symp. Dig.*, pp. 246–252, June 1977.
14. W. H. Emerson and H. B. Sefton, "An Improved Design for Indoor Ranges," *Proc. IEEE*, Vol. 53, pp. 1079–1081, August 1965.
15. J. R. Mentzer, "The Use of Dielectric Lenses in Reflection Measurements," *Proc. IRE*, Vol. 41, pp. 252–256, February 1953.

16. P. A. Beeckman, "Prediction of the Fresnel Region Field of a Compact Antenna Test Range with Serrated Edges," *IEE Proc.*, Vol. 133, Pt. H, No. 2, pp. 108–114, April 1986.
17. H. F. Schluper, "Compact Antenna Test Range Analysis Using Physical Optics," *AMTA Proceedings*, pp. 309–312, Seattle, WA, October 1987.
18. H. F. Schluper, "Verification Method for the Serration Design of CATR Reflectors," *AMTA Proceedings*, pp. 10-9–10-14, Monterey, CA, October 1989.
19. W. D. Burnside, M. C. Gilreath, and B. Kent, "A Rolled Edge Modification of Compact Range Reflectors," Presented at *AMTA Conf.*, San Diego, CA, October 1984.
20. W. D. Burnside, M. C. Gilreath, B. M. Kent, and G. L. Clerici, "Curved Edge Modification of Compact Range Reflector," *IEEE Trans. Antennas Propagat.*, Vol. AP-35, No. 2, pp. 176–182, February 1987.
21. M. R. Hurst and P. E. Reed, "Hybrid Compact Radar Range Reflector," *AMTA Proceedings*, pp. 8-9–8-13, Monterey, CA, October 1989.
22. J. P. McKay and Y. Rahmat-Samii, "Multi-Ring Planar Array Feeds for Reducing Diffraction Effects in the Compact Range," *AMTA Proceedings*, pp. 7-3–7-8, Columbus, OH, October 1992.
23. J. P. McKay, Y. Rahmat-Samii, and F. M. Espiau, "Implementation Considerations for a Compact Range Array Feed," *AMTA Proceedings*, pp. 4-21–4-26, Columbus, OH, October 1992.
24. J. P. McKay and Y. Rahmat-Samii, "A Compact Range Array Feed: Tolerances and Error Analysis," *1993 Int. IEEE/AP-S Symp. Dig.*, Vol. 3, pp. 1800–1803, June 1993.
25. J. P. McKay, Y. Rahmat-Samii, T. J. De Vicente, and L. U. Brown, "An X-Band Array for Feeding a Compact Range Reflector," *AMTA Proceedings*, pp. 141–146, Dallas, TX, October 1993.
26. H. F. Schluper, J. Van Damme, and V. J. Vokurka, "Optimized Collimators—Theoretical Performance Limits," *AMTA Proceedings*, p. 313, Seattle, WA, October 1987.
27. J. D. Huff, J. H. Cook, Jr., and B. W. Smith, "Recent Developments in Large Compact Range Design," *AMTA Proceedings*, pp. 5-39–5-44, Columbus, OH, October 1992.
28. H. F. Schluper and V. J. Vokurka, "Troubleshooting Limitations in Indoor RCS Measurements," *Microwaves & RF*, pp. 154–163, May 1987.
29. V. J. Vokurka, "Seeing Double Improves Indoor Range," *Microwaves & RF*, pp. 71–76, 94, February 1985.
30. T. Harrison, "A New Approach to Radar Cross-Section Compact Range," *Microwave Journal*, pp. 137–145, June 1986.
31. K. W. Lam and V. J. Vokurka, "Hybrid Near-Field/Far-Field Antenna Measurement Techniques," *AMTA Proceedings*, pp. 9-29–9-34, Boulder, CO, October 1991.
32. C. R. Birtcher, C. A. Balanis, and V. J. Vokurka, "RCS Measurements, Transformations, and Comparisons Under Cylindrical and Plane Wave Illumination," *IEEE Trans. Antennas Propagat.*, Vol. AP-42, No. 3, pp. 329–334, March 1994.
33. C. R. Birtcher, C. A. Balanis, and V. J. Vokurka, "Quiet Zone Scan of the Single-Plane Collimating Range," *AMTA Proceedings*, pp. 4-37–4-42, Boulder, CO, October 1991.
34. A. D. Yaghjian, "An Overview of Near-Field Antenna Measurements," *IEEE Trans. Antennas Propagat.*, Vol. AP-34, pp. 30–45, January 1986.
35. P. M. Morse and H. Feshbach, *Methods of Theoretical Physics*, McGraw-Hill, New York, 1953, Chapter 13.
36. A. V. Oppenheim and R. W. Schaffer, *Digital Signal Processing*, Prentice-Hall, Englewood Cliffs, NJ, 1975, Chapter 3.
37. Y. Rahmat-Samii, V. Galindo, and R. Mittra, "A Plane-Polar Approach for Far-Field Construction from Near-Field Measurements," *IEEE Trans. Antennas Propagat.*, Vol. AP-28, No. 3, pp. 216–230, March 1980.
38. L. I. Williams and Y. Rahmat-Samii, "Novel Bi-Polar Planar Near-Field Measurement Scanner at UCLA," 1991 Int. IEEE/AP-S Symp. Dig., London, Ontario, Canada, June 1991.
39. G. D. Bergland, "A Guided Tour of the Fast Fourier Transform," *IEEE Spectrum*, pp. 41–52, July 1969.
40. W. M. Leach, Jr. and D. T. Paris, "Probe Compensated Near-Field Measurements on a Cylinder," *IEEE Trans. Antennas Propagat.*, Vol. AP-21, No. 4, pp. 435–445, July 1973.
41. C. F. Stubenrauch and A. C. Newell, "Some Recent Near-Field Antenna Measurements at NBS," *Microwave J.*, pp. 37–42, November 1980.

42. J. Lemanczyk and F. H. Larsen, "Comparison of Near-Field Range Results," *IEEE Trans. Antennas Propagat.*, Vol. AP-36, No. 6, pp. 845–851, June 1988.
43. J. J. Lee, E. M. Ferren, D. P. Woollen, and K. M. Lee, "Near-Field Probe Used as a Diagnostic Tool to Locate Defective Elements in an Array Antenna," *IEEE Trans. Antennas Propagat.*, Vol. AP-36, No. 6, pp. 884–889, June 1988.
44. Y. Rahmat-Samii and J. Lemanczyk, "Application of Spherical Near-Field Measurements to Microwave Holographic Diagnosis of Antennas," *IEEE Trans. Antennas Propagat.*, Vol. AP-36, No. 6, pp. 869–878, June 1988.
45. A. C. Newell, R. C. Baird, and P. F. Wacker, "Accurate Measurement of Antenna Gain and Polarization at Reduced Distances by an Extrapolation Technique," *IEEE Trans. Antennas Propagat.*, Vol. AP-21, No. 4, pp. 418–431, July 1973.
46. L. H. Hemming and R. A. Heaton, "Antenna Gain Calibration on a Ground Reflection Range," *IEEE Trans. Antennas Propagat.*, Vol. AP-21, No. 4, pp. 532–537, July 1973.
47. R. G. FitzGerrell, "Gain Measurements of Vertically Polarized Antennas over Imperfect Ground," *IEEE Trans. Antennas Propagat.*, Vol. AP-15, No. 2, pp. 211–216, March 1967.
48. R. G. FitzGerrell, "The Gain of a Horizontal Half-Wave Dipole over Ground," *IEEE Trans. Antennas Propagat.*, Vol. AP-15, No. 4, pp. 569–571, July 1967.
49. R. G. FitzGerrell, "Limitations on Vertically Polarized Ground-Based Antennas as Gain Standards," *IEEE Trans. Antennas Propagat.*, Vol. AP-23, No. 2, pp. 284–286, March 1975.
50. E. H. Newman, P. Bohley, and C. H. Walter, "Two Methods for the Measurement of Antenna Efficiency," *IEEE Trans. Antennas Propagat.*, Vol. AP-23, No. 4, pp. 457–461, July 1975.
51. M. Sucher and J. Fox, *Handbook of Microwave Measurements*, Vol. I, Polytechnic Press of the Polytechnic Institute of Brooklyn, New York, 1963.
52. C. G. Montgomery, *Techniques of Microwave Measurements*, Vol. II, MIT Radiation Laboratory Series, Vol. 11, McGraw-Hill, New York, 1947, Chapter 8.
53. ANSI/IEEE Std 148–1959 (Reaff 1971).
54. R. E. Collin, *Foundations for Microwave Engineering*, pp. 248–257, McGraw-Hill, New York, 1992.
55. J. D. Kraus, *Antennas*, McGraw-Hill, New York, 1988.
56. G. Barzilai, "Experimental Determination of the Distribution of Current and Charge Along Cylindrical Antennas," *Proc. IRE (Waves and Electrons Section)*, pp. 825–829, July 1949.
57. T. Morita, "Current Distributions on Transmitting and Receiving Antennas," *Proc. IRE*, pp. 898–904, August 1950.
58. A. F. Rashid, "Quasi-Near-Zone Field of a Monopole Antenna and the Current Distribution of an Antenna on a Finite Conductive Earth," *IEEE Trans. Antennas Propagat.*, Vol. AP-18, No. 1, pp. 22–28, January 1970.
59. H. G. Booker, V. H. Rumsey, G. A. Deschamps, M. I. Kales, and J. I. Bonhert, "Techniques for Handling Elliptically Polarized Waves with Special Reference to Antennas," *Proc. IRE*, Vol. 39, pp. 533–552, May 1951.
60. E. S. Gillespie, "Measurement of Antenna Radiation Characteristics on Far-Field Ranges," Chapter 32 in *Antenna Handbook* (Y. T. Lo and S. W. Lee, eds.), pp. 32-1–32-91, Van Nostrand Reinhold Co., Inc., New York, 1988.
61. G. Sinclair, "Theory of Models of Electromagnetic Systems," *Proc. IRE*, Vol. 36, pp. 1364–1370, November 1948.
62. C. A. Balanis, *Advanced Engineering Electromagnetics* (2nd edition), Chapter 11, John Wiley & Sons, Inc., 2012.

APPENDIX I

$$f(x) = \frac{\sin(x)}{x}$$

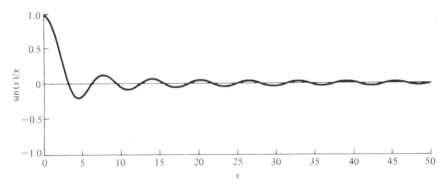

Figure I.1 Plot of $\sin(x)/x$ function.

APPENDIX II

$$f_N(x) = \left| \frac{\sin(Nx)}{N \sin(x)} \right| \quad N = 1, 3, 5, 10, 20$$

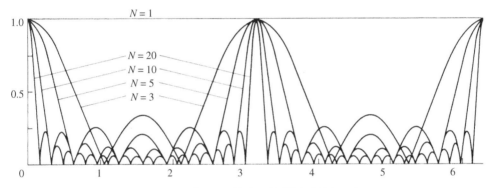

Figure II.1 Curves of $|\sin(Nx)/N\sin(x)|$ function.

APPENDIX III

Cosine and Sine Integrals

$$S_i(x) = \int_0^x \frac{\sin(\tau)}{\tau} d\tau \tag{III-1}$$

$$C_i(x) = -\int_x^\infty \frac{\cos(\tau)}{\tau} d\tau = \int_\infty^x \frac{\cos(\tau)}{\tau} d\tau \tag{III-2}$$

$$C_{in}(x) = \int_0^x \frac{1-\cos(\tau)}{\tau} d\tau \tag{III-3}$$

$$C_{in}(x) = \ln(\gamma x) - C_i(x) = \ln(\gamma) + \ln(x) - C_i(x)$$

$$C_{in}(x) = \ln(1.781) + \ln(x) - C_i(x) = 0.577215665 + \ln(x) - C_i(x) \tag{III-4}$$

Also

$$S_i(x) = \sum_{k=0}^\infty \frac{(-1)^k x^{2k+1}}{(2k+1)(2k+1)!}$$

$$C_i(x) = C + \ln(x) + \sum_{k=1}^\infty (-1)^k \frac{x^{2k}}{2k(2k)!}$$

$$C_{in}(x) = \sum_{k=1}^\infty (-1)^{k+1} \frac{x^{2k}}{2k(2k)!}$$

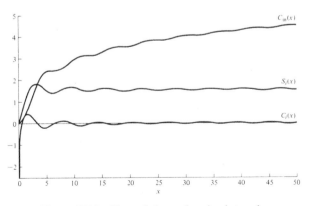

Figure III.1 Plots of sine and cosine integrals.

APPENDIX IV

Fresnel Integrals

$$C_0(x) = \int_0^x \frac{\cos(\tau)}{\sqrt{2\pi\tau}} \, d\tau \tag{IV-1}$$

$$S_0(x) = \int_0^x \frac{\sin(\tau)}{\sqrt{2\pi\tau}} \, d\tau \tag{IV-2}$$

$$C(x) = \int_0^x \cos\left(\frac{\pi}{2}\tau^2\right) d\tau \tag{IV-3}$$

$$S(x) = \int_0^x \sin\left(\frac{\pi}{2}\tau^2\right) d\tau \tag{IV-4}$$

$$C_1(x) = \int_x^\infty \cos(\tau^2) \, d\tau \tag{IV-5}$$

$$S_1(x) = \int_x^\infty \sin(\tau^2) \, d\tau \tag{IV-6}$$

$$C(x) - jS(x) = \int_0^x e^{-j(\pi/2)\tau^2} \, d\tau = \int_0^{(\pi/2)x^2} \frac{e^{-j\tau}}{\sqrt{2\pi\tau}} \, d\tau$$

$$C(x) - jS(x) = C_0\left(\frac{\pi}{2}x^2\right) - jS_0\left(\frac{\pi}{2}x^2\right) \tag{IV-7}$$

$$C_1(x) - jS_1(x) = \int_x^\infty e^{-j\tau^2} \, d\tau = \sqrt{\frac{\pi}{2}} \int_{x^2}^\infty \frac{e^{-j\tau}}{\sqrt{2\pi\tau}} \, d\tau$$

$$C_1(x) - jS_1(x) = \sqrt{\frac{\pi}{2}} \left\{ \int_0^\infty \frac{e^{-j\tau}}{\sqrt{2\pi\tau}} \, d\tau - \int_0^{x^2} \frac{e^{-j\tau}}{\sqrt{2\pi\tau}} \, d\tau \right\}$$

$$C_1(x) - jS_1(x) = \sqrt{\frac{\pi}{2}} \left\{ \left[\frac{1}{2} - j\frac{1}{2}\right] - [C_0(x^2) - jS_0(x^2)] \right\}$$

$$C_1(x) - jS_1(x) = \sqrt{\frac{\pi}{2}} \left\{ \left[\frac{1}{2} - C_0(x^2)\right] - j\left[\frac{1}{2} - S_0(x^2)\right] \right\} \tag{IV-8}$$

Antenna Theory: Analysis and Design, Fourth Edition. Constantine A. Balanis.
© 2016 John Wiley & Sons, Inc. Published 2016 by John Wiley & Sons, Inc.
Companion Website: www.wiley.com/go/antennatheory4e

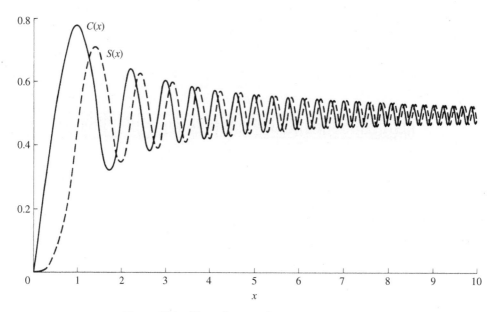

Figure IV.1 Plots of $C(x)$ and $S(x)$ Fresnel integrals.

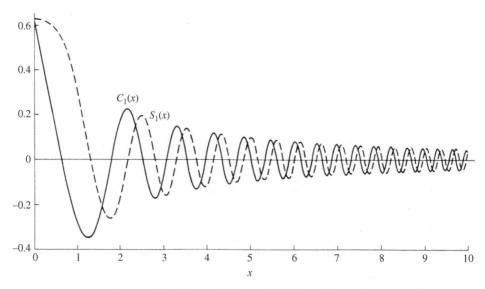

Figure IV.2 Plots of $C_1(x)$ and $S_1(x)$ Fresnel integrals.

APPENDIX V

Bessel Functions

Bessel's equation can be written as

$$x^2 \frac{d^2y}{dx^2} + x\frac{dy}{dx} + (x^2 - p^2)y = 0 \qquad \text{(V-1)}$$

Using the method of Frobenius, we can write its solutions as

$$y(x) = A_1 J_p(x) + B_1 J_{-p}(x), \quad p \neq 0 \text{ or integer} \qquad \text{(V-2)}$$

or

$$y(x) = A_2 J_n(x) + B_2 Y_n(x), \quad p = n = 0 \text{ or integer} \qquad \text{(V-3)}$$

where

$$J_p(x) = \sum_{m=0}^{\infty} \frac{(-1)^m (x/2)^{2m+p}}{m!(m+p)!} \qquad \text{(V-4)}$$

$$J_{-p}(x) = \sum_{m=0}^{\infty} \frac{(-1)^m (x/2)^{2m-p}}{m!(m-p)!} \qquad \text{(V-5)}$$

$$Y_p(x) = \frac{J_p(x)\cos(p\pi) - J_{-p}(x)}{\sin(p\pi)} \qquad \text{(V-6)}$$

$$m! = \Gamma(m+1) \qquad \text{(V-7)}$$

$J_p(x)$ is referred to as the Bessel function of the first kind of order p, $Y_p(x)$ as the Bessel function of the second kind of order p, and $\Gamma(x)$ as the gamma function.

When $p = n =$ integer, using (V-5) and (V-7) it can be shown that

$$J_{-n}(x) = (-1)^n J_n(x) \qquad \text{(V-8)}$$

Antenna Theory: Analysis and Design, Fourth Edition. Constantine A. Balanis.
© 2016 John Wiley & Sons, Inc. Published 2016 by John Wiley & Sons, Inc.
Companion Website: www.wiley.com/go/antennatheory4e

and no longer are the two Bessel functions independent of each other. Therefore a second solution is required and it is given by (V-3). It can also be shown that

$$Y_n(x) = \lim_{p \to n} Y_p(x) = \lim_{p \to n} \frac{J_p(x)\cos(p\pi) - J_{-p}(x)}{\sin(p\pi)} \tag{V-9}$$

When the argument of the Bessel function is negative and $p = n$, using (V-4) leads to

$$J_n(-x) = (-1)^n J_n(x) \tag{V-10}$$

In many applications Bessel functions of small and large arguments are required. Using asymptotic methods, it can be shown that

$$\left. \begin{array}{l} J_0(x) \simeq 1 \\[6pt] Y_0(x) \simeq \dfrac{2}{\pi} \ln\left(\dfrac{\gamma x}{2}\right) \\[6pt] \gamma = 1.781 \end{array} \right\} x \to 0 \tag{V-11}$$

$$\left. \begin{array}{l} J_p(x) \simeq \dfrac{1}{p!}\left(\dfrac{x}{2}\right)^p \qquad x \to 0 \\[10pt] Y_p(x) \simeq -\dfrac{(p-1)!}{\pi}\left(\dfrac{2}{x}\right)^p \quad p > 0 \end{array} \right\} \tag{V-12}$$

and

$$\left. \begin{array}{l} J_p(x) \simeq \sqrt{\dfrac{2}{\pi x}} \cos\left(x - \dfrac{\pi}{4} - \dfrac{p\pi}{2}\right) \\[10pt] Y_p(x) \simeq \sqrt{\dfrac{2}{\pi x}} \sin\left(x - \dfrac{\pi}{4} - \dfrac{p\pi}{2}\right) \end{array} \right\} x \to \infty \tag{V-13}$$

For wave propagation it is often convenient to introduce Hankel functions defined as

$$H_p^{(1)}(x) = J_p(x) + jY_p(x) \tag{V-14}$$

$$H_p^{(2)}(x) = J_p(x) - jY_p(x) \tag{V-15}$$

where $H_p^{(1)}(x)$ is the Hankel function of the first kind of order p and $H_p^{(2)}(x)$ is the Hankel function of the second kind of order p. For large arguments

$$H_p^{(1)}(x) \simeq \sqrt{\dfrac{2}{\pi x}} e^{j[x - p(\pi/2) - \pi/4]}, \qquad x \to \infty \tag{V-16}$$

$$H_p^{(2)}(x) \simeq \sqrt{\dfrac{2}{\pi x}} e^{-j[x - p(\pi/2) - \pi/4]}, \qquad x \to \infty \tag{V-17}$$

A derivative can be taken using either

$$\frac{d}{dx}[Z_p(\alpha x)] = \alpha Z_{p-1}(\alpha x) - \frac{p}{x} Z_p(\alpha x) \tag{V-18}$$

or

$$\frac{d}{dx}[Z_p(\alpha x)] = -\alpha Z_{p+1}(\alpha x) + \frac{p}{x} Z_p(\alpha x) \tag{V-19}$$

where Z_p can be a Bessel function (J_p, Y_p) or a Hankel function [$H_p^{(1)}$ or $H_p^{(2)}$].
A useful identity relating Bessel functions and their derivatives is given by

$$J_p(x) Y_p'(x) - Y_p(x) J_p'(x) = \frac{2}{\pi x} \tag{V-20}$$

and it is referred to as the Wronskian. The prime (′) indicates a derivative. Also

$$J_p(x) J_{-p}'(x) - J_{-p}(x) J_p'(x) = -\frac{2}{\pi x} \sin(p\pi) \tag{V-21}$$

Some useful integrals of Bessel functions are

$$\int x^{p+1} J_p(\alpha x)\, dx = \frac{1}{\alpha} x^{p+1} J_{p+1}(\alpha x) + C \tag{V-22}$$

$$\int x^{1-p} J_p(\alpha x)\, dx = -\frac{1}{\alpha} x^{1-p} J_{p-1}(\alpha x) + C \tag{V-23}$$

$$\int x^3 J_0(x)\, dx = x^3 J_1(x) - 2x^2 J_2(x) + C \tag{V-24}$$

$$\int x^6 J_1(x)\, dx = x^6 J_2(x) - 4x^5 J_3(x) + 8x^4 J_4(x) + C \tag{V-25}$$

$$\int J_3(x)\, dx = -J_2(x) - \frac{2}{x} J_1(x) + C \tag{V-26}$$

$$\int x J_1(x)\, dx = -x J_0(x) + \int J_0(x)\, dx + C \tag{V-27}$$

$$\int x^{-1} J_1(x)\, dx = -J_1(x) + \int J_0(x)\, dx + C \tag{V-28}$$

$$\int J_2(x)\, dx = -2 J_1(x) + \int J_0(x)\, dx + C \tag{V-29}$$

$$\int x^m J_n(x)\, dx = x^m J_{n+1}(x) - (m - n - 1) \int x^{m-1} J_{n+1}(x)\, dx \tag{V-30}$$

$$\int x^m J_n(x)\, dx = -x^m J_{n-1}(x) + (m + n - 1) \int x^{m-1} J_{n-1}(x)\, dx \tag{V-31}$$

$$J_1(x) = \frac{2}{\pi} \int_0^{\pi/2} \sin(x \sin\theta) \sin\theta\, d\theta \tag{V-32}$$

$$\frac{1}{x}J_1(x) = \frac{2}{\pi}\int_0^{\pi/2} \cos(x\sin\theta)\cos^2\theta\, d\theta \tag{V-33}$$

$$J_2(x) = \frac{2}{\pi}\int_0^{\pi/2} \cos(x\sin\theta)\cos 2\theta\, d\theta \tag{V-34}$$

$$J_n(x) = \frac{j^{-n}}{2\pi}\int_0^{2\pi} e^{jx\cos\phi}e^{jn\phi}\, d\phi \tag{V-35}$$

$$J_n(x) = \frac{j^{-n}}{\pi}\int_0^{\pi} \cos(n\phi)e^{jx\cos\phi}\, d\phi \tag{V-36}$$

$$J_n(x) = \frac{1}{\pi}\int_0^{\pi} \cos(x\sin\phi - n\phi)\, d\phi \tag{V-37}$$

$$J_{2n}(x) = \frac{2}{\pi}\int_0^{\pi/2} \cos(x\sin\phi)\cos(2n\phi)\, d\phi \tag{V-38}$$

$$J_{2n}(x) = (-1)^n\frac{2}{\pi}\int_0^{\pi/2} \cos(x\cos\phi)\cos(2n\phi)\, d\phi \tag{V-39}$$

The integrals

$$\int_0^x J_0(\tau)\, d\tau \quad \text{and} \quad \int_0^x Y_0(\tau)\, d\tau \tag{V-40}$$

often appear in solutions of problems but cannot be integrated in closed form. Graphs and tables for each, obtained using numerical techniques, are included.

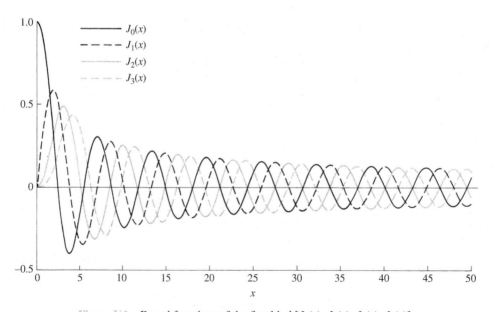

Figure V.1 Bessel functions of the first kind [$J_0(x), J_1(x), J_2(x), J_3(x)$].

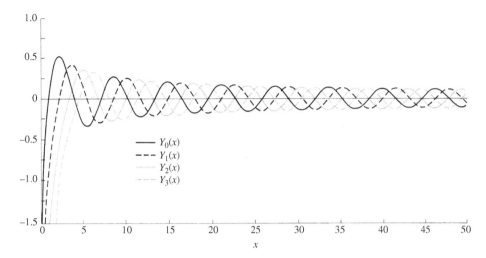

Figure V.2 Bessel functions of the second kind $[Y_0(x), Y_1(x), Y_2(x), Y_3(x)]$.

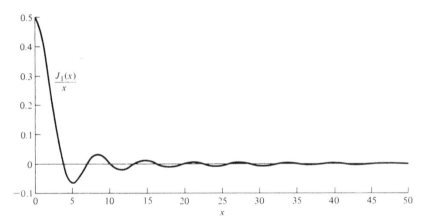

Figure V.3 Plot of $J_1(x)/x$ function.

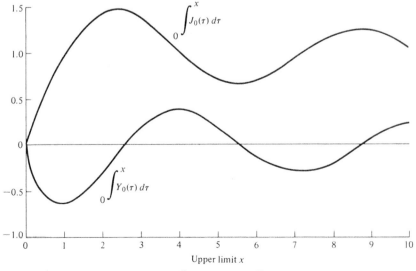

Figure V.4 Plots of $\int_0^x J_0(\tau)d\tau$ and $\int_0^x Y_0(\tau)d\tau$ functions.

APPENDIX VI

Identities

VI.1 TRIGONOMETRIC

1. Sum or difference:
   a. $\sin(x+y) = \sin x \cos y + \cos x \sin y$
   b. $\sin(x-y) = \sin x \cos y - \cos x \sin y$
   c. $\cos(x+y) = \cos x \cos y - \sin x \sin y$
   d. $\cos(x-y) = \cos x \cos y + \sin x \sin y$
   e. $\tan(x+y) = \dfrac{\tan x + \tan y}{1 - \tan x \tan y}$
   f. $\tan(x-y) = \dfrac{\tan x - \tan y}{1 + \tan x \tan y}$
   g. $\sin^2 x + \cos^2 x = 1$
   h. $\tan^2 x - \sec^2 x = -1$
   i. $\cot^2 x - \csc^2 x = -1$

2. Sum or difference into products:
   a. $\sin x + \sin y = 2 \sin \frac{1}{2}(x+y) \cos \frac{1}{2}(x-y)$
   b. $\sin x - \sin y = 2 \cos \frac{1}{2}(x+y) \sin \frac{1}{2}(x-y)$
   c. $\cos x + \cos y = 2 \cos \frac{1}{2}(x+y) \cos \frac{1}{2}(x-y)$
   d. $\cos x - \cos y = -2 \sin \frac{1}{2}(x+y) \sin \frac{1}{2}(x-y)$

3. Products into sum or difference:
   a. $2 \sin x \cos y = \sin(x+y) + \sin(x-y)$
   b. $2 \cos x \sin y = \sin(x+y) - \sin(x-y)$
   c. $2 \cos x \cos y = \cos(x+y) + \cos(x-y)$
   d. $2 \sin x \sin y = -\cos(x+y) + \cos(x-y)$

4. Double and half angles:
   a. $\sin 2x = 2 \sin x \cos x$
   b. $\cos 2x = \cos^2 x - \sin^2 x = 2\cos^2 x - 1 = 1 - 2\sin^2 x$
   c. $\tan 2x = \dfrac{2 \tan x}{1 - \tan^2 x}$

Antenna Theory: Analysis and Design, Fourth Edition. Constantine A. Balanis.
© 2016 John Wiley & Sons, Inc. Published 2016 by John Wiley & Sons, Inc.
Companion Website: www.wiley.com/go/antennatheory4e

d. $\sin\frac{1}{2}x = \pm\sqrt{\frac{1-\cos x}{2}}$ or $2\sin^2\theta = 1 - \cos 2\theta$

e. $\cos\frac{1}{2}x = \pm\sqrt{\frac{1+\cos x}{2}}$ or $2\cos^2\theta = 1 + \cos 2\theta$

f. $\tan\frac{1}{2}x = \pm\sqrt{\frac{1-\cos x}{1+\cos x}} = \frac{\sin x}{1+\cos x} = \frac{1-\cos x}{\sin x}$

5. Series:

   a. $\sin x = \frac{e^{jx} - e^{-jx}}{2j} = x - \frac{x^3}{3!} + \frac{x^5}{5!} - \frac{x^7}{7!} + \cdots$

   b. $\cos x = \frac{e^{jx} + e^{-jx}}{2} = 1 - \frac{x^2}{2!} + \frac{x^4}{4!} - \frac{x^6}{6!} + \cdots$

   c. $\tan x = \frac{e^{jx} - e^{-jx}}{j(e^{jx} + e^{-jx})} = x + \frac{x^3}{3} + \frac{2x^5}{15} + \frac{17x^7}{315} + \cdots$

VI.2 HYPERBOLIC

1. Definitions:
   a. Hyperbolic sine: $\sinh x = \frac{1}{2}(e^x - e^{-x})$
   b. Hyperbolic cosine: $\cosh x = \frac{1}{2}(e^x + e^{-x})$
   c. Hyperbolic tangent: $\tanh x = \frac{\sinh x}{\cosh x}$
   d. Hyperbolic cotangent: $\coth x = \frac{1}{\tanh x} = \frac{\cosh x}{\sinh x}$
   e. Hyperbolic secant: $\text{sech } x = \frac{1}{\cosh x}$
   f. Hyperbolic cosecant: $\text{csch } x = \frac{1}{\sinh x}$

2. Sum or difference:
   a. $\cosh(x+y) = \cosh x \cosh y + \sinh x \sinh y$
   b. $\sinh(x-y) = \sinh x \cosh y - \cosh x \sinh y$
   c. $\cosh(x-y) = \cosh x \cosh y - \sinh x \sinh y$
   d. $\tanh(x+y) = \frac{\tanh x + \tanh y}{1 + \tanh x \tanh y}$
   e. $\tanh(x-y) = \frac{\tanh x - \tanh y}{1 - \tanh x \tanh y}$
   f. $\cosh^2 x - \sinh^2 x = 1$
   g. $\tanh^2 x + \text{sech}^2 x = 1$
   h. $\coth^2 x - \text{csch}^2 x = 1$
   i. $\cosh(x \pm jy) = \cosh x \cos y \pm j \sinh x \sin y$
   j. $\sinh(x \pm jy) = \sinh x \cos y \pm j \cosh x \sin y$

3. Series:

   a. $\sinh x = \frac{e^x - e^{-x}}{2} = x + \frac{x^3}{3!} + \frac{x^5}{5!} + \frac{x^7}{7!} + \cdots$

   b. $\cosh x = \frac{e^x + e^{-x}}{2} = 1 + \frac{x^2}{2!} + \frac{x^4}{4!} + \frac{x^6}{6!} + \cdots$

   c. $e^x = 1 + x + \frac{x^2}{2!} + \frac{x^3}{3!} + \frac{x^4}{4!} + \cdots$

VI.3 LOGARITHMIC

1. $\log_b(MN) = \log_b M + \log_b N$
2. $\log_b(M/N) = \log_b M - \log_b N$
3. $\log_b(1/N) = -\log_b N$
4. $\log_b(M^n) = n \log_b M$
5. $\log_b(M^{1/n}) = \dfrac{1}{n} \log_b M$
6. $\log_a N = \log_b N \cdot \log_a b = \log_b N / \log_b a$
7. $\log_e N = \log_{10} N \cdot \log_e 10 = 2.302585 \log_{10} N$
8. $\log_{10} N = \log_e N \cdot \log_{10} e = 0.434294 \log_e N$

APPENDIX VII

Vector Analysis

VII.1 VECTOR TRANSFORMATIONS

In this appendix we present the vector transformations from rectangular to cylindrical (and vice versa), from cylindrical to spherical (and vice versa), and from rectangular to spherical (and vice versa). The three coordinate systems are shown in Figure VII.1.

VII.1.1 Rectangular to Cylindrical (and Vice Versa)

The coordinate transformation from rectangular (x, y, z) to cylindrical (ρ, ϕ, z) is given, referring to Figure VII.1(b)

$$\left. \begin{array}{l} x = \rho \cos \phi \\ y = \rho \sin \phi \\ z = z \end{array} \right\} \quad \text{(VII-1)}$$

In the rectangular coordinate system, we express a vector $\mathbf{A}$ as

$$\mathbf{A} = \hat{\mathbf{a}}_x A_x + \hat{\mathbf{a}}_y A_y + \hat{\mathbf{a}}_z A_z \quad \text{(VII-2)}$$

where $\hat{\mathbf{a}}_x, \hat{\mathbf{a}}_y, \hat{\mathbf{a}}_z$ are the unit vectors and A_x, A_y, A_z are the components of the vector $\mathbf{A}$ in the rectangular coordinate system. We wish to write $\mathbf{A}$ as

$$\mathbf{A} = \hat{\mathbf{a}}_\rho A_\rho + \hat{\mathbf{a}}_\phi A_\phi + \hat{\mathbf{a}}_z A_z \quad \text{(VII-3)}$$

where $\hat{\mathbf{a}}_\rho, \hat{\mathbf{a}}_\phi, \hat{\mathbf{a}}_z$ are the unit vectors and A_ρ, A_ϕ, A_z are the vector components in the cylindrical coordinate system. The z-axis is common to both of them.

Referring to Figure VII.2, we can write

$$\left. \begin{array}{l} \hat{\mathbf{a}}_x = \hat{\mathbf{a}}_\rho \cos \phi - \hat{\mathbf{a}}_\phi \sin \phi \\ \hat{\mathbf{a}}_y = \hat{\mathbf{a}}_\rho \sin \phi + \hat{\mathbf{a}}_\phi \cos \phi \\ \hat{\mathbf{a}}_z = \hat{\mathbf{a}}_z \end{array} \right\} \quad \text{(VII-4)}$$

Antenna Theory: Analysis and Design, Fourth Edition. Constantine A. Balanis.
© 2016 John Wiley & Sons, Inc. Published 2016 by John Wiley & Sons, Inc.
Companion Website: www.wiley.com/go/antennatheory4e

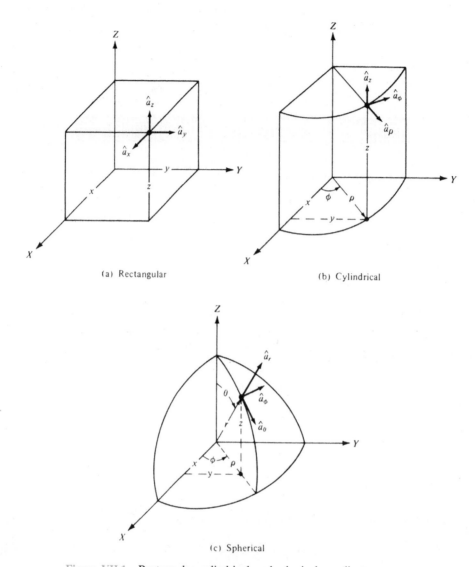

Figure VII.1 Rectangular, cylindrical, and spherical coordinate systems.

Using (VII-4) reduces (VII-2) to

$$\mathbf{A} = (\hat{\mathbf{a}}_\rho \cos\phi - \hat{\mathbf{a}}_\phi \sin\phi)A_x + (\hat{\mathbf{a}}_\rho \sin\phi + \hat{\mathbf{a}}_\phi \cos\phi)A_y + \hat{\mathbf{a}}_z A_z$$

$$\mathbf{A} = \hat{\mathbf{a}}_\rho(A_x \cos\phi + A_y \sin\phi) + \hat{\mathbf{a}}_\phi(-A_x \sin\phi + A_y \cos\phi) + \hat{\mathbf{a}}_z A_z \quad \text{(VII-5)}$$

which when compared with (VII-3) leads to

$$\left.\begin{array}{l} A_\rho = A_x \cos\phi + A_y \sin\phi \\ A_\phi = -A_x \sin\phi + A_y \cos\phi \\ A_z = A_z \end{array}\right\} \quad \text{(VII-6)}$$

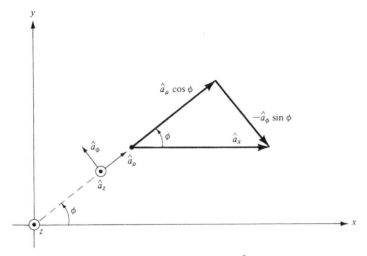

(a) Geometry for unit vector $\hat{a}_x$

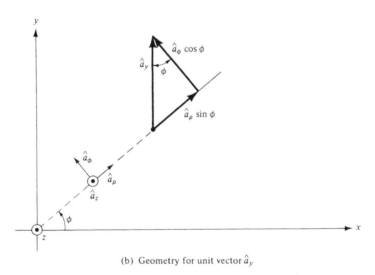

(b) Geometry for unit vector $\hat{a}_y$

Figure VII.2 Geometrical representation of transformations between unit vectors of rectangular and cylindrical coordinate systems.

In matrix form, (VII-6) can be written as

$$\begin{pmatrix} A_\rho \\ A_\phi \\ A_z \end{pmatrix} = \begin{pmatrix} \cos\phi & \sin\phi & 0 \\ -\sin\phi & \cos\phi & 0 \\ 0 & 0 & 1 \end{pmatrix} \begin{pmatrix} A_x \\ A_y \\ A_z \end{pmatrix} \quad \text{(VII-6a)}$$

where

$$[A]_{rc} = \begin{bmatrix} \cos\phi & \sin\phi & 0 \\ -\sin\phi & \cos\phi & 0 \\ 0 & 0 & 1 \end{bmatrix} \quad \text{(VII-6b)}$$

is the transformation matrix for rectangular-to-cylindrical components.

Since $[A]_{rc}$ is an orthonormal matrix (its inverse is equal to its transpose), we can write the transformation matrix for cylindrical-to-rectangular components as

$$[A]_{cr} = [A]_{rc}^{-1} = [A]_{rc}^{t} = \begin{bmatrix} \cos\phi & -\sin\phi & 0 \\ \sin\phi & \cos\phi & 0 \\ 0 & 0 & 1 \end{bmatrix} \quad \text{(VII-7)}$$

or

$$\begin{pmatrix} A_x \\ A_y \\ A_z \end{pmatrix} = \begin{pmatrix} \cos\phi & -\sin\phi & 0 \\ \sin\phi & \cos\phi & 0 \\ 0 & 0 & 1 \end{pmatrix} \begin{pmatrix} A_\rho \\ A_\phi \\ A_z \end{pmatrix} \quad \text{(VII-7a)}$$

or

$$\left. \begin{array}{l} A_x = A_\rho \cos\phi - A_\phi \sin\phi \\ A_y = A_\rho \sin\phi + A_\phi \cos\phi \\ A_z = A_z \end{array} \right\} \quad \text{(VII-7b)}$$

VII.1.2 Cylindrical to Spherical (and Vice Versa)

Referring to Figure VII.1(c), we can write that the cylindrical and spherical coordinates are related by

$$\left. \begin{array}{l} \rho = r\sin\theta \\ z = r\cos\theta \end{array} \right\} \quad \text{(VII-8)}$$

In a geometrical approach similar to the one employed in the previous section, we can show that the cylindrical-to-spherical transformation of vector components is given by

$$\left. \begin{array}{l} A_r = A_\rho \sin\theta + A_z \cos\theta \\ A_\theta = A_\rho \cos\theta - A_z \sin\theta \\ A_\phi = A_\phi \end{array} \right\} \quad \text{(VII-9)}$$

or in matrix form by

$$\begin{pmatrix} A_r \\ A_\theta \\ A_\phi \end{pmatrix} = \begin{pmatrix} \sin\theta & 0 & \cos\theta \\ \cos\theta & 0 & -\sin\theta \\ 0 & 1 & 0 \end{pmatrix} \begin{pmatrix} A_\rho \\ A_\phi \\ A_z \end{pmatrix} \quad \text{(VII-9a)}$$

Thus the cylindrical-to-spherical transformation matrix can be written as

$$[A]_{cs} = \begin{bmatrix} \sin\theta & 0 & \cos\theta \\ \cos\theta & 0 & -\sin\theta \\ 0 & 1 & 0 \end{bmatrix} \quad \text{(VII-9b)}$$

The $[A]_{cs}$ matrix is also orthonormal so that its inverse is given by

$$[A]_{sc} = [A]_{cs}^{-1} = [A]_{cs}^{t} = \begin{bmatrix} \sin\theta & \cos\theta & 0 \\ 0 & 0 & 1 \\ \cos\theta & -\sin\theta & 0 \end{bmatrix} \quad \text{(VII-10)}$$

and the spherical-to-cylindrical transformation is accomplished by

$$\begin{pmatrix} A_\rho \\ A_\phi \\ A_z \end{pmatrix} = \begin{pmatrix} \sin\theta & \cos\theta & 0 \\ 0 & 0 & 1 \\ \cos\theta & -\sin\theta & 0 \end{pmatrix} \begin{pmatrix} A_r \\ A_\theta \\ A_\phi \end{pmatrix}$$ (VII-10a)

or

$$\left.\begin{aligned} A_\rho &= A_r \sin\theta + A_\theta \cos\theta \\ A_\phi &= A_\phi \\ A_z &= A_r \cos\theta - A_\theta \sin\theta \end{aligned}\right\}$$ (VII-10b)

This time the component A_ϕ and coordinate ϕ are the same in both systems.

VII.1.3 Rectangular to Spherical (and Vice Versa)

Many times it may be required that a transformation be performed directly from rectangular-to-spherical components. By referring to Figure VII.1, we can write that the rectangular and spherical coordinates are related by

$$\left.\begin{aligned} x &= r\sin\theta\cos\phi \\ y &= r\sin\theta\sin\phi \\ z &= r\cos\theta \end{aligned}\right\}$$ (VII-11)

and the rectangular and spherical components by

$$\left.\begin{aligned} A_r &= A_x \sin\theta\cos\phi + A_y \sin\theta\sin\phi + A_z \cos\theta \\ A_\theta &= A_x \cos\theta\cos\phi + A_y \cos\theta\sin\phi - A_z \sin\theta \\ A_\phi &= -A_x \sin\phi + A_y \cos\phi \end{aligned}\right\}$$ (VII-12)

which can also be obtained by substituting (VII-6) into (VII-9). In matrix form, (VII-12) can be written as

$$\begin{pmatrix} A_r \\ A_\theta \\ A_\phi \end{pmatrix} = \begin{pmatrix} \sin\theta\cos\phi & \sin\theta\sin\phi & \cos\theta \\ \cos\theta\cos\phi & \cos\theta\sin\phi & -\sin\theta \\ -\sin\phi & \cos\phi & 0 \end{pmatrix} \begin{pmatrix} A_x \\ A_y \\ A_z \end{pmatrix}$$ (VII-12a)

with the rectangular-to-spherical transformation matrix being

$$[A]_{rs} = \begin{pmatrix} \sin\theta\cos\phi & \sin\theta\sin\phi & \cos\theta \\ \cos\theta\cos\phi & \cos\theta\sin\phi & -\sin\theta \\ -\sin\phi & \cos\phi & 0 \end{pmatrix}$$ (VII-12b)

The transformation matrix of (VII-12b) is also orthonormal so that its inverse can be written as

$$[A]_{sr} = [A]_{rs}^{-1} = [A]_{rs}^t = \begin{pmatrix} \sin\theta\cos\phi & \cos\theta\cos\phi & -\sin\phi \\ \sin\theta\sin\phi & \cos\theta\sin\phi & \cos\phi \\ \cos\theta & -\sin\theta & 0 \end{pmatrix}$$ (VII-13)

and the spherical-to-rectangular components related by

$$\begin{pmatrix} A_x \\ A_y \\ A_z \end{pmatrix} = \begin{pmatrix} \sin\theta\cos\phi & \cos\theta\cos\phi & -\sin\phi \\ \sin\theta\sin\phi & \cos\theta\sin\phi & \cos\phi \\ \cos\theta & -\sin\theta & 0 \end{pmatrix} \begin{pmatrix} A_r \\ A_\theta \\ A_\phi \end{pmatrix} \quad \text{(VII-13a)}$$

or

$$\left.\begin{aligned} A_x &= A_r \sin\theta\cos\phi + A_\theta \cos\theta\cos\phi - A_\phi \sin\phi \\ A_y &= A_r \sin\theta\sin\phi + A_\theta \cos\theta\sin\phi + A_\phi \cos\phi \\ A_z &= A_r \cos\theta - A_\theta \sin\theta \end{aligned}\right\} \quad \text{(VII-13b)}$$

VII.2 VECTOR DIFFERENTIAL OPERATORS

The differential operators of gradient of a scalar ($\nabla\psi$), divergence of a vector ($\nabla \cdot \mathbf{A}$), curl of a vector ($\nabla \times \mathbf{A}$), Laplacian of a scalar ($\nabla^2\psi$), and Laplacian of a vector ($\nabla^2\mathbf{A}$) frequently encountered in electromagnetic field analysis will be listed in the rectangular, cylindrical, and spherical coordinate systems.

VII.2.1 Rectangular Coordinates

$$\nabla\psi = \hat{\mathbf{a}}_x \frac{\partial\psi}{\partial x} + \hat{\mathbf{a}}_y \frac{\partial\psi}{\partial y} + \hat{\mathbf{a}}_z \frac{\partial\psi}{\partial z} \quad \text{(VII-14)}$$

$$\nabla \cdot \mathbf{A} = \frac{\partial A_x}{\partial x} + \frac{\partial A_y}{\partial y} + \frac{\partial A_z}{\partial z} \quad \text{(VII-15)}$$

$$\nabla \times \mathbf{A} = \hat{\mathbf{a}}_x \left(\frac{\partial A_z}{\partial y} - \frac{\partial A_y}{\partial z} \right) + \hat{\mathbf{a}}_y \left(\frac{\partial A_x}{\partial z} - \frac{\partial A_z}{\partial x} \right) + \hat{\mathbf{a}}_z \left(\frac{\partial A_y}{\partial x} - \frac{\partial A_x}{\partial y} \right) \quad \text{(VII-16)}$$

$$\nabla \cdot \nabla\psi = \nabla^2\psi = \frac{\partial^2\psi}{\partial x^2} + \frac{\partial^2\psi}{\partial y^2} + \frac{\partial^2\psi}{\partial z^2} \quad \text{(VII-17)}$$

$$\nabla^2\mathbf{A} = \hat{\mathbf{a}}_x \nabla^2 A_x + \hat{\mathbf{a}}_y \nabla^2 A_y + \hat{\mathbf{a}}_z \nabla^2 A_z \quad \text{(VII-18)}$$

VII.2.2 Cylindrical Coordinates

$$\nabla\psi = \hat{\mathbf{a}}_\rho \frac{\partial\psi}{\partial\rho} + \hat{\mathbf{a}}_\phi \frac{1}{\rho}\frac{\partial\psi}{\partial\phi} + \hat{\mathbf{a}}_z \frac{\partial\psi}{\partial z} \quad \text{(VII-19)}$$

$$\nabla \cdot \mathbf{A} = \frac{1}{\rho}\frac{\partial}{\partial\rho}(\rho A_\rho) + \frac{1}{\rho}\frac{\partial A_\phi}{\partial\phi} + \frac{\partial A_z}{\partial z} \quad \text{(VII-20)}$$

$$\nabla \times \mathbf{A} = \hat{\mathbf{a}}_\rho \left(\frac{1}{\rho}\frac{\partial A_z}{\partial\phi} - \frac{\partial A_\phi}{\partial z} \right) + \hat{\mathbf{a}}_\phi \left(\frac{\partial A_\rho}{\partial z} - \frac{\partial A_z}{\partial\rho} \right)$$
$$+ \hat{\mathbf{a}}_z \left(\frac{1}{\rho}\frac{\partial(\rho A_\phi)}{\partial\rho} - \frac{1}{\rho}\frac{\partial A_\rho}{\partial\phi} \right) \quad \text{(VII-21)}$$

$$\nabla^2\psi = \frac{1}{\rho}\frac{\partial}{\partial\rho}\left(\rho\frac{\partial\psi}{\partial\rho}\right) + \frac{1}{\rho^2}\frac{\partial^2\psi}{\partial\phi^2} + \frac{\partial^2\psi}{\partial z^2} \quad \text{(VII-22)}$$

$$\nabla^2\mathbf{A} = \nabla(\nabla \cdot \mathbf{A}) - \nabla \times \nabla \times \mathbf{A} \quad \text{(VII-23)}$$

or in an expanded form

$$\nabla^2 \mathbf{A} = \hat{\mathbf{a}}_\rho \left(\frac{\partial^2 A_\rho}{\partial \rho^2} + \frac{1}{\rho} \frac{\partial A_\rho}{\partial \rho} - \frac{A_\rho}{\rho^2} + \frac{1}{\rho^2} \frac{\partial^2 A_\rho}{\partial \phi^2} - \frac{2}{\rho^2} \frac{\partial A_\phi}{\partial \phi} + \frac{\partial^2 A_\rho}{\partial z^2} \right)$$

$$+ \hat{\mathbf{a}}_\phi \left(\frac{\partial^2 A_\phi}{\partial \rho^2} + \frac{1}{\rho} \frac{\partial A_\phi}{\partial \rho} - \frac{A_\phi}{\rho^2} + \frac{1}{\rho^2} \frac{\partial^2 A_\phi}{\partial \phi^2} + \frac{2}{\rho^2} \frac{\partial A_\rho}{\partial \phi} + \frac{\partial^2 A_\phi}{\partial z^2} \right)$$

$$+ \hat{\mathbf{a}}_z \left(\frac{\partial^2 A_z}{\partial \rho^2} + \frac{1}{\rho} \frac{\partial A_z}{\partial \rho} + \frac{1}{\rho^2} \frac{\partial^2 A_z}{\partial \phi^2} + \frac{\partial^2 A_z}{\partial z^2} \right) \quad \text{(VII-23a)}$$

In the cylindrical coordinate system $\nabla^2 \mathbf{A} \neq \hat{\mathbf{a}}_\rho \nabla^2 A_\rho + \hat{\mathbf{a}}_\phi \nabla^2 A_\phi + \hat{\mathbf{a}}_z \nabla^2 A_z$ because the orientation of the unit vectors $\hat{\mathbf{a}}_\rho$ and $\hat{\mathbf{a}}_\phi$ varies with the ρ and ϕ coordinates.

VII.2.3 Spherical Coordinates

$$\nabla \psi = \hat{\mathbf{a}}_r \frac{\partial \psi}{\partial r} + \hat{\mathbf{a}}_\theta \frac{1}{r} \frac{\partial \psi}{\partial \theta} + \hat{\mathbf{a}}_\phi \frac{1}{r \sin \theta} \frac{\partial \psi}{\partial \phi} \quad \text{(VII-24)}$$

$$\nabla \cdot \mathbf{A} = \frac{1}{r^2} \frac{\partial}{\partial r}(r^2 A_r) + \frac{1}{r \sin \theta} \frac{\partial}{\partial \theta}(A_\theta \sin \theta) + \frac{1}{r \sin \theta} \frac{\partial A_\phi}{\partial \phi} \quad \text{(VII-25)}$$

$$\nabla \times \mathbf{A} = \frac{\hat{\mathbf{a}}_r}{r \sin \theta} \left[\frac{\partial}{\partial \theta}(A_\phi \sin \theta) - \frac{\partial A_\theta}{\partial \phi} \right] + \frac{\hat{\mathbf{a}}_\theta}{r} \left[\frac{1}{\sin \theta} \frac{\partial A_r}{\partial \phi} - \frac{\partial}{\partial r}(rA_\phi) \right]$$

$$+ \frac{\hat{\mathbf{a}}_\phi}{r} \left[\frac{\partial}{\partial r}(rA_\theta) - \frac{\partial A_r}{\partial \theta} \right] \quad \text{(VII-26)}$$

$$\nabla^2 \psi = \frac{1}{r^2} \frac{\partial}{\partial r} \left(r^2 \frac{\partial \psi}{\partial r} \right) + \frac{1}{r^2 \sin \theta} \frac{\partial}{\partial \theta} \left(\sin \theta \frac{\partial \psi}{\partial \theta} \right) + \frac{1}{r^2 \sin^2 \theta} \frac{\partial^2 \psi}{\partial \phi^2} \quad \text{(VII-27)}$$

$$\nabla^2 \mathbf{A} = \nabla(\nabla \cdot \mathbf{A}) - \nabla \times \nabla \times \mathbf{A} \quad \text{(VII-28)}$$

or in an expanded form

$$\nabla^2 \mathbf{A} = \hat{\mathbf{a}}_r \left(\frac{\partial^2 A_r}{\partial r^2} + \frac{2}{r} \frac{\partial A_r}{\partial r} - \frac{2}{r^2} A_r + \frac{1}{r^2} \frac{\partial^2 A_r}{\partial \theta^2} + \frac{\cot \theta}{r^2} \frac{\partial A_r}{\partial \theta} + \frac{1}{r^2 \sin^2 \theta} \frac{\partial^2 A_r}{\partial \phi^2} \right.$$

$$\left. - \frac{2}{r^2} \frac{\partial A_\theta}{\partial \theta} - \frac{2 \cot \theta}{r^2} A_\theta - \frac{2}{r^2 \sin \theta} \frac{\partial A_\phi}{\partial \phi} \right)$$

$$+ \hat{\mathbf{a}}_\theta \left(\frac{\partial^2 A_\theta}{\partial r^2} + \frac{2}{r} \frac{\partial A_\theta}{\partial r} - \frac{A_\theta}{r^2 \sin^2 \theta} + \frac{1}{r^2} \frac{\partial^2 A_\theta}{\partial \theta^2} + \frac{\cot \theta}{r^2} \frac{\partial A_\theta}{\partial \theta} \right.$$

$$\left. + \frac{1}{r^2 \sin^2 \theta} \frac{\partial^2 A_\theta}{\partial \phi^2} + \frac{2}{r^2} \frac{\partial A_r}{\partial \theta} - \frac{2 \cot \theta}{r^2 \sin \theta} \frac{\partial A_\phi}{\partial \phi} \right)$$

$$+ \hat{\mathbf{a}}_\phi \left(\frac{\partial^2 A_\phi}{\partial r^2} + \frac{2}{r} \frac{\partial A_\phi}{\partial r} - \frac{1}{r^2 \sin^2 \theta} A_\phi + \frac{1}{r^2} \frac{\partial^2 A_\phi}{\partial \theta^2} + \frac{\cot \theta}{r^2} \frac{\partial A_\phi}{\partial \theta} \right.$$

$$\left. + \frac{1}{r^2 \sin^2 \theta} \frac{\partial^2 A_\phi}{\partial \phi^2} + \frac{2}{r^2 \sin \theta} \frac{\partial A_r}{\partial \phi} + \frac{2 \cot \theta}{r^2 \sin \theta} \frac{\partial A_\theta}{\partial \phi} \right) \quad \text{(VII-28a)}$$

Again note that $\nabla^2 \mathbf{A} \neq \hat{\mathbf{a}}_r \nabla^2 A_r + \hat{\mathbf{a}}_\theta \nabla^2 A_\theta + \hat{\mathbf{a}}_\phi \nabla^2 A_\phi$ since the orientation of the unit vectors $\hat{\mathbf{a}}_r, \hat{\mathbf{a}}_\theta$, and $\hat{\mathbf{a}}_\phi$ varies with the $r, \theta,$ and ϕ coordinates.

VII.3 VECTOR IDENTITIES

VII.3.1 Addition and Multiplication

$$\mathbf{A} \cdot \mathbf{A} = |\mathbf{A}|^2 \tag{VII-29}$$

$$\mathbf{A} \cdot \mathbf{A}^* = |\mathbf{A}|^2 \tag{VII-30}$$

$$\mathbf{A} + \mathbf{B} = \mathbf{B} + \mathbf{A} \tag{VII-31}$$

$$\mathbf{A} \cdot \mathbf{B} = \mathbf{B} \cdot \mathbf{A} \tag{VII-32}$$

$$\mathbf{A} \times \mathbf{B} = -\mathbf{B} \times \mathbf{A} \tag{VII-33}$$

$$(\mathbf{A} + \mathbf{B}) \cdot \mathbf{C} = \mathbf{A} \cdot \mathbf{C} + \mathbf{B} \cdot \mathbf{C} \tag{VII-34}$$

$$(\mathbf{A} + \mathbf{B}) \times \mathbf{C} = \mathbf{A} \times \mathbf{C} + \mathbf{B} \times \mathbf{C} \tag{VII-35}$$

$$\mathbf{A} \cdot \mathbf{B} \times \mathbf{C} = \mathbf{B} \cdot \mathbf{C} \times \mathbf{A} = \mathbf{C} \cdot \mathbf{A} \times \mathbf{B} \tag{VII-36}$$

$$\mathbf{A} \times (\mathbf{B} \times \mathbf{C}) = (\mathbf{A} \cdot \mathbf{C})\mathbf{B} - (\mathbf{A} \cdot \mathbf{B})\mathbf{C} \tag{VII-37}$$

$$(\mathbf{A} \times \mathbf{B}) \cdot (\mathbf{C} \times \mathbf{D}) = \mathbf{A} \cdot \mathbf{B} \times (\mathbf{C} \times \mathbf{D})$$
$$= \mathbf{A} \cdot (\mathbf{B} \cdot \mathbf{D}\mathbf{C} - \mathbf{B} \cdot \mathbf{C}\mathbf{D})$$
$$= (\mathbf{A} \cdot \mathbf{C})(\mathbf{B} \cdot \mathbf{D}) - (\mathbf{A} \cdot \mathbf{D})(\mathbf{B} \cdot \mathbf{C}) \tag{VII-38}$$

$$(\mathbf{A} \times \mathbf{B}) \times (\mathbf{C} \times \mathbf{D}) = (\mathbf{A} \times \mathbf{B} \cdot \mathbf{D})\mathbf{C} - (\mathbf{A} \times \mathbf{B} \cdot \mathbf{C})\mathbf{D} \tag{VII-39}$$

VII.3.2 Differentiation

$$\nabla \cdot (\nabla \times \mathbf{A}) = 0 \tag{VII-40}$$

$$\nabla \times \nabla \psi = 0 \tag{VII-41}$$

$$\nabla(\phi + \psi) = \nabla\phi + \nabla\psi \tag{VII-42}$$

$$\nabla(\phi\psi) = \phi\nabla\psi + \psi\nabla\phi \tag{VII-43}$$

$$\nabla \cdot (\mathbf{A} + \mathbf{B}) = \nabla \cdot \mathbf{A} + \nabla \cdot \mathbf{B} \tag{VII-44}$$

$$\nabla \times (\mathbf{A} + \mathbf{B}) = \nabla \times \mathbf{A} + \nabla \times \mathbf{B} \tag{VII-45}$$

$$\nabla \cdot (\psi \mathbf{A}) = \mathbf{A} \cdot \nabla \psi + \psi \nabla \cdot \mathbf{A} \tag{VII-46}$$

$$\nabla \times (\psi \mathbf{A}) = \nabla \psi \times \mathbf{A} + \psi \nabla \times \mathbf{A} \tag{VII-47}$$

$$\nabla(\mathbf{A} \cdot \mathbf{B}) = (\mathbf{A} \cdot \nabla)\mathbf{B} + (\mathbf{B} \cdot \nabla)\mathbf{A} + \mathbf{A} \times (\nabla \times \mathbf{B}) + \mathbf{B} \times (\nabla \times \mathbf{A}) \tag{VII-48}$$

$$\nabla \cdot (\mathbf{A} \times \mathbf{B}) = \mathbf{B} \cdot \nabla \times \mathbf{A} - \mathbf{A} \cdot \nabla \times \mathbf{B} \tag{VII-49}$$

$$\nabla \times (\mathbf{A} \times \mathbf{B}) = \mathbf{A}\nabla \cdot \mathbf{B} - \mathbf{B}\nabla \cdot \mathbf{A} + (\mathbf{B} \cdot \nabla)\mathbf{A} - (\mathbf{A} \cdot \nabla)\mathbf{B} \tag{VII-50}$$

$$\nabla \times \nabla \times \mathbf{A} = \nabla(\nabla \cdot \mathbf{A}) - \nabla^2 \mathbf{A} \tag{VII-51}$$

VII.3.3 Integration

$$\oint_C \mathbf{A} \cdot d\mathbf{l} = \iint_S (\nabla \times \mathbf{A}) \cdot d\mathbf{s} \qquad \text{Stoke's theorem} \qquad \text{(VII-52)}$$

$$\oiint_S \mathbf{A} \cdot d\mathbf{s} = \iiint_V (\nabla \cdot \mathbf{A}) dv \qquad \text{Divergence theorem} \qquad \text{(VII-53)}$$

$$\oiint_S (\hat{\mathbf{n}} \times \mathbf{A}) ds = \iiint_V (\nabla \times \mathbf{A}) dv \qquad \text{(VII-54)}$$

$$\oiint_S \psi d\mathbf{s} = \iiint_V \nabla \psi \, dv \qquad \text{(VII-55)}$$

$$\oint_C \psi d\mathbf{l} = \iint_S \hat{\mathbf{n}} \times \nabla \psi \, ds \qquad \text{(VII-56)}$$

APPENDIX VIII

Method of Stationary Phase

In many problems, the following integral is often encountered and in most cases cannot be integrated exactly:

$$I(k) = \int_a^b \int_c^d F(x,y) e^{jkf(x,y)} \, dx \, dy \quad \text{(VIII-1)}$$

where

k = real

$f(x, y)$ = real, independent of k, and nonsingular

$F(x, y)$ = may be complex, independent of k, and nonsingular

Often, however, the above integral needs to be evaluated only for large values of k, but the task is still formidable. An approximate technique, known as the *Method of Stationary Phase*, exists that can be used to obtain an approximate asymptotic expression, for large values of k, for the above integral.

The method is justified by the asymptotic approximation of the single integral

$$I'(k) = \int_a^b F(x) e^{jkf(x)} \, dx \quad \text{(VIII-2)}$$

where

k = real

$f(x, y)$ = real, independent of k, and nonsingular

$F(x, y)$ = may be complex, independent of k, and nonsingular

which can be extended to include double integrals.

Antenna Theory: Analysis and Design, Fourth Edition. Constantine A. Balanis.
© 2016 John Wiley & Sons, Inc. Published 2016 by John Wiley & Sons, Inc.
Companion Website: www.wiley.com/go/antennatheory4e

The asymptotic evaluation of (VIII-1) for large k is based on the following: $f(x, y)$ is a well behaved function and its variation near the stationary points x_s, y_s determined by

$$\left.\frac{\partial f}{\partial x}\right|_{\substack{x=x_s \\ y=y_s}} \equiv f'_x(x_s, y_s) = 0 \tag{VIII-3a}$$

$$\left.\frac{\partial f}{\partial y}\right|_{\substack{x=x_s \\ y=y_s}} \equiv f'_y(x_s, y_s) = 0 \tag{VIII-3b}$$

is slow varying. Outside these regions, the function $f(x, y)$ varies faster such that the exponential factor $\exp[jkf(x, y)]$ of the integrand oscillates very rapidly between the values of $+1$ and -1, for large values of k. Assuming $F(x, y)$ is everywhere a slow varying function, the contributions to the integral outside the stationary points tend to cancel each other. Thus the only contributors to the integral, in an approximate sense, are the stationary points and their neighborhoods. Thus, we can write (VIII-1) approximately as

$$\begin{aligned} I(k) &\simeq \int_{-\infty}^{+\infty} \int_{-\infty}^{+\infty} F(x_s, y_s) e^{jkf(x,y)} \, dx \, dy \\ &= F(x_s, y_s) \int_{-\infty}^{+\infty} \int_{-\infty}^{+\infty} e^{jkf(x,y)} \, dx \, dy \end{aligned} \tag{VIII-4}$$

where the limits have been extended, for convenience, to infinity since the net contribution outside the stationary points and their near regions is negligible.

In the neighborhood of the stationary points, the function $f(x, y)$ can be approximated by a truncated Taylor series

$$f(x, y) \simeq f(x_s, y_s) + \tfrac{1}{2}(x - x_s)^2 f''_{xx}(x_s, y_s) + \tfrac{1}{2}(y - y_s)^2 f''_{yy}(x_s, y_s) \\ + (x - x_s)(y - y_s) f''_{xy}(x_s, y_s) \tag{VIII-5}$$

since

$$f'_x(x_s, y_s) = f'_y(x_s, y_s) = 0 \tag{VIII-6}$$

by (VIII-3a) and (VIII-3b). For convenience, we have adopted the notation

$$\left.\frac{\partial^2 f}{\partial x^2}\right|_{\substack{x=x_s \\ y=y_s}} \equiv f''_{xx}(x_s, y_s) \tag{VIII-7a}$$

$$\left.\frac{\partial^2 f}{\partial y^2}\right|_{\substack{x=x_s \\ y=y_s}} \equiv f''_{yy}(x_s, y_s) \tag{VIII-7b}$$

$$\left.\frac{\partial^2 f}{\partial x \partial y}\right|_{\substack{x=x_s \\ y=y_s}} \equiv f''_{xy}(x_s, y_s) \tag{VIII-7c}$$

For brevity, we write (VIII-5) as

$$f(x, y) \simeq f(x_s, y_s) + A\xi^2 + B\eta^2 + C\xi\eta \tag{VIII-8}$$

where

$$A = \frac{1}{2} f''_{xx}(x_s, y_s) \qquad \text{(VIII-8a)}$$

$$B = \frac{1}{2} f''_{yy}(x_s, y_s) \qquad \text{(VIII-8b)}$$

$$C = f''_{xy}(x_s, y_s) \qquad \text{(VIII-8c)}$$

$$\xi = (x - x_s) \qquad \text{(VIII-8d)}$$

$$\eta = (y - y_s) \qquad \text{(VIII-8e)}$$

Using (VIII-8)–(VIII-8e) reduces (VIII-4) to

$$I(k) \simeq F(x_s, y_s) e^{jkf(x_s, y_s)} \int_{-\infty}^{+\infty} \int_{-\infty}^{+\infty} e^{jk(A\xi^2 + B\eta^2 + C\xi\eta)} \, d\xi \, d\eta \qquad \text{(VIII-9)}$$

We now write the quadratic factor of the exponential, by a proper rotation of the coordinate axes ξ, η to μ, λ, in a diagonal form as

$$A\xi^2 + B\eta^2 + C\xi\eta = A'\mu^2 + B'\lambda^2 \qquad \text{(VIII-10)}$$

related to A, B, and C by

$$A' = \frac{1}{2}[(A+B) + \sqrt{(A+B)^2 - (4AB - C^2)}] \qquad \text{(VIII-10a)}$$

$$B' = \frac{1}{2}[(A+B) - \sqrt{(A+B)^2 - (4AB - C^2)}] \qquad \text{(VIII-10b)}$$

which are found by solving the secular determinant

$$\begin{vmatrix} (A - \zeta) & C/2 \\ C/2 & (B - \zeta) \end{vmatrix} = 0 \qquad \text{(VIII-11)}$$

with $\zeta_1 = A'$ and $\zeta_2 = B'$. Substituting (VIII-10) into (VIII-9) we can write

$$I(k) \simeq F(x_s, y_s) e^{jkf(x_s, y_s)} \int_{-\infty}^{+\infty} \int_{-\infty}^{+\infty} e^{jk(A'\mu^2 + B'\lambda^2)} \, d\mu \, d\lambda$$

$$I(k) \simeq F(x_s, y_s) e^{jkf(x_s, y_s)} \int_{-\infty}^{+\infty} e^{\pm jk|A'|\mu^2} \, d\mu \int_{-\infty}^{+\infty} e^{\pm jk|B'|\lambda^2} \, d\lambda \qquad \text{(VIII-12)}$$

where the signs in the exponents are determined by the signs of A' and B', which in turn depend upon A and B, as given in (VIII-10a) and (VIII-10b). The two integrals in (VIII-12) are of the same form and can be evaluated by examining the integral

$$I''(k) = \int_{-\infty}^{+\infty} e^{\pm jk|\alpha|t^2} \, dt = 2 \int_{0}^{\infty} e^{\pm jk|\alpha|t^2} \, dt \qquad \text{(VIII-13)}$$

where α can represent either A' or B' of (VIII-12). Making a change of variable by letting

$$k|\alpha|t^2 = \frac{\pi}{2}\tau^2 \qquad \text{(VIII-13a)}$$

$$dt = \sqrt{\frac{\pi}{2k|\alpha|}} \, d\tau \qquad \text{(VIII-13b)}$$

we can rewrite (VIII-13) as

$$I''(k) = 2\sqrt{\frac{\pi}{2k|\alpha|}} \int_0^\infty e^{\pm j\frac{\pi}{2}\tau^2} d\tau \qquad \text{(VIII-14)}$$

The integral is recognized as being the complex Fresnel integral whose value is

$$\int_0^\infty e^{\pm j\frac{\pi}{2}\tau^2} d\tau = \frac{1}{2}(1 \pm j) = \frac{1}{\sqrt{2}} e^{\pm j\frac{\pi}{4}} \qquad \text{(VIII-15)}$$

which can be used to write (VIII-14) as

$$I''(k) = 2\sqrt{\frac{\pi}{2k|\alpha|}} \int_0^\infty e^{\pm j\frac{\pi}{2}\tau^2} d\tau = \sqrt{\frac{\pi}{k|\alpha|}} e^{\pm j\frac{\pi}{4}} \qquad \text{(VIII-16)}$$

The result of (VIII-16) can be used to reduce (VIII-12) to

$$I(k) \simeq F(x_s, y_s) e^{jkf(x_s,y_s)} \frac{\pi}{k\sqrt{|A'||B'|}} e^{\pm j\frac{\pi}{4}} e^{\pm j\frac{\pi}{4}} \qquad \text{(VIII-17)}$$

If A' and B' are both positive, then $e^{\pm j\frac{\pi}{4}} e^{\pm j\frac{\pi}{4}} = e^{+j\frac{\pi}{2}} = +j$

If A' and B' are both negative, then $e^{\pm j\frac{\pi}{4}} e^{\pm j\frac{\pi}{4}} = e^{-j\frac{\pi}{2}} = -j$

If A' and B' have different signs, then $e^{\pm j\frac{\pi}{4}} e^{\pm j\frac{\pi}{4}} = 1$

Thus, (VIII-17) can be cast into the form

$$I(k) \simeq F(x_s, y_s) e^{jkf(x_s,y_s)} \frac{j\pi\delta}{k\sqrt{|A'||B'|}} \qquad \text{(VIII-18)}$$

where

$$\delta = \begin{cases} +1 & \text{if } A' \text{ and } B' \text{ are both positive} \\ -1 & \text{if } A' \text{ and } B' \text{ are both negative} \\ -j & \text{if } A' \text{ and } B' \text{ have different signs} \end{cases} \qquad \text{(VIII-18a)}$$

Examining (VIII-10a) and (VIII-10b), it is clear that

a. A' and B' are real (because $A, B,$ and C are real)
b. $A' + B' = A + B$ \qquad (VIII-19)
c. $A'B' = (4AB - C^2)/4$

Using the results of (VIII-19), we reduce (VIII-18) to

$$I(k) \simeq F(x_s, y_s) e^{jkf(x_s,y_s)} \frac{j2\pi\delta}{k\sqrt{|4AB - C^2|}}. \qquad \text{(VIII-20)}$$

To determine the signs of A' and B', let us refer to (VIII-19).

a. If $4AB > C^2$, then A and B have the same sign and $A'B' > 0$. Thus, A' and B' have the same sign.
   1. If $A > 0$ then $B > 0$ and $A' > 0, B' > 0$
   2. If $A < 0$ then $B < 0$ and $A' < 0, B' < 0$

b. If $4AB < C^2$, then $A'B' < 0$, and A' and B' have different signs. Summarizing the results we can write that
   1. If $4AB > C^2$ and $A > 0$, then A' and B' are both positive
   2. If $4AB > C^2$ and $A < 0$, then A' and B' are both negative
   3. If $4AB < C^2$, then A' and B' have different signs

Using the preceding deductions, we can write the sign information of (VIII-18a) as

$$\delta = \begin{cases} +1 & \text{if } 4AB > C^2 \text{ and } A > 0 \\ -1 & \text{if } 4AB > C^2 \text{ and } A < 0 \\ -j & \text{if } 4AB < C^2 \end{cases} \qquad \text{(VIII-21)}$$

in the evaluation of the integral in

$$I(k) \simeq F(x_s, y_s) e^{jkf(x_s, y_s)} \int_{-\infty}^{+\infty} \int_{-\infty}^{+\infty} e^{jk(A\xi^2 + B\eta^2 + C\xi\eta)} \, d\xi \, d\eta$$

$$I(k) \simeq F(x_s, y_s) e^{jkf(x_s, y_s)} \frac{j2\pi\delta}{k\sqrt{|4AB - C^2|}} \qquad \text{(VIII-22)}$$

APPENDIX IX

Television, Radio, Telephone, and Radar Frequency Spectrums

IX.1 TELEVISION

IX.1.1 Very High Frequency (VHF) Channels

Channel number	2 3 4	5 6	7 8 9 10 11 12 13
Frequency (MHz)	54 ↑ 60 ↑ 66 ↑ 72	76 ↑ 82 ↑ 88	174 ↑ 180 ↑ 186 ↑ 192 ↑ 198 ↑ 204 ↑ 210 ↑ 216

IX.1.2 Ultra High Frequency (UHF) Channels*

Channel number	14 15 16 17 18 19 20 ... 82 83
Frequency (MHz)	470 ↑ 476 ↑ 482 ↑ 488 ↑ 494 ↑ 500 ↑ 506 ↑ 512 ... 878 ↑ 884 ↑ 890

For both VHF and UHF channels, each channel has a 6-MHz bandwidth. For each channel, the carrier frequency for the video part is equal to the lower frequency of the bandwidth plus 1.25 MHz while the carrier frequency for the audio part is equal to the upper frequency of the bandwidth minus 0.25 MHz.

Examples: Channel 2 (VHF): $f_0(\text{video}) = 54 + 1.25 = 55.25$ MHz
$f_0(\text{audio}) = 60 - 0.25 = 59.75$ MHz

Channel 14 (UHF): $f_0(\text{video}) = 470 + 1.25 = 471.25$ MHz
$f_0(\text{audio}) = 476 - 0.25 = 475.75$ MHz

*In top ten urban areas in the United States, land mobile is allowed in the first seven UHF TV channels (470–512 MHz).

Antenna Theory: Analysis and Design, Fourth Edition. Constantine A. Balanis.
© 2016 John Wiley & Sons, Inc. Published 2016 by John Wiley & Sons, Inc.
Companion Website: www.wiley.com/go/antennatheory4e

IX.2 RADIO

IX.2.1 Amplitude *M*odulation (AM) Radio

Number of channels: 107 (each with 10-kHz separation)
Frequency range: 535–1605 kHz

IX.2.2 *F*requency *M*odulation (FM) Radio

Number of channels: 100 (each with 200-kHz separation)
Frequency range: 88–108 MHz

IX.3 AMATEUR BANDS

Band	Frequency (MHz)	Band	Frequency (MHz)
160-m	1.8–2.0	2-m	144.0–148.0
80-m	3.5–4.0	—	220–225
40-m	7.0–7.3	—	420–450
20-m	14.0–14.35	—	1215–1300
15-m	21.0–21.45	—	2300–2450
10-m	28.0–29.7	—	3300–3500
6-m	50.0–54.0	—	5650–5925

IX.4 CELLULAR TELEPHONE

IX.4.1 Land Mobile Systems

Uplink: MS to BS (mobile station to base station)
Downlink: BS to MS (base station to mobile station)

System	Uplink (MHz)**	Downlink (MHz)*	Major Covered Areas
CDMA IS-95	824–849	869–894	North America, Korea, China
GSM	890–915	935–960	North America, Europe, China, Japan
Extended-GSM	880–915	925–960	North America, Europe, China, Japan
DCS 1800	1710–1785	1805–1880	Europe
US PCS 1900	1850–1910	1930–1990	North America
WCDMA	1920–1980	2110–2170	Everywhere
CDMA2000	All existing CDMA system frequencies		Everywhere

*Downlink is the channel from the base station to the mobile unit, which is also called forward-link.
**Uplink is the channel from the mobile unit to the base station, which is also called reverse-link.

IX.4.2 Cordless Telephone

United States of America: 46–49 MHz
Digital European Cordless Telecommunications (DECT): 1.880–1.990 GHz

IX.5 RADAR IEEE BAND DESIGNATIONS

HF (High Frequency):	3–30	MHz
VHF (Very High Frequency):	30–300	MHz
UHF (Ultra High Frequency):	300–1,000	MHz
L-band:	1–2	GHz
S-band:	2–4	GHz
C-band:	4–8	GHz
X-band:	8–12	GHz
K_u-band:	12–18	GHz
K-band:	18–27	GHz
K_a-band:	27–40	GHz
Millimeter wave band:	40–300	GHz

Index

adaptive beamforming, 950–953, 951f, 952f, 969–971, 970f, 971f, 972f
amplitude pattern shape, 32–33, 33f
amplitude pattern shape, radiation pattern (antenna pattern), 32–33, 33f
analysis methods, 20–21
antena, radiation mechanism
    dipole, 13, 14f
    E-plane sectoral horn, unbounded medium (te_horn), 15
    infinite line source, unbounded medium (tm_open), 15
    radiation problems, computer animated-visualization, 13–15
    single wire, 7–10, 8f, 10f
    two-wires, 10–11, 11f, 12f
antenna
    current distribution, thin wire antenna, 15–18, 16f, 17f, 18f
    defined, overview, 1
    as transition device, 2f
    transmitting mode, transmission-line Thevenin equivalent, 1, 2f
antenna beamforming
    adaptive beamforming, 950–953, 951f, 952f
    direction-of-arrival (DOA) algorithms, 947–950, 947f, 950f
    optimal beamforming techniques, least mean square (LMS) algorithm, 956–959, 957f, 958f, 959f
    optimal beamforming techniques, minimum mean square error (MMSE) criterion, 955–956
antenna efficiency, 60–61, 61f
antenna elements, 19–20
antenna equivalent areas, 83–86
antenna, historical advancement, 18
    analysis methods, 20–21
    antenna elements, 19–20
    future challenges, 21

antenna measurements
    current measurements, 1014
    directivity measurements, 1010–1012
    impedance measurements, 1012–1014
    overview, 981–982, 982f
    polarization measurements, 1014–1019, 1015f, 1016f, 1017f, 1018f
    radiation efficiency, 1012
antenna measurements, antenna ranges, 982
    compact ranges, 986–987, 986f
    compact ranges, CATR designs, 989–992, 991f, 992f
    compact ranges, CATR performance, 987–989, 988f, 989f, 990f
    free-space ranges, anechoic chambers, 984–986, 985f
    free-space ranges, elevated ranges, 983–984, 984f
    free-space ranges, slant ranges, 984, 984f
    near-field/far-field methods, measurements and computations, 997–1000, 999f
    near-field/far-field methods, modal-expansion method for planar systems, 996–997
    reflection ranges, 983, 983f
antenna measurements, gain measurements, 1003, 1005–1006
    realized-gain measurements, extrapolation method, 1008
    realized-gain measurements, gain-transfer (gain-comparison), measurements, 1009–1010
    realized-gain measurements, ground-reflection range method, 1008–1009
    realized-gain measurements, three-antenna method, 1007–1008
    realized-gain measurements, two-antenna method, 1006, 1007f
antenna measurements, near-field/far-field methods, 992–996, 993f, 994f, 995f

---

Antenna Theory: Analysis and Design, Fourth Edition. Constantine A. Balanis.
© 2016 John Wiley & Sons, Inc. Published 2016 by John Wiley & Sons, Inc.
Companion Website: www.wiley.com/go/antennatheory4e

antenna measurements, radiation patterns,
   1000–1001, 1000f
  amplitude pattern, 1001f, 1003, 1004f, 1005f
  instrumentation, 1001–1003, 1001f, 1002f
  phase measurements, 1003, 1005f
antenna measurements, scale model measurements,
   1019, 1019t
  echo area (RCS) measurements, simulations and
     comparisons, 1021–1024, 1022f, 1023f
  gain (amplitude) measurements, simulations and
     comparisons, 1020–1021, 1020f
antenna miniaturization, 619
  folding, 624–626, 624f, 626f
  metamaterials, 626–627
  monopole antenna, impedance loading, 620–622,
     620f
  monopole antenna, materials loading, 622–624,
     622f
  patch antennas, 626
antenna radar cross section (RCS), 92–96, 92t, 96f,
   97f
antenna radiation efficiency, 80
antenna synthesis, continuous sources
  continuous aperture sources, 417
  continuous aperture sources, circular aperture,
     418–419, 419t
  continuous aperture sources, rectangular aperture,
     418
  discretization of continuous sources, 387, 387f
  Fourier transform method, linear array, 395–398,
     398f
  Fourier transform method, line-source, 392–394,
     395f
  line-source, 386–387, 387f
  line-source phase distributions, 416–417,
     417f
  overview, 385
  Schelkunoff polynomial method, 387–391, 389f,
     390f, 391f, 392f
  Taylor line-source (one-parameter), 408–414,
     411f, 413f, 414f
  Taylor line-source (Tschebyscheff-error), 404–405
  Taylor line-source (Tschebyscheff-error), design
     procedure, 406–407, 408f
  triangular, cosine, cosine-squared amplitude
     distributions, 415–416, 415t
  Woodward-Lawson method, 398
  Woodward-Lawson method, linear array, 403–404
  Woodward-Lawson method, line-source, 399–402,
     402f
antenna temperature, 96, 98–1, 99f
antenna types
  aperture antennas, 3, 4f, 6f
  array antennas, 5, 6f
  lens antennas, 6–7, 8f
  microstrip antennas, 5, 5f, 6f
  reflector antennas, 6, 7f
  wire antennas, 3, 4f, 6f
antenna vector effective length, equivalent areas
  antenna equivalent areas, 83–86
  maximum directivity, maximum effective area,
     86–88, 86f
  vector effective length, 81–83, 81f
aperture antennas, 3, 4f, 6f
  Babinet's principle, 680–684 , 680f, 681f, 682f,
     683f
  directivity, 648
  field equivalence principle, Huygen's principle,
     639–645, 640f, 641f, 645f
  ground plane edge effects, geometrical theory of
     diffraction, 702–703, 703f, 704f, 705–707, 705f,
     706f
  radiation equations, 645–647
aperture antennas, circular apertures, 667–669, 668f
  beam efficiency, 675, 677f
  TE_{11}-mode distribution on infinite ground plane,
     671, 672–673t, 674, 674f, 676f
  uniform distribution on infinite ground plane,
     669–671, 671f
aperture antennas, design considerations, 676
  circular aperture, 678–679, 679t
  rectangular aperture, 677
aperture antennas, Fourier transforms
  aperture admittance, 695–702, 698f, 700f
  asymptotic evaluation of radiated field, 689–694
  dielectiric-covered apertures, 694–695, 696f
  Fourier transforms-spectral domain, 685
  radiated fields, 685–689, 685f
aperture antennas, rectangular apertures, 648–650,
   649f
  beam efficiency, 666–667, 666f, 677f
  TE_{10}-mode distribution on infinite ground plane,
     663, 663f, 664f, 665f
  uniform distribution in space, 658–659t, 661–663,
     662f
  uniform distribution on infinite ground plane,
     650–652, 650f, 653f, 654–657, 654f, 658–659t,
     660
array antennas, 5, 6f
arrays, array design, 285. *See also* mutual coupling
   in arrays
  design considerations, 361–362
  design procedure, 318–319
  linear array design, 967–968, 968f
  planar array design, 968–969
arrays, circular array
  array factor, 363–365, 363f, 367366f
arrays, feed networks
  microstrip and mobile communications antennas,
     832–837, 835f, 836f, 837f, 838f

arrays, *N*-element linear array: directivity, 312
 broadside, end-fire arrays, 313–314, 318*t*
 Hansen-Woodyard end-fire array, 317–318
 ordinary end-fire array, 315–317, 318*t*
arrays, *N*-element linear array: three-dimensional characteristics
 N-elements along *X*- or *Y*-axis, 320–322, 320*f*, 321*f*
 N-elements along *Z*-axis, 319–320
arrays, *N*-element linear array: uniform amplitude/spacing, 293–297, 294*f*
 broadside array, 297–299, 298*f*, 299*f*
 Hansen-Woodyard end-fire array, 304–312, 309*f*, 311*f*, 312*t*, 313*t*
 ordinary end-fire array, 299–300, 300*t*, 301*f*, 303*t* 302*f*, 304*t*
 phased (scanning) array, 302–304, 303*f*, 305*f*, 306*f*
arrays, *N*-element linear array: uniform spacing, nonuniform amplitude, 323–324
 array factor, 325–326, 325*f*
 binomial array, design procedure, 327–329, 328*f*
 binomial array, excitation coeffecients, 326–327
 Dolph-Tschebyscheff array broadside, array design, 331–338, 332*f*, 334*f*, 336*f*, 336*t*, 339*f*
 Dolph-Tschebyscheff array broadside, array factor, 330–331
 Dolph-Tschebyscheff array broadside, beamwidth and directivity, 338, 340–341, 340*f*
 Dolph-Tschebyscheff array broadside, design, 341–343, 343*f*
 Tschebyscheff design, scanning, 344–345, 345*f*
arrays, planar array, 348*f*
 array factor, 348–354, 350*f*, 353*f*, 354*f*, 355*f*, 356*f*
 beamwidth, 354, 357–359, 358*f*
 directivity, 359–350
arrays, rectangular-to-polar graphical solution, 322–333, 324*f*
arrays, superconductivity, 345
 designs with constraints, 346–348
 efficiency, directivity, 346
arrays, two-element array, 286–293, 287*f*, 291*f*, 292*f*, 293*f*

Babinets principle, 680–684
baluns, transformers, 521–523, 522*f*, 523*f*
bandwidth, 65–66
beam efficiency, 65
beamforming, diversity combining, Raleigh-fading, trellis coded modulation, 972–975, 972*t*, 973*f*, 974*f*
beamwidth, 40
broadband antennas. *See also* traveling wave antennas
 electric-magnetic dipole, 559, 560*f*
 helical antenna, 549–550, 550*f*, 551*f*
 Yagi-Uda array of linear elements, 559–561, 560*f*
broadband dipoles, matching techniques. *See also* cylindrical dipole; folded dipole; matching techniques; triangular sheet, bow-tie, and wire simulation of biconical antenna; Vivaldi antenna
 biconical antenna, input impedance for unipole, 491–492
 biconical antenna, input impedance for finite cones, 491
 biconical antenna, input impedance for infinite cones, 490–491, 492*f*
 biconical antenna, radiated fields, 487, 488*f*, 489–490
 overview, 485–487, 486*f*, 487*f*

Cassegrain reflectors, 915–916, 917*f*
 Cassegrain and Gregorian forms, 919–920, 919*f*
 classical form, 917–919, 918*f*
cellular radio systems evolution
 omnidirectional systems, 933, 933*f*
 omnidirectional systems, cell splitting, 934, 934*f*
 omnidirectional systems, sectorized systems, 934–936, 935*f*
 smart-antenna systems, adaptive array systems, 937–938, 937*f*, 938*f*
 smart-antenna systems, spatial division multiple access (SDMA), 938–939, 939*f*
 smart-antenna systems, switch-beam systems, 936–937, 936*f*, 937*f*
circular loop of constant current, loop antennas
 power density, radiation intensity, radiation resistance, directivity, 252–253, 253*f*, 254*f*, 255–259, 257*f*, 258*f*
 radiated fields, 250–252, 251*f*
circular loop of nonuniform current, loop antennas, 259–260, 260*f*, 261*f*, 262–266, 262*f*, 263*f*, 264*f*, 265*f*
 arrays, 266
 design procedure, 267–268
circular patch, microstrip and mobile communications antennas, 815, 815*f*
 conductance, directivity, 821–822, 822*f*, 823*f*
 design, 818–819, 819*f*
 electric and magnetic fields - TM_{mnp}^z, 816–817
 equivalent current densities, fields related, 819–821, 820*f*
 resonant frequencies, 817–818
 resonant input resistance, 822–823
coordinate system for antenna analysis, 25–26, 26*f*
corner reflector, 876–878, 876*f*, 877*f*, 878*f*, 880–884, 881*f*, 882*f*, 883*f*
cross-polarization, reflector antennas (front-fed parabolic), 896–897, 896*f*, 897*f*

1068 INDEX

current distribution method, reflector antennas (front-fed parabolic), 897–901, 898f, 900f
current distribution, thin wire antenna, 15–18, 16f, 17f, 18f
cylindrical dipole. *See also* folded dipole
  bandwidth, 501
  equivalent radii, 504–505, 506t
  input impedance, 501–502, 502f
  radiation patterns, 503–504, 505f
  resonance, ground plane simulation, 503, 503t, 504f

dielectric resonator antennas (DRAs), 847–848
  analysis and design methods, 849–850
  basic geometries, 848, 848f, 849f
  cavity model resonant frequencies (TE, TM modes), 850–852
  hybrid modes: resonant frequencies, quality factors, 852–853, 853f, 854t
  radiated fields, 855–859, 856f, 858f 857f
dipole array design, 608–609, 609f
  computer program, 613–614
  design equations, 609–612, 611f
  design procedure, 611–613, 613f
dipole, *See also* linear wire antennas, dipole array, 13, 14f, 602–605, 603f, 606f, 607–608, 607f, 607t, 608f
  directional patterns, 47–48, 47f, 49f, 50–51, 51t
direction-of-arrival (DOA) algorithms, 947–950, 947f, 950f
directivity, 41–46, 44f, 45f
  aperture efficiency, 898f, 900f, 901–909, 903f, 906f, 908f, 909t, 910f, 911f
  directional patterns, 47–48, 47f, 49f, 50–51, 51t
  omnidirectional patterns, 51–52, 52f, 53f, 54
discone, conical skirt model, 512–513, 512f, 513t
duality theorem, 138–139, 139t

electric and magnetic fields, electric (**J**) and magnetic (**M**) current sources, 131–132
electrically small antennas, fundamental limits, 614–619, 615f, 617f, 618f, 618t
electric-magnetic dipole, 559, 560f
electrostatic charge distribution, 432
  bent wire, 437–439, 438f
  finite straight wire, 433–437, 437f 433f
E-plane sectoral horn, 720f, 721f
  aperture fields, 719–722
  directivity, 728–733, 730f, 731f, 732f
E-plane sectoral horn, unbounded medium (te_horn), 15

equiangular spiral antennas, 593–594
  conical spiral, 598, 599f
  planar spiral, 593–598, 596f, 597f

far-field radiation, 136–137
feed design, reflector antennas (front-fed parabolic), 913–915, 915f
ferrite loop, loop antennas
  ferrite-loading receiving loop, 271–272
  radiation resistance, 270–271, 271f
field regions, 31–33, 32f, 33f
folded dipole, 505–512, 507f, 509f, 511f
4×4 vs 8×8 planar array, 969
fractal antennas, 627–633, 628f, 629f, 630f, 631f, 632f
frequency independent antennas. *See also* antenna miniaturization; electrically small antennas, fundamental limits; equiangular spiral antennas; fractal antennas; log-periodic antennas
  overview, 591–592
  theory, 592–593
Friis transmission equation, 88–90, 88f
future challenges, 21

gain, realized gain, 61–64
ground effects, 203–216

half wavelength dipole, 176–179
Hallén's integral equation, 444–445
helical antenna, 549–550, 550f, 551f
  design procedure, 553–558, 557f
  end-fire mode, 553
  feed design, 558–559, 559f
  normal mode, 550–553, 552f
horizontal electric dipole, 195–203
horn antennas
  aperture-matched horns, 766–769, 767f, 768f
  conical horn, 756–759, 757f, 758f, 759f, 760f, 761t
  corrugated horn, 761–764, 762f, 763f, 765f, 766
  dielectric-loaded horns, 771, 773
  multimode horns, 769–771, 769f, 772–773f
  phase center, 773–774
horn antennas, *E*-plane sectoral horn, 720f, 721f
  aperture fields, 719–722
  directivity, 728–733, 730f, 731f, 732f
horn antennas, *H*-plane sectoral horn, 734f
  aperture fields, 733
  directivity, 738–741, 741f, 742f, 743
  radiated fields, 734–738, 737f, 738f, 739f
horn antennas, pyramidal horn, 743, 744f
  aperture fields, equivalent, radiated fields, 744–748, 746f, 747f, 748f, 749f
  design procedure, 754–756
  directivity, 748–754, 751f, 752f

H-plane sectoral horn, 734*f*
   aperture fields, 733
   directivity, 738–741, 741*f*, 742*f*, 743
   radiated fields, 734–738, 737*f*, 738*f*, 739*f*

important parameters, associated formulas, equation numbers, 101–103*t*
induced EMF method, 458
   near-field of dipole, 458–460, 458*f*
   self-impedance, 460–463, 460*f*, 462*f*, 463*f*
induced EMF method, mutual impedance between linear elements, 467–473, 470*f* 468*f*, 471*f*, 473*f* 472*f*
infinite line source, unbounded medium (tm_open), 15
inhomogeneous vector potential wave solution, 132–136, 133*f*
input impedance, 75–79, 75*f*, 78*f*
integral equation-moment method, 455–457, 457*f*, 457*t*
   Mini-Numerical Electromagnetic Code (MININEC), 467
   mutual impedance between linear elements, 465
   Numerical Electromagnetic Code (NEC), 466
integral equations, 431. *See also* moment method solution
integral equations, finite diameter wires
   Hallén's integral equation, 444–445
   Pocklington's integral equation, 440–444, 440*f*
   source modeling, delta gap, 445, 446*f*, 448*t*
   source modeling, magnetic-frill generator, 446–448, 446*f*, 448*t*
integral equations, Integral Equation (IE) method
   electrostatic charge distribution, 432
   electrostatic charge distribution, bent wire, 437–439, 438*f*
   electrostatic charge distribution, finite straight wire, 433–437, 437*f* 433*f*
   integral equation, 439
inverted-F antenna (IFA), 843–845, 843*f*, 845*f*
isotropic, directional, omnidirectional patterns, 30, 31*f*

least mean square (LMS) algorithm, optimal beamforming techniques, 956–959, 957*f*, 958*f*, 959*f*
lens antennas, 6–7, 8*f*
linear, circular, elliptical polarization, 68–71
linear elements near/on infinite perfect electric conductors (PEC), perfect magnetic conductors (PMC), and electromagnetic band-gap (EBG) surfaces, 179
   ground effects, 203
   ground planes: electric, magnetic, 180–182, 182*f*, 183*f*

   horizontal electric dipole, 195, 196*f*, 197–203, 198*f*, 200*f*, 201*f*, 202*f*, 203*f*
   image theory, 182–183, 184*f*, 185*f*
   mobile communication devices, antennas for mobile communication, 192–195, 195*f*, 196*f*
   rapid calculations and design, approximate formulas, 191–192
   vertical electric dipole, 183–187, 184*f*, 185*f*, 187*f*, 188*f*, 189, 189*f*, 190*f*, 191, 191*f*
linear wire antennas. *See also* linear elements near/on infinite perfect electric conductors (PEC), perfect magnetic conductors (PMC), and electromagnetic band-gap (EBG) surfaces
   computer codes, 216
   dipole in far field, parameters, formulas, equation numbers, 217–218*t*
linear wire antennas, finite length dipole
   current distribution, 164
   directivity, 172–173
   input resistance, 173–175, 174*f*, 176*f*
   power density, radiation intensity, radiation resistance, 166–172, 167*f*, 168*f*, 169*f*, 171*f*
   radiated fields: element, space, and pattern multiplication, 164–166
linear wire antennas, ground effects, 203
   earth curvature, 211–216, 212*f*, 213*f*, 215*f*
   horizontal electric dipole, 205, 207, 207*f*, 208*f*, 209*f*, 210*f*
   PEC, PMC, EBG surfaces, 207, 210
   verticle electric dipole, 204–205, 206*f*
linear wire antennas, half-wavelength dipole, 176–177, 178*f*, 179, 179*t*
linear wire antennas, infinitesimal dipole
   directivity, 154–155, 155*f*
   far-field ($kr \gg 1$) region, 153–154
   intermediate-field ($kr > 1$) region, 152
   near-field ($kr \ll 1$) region, 151–152
   power density, radiation resistance, 148–150
   radian distance, radian sphere, 150–151, 151*f*
   radiated fields, 145–148, 146*f*
linear wire antennas, region separation
   far-field (Fraunhofer) region, 160–162
   radiating near-field (Fresnel) region, 162–163
   reactive near-field region, 163–164
linear wire antennas, small dipole, 155, 156*t*, 157*f*, 158
log-periodic antennas, 598
   dipole array, 602–605, 603*f*, 606*f*, 607–608, 607*f*, 607*t*, 608*f*
   dipole array design, 608–609, 609*f*
   dipole array design, computer program, 613–614
   dipole array design, design equations, 609–612, 611*f*

log-periodic antennas (*Continued*)
  dipole array design, design procedure, 611–613, 613*f*
  planar, wire surfaces, 599–602, 600*f*, 601*f*
long wire
  amplitude patterns, maxima, nulls, 538–539, 539*f*, 540*f*, 541*f*
  input impedance, 541–542
  resonant wires, 542–543
loop antennas, 235, 236*f*
  circular loops ground/earth curvature effects, 268–269, 268*f*, 269*f*
  loop in far field paramaters, formulas, equation numbers, 274–275*t*
  mobile communications applications, 272, 273*f*
  polygonal loop antennas, 269–270
loop antennas, circular loop of constant current
  power density, radiation intensity, radiation resistance, directivity, 252–253, 253*f*, 254*f*, 255–259, 257*f*, 258*f*
  radiated fields, 250–252, 251*f*
loop antennas, circular loop of nonuniform current, 259–260, 260*f*, 261*f*, 262–266, 262*f*, 263*f*, 264*f*, 265*f*
  arrays, 266
  design procedure, 267–268
loop antennas, ferrite loop
  ferrite-loading receiving loop, 271–272
  radiation resistance, 270–271, 271*f*
loop antennas, small circular loop
  equivalent circuit, receiving mode, 249–250, 249*f*
  equivalent circuit, transmitting mode, 247–249, 248*f*
  far-field ($kr \gg 1$) region, 245–246
  near-field ($kr \ll 1$) region, 246
  power density, radiation resistance, 241–245, 244*f*
  radiated fields, 236–241, 237*f*
  radiation intensity, directivity, 246–247
  small loop, infinitesimal magnetic dipole, 241

matching techniques
  baluns, transformers, 521–523, 522*f*, 523*f*
  quarter-wavelength transformer, binomial design, 515–518
  quarter-wavelength transformer, multiple sections, 515
  quarter-wavelength transformer, single section, 514–515
  quarter-wavelength transformer, Tschebyscheff design, 518–522, 518*f*, 520*f*
  stub-matching, 513
maximum directivity, maximum effective area, 86–88, 86*f*
microstrip and mobile communications antennas
  analysis methods, 787–788
  arrays, feed networks, 832–837, 835*f*, 836*f*, 837*f*, 838*f*
  basic characteristics, 784–785, 784*f*, 785*f*
  circular polarization, 830–832, 832*f*, 832*t*, 833*f*, 834*f*
  coupling, 827–830, 828*f*, 829*f*
  dielectric resonator antennas (DRAs), 847–848
  feeding methods, 785–787, 786*f*, 787*f*
  input impedance, 826–827, 826*f*
  inverted-F antenna (IFA), 843–845, 843*f*, 845*f*
  multiband antennas for mobile units, 846–847, 846*f*
  overview, 783–784, 784*t*
  parameters, formulas, equation numbers, 859*t*
  planar inverted-F antenna (PIFA), 838*f*, 839–840, 839*f*, 841*f*
  quality factor, bandwidth, efficiency, 823–826, 825*f*
  slot antenna, 841–843, 842*f*
microstrip and mobile communications antennas, circular patch, 815, 815*f*
  conductance, directivity, 821–822, 822*f*, 823*f*
  design, 818–819, 819*f*
  electric and magnetic fields - TM^z_{mnp}, 816–817
  equivalent current densities, fields radiated, 819–821, 820*f*
  resonant frequencies, 817–818
  resonant input resistance, 822–823
microstrip and mobile communications antennas, rectangular patch
  cavity model, 799, 799*f*
  cavity model, equivalent current densities, 804–807, 805*f*, 806*f*, 807
  cavity model, field configurations (modes) - TM^x, 800–803, 801*f*, 804*f*
  directivity, double slot ($k_0 h \ll 1$), 812–815, 814*f*
  directivity, single slot $k_0 h \ll 1$), 811–812, 812*f*
  nonradiating slots, 810–811
  radiating slots, 807–810, 809*f*
  transmission-line model effective length, conductance, 793–794, 793*f*, 795*f*
  transmission-line model effective length, design, 791–792, 792*f*
  transmission-line model effective length, matching techniques, 796–799, 796*f*, 799*f*
  transmission-line model effective length, resonant frequency, effective width, 790–791, 790*f*
  transmission-line model effective length, resonant input resistance, 794–796
  transmission-line model, fringing effects, 788–789, 789*f*
microstrip antennas, 5, 5*f*, 6*f*
Mini-Numerical Electromagnetic Code (MININEC), 467

mobile ad hoc networks (MANETs)
  MANETs employing smart-antenna systems, protocol, 962–964, 963f
  MANETs employing smart-antenna systems, wireless network, 961–962, 962f
  overview, 960–961, 960f
moment method solution, 448
  basis (expansion) functions, entire-domain functions, 453
  basis (expansion) functions, subdomain functions, 449–452, 450f, 451f, 452f
  weighting (testing) functions, 453, 454f, 455
multiband antennas for mobile units, 846–847, 846f
mutual coupling in arrays
  active element pattern in array, 478–480, 478f, 480f
  infinite regular array coupling, 476–478
  mutual coupling on array performance, 476
  receiving mode coupling, 476
  transmitting mode coupling, 474–476, 475f
mutual impedance between linear elements, 463–465, 464f, 466f. *See also* self-impedance
  induced EMF method, 467–473, 470f 468f, 471f, 473f 472f
  integral equation-moment method, 465
  integral equation-moment method, Mini-Numerical Electromagnetic Code (MININEC), 467
  integral equation-moment method, Numerical Electromagnetic Code (NEC), 466
mutual impedance between linear elements, integral equation-moment method, 465

90° corner reflector, 878–880, 879f
Numerical Equivalent Code (NEC), 466

omnidirectional patterns, 51–52, 52f, 53f, 54
omnidirectional systems, 933, 933f
  cell splitting, 934, 934f
  sectorized systems, 934–936, 935f
optimal beamforming techniques
  least mean square (LMS) algorithm, 956–959, 957f, 958f, 959f
  minimum mean square error (MMSE) criterion, 955–956
optimal beamforming techniques, minimum mean square error (MMSE) criterion, 955–956

phase errors, reflector antennas (front-fed parabolic), 910–913, 914f
planar inverted-F antenna (PIFA), 838f, 839–840, 839f, 841f
planar, wire surfaces, 599–602, 600f, 601f
plane reflector, 875–876, 876f
Pocklington's integral equation, 440–444, 440f
polarization, 66–68, 67f
  linear, circular, elliptical polarization, 68–71
  polarization loss factor (PLF), efficiency, 71–75, 71f, 74f
pyramidal horn, 743, 744f
  aperture fields, equivalent, radiated fields, 744–748, 746f, 747f, 748f, 749f
  design procedure, 754–756
  directivity, 748–754, 751f, 752f

quarter-wavelength transformer
  binomial design, 515–518
  multiple sections, 515
  single section, 514–515
  Tschebyscheff design, 518–522, 518f, 520f

radar cross section (RCS), 92–96, 92t, 96f, 97f
radar range equation, 90–92, 91f
radian, steradian, 33, 34f
radiating far-field (Fraunhofer) region, 32, 32f
radiating near-field (Fresnel) region, 31–32, 32f
radiation integrals, auxilliary potential functions
  electric and magnetic sources, computing fields block diagram, 128f
  overview, 127–128
radiation intensity, 37–38, 39f
radiation pattern (antenna pattern), 25–26, 26f, 27f
  amplitude pattern shape, 32–33, 33f
  coordinate system for antenna analysis, 25–26, 26f
  field regions, 31–33, 32f, 33f
  isotropic, directional, omnidirectional patterns, 30, 31f
  principle patterns, 30–31
  radiation pattern lobes, 26–30, 28f
radiation pattern lobes, 26–30, 28f
radiation patterns, numerical techniques, 55–57, 56f, 58f, 59–60, 60f
radiation power density, 35–37
radiation problems, computer animation-visualization, 13–15
reactive near-field region, 31, 32f
reciprocity, reaction theorems, 138–140
  for antenna radiation patterns, 141–143, 142f
  for two antennas, 140–141, 140f, 141f
rectangular patch, 965–966, 966f, 967f, 968f
rectangular patch, microstrip and mobile communications antennas
  nonradiating slots, 810–811
reflector antennas, 6, 7f
  90° corner reflector, 878–880, 879f
  corner reflector, 876–878, 876f, 877f, 878f, 880–884, 881f, 882f, 883f
  plane reflector, 875–876, 876f
reflector antennas, front-fed parabolic
  aperture distribution method, 890–896, 891f, 893f, 895f
  Cassegrain reflectors, 915–916, 917f

reflector antennas, front-fed parabolic (*Continued*)
  Cassegrain reflectors, Cassegrain and Gregorian forms, 919–920, 919*f*
  Cassegrain reflectors, classical form, 917–919, 918*f*
  cross-polarization, 896–897, 896*f*, 897*f*
  current distribution method, 897–901, 898*f*, 900*f*
  directivity and aperture efficiency, 898*f*, 900*f*, 901–909, 903*f*, 906*f*, 908*f*, 909*t*, 910*f*, 911*f*
  feed design, 913–915, 915*f*
  induced current, 890
  phase errors, 910–913, 914*f*
  surface geometry, 887–890, 888*f*
reflector antennas, parabolic, 884–887, 885*f*, 886*f*
reflector antennas, spherical reflector, 920–922, 920*f*, 922*f*
rhombic antenna geometry, design equations, 549

self-impedance. *See also* mutual impedances between linear elements
  induced EMF method, 458
  induced EMF method, near-field of dipole, 458–460, 458*f*
  induced EMF method, self-impedance, 460–463, 460*f*, 462*f*, 463*f*
  integral equation-moment method, 455–457, 457*f*, 457*t*
slot antenna, 841–843, 842*f*
smart antennas
  antenna, array design, 943
  antenna, linear array, 944–945, 944*f*, 945*f*
  antenna, planar array, 945–946, 946*f*
  beamforming, diversity combining, Raleigh-fading, trellis code modulation, 972–975, 972*t*, 973*f*, 974*f*
  other geometries, 975–976, 975*f*
  signal propagation, 939–941, 940*f*, 941*f*, 942*f*
  smart antennas' benefits, 942–943
  smart antennas' drawbacks, 942–943
  smart antenna analogy, 931–932, 932*f*
smart antennas, antenna beamforming, 946
  adaptive beamforming, 950–953, 951*f*, 952*f*
  direction-of-arrival (DOA) algorithms, 947–950, 947*f*, 950*f*
  optimal beamforming techniques, least mean square (LMS) algorithm, 956–959, 957*f*, 958*f*, 959*f*
  optimal beamforming techniques, minimum mean square error (MMSE) criterion, 955–956
smart antennas, cellular radio systems evolution
  omnidirectional systems, 933, 933*f*
  omnidirectional systems, cell splitting, 934, 934*f*
  omnidirectional systems, sectorized systems, 934–936, 935*f*
  smart-antenna systems, adaptive array systems, 937–938, 937*f*, 938*f*
  smart-antenna systems, spatial division multiple access (SDMA), 938–939, 939*f*
  smart-antenna systems, switch-beam systems, 936–937, 936*f*, 937*f*
smart antennas, mobile ad hoc networks (MANETs)
  MANETs employing smart-antenna systems, protocol, 962–964, 963*f*
  MANETs employing smart-antenna systems, wireless network, 961–962, 962*f*
  overview, 960–961, 960*f*
smart antennas, system design/simulation/results
  4×4 vs 8×8 planar array, 969
  adaptive beamforming, 969–971, 970*f*, 971*f*, 972*f*
  array design, linear array design, 967–968, 968*f*
  array design, planar array design, 968–969
  design process, 964–965
  rectangular patch, 965–966, 966*f*, 967*f*, 968*f*
  single element, microstrip design, 965
stub-matching, 513
surface geometry, reflector antennas (front-fed parabolic), 887–890, 888*f*

transmitting mode, transmission-line Thevenin equivalent, 1, 2*f*
traveling wave antennas, 533–535, 534*f*, 535*f*
  long wire amplitude patterns, maxima, nulls, 538–539, 539*f*, 540*f*, 541*f*
  long wire, input impedance, 541–542
  long wire, resonant wires, 542–543
  rhombic antenna geometry, design equations, 549
  V antenna, 543–547, 544*f*, 545*f*, 546*f*, 547*f*
triangular sheet, bow-tie, and wire simulation of biconical antenna, 492–496, 493*f*, 494*f*, 495*f*, 496*f*
two-wires, 10–11, 11*f*, 12*f*

V antenna, 543–547, 544*f*, 545*f*, 546*f*, 547*f*
vector effective length, 81–83, 81*f*
vector potential **A**, electric current source **J**, 129–130
vector potential **F**, magnetic current source **M**, 130–131
vertical electric dipole, 183–195
Vivaldi antenna, 496–499, 497*f*, 499*f*, 500*f*

wire antennas, 3, 4*f*, 6*f*

Yagi-Uda array of linear elements, 559–561, 560*f*
  computer program and results, 568–571, 569*f*, 570*f*
  design procedure, 575–578, 576*t*, 577*f*, 579*f*
  far-field pattern, 566–567
  impedance and matching techniques, 572, 575, 575*t*
  integral equation-moment method, 562–565, 566*f*
  optimization, 571–572, 572*t*, 573*t*, 574*f*
Yagi-Uda array of loops, 579–580, 579*f*